Lecture Notes in Computer Science 4128

Commenced Publication in 1973
Founding and Former Series Editors:
Gerhard Goos, Juris Hartmanis, and Jan van Leeuwen

Editorial Board

David Hutchison
 Lancaster University, UK
Takeo Kanade
 Carnegie Mellon University, Pittsburgh, PA, USA
Josef Kittler
 University of Surrey, Guildford, UK
Jon M. Kleinberg
 Cornell University, Ithaca, NY, USA
Friedemann Mattern
 ETH Zurich, Switzerland
John C. Mitchell
 Stanford University, CA, USA
Moni Naor
 Weizmann Institute of Science, Rehovot, Israel
Oscar Nierstrasz
 University of Bern, Switzerland
C. Pandu Rangan
 Indian Institute of Technology, Madras, India
Bernhard Steffen
 University of Dortmund, Germany
Madhu Sudan
 Massachusetts Institute of Technology, MA, USA
Demetri Terzopoulos
 University of California, Los Angeles, CA, USA
Doug Tygar
 University of California, Berkeley, CA, USA
Moshe Y. Vardi
 Rice University, Houston, TX, USA
Gerhard Weikum
 Max-Planck Institute of Computer Science, Saarbruecken, Germany

Wolfgang E. Nagel
Wolfgang V. Walter
Wolfgang Lehner (Eds.)

Euro-Par 2006
Parallel Processing

12th International Euro-Par Conference
Dresden, Germany, August 28 – September 1, 2006
Proceedings

Springer

Volume Editors

Wolfgang E. Nagel
TU Dresden
Zentrum für Informationsdienste und Hochleistungsrechnen
Fakultät Informatik, Institut für Technische Informatik
01062 Dresden, Germany
E-mail: wolfgang.nagel@tu-dresden.de

Wolfgang V. Walter
TU Dresden
Fakultät Mathematik und Naturwissenschaften
Institut für Wissenschaftliches Rechnen
01062 Dresden, Germany
E-mail: wolfgang.walter@tu-dresden.de

Wolfgang Lehner
TU Dresden
Fakultät Informatik, Institut für Systemarchitektur
01062 Dresden, Germany
E-mail: wolfgang.lehner@tu-dresden.de

Library of Congress Control Number: 2006930999

CR Subject Classification (1998): C.1-4, D.1-4, F.1-3, G.1-2, H.2

LNCS Sublibrary: SL 1 – Theoretical Computer Science and General Issues

ISSN   0302-9743
ISBN-10   3-540-37783-2 Springer Berlin Heidelberg New York
ISBN-13   978-3-540-37783-2 Springer Berlin Heidelberg New York

This work is subject to copyright. All rights are reserved, whether the whole or part of the material is concerned, specifically the rights of translation, reprinting, re-use of illustrations, recitation, broadcasting, reproduction on microfilms or in any other way, and storage in data banks. Duplication of this publication or parts thereof is permitted only under the provisions of the German Copyright Law of September 9, 1965, in its current version, and permission for use must always be obtained from Springer. Violations are liable to prosecution under the German Copyright Law.

Springer is a part of Springer Science+Business Media

springer.com

© Springer-Verlag Berlin Heidelberg 2006
Printed in Germany

Typesetting: Camera-ready by author, data conversion by Scientific Publishing Services, Chennai, India
Printed on acid-free paper   SPIN: 11823285   06/3142   5 4 3 2 1 0

Preface

Euro-Par Conference Series

Euro-Par is an annual series of international conferences dedicated to the promotion and advancement of all aspects of parallel computing. The major themes can be divided into the broad categories of hardware, software, algorithms and applications for parallel computing. The objective of Euro-Par is to provide a forum within which to promote the development of parallel computing both as an industrial technique and an academic discipline, extending the frontier of both the state of the art and the state of the practice. This is particularly important at a time when parallel computing is undergoing strong and sustained development and experiencing real industrial take-up. The main audience for, and participants in, Euro-Par are researchers in academic departments, government laboratories and industrial organizations. Euro-Par's objective is to be the primary choice of such professionals for the presentation of new results in their specific fields of expertise. Euro-Par is also interested in applications that demonstrate the effectiveness of the main Euro-Par themes.

Previous Euro-Par conferences took place in Stockholm, Lyon, Passau, Southampton, Toulouse, Munich, Manchester, Paderborn, Klagenfurt, Pisa, and Lisbon. The next conference will take place in Rennes. The Euro-Par conference series is traditionally organized in cooperation with the International Federation for Information Processing (IFIP), the Association for Computer Machinery (ACM), and the Institute of Electrical and Electronics Engineers (IEEE) Computer Society, Technical Committee on Parallel Processing (TCPP). Euro-Par has a permanent website where its history and organization are described: http://www.europar.org.

Euro-Par 2006 in Dresden, Germany

Euro-Par 2006, the 12th conference in the Euro-Par series, was chiefly organized by the Center for Information Services and High Performance Computing (ZIH) in collaboration with the Department of Informatics and the Institute of Scientific Computing at the Technische Universität Dresden.

Three prominent workshops were collocated with the conference: the CoreGrid Workshop on Grid Middleware, the UNICORE Summit Workshop, and the Petascale Computational Biology and Bioinformatics Workshop. For the first time at a Euro-Par conference, real-time applications requiring high bandwidth were presented as part of a Grid Village exhibit complementing the Industrial Exhibition. A special Industrial Session was reserved within the main conference program for technical presentations by the exhibitors.

Euro-Par 2006 was able to attract three renowned invited speakers. Their talks highlighted some of the most recent trends:

Dominik Marx (Theoretical Chemistry, Ruhr-Univ. Bochum): *The Virtual Laboratory: Scientific Computing Beyond the Cutting Edge.*
Ajay K. Royyuru (Computational Biology Center, IBM T.J. Watson Research Center): *Deep Computing in Biology: Challenges and Progress.*
Alistair Dunlop (OMII-Europe, the Open Middleware Infrastructure Institute, Univ. of Southampton): *Integrating and Interoperability of Grid Infrastructures in OMII-Europe.*

Euro-Par 2006 Statistics

Compared with the traditional conference format, Euro-Par 2006 introduced two new topics — *High-Performance Bioinformatics* and *Embedded Parallel Systems* — for a total of 18 topics covering a large variety of aspects of parallel and distributed computing. Each topic was supervised by a committee of four: a global chair, a local chair, and two vice-chairs. The call for papers attracted a total of 290 submissions. For most of the submitted papers, at least three and often four individual review reports were collected. A total of 110 full papers were finally accepted, of which 5 received the special honor of distinguished paper. The acceptance rate was thus below 38 %. The submitting authors of accepted papers are from 26 countries, with the four main contributing countries — Germany, Spain, France and the USA — accounting for 67 accepted papers.

Acknowledgments

Euro-Par 2006 was made possible through the generous support and diligent work of many individuals and organizations. At the Technische Universität Dresden, the Center for Information Services and High Performance Computing (ZIH), the Department of Informatics, and the Institute of Scientific Computing were heavily involved in the organizational tasks, contributing their staff and their infrastructure.

Furthermore, a number of institutional and industrial sponsors made contributions and participated in the Industrial Exhibition and the Grid Village. Their names and logos appear on the Euro-Par 2006 website at http://www.tu-dresden.de/europar2006/ . We gratefully acknowledge their support.

At the ZIH, Claudia Schmidt undoubtedly bore the brunt of the administrative and organizational chores, all the while keeping her good humor (and her senses), Stefan Pflüger did a superb job in organizing and setting up the Industrial Exhibition and the Grid Village, and Guido Juckeland took on the difficult task of assembling the present volume of proceedings. The organizers are greatly indebted to them and the numerous other helping hands who donated their valuable time and went to great lengths to make Euro-Par 2006 a success.

We are greatly indebted to José Cunha, the organizer of Euro-Par 2005 in Lisbon, who never failed to give us prompt advice regarding any organizational matters.

We are also grateful to the members of the Euro-Par Steering Committee for their support. We owe special thanks to Christian Lengauer, chairman of the Steering Committee, who was always available to share with us his experience in the organization of Euro-Par, for giving us friendly advice, support, and encouragement. We also thank Luc Bougé, vice-chair, for his vision and contributions to improve Euro-Par conferences.

Specific thanks are due to the authors of all the submitted papers, the 72 members of the 18 topic committees, and the vast number of reviewers, for contributing to the success of this conference.

Last, but not least, we are grateful to Springer for agreeing to publish these proceedings, thus continuing a long and successful Euro-Par tradition.

It was a pleasure and an honor to host Euro-Par 2006 at the Technische Universität Dresden. We hope that we were able to make it a memorable event for all participants.

Dresden, June 2006

Wolfgang E. Nagel
Wolfgang V. Walter
Wolfgang Lehner

Organization

Euro-Par Steering Committee

Chair
Christian Lengauer — University of Passau, Germany

Vice-Chair
Luc Bougé — ENS Cachan, France

European Representatives
José Cunha — New University of Lisbon, Portugal
Marco Danelutto — University of Pisa, Italy
Rainer Feldmann — University of Paderborn, Germany
Christos Kaklamanis — Computer Technology Institute, Greece
Paul Kelly — Imperial College, UK
Harald Kosch — University of Passau, Germany
Thomas Ludwig — University of Heidelberg, Germany
Emilio Luque — Universitat Autònoma de Barcelona, Spain
Luc Moreau — University of Southampton, UK
Rizos Sakellariou — University of Manchester, UK

Non-European Representatives
Jack Dongarra — University of Tennessee at Knoxville, USA
Shinji Tomita — Kyoto University, Japan

Honorary Members
Ron Perrott — Queen's University Belfast, UK
Karl Dieter Reinartz — University of Erlangen-Nuremberg, Germany

Observers
Wolfgang E. Nagel — Technische Universität Dresden, Germany
Anne-Marie Kermarrec — IRISA Rennes, France

Euro-Par 2006 Local Organization

Euro-Par 2006 was organized by the Center for Information Services and High Perfomance Computing (ZIH) of the Technische Universität Dresden.

Conference Chairs

Wolfgang E. Nagel	Center for Information Services and High Performance Computing (ZIH) and Dept. of Computer Science, Inst. for Computer Engineering, Technische Universität Dresden
Wolfgang V. Walter	Dept. of Mathematics, Inst. for Scientific Computing, Technische Universität Dresden
Wolfgang Lehner	Dept. of Computer Science, Inst. for System Architecture, Technische Universität Dresden

General Organization
Claudia Schmidt

Technical Support
Kirsten Kern, Ronny Zschitzschmann

Grid Village, Exhibits
Dietmar Augustin, Stefan Pflüger

Proceedings
Guido Juckeland

Secretariat
Jenny Baumann

Euro-Par 2006 Program Committee

Topic 1: Support Tools and Environments

Global Chair
Bronis R. de Supinski — Lawrence Livermore National Laboratory, Center for Applied Scientific Computing, Livermore, CA, USA

Local Chair
Matthias Brehm — Leibniz-Rechenzentrum, High Performance Computing Group, Munich, Germany

Vice Chairs
    Tomas Margalef   Univ. Autònoma de Barcelona, Computer Architecture and Operating Systems Dept., Barcelona, Spain

    Luiz DeRose   Cray Inc., USA

Topic 2: Performance Prediction and Evaluation

Global Chair
    Jesús Labarta   Technical University of Catalunya (UPC), Barcelona Supercomputing Center (BSC), Barcelona, Spain

Local Chair
    Bernd Mohr   Forschungszentrum Jülich, ZAM, Jülich, Germany

Vice Chairs
    Allan Snavely   San Diego Supercomputer Center, University of California, San Diego, USA

    Jeffrey Vetter   Oak Ridge National Laboratory, Oak Ridge, Tennessee, USA

Topic 3: Scheduling and Load Balancing

Global Chair
    Michael A. Bender   State University of New York at Stony Brook, Department of Computer Science, Stony Brook, NY, USA

Local Chair
    Uwe Schwiegelshohn   University Dortmund, Robotics Research Institute, Dortmund, Germany

Vice Chairs
    Dror Feitelson   The Hebrew University, School of Computer Science and Engineering, Jerusalem, Israel

    Allan Gottlieb   New York University, Computer Science Department, New York, NY, USA

Topic 4: Compilers for High Performance

Global Chair
    William Jalby   University of Versailles Saint Quentin ARPA, France

Local Chair
Oscar Plata — University of Malaga, Department of Computer Architecture, Malaga, Spain

Vice Chairs
Barbara Chapman — University of Houston, Department of Computer Science, Houston, USA

Paul Kelly — Imperial College London, Department of Computing, London, UK

Topic 5: Parallel and Distributed Databases, Data Mining and Knowledge Discovery

Global Chair
Patrick Valduriez — INRIA and LINA, Nantes, France

Local Chair
Wolfgang Lehner — TU Dresden, Dept. of Computer Science, Dresden, Germany

Vice Chairs
Domenico Talia — DEIS, University of Calabria, Rende CS, Italy

Paul Watson — University of Newcastle Upon Tyne, School of Computing Science, Newcastle Upon Tyne, UK

Topic 6: Grid and Cluster Computing: Models, Middleware and Architectures

Global Chair
Domenico Laforenza — Istituto di Scienza e Tecnologie dell'Informazione (ISTI-CNR), Pisa, Italy

Local Chair
Alexander Reinefeld — Zuse Institute Berlin, Berlin, Germany

Vice Chairs
Dieter Kranzlmüller — Johannes Kepler Universität Linz, GUP - Institute of Graphics and Parallel Processing, Linz, Austria

Luc Moreau — University of Southampton, School of Electronics & Computer Science, Southampton, UK

Topic 7: Parallel Computer Architecure and Instruction-Level Parallelism

Global Chair
Eduard Ayguade — Barcelona Supercomputing Center (BSC), Technical University of Catalunya (UPC), Spain

Local Chair
Wolfgang Karl — Universität Karlsruhe (TH), Institut für Technische Informatik, Karlsruhe, Germany

Vice Chairs
Koen De Bosschere — Ghent University, ELIS Department, Ghent, Belgium

Jean-François Collard — Hewlett-Packard Laboratories, Palo Alto, CA, USA

Topic 8: Distributed Systems and Algorithms

Global Chair
Andrzej M. Goscinski — Deakin University, School of Information Technology, Victoria, Australia

Local Chair
Gudula Rünger — TU Chemnitz, Dept. of Computer Science, Chemnitz, Germany

Vice Chairs
Christine Morin — Campus universitaire de Beaulieu, IRISA / INRIA, France

Edgar Gabriel — University of Houston, Computer Science Department, Houston, TX, USA

Topic 9: Parallel Programming: Models, Methods, and Languages

Global Chair
José C. Cunha — New University of Lisbon, Faculty of Sciences and Technology, Department of Informatics, Monte de Caparica, Portugal

Local Chair
Sergei Gorlatch — TU Berlin, Faculty of Electrical Engineering and Computer Science Berlin, Germany

Vice Chairs

Daniel Quinlan Lawrence Livermore National Laboratory, Center for Applied Scientific Computing, Livermore, CA, USA

Peter H. Welch University of Kent, Computing Laboratory, Kent, UK

Topic 10: Parallel Numerical Algorithms

Global Chair

Michel Cosnard INRIA & Université de Nice - Sophia Antipolis, France

Local Chair

Hans-Joachim Bungartz TU München, Dept. of Computer Science, Munich, Germany

Vice Chairs

Efstratios Gallopoulos University of Patras, Dept. of Computer Engineering & Informatics, Patras, Greece

Yousef Saad University of Minnesota, Dept. of Computer Science and Engineering, Minneapolis, USA

Topic 11: Distributed and High-Performance Multimedia

Global Chair

Geoff Coulson Lancaster University, Computing Department, Lancaster, UK

Local Chair

Harald Kosch University of Passau, Department of Mathematics and Informatics, Passau, Germany

Vice Chairs

Frank Seinstra University of Amsterdam, Intelligent Systems Lab Amsterdam ISLA, Amsterdam, The Netherlands

Odej Kao University of Paderborn, Dept. of Computer Science, Paderborn, Germany

Topic 12: Theory and Algorithms for Parallel Computation

Global Chair

Danny Krizanc Wesleyan University, Mathematics and Computer Science, Middletown, CT, USA

Local Chair
    Michael Kaufmann   University of Tuebingen, Wilhelm Schickard Institute of Informatics, Tuebingen, Germany

Vice Chairs
    Pierre Fraigniaud   CNRS, LRI, University of Paris Sud, Paris, France

    Christos Zaroliagis   CTI, and Dept. of Computer Eng. & Informatics, University of Patras, Patras, Greece

Topic 13: Routing and Communication in Interconnection Networks

Global Chair
    Jose A. Gregorio   University of Cantabria, Computer Architecture Group, Santander, Spain

Local Chair
    Bettina Schnor   University Potsdam, Dept. of Computer Science, Potsdam, Germany

Vice Chairs
    Angelos Bilas   ICS-FORTH & University of Crete, Greece
    Olav Lysne   University of Oslo and Simula Research Laboratory, Lysaker, Norway

Topic 14: Mobile and Ubiquitous Computing

Global Chair
    Alois Ferscha   Johannes Kepler University Linz, Dept. of Pervasive Computing, Linz, Austria

Local Chair
    Alexander Schill   TU Dresden, Dept. of Computer Science, Dresden, Germany

Vice Chairs
    Gianluigi Ferrari   University of Pisa, Computer Science Department, Pisa, Italy

    Valérie Issarny   INRIA-Rocquencourt, Domaine de Voluceau,Rocquencourt, France

Topic 15: Peer-to-Peer and Web Computing

Global Chair
Henrique J. Domingos — New University of Lisbon, Department of Informatics, Caparica, Lisbon, Portugal

Local Chair
Anne-Marie Kermarrec — INRIA/IRISA, Campus Universitaire de Beaulieu, Rennes, France

Vice Chairs
Pascal Felber — University of Neuchâtel, Department of Computer Science, Switzerland

Márk Jelasity — Dept. of Computer Science, University of Bologna, Bologna, Italy

Topic 16: Applications of High-Performance and Grid Computing

Global Chair
Simon J. Cox — University of Southampton, School of Engineering Science, Southamton, UK

Local Chair
Thomas Lippert — Research Center Juelich, Central Institute for Applied Mathematics, Juelich, Germany

Vice Chairs
Giovanni Erbacci — CINECA, High Performance Systems Department, Bologna, Italy

Denis Trystram — ID-IMAG, Grenoble, France

Topic 17: High-Performance Bioinformatics

Global Chair
Craig A. Stewart — Indiana University, University Information Technology Services, Indiana, USA

Local Chair
Michael Schroeder — TU Dresden, Biotec, Dresden, Germany

Vice Chairs
Concettina Guerra — University of Padova, Padova, Italy, and Georgia Tech, Atlanta, USA

Konagaya Akihiko — Riken Genomic Sciences Center, Yokohama City, Japan

Topic 18: Embedded Parallel Systems

Global Chair
Stefanos Kaxiras — University of Patras Department of Electrical and Computational Engineering, Patras, Greece

Local Chair
Jürgen Teich — University of Erlangen-Nuremberg, Dept. of Computer Science, Erlangen, Germany

Vice Chairs
Toomas Plaks — South Bank University, School of Computing, Information, Systems and Mathematics, London, UK

Krisztian Flautner — ARM Ltd.

Euro-Par 2006 Referees

Not including members of the Program and Steering Committees.

Pablo Abad
J. H. Abawajy
Maha Abdallah
Ana Paula Afonso
Adnan Agbaria
Marco Aldinucci
Pinar Alper
Martin Alt
Patrick Amestoy
Emmanuelle Anceaume
Peter Andras
Artur Andrzejak
Eric Angel
George Apostolopoulos
Shawki Areibi
Masanori Arita
Kubilay Atasu

David A. Bader
Michael Bader
Rosa M. Badia
R. Badrinath
Arati Baliga

E. Bampis
Cyril Banino-Rokkones
Francoise Baude
Oliver Beaumount
Simon Becher
Christian Becker
Olav Beckmann
Andrzej Bednarski
Ramon Beivide
Domingo Benítez
Petra Berenbrink
Martin Bernreuther
Marin Bertier
Carlo Bertolli
Kristof Beyls
Christoph Biardzki
Florent Blachot
Lars Ailo Bongo
Markus Borschbach
Mireille Bossy
Jacob Bower
Andrea Bracciali
Luis Diego Briceno

Mats Brorsson
Ashley Brown
Ulrich Bruening
Marian Bubak
Rainer Buchty
Marzia Buscemi
Yann Busnel

Qiong Cai
Luis Caires
Rui Camacho
Blanca Caminero
Mario Cannataro
Mauro Caporuscio
Ioannis Caragiannis
Franco Alberto Cardilli
Roberto Speicys Cardoso
Denis Caromel
Laura Carrington
Carmen Carrion
Antonio Caruso
Carlo Cavazzoni
Eduardo César
Eugenio Cesario
Damien Charlet
Guillaume Chazarain
Zheng Chen
Nikolaos Chrysos
Tarik Cicic
Marcelo Cintra
Giovanni Ciriello
Antonio Cisternino
Peter Clayton
Diego Colombo
John Colquhoun
Matteo Comin
Carmela Comito
Antonio Congiusta
George A. Constantinides
Massimo Coppola
Jay Cornwall
Vitor Santos Costa
Adrian Cristal

Marco Danelutto
Karen Devine

Robert Dew
Erik D'Hollander
Marios D. Dikaiakos
Robert Dimond
Gero Dittmann
Ramon Doallo
Mario Döller
Henrique Domingos
Andreas Doms
Rubing Duan
Sergio Duarte
Vitor Duarte
Jose Duato
Jan Dünnweber
Pierre-Francois Dutot

Lieven Eeckhout
Pål Engelstad
Dick Epema
Jocelyne Erhel
Yoav Etsion
Stijn Eyerman

Thomas Fahringer
Weijian Fang
Paolo Ferragina
Brigid Ferrari
Rainer Fischer
José Flich
Dimitris Fotakis
Damien Fournier
Carsten Franke
Manoj Franklin
Manel Fredj
C. F. Freytag
Holger Froening
K. Fürlinger

Giulio Galante
Luis Garcés-Erice
Claudio Garutti
Thierry Gautier
Markus Geimer
Ingolf Geist
Cecile Germain-Renaud
Michael Gerndt

Vladimir Getov
Stein Gjessing
Brighten Godfrey
Roberto Gori
Anastasios Gounaris
Paul Grace
Xavier Granier
Phil Greenwood
Dan Grigoras
Laura Grigori
Paul Groth
Roberto Guanciale
Mitchell Gusat
Eladio Gutierrez

Pål Halvorsen
Frank Hannig
Audun Fosselie Hansen
Laurent Hascoet
Jiahua He
Manfred Hechinger
Frank Heller
Jörg Henkel
Pascal Henon
Andreas Henschel
Wim H. Hesselink
Elisa Heymann
Günther Höbling
Mikael Hoegqvist
J. Hofer
Bo Hong
Thomas Huckle
Danny Hughes
Sascha Hunold
Felix Hupfeld

Hitoshi Iba
Liviu Iftode

David Janovy
Mohamed Jemni
Morris Jette
Sheng Jiang
Josep Jorba

Peter Kacsuk
Marcus Kaiser

Helen Karatza
Sven Karlsson
Darren J. Kerbyson
Arie Keren
Omid Khalili
Thilo Kielmann
Andy King
Dirk Koch
Spyros Kontogiannis
Pierre Kuonen
Amund Kvalbein
Oren Laadan
Tobias Langhammer
Stéphane Lanteri
Isabelle Guérin Lassous
Gregor von Laszewski
Michael Laurenzano
Mario Lauria
Cynthia Bailey Lee
Sang Ik Lee
Arnaud Legrand
Michael H. Leitner
Joachim Lepping
Jean-Yves L`Excellent
Frank Li
Hui Li
Keqin Li
Zhongze Li
Stefan Liske
Jinshan Liu
Angel Llano
Josep Llosa
Luis Lopes
Phillip Lord
Joao Lourenço
Andre Luckow

Scott MacLachlan
Amine Mahjoub
Paolo Malfetti
Andrew Maloney
Anirban Mandal
Paolo Manghi
Manolis Marazakis
Annalisa Marsico
Maxime Martinasso

Tsutomu Maruyama
Gaia Maselli
Carlo Mastroianni
Ludek Matyska
John M. May
Michael O. McCracken
Sally McKee
Pedro Medeiros
Miriam Mehl
Michael Mellor
Rafael Menendez
Andre Merzky
Michael Messig
Norbert Meyer
Klaus Meyer-Wegener
Alessio Micheli
Jose Miguel-Alonso
Simon Miles
Sonia Ben Mokhtar
Burkhard Monien
Aad van Moorsel
Anna Morajko
Jose Moreira
Francesco Moscato
Andreas Moshovos
Achour Mostefaoui
Athanasios Mouchtaris
Irene Moulitsas
Gregory Mounie
Juan Carlos Moure
Gero Muehl
Jens Müller
Bruno Müller-Clostermann
Ralf-Peter Mundani
Steven Munroe

Francesco Nidito
Nils Agne Nordbotten

Adam J. Oliner
Catherine Olschanowsky
Salvatore Orlando
Motonori Ota

Esther Pacitti
Vijay S. Pai

D. K. Panda
Evaggelos Papapetrou
Alexander Papaspyrou
Nikos Parlavantzas
Fanny Pascual
David J. Pearce
Johnatan Pecero Sanchez
Miquel Pericas
Panos Periorellis
Kathrin Peter
Stefan Petri
Antonio Pinizzotto
Steven Plimpton
Eleftherios Polychronopoulos
Francesco Potorti
Nuno Preguiça
Giuseppe Prencipe
Thierry Priol
Valentin Puente
Diego Puppin

Jun Qin

Rolf Rabenseifner
Bruno Raffin
Omer F. Rana
Thomas Rauber
Pierre-Guillaume Raverdy
Sebastian Richly
Fco. Javier Ridruejo
Andreas Riener
Louis Rilling
Etienne Riviere
Yves Robert
Thomas Roeblitz
Jean Roman
Marie-Christine Rousset
Sean Rul

Francoise Sailhan
Rizos Sakellariou
Oliverio J. Santana
Eike Schallehn
Thomas Scheffler
Maraike Schellmann

Olaf Schenk
Isaac D. Scherson
Enrico Schiattarella
Florian Schintke
Thomas Schlichter
Bertil Schmidt
Lars Schneidenbach
Anke Schneidewind
Thorsten Schuett
Andreas S. Schulz
Martin Schulz
Frank Olaf Sem-Jacobsen
Miquel A. Senar
H. J. Siegel
Fernando Silva
Fabrizio Silvestri
Bernd Simeon
Brett Sinclair
Vasilios Siris
Garry Smith
Jim Smith
Peter Sobe
Angela C. Sodan
Thomas Sødring
Masha Sosonkina
Paul Spirakis
Parthasarathy Srinivasan
G. S. Stiles
Heinz Stockinger
Achim Streit
Martin Streubühr
Jaspal Subhlok
Remo Suppi
F. Suter
Bjorn De Sutter
Duane Szafron

Andrea Tagarelli
Victor Tan
Jie Tao
Christian Tenllado
J. Thiyagalingam
Fernando Tinetti
Nicola Tonellotto
Paolo Trunfio
George Tsaggouris

Konstantinos Tserpes
Giovanni Turi

Enrique Vallejo
Fernando Vallejo
Frederik Vandeputte
Hans Vandierendonck
Dora Varvarigou
Luis Veiga
Pierangelo Veltri
Salvatore Venticinque
K. V. Viswanathan
Frédéric Vivien

Thomas Wächter
Marcel Waldvogel
Chris Walshaw
Bin Wang
Rolf Wanka
Tan Tin Wee
Jonathan Weinberg
Matthias Werner
Stefan Wesner
Sebastian Winkel
Christoph Winter
Felix Wolf
Adam K.L. Wong
Brian J.N. Wylie

Polychronis Xekalakis
Gaogang Xie
Cheng-Zhong Xu

Ramin Yahyapour
Yoshiki Yamaguchi
Robert K. Yates
Han Yu

Serafeim Zanikolas
Apostolos Zarras
Rong Zheng
Yunkai Zhou
Fujun Zhu
Wolfgang Ziegler
Stefan Zimmer
Wolf Zimmermann
Martina Zitterbart

Table of Contents

Topic 1: Support Tools and Environments

Introduction.. 1
  Bronis R. de Supinski, Matthias Brehm, Luiz DeRose,
  Tomás Margalef (Topic Chairs)

IOAgent: A Parallel I/O Workload Generator........................ 3
  Sergio Gómez-Villamor, Victor Muntés-Mulero,
  Marta Pérez-Casany, John Tran, Steve Rees, Josep-L. Larriba-Pey

TDP_SHELL: An Interoperability Framework for Resource Management
Systems and Run-Time Monitoring Tools 15
  Vicente Ivars, Ana Cortes, Miquel A. Senar

Supporting Cache Locality Optimization with a Toolset............. 25
  Jie Tao, Wolfgang Karl

Model-Based Performance Diagnosis of Master-Worker Parallel
Computations 35
  Li Li, Allen D. Malony

Specification of Inefficiency Patterns for MPI-2 One-Sided
Communication.. 47
  Andrej Kühnal, Marc-André Hermanns, Bernd Mohr, Felix Wolf

Topic 2: Performance Prediction and Evaluation

Introduction... 63
  Jesús Labarta, Bernd Mohr, Allan Snavely, Jeffrey Vetter
  (Topic Chairs)

Hierarchical Model Validation of Symbolic Performance Models of
Scientific Kernels 65
  Sadaf R. Alam, Jeffrey S. Vetter

Tuning Application in a Multi-cluster Environment 78
  Eduardo Argollo, Adriana Gaudiani, Dolores Rexachs,
  Emilio Luque

Analyzing the Interaction of OpenMP Programs Within
Multiprogramming Environments on a Sun Fire E25K System
with PARbench... 89
  Rick Janda, Wolfgang E. Nagel, Bernd Trenkler

Early Experiences with KTAU on the IBM BG/L 99
  Aroon Nataraj, Allen D. Malony, Alan Morris, Sameer Shende

PAM-SoC: A Toolchain for Predicting MPSoC Performance 111
  Ana Lucia Varbanescu, Henk Sips, Arjan van Gemund

Analysis of the Memory Registration Process in the Mellanox
InfiniBand Software Stack 124
  *Frank Mietke, Robert Rex, Robert Baumgartl, Torsten Mehlan,
  Torsten Hoefler, Wolfgang Rehm*

Optimization of Dense Matrix Multiplication on IBM Cyclops-64:
Challenges and Experiences 134
  Ziang Hu, Juan del Cuvillo, Weirong Zhu, Guang R. Gao

Optimizing OpenMP Parallelized DGEMM Calls on SGI Altix 3700 145
  *Daniel Hackenberg, Robert Schöne, Wolfgang E. Nagel,
  Stefan Pflüger*

Topic 3: Scheduling and Load Balancing

Introduction 155
  *Michael Bender, Dror Feitelson, Allan Gottlieb, Uwe Schwiegelshohn
  (Topic Chairs)*

The Price of Approximate Stability for Scheduling Selfish Tasks on
Two Links 157
  Eric Angel, Evripidis Bampis, Fanny Pascual

Master-Slave Tasking on Asymmetric Networks 167
  Cyril Banino-Rokkones, Olivier Beaumont, Lasse Natvig

Using On-the-Fly Simulation for Estimating the Turnaround Time
on Non-dedicated Clusters 177
  *Mauricio Hanzich, Josep L. Lérida, Matías Torchinsky,
  Francesc Giné, Porfidio Hernández, Emilio Luque*

An Adaptive Scheduling Method for Grid Computing 188
  Salah-Salim Boutammine, Daniel Millot, Christian Parrot

On the Placement of Reservations into Job Schedules 198
  Thomas Röblitz, Krzysztof Rzadca

A Practical Approach of Diffusion Load Balancing Algorithms 211
  Emmanuel Jeannot, Flavien Vernier

Fast Diffusion Load Balancing Algorithms on Torus Graphs 222
  Gregory Karagiorgos, Nikolaos M. Missirlis, Filippos Tzaferis

A Parallel Shape Optimizing Load Balancer 232
  Henning Meyerhenke, Stefan Schamberger

Improvement of the Efficiency of Genetic Algorithms for Scalable
Parallel Graph Partitioning in a Multi-level Framework 243
  Cédric Chevalier, François Pellegrini

Probablistic Self-Scheduling 253
  Milind Girkar, Arun Kejariwal, Xinmin Tian,
  Hideki Saito, Alexandru Nicolau, Alexander Veidenbaum,
  Constantine Polychronopoulos

Data Sharing Conscious Scheduling for Multi-threaded
Applications on SMP Machines 265
  Shlomit S. Pinter, Marcel Zalmanovici

Topic 4: Compilers for High Performance

Introduction... 277
  William Jalby, Oscar Plata, Barbara Chapman, Paul Kelly
  (Topic Chairs)

Compiler Technology for Blue Gene Systems 279
  Stefan Kral, Markus Triska, Christoph W. Ueberhuber

SCAN: A Heuristic for Near-Optimal Software Pipelining 289
  Florent Blachot, Benoît Dupont de Dinechin, Guillaume Huard

Code Generation for STA Architecture 299
  J. Guo, T. Limberg, E. Matus, B. Mennenga, R. Klemm,
  G. Fettweis

Multi-dimensional Kernel Generation for Loop Nest Software
Pipelining... 311
  Alban Douillet, Hongbo Rong, Guang R. Gao

Towards a Versatile Pointer Analysis Framework 323
  R. Castillo, A. Tineo, F. Corbera, A. Navarro, R. Asenjo,
  E.L. Zapata

Topic 5: Parallel and Distributed Databases, Data Mining and Knowledge Discovery

Introduction... 335
  Patrick Valduriez, Wolfgang Lehner, Domenico Talia, Paul Watson
  (Topic Chairs)

Dynamic and Distributed Reconciliation in P2P-DHT Networks 337
  Vidal Martins, Esther Pacitti

HyParSVM – A New Hybrid Parallel Software for Support Vector
Machine Learning on SMP Clusters 350
  Tatjana Eitrich, Wolfgang Frings, Bruno Lang

Supporting a Real-Time Distributed Intrusion Detection Application
on GATES 360
  Qian Zhu, Liang Chen, Gagan Agrawal

On the Use of Semantic Annotations for Supporting Provenance
in Grids 371
  Liming Chen, Zhuoan Jiao, Simon J. Cox

Topic 6: Grid and Cluster Computing: Models, Middleware and Architectures

Introduction 381
  *Domenico Laforenza, Alexander Reinefeld, Dieter Kranzlmüller,
  Luc Moreau (Topic Chairs)*

Supporting Efficient Execution of MPI Applications Across
Multiple Sites 383
  Enol Fernández, Elisa Heymann, Miquel Àngel Senar

Private Virtual Cluster: Infrastructure and Protocol for Instant
Grids 393
  Ala Rezmerita, Tangui Morlier, Vincent Neri, Franck Cappello

Reducing Communication Overhead and Page Faults in SDSM
Platforms 405
  Artemis A. Christopoulou, Eleftherios D. Polychronopoulos

Flexible I/O Support for Reconfigurable Grid Environments 415
  *Marc-André Hermanns, Rudolf Berrendorf, Marcel Birkner,
  Jan Seidel*

Storage Exchange: A Global Trading Platform for Storage Services .. 425
  Martin Placek, Rajkumar Buyya

Vigne: Towards a Self-healing Grid Operating System 437
  Louis Rilling

Problems for Resource Brokering in Large and Dynamic Grid
Environments 448
  Catalin L. Dumitrescu

Topic 7: Parallel Computer Architecture and Instruction Level Parallelism

Introduction 459
  Eduard Ayguadé, Wolfgang Karl, Koen De Bosschere,
  Jean-Francois Collard (Topic Chairs)

Optimal Integrated VLIW Code Generation with Integer Linear
Programming 461
  Andrzej Bednarski, Christoph Kessler

Speeding-Up Synchronizations in DSM Multiprocessors 473
  A. de Dios, B. Sahelices, P. Ibáñez, V. Viñals, J.M. Llabería

Design and Effectiveness of Small-Sized Decoupled Dispatch Queues 485
  Won W. Ro, Jean-Luc Gaudiot

Sim-async: An Architectural Simulator for Asynchronous Processor
Modeling Using Distribution Functions 495
  J.M. Colmenar, O. Garnica, J. Lanchares, J.I. Hidalgo,
  G. Miñana, S. Lopez

A Hybrid Hardware/Software Generated Prefetching Thread
Mechanism on Chip Multiprocessors 506
  Hou Rui, Longbing Zhang, Weiwu Hu

Topic 8: Distributed Systems and Algorithms

Introduction 517
  Andrzej Goscinski, Gudula Rünger, Edgar Gabriel,
  Christine Morin (Topic Chairs)

Distributed Approximation Allocation Resources Algorithm for
Connecting Groups 519
  Fabien Baille, Lelia Blin, Christian Laforest

Rollback-Recovery Protocol Guarantying MR Session Guarantee
in Distributed Systems with Mobile Clients 530
  Jerzy Brzeziński, Anna Kobusińska, Michał Szychowiak

A Practical Single-Register Wait-Free Mutual Exclusion Algorithm
on Asynchronous Networks 539
  Hyungsoo Jung, Heon Y. Yeom

Optimal and Practical WAB-Based Consensus Algorithms 549
  Lasaro Camargos, Edmundo R.M. Madeira, Fernando Pedone

Self-stabilizing Deadlock Detection Under the OR Requirement
Model 559
  Christian F. Orellana, Cristian Ruz, Yadran Eterovic S.

Incremental Distributed Garbage Collection Using Reverse
Reference Tracking 571
    M. Schoettner, R. Goeckelmann, S. Frenz, M. Fakler, P. Schulthess

Run-Time Switching Between Total Order Algorithms 582
    José Mocito, Luís Rodrigues

On Greedy Graph Coloring in the Distributed Model 592
    Adrian Kosowski, Lukasz Kuszner

Topic 9: Parallel Programming: Models, Methods and Languages

Introduction... 603
    José C. Cunha, Sergei Gorlatch, Daniel Quinlan, Peter H. Welch
    (Topic Chairs)

Surrounding Theorem: Developing Parallel Programs for
Matrix-Convolutions 605
    Kento Emoto, Kiminori Matsuzaki, Zhenjiang Hu, Masato Takeichi

Dynamic Task Generation and Transformation Within a Nestable
Workpool Skeleton 615
    Steffen Priebe

Data Parallel Iterators for Hierarchical Grid and Tree Algorithms 625
    Gerhard Zumbusch

Implementing Irregular Parallel Algorithms with OpenMP 635
    Michael Süß, Claudia Leopold

Toward Enhancing OpenMP's Work-Sharing Directives 645
    Barbara M. Chapman, Lei Huang, Haoqiang Jin, Gabriele Jost,
    Bronis R. de Supinski

Toward a Definition of and Linguistic Support for Partial
Quiescence 655
    Billy Yan-Kit Man, Hiu Ning (Angela) Chan, Andrew J. Gallagher,
    Appu S. Goundan, Aaron W. Keen, Ronald A. Olsson

Tying Memory Management to Parallel Programming Models 666
    Ioannis E. Venetis, Theodore S. Papatheodorou

Topic 10: Parallel Numerical Algorithms

Introduction... 677
    Michel Cosnard, Hans-Joachim Bungartz, Efstratios Gallopoulos,
    Yousef Saad (Topic Chairs)

Parallel LOD Scheme for 3D Parabolic Problem with Nonlocal
Boundary Condition... 679
    Raimondas Čiegis

Online Checkpointing for Parallel Adjoint Computation in PDEs:
Application to Goal-Oriented Adaptivity and Flow Control............. 689
    Vincent Heuveline, Andrea Walther

Parallel Fault Tolerant Algorithms for Parabolic Problems 700
    Hatem Ltaief, Marc Garbey, Edgar Gabriel

Parallel Solution of Large-Scale and Sparse Generalized Algebraic
Riccati Equations 710
    José M. Badía, Peter Benner, Rafael Mayo,
    Enrique S. Quintana-Ortí

Applicability of Load Balancing Strategies to Data-Parallel Embedded
Runge-Kutta Integrators 720
    Matthias Korch, Thomas Rauber

A Software Framework for the Portable Parallelization
of Particle-Mesh Simulations 730
    I.F. Sbalzarini, J.H. Walther, B. Polasek, P. Chatelain, M. Bergdorf,
    S.E. Hieber, E.M. Kotsalis, P. Koumoutsakos

Parallelization of a Discrete Radiosity Method 740
    Rita Zrour, Pierre Chatelier, Fabien Feschet, Rémy Malgouyres

Parallelising Matrix Operations on Clusters for an Optimal
Control-Based Quantum Compiler 751
    T. Gradl, A. Spörl, T. Huckle, S.J. Glaser, T. Schulte-Herbrüggen

Topic 11: Distributed and High-Performance Multimedia

Introduction.. 763
    Geoff Coulson, Harald Kosch, Odej Kao, Frank Seinstra
    (Topic Chairs)

Supporting Reconfigurable Parallel Multimedia Applications............ 765
    Maik Nijhuis, Herbert Bos, Henri E. Bal

Providing VCR in a Distributed Client Collaborative Multicast
Video Delivery Scheme 777
    X.Y. Yang, P. Hernández, F. Cores, A. Ripoll, R. Suppi, E. Luque

Linear Hashtable Motion Estimation Algorithm for Distributed
Video Processing... 788
    Yunsong Wu, Graham Megson

Topic 12: Theory and Algorithms for Parallel Computation

Introduction.. 799
Danny Krizanc, Michael Kaufmann, Pierre Fraigniaud, Christos Zaroliagis (Topic Chairs)

A Hierarchical CLH Queue Lock 801
Victor Luchangco, Dan Nussbaum, Nir Shavit

Competitive Freshness Algorithms for Wait-Free Data Objects 811
Peter Damaschke, Phuong Hoai Ha, Philippas Tsigas

A Parallel Algorithm for the Two-Dimensional Cutting Stock Problem 821
Luis García, Coromoto León, Gara Miranda, Casiano Rodríguez

A BSP/CGM Algorithm for Finding All Maximal Contiguous Subsequences of a Sequence of Numbers 831
Carlos Eduardo Rodrigues Alves, Edson Norberto Cáceres, Siang Wun Song

On-Line Adaptive Parallel Prefix Computation 841
Jean-Louis Roch, Daouda Traoré, Julien Bernard

Topic 13: Routing and Communication in Interconnection Networks

Introduction.. 851
Jose A. Gregorio, Bettina Schnor, Angelos Bilas, Olav Lysne (Topic Chairs)

A Model for the Development of AS Fabric Management Protocols 853
Antonio Robles-Gómez, Eva M. García, Aurelio Bermúdez, Rafael Casado, Francisco J. Quiles

On the Influence of the Selection Function on the Performance of Fat-Trees 864
F. Gilabert, M.E. Gómez, P. López, J. Duato

Scalable Ethernet Clos-Switches 874
Norbert Eicker, Thomas Lippert

Towards a Cost-Effective Interconnection Network Architecture with QoS and Congestion Management Support 884
A. Martínez, P.J. García, F.J. Alfaro, J.L. Sánchez, J. Flich, F.J. Quiles, J. Duato

Topic 14: Mobile and Ubiquitous Computing

Introduction... 897
   Alois Ferscha, Alexander Schill, Gianluigi Ferrari, Valerie Issarny (Topic Chairs)

Multi-rated Packet Transmission Scheme for IEEE 802.11 WLAN
Networks.. 899
   Namgi Kim

Comparison of Different Methods for Next Location Prediction......... 909
   Jan Petzold, Faruk Bagci, Wolfgang Trumler, Theo Ungerer

SEER: Scalable Energy Efficient Relay Schemes in MANETs............ 919
   Lin-Fei Sung, Cheng-Lin Wu, Yi-Kai Chiang, Shyh-In Hwang

Multicost Routing over an Infinite Time Horizon in Energy and
Capacity Constrained Wireless Ad-Hoc Networks........................ 931
   Christos A. Papageorgiou, Panagiotis C. Kokkinos, Emmanouel A. Varvarigos

An Adaptive Self-organization Protocol for Wireless Sensor
Networks.. 941
   Kil-Woong Jang, Byung-Soon Kim

COPRA – A Communication Processing Architecture for Wireless
Sensor Networks.. 951
   Reinhardt Karnapke, Joerg Nolte

DAEDALUS – A Peer-to-Peer Shared Memory System for
Ubiquitous Computing... 961
   Peter Ibach, Vladimir Stantchev, Christian Keller

Context Awareness: An Experiment with Hoarding..................... 971
   João Garcia, Luís Veiga, Paulo Ferreira

A Client-Server Approach to Enhance Interactive Virtual
Environments on Mobile Devices over Wireless Ad Hoc Networks....... 981
   Azzedine Boukerche, Richard Werner Nelem Pazzi, Tingxue Huang

Topic 15: Peer-to-Peer and Web Computing

Introduction... 993
   Henrique J. Domingos, Anne-Marie Kermarrec, Pascal Felber, Mark Jelasity (Topic Chairs)

Top k RDF Query Evaluation in Structured P2P Networks............. 995
   Dominic Battré, Felix Heine, Odej Kao

Roogle: Supporting Efficient High-Dimensional Range Queries
in P2P Systems 1005
    Di Wu, Ye Tian, Kam-Wing Ng

Creating and Maintaining Replicas in Unstructured Peer-to-Peer
Systems 1015
    Elias Leontiadis, Vassilios V. Dimakopoulos, Evaggelia Pitoura

DOH: A Content Delivery Peer-to-Peer Network 1026
    Jimmy Jernberg, Vladimir Vlassov, Ali Ghodsi, Seif Haridi

Topic 16: Applications of High-Performance and Grid Computing

Introduction 1041
    Simon J. Cox, Thomas Lippert, Giovanni Erbacci, Denis Trystram (Topic Chairs)

Task Pool Teams Implementation of the Master Equation Approach for
Random Sierpinski Carpets 1043
    K.H. Hoffmann, M. Hofmann, G. Rünger, S. Seeger

A Preliminary Out-of-Core Extension of a Parallel Multifrontal
Solver 1053
    Emmanuel Agullo, Abdou Guermouche, Jean-Yves L'Excellent

A Parallel Adaptive Cartesian PDE Solver Using Space–Filling
Curves 1064
    Hans-Joachim Bungartz, Miriam Mehl, Tobias Weinzierl

Load Balanced Parallel Simulated Annealing on a Cluster of SMP
Nodes 1075
    Agnieszka Debudaj-Grabysz, Rolf Rabenseifner

A Grid Computing Based Virtual Laboratory for Environmental
Simulations 1085
    I. Ascione, G. Giunta, P. Mariani, R. Montella, A. Riccio

Exploiting Throughput for Pipeline Execution in Streaming Image
Processing Applications 1095
    F. Guirado, A. Ripoll, C. Roig, A. Hernàndez, E. Luque

dCache, Storage System for the Future 1106
    Patrick Fuhrmann, Volker Gülzow

Computing the Diameter of 17-Pancake Graph Using a PC Cluster ... 1114
    Shogo Asai, Yuusuke Kounoike, Yuji Shinano, Keiichi Kaneko

Topic 17: High-Performance Bioinformatics

Introduction 1125
*Craig A. Stewart, Michael Schroeder, Concettina Guerra,
Konagaya Akihiko (Topic Chairs)*

Multidimensional Dynamic Programming for Homology Search on
Distributed Systems 1127
*Shingo Masuno, Tsutomu Maruyama, Yoshiki Yamaguchi,
Akihiko Konagaya*

Load Balancing and Parallel Multiple Sequence Alignment
with Tree Accumulation 1138
Guangming Tan, Liu Peng, Shengzhong Feng, Ninghui Sun

ZIB Structure Prediction Pipeline: Composing a Complex Biological
Workflow Through Web Services 1148
Patrick May, Hans-Christian Ehrlich, Thomas Steinke

Evaluation of Parallel Paradigms on Anisotropic Nonlinear Diffusion 1159
S. Tabik, E.M. Garzón, I. García, J.J. Fernández

Improving the Research Environment of High Performance
Computing for Non-cluster Experts Based on Knoppix
Instant Computing Technology 1169
*Fumikazu Konishi, Manabu Ishii, Shingo Ohki, Yusuke Hamano,
Shuichi Fukuda, Akihiko Konagaya*

Topic 18: Embedded Parallel Systems

Introduction 1179
*Jürgen Teich, Stefanos Kaxiras, Toomas Plaks, Krisztián Flautner
(Topic Chairs)*

Efficient Realization of Data Dependencies in Algorithm Partitioning
Under Resource Constraints 1181
Sebastian Siegel, Renate Merker

FPGA Implementation of a Prototype Hierarchical Control Network
for Large-Scale Signal Processing Applications 1192
Jérôme Lemaitre, Ed Deprettere

An Embedded Systems Programming Environment for C 1204
Bernd Burgstaller, Bernhard Scholz, Anton Ertl

Author Index 1217

Topic 1: Support Tools and Environments

Bronis R. de Supinski, Matthias Brehm, Luiz DeRose, and Tomás Margalef

Topic Chairs

Support tools and environments are vital to the production of efficient parallel and distributed applications. This year, eleven papers were submitted to this topic area, from which five were accepted as full papers.

As in previous years, many submitted papers focused on the internal behaviour of applications, including how to enhance cache locality, how to detect inefficiency patterns in MPI applications and how to profile performance effectively. The infrastructure of the tools ranges from profiling, analysis, and simulation to visualization tools for graphical representation and code development.

Automatic performance analysis that can detect bottlenecks related to code regions or to data structures were of particular interest. Pattern-based performance diagnosis is not only applied to master-worker computations, but also, in a distinguished paper, to patterns that arise from inefficiencies in one-sided communication. Another paper presents an automatic synthetic I/O workload in which multiple factor analysis describes the behaviour of the I/O subsystem.

Other submitted papers this year focused on tools for grid and P2P infrastructures, particularly on resource management. An accepted paper with this focus presents a generic framework that enables run-time monitoring tools to be launched under the control of a resource management system. The crucial issues of process creation, tool creation, process monitoring and control and front-end/back-end coordination are addressed.

The qualified papers demonstrate significant tool improvement and maturation. We look forward to the stimulating discussions during session meetings that they will engender.

IOAgent: A Parallel I/O Workload Generator

Sergio Gómez-Villamor[1], Victor Muntés-Mulero[1], Marta Pérez-Casany[1],
John Tran[2], Steve Rees[2], and Josep-L. Larriba-Pey[1,*]

[1] DAMA-UPC
Computer Architecture Dept., Universitat Politècnica de Catalunya
Jordi Girona 1-3, Campus Nord-UPC, Modul C6, 08034 Barcelona
{sgomez, vmuntes, larri}@ac.upc.edu, marta.perez@upc.edu
http://www.dama.upc.edu
[2] IBM Toronto Lab.
8200 Warden Ave., Markham, ON L6G1C7, Canada
{jbtran, srees}@ca.ibm.com

Abstract. The purpose of this paper is twofold. First, we present IOAgent, a tool that allows to generate synthetic workloads for parallel environments in a simple way. IOAgent has been implemented for Linux and takes into account different I/O characteristics like synchronous and asynchronous calls, buffered and unbuffered accesses, as well as different numbers of disks, intermediate buffers and number of agents simulating the workload. Second, we propose statistical models that help us to analyze the I/O behaviour of an IBM e-server OpenPower 710, with 4 SCSI drives. The observations used to build the model have been obtained using IOAgent.

Keywords: parallel I/O, synthetic workload generator, Linux, performance evaluation, statistical modelling.

1 Introduction

The quest for tools that generate workloads for the evaluation of parallel I/O comes from long ago [1]. It is a need both as an aid for the optimum configuration of complex applications on complex computer architectures, and for the evaluation of research on Operating Systems and storage performance. The reason for such tools to exist is that they can emulate the behaviour of complex applications, avoiding the use of such applications for evaluation.

Bonnie [2], LMbench [3] and FileBench [4] are examples of tools that allow this type of workload emulation for Unix-like environments. However, such tools are restricted from many different points of view, like the number of threads and the type of I/O operating system calls that they can trigger.

---

[*] The authors at UPC thank Generalitat de Catalunya for its support through grant GRE-00352. Sergio Gómez-Villamor thanks IBM for its support through a CAS grant.

In order to give access to a generic workload generator for the evaluation of parallel I/O subsystems, we present IOAgent. Our tool sits on top of the Operating System in the same way as an application would do, and allows for the generation of workloads tailored to the needs of the system to evaluate. IOAgent allows users to generate synthetic application-level requests and defines a considerable number of behaviour-variables that allow to mimic an application. Our tool has been tested in large environments with up to 24 disks to emulate the accesses caused by a DBMS with a transactional processing workload [5].

In addition, as an example of the use of IOAgent, we evaluate an Open-Power 710 I/O subsystem. For the evaluation, we generate different workloads that stress the I/O subsystem in different ways and execute them to obtain more than 10 thousand execution time measures. With all those data, we propose a statistical model of the system using the Analysis of Variance (ANOVA). We show that using statistical tools it is easier and more reliable to extract conclusions.

The rest of this paper is organized as follows. In Section 2 we explain IOAgent. Then, in Section 3, we describe the environment and tests performed. In Section 4 we describe the statistical models obtained and in Section 5 we discuss the results with the help of the models. In Section 6 we give a short overview of the literature on the topic and, finally, we conclude.

2 IOAgent

IOAgent offers the possibility to mimic the stress imposed by an application on the I/O subsystem. This is done by simulating processes that exercise read and write predefined patterns on the accessed devices. Thus, IOAgent allows to evaluate parallel environments, both from the processing and the I/O points of view.

Currently, most OSs provide different system calls to execute I/O operations in a *synchronous* or *asynchronous* manner. Also, depending on whether the data blocks involved in I/O operations are mapped onto kernel buffers we can distinguish between *buffered* or *unbuffered* I/O. IOAgent allows for all those possibilities.

At this moment, IOAgent is implemented for Linux but could be ported to any other OS with little effort. IOAgent can be freely accessed in [6].

2.1 Basic Structures

The synthetic workload configuration of IOAgent is set by means of (i) a set of per-thread access-patterns, also called *agents*, and (ii) a set of pseudo-devices, also called *files*.

Agents are responsible for performing the desired stress. Every *agent* is a thread which performs one access pattern. The mixed execution of different *agents* will determine the desired global workload. Each *agent* in a simulation will have specific values for the following fields:

- **File.** The *file* used to perform the I/O operations.
- **Operation mode.** There are four simultaneously compatible operation modes:
    - **Synchronous or asynchronous.** Synchronous I/O operations block a process until the I/O is performed while asynchronous I/O operations do not block the process.
    - **Number of buffers.** IOAgent associates each I/O operation to a buffer. Synchronous operations are performed sequentially, thus, IOAgent requires only one buffer for synchronous agents. Asynchronous operations can be performed concurrently, thus, asynchronous agents require as many buffers as I/O operations on the fly. The buffers of IOAgent emulate those provided by the application it is emulating.
    - **Buffered or unbuffered.** With buffered I/O, IOAgent uses its own buffers, and indirectly those provided by the OS as an intermediate step for the I/O operations. With unbuffered I/O, IOAgent only uses its own buffers. Unbuffered I/O can be specified using raw devices or the *Direct I/O* mode, which can be associated to a block device interface or a file of some file systems (*i.e.* some file systems do not allow *Direct I/O* mode).
    - **Read or write.** Each agent will perform exclusively read or write operations.
    - **Sequential or random.** Sequential accesses perform a set of consecutive or strided I/O operations. Random accesses perform a number of read or writes over each position of the file following a certain probability distribution. At present, it is possible to choose among *Uniform* or *Poisson* distributions.
- **Operation size.** The size of the I/O operations is fixed for an *agent*.
- **Inter-arrival times.** We fix this per-*agent* value as the number of I/O operations per unit of time that each *agent* must generate.

Files are used in our environment to encapsulate the different storage capabilities of a system. Therefore, every *agent* is associated to a *file* which represents where the *agent* performs its I/O operations. *Files* are devices or common files on which I/O operations are executed. Regarding the storage system support, IOAgent allows for:

- **File systems.** The most common way to access hard disks (or its logical partitions) is through files of a built-in file system (*e.g. ext3, ext2, reiserfs, xfs, jfs*, etc.).
- **Block devices interface.** Different block devices (*e.g.* hard disks) of UNIX-type systems can be accessed through the /dev interface.
- **Raw devices interface.** A raw device can be bound to an existing block device (*e.g.* a disk) and can be used to perform raw I/O with that existing block device. Such raw I/O bypasses the caching that is normally associated with block devices.

For a comprehensive explanation of other parameters not used in this paper, and how to configure a workload generation, we refer the reader to [5].

Table 1. OpenPower 710 configuration

Processor	2 Power5 at 1.65 GHz
Memory	4 GBytes, DDR-I ECC at 266 MHz
L1 data cache	4-way set associative LRU
L2 cache	10-way set associative 1.9 MBytes
L3 cache	36 MBytes
Storage	4 146.8 GBytes drives at 10 Krpm 2 channel Ultra320 SCSI controller 320 MBps peak transfer

3 Evaluation Setup

In order to show the use of IOAgent, we perform an analysis of the I/O performance characteristics of an IBM e-server OpenPower 710 [7], with 4 SCSI drives, and Red Hat Enterprise Linux AS v4 for 64-bit IBM Power based on the 2.6 Kernel. The configuration of the OpenPower 710 evaluated in this paper is as shown in Table 1.

We run extensive executions of IOAgent on the system. The executions added up to 10,368 performance measures (two weeks of executions). For the evaluation, we have set up the parameters of IOAgent shown in Table 2.

Buffered and unbuffered I/O tests were both performed on 4 GBytes files, one per disk used. For buffered I/O tests, we used files mounted with an *ext3* file system and to allow buffered file system asynchronous I/O we patched a 2.6.12 kernel version [8]. Asynchronous I/O is a very recent feature in the Linux kernel, therefore we focused our studies on this new feature. For unbuffered I/O we used a 2.6.9 kernel version.

In order to analyze the system under maximum stress, we fix the inter-arrival time to zero for all the agents.

4 Statistical Modelling

We study five different categorical variables (factors) related to the I/O subsystem performance, namely, the use of the OS buffers, the number of disks accessed by the application, the number of agents accessing the disks, the number of intermediate application buffers used to store the data managed during the I/O operations, and the size of those intermediate application buffers. As a response variable, we study the average transfer rate from disk for four different access patterns: sequential reads, sequential writes, random reads and random writes. We propose two models, a general one useful for the four access patterns, and a simplified specific model for random reads, based on the five factors mentioned above.

The models we propose provide a way to analyze the data collected and allow us to find out which parameters are most significant.

Table 2. Parameters used for the evaluation

Factor	Name	Levels	#levels
Buffered	O	Buffered and Unbuffered	2
#disks	D	1, 2 and 4	3
#agents/disk	A	1, 2, 4, 8, 16 and 32	6
#buffers/agent	B	1, 2, 4, 8, 16 and 32	6
buffer size (KBytes)	S	8, 32 and 128	3

4.1 The Models

We use the Analysis of Variance (ANOVA) because it is the classical statistical technique to describe the behaviour of a response variable as a function of some factors [9]. Conversely, regression plays the same role for continuous variables. The factors considered in the models and their levels are summarized in Table 2. All of them are fixed effect factors, which means that the levels are considered constants. This case is opposed to the random effects case, in which they are considered observations from a random variable.

First, we try to model the transfer rate as a function of the main effects of the factors without interactions. The results are not satisfactory since the errors (difference between the observed and the predicted values) do not satisfy the hypothesis of independence, normality and equal variance, required for the ANOVA. The problem disappears by transforming the response variables by means of a logarithm, and considering some interactions in the model. Following the Principle of Parsimony [9], we have finally accepted the following model for the four access patterns analyzed in this paper:

$$y_{ijkml} = \mu + O_i + D_j + A_k + S_l + B_m + \\ + (OD)_{ij} + (OA)_{ik} + (OS)_{il} + (OB)_{im} + (DA)_{jk} + \\ + (AS)_{kl} + (AB)_{km} + (SB)_{lm} + e_{ijklm} \quad (1)$$

for $i = 1..2$ (2 levels for factor O), $j = 1..3$ (3 levels for factor D), $k = 1..6$ (6 levels for factor A), $l = 1..3$ (3 levels for factor S), and $m = 1..6$ (6 levels for factor B) where,

1. y_{ijkml} is the logarithm of the average transfer rate of 4 executions of the application, that have been run under conditions i, j, k, m, l of the factors.
2. μ is known as the general average. In our case, it represents the mean value of the logarithm of the average transfer rate expected if the conditions under which the observation has been obtained are unknown.
3. O_i, D_j, A_k, S_l and B_m correspond to the main effects of the five factors explained before. Specifically, O_i corresponds to the effect of the ith level of O, D_j corresponds to the effect of the jth level of D, and so on. For being a fixed effects model, the conditions $\sum_i O_i = \sum_j D_j = \sum_k A_k = \sum_l S_l = \sum_m B_m = 0$ must be satisfied.

4. $(OD)_{ij}$ corresponds to the interaction of the ith level of O with the jth level of D. Those constants must verify that $\forall i, \sum_j (OD)_{ij} = 0$ and $\forall j$, $\sum_i (OD)_{ij} = 0$. Analogously, $(OA)_{ik}$, $(OS)_{il}$, $(OB)_{im}$, $(DA)_{jk}$, $(AS)_{kl}$, $(AB)_{km}$ and $(SB)_{lm}$ correspond to the different interactions between the levels of the corresponding factors, and the corresponding analogous restrictions must be verified.
5. e_{ijklm} corresponds to the experimental error and contains the information in the data which is not explained by the considered factors.

However, for random reads, the model above can be simplified since some interactions are not statistically significant. The simplified model is:

$$y_{ijkml} = \mu + O_i + D_j + A_k + S_l + B_m + \\ +(OD)_{ij} + (OS)_{il} + (OB)_{im} + (AB)_{km} + e_{ijklm} \qquad (2)$$

The R-Squares of the four response variables modeled (sequential reads and writes, and random reads and writes) are 0.81, 0.95, 0.99 and 0.99 respectively, using model (1) for all the cases except for random reads, where we use model (2). This means that the models explain the corresponding percentage of the total variability in the data (*i.e.* 81% for 0.81). The error terms for each model are independent and follow a normal distribution with zero mean and constant variance. Therefore we can accept model (1) for sequential I/O activity and random writes and model (2) for random reads.

5 Discussion

In the following paragraphs we dissect the general and specific characteristics of the four types of accesses that we exercised: sequential reads and writes, and random reads and writes. Although the models characterize the logarithm of the transfer rate, we always refer to the transfer rate of the I/O subsystem for simplicity in the text. All the plots show averages obtained from real executions.

5.1 Single Factor Analysis

First of all, it is remarkable that the models show an important difference between the levels of factor O, OS buffered/unbuffered accesses. While OS buffered accesses work better in sequential reads, unbuffered accesses work better in the rest of the cases. This can be understood from the fact that the OS monitors sequential read access patterns and prefetches data blocks in those cases.

A general characteristic for factor A is that, in random accesses, the larger the number of agents, the better, while in sequential accesses, the smaller the number of agents, the better. An explanation to this follows. One of the strategies to maximize throughput is to sort disk accesses to reduce the number of backward and forward disk arm movements [10]. Therefore, in random accesses, a small number of agents may cause the disk arms to move backward and forward constantly. A larger number of random accesses (*i.e.* a larger number of agents) causes more chance to have accesses to be on the route between two far away accesses, improving the usage of the resource. On the other hand, in sequential accesses, there is a trade off between the number of accesses and the randomness introduced by having several sequential accesses along a significant lapse of time.

Factors D and S increase the transfer rate of the I/O subsystem as their values increase. Finally, although factor B behaves in the same way, there are cases where its influence is unnoticeable.

5.2 Multiple Factor Analysis

Now we analyze the most significant interactions between pairs of factors for the different access patterns.

Sequential Reads. Among the interactions modeled for this type of I/O accesses, we found that the most interesting were the two shown in Figure 1: (top chart) between the OS buffered/unbuffered I/O, factor O, and the size of the application buffers, factor S, and (bottom chart) between OS buffered/unbuffered I/O, factor O, and the number of application buffers used, factor B.

As shown in Figure 1, OS buffered I/O always behaves better than unbuffered I/O for sequential reads. Also, the interactions show that both increasing the number of application buffers (factor B) or their size (factor S) benefits unbuffered I/O while, in general, it is less significant for OS buffered I/O throughput. This makes sense since the OS prefetches or reads data ahead when it detects sequential access patterns. On the other hand, unbuffered I/O improves with larger number and size of buffers (factors B and S respectively) because the bandwidth of the I/O is exercised more intensely either with more agents in parallel or with larger data sets accessed sequentially that, in both cases, allow for better throughput.

Sequential Writes. The interactions that we chose in this case for their significance are shown in Figure 2: (top chart) between the number of agents, factor A, and the OS buffered/unbuffered I/O, factor O, and (bottom chart) between the number of agents, factor A, and the number of buffers per agent, factor B.

The first interaction shows how increasing from one agent to two decreases the performance of the I/O subsystem in OS buffered accesses. This is caused by the randomness added by one more agent accessing a different set of data. The performance for two agents is sustained for more agents as shown in the same plot. Note that if we increase the size of the buffer (not shown in the plots), we improve the performance, showing that increasing the sequentiality benefits performance.

Turning to the plot at the bottom, we can see that the number of buffers used per agent saturates the performance at a certain point with 8 or 16 buffers per agent for one agent and more for more agents. This shows that having a larger number of buffers allows for a better planning of the I/O activity when an agent is writing data to the disks. This is also true when we increase the number of agents but, in those cases, the randomness and the number of context switches introduced

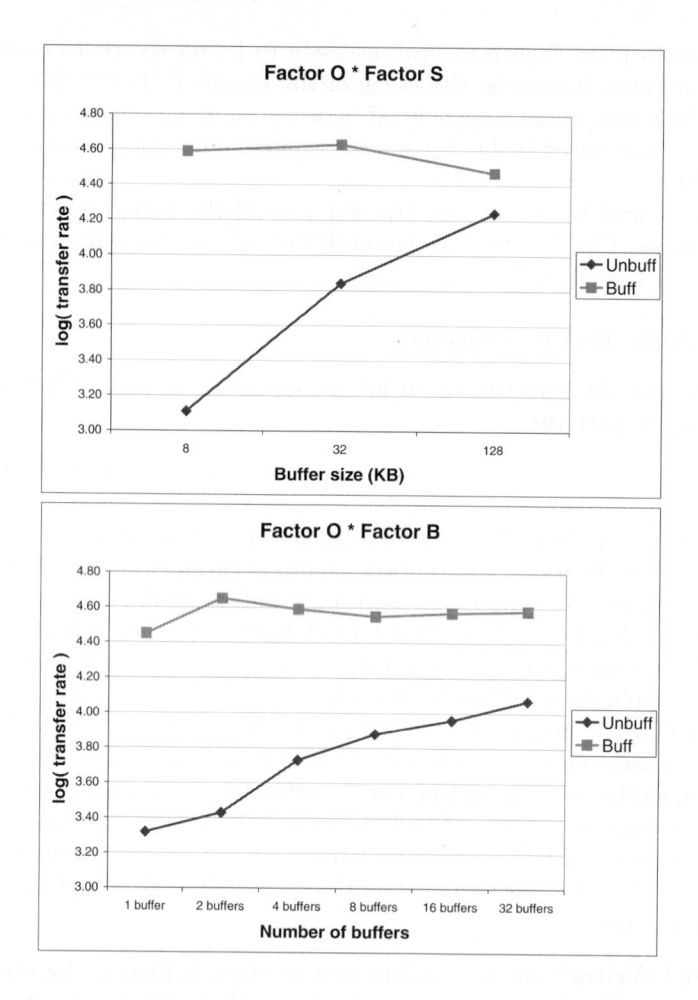

Fig. 1. (top) Interaction between OS buffered I/O, factor O, and the application buffer size, factor S, and (bottom) interaction between OS buffered I/O, factor O, and the number of application buffers, factor B, for sequential data reads

due to a larger number of agents reduces significantly the performance of the system. Therefore, the performance decreases when the number of agents increases.

In general, we can say that it is beneficial to have a large number of buffers per agent. In particular, the performance starts to saturate at 16 to 32 buffers for the case of one agent.

Finally, not in the plots, we observe that the interaction between the number of buffers per agent (factor B) and their size (factor S) is also significant. In this case, the smaller the size of the buffer, the more beneficial it is to have a larger number of buffers. In any case, the combination of large buffers with large number of buffers is the best, even though it is more important to have large buffers than a large number of them.

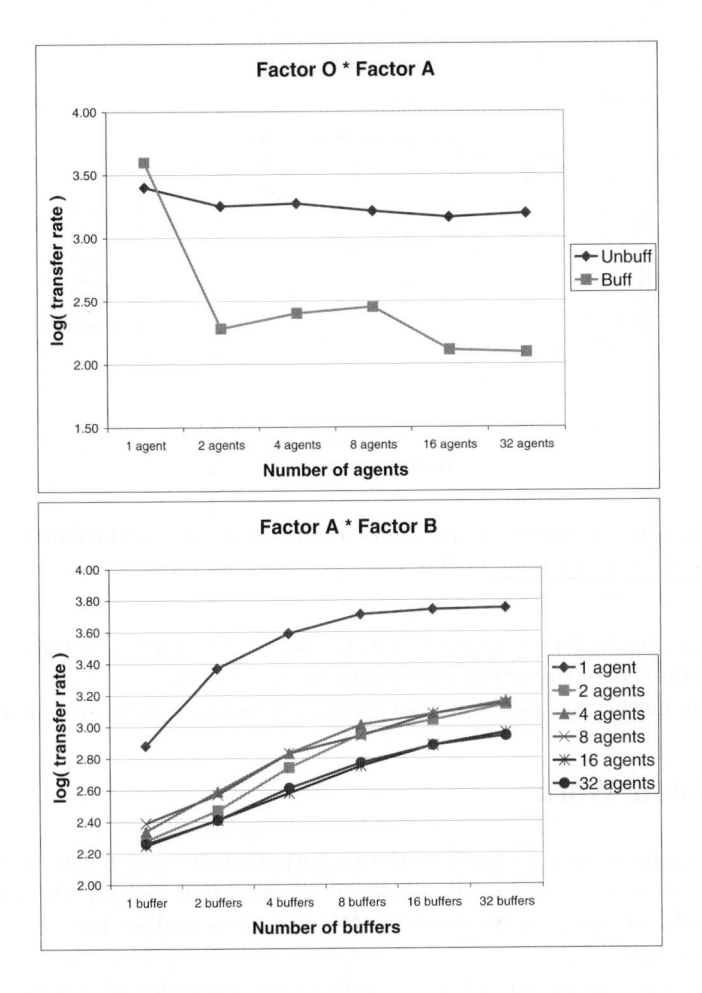

Fig. 2. (top) Interaction between OS buffered I/O, factor O, and the number of agents, factor A, and (bottom) interaction between the number of agents, factor A, and the number of buffers per agent, factor B, for sequential data writes

Random Reads and Random Writes. As a general observation, in the plot of Figure 3 we observe that a larger number of agents is beneficial for random operations, as opposed to sequential operations (plot at the bottom in Figure 2), where it was better to have less agents.

More specifically, in the plot of Figure 3, we show the interaction between the number of buffers per agent (factor B) and the number of agents (factor A) for random reads (the behaviour for random writes is similar although the model for random reads is simpler). The plot shows a clear tendency to converge to an asymptote, with larger numbers of agents converging faster. This is so because there is

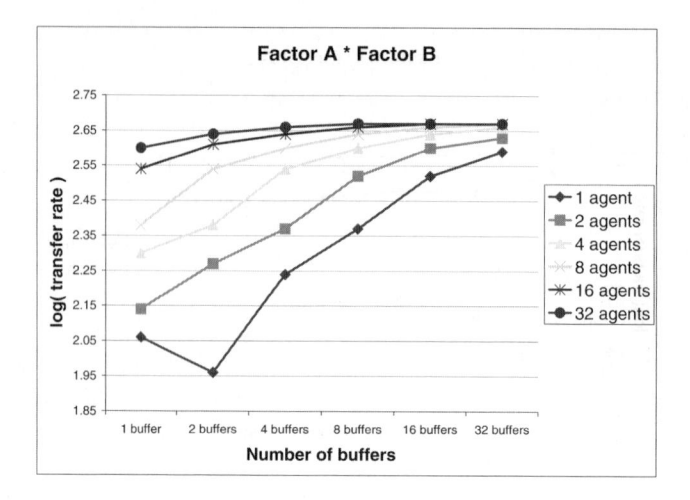

Fig. 3. Interaction between the number of agents, factor A, and the number of buffers per agent, factor B, for random data reads

a better planing of the resources when the number of agents and buffers is large. However, the size of the buffer (not shown in the plot) does not have a significant interaction neither on the number of agents nor on the number of buffers.

6 Related Work

There is a significant amount of work on synthetic workload generation tools for Unix-like systems as *Bonnie* [2], *LMbench* [3] and *FileBench* [4]. However, those softwares do not have all the features IOAgent offers and we have used here, as explained above.

On the other hand, we are not aware of the existence of a comprehensive study of the performance of the Linux asynchronous I/O on parallel devices before, or the Linux I/O performance on Power-based architectures. Moreover, our characterization can be regarded as the first one to use statistical methods for the analysis of the results. However, we want to mention the following pieces of work related to Linux I/O characterization, as examples. Ram Pai *et al.* [11] used *iozone* to study the performance improvement through readahead opti-

Table 3. Recommended configurations. SNS stands for Statistically Not Significant

Access pattern	Buffered mode	Number of agents	Buffer size	Number of buffers
Sequential reads	Buffered	small	large	SNS
Sequential writes	Unbuffered	small	SNS	large
Random reads and writes	Unbuffered	large	SNS	large

mization. Unfortunately, the tool used did not allow to test the optimizations under the Linux native asynchronous I/O. Also, Suparna *et al.* [12] analyzed the performance and robustness of the new Linux asynchronous I/O for enterprise workloads. The main difference between the evaluation done in this paper and that in [12] is that we give a more generic view and make extensive use of statistical infrastructure to validate the model presented.

7 Conclusions

This paper shows the importance of using a solid software and mathematical infrastructure for the evaluation of hardware/software systems. Our most important conclusion is that using a workload generator and a statistical methodology, we can extract solid and sound conclusions about the interaction of a simulated application, the OS and the hardware at use, in particular, the I/O subsystem.

We have generated test cases for the evaluation of the I/O subsystem of an IBM OpenPower 710. With more than 10 thousand execution results, we have built a statistical model that describes and fits accurately the behaviour of the I/O subsystem. Table 1 summarizes the best configurations for the four scenarios analyzed in this paper.

From the point of view of the use of IOAgent in a parallel environment, we show that the benefits obtained with an increase in the degree of parallelism is not straight forward.

References

1. Greg Ganger. Generating Representative Synthetic Workloads: An Unsolved Problem. *Proceedings of the Computer Measurement Group (CMG) Conference*, 1995.
2. Tim Bray. Bonnie Benchmark for UNIX Filesystem Operations. *http://www.textuality.com/bonnie*.
3. Larry McVoy and Carl Staelin. LMbench: Portable Tools for Performance Analysis. *USENIX Annual Technical Conference, San Diego, California, USA*, 1996.
4. Richard McDougall, Joshua Crase, and Shawn Debnath. FileBench: File System Microbenchmark. *http://www.solarisinternals.com/si/tools/filebench*.
5. Sergio Gómez-Villamor, John Tran, Steve Rees, Victor Muntés-Mulero, and Josep-L. Larriba-Pey. IOAgent: Leveraging the Application Analysis of Workload Effects. *Technical Report UPC-DAC-RR-2005-49, Department of Computer Architecture, Universitat Politecnica de Catalunya*, 2005.
6. DAMA-UPC. Data Management group at Universitat Politècnica de Catalunya. *http://www.dama.upc.edu/en/research/ioagent.html*.
7. *IBM e-server OpenPower 710 Technical Overview and Introduction*.
8. Suparna Bhattacharya. Patches for buffered file system AIO in 2.6 Linux kernel. *http://www.kernel.org/pub/linux/kernel/people/suparna/aio*.
9. Douglas C. Montgomery. *Design and analysis of experiments*. John Wiley, New York, fifth edition, 2001.

10. Bruce L. Worthington, Gregory R. Ganger, and Yale N. Patt. Scheduling Algorithms for Modern Disk Drives. *SIGMETRICS Perform. Eval. Rev.*, 1994.
11. Ram Pai, Badari Pulavarty, and Mingming Cao. Linux 2.6 Performance Improvement Through Readhead Optimization. *Proceedings of the Linux Symposium, Ottawa, Canada*, 2004.
12. Suparna Bhattacharya, John Tran, Mike sullivan, and Chris Mason. Linux AIO Performance and Robustness for Enterprise Workloads. *Proceedings of the Linux Symposium, Ottawa, Canada*, 2004.

TDP_SHELL: An Interoperability Framework for Resource Management Systems and Run-Time Monitoring Tools*

Vicente Ivars, Ana Cortes, and Miquel A. Senar

Departament d'Arquitectura d'Ordinadors i Sistemes Operatius
Universitat Autònoma de Barcelona
Barcelona, Spain
vicente@aomail.uab.es, {ana.cortes, miquelangel.senar}@uab.es

Abstract. Resource management systems and tool support are two important factors for efficiently developing applications in large clusters. On the one hand, management systems (in the form of batch queue systems) are responsible for all issues related to executing jobs on the existing machines. On the other hand, run-time tools (in the form of debuggers, tracers, performance analyzers, etc.) are used to guarantee the correctness and the efficiency of execution. Executing an application under the control of both a resource management system and a run-time tool is still a challenging problem in most cases. Using run-time tools might be difficult or even impossible in usual environments due to the restrictions imposed by resource managers. We propose TDP-Shell as a framework for providing the necessary mechanisms to enable and simplify using run-time tools under a specific resource management system. We have analyzed the essential interactions between common run-time tools and resource management systems and implemented a pilot TDP-Shell. The paper describes the main components of TDP-Shell and its use with some illustrative examples.

1 Introduction

Large distributed clusters are becoming a common platform for running compute-intensive applications. Developing applications that run in these environments is still a difficult task, even after several decades of intense research on methodologies and supporting tools. The intrinsic complexity of distributed systems and the continuous changes in hardware, operating systems and middleware platforms contribute to this complexity. Run-time tools play an important role in program development and we can find tools of various types to help programmers to develop, optimize and maintain code [1][2]. These tools can be used for debugging, performance analysis, program tracing, program flow visualization and computational steering.

While these tools are readily available in the programmer's desktop computer, when programs move into a distributed environment their usage and availability

---

* This work was supported by MEyC-Spain under contract TIN 2004-03388, and partially supported by NATO under contract EST.EAP.CLG 981032.

become more difficult. On one hand, the required components of the run-time tools may not be in place and, on the other hand, the run-time tool may not be able to carry out some necessary actions to control and monitor the application. These problems generally arise due to conflicts between executing run-time tools and the existence of resource management systems that schedule access to the distributed resources. Resource management software is responsible for accessing resources and providing the resources needed to run a job, as well as monitoring the job's execution and retrieving any results produced by the job. It plays a crucial role in any distributed cluster because it guarantees that applications are executed seamlessly and securely. Common resource management systems are used in local clusters in the form of batch queuing environments (e.g. Condor [3] or PBS[4]).

Using run-time tools in a distributed environment is difficult because of complex interactions between the application program, operating system, and layers of job and resource management software. This leads to situations where executing a run-time tool to monitor an application is incompatible with the existence of a resource management system unless the run-time tool is individually ported and adapted to run under each particular resource management system. We refer to this problem as a problem of interoperability between run-time tools and resource management systems. By interoperability, we mean the ability of different tools and resource managers to co-operate in controlling user applications by using common services and communication mechanisms.

A different approach to the interoperability problem was also considered in the literature when the concept of interoperable tools was used by some authors to refer to using more than one run-time tool in a user's program [5][6][7][8]. For instance, using a debugger and a performance analyzer concurrently is an example of two interoperable tools. In this case, the problem of interoperability applies to sets of tools that work at the same level, while in our work we consider resource managers and run-time tools, which have a different hierarchy in the system. Run-time tools have to run under the control and supervision of resource managers.

There are few references to this case of the interoperability problem in the literature on development tools for distributed systems. The TDP (Tool Daemon Protocol) [9] constitutes a recent contribution aimed at providing a standard interface that codifies the essential interactions between the run-time tool, the resource manager and the application program in order to make them interoperable. Using TDP-enabled tools and resource managers significantly simplifies the interoperability problem because it reduces the porting effort. Obviously, this requires modifications in existing tools and resource management systems in order to include the necessary TDP pieces. This limitation of TDP has motivated the continuation of this work towards TDP-Shell, a framework intended to provide interoperability in a flexible and easy way, which does not require changes in run-time tools or in resource managers. TDP-Shell provides a scripting-like language that allows programmable steps to be specified to guarantee their cooperation while executing a user application with maximum transparency and portability. It also uses agents interposed between the resource manager and the run-time tool that execute all these programmed steps.

The rest of the paper is organized as follows. Next, in Section 2, we discuss the problem background and we review the main characteristics of TDP, which has been used as a basis for TDP-Shell. Section 3 describes the general architecture of

TDP-Shell and Section 4 illustrates usage examples of TDP-Shell. We present conclusions and future work in Section 5.

2 Problem Background

Analysis of run-time tools and resource management environments shows a common architecture based on the existence of two main pieces: a front-end part that runs on the user's desktop computer and a back-end part that runs on the remote host where the user's application is also running. This common configuration requires a communication channel (typically a TCP/IP connection) that is established between the front-end and the back-end processes.

There are several crucial issues that must be addressed when an application running under the control of a resource management system also has to be monitored by a run-time tool.

- *Process creation:* This operation can be in conflict with a tool such as a debugger or profiler that also expects to launch the process. While most sophisticated run-time tools have the ability to attach to a running process, they cannot handle the case when the tool wants to attach to the process before it starts execution. There might be scenarios in which an application process is created but does not start, so the run-time tool attaches to the process and performs its initial processing, and then starts the application. Tools such as gdb, Totalview, and Paradyn use this technique. The appropriate information must be provided to the run-time tool so that it can find and operate on the application program.

- *Tool creation:* Similarly to the application, the resource manager is responsible for launching the run-time tool. This action also implies that configuration and data files needed by the run-time tool are transferred to the execution nodes. The run-time tool might be launched before the application is created (as above) or launched afterwards. In this second case, the resource manager must provide the appropriate information to the run-time tool so that it can attach to and operate on the application.

- *Process monitoring and control:* In the course of normal operation, the resource manager may pause and resume or vacate the application process. All these actions should also affect the run-time tool back-end, while the tool front-end is notified somehow.

- *Front-end/back-end coordination:* The front-end and the back-end of the run-time tool need to communicate and this communication is typically done with TCP/IP sockets. This communication can generally be established by a host/port number pair that must be provided either to the front-end or to the back-end when the other party has started execution. Executing a back-end controlled by the resource manager implies that in most cases this information is not known before hand and the front-end must wait until the resource manager allocates a particular resource and starts the back-end there. Therefore, a synchronization and coordination mechanism is required to guarantee a proper connection between the front–end and the back-end.

TDP-Shell is specifically targeted at the above mentioned issues, which are related to the problem of interoperability between a resource manager and a run-time tool. Other significant works have studied the problem of interoperability between run-time tools and the issues involved in coordinating the interactions between multiple run-time tools. While TDP-Shell is designed to allow multiple tools to be launched for a given application, the interactions between these tools must be coordinated by the tools themselves.

To the best of our knowledge, no other significant works have been proposed to deal with the interoperability problem as considered in this paper. There are isolated solutions (such as Totalview [10] running under MPICH) but only the Tool Daemon Protocol (TDP) mentioned in the introduction addresses the problem in a general way by proposing an interface that codifies the essential interactions between run-time tools and resource managers.

2.1 The Tool Daemon Protocol (TDP) Library

Our work on TDP-Shell uses the basic functionality provided by the Tool Daemon Protocol library. The TDP library provides three main groups of services: process management, inter-daemon communication interface and event notification. We outline below the main features related to these three groups. The core component of TDP consists in an Attribute Space that is used as a medium for data exchanging, for process synchronization and for event notification. The Attribute Space has many similarities with a Linda tuple space [11]. The two basic Attribute Space primitives are *tdp_get* and *tdp_put*. Information in the shared environment space is kept in the form of (attribute, value) pairs, where both the attribute and value are constrained to only being null terminated strings. An attribute consists simply in a character string that names data in the shared space. In the current implementation, the attribute space is supported by a Central Attribute Space Service that runs on the front-end machine. Any process using the TDP library can access the attribute space of the CASS.

The basic functions to work with the attribute space are the following:

- *tdp_get*: obtains the value of a given attribute from the attribute space
- *tdp_put*: inserts a new (attribute, value) pair into the attribute space.
- *tdp_del*: obtains and removes a given attribute from the space.

These operations block forms of communication between a daemon and the C*ASS*. Asynchronous versions for retrieving and storing information from the shared space are also available: (*tdp_async_get, tdp_async_put and tdp_aync_del*).

Finally, the TDP library provides several process management functions that are used to create, destroy, attach to, detach from, suspend or resume processes. While similar process management functions are present in common operating systems, TDP provides its own interfaces that are OS neutral.

3 The TDP-Shell Architecture

TDP-Shell is a framework based on two process agents. One agent runs on the front-end machine (referred to as TDP-SC:TDP-Shell Console) and the other agent runs on the back-end machine (referred to as TDP-SA: TDP-Shell Agent). Communication between the TDP-SC and the TDP-SA is based on the TDP attribute space. The framework also includes a scripting-like language that is used to describe actions that must be carried out both at the front-end machine and the back-end machine in order to start the run-time tool components and the process application in the right order. Both TDP-SC and TDP-SA execute a command file written in the above mentioned scripting language. Most of the available commands consist of wrappers to the TDP library functions mentioned above. Additionally, the command file may include simple assignment and control flow statements. Figure 1 shows the main components of the framework and their connection to the resource management and the run-time tool daemons.

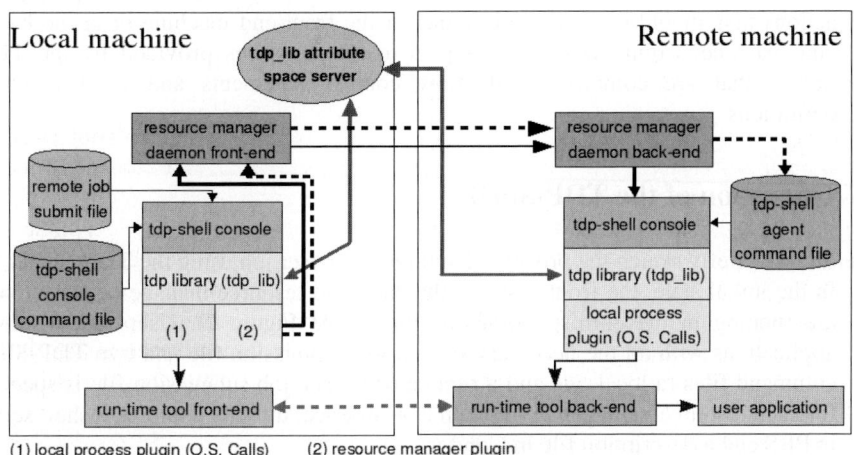

Fig. 1. Architecture of the TDP-Shell framework

1. TDP-Shell Console (TDP-SC): this is the process that runs at the front-end machine. It receives a job submit command file that specifies all the actions that should be carried out in the front-end machine. Among others, the job submit command file contains a set of commands to submit the application process to the resource manager, commands to start the run-time tool front-end and commands for the subsequent steps required to synchronize the run-time tool front-end with the run-time tool back-end. TDP-SC also contains the specific logic that guarantees that all the necessary components of the run-time tool are transferred to the remote machine. It does this by generating an appropriate submit file that is later submitted to the resource manager. A specific plug-in for different resource managers is used in TDP-SC to generate these submission files. TPD-SC is also

responsible for starting the central attribute space and it may run as an interactive process or in the background.

2. TDP-Shell Agent (TDP-SA): this is the process that runs in the back-end machine. It is started by the back-end daemon of the resource manager and, similarly to the TDP-SC, it executes a command file that specifies the actions that must be carried out to start the application process and the run-time tool back-end daemon. The command file also contains communication and synchronization actions between TDP-SA and TDP-SC through the attribute space provided by the TDP library. TDP-SA sits between the resource manager daemon and the application process and therefore it is also responsible for forwarding to the latter all control actions generated by the former.

3. TDP-Shell command file: this file contains the set of commands that specify the actions that should be carried out either at the front-end machine or at the back-end machine. Commands are wrappers to the functions provided by the TDP library that are combined with flow control statements and local function definitions.

4 Operation of the TDP-Shell

Below, we briefly sketch the process of submitting a user job using the TDP-Shell.

1. In the initial state, the front-end and the back-end resource management daemons are running in the corresponding machines (see Figure 2). Users supply their applications with all the necessary files, a job submission file and two TDP-Shell command files (a local one and a remote one). The job submission file is specific for each resource management system (for instance, this file could be a shell script in PBS and a description file in Condor).

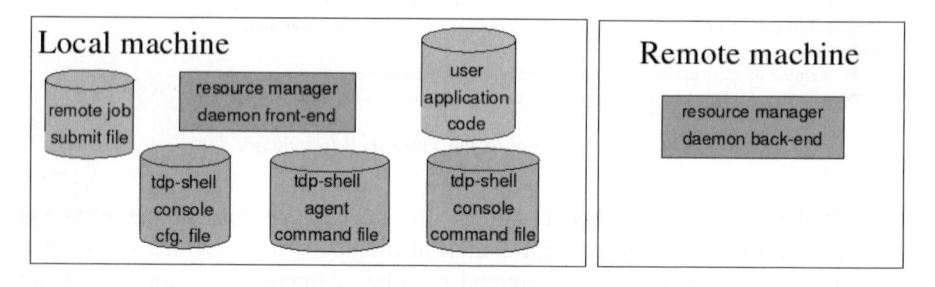

Fig. 2. TDP-Shell operation: initial state

2. Users create the TDP-Shell Console (using a *tdp-sc* command). The corresponding TDP-Shell command file is provided as a complement to TDP-SC (Figure 3 shows an example of the command file that uses *gdb* as a run-time tool). The TDP-SC starts the execution of the TDP-Shell command file. Following the example in Figure 3, the attribute space is created once the **tdp_init** command is called (see Figure 4).

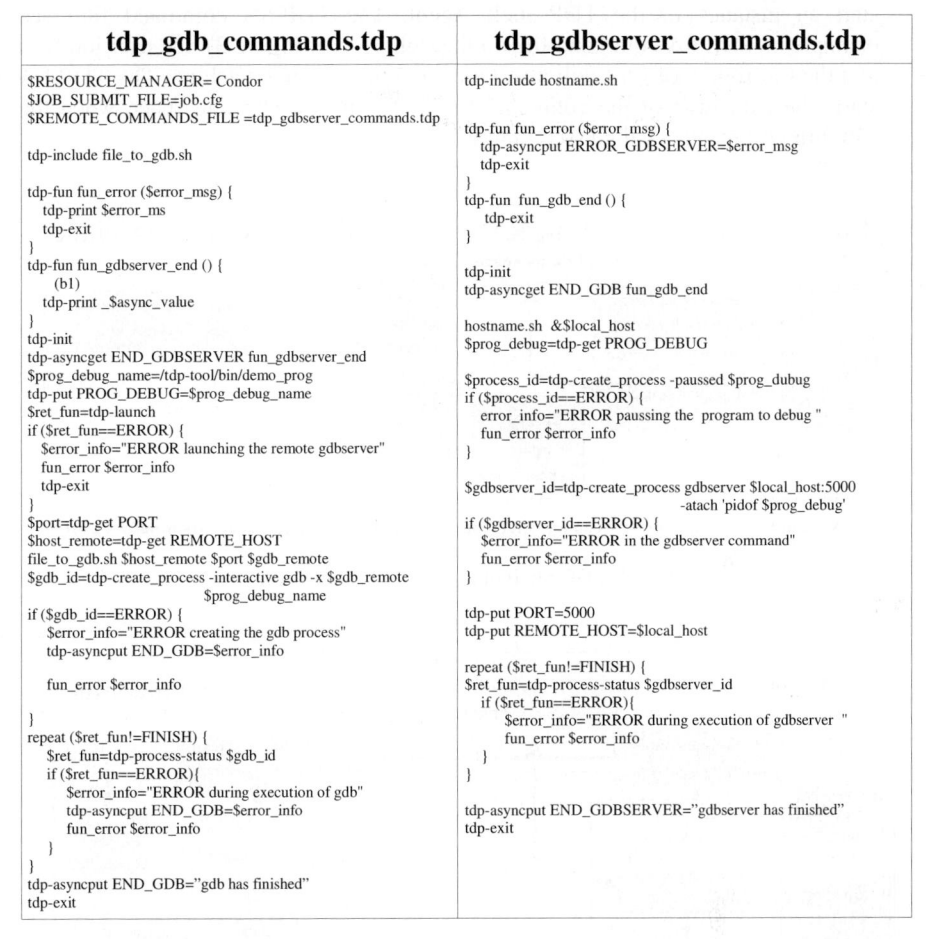

tdp_gdb_commands.tdp	**tdp_gdbserver_commands.tdp**

$RESOURCE_MANAGER= Condor
$JOB_SUBMIT_FILE=job.cfg
$REMOTE_COMMANDS_FILE =tdp_gdbserver_commands.tdp
 	
tdp-include file_to_gdb.sh
 	
tdp-fun fun_error ($error_msg) {
  tdp-print $error_ms
  tdp-exit
}
tdp-fun fun_gdbserver_end () {
    (b1)
  tdp-print _$async_value
}
tdp-init
tdp-asyncget END_GDBSERVER fun_gdbserver_end
$prog_debug_name=/tdp-tool/bin/demo_prog
tdp-put PROG_DEBUG=$prog_debug_name
$ret_fun=tdp-launch
if ($ret_fun==ERROR) {
  $error_info="ERROR launching the remote gdbserver"
  fun_error $error_info
  tdp-exit
}
$port=tdp-get PORT
$host_remote=tdp-get REMOTE_HOST
file_to_gdb.sh $host_remote $port $gdb_remote
$gdb_id=tdp-create_process -interactive gdb -x $gdb_remote
                  $prog_debug_name
if ($gdb_id==ERROR) {
  $error_info="ERROR creating the gdb process"
  tdp-asyncput END_GDB=$error_info
 	
  fun_error $error_info
 	
}
repeat ($ret_fun!=FINISH) {
  $ret_fun=tdp-process-status $gdb_id
  if ($ret_fun==ERROR){
    $error_info="ERROR during execution of gdb"
    tdp-asyncput END_GDB=$error_info
    fun_error $error_info
  }
}
tdp-asyncput END_GDB="gdb has finished"
tdp-exit
	tdp-include hostname.sh
	
	tdp-fun fun_error ($error_msg) {
	  tdp-asyncput ERROR_GDBSERVER=$error_msg
	  tdp-exit
	}
	tdp-fun fun_gdb_end () {
	  tdp-exit
	}
	
	tdp-init
	tdp-asyncget END_GDB fun_gdb_end
	
	hostname.sh &$local_host
	$prog_debug=tdp-get PROG_DEBUG
	
	$process_id=tdp-create_process -paussed $prog_dubug
	if ($process_id==ERROR) {
	  error_info="ERROR paussing the program to debug "
	  fun_error $error_info
	}
	
	$gdbserver_id=tdp-create_process gdbserver $local_host:5000
	                            -atach 'pidof $prog_debug'
	if ($gdbserver_id==ERROR) {
	  $error_info="ERROR in the gdbserver command"
	  fun_error $error_info
	}
	
	tdp-put PORT=5000
	tdp-put REMOTE_HOST=$local_host
	
	repeat ($ret_fun!=FINISH) {
	$ret_fun=tdp-process-status $gdbserver_id
	  if ($ret_fun==ERROR){
	    $error_info="ERROR during execution of gdbserver "
	    fun_error $error_info
	  }
	}
	tdp-asyncput END_GDBSERVER="gdbserver has finished"
	tdp-exit

Fig. 3. TDP-Shell command file example. Left: TDP-Shell Console part; Rigth: TDP-Shell Agent part

3. A job is submitted to the resource manager when the *tdp_launch* command is found. The TDP-Shell Console creates a special submission file taking into account the particular resource manager used in the system (in our example, $RESOURCE_MANAGER = Condor), the original job submission file ($JOB_SUBMIT_FILE = job.cfg) and if necessary the binaries and other additional files required by the back-end daemon of the run-time tool (in our example, we assume that these binaries are available in every remote machine, so they don't need to be transferred). A specific plug-in for each resource management system (see Figure 1) is responsible for generating the special submission file. Actually, TDP-SC submits a job that consists in the TDP-Shell Agent. The information about the original user's job and the run-time tool are combined in this TDP-SA job automatically and transparently. As a consequence, when the resource manager finds a suitable machine for the job, it will actually

start an instance of the TDP-Shell Agent. The TDP-SA command file (see example of TDP-SA command file on the right side of Figure 3), the user job files and the run-time tool files will also be copied in the same machine. The TDP-SA starts the execution of the command file and joins the attribute space when the **tdp_init** command is called (see Figure 5).

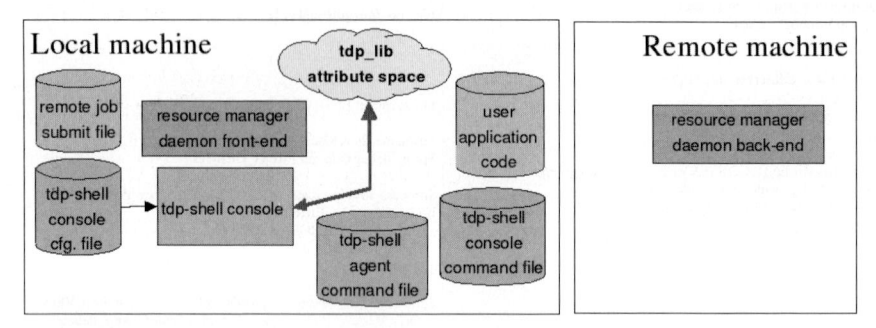

Fig. 4. TDP-Shell operation: job submission

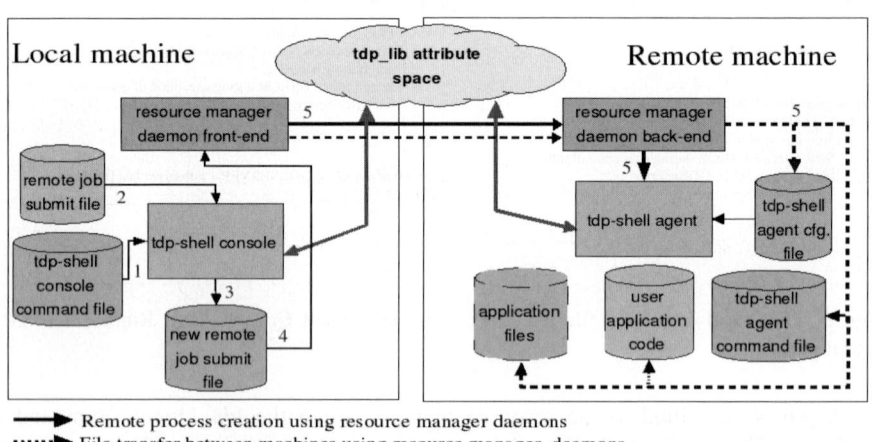

Fig. 5. TDP-Shell operation: TDP-SA start up

4. The TDP-SC and the TDP-SA execute their commands concurrently and they synchronize or exchange information using the attribute space provided by the TDP library. Specifically, the TDP-SC will start the run-time tool front-end daemon and the TDP-SA will create the user job and the run-time tool-back end daemon (see Figure 6).

The order for creating these elements depends strongly on the run-time tool's characteristics. Some tools require the front-end daemon to start before the back-end daemon (i.e. Paradyn [12]), while others require the opposite order (i.e. Totalview [10] or gdb [13]). In order to accommodate the requirements of any particular run-time tool, users can control the order of creation by using the appropriate synchronization between the TDP-SC and the TDP-SA. In our example in Figure 3, the TDP-SA first creates the gdbserver (back-end daemon) and puts two attributes in the attribute space (PORT and REMOTE_HOST), which are needed in the TDP-SC to start the gdb front-end. The TDP-SC is blocked until these two attributes are put in the attribute space and then it starts the gdb front-end. In general, most run-time tools publish some information that is required to establish the connection between its back-end and front-end daemons properly. Unfortunately, there is no common and easy way to obtain this information. For instance, the Paradyn front-end publishes its two connection ports in an external file, the gdb front-end must know the host name of the machine where the gdbserver has been starter, and so on. An external user function provided by the user can be used to obtain this information. This external function (shell script) must publish the information in the *stdout*; it is declared with the *tdp_include* statement and invoked within the TDP-Shell command file. Finally, the application can be created either in a paused or non-paused way. The choice depends on the ability of the tool to attach to the application later and continue it.

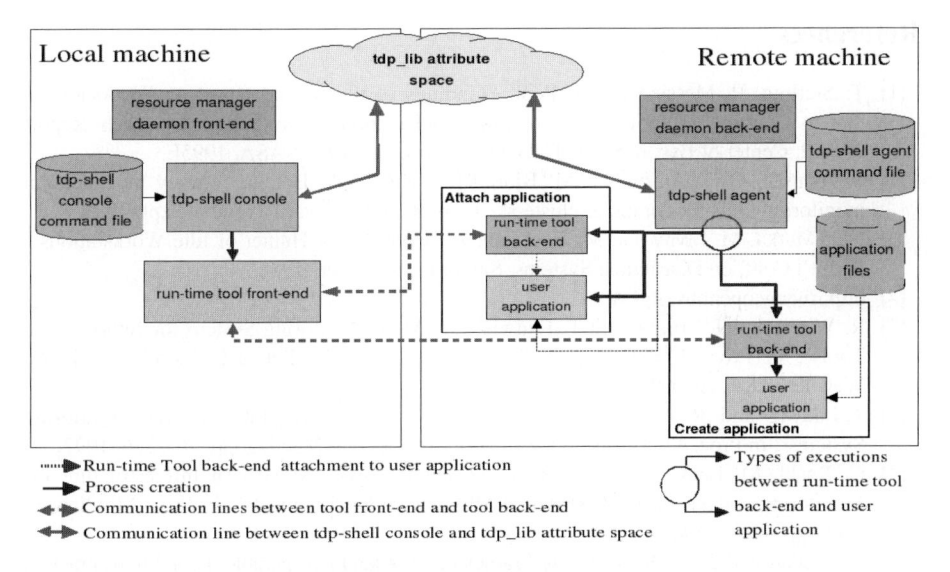

Fig. 6. TDP-Shell operation: run-time tool and application start-up

Once the application and the run-time tools have been successfully created, they continue their execution in a normal way. Users may interact with the run-time tool front-end and control the execution of the application as usual. TDP-SA detects the finalization of the application or the run-time back-end daemon. It will carry out any programmed action included in the command file, which includes notifications through the attribute space (the example in Figure 3 contains some examples of error control statements that are invoked asynchronously).

5 Conclusions and Future Work

Large-scale distributed environments imply a new scenario that requires new mechanisms that enable run-time monitoring tools to be launched under the control of a resource management system. We have developed TDP-Shell as a generic framework that is able to deal with a wide range of different run-time tools and resource managers. TDP-Shell uses a simple and easy notation mechanism, similar to the one exhibited by most OS shells, to specify the interactions between the run-time tool and the user application when executed by a given resource management system. TDP-Shell is based on two agents that have little impact on the normal execution of the application and introduce minimum overhead (mostly at the application launching time). Our current prototype includes two resource manager plug-ins (one for PBS and one for Condor) and it has been successfully used to submit sequential applications to these two batch systems and monitor them using gdb and Paradyn. Currently, a user is still limited to specifying the usage of a run-time tool at the submission time and it is not possible to start a run-time tool on-the-fly if it was not specified when the application was submitted. Our future aims are to overcome this limitation and also to support parallel applications (based on MPI).

References

[1] T. Sterling. P. Messina and J. Pool. "Findings of the second Pasadena Workshop on system software and tools for high performance computing environments", Tech. Report 95-162, Center of Exc. in Space Data and Inform. Sciences, NASA, 1995E.
[2] S. Johnsrun, O. J. Anshus, J. M. Bjørndalen, and L. A. Bongo, "Survey of execution monitoring tools for computer clusters", Tech. Report, Univ. of Tromso, Sept. 2003.
[3] M.J. Mutka, M. Livny and M. W. Litzkow, "Condor – A Hunter of Idle Workstations", 8th Int'l Conf. on Distributed Systems, San Francisco, June, 1988.
[4] http://www.openpbs.org/
[5] R. Wismuller, J. Trinitis and T. Ludwig, "OCM-A Monitoring System for Interiperable Tools", in Proc. 2nd SIGMETRICS Symposium on Parallel and Distrubuted Tools, Welches, USA, August 1998.
[6] T. Ludwig and R. Wismüller, "OMIS 2.0 -- A Universal Interface for Monitoring Systems", in Proc. 4th European PVM/MPI Users' Group Meeting, pp. 267-276, 1997.
[7] G. Rackl, M. Lindermeier, M. Rudorfer and B. Süss, "MIMO-An Infraestructure for Monitoring and Managing Distributed Middleware Environments", in Proc. Middleware 2000, pp. 71-87, 2000.
[8] R. Prodan and J. M. Kewley, "A Framework for an Interoperable Tool Environment", Proc. of EuroPar 2000, Lecture Notes in Computer Science, vol. 1900, pp. 65-69, 2000.
[9] B. Miller, A. Cortes, M. A. Senar, and M. Livny, "The Tool Daemon Protocol (TDP)", Proc. SuperComputing, November, 2003.
[10] Etnus LLC, "TotalView User's Guide", Document version 6.0.0-1, January 2003. http://www.etnus.com
[11] N. Carriero and D. Gelernter, "Linda in Context", Comm. of the ACM, 32, 4, pp. 444-458, 1989.
[12] B.P. Miller, et alter, "The Paradyn Parallel Performance Measurement Tools", IEEE Computer 28, 11, 1995.
[13] http://www.gnu.org/software/gdb/

Supporting Cache Locality Optimization with a Toolset

Jie Tao and Wolfgang Karl

Institut für Technische Informatik
Universität Karlsruhe (TH), 76128 Karlsruhe, Germany
{tao, karl}@ira.uka.de

Abstract. Cache performance significantly influences the computation power of modern processors. With the trend of microprocessor design for both general use and embedded systems towards chip-multiple, cache performance becomes more important because an off-chip access is rather expensive in comparison with on-chip references. This means cache locality optimization remains a hot research area for the next generation of computer architectures.

In this paper we present a tool environment aiming at providing the programmers sufficient support in the task of optimizing source codes for better runtime cache behavior. This environment contains a set of tools ranging from profiling, analysis, and simulation tools for gathering performance data, to visualization tools for graphical presentation and platforms for program development. Together, these tools establish a feedback loop for tuning cache performance on current and emerging uniprocessor and multiprocessor systems.

1 Introduction

Processor speed is growing at an exponential rate. In contrast, the increase of memory speed is rather lower. This results in an ever widening gap and the continuously growing memory wall. Uniprocessor chips usually rely on large cache size to overcome the problem, however, this solution is often not efficient due to the complexity in both application access pattern and the memory system. In addition, such solutions can not be commonly applied to chip-multiprocessor machines, while the increase in number of on-chip processors usually results in the decrease of per-processor caches.

For a deeper insight into the influence of cache memories we examine a simple, easy-to-understand example. Suppose a program generate 180 millions memory accesses during execution. If each memory access takes 50ns, 9 seconds are needed to access the memory. Would 20% of the data be found in cache, which is 20-fold faster than the main memory, the time for memory operation takes 7.29 seconds, forming an improvement of 19%. If the cache access behavior could be optimized and 80% of the data was acquired from the cache, the time for data accesses would only account for 4.95 seconds, which is a 56% improvement to the no-cache version.

This example depicts that cache performance can significantly affect the overall execution behavior of an application. It also demonstrates the necessity of code optimization with respect to cache performance. A prerequisite for this kind of optimization is the cache access behavior and the access pattern of applications. For this knowledge, programmers must rely on the help of tools, since static analysis of the source code usually show inefficiency and tedious work.

In this case, a visualization tool is needed to present such access pattern and the runtime behavior in user understandable manner. This tool, in turn, requires performance data, potentially not at high-level but in detail and covering various aspects.

Actually, modern processors offer a set of performance counters to collect specific events including those about caches. This helps to locate performance bottlenecks, however, does not suffice for a comprehensive understanding of the access pattern and behavior. For example, it does not present the reason of misses; it also say nothing about the affinity among data accesses. Therefore, due to this limit, most existing visualization tools are only capable of providing an environment for programmers to analyze the problem of the code and to detect the regions or data structures responsible for the poor performance. It is difficult or even impossible for users to detect an appropriate optimization scheme with their help.

In this case we implemented a set of tools with the goal of efficiently helping the users in locality optimization by presenting both the problem and hints for solutions. This toolset contains the following tools:

– A visualization tool that displays the performance data in user-understandable way. In contrast to existing tools, this cache visualizer not only shows the updates of cache contents but also, and with more endeavor, depicts the reason of cache misses enabling the detection of optimization strategies.
– A profiling tool that utilizes performance counters to collect information about individual events like cache hit/miss, total memory access, and access latency.
– A pattern analyzer that detects repeated address sequence and access stride. The former helps to allocate data with spatial locality in the same cache block, while the latter can be used to guide software prefetching.
– A cache simulator that models the runtime cache activities and analyzes the feature of cache misses.
– A program development environment that establishes a platform for application design.

All these tools work on both sequential and parallel applications, hence can be used to optimize the data locality on both uniprocessor and multiprocessor architectures.

In the following, Section 2 depicts the whole infrastructure of the toolset. This is followed by a detailed description of individual tools, first the visualization tool in Section 3, then the tools for data acquisition in Section 4, and lastly the development environment in Section 5. In Section 3 we also demonstrate how to apply the visualized information to optimize applications towards better runtime cache performance. The paper concludes in Section 6 with some current and future directions.

2 Infrastructure of the Toolset

The infrastructure of the toolset is depicted in Figure 1. As shown on the most right side of the figure, the program development environment is the core of this infrastructure. From there execution commands can be issued and performance profiling, pattern analyzing, cache simulation, and visualization can be started. The profiling has to be done during the execution of an application because it relies on the performance counters to acquire performance data. A memory reference trace is also generated at the runtime. This trace records all memory accesses performed during the program run. The trace is used as input of both the pattern analyzer and the cache simulator which in turn produce access patterns and cache miss information individually. Performance data from the profiler and the cache simulator is displayed by the visualization tool, where the profiler contributes statistical numbers, while the simulator provides cache miss analysis and other detailed information about the cache behavior like the runtime cache operations. The visualization of the output from the Analyzer, however, is integrated in the development environment due to the combination with source codes.

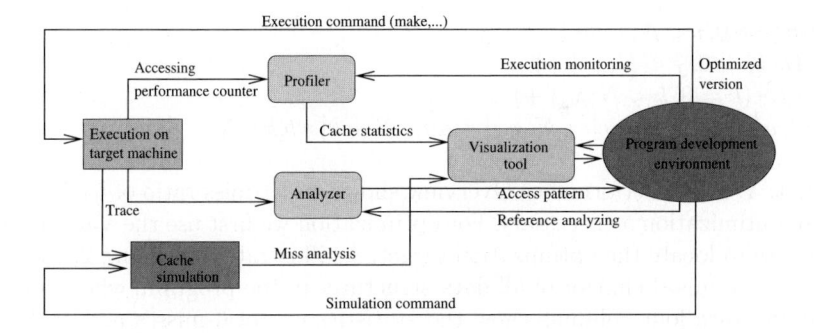

Fig. 1. Infrastructure of the toolset

Figure 1 also demonstrates the feedback loop for tuning cache performance. This loop contains four steps: running the program, gathering performance data, cache visualization, and optimization in source code. After the optimization, users can execute the optimized version of the program and examine the optimization result using the development environment. If cache problems still exist, another feedback loop may be started and further optimizations can be conducted.

3 Cache Visualization

Core of the tool infrastructure is a visualization tool, called YACO [6], which shows both bottlenecks and the reason for them. Following the common optimization process, YACO uses a top-down approach to direct the user step-by-step to detect the problem and the solution. First, users acquire an overview about the cache access behavior shown by the chosen program. Based on this overview, users can determine whether an optimization is essential. In the next step, the access hotspots can be located. After that, the reasons and interrelations between memory references can be detected using the presented information about runtime cache operations. This information also enables the user to select appropriate optimization scheme and related parameters.

Specific feature of YACO, in comparison with existing cache visualization tools like CACHEVIZ [11], lies in its ability of presenting the miss reason. This information can guide the user to deploy effective optimization schemes. Actually, there exist a number of strategies for optimizing the cache locality. However, individual scheme is usually effective to certain cache miss. For example, array padding can tackle conflict miss, but not capacity miss. Hence, in order to adequately apply theses techniques users have to know the reason of cache misses. For this, YACO depicts the runtime activities of both data structures and the caches enabling the detection of the cache miss reason.

Now we use a matrix multiplication code to demonstrate how YACO helps the programmers to perform data locality optimization. The code contains mainly the following loop for computation:

$$for(i = 0; i < N; i {++})$$
$$for(j = 0; j < N; j {++})$$
$$for(k = 0; k < N; k {++})$$
$$a[i * N + j] = c[i * N + j] + a[i * N + k] * b[k * N + j];$$

First, YACO's Performance Overview shows an L1 miss ratio of 54%, rendering an optimization as necessary. For optimization we first use the Variable Miss Overview to locate the optimization target. As illustrated in Figure 2, this view presents the miss behavior of all data structures in the program, where for each data structure four columns show the statistics on total misses and each miss category. Absolute numbers of misses are depicted at the left bottom.

It can be seen that all three matrices introduce cache misses, especially matrix b. It can be also seen that most misses with a and b are capacity miss, while with c mapping conflict is the main miss reason.

For a and b this view indicates an insufficient number of cache lines for holding all the required data within an iteration. Examining the code region above, it can be observed that each k loop calculates a single element of matrix c and for this it needs a whole row of a and a whole column of b. More importantly, the row of a is reused for computing the next element. The capacity miss with a means that these elements have to be evicted from the cache before being reused. This problem can be tackled with loop blocking, an efficient approach

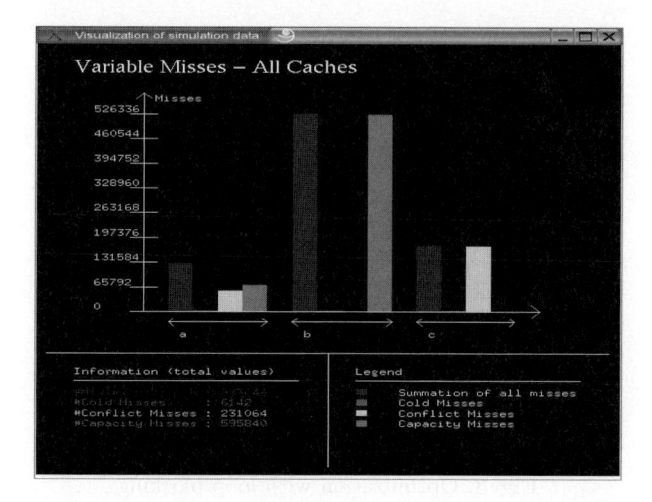

Fig. 2. Variable Miss Overview

for reducing capacity miss [8]. With blocking we generate the following code:

$for(block = 0; block < N; block += N/2)$
$\quad for(i = 0; i < N; i ++)$
$\quad \quad for(j = 0; j < N; j ++)$
$\quad \quad \quad for(k = block; k < block+N/2; k ++)$
$\quad \quad \quad \quad c(i * N + j) = c[i * N + j] + a[i * N + k] * b[k * N + j];$

The difference with the original code lies in that the innermost loop (loop k) does not perform the whole work for generating an element of matrix c; rather it does only a half of the whole work. The additional loop with *block* guarantees the whole work to be covered.

This optimization introduces a significant performance gain, which can be observed in Figure 3, which illustrates two Performance Overview of YACO showing the overall cache performance of both unoptimized version and the version with blocking. Observing the second column of both diagrams it can be noted that the number of cache hits is significantly increased with the optimization. As a result, the cache hit ratio raises from 46% to 62%.

For matrix c, however, a further study on conflict is needed in order to decide how to optimize the code. For this, we examine the Cache Set view of YACO, which shows the content update in the cache.

As depicted in Figure 4, this view contains several horizontal blocks, in this case 2 blocks, which corresponds to the cache lines in a cache set. These blocks demonstrate the operations and content update in the specific set. The operations are presented in chronological order with the right followed by the left. Figure 4 shows the operations of initializing the last few elements of matrix b and the begin of calculating the first element of matrix c.

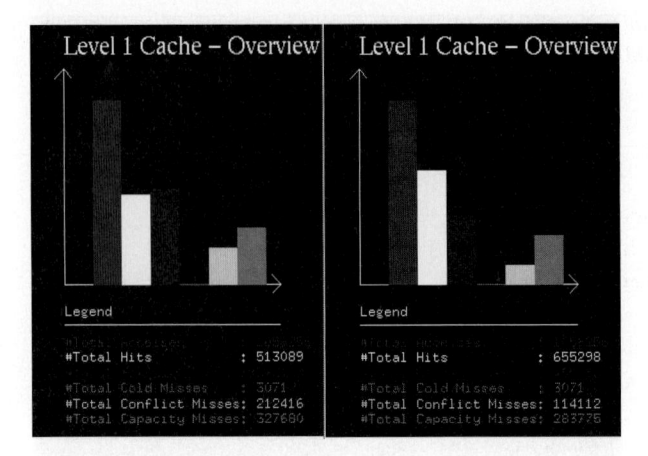

Fig. 3. Optimization with loop blocking

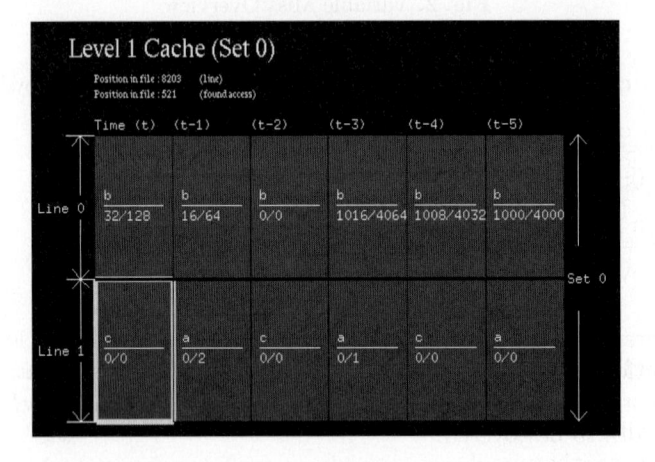

Fig. 4. The Cache Set view of set 0 in L1

For each operation, the operation type, which can be a load, a replacement, or a hit, and the operation target, i.e. a variable or an element in an array, are presented. Examining line 1 it can be seen that the same element of matrix c, here $c\ 0/0$ in the form of block/element, is reused but replaced by elements of matrix a every time after the access. Actually, this corresponds to the computing process where first a multiplication is performed and then the result is accumulated to the c element. The multiplication is performed on different elements of a and b, but the accumulation targets on the same element of c. Hence, better performance can be achieved if the elements of c can be kept in cache.

An efficient approach to tackle conflict miss is padding [5]. The idea of this approach is to additionally allocate a memory buffer to change the mapping behavior of data structures in the cache. Following this approach we add a pad of one cache line between matrix b and c like this:

$$a = (double*)malloc(sizeof(double) * N * N)$$
$$b = (double*)malloc(sizeof(double) * N * N)$$
$$d = (double*)malloc(sizeof(double) * 4)$$
$$c = (double*)malloc(sizeof(double) * N * N)$$

From the Cache Set views for this code version we see that with the padding the mapping of matrix c has been changed to set 1 rather than set 0. This introduces cache hits not only with c but also with a. As a result, this optimization shows a 40% of cache miss reduction which results in a raise of 23% in cache hit ratio.

All these performance gains have to be a contribution of the visualization tool that allows the user to detect the miss reason and further the optimization strategy.

4 Tools for Gathering Cache Performance Data

A prerequisite for any visualization is cache performance data. We currently rely on three tools to acquire this data with one extending another. Fist we deploy performance counters to collect information of global cache events. This information can be used to detect access bottlenecks. We then use a pattern analyzer to acquire access patterns, such as spatial relationship among references and access strides. This information shows the user how to accurately use certain optimization schemes, such as grouping and prefetching, to reduce cache misses. Finally, we deploy a cache simulator to deliver information about miss feature and runtime cache activities. This information enables the application of more efficient optimization schemes like code transformation and array padding.

Data Structure Profiling with Performance Counters. Performance counters are specially provided by modern architectures to gather performance information with limited overhead. This feature has to be utilized to first detect access hotspots where a large number of cache misses occur. For user-level optimization, however, the hotspots have to be related to code regions and data structures. This requires the counters to deliver access addresses together with the miss report. In this case, we developed a novel data profiler on Itanium 2 [4]. However, this profiler can be ported to other architectures with the requested counter feature.

Based on the *pfmon* kernel interface [3], we are able to control and access the performance counters, and generate a trace file for captured cache miss events. For each event, this file stores the address of the instruction issuing this access, the data address, and the latency for acquiring the missing data. For mapping the addresses to variables in the source code, we rely on the debugging information in the binary for static data structures and a self-made library for dynamic data structures. This library is used to instrument the *malloc* calls and collect the mapping information. Based on this mapping information and also the source code, miss events in the trace file are ordered to individual data structures and as a result miss statistics on them are generated. These statistics can cover the whole problem, a single function, or even a code line, allowing hence to exactly locate the concrete access hotspots.

Pattern Analysis. The second tool for collecting performance data is a pattern analyzer [10] that provides address groups, access strides, and push back distances. The base of this tool is a memory reference trace, which is acquired with a code instrumentor. The instrumentor takes an assembly code as input, marks all memory references, inserts calls for acquiring thread IDs, and generates an extended version of the code. The execution of this version results in a creation of the reference trace.

Based on the trace, the analyzer applies appropriate algorithms to detect the regularity among the references and to give hints about optimization possibilities. For detecting address groups it applies Teiresias [7], an algorithm often used for pattern recognition in Bioinformatics, to find accesses which repeatedly occur together but target different memory locations. This helps to pack the target addresses of these accesses in the same data block so that the needed data can be loaded into the cache with the data for the first access, guaranteeing that the rest accesses are all cache hit. This kind of optimization is called grouping, another useful technique to enhance the cache performance. However, for deploying this scheme the knowledge about reference affinity is necessary. In comparison with common-used compiler-based approach [9], our analyzer is simple and based on runtime references.

For detecting access strides the analyzer uses an algorithm similar to that described in [2] to find the access regularity within a data array. The results are a set of records that describe each detected stride with three parameters: start address, access distance, and number of repeating. This information can be used for efficient prefetching, because it shows the address of the next required data.

Additionally, this analysis tool determines for each memory access whether it is a cache hit or a cache miss by computing the reuse distance and set reuse distance. For a cache miss, it also calculates the push back distance which shows how many steps a miss access must be shifted in order to achieve a hit.

The Cache Simulator. Simulation has been generally used to evaluate architecture design, to understand the behavior of applications and target machines, to deliver state information, and to optimize system behavior including the cache performance [1]. Similarly, we developed a cache simulator.

The cache simulator models the runtime cache operations and gathers information about cache activities and information exhibiting cache miss reason. It also uses the memory reference trace as input. For each record in the trace, it simulates the search process in a real multiprocessor cache hierarchy and stores the result in an operation sequence file.

For a cache miss, the simulator analyses the reason for it, i.e. whether it is a cold miss, a conflict miss, or a capacity miss. While cold miss can be simply identified by examining if an access is a first reference, identifying capacity and conflict miss has to rely on reuse and set reuse distance, which need specially designed algorithms to compute efficiently.

Besides, the cache simulator provides specific functionality for multiprocessor systems. For example, it models a variety of cache coherence protocols and can detect false-sharing across processors.

5 Program Development Environment

In order to simplify the user's task in code optimization, we implemented a programming environment which builds a platform for code development and also for integrating all tools.

This environment is mainly composed of three components: window for code development, window for visualization of access patterns and for comparison of cache performance by different runs, and window for information like runtime output.

The first component offers a platform for modifying, compiling, analyzing, and executing an application. It uses a menu concept to enable the issuing of execution command or the starting of those tools for data collection and visualization. The second component provides a platform for displaying the access pattern, in this case the address group and access stride, the hit/miss behavior of each memory reference, and the execution comparison between transparent and optimized version of the same program. Address groups are presented using virtual addresses. However, by clicking an individual group the corresponding variables to the addresses in the group are immediately marked in the source code. Similarly, for the view showing the hit/miss behavior users can select any reference and in the source code all occurrence of the corresponding variable is marked and in case of a cache miss the position is highlighted, where the reference has to be issued in order to achieve a cache hit for it. The access stride, however, is directly presented within the arrays that are depicted using a diagram showing the array elements holding this stride. In addition, the hit/miss behavior of the access to the elements is also presented.

6 Conclusions

In this paper we introduce a set of tools developed for supporting users in the task of cache locality optimization. This toolset contains components for acquiring cache performance data, tools for behavior visualization, and platform for program development. Based on theses facilities, users can detect cache problem and optimization strategies, and further perform the optimization directly in the source code.

A limitation of this toolset is the workload size of examined applications because both the simulator and the pattern analyzer use a reference trace as input. This trace could be rather large for realistic applications. Currently we are working on different approaches to reduce the trace size. First, we use smaller workload size to detect optimization strategies and then deploy these strategies to realistic applications. This means also a corresponding reduction of the cache size, which indicates additional burden to users. The second approach is to first use counter information to find critical functions and then only instrument these functions. However, this can still generate large trace file, while most applications show considerate accesses even within a function. Hence, our third approach is to generate a reduced trace which only stores the accesses in the first few iterations of all loops. All theses approaches have to be examined in detail with respect to accuracy and the trace size.

We also intend to investigate compiler-level optimization. This means the gathered cache information is delivered to compilers which based on this information transparently perform the optimization during the generation of executables. This will introduce better performance than approaches using heuristic analysis, because we provide the runtime access pattern for compilers to make optimization decisions.

References

1. L. DeRose, K. Ekanadham, J. K. Hollingsworth, and S. Sbaraglia. SIGMA: A Simulator Infrastructure to Guide Memory Analysis. In *Supercomputing'02: Proceedings of the 2002 ACM/IEEE conference on Supercomputing*, pages 1–13, 2002.
2. T. Mohan et. al. Identifying and Exploiting Spatial Regularity in Data Memory References. In *Supercomputing 2003*, Nov. 2003.
3. HP. Perfmon Project Web Site. available at http://www.hpl.hp.com/research/linux/perfmon/.
4. Intel Corporation. *Intel Itanium Architecture Software Developer's Manual*, volume 1–3. 2002. available at http://developer.intel.com/design/itanium/manuals/iiasdmanual.htm.
5. K. Ishizaka, M. Obata, and H. Kasahara. Cache Optimization for Coarse Grain Task Parallel Processing Using Inter-Array Padding. In *Languages and Compilers for Parallel Computing: 16th International Workshop, LCPC 2003*, volume 2958 of *Lecture Notes in Computer Science*, pages 64–76, 2003.
6. B. Quaing, J. Tao, and W. Karl. YACO: A User Conducted Visualization Tool for Supporting Cache Optimization. In *High Performance Computing and Communcations: First International Conference, HPCC 2005. Proceedings*, volume 3726 of Lecture Notes in Computer Science, pages 694–703, Sorrento, Italy, September 2005.
7. I. Rigoutsos and A. Floratos. Combinatorial Pattern Discovery in Biological Sequences: the TEIRESIAS Algorithm. *Bioinformatics*, 14(1):55–67, January 1998.
8. G. Rivera and C. Tseng. Tiling Optimizations for 3D Scientific Computations. In *Proceedings of Supercomputing'2000*, 2000.
9. X. Shen, Y. Gao, C. Ding, and R. Archambault. Lightweight Reference Affinity Analysis. In *ICS '05: Proceedings of the 19th annual international conference on Supercomputing*, pages 131–140, New York, NY, USA, 2005.
10. J. Tao, S. Schloissnig, and W. Karl. Analysis of the Spatial and Temporal Locality in Data Accesses. In *Proceedings of ICCS 2006*, number 3992 in Lecture Notes in Computer Science, pages 502–509, May 2006.
11. Y. Yu, K. Beyls, and E.H. D'Hollander. Visualizing the Impact of the Cache on Program Execution. In *Proceedings of the 5th International Conference on Information Visualization (IV'01)*, pages 336–341, July 2001.

Model-Based Performance Diagnosis of Master-Worker Parallel Computations

Li Li and Allen D. Malony

Performance Research Laboratory
Department of Computer and Information Science
University of Oregon, Eugene, OR, USA
{lili, malony}@cs.uoregon.edu

Abstract. Parallel performance tuning naturally involves a diagnosis process to locate and explain sources of program inefficiency. Proposed is an approach that exploits parallel computation patterns (models) for diagnosis discovery. Knowledge of performance problems and inference rules for hypothesis search are engineered from model semantics and analysis expertise. In this manner, the performance diagnosis process can be automated as well as adapted for parallel model variations. We demonstrate the implementation of model-based performance diagnosis on the classic Master-Worker pattern. Our results suggest that pattern-based performance knowledge can provide effective guidance for locating and explaining performance bugs at a high level of program abstraction.

Keywords: Performance diagnosis, parallel models, master-worker, measurement, analysis.

1 Introduction

Performance tuning (a.k.a. *performance debugging*) is a process that attempts to find and to repair performance problems (performance bugs). For parallel programs, performance problems may be the result of poor algorithmic choices, incorrect mapping of the computation to the parallel architecture, or a myriad of other parallelism behavior and resource usage problems that make a program slow or inefficient. Expert parallel programmers often approach performance tuning in a systematic, empirical manner by running experiments on a parallel computer, generating and analyzing performance data for different parameter combinations, and then testing performance hypotheses to decide on problems and prioritize opportunities for improvement. Implicit in this process is the expert's knowledge of the program's code structure, its parallelism approach, and the relationship of application parameters with performance factors. We can view performance tuning as involving two steps: detecting and explaining performance problems (a process we call *performance diagnosis*), and performance problem repair (commonly referred to as *performance optimization*). This paper focuses on *parallel performance diagnosis* and how it can be supported as an automated knowledge-based process in performance analysis tools.

Performance diagnosis, as a process, is best based on understanding of how expert parallel programmers debug performance problems. That is, we should regard performance diagnosis as an intelligent system wherein we capture *knowledge* about performance problems and how to detect them, and then apply this knowledge in a diagnosis framework. The key idea is to extract performance knowledge from parallel computational models that represent structural and communication pattern of a program. The models provide semantically rich descriptions that enable better interpretation and understanding of performance. The goal is to engineer the performance knowledge to support bottom-up inference of performance causes effectively. A diagnosis system can then use the performance knowledge for performance problem search and reasoning. The problem we focus on in this paper is the knowledge engineering required for model-based performance diagnosis. We will show in a particular scenario, the classic *Master-Worker* parallel model, that the performance knowledge derived from parallel models provides a sound basis for automating performance diagnosis processes and can explain performance loss from high-level computation semantics.

In Section §2, we more formally discuss performance diagnosis as a general intelligent process and provide background on why we advocate a model-based diagnosis approach. From this perspective, Section §3 describes our approach to engineering performance knowledge and problem inference. A prototype diagnosis system, *Hercule*, was developed based on this approach and is presented. The Master-Worker pattern is illustrated in section §4 to demonstrate how performance knowledge is engineered in Hercule and to show automatic performance diagnosis in action. Section §5 highlights related research and Section §6 concludes with observations and future work.

2   Model-Based Performance Diagnosis

Performance diagnosis is the process of locating and explaining sources of performance loss in a parallel execution. Expert parallel programmers often improve program performance by iteratively running their programs on a parallel computer, then interpreting the experiment results and performance measurements to suggest changes to the program. Specifically, the process involves:

Designing and running performance experiments. Parallel computing researchers have developed integrated measurement systems to facilitate performance analysis [18,15,2]. The performance experiments specify input data, number of processors, and other parameters. The experiments also decide on points of instrumentation and what information to capture. Performance data are then collected from experiment runs.

Finding symptoms. We define a *symptom* as an observation that deviates from performance expectation. General metrics for evaluating performance includes execution time, parallelization overhead, speedup, efficiency, and cost. By comparing the metric values computed from performance data with what is expected, we can find symptoms such as low scalability, poor efficiency, and high overhead.

Inferring causes from symptoms. *Causes* are explanations of observed symptoms. Expert programmers interpret performance symptoms at different levels of abstraction. They may explain symptoms by looking at more specific performance properties [9], such as load balance and communication cost, or tracking down specific source code fragments that are responsible for major performance loss. Performance analysis expertise and knowledge about code structure and parallelization design can help form performance hypotheses, capture supporting performance information, synthesize raw performance data to test the hypotheses, and iteratively refine hypotheses toward higher-level abstractions until some cause is found.

A parallel computational model [13,14], also called a parallel pattern [16] or programming paradigm [6] in the literature, is a recurring parallel solution to a class of problems. Typical models include master-worker, pipeline, divide-and-conquer, and domain decomposition [13]. Models usually describes computational components and their behaviors (semantics) and how multiple threads of execution interact and collaborate in a parallel solution (parallelism). Parallel programming models abstract parallelism common in realistic parallel applications and serve as a computational basis for parallel program development. It is possible to extract from them a performance knowledge foundation based on which we are able to derive performance diagnosis processes tailored to realistic program implementations. Specifically, we envision that computational models can play an active role in the following aspects of performance diagnosis:

Selective instrumentation. Performance diagnosis naturally involves mapping low-level performance details to higher-level program designs, which raises the problems of what low-level information to collect and how to specify experiment to generate the information. Parallel models identify major computational components in a program, and can therefore guide the code instrumentation and organize performance data produced.

Detection and interpretation of performance bugs. In a parallel program, a significant portion of performance inefficiencies is due to process interactions arising from data/control dependency. Parallel models capture information about computational structures and process coordination patterns generic to a broad range of parallel applications. This information provides a context for describing performance properties and associated behaviors.

Expert analysis of performance problems. There is a collection of commonly-used parallel models for constructing parallel applications. Expert knowledge about the model performance includes typical performance properties and corresponding factors at the level of program/algorithm design. If we can represent and manage the performance knowledge in a proper manner, they will effectively drive diagnosis process with little or no user intervention.

3 Performance Diagnosis Engineering and Hercule

To build a performance diagnosis system, we need to generate performance knowledge from computational models and represent it in a knowledge base for

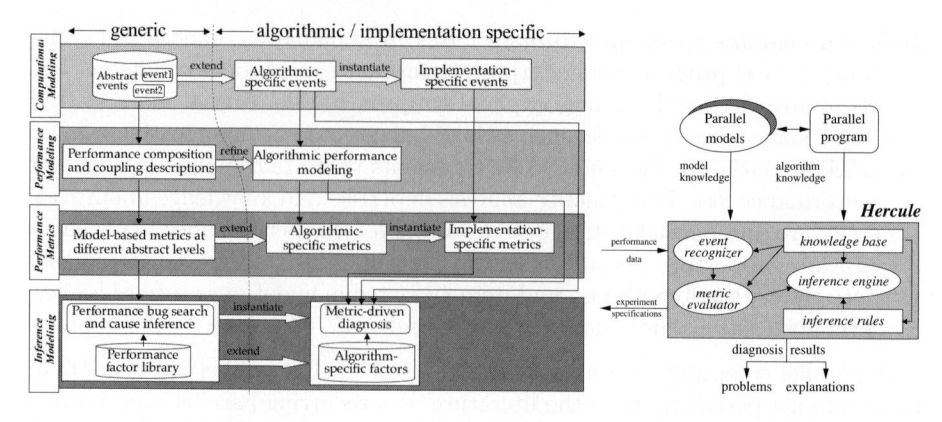

Fig. 1. Generating performance knowledge from models. The dashed line draws a boundary between model-based and algorithm/implementation specific knowledge generation.

Fig. 2. Hercule Framework

use in experimentation and problem discovery. Extracting performance knowledge from parallel computational models involves four types of actions, which are shown in Figure 1. The *computational modeling* captures knowledge of program execution semantics as behavioral patterns represented by a set of abstract events at varying detail levels, depending on the complexity of the model and diagnosis needs. The purpose of the abstract events in the diagnosis system is to give contextual informaton for performance modeling, metric analysis, and diagnosis inferencing.

Performance modeling is carried out based on structural information in the abstract events. The modeling identifies performance attributes with respect to the behavior patterns represented by abstract events and model-specific performance overhead categories. *Performance metrics* are then defined, in terms of performance attributes in related abstract events, to evaluate the performance properties for problem interpretation. *Inference modeling* (i.e., performance bug search) is driven by the metric and property evaluation. Cause inference tries to explain found performance problems with performance-critical program design factors, as specific to the particular computational model. The performance problem analysis and cause refinement are captured in the form of an inference tree linking symptoms to sources.

Algorithmic implementations of a computational model may introduce new performance knowledge with regard to behavioral models, performance properties, performance-critical design factors, or cause inference. Following our four-step knowledge extraction approach, the new knowledge can be generated by the users in the form of refinements or extensions of the generic model knowledge, as shown on the right hand part of Figure 1. In our design and implementation of a model-based performance diagnosis system, we will allow the expression of algorithmic features and incorporating it into the inference system that is initially based on generic model knowledge.

We have built a prototype automatic performance diagnosis system called *Hercule*[1], which implements the model-based performance diagnosis approach discussed above; see Figure 2. The Hercule system operates as an expert system within a parallel performance measurement and analysis toolkit, in this case, the TAU [2] performance system. Hercule includes a knowledge base composed of an abstract event library, metrics set, and performance factors for individual parallel models. Below, we describe in more detail how the performance diagnosis engineering is accomplished in Hercule.

The *abstract event description* used in EBBA [4] is adapted in Hercule to describe behavioral characteristics of a target computational model. The description of each abstract event type consists of one required component, expression, and four optional components, constituent event format, associated events, constraints, and performance attributes. An abstract event usually represents a sequence of constituent events. A constituent event can be a primitive event presenting an occurrence of a predefined action in the program (e.g., interprocess communication or regular routine invocations), or an instance of other abstract event type. The *expression* is a specification that names the constituent events and enforces their occurrence order using event operators. The order can be *sequential* ($\circ$), *choice* ($|$), *concurrent* (Δ), *repetition* (+ or *), and *occur zero or one time* ($[]$). *Constituent event format* specifies the format and/or types of the constituent events. For primitive events, the format often takes the form of an ordered tuple that consists of the event identifier, the timestamp when the event occurred, the event location, etc. For constituent abstract events, their types are specified. *Associated events* are a list of related abstract event types, such as a matching event on a collaborating process or the successive event on the same process. *Constraints* indicate what attribute values an instance of an abstract event type must possess to match its corresponding expression members and associated events. *Performance attributes* present performance properties of the behavior model an abstract event type represents and computing rules to evaluate them. Figure 3 in the next section shows an example abstract event for the Master-Worker (M-W) computational model.

Hercule implements the abstract event representation in a Java class library. The *event recognizer* in Hercule fits event instances into abstract event descriptions as performance data stream flows through it. It then feeds the event instances into Hercule's performance model evaluator - *metric evaluator*. Performance models in Hercule are coded as Java classes used to represent model-specific metrics and associated performance formulations. The performance metrics will be evaluated based on the related abstract event instances. The event recognizer and metric evaluator can incorporate algorithm-specific abstract event definitions and metric computing rules.

Perhaps the most interesting part of the Hercule knowledge engineering is the cause inferencing system. The expert knowledge used to reason about performance problems based on model symptoms can be structured as *inference trees*

---

[1] The name was chosen in the spirit of our earlier performance diagnosis project, *Poirot* [12].

where the root is the symptom to be diagnosed, the branch nodes are intermediate observations obtained so far, and the leaf nodes are an explanation of the root symptom in terms of high-level performance factors. We encode inference trees with production rules. A production rule consists of one or more performance assertions and performance evidences that must be satisfied to prove the assertions. Hercule makes use of syntax defined in the CLIPS [1] expert system building tool to describe production rules, and the CLIPS inference engine for operation. The inference engine provided in CLIPS is particularly helpful in performance diagnosis because it can repeatedly fire rules with original and derived performance information until no more new facts can be produced, thereby realizing automatic performance experiment generation and cause reasoning. Due to the limitation of space, we refer readers to elsewhere [11] for details of encoding performance knowledge with CLIPS.

The effort involved in implementing performance knowledge base for a computational model consists of two parts: acquiring knowledge with the approach presented above and encoding the knowledge with abstract event specification, performance formulation, and production rules. Work time needed for a performance analyst to generate knowledge varies depending on computational complexity of the model and desired detail level of the targeting inference tree. When using the knowledge base to diagnose a parallel application based on a parallel model, the developer may need to express the programatic or algorithm variations with respect to abstract event descriptions, metric computing specifications, and corresponding inference tree. Because the generic knowledge base is inherited, additional efforts are reduced to adding knowledge specialization.

4 Master-Worker Parallel Pattern and Diagnosis

A widely used parallel computation pattern is the classic Master-Worker (M-W) model. Here we use the M-W model to demonstrate the performance diagnosis methodology above and show how it is implemented in the Hercule framework. Master-Worker models a computation that is decomposed into a number of independent tasks of variable length. A *master* is responsible for assigning the tasks to a group of *workers*. Communications are required between the master and workers before and after processing each task. The workers are independent to one another. The master usually employs certain task scheduling algorithms to achieve load balance and minimize workspan. M-W performance factors we identified, through performance observation of M-W codes and knowledge obtained from expert performance analysts, are:

- *Inherent sequential code fragments in the master.*
- *Number and complexity of tasks assigned to the workers.*
- *Task setup costs in the master and the task scheduling method.*
- *Number of worker processors.*
- *Task scheduling strategy*

In a M-W program, an independent task assigned to a worker process has a well-defined life cycle: first the worker sends a task request to the master, the

```
AbstractEvent TaskLifeCycle (id, pad) {
    Expression
        (WorkerSendReq ∆ MasterRecv) ο ((MasterSetupTask ο MasterSendTask) ∆ WorkerRecv) ο WorkerCompute
    Constituent Event Format
        (EventComponent <name><pid><entering_time><execution_time>[source][dest])
    Associated Events
        TaskLifeCycle preTask, nextTask
    Constraints
        WorkerSendReq.name == "MPI_Send";
        MasterRecv.name == "MPI_Recv";
        MasterRecv.source == WorkerSendReq.pid;
        MasterSetupTask.name == "setup";
        MasterSendTask.name == "MPI_Send";
        WorkerRecv.name == "MPI_Recv";
        WorkerRecv.pid == WorkerSendReq.pid;
        ...
    Performance Attributes
        IsWorkerSendLate := { true   if WorkerSendReq.entering_time > MasterRecv.entering_time;
                              false  otherwise
        IsMasterSendLate := { true   if MasterSendTask.entering_time > WorkerRecv.entering_time;
                              false  otherwise
        WorkerWaitingTimeForTheTask :=
            { MasterSetupTask.execution_time, if IsWorkerSendLate;
              MasterRecv.entering_time - WorkerSendReq.entering_time + MasterSetupTask.execution_time, otherwise
        ...
}
```

Event expression — (Expression line)
Constituent event descriptor — (Constituent Event Format line)
Related abstract event list — (Associated Events line)
Constraint clauses — (Constraints block)
Performance attribute descriptor — (Performance Attributes block)

∆ concurrent
ο sequence

define conditions for abstract events

use later for performance metric evaluation and inference

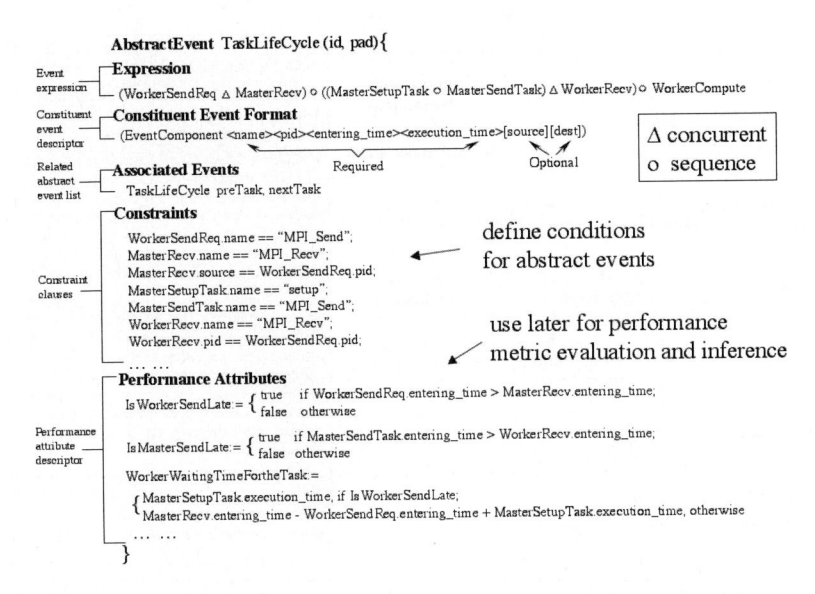

Fig. 3. An abstract event description of Master-Worker model

master receives the request and sets up a task, it then transfers the data and task specification to the requesting worker, and the worker processes the task until finished. At that time, that worker returns the result to the master and the cycle continues until the worker is instructed to terminate. We specify the program behaviors and performance properties associated with a task life cycle by an abstract event type **TaskLifeCycle**, as shown in Figure 3.

Given the program behavior, we can formulate M-W performance models. For instance, a worker's total elapsed time t_{worker} consists of t_{init} (initialization cost), t_{comp} (the amount of time spent computing tasks), t_{comm} (the amount of time spent communicating with the master), t_{wait} (the amount of time spent waiting for task assignment or synchronizing with other workers before finalization, excluding communication overhead), and t_{final} (finalization cost):

$$t_{worker} = t_{init} + t_{comp} + t_{comm} + t_{wait} + t_{final} \qquad (1)$$

Whenever we refer to communication time, we mean effective message passing time that excludes time loss due to communication inefficiencies such as late sender or late receiver in MPI applications. Rather, waiting time accounts for the communication inefficiencies with the purpose of making explicit performance losses attributed to mistimed processor concurrency.

Performance coupling of a worker with the master and the rest of peer workers manifests four performance overheads – t_{seq} (the master initialization and finalization costs translated to idle overhead in the worker), $t_{w-setup}$ (master task setup time), t_{w-bn} (blocking time in master bottlenecks), and $t_{w-final}$ (the cost of synchronization with other workers for finalization).

$$t_{wait} = t_{seq} + t_{w-setup} + t_{w-bn} + t_{w-final} \qquad (2)$$

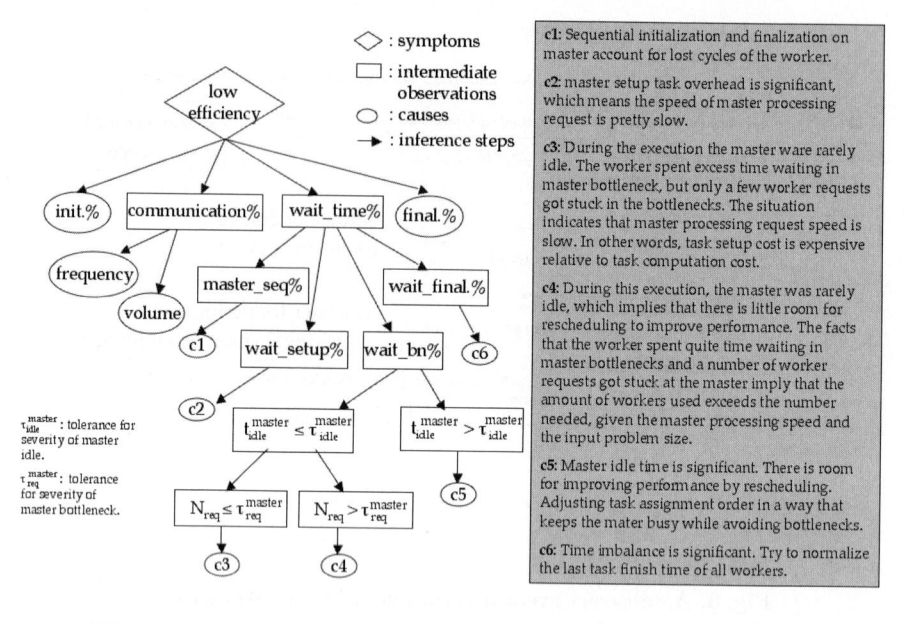

Fig. 4. Inference Tree for Performance Diagnosis of M-W programs

The above performance models enable us to define performance metrics specifically tailored to M-W programs. We start with evaluating individual *worker efficiencies* to detect a top-level symptom because efficiency is a reflection of total worker scalability.

$$\textbf{worker efficiency} := \frac{t^{worker}_{comp}}{t_{worker}} \qquad (3)$$

Refining each item in model (2), we obtain metrics of worker wait time:

$$t_{seq} := \max\{t^{master}_{init} - t^{worker}_{init}, 0\} + \max\{t^{master}_{final} - t^{worker}_{final}, 0\}$$

$$t_{w-setup} := \sum_{i=1}^{M} t^i_{setup}, \qquad t_{w-bn} := \sum_{i=1}^{M} t^i_{w-bn} = \sum_{i=1}^{M} (t^i_{wait} - t^i_{setup})$$

$$t_{w-final} := \max_{all\ workers}\{T_{fin}\} - T_{fin}$$

where M is the number of tasks the worker processes altogether, t^i_{setup} the amount of time for setting up task i, t^i_{w-bn} is the waiting time due to master bottleneck when requesting the ith task, t^i_{wait} is the total amount of worker idle time between sending out request and receiving task i, $\max_{allworkers}\{T_{fin}\}$ is the finish timestamp of the last task computed, and T_{fin} the last task finish timestamp of the observed worker processor.

Now we can incorporate these performance factors and metrics in diagnosis inference rules. An inference tree is created for every symptom. The inference tree for explaining low efficiency of a worker process, for instance, is shown

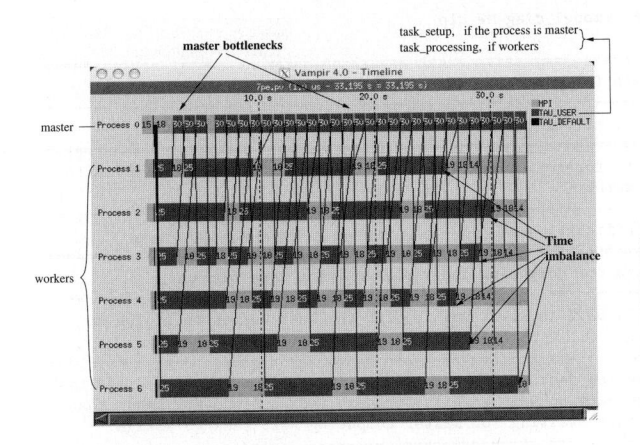

Fig. 5. Vampir timeline view of an example M-W program execution

Fig. 6. Metric values of the run

Metric name	Performance loss%
t_{w-bn}	39.2%
$t_{w-setup}$	34.3%
$t_{w-final}$	14.8%
t_{comm}	6.2%

in Figure 4. The root is the symptom to be diagnosed, the branch nodes are intermediate observations obtained so far (i.e., a performance evaluation with respect to a performance metric, such as waiting time is a *significant* percentage of total elapsed time), and need further performance evidences to explain, and the leaf nodes are an explanation of the root symptom in terms of high-level performance factors. It is interesting to note that nodes at different inference tree levels may enforce varying experiment specifications. Our diagnosis system can construct the experiments according to the abstract event descriptions from which the metrics derive.

We tested Hercule's performance diagnosis capability for the M-W pattern using a synthetic M-W application. This allowed us to introduce various known performance problems (i.e., performance faults) and evaluate whether Hercule would be able to discover them. All experiments were run on an distributed memory Pentium Xeon cluster running Linux. The M-W synthetic program was implemented using MPI.

For the results discussed below, we introduced in the M-W program a performance fault targeting the impact of master-request-processing speed on overall performance. Figure 5 presents a Vampir [3] timeline view of a parallel execution with one master and six workers. The event trace and profiles are generated by TAU [2] with only major model components being instrumented. The red regions in the figure represent task setup periods at the master and task processing periods at the workers. Light blue regions represent MPI function calls. Note that blocking (waiting) time of processors is implicitly included in the elapsed time of blocked **MPI_Send**, **MPI_Recv** and **MPI_Finalize** operations.

Given the program and performance knowledge associated with M-W model, Hercule will automatically request three experiments during the diagnosis of this problem. The inference process and diagnosis results of these experiments are presented in Figure 7. The first experiment collects data for computing efficiencies of each worker. The measurement data shows that worker 3 performs worst.

```
dyna6-166:~/PerfDiagnosis lili$ ./model_diag MW.clp
Begin diagnosing ...
================================================================================
Level 1 experiment - collect data for computing worker efficiencies.
--------------------------------------------------------------------------------
Worker 3 is least utilized, whose efficiency is 0.385.
================================================================================
Level 2 experiment - collect data for computing initialization, communication, finalization
costs, and wait (idle) time of worker 3.
--------------------------------------------------------------------------------
Waiting time of worker 3 is significant.
================================================================================
Level 3 experiment - collect data for computing individual waiting time fields.
--------------------------------------------------------------------------------
Among lost cycles of worker 3, 14.831% is spent waiting for the last worker to finish up
(time imbalance).
--------------------------------------------------------------------------------
Master processing time for assigning task to workers is significant relative to average task
processing time, which causes workers to wait a while for next task assignment. Among lost
cycles of worker 3, 34.301% is spent waiting for master computing next task to assign.
--------------------------------------------------------------------------------
Among lost cycles of worker 3, 39.227% is spent waiting for the master to process other
workers' requests in bottlenecks. This is because master processing time for assigning
task is expensive relative to average task processing time, which causes some workers to
queue up waiting for task assignment.
================================================================================
Diagnosing finished...
```

Fig. 7. Diagnosis result output from Hercule of the M-W test program

Then Hercule investigates the performance loss of worker 3 (of course, any worker can be identified for additional study), and issues the second experiment to evaluate individual overheads in equation (1). Waiting time cost stands out as a result of this inference step. The third experiment then targets performance loss categories in equation (2). Figure 6 presents model-specific metrics computed during the diagnosis in the form of percentage that each overhead category contributes to the overall performance loss (i.e., total elapsed execution time minus effective task processing time). It is important to note that diagnosis results can be encoded to present output in a manner close to programmer's reasoning and understanding of the M-W computation model.

5 Related Research

There are several related projects to our work. Paradyn [15] is a performance analysis system that automatically locates bottlenecks using the W^3 search model. According to the model, searching for a performance problem is an iterative process of refining the answer to three questions: *why* is the application performing poorly, *where* is the bottleneck, and *when* does the problem occur. Unlike Hercule, the performance bugs Paradyn targets are not in direct relation to parallel program design and not intended for explanation of high-level causes. In [10], a cause-effect analysis approach is proposed to explain inefficiencies in distributed programs. It interprets performance losses by comparing earlier execution paths of behaviorally inconsistent processes. Similarly, [17] looks for cause of communication inefficiencies in message passing programs by classification. They train decision trees with real trace data in order to classify individual communication operations and find inefficient behaviors automatically.

Several research work use parallel computational models in performance modeling and evaluation. [5] evaluates the performance of parallel programs coded in algorithmic skeletons with process algebras. POETRIES [7] is a performance tuning tool that takes advantage of the knowledge about the high-level structure of the application to detect and correct performance drawbacks. It builds analytical models based on the structures and attributes performance degradation to parameters composing the models. Hercule differs to POETRIES in that, first, it targets performance explanation and, second, it features a knowledge-based inference system that diagnoses performance in an automated manner.

The project closest to Hercule is Kappa-Pi [8]. This is an automatic performance analysis tool that encodes knowledge about commonly-seen performance problems into deduction rules at various abstraction levels. It explains the problem found by building an expression of the highest-level deduced fact which includes the situation found, the importance of such a problem, and the program elements involved in the problem. Kappa-Pi has been applied to the Master-Worker problem with excellent success. Our work builds on the Kappa-Pi objectives by proposing a systematic approach to extracting knowledge from high-level parallel design patterns.

6 Conclusions and Future Directions

This paper describes a systematic approach to generating and representing performance knowledge for the purpose of automatic performance diagnosis. The methodology makes use of operation semantics and parallelism found in parallel models as a basis for performance bug search and explanation. In order to generate performance knowledge from computational models and apply it to diagnosing realistic parallel programs, we specifically identify methods for behavioral model representation, performance modeling, metric definition, and performance bug search and interpretation. The methods address not only performance cause interpretation at high-level program abstractions, but adaptivity to allow algorithm and implementation variants.

The Hercule framework offers a prototype performance diagnosis system based on computational patterns. We demonstrated the use of Hercule on the Master-Worker pattern to validate the approach. However, there is still much work to be done for further improvement and application of model-based diagnosis. First, we are encoding additional parallel patterns, such as *Wavefront* and *Divide-and-Conquer*. As parallel applications can use a combination of parallel paradigms, an important target for our future work is the inclusion of compositional patterns that allow hierarchical reasoning about performance problems.

References

1. Clips. http://www.ghg.net/clips/CLIPS.html.
2. Tau tuning and analysis utilities. http://www.cs.uoregon.edu/research/paracomp/tau/tautools/ .
3. Vampir. http://www.pallas.com/e/products/index.htm.
4. P. Bates. Debugging heterogeneous distributed systems using event-based models of behavior. *ACM Trans. on Computer Systems*, 13(1):1–31, 1995.
5. A. Benoit, M. Cole, S. Gilmore, and J. Hillston. Evaluating the performance of skeleton-based high level parallel programs. In *M. Bubak, D. van Albada, P. Sloot, and J. Dongarra, editors, The International Conference on Computational Science (ICCS 2004), Part III, LNCS*, pages 299–306. Springer Verlag, 2004.
6. N. Carriero and D. Gelernter. How to write parallel programs: A guide to perplexed. *ACM Computing Surveys*, 21(3):323–357, 1989.
7. E. Csar, J. G. Mesa, J. Sorribes, and E. Luque. Modeling master-worker applications in poetries. In *HIPS 2004*.
8. A. Espinosa. *Automatic Performance Analysis of Parallel Programs*. PhD thesis, Computer Science Department, University Autonoma de Barcelona, Spain, 2000.
9. T. Fahringer and C. Seragiotto. Automatic search for performance problems in parallel and distributed programs by using multi-experiment analysis. In *International Conference On High Performance Computing (HiPC 2002), Bangalore, India, December 2002*. Springer Verlag.
10. W. M. Jr., T. Leblanc, and V. Almeida. Using cause-effect analysis to understand the performance of distributed programs. In *SIGMETRICS Symposium on Parallel and Distributed Tools*, 1998.
11. L. Li and A. Malony. Knowledge engineering for automatic parallel performance diagnosis. submitted to *Concurrency and Computation: Practice and Experience*.
12. A. Malony and B. Helm. A theory and architecture for automating performance diagnosis. *Future Generation Computer Systems*, 18, 2001.
13. B. Massingill, T. Mattson, and B. Sanders. Patterns for parallel application programs. In *6th Pattern Languages of Programs Workshop*, 1999.
14. B. Massingill, T. Mattson, and B. Sanders. Some algorithm structure and support patterns for parallel application programs. In *9th Pattern Languages of Programs Workshop*, 2002.
15. B. Miller and et al. The paradyn parallel performance measurement tool. *IEEE Computer*, 28(11):37–46, 1995.
16. F. Rabhi and S. Gorlatch. *Patterns and Skeletons for Parallel and Distributed Domputing*. Springer-Verlag, 2003.
17. J. Vetter. Performance analysis of distributed applications using automatic classification of communication inefficiencies. In ACM International Conference on Supercomputing 2002.
18. J. Yan. Performance tuning with AIMS — an Automated Instrumentation and Monitoring System for multicomputers. In *Proc. 27th Hawaii International Conference on System Sciences*, pages 625–633, 1994.

Specification of Inefficiency Patterns for MPI-2 One-Sided Communication

Andrej Kühnal, Marc-André Hermanns, Bernd Mohr, and Felix Wolf

Forschungszentrum Jülich,
Zentralinstitut für Angewandte Mathematik,
52425 Jülich, Germany
{a.kuehnal, m.a.hermanns, b.mohr, f.wolf}@fz-juelich.de

Abstract. Automatic performance analysis of parallel programs can be accomplished by scanning event traces of program execution for patterns representing inefficient behavior. The temporal and spatial relationships between individual runtime events recorded in the event trace allow the recognition of wait states as a result of suboptimal parallel interaction. In our earlier work [1], we have shown how patterns related to MPI point-to-point and collective communication can be easily specified using common abstractions that represent execution-state information and links between related events. In this article, we present new abstractions targeting remote memory access (also referred to as one-sided communication) as defined in the MPI-2 standard. We also describe how the general structure of these abstractions differs from our earlier work to accommodate the more complicated sequence of data-transfer and synchronization operations required for this type of communication. To demonstrate the benefits of our methodology, we specify typical performance properties related to one-sided communication.

1 Introduction

Remote memory access (RMA) describes the ability of a process to access a part of the memory of a remote process directly without explicit participation of the remote process in the data transfer. Since all parameters for the data transfer are determined by one process, it is also called *one-sided* or *single-sided* communication. One-sided communication is often made available to the programmer in the form of platform or vendor-specific libraries, such as SHMEM (Cray/SGI) or LAPI (IBM). In 1997, one-sided communication was added to the portable MPI standard version 2 [2].

On platforms with special hardware providing RMA support, one-sided communication can be used to improve the efficiency of parallel applications. For example, NASA researchers reported a 39% improvement in throughput after replacing MPI-1 non-blocking with MPI-2 one-sided communication in a global atmosphere simulation program [3]. As more and more scientists adopt this new paradigm to better utilize the underlying hardware, the demand for performance tools supporting RMA communication will increase. This is especially important in view of the complicated sequences of data transfer and synchronization operations involved and the fact that the MPI specification leaves a large degree of freedom to implementors regarding the blocking behavior of corresponding operations.

A performance-analysis technique successfully applied to traditional message-passing applications is event tracing. An event trace records performance-relevant runtime events, such as routine entries or exits or as sending and receiving point-to-point messages. The KOJAK tool [4] uses the temporal and spatial relationships between individual runtime events reflected in the event trace to recognize patterns that occur as a result of suboptimal parallel interaction. These patterns are specified as compound events to (i) allow the classification of inefficient behavior by describing the exact circumstances causing it and to (ii) enable the quantification of wait times incurred.

Compound events consist of multiple primitive events, as recorded in the trace file, and are connected by relationships, such as message transfers, that are often specific to a particular parallel programming model, such as MPI. They may be further characterized by constraints imposing, for example, a certain temporal order of events. To keep the pattern specifications as simple as possible and also to make the simultaneous search for different patterns more efficient, KOJAK includes a separate layer with common abstractions representing execution-state information and links between related events in terms of which the actual patterns are described.

In our earlier work [1], we have defined abstractions along with typical patterns describing performance properties in the context of traditional message passing (MPI-1) and shared-memory (OpenMP) programming. In [5], we provided informal descriptions of patterns suitable to diagnose inefficiencies related to one-sided communication. In this article, we outline the formal specification of these new abstractions and patterns. Compared to the previous ones related to MPI-1 and OpenMP, the abstractions presented here are more complicated to accommodate the complex sequences of data-transfer and synchronization operations involved in MPI-2 one-sided communication and to reflect the intricate inter-process relationships established by groups denoting potential origins or targets during communication epochs. The new patterns have been incorporated into the KOJAK tool, taking advantage of the recently added measurement infrastructure for one-sided communication events reported in [6]. As part of this effort, we have also specified abstractions and patterns related to SHMEM, which, however, are beyond the scope of this paper.

The outline of this article is as follows: We start with a short description of MPI-2 RMA communication and synchronization functions in Section 2. In Section 3, we give a brief overview of related work. In Section 4, we introduce the idea of creating suitable abstractions on top of which inefficiency patterns can be specified and explain their general structure. After that, we define abstractions for MPI-2 one-sided communication in Section 5. To demonstrate the usefulness of our methodology, Section 6 specifies several example MPI-2 patterns containing wait states the application developer may wish to identify. In Section 7, we conclude our paper and give an outlook on future work.

2 MPI-2 One-Sided Communication

The interface for RMA operations defined by MPI-2 differs from the vendor-specific APIs in many respects. This is to ensure that it can be efficiently implemented on a wide variety of computing platforms even if a platform does not provide any direct hardware support for RMA. The design behind the MPI-2 RMA API specification is similar to that of weakly coherent memory systems: correct ordering of memory accesses has to be specified by the user with explicit synchronization calls; for efficiency, the implementation can delay communication operations until the synchronization calls occur.

MPI does not allow RMA operations to access arbitrary memory locations. They can access only designated parts of a process' memory, which are called *windows*. Windows must be explicitly initialized (with a call to MPI_Win_create) and released (with MPI_Win_free) by all processes that either provide memory or want to access this memory. These calls are *collective* between all participating partners and include an internal barrier operation. By *origin* MPI denotes the process that performs an RMA read or write operation, and by *target* the process in which the memory is accessed.

There are three RMA communication calls in MPI: MPI_Put transfers data from the caller's memory to the target memory (*remote write*); MPI_Get transfers data from the target to the origin (*remote read*); and MPI_Accumulate[1] updates locations in the target memory, for example, by replacing them with sums or products of the local and remote data values (*remote update*). These operations are *nonblocking*: the call initiates the transfer, but the transfer may continue after the call returns. The transfer is completed, both at the origin and the target, only when a subsequent synchronization call is issued by the caller on the involved window object. Only then are the transferred values (and the associated communication buffers) available to the program. RMA communication falls in two categories: *active target* and *passive target* communication. In both modes, the parameters of the data transfer are specified only at the origin, however in active mode, both origin and target processes have to participate in the synchronization of the RMA accesses. Only in passive mode is the communication and synchronization completely one-sided.

RMA accesses to locations inside a specific window must occur only within an *access epoch* for this window. Such an access epoch starts with an RMA synchronization call, proceeds with any number of remote read, write, or update operations on this window, and finally completes with another (matching) synchronization call. Additionally, in active target communication, a target window can only be accessed within an *exposure epoch*. RMA operations issued by an origin process for a target window will access that target window during the same exposure epoch if and only if they were issued during the same access epoch. Distinct epochs for a window at the same process must be disjoint. However, epochs pertaining to different windows may overlap.

MPI provides three RMA synchronization mechanisms:

Fences: The MPI_Win_fence collective synchronization call is used for active target communication. An access epoch at an origin process or an exposure epoch at a target process are started and completed by such a call. All processes who participated in the creation of the window synchronize, which in most cases includes a barrier. The data transfered is only accessible to user code after the fence.

General Active Target Synchronization (GATS): Here, synchronization is reduced: only pairs of communicating processes synchronize, and they do so only when needed to correctly order accesses to a window with respect to local accesses to

---

[1] In our model, we consider an accumulate operation as a special version of a put operation and, therefore, distinguish only between get and put in the remainder.

that window. An access epoch is started at an origin process by `MPI_Win_start` and is terminated by a call to `MPI_Win_complete`. The start call specifies the group of targets for that epoch. An exposure epoch is started at a target process by `MPI_Win_post` and is completed by `MPI_Win_wait` or `MPI_Win_test`. Again, the post call specifies the group of origin processes for that epoch. Data written is only accessible after the wait (or test) call, however data can only be read after the complete call.

Locks: Finally, shared and exclusive locks are provided through the `MPI_Lock` and `MPI_Unlock` calls. They are used for passive target communication. In addition, they also define the access epoch for this window at the origin. Data read or written is only accessible from user code after the unlock operation has completed.

It is implementation-defined whether some of the described calls are blocking or non-blocking; for example, in contrast to other shared memory programming paradigms, the lock call must not be blocking. For a complete description of MPI-2 RMA communication see [2].

3 Related Work

Currently, there are only very few tools that support the measurement and analysis of one-sided communication and synchronization on a wide range of platforms. The well-known Paradyn tool which performs an automatic on-line bottleneck search, was recently extended to support several major features of MPI-2 [7]. For RMA analysis, it collects basic, process-local statistical data (i.e., transfer counts and execution time spent in RMA functions). It does not take inter-process relationships into account nor does it provide detailed trace data. Also, it does not support the analysis of SHMEM programs. The very portable TAU performance analysis tool environment [8] supports profiling and tracing of MPI-2 and SHMEM one-sided communication. However, it only monitors the entry and exit of the RMA functions; it does not provide RMA transfer statistics nor are the transfers recorded in tracing mode. The commercial Intel Trace Collector tool (formerly known as VampirTrace) [9] records MPI execution traces. When used with MPI-2, it does not measure the routines of the general active target synchronization, creating the wrong impression that useful user calculations are done instead. Also, message lines show the RMA transfer as completed by the end of the put or get operation, which does not reflect the user-visible behavior, as specified by the MPI-2 standard. Finally, it does not record the collective nature of MPI-2 window functions. Besides these there are also some non-portable vendor tools with similar limitations.

4 State Sequences and Pointer Attributes

Event tracing models the execution of a program as a sequence of events representing actions relevant to the purpose of the observation. Therefore, the selection of event types to be observed defines the view of program execution an event trace can provide. An *event model* defines the formal properties of that view. It comprises a set of event types with an associated set of attributes and constraints defining correct event ordering. Each event has a location attribute as well as a wall-clock time stamp. The event location is an abstraction usually referring to the process or the thread generating an event. Since the following discussion only considers pure MPI applications, the location of an event can be regarded as equivalent to the MPI process, as identified by the rank in MPI_COMM_WORLD. Another attribute denotes the event type. An *event trace* is a finite indexed set of events $E := \{e_1, \ldots, e_{n_e}\}$. The indexing reflects the time-sequenced order of event records in the trace file.

To be able to express complex relationships among the constituents of a compound event, the event model of system observation can be extended by creating instances of two different categories of abstractions: (i) state sequences and (ii) pointer attributes. The process of creating these abstractions is called *event model enhancement* because it enhances the model's capabilities to describe complex situations of execution behavior. We summarize the key concepts below. The interested reader may refer to [10] for more details.

State sequences. An event happening in a parallel system indicates a change in its state, thus, events can be regarded as state transitions. An event trace can be seen as a sequence of state transitions starting at an initial state and changing into the next state, event by event, until a final state is reached after the last event. The state entered as the result of an event is a useful abstraction when specifying compound events that represent inefficient behavior.

The overall state of a parallel system is characterized by different aspects. For example, one aspect might be the set of messages being transferred at a given moment, another aspect might be the dynamic call stack of a process or thread. Such a state aspect can be conveniently characterized in terms of the events that caused that aspect's state. For each of these aspects we can define a *state sequence* that describes the evolution of that aspect over time. A state sequence is inductively defined by a transition operator. The transition operator is applied to the current state and the next event to compute the next state in the sequence. Since a state sequence describes only one aspect of the system, we can combine all state sequences into a vector of state sequences to obtain the *overall-state sequence*.

In our earlier work, a state sequence has been defined as a sequence of event sets. Starting with the empty set, the transition operator either added the current event or removed events related to the current event, changing the event set describing an aspect of the overall system. To conveniently retrieve event sets of interest during trace analysis, we have defined auxiliary functions that can be applied to individual states of a sequence. For example, a scheme that proved to be useful to identify individual collective-operation instances was to collect all events belonging to an instance and retrieving it using an auxiliary function upon its completion if needed. Immediately after this point, the transition operator removes the instance from the set. Later, we will see that simple event sets are inconvenient to describe patterns involving intertwined steps of communication and synchronization, such as occur in one-sided communication, and that a hierarchical grouping of events becomes necessary.

Pointer attributes. Another useful abstraction is a link connecting related events, so that one can navigate from one event to another related event. An example is a link from the event of receiving a message back to the corresponding event of sending it.

This mechanism permits navigation along a path of related events and the definition of relationships among the constituents of a compound event using such paths. A natural way of representing such links is to provide event attributes with pointer semantics, which we call *pointer attributes*.

5 One-Sided Abstractions

In this section, we describe abstractions suitable as building blocks for the specification of inefficiency patterns related to MPI-2 one-sided communication. For reasons of brevity, we refrain from presenting the unabridged formalism underlying our abstractions and try to restrict ourselves to key concepts explained in natural language as far as possible. See [11] for a complete specification.

The state sequences and pointer attributes presented in this article apply to the underlying KOJAK event model, whose relevant portions are summarized in Table 1.

Table 1. Event types in KOJAK relevant to MPI-2 RMA analysis

Abstraction	Event type	Type specific Attributes
Entering / leaving a region	ENTER	region id
(e.g., a function)	EXIT	region id
Leaving an MPI collective function	MPICEXIT	region id, comm id, root loc, sent, recvd
Start / end / origin of RMA	PUT_1TS	window id, rma id, length, dest loc
one-sided transfers	PUT_1TE	window id, rma id, length, src loc
	GET_1TO	window id, rma id
	GET_1TS	window id, rma id, length, dest loc
	GET_1TE	window id, rma id, length, src loc
Leaving an MPI GATS function	MPIWEXIT	window id, region id, group id
Leaving an MPI collective RMA function	MPIWCEXIT	window id, region id, comm id
Locking / unlocking an MPI window	WLOCK	window id, lock loc, type
	WUNLOCK	window id, lock loc

The table lists type-specific attributes that are added to the location attribute and the timestamp mentioned in Section 4. For entries and exits of regions and ,in particular, MPI functions, we record which region was entered or left. In the case of collective MPI functions, instead of "normal" EXIT events, special collective events are used to capture the attributes of the collective operation. This is the communicator, the root process, and the amounts of data sent and received during this operation. Start and end of RMA one-sided transfers are marked with PUT_1TS and PUT_1TE (for remote writes and updates) or with GET_1TS and GET_1TE (for remote reads). For these events, we collect the source and destination and the amount of data transferred, as well as a unique RMA operation identifier which allows an easier mapping of #_1TE to the corresponding #_1TS events in the analysis stage later on. For all MPI RMA communication and synchronization operations we also collect an identification for the window on which the operation was performed. Exits of MPI-2 functions related to general active target synchronization (GATS) are marked with a MPIWEXIT event which also captures the groups of origin or target processes. For collective MPI-2 RMA functions, we use an MPIWCEXIT event and record the communicator that defines the group of processes participating in the collective operation. Finally, MPI window lock and unlock operations are represented by WLOCK and WUNLOCK events. A more detailed description of the MPI-2-specific events and their implementation in KOJAK can be found in [6].

Collective operations. In active target mode, access and exposure epochs may be enclosed in collective fence synchronization operations. The synchronizing character of these operations may result in wait times when processes reach the fence at different points in time. The same applies to functions to create and destroy windows. To detect wait states resulting from collective synchronization, we have defined a state sequence modeling the progress of collective operations on RMA windows - similar to the one for MPI-1 collective communication defined in our previous model.

Since the structure of the RMA-collective sequence is nearly identical to the sequence used in our previous model, we have introduced the concept of generic meta-sequences that can be instantiated with a type argument to simplify the formulation of sequences describing arbitrary collective operations. We have created a meta-sequence $\mathfrak{C}^g < T >$ collecting the exit events of collective operations carried out by members of a group g of processes. Depending on the type T of these exit events, this group is identified either by an MPI communicator, an OpenMP team, or an RMA window. Once all events of type T belonging to a collective operation instance are present, the complete instance is removed upon the next event applied to the set. An auxiliary function $complete < T > (e)$ is provided to query for instances completed by an event e, which is useful to measure waiting times. The state sequence for collective window operations is created by instantiating $\mathfrak{C}^g < MPIWCExit >$. Note that this abstraction can also be used for SHMEM collective operations.

Data transfers. Data transfers are modeled as pairs of events: (i) a start event initiating the transfer (i.e., PUT_1TS or GET_1TS) and (ii) an end event completing the transfer (i.e., PUT_1TE or GET_1TE). KOJAK's event model observes the MPI-2 synchronization semantics and, therefore, reflects the user-visible behavior of MPI-2 RMA operations. Figure 1 shows the model for the three different synchronization methods defined by MPI-2. The transfer line shown in the picture is not part of the model and is only shown for clarity.

The end GATS calls is modeled with MPIWEXIT events, the end of fences with MPIWCEXIT events to capture their collective nature. The transfer-start event is placed at the source process immediately after the begin of the corresponding communication function. However, the transfer-end event is placed at the destination process shortly before the exit of the RMA synchronization function which completes the transfer according to the MPI-2 standard rules. Unfortunately, this has an undesired side effect. As one can see in the figure, this results in a separation of the data transfer for remote reads from the corresponding `MPI_Get` function. To rectify this situation, we have introduced a new event GET_1TO indicating time and location of the transfer's origin.

To access all events belonging to the same data transfer, we have defined pointer attributes *startptr* and *originptr*, which connect the end event with its corresponding

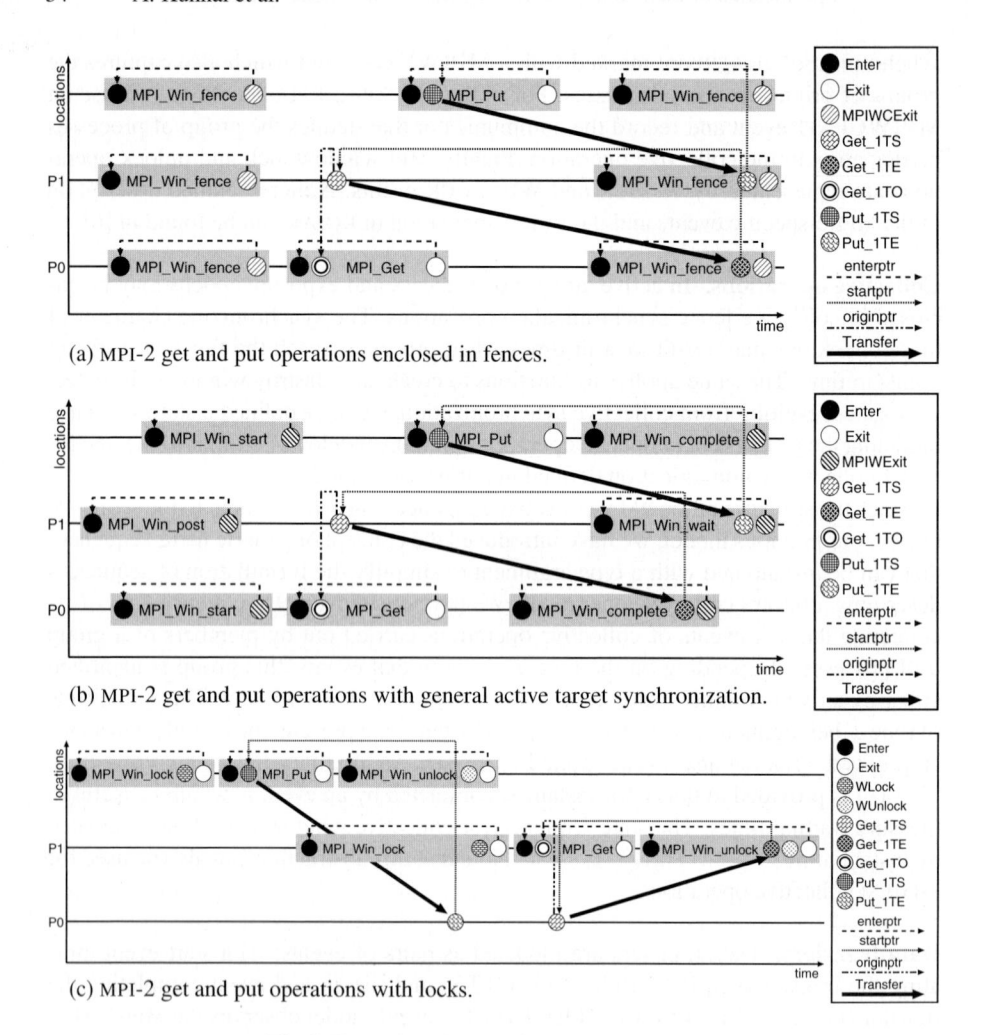

Fig. 1. Examples of KOJAK's MPI-2 event model

start event and the start event with its corresponding origin event, respectively. Their definition is based on state sequences collecting transfer events separately for each location (i.e., process) - similar to the queue for point-to-point messages defined in our earlier model. The identification of events belonging to the same transfer is based on the *rma id* attribute assigned during trace generation. Subsequently, we use these pointer attributes to reach start and origin events for given transfer-end events. Beyond that, these pointer attributes can be useful to calculate matrices with amounts of data transferred between processes.

Access and exposure epochs in general active target synchronization. The most challenging part of analyzing MPI-2 one-sided communication is GATS synchronization. To facilitate cross-process analysis in GATS mode, it is necessary to identify corresponding access and exposure epochs. Here, we present a multi-step method to recognize all access epochs belonging to a given exposure epoch (and vice versa) with the goal of providing all data needed for their analysis. This is the most intricate part of our model as it requires considering sets of sets of events to reflect the hierarchical grouping of the events involved. Note that this constitutes an important difference to the abstractions defined in our earlier work. We start with an introduction of the overall structure of event sets related to GATS communication:

Data transfer. A put or get operation.

Put operation. A PUT_1TE and its corresponding PUT_1TS event connected by the $startptr$ attribute.

Get operation. A GET_1TE and its corresponding GET_1TS and GET_1TO events connected by $startptr$ and $originptr$ attributes.

Epoch. An access or exposure epoch.

Access epoch. Includes two MPIWCEXIT events, one for each call to MPI_Win_start and MPI_Win_complete, plus all GET_1TE events in between at the same location and referencing the same window to represent all get operations belonging to this epoch. Note that put operations are represented by their respective PUT_1TE events inside the exposure epoch.

Exposure epoch. Includes two MPIWCEXIT events, one for each call to MPI_Win_post and MPI_Win_wait, plus all PUT_1TE events in between at the same location and referencing the same window to represent all put operations belonging to this epoch. Note that get operations are represented by their respective GET_1TE events inside the access epoch.

Epoch pair. Union of an access epoch at location l with a corresponding exposure epoch at location k but without any communication events not related to communication between l and k.

Access transaction. Union of an access epoch at location l with all corresponding exposure epochs at locations $k_1, \ldots, k_n$, but without any communication events not related to communication between l and $k_1, \ldots, k_n$. Figure 2 (left) shows an access transaction involving one access and two exposure epochs.

Exposure transaction. Union of an exposure epoch at location l with all corresponding access epochs at locations $k_1, \ldots, k_n$, but without any communication events not related to communication between l and $k_1, \ldots, k_n$. Figure 2 (right) shows an exposure transaction involving one exposure and two access epochs.

Matching GATS-based patterns requires the recognition of the above structures in the event trace. For this purpose, we have defined a hierarchical system of state sequences

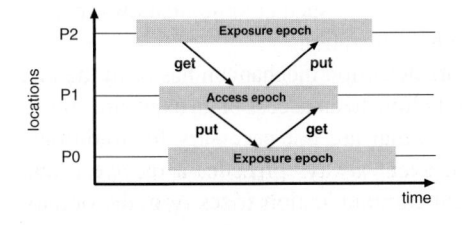

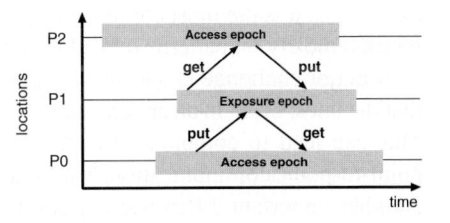

Fig. 2. An access transaction (left) and an exposure transaction (right)

that detects higher-level structures step-by-step based on lower-level structures already detected.

At the bottom, there are two state sequences $\mathfrak{A}^{l,w}$ and $\mathfrak{E}^{l,w}$ responsible for collecting all events belonging to an access or exposure epoch taking place at location l and referring to window w. The separation by window ensures that epochs belonging to the same window do not overlap in time at the same location. Once the event set describing an epoch is complete, the state is cleared upon the occurrence of the next event.

Completed epochs are combined into epoch pairs by a state sequence $\bar{\mathfrak{P}}^{k,l,w}$, which is defined for a target location k, an origin location l, and a window w. Before combining the two epochs, however, all events not related to communication between the two sides of the pair are removed. Again, after completion of the whole pair, the state is cleared. Different from our earlier sequence, the states of this sequence contain sets of event sets. This is necessary to express the hierarchical grouping of events typical for GATS transactions that consist of zero or more data transfer events enclosed by synchronization operations at each participating location. The auxiliary function $epoch\_pair(e, l)$ extracts a complete epoch pair as a flat set if e constitutes the last event of a pair with l being the location of the counter epoch.

The next level of composition is achieved through an auxiliary function $expta(e, \bar{P})$ that can be applied to an event e and a set of epoch pairs $\bar{P}$ and that returns all epoch pairs belonging to an exposure transaction if e constitutes the last event of this transaction. Using this and the function above, we have defined a state sequence $\bar{\mathfrak{E}}^{l,w}$ for a location l and a window w that successively adds epoch pairs as they are finished until a full exposure transaction has been completed, which then can be extracted using $expta(e, \bar{P})$. l denotes the location of the access epoch. Similarly, we have defined a function $accta(e, \bar{P})$ and a state sequence $\bar{\mathfrak{A}}^{l,w}$ to identify whole access transactions for later performance analysis.

6 One-Sided Patterns

Now, we use the abstractions defined in the previous section to specify complex inefficiency patterns spanning more than one process as a prerequisite for their automatic detection in event traces. The general structure of a pattern consists of a *root event* described by a simple test condition and zero or more constituents that can be located from the root event using the abstractions. The root event is the latest constituent event because the search for the remaining ones occurs backwards for efficiency reasons. An additional rule specifies how to quantify the pattern's performance impact (i.e., the time lost). Since it is the most complicated part of MPI-2 one-sided communication, we have focused mostly on patterns related to GATS synchronization.

A major challenge in specifying appropriate detection mechanism has been the fact that the latest event in an epoch pair can either belong to an access or an exposure epoch. This can lead to complicated case distinctions that are not necessary for traditional point-to-point communication, where a send event always precedes a receive event. Another important difference to point-to-point communication arises from the one-to-many relationships existing between access and exposure epochs involving more than two processes. For example, during an exposure epoch, a window may be accessed

Specification of Inefficiency Patterns for MPI-2 One-Sided Communication 57

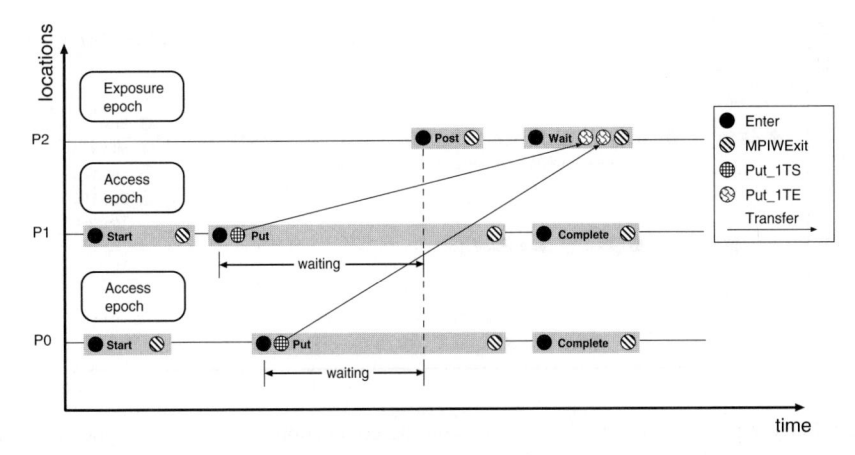

Fig. 3. Early Transfer: MPI_Get/Put() blocks during an access epoch until the related exposure epoch is started with MPI_Win_post()

by multiple processes each passing through a separate access epoch according to our definition above.

Early Transfer. This pattern describes a situation that may happen when communicating in GATS mode. MPI_Get/Put() blocks during an access epoch until the related exposure epoch is started with MPI_Win_post() (Figure 3). Recognizing the pattern requires considering epoch pairs. The root event is the last event of an epoch pair and is of type MPIWEXIT. It either completes the access or the exposure epoch and, therefore, either belongs to MPI_Win_complete() or to MPI_Win_wait/test().

The complete set of epoch pairs finished by the root event is determined by calculating $epoch\_pair(root, l)$ for every location l being a member of the partner group recorded with the root event. If the root event belongs to an exposure epoch, the pattern covers all corresponding access epochs already finished.

The waiting time is counted from the start of an access operation within the access epoch until the corresponding post operation has been issued during the matching exposure epoch. The begin of the access operation is identified using the pointer attributes $startptr$ and $originptr$ in the case of a get operation.

Early Wait. This pattern represents the premature request to finish an exposure epoch using MPI_Win_wait() and is depicted in Figure 4. We consider the request as premature if it was posted before the last access epoch's closure has been requested using MPI_Win_complete() within the same exposure transaction.

The recognition of this pattern requires the recognition of an exposure transaction. Two cases must be distinguished: (i) the transaction is completed by an exposure epoch or (ii) the transaction is completed by an access epoch. In the first case, the root event is the MPIWEXIT event of the wait operation and the full transaction is easily obtained by applying $expta()$ to the root event and $\bar{\mathfrak{E}}^{l,w}$ with l being the location of the root event.

In the second case, the root event is the MPIWEXIT event of a complete operation and, therefore, finishes an access epoch. Now, the detection mechanism needs to

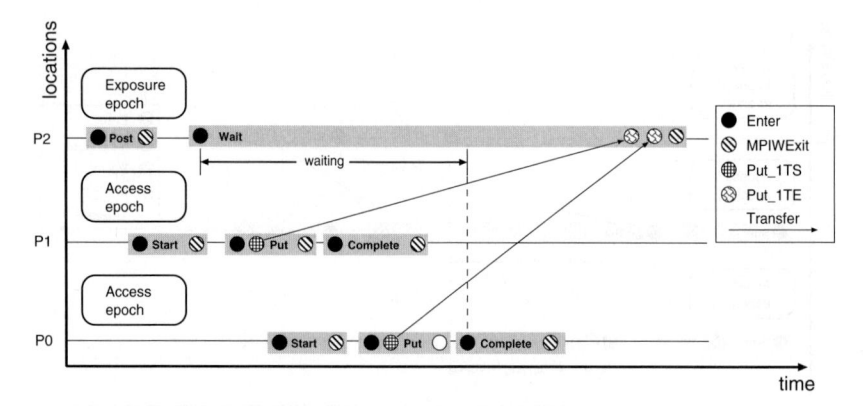

Fig. 4. Early Wait: `MPI_Win_wait()` blocks during completion of an exposure epoch until all related access epochs are completed

find all exposure transactions finished with this access epoch. This is accomplished by iterating over all exposure epochs belonging to epoch pairs completed by the root event and extracting completed exposure transactions from $\bar{\mathfrak{E}}^{k,w}$ using the $expta()$ function. The exposure epochs are found by means of $\bar{\mathfrak{P}}^{k,l,w}$ with l being the location of the root event and k being a location in the root event's partner group. Since the exposure transactions we are looking for have been completed by the root event, we need to consider $\bar{\mathfrak{E}}^{k,w}$ at the time of the root event. The waiting time is the period between the start of the wait until the beginning of the latest complete operation in the transaction.

Late Complete. If a process delays the completion of an access epoch by performing work between the last access and the complete operation and the wait operation has already been posted, a situation named Late Complete occurs (Figure 5). It is actually a sub-property of Early Wait. This pattern considers an exposure transaction and

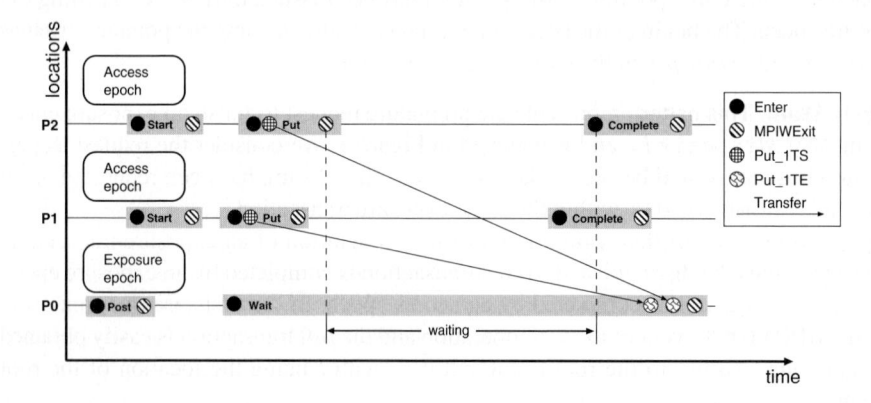

Fig. 5. Late Complete: Describes the time wasted between the last access and the call to the corresponding `MPI_Win_complete()` operation

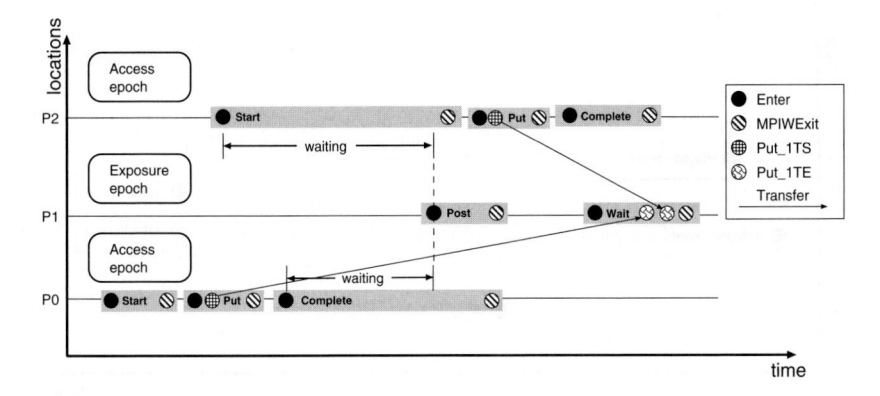

Fig. 6. Late Post: MPI_Win_start() or MPI_Win_complete() block because the corresponding exposure epoch has not started yet

measures the time spent in the wait operation between exiting the last put or get and entering the corresponding complete (or the latest complete if the last get/put is not unique). The recognition of the exposure transaction is similar to Early Wait.

Late Post. Refers to access-sided synchronization operations that block until access is granted by an exposing process (Figure 6). Depending on the MPI implementation, this may happen either during MPI_Win_start() or during MPI_Win_complete(). Since the exact blocking semantics are usually not known to a performance tool, our pattern counts time spent in both operations before the earliest post call within the same access transaction is issued in the case that MPI_Win_start() does not block. Then, however, the time spent in the start operation will be small and the resulting inaccuracy negligible. Whereas the semantics of the pattern are closer to Early Transfer, its recognition is very similar to Early Wait, only that it requires the recognition of an access transaction using $\bar{\mathfrak{A}}^{l,w}$. Like Early Wait, this pattern needs to distinguish two cases: (i) the root event finishes an access epoch or (ii) the root event finishes an exposure epoch, in which case the access transactions have to be identified by iterating over all related access epochs.

Wait at Fence. Whereas the previous patterns all refer to GATS synchronization, this pattern covers the simpler case of synchronization with MPI_Win_fence(). Since fence normally[2] implies a barrier, waiting times occur if the fence is not reached simultaneously by all participating processes (Figure 7). Early processes have to wait for the latest one. The recognition of Wait at Fence is accomplished using $\mathfrak{C}^g < MPIWCExit >$, which collects collective window operation instances. After retrieving such an instance using $complete < MPIWCExit > ()$ and identifying the latest entry into the operation, the waiting times of different processes can be easily determined.

---

[2] The internal barrier can be avoided by passing additional "hints" to the fence call as a second parameter.

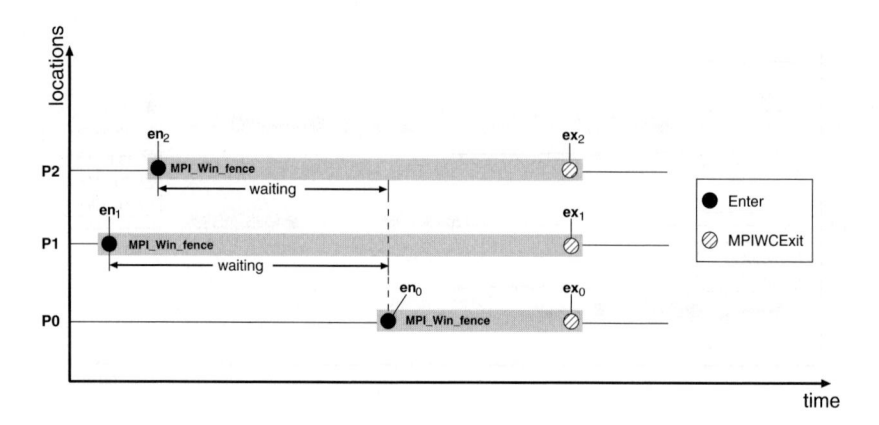

Fig. 7. Wait at Fence: Time spent waiting in front of a synchronizing `MPI_Win_fence()` operation

7 Conclusion

To the best of our knowledge, this is the first systematic approach of automatically identifying wait states related to MPI-2 one-sided communication in event traces. Building upon our earlier framework to identify wait states in traditional two-sided and collective communication, we have defined new abstractions representing higher-level events related to one-sided operations. These abstractions serve as a useful prerequisite to specify inefficiency patterns in a way facilitating their automatic detection in the event stream.

A major difficulty that has been solved within our new framework is the fact that one-sided communication is accomplished in complex sequences of synchronization and communication, where the notion of send and receive operations is replaced by the notion of access and exposure epochs comprising both synchronization and access operations. Also, a single epoch may perform communication with an entire group of processes, which requires the recognition of all counter epochs performed by members of this group. In addition, the root event from where the constituents of a pattern may be located may reside on either side of an epoch pair, which involves complex case distinctions on the side of the detection mechanism.

To demonstrate the usefulness of our framework, we have specified several complex patterns of inefficient behavior targeting, in particular, general active target synchronization, which can be challenging for programmers. Meanwhile, we have completed the implementation of all KOJAK modules necessary for the instrumentation, measurement, conversion, and analysis of parallel applications based on MPI-2 RMA and we have a prototype version for SHMEM programs. Figure 8 shows a summary of the currently implemented pattern hierarchy. We have also extended our internal test suite to cover one-sided communication and used it to verify our implementation. As a next step, we need to evaluate the relevance of these patterns using real-world applications.

Finally, we hope that some of the complexity in the analysis can be avoided, when transferring this approach to the new parallel analyzer architecture developed in the SCALASCA [12] project. By exploiting distributed memory and parallel processing

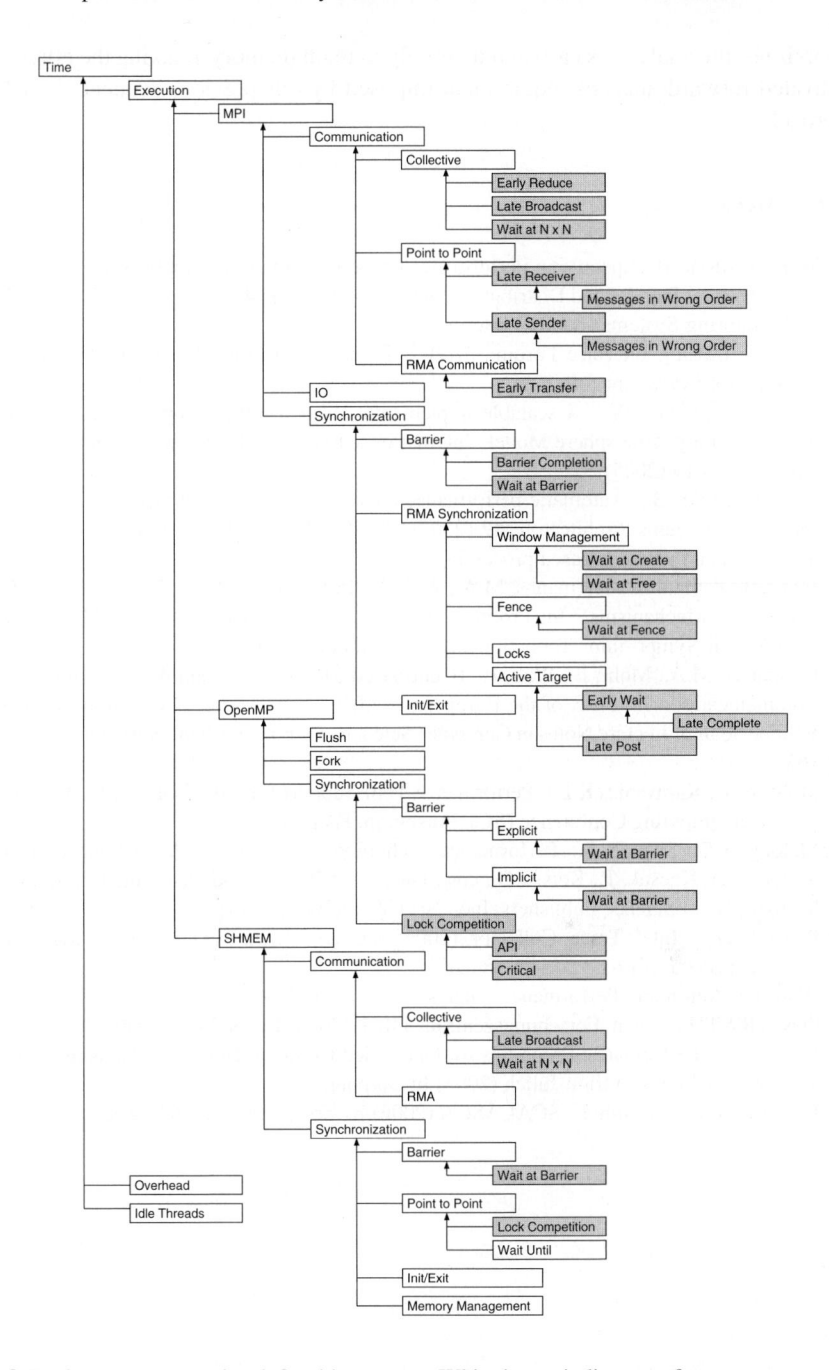

Fig. 8. Performance properties defined by KOJAK. White boxes indicate performance properties based on summary information which could also be provided by a profiling tool. However, the second type, indicated by gray boxes, involves idle times that can only be determined by comparing the chronological relation between individual events.

capabilities, the analysis is carried out entirely in main memory, relaxing the efficiency-motivated forward-analysis requirement imposed by our previous sequential analysis approach.

References

1. Wolf, F., Mohr, B.: Specifying Performance Properties of Parallel Applications Using Compound Events. Parallel and Distributed Computing Practices **4** (2001) 301–317 Special Issue on Monitoring Systems and Tool Interoperability.
2. Message Passing Interface Forum: MPI-2: Extensions to the Message-Passing Interface (1997) http://www.mpi forum.org.
3. Mirin, A., Sawyer, W.: A scalable implementation of a finite volume dynamical core in the Community Atmosphere Model. International Journal of High Performance Computing Applications **19** (2005) 203–212
4. Wolf, F., Mohr, B.: Automatic performance analysis of hybrid MPI/OpenMP applications. Journal of Systems Architecture **49** (2003) 421–439 Special Issue "Evolutions in parallel distributed and network-based processing".
5. Mohr, B., Kühnal, A., Hermanns, M.A., Wolf, F.: Performance Analysis of One-sided Communication Mechanisms. In: Proceedings of Parallel Computing (ParCo), Malaga, Spain (2005) Mini-Symposium "Tools Support for Parallel Programming".
6. Hermanns, M.A., Mohr, B., Wolf, F.: Event-based Measurement and Analysis of One-sided Communication. In: Proc. of the European Conference on Parallel Computing (Euro-Par). Volume 3648 of Lecture Notes in Computer Science., Lisboa, Portugal, Springer (2005) 156–165
7. Mohror, K., Karavanic, K.L.: Performance Tool Support for MPI-2 on Linux. In: Proc. of the Supercomputing Conference (SC), Pittsburgh, PA (2004)
8. Malony, A.D., Shende, S.: Performance Technology for Complex Parallel and Distributed Systems. In Kacsuk, P., Kotsis, G., eds.: Quality of Parallel and Distributed Programs and Systems, Nova Science Publishers, Inc., New York (2003) 25–41
9. Pallas/Intel: Intel Trace Collector (2006) http://www.intel.com/software/products/cluster/tcollector/
10. Wolf, F.: Automatic Performance Analysis on Parallel Computers with SMP Nodes. PhD thesis, RWTH Aachen, Forschungszentrum Jülich (2003) ISBN 3-00-010003-2.
11. Kühnal, A.: Performance Properties for One-Sided Communication Mechanisms. Diploma Thesis. Forschungszentrum Jülich (2005) In German.
12. Forschungszentrum Jülich: SCALASCA (2006) http://www.scalasca.org.

Topic 2: Performance Prediction and Evaluation

Jesús Labarta, Bernd Mohr, Allan Snavely, and Jeffrey Vetter

Topic Chairs

Parallel computing enables solutions to computational problems that are impossible on sequential systems due to their limited performance. To meet this objective, it is critical that users can both measure performance on a given system and predict the performance for other systems. Achieving high performance on parallel computer systems is the product of an intimate combination of hardware architecture (processor, memory, interconnection network), system software, runtime environment, algorithms, and application design. Performance evaluation is the science of understanding these factors that contribute to the overall expression of parallel performance on real machines and on systems yet to be realized. Benchmarking and performance characterization methodologies and tools provide an empirical foundation for performance evaluation. Performance prediction techniques provide a means to model performance behaviors and properties as system, algorithm, and software features change, particularly in the context of large-scale parallelism. These two areas are closely related since most prediction requires data to be gathered from measured runs of a program, to identify application signatures or to understand the performance characteristics of current machines.

A total of eighteen papers were submitted to the performance prediction and evaluation topic area. The submissions covered a broad range of prediction and evaluation topics, and reflect a high level of current interest in the parallel computing community. The eight papers accepted (44state-of-the-art results from leading parallel performance researchers in the field today. The papers cover two general themes in performance prediction and evaluation. The first theme considers methods to explore performance properties from different evaluation contexts: data access, processor, and interconnect. Three papers investigate performance issues on shared-memory machines (IBM Cyclops-64, SGI Altix 3700, and Sun Fire E25K). Another three articles center around the analysis of applications on distributed memory architectures (IBM BlueGene/L, Linux Multiclusters, Clusters with Infiniband interconnect). The second theme concerns advances in performance prediction with two papers about tools for predicting multi-processor system on a chip (MPSoC) performance and system for hierarchical model validation.

Finally, we would like to thanks all contributing authors as well as all reviewers for their work.

Hierarchical Model Validation of Symbolic Performance Models of Scientific Kernels

Sadaf R. Alam and Jeffrey S. Vetter

Oak Ridge National Laboratory
Oak Ridge, TN 37831, U.S.A.
{alamsr, vetter}@ornl.gov

Abstract. Multi-resolution validation of hierarchical performance models of scientific applications is critical primarily for two reasons. First, the step-by-step validation determines the correctness of all essential components or phases in a science simulation. Second, a model that is validated at multiple resolution levels is the very first step to generate predictive performance models, for not only existing systems but also for emerging systems and future problem sizes. We present the design and validation of hierarchical performance models of two scientific benchmarks using a new technique called the modeling assertions (MA). Our MA prototype framework generates symbolic performance models that can be evaluated efficiently by generating the equivalent model representations in Octave and MATLAB. The multi-resolution modeling and validation is conducted on two contemporary, massively-parallel systems, XT3 and Blue Gene/L system. The workload distribution and the growth rates predictions generated by the MA models are confirmed by the experimental data collected on the MPP platforms. In addition, the physical memory requirements that are generated by the MA models are verified by the runtime values on the Blue Gene/L system, which has 512 MBytes and 256 MBytes physical memory capacity in its two unique execution modes.

1 Introduction

Performance models of scientific applications have been generated using analytical techniques and measurement-based techniques. Analytical techniques like the one presented by Almasi *et. al.* [2] provide detailed information about the application structure and underlying algorithms but do not capture the computation and workload characteristics in detail that is essential to carry out performance prediction studies on a given target architecture. The measurement-based techniques [7, 9], for instance, techniques based on collecting detailed memory tracing data on target systems provide detailed system-specific performance characteristics of an application [7]. However, these approaches do not capture the algorithmic and problem resolution metrics of scientific applications in the performance models. Thus, the applicability is limited if an underlying algorithm or the target architecture characteristics are modified. For instance, the prediction error rates can change dramatically if the memory hierarchy of a target system varies from the architecture on which the measurements are taken. A successful performance modeling and prediction effort is presented by Kerbyson et. al. [5] for a large-scale scientific application and its kernels; however, this scheme requires an expert understanding of the application and underlying algorithms as well as detailed information about the features of the target parallel platform.

We have proposed a portable and extensible approach for developing performance models of scientific applications called modeling assertions (MA) [1]. Our approach encapsulates an application's key input parameters as well as the workload parameters including the computation and the communication characteristics of the modeled applications. The MA scheme requires an application developer to describe the workload requirements of a given block of code using the MA API in form of code annotations. These code annotations are independent of the target platforms. Moreover, the MA scheme allows multi-resolution modeling of scientific applications. In other words, a user can decide which functions are critical to a given application, and can annotate and subsequently develop detailed performance models of the key functions. Depending on the runtime accuracy of the model, a user can develop hierarchical, multi-resolution performance models of selected functions, for instance, models of critical loop blocks within a time-consuming function. MA models can capture the control structure of an application. Thus, not only an aggregated workload metric is generated but also the distribution of a given workload over an entire execution cycle can be modeled using the MA framework.

In this paper, we present the step-by-step validation of the MA performance models for two scientific kernels, the NAS parallel, message passing (MPI) CG and SP benchmarks [3]. The runtime measurement has been conducted on two distributed memory, teraflop/s scale systems, Cray XT3 [8] and IBM Blue Gene/L [6]. The XT3 systems is based on a 2.4 GHz Opteron processor connected with a high-speed Hypertransport link. A single processor is capable of delivering 4.8 gigaFLOP/s and provides up to ~6200 Mbytes/s memory bandwidth. The Blue Gene/L system has a unique memory hierarchy because of the two modes of execution namely co-processor mode and the virtual-node mode [6]. The Blue Gene/L system has a small physical memory per processor, 512 Mbytes and in the virtual-node execution mode, only half (256 Mbytes) is available to the user processes. The peak performance of a Blue Gene/L processing core is 1.4 gigaFLOPS/s and its main memory bandwidth is ~3200 Mbytes/s, which is shared between two processors in the virtual-node mode.

We developed workload models for the two NAS MPI benchmarks, one each for the floating-point computation, memory operations, memory capacity and sizes and patterns of MPI operations [1]. The model predictions are validated with the runtime data by altering the key input parameter values and by running a fix size application in strong-scaling mode. The MA models capture the workload distribution that is represented as a function of key input parameters. These workload requirements are validated with the runtime performance data measured on two parallel systems. In addition to workload requirements, the MA models of the two NAS benchmarks can generate the growth rates for the workload distribution as a function of input parameters. The growth rates and sensitivity studies are also validated with the runtime data.

The outline of the paper is as follows: section 2 presents the design of the MA models using the MA framework. A brief description of NAS CG and SP models of computation and communication is presented in section 3. The step-by-step process of model validation at multiple-resolutions is provided in section 4. Section 4 also presents the validation results for the MA models that show the workload growth rate as a function of input parameters. Section 5 gives a summary of the MA approach and future research directions.

2 The MA Framework

In order to evaluate our approach of developing symbolic models with MA, we have designed a prototype framework [1]. This framework has two main components: an API and a post-processing toolset. Figure 1 shows the components of the MA framework. The MA API is used to annotate the source code. As the application executes, the runtime system captures important information in trace files. These trace files are then post-processed to validate, analyze, and construct models. The post-processing toolset is a collection of tools or Java classes. The post-processor currently has three main classes: model validation, control-flow model creation and symbolic model generation classes. The symbolic model shown in the Figure 1 is generated for the MPI send volume. This symbolic model can be evaluated and is compatible with MATLAB [10] and Octave [11].

The MA API provides a set of functions to annotate a given FORTRAN or C code with MPI message-passing communication library. For example, `ma_loop_start`, a MA API function, can be used to mark the start of a loop. Upon execution, the code instrumented with MA API functions generates trace files. For parallel applications, one trace file is generated for each MPI task. The trace files contain traces for `ma_xxx` calls and MPI communication events. Most MA calls require a pair of `ma_xxx_start` and a `ma_xxx_end` calls. The `ma_xxx_end` traces are primarily used to validate the modeling assertions against the runtime values. The assertions for hardware counter values, ma_flop_start/stop, invoke the PAPI hardware counter API [4]. The `ma_mpi_xxx` assertions on the other hand are validated by implementing MPI wrapper functions (PMPI) and by comparing `ma_mpi_xxx` traces to `PMPI_xxx` traces. Additional functions are provided in the MA API to control the tracing volume, for example, the size of the trace files, by enabling and disabling the tracing at compile time and also at runtime. At runtime, the MA runtime system (MARS) tracks and captures the actual instantiated values as they execute in the application. MARS creates an internal control flow representation of the calls to the MA library as they are executed. It also captures both the symbolic values and the actual values of the expressions. Multiple calls to the same routines with similar parameters maps onto the same call graph, therefore, the data volume is manageable.

The validation of an MA performance model is a two-stage process. When a model is initially being created, validation plays an important role in guiding the resolution of the model at various phases in the application. Later, the same model and validation technique can be used to validate against historical data and across the parameter space.

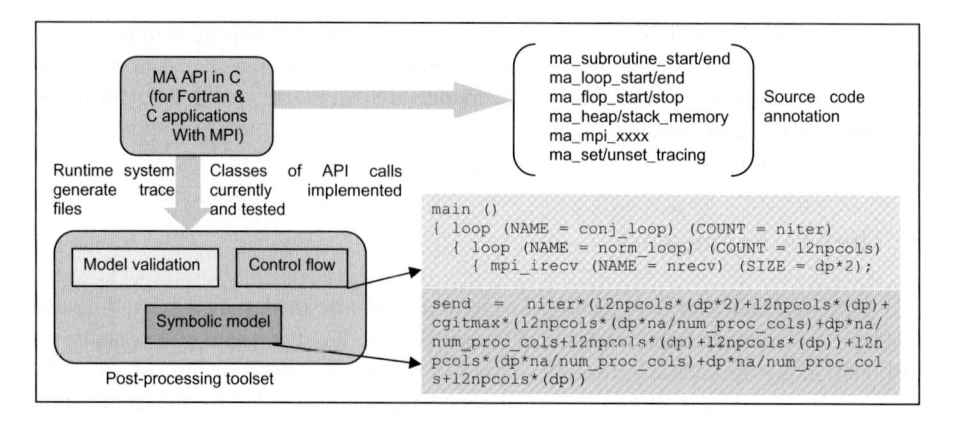

Fig. 1. Design components of the Modeling Assertion (MA) framework: The MA API, which is written in C and the extensible, post-processing toolset classes in Java. The MA API is available for C and FORTRAN code.

3 Evaluation of Symbolic Models

NAS CG computes an approximation to the smallest eigenvalue of a large, sparse, symmetric positive definite matrix, which is characteristic of unstructured grid computations. The main subroutine is `conj_grad`, which is being a called `niter` time. The first step was to identify the key input parameters, `na`, `nonzer`, `niter` and `nprocs` (number of MPI tasks); the MA symbolic models for floating-point, load-store, physical memory and communication volume requirements are generated in terms of these four input parameters.

The NAS parallel benchmarks provide different problem sizes or classes where class S is the smallest problem size. The MA model for SP is represented in terms of one input parameter, `problem_size`. In addition, the number of MPI tasks determines some derived parameters like the $\log_2$ of MPI tasks in CG and the square-root of number of processors in the SP benchmark to simplify model representations. Both CG and SP benchmarks follow a Single Program Multiple Data (SPMD) programming paradigm. Hence, the workload and memory mapping and distribution per processor not only depend on the key input parameters but also on the number of MPI tasks.

Upon termination of a runtime experiment, MA outputs a control flow model representation, an intermediate file and symbolic models for number of floating-point, load-store and communication operations. The control flow model representation is similar to the actual code annotations; that is, it is a high level, visual flow of the annotated parts of the application. The intermediate representation serves as an input to develop symbolic models for user-defined characteristics or relative quantities like memory byte-to-flop ratio. For instance, a user can create models for load/store-to-flop ratios using the intermediate representation. The symbolic models generated by the MA framework are compatible with Matlab and Octave script format. Figure 2 shows symbolic model representation for SP communication operations. Only three

```
ncells = sqrt (no_nodes)
elems = problem_size / ncells
dp = sizeof (double)
niter * (elems ^ 2 * (ncells - 1) * 10 * dp
+ elems ^ 2 * (ncells - 1) * 10 * dp
+ . . .
+ ncells * (22 * (elems - 1) ^ 2 * dp)
+ ncells * (10 * (elems - 1) ^ 2 * dp)
+ . . .
```

Fig. 2. A MATLAB/Octave compatible symbolic model generated by the MA framework

input parameters are required to evaluate this model, no_nodes (number of MPI tasks), niter (number of time step iterations) and problem_size (application input parameter). Our target is to be able to generate symbolic models that represent the architecture independent requirements of an application and that can be evaluated efficiently by existing mathematical software frameworks.

One of the aims of creating the models of scientific applications is to be able to predict the application requirements for the future problem configurations. We used our MA models to understand the sensitivity of floating-point operations, memory requirements per processor, and message volume to applications' input parameters. We begin experiments with a validated problem instance, Class C, for both the NAS CG and SP benchmarks, and scale the input parameters linearly. Note that the MA framework has a post-processing toolset that allows validation of MA model annotations with the runtime values. For instance, the PAPI_FP_OPS (number of floating-point operations) data was compared with the ma_flop runtime value. The validated problem instances, Class C, have na=150000, nonzer=15, for CG Class C benchmark with 128 MPI tasks. We increase the value of na linearly and generate the floating-point and load-store operation count using the MA symbolic models of the NAS CG benchmark. Figure 3 shows that the floating-point and load-store cost in the CG experiments increase linearly with the na parameter value.

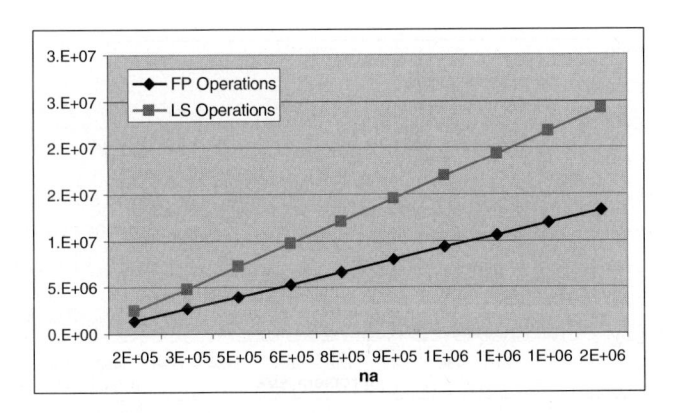

Fig. 3. Sensitivity of the number of Floating-point (FP) and load-store (LS) operations per processor in a 128 processor experiment by increasing the array size parameter: na

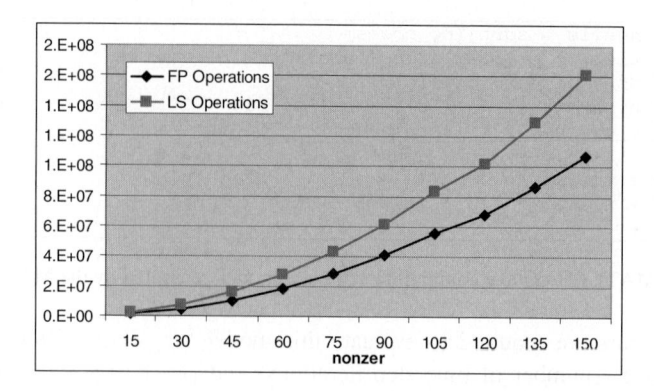

Fig. 4. Sensitivity of FP and LS by increasing the number of non-zero elements parameter: nonzer

Similarly, we generated the growth rates for the floating-point and load-store operation cost for the other input parameter, `nonzer`. Results in Figure 4 show that the floating-point and load-store operation cost in CG is relatively more sensitive to the increase in the number of `nonzer` elements in the array than the array size: `na`.

The NAS SP benchmark has a single application parameter, `problem_size`, which we have used to represent the workload requirements (floating-point, load-store, memory and communication) in the MA symbolic models. Figure 5 shows the increase in the floating-point and load-store operation count by increasing the `problem_size` linearly. Note that like CG, the initial set of experiments (Class S, W, A, B, C and D) are validated on the target MPP platforms. Figure 5 shows that the floating-point operation cost increases at a very high rate by increasing the `problem_size`.

Using the MA models, we can not only generate the aggregated workload requirements shown earlier, but also get an insight into the scaling behavior of the

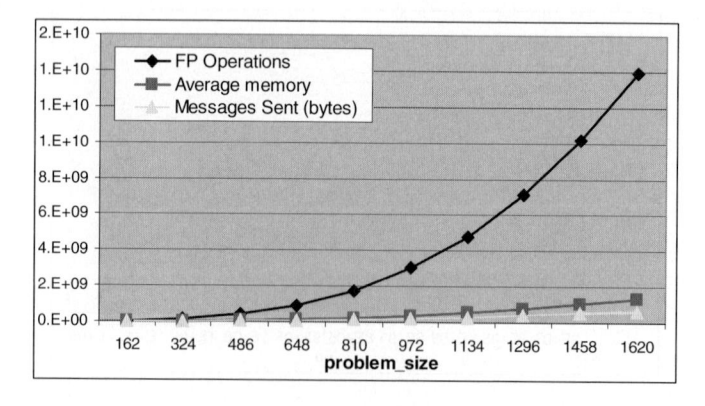

Fig. 5. Sensitivity of workload requirements with respect to the SP input parameter: `problem_size`

workload requirements within an application as a function of the `problem_size` parameter. Figure 6 shows contribution of different functions in total floating-point operation count in SP time step iterations. The results shown in Figure 6 are generated for a fix number of MPI tasks and by increasing the `problem_size` parameter linearly. The floating-point workload requirements generated by the MA model show that the `z_solve` is the most expensive function for runs with large number of processors. The cost of `x_solve` and `y_solve` are identical and consistent. Moreover, based on the MA model results shown in Figure 6, we can safely ignore cost of `txinvr` and `add` functions in the further analysis.

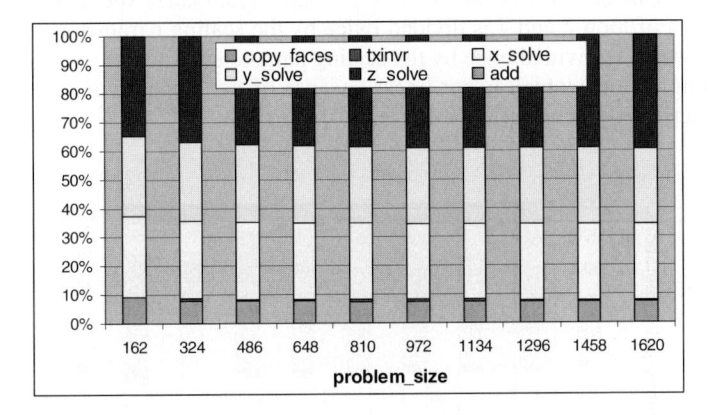

Fig. 6. Impact of individual functions on the overall increase in the number of floating-point operations by increasing the input parameter `problem_size` on 1024 processors

4 Multi-resolution Validation of MA Models

The model verification output enables us to identify the most floating-point intensive loop block of the code in the CG benchmark. This loop block is shown in Figure 7, which is called twice during a conjugate gradient calculation in the CG benchmark. The symbolic floating point operation cost of the loop is approximately $2*na/(num\_proc\_cols*nonzer*ceiling(nonzer/nprows))$.

Using the MA models, we generated the scaling of the floating-point operation cost of the loop block in Figure 7 with the other loop blocks within a conjugate gradient

```
do j=1,lastrow-firstrow+1
    sum = 0.d0
    do k=rowstr(j),rowstr(j+1)-1
        sum = sum + a(k)*p(colidx(k))
    enddo
    w(j) = sum
enddo
```

Fig. 7. The partition submatrix-vector multiply

iteration. The model predictions are shown in Figure 8. The total cost of two invocations of the submatrix vector multiply operation contributes to a large fraction of the total floating-point operation cost. l1 is the first loop block in the CG timestep iteration and l2 is the second. Figure 8 shows that the workload is not evenly distributed among the different loop blocks (or phases of calculations), and the submatrix vector multiply loop can be serious bottleneck. Furthermore, as we scale the problem to a large number of processors, we begin to identify loop that are either the Amdahl's proportions of the serial code or their loop count is directly propotional to the number of MPI tasks in the system. We found that the loop count of loop number 3 and 8 depend on the number of MPI tasks ($\log_2(\log_2(MPI\_Tasks))$), while loop 1 and 8 scale at a slower rate than loop 2 and 7 (submatrix vector multiply loop), since the cost of loop 2 and 7 is divided twice by the scaling parameters as compared to 1 and 8, which is divided once by the scaling parameter. Another interesting feature is the scaling pattern, which is not linear because of the mapping and distribution of workload depends on ceiling($\log_2(MPI\_tasks)$).

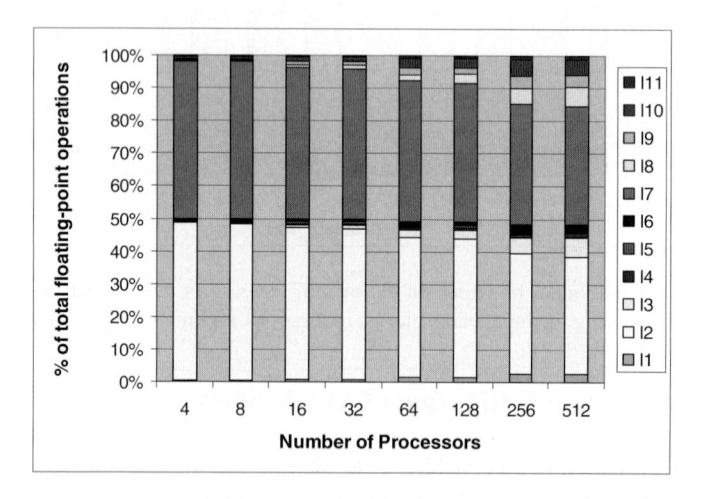

Fig. 8. Distribution of floating-point operation cost within a time step iteration in the NAS CG benchmark. Default `na` and `nonzer` parameter values for the Class C problem instance are used for the experiments. l1 is the first loop block and l2 is the second loop block in the conjugate gradient iterations. l2 and l7 perform calculations shown in Figure 7.

We collected the runtime data for the loops blocks in CG time step iterations on XT3 and Blue Gene/L processors to validate our workload distribution and scaling patterns. Figure 9 shows the percentage of runtime spent in individual loop blocks. Comparing it with the workload distribution in Figure 8, we observe not only a similar workload distribution but also a similar scaling pattern. Note that the message passing communication times are not included in these runtime measurements. We collected data for the Class D CG benchmark on the XT3 system, which also validate the floating-point message count distribution and scaling characteristics that are generated by the symbolic MA models.

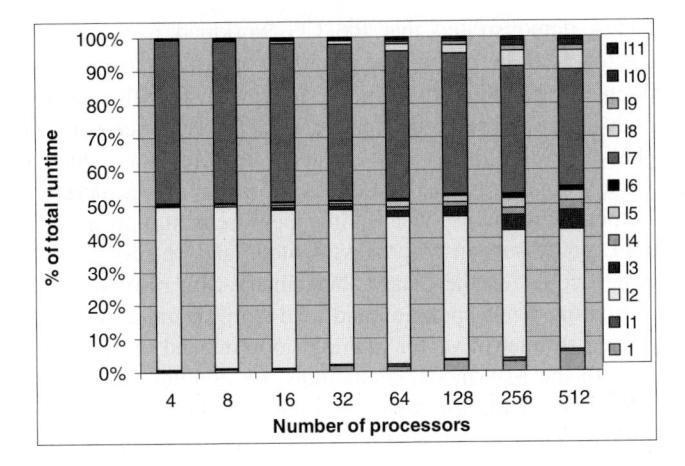

Fig. 9. Percentage of total runtime spent in individual loop blocks in a CG iteration. These runtime are measured in the co-processor mode. Default `na` and `nonzer` parameter values for the Class C problem instance are used for the experiments.

On Blue Gene/L we collected data in the two execution modes to investigate the effect of reduced memory bandwidth. The memory bandwidth in the virtual node mode is shared by two Blue Gene/L processors, while in the co-processor mode, a single compute processor accesses the main memory during the computation. In the co-processor mode, the other processor, the communication processors, is typically associated with MPI communication data movement. We did not observe a significant increase in runtime in the virtual-node mode experiments on the Blue Gene/L system as shown in Figure 10. Although the runtime increases for the most time-consuming loops, we conclude that the memory traffic is not a major issue in the most time-consuming block of the code. We expect the memory bandwidth to be an issue when workload sizes per processor are increased significantly.

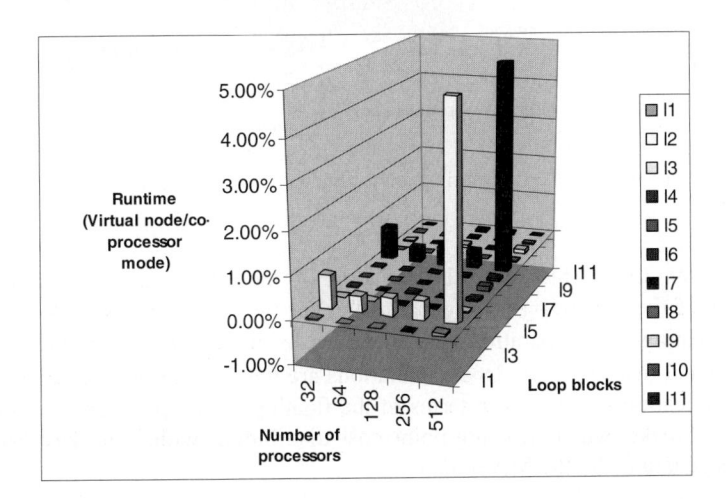

Fig. 10. Percentage increase in the runtime values in the virtual-node mode experiments

Figure 3 and 4 demonstrated that the CG workload is more sensitive to the `nonzer` parameter. In order to validate our MA performance model predictions, we ran the experiments by doubling the `nonzer` parameter. On Blue Gene/L, we ran the experiments both in the virtual node mode and in the co-processor mode. First, we validate the MA model for the physical memory requirements. Our model predicts that the sizes of all large arrays depend on the `nonzer` value, therefore, the overall memory requirement will double. The runtime measurements confirm that a minimum of 8 Blue Gene/L processors in co-processor mode and 16 processors in the virtual node mode are needed to run the Class C benchmark with `nonzer=30`. Second, the MA models of floating-point operation and load-store operation count predicted that only the cost of the submatrix vector multiply loop depend on the nonzer parameter value. According to the MA model predictions, by doubling the value of `nonzer` (Class C problem instance), the floating-point and the load-store operation cost increases by ~300%. The runtime data in Figure 11 confirm the cost distribution and scaling by doubling the `nonzer` parameter as predicted by the MA model. We also validated our MA model for up to five times increase in the `nonzer` value with the runtime data.

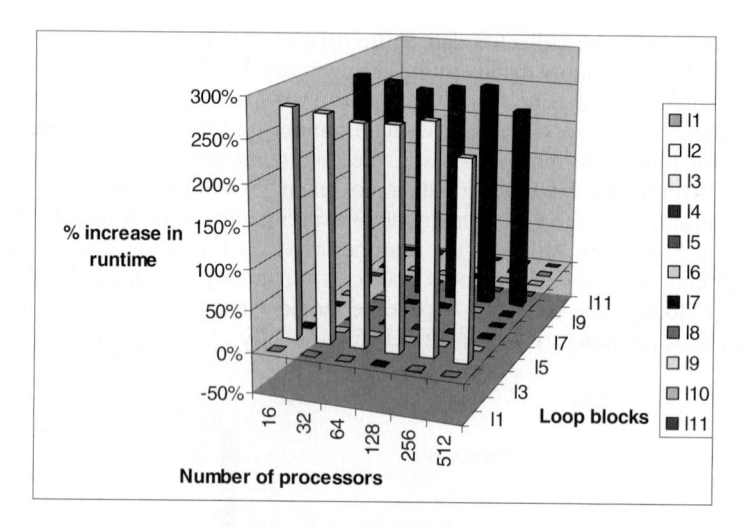

Fig. 11. Percentage increase in runtime of individual loop blocks by doubling the nonzer parameter

In addition to the hierarchical validation of the CG model, we validated the sensitivity of the `problem_size` parameter for the NAS SP model. We identified that the floating-point operation cost increases by increasing the `problem_size` parameter and that the `z_solve` calculations are the most expensive calculations in the SP application simulation in terms of the floating-point operation cost. Figure 12 shows the breakdown of floating-point cost distribution within the key calculation phases as generated by the MA model.

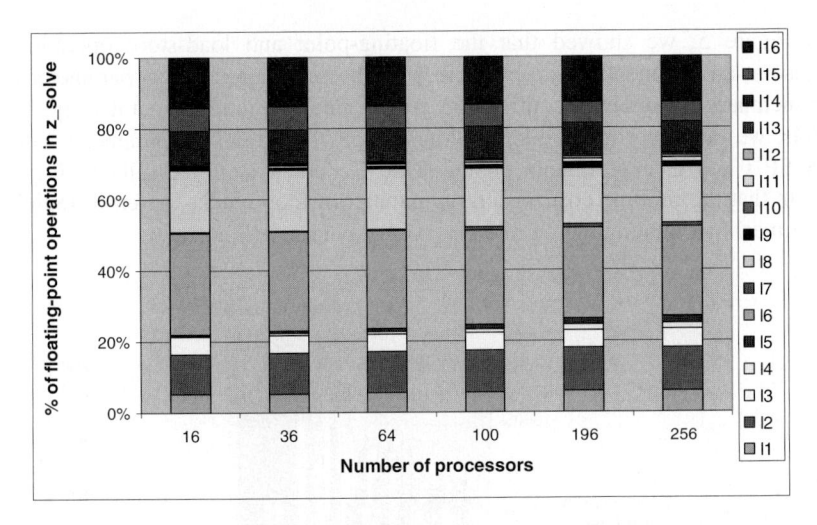

Fig. 12. Distribution of the floating-point operation cost in different code blocks in the z_solve method of the NAS SP benchmark. These distributions are generated by the MA models.

We conducted fine-grain measurements on the z_solve operation and collected runtime data on the XT3 and Blue Gene/L processors. Figure 13 shows the runtime data collected on the Blue Gene/L processor for different phases of calculation in the z_solve operation. The distribution of the runtime cost in the z_solve function confirm the workload distribution and scaling pattern that was generated by the MA model for the NAS SP benchmark (Figure 12).

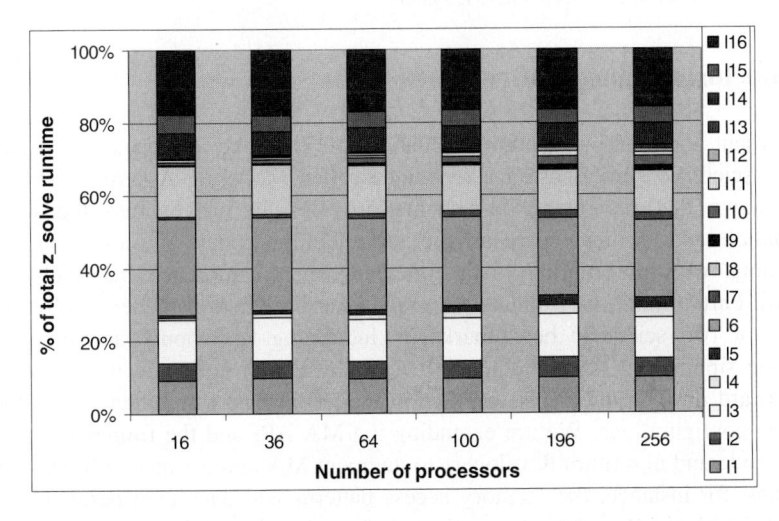

Fig. 13. The distribution of the runtime cost measured on the Blue Gene/L processor for the Class C problem instance of the NAS Parallel benchmark

In Figure 5, we showed that the floating-point and load-store operation cost increase at an exponential rate by increasing the `problem_size` parameter in the SP benchmark. In order to confirm the growth rate prediction generated by the MA models, we ran experiments by doubling the `problem_size` parameter value for NAS SP Class C experiments on the Blue Gene/L system. Results in Figure 14 confirm that the runtime cost for a large number of loop blocks increases rapidly (up to 10 times) by doubling the `problem_size` parameter.

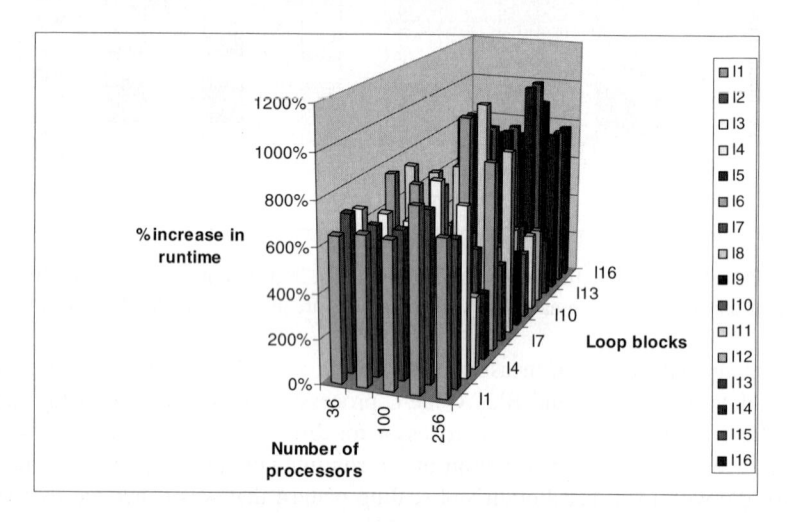

Fig. 14. Percentage increase in runtime by doubling the problem_size parameter in NAS SP Class C experiments on the Blue Gene/L system

5 Conclusions and Future Work

We present the multi-resolution validation of symbolic performance models of parallel scientific kernels using a technique called Modeling Assertions (MA). We have shown that our modeling scheme provides an insight into the workload distribution and scaling characteristic of scientific codes by comparing model predictions with the runtime data collected on contemporary massively-parallel systems. Furthermore, we validate the growth rates predictions generated by the MA models of two scientific benchmarks by increasing key input parameters of the scientific simulations. Development of hierarchical MA symbolic models is a first step toward developing precise prediction model on target architectures and future problem configurations. We are extending the MA API and the framework that will enable code and algorithm developers to augment MA annotations with performance attributes, for instance, the memory access patterns and data-level parallelism for a given loop block. We also plan to introduce a set of modeling attributes that can represent the unique performance enhancing features of emerging architectures.

Acknowledgements

This research was sponsored by the Office of Mathematical, Information, and Computational Sciences, Office of Science, U.S. Department of Energy under Contract No. DE-AC05-00OR22725 with UT-Batelle, LLC. Accordingly, the U.S. Government retains a non-exclusive, royalty-free license to publish or reproduce the published form of this contribution, or allow others to do so, for U.S. Government purposes.

References

1. S. Alam, and J. Vetter, A Framework to Develop Symbolic Performance Models of Parallel Applications, 5th International Workshop on Performance Modeling, Evaluation, and Optimization of Parallel and Distributed Systems (PMEO-PDS 2006). Held in conjunction with IPDPS 2006.
2. George S. Almasi, Calin Cascaval, José G. Castaños, Monty Denneau, Wilm E. Donath, Maria Eleftheriou, Mark Giampapa, C. T. Howard Ho, Derek Lieber, José E. Moreira, Dennis M. Newns, Marc Snir and Henry S. Warren Jr, Demonstrating the scalability of a molecular dynamics application on a Petaflop computer, Proceedings of Int'l Conf. Supercomputing, 2001.
3. D. Bailey, E. Barszcz et al., The NAS Parallel Benchmarks (94), NASA Ames Research Center, RNR Technical Report RNR-94-007, 1994, http://www.nas.nasa.gov/Pubs/TechReports/RNRreports/dbailey/RNR-94-007/RNR-94-007.html.
4. S. Browne, J Dongarra, N Garner, G. Ho, P Mucci, A Portable Programming Interface for Performance Evaluation on Modern Processors, The International Journal of High Performance Computing Applications, Volume 14, number 3, Fall 2000.
5. Darren J. Kerbyson, Henry J. Alme, Adolfy Hoisie, Fabrizio Petrini, Harvey J. Wasserman, M. Gittings: Predictive performance and scalability modeling of a large-scale application. Proceedings of Int'l Conf. Supercomputing, 2001.
6. M. Ohmacht, R. A. Bergamaschi, S. Bhattacharya, A. Gara, M. E. Giampapa, B. Gopalsamy, R. A. Haring, D. Hoenicke, D. J. Krolak, J. A. Marcella, B. J. Nathanson, V. Salapura, M. E. Wazlowski, Blue Gene/L compute chip: Memory and Ethernet subsystem, IBM Journal of Research and Development, Vol. 49, No. 2/3, 2005.
7. A. Snavely, L. Carrington, N. Wolter, J. Labarta, R. Badia A. Purkayastha., A Framework for Performance Modeling and Prediction, Proceedings of Int'l Conf. Supercomputing (electronic publication), 2002.
8. J. Vetter, S. Alam, T. Dunigan, M. Fahey, P. Roth and P. Worley, Early Evaluation of the Cray XT3, 20th IEEE International Parallel & Distributed Processing Symposium (IPDPS), 2006.
9. T. Yang, X. Ma and F. Mueller, "Predicting Parallel Applications' Performance Across Platforms using Partial Execution," ACM/IEEE Supercomputing Conference, 2005.
10. MATLAB, http://www.mathworks.com/products/matlab/.
11. Octave, http://www.gnu.org/software/octave/.

Tuning Application in a Multi-cluster Environment

Eduardo Argollo[1], Adriana Gaudiani[2], Dolores Rexachs[1], and Emilio Luque[1]

[1] Computer Architecture and Operating System Department,
Universidad Autónoma de Barcelona. 08193 Barcelona, Spain
eduardo.argollo@aomail.uab.es,
{Dolores.Rexachs, Emilio.Luque}@uab.es
[2] Instituto de Ciências, Informática, Universidad Nacional de General Sarmiento,
Buenos Aires, Argentina
agaudi@ungs.edu.ar

Abstract. The joining of geographically distributed heterogeneous clusters of workstations through the Internet can be a simple and effective approach to speed up a parallel application execution. This paper describes a methodology to migrate a parallel application from a single-cluster to a collection of clusters, guaranteeing a minimum level of efficiency. This methodology is applied to a parallel scientific application to use three geographically scattered clusters located in Argentina, Brazil and Spain. Experimental results prove that the speedup and efficiency estimations provided by this methodology are more than 90% precision. Without the tuning process of the application a 45% of the maximum speedup is obtained whereas a 94% of that maximum speedup is attained when a tuning process is applied. In both cases efficiency is over 90%.

1 Introduction

The usage of heterogeneous clusters of computers to solve complex scientific problems became ubiquitous in universities departments around the globe. These systems represent a cost-effective tool to achieve data intensive computation. The joint of these clusters for a parallel application execution could enhance the application's problem study by achieving faster results or by increasing its dimension.

The evolution of Internet made usual the interconnection and organization of distributed clusters in computational grids [1]. Although it is not a trivial matter to reach efficiency in heterogeneous clusters [2] and, despite the simplicity of physically interconnecting them, this complexity is increased in a multi-cluster environment [3]. It has been shown that parallel applications written for a single cluster do not run efficiently without modifications on multi-cluster systems [4].

The difficulties are even magnified when Internet represents the multi-cluster interconnection network because of Internet's unpredictable latency, throughput and performance limitations. The study of workload distribution policies is then crucial for obtaining applications speedup in such a heterogeneous and unstable environment.

This paper describes a methodology to migrate master-worker parallel applications from their original cluster to a multi-cluster environment. The proposed methodology targets to decrease the execution time in the multi-cluster environment guaranteeing a pre-established threshold level of efficiency. The methodology "inputs" are the desired efficiency, some characteristics of the application (computation and communication volume) and multi-cluster system (computers performance and networks throughputs).

The methodology described in this paper is based in a developed hierarchical master-worker system architecture and an analytical system model [5]. Hierarchical approaches proved to be a viable alternative for handling the communication heterogeneity in a multi-cluster [6, 7].

The system architecture enables the collection of clusters to be seen as a single entity, allowing the transparent transition of a master-worker application to the multi-cluster environment, overcoming possible problems with Internet communication [8]. The analytical system model is based on the computation-communication analysis and evaluates which level of collaboration is possible (if possible) between clusters. Before the methodology is used, the application is adapted to the architecture.

The methodology can be divided in three basic phases: Local Cluster Analysis, Multi-Cluster Analysis, and Application Tuning. The Local Cluster Analysis evaluates the possibility for the application in the original cluster to get performance contribution from external clusters.

The Multi-Cluster Analysis evaluates the speedup that each available external cluster can add to the local one executing the specific application. Each cluster's resources are then selected to reach the evaluated speedup respecting the target efficiency.

The Application Tuning phase evaluates the possibilities of tuning the application for improving the reachable performance. For doing this the methodology provides some guide and parameters values to the application developer.

We present our methodology applying it to the execution of a complex scientific application migrated to a multi-cluster system composed of three clusters located in Argentina, Brazil and Spain. The selected application was described in [9] and it studies the problem of short-term memory storage in the central nervous system through simulations of a ring of bistable oscillators. The phenomenon is called Stochastic Resonant Memory Storage Device (SRMSD).

Experiments to validate the methodology by checking the predicted and the real obtained efficiency and speedup are also shown aside the methodology explanation. The experiments proved that for this application it is possible, without the application tuning changes, to reach 45% of the maximum speedup with a level of efficiency over 90%. After the tuning process, based on the Application Tuning recommendations, 94% of the maximum speedup was achieved, maintaining the efficiency over 90%.

The following sections present our study in further detail. Section 2 describes the methodology phases. The SRMSD problem, the system architecture and the basic adaptation of the application to the proposed architecture are explained in section 3. In section 4 the three phases methodology is applied to the selected application, to show the application tuning process for the above mentioned multi-cluster system. Finally, conclusions and further work are presented in section 5.

2 Migrating Form a Single Cluster to a Multi-cluster System

The methodology to adapt a parallel application to a heterogeneous multi-cluster environment has as basic goals the *decrease of the execution time*, maintaining the *efficiency* over *a selected threshold*.

The methodology is divided in three basic phases providing an estimated efficiency and speedup, and the selected resources to be used in each cluster. The methodology flow diagram is shown in Fig.1.

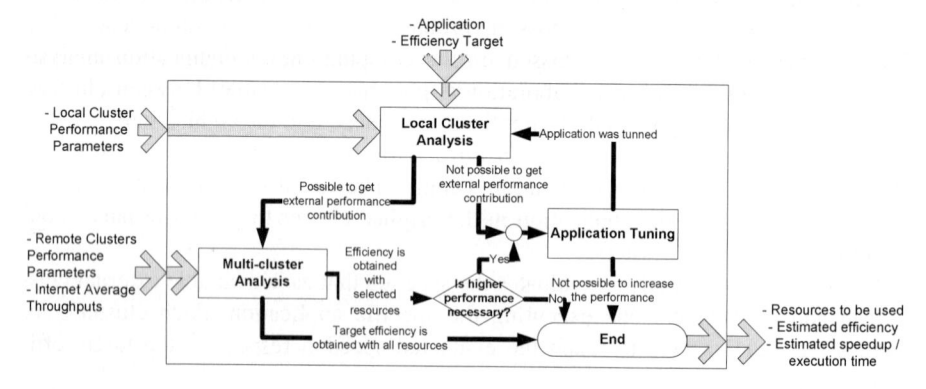

Fig. 1. A flow diagram of the multi-cluster tuning methodology

The speedup is the metric for measuring the performance improvement. It is defined as the ratio between the application original execution time in the local base cluster and the execution time in the multi-cluster system.

For parallel applications in homogeneous clusters the efficiency is defined as the ratio between the execution speedup and the number of computers. This ratio is not directly appropriate when heterogeneous computers are used [10]. For our work efficiency is redefined as the ratio between the obtained and available performances.

The first phase is the Local Cluster Analysis that requires the algorithms communication volume and the local cluster *performance parameters* (the performance of each computer running the application and the local area network throughput) to evaluate if the application is apt to receive external clusters' performance collaboration.

When the remote clusters can collaborate raising up the speedup then the Multi-Cluster Analysis is the following phase. On the contrary, the application should be tuned to be able to use the multi-cluster resources. The Multi-Cluster Analysis has as "inputs" the *performance parameters* of all the remote clusters and the average value of Internet throughput between the clusters.

The Multi-Cluster Analysis identifies for each remote cluster the resources that can collaborate with the local cluster, keeping the efficiency threshold. This analysis also provides the attainable speedup for the application execution using these resources.

There are two possible outputs from the Multi-Cluster Analysis. The first output, when the obtainable performance is satisfactory, leads to the end of the process, the other output, for improving the performance, is the Application Tuning phase.

The Application Tuning evaluates the possibilities of adapting the application data distribution strategy for improving the collaborative performance. Some *performance recommendations* are presented to guide the possible application changes. If the application is changed the application needs to be re-evaluated.

Detailed explanations of the methodology phases are provided in section 4.

3 The Testbed Description: The Application and the Architecture

3.1 The SRMSD Problem and Simulation Application

The Stochastic Resonant Memory Storage Device SRMSD application [9] represents a numerical simulation to study the response of a system of bistable oscillators coupled unidirectional driven by a source of external noise and a periodic and temporal stimulus.

The physical phenomena underlying such short-time storage is that of stochastic resonant (SR) [11], that is, the presence of external noise is essential in sustaining the stored information for an appreciable time once the external stimulus has disappeared. Under this condition the set up acts as a SRMSD.

The application is used to perform several numerical simulations involving rings with different numbers of links and delay times in the coupling. The result is the evaluation of the power-spectral density of the first oscillator during the travelling signal loops, averaging it over a suitable ensemble of N initial conditions in order to hinder fluctuations. The windowed Fourier transform is used to provide a picture of the decaying process.

The SRMSD model and its simulation in a single-cluster environment were previously developed by the Argentinean research group using the master-worker paradigm [9]. The master distributes to all workers the command to execute one simulation. The worker then generates a random set of initial conditions and simulates the traveling signal, sending back to the master a tri-dimensional matrix representing the simulation result. When the master receives one result back, it sends to this worker another simulation command, reaching a dynamic load balancing in the heterogeneous cluster.

When all the N simulations matrices results are received and added by the master, the master calculates the average value of these tri-dimensional matrices, writes it in a file and terminates the program execution.

3.2 Multi-cluster Architecture

To interconnect clusters in such way that it is possible to reach efficient collaboration in a multi-cluster environment a run-time architecture was created. This architecture is a hierarchical master-worker on which the remote clusters are called sub-clusters, with their sub-masters and sub-workers.

The architecture supports the overlapping of computation and communication in the workers and a dynamic on-demand tasks distribution policy. To isolate the local network and Internet, overcoming the problems on the Internet communication between clusters, the architecture includes a module called Communication Manager (CM) [8].

The required modifications in the single-cluster SRMSD application to be executed in the multi-cluster system were to add the CM module and to allow the sub-master to execute receiving its amount of work from the main cluster. The integration of the CM simply implies the creation of a special worker process that communicates through Internet. For the master, this "special worker" acts like any worker with the difference that it has a higher performance (correspondent to the whole remote cluster) and a higher latency (because of the use of Internet). Fig. 2 shows the three testbed clusters.

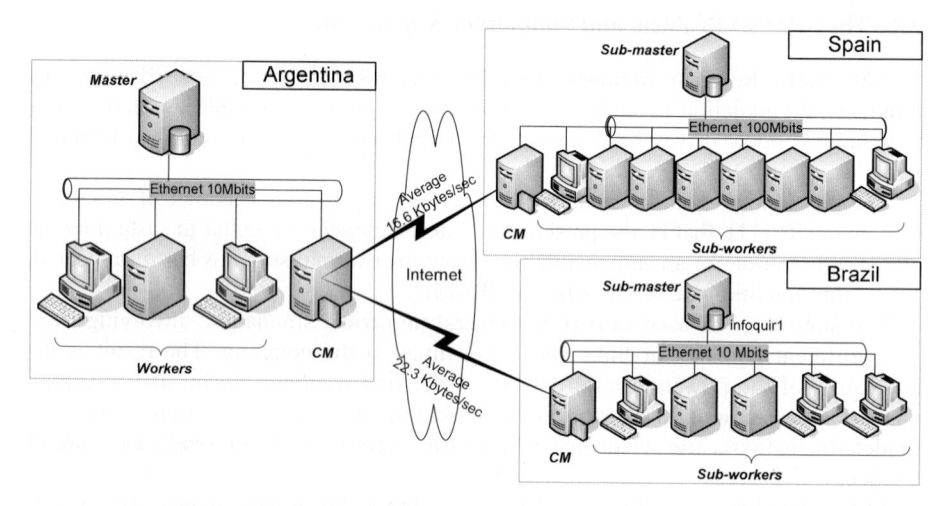

Fig. 2. Multi-cluster testbed system with the local cluster in Argentina

4 Applying the Methodology to the SRMSD

Throughout this section the methodological phases are applied to the SRMSD application migrating it from its original single-cluster version (Argentina) to the multi-cluster (Argentina, Brazil and Spain) scenario. The target threshold efficiency is fixed in 85% and the SRMSD application is executed for a set of N=500 initial conditions.

The Fig. 3 shows the methodology flow diagram with the internal structure of its phases – Local Cluster Analysis (LCA), Multi-Cluster Analysis (MCA) and Application Tuning (AT).

4.1 Local Cluster Analysis

The main process inside the LCA is to apply the computation-communication analysis [5] to the local cluster. This analysis is oriented to the estimation of the speedup and efficiency that can be obtained in the parallel application execution.

If the analysis' concludes that the application is not able to obtain the whole *Available Performance* in the local cluster then the application should be tuned (AT phase). Otherwise the next phase is the MCA.

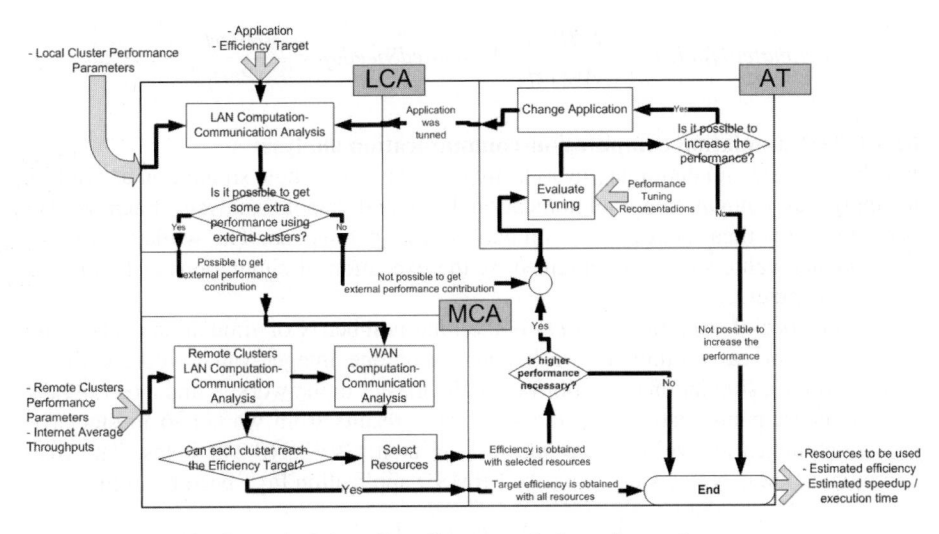

Fig. 3. Methodology flow diagram and phases internal structure

a) General computation-communication analysis

Considering the master does not represent the bottleneck, a parallel application execution time is either limited by the computers performance (computation-bounded) or by the network throughput (communication-bounded). The *Maximum Performance* (*MaxPerf*) for a specific execution is achieved when the execution is computation-bounded. This happens when the application's computation time is greater than or equal to its communication time.

For a worker task running on a processor, the *computation time* (T_{Cpt}) is defined as the ratio between the task number of *operations* (*Oper*) and the processor *performance* (*Perf*): $T_{Cpt}=Oper/Perf$. The *communication time* (T_{Comm}) is the ratio between the volume of *data communication* (*Comm*) (worker task data from and to the master) and the *network throughput* (*TPut*): $T_{Comm}=Comm/TPut$. The *MaxPerf* is the performance that can be obtained when $T_{Cpt} \geq T_{Comm}$ (Eq. 1).

In a multi-cluster environment there are two communication levels: intra-cluster and inter-clusters. To calculate the *MaxPerf* for the intra-cluster (*MaxPerf$_{intra}$*) it is necessary to consider the local-area throughput and for the *MaxPerf* for the inter-cluster (*MaxPerf$_{inter}$*) the Internet average throughput between the clusters.

The *Available Performance* (*AvPerf*) is the addition of each worker performance executing standalone the application tasks. The *Estimated Performance* (*EstPerf*) that a cluster can provide to the multi-cluster system is then minimum value between its AvPerf, *MaxPerf$_{LAN}$* and *MaxPerf$_{Inet}$* (Eq.2). The *Estimated Efficiency* for each cluster is the ratio between the cluster *EstPerf* and *AvPerf* while the *Estimated Speedup* is the ratio between the cluster *EstPerf* and the local cluster *EstPerf* (*EstPerf$_{LC}$*) (Eq. 3).

$$T_{Cpt} \geq T_{Comm} \Rightarrow MaxPerf = Perf \leq \frac{Oper*TPut}{Comm} \quad (1)$$

$$EstPerf = \min(AvPerf, MaxPerf_{LAN}, MaxPerf_{Inet}) \quad (2)$$

$$EstimatedEfficiency = \frac{EstPerf}{AvPerf} \; ; \; EstimatedSpeedup = \frac{EstPerf}{EstPerf_{LC}} \qquad (3)$$

b) SRMSD simulation computation-communication analysis

For the SRMSD simulation application we consider each single simulation as the task to apply our computation-communication local and remote analysis. Each worker performance is then measured in simulation tasks per second. The worker *available performance* value should be obtained by the execution of simulations tasks in each cluster computer.

The number of operations (*Oper*) is the total number N of simulations. The communication for each simulation task consists of one integer value (representing a command for a simulation execution) from the master to the worker, and a 601x31x31 float-point elements matrix with the simulation results from worker to master. The total communication volume for one task is then CV=2,310,248 bytes. The total communication is N times CV. For the SRMSD application Eq.1 become in Eq.4.

$$MaxPerf = \frac{N*Tput}{N*CV} = \frac{Tput}{CV} \qquad (4)$$

In this Local Cluster Analysis step the efficiency and speedup are evaluated for the local cluster (Argentina). Table 1 shows the testbed parameters, and the estimated and experimentally obtained values for the Speedup and the Efficiency in the Argentinean Cluster. The testbed parameters are the number of computers, the LAN throughput and the *Available Performance* in tasks/second. The Eq. 3 were used to estimate the speedup and efficiency. The comparison from the experimental speedup and the estimated one is our estimation precision.

Table 1. Testbed parameters, estimated and experimental Speedup and Efficiency, and estimation precision for the local cluster stand alone execution of the application

Cluster	#Com	LAN TPut (Mbytes/sec)	AvPerf (10^{-3} Tasks/sec)	Estimated Speedup	Estimated Efficiency	Experimental Speedup	Experimental Efficiency	Estimation Precision
Argentina	4	1	1.754	1.000	100%	0.918	92%	92%

The experimental data shows that a high-level of efficiency is possible although the theoretical peak could not be achieved. The differences between the estimated and experimental efficiency values are mainly caused by light load-imbalance. Since the application is able to execute efficiently in the local cluster, the next phase is the MCA.

4.2 Multi-cluster Analysis

The first step of the MCA is to apply the computation-communication analysis to each remote cluster. The MCA parameters, the estimated and experimental Speedup and Efficiency values for the clusters are displayed in Table 2.

Table 2. Testbed parameters, estimated and experimental Speedup and Efficiency, and estimation precision for the multi-cluster execution of the application with all resources

Cluster	#Workers	Inet Tput (Kbytes/sec)	AvPerf (10^{-3} Tasks/sec)	Estimated Speedup	Estimated Efficiency	Experimental Speedup	Experimental Efficiency	Estimation Precision
Argentina	3		1.754	1.000	100%	0.886	89%	89%
Brazil	5	25.3	3.106	1.771	100%	1.621	92%	92%
Spain	8	21.0	23.570	5.307	39%	5.795	43%	92%
Total	16		28.430	8.077	50%	8.302	51%	97%

From the Table 2 we can infer that it is possible to get a speedup of 8 for the application execution using the external clusters. The experimental speedup of the Spanish cluster is higher than estimated because during the experiments the real Internet average throughput was 23.0 Kbytes/sec, greater than the estimated average used in the analysis.

The estimation data in Table 2 shows that the target threshold efficiency for the cluster in Spain is not reachable. To attain the efficiency threshold it is necessary to select the resources to be used in this cluster. This adjustment is done by selecting the workers for which the addition of *Available Performance* is approximate to the *Estimated Performance* (Eq. 3). Table 3 shows the data using just the selected computers of the Spanish cluster.

This experiment shows that the speedup decreased from 8.3 to 7.3 while the efficiency of the Spanish cluster increased from 39% to 96%. This means a better utilization of the resources or a lower cost, in case of paying for remote resources utilization.

Even though the execution speedup can be 8, comparing the total *AvPerf* and the Argentinean cluster *AvPerf* in Table 2, we can infer that potential speedup of the multi-cluster is 16.2. To try to get the maximum speedup it is necessary to apply the Application Tuning phase.

Table 3. Testbed parameters, estimated and experimental Speedup and Efficiency, and estimation precision for the multi-cluster execution of the application with selected resources

Cluster	#Workers	Inet Tput (Kbytes/sec)	AvPerf (10^{-3} Tasks/sec)	Estimated Speedup	Estimated Efficiency	Experimental Speedup	Experimental Efficiency	Estimation Precision
Argentina	3		1.754	1.000	100%	0.858	86%	86%
Brazil	5	25.3	3.106	1.771	100%	1.621	92%	92%
Spain	3	21.0	8.885	5.065	100%	4.863	96%	96%
Total	11		13.745	7.836	100%	7.341	94%	94%

4.3 Application Tuning

At the MCA we concluded that, without modifications, the SRMSD application can not obtain the whole of the Spanish cluster performance. Based on the data in Table 2, the value for the Spanish cluster $MaxPerf_{inter}$ (Eq.4) is of 9.30×10^{-3} tasks/sec. Comparing this value with the Spanish cluster *Available Performance* (23.57x10^{-3} tasks/sec), we can conclude that Internet communication between Spain and Argentina is saturated.

In this case, the Application Tuning (AT) phase intends to improve the application computation-communication ratio to enhance the obtained performance. It is necessary then to evaluate the increase in the inter-cluster granularity, executing more tasks with the necessity of less communication.

The final operation at the SRMSD application is to average the different simulations results. The granularity could be changed if sub-masters perform the addition of workers partial results and just communicates to the local master the result of a certain number S of additions. This granularity change means that each result simulation communicated (CV bytes) represents S simulation tasks. The number S of simulations that need to be aggregated before sending the results depends on the Internet throughput. S should be small enough to avoid load imbalance and big enough to reach the desired performance.

Once the tuning strategy is set, the application needs to be changed and re-evaluated. Nothing was changed in the local execution and the LCA can be skipped. On the remote execution, for N simulation tasks (*oper*), (N/S)*CV bytes are communicated (*comm*). Applying it in the Eq. 2, the new $MaxPerf_{Inet}$ is Eq. 5.

$$MaxPerf_{Inet} = \frac{S * Tput}{CV} \qquad (5)$$

The Spanish cluster has an *Available Performance* of 23.57 x10^{-3} tasks/sec. Using this as the $MaxPerf_{Inet}$ in the Eq. 5 we can conclude the number of aggregated simulations S should be 2.53. The amount of simulations needs to be an integer and for choosing its value it is necessary to evaluate the impact of this number in the balance of the load.

For S=2 we would not achieve the totality of the Spanish performance. For S=3 the load-imbalance is increased. Since the value 3 is still not significant when compared to the total number of tasks (N=500) we choose this value for the application tuning.

Table 4 shows the testbed parameters, the estimated and experimental Speedup and Efficiency for the tuned application. This table shows that with the AT the speedup was increased from 7.34 to 15.33. The efficiency in all clusters was 95% and, since all the clusters' resources were used, this means 95% of the total *AvPerf*.

Table 4. Testbed parameters, estimated and experimental Speedup and Efficiency, and estimation precision for the multi-cluster execution of the tuned application

Cluster	Testbed parameters		Theoretical peak performance		Real performance		Estimation Precision
	Inet Tput (Kbytes /sec)	AvPerf (10^{-3} Tasks /sec)	Estimated Speedup	Estimated Efficiency	Experimental Speedup	Experimental Efficiency	
Argentina		1.754	1.000	100%	0.889	89%	89%
Brazil	25.3	3.106	1.771	100%	1.594	90%	90%
Spain	21.0	23.570	13.437	100%	12.848	96%	96%
Total		28.430	16.208	100%	15.332	95%	95%

Table 5 presents a summary of the predicted values and the experimental results obtained as results of the different phases of the proposed methodology.

Table 5. Estimation and experiments speedup, efficiency and execution time comparison for different steps along the methodology

Description	Estimated Indexes Values		Experimental Indexes Values		Prediction Precision
	Speedup	Efficiency	Speedup	Efficiency	
Original application single cluster execution.	1	100%	0.918	92%	92%
Intermediate evaluation in the MCA phase: All clusters with all resources.	8.077	50%	8.302	51%	97%
After the MCA phase: All the clusters with selected resources.	7.836	100%	7.341	94%	94%
After the AT phase: All the clusters with all resources **with tuned application**	16.208	100%	15.332	95%	95%

5 Conclusions and Future Work

A multi-cluster environment using Internet as inter-communication network represents a cost-effective way to speedup the execution of scientific applications. This paper presented a methodology for adapting a parallel application from a single cluster to a multi-cluster environment, keeping a threshold level of efficiency. The methodology phases were presented and an example application was used to follow the methodological steps.

To validate the methodology, experiments were done in different stages. The results proved that the estimation is over 90% precision.

The application execution time was reduced to 12% its original single cluster time just with the adaptation to the proposed architecture. When the application was tuned the execution time was reduced to 6% its original value. The efficiency of the system was kept over 90%.

Our methodology should be considered as a low level, direct resources utilization of computational grid environments. Therefore it provides an upper limit of the environment utilization, providing a good prediction of the maximum available speedup guarantying a defined resources utilization efficiency.

Future lines are to extend the multi-cluster computation-communication model to other parallel programming paradigms and to include in the model other system parameters (memory size, cache size, ...).

References

[1] I. Foster. "The grid: A new infrastructure for 21st century science". Physics Today, 55(2), pp 42–47, 2002.
[2] Olivier Beaumont, Arnaud Legrand, and Yves Robert. "The master-slave paradigm with heterogeneous processors". IEEE Trans. Parallel Distributed Systems, 14(9):897-908, 2003.

[3] B. Javadi, M. Akbari and J. Abawajy. "Performance analysis of heterogeneous multi-cluster systems". ICPP 2005, pp 493-500, 2005.
[4] Bal, H.E.; Plaat, A.; Bakker, M.G.; Dozy, P.; Hofman, R.F.H., "Optimizing parallel applications for wide-area clusters". Proceedings of IPPS/SPDP 98, pp.784-790, 1998.
[5] E. Argollo, D. Rexachs, F. Tinetti and E. Luque. "Efficient Execution of Scientific Computation on Geographically Distributed Clusters". In LNCS vol. 3732, 2004.
[6] Aida, K.; Natsume, W. & Futakata, Y. "Distributed computing with hierarchical master-worker paradigm for parallel branch and bound algorithm" CCGrid 2003, pp. 156-163, 2003.
[7] Nieuwpoort, R.V.; Kielmann, T. & Bal, H.E. "Efficient load balancing for wide-area divide-and-conquer applications". PPoPP '01, ACM Press, 34-43, 2001
[8] E. Argollo, J. R. de Souza, D. Rexachs, and E. Luque. "Efficient Execution on Long-Distance Geographically Distributed Dedicated Clusters". In D. Kranzlmüller et al. editors, Proceedings of the 11th Euro PVM/MPI 2004, LNCS vol. 3241, pages 311--319, 2004.
[9] M.F. Carusela, R.P.J.Perazzo and L. Romanelli. "Stochastic resonant memory storage device". Physical Review, 64(3 pt 1):031101, 2001.
[10] L. Colombet, and L. Desbat. "Speedup and efficiency of large-size applications on heterogeneous networks". Theoretical Computer Science, 196 ,pp 31-44, 1998.
[11] Bruce McNamara and Kurt Wiesenfeld. "Theory of stochastic resonance". Physical Review, A 39, pp 4854 – 4869, 1989.

Analyzing the Interaction of OpenMP Programs Within Multiprogramming Environments on a Sun Fire E25K System with PARbench

Rick Janda, Wolfgang E. Nagel, and Bernd Trenkler

Center of Information Services and
High Performance Computing
Dresden University of Technology
01162 Dresden, Germany
`rick.janda@zih.tu-dresden.de`, `bernd.trenkler@tu-dresden.de`,
`wolfgang.nagel@tu-dresden.de`

Abstract. Nowadays, most high performance computing systems run in multiprogramming mode with several user programs simultaneously utilizing the available CPUs. Even though most current SMP systems are implemented as ccNUMA to reduce the bottleneck of main memory access, the user programs still interact as they share other system resources and influence the scheduler decisions with their generated load. PARbench was designed to generate complete load scenarios based on synthetic jobs and to measure the job behavior during the execution of these scenarios. The E25K is a ccNUMA system with up to 72 dual core CPUs and a crossbar-based connection network. This paper describes the results of the examination of such a Sun Fire E25K system with PARbench. First, PARbench was used to investigate the performance impact caused by the interactions of jobs on fully loaded and overloaded machines. Second, the impact of operating system tasks to the performance of OpenMP parallelized programs in scenarios of full load as created by the cluster batch engine is quantized, especially when these system tasks are not considered in the initial load calculation. Additionally, the generated scenarios were used for a statistical analysis of the scheduling of OpenMP programs, focusing on data locality and migration frequency.

1 Introduction

Current installations of high performance computing systems often contain systems with several hundred processors. Not all user programs need this huge amount of CPUs. Thus, the systems run several user jobs simultaneously in multiprogramming mode. While these jobs can often use a subset of the available CPUs almost exclusively, they nevertheless share common resources like data connections, caches, or the I/O subsystem. The scheduler may also migrate jobs to other CPUs in order to assign resources equitably. Such migrations and saturated memory connections are major causes of performance degradations. The performance impact becomes more and more substantial as the processor/memory speed gap widens.

W.E. Nagel et al. (Eds.): Euro-Par 2006, LNCS 4128, pp. 89–98, 2006.
© Springer-Verlag Berlin Heidelberg 2006

OpenMP is a widely used approach to parallelize calculations on SMP systems. In the past a lot of work has been presented that assesses the performance of OpenMP programs on dedicated SMP systems or measures the runtime of different OpenMP directives. However, these benchmarks assess the machine's performance under quite favorable circumstances and say nothing about the interaction of several user jobs in production environments. They also do not consider the load generated by the operating system itself, that may be much higher in real production environments than in benchmark situations.

PARbench was designed to address exactly this issue. It permits the user to generate synthetic jobs with various characteristics based on sequences of simple benchmark kernels. After that, several of these jobs can be executed simultaneously. This permits the composition of almost arbitrary workload scenarios. During the execution of the whole scenario the runtime and the CPU time of each job is measured. Thereby, PARbench can not only generate sequential jobs but even tightly coupled parallel programs based on OpenMP. The first version of PARbench had been designed from 1988-1990 to measure the interaction of several programs during execution in multiprogramming mode [1], [2]. In 2001 PARbench was ported to the SGI Origin 3800 by Sebastian Boesler and the use of OpenMP as a standard for parallelization was introduced [3]. Meanwhile, the vector systems NEC SX-4 and SX-5 and the IBM p690 series were also analyzed [4], [5].

Object of this investigation were the Sun Fire E25Ks of the RWTH Aachen with 72 dual core UltraSPARC IV CPUs each, running on Solaris 9. The PARbench code was compiled with the Sun Studio 9 Compiler Collection. A comprehensive view of the architecture of the Sun Fire E25K can be found in [6]. More details about the UltraSAPRC IV CPU can be found in [7].

The work is parted in the following sections: The first part of the investigation will be a general assessement of the scalability of the crossbar-based ccNUMA architecture with dual core CPUs. The second part will examine the operation mode driven by the cluster batch engine (Sun N1 Grid Engine 5.3). The performance impact to OpenMP programs due to operating system tasks, which are not considered in the load calculation of the batch engine, will be elaborated. The last part is dedicated to a statistical analysis of the scheduling and will reveal some weaknesses in the treatment of OpenMP programs.

2 Performance of Sequential Jobs Under Full Load and Overload

The main reason for applying the ccNUMA for large symmetric multiprocessor systems is to effectively widen the memory access bottleneck in comparison to UMA systems. Therefore, ccNUMA systems should scale significantly better for a larger number of CPUs. The first question, that was to be answered by this investigation, is how well the E25Ks do scale. The amount of interaction between the user jobs, due to sharing system resources and maintaining cache coherency in the system, was examined.

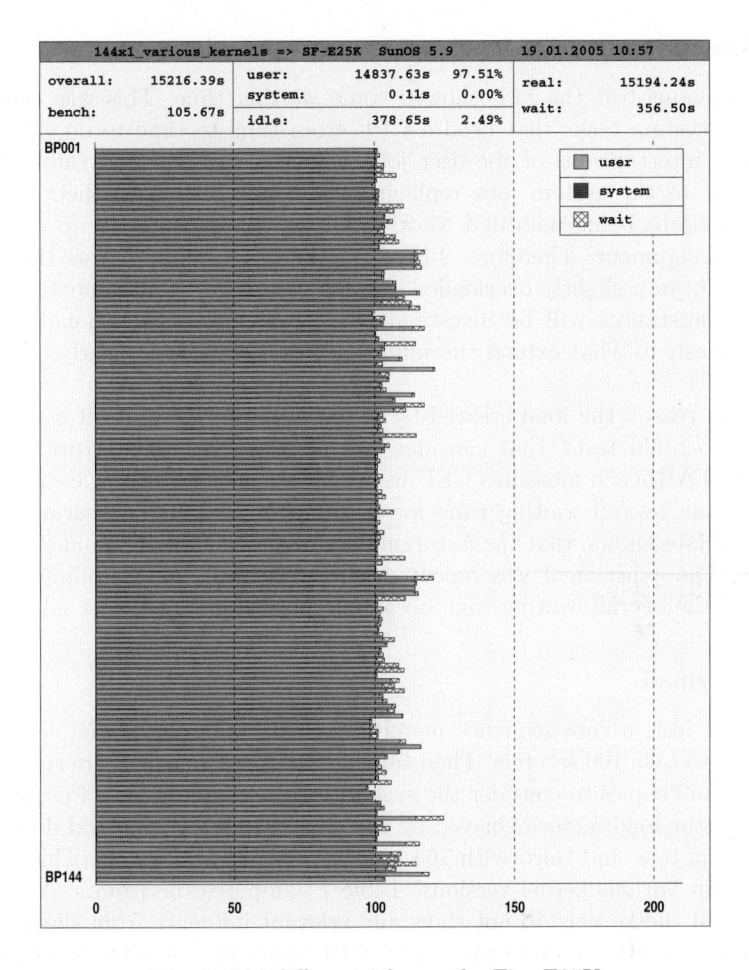

Fig. 1. 144 different jobs on the Fire E25K

2.1 Full Load

To obtain a first overview about the scalability of the architecture, 36 jobs with different kernel numbers and 100 seconds runtime each were generated. These kernels covers all of the 17 internally used math cores, do not perform any I/O operations and span as much as possible of the available data matrices to run out of the caches and stress the memory subsystem. After the generation all generated jobs were executed simultaneously with four copies each to achieve 144 jobs. The result is shown in fig. 1. Some jobs clearly distinguish from other by their increased CPU usage but none of the jobs suffers from a crucial performance impact. Even the most affected job took only 120 seconds which correspond to a relative increase of just 20%. From this point of view the hardware scales very well.

2.2 Specifying the OS Load

In the previous test the jobs showed some waiting time. This was caused by operating system tasks that need a CPU from time to time to do their work. The short interruptions of the user jobs may lead to increased rates of cache misses due to the system jobs replacing the cache lines with their own data or the user jobs being migrated to other CPUs by the scheduler to assure fair resource assignment. Therefore, 144 user jobs at once as well as the system tasks results in a slightly overloaded system. The impact on sequential jobs by these circumstances will be investigated in a next step. Additionally it may be of interest, to what extend the influence increases within clearly overloaded systems.

For this reason the load created by the operating system itself is quantified in order to build tests, that consider this load and avoid interruption of the user jobs. PARbench measures CPU usage time and real time for every job and calculate the overall waiting time for a scenario. A simple scenario with 144 sequential jobs shows, that the jobs remained without CPU in about 2% of their runtime. The experiment was repeated with gradually reduced number of user jobs until the overall waiting time became almost zero at 140 user jobs.

2.3 Overload

For a next test, a core sequence merely based on only one kernel version was generated to take 100 seconds. Then several copies of this jobs were run at once, first with 140 copies to consider the system tasks, second with 144 copies which comply to the load factor achieved by the cluster batch engine and do not consider system task and third with 164 jobs to overload the system. This test was repeated for various kernel versions. Table 1 compares the results. As one can see, most of the kernels do not show any relevant influence from slight or conspicuous overload in comparison to their CPU usage time in a full loaded system with 140 user jobs plus system jobs. Only version 241, 281 and 301 consume recognizable more CPU time on the clearly overloaded system. For a analysis of this behavior, some additional data from the jobs, like cache usage and percentage of write-accesses, is needed. This can be achieved with the performance counters of the UltraSPARC IV CPU but is not yet included in PARbench. The code of the math cores, however, reveals that these cores contains some really odd access patterns to confuse the cache usage and stress the memory subsystem. Hence, their generated load in the memory subsystem is not typical for scientific computations. In the outcome sequential jobs are practical not affected by the system tasks.

3 Influence of Operating System Processes on OpenMP Programs

The cluster batch engine of the RWTH Aachen uses a simple scheme to avoid overloading the systems. The users have to specify the number of processes or threads

Table 1. Relative increase in CPU usage time with raising load factor for different kernel versions

Kernel version	MREFS	FLOPS	Relative average CPU time			
			single	140x	144x	164x
126	765.7	770.1	100%	100.1%	100.1%	100.1%
111	406.9	638.6	100%	100.2%	100.4%	100.2%
101	451.3	702.3	100%	100.4%	100.3%	100.4%
151	919.8	893.2	100%	100.4%	100.3%	100.4%
81	205.2	410.6	100%	102.5%	102.6%	102.4%
226	669.0	79.0	100%	104.5%	104.5%	103.0%
61	71.5	286.0	100%	106.7%	106.5%	106.5%
246	311.2	1.4	100%	107.4%	107.4%	107.5%
221	485.1	54.1	100%	111.5%	113.0%	108.4%
121	309.5	309.6	100%	109.1%	108.9%	108.8%
141	130.7	126.7	100%	118.6%	118.6%	118.4%
281	8.5	0.0	100%	**117.6%**	**120.0%**	**122.0%**
241	325.2	0.1	100%	**109.1%**	**109.1%**	**124.7%**
286	75.1	0.6	100%	124.6%	124.7%	124.8%
201	593.9	119.3	100%	125.3%	125.3%	124.8%
301	63.3	31.7	100%	**124.4%**	**127.0%**	**135.2%**

that they want to use for their MPI or OpenMP jobs. The batch engine then executes only as many jobs simultaneously on the according system that each MPI instance and each thread of the OpenMP programs can theoretically use a CPU exclusively. For the E25K systems the batch engine adjust the number of user programs and threads to 144, which is the number of CPU cores. However, operating system tasks are not considered at all, thus, the system runs slightly overloaded. As it could be seen in the previous section, sequential programs do not suffer much from the short interruptions caused by system tasks. On the other hand, parallelized programs, especially fine-grained parallelized programs like OpenMP programs, may experience much more impact from these short interruptions. To ensure data validity, OpenMP programs contain a lot of synchronization barriers, that have to be reached by all worker threads before the calculation can proceed. In order to obtain efficiently parallelized programs, the work will be spread to every worker thread equally. If every thread can use a CPU exclusively, all worker threads will reach the barriers at the same time and the next part of the work is spread to the worker threads immediately. But if some worker threads are interrupted by system tasks, all other threads of the OpenMP program will have to wait on the next barrier for the interrupted thread. Part of the investigation was to quantify these impact on the performance of OpenMP programs.

For this purpose a job with 200 seconds sequential runtime and no I/O was created. This job was generated to average 500 MREFS and 500 MFLOPS and contains about 5000 kernels, which means 5000 implicit barriers for the OpenMP parallelization. The scenario contains nine copies of this job which were parallelized to eight threads each. First these jobs were executed with 72 sequential

Table 2. Average CPU usage time (user+system) and runtime of the parallelized jobs (9x8) with spinning threads (busy waiting) and different background load in the system

Total number of user threads	Test setup	CPU usage time [s] of the par. jobs	Runtime [s] (average)
8	1x8	232.4	29.1
72	9x8	234.7	29.3
138	66x1 9x8	238.7	30.0
140	68x1 9x8	239.5	30.1
142	70x1 9x8	248.2	31.4
144	72x1 9x8	**272.7**	**34.9**

Table 3. Runtime differences of the parallelized jobs between considering operating system tasks and not

Threads/job	Average runtime of the par. jobs [s] for total number of user threads		Rel. extension [%]
	140	144	
8	30.1	34.9	15.9%
16	19.3	26.3	36.3%
24	14.1	22.3	58.2%
32	10.7	20.6	92.5%

jobs simultaneously and then with only 68 sequential jobs in parallel to ensure four free CPUs for the system tasks. Fig. 2 and fig. 3 show the results. Some of the OpenMP jobs ran notably longer without free CPUs for system jobs. To ensure that the impact does not originate from other influences, the measurement was also repeated without the simultaneous sequential jobs to compare CPU usage and runtime with these values. Table 2 contains the average runtime and the average CPU usage for all of the parallelized jobs and the different scenarios. As one can see, the runtime does not differ much for lower system loads but does notably increase with more than 140 user threads in the system[1]. But then again, an increase to 35 seconds compared to 30 seconds is practically neglectable.

After that the examination was expanded to a higher degree of parallelization with up to 32 threads. Again the OpenMP jobs were executed with some sequential jobs simultaneously to fill the system and then with four sequential jobs lesser to avoid interruption by system tasks. Table 3 compares the average runtime of the parallel jobs in the two scenarios for the an increasing number of parallel threads. One can see, that the impact increases rapidly with higher parallelization. The slight overload caused by the operating system has a dramatic impact on the performance of the OpenMP programs with higher parallelization. OpenMP jobs with 32 threads will already run almost twice as long, if the operating system tasks are not considered in the load calculation and slightly overload the system.

---

[1] Sequential programs are counted as one thread.

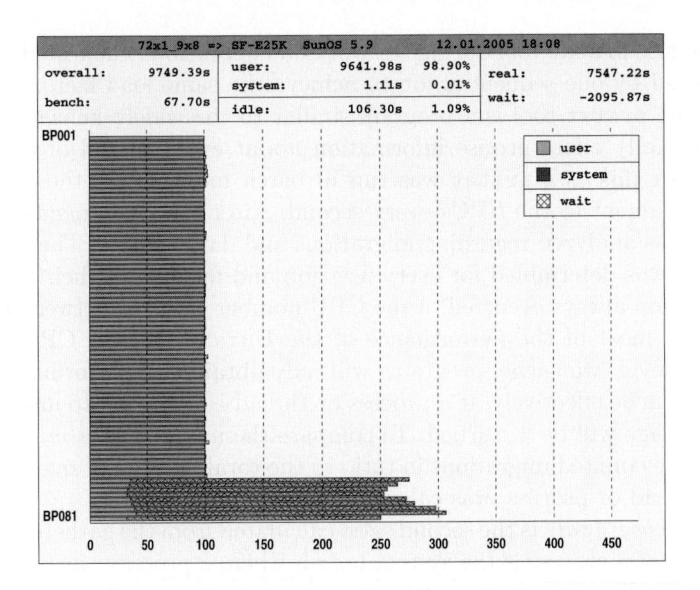

Fig. 2. 144 user threads within the SF-E25K (144 CPUs)

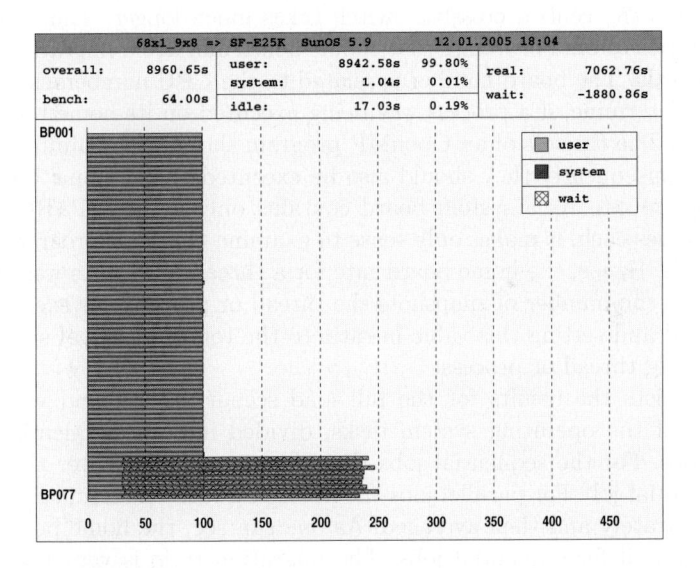

Fig. 3. 140 user threads within the SF-E25K (144 CPUs)

4 Scheduling Analysis

PARbench allows assessing the performance of jobs in different multiprogramming scenarios. The measured interaction, however, may caused by hardware limitation as well as by unfavorable scheduling. To obtain a direct access to the decisions of the scheduler the PARbench startup script was extended in order to run the Solaris `prstat` tool during the workload execution. The tested workloads were reduced by one sequential job to achieve the same load factor as without `prstat`. The `prstat` tool is a program similar to the widely known `top` utility but additionally offers precise information about every thread of the running programs. In this case `prstat` was run in batch mode to log the process and thread assignment to the CPUs every second. After the experiments, the gathered data was analyzed regarding migrations and data locality. The sequence of used CPUs was determined for every user job and for each of their threads.

A migration always occurred, if the CPU number changed between two snapshots. Since most of the performance of the UltraSPARC IV CPU is related to its huge level two cache, programs will only obtain this performance, if they utilize this cache effectively. If a process or thread is migrated to another CPU, the cache usage will be disturbed. To compare the migrations, a *migration rate*, which set the counted migrations in ratio to the total number of snapshots made for each thread or process, was calculated.

The *home board rate* is the second ratio calculated from the gathered data. The home board of a process is the system board, where a process was first executed and where it allocated its data structures in main memory (first touch policy). Within a system board, the latency to main memory is almost the same but data access across the central crossbar switch takes much longer. Thus, the Solaris scheduler tries to bind processes to its home board and avoid migrations to other system boards. The board number is related to the CPU numbers, so it is quite simple to determine, if a process was being executed on its home board during a snapshot. The threads of an OpenMP program should also gain fast access to the data, consequently they should also be executed on the home board of the according process. As a system board contains only 4 UltraSPARC IV CPUs with two cores each, it makes only sense to examine OpenMP programs with up to 8 threads. Hence, the home board rate for a thread or process was calculated by counting the number of snapshots the thread or process was executed on its home board and setting this value in ratio to the total number of snapshots for the according thread or process.

Table 4 lists the results for the full load scenarios with and without consideration of the operating system tasks, divided into the sequential and the parallel jobs. For the sequential jobs the average was taken over the values of each sequential job. For parallel jobs the according rate was calculated for every thread separately and then averaged. As one can see, the home board binding works very well for sequential jobs. The migration ratio is very low, too, and

Table 4. Average home board rate and thread migrations rate separated for sequential an parallelized jobs

Total number of user threads	Average home board rate [%]		Average migration rate [%]	
	sequential (x1)	parallel (x8)	sequential (x1)	parallel (x8)
140	98.8	**38.4**	2.6	8.2
142	96.6	**36.3**	3.1	11.3
144	**95.7**	**38.2**	**5.7**	**16.5**

Table 5. Average executions ratio on the home board separated for the thread numbers

Total number of user theads	Test setup	Home board rate [%] for thread							
		1 (master)	2	3	4	5	6	7	8
140	67x1, 9x8, prstat	**98.7**	83.2	44.4	33.3	**0.0**	11.1	19.1	11.5
142	69x1, 9x8, prstat	**96.7**	62.2	56.3	12.2	**11.5**	22.2	0.0	22.2
144	71x1, 9x8, prstat	**95.7**	77.3	67.6	29.6	**10.1**	21.6	9.3	3.4

does only slightly increase with the small overload caused by the system tasks in the scenario with 144 user threads. The tide turns for parallelized jobs. All threads of the OpenMP programs were only executed one third of the time on the home board and suffer from significantly more migrations than the sequential jobs. OpenMP programs compiled with the Sun Studio 9 compilers allocate all threads at the program's start. In order to save the time for stopping and starting the threads over and over again, temporarily idle threads do busy waiting per default in order not to loose their CPU. This approach is reasonable, if the system does not become overloaded and no other jobs are available to use the freed CPUs. Accordingly, starting and stopping of threads is not the reason for the higher migration rate.

The assumption was that the scheduler does not distinguish between a sequential program and the master thread of an OpenMP program but has no favorable strategy for further threads. Thus, the same data was analyzed again but the values were grouped by the thread number. The values of the sequential jobs and the values of the master thread of the parallel jobs become one group for averaging, the second group contains the second thread of each parallel program and so on. Table 5 lists these average home board rates.

The values for the first thread resemble the values of the sequential jobs in the previous table, which supports the assumption. The binding to the home board rapidly decrease for larger thread numbers. Starting with thread five the threads are practically no more subject to any home board binding and will suffer from slow remote access to their data in the main memory of another CPU board.

5 Conclusions

In this paper, the performance of the Sun Fire E25K in multiprogramming mode was evaluated. The tests with workloads containing different sequential jobs indicate that the hardware scales very well and the ccNUMA architecture with crossbar-based connection network provide sufficient throughput to effectively eliminate the memory access bottleneck.

In contrast to sequential programs, that are not very influenced by the interruptions caused by system jobs, tightly coupled parallel programs suffer much more from those CPU losses. The performance impact increases rapidly with larger number of threads and it was shown how the consideration of the system tasks in the load calculation already reduces the runtime of OpenMP jobs with 32 threads in half. Thus, system tasks could not be ignored in the batch engine's load calculation. On the E25Ks the batch engine should limit the number of user processes and threads to 140 instead of 144, which is the current configuration.

The statistical analysis of the thread scheduling reveals some weaknesses. The first thread, which is the only thread for sequential jobs, is tightly coupled to the board, where the program data was allocated. This ensures fast access to the data and reduce the usage of the central crossbar switch. Further threads are not subject to this tight home board binding and suffer from higher memory access latency for that reason. They also show higher migration rates, which reduce cache utilization. Better thread scheduling could offer some performance increases here. Sun promises much improved thread handling with Solaris 10, which will be subject for further research.

Acknowledgments

The research described in this work was developed using the resources of the computing center of the RWTH Aachen. We would like to thank Dieter an Mey, Andrea Lorenz and Hans-Jürgen Schnitzer as members of the HPC team for providing useful advices and technical assistance.

References

1. Linn, M.A.: Eine Programmierumgebung zur Messung der wechselseitigen Einflüsse von Hintergrundlast und parallelem Programm. Techn. Report Jül-2416, Forschungszentrum Jülich (1990)
2. Nagel, W.E.: Performance evaluation of multitasking in a multiprogramming environment. Techn. Report KF-ZAM-IB-9004, Forschungszentrum Jülich (1990)
3. Boesler, S.: Performance-Analyse von Hochleistungsrechnern im Multiprogramming-Betrieb: Untersuchungen auf der SGI Origin. Diplomarbeit, Center for High Performance Computing, Dresden University of Technology (2001)
4. Kowarz, A.: Performance-Untersuchungen mit dem PARbench-System auf unterschiedlichen Parallelrechnern. Diplomarbeit, Zentrum für Hochleistungsrechnen, Technische Universität Dresden (2003)
5. Dietze, H.: Das PARbench-System: Untersuchungen zum Scheduling von parallelen Programmen auf der IBM p690. Diplomarbeit, Zentrum für Hochleistungsrechnen, Technische Universität Dresden (2004)
6. Sun Microsystems, Inc.: Sun Fire E25K/E20K Systems Overview. (2004)
7. Sun Microsystems, Inc.: UltraSPARC IV Whitepaper. (2004)

Early Experiences with KTAU on the IBM BG/L

Aroon Nataraj, Allen D. Malony, Alan Morris, and Sameer Shende

Performance Research Laboratory, Department of Computer and Information Science
University of Oregon, Eugene, OR, USA
{anataraj, malony, amorris, sameer}@cs.uoregon.edu

Abstract. The influences of OS and system-specific effects on application performance are increasingly important in high performance computing. In this regard, OS kernel measurement is necessary to understand the interrelationship of system and application behavior. This can be viewed from two perspectives: *kernel-wide* and *process-centric*. An integrated methodology and framework to observe both views in HPC systems using OS kernel measurement has remained elusive. We demonstrate a new tool called KTAU (Kernel TAU) that aims to provide parallel kernel performance measurement from both perspectives. KTAU extends the TAU performance system with kernel-level monitoring, while leveraging TAU's measurement and analysis capabilities. As part of the ZeptoOS scalable operating systems project, we report early experiences using KTAU in ZeptoOS on the IBM BG/L system.

Keywords: Kernel, performance, measurement, analysis.

1 Introduction

As High Performance Computing (HPC) moves towards ever larger parallel environments, the influence of OS and system-specific effects on application performance are increasingly important to include in a comprehensive performance view. These effects have already been demonstrated ([10], [16]) to be potential bottlenecks, but an integrated methodology and framework to observe their influence relative to application activities and performance has yet to be fully developed. Such an approach will require OS performance monitoring. OS performance can be observed from two different perspectives. One way is to view the entire kernel operation as a whole, aggregating performance data from all active processes in the system and including the activities of the OS when servicing system-calls made by applications as well as activities not directly related to applications (e.g. servicing interrupts or keeping time). We will refer to this as the *kernel-wide* perspective, which is helpful to understand OS behavior and to identify and remove kernel hot spots. Unfortunately, this view does not provide insight into what parts of an application spend time inside the OS and why.

Another way to view OS performance is within the context of an application's execution. Application performance is affected by the interaction of user-space behavior with the OS, as well as what is going on in the rest of the system. By looking at how the OS behaves in the context of individual processes (instead of aggregate performance) we can provide a view that details interactions between programs, daemons, and system services. This *process-centric* perspective is helpful in tuning the OS for a specific workload, tuning the application to better conform to the OS configuration, and in exposing the source of performance problems (in the OS or the application).

Both these perspectives are important in OS performance measurement and analysis on HPC. The challenge is how to support both while providing a monitoring infrastructure that provides detailed visibility of kernel actions and that is easy to use by different tools. For example, the interactions between applications and the OS mainly occur through five different mechanisms: system-calls, exceptions, interrupts, scheduling, and signals. It is important to understand all forms of interactions as the application performance is influenced by each. However, some are easier to observe than others.

Our approach to the challenges above is the development of a new Linux Kernel performance measurement facility called KTAU, which extends the TAU [4] performance system with kernel-level monitoring. KTAU allows both a *kernel-wide* perspective of the OS performance as well as a *process-centric* perspective which merges kernel-space measurements with user-space performance data measured by TAU. KTAU is part of the ZeptoOS [5] research project to study operating systems for petascale systems and is included in the ZeptoOS distribution for the IBM BG/L I/O nodes. Below we describe the design of KTAU in Section 2, the BG/L in Section 3, KTAU's integration in the ZeptoOS and our early experiences on the IBM BG/L platform in Section 4, Related work is given in Section 5. Section 6 offers final remarks and future directions.

2 KTAU Design and Implementation

Kernel Tuning and Analysis Utilities (KTAU) is a toolkit for profiling and tracing the Linux Kernel. The toolkit is unique in its ability to produce both a kernel-specific and process-specific view of performance. Its main strength is in analyzing program behavior within the context of the kernel. KTAU generates performance data compatible with the TAU performance system [4], allowing TAU's analysis tools to be used.

As shown in Figure 1, KTAU consists of five distinct parts, namely,

- the Kernel Instrumentation
- the Kernel Infrastructure
- the KTAU proc filesystem
- the libKtau user-space library
- and clients of KTAU including the integrated TAU framework and daemons

2.1 Kernel Source Instrumentation

The kernel instrumentation is composed of easy to use C macros and functions. The instrumentation allows KTAU to intercept the kernel execution path and

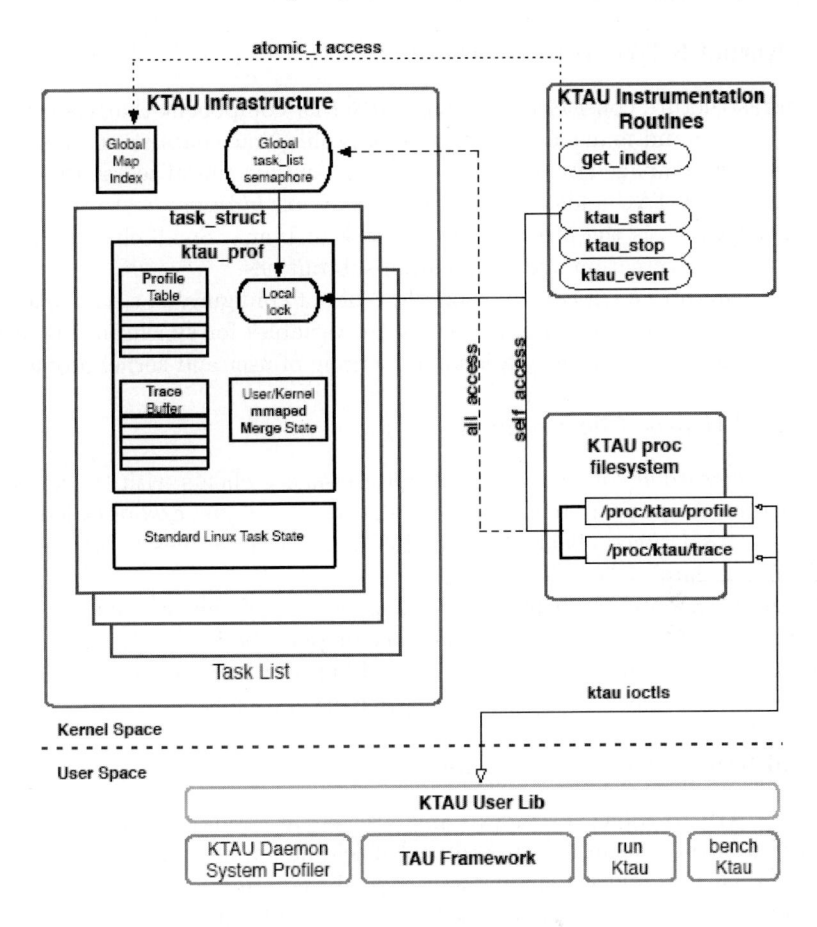

Fig. 1. KTAU Architecture

at each point measurement data is recorded. Profiling and tracing share the same instrumentation points. Instrumentation points are coarsely grouped based on various aspects of the kernel such as the subsystem in which they occur (e.g. scheduling, networking) or in what contexts they occur (e.g. system calls, interrupt, bottom-half handling). Compile-time configuration options and boot-time parameters control which groups of instrumentation points are turned on.

Three different types of macros are provided, namely *mapping*, *timer* and *event* macros. The *mapping* macro is necessary in every function that contains the other two types of instrumentations. It performs the function of providing identities to instrumentation points and mapping performance data to the instrumentation points. The next macro is a start-stop *timer* that calculates time elapsed between an entry and exit point. To obtain high-granularity timing resolution, low-level hardware timers are used. Lastly the *event* macro is used for events that do not conform to the *entry/exit* semantics or for events with non-monotonically increasing values.

2.2 Kernel KTAU Infrastructure

The infrastructure consists of the main in-kernel component that collects performance data and manages the lifecycle of profile/trace data structures. These structures are managed on a per-process basis and are initialized at the process creation time. KTAU adds a member of type *struct ktau_prof* ∗ to the process' *task_struct* which is the process control block in Linux. Small changes are also made to process creation and termination subroutines.

struct ktau_prof comprises of a profile table of configurable size, a fixed-size circular trace buffer and various other state variables for synchronizing access, inclusive/exclusive time calculation and merging of user and kernel profile data.

2.3 KTAU *proc* Filesystem

The *proc filesystem* component interfaces user-space clients with the kernel infrastructure component. It exposes two entries under */proc/ktau* called *profile* and *trace*. User-space clients perform IOCTLs on these files to access kernel performance data.

To accommodate frequent use clients, such as daemons that repeatedly retrieve data at small intervals, a memory-mapped buffer strategy (using *get_user_pages* and *vmap()* kernel routines in Linux) is being experimented with. This removes the need to repeatedly copy to/from userspace.

2.4 libKtau – User Library and API

The User API to KTAU provides a small set of easy to use functions that hide the details of the *proc* filesystem protocol, shielding applications from changes to KTAU kernel components. It provides control (merging, overhead calculation), data retrieval, data conversion (ASCII/binary) and formatted output.

All requests from user-space are grouped into three accessing schemes namely 'self', 'others' and 'all', referring to the set of processes of interest. For 'self' requests, kernel-mode performance data of the same process as the user-space client making the request is accessed. 'other' requests are for a set of processes (one or more) explicitly named in the request. And 'all', just an extension of 'other', is a convenience sparing clients from having to find the pids of all the processes on the system. The reason for having different modes has to do with reducing or removing the need for locking(a process accessing its own kernel-level profile does not need synchronization as it cannot race against itself).

2.5 KTAU Clients

KTAUD - KTAU Daemon. KTAUD periodically extracts performance data from the kernel. It can be configured to extract information for all or a subset of processes. Although it supports extracting profile data, its periodic nature suits it to dumping trace data, as trace buffers within the kernel can become full.

Integrated TAU Framework. The TAU measurement framework has been integrated with KTAU and is a client of libKtau. Applications instrumented with TAU automatically have access to kernel performance data of their own process (i.e. they will be self-profiling clients, using the 'self' mode of libKtau). When enabled through configuration, TAU will generate merged user/kernel profile information. This is supported only on KTAU-patched Linux OSes.

runKtau. Another type of client is similar to how the 'time' command under UNIX works. 'time' spawns a child process, executes the required job within that process, and then gathers rudimentary performance data by doing a waitpid on the child's pid. Similarly 'runktau' extracts the process' detailed KTAU profile.

3 The BG/L and Its Performance Observation

3.1 Brief Description of the Blue Gene/L

The Blue Gene L(BG/L) is a recent massively parallel supercomputer system from IBM, developed in collaboration with LLNL, that scales to 65,536 compute nodes [13]. Its architecture includes dual-processor Compute Nodes, I/O Nodes, front-end nodes and a Service node. Five different interconnect networks provide file I/O, control, debugging and interprocessor communication. Our focus is currently toward the compute and I/O nodes. We provide only a brief description of those aspects relevant to the current work as other work describes all aspects of the BG/L architecture and system software in detail([11],[6]).

The dual-processor (700 Mhz PPC 440) compute nodes, running a proprietary IBM operating system called the Compute Node Kernel (CNK) act as the main computation engines. The CNK is a small, light-weight OS, written in C++, without multitasking or virtual memory support allowing as many cycles as possible to application processing. File I/O is not directly implemented by the CNK and instead is call-forwarded to dedicated I/O nodes.

The I/O nodes serve two purposes. They participate in control of compute nodes including initialization, program launch and termination. They also perform all File I/O processing on behalf of the compute nodes. Multiple Compute nodes share a single I/O node. The ratio of I/O to compute nodes is configurable during system setup with values ranging from 1:8 to 1:128. The I/O node along with the compute nodes connected to it, forms a partition set (or pset).

The IO node OS is a modified Linux kernel (called the IO Node Kernel or INK). Modifications include patches to change interrupt/exception handling, add device drivers and floating point unit support. Due to the nodes being disk-less, a ramdisk containing shells and utilities is loaded into memory during bootup and forms the root filesystem. After bootup remote filesystems can be mounted onto the I/O Node based on the configuration of the ramdisk.

The Control and Input/Output daemon (CIOD) running on the I/O Node manages the control and file I/O of compute nodes in its pset. It listens for and accepts requests for processing of forwarded I/O system calls from the compute node applications. The file-I/O is blocking on the application side. The CIOD re-issues the system-calls through the VFS (Virtual File system) on the I/O Node which in turn is implemented by the file-system in use (e.g. NFS or PVFS2).

A tree network connects the Compute Nodes in a pset to their corresponding I/O node. The tree is used for collective operations as well as file I/O.

3.2 The Performance Measurement Problem

The BG/L system involves different nodes and multiple interacting software components within the nodes. Our approach is to place a unified measurement framework, based on TAU and KTAU, that can observe the applications, the system software and the operating systems and can correlate the performance data from the disparate sources.

The compute and I/O nodes tend to influence application performance the most. The I/O node being a shared resource within a pset, it is important to understand aspects of sharing such as fairness and utilization. Key questions regarding configuration of CIOD and the kernel, kernel version, choice of filesystem (NFS, PVFS2, Lustre etc) need to be answered with reference to performance data. The problem can be further split as follows:

1. Measurement of the Compute Node applications and the CNK and correlation of both performance results across compute nodes. This is (partially) tackled by using applications instrumented with TAU. TAU instrumentation allows tracing/profiling the applications and libraries such as MPI (using the MPI profiling interface). The CNK is closed and proprietary (akin to a black-box) and hence cannot be profiled or traced. But it is considered to be light-weight and built to stay out of the way of the application. We do not go into the details of using the TAU system on the compute nodes any further in this paper.

2. Measurement and correlation of performance of the CIOD, system daemons and IO Node Kernel. This is the focus of out current work including the integration with the ZeptoOS IO Node kernel on the Blue Gene/L at Argonne National Labs and our early experiences and results in using KTAU on the IO Nodes. The black-box nature of the CIOD is one of the challenges in measuring I/O node performance. The next section describes our approach.

3. Correlation of performance between the compute and I/O nodes within a pset and across multiple psets. This faces the problem of two black-boxes (CNK and CIOD) obscuring the call flow between the Compute and I/O-nodes, making correlation of events challenging. This is the target of future work.

4 KTAU On BG/L

Figure 2 shows a simplified view of the architecture of the BG/L's I/O and Compute Nodes within a pset and the main software components on those nodes. The ZeptoOS Linux kernel on the IO Node has been patched with KTAU support and KTAUD has been added to the IO Node utilities. On the compute node side, the applications can be instrumented with TAU to generate profiles or traces.

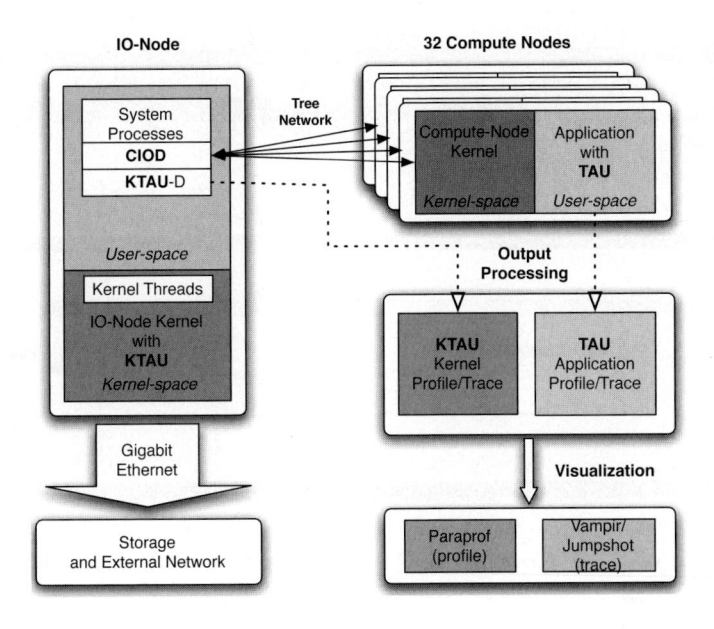

Fig. 2. KTAU on IBM BG/L

KTAU's process-centric kernel level profiling and tracing support is used to observe the CIOD despite it being closed-source. Its interactions at the kernel-level are the only way to peer into it. While utilities such as 'strace' are being used for this purpose, the perturbation caused by strace's trap-and-signal mechanism can cause significant differences between observed and 'real' behavior.

The integration with ZeptoOS (see [1] for details) includes patching and configuration changes to the Kernel source and the ramdisk image.

4.1 Performance Observation Capabilities Demonstrated

The following experiments aim to show KTAU's fine-grained performance observation capabilities on the BG/L IO Nodes. The experimental setup consists of a single pset (a single I/O node with 1 to 32 compute nodes). The compute nodes run a MPI I/O benchmark called *iotest* (used at ANL) which produces aggregate bandwidth numbers over varying block-sizes, number of processors and iterations. While the benchmark is run on the compute nodes, the CIOD services the read and write calls on the I/O node. On the I/O node while KTAU captures the kernel interactions, KTAUD periodically queries and saves the KTAU trace or profile information. Paraprof and Vampir [17] are used for visualization.

1. Fine Grained CIOD Tracing: We first show a trace fragment of the CIOD along with a zoomed in view in Figure 3. The groups of kernel functions shown in the upper image include TCP/IP, Socket calls, Interrupts, bottom half handling and scheduling. The lower zoomed-in image shows a typical CIOD kernel interaction when servicing a write system call from the compute node. The CIOD

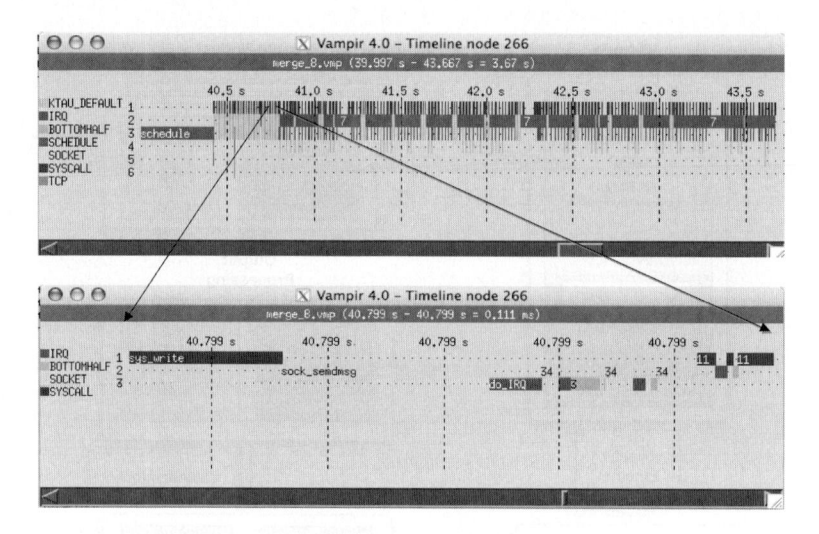

Fig. 3. KTAU Trace of CIOD – Bottom is Zoomed-In view of Top

calls sys_write on the IO node which in turn gets translated into socket and bottom half handling. UDP, instead of TCP, is used due to the filesystem being NFS (but the UDP activity is not shown as those instrumentation points were not enabled). The fine grained detail also shows interrupts occurring in between (do_IRQ). The 'node 266' in the title refers to process-id 266 (of the CIOD).

2. Effect of Increasing Compute Jobs on CIOD: As the I/O Node is a shared resource among the compute nodes in the pset, the load on the CIOD will no-doubt increase as the iotest benchmark is run in parallel over multiple compute nodes. Using KTAU we are able to capture the effect on the CIOD through its increasing interactions with the INK as it tries to service all of the compute nodes. Figure 4 shows five runs each with varying number of compute jobs.

3. Correlating behavior of Daemons and Kernel threads: It is possible to loosely correlate activity between different interacting daemons and kernel threads using traces collected from them. While no actual causal relationships can be directly deduced from the traces, intelligent guesses can be made based on timestamps and functionality of the processes. Figure 5 shows trace fragments from two processes on the IO node namely the CIOD and the RPCIOD (RPC I/O daemon). CIOD can be seen to repeatedly make sys_write calls and then be scheduled-out. At the same time, RPCIOD can be seen to be scheduled-in and perform sock_sendmsg followed by bottom-half handling. This behavior can be explained by the fact that with the underlying filesystem being NFS, the CIOD's sys_write calls are being handled by the RPCIOD.

4. Effect of Filesystem choice on CIOD: The backend filesystem mounted by the IO Node after bootup can be varied. By default on the ANL BG/L this is NFS, but a PVFS2 based configuration is also supported. The choice and configuration parameters of the backend filesystem can significantly influence I/O performance. To demonstrate KTAU's characterization of this effect two

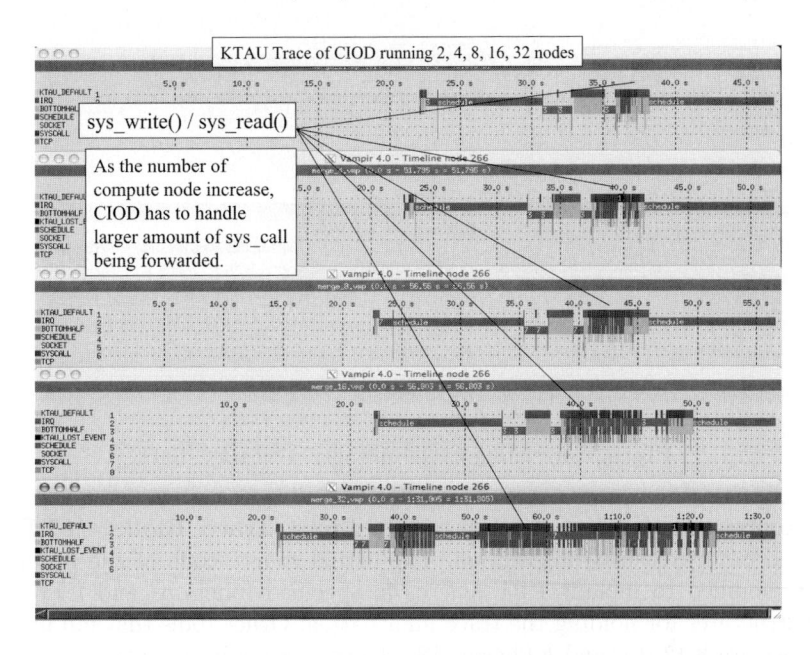

Fig. 4. KTAU Trace of CIOD - Effect of Increasing Compute Nodes from 2 to 32

runs of the benchmark were conducted, one under NFS and the under PVFS2. The traces[1] show clear differences in behavior (including obvious ones such as the use of TCP in the PVFS client versus the use of UDP in the RPCIOD).

5. Profiling support on the IO Node: Profiling support is important to provide as it can quickly help locate performance bottlenecks. While trace information can be post-processed to provide profiles, direct online profiling has much lower overheads. KTAU can collect profiling data, in addition to above mentioned traces, that can be visualized in Paraprof. The profiles[1] show inclusive and exclusive times taken by various kernel routines in the context of the CIOD.

5 Related Work

It is interesting to compare KTAU to other kernel instrumentation and measurement projects. We discuss below a few of the tools presented in Table 1.

KTAU is clearly distinguished from tools that use dynamic instrumentation rather than modify the kernel source. In **KernInst** [8] and **DTrace** [9] kernel routines are instrumented by *splicing* in measurement code dynamically at runtime. While kernel measurement can be modified during execution, the overhead of changing instrumentation can be greater than direct code instrumentation. KernInst, by itself, does not support merging user and kernel related performance information. Dtrace's user-level instrumentations also trap into the kernel, making it a costly choice for measurement of parallel HPC codes.

---

[1] Not shown here due to lack of space.

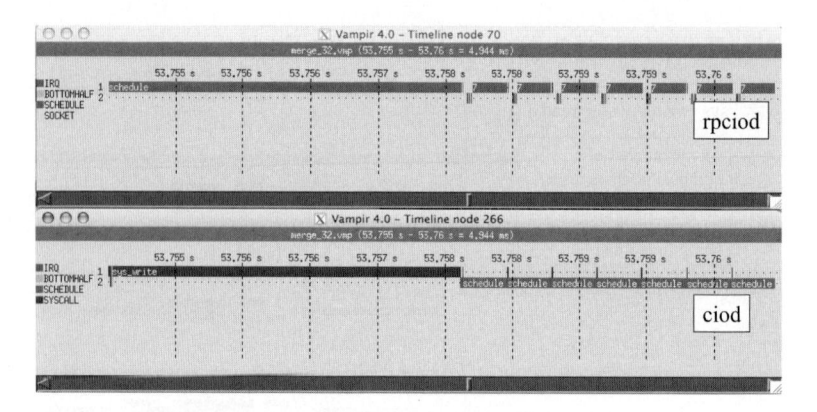

Fig. 5. Correlating CIOD Activity with RPC-IOD

In contrast to KernInst and DTrace, the **Linux Trace Toolkit (LTT)** [12] is based on source instrumentation. The actual source code of the Linux kernel is modified to include LTT macros in specified functions and LTT based data structures for holding the trace information. Other tools that fall into this category are K42 [14] and KLogger [18]. All of these only provide tracing.

5.1 Measured and Statistical Profiling Tools

SGI's KernProf [3] (under call-graph modes) is an example of measured-profiling tools. It uses compiler (gcc -pg option) generated profiling support. Every function is instrumented at compile-time with code to track a call-count and which functions called it and which functions were called from it. This is used to generate a call-graph of the kernel.

Oprofile [2], a statistical profiler, is meant to be a type of *continuous* profiler for Linux meaning always turned on. It performs both user-mode and kernel-mode profiling across the system providing merged user/kernel information. Its shortcomings include its inability to provide online information and the costly requirement of a daemon. SGI KernProf's flat-profile mode also uses sampling.

5.2 Merged User-Kernel Performance Analysis Tools

These tools explicitly provide support to merge the performance information between the application and the kernel. This enables understanding program-OS interaction and being able to pin-point bottleneck location in overall program/OS stack. It may also allow identifying intrusive effects such as excessive scheduling or interrupts that can steal cycles from applications.

CrossWalk [7] is a tool that *walks* a merged call-graph across the user-kernel boundary in search of the real source of performance bottlenecks. Using a specified performance threshold, it tries to find routines that take longer than the threshold starting its search from the *main()* function and examining the entire user/kernel control flow. **DeBox** [19] and [15] are also merged performance measurement tools. All three suffer from the fundamental problem of not providing merged support, unlike KTAU, for interrupts, exceptions and scheduling.

5.3 Discussion

The tools mentioned above are unable to produce valuable merged information for all aspects of program-OS interaction. In addition, online OS performance information and ability to function without a daemon is not widely available. Most of the tools do not provide explicit support for collecting, analyzing and visualizing parallel performance data. KTAU aims to explicitly support online merged user/kernel performance analysis for all program-OS interactions in parallel HPC execution environments while using existing visualization tools.

Table 1. Related Work (N/E stands for 'No Explicit support')

Tool	Instr.	Measurement	User+Kernel	Parallel	SMP	OS
KernInst	Runtime	Flexible	N/E	N/E	Yes	Sun
DTrace	Runtime	Flexible	Trap into OS	N/E	Yes	Sun
KLogger	Static Src.	Trace	N/E	N/E	Yes	Linux
LTT	Static Src.	Trace	N/E	N/E	Yes	Linux
OProfile	Not App.	Stat. Prof.	Partial	N/E	Yes	Linux
KernProf(Flat)	Not App.	Stat. Prof.	N/E	N/E	Yes	Linux
KernProf(C Path)	gcc -pg	Call Path	N/E	N/E	Yes	Linux
LACSI'05	Static Src.	Trace	Syscall Only	N/E	No	Linux
CrossWalk	Runtime	Flexible	Syscall Only	N/E	Yes	Sun
DeBox	Static Src.	Meas. Prof., Trace	Syscall Only	N/E	Yes	Linux
KTAU+TAU	Static Src.	Meas. Prof., Trace	Full	Explicit	Yes	Linux

6 Conclusions and Future Work

The desire for a kernel monitoring infrastructure that can provide both a *kernel-wide* and *process-centric* performance perspective led us to the design and development of KTAU. KTAU is unique in its ability to measure the complete program-OS interaction, its support for joint daemon and program access to kernel performance data, and its integration with a robust application performance measurement and analysis system, TAU. In this paper, we described using KTAU as part of a performance measurement framework on Argonne's IBM BG/L system within the scope of the ZeptoOS project. Our early experiences indicate that KTAU along with TAU can be used to perform fine-grained performance measurement across the system. We demonstrated KTAU's measurement capabilities showing tracing/profiling of the I/O Node processes.

The I/O Node and the BG/L system as a whole provide many interesting performance questions that KTAU can be used to answer. We intend to deepen these early efforts through full-fledged experiments to study BG/L I/O performance and scaling under different backend filesystems, application loads and IO

Node (and CIOD) configurations. Another area we intend to explore is correlating performance observations between the Compute and I/O nodes. As the ZeptoOS project matures, KTAU will be also be used to provide kernel performance measurement and analysis for dynamically adaptive kernel configuration.

References

1. KTAU / ZeptoOS Integration. http://www.cs.uoregon.edu/research/ktau/docs.php.
2. Oprofile. http://sourceforge.net/projects/oprofile/.
3. Sgi kernprof. http://oss.sgi.com/projects/kernprof/.
4. TAU: Tuning and Analysis Utilities. http://www.cs.uoregon.edu/research/paracomp/tau/.
5. ZeptoOS: The Small Linux for Big Computers. http://www.mcs.anl.gov/zeptoos/.
6. A. Gara et. al. Overview of the Blue Gene/L system architecture. *IBM Journal of Research and Development*, 49(2/3):195–212, 2005.
7. A. Mirgorodskiy et. al. Crosswalk: A tool for performance profiling across the user-kernel boundary.
8. A. Tamches et. al. Fine-grained dynamic instrumentation of commodity operating system kernels. In *Operating Systems Design and Implementation*, 1999.
9. B. M. Cantrill et. al. Dynamic instrumentation of production systems. In *Proceedings of the 2004 USENIX Annual Technical Conference*, Boston, MA, USA.
10. F. Petrini et. al. The case of the missing supercomputer performance: Achieving optimal performance on the 8,192 processors of ASCI Q. In *SC '03: Proceedings of the 2003 ACM/IEEE conference on Supercomputing*, Washington, DC, USA, 2003.
11. J. E. Moreira et. al. Blue Gene/L programming and operating environment. *IBM Journal of Research and Development*, 49(2/3):367–376, 2005.
12. K. Yaghmour et. al. Measuring and characterizing system behavior using kernel-level event logging. In *USENIX '00: Proceedings of the 2000 USENIX Annual Technical Conference*, Boston, MA, USA, 2000.
13. N.R. Adiga et. al. An overview of the Blue Gene/L supercomputer. In *SC '02: Proceedings of the 2002 ACM/IEEE conference on Supercomputing*, 2002.
14. R. W. Wisniewski et. al. Efficient, unified, and scalable performance monitoring for multiprocessor operating systems.
15. S. Sharma et. al. A Framework for Analyzing Linux System Overheads on HPC Applications. In *LACSI '05: Proceedings of the 2005 Los Alamos Computer Science Institute Symposium*, Santa Fe, NM, USA, 2005.
16. T. Jones et. al. Improving the scalability of parallel jobs by adding parallel awareness to the operating system. In *SC '03: Proceedings of the 2003 ACM/IEEE conference on Supercomputing*, Washington, DC, USA, 2003.
17. W. E. Nagel et. al. VAMPIR: Visualization and analysis of MPI resources. *Supercomputer*, 12(1):69–80, 1996.
18. Y. Etsion et. al. Fine Grained Kernel Logging with KLogger: Experience and Insights.
19. Y. Ruan et. al. Making the "Box" Transparent: System Call Performance as a First-class Result. In *USENIX '04: Proceedings of the 2004 USENIX Annual Technical Conference*, Boston, MA, USA, 2004.

PAM-SoC: A Toolchain for Predicting MPSoC Performance*

Ana Lucia Varbanescu, Henk Sips, and Arjan van Gemund

Department of Computer Science, Delft University of Technology, The Netherlands
A.L.Varbanescu@tudelft.nl

Abstract. In the past, research on Multiprocessor Systems-on-Chip (MPSoC) has focused mainly on increasing the available processing power on a chip, while less effort was put into specific system-level performance analysis, or into behavior prediction. This paper introduces PAM-SoC, a light-weight performance predictor for MPSoC system-level performance. Being based on PAMELA, a static performance predictor for parallel applications, PAM-SoC can compute its prediction in seconds for cases when cycle-accurate simulation takes tens of minutes. The paper includes a set of PAM-SoC validation experiments, as well as two sets of experiments to show how PAM-SoC can be used for either application tuning or MPSoC platform tuning in early system design phases.

1 Introduction

Systems-on-Chips (SoCs) are built to answer the increased processing power requirements of real-time embedded applications by integrating (most of) the functions of a complete electronic system on a single chip [1]. Multiprocessor SoCs (MPSoCs) are SoCs that integrate several programmable processors, adding more flexibility and programmability to these devices. Currently, consumer electronics, automotive and dedicated industrial control systems are foreseen as the main consumers of MPSoC technology.

So far, MPSoC research was focused mainly on hardware issues, allowing designers to prove their skills in squeezing as much processing power as possible on a single chip. As a result, many different platforms have emerged [2,3]: IXP2850 Network Processor (Intel), OMAP (Texas Instruments), Nexperia[TM](Philips), Nomadik[TM](STMicroelectronics), or Cell (IBM/Sony/Toshiba).

Currently, MPSoCs face a difficult challenge: predictable programmability in terms of performance. Unfortunately, integrating more resources on the same chip does not directly increase performance. It does, however, increase the analysis complexity and the software design time. On top of this complexity, the inherent *hardware imbalance* between the almost unlimited available processing power and the severely limited on-chip memory may induce performance gaps

---

* The research is part of the Scalp project http://scalp.ewi.tudelft.nl funded by STW-Progress.

that have not been foreseen during design. Furthermore, MPSoCs lack dedicated performance analysis methodologies. Most of the current analysis is based on simulations, a time-consuming solution that also requires specific well-defined benchmarks (not yet available), and specific programming models (emerging [4]), in order to provide meaningful conclusions. In other words, current performance analysis is expensive and cumbersome, thus difficult to use in any system design flow feedback loops.

This paper presents PAM-SoC, a light semi-static performance prediction toolchain that computes system-level performance estimations for applications running on MPSoCs. PAM-SoC is based on PAMELA[5], a static performance prediction methodology for general purpose parallel platforms (GPPPs). By coupling an application model with the target machine model, PAMELA computes the lower bound of the execution time of the application on the target architecture. To address the specifics of MPSoCs, PAM-SoC includes new techniques for machine modeling and tools for gathering memory behavior statistics. In its prediction, PAM-SoC trades accuracy for estimation speed: in cases when cycle-accurate simulation would take tens of minutes, application behavior can be estimated in tens of seconds. Thus, PAM-SoC can be part of the MPSoC design flow, for either application or architecture tuning.

The paper is organized as follows: Section 2 briefly presents the PAMELA methodology. Section 3 introduces the PAM-SoC predictor, discussing the application and machine modeling in detail. Section 4 presents the validation process of PAM-SoC and two interesting usage scenarios. Section 5 presents related work, while Section 6 draws the conclusions and presents future work directions.

2 PAMELA Methodology

PAMELA (PerformAnce ModEling LAnguage) [6] is a performance simulation formalism that facilitates symbolic cost modeling, featuring a modeling language, a compiler, and a performance analysis technique. The PAMELA model of a Series-Parallel (SP) program[7] is a set of explicit, algebraic performance expressions in terms of program parameters (e.g., problem size), and machine parameters (e.g., number of processors). These expressions are automatically *compiled* into a *symbolic cost model*, that can be further reduced and compiled into a *time-domain cost model* and, finally, evaluated into a *time estimate*. Note that PAMELA models trade prediction accuracy for the lowest possible solution complexity. Fig. 1 presents the PAMELA methodology.

Modeling. The PAMELA modeling language is a process-oriented language designed to capture concurrency and timing behavior of parallel systems. Data computations from the original source code are modeled into the *application model* in terms of their resource requirements and workload. The available resources and their usage policies are specified by the *machine model*.

Any PAMELA model is written as a set of process equations, composed from `use` and `delay` basic processes, using sequential, parallel, and conditional

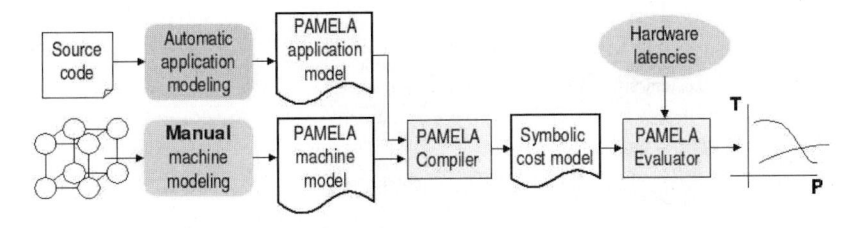

Fig. 1. The PAMELA symbolic cost estimation

composition operators. The construct `use(Resource,t)` stands for exclusive acquisition of `Resource` for t units of (virtual) time. The construct `delay(t)` stalls program execution for t units of (virtual) time. A *machine model* is expressed in terms of available resources and an abstract instruction set (AIS) for using these resources. The *application model* of the parallel program is implemented using an (automated) translator from the source instruction set to the machine AIS. The example below illustrates the modeling of a block-wise parallel addition computation $y = \sum_{i=1}^{N} x_i$ on a machine with P processors and shared memory mem:

```
// application model:                // machine model
    par (p=1,P) {                        load=use(mem,t_access)
        seq (i=1,N/P) { load ; add } ;   add=delay(t_add)
        store }
```

Symbolic Compilation and Evaluation. A PAMELA model is translated into a *time-domain performance model* by substituting every process equation by a numeric equation that models the execution time associated with the original process. The result is a new PAMELA model that only comprises numeric equations, as the original `process` and `resource` equations are no longer present. The PAMELA compiler can further reduce and evaluate this model for different numerical values of the parameters, computing the lower bound of the application execution time. The analytic approach underlying the translation, together with the algebraic reduction engine that drastically optimizes the evaluation time, are detailed in [8].

3 The PAM-SoC Toolchain

Using PAMELA for MPSoC performance predictions is quite difficult because of the architecture and application modeling efforts required. Details that can be safely ignored for GPPP models, as they do not have a major influence on the overall performance, may have a significant influence on MPSoC behavior. As a consequence, for correct modeling of MPSoC applications and architectures, we have extended PAMELA with new techniques and additional memory behavior tools. The resulting PAM-SoC toolchain is presented in Fig. 2. In this section we will further detail its specific components.

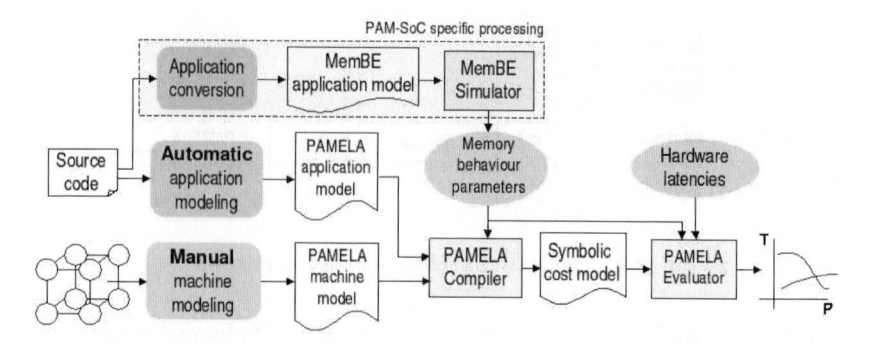

Fig. 2. The PAM-SoC toolchain

3.1 MPSoC Machine Modeling

The successfull modeling of a machine architecture starts with accurate detection of its important *contention points*, i.e., system resources that may limit performance when used concurrently. Such resources can be modeled at various degrees of detail, i.e., *granularities*, by modeling more or less from their internal sub-resources. The model granularity is an essential parameter in the speed-to-accuracy balance of the prediction: a finer model leads to a more accurate prediction (due to better specification of its contention points), but it is evaluated slower (due to its increased complexity). Thus, a model *granularity boundary* should be established for any architecture, so that the prediction is is still fast and sufficiently accurate. This boundary is usually set empirically and/or based on a set of validation experiments.

Previous GPPPs experiments with PAMELA typically used coarse models, based on three types of system resources: the processing units, the communication channels and the memories. For MPSoC platforms, we have established a new, extended set of resources to be included in the machine model, as seen in Fig. 3. The new granularity boundaries (the leaf-level in the resource tree in Fig. 3) preserve a good speed-to-accuracy balance, as proved by various experiments we did [9], while allowing drastic simplification of the MPSoC machine modeling procedure. Some additional comments with respect to the machine modeling are the following:

- When on-chip programmable processors have subunits able to work in parallel, they should be modeled separately, especially when analyzing applications that specifically stress them.
- The communication channels require no further detailing for shared-bus architectures. For more complex communication systems, involving networks-on-chips or switch boxes, several channels may be acting in parallel. In this case, they have to be detected and modeled separately.
- The memory system is usually based on individual L1's, an L2 shared cache (maybe banked) and off-chip memory (eventually accessed by dedicated L2 Refill and L2 Victimize engines). If hardware snooping coherency is enforced,

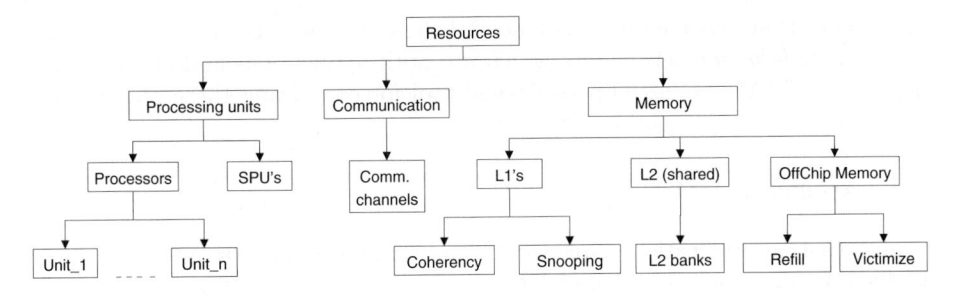

Fig. 3. The extended set of resources to be included in a PAM-SoC machine model

two more dedicated modules become of importance: the Snooping and the Coherency engines. Any of these components present in the architecture must be also included in the model.

After identifying model resources, the AIS has to be specified as a set of rules for using these resources, requiring (1) in-depth knowledge on the functionality of the architecture, for detecting the resource accesses an instruction performs, and (2) resource latencies. As an example, Table 1 presents a snippet from a (possible) AIS, considering an architecture with: several identical programmable processors (`Procs(p)`), each one having parallel arithmetic (`ALU(p)`) and multiplication (`MUL(p)`) units and its own `L1(p)` cache; several Specialized Functional Units (`SFU(s)`); a shared L2 cache, banked, with dedicated `Refill` and `Victimize` engines; virtually infinite off-chip memory (`mem`).

The cache hit/miss behavior cannot be evaluated using the cache (directory) state, because PAMELA, being algebraic, is a state-less formalism. Thus, we compute a probabilistic average cache latency, depending on the cache hit ratio, h^{ratio}, and on the hit/miss latencies, t^{hit} and t^{miss}. Also the `if` branches in the `READ(addr)` model are addressed in a probabilistic manner. For example, `if(missL1)` is replaced by a quantification with $(1 - h_{L1}^{ratio})$, which is the

Table 1. Snippet from an AIS for a generic MPSoC

Operation	Model
p: ADD	use(ALU(p), t_{ADD})
p: MUL	use(MUL(p), t_{MUL})
s: EXEC	use(SFU(s), t_{SFU})
p: accessL1(addr)	use(L1(p), $t_{L1}^{hit} * h_{L1}^{ratio} + t_{L1}^{miss} * (1 - h_{L1}^{ratio})$)
p: accessL2(addr)	use(L2(bank(addr)), $t_{L2}^{hit} * h_{L2}^{ratio} + t_{miss} * (1 - h_{L2}^{ratio})$)
p: refillL2(addr)	use(Refill, t_{Mem}^{RD})
p: victimizeL2(addr)	use(Victimize, $victimization^{ratio} * t_{Mem}^{WR}$)

p: READ(addr)	accessL1(addr);
	if (missL1) { accessL2(addr);
	    if (missL2) {
	        if (victimize) victimizeL2(addr);
	        refillL2(addr)}}

probability that this condition is true. All these probabilistic models are based on *memory behavior parameters* which are both application- and architecture-dependent. PAM-SoC uses an additional tool for computing these parameters, which is presented in Section 3.3.

3.2 Application Modeling

Translating an application implemented in a high-level programming language to its PAMELA application model (as well as writing a PAMELA model from scratch) implies two distinct phases: (1) modeling the application as a series-parallel graph of processes, and (2) modeling each of the processes in terms of PAMELA machine instructions. However, modeling an existing application to its PAMELA model is a translation from one instruction set to another, and it can be automated if both instruction sets are fully specified as exemplified in [5].

3.3 The Memory Statistics

For computing its prediction, PAM-SoC uses two types of numerical parameters: (1) the hardware latencies (measured under no-contention conditions), and (2) the memory statistics. While the former have been also required by GPPP models, the latter become of importance mainly for modeling MPSoC platforms.

The hardware latencies are fixed values for a given architecture and can be either obtained from the hardware specification itself (i.e., theoretical latencies) or by means of micro-benchmarking (i.e., measured latencies). We have based our experiments on the theoretical latencies.

The memory statistics are both machine- and application-dependent, and they have to be computed/evaluated on a per-application basis. For this, we have built MemBE, a custom light-weight *Memory system Behavior Emulator* able to obtain memory statistics like cache hit ratios, snooping success ratios, or vitimization ratios, with good speed and satisfactory accuracy. MemBE is built as a multi-threaded application that permits the (re)configuration of a custom memory hierarchy using the memory components supported by PAM-SoC. MemBE emulates the memory system of the target architecture and executes a memory-skeleton version of the analyzed application[1]. The memory skeleton is stripped of any data-processing, which allows MemBE to run faster and to focus exclusively on monitoring the application data-path.

4 Experiments and Results

In this section we present the validation experiments for our PAM-SoC toolchain, as well as two sets of design-space exploration experiments, one for architecture tuning and one for application tuning, respectively.

---

[1] Currently, the application simplification from the source code to the memory-skeleton is done by hand. In principle, we believe that PAMELA and MemBE can both start from a common, autmatically-generated application model.

4.1 Validation Experiments

The validation process of PAM-SoC aims to prove its abilities to correctly predict application behavior on a given MPSoC platform. For these experiments, we have modeled the Wasabi platform, one tile of the CAKE architecture from Philips [10,11]. A Wasabi chip is a shared-memory MPSoC, having 1-8 programmable processors, several SFUs, and various interfaces. The tile memory hierarchy has three levels: (1) private L1's for each processor, (2) one shared on-chip L2 cache, available to all processors, and (3) one off-chip memory module. Hardware consistency between all L1's and L2 is enforced. For software support, Wasabi runs eCos[2], a modular open-source Real-Time Operating System (RTOS), which has embedded support for multithreading. Programming is done in C/C++, using eCos synchronization system calls and the default eCos thread scheduler.

The simulation experiments have been run on Wasabi's configurable cycle-accurate simulator, provided by Philips. For our experiments, we have chosen a fixed memory configuration (L1's are 256KB, L2 is 2MB, and the off-chip memory is 256MB) and we have used up to 8 identical Trimedia processors[3].

For validation, we have implemented a set of six simple benchmark applications, each of them being a possible component of a more complex, real-life MPSoC application. These applications are: (1) **element-wise matrix addition** - memory intensive, (2) **matrix multiplication** - computation intensive, (3) **RGB-to-YIQ conversion** - a color-space transformation filter, from the EEMBC Consumer suite[4], (4) **RGB-to-Grey conversion** - another color-space transformation, (5) **high-pass Grey filter** - an image convolution filter, from the EEMBC Digital Entertainment suite[5], and (6) **filter chain** - a chain of three filters (YIQ-to-RGB, RGB-to-Grey, high-pass Grey) successively applied on the same input data. All benchmark applications have been implemented for shared-memory (to comply with the Wasabi memory system), using the SP-programming model and exploiting data-parallelism only (no task parallelism, which is a natural choice for the case of these one-task applications).

The results of PAM-SoC prediction and Wasabi simulation for matrix addition and RGB-to-Grey transformation are presented together in Fig. 4. Due to space limitation, we have only included these two graphs, for the two applications that clearly exhibit memory contention and therefore show the prediction abilities of PAM-SoC. The complete set of graphs (for all the applications) and experiment results are presented in [9]. For all the applications, the behavior trend is correctly predicted by PAM-SoC. The average error between simulation and prediction is within 19%, while the maximum is less than 25% [9]. These deviations are due to (1) the differences between the theoretical Wasabi latencies and the ones implemented in the simulator (50-70%), (2) the averaging of the memory behavior data, and (3) the PAMELA accuracy-for-speed trade.

---

[2] http://ecos.sourceware.org/
[3] TriMedia is a family of Philips VLIW processors optimized for multimedia processing.
[4] http://www.eembc.org/benchmark/consumer.asp
[5] http://www.eembc.org/benchmark/digital_entertainment.asp

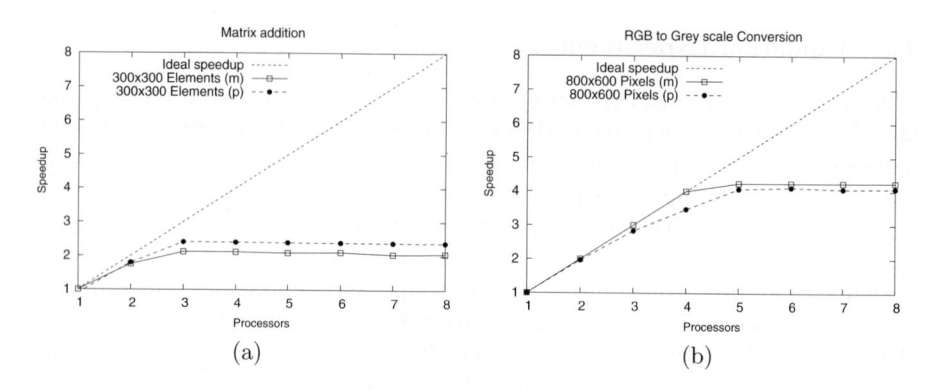

Fig. 4. Predicted and measured speedup for (a) matrix addition and (b) RGB-to-Grey conversion

Table 2. Simulation vs. prediction times [s]

Application	Data size	T_{sim}	T_{MemBE}	T_{Pam}	T_{PAMSoC}	Speed-up
MADD	3x1024x1024 words	94	2	1	3	31.3
MMUL	3x512x512 words	8366	310	2	312	26.8
RGB-to-YIQ	6x1120x840 bytes	90	7	4	11	8.1
RGB-to-Grey	4x1120x840 bytes	62	3	1	4	15.5
Grey-Filter	2x1120x840 bytes	113	6	4	10	11.3
Filter chain	8x1120x840 bytes	347	20	12	32	10.8

While Fig. 4 demonstrates how PAM-SoC is accurate in terms of application behavior, Table 2 emphasizes the important speed-up of PAM-SoC prediction time ($T_{PAMSoC} = T_{MemBE} + T_{Pam}$) compared to the cycle-accurate simulation time, T_{sim}, for the considered benchmark applications and the largest data sets we have measured. While a further increase of the data set size leads to a significant increase for T_{sim} (tens of minutes), it has a minor impact on T_{Pam} (seconds) and leads to a moderate increase of T_{MemBE} (tens of seconds up to minutes). Because MemBE is at its first version, there is still much room for improvement, by porting it on a parallel machine and/or by including more aggressive optimizations. In the future, alternative cache simulators or analytical methods (when/if available) may even replace MemBE for computing memory statistics.

4.2 Design Space Exploration

PAM-SoC can be successfully used for early design space exploration, both for *architecture tuning*, where architectural choices effects can be evaluated for a given application, and for *application tuning* where application implementation choices effects can be evaluated for a given architecture.

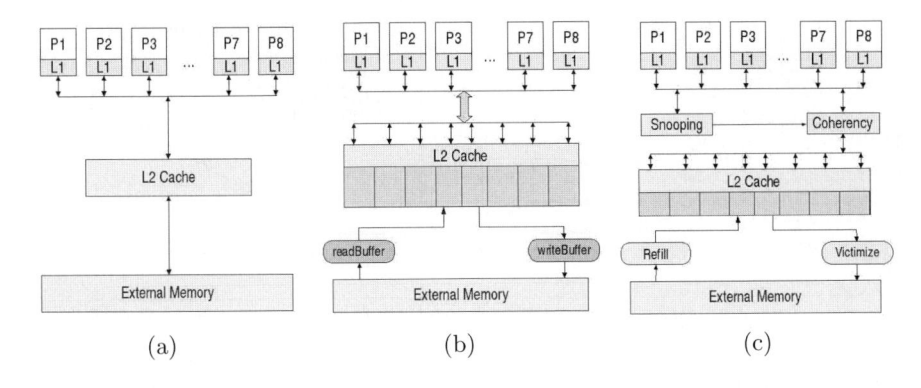

Fig. 5. The hypothetical MPSoCs: (a) **M1**, (b) **M2**, (c) **M3**

Architecture tuning. For *architecture tuning*, we have considered a hypothetical system with up to eight identical programmable processors, each processor having its own L1 cache, a fast shared on-chip L2 cache and practically unlimited external memory (i.e., off-chip). We have modeled three variants of this hypothetical system, named **M1**, **M2**, and **M3**, and we have estimated the performance of a benchmark application (matrix addition, implemented using 1-dim block distribution) on each of them. In this experiment we have chosen to tune the memory configuration of the architecture for these different models, but different processing and/or communication configurations can be evaluated in a similar manner.

M1 is a basic model presented in Fig. 5(a), being a good starting point for modeling any MPSoC architecture. Its key abstraction is the correct approximation of the off-chip memory latencies - for both READ and WRITE operations.

M2 is an improved version of **M1**, presented in Fig. 5(b). It has multiple interleaved banks for the L2 cache (providing concurrent access to addresses that do not belong to the same bank) and buffered memory access to the external memory. Due to these changes, we expect the execution of an application to speed-up on **M2** compared to **M1**. The **M2** model can be adapted to suite any MPSoC architecture with shared banked on-chip memory, if the number of banks, the sizes of the on-chip buffers, and the banking interleaving scheme are adapted for the target machine.

M3, presented in Fig. 5(c), has hardware-enforced *cache coherency*, based on a snooping protocol. The snooping mechanism may increase performance for successful snoopings, when an eventual L1-to-L1 transfer is replacing a slower L2-to-L1 fetch. On the other hand, the L1-to-L2 coherency writes may slow down the application execution. Furthermore, due to the L2-to-Memory transfers performed by the Victimize (for WRITE) and Refill (for READ) engines, two new contention points are added. Overall, because matrix addition has almost no successfull snoopings, the application execution on **M3** is slowed down compared to its run on **M2**. **M3** covers the most complex variants of a three-level memory

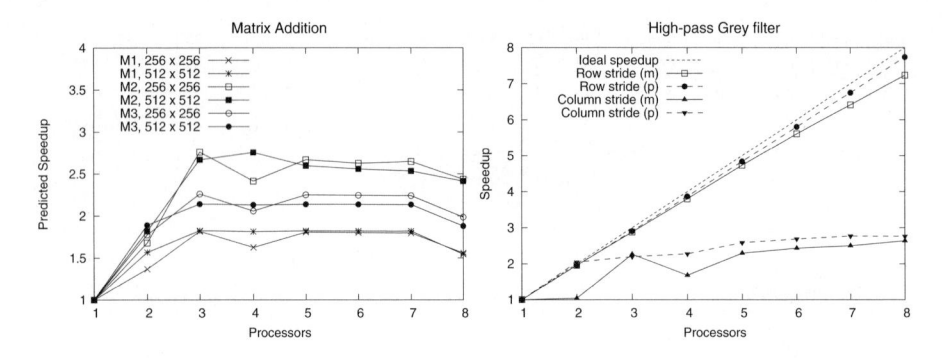

Fig. 6. Architecture tuning results (predicted)

Fig. 7. Application tuning results (predicted vs. measured)

hierarchy to be found in shared memory MPSoCs. Fig. 6 shows that PAM-SoC is able to correctly (intuitively) predict the behavior of the given application on these three models. Unfortunately, we could not run validation experiments for the same data sets because these would require cycle-accurate simulators for M1, M2, and M3[6], which are not available. However, previous PAMELA results [5] provide ample evidence that PAM-SoC will not predict a wrong relative order in terms of performance: it either correctly identifies the best architecture for the given application, or it cannot distinguish one single best architecture.

Application tuning. The aim of *application tuning* experiments is to try evaluate several possible implementations of the same application and choose the best one. To prove the use of PAM-SoC for application tuning, we have used the model of the Wasabi platform and we have implemented the high-pass Grey filter mentioned in Section 4.1 using row and column stride distribution. For this example, based on the PAM-SoC predictions (which are validated by the simulation results), we can decide that row-stride distribution is the best for the Wasabi architecture. Similarly, the experiments like, for example, matrix addition (see Fig. 4(a)), can detect the maximum number of processors to be used for the computation. Fig. 7 shows how PAM-SoC correctly detects the application implementation effects on the given architecture. The simulation results that validate the predictions are also included in the graph.

5 Related Work

Because performance prediction and analysis specifically targeted to MPSoCs is still young, we relate our work to more mature adjacent fields, namely (1) performance prediction for parallel applications, (2) MPSoC performance evaluation and design space exploration techniques, and (3) methods for estimating embedded systems performance.

---

[6] Wasabi is a variant of M3, but its simulator is implemented as a combination of M2 and M3.

For static performance prediction of (scientific) parallel applications, we direct the reader to PAMELA and its related work, to be found in [5,12,13]. Most of these methodologies are difficult to adapt for MPSoCs and their applications, mainly because of the sheer complexity of the hardware platforms. On the practical side, we name the PERC framework [14], close to PAM-SoC in both objectives and realization, as it aims to estimate parallel application behavior by combining machine and application models with the aid of behavior statistics; however, because PERC is intended for scientific applications running on high-performance computers, its analysis granularity is too coarse for direct applicability to MPSoCs.

MPSoC performance analysis is still relying heavily on simulation. MPSoC designers and producers deliver proprietary toolchains, while generic frameworks, like [15,16], provide complex solutions for hardware/software co-simulation and integrated performance estimation. Although very accurate, these simulations are still expensive in terms of computation time. Many hybrid performance estimation techniques have been developed in the context of design space exploration of both MPSoC-specific applications and architectures [17,18]. For example, an interesting hybrid co-simulation solution, similar to PAM-SoC in combining performance estimation with simulation techniques, is presented in [19]. The reduction in complexity is obtained by simulating the architecture at the functional level, with approximate timing behavior which, compared to PAM-SoC, is much coarser. Other MPSoC performance analysis methods are dedicated to estimate the performance of MPSoC components, such as on-chip communication [20,21] or memory systems [22]. An available solution for formal system-level performance verification of MPSoCs is presented in [23], but their approach aims to verify the performance of the hardware system, not to estimate its application-specific behavior.

Although there is no clear border between embedded systems and MPSoC platforms, most of the existing performance evaluation methods for embedded systems, like those presented in [24,25], have not been tested/adapted for MPSoCs, so there are no clear results in this direction.

6 Conclusions and Future Work

In this paper we have presented PAM-SoC, the first toolchain (to the best of our knowledge) for MPSoCs semi-static performance prediction.

Static performance prediction for MPSoCs is motivated by its reduced cost, which allows it to be a part of the design loop. Even though static performance predictors trade accuracy for estimation cost, a behavior estimation within minutes is more valuable, in the early design phases, than an hours-long simulation with very precise results. We have shown how PAM-SoC is used to predict the performance of MPSoC platforms. To validate the methodology, we have modeled a real MPSoC platform and compared the PAM-SoC prediction results for a set of six benchmark applications with the simulation results. Furthermore, we have presented the successful results of PAM-SoC in two early design exploration use-cases, namely application tuning and architecture tuning. For the future, we plan to enhance PAM-SoC by exploring three directions: (1) to model and test more complex applications, for further validation/improvement of PAM-SoC, (2) to make the modeling process as automatic as possible, and (3) to investigate how can PAM-SoC become a truly *static* performance predictor.

Acknowledgements. We would like to thank Paul Stravers and Philips Research for providing the Wasabi simulator, and for the help and support we got for understanding the details of the architecture, support that allowed us to properly model the system.

References

1. Jerraya, A., Wolf, W.: Multiprocessor Systems-on-Chips. Morgan Kaufmann Publishers (2004)
2. Wolf, W.: The future of multiprocessor systems-on-chips. In: Proc. DAC'04, ACM Press (2004) 681–685
3. Kahle, J.A., Day, M.N., Hofstee, H.P., Johns, C.R., Maeurer, T.R., Shippy, D.: Introduction to the Cell multiprocessor. IBM Journal of Research and Development **49**(4/5) (2005)
4. Nijhuis, M., Bos, H., Bal, H.: Supporting reconfigurable parallel multimedia applications. In: EuroPAR'06. (2006)
5. van Gemund, A.: Symbolic performance modeling of parallel systems. IEEE TPDS **14**(2) (2003) 154–165
6. van Gemund, A.: Performance Modeling of Parallel Systems. PhD thesis, Delft University of Technology (1996)
7. Gonzalez-Escribano, A.: Synchronization Architecture in Parallel Programming Models. PhD thesis, Dpto. Informatica, University of Valladolid (2003)
8. Gautama, H., van Gemund, A.: Static performance prediction of data-dependent programs. In: Proc. WOSP'00, ACM (2000) 216–226
9. Varbanescu, A.L.: PAM-SoC experiments and results. Technical Report PDS-2006-001, (Delft University of Technology, http://www.pds.twi.tudelft.nl/reports/)
10. Stravers, P., Hoogerbrugge, J.: Homogeneous multiprocessing and the future of silicon design paradigms. In: Proc. VLSI-TSA'01. (2001)
11. Borodin, D.: Optimisation of multimedia applications for the Philips Wasabi multiprocessor system. Master's thesis, TU Delft (2005)
12. Adve, V.S.: Analyzing the behavior and performance of parallel programs. PhD thesis, Dept. of Computer Sciences, University of Wisconsin-Madison (1993)
13. Adve, V.S., Vernon, M.K.: A deterministic model for parallel program performance evaluation. Technical Report TR98-333 (1998)
14. Snavely, A., Carrington, L., Wolter, N., Labarta, J., Badia, R., Purkayastha, A.: A framework for performance modeling and prediction. In: Proc. Supercomputing '02, IEEE Computer Society Press (2002) 1–17
15. Mahadevan, S., Storgaard, M., Madsen, J., Virk, K.: ARTS: A system-level framework for modeling MPSoC components and analysis of their causality. In: Proc. MASCOTS '05, IEEE Computer Society (2005) 480–483
16. Pimentel, A.D.: The Artemis workbench for system-level performance evaluation of embedded systems. Int. Journal of Embedded Systems **1**(7) (2005)

17. Gries, M.: Methods for evaluating and covering the design space during early design development. Technical Report UCB/ERL M03/32, Electronics Research Lab, University of California at Berkeley (2003)
18. Kienhuis, B.: Design Space Exploration of Stream-based Dataflow Architectures. PhD thesis, Delft University of Technology (1999)
19. Baghdadi, A., Zergainoh, N.E., Cesario, W.O., Jerraya, A.A.: Combining a performance estimation methodology with a hardware/software codesign flow supporting multiprocessor systems. IEEE TSE **28**(9) (2002) 822–831
20. Loghi, M., Angiolini, F., Bertozzi, D., Benini, L., Zafalon, R.: Analyzing on-chip communication in a MPSoC environment. In: Proc. DATE '04, IEEE Computer Society (2004) 20752
21. Pande, P.P., Grecu, C., Jones, M., Ivanov, A., Saleh, R.: Performance evaluation and design trade-offs for Network-on-Chip interconnect architectures. IEEE TC **54**(8) (2005) 1025–1040
22. Loghi, M., Poncino, M.: Exploring energy/performance tradeoffs in shared memory MPSoCs: Snoop-based cache coherence vs. software solutions. In: Proc. DATE'05, IEEE Computer Society (2005) 508–513
23. Richter, K., Jersak, M., Ernst, R.: A formal approach to MpSoC performance verification. IEEE Computer **36**(4) (2003) 60–67
24. Lazarescu, M., Bammi, J., Harcourt, E., Lavagno, L., Lajolo, M.: Compilation-based software performance estimation for system level design. In: Proc. IEEE HLDVT'00. (2000)
25. Thiele, L., Wandeler, E.: Performance Analysis of Embedded Systems. In: The Embedded Systems Handbook. CRC Press (2004)

Analysis of the Memory Registration Process in the Mellanox InfiniBand Software Stack

Frank Mietke, Robert Rex, Robert Baumgartl,
Torsten Mehlan, Torsten Hoefler, and Wolfgang Rehm

Department of Computer Science
Chemnitz University of Technology, Germany
{firstname.surname}@informatik.tu-chemnitz.de

Abstract. To leverage high speed interconnects like InfiniBand it is important to minimize the communication overhead. The most interfering overhead is the registration of communication memory.

In this paper, we present our analysis of the memory registration process inside the Mellanox InfiniBand driver and possible ways out of this bottleneck. We evaluate and characterize the most time consuming parts in the execution path of the memory registration function using the Read Time Stamp Counter (RDTSC) instruction. We present measurements on AMD Opteron and Intel Xeon systems with different types of Host Channel Adapters for PCI-X and PCI-Express. Finally, we conclude with first results using Linux hugepage support to shorten the time of registering a memory region.

1 Introduction

High speed interconnects like InfiniBand [4] or Myrinet [11] use DMA engines in conjunction with user level communication protocols to achieve high bandwidth, low latency and a low CPU utilization. That is the user level application (Consumer in InfiniBand Architecture) just creates a communication request including the relevant information like starting address and length of the communication buffer. This communication request is then transmitted to the network adapter (Host Channel Adapter in InfiniBand) through a simple user level API function call. For a normal send operation the HCA takes the request to create the appropriate packet structure and programs the DMA engine to get the user data. After this the packet is immediately transferred to the other communication partner. This process is depicted in figure 1.

The DMA engine responsible for transferring the data from main memory to the network adapter handles only physical addresses. Thus the virtual addresses of the communication buffer have to be translated into a physical one. Furthermore it is important to ensure that every page of the communication buffer is pinned to prevent swapping. This process of pinning and address translation is called memory registration. Every communication operation of InfiniBand needs registered memory except the inline send operation where the data is directly transferred to the network adapter inside the communication request.

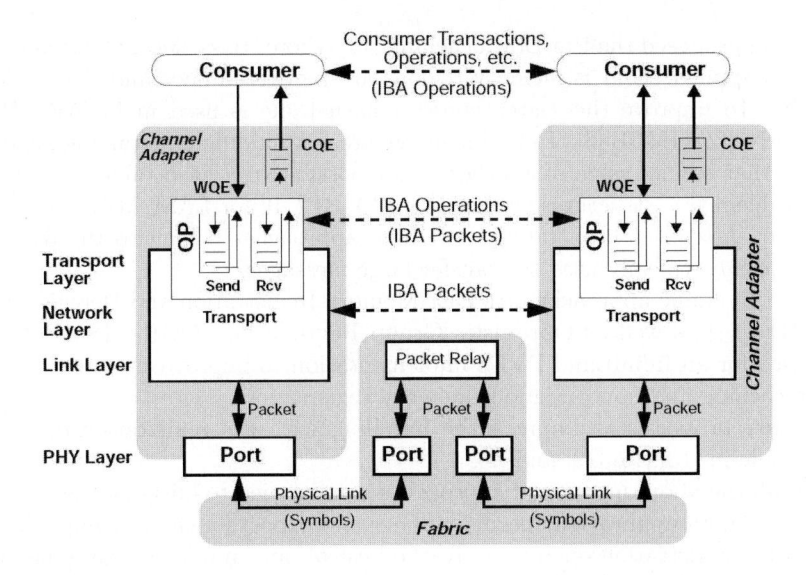

Fig. 1. InfiniBand Architecture Communication Stack ([4])

To avoid the expensive registration costs several approaches were investigated and integrated. We present them in the next section. In section 3 we give a detailed description of the memory registration function of the Mellanox InfiniBand software stack and how we measured the single components inside this function. This section includes also the appropriate measurements and presents the results of our first approach to shorten the registration time. We summarize and conclude in section 4.

2 Related Work

Due to the costs of memory registration several approaches try to reduce the impact of this operation on middleware or application level. These approaches can roughly be categorized in two classes static and dynamic. Static means that every memory area is registered in advance or it is hidden in a memory allocation call. Dynamic means that a memory area is registered on the fly in the communication path. Typically – to complicate there work further – the amount of pinned memory pages and the number of registered memory regions may be limited.

Avoiding the registration operation by using memory copies in conjunction with pre-registered memory regions belongs to the static class. This is typically used if only small messages are sent or received to improve the latency behaviour. Application examples are several InfiniBand MPI implementations like MVAPICH [8], MVAPICH2 [7] and MPICH2-CH3-IB [3].

Registration of the whole physical memory or parts thereof in advance is another approach in the static class. A call to malloc allocates then already registered memory for the application. DSM systems like [6] use this approach.

Tezuka proposed the Pin-Down caching [14] where a lazy deregistration mechanism is applied. That is memory regions are registered once and then hold in a cache. To improve the search speed a hash table is used in MVAPICH [8] and MPICH2-CH3-IB [3]. To find memory areas with different starting page addresses that reside inside of another is not possible in a hash table. To remedy this problem tree structures are used in VIA-RPI [9] for LAM/MPI and Open-MPI [16] instead of hash tables. All these approaches belong to the dynamic class and are typically used to transfer large messages.

Other dynamic approaches are Fast Memory Registration and Deregistration (FMRD) [17] as well as Optimistic Group Registration (OGR) [18]. Both are proposed for an InfiniBand PVFS implementation to improve the speed of Pin-Down caching and noncontiguous memory registration.

Further proposals to improve the handling with the registration operation were made in [13],[15],[19] and [2].

But all the above mentioned approaches merely tried to mitigate the registration costs in an application specific manner and expect an efficient implementation of the registration operation. To the best of our knowledge there has been no detailed analysis which went underneath the registration call.

3 Memory Registration Analysis

It has been observed by several researchers that registering memory for communication is very time-consuming. Table 1 compares the best case (no registration at all) and worst case (every buffer must be registered) scenario running the SendRecv test of the Intel MPI Benchmark [5] suite between two Opteron testsystems each hosting a PCI-express InfiniBand HCA. This comparison clearly shows how big the influence on communication performance is. A detailed analysis regarding the impact on applications is done in [10].

The main goal of the work described here is to obtain a precise understanding of the execution timing of all InfiniBand driver functions contributing to memory registration. We aimed at identifying potential performance bottlenecks and entry points for optimization.

Table 1. Factor of improvement (FOI) when there are no registration costs

Msg size	Bandwidth when Registration necessary	FOI if No Registration
32kB	270MB/s	3.22
64kB	457MB/s	2.55
128kB	701MB/s	2.00
256kB	892MB/s	1.74
512kB	1058MB/s	1.56
1024kB	1217MB/s	1.39
2048kB	1295MB/s	1.33
4096kB	1332MB/s	1.31

3.1 Profiling the Driver

Prior to data transfer, the following main functions are performed sequentially by the driver:

- pin the requested quantity of memory pages for subsequent DMA transfers by the IB controller using `mlock()`,
- translate the virtual addresses of the pinned pages into physical addresses,
- transmit the obtained physical addresses to the IB controller.

It is irrelevant, whether a send or a receive operation follows that preparation. The sequence constitutes the *memory registration*.

The first experiment focused on profiling this sequence. We instrumented the relevant driver functions (mainly `VAPI_register_mr()`) of the Mellanox Infini-Band driver API with `rdtsc` machine instructions and code to write the obtained time stamps into the kernel log. This writing needed approximately 2000-5000 clock cycles which is two to three orders of magnitude smaller than the functions profiled. Therefore we could safely neglect that measurement error.

Two different situations concerning `mlock()` can be distinguished:

a) All or most of the pages to be pinned are present in main memory.
b) The pages are not present in memory. `mlock()` generates page faults and its performance degrades.

Situation b) typically occurs when: Either the buffer is allocated and registered for the very first time or the pages have been swapped out due to tight memory. That is, the former case usually occurs when receiver memory is registered, or sender memory is pre-registered during the init stage. The latter case should be avoided at all costs, e.g, by fitting a maximum of physical main memory into the machine. We conducted our experiments for both situations.

The experimental setup consisted of an AMD Opteron 244 Dual Processor Machine clocked at 1.8 GHz and equipped with 2 GBytes of RAM. We used the PCI-Express InfiniBand Adapter MT25208 InfiniHost III Ex with 256MB RAM and the MemFree version respectively. All experiments were conducted with the Mellanox InfiniBand Gold Edition Package (IBGD), versions 1.7.0 and 1.8.0. The operating system was a standard Linux kernel, version 2.6.10UP and SuSe 2.6.11.4-20a-smp (MemFree HCA). One observation that can be made between the UP and SMP kernel is a slightly bigger overhead due to the spinlock insertion in SMP kernels. All figures are in doubly logarithmic representation.

Figure 2 and 3 depicts the execution timings for the individual steps and the overall memory registration for different communication buffer sizes when the pages of the buffer to be registered are present in memory.

The most time-consuming factor is the address transfer to the IB controller. Pinning and address translation contribute to overall timing only marginally with a slightly larger influence for large buffers. Unfortunately, the communication with the IB controller does not exhibit much optimization potential, because it is bound by the controller's reaction time.

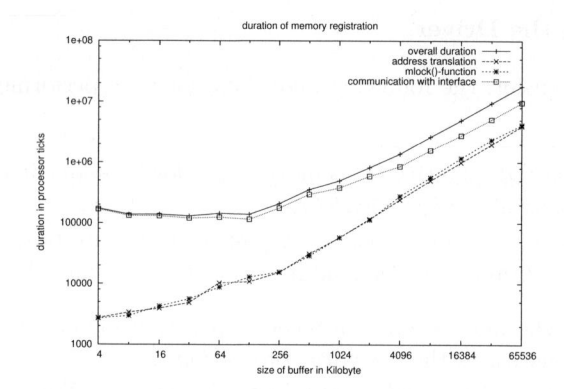

Fig. 2. Memory Registration Performance, Pages Present

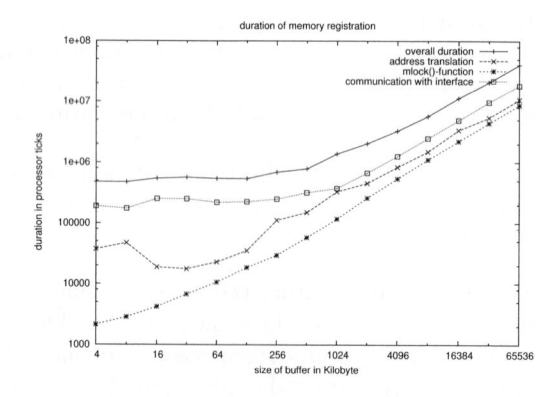

Fig. 3. Memory Registration Performance, Pages Present (MemFree HCA)

Figure 4 and 5 depicts the same execution timings with the pages of the buffer not present in memory.

Several observations can be made here: For buffers larger than 256 kBytes registration time is almost completely dominated by pinning whereas for small buffer sizes the communication with the adapter is the most influential factor. This is not surprising, because the number of occuring page faults increases with buffer size. Virtual-to-Physical address translation is almost not influencing the registration timing. Registration of small buffers has almost a constant timing overhead regardless of the exact buffer size.

We repeated both experiments on an Intel Xeon SMP system hosting 2 CPUs at 2.4 GHz and 2 GBytes of memory and a PCI-X InfiniBand HCA. The software used was the same as on the Opteron System with the 256MB HCA mentioned above. We used a slightly modified methodology due to some driver peculiarities and obtained very similar timing proportions as one can see in figure 6 and 7.

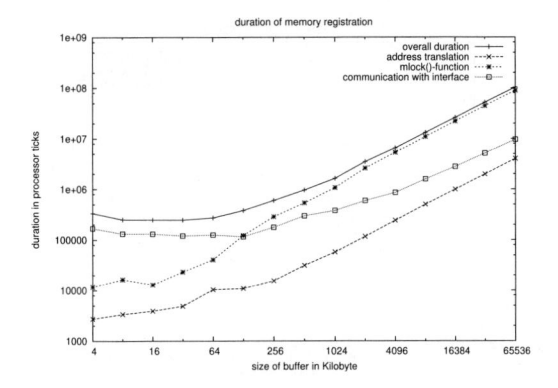

Fig. 4. Memory Registration Performance, Pages not Present

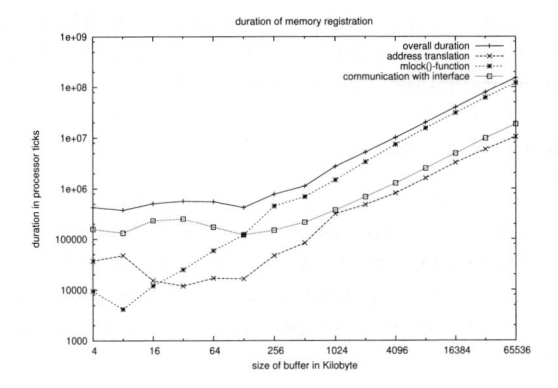

Fig. 5. Memory Registration Performance, Pages not Present (MemFree HCA)

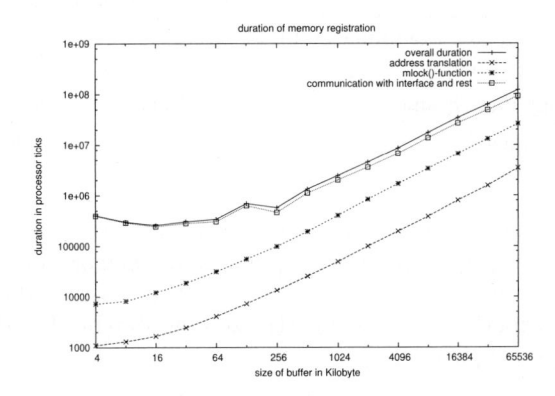

Fig. 6. Memory Registration Performance, Pages Present (PCI-X HCA)

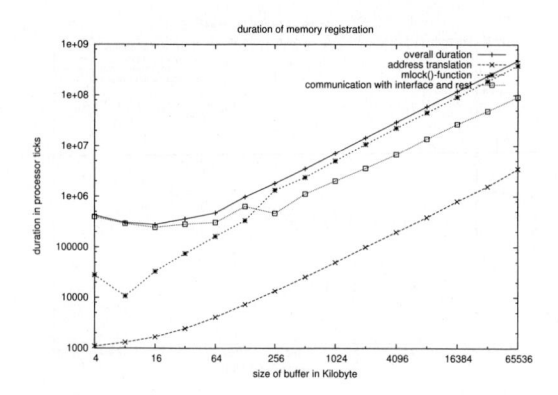

Fig. 7. Memory Registration Performance, Pages not Present (PCI-X HCA)

But the quantitative values are worse compared to AMD Opterons due to the different locations of the memory controller and PCI-X vs. PCI-express (Opteron) HCAs. More details can be found in [12].

3.2 Performance of `mlock()`

Because `mlock()` performance seemed relevant if pages are not present, we next concentrated on profiling it. The table lists the obtained timings for pinning a single page of 4kBytes on the AMD Opteron and the Intel Xeon processor when the page is not present in memory. The shown profile is a typical case when registering receiver memory or preregister sender memory. All times are in processor cycles.

Table 2. Timing Profile for mlock()

Function	AMD Opteron	Intel Xeon
Search for Free Page Frame and Update Page Table	3500	9000
Zero out Page Frame	1000	2000
Pin the Page	1800	5400

Even if you normalize the numbers of the timings the Xeon system needs almost twice the time to execute `mlock()` due to its memory subsystem. As you can see the pinning itself is now only a fraction of the costs of the `mlock()` call. We tried in some experiments to remove the zeroing step but failed with libraries which presume zeroed pages.

3.3 Using Large Pages

Most modern processors like Intel Xeon or AMD Opteron support different page sizes. The most obvious improvement of registration time could be the usage of larger pages. The current 2.6 Linux kernels [1] provide the hugetlbfs to use these different page sizes simultaneously. Apparent advantages using larger page sizes for registering memory are:

- `mlock()` has to pin less pages
- there are less address translations
- and thus less translations has to be transferred to the HCA
- the Mellanox driver can already use large page sizes

Table 3. Comparison of registration time for 4kB and 2MB page sizes

Buffer Size in kB	Registration Time 4kB (ms)	Registration Time 2MB (ms)
2048	1.8	1.5
4096	3.7	2.9
8192	7.4	5.7
16384	14.7	11.3
32768	28.8	22.5
65536	57.9	45.0

To use the hugetlbfs it is necessary to utilize `mmap()` or shared memory system calls. In table 3 the registration times are shown using 4kB and 2MB page sizes. The timings in the 4kB column correspond to the values of figure 4. To be comparable the timings in the 2MB column include `mmap()` and the register call. By using hugetlbfs one attains improvements of 15% up to 25%.

4 Summary and Conclusions

With this paper we have given a quantitative analysis of the execution timing of the memory registration inside the Mellanox InfiniBand driver. We showed that in the case where the pages are not present the `mlock()` call is the dominant factor. Otherwise the communication with the adapter to communicate the address translations is the dominant part. Furthermore, we showed that the AMD Opteron has a much better timing behaviour of the `mlock()` call than the Intel Xeon.

Finally we presented our first results using larger page sizes and showed that improvements of 15% up to 25% are attainable using the mmap approach.

To improve the behaviour of `mlock()` when pages are not present, a seperate kernel thread could fill the pages with zeros when the kernel has time. This could drastically reduce the amount of work which `mlock()` does. To avoid the address translation and thus the communication with the HCA, one would have to change the behaviour of the HCA that it can handle virtual addresses and has access to the kernel page tables. Finally to better utilize hugetlbfs we have to provide a malloc/free library that supports multiple page sizes simultaneously. Therefore also the communication protocol to convey the address translation in the InfiniBand driver has to be changed. Then the applications can transparently make use of this kernel feature in a memory footprint efficient manner.

All these propositions will be investigated in the future.

References

[1] L. K. Archives. Website. http://www.kernel.org.
[2] C. Bell and D. Bonachea. A New DMA Registration Strategy for Pinning-Based High Performance Networks. In *In Proceedings of Int'l Parallel and Distributed Processing Symposium (IPDPS 03)*, April 2003.
[3] R. Grabner, F. Mietke, and W. Rehm. Implementing an MPICH-2 Channel Device over VAPI on InfiniBand. In *Proceedings of the 18th Int'l Parallel and Distributed Processing Symposium, IPDPS*, 2004.
[4] InfiniBand Trade Association. *InfiniBand Architecture Specification 1.2*, 2004.
[5] Intel GmbH, Hermlheimer Str. 8a, D-50321 Brhl, Germany. *Intel MPI Benchmarks – Users Guide and Methodology Description*.
[6] L. Liss, Y. Birk, and A. Schuster. In-Kernel Integration of Operating System and Infiniband Functions for High Performance Computing Clusters: A DSM Example. *IEEE Transactions on Parallel and Distributed Systems*, 16(9), September 2005.
[7] J. Liu, W. Jiang, P. Wyckoff, D. K. Panda, D. Ashton, D. Buntinas, W. Gropp, and B. Toonen. Design and Implementation of MPICH2 over InfiniBand with RDMA Support. In *In Proceedings of Int'l Parallel and Distributed Processing Symposium (IPDPS 04)*, April 2004.
[8] J. Liu, J. Wu, S. P. Kini, P. Wyckoff, and D. K. Panda. High Performance RDMA-Based MPI Implementation over InfiniBand. In *In the Proceedings of 17th Annual ACM International Conference on Supercomputing*, June 2003.
[9] T. Mehlan, W. Rehm, R. Engler, and T. Wenzel. Providing a High-Performance VIA-Module for LAM/MPI. In *In Proceedings of IEEE International Conference on Parallel Computing in Electrical Engineering (PARELEC'04)*, September 2004.
[10] F. Mietke, R. Rex, T. Mehlan, T. Hoefler, and W. Rehm. Reducing the Impact of Memory Registration in InfiniBand. In *Proceedings of the 1. Workshop Kommunikation in Clusterrechnern und Clusterverbundsystemen (KiCC)*, 2005.
[11] Myrinet. Myrinet Inc. http://www.myri.com.
[12] R. Rex. Analysis and Evaluation of Memory Locking Operations for High-Speed Network Interconnects. Student Project, Chemnitz University of Technology, October 2005.
[13] S. Sur, U. Bondhugula, A. Mamidala, H.-W. Jin, and D. K. Panda. High Performance RDMA Based All-to-all Broadcast for InfiniBand Clusters. In *In Proceedings of International Conference on High Performance Computing (HiPC 2005)*, December 2005.
[14] H. Tezuka, F. O'Carroll, A. Hori, and Y. Ishikawa. Pin-down Cache: A Virtual Memory Management Technique for Zero-copy Communication. In *In Proceedings of 12th Int. Parallel Processing Symposium*, March 1998.
[15] V. Tipparaju, G. Santhanaraman, J. Nieplocha, and D. K. Panda. Host-Assisted Zero-Copy Remote Memory Access Communication on InfiniBand. In *In Proceedings of Int'l Parallel and Distributed Processing Symposium (IPDPS 04)*, April 2004.

[16] O. M. Website. A High Performance Message Passing Library. http://www.open-mpi.org.
[17] J. Wu, P. Wyckoff, and D. K. Panda. PVFS over InfiniBand: Design and Performance Evaluation. In *In Proceedings of International Conference on Parallel Processing (ICPP 03)*, October 2003.
[18] J. Wu, P. Wyckoff, and D. K. Panda. Supporting Efficient Noncontiguous Access in PVFS over InfiniBand. In *In Proceedings of IEEE International Conference on Cluster Computing (Cluster'2003)*, December 2003.
[19] J. Wu, P. Wyckoff, D. K. Panda, and R. Ross. Unifier: Unifying Cache Management and Communication Buffer Management for PVFS over InfiniBand. In *In Proceedings of IEEE/ACM International Symposium on Cluster Computing and the Grid (CCGrid 04)*, April 2004.

Optimization of Dense Matrix Multiplication on IBM Cyclops-64: Challenges and Experiences

Ziang Hu, Juan del Cuvillo, Weirong Zhu, and Guang R. Gao

Department of Electrical and Computer Engineering
University of Delaware
Newark, Delaware 19716, U.S.A.
{hu, jcuvillo, weirong, ggao}@capsl.udel.edu

Abstract. This paper presents a study of performance optimization of dense matrix multiplication on IBM Cyclops-64(C64) chip architecture. Although much has been published on how to optimize dense matrix applications on shared memory architecture with multi-level caches, little has been reported on the applicability of the existing methods to the new generation of multi-core architectures like C64. For such architectures a more economical use of on-chip storage resources appears to discourage the use of caches, while providing tremendous on-chip memory bandwidth per storage area.

This paper presents an in-depth case study of a collection of well known optimization methods and tries to re-engineer them to address the new challenges and opportunities provided by this emerging class of multi-core chip architectures. Our study demonstrates that efficiently exploiting the memory hierarchy is the key to achieving good performance. The main contributions of this paper include: (a) identifying a set of key optimizations for C64-like architectures, and (b) exploring a practical order of the optimizations, which yields good performance for applications like matrix multiplication.

1 Introduction

Cyclops-64 (C64) [1,2] is a petaflop supercomputer project under development at IBM. As shown in Figure 1(a), a C64 system is built from thousands of C64 chips that employ a unique multiprocessor-on-a-chip design. Each chip consists of 160 thread units and the same number of SRAM memory banks connected by an on-chip crossbar network (see Figure 1(b)). C64 chip architecture features massive intra-chip parallelism and on-chip memory bandwidth (320GB/s). Given such a novel architecture, the challenge is how to use these two features to obtain high sustained performance for scientific and engineering applications.

During the past two decades, there has been a considerable amount of work on how to optimize dense matrix applications on shared memory architectures with multi-level caches. However, it is not clear whether the existing methods are applicable to the new generation of multi-core architectures, such as C64.

This paper presents an in-depth case study of how a collection of well known optimization methods can be applied to address the new challenges and opportunities that the emerging class of multi-core chip architectures may present. The *phase ordering* of different optimizations has long been challenging but interesting research problem

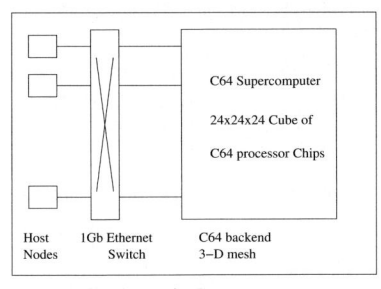

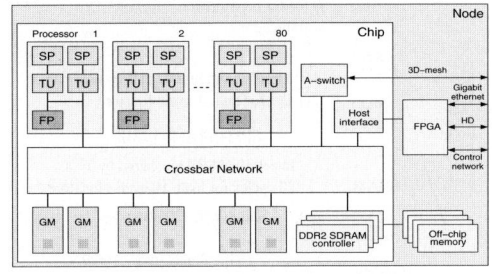

(a) Cyclops-64 Supercomputer   (b) Cyclops-64 Chip Architecture

Fig. 1. Cyclops-64 Architecture

that still remains open [3]. Furthermore, previous work [4], which has established the optimization order for cache-based architectures, may or may not be applicable to a cacheless architecture like C64. In this work, we apply specific optimizations following the order dictated by our experience and knowledge of the problem at hand. However, we do not in any way claim that this order is optimal. Our goal is to demonstrate that overall, for a given dense matrix operation, it is possible to derive a good order of optimization. We hope that the experience reported in this paper will prove to be useful for developers, in designing compilers and runtime systems for C64-like multi-core architectures.

2 Cyclops64 Chip Architecture

The work described in this paper focuses on a single C64 chip [1,2], the main component of a C64 node (see Figure 1(b)). Within a C64 chip there are 80 processors, each consisting of two thread units, a floating-point unit, and two SRAM memory banks of 32KB each. Hence, the total on-chip memory is approximately 5MB. A 32KB instruction cache, not shown in the figure, is shared among five processors.

At boot time, SRAM banks are partitioned into two segments. One segment contributes to the globally shared interleaved on-chip memory. Processors and interleaved memory are logically arranged in a dancehall configuration with processors and memory banks on opposite sides connected by a one-level crossbar switch. The other segment, called scratchpad memory (SPM), is regarded as local memory since the corresponding thread unit has fast access to its own SPM. The C64 architecture also provides four DRAM controllers. Each one is attached to a 256MB bank, hence a C64 node features 1GB off-chip DRAM. As a summary, Figure 2(a) reflects the current size, latency (when there is no contention) and bandwidth of each level of the memory hierarchy. The C64 instruction set architecture incorporates efficient support for thread level execution, hardware barriers, and atomic in-memory operations.

3 The Problem and Experimental Method

This paper is a case study of square matrix multiplication (MM), which is a widely used computation kernel for scientifc and engineering applications. For our baseline,

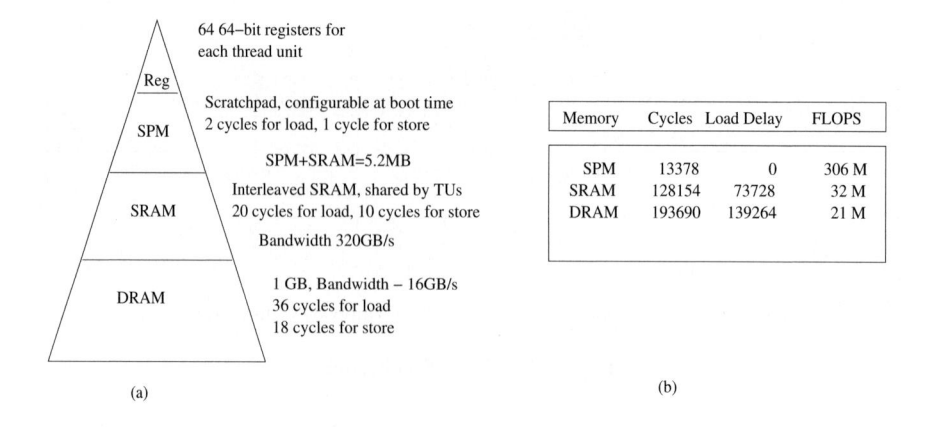

Fig. 2. Cyclops-64 Memory Segments and Baseline Numbers

we choose a straightforward implementation of the sequential algorithm. To parallelize matrix multiplication, we partitioned the three matrices into t^2 blocks and we assign each thread unit the computation of a number of such blocks. The computation of a $\mathbf{C}_{m,n}$ block requires t block multiplications and additions according to the following expression:

$$\mathbf{C}_{m,n} + = \sum_{k=0}^{t-1} \mathbf{A}_{m,k} \times \mathbf{B}_{k,n} \quad (1)$$

To exploit spatial locality, it is best to assign the calculation of $\mathbf{C}_{m,n}$ to a single thread, as the resultant matrix block does not need to move around.

To study matrix multiplication on the C64 architecture we used the FAST simulator [5]. FAST is an functionally-accurate simulator that, among other features, models the memory hierarchy of C64 architecture, including the latencies and bandwidth of each memory segment.

4 Evolutionary Performance Tuning

In this paper, 128×128 and 256×256 matrix multiplications are simulated on up to 68 thread units. The former ensures the matrix fits into on-chip memory, the latter forces some blocks of the matrix to be stored in DRAM. However, the results hereby presented can be extrapolated to larger matrices.

The study begins with the sequential version, where the code always resides in off-chip DRAM, and data is placed in each of the three memory segments one at a time. We compare the performance and memory latencies of the three cases. Then a straightforward parallel version of the MM is introduced, with data stored in SRAM and DRAM, respectively. In the following sections, we improve the performance of the parallel implementation and measure the effectiveness of various optimizations.

4.1 Sequential Matrix Multiplication

We start by comparing the performance achieved by the sequential implementation, with matrices placed in SPM, interleaved SRAM, and DRAM, respectively. SPM size is quite limited. In addition it holds both the runtime stack and thread-private data. Hence, the maximum size allowed for each matrix is 16×16 only. The results from this experiment are shown in Figure 2(b).

It is apparent that the performance difference comes from the latency incurred by load operations accessing different memory segments. We may conclude that data should always be loaded into SPM first before starting the computation. However, data needs to be loaded from SRAM/DRAM into registers first and stored into SPM afterwards on this architecture. If data reuse rate is low, it is not worth performing this "prefetching". Therefore, data reuse is a key issue for achieving high performance on C64. Matrix multiplication has the potential for high data reuse as the memory size is $O(n^2)$ and computation is $O(n^3)$.

4.2 Matrix Multiplication Parallelization in On-Chip SRAM

We implemented a straightforward parallel version, which places three 128×128 matrices into interleaved SRAM. This version will be used as the baseline version for performance comparison. The matrices are partitioned into 8^2 blocks, each with 16×16 size. At most 64 thread units are used in this experiment, as there are 64 blocks in total. Thus, it is natural to assign one resultant matrix block to each thread, as well as all the computation for that block. We encapsulate the computation for one resultant block into one task. Also notice that the resultant block can be reused 8 times while the other blocks are used only once for each task. A task array is employed to store the tasks. Each task consists of a pointer to the resultant block, and two arrays of pointers that point to 8 pairs of source blocks.

Table 1. Baseline Parallel Version

Num of Threads	Cycles	FLOPS	Speedup
1	93,435,509	22.5M	1.00
2	46,750,840	44.9M	2.00
4	23,413,382	89.6M	3.99
8	11,783,500	178.0M	7.93
16	5,942,832	352.9M	15.72
32	3,207,410	653.9M	29.13
64	1,627,767	1.3G	57.40

Each thread tries to obtain the next available task from the task pool. When successful, it performs the computations, writes the resultant block back, and attempts to get a new task until the task pool is empty. The result is shown in Table 1. Although we get near linear speed up, the overall performance is still low - up to 1.3GFLOPS for 64 threads - 4% of the peak performance (32GFLOPS with 64 thread units).

Next, we will study a sequence of optimizations to improve the parallel performance.

Using SPM. The next step is to use the SPM as a high speed buffer to accelerate the corresponding thread unit in the computation. We still perform the 16 × 16 matrix multiplication in SPM. The matrices are copied into SPM block by block. The computation is conducted and the result is stored back into SRAM. It is worth copying the resultant block into SPM as it will be used 8 times. Since the two source matrices are only used once, they are not copied into the SPM. Implementing this yielded 1.79GFLOPS. This represents a 38% performance improvement over the base version (See "Using SPM" in Table 2).

Table 2. 128x128 MMM Incremental Optimizations in SRAM

Optimizations	GFLOPS	Speedup Over Baseline Parallel Version	Speedup Over Sequential Version	Incremental
Baseline	1.29	1.00	40.31	0%
Using SPM	1.79	1.38	55.94	38%
Tiling+Unrolling	2.77	2.15	88.56	55%
Reg. Tiling	5.05	3.91	157.81	82%
Inst. Sched.	10.02	7.77	313.12	99%
Reg. Alloc.	11.03	8.55	344.69	10%
Sync. Opt.	13.70	10.61	428.12	24%

Loop Tiling and Unrolling. Loop tiling is a very effective optimization for architectures with caches. The tile size is chosen to allow all the data accessed by the inner most tile to fit into the cache. For matrix multiplication, the 16 × 16 matrix is split into two levels of 4 × 4 tiles.

A simple tiling does not bring performance gain as the number of branch instructions and code size are increased. By unrolling the next level of inner loops, 2.77GFLOPS, which is a 55% improvement over "Using SPM", is achieved.

Register Tiling (Manually). For the inner most 3 loop nests, there are total $4 \times 4 \times 3 = 48$ data elements that can fit into 64 registers of C64. The data reuse rate is 4 for each element of A and B, and 32 for C.

Because of the current limitation in the compiler, we manually did the register tiling by allocating registers properly to the data elements of the 3 matrices, as well as other index variables. Those elements are used in the 2 inner most loop nests, with A and B inside and C one level outside. After manually performing register tiling and allocation, the optimized code achieved 5.05GFLOPS, which is an 82% improvement over the simple tiling plus unrolling.

Instruction Scheduling (Manually). After register tiling, by properly scheduling the instructions in the innermost loop, we can hide the latencies of most memory and floating point operations and achieve 10.02GFLOPS - another 99% improvement over the register tiling. By moving accesses to C outside of the inner most loop, the performance reaches 11.03GFLOPS.

A good instruction scheduler is very important to the MM application as well as other programs. The key issue is that the scheduler should be aware of the different latencies when accessing different memory segments (SPM, SRAM and DRAM). Most existing compilers assume cache latency when they do instruction scheduling. For this architecture, there is no data cache and each load/store may have different latency depending on the target memory segment. Explicit multi-level memory hierarchy aware instruction scheduling is a key optimization for the C64 architecture. In fact, loop tiling, register tiling and instruction scheduling have to be tightly coupled, and the aggregation of the 3 optimizations is the key to generate optimal code for even a simple matrix multiplication.

Remove Unnecessary Synchronization. In all the above experiments, mutex is used to control the access to the task pool. When one thread is getting a task from the task pool and updating the status of the allocated task, all other threads have to wait for the release of the mutex lock.

Since MM is a regular application, an alternative approach is to statically assign workload, i.e., each thread is assigned to a fixed number of tasks. As a result, the mutex lock is not needed. After removing the mutex, we get 13.70GFLOPS, which is 42.8% of the potential peak performance (32GFLOPS for 64 threads).

All of the above results are based on the assumption that 3 matrices are stored into on-chip SRAM. The memory bandwidth (320GB/s) is enough to sustain the computation. However, when the matrices become larger and larger such that they cannot be stored into on-chip SRAM, bandwidth of DRAM becomes a major issue. In the next section, we are going to investigate bandwidth optimizations to bring high performance to the algorithm assuming that data resides in off-chip DRAM.

4.3 Parallelizing Matrix Multiplication in DRAM

Off-chip DRAM is the largest memory resource of the C64 architecture. Most data and code will be stored there for real applications. On-chip SRAM and SPM are smaller and more expensive resources, and should be used more carefully.

To demonstrate the optimizations, we use 256×256 matrices that need to be stored in DRAM with 128×128 sub-banks buffered in SRAM. Therefore, the application has to move data between DRAM and SRAM. In this section we study the impact of DRAM bandwidth limitation on the application's performance and how to tackle this problem by hiding the communication latency between DRAM and SRAM with computation. A nice feature of C64 is that thread units are not expensive - there are very many of them. On-chip memory resources are more expensive. We can use a set of thread units to do the computation and another group of thread units to move data between DRAM and SRAM. In this case study, we use two sets of SRAM banks (double buffering). One set for computation and another set for preloading, and switch between them during the computation.

DRAM Bandwidth. For the first version of C64 chip design, the DRAM can transfer at most 32 bytes every cycle. Hence, the total DRAM bandwidth is 16GB/s.

To make the best utilization of the DRAM bandwidth, load multiple and store multiple (of 8 doublewords or 64 bytes) instructions should be used and the starting address should be 64 byte aligned.

Bandwidth limitation is the major challenge here. For 128×128 matrix multiplication, the total number of memory accesses is $128 \times 128 \times 128 \times 8 \times (3 + 1)$ bytes (3 loads, 1 store), or $67,108,864$ bytes. Then, the ideal access to memory time is $67,108,864/32$, or $2,097,152$ cycles. Even excluding load/store conflicts and ignoring other instructions, the peak performance can only be 1GFLOPS.

We may assume the C array is loaded and stored in the second innermost loop. The total bytes to be accessed becomes $128 \times 128 \times 128 \times 8 \times 2 + 128 \times 128 \times 8 \times 2$, or $33,816,576$ bytes. In this case, the ideal performance increases to 1.98 GFLOPS. But we are still far from the peak performance (32GFLOPS for 64 threads).

This means that we have to use on-chip SRAM and/or SPM to buffer matrix blocks, perform the computation in SRAM/SPM, and store the results back to off-chip DRAM. In other words, we have to reduce the DRAM bandwidth requirements via the on-chip data reuse.

Using LDM and STM. One optimization is to use LDM and STM instructions to aggregate multiple memory accesses. Four LDD (load doubleword) are combined into one LDM and four STD are combined into one STM. Hence, DRAM requests are effectively reduced to 1/4 of its original number, and DRAM bandwidth has been better utilized here. The best case is to combine 8 LDD into one LDM and 8 STD into one STM. But for register tiling, 4x4 is a better choice. If we do 8x8, although we can load sub-blocks into registers, we cannot consume them and have to store them into on-chip memory. This is not good for matrices A and B.

Using On-chip Memory. To reduce the bandwidth requirement to DRAM, we try to move sub-blocks of matrices into SRAM, and move intermediate results back to DRAM whenever it is necessary. We also pipeline the process by using two SRAM blocks for each matrix: one for computation and the other for load/store.

In this study, we assume the original size of the three matrices is 256×256 and they reside in DRAM. The on-chip block size is 128×128. Each matrix has two blocks in SRAM and half of each is loaded into SRAM. We assume $c1$ and $c2$ for matrix C, $a1$ and $a2$ for A, and $b1$ and $b2$ for B. While one set of SRAM blocks is used for computation, the other set can be used to load or store. The pipeline is designed as follows:

The total DRAM accesses: $128 \times 128 \times 8 \times$ (4 loads of C + 4 stores of C + 4 loads of A + 6 loads of B) $= 2,359,296$ bytes. The ideal DRAM access time in this case is $73,728$ cycles, which is equivalent to 56.9GFLOPS without considering other computations.

Synchronization Overhead. To implement the above pipelined scheme, a barrier is inserted at the end of each step. There are 12 barrier invocations in the implementation. This guarantees that computation happens after loading all the required data, and storing follows the corresponding computation stage. C64 has hardware barrier support with low cost. A barrier can be completed in as little as dozens of cycles.

Computation Threads	Memory Access Threads
	load c00 (to c1), a00(a1),b00 (b1)
compute c00/a00/b00 in c1/a1/b1	load a01(to a2) b10 (to b2)
compute c00/a01/b10 in c1/a2/b2	load c01(to c2) b01 (to b1)
	store c00
compute c01/a00/b01 in c2/a1/b1	load b11(to b2)
compute c01/a01/b11 in c2/a2/b2	load c11(to c1) a10 (to a1)
	store c01
compute c11/a10/b01 in c1/a1/b1	load a11(to a2)
compute c11/a11/b11 in c1/a2/b2	load c10(to c2) b00 (to b1)
	store c11
compute c10/a10/b00 in c2/a1/b1	load b10(to b2)
compute c10/a11/b10 in c2/a2/b2	
	store c10

Fig. 3. Execution Steps When the Matrices are in DRAM

Optimized memcpy(). The standard C library features an optimized version of memcpy(), which is up to 20 times faster than the initial straightforward implementation. It takes into account possible unalignment at the source and destination, as well as different copy lengths. It is also capable of pipelining the three basic stages: loading from the source array, address computation and storing into the destination array.

Using More Threads for Load/Store. In previous sections, only one thread handles the work of loading and storing. To further improve the performance, we assign three more threads, four in total. Three threads are responsible for preloading each of the three matrices, and the main thread handles the task pool creation and stores the resulting sub matrices back to DRAM.

The final result we achieve is $1,206,048$ cycles and 13.9 GFLOPS for a 256×256 problem size, which is 43.4% of the peak performance (we use 68 threads in this case: 64 threads for computation, 4 threads for load/store).

Table 3. Optimizations for Matrices in DRAM

Optimizations	Size	Cycles	Mem/Delay	FLOPS	Speedup
No Opt	128	6,499,276	5,401,783	322.7M	
No Opt	256	42,078,325	35,060,687	398.7M	1.00
LDM/STM	128	1,745,340	1,439,301	1.2G	
LDM/STM	256	13,996,754	11,652,068	1.2G	3.00
All Opt	256	1,206,048	810,997	13.9G	34.86

5 Conclusions

Our results demonstrate that efficiently exploiting the multi-level memory hierarchy is the key to achieve good performance on C64. When data fits into SRAM, tiling, loop unrolling, register allocation, and instruction scheduling are the most important optimizations. SPM can also be used to buffer frequently accessed data. When data does not fit in SRAM, DRAM bandwidth becomes the bottleneck. To overcome this issue, first we use SRAM to buffer blocks of DRAM data, which additionally reduces the bandwidth requirements to DRAM. Second, we overlap DRAM accesses with computation in SRAM to dramatically improve the performance.

For compiler designers, inner most register tiling is very important. The instruction scheduler should be aware of the latency for each memory segment. High level loop optimization should be able to automatically choose SPM buffers for SRAM data and/or SRAM buffers for DRAM data.

6 Related and Future Work

Locality optimizations have been studied by numerous researchers which resulted in many publications on cache-based architectures. Loop transformations have been investigated to exploit computation parallelism and data locality for scientific applications [6,7,8,9,10,11]. Loop tiling is a well known loop transformation to increase cache locality (see [12,7,9,13,14] and their references). We use loop tiling (and register tiling) to map a matrix block into the register file, SPM, and SRAM. Bandwidth optimization has also been extensively explored in [15,16,17,18,19,20,21] and their references. Indeed, we have shown that an efficient utilization of the memory bandwidth is critical for C64 when data is stored in DRAM. Phase order problem has been studied in [4,3] and their references. We identify a set of useful optimizations for C64-like architectures. Moreover, we explore a practical sequence order of optimizations for the matrix multiplication that yields 14GFLOPS.

As future work we intend to the study other representative benchmarks. The identified optimizations will be implemented in the C64 compiler. Traditional loop optimizations may be extended to support automatic storage and thread unit management by allocating SPM and SRAM to the hot data at certain computation phases, and automatically overlap memory transfer with computation.

Acknowledgments

We acknowledge support from IBM, in particular, Monty Denneau, Henry Warren, José Castaños, and Christos Georgiou, ETI, the Department of Defense, the Department of Energy (DE-FC02-01ER25503), the National Science Foundation (CNS-0509332), and other government sponsors. Thanks to many CAPSL members for helpful discussions, in particular, Fei Chen, Brice Dobry, Geoff Gerfin, Ge Gan, Wesley Toland, and John Tully.

References

1. Denneau, M., Warren, Jr., H.S.: 64-bit Cyclops principles of operation part I. Technical report, IBM Watson Research Center, Yorktown Heights, NY (2005)
2. Denneau, M., Warren, Jr., H.S.: 64-bit Cyclops principles of operation part II: Memory organization, the A-switch, and SPRs. Technical report, IBM Watson Research Center, Yorktown Heights, NY (2005)

3. Almagor, L., Cooper, K.D., Al., E.: Finding effective compilation sequences. In: LCTES'04, Wahsington, DC, USA (2004)
4. Wolf, M.E., Maydan, D.E., Chen, D.K.: Combining loop transformations considering caches and scheduling. In: Proceedings of the 29th Annual International Symposium on Microarchitecture, Paris, IEEE-CS TC-MICRO and ACM SIGMICRO (1996) 274–286
5. del Cuvillo, J., Zhu, W., Hu, Z., Gao, G.R.: Fast: A functionally accurate simulation toolset for the cyclops-64 cellular architecture. In: Workshop on Modeling, Benchmarking and Simulation (MoBS'05) of ISCA'05, Madison, Wisconsin (2005)
6. Allen, R., Kennedy, K.: Optimizing Compilers for Modern Architectures: A Dependence-based Approach. Morgan Kauffmann Publishers (2001)
7. Wolf, M.E., Lam, M.S.: A data locality optimizing algorithm. In: Proceedings of the ACM SIGPLAN '91 Conference on Programming Language Design and Implementation, Toronto, Ontario (1991) 30–44 *SIGPLAN Notices*, 26(6), June 1991.
8. Carr, S., McKinley, K.S., Tseng, C.W.: Compiler optimizations for improving data locality. In: Proceedings of the Sixth International Conference on Architectural Support for Programming Languages and Operating Systems, San Jose, California, ACM SIGARCH, SIGOPS, SIGPLAN, and the IEEE Computer Society (1994) 252–262 *Computer Architecture News,* 22, October 1994; *Operating Systems Review,* 28(5), December 1994; *SIGPLAN Notices,* 29(11), November 1994.
9. Anderson, J.M., Lam, M.S.: Global optimizations for parallelism and locality on scalable parallel machines. In: Proceedings of the ACM SIGPLAN '93 Conference on Programming Language Design and Implementation, Albuquerque, New Mexico (1993) 112–125 *SIGPLAN Notices,* 28(6), June 1993.
10. Wolfe, M.J.: High Performance Compilers for Parallel Computing. Addison-Wesley Longman Publishing, Boston, MA (1995)
11. Lim, A.W., Lam, M.S.: Maximizing parallelism and minimizing synchronization with affine transforms. In: Conference Record of POPL'97: The 24th ACM SIGPLAN-SIGACT Symposium on Principles of Programming Languages, Paris (1997) 201–214
12. Wolfe, M.: Iteration space tiling for memory hierarchies. (SIAM) Parallel Processing for Scientific Computing (1987) 36–361
13. Andonov, R., Bourzoufi, H., Rajopadhye, S.: Two-dimensional orthogonal tiling: from theory to pratice. In: HiPC 1996, Trivandrum, India (1996)
14. Xue, J.: Loop Tiling for Parallelism. Kluwer Academic Publishers (2000)
15. Calder, B., Krintz, C., John, S., Austin, T.: Cache-conscious data placement. In: Proceedings of the Eighth International Conference on Architectural Support for Programming Languages and Operating Systems, San Jose, California, ACM SIGARCH, SIGOPS, SIGPLAN, and the IEEE Computer Society (1998) 139–149 *Computer Architecture News,* 26, October 1998; *Operating Systems Review,* 32(5), December 1998; *SIGPLAN Notices,* 33(11), November 1998.
16. Ding, C., Kennedy, K.: Improving cache performance in dynamic applications through data and computation reorganization at run time. [22] 229–241 *SIGPLAN Notices,* 34(5), May 1999.
17. Kennedy, K., Kremer, U.: Automatic data layout for distributed memory machines. ACM Transactions on Programming Languages and Systems **20**(4) (1998)
18. Chilimbi, T.M., Davidson, B., Larus, J.R.: Cache-conscious structure definition. [22] 13–24 *SIGPLAN Notices,* 34(5), May 1999.
19. N.Gloy, Smith, M.D.: Procedure placement using temporal-ordering information. ACM Transactions on Programming Languages and Systems **21**(6) (1999)
20. Ding, C., Kennedy, K.: Improving effective bandwidth through compiler enhancement of global cache reuse. Parallel and Distributed Computing **64**(1) (2004)

21. Ding, C., Orlovich, M.: The potential of computation regrouping for improving locality. In: SuperComputing 2004, Pittsburgh, PA. (2004)
22. Proceedings of the ACM SIGPLAN '99 Conference on Programming Language Design and Implementation. In: Proceedings of the ACM SIGPLAN '99 Conference on Programming Language Design and Implementation, Atlanta, Georgia (1999) *SIGPLAN Notices,* 34(5), May 1999.

Optimizing OpenMP Parallelized DGEMM Calls on SGI Altix 3700

Daniel Hackenberg, Robert Schöne, Wolfgang E. Nagel, and Stefan Pflüger

Technische Universität Dresden,
Center for Information Services and High Performance Computing (ZIH),
01062 Dresden, Germany
{daniel.hackenberg, robert.schoene, wolfgang.nagel, stefan.pflueger}
@zih.tu-dresden.de
http://www.tu-dresden.de/zih

Abstract. Using functions of parallelized mathematical libraries is a common way to accelerate numerical applications. Computer architectures with shared memory characteristics support different approaches for the implementation of such libraries, usually OpenMP or MPI.

This paper's content is based on the performance comparison of DGEMM calls (floating point matrix multiplication, double precision) with different OpenMP parallelized numerical libraries, namely Intel MKL and SGI SCSL, and how they can be optimized. Additionally, we have a look at the memory placement policy and give hints for initializing data. Our attention has been focused on a SGI Altix 3700 Bx2 system using BenchIT [1] as a very convenient performance measurement suite for the examinations.

1 Measurement Environment

For a detailed analysis of a system architecture by parameter studies, the choice of a suitable measuring framework is an important decision. To benchmark the DGEMM calls we use BenchIT. This performance measurement suite helps to compare different algorithms, implementations of algorithms, features of the software stack, and hardware details of whole systems. It has been designed to run many microbenchmarks on every POSIX 1.003 compatible system in a very user-friendly way. BenchIT has been developed at the Center for Information Services and High Performance Computing (ZIH) at the Technische Universität Dresden and was previously mentioned at [2,3,4]. Sources and results are freely available at [1].

2 The SGI Altix 3700 Bx2 System

2.1 System Architecture

The SGI [5] Altix 3700 Bx2 is a ccNUMA shared memory system based on Intel Itanium 2 processors and SGI's scalable node architecture SN2. In developing this, special attention has been paid to building a highly scalable computer with large bandwidths on all data paths. It provides cache coherency in one coherent sharing domain (CSD), which runs a Linux kernel and can scale up to 512 processors each. A single processor is operating at 1.5 GHz and therefore a maximum floating point performance of 6 GFLOPS can be reached. More information, especially about the SGI bricks and application benchmarks, can be found at [6].

2.2 First Touch Policy

In contrast to former SGI systems like the Origin 3800, the Altix does not move data near the processor which is using it most. Instead, it uses the so-called first touch policy which means that data is placed next to the processor that writes to it first. This may have no effect if the application was parallelized with MPI, and the data was spread manually. Multithreaded programs usually don't spread data because all addresses can be accessed from every thread.

In the worst case, all data is placed in just one memory module when all other OMP threads want to access it. The remaining bandwidth for each thread would shrink to b/p, where b is the bandwidth of a single memory module and p the number of participating processors and threads respectively.

3 Optimizing the DGEMM Call

The cblas_dgemm call is defined as shown in Listing 1.1. A description of the parameters can be found at [7].

```
void cblas_dgemm (
    const enum CBLAS_ORDER Order,
    const enum CBLAS_TRANSPOSE TransA,
    const enum CBLAS_TRANSPOSE TransB,
    const int M, const int N,
    const int K, const double alpha,
    const double *A, const int lda,
    const double *B, const int ldb,
    const double beta,
    double *C, const int ldc);
```

Listing 1.1. cblas_dgemm call declaration

A simple cblas_dgemm call for a matrix multiplication for two square matrices of order *size* would look like this:

```
cblas_dgemm (CblasRowMajor, CblasNoTrans, CblasNoTrans,
    size, size, size, 1.0, A, size, B,
    size, 1.0, C, size);
```

Listing 1.2. Simple matrix multiplication using cblas_dgemm

The following sections will explain how these DGEMM calls can be parallelized and optimized to scale well up to at least 124 processors on the SGI Altix 3700.

3.1 The BenchIT DGEMM Kernel

As previously mentioned, BenchIT is a measurement environment which helps to examine a system with microbenchmarks. Some of these execute a sequential matrix multiplication with different libraries. During each measuring run, which means one performance measurement for one problem size, the matrices are allocated and filled with variables. After the measurement the allocated memory is being released. Therefore, each measurement consists of three steps: initializing the data, recording the duration of processing, and destroying data. The problem sizes that are to be measured can be set as parameters for the microbenchmark.

The original hardware vendor mathematical libraries available on SGI Altix 3700 are the Intel Math Kernel Library - MKL 8.0 [8] and the SGI Scientific Computing Software Library - SCSL 1.6.1.0 [9]. They both nearly reach peak performance on a single processor (Fig. 1).

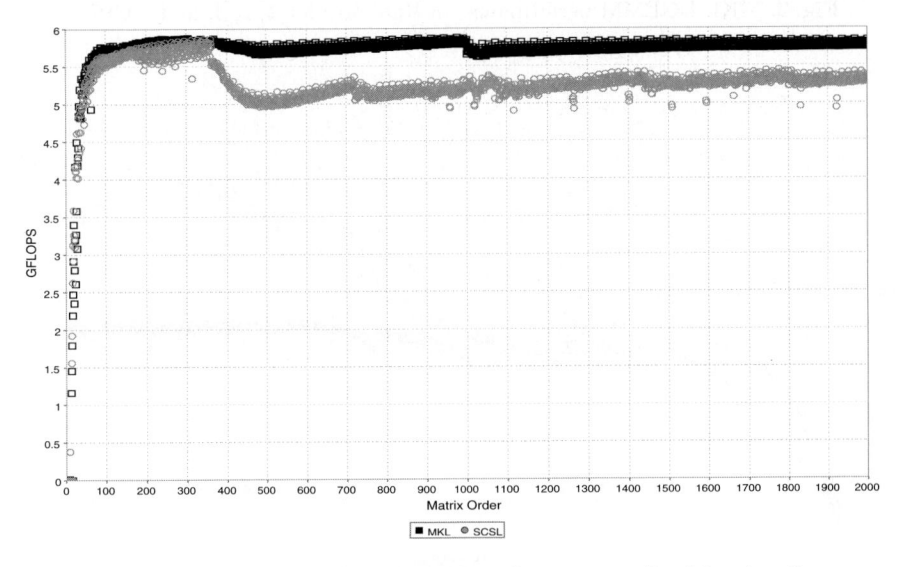

Fig. 1. MKL and SCSL DGEMM performance on Intel Itanium 2

3.2 Library Provided OpenMP Parallelization

The DGEMM calls can easily be parallelized by setting the environment variable OMP_NUM_THREADS. This implicit parallelization is provided by both the MKL and SCSL, using OpenMP to handle different threads in order to calculate the matrix multiplication faster.

Fig. 2 shows a nearly linear speedup for 2, 4, and 8 processors. The graph for 16 CPUs in Fig. 3 also shows acceptable performance, but using 32 or 64

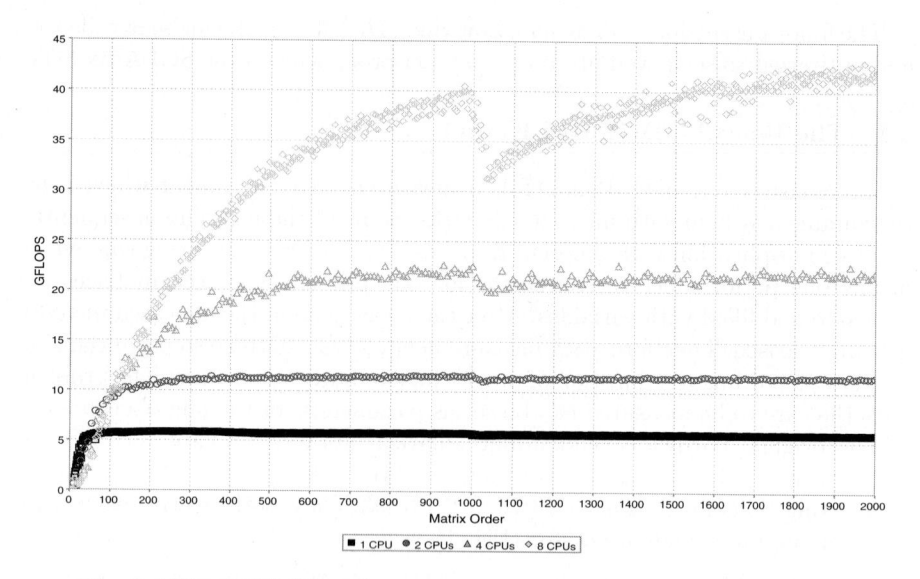

Fig. 2. MKL DGEMM performance on Itanium 2 for 1, 2, 4, and 8 CPUs

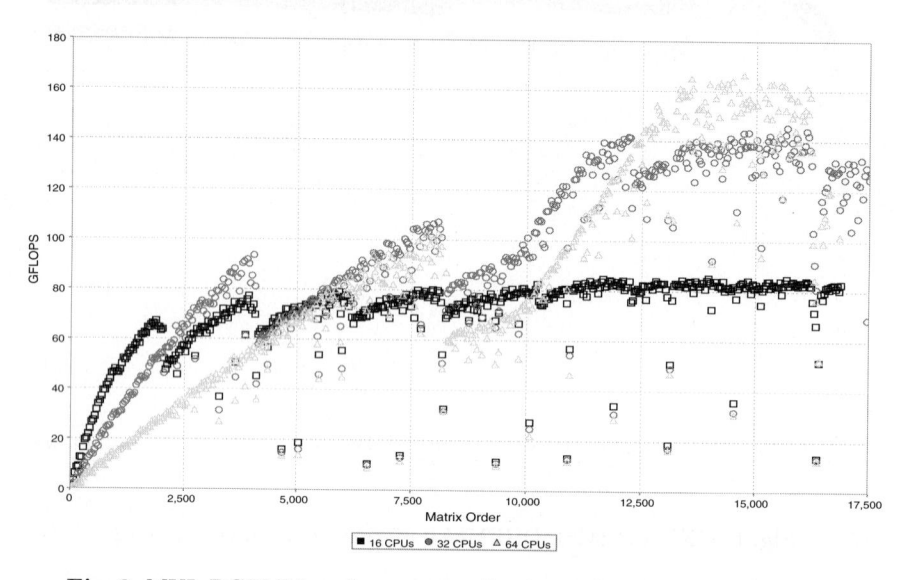

Fig. 3. MKL DGEMM performance on Itanium 2 for 16, 32, and 64 CPUs

processors will need very large matrices to achieve at least a small speedup, and can even be slower for small matrices.

The SCSL shows equally discouraging results as the MKL does in Fig. 3, which is not surprising, as the limiting factor for more than 16 processors is just the memory bandwidth due to the bad data distribution. The next subsection will describe how this performance degradation can easily be corrected for both libraries.

3.3 Parallelizing the Data Initialization

With the first touch policy (see 2.2) in mind, a better data distribution on the machine should be implemented. This can be achieved by surrounding the initialization loop that fills the matrices with an OpenMP pragma. Listing 1.3 shows the corresponding C code. Please note that the data is distributed on the machine by different threads, each of them running on a different processor. Therefore the matrices are spread row-wise over the memory modules. Special attention should be paid to this parallelization in order not to scatter the cache lines. We did not find a better data distribution by using, for example, different chunksizes or other scheduling strategies such as dynamic scheduling.

Without having extended knowledge of the DGEMM library, it is impossible to know which processor will predominantly use which data to further improve the code in listing 1.3. Thus we can only spread the data over the memory modules in order to use the whole memory bandwidth of the system as efficiently as possible without providing a perfect data distribution. However, when using OpenMP-parallized calculations in a specific program, the data distribution should be adjusted accordingly.

```
#pragma omp parallel for schedule(static,1) \
  private(x,index,max) shared(A,B,C,size)
for(x = 0; x < size; x++) {
    index = x * size;
    max = index + size;
    for(index; index < max; index++) {
        A[index] = 30.0;
        B[index] = 0.01;
        C[index] = 0.0;
    }
}
```

Listing 1.3. Surrounding pragma for initialization loop

As shown in Fig. 4, the improved data distribution speeds up the matrix multiplication for a large number of processors significantly. The 64 processor measurement now peaks at about 330 GFLOPS instead of 170 GFLOPS. As we have noticed for example for `getrf` (LAPACK) measurements, the effect can be much larger for algorithms which are not as cache efficient as DGEMM but depend more on memory bandwidth. As a welcome side effect, the improved data distribution has resolved the issue of the striking variability in the performance of the MKL calls in Fig. 3.

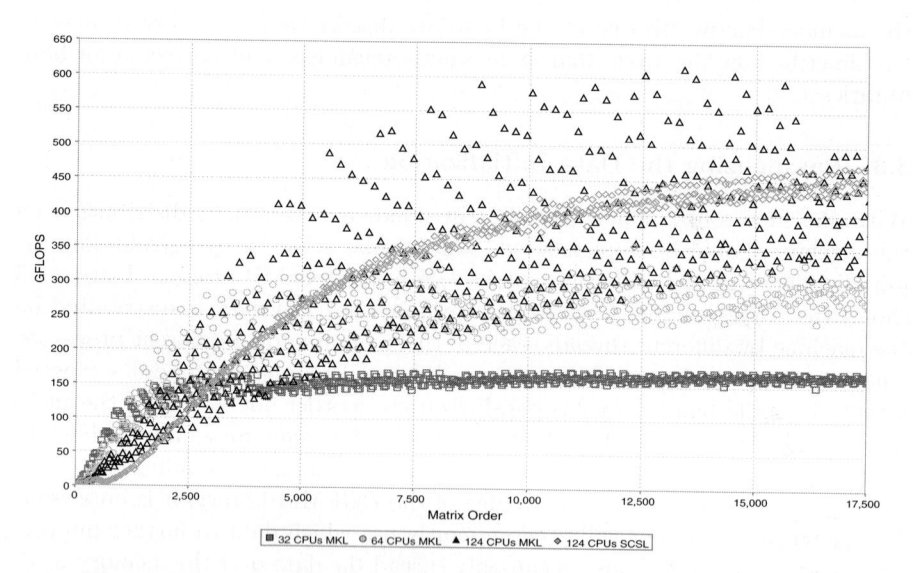

Fig. 4. Optimized MKL and SCSL DGEMM performance on Itanium 2

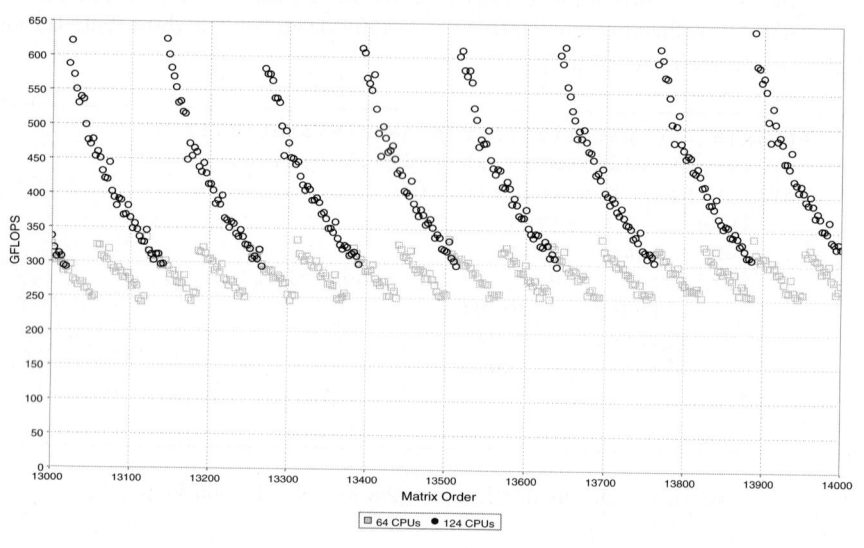

Fig. 5. Zoom on MKL DGEMM performance for 64 and 124 CPUs

Despite the obvious performance improvement, Fig. 4 also shows that MKL DGEMM calls with 64 and more processors are still not perfect. Instead, they show a very unsteady behavior. A more detailed view on the results (Fig. 5) shows that MKL scales best when the problem size is a multiple of the number of processors used. For a matrix multiplication running on 124 processors a difference of one as matrix order can decide whether the achieved performance is about 300 or 600 GFLOPS.

The SCSL on the other hand does not show this unsteady behavior as the plot for 124 CPUs represents in Fig. 4. However, the MKL peak performance is significantly higher, which suggests a further improvement of the MKL DGEMM call.

3.4 Partitioning the DGEMM Call

It was shown that a very good but not constant performance for a DGEMM call was reached by the MKL. For further optimization, we split up the DGEMM call to limit the computation of the resulting matrix in the number of columns and rows to a multiple of the number of processors and threads respectively. The remaining parts of the resulting matrix are calculated later on in separate DGEMM calls. Fig. 6 shows how this partitioning is done for a square matrix. There are four separate matrix multiplications and four corresponding DGEMM calls to be calculated. *optsize* equals the original matrix order (*size*) truncated to a multiple of the number of processors (p) and *diff* represents the overlapping part. Or as a formula: $diff = size \bmod p$ and $optsize = size - diff$.

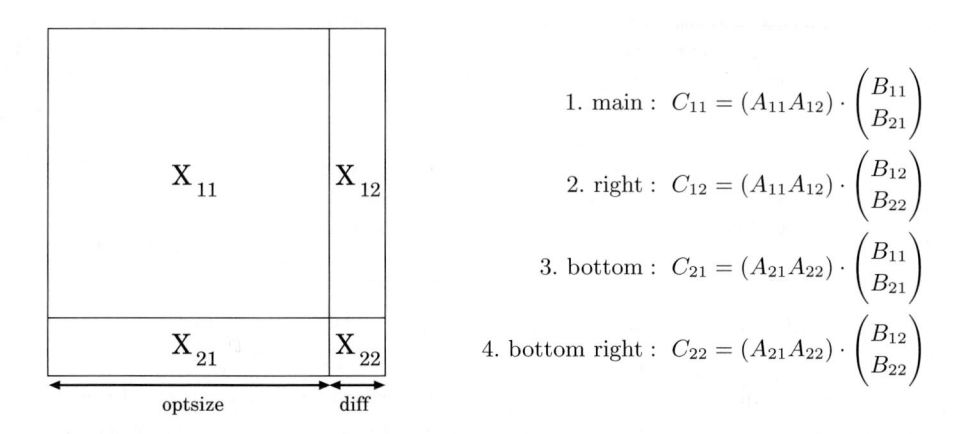

1. main : $C_{11} = (A_{11} A_{12}) \cdot \begin{pmatrix} B_{11} \\ B_{21} \end{pmatrix}$

2. right : $C_{12} = (A_{11} A_{12}) \cdot \begin{pmatrix} B_{12} \\ B_{22} \end{pmatrix}$

3. bottom : $C_{21} = (A_{21} A_{22}) \cdot \begin{pmatrix} B_{11} \\ B_{21} \end{pmatrix}$

4. bottom right : $C_{22} = (A_{21} A_{22}) \cdot \begin{pmatrix} B_{12} \\ B_{22} \end{pmatrix}$

Fig. 6. Partitioning the matrix multiplication

By splitting up the DGEMM calls we make sure that the largest part of the calculation (C_{11} in Fig. 6) is done with optimal performance. Considering the cubic complexity of the matrix multiplication, the computing effort for the remaining parts is, especially for larger matrices, very small.

The optimized C code for square matrices is shown in Listing 1.4. The code for non-square matrices is similar but slightly more sophisticated, as there are three optimal sizes that need to be calculated.

With this implementation, a maximal performance of about 600 GFLOPS for 124 processors is reached and stays nearly at the very same level. This means a speedup of about 102 in comparison to a single Itanium 2 processor for DGEMM calls. The performance of the new implementation is dominating the old one which means that there is nearly no performance loss. A speedup of 2 is reached for problem sizes N, when $N = p * n - 1$, where p is the number of processors and n is a large natural number. The lowest performance improvement is visible for problem sizes $N = p*n$, as they already had a good performance before the optimization. The overhead for these problem sizes is very small as there are only three additional integer calculations to be executed.

```
diff = size % omp_get_max_threads();
optsize = size - diff;
cblas_dgemm(CblasRowMajor,CblasNoTrans,CblasNoTrans,
   optsize, optsize, size, 1.0,
   A, size, B, size, one, C, size);      /* main part */
if (diff > 0)                /* calculate remaining parts */
{
  /* right */
  cblas_dgemm(CblasRowMajor,CblasNoTrans,CblasNoTrans,
     optsize, diff, size, 1.0, A, size, &(B[optsize]),
     size, one, &(C[optsize]), size);
  /* bottom */
  cblas_dgemm(CblasRowMajor,CblasNoTrans,CblasNoTrans,
     diff, optsize, size, 1.0, &(A[size*optsize]),
     size, B, size, one, &(C[size*optsize]), size);
  /* bottom right */
  cblas_dgemm(CblasRowMajor,CblasNoTrans,CblasNoTrans,
     diff, diff, size, 1.0, &(A[size*optsize]),
     size, &(B[optsize]), size, one,
     &(C[size*optsize+optsize]), size);
}
```

Listing 1.4. Optimized DGEMM call for MKL on SGI Altix

The overall speedup for the optimization described in this section is between one and two. Figure 7 compares the DGEMM performance on 124 processors for SCSL and MKL with improved data distribution according to 3.3 and MKL with additional DGEMM partitioning.

However, a direct optimization within the Intel MKL library might deliver even higher performance than the implementation described above.

3.5 Reinitializing the Data

Further examinations have revealed that beyond the DGEMM call a lot of time is used for initializing the matrices, even though this was parallelized. In fact, the parallel initialization of the three matrices with an order of 13000 takes about nine seconds on 124 processors. The corresponding write bandwidth is about 31 MByte/s. According to SGI [10], the glibc `malloc/free` calls consume significant system time due to memory management overhead. In order to prevent glibc from using `mmap`, the environment variables `MALLOC_TRIM_THRESHOLD_=-1`

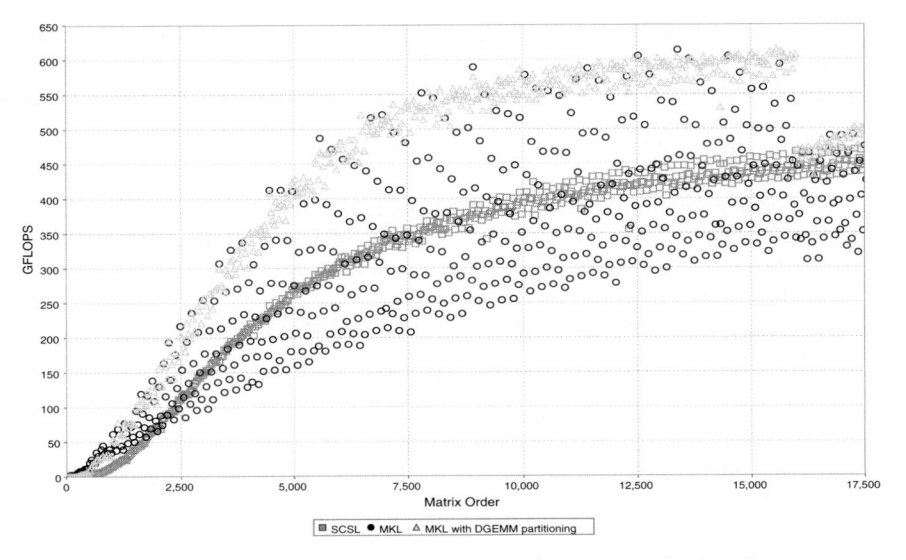

Fig. 7. Optimized MKL and SCSL DGEMM performance for 124 CPUs

and `MALLOC_MMAP_MAX_=0` have to be set. Thereby, repeated matrix reinitializations as in our BenchIT performance measurement runs are about one order of magnitude faster. However, these time values are unsteady and therefore further examination is planned.

4 Conclusion

The Altix 3700 combined with the Intel MKL or SGI SCSL provides a fast computation of DGEMM calls. These calls can easily be parallelized, but special attention should be paid to the first touch policy. For matrices of order 5000 or higher, 64 or more CPUs can be used efficiently but native MKL DGEMM calls show unsteady performance. We have described how to optimize these calls to offer steady performance and clearly outperform SCSL, mostly independent of the matrix order and the number of OpenMP threads used.

References

1. BenchIT: Homepage http://www.benchit.org
2. Juckeland, G., Börner, S.; Kluge, M., Kölling, S., Nagel, W.E., Pflüger, S., Röding, H., Seidl, S., William, T., Wloch, R.: ParCo 2003: BenchIT - Performance Measurement and Comparison for Scientic Applications http://www.benchit.org/DOWNLOAD/DOCUMENTS/parco_paper.pdf
3. Juckeland, G., Kluge, M., Nagel W.E., Pflüger, S.: QEST, IEEE Computer Society, 2004.: Performance Analysis with BenchIT: Portable, Flexible, Easy to Use ISBN 0-7695-2185-1, S. 320-321

4. Schöne, R., Juckeland, G., Nagel W.E., Pflüger, S., Wloch, S.: Parco 2005: Performance comparison and optimization: Case studies using BenchIT http://www.benchit.org/downloads/documents/parco_05_abstract.pdf
5. Silicon Graphics Inc.: Homepage http://www.sgi.com
6. Oak Ridge National Laboratoy: Evaluation of the Altix 3700 at Oak Ridge National Laboratoy
   http://www.gelato.unsw.edu.au/archives/linux-ia64/0409/10993.html
7. University of Tennessee: Basic Linear Algebra Subprograms Technical (BLAST) Forum http://www.netlib.org/utk/papers/blas-report.ps
8. Intel: Intel Math Kernel Library 8.0
   http://www.intel.com/cd/software/products/asmo-na/eng/perflib/mkl/index.htm
9. Silicon Graphics Inc.: Scientific Computing Software Library
   http://www.sgi.com/products/software/scsl.html
10. Silicon Graphics Inc.: Linux Application Tuning Guide
    http://techpubs.sgi.com/library/manuals/4000/007-4639-004/pdf/007-4639-004.pdf

Topic 3: Scheduling and Load Balancing

Michael Bender, Dror Feitelson, Allan Gottlieb, and Uwe Schwiegelshohn

Topic Chairs

An increasing variety of parallel and distributed systems is being developed throughout the world. Parallelism is available at granularities ranging from multi-core processors, via SMPs and clusters, to Grids. However, much of the computing power in these systems remains unusable because adequate algorithms and software for managing these resources are unavailable.

Topic 3 is devoted to scheduling and resource management. The broad objective of our topic is to develop the theory and technology so that the vast computing power of parallel and distributed systems can be used more efficiently. The subjects presented in Topic 3 cover many aspects of scheduling and load-balancing, including system-level techniques, theoretical foundations, applications, and scheduling tools. New trends and emerging models are also discussed.

Scheduling is the common term for the allocation of resources over time to computational tasks. In parallel computing, these resources include the computing devices (processors, computers), the communication medium (networks, buses, etc), and memory (disk, registers, etc). New parameters, such as heterogeneity, hierarchical character of memory or communication links, and large-scale computing, are also considered. As parallel and distributed computing evolves, classical models and techniques sometimes apply, but often new ones must be proposed, implemented, and validated.

There were 27 papers submitted to the Topic 3 track. Each submitted paper has been reviewed by 4 reviewers, and finally 11 papers were chosen for inclusion in the final program. They reflect the synergy between theoretical approaches (models, analysis of algorithms, complexity, approximability results, multi-criteria analysis) and practical realizations and tools (new methods, simulation results, experiments, specific tuning for an application).

We would like to thank to our colleagues, experts in the field, who helped in the reviewing process.

– Michael, Dror, Allan, and Uwe

The Price of Approximate Stability for Scheduling Selfish Tasks on Two Links

Eric Angel, Evripidis Bampis, and Fanny Pascual

Université d'Évry-Val d'Essonne, IBISC CNRS FRE 2873,
523 place des Terrasses, 91000 Évry, France
{angel, bampis, fpascual}@lami.univ-evry.fr

Abstract. We consider a *scheduling game*, where a set of selfish agents (traffic loads) want to be routed in exactly one of the two parallel links of a system. Every agent aims to minimize *her own completion time*, while the social objective is the *makespan*, i.e. the time at which the last agent finishes her execution. We study the problem of optimizing the makespan under the constraint that the obtained schedule is a (pure) Nash equilibrium, i.e. a schedule in which no agent has incentive to unilaterally change her strategy (link). We consider a relaxation of the notion of *equilibrium* by considering α-approximate Nash equilibria where an agent does not have *sufficient* incentive (w.r.t. the value of α) to unilaterally change her strategy. Our main contribution is the study of the *tradeoff* between the *approximation ratio* for the makespan and the value of α. We first give an algorithm which provides a solution with an approximation ratio of $\frac{8}{7}$ for the makespan and which is a 3-approximate Nash equilibrium, provided that the local policy of each link is *Longest Processing Time* (LPT). Furthermore, we show that a slight modification of the classical *Polynomial Time Approximation Scheme* (PTAS) of Graham allows to obtain a schedule whose makespan is arbitrarily close to the optimum while keeping a constant value for α. Finally, we give bounds establishing relations between the value of α and the best possible value of the approximation ratio, provided that the local policies of the links are LPT.

1 Introduction

The scheduling setting that we consider in this paper is the following one: we are given a simple network of two parallel links and a set of n selfish agents. Each agent has some (positive) traffic load and wants to use exactly one of the parallel links to route it through. Equivalently, every agent can be viewed as a task, and each link as a machine. Every agent aims to maximize her own profit, and there are two basic models depending on what is considered as the individual profit of the agents:

- In [2], the *profit* of an agent is the inverse of the completion time of the machine on which she is assigned to. This model is known as the KP model.
- In [1], the *profit* of an agent is the inverse of her completion time. This model is known as the CKN model.

In both cases, the *social optimum (or global objective function)* is the *makespan*, i.e. the time at which the last agent terminates her execution.

Our aim is to obtain a schedule which minimizes the makespan and which at the same time is *stable*, i.e. such that no agent has incentive to unilaterally change her link. More precisely, we assume that each processor uses a local policy (known by all the agents) in order to schedule the tasks assigned to it, and if a task decides to leave, in the proposed schedule, its current machine to go on another machine, knowing its local policy, it can calculates its new completion time. Therefore, the solution we seek is a *best Nash equilibrium* w.r.t. the global objective function. A new measure for evaluating the impact of searching a stable solution, i.e. a Nash equilibrium, has been introduced by Schultz et al. [3] and by Anshelevich et al. [4]: the *price of stability* is defined as the ratio of the objective function in a *best* Nash equilibrium and a global optimum schedule (this maximum is taken over all solutions). This measure can be viewed as the *optimistic* price of anarchy [2].

In the KP model there always exists a pure Nash equilibrium which is an optimal solution w.r.t. the makespan, and so the price of stability is 1 for this model. This result is a direct corollary of the following fact: it is always possible in the KP model to transform (*nashify*) an initial solution into a pure Nash equilibrium without increasing the value of the makespan [7]. It is then natural to ask if this is also possible for the CKN model. Notice that the the price of stability depends on the local policies of the links. We assume that each link schedules the tasks assigned to it using the Longest Processing Time (LPT) order, i.e. each link schedules its tasks from the largest one to the smallest one. In that case, it is not difficult to see that there is only one pure Nash equilibrium and that this schedule can be obtained using the classical LPT list scheduling algorithm. This shows that the price of stability in that case is 7/6, i.e. the approximation ratio of the LPT algorithm [6].

A natural way to improve this ratio, is to relax the definition of a "stable" schedule. We consider that a schedule is stable if it is an α-*approximate Nash equilibrium* ($\alpha \geq 1$), i.e. a schedule in which no agent has *sufficient incentive* to unilaterally change its behavior. We say that an agent does not have *sufficient incentive* to unilaterally leave the link on which it is scheduled, if and only if this does not increase its profit by more than α times its current profit[1]. We can define now the *price of α-approximate stability* as the ratio of the objective function in a *best* α-approximate Nash equilibrium and a global optimum schedule (this maximum is taken over all solutions). If $\alpha = 1$ we get the price of stability defined before. Thus, if the price of α-approximate stability is γ, it means that no algorithm (even an exponential one) with an approximation ratio smaller than γ can return schedules which are always α-approximate Nash equilibrium.

---

[1] Note that this definition is different from the definition of an ϵ-Nash equilibrium in [5]: in [5], if a solution is an ϵ-Nash equilibrium, then if an agent unilaterally changes strategy, her profit should be smaller than or equal to her current profit *plus* ϵ (and not *times* α as in our definition).

Example: Let us consider two machines using LPT policy, and the following tasks: two tasks t_1 and t_2 of length 3 and 3 tasks t_3, t_4 and t_5 of length 2. The only pure Nash equilibrium is the schedule where the tasks of length 3 are scheduled at time 0, and are followed by the tasks of length 2. This solution has a makespan of 7, whereas the makespan of the optimal solution is 6. In the optimal solution, let us consider the task of length 3 which starts at time 3. By going on the second machine, this task would be the largest one of this machine and would then be scheduled in the first position. Its completion time would then be 3 instead of 6. Since this task can reduce its completion time by a factor of 2 by changing machine, we will say that this task is *2-approximate*. Each task of this schedule can reduce its completion time by at most 2 by changing machine, so this schedule is a 2-approximate Nash equilibrium.

Our contribution: We give relations between α and the approximation ratio for the makespan, provided that the policies of the links are LPT. We show that the price of α-approximate stability is at least $\frac{8}{7}$ for all $\alpha < 2.1$, and at most $\frac{8}{7}$ for all $\alpha \geq 3$. We give in Section 2 an algorithm which shows this last bound: it achieves an approximation ratio of $\frac{8}{7}$ for the makespan and returns a 3-approximate Nash equilibrium. We show that the price of α-approximate stability is larger than or equal to $1 + \varepsilon$ for any $\alpha \leq k$, where k is a certain constant in $\Theta(\varepsilon^{-1/2})$, and that it is smaller than or equal to $1 + \varepsilon$ for any $\alpha \geq \frac{1}{\varepsilon}$ (see Section 3). This last bound is obtained by analyzing the PTAS of Graham (slightly modified) [6]. Section 4 shows a summary of the negative and positive results of this paper.

2 A Variant of LPT

For simplicity in the sequel, we adopt the classical terminology of scheduling theory. We want to schedule n tasks $\{t_1, \ldots, t_n\}$ on 2 identical machines, and we want to minimize the maximum completion time of the last task scheduled, i.e. the *makespan*. We denote by P_1 the machine with maximum load, and by P_2 the other machine. Let n_i be the number of tasks scheduled on P_i. Let x_i be the ith task on P_1, and y_i the ith task on P_2. We denote by $l(t_i)$ the execution time (or length) of task t_i.

We represent a schedule by two sets A and B, where A (respectively B) is the set of the tasks scheduled on P_1 (respectively P_2). On each machine, the tasks are scheduled in the decreasing order of execution times. Let $\xi = (A, B)$ be a schedule of the n tasks on the two machines. Let a (respectively b) be a subset of tasks scheduled on P_1 (respectively P_2). We denote by $swap(\xi, a \leftrightarrow b)$ the schedule $((A \setminus a) \cup b, (B \setminus b) \cup a)$. In this new schedule, each machine still executes its tasks using the LPT policy (if two tasks have the same lengths, the one with the smallest identification number is scheduled first).

Let us now consider the following algorithm LPT_{swap}.

Theorem 1. *The algorithm LPT_{swap} achieves an approximation ratio of $\frac{8}{7}$ for the makespan.*

```
Order tasks by non increasing execution times. At each step i, for 1 ≤ i ≤ n,
schedule the current task t_i on the machine which has the smallest completion
time. Let LPT denote the schedule obtained in this way.
Let S = Σ_{i=1}^n l(t_i).
if n_2 ≥ 2 and ((n_1 = 3 and l(x_1) + l(x_2) + l(x_3) > (4/7)S) or (n_1 = 4)) then
    if n_1 = 3 then
        Let ξ_1 = swap(LPT, {x_1} ↔ {y_2}), ξ_2 = swap(LPT, {x_2} ↔ {y_2}) and
        ξ_3 = swap(LPT, {x_1} ↔ {y_1}).
        Return a schedule which minimizes the makespan among the schedules:
        LPT, ξ_1, ξ_2 and ξ_3.
    end
    if n_1 = 4 then
        Let ξ_4 = swap(LPT, {x_3, x_4} ↔ {y_2}).
        Return a schedule which minimizes the makespan among the schedules:
        LPT and ξ_4.
    end
end
else
 |  Return LPT.
end
```

Algorithm LPT_{swap}

Proof: Let us assume that ξ is a LPT schedule which does not achieve a $\frac{8}{7}$-approximation ratio. We show that the algorithm LPT_{swap} transforms this schedule into a schedule which achieves a $\frac{8}{7}$-approximation ratio.

Let t_{max} be the last task scheduled on P_1 (the machine with maximum load). We say that a task is *large* if its execution time is greater than or equal to the execution time of t_{max}. A task is *small* if its execution time is smaller than the one of t_{max}.

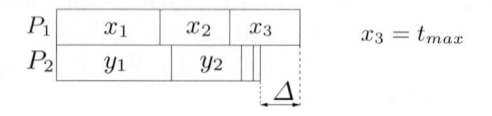

$$\begin{array}{|c|c|c|c|} \hline P_1 & x_1 & x_2 & x_3 \\ \hline P_2 & y_1 & y_2 & \\ \hline \end{array} \quad x_3 = t_{max}$$

Fig. 1. A LPT schedule

Claim: There are 3 or 4 tasks on P_1 and exactly 2 large tasks on P_2.
Let us show that we have exactly two large tasks on

Let us call Δ the difference between the completion time of the last task of P_1 and the last task of P_2 (see Figure 1). The makespan of the schedule ξ is $\frac{\sum_{i=1}^n l(t_i) + \Delta}{2}$. Since ξ does not achieve a $\frac{8}{7}$-approximation ratio, we have $\frac{\sum_{i=1}^n l(t_i) + \Delta}{2} > \frac{8}{7}OPT$. Since the makespan OPT of an optimal solution is at least $\frac{\sum_{i=1}^n l(t_i)}{2}$, we have $\frac{\Delta}{2} > (\frac{8}{7} - 1)OPT$, and so $\Delta > \frac{2}{7}OPT$. Let $\epsilon > 0$ be such that $\Delta = (\frac{2}{7} + \epsilon)OPT$.

We know that $l(t_{max}) \geq \Delta$ (because in a LPT schedule, at each step, we add a task on the machine with the smallest load), so the minimum execution time of a large task is Δ. We also know that the completion time of the last task on P_2 is smaller than or equal to $OPT - (\frac{1}{7} + \epsilon_1)OPT$, with $\epsilon_1 > 0$, because otherwise the schedule ξ would achieve a $\frac{8}{7}$-approximation ratio. So the maximum number of large tasks on P_2 is equal to $\lfloor \frac{OPT - (\frac{1}{7} + \epsilon_1)OPT}{\Delta} \rfloor = \lfloor \frac{1 - (\frac{1}{7} + \epsilon_1)}{\frac{2}{7} + \epsilon} \rfloor = 2$.

Moreover we can deduce that there are at least two large tasks on P_2 because otherwise the schedule ξ would be an optimal schedule.

Let us show now that there are either 3 or 4 tasks on P_1. There are at least 3 tasks on P_1 because otherwise the schedule ξ would be optimal. Indeed, if there is only one task on P_1 it is trivial that ξ is optimal. If there are two tasks on P_1 and $l(x_1) \leq l(y_1)$ then we would not decrease the makespan by putting x_2 and y_1 on the same machine (because P_1 is the machine which has the largest load). If there are two tasks on P_1 and $l(x_1) > l(y_1)$, then $l(y_2) \geq l(x_2)$ (schedule LPT): if we add x_2 to the large tasks of P_2 (y_1 and y_2) we do not decrease the makespan (x_2 starts before the completion time of y_2), and if we add to x_1 one of the large tasks of P_2 (i.e. we exchange x_2 with y_1 or y_2), we do not decrease the makespan because the execution time of each of these tasks (y_1 or y_2) is greater than or equal to the execution time of x_2.

Let us now show that there are at most 4 tasks on P_1. We know that $l(t_{max}) \geq \Delta > \frac{2}{7}OPT$, and each task on P_1 is larger than or equal to t_{max}. Since the maximum makespan of a LPT schedule on two machines is at most $\frac{7}{6}OPT$ [6] then the maximum number of tasks on P_1 is $\lfloor \frac{(\frac{7}{6})OPT}{\Delta} \rfloor = \lfloor \frac{(\frac{7}{6})OPT}{(\frac{2}{7} + \epsilon)OPT} \rfloor = \lfloor \frac{49}{12} \rfloor = 4$.

We have shown that if a LPT schedule does not achieve a $\frac{8}{7}$-approximation ratio, then it has 3 or 4 tasks on P_1, and at least 2 tasks on P_2. Moreover, $\sum_{t_i \in P_1} l(t_i) > (\frac{8}{7})OPT \geq \frac{8}{7}(\frac{\sum_{i=1}^n l(t_i)}{2}) = \frac{4}{7} \sum_{i=1}^n l(t_i)$. Thus all the conditions to enter in the first "if" of the algorithm LPT_{swap} are fulfilled. Due to space limitation, the sequel of the proof which proceeds by cases analysis, is omitted here. □

Theorem 2. *The schedule returned by LPT_{swap}, on two machines whose policies are LPT, is a 3-approximate-Nash equilibrium.*

Proof: Let ξ be a LPT schedule of the n tasks and ξ_s be the schedule returned by LPT_{swap}. Let m_ξ denote the makespan of ξ, and m_{ξ_s} the makespan of ξ_s. If $\xi_s = \xi$, then ξ_s is a LPT schedule and it is a Nash equilibrium. Else $\xi_s \neq \xi$, and LPT_{swap} has done a swap: ξ_s is then equal to ξ_1, ξ_2, ξ_3 or ξ_4. Let us show that each task of ξ_s does not have incentive to go on the other machine.

First of all, x_3 and all the small tasks (the tasks whose lengths are smaller than the one of t_{max}) do not have incentive to go on the other machine because if they change they will always be started after at least two large tasks, that is a length greater than $\frac{m_{\xi_s}}{3}$, and without changing they are always completed before or at m_{ξ_s}. Indeed, $l(x_3) \leq \frac{m_{\xi_s}}{3}$ and if a small task (or x_3) changes, it will be scheduled at the earliest at $m_{\xi_s} - l(x_3) \geq m_{\xi_s} - \frac{m_{\xi_s}}{3} > \frac{m_{\xi_s}}{3}$. Likewise, it is trivial that the tasks which are scheduled at the beginning of the schedule (e.g. x_1 and y_1 in ξ_2) do not have incentive to swap.

We then prove by case analysis that every large task of ξ_1, ξ_2, ξ_3 or ξ_4 does not have incentive to change machine. Due to space limitations, the end of the proof is omitted here. □

Corollary 1. *The price of α-approximate stability is at most $\frac{8}{7}$, for all $\alpha \geq 3$.*

Theorem 3. *Let ε be any small constant such that $\varepsilon > 0$. The price of α-approximate stability is at least $\frac{8}{7}$, for all $\alpha \leq 2.1 - \varepsilon$.*

Proof: Let ε be a small number such that $\varepsilon > 0$. Consider the following tasks: a task of length $3.3 - \varepsilon$, a task of length $3 + \varepsilon$, and three tasks of length 2.1. The optimal schedule (for the makespan) of these tasks is achieved if and only if the tasks of length 2.1 are on the same machine, and the other two tasks are on the other machine (see Figure 2 *Left*). The makespan of this schedule is $OPT = 6.3$. In this schedule, the completion time of the task of length $3 + \varepsilon$ is 6.3 but this task could be on the first position if it goes on the other machine (because the policies of the machines are LPT), and its completion time would then be $3 + \varepsilon$. So this schedule is $\frac{6.3}{3+\varepsilon} > (2.1 - \varepsilon)$-Nash approximate.

P_1	$3.3 - \varepsilon$	$3 + \varepsilon$	
P_2	2.1	2.1	2.1

Makespan = 6.3

P_1	$3.3 - \varepsilon$	2.1	
P_2	$3 + \varepsilon$	2.1	2.1

Makespan = $7.2 + \varepsilon$

Fig. 2. *Left:* Optimal schedule for the makespan. This is a $\frac{6.3}{3+\varepsilon}$-approximate Nash equilibrium. *Right:* Approximate schedule for the makespan.

Notice that all the other schedules have a makespan of at least $7.2 + \varepsilon > \frac{8}{7}OPT$. So if we want a schedule which is $\frac{8}{7}$-approximate, this schedule will not be an α-approximate Nash equilibrium, with $\alpha < 2.1 - \varepsilon$. Thus the price of α-approximate stability is at least $\frac{8}{7}$, for all $\alpha \leq 2.1 - \varepsilon$. □

An interesting question concerns the relation between the approximation ratio and α, i.e. what happens if we consider other values of the approximation ratio we wish to obtain, or other values of α? In particular, does there exist algorithms with smaller approximation ratios and which return α-Nash approximate equilibria, with α bounded (and as small as possible)? The following section gives an answer to this question.

3 A Variant of Graham's Algorithm

We are now interested in $(1 + \varepsilon)$-approximate schedules, for any $\varepsilon > 0$. We show that the price of α-approximate stability is smaller than $(1 + \varepsilon)$ if α is at least equal to k, where k is a constant smaller than $\frac{1}{\varepsilon}$.

We consider the algorithm of Graham [6], slightly modified:
(i) Let k be some specified and fixed integer.
(ii) Obtain an optimal schedule for the k longest tasks, such that:

- Once tasks are assigned to each machine, they are scheduled on their machines in order of decreasing lengths (i.e. for a given machine, tasks are scheduled from the largest one to the smallest one).
- If two tasks have the same length, the one which has the smallest identification number is scheduled the first.

(iii) Schedule the remaining $n - k$ tasks using the LPT rule.

This algorithm is a polynomial time approximation scheme (PTAS), and its approximation ratio is $1 + \varepsilon$, where ε is equal to $\frac{1}{2 + 2\lfloor \frac{k}{2} \rfloor}$, if the k largest tasks of the schedule are optimally scheduled [6]. Let us now show that this algorithm, denoted by OPT-LPT(k), returns α-approximate-Nash equilibria, with $\alpha < k-2$.

Theorem 4. *The schedules returned by algorithm OPT-LPT(k) are α-approximate-Nash equilibria, with $\alpha < k - 2$.*

Proof: Let us show that each task of an OPT-LPT(k) schedule either does not have incentive to change machine, or does not decrease its completion time by a factor larger than or equal to $k - 2$, by going on the other machine. The $n - k$ smallest tasks of the schedule are scheduled using the LPT rule, so they do not have incentive to change machine. Thus we consider the case of the k largest tasks. Let OPT be the optimal solution of these tasks, such as computed by OPT-LPT(k). We now consider three cases.

• In the first case, there are, in OPT, only one task on a machine (w.l.o.g. on P_1), and $k - 1$ tasks on the other machine. Since this schedule is an optimal solution, the task on P_1 is necessarily the largest task on the schedule, and this schedule is a LPT schedule. So no task has incentive to change machine in this case.

• Let us now consider the case where there are exactly two tasks on a machine (w.l.o.g. on P_1) in OPT. The others $k - 2$ tasks are then on P_2.

We first show that no task scheduled on P_2 has incentive to go on P_1. By construction, we know that $l(x_1) + l(x_2)$ is larger than or equal to the sum of the lengths of the $k - 3$ first tasks of P_2, $\sum_{j=1}^{k-3} y_j$. Let i be the largest number such that $l(x_1) \geq \sum_{j=1}^{i} y_j$: the $i + 1$ first tasks of P_2 (i.e. the tasks who start at the latest at the end of x_1) do not have incentive to go on P_1, otherwise they would be scheduled after P_1 and would not decrease their completion times. Moreover, we know that $l(x_2) \geq \sum_{j=i+2}^{k-3} y_j$: thus the tasks from y_{i+2} to y_{k-3} do not have incentive to change machine. Likewise, y_{k-2} does not have incentive to change: if it is smaller than x_2, then it would be scheduled on P_1 after x_2, and would not decrease its completion time, since OPT is an optimal solution. If y_{k-2} is larger than x_2, then y_{k-2} starts to be executed before (or at the same time as) x_2. Thus, if it goes on P_1, y_{k-2} will be scheduled after x_1, and then will not decrease its completion time.

The only task which may have incentive to change machine is x_2. If x_2 is smaller than all the other tasks, then it does not have incentive to change. Otherwise, since OPT is an optimal solution, we know that at least a task of P_2 starts at the same time or after x_2. In the best case, x_2 can go to the first position on P_2: by doing this, it starts on P_2 before at most $k-3$ tasks which started before it when it was on P_1. These $k-3$ tasks are smaller than x_2: the sum of their completion time, S, is thus smaller than $(k-3)\,l(x_2)$. The completion time of x_2 decreases with this change, from $S + l(x_2) < (k-2)\,l(x_2)$ to $l(x_2)$. Thus x_2 is, in OPT, α-Nash-approximate, with $\alpha < k-2$.

- Let us now consider the case where there are exactly $a < k-2$ tasks on P_1, and $b < k-2$ tasks on P_2. Let t be a task on P_1 (respectively P_2) which has incentive to change machine. When it changes machine, t overtakes p tasks of P_2 (respectively P_1), i.e. it starts to be executed before p tasks which started to be executed before t before the change. We know that p is smaller than $k-2$ because there are less than $k-2$ tasks on each machine. Moreover, these tasks have a length smaller than the one of t, otherwise t would not overtake them. Thus, in the best case, t overtakes $k-3$ tasks of length almost equal to $l(t)$, and the completion time of t decreases from a value smaller than $(k-2)\,l(t)$ to $l(t)$. Thus t is α-Nash-approximate, with $\alpha < k-2$. □

Theorem 5. *Let ε be any small number such that $0 < \varepsilon < 1$. OPT-LPT(k) can return α-approximate Nash equilibria, with $\alpha \geq k - 2 - \varepsilon$ and $k \geq 5$.*

Proof: Let $\varepsilon' = \frac{\varepsilon}{k-2-\varepsilon}$, and let us consider the following instance: a task of length $k - 3 - \varepsilon'$, a task of length $1 + \varepsilon'$ denoted by t, and $k-2$ tasks of length 1. In the only optimal solution for this instance, it can be shown that t is $(k-2-\varepsilon)$-approximate. The schedule returned by OPT-LPT(k) on this instance is an α-approximate Nash equilibrium, with $\alpha \geq k - 2 - \varepsilon$. □

We can deduce from Theorem 4, and from the fact that the approximation ratio of OPT-LPT(k) is $\frac{1}{2+2\lfloor \frac{k}{2} \rfloor}$, the following result:

Corollary 2. *Let k be any integer larger than or equal to 5. The price of α-approximate stability is at most $1 + \varepsilon$, where $\varepsilon = \frac{1}{4+2\lfloor \frac{k}{2} \rfloor} < \frac{1}{k}$, for all $\alpha \geq k$.*

Note that if we want to get an algorithm $\frac{8}{7}$-approximate which returns solutions as stable as possible, then LPT_{swap} is better than OPT-LPT(k): indeed, the solution of LPT_{swap} can be found faster (in OPT-LPT(k) we have 64 schedules to compare, whereas in LPT_{swap} there are at most 4 schedules to compare), and the OPT-LPT(k) returns 4-approximate-Nash equilibria (versus 3-approximate Nash equilibria for LPT_{swap}). On the other hand, OPT-LPT(k) is useful if we wish algorithms with smaller approximation ratios, since OPT-LPT(k) is a PTAS.

4 Tradeoffs

We first show that if we want a price of α-approximate stability smaller than or equal to $(1+\varepsilon)$, then α must be larger than a certain constant in $\Theta(\varepsilon^{-1/2})$.

Theorem 6. *Let $\varepsilon > 0$ and $k > 0$ such that $\varepsilon = \frac{1}{k(k+1)}$. Then to get a price of α-approximate stability smaller than $1 + \varepsilon$, we have to set $\alpha \geq k$.*

Proof: Consider the following instance: a task of length $k-1$, a task of length 1, and $k+1$ tasks of length $\frac{k}{k+1}$. The optimal schedule (for the makespan) of these tasks is achieved if and only if the tasks of length $\frac{k}{k+1}$ are on the same machine, and the other two tasks are on the other machine (see Figure 3 *Left*). The makespan of this schedule is $OPT = k$. This schedule is a k-approximate Nash equilibrium. Indeed, the completion time of the task of length 1 is k but this task could be on the first position if it goes on the other machine (because the policies of the machines are LPT), and its completion time would then be 1 (which is n times smaller than its current completion time).

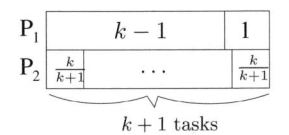

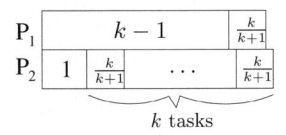

Fig. 3. *Left:* Optimal schedule for the makespan. This is a k-approximate Nash equilibrium (the policies of the machines are LPT). *Right:* Approximate schedule for the makespan.

Figure 3 *Right* shows the second smallest makespan schedule with these tasks: all the schedules which are not the optimal one have a makespan greater than or equal to the makespan of this schedule. This makespan is $k - (\frac{k}{k+1}) + 1$. So the ratio between this makespan and OPT is $\frac{k-(\frac{k}{k+1})+1}{k} = 1 + \frac{1}{k} - \frac{1}{k+1} = 1 + \frac{1}{k(k+1)}$. Thus if we want a schedule which is $(1+\varepsilon)$-approximate, with $\varepsilon < \frac{1}{k(k+1)}$, this schedule will not be an α-approximate Nash equilibrium, with $\alpha < k$. □

We can also prove:

Theorem 7. *The price of α-approximate stability is at least 1.1 (respectively $\frac{9}{8}$), for all $\alpha \leq \frac{10}{3}$ (respectively $\alpha \leq \frac{8}{3}$).*

Figure 4 illustrates the tradeoffs between the approximation ratio of an algorithm, and the stability of the schedules it can returns. The dark grey zone illustrates the results showed in Theorem 6 (for $2 \leq k \leq 10$), Theorem 3, and Theorem 7. Each point (x, α) belonging to the dark grey zone represents a negative result, i.e. it means that there is no x-approximate algorithm which returns α-Nash equilibria. The light grey zone illustrates the results showed by

Theorem 4 and Theorem 1: each point (x, α) belonging to the light grey zone represents a positive result, i.e. it means that there is an x-approximate algorithm which returns α-Nash equilibria (this algorithm is either OPT-LPT(k) or LPT_{swap}). If a point (x, α) belongs to the white zone, then it means that we do not have any algorithm corresponding to this point, nor any impossibility result.

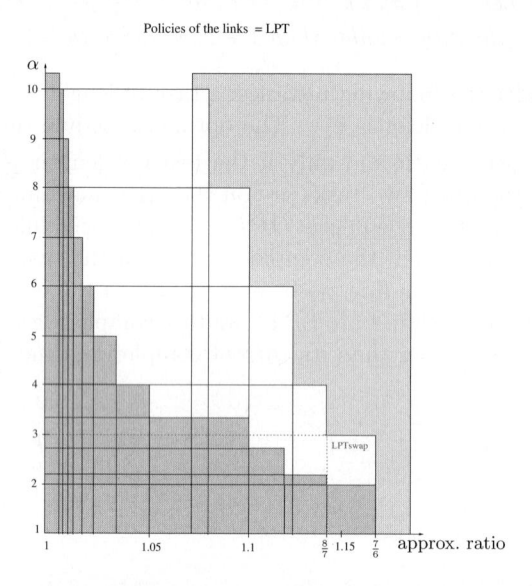

Fig. 4. *Light grey* (respectively *Dark grey*): Relation between α and approximation ratios that it is possible (respectively it is not possible) to obtain if we wish an algorithm which returns α-approximate Nash equilibria

References

1. G. Christodoulou, E. Koutsoupias, A. Nanavati *Coordination mechanisms*. In Proc. of ICALP 2004, LNCS 3142, pages 345-357.
2. E. Koutsoupias, C. H. Papadimitriou *Worst-case equilibria*. In Proc. of STACS 1999, LNCS 1563, pages 404-413.
3. A. S. Schulz, N. Stier Moses *On the performance of user equilibria in traffic networks*. In Proc. of SODA 2003, pages 86-87.
4. E. Anshelevich, A. Dasgupta, J. Kleinberg, É. Tardos, T. Wexler, T. Roughgarden *The price of stability for network design with fair cost allocation*. In Proc. of FOCS 2004, pages 295-304.
5. R. Lipton, E. Markakis, A. Mehta *Playing Large Gamed Using Simple Strategies*. ACM Conference on Electronic Commerce, pp. 36-41, 2003.
6. R. Graham *Bounds on Multiprocessing Timing Anomalies*. SIAM Jr. on Appl. Math., 17(2), pp.416-429, 1969.
7. D. Fotakis, S. Kontogiannis, E. Koutsoupias, M. Mavronicolas, P. Spirakis, *The structure and complexity of nash equilibria for a selfish routing game*. In Proc. of ICALP 2002, LNCS 2380, pages 123-134.

Master-Slave Tasking on Asymmetric Networks

Cyril Banino-Rokkones[1], Olivier Beaumont[2], and Lasse Natvig[1]

[1] Norwegian University of Science and Technology, NO-7491 Trondheim, Norway
{Cyril.Banino, Lasse.Natvig}@idi.ntnu.no
[2] LaBRI, UMR CNRS 5800, Domaine Universitaire, 33405 Talence Cedex, France
Olivier.Beaumont@labri.fr

Abstract. This paper presents new techniques for master-slave tasking on tree-shaped networks with fully heterogeneous communication and processing resources. A large number of independent, equal-sized tasks are distributed from the master node to the slave nodes for processing and return of result files. The network links present bandwidth asymmetry, i.e. the send and receive bandwidths of a link may be different. The nodes can overlap computation with at most one send and one receive operation. A centralized algorithm that maximizes the platform throughput under static conditions is presented. Thereafter, we propose several distributed heuristics making scheduling decisions based on information estimated locally. Extensive simulations demonstrate that distributed heuristics are better suited to cope with dynamic environments, but also compete well with centralized heuristics in static environments.

1 Introduction

In this paper, we consider the allocation of a large number of independent equal-sized tasks onto a tree platform. We concentrate on tree-shaped platforms since they represent a natural framework for master slave tasking. More importantly, administrative organizations often rely on tree-shaped networks to interconnect computing resources [1]. Initially, the root of the tree (master node) holds a large bunch of tasks. Those tasks will be either processed by the master node or transmitted to its child nodes (also called slave nodes). Then, in turn, the child nodes face the same allocation problem (either processing the tasks locally or forwarding them to their child nodes). We consider the case where slave processors need to send back a file of results after processing each task. Even if this is the most natural situation, it is worth noting that most of the papers on independent tasks scheduling or Divisible Load Theory (DLT) do not consider those return communications. Targeted platforms are fully heterogeneous, i.e. both the processing resources and the communication resources have different capacities in terms of processing power and bandwidth. Moreover, the network links present bandwidth asymmetry in the sense that the bandwidth for sending tasks down the tree may be different from the bandwidth for returning results up the tree.

We concentrate on the influence of dynamic resource characteristics on the allocation scheme. In shared and unstable environments such as grids and peer to peer systems, the performance of the resources may well change during the execution of the whole process. In this context, it is not realistic to assume that one of the nodes knows at any time step the exact performance of all resources and is able to make optimal scheduling decisions [1]. Therefore, the main question consists in determining whether the allocation scheme can make use of some static knowledge about the platform (for instance, the optimal solution computed from an initial snapshot of the platform), or whether we need to rely on fully dynamic scheduling schemes. In order to answer this question, we first derive optimal scheduling algorithms (with respect to throughput maximization). Then we present several heuristics. Some of them make their scheduling decisions using the optimal scheduling policy, computed using a snapshot of resource performance characteristics. Those heuristics may lead to optimal scheduling decisions in static environments. On the other hand, we propose a set of fully dynamic allocation heuristics that make their scheduling decisions only according to information measurable locally. Those heuristics may give poor results in static environments, but their performances are expected to be more robust in dynamic environments. We compare all those heuristics through extensive simulations using the SimGrid toolkit [2]. We rely on simulations rather than direct experiments in order to make a fair comparison between proposed heuristics. Indeed, simulation enables running of the different tests on computing platforms having exactly the same dynamic behavior. Moreover, SimGrid enables to define the trace of performance data over time for each processing or communication resource. Therefore, it is possible to compute (off-line) the optimal solution at any given time step and it is therefore possible to compare the performances of the different heuristics between them and against the optimal ideal solution.

The rest of the paper is organized as follows. Section 2 is devoted to a survey of related work, both DLT studies, independent tasks scheduling and on dynamic scheduling. Then, we present our platform model in Section 3 and how to find the optimal solution, in presence of return messages, in Section 4. Section 5 states the main Theorem of this paper, which provides a mean to optimize the nodes bandwidth utilization. Section 6 presents a task-flow control mechanism that regulates the amount of tasks and results buffered by the nodes throughout the execution. The set of centralized and distributed heuristics are described in Section 7. The methodology and results of the simulations are discussed in Section 8. Finally, we give some remarks and conclusions in Section 9. Due to space limitation, many of the technical details have been omitted, but can be found in the extended version of this paper [3].

2 Related Work

The problem of master-slave tasking on heterogeneous tree platforms has already been widely studied, both in the context of Divisible Load Theory (DLT) and independent tasks scheduling. A divisible load is a perfect parallel task that can be arbitrarily split and allocated to slave processors, without processing overhead. The overall load is first split at the master node in order to minimize the total execution time. Tasks are distributed in one round to the slaves, so that the master node makes the decisions about the set of slaves to be used, the amount of data to be sent to each slave, and the communication ordering [4,5,6]. When return messages are taken into account, two permutations must then be determined (one for tasks distribution and one for results collection) [7,8]. Although the complexity of this problem is still open, Rosenberg et al. [9] proved that in the case of a homogeneous single-level tree, the optimal schedule for both outgoing and incoming messages can be determined, and the optimal LIFO and FIFO orderings are given in [10] for heterogeneous single-level trees.

On the other hand, when considering independent tasks scheduling, the master node faces the allocation problem for each task and the communications with its child nodes may well be split into several rounds [11,12,13]. Recently research studies have focused on steady-state scheduling, i.e. throughput maximization [11, 14, 15]. The steady-state scheduling approach has been pioneered by Bertsimas and Gamarnik [16] who considered packet routing and proposed to concentrate first on resource occupation rather than scheduling. The optimal solution for resource occupation, given link capacities, is obtained via a linear program. Then, an algorithm based on super-steps is proposed for building the actual schedule of packets. This idea has been adapted in [14] to the distribution of independent tasks on static platforms. Results collection was not considered in [14], but the linear program presented in Section 4 is a direct adaptation of the solution proposed in [14].

Dynamic scheduling of independent tasks has not been widely studied. Recently, Hong et al. [1, 15, 17] proposed a very nice algorithm, based on decentralized versions of flow algorithms. It is worth noting that this algorithm assumes a strongly different communication model than the one presented in this paper, and consequently cannot be easily adapted to our model. Here again, the results collection has not been considered.

3 Platform Model

The model considered in this paper is based on the model proposed in [14] that we augment by introducing communication weights for returning computation results back to the master. Processing nodes are assumed to be connected via a node-weighted edge-weighted tree $T = (V, E, w, c, c')$ as depicted in Figure 1.

Each node $P_i \in V$ represents a computing resource of weight w_i, meaning that node P_i requires w_i units of time to process one task. Each edge corresponds to a communicating resource and is weighted by two values: c_i which represents the time needed by a parent node to send one task to its child P_i, and c'_i which represents the time needed by the child P_i to send one result back to its parent. All the w_i's are assumed to be positive rational numbers since they represent node processing times. We disallow $w_i = 0$ since it would permit node P_i to perform an infinite number of tasks. Similarly, we assume that all c_i's and c'_i's are positive rational numbers since they correspond to the communication times between two processors. A node can perform three kinds of activity simultaneously: (i) it can process a task, (ii) it can receive a task file from its parent or a result file from one of its children, and (iii) it can send a result file to its parent or a task file to one

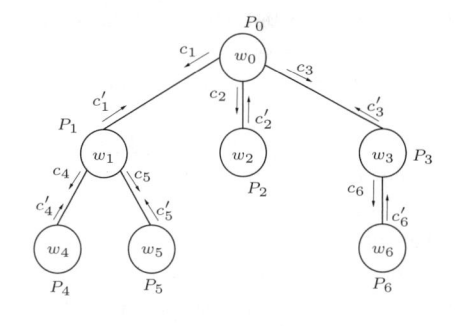

Fig. 1.

of its children. This model is known under the name *full overlap, bidirectional-single-port* model [14, 11]. At any given time-step, a node may overlap computation with only two connections, one for incoming communications and one for outgoing communications. Computation and communication are assumed to be atomic operations, i.e. once initiated they cannot be preempted. Finally the communication model works in a store-and-forward fashion.

4 Maximizing the Throughput

Given the resources of a weighted tree T operating under the *full overlap, bidirectional-single-port* model, we aim at maximizing the number of tasks processed per time unit. Let $\mathcal{C}_i$ denote the set of P_i's children. During one time unit, let α_i be the fractional number of tasks processed by P_i, and β_i be the fractional number of tasks received by P_i from its parent. Equivalently, α_i and β_i correspond respectively to the fractional number of results produced by P_i, and to the fractional number of results sent by P_i to its parent. The optimal throughput is obtained by solving the following linear programming problem (LPP), whose objective function is to maximize the number of tasks processed per time unit.

Maximize $n_{task}(T) = \sum_i \alpha_i$
subject to
$$\begin{cases} \forall i, & 0 \leq \alpha_i \leq \frac{1}{w_i} \\ \forall i \neq m, & 0 \leq \beta_i \\ \forall i \neq m, & \beta_i = \alpha_i + \sum_{j \in \mathcal{C}_i} \beta_j \\ \forall i, & \sum_{j \in \mathcal{C}_i} c_j \beta_j + c'_i \beta_i \leq 1 \\ \forall i, & \sum_{j \in \mathcal{C}_i} c'_j \beta_j + c_i \beta_i \leq 1 \end{cases}$$

The first set of constraints states that computation resources are limited. The second set of constraints confines the variables β_i within non-negative values. Note that the master P_m does not have a parent, so that we let $\beta_m = 0$. The third set of constraints deals with *conservation laws*. For each node P_i (except the master), the number of tasks received by P_i, should be equal to the number of tasks that P_i processes locally, plus the number of tasks forwarded to the children of P_i. Equivalently, the number of results sent by P_i to its parent, should be equal to the number of results produced locally by P_i, plus the number of results received from its children. The last constraints account for the single-port model. The send and receive operations performed by the nodes are assumed to be sequential.

Since we are looking for a solution of the LPP into rational numbers, optimal rational values for all variables can be obtained in polynomial time. However, the solution of the above LPP is in general not unique and some solutions might be more interesting than others in our context. In particular, *compact* solutions, i.e. that utilize nodes close to the root in priority, are more preferable than *stretched* solutions (that utilize nodes far away from the root). Indeed, start-up time (required to enter the steady-state) and wind-down time (required to gather the last results to the root) will be longer for stretched solutions than for compact ones. In order to obtain compact solutions, we first need to solve the initial LPP to derive the optimal throughput $n_{task}(T)$ of the tree. The objective function of the second LPP becomes the minimization of all the communications, under the aforementioned constraints plus an additional one that states the conservation of the optimal throughput obtained by the former LPP. Minimizing the amount of communications while maintaining the optimal number of tasks processed implicitly enforces compact solutions. We hence add the following constraint: $\sum \alpha_i = n_{task}(T)$. And the objective function of the second LPP becomes: **Minimize** $\sum_i \beta_i$. Once a solution has been obtained, one needs to construct a schedule that (i) ensures that the optimal throughput is achieved and (ii) exhibits a correct orchestration of communication events, i.e. where simultaneous communications involve disjoint pairs of senders and receivers. We can obtain a time period Γ by taking the least common multiple (lcm) of all the denominators of the variables α_i. Then, the integer number of tasks γ_i that must be communicated to P_i during each time period Γ is obtained by $\gamma_i = \beta_i \Gamma$.

Proposition 1. *Sending and receiving files by bunches of γ_i in a round robin fashion generates an optimal steady-state schedule where single-port constraints are satisfied.*

Proof. The proof is done by induction over h, the height of the tree T [3]. □

Initially, nodes do not dispose of tasks nor results buffered locally to comply with Proposition 1. Therefore an initialization phase must take place before entering steady-state. During start-up, nodes will act as if they were in steady-state, at the difference that fake results will be sent to the parents if not enough results are available. Thus, tasks will be propagated down the tree, while fake results will be propagated up the tree. The fake results received by parents nodes are simply discarded. Once the first bunch of results processed by all the deepest nodes used in the schedule have been transmitted to the root node, then steady-state has been reached.

5 Bandwidth Optimization

A simple scheduling principle is presented in [14] when returning results is neglected. This scheduling algorithm was termed *bandwidth-centric* because priorities do not depend on the children processing capabilities, but only on their communication capabilities. The bandwidth-centric principle is extended to our problem as follows. First, observe that for each task that a node P_i delegates to a child P_j, P_i must first receive the task from its parent, then forward it to P_j, receive the associated result back, and finally send the result to its parent. Consequently, P_i will spend $x_j = c_j + c'_i$ time units sending data, and $y_j = c'_j + c_i$ time units receiving data. Since the master P_m does not have a parent, we let $x_m = c_m$ and $y_m = c'_m$. The bandwidth utilization of a node P_i can be sketched within the Cartesian plane, where the X and Y axes represent the time spent in emission and reception respectively. Hence, allocating a task to child P_j corresponds to a displacement in the Cartesian plane along vector v_j of components (x_j, y_j).

Theorem 1. *In steady-state, the bandwidth utilization of a parent node is optimized when using at most 2 children (if processing capabilities are not taken into account).*

Proof. The proof is done by induction over n, the number of children that are utilized by a parent in addition to the two nodes mentioned in Theorem 1. Consider the case where

$n = 1$, i.e. when a parent delegates α_1, α_2 and α_3 tasks per time unit to three children P_1, P_2 and P_3 respectively (see Figure 2). Displacements OA_1, A_1A_2 and A_2A_3 stand for delegating α_1, α_2 and α_3 tasks to the children P_1, P_2 and P_3 respectively.

Consider the triangle A_1A_2P where the displacements A_1P and PA_2 amount to allocate j_1 and j_3 tasks to P_1 and P_3 respectively. Consider now both quantities $(j_1 + j_3)$ and α_2. If $(j_1 + j_3) \geq \alpha_2$, it means that it is more profitable to spend the bandwidth time assigned to P_2 by allocating more tasks to P_1 and P_3. As a consequence, P_2 should not be used. But if $(j_1 + j_3) < \alpha_2$, then consider the triangle ORA_1, where the displacements OR and A_1R amount to allocate k_2 and k_3 tasks to P_2 and P_3 respectively. Since both triangles A_1A_2P and ORA_1 are equal (since their internal angles are equal), if $(j_1+j_3) < \alpha_2$ then $(\alpha_1 + k_3) < k_2$. In that case, it becomes more profitable to assign k_2 tasks to P_2 instead of α_1 tasks to P_1 and j_3 tasks to P_3, and P_1 should not be used. Assume now that Theorem 1 is true for rank n, and let us prove that it holds also for rank $n + 1$. Consider a parent utilizing $n + 3$ children. Extract 3 of the $n + 3$ children and apply the aforementioned geometric transformation. One then utilizes only $n + 2$ children without degrading the initial throughput. □

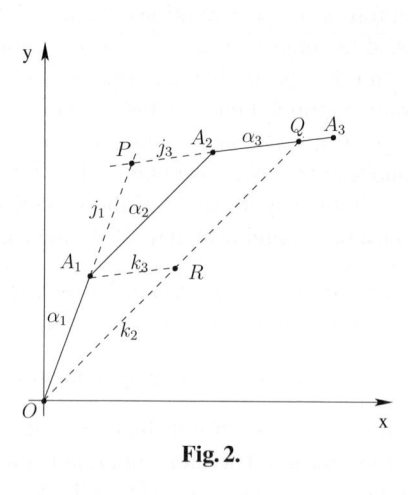

Fig. 2.

Theorem 1 assumes that nodes can provide as much computing power as necessary which contravenes the fact that computing resources are limited. Nonetheless, it allows identifying the way to optimize the bandwidth of any node P_i in using at most two children. Furthermore, we show in [3] that if such a pair of children exists, then the emission and reception bandwidth of P_i are equally utilized.

6 Task-Flow Control

In order to regulate the number of tasks and results that nodes are allowed to buffer locally throughout the execution, a threshold value θ_i is introduced for each node $P_i, i \neq m$. On the one hand, if the number of tasks buffered locally by P_i is beneath the threshold, then P_i will request more tasks in order to prevent starvation. On the other hand, if the number of results buffered locally by P_i is larger than the threshold, then P_i will not request additional tasks in order to hinder a monotonic accumulation of results. Initially, $\theta_i = 1, \forall i \neq m$. Since we search for compact solutions, parent nodes will try to process as many tasks as possible. If additional tasks arrive while a node is busy processing, then the task will be forwarded down the tree. During the execution, nodes are allowed to increase their local thresholds θ_i only when (i) they are starving and (ii) if they recently succeeded to accumulate θ_i tasks locally (to ensure that the current threshold is not sufficient) and (iii) if the number of results buffered locally is strictly lower than θ_i. This mechanism allows nodes to collect enough tasks locally to feed their sub-trees, while ensuring that results do not accumulate monotonically locally. On the other hand, nodes must decrease their local thresholds whenever the number of results buffered locally exceeds the threshold. This threshold growth mechanism provides a mean to adapt to the platform dynamics.

7 Scheduling Heuristics

Round Robin. (RR) This heuristic implements Proposition 1. Once all the α_i are known, the period Γ is estimated as follows. Let us set $x = \lfloor \log_{10}(\max_i \alpha_i) \rfloor$. If $x \leq 0$ then $\Gamma = 10^{|x|+1}$, $\Gamma = 10^x$ otherwise. The aim is to obtain a compromise between a short time period, and an approximation close to the optimal solution. Then we get the number of tasks computed by each node P_i by rounding $\Gamma \alpha_i$ to the nearest integer.

On the Fly. (OTF) This heuristic makes use of the centralized knowledge. Once all the β_i's are known, each node maintains a table $tasks\_given[j]$, which records the number of tasks delegated to child P_j so far. The child node that has the lowest $\frac{tasks\_given[j]}{\beta_j}$ ratio is served in priority.

FIFO. Tasks are delegated in a first-come first-served basis.

Bandwidth-Centric. (BC) Let $r_j = \min\{\frac{1}{x_j}, \frac{1}{y_j}\}$ denote the maximum amount of tasks that P_i can delegate to child P_j per time unit. The child which has the highest r_j is served in priority.

Geometric. (Geo) This heuristic makes use of Theorem 1, but starts by applying the bandwidth-centric heuristic, in order to determine which child obtains the highest r_j. Then, it inspects if a pair of children can improve that rate. If such a pair of children exists, one must decide which child should be served. In order to make the right decision, we use a variable Δ which works much like a pair of scales. At start, $\Delta = 0$. Each time a child node P_j is served, we put x_j in one scale, and y_j in the other, which amounts to $\Delta = \Delta + x_j - y_j$. When a pair of children nodes is elected, then the child which brings Δ closest to 0 is serve. The aim is to utilize equally the emission and reception bandwidths of the parent nodes. Such strategy will optimize the bandwidth utilization of the nodes, while naturally adapting to the platform dynamics.

8 Simulations Results

To evaluate our heuristics, we simulate the execution of an application on different random trees. Since a sub-tree can be reduced to a single super-node of equivalent processing power [14], it is not necessary to employ thousands of nodes to simulate large-scale systems [15]. We arbitrarily limited the number of nodes in a tree to 100. Each node was arbitrarily restricted to have at most 10 child nodes. A random tree is generated as follows. Each node is numbered with an ID number between 0 and 99. Then, each node $P_i, i \in [1, 99]$ is connected randomly to a node $P_j, j \in [0, i-1]$. The links have static performance values comprised between c_{min} and c_{max} and the nodes between w_{min} and w_{max}. All random distributions are uniform. The dynamic environments used in our simulations were generated as follows. Each resource R_i (node or link) has a cyclic behavior, i.e. its performance changes n_i times per cycle. The number of changes n_i

per cycle is randomly taken within the interval [5, 15]. Resource performance changes will occur every 25 treated tasks in average. We do not claim that these arbitrary decisions correspond to realistic network conditions. Our aim is to compare our heuristics on a set of different tree configurations. Inspired by Kreaseck et al. [11], we determine the throughput rate by using a growing window. The execution time is divided into 100 equal-sized time slots. Then, the window increases in size by step of one time slot, and the throughput rate delivered within the window time-frame is computed. The throughput rates delivered by the trees have been normalized to the maximum steady-state rates obtained with the LPP in static environments. However, throughput rates obtained in dynamic environments have been scaled up by a *dynamic factor* that accounts for the performance loss incurred by the platform dynamics. The dynamic factors have been obtained by successively solving LPPs of static platforms and comparing them to their homologous LPPs where some dynamism have been introduced (i.e. with the same platform topologies but with scaling down resource performances). More details about our methodology as well as a broader set of simulation can be found in [3].

In this paper, we report the simulation of an independent-task application of 2500 tasks on 50 trees where $c_{min} = 1$, $c_{max} = 10$, $w_{min} = 20$ and $w_{max} = 200$. Two scenarios for the data volume associated to the tasks and results were considered: (i) task data are 1000 times larger than result ones ($\frac{t}{r} = 1000$), and (ii) task and result data have the same size ($\frac{t}{r} = 1$). Figure 3 plots an average of the 50 throughput rates (associated to the 50 trees) over time. Figure 3 (a) and (b) correspond to static environments, while Figure 3 (c) and (d) correspond to fully dynamic environments, i.e. where resource performances can degrade down to 1% of the static value. The RR heuristic has been simulated with more than 2500 tasks in order to overcome the long start-up time required to enter steady-state. Still, RR does not outperform the other heuristics in static environments, certainly due to the truncating and rounding operations that occurred when computing Γ and the γ_i's. Not only the integer number of tasks intended to each node may be sub-optimal, but also the schedule of communications gets disturbed. The centralized heuristics (RR and OTF) are the highest performers in static environments, but the lowest ones in dynamic environments. Indeed, the information on which they rely throughout the execution becomes misleading in dynamic settings. As expected, the BC heuristic works very well when result data are small, while Geo only departs from BC when result data become significant.

Interestingly, when result data become significant, the performance of the best heuristics decrease, whereas the performance of FIFO increases. On the one hand, the decline of the best heuristics can be explained by the scheduling problem becoming more complicated. Returning results up the tree taking as long as sending tasks down the tree, parent nodes may sometimes have to stall a long time, waiting for a child to become available in reception. On the other hand, the performance increase of FIFO is a direct consequence of the task-flow control mechanism. When returning results takes a long time, local accumulations of results will arise, hindering the ineffective nodes to request for additional tasks. In contrast, when returning results is quick, no local results accumulations take place, increasing the margin to make wrong scheduling decisions.

Finally, it is worth noticing that BC and Geo compete well with the centralized heuristics even in static environments. See [3] for further details and interpretations.

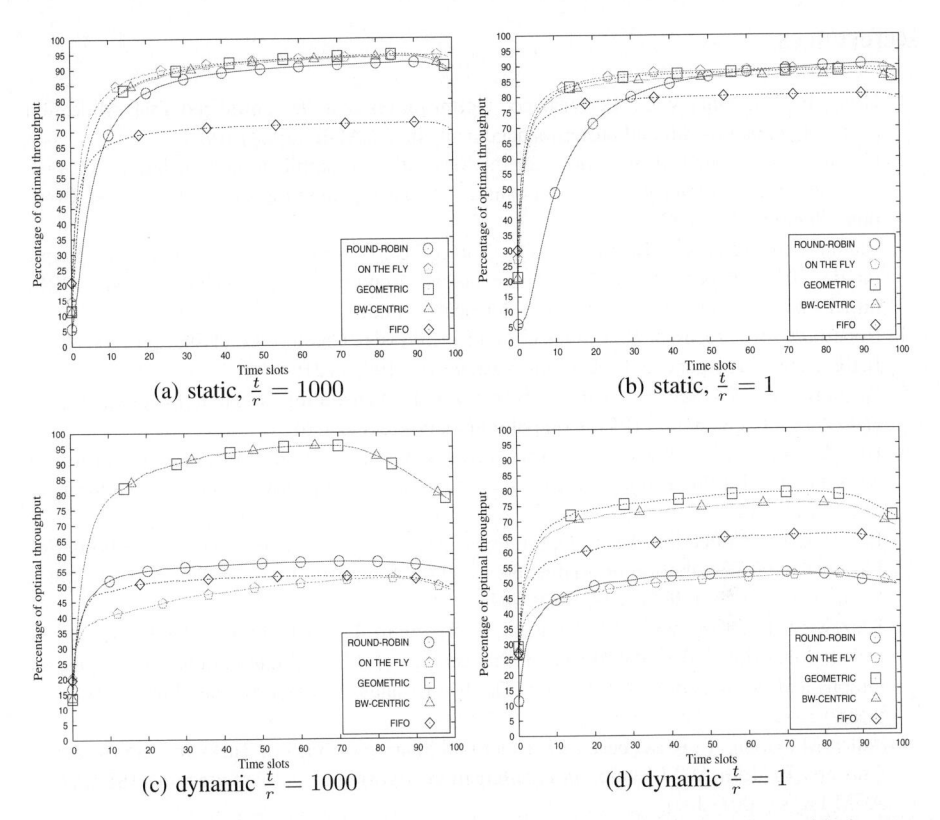

Fig. 3. Average of the 50 throughput rates (associated to the 50 trees) over time, with the computation to communication ratio $\frac{w_i}{c_i} = 20$. In the dynamic environments, resource performances can degrade arbitrarily without failing, i.e. down to 1% of the static performance value.

9 Conclusion and Future Work

The problem of distributing a large number of independent tasks onto heterogeneous tree-shaped platforms with bandwidth asymmetry was considered. In contrast with most previous studies, the cost of returning results to the master node was represented in the problem formulation. We provided theoretical results that were embedded into autonomous heuristics. Simulations results showed that the autonomous heuristics put together with the task-flow control mechanism not only behaved very well in dynamic environments, but also compete well with centralized heuristics in static environments.

The scope of this paper was restricted to tree-shaped networks. However, at the backbone level, various geographically organizations are connected via the Internet resulting in a graph topology. Adapting the theoretical results presented in this paper to graph-shape platforms is a natural continuation of this work, albeit graph topology introduces routing problems. Another direction is to consider master-slave tasking in the presence of multiple masters. This situation arises naturally when several applications share the same platform, or when multiple masters collaborate on a single application.

References

1. Hong, B., Prasanna, V.K.: Performance Optimization of a De-centralized Task Allocation protocol via bandwidth and buffer management. In: CLADE. (2004) 108
2. Casanova, H.: SimGrid: A Toolkit for the Simulation of Application Scheduling. In: Proceedings of the 1st International Symposium on Cluster Computing and the Grid, IEEE Computer Society (2001) 430
3. Banino-Rokkones, C., Beaumont, O., Natvig, L.: Master-Slave Tasking on Asymmetric Tree-Shaped Networks. Technical Report 02/06, NTNU (2006) URL: http://www.idi.ntnu.no/~banino/research/research.html.
4. Robertazzi, T.: Processor Equivalence for a Linear Daisy Chain of Load Sharing Processors. IEEE Trans. Aerospace and Electronic Systems **29** (1993) 1216–1221
5. Bharadwaj, V., Ghose, D., Mani, V., Robertazzi, T.: Scheduling Divisible Loads in Parallel and Distributed Systems. IEEE Computer Society Press (1996)
6. Drozdowski, M., Wolniewicz, P.: Experiments with scheduling divisible tasks in clusters of workstations. In: Proceedings of Euro-Par 2000: Parallel Processing. LNCS 1900, Springer (2000) 311–319
7. Barlas, G.D.: Collection-Aware Optimum Sequencing of Operations and Closed-Form Solutions for the Distribution of a Divisible Load on Arbitrary Processor Trees. IEEE Trans. Parallel Distrib. Syst. **9**(5) (1998) 429–441
8. Blazewicz, J., Drozdowski, M., Guinand, F., Trystram, D.: Scheduling a Divisible Task in a Two-dimensional Toroidal Mesh. In: Proceedings of the third international conference on Graphs and optimization, Amsterdam, The Netherlands, Elsevier Science Publishers B. V. (1999) 35–50
9. Adler, M., Gong, Y., Rosenberg, A.L.: Optimal Sharing of Bags of Tasks in Heterogeneous Clusters. In: 15th ACM Symp. on Parallelism in Algorithms and Architectures (SPAA'03), ACM Press (2003) 1–10
10. Beaumont, O., Marchal, L., Robert, Y.: Scheduling Divisible Loads with Return Messages on Heterogeneous Master-Worker Platforms. In: International Conference on High Performance Computing HiPC'2005. LNCS, Springer Verlag (2005) 123–132
11. Kreaseck, B., Carter, L., Casanova, H., Ferrante, J.: Autonomous Protocols for Bandwidth-Centric Scheduling of Independent-Task Applications. In: IPDPS '03: Proceedings of the 17th International Symposium on Parallel and Distributed Processing, Washington, DC, USA, IEEE Computer Society (2003) 26.1
12. Dutot, P.F.: Complexity of Master-slave Tasking on Heterogeneous Trees. European Journal on Operationnal Research **164**(3) (2005) 690–695
13. Rosenberg, A.L.: Sharing Partitionable Workloads in Heterogeneous NOWs: Greedier is not Better. In: Cluster Computing 2001, IEEE Computer Society Press (2001) 124–131
14. Banino, C., Beaumont, O., Carter, L., Ferrante, J., Legrand, A., Robert, Y.: Scheduling Strategies for Master-Slave Tasking on Heterogeneous Processor Platforms. IEEE Transactions on Parallel and Distributed Systems **15**(4) (2004) 319–330
15. Hong, B., Prasanna, V.K.: Distributed Adaptive Task Allocation in Heterogeneous Computing Environments to Maximize Throughput. In: International Parallel and Distributed Processing Symposium IPDPS'2004, IEEE Computer Society Press (2004) 52b
16. Bertsimas, D., Gamarnik, D.: Asymptotically optimal algorithm for job shop scheduling and packet routing. Journal of Algorithms **33**(2) (1999) 296–318
17. Hong, B., Prasanna, V.K.: Bandwidth-Aware Resource Allocation for Heterogeneous Computing Systems to Maximize Throughput. In: ICPP. (2003) 539–546

Using On-the-Fly Simulation for Estimating the Turnaround Time on Non-dedicated Clusters*

Mauricio Hanzich[2], Josep L. Lérida[1], Matías Torchinsky[2], Francesc Giné[1], Porfidio Hernández[2], and Emilio Luque[2]

[1] Dept. Computer Science, University of Lleida, Spain
{sisco, jlerida}@diei.udl.es
[2] Dept. Computer Architecture and Operating Systems, University Autònoma of Barcelona, Spain
{porfidio.hernandez, emilio.luque}@uab.es,
{mauricio, matias}@aomail.uab.es

Abstract. The computation capacity of the workstations of an open laboratory in almost every university is enough to execute not only the local workload but some distributed computation. Unfortunately, the local workload introduces a big uncertainty into the predictability of the system, which hinders the applicability of the job scheduling strategies.

In this work, we introduce into our job scheduling system, termed CISNE, a simulator, which allows its scheduling decisions to be enhanced by estimating the future cluster state. This process of estimation is backed by analytic procedures which are also described in this study. Likewise, the simulation let us assure some limit to the turnaround time for the parallel user. This paper analyses the performance of the simulation process in relation to different scheduling policies. These results reveal that those policies that respect an FCFS order for the waiting jobs are more predictable than those that alter the job ordering, like Backfilling.

1 Introduction

Several studies [1] have revealed that a high percentage of computing resources (CPU and memory) in a Network Of Workstations (NOW/Cluster) are idle. The possibility of using this computing power to execute distributed applications with a performance equivalent to a Massively Parallel Processor (MPP) and without perturbing the performance of the local users applications on each workstation has led to a proposal for new resource management environments [2,3].

With the aim of taking advantage of these idle computing resources (CPU and memory) available across the cluster, we have developed a new scheduling environment, named CISNE [3], which combines space sharing and time sharing scheduling techniques. The space sharing scheduling component of CISNE is a job scheduler, named LoRaS (Long Range Scheduler). When a parallel job is submitted to the LoRas, the job waits in a queue until it is scheduled and

---

* This work was supported by the MEyC-Spain under contract TIN 2004-03388.

executed. Thus, LoRaS must deal with the *Job Selection* process from a waiting queue, together with the problem of selecting the best set of nodes for executing a job (*Node Selection* policies). This is performed by taking into account the state of the cluster system together with the characteristics of the local and parallel workload. Based on those considerations, different policies are implemented to assign jobs to processors in the LoRaS system.

Once a parallel job is executed, the time sharing component of CISNE, named CCS (Cooperating CoScheduling) [4], takes control of the progression of each parallel job. CCS provides an execution environment where the parallel applications can be dynamically coscheduled. It means that the tasks belonging to the same parallel job are coscheduled according to its communication requirements [5]. In addition, the resources given to parallel tasks are balanced and the interactive responsiveness of the local applications (local workload) is fully preserved by means of a job interaction mechanism, and even when using a MultiProgramming Level of parallel jobs (MPL_{paral}) greater than one [6].

In order to take better scheduling decisions, the CISNE system needs to forecast the future state of the cluster [7]. Likewise, this prediction capacity could help to guarantee some limit in the turnaround time for the applications of any parallel user.

These considerations have stimulated some works focused on the estimation process. The most evaluated alternative is to use a historical system that records the past executions of an application [8,9]. This kind of system normally looks for a state that is similar to the current one by defining a comparison function that determines how similar one state is to another. This focus is valid when the state is defined by a small set of variables, but in our case we have to consider a more complex cluster state (i.e. more variables), due to the non-dedicated characteristic of the environment. Besides, a historical system needs a learning phase to become accurate and, the larger the set of variables to consider, the longer the time needed to achieve precision. Following studies like [10,11], we decided to use a simulation approach to represent our scheduling system. However, in such studies the authors use a historical system for predicting the execution time of each parallel application, while we focus on an analytic schema. The results obtained from those studies can reach an error of around 37% for the execution time prediction of any single application. Our analytic results give an error of around 41%, but with an estimation time that remains constant whatever the cluster state, while a historical system needs a linear time (depending on the number of cases to be studied), for estimating the same value. It should be noted that the number of cases to be evaluated by the historical system in a non-dedicated environment is much greater than those available in studies such as [10,11].

Unfortunately, different policies for distributing the parallel applications may have different effects on the estimation methods and vice-versa. This could make some predicting schemes more suitable for some specific policy but not for others. This has motivated us to propose some estimation methods oriented to non-dedicated clusters and evaluate their performance in relation to different

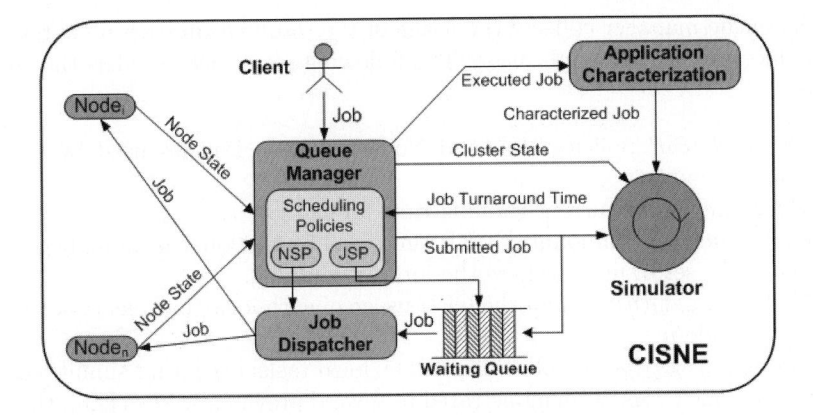

Fig. 1. The integration of the simulation into the CISNE system

distribution policies. With this aim, a simulation tool has been implemented in the CISNE system. In this framework, we have observed that those policies that distribute the resources in a more balanced way are more predictable reaching an accuracy of 12%. On the other hand, those policies that could alter the order of the waiting jobs, such as backfilling, are inherently more inaccurate.

The outline of this work is as follows, section 2 depicts our simulation method. Some proposals for the estimation process are explained in section 3. Next, the experimental results are analyzed in section 4. Finally, some conclusions and the future work are explained.

2 The Simulation Process on the CISNE System

As stated above, the simulation process is integrated into the CISNE system [3]. In order to deal with this estimation in a non-dedicated environment, CISNE needs information from two different inputs: the characterization of the parallel applications and the modelling of the current cluster state, including the local load activity. As we can see in Fig. 1, this information is provided by the queue manager and application characterization block, respectively. At the end of the simulation, the estimated job turnaround time is returned to the queue manager.

The behavior of the parallel applications is obtained by means of running the parallel application in isolation. This is preferred over the information given by the user about the resources used by the applications, which is normally an inaccurate method [12]. For a fixed number of processors per application (n), CISNE collects:

- $ExeTime_{tot}(J)$: execution time of the job J.
- $CPUtime_{tot}(J)$: amount of CPU time used by the job J.
- $CPU(J)$: CPU percentage ($CPUtime_{tot}(J)/ExeTime_{tot}(J)$) used by the job J.

The queue manager collects the usage of the resources in each node together with the state of each application. The following set of data models the cluster state:

- *JSP and NSP* policies: Job and Node Selection Policies used by CISNE, respectively.
- $ExeTime_{cur}(J)$: current running time for the job J.
- $CPUtime_{cur}(J)$: amount of CPU time used by the job J from its beginning.
- $nodes(J)$: set of nodes where the job J is running.
- $CPU_{local/paral}(n)$: sum of the CPU usage of each local/parallel task running in the node n.
- $MPL_{paral/local}(n)$: number of parallel/local tasks executing simultaneously in the node n. As we demonstrated in several previous studies [3,6], the time sharing component of CISNE allows the execution of more than one parallel application in the same set of nodes, whenever it does not disturb the local user.
- $tasks(n)$: set of parallel tasks running in the node n.

Once all the needed elements are collected, CISNE is ready to start the simulation process described in the next section.

3 Turnaround Time Prediction by Simulation

The simulation process is triggered whenever a new application arrives to the CISNE system. Every time that the simulation is started, the turnaround time for each application, either running or waiting, is estimated and adjusted, giving some extra information to the job scheduler about the future cluster state. If the simulator is working when a new application arrives, the whole process has to be restarted considering the new job to be executed. Alg. 1 depicts our simulation method.

The core of the simulation algorithm relies on a *while* that loops as long as any parallel application is running (Alg. 1:3-16). For each iteration, the algorithm estimates the *Remaining Execution Time (RemainTime)* of every job in the running queue (DRQ), selects the next job that will finish (J_i) and removes it from DRQ (Alg. 1:4-6). After that, the $CPUtime_{cur}$ used by each of the remaining jobs in DRQ is calculated (Alg. 1:7) to be used in the next simulation step (Alg. 1:3 loop). Next, another loop tries to execute some waiting jobs using the system scheduling policy, the available resources and those resources released by J_i (Alg. 1:8). Finally, the waiting time for every job in DQ is updated (Alg. 1:14) and the simulation step advances to t_i (Alg. 1:15).

In order to carry out this simulation, we need a pair of extra functions which define the estimation process. The first is the *RemainTime* (Alg. 1:4), which estimates the *remaining execution time* for a given application considering the current cluster and application state. The second tries to predict the *CPU time* ($CPUtime_{cur}$) that the application has used in the past (Alg. 1:7). Our approaches to solving both functions are depicted in the following subsections.

Algorithm 1. Simulation process

1: Duplicate the system state in a *dummy* system state: DQ as a copy of the jobs waiting queue, DRQ as a copy of the running jobs queue and CL_{sim} as a copy of the cluster nodes with their state
2: Store the current time (t_0), as the moment when the simulation has begun.
3: **while** ($\exists\ J$ in DRQ) **do**
4:    **forall** (J in DRQ) Calculate the $RemainTime(J)$.
5:    Assume that the application J_i is the next one to finish in time t_i.
6:    Update the estimated $ExeTime_{tot}(J_i)$ to t_i and remove J_i from DRQ.
7:    **forall** (J in DRQ) Calculate the $CPUtime_{cur}(J)$ in $[t_0, t_i]$.
8:    **while** ($\exists$ usable resources in Cl_{sim} and any job waiting in DQ) **do**
9:      Look for an application J_x in DQ that could be executed in the Cl_{sim} state.
10:      Select the best subset of Cl_{sim} for executing J_x, using the system policy.
11:      Execute the application J_x in the selected subset of Cl_{sim} and add it to DRQ.
12:      Increment the estimated $WaitingTime(J_x)$ in $[t_0, t_i]$.
13:    **end while**
14:    **forall** (J in DQ) Increment the estimated $WaitingTime(J)$ in $[t_0, t_i]$.
15:    Set t_0 to t_i.
16: **end while**

3.1 Remaining Execution Time Approaches

The easiest estimation is to think that the future will be similar to the past. With this in mind, the remaining execution time of a job J, denoted as $RemainTime(J)$, is calculated according to the following equation:

$$RemainTime(J) = \frac{ExeTime_{cur}(J) \times (CPUtime_{tot}(J) - CPUtime_{cur}(J))}{CPUtime_{cur}(J)} \quad (1)$$

Note that Eq.1 assumes that the CPU time ($CPUtime_{tot}(J)$) used by the job J during its complete execution ($RemainTime(J) + ExeTime_{cur}(J)$) is proportional to the CPU time ($CPUtime_{cur}(J)$) used during the current execution time ($ExeTime_{cur}(J)$).

The second proposal considers both the past and current states. It starts by calculating the remaining execution time that the job J would need if it were executed in isolation ($RemainTime_{isol}(J)$). This value, following the same reasoning as Eq. 1, is calculated as follows:

$$RemainTime_{isol}(J) = \frac{ExeTime_{tot}(J) \times (CPUtime_{tot}(J) - CPUtime_{cur}(J))}{CPUtime_{tot}(J)} \quad (2)$$

Next, the maximum MPL ($MPL_{max}(J)$) is defined as:

$$MPL_{max}(J) = max(MPL_{paral}(n) + MPL_{local}(n) \mid n \in nodes(J)). \quad (3)$$

It is worth pointing out that Eq. 3 returns the maximum number of tasks, both local ($MPL_{local}(n)$) and parallel ($MPL_{paral}(n)$), executing concurrently with the job J among the nodes where it is running $(nodes(J))$. Taking $MPL_{max}(J)$ and $RemainTime_{isol}(J)$ into account, the remaining execution time of job J is calculated according to the following equation:

$$RemainTime(J) = RemainTime_{isol}(J) \times MPL_{max}(J) \qquad (4)$$

Our last approach considers not only the number of tasks executing concurrently (MPL) but also the *CPU requirements* of those tasks (in percentage). According to this, the $RemainTime(J)$ is calculated as follows:

$$RemainTime(J) = RemainTime_{isol}(J) \times \frac{CPU(J)}{CPU_{feas}(J)} \qquad (5)$$

where:

$$CPU_{feas}(J) = min(CPU(J), \frac{CPU(J)}{CPU_{max}(J)}), \qquad (6)$$

is the maximum CPU usage (in percentage) that we expect the job J could use, and:

$$CPU_{max}(J) = max(CPU_{paral}(n) + CPU_{local}(n) \mid n \in nodes(J)) \qquad (7)$$

is the maximum CPU usage requirements (in percentage) among the nodes where the job J is running $(nodes(J))$.

It is important to emphasize that no matter which the chosen approach is, the value for $CPUtime_{cur}(J)$ is accurate only at the beginning of the simulation process (Alg. 1:3), but for each simulation step it is necessary to estimate this value again (Alg. 1:7). Therefore, in the next subsection we describe some proposals for estimating this value.

3.2 Used *CPU time* Proposals

This section describes two different proposals for estimating the current CPU time for a given job J at a specific moment (t_i), denoted as $CPUtime_{cur}(J, t_i)$, considering that this value has been measured in the past $(CPUtime_{cur}(J, t_{i-1}))$.

In our first approach, we assume that the application CPU usage is proportional to the maximum MPL calculated in Eq. 3. The following expression represents this proposal.

$$CPUtime_{cur}(J, t_i) = CPUtime_{cur}(J, t_{i-1}) + \frac{(t_i - t_{i-1}) \times CPUtime_{tot}(J)}{MPL_{max}(J) \times ExeTime_{tot}(J)} \qquad (8)$$

Our second proposal is based on the same idea used for the remaining time in Eq. 5, but applied to the *CPUtime*. The following equation represents that idea.

$$CPUtime_{cur}(J, t_i) = CPUtime_{cur}(J, t_{i-1}) + (t_i - t_{i-1}) \times CPU_{feas}(J) \qquad (9)$$

4 Experimentation

In order to carry out the experimentation process, we need two different kinds of workload. On one hand, we need to simulate the local user activity and, on the other hand, we need some parallel applications that arrive at some interval.

The local user activity is represented by a benchmark that could be parameterized in such a way that it uses a percentage of CPU, memory and network. To parameterize this benchmark realistically, we measure our open laboratories for a couple of weeks and use the collected values to run the benchmark (15% CPU, 35% Mem., 0,5KB/sec LAN).

The parallel workload was a list of 30 PVM NAS parallel jobs (CG, IS, MG, BT) with a size of 2, 4 or 8 tasks that entered the system following a Poisson distribution of inter-arrival times with mean=15s. These jobs were merged so that the entire workload had a balanced requirement for computation and communication. It is important to mention that the MPL_{paral} reached for the workload depends on the system state at each moment, but in no case will surpass an $MPL_{paral} = 4$ [6].

This workload was executed with different combination of Job Selection (JSP) and Node Selection policies (NSP). Regarding the JSP policies, *FCFS* (First-Come-First-Served) and Backfilling techniques were tested. A backfilling policy consists of executing a job, not at-the-head of the FCFS queue, whenever this does not delay the start of the job at the head. The set of nodes onto which the selected job will be launched, was chosen according to two different *NSP* policies. The first one, termed *Normal,* selects the nodes for executing a parallel application considering only the resource usage level throughout the cluster, so it does not overload any node in detriment of the local user interactiveness. An example can be observed in Fig. 2.b, where the J_3 shares its nodes with J_1 and J_2. The second approach, called *Uniform,* selects the nodes respecting not only the resource usage but also the job distribution. In this case, the policy executes a pair of jobs of the same size in the same set of nodes whenever possible. Besides, the system tries to execute tasks of different oriented applications (i.e. communication bound vs. computation bound), in the same set of nodes trying to enhance the underlying Time-Sharing schema of our CISNE system [3]. An example can be seen in Fig. 2.a, where J_3 only shares its nodes with J_2. Finally, and for the purpose of comparison, we include a *Basic* policy made up of a *Normal+FCFS* policy with the $MPL_{paral} = 1$, rather than 4 as in the other policies.

The whole system was evaluated in an Linux cluster using 16 P-IV (1,8GHz) nodes with 512MB of memory and a fast Ethernet interconnection network.

4.1 Experimental Results

In this subsection, we present some results showing the effect of the different JSP and NSP policies over the different mixes for the *Remaining Execution Time* (*RemainTime*) and the *Used CPU Time* (*CPUtime*) estimation methods. Fig. 3 shows the estimated turnaround deviation (in %) from the real turnaround time

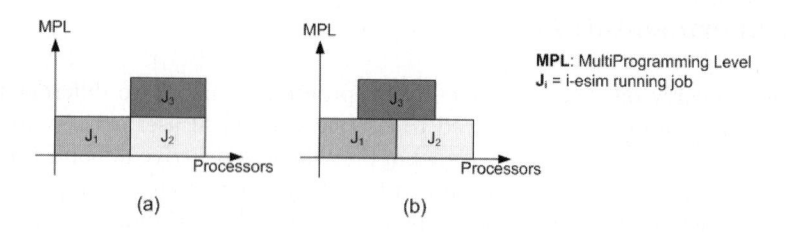

Fig. 2. *Uniform* (a) and *Normal* (b) Node Selection Policy

for the different policies and estimation methods. These results were obtained considering that 25% of the nodes had some kind of local activity.

From Fig. 3 and considering the *RemainTime* methods, we can see that a *Uniform* policy favors the predictability of the system over a *Normal* policy because it tries to balance the resources given to the parallel applications. In such a case, the tasks forming a parallel application have the same resources and then they can evolve jointly letting the estimation methods be more accurate. Likewise, Fig. 3 shows that a Basic policy performs worse than *Normal* or *Uniform* for some cases. This is mainly due to the *Basic* restriction of the MPL_{paral} ($MPL_{paral}=1$). In such a scenario, the waiting queue length increases and hence the waiting time prediction become less accurate. As a consequence, the turnaround time prediction is worse. In addition, this figure reflects that a scheduling policy that includes a backfilling scheme is more unpredictable. This is due to the difficulty of tracking the variation in the order of the elements in the waiting queue. However, and even considering that the resulting estimations are not as good as for those policies without a backfilling scheme, the results are almost always pessimistic, due to the possibility of backfilling some of them, and hence reducing their waiting time. This means that the parallel user has a turnaround time that in the worst case, is overestimated, but never underestimated. Finally, and as was to be expected, the *Proportional* (Prop., Eq. 1) *RemainTime* method performs badly. This is due to the assumption that the future behaves like the past. This assumption is not true when the environment state changes continuously due to the local and parallel loads.

From the same Fig. 3, and focusing now on the *Used CPU Time* methods, it is clear that the estimation via the *CPU* usage (right columns, Eq. 9) is more reliable in most of the cases, compared with the estimation through the *MPL* (left columns, Eq. 8) method. This happens because the *CPUtime* method represents reality better by considering the real percentage of CPU consumed by each task, while the *MPL* method assumes that every task consumes the same percentage of CPU. There is a special case for the *Basic* policy, where these results are apparently contradictory. However, these results are due to the light parallel load imposed on the system (on Basic, the parallel MPL is at most 1), that results in a longer, and hence more unpredictable, waiting queue.

In addition, we want to analyze the influence of the local load on the estimation methods and scheduling policies. Table 1 shows the turnaround deviation for two different combinations to estimate the *RemainTime* and *CPUtime*

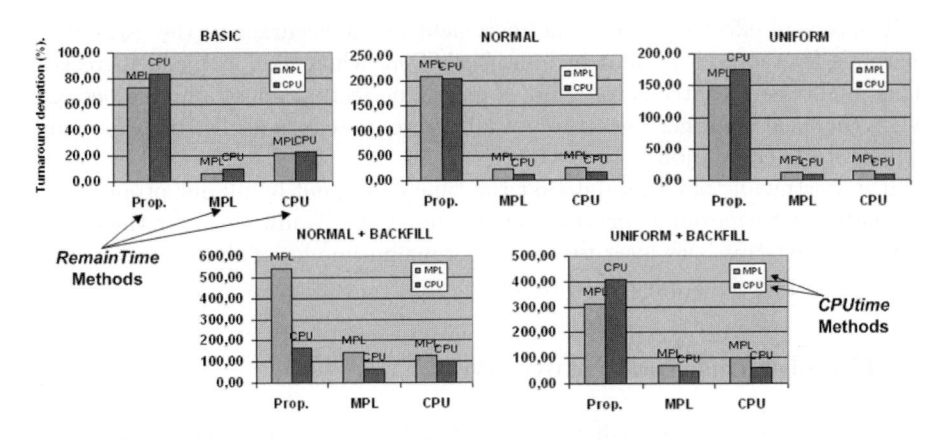

Fig. 3. Turnaround deviation (%) for the different *RemainTime* and *CPUtime* mixes for the scheduling policies evaluated

Table 1. Turnaround deviation (%) for the different *RemainTime* and *CPUtime* mixes for the evaluated policies and different local loads

Local Load	Norm+BF		Unif+BF		Basic		Normal		Uniform	
	MPL	CPU	MPL	CPU	MPL	CPU	MPL	CPU	MPL	CPU
25%	144,33	99,67	69,67	63,33	6,33	23,00	23,00	15,33	12,67	8,67
50%	58,67	59,67	55,00	31,33	14,00	18,67	20,50	21,33	22,00	21,67
100%	46,33	45,00	54,00	46,67	34,33	8,67	16,67	28,33	25,33	18,33

values respectively: *MPL-MPL* (MPL in table 1) and *CPU-CPU* (CPU in table 1). Both combinations were tested varying the number of nodes with local load from 25% to 100% of nodes. From this table, we can see that the CPU-CPU estimation method gives us a better estimate than the MPL-MPL method in most of the cases. This agrees with our expectations, because an estimation process that considers the CPU consumption of each task instead of assuming that every task uses the same amount of CPU, as the MPL-MPL method does, tends to be more accurate. There are, however, some cases where the CPU-CPU method does not give us the best results. One of them was found for the *Normal* policy and 100% of nodes loaded with local tasks. In this case, these bad results are due to an unbalanced distribution of the resources throughout the cluster, which complicates the CPU-CPU method capacity for tracking the CPU usage of the parallel and local loads. The other case that it is not favorable to the CPU-CPU method is for the *Basic* policy. In this case, when the system is unloaded (local load of 25%) and the MPL is at most 1, the CPU usage calculation is misleading because the whole set of tasks could evolve faster than the estimation process assume. However, in such a situation, the estimation process is always pessimistic, so that the application will always finish before the estimated finish time.

The second effect to note is the increment in the accuracy of the estimation methods for the policies that include *Backfilling* when the local load increases. This might seem contradictory, but is in fact perfectly understandable because when the local load increases, the available resources decrease, as do the opportunities to backfill a job.

It is worthwhile pointing out that the time cost spent by all our proposals to estimate the turnaround time of a single application is always lower than $4ms$. It means that this is at least two orders of magnitude lower than the execution time of the parallel applications (minutes).

5 Conclusions and Future Work

In order to improve the prediction capacity of a resource management environment over a non-dedicated cluster, this work presents a simulation algorithm that merges different estimation methods for predicting the turnaround time of parallel applications. The proposed estimation methods focus on two different goals. The first set tries to predict the *Remaining Execution Time* for a given running application and a defined cluster state, while the second set of methods estimates the *CPU usage* that an application could absorb for a given time interval and cluster state. The relationship between these estimation methods and different job scheduling policies are evaluated. We conclude that those methods that consider not only the MultiProgramming Level (MPL) of the parallel and local tasks in each node but also the CPU consumption of each one, are more accurate in the general case. Besides, those policies including a backfilling scheme are inherently more difficult to estimate accurately due to the possibility of altering the job ordering in the queue. However, this estimation always tends to be pessimistic and hence the jobs finish before the predicted finish time. Another effect that could be observed was the influence of the job distribution on the estimation process. For those policies that balance the resources it is easier to generate a more accurate estimation, reaching an accuracy of 12%.

In the future, we want to introduce an hybrid system that merges our simulator with a *Remaining Execution Time* method that uses a historical system. This way it will be possible to generate an estimation method that becomes more and more accurate over time, but without the cost of assuming a whole historical prediction system that has to manage a lot of variables.

References

1. Acharya, A., Setia, S.: Availability and utility of idle memory in workstation clusters. In: Proceedings of the ACM SIGM./PERF.'99. (1999) 35–46
2. Litzkow, M., Livny, M., Mutka, M.: Condor- a hunter of idle workstations. 8th Int'l Conference of Distributed Computing Systems (1988)
3. Hanzich, M., Giné, F., Hernández, P., Solsona, F., Luque, E.: Cisne: A new integral approach for scheduling parallel applications on non-dedicated clusters. EuroPar 2005, Parallel Processing. LNCS **3648** (2005) 220–230

4. Giné, F., Solsona, F., Hernández, P., Luque, E.: Cooperating coscheduling in a non-dedicated cluster. EuroPar 2003 Parallel Processing, LNCS **2790** (2003) 212–218
5. Sobalvarro, P., Pakin, S., Weihl, W., Chien, A.: Dynamic coscheduling on workstation clusters. Job Scheduling Strategies for Parallel Processing, LNCS **1459** (1998) 231–256
6. Hanzich, M., Giné, F., Hernández, P., Solsona, F., Luque, E.: Coscheduling and multiprogramming level in a non-dedicated cluster. EuroPVM/MPI 2004, LNCS **3241** (2004) 327–336
7. Smith, W., Taylor, V., Foster, I.: Using run-time predictions to estimate queue wait times and improve scheduler performance. Job Scheduling Strategies for Parallel Processing, LNCS **1659** (1999) 202–219
8. Lafreniere, B.J., Sodan, A.C.: Scopred—scalable user-directed performance prediction using complexity modeling and historical data. Job Scheduling Strategies for Parallel Processing, LNCS **3834** (2005) 62–90
9. Downey, A.B.: Predicting queue times on space-sharing parallel computers. In: IPPS '97: Proceedings of the 11th International Symposium on Parallel Processing, Washington, DC, USA, IEEE Computer Society (1997) 209–218
10. Smith, W., Wong, P.: Resource selection using execution and queue wait time predictions. NAS Technical Reports (2002)
11. Li, H., Groep, D., Templon, J., Wolters, L.: Predicting job start times on clusters. 4th IEEE/ACM International Symposium on Cluster Computing and the Grid (CCGrid2004) (2004)
12. Mu'alem, A.W., Feitelson, D.G.: Utilization, predictability, workloads, and user runtime estimates in scheduling the ibm sp2 with backfilling. IEEE Transaction on Parallel & Distributed Systems **12(6)** (2001) 529–543

An Adaptive Scheduling Method for Grid Computing

Salah-Salim Boutammine, Daniel Millot, and Christian Parrot

GET / INT
Département Informatique
91011 Évry, France
{Salah-Salim.Boutammine, Daniel.Millot, Christian.Parrot}@int-evry.fr

Abstract. This paper presents an adaptive scheduling method, which can be used for parallel applications whose total workload is unknown a priori. This method can deal with the unpredictable execution conditions commonly encountered on grids. To address this scheduling problem, parameters which quantify the dynamic nature of the execution conditions had to be defined. The proposed scheduling method is based on an on-line algorithm so as to be adaptable to the varying execution conditions, but avoids the idle periods inherent to this on-line algorithm.

Keywords: scheduling, parallel application, grid, master-worker, on-line, multi-round, heterogeneity, dynamicity.

1 Introduction

In this paper, we present an adaptive method for scheduling parallel applications, that can be used in the dynamic context of grids and when some of the information traditionally used by scheduling algorithms is lacking. This method is based on an on-line algorithm from Drozdowski [1]. We assume that a set of grid resources has been identified and tackle the problem of distributing optimally the tasks of a parallel application on this set of resources, so that the application terminates as soon as possible. Precisely, we consider applications that process a finite –but a priori unknown– amount of data independently. The total workload of the application is supposed arbitrarily divisible in any number of chunks where each chunk consists of some amount of data. The same computation is performed on each chunk, producing its own result without any communication. Such applications are suitable for the master-worker programming model, with the master distributing chunks to the workers, then collecting the corresponding results from them. Clearly, for such a parallelization to be useful, the processing cost for a chunk by a worker must dominate the corresponding communication costs between master and worker in a certain sense that will be stated later on, when appropriate notations have been introduced (see inequality (2)).

It has to be noted that although we consider so called divisible load, the DLT (Divisible Load Theory [2, 3, 4, 5]) cannot be straightforwardly applied in our case, as we suppose that the total workload of the application is not known a priori. For this reason, we shall be bound to use on-line algorithms to address our scheduling problem.

We adopt a one-port communication model [6] without contention, which means that for a fixed node neither two emissions nor two receptions can overlap each other, whereas one emission can overlap one reception, and computation can overlap communication. Furthermore, in the context of grid computing, computation and communication latencies must both be considered.

This paper is organized as follows. Section 2 defines precisely the scheduling problem we consider. Section 3 describes the on-line algorithm which our scheduling method is based on and introduces some notations. Section 4 presents the new method itself. It first gives an overview of the approach then successively states the conditions for the method to succeed, details the various computations of the proposed scheduling algorithm and finally compares it with the initial on-line algorithm. Section 5 concludes the paper and outlines future work.

2 The Scheduling Problem

We consider a master-worker model for which the data to be processed are continuously received by the master in an input buffer until the final item is obtained. It is only when the master acquires this last item that the total workload of the application happens to be known. We want to minimize the makespan of the application on a set of grid resources. As this problem is NP-complete when latencies are considered [7], it can only be heuristically dealt with.

Execution parameters on a grid, such as available computing power or network bandwidth, vary both in space (heterogeneity) and time (dynamicity). We assume that we know all past values of these parameters and are unaware of the future ones. Because of the one-port communication model, the workers cannot start their work simultaneously: the master has to finish the emission of some chunk to one worker before being able to begin to send a chunk to another one.

In this paper, we do not consider the fundamental problem of choosing the nodes to be used and the order in which they are served. To terminate the execution of the application as soon as possible, the computation should start as soon as possible on all the selected worker nodes, which should then be sent small initial amounts of work in order to quickly start their computation.

When each worker has received a first chunk, the execution enters the steady-state phase [8]. The main characteristic of this phase is that the total workload is still unknown. If the choice of the computing resources is optimal (i.e. optimal nodes are chosen in optimal proportion), then keeping the selected nodes active minimizes the makespan. When the master gets the final data item to be processed, the steady-state phase ends and the clean-up phase begins.

From this time instant, the problem of scheduling the remaining load is suitable for DLT, as now the total workload is known: namely the amount of data still present in the master input buffer. So, according to the optimality principle, we can try and minimize the makespan by synchronizing the termination of the computation of all the workers. This, being possible only if the master does not overload any worker too much during the steady-state phase, which would cause a discrepancy too large for late workers to catch up, thus preventing a synchronous termination of all workers. In the following, we focus on scheduling during the steady-state phase.

3 Drozdowski's On-Line Scheduling Method

Our scheduling method is based the On-Line method presented by Drozdowski in [1], denoted "OL" thereafter. OL proceeds incrementally, computing the size $\alpha_{i,j}$ of the chunk to be sent to a worker N_i for each new round j, in order to try and maintain a constant duration τ for the different rounds and thus avoid contention at the master.

OL determines $\alpha_{i,j}$ so as to make the distribution asymptotically periodic with period τ, an arbitrarily fixed value, for all the workers. For worker N_i, let $\sigma_{i,j-1}$ be the elapsed time between the begining of the emission of the chunk of its $(j-1)^{th}$ round and the end of the reception of the result corresponding to this chunk. OL determines the value of $\alpha_{i,j}$ as follows:

$$\alpha_{i,j} = \alpha_{i,j-1} \cdot \frac{\tau}{\sigma_{i,j-1}}. \tag{1}$$

That is it allocates comparatively bigger (resp. smaller) chunks to workers with higher (resp. lower) performance. Hence, this method can take the heterogeneous nature of computing and communication resources into account, without explicit knowledge of execution parameters (as equality (1) shows); as Drozdowski states, "the application itself is a good benchmark" [1] (actually the best one).

Lemma 6.1 in [1] shows that, in a static context, with affine cost models for communication, the way $\alpha_{i,j}$ is computed using equation (1) ensures the convergence of $\sigma_{i,j}$ to τ when j increases indefinitely.

Being an estimation of the asymptotic period used for task distribution, τ is also an upper-bound on the discrepancy between workers. Being able to control this bound makes it possible to minimize the makespan during the clean-up phase.

The following notations are used throughout the rest of the paper:

- N number of workers,
- γ_i start-up time for a computation by worker N_i,
- $w_{i,j}$ computation cost for a chunk of size 1 of the jth round by worker N_i,
- β_i (resp. β'_i) start-up time for a communication from the master to N_i (resp. from N_i to the master),
- $c_{i,j}$ (resp. $c'_{i,j}$) transfer cost for a data (resp. result) chunk of size 1 of the jth round from the master to worker N_i (resp. from N_i to the master).

It should be noted that, unlike previous work [1, 9], this paper introduces computation start-up times in order to be more realistic when considering grids. As suggested in section 2, the values of the execution parameters of any worker N_i — here $w_{i,j}$, $c_{i,j}$ and $c'_{i,j}$ — depend on the round. We assume that costs are roundwise affine in the size of chunks. Hence, for a chunk of strictly positive size α (i.e. $\alpha \in \mathbb{R}^{+*}$) of the jth round, we define the cost of:

- sending the chunk to worker N_i $\qquad\qquad \alpha \cdot c_{i,j} + \beta_i,$
- processing the chunk on worker N_i $\qquad\qquad \alpha \cdot w_{i,j} + \gamma_i,$
- receiving the corresponding result from worker N_i $\qquad \alpha \cdot c'_{i,j} + \beta'_i.$

We indicated in section 1 that the processing cost for a chunk should dominate its communication costs in a certain sense. We choose to formulate this assumption as:

$$\forall \alpha \in \mathbb{R}^{+*},\ \gamma_i + \alpha \cdot \min_{j \in \mathbb{N}^*} w_{i,j} \geq \left(\alpha \cdot \max_{j \in \mathbb{N}^*} c_{i,j} + \beta_i \right) + \left(\alpha \cdot \max_{j \in \mathbb{N}^*} c'_{i,j} + \beta'_i \right) \quad (2)$$

$for\ i = 1, N$.

Equation (2) ensures that sending chunks of any size α to a worker N_i and receiving the corresponding results cost less than processing these chunks.

The problem with OL is that computation never overlaps communication in any worker node, as the emission of the chunk of the next round is at best triggered by the return of the result of the previous one, with no possible anticipation.

4 The OLMR Method

4.1 Overview of the Method

Our method is based on OL, but avoids idle time with respect to computing. When the total load is important compared to the available bandwidth between master and workers, the workload should be delivered in multiple rounds [10, 11, 12]. Therefore we will have each worker receive its share of the load through multiple rounds, hence the name On-Line Multi-Round method [9], denoted "OLMR" thereafter. OLMR divides the chunk sent to N_i for each round j into two subchunks "I" and "II" of respective sizes $\overline{\alpha}_{i,j}$ and $\alpha_{i,j} - \overline{\alpha}_{i,j}$. Dividing the chunks in two parts is enough in order to apply the principle, and the division allows the computation to overlap the communications as can be seen in figure FIG.1. In order to compute $\alpha_{i,j}$, we use a value of $\sigma_{i,j-1}$ derived from the measurement of the elapsed time (including both communications and computation) for subchunk I of the previous round: $\overline{\sigma}_{i,j-1}$. We will show that, thanks to this anticipation (compared to OL) in the computation of $\alpha_{i,j}$, we can avoid the inter-round starvation.

Figure FIG.2 gives the OLMR scheduling algorithm. The OLMR scheduler computes $\alpha_{i,j}$ in the same way as the OL scheduler does, and the values of $\sigma_{i,j-1}$ and $\overline{\alpha}_{i,j}$ as detailed later in the next subsections.

Unfortunately, while attempting to deal with the inter-round starvations inherent to OL, there is a risk of creating intra-round starvation between subchunks I and II (see on figure FIG.3 the idle period Δ). We explain below how to prevent both risks.

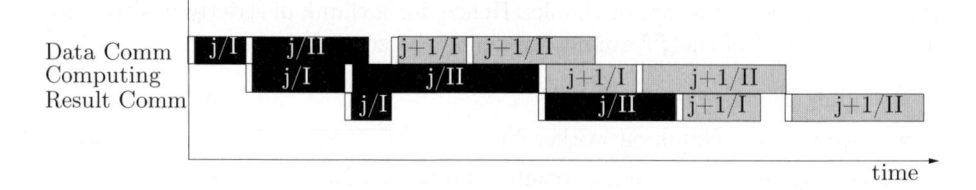

```
Data Comm     | j/I  | j/II  | j+1/I | j+1/II |       |        |
Computing     |      | j/I   | j/II  |       | j+1/I | j+1/II |
Result Comm   |      |       | j/I   | j/II  | j+1/I | j+1/II |
```
→ time

Fig. 1. Overlapping between communication and computation with OLMR

---

while (the last data item has not been acquired) **do**
  if (Reception from N_i of the result of subchunk I of its $(j-1)^{th}$ round) **then**
    • Get $\overline{\sigma}_{i,j-1}, \omega_{i,j-1}, c'_{i,j-1}$ (and γ_i for the first result from N_i)
    • Compute $\sigma_{i,j-1}$ .... (cf. (8))
    • Compute $\alpha_{i,j}$ .... (cf. (1))
    • Compute $\overline{\alpha}_{i,j}$ .... (cf. (7))
    • Send a subchunk of size $\overline{\alpha}_{i,j}$ to N_i as subchunk I of its jth round
    • Send a subchunk of size $(\alpha_{i,j} - \overline{\alpha}_{i,j})$ to N_i as subchunk II of its jth round
  end if
end while

Fig. 2. OLMR scheduler

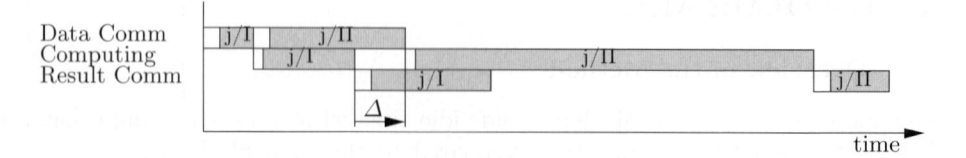

Fig. 3. Example of intra-round starvation with OLMR

As we assume that (2) holds, intra-round starvation can be avoided if $\overline{\alpha}_{i,j}$ is large enough for the processing of subchunk I to overlap the sending of subchunk II of size $\alpha_{i,j} - \overline{\alpha}_{i,j}$. There is no intra-round starvation if and only if

$$\overline{\alpha}_{i,j} \geq \frac{\beta_i - \gamma_i + \alpha_{i,j} \cdot c_{i,j}}{w_{i,j} + c_{i,j}}. \qquad (3)$$

Inter-round starvation between the jth and $(j+1)^{th}$ rounds of N_i could occur if subchunk I happens to be too large compared to subchunk II (see figure FIG.4). Let $\nu_{i,j}$ be some real number dominating $\overline{\alpha}_{i,j+1}$: $\nu_{i,j} \geq \overline{\alpha}_{i,j+1}$. Figure FIG.4 shows that, when N_i is given a subchunk I of size $\nu_{i,j}$ for its $(j+1)^{th}$ round, there is no inter-round starvation if and only if

$$\overline{\alpha}_{i,j} \leq \frac{\alpha_{i,j} \cdot w_{i,j} - \nu_{i,j} \cdot c_{i,j+1} + \gamma_i - (\beta'_i + \beta_i)}{c'_{i,j} + w_{i,j}}. \qquad (4)$$

If inequality (4) holds, then the necessary constraint $\overline{\alpha}_{i,j} < \alpha_{i,j}$ holds too, as soon as $(\beta'_i + \beta_i) > \gamma_i$.

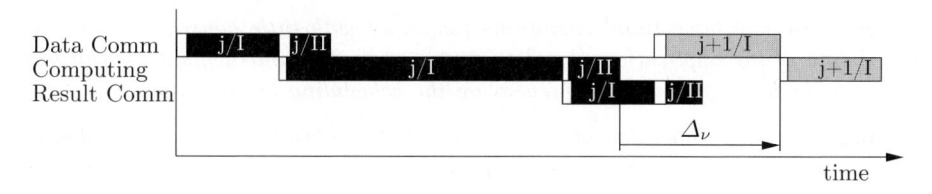

Fig. 4. Example of inter-round starvation with OLMR

Relying on inequations (3) and (4), we can choose $\overline{\alpha}_{i,j}$ so as to avoid idle periods of N_i. Finally, nothing remains but to determine $\sigma_{i,j-1}$ and $\overline{\alpha}_{i,j}$.

4.2 Determining $\overline{\alpha}_{i,j}$

In order to fix the value of $\overline{\alpha}_{i,j}$ according to constraint (4), we need a value for $\nu_{i,j}$. We can decide such a value by extrapolating an upper bound for $\overline{\alpha}_{i,j+1}$ from the values of $\overline{\alpha}_{i,k}$ for the previous rounds, $(\overline{\alpha}_{i,k})_{k=1,j}$. So long as inequalities (3) and (4) hold, an inaccuracy in the value of $\nu_{i,j}$ does not have any dramatic consequence on the course of the method. That is, if inequalities (3) and (4) are compatible, then starvation risks can be avoided.

As the amount of data processed during the steady-state phase is finite, there necessarily exists a real number λ_i ($\lambda_i \geq 1$) for each N_i such that:

$$\overline{\alpha}_{i,j+1} \leq \lambda_i \cdot \overline{\alpha}_{i,j} \qquad \forall j \in \mathbb{N}^*.$$

λ_i characterizes the amplitude of the fluctuations of $\overline{\alpha}_{i,k}$ between two successive rounds. If λ_i can be estimated (see Remarks 2 and 3 for hints), then we have an upper-bound $\nu_{i,j}$ for $\overline{\alpha}_{i,j+1}$:

$$\nu_{i,j} = \lambda_i \cdot \overline{\alpha}_{i,j}. \tag{5}$$

The following Theorem proposes a way to set the value of $\overline{\alpha}_{i,j}$ so that constraints (3) and (4) are both satisfied (see [9] for similar proof).

Theorem 1. *Given $\alpha_{i,j}$, if γ_i, $w_{i,j}$, $c_{i,j}$, β_i, $c'_{i,j}$ and β'_i satisfy (2) and*

$$(\alpha_{i,j} - (\lambda_i + 1)) \cdot w_{i,j} + (\lambda_i + 1) \cdot \gamma_i \geq (\lambda_i \cdot \alpha_{i,j} + (\lambda_i + 1)) \cdot c_{i,j} + (\lambda_i + 1) \cdot \beta_i \tag{6}$$

for i=1,N.
Then, taking

$$\overline{\alpha}_{i,j} = \frac{\alpha_{i,j}}{\lambda_i + 1}, \tag{7}$$

constraints (3) and (4) are satisfied. Therefore, the workers will compute without any idle period during the steady-state phase.

Remark 1. *Parameters τ and λ_i are characteristic of the evolution of the execution parameters. On the one hand, τ characterizes their speed of evolution. Practically, it is the period that should be used for reconsidering their value. On the other hand, λ_i measures the amplitude of their variations on such a period. The obvious dependence between τ and λ_i can take on the most varied forms. For* instance, we can have rapid variations (small τ) with little consequence on the scheduling of the application (λ_i close to 1), or on the contrary slow variations (large τ) with important consequences on the scheduling (λ_i far from 1).

Remark 2. *The knowledge of $\nu_{i,j}$ is implicitly the result of some extrapolation of the values $(\overline{\alpha}_{i,k})_{k=1,j}$ to get an upper-bound of $\overline{\alpha}_{i,j+1}$. If the variations are slight, one can use the quasi-stationary approximation of $\overline{\alpha}_{i,j+1}$ by $\overline{\alpha}_{i,j}$. In this case, we have $\overline{\alpha}_{i,j+1} = \overline{\alpha}_{i,j}$. Then we only have to apply Theorem 1 with $\lambda_i = 1$. More generally, considering a polynomial interpolation of degree (p - 1) for the value of $\overline{\alpha}_{i,j+1}$, we have $\overline{\alpha}_{i,j+1} = p \cdot \overline{\alpha}_{i,j} - \sum_{k=1}^{p-1} \overline{\alpha}_{i,k}$. In this case, it suffices to apply Theorem 1 with $\lambda_i = p$.*

Remark 3. *The satisfaction of the hypotheses of Theorem 1 guarantees the absence of idle time for the workers but requires the knowledge of $(\lambda_i)_{i=1,N}$. Nevertheless, OLMR may still be used when these values (which characterize the dynamicity of execution parameters) are not known. Starting with arbitrary values (e.g. $\lambda_i = 1$ corresponding to a stability assumption) the scheduler could, if necessary, adjust λ_i values at any round according to information provided by the workers. Actually an inappropriate value of λ_i used for some round will lead to an intra- or inter-round starvation observable by the corresponding worker. The scheduler could then adjust this value for the next round, according to the type of starvation observed by the worker.*

Remark 4. *Although different, hypotheses (2) and (6) both make the assumption that processing should dominate communications. Recall that hypothesis (2) ensures an efficient usage of the master-worker paradigm.*

4.3 Determining $\sigma_{i,j-1}$

In order to determine the size of the chunk to be sent for the next round without waiting for the result of the currently processed chunk, it suffices to replace the measured value $\sigma_{i,j-1}$ in expression (1) by some computed value derived from $\overline{\sigma}_{i,j-1}$. But we only know the values of the execution parameters for the data whose result have been received by the master. We choose to get these parameters just after the master has reveived the result for subchunck I of round $j-1$ (see tag "Snapshot" on FIG.5). It is another extrapolation problem. In order to solve it, we assume that the time taken by N_i to process some amount of data during its $(j-1)^{th}$ round is the same for both subchunks I and II.

With the help of figure FIG.5, and omitting the cost of the scheduling algorithm itself, we have

$$\sigma_{i,j-1} = \overline{\sigma}_{i,j-1} + A + B - C,$$
$$\sigma_{i,j-1} = \overline{\sigma}_{i,j-1} + (\alpha_{i,j-1} - \overline{\alpha}_{i,j-1}) \cdot \omega_{i,j-1} + (\alpha_{i,j-1} - 2 \cdot \overline{\alpha}_{i,j-1}) \cdot c'_{i,j-1} + \gamma_i. \quad (8)$$

Remark 5. *The values of $\omega_{i,j-1}$, γ_i and $c'_{i,j-1}$ can be estimated easily by the master with the help of N_i. There is no need to know the value of either the communication start-up times β_i and β'_i or that of $c_{i,j-1}$ in order to compute $\sigma_{i,j-1}$ by means of equation (8).*

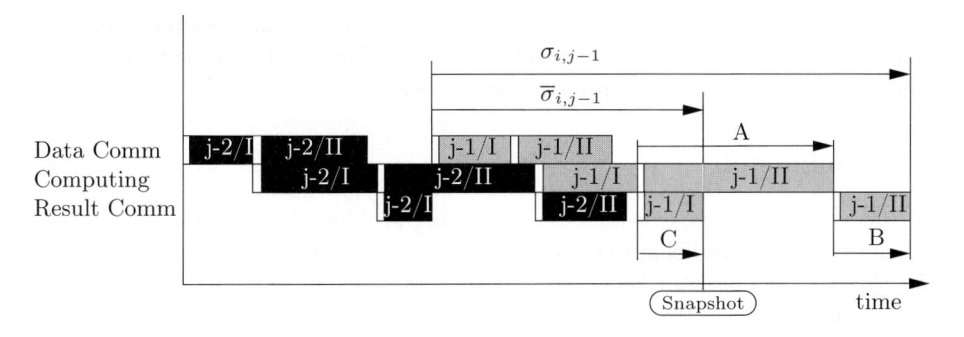

Fig. 5. From the measurement of $\overline{\sigma}_{i,j-1}$ to the computation of $\sigma_{i,j-1}$

4.4 Comparing OL and OLMR

In this section, we compare OL and OLMR, and quantify the benefit of using OLMR compared to OL. We study their behaviour in identical settings: a static context.

Using OLMR requires that the hypotheses of Theorem 1 be satisfied. Lemma 6.1 in [1] sets the context of OL as static. Let us denote c_i, w_i and c'_i the value of $c_{i,j}$, $w_{i,j}$ and $c'_{i,j}$ for any round j as they do not depend on the round, due to the static nature of the execution environment. Under these conditions, both methods send chunks of the same size $\alpha_{i,j}$ to N_i for any round j; for the same value of τ. So processing a workload of size M by both methods requires the same number of rounds δ_M. The gain G_M of OLMR over OL when processing this workload can thus be estimated as :

$$G_M = (\delta_M - 1) \cdot (\beta_i + \beta'_i) - \delta_M \cdot \gamma_i + (M - \overline{\alpha}_{i,1}) \cdot c_i + (M - (\alpha_{i,\delta_M} - \overline{\alpha}_{i,\delta_M})) \cdot c'_i.$$

This gain is the direct consequence of overlapping computation and communications (see figure FIG.6).

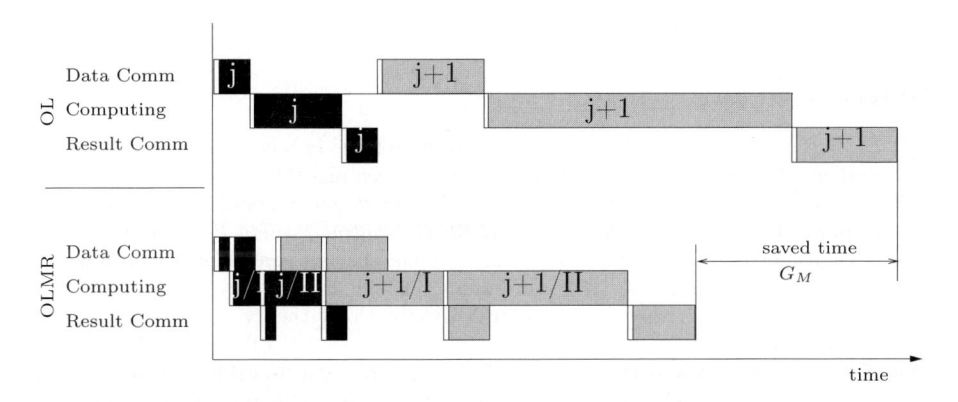

Fig. 6. Comparison between OL and OLMR

Given hypotheses of Theorem 1 and an optimal choice of resources, competitive analysis [13] of OLMR method with an off-line method is not necessary; due to the full use of the computing resources.

5 Conclusion

In this paper, we have considered a scheduling problem that we think is realistic when executing parallel applications on shared resources such as those of a grid. To the best of our knowledge, this scheduling problem has not received much attention up to now. We have presented an adaptive scheduling method, OLMR, to optimize the workload distribution, which can deal with the heterogeneity and dynamicity of the grid if our modelisation hypotheses are realistic; it can also be used when the information that scheduling algorithms traditionally need is lacking. Sufficient conditions have been stated for full usage of the computing resources by means of avoiding idle time.

In order to design the OLMR method, we had to consider the characterization of the dynamicity of the execution conditions. This led us to define $N+1$ parameters: τ and $(\lambda_i)_{i=1,N}$ (see Remark 1). But the improvement made by OLMR to the on-line method presented in [1] has been quantified in a static execution context only.

This approach of scheduling is susceptible to numerous developments, either tending to confirm the results of this paper or aiming at enlarging the potentialities of the OLMR method. First of all, it is useful to check experimentally that, under the hypotheses of our model, the method gives the expected results. For that, we are currently developping simulation programs, using the SimGrid toolkit [14] in order to study OLMR behavior in various conditions and make comparisons with other methods. Furthermore, OLMR could be adapted in different ways: in this paper, τ and λ_i have implicitly been considered as constant throughout all the rounds, but this hypothesis restricts the degree of approximation (order one) of the dynamicity that the scheduler takes into account. From one round to the next, the value of τ could be adapted in order to take further account of the evolution of heterogeneity and dynamicity that would be noticed.

References

[1] M. Drozdowski. *Selected problems of scheduling tasks in multiprocessor computing systems.* PhD thesis, Instytut Informatyki Politechnika Poznanska, Poznan, 1997.
[2] V. Bharadwaj, D. Ghose, V. Mani, and T.G Robertazzi. Scheduling divisible loads in parallel and distributed systems. *IEEE Computing Society Press*, 1996.
[3] T.G. Robertazzi, J. Sohn, and S. Luryi. Load sharing controller for optimizing monetary cost, March 30 1999. US patent # 5,889,989.
[4] T.G. Robertazzi. Ten reasons to use divisible load theory. *IEEE Computer*, 36(5)(63-68), 2003.
[5] K. van der Raadt and H. Casanova Y. Yang. Practical divisible load scheduling on grid platforms with apst-dv. In *Proceeding of the 19th International Parallel and Distributed Processing Symposium (IPDPS'05)*, volume 1, page 29b, IEEE Computing Society Press, April 2005.

[6] O. Beaumont. Nouvelles méthodes pour l'ordonnancement sur plates-formes hétérogènes. Habilitation à diriger des recherches, Université de Bordeaux 1 (France), December 2004.
[7] A. Legrand, Y. Yang, and H. Casanova. Np-completeness of the divisible load scheduling problem on heterogeneous star platforms with affine costs. Technical Report CS2005-0818, UCSD/CSE, March 2005.
[8] L. Marchal, Y. Yang, H. Casanova, and Y. Robert. Steady-state scheduling of multiple divisible load applications on wide-area distributed computing platforms. *Int. Journal of High Performance Computing Applications*, 2006, to appear.
[9] S. Boutammine, D. Millot, and C. Parrot. A runtime scheduling method for dynamic and heterogeneous platforms. In *Proceedings of the 2006 International Conference on Parallel Processing Workshops (ICPPW'06)*, IEEE Computing Society Press, 2006.
[10] V. Bharadwaj, D. Ghose, and V. Mani. Multi-installment load distribution in tree networks with delays. *IEEE Transactions on Aerospace and Electronic Systems*, 31(2):555–567, 1995.
[11] Y. Yang and H. Casanova. *UMR: A Multi-Round Algorithm for Scheduling Divisible Workloads*. IEEE Computing Society Press, April 2003.
[12] O. Beaumont, A. Legrand, and Y. Robert. Optimal algorithms for scheduling divisible workloads on heterogeneous systems. Technical Report 4595, INRIA, Le Chesnay(France), October 2002.
[13] K. Pruhs, J. Sgall, and E. Torng. *Online scheduling, Handbook of Scheduling: Algorithms, Models, and Performance Analysis*. J. Leung, ed., CRC Press, 2004.
[14] H. Casanova, A. Legrand, and L. Marchal. Scheduling distributed applications: the simgrid simulation framework. In *Proceedings of the 3th International Symposium on Cluster Computing and the Grid (CCGrid03)*, IEEE Computing Society Press, 2003.

On the Placement of Reservations into Job Schedules

Thomas Röblitz[1,4] and Krzysztof Rzadca[2,3,4]

[1] Zuse Institute Berlin, D-14195 Berlin-Dahlem, Germany
[2] Laboratoire ID-IMAG, Grenoble, France
[3] Polish-Japanese Institute of Information Technology, Warsaw, Poland
[4] CoreGRID Institute on Resource Management and Scheduling

Abstract. We present a new method for determining placements of flexible reservation requests into a schedule. For each considered placement the *what-if* method inserts a placeholder into the schedule and simulates the processing of batch jobs currently known to the system. Each placement is evaluated wrt. well-known scheduling metrics. This information may be used by a Grid reservation service to choose the most likely successful placement of a reservation. According to the results of extensive simulations, the *what-if* method grants more reservations and improves the performance of local jobs compared to our previously used *load* method.

1 Introduction

Reserving resources is an accepted technique for delivering Quality-of-Service (QoS) [1]. Without support for reservations, QoS-levels may be achieved by cancelling conflicting tasks [2], or by using "best effort" strategies like assigning higher priority to QoS-critical tasks [3] or predicting the future utilization of resources [4,5].

In space-sharing resource management systems the diversity of scheduling policies, the inaccurate estimates of job execution times and unknown future job submissions make predictions of future utilization imprecise. These predictions are clearly not sufficient for supporting QoS formed e.g. by Service Level Agreements.

Furthermore, the autonomy of individual Grid resources makes it difficult to coordinate complex requests such as multi-site jobs or workflows. Such coordination could be managed in a peer-to-peer manner, i.e. the sites' local resource management systems agree on the start time of the parts of a complex request. While the complexity of a single peer is small, the control of the overall behavior of multiple peers is difficult. Thus, we favor another approach which is based on the capability to reserve resources at Grid sites in advance. The coordination is therefore achieved at the Grid level by a reservation service which implicitly solves conflicts between individual sites.

As discussed in our previous work [6] a reservation request should allow fuzzy parameters, such as the requested QoS-level, the requested duration and the start and end times. Fuzzy parameters reduce the communication between the requestor and the reservation service when certain parameters may not be matched, because the reservation service may try alternative configurations automatically. In addition, the reservation scheme may let the Grid site's local resource management system express its preferences on the parameters of a request, i.e. to enforce local scheduling policies and utilization goals.

This paper presents a superior method (*what-if*) to calculate such preferences from the point of view of a Grid site's local resource management system. The *what-if* method calculates different placements of a reservation into a schedule by simulating the current workload including a placeholder for the reservation and measuring well-known scheduling metrics. Using discrete event simulations based on a workload log from the *SDSC Blue Horizon* supercomputer, we found that the *what-if* method performs better than our previously best – the *load* method.

Algorithms for placing reservations have been studied in previous work. Ernemann et al. [7] describe a method for determining available slots at Grid resources. With our method a Grid site does not only determine a list of available slots, but also calculates its preferences for them. [7] also assumes that all jobs have fixed start and end times. Therefore all jobs are in fact reservations. Our method works in situations when users submit both reservations and normal (movable) jobs. Heine et al. [8] propose two schemes for processing rigid reservations requests (with a fixed start time): (1) denying requests conflicting with jobs and (2) delaying jobs to admit more requests. In our approach requests are flexible wrt. the start time. Also, the *what-if* method allows to adjust the admission policy between the two possibilities proposed by [8]. Thus, it is easier to adapt to the systems' need. Smith et al. [9] study the use of advance reservations for co-scheduling multi-site jobs in the Grid. The reservation request specifies a start time for which the algorithm tries to make a reservation. If it fails, the next available start time is taken. As in [7], a resource does not specify its preferences for alternative start times. Through evaluation [9] concludes that backfilling, stopping and restarting jobs and more accurate execution times decrease the impact of reservations on jobs. In our work, we do not consider stopping and restarting of jobs, because this feature is not available on all systems.

The remainder of this paper is structured as follows. In Section 2 we describe basic entities in our context. Then, we briefly present an architecture for processing reservation requests in Section 3. In Section 4 we describe the methods to calculate the availability of a Grid site for placements of a reservation. We experimentally evaluate these methods by simulating a single cluster with normal jobs and reservations in Section 5. We conclude in Section 6.

2 Modelling Resources, Non-reservation Jobs and Reservations

A *resource* R is described by its number of processors R_N. We assume that the Local Resource Management System (LRMS) of a resource uses *first-come-*

first-serve *(FCFS)* scheduling with *EASY backfilling* [10] for scheduling jobs to available processors.

A *non-reservation job* (short: job) is described by its submission time j_{sbt}, its estimated execution time j_{eet} and its requested number of processors j_{np}. The LRMS has no knowledge about a job before it is submitted to the waiting queue at time j_{sbt}. A job is started by the scheduler at some time j_{stt} (start time). Usually j_{eet} is overestimated and a job finishes sooner. We denote the actual execution time of a job as j_{aet}.

The request parameters of a *reservation* are its submission time r_{sbt}, its duration r_{dur}, its earliest start time r_{est}, its latest end time r_{let} and its number of requested processors r_{np}. The reservation algorithm may place a reservation in the time interval $[r_{est}, r_{let}]$. Because there may be multiple feasible parameter sets satisfying a request, a user may specify its preferences r_{pref} to let a reservation system automatically decide which set should be chosen. When a reservation was granted, its start time and end time are denoted by r_{stt} and r_{edt}, respectively.

A scheduling event occurs when a new job or a reservation is submitted or an existing one is completed. On each such event, the LRMS schedules jobs and reservations in the following order:

1. The LRMS assigns the earliest possible start time to the first job in the waiting queue, such that it does not conflict with running jobs and existing (granted) reservations, and locks the requested processors j_{np} for the estimated execution time of the job j_{eet}.
2. All submitted (but not yet granted) reservation requests are handled in their submission order. Reservations are granted if they do not conflict with already scheduled workload and if they do not delay too much the remaining waiting jobs.
3. The LRMS assigns start times to the remaining waiting jobs using FCFS scheduling and EASY backfilling.

3 Architecture and Mechanism for Processing Reservation Requests

In this section, we briefly introduce the architecture and the reservation procedure as it was proposed in our previous work [6]. Figure 1 shows the involved components and their interaction.

The *Grid Reservation Service (GRS)* provides an interface to the clients, coordinates the processing of a request and selects the best time slot to be reserved. The *Grid Information Service (GIS)* stores information about resources in the Grid, such as the operating system, the total number of processors, the number of running jobs, etc. The *Cluster Reservation Service (CRS)* provides methods for *probing* status information and for *reserving* a time slot.

Upon reception of a request (step ①), the GRS queries the GIS for appropriate candidate resources (steps ② and ③). Then the GRS sends *probe* requests to these

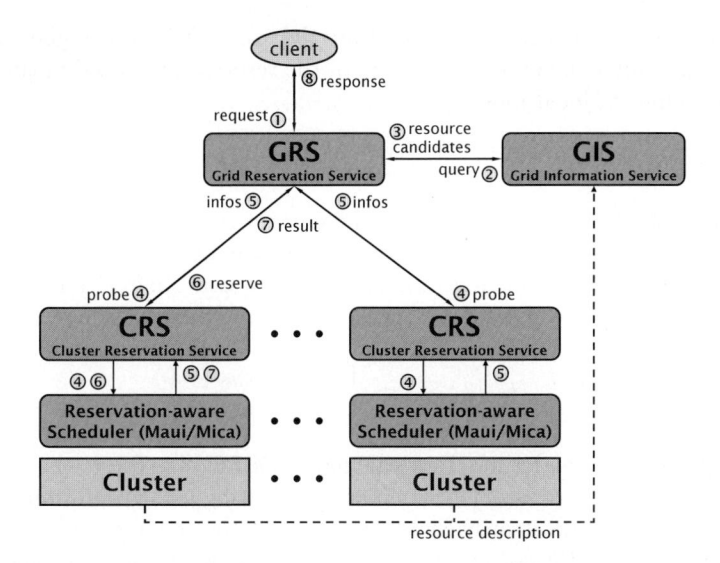

Fig. 1. Architecture and mechanism for processing reservation requests

candidates to obtain detailed information about their availability (step ④). In each candidate resource, the CRS processes the probe and returns a list of time slots including the requested status information to the GRS (step ⑤). When the GRS has received the responses from the candidates, it orders the time slots according to the user preferences. Next, it tries to reserve the best time slot by sending a *reserve* request to the corresponding CRS (step ⑥). If the request fails (step ⑦), the time slot is removed and the processing continues with step ⑥. If any *reserve* request succeeded or all failed (step ⑦), the GRS sends a response including the result to the client (step ⑧). In case of success, the result contains the agreed start time of the reservation. Otherwise, the result may indicate the reason for the failure.

4 Methods for Placing Reservations

When the CRS receives a *probe* request, it determines a limited number of time slots distributed over the time interval $[r_{est}, r_{let}]$. The number is limited by a configurable constant and a constraint on the minimum time gap between two succeeding time slots [6]. For each time slot the CRS calculates properties such as the availability of the resource, the cost for reserving, etc. In this work we concentrate on methods for determining the availability.

In our previous work [6], we presented three methods for calculating the availability of a time slot. The *load* method performed best in all experiments. Thus, we only use the *load* method in the comparison with the newly developed *what-if* method.

Generally, both the *load* method and the *what-if* method work by defining a function, respectively p_{res}^L and p_{res}^W, which ranks each possible slot for a reservation by assigning a real number from the range [0, 1]. The higher the value assigned, the minor the expected impact of the reservation placed in this slot on the performance of local jobs.

4.1 Load Method

The *load* method uses information on the current state of the system: running and waiting jobs and active or pending (but already granted) reservations. This information is used to calculate an approximate end time T for the current workload. Given a time slot $\langle ts_{begin}, ts_{end} \rangle$, we define the function p^L_{res} as follows

$$p^L_{res}(ts_{begin}, T) := \begin{cases} 1, & \text{if } ts_{begin} \geq T \\ 0, & \text{if } ts_{begin} < T \end{cases} \quad (1)$$

The approximate end time T is calculated as follows. First, the remaining execution time (per processor) for running and waiting jobs is determined. Then, it is multiplied by an arbitrarily chosen accuracy factor of 0.5 to take into account the overestimation of jobs' execution times. Next, we iteratively increase the temporary T for reservations which may be active between the current time ct and T. For each reservation r, we add to the time T the area occupied by this reservation $r_{np} \cdot (r_{edt} - \max(r_{stt}, ct))$ divided by the total number of processors in the system R_N.

Although the value for T is only a rough approximation, the method proved to be reasonable in our previous experiments. However, time slots between the current time ct and T are not considered for reservations. Existing reservations starting later than T are also not taken into account. The latter may lead to failing reservation attempts in the presence of existing advance reservations.

4.2 What-If Method

The basic idea of the *what-if* method is to let the availability reflect the impact of a reservation on the non-reservation jobs. For this purpose, the local scheduling system must be able to construct execution plans for jobs without executing them. This requirement is, however, not very restricting, as commonly-used cluster-level schedulers, such as Maui [11] or OAR [12], either provide a simulation mode or can operate in planning mode.

The *what-if* method uses three kinds of execution plans – *original (ORG)*, with *reservation placeholder (RSV)* and with *job placeholder (JOB)*.

Original: The execution plan P^{ORG} for the current workload.
Reservation placeholder: Given a time slot $[ts_{begin}, ts_{end}]$ and a requested number of processors r_{np}, it places a temporary reservation for r_{np} processors from ts_{begin} until ts_{end} into the system. Then it determines the execution plan $P^{RSV/ts_{begin}}$. This procedure is repeated for all time slots.
Job placeholder: The execution plan determines the time when a job with the same requirements would have been started. Therefore a temporary job with an estimated execution time j_{eet} equal to the duration of the reservation

request r_{dur} and the same number of requested processors is submitted to the system. The resulting execution plan P^{JOB} defines a new time slot by setting its properties as follows: $ts_{begin} := j_{stt}, ts_{end} := j_{stt} + j_{eet}$.

Next, the algorithm calculates the availability p_{res}^W for the execution plans $P^{\{JOB,RSV/ts_{begin}\}}$. The availability represents the quality of an execution plan wrt. the well-known scheduling metrics. In this work we measure the makespan and the average completion time of the jobs, which are normalized to map onto the interval $[0,1] \subset \mathbb{R}$. The execution plan P^{ORG} is used as reference in the normalization, possibly defining optimal values for the scheduling metrics. Other well-known scheduling metrics such as slowdown or resource utilization can be easily added if needed. If a time slot cannot be reserved by a reservation placeholder – because it conflicts with running jobs, the first job at the head of the waiting queue (EASY backfilling) or existing reservations – its availability is set to zero.

Let K be the number of jobs in the current workload and P be one of the above execution plans. The start time of job j^i in the execution plan P is denoted by the term $stt(P, j^i)$ where $1 \le i \le K$.

In order to assess the quality of a simulated execution plan, the makespan and the average completion time is computed. The makespan $C_{max}(P)$ of an execution plan P is defined as

$$C_{max}(P) := \max_{1 \le i \le K}(stt(P, j^i) + j^i_{eet}) \qquad (2)$$

C_{max}, stating how long the resource will be occupied, ranks proposed execution plans from the point of view of the owner. Let C^*_{max} denote the minimum makespan for all considered execution plans $P^{\{JOB,RSV/ts_{begin}\}}$.

The average completion time $C_{avg}(P)$ of the jobs in an execution plan P is defined as

$$C_{avg}(P) := \frac{1}{K} \sum_{1 \le i \le K} (stt(P, j^i) + j^i_{eet} - j^i_{sbt}) \qquad (3)$$

C_{avg} expresses how fast on average the jobs are completed and thus rates plans from the point of view of resource's users. Let C^*_{avg} denote the minimum average completion time for all considered execution plans $P^{\{JOB,RSV/ts_{begin}\}}$.

Let $\omega_{C_{max}}$ denote the weight for the makespan ($\omega_{C_{max}} \ge 0, \omega_{C_{max}} \in \mathbb{R}$) and $\omega_{C_{avg}}$ denote the weight for the average completion time ($\omega_{C_{avg}} \ge 0, \omega_{C_{avg}} \in \mathbb{R}$). We require that $\omega_{C_{max}} + \omega_{C_{avg}} = 1$. Considering a time slot $\langle ts_{begin}, ts_{end} \rangle$, the availability $p_{res}^W(ts_{begin}, P)$ of an execution plan is computed as:

$$p_{res}^W(ts_{begin}, P) := \omega_{C_{max}} \cdot \frac{C^*_{max}}{C_{max}(P)} + \omega_{C_{avg}} \cdot \frac{C^*_{avg}}{C_{avg}(P)} \qquad (4)$$

The unweighted parts, both for the makespan and for the average completion time are normalized to fit into the interval $[0,1] \subset \mathbb{R}$. Thus the availability is a number in the interval $[0,1] \subset \mathbb{R}$ too. The more a reservation at ts_{begin} delays the execution of the local jobs, the lower is the availability p_{res}^W.

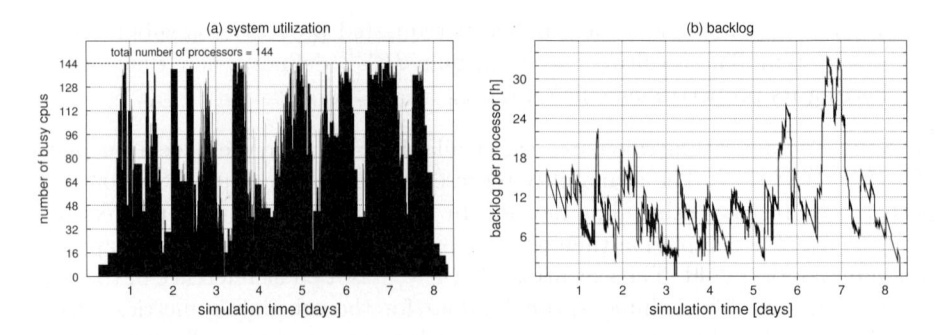

Fig. 2. Non-reservation workload characteristics: (a) system utilization, (b) backlog

5 Experimental Evaluation

We evaluated the presented methods with a simulation based on a workload log of the *SDSC Blue Horizon* supercomputer published in [13]. Because we found it difficult to implement the *what-if* simulations in Maui, we developed our own scheduler Mica (MICrotus Arvalis, field mouse). In the following sections we will briefly describe the simulation environment including Mica. Then, we describe the workloads used in the experiments. Last, we present the results in detail.

5.1 Simulation Settings

In order to measure the performance of reservations accurately, there was one GRS and one CRS in our testing environment (cf. Figure 1). Instead of interfacing a real system, the CRS communicated with the simulation scheduler Mica. Mica is a simple scheduler capable of FCFS scheduling with EASY backfilling. Mica allows users to submit reservations and normal, parallel jobs. In addition, it can perform what-if simulations for both job and reservation requests.

We used the same workloads as in our previous work [6]. Thus, we were able to compare the behavior of the Mica and Maui schedulers. These workloads were generated as follows. We extracted the first 2000 sound jobs of a workload log of the *SDSC Blue Horizon* supercomputer published in [13]. The workload log contains non-reservation jobs only. Fig. 2 shows (a) the utilization of the processors during the simulation and (b) the backlog[1] of the jobs (right). The complete workload lasts about 12.5 simulation days. Because there are only 5 jobs during the first 4.5 days (388800 s), we only show the data for jobs processed after simulation time 370000 s.

The reservation workload was generated by converting 10% of the jobs into reservations. We split the 2000 jobs into blocks of ten jobs (based on consecutive submission times). From each block we selected a single job, removed it from the

---

[1] The backlog is defined as $\left(\sum_{j \in RUN} (j_{ret} \cdot j_{np}) + \sum_{j \in WAIT} (j_{eet} \cdot j_{np}) \right) / R_N$ with $j_{ret} := j_{eet} - (ct - j_{stt})$ being the remaining execution time of a running job.

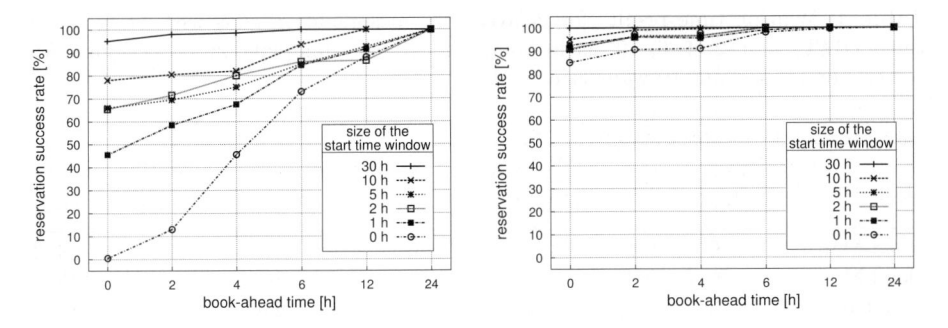

Fig. 3. The reservation success rate for the methods *load* (left) and *what-if* (right)

non-reservation workload and converted it into a reservation request. In such a request r_{sbt}, r_{np} were copied from the original job. The duration r_{dur} was set to the job's actual execution time j_{aet}. An important parameter of a reservation request is the book-ahead time or advance notice time which is derived as $r_{est} - r_{sbt}$. In order to measure the impact of the reservation's parameters on the algorithm, within a particular workload all reservation requests had the same book-ahead time $bat \in \{0, 2, 4, 6, 12, 24\}$ hours and the same size of the start time window $stw \in \{0, 1, 2, 5, 10, 30\}$ hours. For each request, we set its earliest start time as $r_{est} = r_{sbt} + bat$ and its latest end time as $r_{let} = r_{est} + r_{dur} + stw$. We tested all 36 combinations for each reservation placement method.

5.2 The Reservation Success Rate

The reservation success rate can be defined as a percentage of submitted reservations which were granted by the system. Fig. 3 shows the results in detail for the *load* method (left) and the *what-if* method (right). The average success rate for all workloads is 97% with the *what-if* and 80% with the *load* method. When reservation requests have small start time windows and book-ahead times (in scenarios when both parameters are up to 2 hours), the *what-if* method grants 92% of submitted reservations. The *load* method's success rate is only 38%.

Similarly, we observed that the *what-if* method yields significantly better results than the *load* method when the system load is high. For each reservation request r we have defined the load L_r as the backlog of jobs and reservations at the time r_{sbt}. We have ordered the reservations according to the load L_r. Considering the top 20% reservation requests (i.e. submitted at the highest load situations), the *what-if* method granted 92% of such requests, whereas *load* method – only 58%.

5.3 The Impact of Reservations on Jobs

We use two metrics to demonstrate the impact of reservations on jobs: the makespan (more system oriented) and the waiting time of jobs (more user oriented). The makespan resulting from admitting reservations scheduled by either method was extended by 1% to 8% (approx. 60000 s), compared with the

Table 1. Waiting time results for delayed jobs in simulation runs with book-ahead times (BAT) = 0, 2, 4, 6, 12, 24 in hours and a start time window of 30 hours. All reservation requests (200) were successful.

Experiment	BAT [h]	#jobs	⌀ Waiting time [s] original	⌀ Waiting time [s] affected	BAT [h]	#jobs	⌀ Waiting time [s] original	⌀ Waiting time [s] affected
load-Maui	0	186	3191	9296	2	276	2225	6442
load-Mica	0	284	2532	7853	2	341	2138	11572
what-if-Mica	0	172	4155	8112	2	280	3594	8744
load-Maui	4	408	1525	7514	6	356	1939	8436
load-Mica	4	372	1782	11167	6	357	1984	11036
what-if-Mica	4	319	3491	9193	6	429	2426	11081
load-Maui	12	383	1901	8957	24	422	1872	17051
load-Mica	12	390	2244	15405	24	411	1638	12832
what-if-Mica	12	393	1926	16649	24	407	1646	13011

makespan of the workload without any reservation. We found that the *what-if* method performed better than the *load* method in most combinations of book-ahead time and start time window. We do not present the full results because of space limitations.

The waiting time of jobs expresses the performance of the system from the users' point of view. The more jobs are delayed wrt. to their waiting time if no reservations were admitted, the lower is the acceptance of reservations by the users of a system. We only present results for the workloads with a start time window of 30 hours (and different book-ahead times), because all reservation requests were granted with these parameters.

Table 1 shows the number of delayed jobs (#jobs), the average of the original waiting time for these jobs and the average of the affected waiting time. For each book-ahead time we show the results for the experiments load-Maui, load-Mica and what-if-Mica.

The results for the schedulers Maui and Mica using the *load* method differ significantly. Both schedulers show the tendency of delaying more jobs with larger book-ahead times. While the average original waiting times are similar, the difference in the average affected waiting times is much larger. We account the simplified nature of Mica for this.

Comparing the results for the experiments load-Mica and what-if-Mica, the *what-if* method performs better for small book-ahead times (0, 2 and 4 hours). Both the number of delayed jobs and the additional waiting time for these are significantly smaller. With a book-ahead time of 6 hours, more jobs were delayed by the *what-if* method, but the additional waiting time per job was similar to the *load* method. The *load* method performed slightly better than *what-if* in the experiment with 12 hours book-ahead time. With a book-ahead time of 24 hours the results were nearly identical for both experiments.

5.4 Detailed Analysis of the Job Delays

Fig. 4 shows the cumulated additional waiting time of the delayed jobs for the methods *load* (left) and *what-if* (right) using the same experimental settings as in Section 5.3. For each job the additional waiting time was cumulated at the start time of the job.

The sharp increases of the curves show situations in which blocking jobs or reservations finished and many waiting jobs could start in parallel. Studying the logs of the experiments we found a specific pattern for these situations illustrated in Fig. 5. At the current time 'now' the jobs RJ_1, RJ_2, RJ_3 and RJ_4 are being executed. Already granted (advance) reservations are shown along the horizontal axis (boxes with 'RSV'). These reservations were small in their duration and the number of processors. The waiting jobs (WJ_1, WJ_2 and WJ_3) are planned to start after the currently last reservation.

Note, in particular, the very large job WJ_1 which is planned to start at the end of the last reservation. The number of processors this job requested is close to the total available number of processors R_N. Therefore the job is either blocked

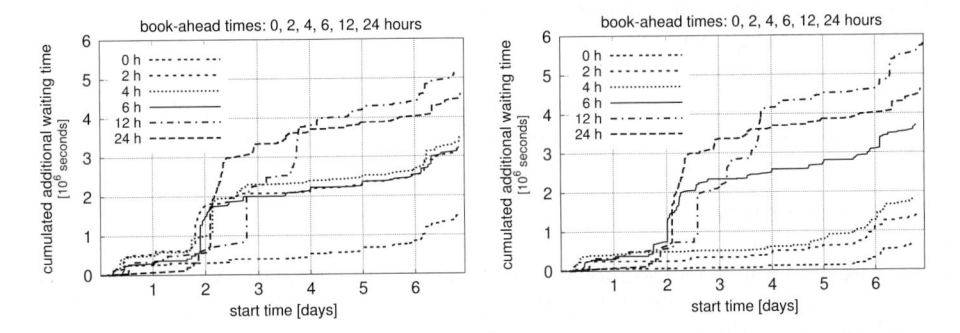

Fig. 4. Cumulated additional waiting time of delayed jobs for the methods *load* (left) and *what-if* (right). The additional waiting time of a job is cumulated at its start time.

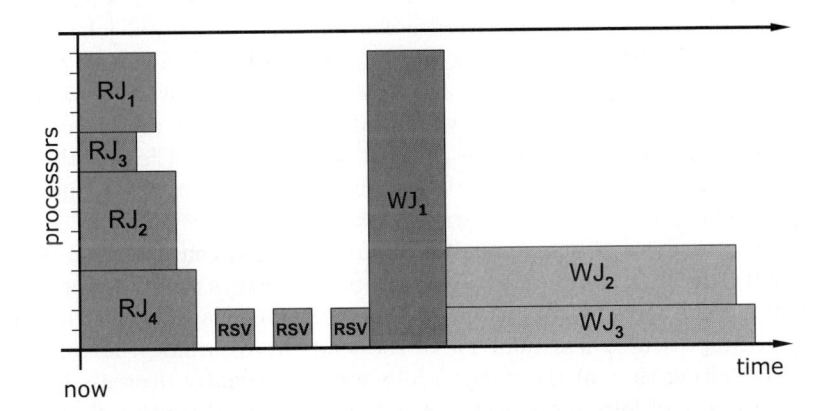

Fig. 5. Pattern causing significant delays of jobs

by already running jobs or by the sequence of reservations. With EASY backfilling, the job is planned into the schedule to prevent its starvation, after it has advanced to the head of the waiting queue (cf. Section 2). The utilization decreases as the time approaches the scheduled start of the large job. The remaining waiting jobs require more processors than the large job leaves available and their estimated execution time is larger than the time remaining to the start of the large job. Therefore, if they were executed before the large job, the large job would be delayed.

Generally, the further in the future reservations may be placed into the system (latest time is given by the book-ahead time plus the size of the start time window), the later a large job will be planned in. Thus, the unused area in the front of a large job will increase as well. The situation may be improved by the following means:

Job size limitation: We repeated the experiments with changed non-reservation workloads. We restricted all jobs to request less than 89 processors. The job delays were significantly improved and the limitation affected less than 10 jobs (out of 1800).

Improving execution time estimates: The influence of inaccurate execution time estimates was extensively studied in other work [14]. Obviously, jobs with more accurate execution time estimates could fit in the hole before a large job. The situation observed will not disappear completely, but the number of delayed jobs and the extent of their delays could be reduced.

6 Conclusion

We presented a new method for placing reservations into a schedule. The *what-if* method simulates the schedule with different placements of the reservation and ranks them according to well-known scheduling metrics, such as makespan and average completion time.

The performance of the *what-if* method was compared with the best method from our previous work, the *load* method, wrt. the reservation success rate, the number of delayed jobs, the average waiting time of delayed jobs and the distribution of the delays. In general, the *what-if* method performs better than the *load* method, except for large book-ahead times and start time windows, where both methods perform equally well. More reservations are granted and, at the same time, the performance of local jobs is better. Clearly, the main advantage of the *what-if* method is the ability to place reservations into holes in the schedule.

A detailed analysis of the simulation data revealed a characteristic pattern when a job requesting a large fraction of processors delays a number of smaller jobs. This pattern accounts for a large fraction of the job delays. When the largest jobs (which represented less than 1% of the load) were shrank, the cumulated additional waiting time of the delayed jobs was significantly decreased. Beside shrinking the size of jobs, more accurate estimates for the execution time of jobs could lower the impact generated by this blocking pattern.

Acknowledgments

The work of the first author was partly funded by German BMBF project AstroGrid-D. Krzysztof Rzadca was partly funded by French CNOUS grant 20045874. This collaborative research work started in the CoreGRID Institute on Resource Management and Scheduling funded by the European Commission (Contract IST-2002-004265). We thank Alexander Reinefeld and the anonymous reviewers for their helpful comments to improve the quality of the paper.

References

1. Foster, I., Kesselman, C., Lee, C., Lindell, R., Nahrstedt, K., Roy, A.: A Distributed Resource Management Architecture that Supports Advance Reservations and Co-Allocation. In: Proceedings of the International Workshop on Quality of Service, IEEE Press: Piscataway, NJ (1999) 27–36
2. Burchard, L.O., Heiss, H.U., Linnert, B., Schneider, J., Kao, O., Hovestadt, M., Heine, F., Keller, A.: The Virtual Resource Manager: Local Autonomy versus QoS Guarantees for Grid Applications. In: Future Generation Grids. Volume 2 of CoreGrid. (2006) 83–98
3. Blake, S., Black, D., Carlson, M., Davies, E., Wang, Z., Weiss, W.: An Architecture for Differentiated Service. RFC 2475 (Informational) (1998) Updated by RFC 3260.
4. Dinda, P.A.: A Prediction-Based Real-Time Scheduling Advisor. In: IPDPS '02: Proceedings of the 16th International Parallel and Distributed Processing Symposium, Washington, DC, USA, IEEE Computer Society (2002) 10–17
5. Downey, A.B.: Using Queue Time Predictions for Processor Allocation. In: IPPS '97: Proceedings of the Job Scheduling Strategies for Parallel Processing, London, UK, Springer-Verlag (1997) 35–57
6. Röblitz, T., Schintke, F., Reinefeld, A.: Resource Reservations with Fuzzy Requests. Concurrency and Computation: Practice and Experience (to appear)
7. Ernemann, C., Yahyapour, R.: Applying Economic Scheduling Methods to Grid Environments. In: Grid Resource Management - State of the Art and Future Trends, Kluwer Academic Publishers (2003) 491–506
8. Heine, F., Hovestadt, M., Kao, O., Streit, A.: On the Impact of Reservations from the Grid on Planning-Based Resource Management. In: Proc. of the International Workshop on Grid Computing Security and Resource Management (GSRM 2005). Volume 3516 of Lecture Notes in Computer Science., Atlanta, USA, Springer-Verlag (2005) 155–162
9. Smith, W., Foster, I., Taylor, V.: Scheduling with Advanced Reservations. In: Proceedings of the 14th International Symposium on Parallel and Distributed Processing, Cancun, Mexico, Washington, DC, USA, IEEE Computer Society (2000) 127–132
10. Lifka, D.A.: The ANL/IBM SP Scheduling System. In: IPPS '95: Proceedings of the Workshop on Job Scheduling Strategies for Parallel Processing, London, UK, Springer-Verlag (1995) 295–303
11. Jackson, D.B., Snell, Q., Clement, M.J.: Core Algorithms of the Maui Scheduler. In Feitelson, D.G., Rudolph, L., eds.: JSSPP. Volume 2221 of Lecture Notes in Computer Science., Springer-Verlag (2001) 87–102

12. Capit, N., Costa, G.D., Georgiou, Y., Huard, G., Martin, C., Mouni, G., Neyron, P., Richard, O.: A batch scheduler with high level components. In: Proceedings of the IEEE International Symposium on Cluster computing and Grid 2005 (CCGrid05). Volume 2. (2005) 776–783
13. Feitelson, D.G.: Parallel Workloads Archive. http://www.cs.huji.ac.il/labs/parallel/workload/ (2006)
14. Chiang, S.H., Arpaci-Dusseau, A.C., Vernon, M.K.: The Impact of More Accurate Requested Runtimes on Production Job Scheduling Performance. In: JSSPP '02: Revised Papers from the 8th International Workshop on Job Scheduling Strategies for Parallel Processing, London, UK, Springer-Verlag (2002) 103–127

A Practical Approach of Diffusion Load Balancing Algorithms

Emmanuel Jeannot[1] and Flavien Vernier[2]

[1] INRIA - LORIA, campus scientifique, BP 239, 54506 Vandoeuvre les Nancy, France
emanuel.jeannot@loria.fr
[2] UHP Nancy-I - LORIA, campus scientifique, BP 239, 54506 Vandoeuvre les Nancy, France
flavien.vernier@loria.fr

Abstract. In this paper, a practical approach of diffusion load balancing algorithms and its implementation are studied. Three problems are investigated. The first one is the determination of the load balancing parameters without any global knowledge. The second problem consists in estimating the cost and the benefit of a load exchange. The last one studies the convergence detection of the load balancing algorithm. For this last point we give an algorithm based on simulated annealing to reduce the convergence towards a load repartition in steps that can be done with discrete loads. Several simulations close this paper and illustrate the impact of the various methods and algorithms introduced.

1 Introduction

One of the most important problems in distributed processing consists in balancing the work load among all processors. The purpose of load (work) balancing is to achieve better performances of distributed computations, by improving load allocation. The load balancing problem was studied by several authors from different points of view [1,2,3,4,5,6,7].

In this paper we focus our study on the iterative load balancing algorithms introduced in [1]. These kinds of algorithms assume that a node manages its load only with its nearest neighbors. They are generic algorithms, useful when the system is decentralized or when some nodes cannot directly communicate with all the other nodes. However these algorithms face several problems. Firstly, the majority of studies about these algorithms use a global knowledge, like the network or the nodes properties, to determine the load balancing parameters. Secondly, most of these algorithms assume that balancing the load is always beneficial and leads to a reduction of the execution time. Thirdly, since the load is not infinitely divisible, the final load balancing (after convergence of the algorithm) can face a *step* problem.

In this paper we propose a practical approach of load balancing that solves the 3 above problems. They are 3 main problems that can appear in an implementation of these load balancing algorithms. To the best of our knowledge no load balancing algorithm of the literature can deal with these 3 issues at the same time. We give methods to determine the diffusion parameters without any global knowledge. We propose an analysis of the cost and benefit of load exchange in order to determine when it is worth exchanging some load. The convergence of the load balancing algorithms with no infinitely divisible loads is also studied in this paper. Finally, the given methods are efficient and easy to implement.

It is important to note that, in this work, we make very few assumptions. We can deal with either static or dynamic load. The network topology can be of any type as long as it is connected. Nodes and networks can be homogeneous or heterogeneous. The notion of load is very abstract, it can be anything that just requires some time to be processed (data, etc.). The proposed methods deal with static networks but the adaptation to dynamic networks is straight forward. Finally, no global knowledge is required to process the algorithm. The knowledge is limited to the neighborhood. The obtained results give in the better case, performance gains greater than 100% and the algorithm does not always use all the available resources: it is able to find the right amount of resources that gives a good speed-up.

This paper is organized as follows. Section 2 presents the related works, we review the diffusion on any static network. In Section 3.1, we study the problem of the connection links heterogeneity. Section 3.2 presents a decentralized method to compute the load balancing parameters. Section 3.3 is dedicated to the not infinitely divisible loads and to the detection of the convergence of the load balancing algorithm. In Section 4 we illustrate the behavior of the load balancing algorithm according to the methods that we give by some experimentation.

2   Related Work

The studied algorithms are generally dedicated to static networks. A static network topology is classically represented by a simple undirected connected graph $G = (V, E)$, where V is the set of vertexes and E is the set of edges, $E \subseteq V \times V$. Each processor is a vertex of the graph and each communication link between two processors i, j is the edge $(i, j) \in E$ between the two vertexes i and j ($i, j \in V$). Vertexes are labeled from 1 to n where n is the number of processors, hence $|V| = n$. Let m be the number of communication links ($|E| = m$). Let F be the vector of edge-weight and let us note $f_{i,j}$ the weight of edge (i, j) ($f_{i,j} = F_k | E_k = (i, j)$). Let C_n be the vector of node-weight such that the average of C_{n_i} is normalized $\frac{\sum_i C_{n_i}}{n} = 1$.

In [1], Cybenko introduced a distributed load balancing (LB) algorithm for static networks called the diffusion algorithm or FOS (First Order Scheme). It assumes that a process i balances its load simultaneously with all its neighbors. To balance the load, a ratio α_{ij} of the load difference between the process i and j is swapped between i and j. In the general case - on heterogeneous networks - the LB step of a process i with all its neighbors is given by Equation (1) where $w_i^{(t)}$ is the work load done by process i at time t.

$$w_i^{(t+1)} = w_i^{(t)} - \sum_j \alpha_{i,j} \cdot f_{i,j} \cdot \left(\frac{w_i^{(t)}}{C_{n_i}} - \frac{w_j^{(t)}}{C_{n_j}} \right) \quad (1)$$

Equation (1) is linear and thus it can be re-written in matrix form: $W^{(t+1)} = M^T W^{(t)}$, where $W^{(t)}$ is the vector $(w_i^{(t)})$ and M is the diffusion matrix defined by

$$m_{ij} = \begin{cases} \frac{\alpha_{ij} f_{i,j}}{C_{n_j}} & \text{if } (i,j) \in E \wedge i \neq j, \\ 1 - \sum_k m_{ik} & \forall k | (i,k) \in E \wedge i = j, \\ 0 & otherwise. \end{cases}$$

This algorithm has often been studied and derived - Dimension Exchange Algorithm [1,2,8], Second Order Scheme [9,4], dynamic networks [10,11,12] ...

In the literature, various methods can be found that determine these parameters $\alpha_{i,j}$ or $f_{i,j}$. There are three classical methods to compute α: Cybenko Choice [1], Boillat Choice [5] or optimal Choice [13]. The optimal Choice and the Cybenko Choice need a global knowledge of the network and the Boillat Choice only needs a knowledge of neighbors degree to determine α: $\alpha_{ij} = \frac{1}{\max(d(i),d(j))+1}$, where $d(i)$ is the degree of node i at time t. The parameter $f_{i,j}$ must be determined according to the constraints of the diffusion matrix M, M must be stochastic, irreducible and aperiodic [1].

3 A Decentralized Practical Approach

3.1 Cost and Benefit of Load Balancing

Let us start by defining the cost and the benefit of a LB algorithm. The cost is the time lost by exchanging the load, it is generally due to communication. The benefit is the time gained by exchanging the load, it is due to a better balance. In Equation (1) the parameter $f_{i,j}$ corresponds to the weight of edge (i,j). Hence, the parameter $f_{i,j}$ must be determined such that the cost of the LB algorithm is lower than the benefit given by the exchange of load. In our practical approach, $f_{i,j}$ is in $\{0,1\}$. If the cost of an exchange $L_{ij}^{(t)}$ between i and j is greater than its benefit, then $f_{i,j}$ is set to 0 and there is no exchange between i and j, otherwise $f_{i,j}$ is set to 1. It can be noted that by this definition $f_{i,j}$ depends on the time, hence it becomes $f_{i,j}^{(t)}$ and its corresponding vector F becomes $F^{(t)}$.

The cost and the benefit of an exchange depends on the size of this exchange. To determine the cost of an exchange we give the following equation,

$$\text{Cost}(L_{ij}^{(t)}) = \text{PreExcCost}(|L_{ij}^{(t)}|) + \text{ExcCost}(|L_{ij}^{(t)}|) + \text{PostExcCost}(|L_{ij}^{(t)}|).$$

The cost of a load exchange $L_{ij}^{(t)}$ is the time to prepare this load for the exchange ($\text{PreExcCost}(|L_{ij}^{(t)}|)$), plus the time of the exchange ($\text{ExcCost}(|L_{ij}^{(t)}|)$), plus the time to integrate it on the receiver ($\text{PostExcCost}(|L_{ij}^{(t)}|)$). PreExcCost and PostExcCost completely depend on the application. ExcCost only depends on the load $L_{ij}^{(t)}$ and on the edge (i,j), a good estimation of this cost can be: ExcCost($|L_{ij}^{(t)}|$) = Lat$_{ij}$ + $\frac{|L_{ij}^{(t)}|}{\text{Bw}_{ij}}$, where Lat$_{ij}$ and Bw$_{ij}$ are respectively the latency and the bandwidth of edge (i,j). Let us note that the communication can always be hidden some computation.

The benefit given by the exchange of $L_{ij}^{(t)}$ can be estimated by the computation time on i and j without exchange minus the computation time on i and j after this exchange. Intuitively the benefit of a load exchange must be positive if the computation time is reduced by this exchange and negative in the other case. Let us recall that the computation time on i and j is given by the maximum between the computation time on i and the computation time on j. The following equation gives the benefit for the cases - $L_{ij}^{(t)} > 0$ and $L_{ij}^{(t)} < 0$.

$$\text{Benefit}(L_{ij}^{(t)}) = \begin{cases} \max(\text{Cp}(w_i^{(t)}), \text{Cp}(w_j^{(t)})) \\ \quad - \max(\text{Cp}(w_i^{(t)} - L_{ij}^{(t)}), \text{Cp}(w_j^{(t)} + L_{ij}^{(t)})) & \text{if } L_{ij}^{(t)} > 0, \\ \max(\text{Cp}(w_i^{(t)}), \text{Cp}(w_j^{(t)})) \\ \quad - \max(\text{Cp}(w_i^{(t)} + L_{ij}^{(t)}), \text{Cp}(w_j^{(t)} - L_{ij}^{(t)})) & \text{if } L_{ij}^{(t)} < 0, \end{cases} \quad (2)$$

where $\text{Cp}(w_i^{(t)} + ...)$ is the computation time of $(w_i^{(t)} + ...)$ on i.

In the iterative LB algorithms, the benefit of an exchange of load at a given iteration can increase with the next iterations. The estimation of the benefit that we give in Equation (2) is evaluated on only one iteration. Hence, a parameter k is introduced to estimate the benefit on the k successive iterations after an exchange. Indeed, $f_{i,j}^{(t)}$ is equal to 1 if and only if $\text{Cost}(L_{ij}^{(t)}) < k*\text{Benefit}(L_{ij}^{(t)})$. The parameter k can be constant or not (in Section 4 the impact of both cases are compared).

One limit of this cost/benefit system appears when the algorithm converges to a load repartition in step. This problem is studied in Section 3.3.

3.2 Parameter Computation

From the general equation of FOS we have determined the parameter $f_{i,j}^{(t)}$ in the previous section. Now, let us study the parameters $\alpha_{i,j}$ and C_{n_i}. In Section 2 we have seen that only the Boillat Choice does not need a global knowledge to compute $\alpha_{i,j}$, but this method is limited to homogeneous networks.

In this section, a method that only needs a local knowledge is given to determine the relation $\frac{\alpha_{i,j}}{C_{n_i}}$. Let us denote C the vector of the processors speeds. Let C_r be the matrix of relative speeds defined by $C_{r_{i,j}}$, the relative speed of j compared to i:

$$C_{r_{i,j}} = \begin{cases} \frac{C_j}{C_i + C_j} & (i,j) \in E, j \neq i \\ 0 & \text{otherwise}. \end{cases}$$

Thus the unit of C is not important, it can be MHz, Mflops or any other. With this definition of a relative speed matrix a diffusion matrix that we denote M_r can be given. M_r is defined such that:

$$M_{r_{ij}} = \begin{cases} \min(\delta_i C_{r_{i,j}}, \delta_j C_{r_{i,j}}) & j \neq i \\ 1 - \sum_{j(j \neq i)} M_{r_{ij}} & j = i \end{cases}$$

with $\delta_i = \frac{1}{\sum_j C_{r_{i,j}}}$. By construction, it is easy to show that $M_{r_{ij}} \geq 0$ and $\sum_j M_{r_{ij}} = 1$, in other words the matrix M_r is stochastic.

Theorem 1. *The diffusion LB algorithm with M_r as diffusion matrix converges toward a load distribution relative to the node speed if and only if M_r is irreducible and aperiodic - the graph G must be connected and non-bipartite.*

Proof. If M_r is stochastic, irreducible and aperiodic, thus the Perron-Frobenius Theorem can be applied, i.e. $\exists \mu$ (μ is a fixed point vector) such that $M_r^T \mu = \mu$. By construction of M_r is stochastique and $M_r^{T\infty}$ tends to the matrix in which each column is $\frac{1}{\sum_i C_i} C$, thus $\mu = hC$ where h is such that $\sum_i w_i^{(0)} = h \sum_i C_i$. Thus for a given $W^{(0)}$ the invariant distribution μ is proportional to C.

As shown by Theorem 1, the LB algorithm converges if the network G is connected and non-bipartite. The connectivity of the network depends on the set E and the network is not bipartite if M_r is well constructed. The previous method does not ensure that the network is not bipartite, to ensure that we can use the following definition to compute $C_{r_{i,j}}$:

$$C_{r_{i,j}} = \begin{cases} \frac{C_j}{C_i + C_j} & (i,j) \in E, j \neq i \\ \frac{C_i}{2C_i} & j = i \\ 0 & otherwise. \end{cases}$$

To build the diffusion matrix M with one of these two methods and the cost/benefit defined in Section 3.1, the vector $L_r^{(t)}$ of load exchange prediction must be defined to compute $f_{ij}^{(t)}$. $L_r^{(t)}$ is given by $L_{r_{ij}}^{(t)} = M_{r_{ij}} w_i^{(t)} - M_{r_{ji}} w_j^{(t)}$. With $F^{(t)}$ and M_r defined, the diffusion matrix $M^{(t)}$ - as F depends on t, M depends on t - is given by:

$$m_{ij}^{(t)} = \begin{cases} f_{ij}^{(t)} M_{r_{ij}} & j \neq i, \\ 1 - \sum_{k(k \neq i)} f_{ik}^{(t)} M_{r_{ik}} & j = i. \end{cases}$$

3.3 Convergence Detection with Unit Size Tokens

The last step that we study in this paper is the termination of the LB algorithm. This step consists in detecting the end of the LB algorithm to stop it and avoid the cost of exchange of information done by the LB algorithm. This cost can be important if the network is slow and if the number of neighbors is high.

The main problem to detect the convergence is that the load is not infinitely divisible for the real applications. This implies that the LB algorithm cannot always reach a uniform load distribution, hence it does not always reach the convergence point. Some steps of load can appear in the system that can block the LB algorithm.

The unit size tokens problem. Let us start by eliminating this step problem. In the literature, the LB problem of indivisible unit-size tokens is studied in [9] where the authors introduced the "I Owe You" (IOU) unit on each edge, and in [14] where the authors introduced a randomized algorithm that deals with heterogeneous networks. In this section, a new approach based on simulated annealing algorithms is used. The objective is to shake the system to move the load of the most loaded nodes toward the least loaded nodes when the classical LB algorithm is blocked. Hence, the algorithm operates as follows: if a node i is unbalanced with its neighbor j and no load is exchanged between these two nodes, a random value denoted alea is drawn between 0 and 1 (0 <alea< 1), and if alea $< e^{(-\kappa * U_{ij})}$, a part of load is exchanged. U_{ij} denotes the number of successive LB iterations during which the neighbors nodes i and j are unbalanced and do not exchange load. The parameter κ defines the probability to exchange load and can be defined by $\kappa = \frac{ln(p)}{\tau}$ where p is the probability to exchange load at the iteration τ of U_{ij}. For example if 50% of probability to exchange is wanted at the second iteration $\kappa = \frac{ln(0.5)}{2}$. Let us note that this method does not ensure to reach the uniform load distribution but it can reduce the unbalance.

Convergence detection problem. Let us recall that the first problem presented in this section is the convergence detection of the LB algorithm. Hence, we must detect that no more load is exchanged in the network. In [15] the authors give a decentralized convergence detection algorithm dedicated to parallel iterative asynchronous algorithms. This algorithm is based on the leader election on the IEEE-1394 (FireWire) protocol, and this base can be used to detect a global state in synchronous algorithms without any centralization. These algorithms operate on a tree, hence a spanning tree of the network must be defined [16,17,18,19].

For the LB algorithm, an adaptation of algorithm given in [15] is used. This adaptation is synchronous and dedicated to binary state detection. The idea of this algorithm is as follows: each node i defines k channels where k is the number of neighbors of i. In the first stage of the algorithm, if a node has only one channel that is not associated to a neighbor, it associates this channel to its neighbor that has no channel and defines this neighbor as its father and sends to its father the state of its sub-tree. If a node receives the state of a sub-tree from a neighbor, it associates a channel to this neighbor and defines this neighbor as one of its children. It is obvious that the leaf nodes of the spanning tree have exactly one such channel at the start of the protocol. Hence, the algorithm is started by a leave that sends its state to its father. In the second and last stage of the algorithm, when a node i has all its channels associated to all its neighbors and that all its neighbors are its children, this node i is the root of the tree. Hence, the state of its sub-tree is the state of the tree, in other words, this node detects the global state of the system. It sends this global state to its children and they do likewise with their children and so on. Thus the information of the global state goes through all the network.

To finish with, if the convergence is detected, the LB algorithm can be stopped if the load is static. In the other case - dynamic load, dynamic networks or other ...

- the convergence detection algorithm can be used to reduce the frequency of LB steps if the system became stable or increase it if it became unbalanced.

4 Simulations

The following simulations are realized with SimGrid [20]. The application that is balanced is represented by an integer that corresponds to the load of the application. Let us recall that the load is static to illustrate the convergence of the algorithm and it is considered homogeneous.

The behavior of the FOS algorithm is studied on the worse configuration - a line topology with all the load on the first node - with 64 homogeneous nodes (2000MFlops). The program that is balanced can be viewed as a parallel and iterative numerical solver that computes 1000 iterations where the topology is virtual and depends on the data dependency - communication for data dependency are simulated. This study is realized for two cases, a first one when the network is a LAN and a second one when the network is a DSL.

4.1 Fast Network

In the former, a bandwidth of 100Mb/s is used with 0.15ms of latency on each edge. Figure 1 shows the gain given by the FOS algorithm with the cost/benefit system and with convergence detection ($Algo2$) compared to the FOS algorithm without cost/benefit system and without convergence detection ($Algo1$). Let us note that in $Algo2$ the cost/benefit parameter k is given by $k^{(t+1)} = k^{(t)} - 1$ with $k^{(0)} = 1000$. The gain is given by $\frac{T1-T2}{T1}$, where $T1$ and $T2$ are the computation time of $Algo1$ and $Algo2$, respectively. In this figure, the gain depends on the load average w^* - the global load is given by $64 \times w^*$ - and on the number of LB steps. The results on Figure 1 show that the gain is significant when w^* is low and also show that the gain is null when w^* is high. This is due to the cost of

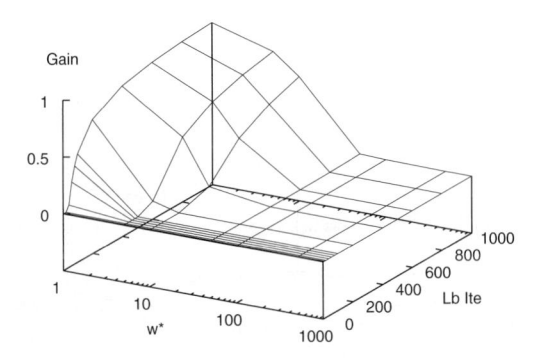

Fig. 1. Gain given with the cost/benefit parameter: $k^{(t+1)} = k^{(t)} - 1$ and with the convergence detection algorithm on a LAN network

the LB algorithm itself: when it has converged, its cost is constant and it only depends of the network. Hence, when the computation time is low - when w^* is low - the cost of load balancing is relatively high and when the computation time is high - when w^* is high - it is negligible. If the cost of load balancing is negligible, the cost/benefit system and the convergence detection are not useful but it can be noted that they are not costly with a LAN network: the gain on Figure 1 is never negative when w^* is high.

4.2 Slow Network

In the latter, the same problem is deployed on a DSL network where the bandwidth is 1Mb/s and the latency is 40ms. Figures 2 and 3 show the program computation times depending on the load average w^* and on the number of LB steps.

Figure 2 corresponds to the program with a classical FOS algorithm without cost/benefit system and without convergence detection algorithm. Here, we can see that the first iterations of the LB algorithm give a gain and that after some iterations of load balancing, the computation time increases and the time to

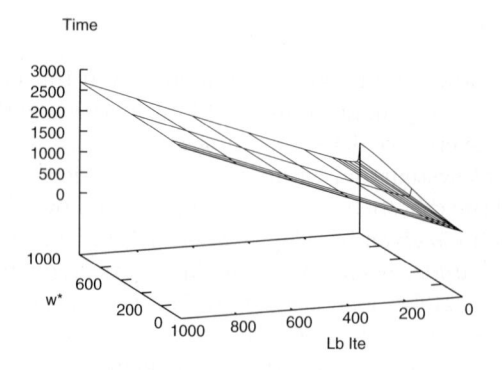

Fig. 2. Classical load balancing

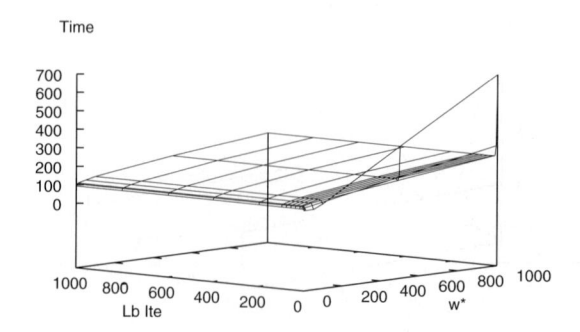

Fig. 3. Load balancing with convergence detection and cost/benefit system with k depends on the computation time of an iteration

compute the program becomes much greater than a sequential computation. This problem has two complementary reasons: the cost of load exchanges and the cost of information exchanges after the convergence of the algorithm. Hence, the cost/benefit system and the convergence detection algorithm can be interesting, in particular also for for small w^*.

An implementation of the convergence detection algorithm and the cost/benefit system as in the LAN configuration - k defined by $k^{(t+1)} = k^{(t)} - 1$ - showed us that this definition of k is not effective in a DSL network.

Figure 3 shows the results obtained with the convergence detection algorithm and the cost/benefit system with k depending on the computation time of an iteration. For a given node, when the computation time of its iteration is greater than the computation time of its previous iteration, it divides its value of k by 2. Figure 3 shows that with this system, the LB algorithm is stopped after a few iterations in which the computation time has increased. Thus the LB algorithm is beneficial to the program in quasi all configurations. When the global load is small - when a parallel computation is costlier than a sequential one - the LB algorithm is not beneficial but it is stopped fast enough for its cost to be negligible. Moreover, it can be noted that with this extreme configuration the LB algorithm with this cost/benefit system does not use all the processors, see Table 1. The optimal value is the number of processors to reach the minimum

Table 1. This table shows for a given load, in line 2 and in line opt, the number of nodes used and the optimal number of nodes with the cost/benefit system

load nxw^*		64x1	64x5	64x10	64x50	64x100	64x500	64x1000
number of	used	3/64	5/64	6/64	7/64	8/64	9/64	10/64
	opt	1/64	1/64	1/64	1/64	3/64	5/64	10/64

computation time with the cost/benefit system. This optimal value is computed using a global knowledge. We see that without global knowledge, we find a result close to the optimal.

5 Conclusion

In this paper we have studied a practical approach of diffusion load balancing. We have proposed an analysis of the cost and benefit of a load exchange. Based on this analysis we are able to decide wherever or not to exchange the load. This cost and benefit mechanism increases the well-known *step* problem. In order to tackle this problem, we propose a new feature based on simulated annealing that shakes the load when required. Finally, we have enhanced the classical convergence detection to take into account these new elements.

In this work very few assumptions are made. We can deal with static or dynamic load, with any kind of network topology, with heterogeneous nodes and networks and with any type of load. Furthermore, no global knowledge is required to perform the algorithm.

Results show that the proposed features do not degrade the performance of the load balancing algorithm and can lead (in the best case) to 100% of performance increase. Furthermore, in case of slow networks, the algorithm does not use all the available resources in order to give a good speed-up.

References

1. G. Cybenko. Dynamic load balancing for distributed memory multiprocessors. *Journal of Parallel and Distributed Computing*, 7:279–301, 1989.
2. S.H. Hosseini, B. Litow, M. Malkawi, J. McPherson, and K. Vairavan. Analysis of a graph coloring based distributed load balancing algorithm. *Jour. of Para. and Dist. Comp.*, 10:160–166, 1990.
3. B. Litow, S.H. Hosseini, K. Vairavan, and G.S. Wolffe. Performance characteristics of a load balancing algorithm. *Jour. of Para. and Dist. Comp.*, 31:159–165, 1995.
4. R. Diekmann, A. Frommer, and B. Monien. Efficient schemes for nearest neighbor load balancing. *Parallel Computing*, 25:289–313, 1998.
5. J.E. Boillat. Load balancing and poisson equation in a graph. *Concurrency: Practice and Experience.*, 2(4):289–313, 1990.
6. J.M. Bahi and J. Gaber. Load balancing on networks with dynamically changing topology. In *Europar 2001 conference, Lecture Notes on Computer Science*, pages 175–182, Manchester, UK, 2001.
7. D.P. Bertsekas and J.N. Tsitsiklis. *Parallel and Distributed Computation: Numerical Methods*. nglewood Cliffs NJ, Prentice-Hall, 1989.
8. C.Z. Xu and F.C.M. Lau. Analysis of the generalized dimension exchange method for dynamic load balancing. *Journal of Parallel and Distributed Computing*, 16(4):385–393, 1992.
9. B. Ghosh, S. Muthukrishnan, and M.H. Schultz. First and second order diffusive methods for rapid, coarse, distributed load balancing. In *Proc. of the 8th Annual ACM Sympo. on Para. Algo. and Archi.*, pages 72–81. ACM NY, 1996.
10. J.M. Bahi, R. Couturier, and F. Vernier. Synchronous distributed load balancing on dynamic networks. *Journal of Parallel and Distributed Computing*, 65(11):1397–1405, 2005.
11. R. Elsässer, B. Monien, and S. Schamberger. Load balancing in dynamic networks. In *Proc. 7th Inter. Sympo. on Para. Archi., Algo. and Net.*, 2004.
12. F. Vernier. *Algorithmique itérative pour l'équilibrage de charge dans les réseaux dynamiques*. PhD thesis, Univ. de Franche-Comté (France), 2004.
13. C.Z. Xu, B. Monien, R. Lüling, and F.C.M. Lau. An analytical comparison of nearest neighbor algorithms for load balancing in parallel computers. In *Proc. of 9th Inter. Para. Proc. Sympo.*, pages 472–479. IEEE CSP, 1995.
14. R. Elssser, B. Monien, and S. Schamberger. Load balancing of indivisible unit size tokens in dynamic and heterogeneous networks. In *Proc. of 12th Annual Euro. Symp. (ESA'04)*, volume 3221, page 640. Springer, 2004.
15. J.M. Bahi, Contassot-Vivier S., R. Couturier, and F. Vernier. A decentralized convergence detection algorithm for asynchronous parallel iterative algorithms. *IEEE Trans. on Para. and Dist. Sys.*, 16(1):4–13, 2005.
16. R.G. Gallager, P.A. Humblet, and P.M. Spira. A distributed algorithm for minimum-weight spanning trees. *ACM Transactions on Programming Languages and Systems*, 5(1):66–77, 1983.

17. I. Lavallee and G. Roucairol. A fully distributed (minimal) spanning tree algorithm. *Information Processing Letters*, 23(2):55–62, 1986.
18. B. Awerbuch. Optimal distributed algorithms for minimum weight spanning tree, counting, leader election, and related problems. In *The 19th Annual ACM Conf. on Theo. of Comp.*, pages 230–240. ACM NY, 1987.
19. I. Lavallee and C. Lavault. Yet another distributed election and (minimum-weight)spanning tree algorithm. Rr-1024, INRIA - Rocquencourt, 1989.
20. H. Casanova, A. Legrand, and L. Marchal. Scheduling Distributed Applications: the SimGrid Simulation Framework. In *Proc. of the 3rd IEEE Inter. Sympo. on Clust. Comp. and the Grid (CCGrid'03)*. IEEE CSP, may 2003.

Fast Diffusion Load Balancing Algorithms on Torus Graphs

Gregory Karagiorgos, Nikolaos M. Missirlis, and Filippos Tzaferis

Department of Informatics and Telecommunications
Panepistimiopolis 157 84, Athens, Greece
{greg, nmis, ftzaf}@di.uoa.gr
National and Kapodistrian University of Athens

Abstract. In this paper, we consider the application of accelerated methods in order to increase the rate of convergence of the diffusive iterative load balancing algorithms. In particular, we compare the application of Semi-Iterative, Second Degree and Variable Extrapolation techniques on the basic Diffusion method and the Extrapolated Diffusion method for torus graphs. It is shown that our methods require approximately 30% less iterations to reach the balanced state compared to the existed ones.

Keywords: Iterative load balancing; Diffusion algorithms; Distributed processor network; accelerated techniques.

1 Introduction

We consider the following abstract distributed load balancing problem. We are given an arbitrary, undirected, connected graph $G = (V, E)$ in which node $v_i \in V$ contains a number u_i of current workload. The goal is to determine a schedule to move an amount of workload across edges so that, the weight on each node is equal. Communication between non-adjacent nodes is not allowed. This problem describes load balancing in synchronous distributed processor networks and parallel machines when we associate a node with a processor, an edge with a communication link of unbounded capacity between two processors, and the weight as infinitly divisible independent tasks. Diffusion algorithms assume that a node of the graph is able to send and receive messages to/from all its neighbours simultaneously.

The performance of a balancing algorithm can be measured in terms of number of iterations to reach a balanced state and the amount of load moved over the edge of the graph. The original algorithm described by Cybenko [2] and, independently, by Boillat [1] lacks in performance because of its very slow convergence to the balanced state [14]. Recently, diffusive algorithms have been proposed [3,6,8,10,12] to speed up the iteration process by an order of magnitude. All commonly used diffusion schemes generate the unique l_2-minimal flow [3,4,8]. Most of the existing iterative dynamic load balancing algorithms [3,8,14] involve two steps:

- Flow calculation: Calculating the amount of workload to be migrated between neighbouring processors such that a uniform load distribution is achieved when the migration is carried out to satisfy the flow.
- Task selection: Deciding which particular tasks are to be migrated, and scheduling these tasks to the appropriate neighbouring processors.

In practice, the diffusion iteration is used for the flow calculation. The real movement of task is complex and the exact details depend on the applications and the data structures used. This paper is concerned with algorithms for the first step. We consider the application of accelerated methods in order to increase the rate of convergence of the diffusive iterative load balancing algorithms for torus graphs. In particular, we apply Semi-Iterative (SI), Second Degree (SD) and Variable Extrapolation (VE) techniques [13,15] on the Extrapolated Diffusion (EDF) method [11] and compare their performances with the Diffusion (DF) method. It is shown that our methods require approximately 30% less iterations to reach the balanced state compared to the existed ones [4,6,8].

The paper is organized as follows. Section 2 presents the extrapolated diffusion method. Section 3 adapts the aforementioned accelerated techniques to the extrapolated diffusion method. These methods increase the rate of convergence of the basic iterative scheme by an order of magnitude. Section 4 presents our results and conclusions.

2 The Extrapolated Diffusion Method

The Extrapolated Diffusion (EDF) method for the load balancing has the form [1,2]

$$u_i^{(n+1)} = u_i^{(n)} - \tau \sum_{j \in A(i)} c_{ij} \left(u_i^{(n)} - u_j^{(n)} \right), \qquad (1)$$

where c_{ij} are diffusion parameters, $A(i)$ is the set of the nearest neighbors of node i of the graph $G = (V, E)$, $u_i^{(n)}$, $i = 0, 1, 2, \ldots, |V|$ is the load after the n-th iteration on node i and $\tau \in \mathbb{R}\setminus\{0\}$ is a parameter that plays an important role in the convergence of the whole system to the equilibrium state. The overall workload distribution at step n, denoted by $u^{(n)}$, is the transpose of the vector $(u_1^{(n)}, u_2^{(n)}, \ldots, u_{|V|}^{(n)})$ and $u^{(0)}$ is the initial workload distribution. In matrix form (1) becomes

$$u^{(n+1)} = M u^{(n)}, \qquad (2)$$

where M is called the *diffusion matrix*. The elements of M, m_{ij}, are equal to τc_{ij}, if $j \in A(i)$, $1 - \tau \sum_{j \in A(i)} c_{ij}$, if $i = j$ and 0 otherwise. With this formulation, the features of diffusive load balancing are fully captured by the iterative process (2) governed by the diffusion matrix M. Also, (2) can be written as $u^{(n+1)} = (I - \tau L)u^{(n)}$, where $L = BWB^T$ is the *weighted Laplacian* matrix of the graph, W is a diagonal matrix of size $|E| \times |E|$ consisting of the coefficients c_{ij} and B is the vertex-edge incident matrix. At this point, we note that if $\tau = 1$, then we obtain the DF method proposed by Cybenko [2] and Boillat [1], independently. If $W = I$, then we obtain the special case of the DF method with a single parameter τ *(unweighted Laplacian)*. In the unweighted case and for network topologies such as chain, 2D-mesh, nD-mesh, ring, 2D-torus, nD-torus and nD-hypercube, optimal values for the parameter τ that maximize the convergence rate have been derived by Xu and Lau [14]. Next, we consider the weighted case.

The diffusion matrix of EDF can be written as

$$M = I - \tau L, \quad L = D - A, \tag{3}$$

where $D = diag(L)$ and A is the weighted adjacency matrix. Because of (3), (2) becomes $u^{(n+1)} = (I - \tau D) u^{(n)} + \tau A u^{(n)}$ or in component form

$$u_i^{(n+1)} = \left(1 - \tau \sum_{j \in A(i)} c_{ij}\right) u_i^{(n)} + \tau \sum_{j \in A(i)} c_{ij} u_j^{(n)}, i = 1, 2, \ldots, |V|. \tag{4}$$

The diffusion matrix M must have the following properties: nonnegative, symmetric and stochastic [2,1]. The eigenvalues of L are $0 = \lambda_1 < \lambda_2 \leq \ldots \leq \lambda_n$. In case $c_{ij} = $ constant, the optimum value of τ is attained at [13,15]

$$\tau_o = \frac{2}{\lambda_2 + \lambda_n} \tag{5}$$

and the corresponding minimum value of the convergence factor

$$\gamma(M) = \max\{|1 - \tau \lambda_n|, |1 - \tau \lambda_2|\} \tag{6}$$

is given by

$$\gamma_o(M) = \frac{P(L) - 1}{P(L) + 1}, \text{ where } P(L) = \frac{\lambda_n}{\lambda_2}, \tag{7}$$

which is the P-condition number of L. Note that if $P(L) \gg 1$, then the rate of convergence of the DF method is given by

$$R(M) = -\log \gamma_o(M) \simeq \frac{2}{P(L)}, \tag{8}$$

which implies that the rate of convergence of the DF method is a decreasing function of $P(L)$. In the sequel, we will express the optimum values of the parameters involved, in each considered iterative scheme, using the second minimum and maximum eigenvalues λ_2, λ_n, respectively of the Laplacian matrix. A first advantage of EDF is that it converges for any positive, real values of the parameters c_{ij} if $\tau \in (0, 1/||A||_\infty)$ [11], whereas in DF it is required that c_{ij} must satisfy the conditions $\sum_{j \in A(i)} c_{ij} < 1$ for at least one i. The problem of determining the diffusion parameters c_{ij} such that EDF attains its maximum rate of convergence is an active research area [3,5,11]. Introducing the set of parameters $\tau_i, i = 1, 2, \ldots, |V|$, instead of a fixed parameter τ in 4, the problem moves to the determination of the parameters τ_i in terms of c_{ij}. By considering local

Fourier analysis [9,11] we were able to determine good values (near the optimum) for τ_i. These values become optimum in case the diffusion parameters are constant in each dimension and satisfy the relation $c_j^{(2)} = \sigma_2 c_i^{(1)}$, $i = 1, 2, \ldots, N_1$, $j = 1, 2, \ldots, N_2$, where $\sigma_2 = \frac{1-\cos\frac{2\pi}{N_1}}{1-\cos\frac{2\pi}{N_2}}$ and $c_i^{(1)}$, $c_j^{(2)}$ are the row and column diffusion parameters, respectively, of the torus (see Fig. 1).

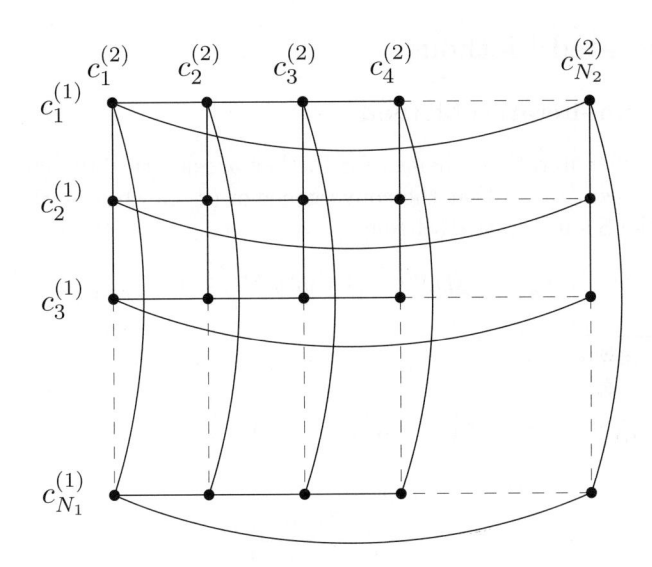

Fig. 1. The diffusion parameters in a 2-D torus

At the optimum stage EDF is twice as fast as DF for stretched torus, that is a torus with either $N_1 \gg N_2$ or $N_2 \gg N_1$ [11]. Apart from the fact that our approach produces a monoparametric set of optimum values for the diffusion parameters it also has the advantage of determining a closed form formula for the involved parameter τ and the convergence factor γ (see Table 1). These facts have two consequences. First, we avoid the computation of the second smallest and largest eigenvalue of the Laplacian matrix for the determination of the optimum

Table 1. Formulae for the optimum τ_o and $\gamma_o(M)$

N_1	N_2	Case	τ_o	$\gamma_o(M)$
Even	Even	1	$\left[3 + 2\sigma_2 - \cos\frac{2\pi}{N_1}\right]^{-1}$	$\frac{1+2\sigma_2+\cos\frac{2\pi}{N_1}}{3+2\sigma_2-\cos\frac{2\pi}{N_1}}$
Odd	Odd	2	$\left[2 + \sigma_2(1+\cos\frac{\pi}{N_2}) + \cos\frac{\pi}{N_1} - \cos\frac{2\pi}{N_1}\right]^{-1}$	$\frac{\cos\frac{\pi}{N_1}+\cos\frac{2\pi}{N_1}+\sigma_2(1+\cos\frac{\pi}{N_2})}{2+\sigma_2(1+\cos\frac{\pi}{N_2})+\cos\frac{\pi}{N_1}-\cos\frac{2\pi}{N_1}}$
Even	Odd	3	$\left[3 - \cos\frac{2\pi}{N_1} + \sigma_2(1+\cos\frac{\pi}{N_2})\right]^{-1}$	$\frac{1+\cos\frac{2\pi}{N_1}+\sigma_2(1+\cos\frac{\pi}{N_2})}{3-\cos\frac{2\pi}{N_1}+\sigma_2(1+\cos\frac{\pi}{N_2})}$
Odd	Even	4	$\left[2 + 2\sigma_2 + \cos\frac{\pi}{N_1} - \cos\frac{2\pi}{N_1}\right]^{-1}$	$\frac{2\sigma_2+\cos\frac{\pi}{N_1}-\cos\frac{2\pi}{N_1}}{2+2\sigma_2+\cos\frac{\pi}{N_1}-\cos\frac{2\pi}{N_1}}$

value for τ. This is a time consuming process and was an open problem as far as we know. Second, we were able to study the convergence behaviour of the EDF method and predict its performance using the expression for γ. In order to further improve, by an order of magnitude, the rate of convergence of EDF we can apply accelerated techniques (Semi-Iterative, Second-Degree and Variable Extrapolation) following [15,13,7,6,12,10].

3 Accelerated Methods

3.1 The Semi-iterative Method

We now consider iterative schemes for further accelerating the convergence of EDF. It is known [13,15] that the convergence of (2) can be greatly accelerated if one uses the Semi-Iterative scheme

$$u^{(n+1)} = \rho_{n+1}\left[\bar{\rho}Mu^{(n)} + (1-\bar{\rho})u^{(n)}\right] + (1-\rho_{n+1})u^{(n-1)} \qquad (9)$$

with

$$\bar{\rho} = \frac{2}{2-(\beta+\alpha)}, \rho_1 = 1, \rho_2 = \left(1-\frac{\sigma^2}{2}\right)^{-1}, \rho_{n+1} = \left(1-\frac{\sigma^2}{4}\rho_n\right)^{-1}, \ n=2,3,\ldots, \qquad (10)$$

where

$$\sigma = \frac{\beta-\alpha}{2-(\beta+\alpha)}, \qquad (11)$$

with

$$\alpha \leq \mu_i \leq \beta, \qquad (12)$$

where μ_i are the eigenvalues of M. Because of (3), the eigenvalues of M and the Laplacian matrix L are related via the following relationship

$$\mu_i = 1 - \tau\lambda_i, \quad i = 1, 2, \ldots, n, \qquad (13)$$

hence

$$\alpha = 1 - \tau\lambda_n \quad \text{and} \quad \beta = 1 - \tau\lambda_1, \qquad (14)$$

since $\tau > 0$. Expressing $\bar{\rho}$ and σ (see (10) and (11)) in terms of λ_1 and λ_n, with the use of (14), we find that

$$\bar{\rho} = \frac{2}{\tau(\lambda_1+\lambda_n)} \quad \text{and} \quad \sigma = \frac{P(L)-1}{P(L)+1}, \qquad (15)$$

with

$$P(L) = \frac{\lambda_n}{\lambda_1}. \qquad (16)$$

But the optimum value of τ, τ_o is given by (5) which on substitution in the expression of $\bar{\rho}$ in (15) yields

$$\bar{\rho} = 1. \qquad (17)$$

Moreover, for $\tau = \tau_o$, (6) becomes

$$\gamma_o(M) = \frac{P(L) - 1}{P(L) + 1}. \tag{18}$$

By (18) and (15) we have that

$$\sigma = \gamma_o(M). \tag{19}$$

Thus expressing (9) in terms of the Laplacian matrix L, using (3) and (17), we obtain

$$u^{(n+1)} = \rho_{n+1}(I - \tau_o L)u^{(n)} + (1 - \rho_{n+1})u^{(n-1)}, \tag{20}$$

where $\rho_{n+1}, \sigma, \tau_o$ are given by (10), (19) and (5), respectively. It is worth noting that σ is equal to $\gamma_o(M)$ (see (19)), which is the minimum value of the convergence factor of EDF. In addition, $\gamma_o(M)$ and τ_o, for EDF, are given by the expressions of Table 1 for the corresponding values of N_1 and N_2. It can be shown [7,13,15] that

$$\gamma(P_n(M)) = \frac{2r^{n/2}}{1 + r^n}, \tag{21}$$

where $P_n(M)$ is a certain polynomial in M (which is related to Chebyshev polynomials) and

$$r^{1/2} = \frac{\sigma}{1 + \sqrt{1 - \sigma^2}} = \frac{\sqrt{P(L)} - 1}{\sqrt{P(L)} + 1}.$$

In addition, for $P(L) \gg 1$, we have

$$r \simeq 1 - \frac{4}{\sqrt{P(L)}}, \tag{22}$$

thus the asymptotic average rate of convergence for the Semi-Iterative EDF (SI-EDF) method is given by

$$R_\infty(P_n(M)) = -\frac{1}{2}\log r \simeq \frac{2}{\sqrt{P(L)}} \tag{23}$$

as $n \to \infty$. From (8) and (23) the following relationship holds between the reciprocal rates of convergence[1] of SI-EDF and EDF

$$RR_\infty(P_n(M)) \simeq \frac{\sqrt{RR(M)}}{2}. \tag{24}$$

Therefore, the use of Semi-Iterative techniques results in an order of magnitude improvement in the reciprocal rate of convergence of EDF and in turn in the number of iterations.

---

[1] $RR(.) = \frac{1}{R(.)}$.

3.2 The Second Degree Method

An accelerated scheme similar to (9) can be produced by considering constant iteration parameters throughout the process. It is known as the Second Degree (SD) method and is given by [15]

$$u^{(n+1)} = u^{(n)} + (\hat{\omega}_o - 1)(u^{(n)} - u^{(n-1)}) + \hat{\omega}_o(Mu^{(n)} - u^{(n)}), \quad (25)$$

where $\hat{\omega}_o = \frac{2}{1+\sqrt{1-\sigma^2}}$ with σ given by (19). Expressing (25) in terms of the Laplacian matrix L, we obtain

$$u^{(n+1)} = \hat{\omega}_o(I - \tau_o L)u^{(n)} + (1 - \hat{\omega}_o)u^{(n-1)}. \quad (26)$$

If $\hat{M}$ is the iteration matrix of the SD method, then [15]

$$\gamma(\hat{M}) = (\hat{\omega}_o - 1)^{1/2} = r^{1/2}, \quad (27)$$

thus the rate of convergence of the Second Degree EDF (SD-EDF) method is $R(\hat{M}) = -\frac{1}{2}\log r$, which is comparable with the one obtained by semi-iterative techniques. Also, by (23) and (27) we conclude that the rate of convergence of semi-iterative and second degree methods depends on the same quantity r. This implies that (23) and (24) hold also for the SD method.

3.3 The Variable Extrapolation Method

In the previous sections, it was shown how we can find effective iterative processes. Note that in the new procedures each vector $u^{(n+1)}$ requires the computation of the two previous vectors $u^{(n)}$ and $u^{(n-1)}$. In case we face memory limitation problems we can consider another iterative scheme (sometimes called the Richardson method [13,15]) of accelerating the EDF method, where $u^{(n+1)}$ is computed using $u^{(n)}$ only. This can be achieved by applying the following iterative scheme

$$u^{(n+1)} = \theta_{n+1} Mu^{(n)} + (1 - \theta_{n+1})u^{(n)}, \quad (28)$$

where $\theta_{n+1} = \frac{2}{2-(\beta-\alpha)\cos\frac{(2n-1)\pi}{2m}-(\beta+\alpha)}$. The iteration parameters θ_{n+1} are selected in the cyclic order $\theta_1, \theta_2, \ldots, \theta_m, \theta_1, \theta_2, \ldots, \theta_m$, where m is an integer. Expressing (28) in terms of L we have

$$u^{(n+1)} = (I - \hat{\theta}_{n+1} L)u^{(n)}, \quad (29)$$

where $\hat{\theta}_{n+1} = \theta_{n+1}\tau_o = \frac{\tau_o}{1-\sigma\cos\frac{(2n-1)\pi}{2m}}$ with σ given by (19). The spectral radius of the Variable Extrapolation EDF (VE-EDF) method is given by [15]

$$\gamma(P_{\ell m}(M)) = \left(\frac{2r^{m/2}}{1+r^m}\right)^\ell, \quad (30)$$

where ℓ is an integer determining the number of cycles. It can be seen from (21) and (30) that as m increases, then the rapidity of convergence tends to the one given by the semi-iterative method. However, numerical experiments [15] show that, for large m, numerical instability may occur. Also, it is undesirable to select m very large because convergence is expected after ℓm iterations.

4 Numerical Results and Conclusions

The purpose of the paper is to compare the application of accelerated techniques to the DF and EDF methods. As expected, both methods produce the same results in case of square torus [11], whereas for rectangular torus EDF tends to achieve twice as fast convergence rate compared to DF. In particular,

$$RR_\infty(EDF) \simeq \frac{1}{2} RR_\infty(DF) \tag{31}$$

when $N_1 \gg N_2$ or $N_1 \ll N_2$ (stretched torus) [11]. This is shown in columns DF and EDF of Tables 2 and 3, respectively, where we present the number of iterations for both schemes to converge using the same criterion. The convergence criterion for both schemes was $\sum_{i=1}^{|V|} \left(u_i^{(u)} - \bar{u}\right)^2 < \epsilon$, where $\bar{u} = \left(\sum_{i=1}^{|V|} u_i\right)/|V|$ and $\epsilon = 10^{-6}$, while the initial load $u_i^{(0)} = u_i$ was randomly distributed on the nodes of the graph. For all cases we used the optimum values for the parameters involved. These values are τ_o and $\sigma = \gamma_o(M)$, which were obtained by the formulae of Table 1.

Table 2. Number of iterations for DF, SI-DF, SD-DF and VE-DF methods

N1 × N2	τ_o	$\gamma_o(M)$	DF	SI-DF	SD-DF	VE-DF	m	ℓ
5 × 5	0.232	0.679	40	16	18	18	9	2
5 × 11	0.254	0.919	174	38	41	42	14	3
5 × 21	0.260	0.977	605	74	79	81	27	3
5 × 51	0.262	0.996	3375	182	194	178	30	6
5 × 101	0.262	0.999	13102	366	366	597	30	20
6 × 6	0.222	0.778	60	21	23	22	10	2
6 × 10	0.200	0.908	184	55	38	38	19	2
6 × 20	0.246	0.976	572	73	76	81	27	3
6 × 50	0.249	0.996	3375	182	192	232	29	8
6 × 100	0.249	0.999	12799	366	361	575	34	20

By (24) we have

$$RR_\infty(SI - EDF) \simeq \frac{1}{2}(RR_\infty(EDF))^{1/2}, \tag{32}$$

or using (31),

$$RR_\infty(SI - EDF) \simeq \frac{1}{2\sqrt{2}}(RR_\infty(DF))^{1/2}. \tag{33}$$

But (24) holds also for the DF method, this means that

$$RR_\infty(SI - DF) \simeq \frac{1}{2}(RR_\infty(DF))^{1/2}. \tag{34}$$

Therefore, (33), because of (34), yields

$$RR_\infty(SI - EDF) \simeq \frac{1}{\sqrt{2}} RR_\infty(SI - DF) \tag{35}$$

which indicates that the number of iterations of SI-EDF will be approximately 30% less than the number of iterations of SI-DF in case of stretched torus. Clearly, this result holds for the other two accelerated methods (SD-EDF and VE-EDF) as both these methods tend to obtain the same rate of convergence as SI-EDF (see (27) and (30)). In Tables 2 and 3 we also present the number of iterations for the accelerated versions of DF and EDF methods, respectively. For the VE version of both methods the value of m was determined experimentaly such that the number of iterations is minimum. The results of Tables 2 and 3 clearly show that fixing one dimension of a torus and increasing the other, the number of iterations of the accelerated versions of EDF (SI-EDF, SD-EDF, VE-EDF) is 30% less than the number of iterations of the corresponding versions of DF. Similar results were also obtained in the odd/even cases. Therefore, our theoretical expectation, which is expressed by (35), is verified. Finally, comparing the accelerated versions of EDF we note that SI and SD have similar behaviour, which is better than the VE version (Table 3).

Table 3. Number of iterations for EDF, SI-EDF, SD-EDF and VE-EDF methods

N1 × N2	τ_{opt}	γ_{opt}	EDF	SI-EDF	SD-EDF	VE-EDF	m	ℓ
5 × 5	0.232	0.679	40	16	18	18	9	2
5 × 11	0.091	0.874	113	30	32	30	30	1
5 × 21	0.029	0.958	348	54	59	58	29	2
5 × 51	0.005	0.992	1966	133	142	161	27	6
5 × 101	0.001	0.998	7176	264	269	387	30	13
6 × 6	0.222	0.777	60	21	23	22	10	2
6 × 10	0.129	0.870	109	29	32	30	30	1
6 × 20	0.043	0.956	328	53	56	58	29	2
6 × 50	0.007	0.992	1770	130	137	150	30	5
6 × 100	0.001	0.998	6824	261	260	385	26	13

Acknowledgements

We would like to thank the anonymous referees for their positive comments which resulted in the present form of the paper.

The project is co-funded by the European Social Fund and National Resources (EPEAK II) Pythagoras.

References

1. J. E. Boillat, Load balancing and poisson equation in a graph, *Concurrency: Practice and Experience* **2**, 1990, 289-313.
2. G. Cybenko, Dynamic load balancing for distributed memory multi-processors, *Journal of Parallel and Distributed Computing* **7**, 1989, 279-301.
3. R. Diekmann, S. Muthukrishnan, M. V. Nayakkankuppam, Engineering diffusive load balancing algorithms using experiments, *IRREGULAR'97*, **LNCS 1253**, 1997, 111-122.

4. R. Diekmann, A. Frommer, B. Monien, Efficient schemes for nearest neighbour load balancing, *Parallel Computing* **25**, 1999, 789-812.
5. R. Elsässer, B. Monien, S. Schamberger, G. Rote, Toward optimal diffusion matrices, *International Parallel and Distributed Processing Symposium*, IEEE Computer Society Press, 2002.
6. B. Ghosh, S. Muthukrishnan and M. H. Schultz, First and second order diffusive methods for rapid, coarse, distributed load balancing, *8thACM Symposium on Parallesim in Algorithms and Architectures*, 1996, 72-81.
7. G. H. Golub and R. S. Varga, Chebyshev semi-iterative methods, successive over-relaxation iterative methods, and second-order Richardson iterative methods, *Numer. Math. Parts I and II* **3**, 1961, 147-168.
8. Y. F. Hu and R. J. Blake, An improved diffusion algorithm for dynamic load balancing, *Parallel Computing* **25**, 1999, 417-444.
9. G. Karagiorgos and N. M. Missirlis, Fourier analysis for solving the load balancing problem, *Foundations of Computing and Decision Sciences* **27**, No 3, 2002.
10. G. Karagiorgos and N. M. Missirlis, Accelerated diffusion algorithms for dynamic load balancing, *Information Processing Letters* **84**, 2002, 61-67.
11. G. Karagiorgos and N. M. Missirlis, Local convergence analysis for the diffusion load balancing methdod in torus *(submitted)*.
12. S. Muthukrishnan, B. Ghosh and M. H. Schultz, First and second order Diffusive methods for rapid, coarse, distributed load balancing, *Theory of Computing Systems* **31**, 1998, 331-354.
13. R. Varga, Matrix iterative analysis, *Prentice-Hall*, Englewood Cliffs, NJ, 1962.
14. C. Z. Xu and F. C. M. Lau, Load balancing in parallel computers: *Theory and Practice*, Kluwer Academic Publishers, Dordrecht, 1997.
15. D. M. Young, Iterative solution of large linear systems, *Academic Press*, New York, 1971.

A Parallel Shape Optimizing Load Balancer[*]

Henning Meyerhenke and Stefan Schamberger

Universität Paderborn,
Fakultät für Elektrotechnik, Informatik und Mathematik
Fürstenallee 11, D-33102 Paderborn
{henningm, schaum}@uni-paderborn.de

Abstract. Load balancing is an important issue in parallel numerical simulations. However, state-of-the-art libraries addressing this problem show several deficiencies: they are hard to parallelize, focus on small edge-cuts rather than few boundary vertices, and often produce disconnected partitions.

We present a distributed implementation of a load balancing heuristic for parallel adaptive FEM simulations. It is based on a disturbed diffusion scheme embedded in a learning framework. This approach incorporates a high degree of parallelism that can be exploited and it computes well-shaped partitions as shown in previous publications. Our focus lies on improving the condition of the involved matrix and solving the resulting linear systems with local accuracy. This helps to omit unnecessary computations as well as allows to replace the domain decomposition by an alternative data distribution scheme reducing the communication overhead, as shown by experiments with our new MPI based implementation.

Keywords: Load balancing, graph partitioning, parallel adaptive FEM computations.

1 Introduction

Finite Element Methods (FEM) play a very important role in engineering for analyzing a variety of physical processes that can be expressed via Partial Differential Equations (PDE). The domain on which the PDEs have to be solved is discretized into a mesh, and the PDEs are transformed into a set of equations defined on the mesh's elements (see e.g. [5]). Due to the sparseness of the discretization matrices these equations are typically solved by iterative methods such as Conjugate Gradient (CG) or multigrid.

Since an accurate approximation of the original problem requires a very large number of elements, this method has become a classical application for parallel computers. The parallelization of numerical simulation algorithms usually

---

[*] This work is supported by German Science Foundation (DFG) Research Training Group GK-693 of the Paderborn Institute for Scientific Computation (PaSCo) and DFG Collaborative Research Centre SFB-376.

follows the Single-Program Multiple-Data paradigm: Each of the P processors executes the same code on a different part of the data. Thus, the mesh has to be split into sub-domains, each being assigned to one processor. To minimize the overall computation time, all processors should roughly contain the same number of elements. Furthermore, since iterative solution algorithms perform mainly local operations, the parallel algorithm mostly requires communication at the partition boundaries. Hence, these should be as small as possible due to the very high communication costs involved.

Depending on the application, some areas of the simulation space require higher resolutions and therefore more elements. Since in many cases the location of these areas varies over time, the mesh is refined and coarsened during the computation. Yet, this can cause imbalance between the processor loads and therefore delay the simulation. To avoid this, the element distribution needs to be rebalanced during runtime. For this, the application is interrupted and the repartitioning problem is solved. Although this interruption should be as short as possible, it is also important to find a new balanced partitioning with small boundaries that does not cause too many elements to change their processor. Migrating elements can be extremely costly since large amounts of data have to be sent over communication links and stored in complex data structures.

In previous work [19,15] we have shown that (re-)partitioning heuristics focusing on the shape of partitions are able to find solutions with a small number of boundary vertices while also causing little migration. There, we have compared our method to the state-of-the-art libraries Metis [11] and Jostle [21] regarding solution quality and runtime. It turns out that while the solution quality of the shape-optimizing approach is usually the best, its main drawback is its long runtime and high memory consumption. Therefore, in this paper we present a new parallel implementation based on the message-passing interface MPI that incorporates several improvements addressing these problems.

The remaining part of the paper is organized as follows. In the next Section we recapture related work and explain the shape optimizing bubble framework. Section 3 describes the diffusion scheme applied within this framework as its growth mechanism and an enhancement to the condition of the involved matrix. Our parallel MPI based implementation is presented in Section 4. The new concept of solving the linear systems with local accuracy reduces the computation time as well as the memory requirements. Additionally, it facilitates a new data distribution scheme decreasing communication. Subsequently, we present some of our experiments in Section 5 before we give a short conclusion.

2 Related Work

2.1 Graph Partitioning and Load Balancing Heuristics

Balancing an FEM mesh can be expressed as a graph (re-)partitioning problem. The mesh is transformed into a graph whose vertices represent the computational work and the edges their interdependencies. Due to its complexity existing libraries for this problem are based on heuristics. State-of-the-art implementations like Metis [11], Jostle [21] or Party [17] follow the multilevel scheme [7] with a local improvement heuristic based on exchanging vertices between partitions. This heuristic reduces the number of cut-edges or the boundary size as well as balances the partition sizes. Hence, the final solution quality mainly depends on this method. Implementations are mostly based on the Kernighan-Lin (KL) heuristic [12], while the local refinement in Party is derived from theoretical analysis with Helpful-Sets (HS) [8]. To address the load balancing problem during parallel computations, distributed versions of the libraries Metis [20] and Jostle [22] have been developed. However, due to the sequential nature of the KL heuristic, their parallelization is difficult. This situation is even worse with the HS heuristic in Party due to the large overhead for exchanging large vertex sets.

While the global edge-cut is the classical metric that most graph partitioners optimize, it is not necessarily the best metric to follow [6] because it does not model the real communication and runtime costs of FEM computations. Hence, different metrics have been implemented to model the real objectives more closely [16,11]. As an example, since the convergence rate of the CGBI solver in the PadFEM environment depends on the geometric shape of a partition, its load balancer iteratively decreases the partitions' aspect ratios by applying the algorithm "Bubble" [3], whose basic idea appeared already in [23]. Yet, its implementation contains a strictly sequential part and suffers from some other difficulties described in [18]. Details about this algorithm and how to overcome its issues are discussed in the following.

2.2 The Bubble Framework

The bubble framework is related to the k-means algorithm well-known in cluster analysis [13] and transfers its ideas to graphs: First, one chooses randomly for each partition one vertex as its center vertex. With this initial set of seed vertices at hand, all remaining vertices are assigned to their closest seed based on some distance measure. (This resembles the simultaneous growth of soap bubbles starting at the seed vertices and colliding at common borders.) After all vertices of the graph have been assigned this way, each sub-domain computes its new center, which acts as the seed in the next iteration. This can be repeated until a stable state is reached. Fig. 1 illustrates the three main operations. This framework can be implemented in various ways, but many approaches show some major disadvantages for our given problem (cf. [14] for a broader discussion). They can be overcome by the growth mechanism explained next.

3 The Diffusion Based Growth Mechanism

Our implementation of the bubble operations is based on solving diffusion problems on the input graph. This is due to the fact that diffusion prefers densely

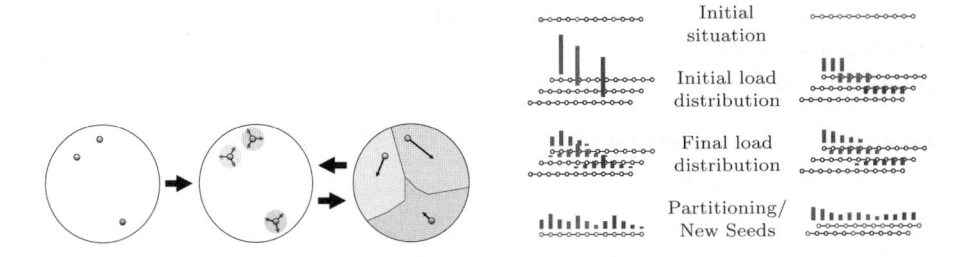

Fig. 1. The main bubble framework operations: Determine initial seeds for each partition (left), grow partitions around the seeds (middle), move seeds to the partition centers (right)

Fig. 2. Schematic view: Placing load on single vertices (left) or a partition (right), the diffusion process, and the mapping of vertices to the partitions according to the load

connected regions of the graph. Thus, one can expect to identify vertex sets that tend to possess a small number of boundary vertices.

3.1 The FOS/C Diffusion Scheme

Generally speaking, a diffusion problem consists of distributing load from some given seed vertex (or vertices) into the whole graph by load exchanges between neighbor vertices. Standard diffusion schemes like FOS [1] converge to fully balanced load distributions. This is undesired here because the amount of load should represent a distance between vertices. Hence, we disturb FOS to obtain a hill-like load distribution with meaningful diffusion distances between vertices.

How this hill-like distribution is interpreted as distance values is illustrated in figure 2. Given a seed vertex for each partition (left), we place load on the respective seed and use a diffusive process to have it spread into the graph. This is performed independently for every partition. After the load is distributed, we assign each vertex to that partition it has obtained the highest load amount from (highest load means shortest distance). The next step (right) does not place load on a single seed vertex only, but distributes it evenly among all vertices of the given partition. After performing the diffusion process, the resulting load distribution can either be used as an optional consolidation or for contracting the partitions to the seed vertices of the next iteration. A consolidation again assigns the vertices to partitions according to the highest load as in the previous step. This further improves the partition shapes. During a contraction, for each partition the vertex containing the highest load becomes its new seed.

We now restate some important properties of the diffusion scheme applied (for details cf. [14]). Let $\mathbf{L}$ be the Laplacian matrix of the unweighted, connected input graph $G = (V, E)$. Shifting a small load amount δ (drain) from each vertex back to the seed vertex/vertices (comprised in the set $S \subset V$) in each iteration by the drain vector d leads to the desired disturbed diffusion scheme called FOS/C with the matrix/vector notation $w^{(i+1)} = \mathbf{M}w^{(i)} + d$, where $\mathbf{M} = \mathbf{I} - \alpha\mathbf{L}$ is the diffusion matrix with some suitable constant $\alpha > 0$, $w^{(i)}$ is the load vector at iteration i and d the drain vector, whose vector sum is 0, so that $d \in range(\mathbf{L})$.

Theorem 1 (Convergence of FOS/C). *[14] The FOS/C scheme converges for any arbitrary initial load vector $w^{(0)}$.*

Corollary 1. *[14] The convergence state $w^{(*)}$ of FOS/C can be characterized as $w^{(*)} = \mathbf{M}w^{(*)} + d \Leftrightarrow (\mathbf{I} - \mathbf{M})w^{(*)} = d \Leftrightarrow \alpha\mathbf{L}w^{(*)} = d$. Hence, the convergence state can be determined by solving the linear system $\mathbf{L}w = d$, where $w = \alpha w^{(*)}$.*

The resulting load vector w represents the hill-like distribution we need in order to compute diffusion distances between a seed and some other vertex. Therefore, we have the choice to compute this vector by local operations (e. g. the second order diffusion scheme [4]) or by generally faster solvers using global knowledge (such as CG or multigrid), whichever is more appropriate.

3.2 Improving the Matrix Condition

Observe that if load diffuses faster into dedicated regions, then the flow over the edges directing there must be higher than the flow over edges pointing elsewhere. Due to [9] and [2] we know that the solution of the FOS/C diffusion problem is equivalent to a $\|\cdot\|_2$-minimal flow over the edges of the graph. The diffusion problem can therefore be regarded as a flow problem, too. To make the sink of the flow unique, we insert an extra vertex into the input graph G of n nodes as in [15]. This new vertex is connected to every other vertex in G by an edge of weight $\phi > 0$, which leads to a modified Laplacian matrix $\mathbf{L}_\phi$ having one additional row and column whose off-diagonal entries are all $-\phi$. The diagonal of $\mathbf{L}_\phi$ contains for each row the weighted degree of the corresponding vertex, so that it is symmetric positive-semidefinite (spsd) and of rank n. The resulting linear system is denoted by $\mathbf{L}_\phi w_\phi = d_\phi$ with the following drain vector d_ϕ:

$$d_\phi(v) = \begin{cases} \delta \cdot |V|/|S| & : v \in S \\ -\delta \cdot |V| & : v \text{ is the extra vertex} \\ 0 & : \text{otherwise} \end{cases}$$

Solving this spsd system by iterative methods can be made faster and more robust to numerical imprecision by fixing entries (as many as the dimension of the null space of $\mathbf{L}_\phi$) of w_ϕ and deleting their corresponding rows and columns from the matrix [10]. Hence, we improve on previous work [15,14] by fixing the value of the extra vertex to be zero and delete the row and column appended to $\mathbf{L}$ before. What remains is the addition of ϕ to the diagonal values of $\mathbf{L}$. This results in a symmetric positive-definite (spd) matrix whose condition can be controlled by the parameter ϕ (therefore we actually solve $\mathbf{L}'w' = d'$, where $\mathbf{L}' = \mathbf{L} + \phi \mathbf{I}$ and d' (resp. w') equals d_ϕ (resp. w_ϕ) without the entry for the extra vertex). Note that this simple preconditioning is well-defined by the notion of the extra vertex.

Using $\mathbf{L}'$ has even more advantages than improving the convergence and robustness of iterative solvers: the distributions of different seeds are comparable without post-processing because the extra vertex acts as a common reference point. Moreover, unlike in [15], the extra vertex is eliminated from the actual solution process, which makes the use of multigrid/multilevel methods easier and further speeds up computations.

4 Parallel Implementation of Bubble-FOS/C

In this section we present the Bubble-FOS/C algorithm, its new MPI based implementation, and show improvements to the algorithm in terms of runtime and memory consumption. For sake of simplicity we denote the linear system of our diffusion/flow problem from now on $\mathbf{L}w = d$, although it has the structure of $\mathbf{L}'w' = d'$ from the previous Section. A specific system corresponding to partition p is denoted by $\mathbf{L}w_p = d_p, p \in \{1, ..., P\}$.

4.1 The Bubble-FOS/C Heuristic

```
Algorithm Bubble-FOS/C(G, π, l, i)
01  in each loop l
02    if π is undefined π = determine-seeds(G)
03    else parallel for each partition p
04      centers = Contraction(G, π)
05    parallel for each partition p
06      π = AssignPartition(G, centers)
07    in each iteration i
08      parallel for each partition p
09        π = Consolidation(G, π)
10      π = scale-balance(π)
11      π = greedy-balance(π)
12  return smooth(π)
```

Fig. 3. Sketch of the algorithm

Incorporating FOS/C into the bubble framework results in the algorithm sketched in Figure 3. It can be invoked with or without a valid partitioning π. In the latter case, we determine initial seeds randomly (line 2). Otherwise, we contract the given partitions (lines 3-4) by applying the proposed mechanism based on solving the P linear systems $\mathbf{L}w_p = d_p$. Then, we determine a partitioning (lines 5-6) before performing optional consolidations (lines 7-9). These consolidations can also be used for balancing by scaling the vectors w_p (line 10). This approach can quickly find almost balanced solutions in most cases. If necessary, we perform an additional greedy balancing operation (line 11) to guarantee a certain partition size.

Depending on the quality of the initial solution, it is advisable to repeat the learning process several times. (A multilevel scheme can help to keep the number of repetitions small [14].) Before returning the partitioning π, vertices can be migrated optionally if the number of their neighbors in another partition is larger than the number in their own partition (line 12). This further smooths the partition boundaries but might lead to a slightly higher imbalance.

4.2 Partial Graph Coarsening

As explained above, one needs to solve P linear systems $\mathbf{L}w_p = d_p$ with the same matrix $\mathbf{L}$ and a different right-hand side d_p for each bubble operation based on FOS/C. However, since a vertex is assigned according to the maximum load value, we notice that only a part of the solution is relevant to the vertex assignment due to its hill-like manner. Hence, it is not necessary to compute the exact solution for all vertices of the graph, but only in the important areas surrounding the respective partition, and an approximation elsewhere. This observation can be exploited to both speed up the computations and reduce the memory requirements in a parallel implementation:

Before the first computation, each domain creates a local level hierarchy. Similar as in state-of the-art graph partitioning libraries, this is achieved by calculating a 2-approximation of a maximum weighted matching restricted to edges connecting local vertices. These are then combined to form the vertices of the next level. After that, the implemented data structure allows us to solve linear systems that are composed of different levels of the hierarchy, reflecting the different solution accuracies on the domains.

To solve a linear system, we project the drain vector onto the respective vertices of the lowest hierarchy level and first compute load values there. Figure 4 (left) illustrates a solution for one partition on the lowest levels. One can see that the highest solution values can be found close to the originating domain. Since the matching process preserves the graph structure, the solution on the lowest level is similar to the expected load distribution in the original graph. Hence, we are able to use it to determine the most relevant parts of the solution. Important domains will be switched to a higher hierarchy level while the unimportant ones remain on the lowest one.

The approximate solution is then interpolated to higher levels where necessary and the system is solved again more accurately. Figure 4 (middle) gives an example of a load distribution that has been calculated with varying accuracy. In the important regions of the graph, the linear system is solved on the highest hierarchy level, that is the original graph, while in areas further away from the respective domain lower levels are used.

Although we now have to solve two linear systems per partition, a small one on the lowest hierarchy levels and the second one on the mixed levels, less runtime

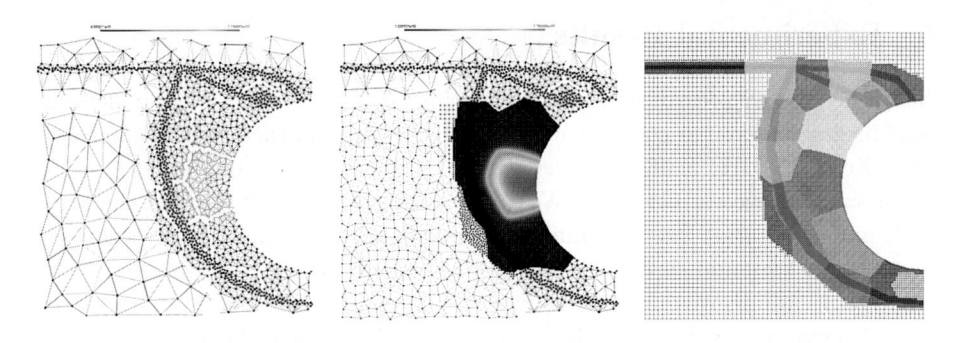

Fig. 4. Vertex loads on the lowest levels (left) and the final solution with local accuracy on the respective levels (middle). The shown solution has been computed for the pink domain leading to the displayed partitioning (right). Edges between vertices of different domains (the initial partitioning) are cut.

is required in total compared to solving one system on the original graph. The lowest levels of the hierarchy are very small and can be processed quickly. The additional time spent in this computation is compensated by the reduction of the system sizes in the second computation.

Note that for each of the P linear systems a different part of the graph is important. Hence, on each domain a number of hierarchy levels contribute to the respective solutions. In our implementation, all systems are solved simultaneously with a standard CG solver. Therefore, we are able to combine the data sent by all P instances and reduce the number of necessary messages.

4.3 Domain Decomposition vs Domain Sharing

Usually a domain decomposition is applied to distribute a graph on a parallel computer. Following this practice, the implemented CG solver requires three communications per iteration, one matrix communication that updates the halo values and two scalar products. Hence, the number of messages is proportional to the number of iterations, which typically grows with the system size.

Since we solve P linear systems concurrently, a second possibility to distribute the computations onto the processing nodes exists. Instead of letting every node process the chosen hierarchy level of its own domain for each of the P systems, it is possible to assemble one complete linear system on each processor. The systems are then solved locally without any communication, and finally the solution is sent back to the domains. We call this approach *domain sharing*.

Domain sharing requires copies of all domains on every other node which usually is impossible due to the involved memory requirements. However, we have seen that an accurate solution is not required in many areas of the graph, especially if the number of partitions is large. Hence, mainly lower levels of the hierarchy have to be copied which reduces the memory requirements significantly.

5 Experimental Results

In this Section we present some of our experiments executed on a Fujitsu Siemens hpcLine2. This system consists of 200 computing nodes, each of which has two Intel Xeon 3.2 GHz EM64T processors and 4 GB RAM. In our tests, we only use a single processor per node. We apply the Intel Compiler 8.1 and the Scali MPI implementation via the Infiniband interconnection. The test set comprises a number of two- and three-dimensional FEM graphs of different sizes. Since the results are similar, we only include five of them here.

As mentioned, it has already been shown [15,14] that the Bubble-FOS/C algorithm is able to produce partitionings with few boundary vertices. Since the solution quality varies only little in the settings, we focus our attention on the run-time improvements here.

Table 1 displays the recorded run-times for the five selected graphs. The first column contains the values for the classical domain decomposition approach without constructing a level hierarchy. Note that with the number of partitions the number of linear systems doubles, as well as the number of CPUs. Hence, in the optimal case of this setting all run-times were roughly the same, slightly varying due to the different right hand sides of the linear systems and the resulting number of CG iterations. Of course, the communication overhead prohibits this.

The middle column lists the run-times applying the level approach and domain decomposition. Though we construct the hierarchy and solve the additional small systems on the lowest levels, usually comparable run-times for 8 processors can be achieved. Note that with 8 processors it is very likely that every part of the hierarchy has to be solved on the highest level, especially for three-dimensional graphs, since almost all domains share a common border. If the number of partitions increases, we notice some run-time reduction in contrast to the approach

Table 1. Running times (s) for Bubble-FOS/C algorithm without the level approach (nolevel, domain decomposition by default), with hierarchy and domain decomposition (DD), and with hierarchy and domain sharing (DS). The shock9 ($|V| = 36476, |E| = 71290, \phi = 0.008$), ocean ($|V| = 143437, |E| = 409593, \phi = 0.06$), wave ($|V| = 156317, |E| = 1059331, \phi = 0.07$), auto ($|V| = 448695, |E| = 3314611, \phi = 0.125$), and hermes ($|V| = 320194, |E| = 3722641, \phi = 0.12$) graphs have been repartitioned on 8, 16, 32, and 64 processors respectively.

Graph	nolevel (DD) 8	16	32	64	DD 8	16	32	64	DS 8	16	32	64
shock9	2.68	2.97	3.53	4.17	2.49	2.26	1.53	1.74	1.96	1.34	0.99	1.01
ocean	7.27	8.43	8.90	9.61	8.79	6.67	4.90	8.17	8.84	4.75	3.17	3.02
wave	17.93	18.48	20.75	37.02	31.90	25.76	26.79	25.26	43.67	27.40	18.16	10.22
auto	37.92	40.84	48.39	53.08	126.15	106.04	79.08	52.19	141.93	77.88	42.30	24.53
hermes	70.89	74.35	77.63	112.93	191.71	144.11	95.17	87.26	165.75	105.31	54.08	29.08

without levels. However, for 64 processors the run-times increase again for the two small graphs, which can be explained with the increasing communication overhead.

The results of the domain sharing approach can be found in the right column. Although large messages are sent before and after solving the linear systems, it turns out that avoiding communication inside the CG solver speeds up the calculation significantly. This advantage becomes larger with a growing number of partitions, because the fraction of vertices where no exact solution is required increases as well. Hence, for larger number of processors (32 and more) this new scheme shows a clear improvement to the original method regarding run-time.

6 Conclusion

We have presented the parallel load balancing heuristic Bubble-FOS/C and significant improvements concerning a parallel implementation. By introducing an extra vertex, we are able to improve the condition of the involved matrices and therefore the numerical stability and complexity, without changing the matrix structure. Constructing local hierarchies and solving the linear systems with partial accuracy reduces the problem size and therefore the memory requirements. This allows us to solve the linear systems locally and avoid high latency communication inside the solver, which leads to a significant run-time reduction in case of a larger number of partitions.

In the future, it would be interesting to replace the Conjugate Gradient method and combine the presented hierarchical approach with a faster algebraic multigrid solver instead. Note that the latter is based on hierarchy levels by default.

References

[1] G. Cybenko. Dynamic load balancing for distributed memory multiprocessors. *Parallel and Distributed Computing*, 7(2):279–301, 1989.
[2] R. Diekmann, A. Frommer, and B. Monien. Efficient schemes for nearest neighbor load balancing. *Parallel Computing*, 25(7):789–812, 1999.
[3] R. Diekmann, R. Preis, F. Schlimbach, and C. Walshaw. Shape-optimized mesh partitioning and load balancing for parallel adaptive FEM. *J. Parallel Computing*, 26:1555–1581, 2000.
[4] R. Elsässer, B. Monien, and R. Preis. Diffusion schemes for load balancing on heterogeneous networks. *Theory of Computing Systems*, 35:305–320, 2002.
[5] G. Fox, R. Williams, and P. Messina. *Parallel Computing Works!* Morgan Kaufmann, 1994.
[6] B. Hendrickson. Graph partitioning and parallel solvers: Has the emperor no clothes? In *Irregular'98*, number 1457 in LNCS, pages 218–225, 1998.
[7] B. Hendrickson and R. Leland. A multi-level algorithm for partitioning graphs. In *Supercomputing'95*, 1995.

[8] J. Hromkovic and B. Monien. The bisection problem for graphs of degree 4. In *Math. Found. Comp. Sci. (MFCS '91)*, volume 520 of *LNCS*, pages 211–220, 1991.

[9] Y. F. Hu and R. F. Blake. An improved diffusion algorithm for dynamic load balancing. *Parallel Computing*, 25(4):417–444, 1999.

[10] E. F. Kaasschieter. Preconditioned conjugate gradients for solving singular systems. *J. of Computational and Applied Mathematics*, 24(1-2):265–275, 1988.

[11] G. Karypis and V. Kumar. *MeTis: A Software Package for Partitioning Unstrctured Graphs, Partitioning Meshes, [...], Version 4.0*, 1998.

[12] B. W. Kernighan and S. Lin. An efficient heuristic for partitioning graphs. *Bell Systems Technical Journal*, 49:291–308, 1970.

[13] J. B. MacQueen. Some methods for classification and analysis of multivariate observations. In *Proc. of 5th Berkeley Symposium on Mathematical Statistics and Probability*, pages 281–297. Berkeley, University of California Press, 1967.

[14] H. Meyerhenke, B. Monien, and S. Schamberger. Accelerating shape optimizing load balancing for parallel FEM simulations by algebraic multigrid. In *Proc. 20th IEEE Intern. Parallel and Distributed Processing Symposium, (IPDPS'06)*, page 57 (CD). IEEE Computer Society, 2006.

[15] H. Meyerhenke and S. Schamberger. Balancing parallel adaptive fem computations by solving systems of linear equations. In *Proc. Euro-Par 2005*, number 3648 in LNCS, pages 209–219, 2005.

[16] L. Oliker and R. Biswas. PLUM: Parallel load balancing for adaptive unstructured meshes. *J. Par. Dist. Comp.*, 52(2):150–177, 1998.

[17] S. Schamberger. Graph partitioning with the Party library: Helpful-sets in practice. In *Comp. Arch. and High Perf. Comp.*, pages 198–205, 2004.

[18] S. Schamberger. On partitioning FEM graphs using diffusion. In *HPGC, Intern. Par. and Dist. Processing Symposium, IPDPS'04*, page 277 (CD), 2004.

[19] S. Schamberger. A shape optimizing load distribution heuristic for parallel adaptive FEM computations. In *Parallel Computing Technologies, PACT'05*, number 2763 in LNCS, pages 263–277, 2005.

[20] Kirk Schloegel, George Karypis, and Vipin Kumar. Multilevel diffusion schemes for repartitioning of adaptive meshes. *J. Par. Dist. Comp.*, 47(2):109–124, 1997.

[21] C. Walshaw. *The parallel JOSTLE library user guide: Version 3.0*, 2002.

[22] C. Walshaw and M. Cross. Parallel optimisation algorithms for multilevel mesh partitioning. *J. Parallel Computing*, 26(12):1635–1660, 2000.

[23] C. Walshaw, M. Cross, and M. G. Everett. A Localised Algorithm for Optimising Unstructured Mesh Partitions. *Intl. J. Supercomputer Appl.*, 9(4):280–295, 1995.

Improvement of the Efficiency of Genetic Algorithms for Scalable Parallel Graph Partitioning in a Multi-level Framework

Cédric Chevalier and François Pellegrini

LaBRI and INRIA Futurs
Université Bordeaux I
351, cours de la Libération, 33405 TALENCE, France
{cchevali, pelegrin}@labri.fr

Abstract. Parallel graph partitioning is a difficult issue, because the best sequential graph partitioning methods known to date are based on iterative local optimization algorithms that do not parallelize nor scale well. On the other hand, evolutionary algorithms are highly parallel and scalable, but converge very slowly as problem size increases. This paper presents methods that can be used to reduce problem space in a dramatic way when using graph partitioning techniques in a multi-level framework, thus enabling the use of evolutionary algorithms as possible candidates, among others, for the realization of efficient scalable parallel graph partitioning tools. Results obtained on the recursive bipartitioning problem with a multi-threaded genetic algorithm are presented, which show that this approach outperforms existing state-of-the-art parallel partitioners.

1 Introduction

Graph partitioning is an ubiquitous technique which has applications in many fields of computer science and engineering, such as workload balancing in parallel computing, database storage, VLSI design or bio-informatics. It is mostly used to help solving domain-dependent optimization problems modeled in terms of weighted or unweighted graphs, where finding good solutions amounts to computing, eventually recursively in a divide-and-conquer framework, small vertex or edge cuts that balance evenly the weights of the graph parts.

For instance, the obtainment of small and balanced bipartitions is essential to the reordering of sparse matrices by nested dissection [5]. This method consists in computing a small vertex set that separates the adjacency graph of the sparse matrix in two parts, ordering the separator vertices with the highest indices available, then proceeding recursively on the two separated subgraphs until their size is smaller than some specified threshold. The smaller and more balanced the separators are, the smaller the fill-in incurred at the factorization stage, and thus the number of operations required to factor the matrix (referred to as the operation count, or OPC), is likely to be.

Currently, general-purpose sequential ordering software such as SCOTCH [12] or METIS [9] can handle graphs of about ten million vertices on an average workstation. However, as the power of parallel machines increases, so does the size of the problems to handle, and since the large graphs which model these problems cannot be processed on a single computer without incurring swapping, it is necessary to resort to parallel graph ordering tools, based on parallel graph bipartitioning algorithms. Several such tools have already been developed [9], but their outcome is mixed. In particular, they do not scale well, as partitioning quality tends to decrease, and thus fill-in to increase much, when the number of processors which run the program increases.

The purpose of the PT-SCOTCH software ("*Parallel Threaded* SCOTCH", an extension of the sequential SCOTCH software), developed at LaBRI within the SCALAPPLIX project of INRIA Futurs, is to provide efficient parallel tools to partition graphs with sizes up to a billion vertices, distributed over a thousand processors. Among our target applications is the parallel ordering of large graphs.

PT-SCOTCH is still under development, but several results have already been achieved. Section 2 presents a constrained banding technique which, based on the characteristics of the local optimization algorithms that are used to refine the partitions, reduces considerably the size of the problem space without loss of quality, already allowing one to develop semi-parallel programs that can compute efficient bipartitions of graphs having a billion nodes. Section 3 describes how this reduction enables us to use genetic algorithms, which are highly scalable but slow to converge, in a practical way. Some graph ordering results are presented, using a multi-threaded shared-memory genetic algorithm, which illustrate the quality of the orderings that can be produced. Then comes the conclusion.

2 Reducing Problem Space in a Multi-level Framework

Experience has shown that best partition quality is achieved when using a multi-level framework. This method, which derives from the multi-grid algorithms used in numerical physics, repeatedly reduces the size of the graph to partition by finding matchings that collapse vertices and edges, computes an initial partition for the coarsest graph obtained, and projects the result back to the original graph [2,6,8]. It is most often combined with greedy iterative algorithms, such as Kernighan-Lin [10] or Fiduccia-Mattheyses [4] (FM), to refine the projected partitions at every level, so that the granularity of the solution is the one of the original graph and not the one of the coarsest graph.

Because of the local nature of both the FM and the uncoarsening algorithms, it is most likely that the refined partition computed at any level will not differ much from the partition that was projected back to this level, as this latter is itself the projection of a partition that was a local optimum in the coarser levels. Therefore, to refine a partition, FM-like algorithms may not need to know more of the graph topology than a small "band" around the boundary of the projected partition. The locality of the optimization process is already exploited in many implementations of FM-like algorithms which, in order to save time and memory,

Table 1. Some of the test graphs that we use

Graph	Size ($\times 10^3$) V	E	Average degree	Graph	Size ($\times 10^3$) V	E	Average degree
598a	111	742	13.37	audikw1	944	38354	81.28
aatken	43	88	4.14	b5tuer	163	3874	47.64
auto	449	3315	14.77	bmw32	227	5531	48.65
bcsstk29	14	303	43.27	bmwcra1	149	5248	70.55
bcsstk30	29	1007	69.65	crankseg2	64	7043	220.64
bcsstk32	45	985	44.16	inline1	504	18156	72.09
body	45	164	7.26	mt1	98	4828	98.96
bracket	63	367	11.71	oilpan	74	1762	47.77
coupole8000	1768	41657	47.12	ship001	35	2305	132.00
m14b	215	1679	15.64	shipsec5	180	4967	55.23
ocean	143	410	5.71	thread	30	2220	149.32
pwt	37	145	7.93	x104	108	5030	92.81
rotor	100	662	13.30	altr4	26	163	12.50
s3dkq4m2	90	2365	52.30	chanel1m	81	527	13.07
tooth	78	453	11.58	conesphere1m	1055	8023	15.21

compute and update vertex swapping gains only for vertices that have to be considered, that is, the ones that are in the immediate vicinity of vertices that currently belong to the separator. However, these vertices cannot be known in advance. Our idea is that, since the FM algorithm is local, we can constrain it to operate on a small, predefined band of graph vertices without changing significantly its outcome.

To validate this assumption, we have instrumented our SCOTCH sequential partitioning software in order to measure how much refined partitions differ from projected partitions. Since our current target application requires vertex separators, we have focused on them for these experiments, but the same kind of measures could be obtained from edge separation routines as well. The test graphs we have used in all of our experiments are well-known cases of various sizes, listed in Table 1.

For every separator computed in a nested dissection process (which stops when subgraphs are of sizes of about a hundred vertices), we accumulate the numbers of refined separator vertices that end up at a given distance from the projected separators. These results are presented in Table 2.

As expected, the overwhelming majority of refined separator vertices is not located at a distance greater than three from the vertices of the projected separators. Therefore, it can be assumed that the quality of partitions should not be impacted if refined partitions are computed on band graphs only. In order to validate this second assumption, we have developed in SCOTCH a partitioning method which extracts a band subgraph of given width from a given graph and its given initial separator, applies a FM separator refinement method to the initial separator of the band subgraph, and projects back the refined band separator to the full graph. We have then replaced all of our calls to the FM refinement algorithm by calls to this band FM refinement algorithm.

Table 2. Distance histogram (in % of the number of separator vertices) of the location of refined separator vertices with respect to projected separators. These statistics have been collected over all separators when performing nested dissection on the given graphs.

Graph	Distance					Graph	Distance				
	0	1	2	3	≥4		0	1	2	3	≥4
598a	76.23	23.45	0.32	0.00	0.00	audikw1	91.44	8.55	0.01	0.00	0.00
aatken	77.00	20.45	2.27	0.24	0.04	b5tuer	74.18	22.96	1.85	0.42	0.59
auto	77.89	21.89	0.22	0.00	0.00	bmw32	80.98	18.31	0.50	0.08	0.14
bcsstk29	82.04	17.66	0.30	0.00	0.00	bmwcra1	91.29	8.58	0.13	0.00	0.00
bcsstk30	87.17	12.53	0.29	0.01	0.00	crankseg2	95.80	4.17	0.01	0.02	0.00
bcsstk32	81.91	17.80	0.23	0.03	0.02	inline1	87.57	12.35	0.08	0.00	0.00
body	67.49	30.20	2.08	0.20	0.04	mt1	84.79	14.00	0.93	0.25	0.04
bracket	72.47	26.19	1.08	0.16	0.10	oilpan	77.60	20.54	1.20	0.17	0.49
coupole8000	90.23	9.74	0.03	0.00	0.00	ship001	91.43	8.51	0.05	0.00	0.00
m14b	78.65	21.17	0.18	0.00	0.00	shipsec5	82.29	17.28	0.41	0.03	0.00
ocean	60.43	32.86	4.58	1.29	0.84	thread	91.40	8.53	0.06	0.00	0.00
pwt	54.35	37.31	6.10	1.56	0.69	x104	86.64	12.81	0.51	0.03	0.00
rotor	77.09	21.99	0.75	0.11	0.06	altr4	74.19	24.89	0.80	0.12	0.00
s3dkq4m2	78.72	20.34	0.89	0.04	0.00	chanel1m	74.65	24.09	1.16	0.10	0.00
tooth	69.90	26.82	2.42	0.63	0.24	conesphere1m	82.16	17.67	0.17	0.00	0.00

The quality criterion that we have chosen is the operation count (OPC) required to factor the reordered matrix using a Cholesky method; it is an indirect measurement of the overall quality of all bipartitions, in the practical context of nested dissection ordering. The results that we obtain for all of our test matrices, using band graphs with a width of three, show only marginal differences in OPC compared to the original FM refinement algorithm, and no difference on average. An explanation to this is that, even if the separator cannot move more than three vertices away at any level, it has the ability to move again at the next levels to reach its local optimum, therefore compensating on several levels for the moves it could not do on a single level.

An interesting feature of band FM refinement is that is seems to be more stable than the classical FM algorithm. In the production version of SCOTCH, two runs of multi-level bipartitioning were performed for each subgraph, and then the best separator of the two was kept. When using band FM refinement, equivalent results are obtained with only one run, as presented in Table 3. Most of the time, the quality of band FM lies between the one exhibited by one and two runs of the classical FM method. In terms of time, we can evidence a moderate over-cost with respect to a single run of classical FM, because of the computation of the band graph. It seems that, by "amortizing" the move of the frontier, the band FM algorithm prevents it from exploring local minima that differ too much from the "pseudo-global" solution computed at the coarsest level and in which it could be trapped afterwards. Further experiments are required to investigate this.

Table 3. Comparison between band FM and classical FM. Tests have been run on a 375MHz IBM SP3.

Graph	Band FM (1 run)		FM (2 runs)		FM (1 run)	
	OPC	Time (s)	OPC	Time (s)	OPC	Time (s)
aatken	1.72e+11	6.17	**1.70e+11**	10.79	1.73e+11	5.38
auto	5.14e+11	47.09	**4.98e+11**	75.00	5.27e+11	39.40
bcsstk32	1.40e+9	1.16	**1.28e+9**	1.65	1.40e+9	1.02
coupole8000	7.57e+10	210.15	**7.48e+10**	346.81	7.57e+10	183.72
m14b	6.27e+10	21.4	6.31e+10	33.42	**6.03e+10**	17.56
tooth	**6.50e+9**	5.66	6.51e+9	9.01	6.71e+9	4.64
audikw1	5.58e+12	59.32	**5.48e+12**	86.78	5.64e+12	50.33
bmw32	3.15e+10	4.52	**2.75e+10**	6.51	3.07e+10	4.08
oilpan	2.92e+9	0.73	**2.74e+9**	0.95	2.99e+9	0.69
thread	4.17e+10	1.62	**4.14e+10**	2.30	4.17e+10	1.44
x104	1.84e+10	1.97	**1.64e+10**	2.60	1.80e+10	1.83
altr4	3.68e+8	1.55	**3.65e+8**	2.52	3.84e+8	1.32
conesphere1m	**1.83e+12**	122.03	1.85e+12	192.27	1.88e+12	100.19

By using this limitation of problem space, we can already devise a way to compute high-quality partitions of distributed 3D mesh graphs of up to a billion vertices: since the expected size of the separator of a n-vertex 3D mesh graph is in $O(n^{2/3})$ [14], the order of magnitude of the first separator of a 3D graph of about a billion vertices should be of about a million vertices, which can be handled by a sequential computer. Therefore, basing on existing parallel coarsening algorithms such as the one of [13], one can coarsen a distributed graph so as to get a coarsened graph that fits in the memory of a sequential computer, compute an initial bipartition of this coarse graph using existing sequential partitioners, and project back this partition as follows. During each uncoarsening step, once the separator has been projected back to the finer distributed graph, a centralized copy of the distributed band graph surrounding the projected separator is gathered on every processor. All of the processors can then run independently a classical sequential FM algorithm on their centralized band graph, leading to a better exploration of the reduced problem space, after which the best refined separator found is projected back to the finer distributed graph. This uncoarsening process is repeated up to obtain a distributed bipartition of the original graph. Recursive bipartitioning can then take place on the two parts created, with separators of smaller sizes.

The above scheme, which may be useful to handle large graphs at the expense of quite little work on top of existing software, is clearly not fully satisfactory, since the refinement of the partitions is sequential in nature, and thus not scalable. In fact, local optimization algorithms are not well suited, because of their iterative nature, while global heuristics, although more scalable, are usually not considered as good candidates because of the size of the problem spaces to explore. However, taking advantage of the reduction of problem space that we have evidenced, they could be, as described in the following.

3 Using Genetic Algorithms in the Reduced Problem Space

Currently, there exist only few software that do graph ordering in parallel, and their quality is not equivalent to the one of sequential algorithms. For instance, PaRMeTiS [9] implements a parallel version of a FM algorithm to refine its bipartitions but, in order to relax the strong sequentiality constraint of the algorithm when moving vertices that have neighbors on other processors, only such moves that improve the quality of the solution are accepted, therefore limiting the hill-climbing feature of the FM algorithm and reducing further the quality of the solutions as the number of processors (and thus, of potential distant neighbors) increase.

To avoid this intrinsic sequentiality problem, we have decided to turn to a completely different class of algorithms. Genetic algorithms (GA) are highly scalable meta-heuristics which allow to solve multi-criteria optimization problems using an evolutionary method. It is an iterative method that consists in simulating the evolution of a population of individuals which represent solutions to the problem, selecting best-fitting individuals as candidates for breeding the next generation. GA are known to converge very slowly and cannot therefore be applied to large graphs [1,3], but might be of use in the reduced problem spaces of band graphs. In the graph separation problem, every vertex can belong to three different domains: the separator, or any of the two separated parts. Therefore, every individual in the population is implemented as a linear array, similar in principle to a chromosome, which associates a number between 0 and 2 to any graph vertex index.

The reproduction operator is a classical multi-points cross-over operator, which is applied at a randomly-selected position of two mated individuals, and swaps one part of their arrays to produce two descendants. The mutation operator consists in swapping the part of randomly chosen vertices on some individuals. Since these naive operators cannot enforce that the crossed-over and mutated individuals be valid solutions, they are post-processed with a consistency-checking phase which adds vertices to the separator whenever necessary, and removes unneeded separator vertices.

Individuals are evaluated by means of a fitness function, which linearly combines dimensionless numbers such as the ratio of graph vertices that belong to the separator, the imbalance between the two parts, and the ratio of graph edges that link separator vertices. The first generation is made up of individuals that are mutations of the projected partition, plus some entirely random individuals which provide genetic diversity. To select and mate individuals, we have implemented several classical algorithms [7,11]. Although all methods behave quite similarly, best results were achieved with a mix of the elitism and roulette methods: the 5% best individuals are kept unconditionally, and each of the remaining ones is kept with a probability proportional to its fitness. Then, individuals are mated by pairs of descending fitness, and bred so as to keep constant population.

In order to increase concurrency in the GA algorithm, all of the individuals that are located on the same processor are considered as an isolated population

Table 4. OPC of the reordered *bcsstk29* matrix when multi-level band GA is used for all levels of nested dissection. Classical multi-level FM yields an OPC of $3.43e+8$ in $0.74s$.

Deme size	# Demes	Generations	OPC	Time (s)
40	1	25	5.322334e+08	4.05
80	1	25	5.370016e+08	7.95
80	1	100	4.355475e+08	25.72
40	2	25	4.653384e+08	6.61
40	2	100	4.569806e+08	20.17
80	8	100	3.751443e+08	50.90

(also called "deme") living on an island [15]. Only occasionally can a few "champions" move from one island to another, to propagate their successful chromosomes into other populations which can have been trapped in local optima. In our current sequential implementation, every deme is handled by a different thread. Migration is performed when the variety of the population in some deme decreases, *i.e.* when individuals are too similar to their local champion.

To evaluate the convergence speed of our GA algorithm, we have computed nested dissection orderings of several test graphs with our multi-level band GA method. All of our tests were run on the M3PEC machine of the Université Bordeaux I, an eleven-node IBM machine with eight 1.5 MHz dual-core processors and 32 GB of memory per node. Since our current implementation is thread-based only, timings of tests involving more than sixteen threads (written between parentheses) are estimated: these tests are still run on a single SMP node, with as many threads per core as necessary, and the running time is divided by the appropriate ratio. PARMETIS, however, uses MPI, and runs fully in parallel.

Table 4 provides some results for graph *bcsstk29*. These results show that GA converges quite well, and that quality can be improved by increasing computation time and/or population size. As expected, running times are high, but GA are highly scalable, so that computation time can be reduced by adding processors, and partitioning quality can be increased by giving more time.

The second class of experiments that we have run aimed at evaluating the scalability of our method in terms of quality and running time. In order to compare our ordering software to PARMETIS in similar conditions, we ran our method on numbers of processors p that are powers of two (while our method does not require it), and performed band GA on the first $\log_2(p)$ levels only, using band FM afterwards; we will refer to this method as "limited GA" (LGA) in all of the following. When running GA, the population is evenly spread on all of the threads, with at least 100 individuals on the whole and at least 25 individuals per deme; therefore, above 4 threads, the population doubles along with the number of threads.

Our results, which are summarized in Table 5, are extremely encouraging. First of all, partitioning quality is not degraded too much when the number of processors increases: on our worst case, *bmw32*, we loose about 60% in OPC quality between 1 and 64 processors, and the quality is almost constant for *coupole8000*. Above

Table 5. Comparison between PARMETIS (PM) and our multi-level limited band genetic algorithm (LGA) for several graphs. C_{LGA} and C_{PM} are the OPC for LGA and PM, respectively. Dashes indicate abortion due to memory shortage. LGA timings between parentheses are extrapolated times for cases requiring more than 16 threads, as we had to run several threads per core on a single SMP node. Timings for PARMETIS are provided for graph *altr4* to give an idea of its speed, but t_{PM} and t_{LGA} cannot be compared, because PM is a fully parallel program, while our LGA testbed is the purely sequential nested dissection routine of SCOTCH, which has been parametrized so as to run the multi-threaded LGA algorithm only during the uncoarsening phases of the first $\log_2(p)$ stages of the nested dissection process.

Test case	Number of processors or threads						
	1	2	4	8	16	32	64
bcsstk32							
C_{LGA}	1.60e+9	1.55e+9	1.67e+9	**1.82e+9**	**1.83e+9**	1.53e+9	2.07e+9
C_{PM}	**1.29e+9**	1.55e+9	**1.62e+9**	3.09e+9	4.11e+9	5.85e+9	4.01e+9
t_{LGA}	0.42	0.88	0.84	0.97	2.07	(2.86)	(4.06)
audikw1							
C_{LGA}	**5.68e+12**	**5.91e+12**	**5.70e+12**	**5.82e+12**	**5.99e+12**	**6.44e+12**	**6.02e+12**
C_{PM}	–	–	–	7.78e+12	8.88e+12	8.91e+12	1.07e+13
t_{LGA}	19.78	22.77	29.55	32.89	60.24	(74.64)	(91.78)
bmw32							
C_{LGA}	3.04e+10	3.44e+10	**3.75e+10**	**4.13e+10**	**4.64e+10**	**4.57e+10**	**5.01e+10**
C_{PM}	**2.84e+10**	**3.22e+10**	4.09e+10	5.11e+10	5.61e+10	5.74e+10	6.31e+10
t_{LGA}	1.69	1.79	2.48	2.36	3.67	(5.11)	(7.80)
altr4							
C_{LGA}	**3.46e+8**	**3.71e+8**	**4.23e+8**	**4.06e+8**	**4.31e+8**	**4.92e+8**	**4.71e+8**
C_{PM}	4.25e+8	4.20e+8	4.49e+8	4.46e+8	4.64e+8	5.03e+8	5.16e+8
t_{LGA}	0.65	1.78	2.25	1.95	3.36	(5.43)	(7.20)
t_{PM}	0.58	0.31	0.20	0.13	0.11	0.27	0.31
conesphere1m							
C_{LGA}	**1.90e+12**	**1.92e+12**	**1.99e+12**	**2.37e+12**	**2.34e+12**	**2.53e+12**	**2.63e+12**
C_{PM}	2.04e+12	2.20e+12	2.46e+12	2.78e+12	2.96e+12	2.99e+12	3.29e+12
t_{LGA}	44.03	69.66	86.47	90.44	120.87	(134.85)	(158.07)
coupole8000							
C_{LGA}	**7.64e+10**	**7.64e+10**	**7.62e+10**	**7.65e+10**	**7.66e+10**	**7.68e+10**	**7.66e+10**
C_{PM}	–	–	–	8.17e+10	8.26e+10	8.58e+10	8.71e+10
t_{LGA}	125.69	75.40	55.19	49.16	52.59	(61.93)	(77.26)
thread							
C_{LGA}	4.10e+10	3.99e+10	**4.41e+10**	**4.64e+10**	**4.43e+10**	**4.59e+10**	**5.19e+10**
C_{PM}	**3.65e+10**	**3.98e+10**	6.60e+10	1.03e+11	1.24e+11	1.53e+11	1.28e+11
t_{LGA}	0.56	2.33	3.10	2.93	4.22	(5.02)	(5.92)

8 processors, our results clearly outperform the ones of PARMETIS, by a factor greater than two for *thread*. As expected, the higher the degree of the graph is, the bigger the difference is, because PARMETIS can only do gradient local optimizations on nodes which have neighbors on other processors.

Partitioning times are very good, too. Although the running time of a single sequential band GA refinement algorithm is between 30 and 80 times higher than the one of its sequential band FM counterpart, the overall running time of our LGA ordering program does not increase too much when the number of processors increase. While a doubling of the number of processors implies the turning of a whole level of band FM refinements into band GA refinements, the running time of LGA increases reasonably along with the number of threads, because when the number of processors increases it is levels of smaller subgraphs that are passed to the GA, which only results in a limited increase in the overall running time compared to the time taken by the first GA levels. Much hope is therefore placed in the development of a fully parallel, distributed-memory LGA algorithm.

4 Conclusion and Future Work

In this paper, we have presented a constrained banding approach which dramatically decreases problem size during the refinement phase of multi-level partitioning schemes. This method, which can be used with any refinement algorithm, allows us to take advantage of heuristics which are usually too expensive to be considered, such as genetic algorithms. We have implemented a shared memory multi-threaded GA, and tried it on numerous test cases. Although our GA is slower than distributed FM-like algorithms, it is scalable and provides better results, and its quality can be parametrized more easily (in terms of population size and of number of generations) to account for eventual time or quality constraints.

We are currently developing a distributed memory version of our GA algorithm, based on MPI, which will allow us to run tests on a larger number of processors, and to investigate the limits of using GA as a band refinement method for very large graphs. Since the testbed that we will use for this new version will be the parallel ordering routine of PT-SCOTCH, we will be able to compare its running time with the one of other parallel ordering software. Moreover, in order to have a reference for the quality of orderings, we are also currently completing the coding in PT-SCOTCH of the centralized band FM refinement algorithm described at the end of Section 2, which will allow us to compute, in a semi-parallel fashion, high quality orderings of very large graphs.

References

1. S. Areibi and Zeng Y. Effective memetic algorithms for VLSI design automation = genetic algorithms + local search + multi-level clustering. *Evolutionary Computation*, 12(3):327–353, 2004.
2. S. T. Barnard and H. D. Simon. A fast multilevel implementation of recursive spectral bisection for partitioning unstructured problems. *Concurrency: Practice and Experience*, 6(2):101–117, 1994.
3. T. N. Bui and B. R. Moon. Genetic algorithm and graph partitioning. *IEEE Trans. Comput.*, 45(7):841–855, 1996.

4. C. M. Fiduccia and R. M. Mattheyses. A linear-time heuristic for improving network partitions. In *Proc. 19th Design Automat. Conf.*, pages 175–181. IEEE, 1982.
5. A. George and J. W.-H. Liu. *Computer solution of large sparse positive definite systems*. Prentice Hall, 1981.
6. B. Hendrickson and R. Leland. A multilevel algorithm for partitioning graphs. In *Proceedings of Supercomputing*, 1995.
7. J. Horn, N. Nafpliotis, and D. E. Goldberg. A niched Pareto genetic algorithm for multiobjective optimization. In *IEEE World Congress on Computational Intelligence*, volume 1, pages 82–87, 1994.
8. G. Karypis and V. Kumar. A fast and high quality multilevel scheme for partitioning irregular graphs. *SIAM J. on Scientific Computing*, 20(1):359–392, 1998.
9. MeTiS: Family of multilevel partitioning algorithms. http://glaros.dtc.umn.edu/gkhome/views/metis.
10. B. W. Kernighan and S. Lin. An efficient heuristic procedure for partitioning graphs. *Bell System Technical Journal*, 49:291–307, February 1970.
11. P. Moscato. On evolution, search, optimization, genetic algorithms and martial arts, towards memetic algorithms. Technical Report 826, California Intitute of Technology, Pasadena, CA 91125, U.S.A., 1989.
12. Scotch: Static mapping, graph partitioning, and sparse matrix block ordering package. http://www.labri.fr/~pelegrin/scotch/.
13. K. Schloegel, G. Karypis, and V. Kumar. Parallel multilevel algorithms for multi-constraint graph partitioning. In *Proceedings of EuroPar*, pages 296–310, 2000.
14. H. D. Simon and S.-H. Teng. How good is recursive bipartition. *SIAM J. Sc. Comput.*, 18(5):1436–1445, 1995.
15. D. Whitley, S. Rana, and R. B. Heckendorn. The island model genetic algorithm: On separability, population size and convergence. *Journal of Computing and Information Technology*, 7:33–47, 1999.

Probablistic Self-Scheduling

Milind Girkar[1], Arun Kejariwal[2],
Xinmin Tian[1], Hideki Saito[1], Alexandru Nicolau[2],
Alexander Veidenbaum[2], and Constantine Polychronopoulos[3]

[1] Intel Corporation
3600, Juliette Lane
Santa Clara, CA, 95050, USA
[2] Center for Embedded Computer Systems
University of California at Irvine
Irvine, CA 92697, USA
[3] Center for Supercomputing Research and Development
University of Illinois at Urbana-Champaign
Urbana, IL 61801, USA

Abstract. Scheduling for large parallel systems such as clusters and grids presents new challenges due to multiprogramming/polyprocessing [1]. In such systems, several jobs (each consisting of a number of parallel tasks) of multiple users may run at the same time. Processors are allocated to the different jobs either statically or dynamically; further, a processor may be taken away from a task of one job and be reassigned to a task of another job. Thus, the number of processors available to a job varies with time. Although several approaches have been proposed in the past for scheduling tasks on multiprocessors, they assume a dedicated availability of processors. Consequently, the existing scheduling approaches are not suitable for multiprogrammed systems. In this paper, we present a novel probabilistic approach for scheduling parallel tasks on multiprogrammed parallel systems. The key characteristic of the proposed scheme is its self-adaptive nature, i.e., it is responsive to systemic parameters such as number of processors available. Self-adaptation helps achieve better load balance between the different processors and helps reduce the synchronization overhead (number of allocation points). Experimental results show the effectiveness of our technique.

1 Introduction

Scheduling for parallel systems is done at two levels. At the first level, jobs (of different users) are scheduled such that each job receives a fair share of the resources. On the other hand, tasks of a job are scheduled on different processors such that the overall completion time (also known as *makespan*) is minimized. In context of dynamic scheduling schemes, this also involves minimizing the runtime scheduling overhead. Processors may be allocated (to a job) either statically or dynamically. Further, a processor may be taken away from a task of one job and be reassigned to a task of another job. As a consequence, the number of

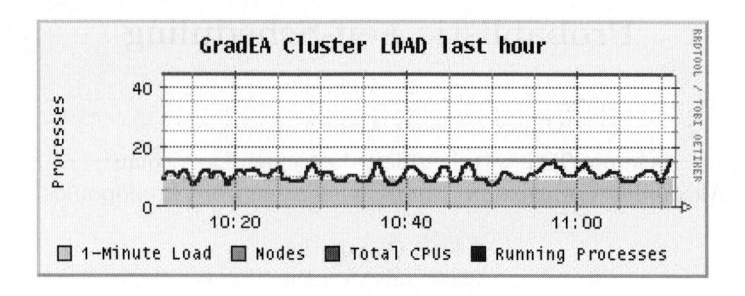

Fig. 1. Variation in the number of processes over an hour on a real multiprogramming system with 15 nodes (44 CPUs) (http://www.gradea.uci.edu)

processors available to a job varies with time. To validate this, we recorded the number of user-level processes on a real multiprogrammed system for an hour (see Figure 1). Clearly, the number of idle processors varies with time, by as much as 10%, which has a direct effect on the performance of a scheduling policy. Therefore, a scheduling policy should be designed such that it is aware of such systemic variations.

In this paper, we address the problem of scheduling parallel tasks of a given job. Without loss of any generality, we focus on scheduling iterations of a DOALL loop [2]; note that the proposed technique is general in nature, e.g., it can also be used for scheduling coarse-grain (function-level) parallel independent tasks. We model the problem as a task allocation problem wherein at any scheduling step, given a set of idle processors, one or more iterations are allocated to each processor. The key consideration in task allocation is the selection of the task size, i.e., the number of iterations constituting a task. While a small task size incurs significant scheduling overhead, a large task size results in load imbalance. Thus, the task allocation problem naturally reduces to determining the optimal task size in order to minimize the total execution time. For this, several self-scheduling techniques have been proposed for scheduling parallel loops [3]. However, none of the existing techniques account for the variation in the number of available processors with time. For this, we propose a novel approach, referred to as *Probabilistic Self-Scheduling* (PSS), for scheduling of (nested) parallel loops on multiprogrammed parallel systems. At any scheduling step, the number of iterations allocated to an idle processor is determined based on the number of remaining iterations and the number of processors expected to be available in future. The latter is determined based on the the number of processors available in the past. The proposed approach is compatible with the environment established by auto-scheduling compilers [4].

The rest of the paper is organized as follows. In the next section, we present a motivating example. Section 3 presents our approach PSS. Experimental setup and results are presented in Section 4. Previous work is discussed in Section 5. Finally, in Section 6, we conclude with directions for future research.

2 A Motivating Example

In this section we illustrate the intuitive idea behind our approach (PSS) with the help of an example. For comparison purposes, we consider two well known self-scheduling techniques, viz., *guided self-scheduling* GSS(1) [5] and *factoring* [6]. Assuming identical processors, at a given scheduling step GSS(1) assigns $\frac{1}{\mathbf{P}}$ of the remaining iterations to an idle processor, where $\mathbf{P}$ is the total number of processors; factoring assigns iterations to the processors in batches of $\mathbf{P}$ chunks, where the batch size is half the number of remaining iterations, for example, given 100 iterations and 4 processors, the initial batch size is 50 ($= 100/2$) and the chunk size of the first four chunks is 13 ($= \lceil 50/4 \rceil$). Consider a multiprogrammed system consisting of 4 processors. Let processors P_1 and P_2 be available for time $t \geq 0$ and $t \geq 22$ respectively and let processors P_3 and P_4 be busy serving other jobs in the system. For a `DOALL` loop with 100 iterations (for simplicity of exposition, we assume that each iteration has a workload of 1 unit), the chunk sizes for GSS(1) and factoring are shown in Table 1. From the table we note that GSS(1) and factoring incur large synchronization overhead due to large number of allocation points. This can be attributed to the fact that GSS(1) and factoring are oblivious of the number of processors available. In contrast, PSS assigns $\frac{1}{E[P]}$ of the remaining iterations to an idle processor, where $E[P]$ is the average number of processors available to the job under consideration. In the current context, $E[P] = 2$, as processors P_3 and P_4 are never available for scheduling. The chunk sizes for PSS is shown in Table 1. From the table we see that PSS reduces the synchronization overhead by 50% w.r.t. GSS(1) and by 65% w.r.t. factoring. Clearly, PSS yields better performance than GSS(1) and factoring as it incurs far less synchronization overhead.

Table 1. Total number of iterations = 100, $\mathbf{P} = 4$

Scheme	Chunk Sizes	# of Allocation Points
GSS(1)	25 19 14 11 8 6 5 3 2 2 2 1 1 1	14
Factoring	13 13 13 13 6 6 6 6 3 3 3 3 2 2 2 2 1 1 1 1	20
PSS	50 25 13 6 3 2 1	7

3 The Approach

In this section we present the algorithm for our approach - *Probabilistic Self-Scheduling*. Although several models have been proposed, viz., global, local and hybrid, for work queues in context of self-scheduling, we adopt the model proposed by Polychronopoulos and Kuck in [5] owing to its simplicity. Note that model selection is orthogonal to the concerns we address in this paper. The algorithm is designed for non-preemptive scheduling, whereby a task once assigned to a processor may not be removed until it has finished execution. The rest of the section describes the different phases of our scheduling algorithm.

3.1 Expected Processor Availability

As discussed in the previous section, the presence of other jobs in a multiprogramming environment has direct impact on the performance of a self-schedule. In order to address the above, at a given scheduling step t, PSS computes the chunk size (discussed further in subsection 3.2) based on the number of remaining iterations and the expected value of the number of processors available after step t assuming that it would be the same as the average number of processors available in the past. Before discussing how to compute the above, we defines some terms of probability (for a detailed discussion, the reader is referred to the book by Meyer [7]).

Preliminaries

Let X be a discrete random variable and its range space, denoted by R_X, consist of a countably infinite number of values, $x_1, x_2, \ldots$. With each possible outcome x_i, we associate a number $p(x_i) = P(X = x_i)$, called the probability of x_i. The numbers $p(x_i), i = 1, 2, \ldots$ must satisfy the following:

$$p(x_i) \geq 0, \quad \forall i$$

$$\sum_{i=1}^{\infty} p(x_i) = 1.$$

The function p defined above is called the *probability function* of the random variable X. The collection of pairs $(x_i, p(x_i))$, for $i = 1, 2, \ldots$ is called the *probability distribution* of X.

Definition 1. *The expected value of a discrete random variable X, denoted by $E(X)$, is defined as:*

$$E(X) = \sum_{i=1}^{\infty} x_i p(x_i) \tag{1}$$

if the series $\sum_{i=1}^{\infty} x_i p(x_i)$ converges absolutely, i.e., if $\sum_{i=1}^{\infty} |x_i| p(x_i) < \infty$. $E(X)$ is also referred to as the mean value of X.

Processor availability

We model the number of processor available at each scheduling step as a discrete random variable P. At each scheduling step t, we record the number of available processors. Also, we determine the expected value of the number of processors available subsequently. For this, we define a window of width w to compute the above. The processor availability during this window can be represented as a histogram, as illustrated in Figure 2(a). From this histogram, the probability distribution of processor availability is computed [7]. For example, for the window shown in Figure 2(a), $p(x_i = 8) = 4/16 = 0.25$, as shown in Figure 2(b). Finally, the expected value is computed using Equation 1.

The window width is parameterized. A larger width increases the accuracy of the update process, however, it incurs more overhead. It has been shown that

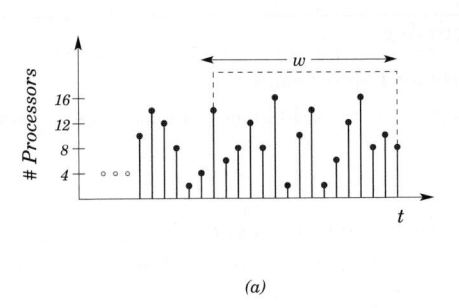

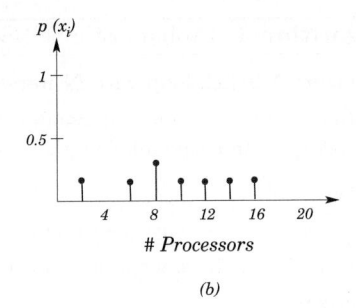

Fig. 2. An illustration of how to determine the probability distribution from the processor availability record. a) For a given time t, processor availability in the past; b) probability distribution of processor availability during the window w. (Total number of processors in the multiprogrammed system $\mathbf{P} = 20$)

run-time performance measurement via use of hardware performance counters incurs minimal scheduling overhead [8]. Note that the expected value computed above is not fixed. This is due to the fact that the processor availability profile in two different windows need not be the same, as evident from Figure 1. Also, the expected value cannot be determined statically as the processor availability profile in a given window is non-deterministic. Hence, under PSS, the expected value is "updated" at every scheduling step.

3.2 Chunk Size

Under GSS, at a given scheduling step, the chunk size (denoted by Λ) is determined as follows:

$$\Lambda = \left\lceil \frac{W_R}{\mathbf{P}} \right\rceil \qquad (2)$$

where, W_R is the number of remaining iterations. However, as discussed in [6], the above may result in allocation of too much work to early chunks; specifically, two-thirds of the work is assigned to first $\mathbf{P}$ chunks in case of identical processors. It has been shown that 50% of the total number of iterations is sufficient to even out the finishing times of the processors [6]. Therefore, we introduce a correction factor to "relax" the exponential decay of chunk size. Assuming identical processors, the number of iterations remaining after $\mathbf{P}$ allocations can be approximated as $(1 - \frac{1}{\eta \mathbf{P}})^{\mathbf{P}} W_R$, where η is the correction factor and $\mathbf{P}$ is the number of processors. From the above, η must satisfy the following:

$$\lim_{\mathbf{P} \to \infty} \left(1 - \frac{1}{\eta \mathbf{P}}\right)^{\mathbf{P}} = 0.5$$

Therefore, $\eta = 1.5$. The modified formula for the function Λ is as follows:

$$\Lambda = \left\lceil \frac{W_R}{1.5\,\mathbf{P}} \right\rceil \qquad (3)$$

Algorithm 1. Probabilistic Self-Scheduling

Input: A DOALL loop with **N** iterations and **P** processors.
Output : A near-optimal dynamic schedule w.r.t. load balance amongst the different processors and schedule length
$\mathbf{W_R} \leftarrow \mathbf{N}$
/* Generate the schedule (assuming implicit loop coalescing [10][1]) */
Let $P_{\text{idle}} \subseteq \mathbf{P}$ be a set of idle processors at any given scheduling step
repeat
  if $|P_{\text{idle}}| \: != 0$ **then**
    Determine $E[P]$
    for all $p_i \in P_{\text{idle}}$ **do**
      /* Compute the chunk size */

$$\Lambda = \max\left(W_{\min}, \left\lceil \frac{\mathbf{W_R}}{1.5 E[P]} \right\rceil\right) \quad (5)$$

    Compute index range for each processor
    Allocate the iterations corresponding to index range to p_i
  end for
end if
$\mathbf{W_R} \leftarrow \mathbf{W_R} - |P_{\text{idle}}| \times \Lambda$
until $\mathbf{W_R} > 0$

where, $W_{\min}$ is the minimum chunk size (pre-specified by the user).

Based on our discussion in the previous subsection, we adapt Equation 3 for multiprogramming systems as follows:

$$\Lambda = \left\lceil \frac{W_R}{1.5 \: E[P]} \right\rceil \quad (4)$$

So far, the chunk size is computed oblivious of the variation in the number workload (execution time) of the different iterations. To account for this, Equation 4 can be further refined as proposed in [9]. A detailed discussion of an integrated approach is beyond the scope of the paper.

3.3 The Algorithm

In this section we present a formal description of the algorithm for PSS. At each scheduling step, Algorithm 1 first determines the expected number of available processors (refer to subsection 3.1). Subsequently, it determines the chunk size Λ (given by Equation 5), i.e., the number of iterations to be allocated to an idle processor p_i. Next, it determines the range of the iterations to be mapped to processor and maps the corresponding iterations on to processor p_i. Note that

PSS is an online algorithm as the chunk size is determined at run-time based on $\mathbf{W_R}$ and $E[P]$.

4 Experiments

We obtained traces of processor availability on a real multiprogramming system with 15 nodes (44 CPUs) (http://www.gradea.uci.edu). Also, we extracted several kernels (DOALL loops $L_1, L_2, \ldots, L_{10}$) from SPEC OMP 2001M [11] and other scientific applications such as LAMMPS [12] and DAKOTA [13]. We used the above two as inputs to our simulator [14] to compare the performance of PSS with *adaptive self-tuning scheduling* [15] (referred to as HLS in the rest of the paper). For consistency purposes (w.r.t. the task granularity), we only consider the "upper algorithm" of HLS which does scheduling at the iteration level. HLS samples the performance of a number of self-scheduling techniques, such as guided self-scheduling, factoring, trapezoidal self-scheduling et cetera, at runtime to determine the best scheme for each loop in a given application program. Thus, HLS is in essence the best of all the self-scheduling techniques proposed so far. Due to this, we demonstrate the effectiveness of our approach over HLS only.

4.1 Results

We conducted two sets of experiments: (i) First, we evaluated the effectiveness of our approach and compare it with HLS for a small multiprogramming system (http://www.gradea.uci.edu) with 15 nodes (44 CPUs); (ii) Second, assuming random processor availability, we evaluated the effectiveness of PSS for number of processors — 2,068 as in the Bigben [16] and 10,240 processors such as in the Columbia supercomputer [17]. Note that the applicability of our approach is not restricted to any particular processor configuration.

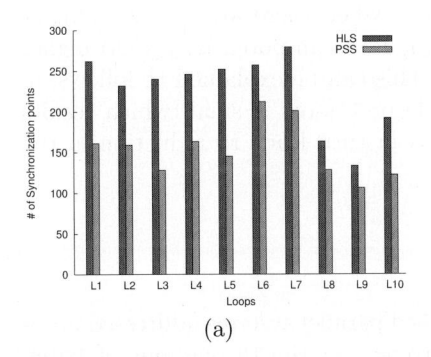

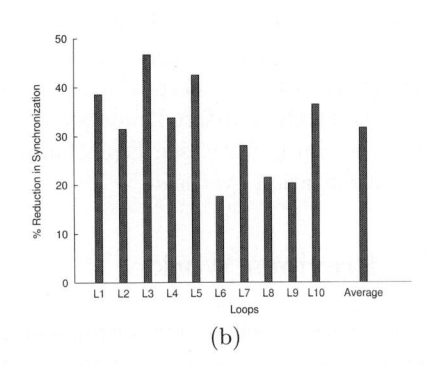

Fig. 3. a) Number of synchronization points for the different kernels; b) % Reduction in synchronization

---

[1] Loop coalescing transforms multiply nested DOALL loops into singly nested loops.

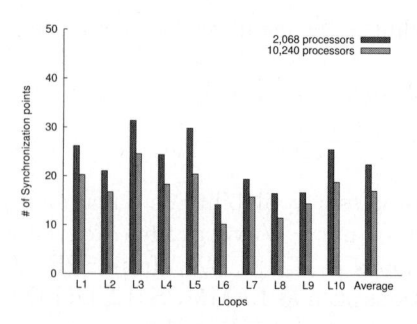

Fig. 4. % Reduction in synchronization (w.r.t. HLS) on systems with 2,068 and 10,240 processors

Figure 3 presents a performance (number of synchronization points) comparison of PSS with HLS. In order to minimize the effect of uneven start times of the processors, the number of synchronization points required was computed as an average of 10 simulation runs. We observe that PSS reduces the synchronization overhead by a maximum of 46.67% and by 31.67% on an average. The decrease in synchronization directly increases performance at the application level. The better allocation of the processors will also tend to increase the performance at the system level. The latter can be attributed to the reduced contention for accessing the interconnection network which yields higher throughput.

Next, we evaluated the performance of PSS for large parallel systems, such as the Bigben [16] and the Columbia supercomputer [17]. Since we did not have processors availability traces for such systems, we simulated the same using a random number generator, as in [5]. Figure 4 presents the results for the performance (% reduction in synchronization) of PSS w.r.t. HLS for 2,068 and 10,240 processors. From the figure, we see that PSS reduces synchronization overhead by a maximum of 29.86% and 22.55% for a system consisting of 2,068 and 10,240 processors respectively and by 22.54% and 17.14% on an average respectively. In case of heavy workloads (i.e., when there are a large number of jobs of other users running on the system) PSS can potentially yield higher reduction in the synchronization overhead. This can be explained as follows: in such cases the expected number of available processors is small which results in large chunk sizes, see Equation 4. This is in turn leads to reduction in the synchronization overhead.

5 Previous Work

Early work on scheduling for multiprogrammed parallel systems addressed problems such as the effect of program concurrency on the throughput of batch processing systems [18,19]. Later, Ousterhout proposed *co-scheduling*, where groups of cooperating processes are assigned processors at the same time [20].

In order to facilitate sharing of the multiprocessor system amongst several groups of processors, a group of cooperating processes would execute on the processors in time-multiplexed fashion. Rommel et al. analyzed the processor sharing discipline in context of parallel jobs running on uniprocessor systems [21]. Approaches for multiprogramming distributed memory systems are discussed in [22,23]. Policies for processor allocation in multiprogrammed environments are discussed in [24,25]. Program characterization and performance evaluation of scheduling algorithms in multiprogrammed systems is discussed in [26,27,28].

Probabilistic scheduling approaches have been proposed in several different fields of research. In [29], Chandy and Reynolds proposed an approach for scheduling partially ordered tasks with probabilistic execution times. Bruno and Downey studied the probabilistic bounds on list scheduling in [30]. Tongsima et al. [31] proposed confidence-based probabilistic scheduling of data-flow graphs. In [32], Som et al. presented a probabilistic event scheduling policy for optimistic parallel discrete event simulation. Burns et al. [33] proposed a scheduling policy based on probabilistic guarantees for fault-tolerant real-time systems. In [34], Fujita and Zhou proposed a multiprocessor scheduling problem with probabilistic execution costs. Li and Pan presented a probabilistic analysis of scheduling precedence constrained parallel tasks on multicomputers with contiguous processor allocation [35]. Moulin [36] proposed a probabilistic approach for split-proof[2] scheduling of parallel jobs to ensure fairness between the different users. Özsoy [37] investigated the effect of coordinated splitting by several users and proposed that the *uniform rule* — given n jobs, choose each ordering of the n jobs (for scheduling) with an equal probability of $1/n!$ — is the only rule immnue to coordinated splitting. Recently, Glatard et al. [38] proposed a probabilistic approach for job partitioning and scheduling on a grid infrastructures. The problem addressed in each of the aforementioned works is orthogonal to the problem addressed in this paper (load balancing between the different processors). Furthermore, the techniques proposed in prior work assume that a "fixed" number of processors are available for scheduling each job. This assumption is not representative of the multiprogrammed systems thereby restricting their applicability.

6 Conclusion

In this paper we presented an algorithm for self-scheduling of parallel tasks in multiprogrammed systems. The key characteristic of our approach is the dynamic adaptation of the chunk size based on the variation in the number of available processors. The approach achieves dual objectives: (i) it achieves load balance between different processors; and (ii) reduces the synchronization overhead by reducing the number of allocations points. As future work, we would like to extend our approach to address other issues such as minimizing maximum tardiness.

---

[2] Under *splitting*, a user breaks down his job into multiple smaller jobs under different aliases. This can potentially reduce the expected wait time if the shortest jobs are served first.

References

1. C. D. Polychronopoulos. Multiprocessing vs multiprogramming. In *Proceedings of the 1989 International Conference on Parallel Processing*, pages II–223–II–230, August 1989.
2. S. Lundstrom and G. Barnes. A controllable MIMD architectures. In *Proceedings of the 1980 International Conference on Parallel Processing*, St. Charles, IL, August 1980.
3. A. Kejariwal and A. Nicolau. Reading list of self-scheduling of parallel loops. http://www.ics.uci.edu/~akejariw/SelfScheduleReadingList.pdf.
4. C. D. Polychronopoulos. Towards autoscheduling compilers. *Journal of Supercomputing*, 2(3):297–330, 1988.
5. C. D. Polychronopoulos and D. J. Kuck. Guided self-scheduling: A practical scheduling scheme for parallel supercomputers. *IEEE Transactions on Computers*, 36(12):1425–1439, 1987.
6. S. F. Hummel, E. Schonberg, and L. E. Flynn. Factoring: a method for scheduling parallel loops. *Communications of the ACM*, 35(8):90–101, 1992.
7. P. L. Meyer. *Introductory Probability and Statistical Applications*. Reading, MA, 1970.
8. A. B. Downey and D. G. Feitelson. The elusive goal of workload characterization. *SIGMETRICS Performance Evaluation Review*, 26(4):14–29, 1999.
9. A. Kejariwal, A. Nicolau, and C. D. Polychronopoulos. Feedback-based guided self-scheduling. In *Proceedings of the 12th SIAM Conference on Parallel Processing for Scientific Computing*, San Francisco, CA, February 2006.
10. C. Polychronopoulos. Loop coalescing: A compiler transformation for parallel machines. In *Proceedings of the 1987 International Conference on Parallel Processing*, pages 235–242, August 1987.
11. SPEC OMP. http://www.spec.org/omp.
12. LAMMPS. http://www.cs.sandia.gov/~sjplimp/lammps.html.
13. DAKOTA. http://endo.sandia.gov/DAKOTA/software.html.
14. A. Kejariwal, A. Nicolau, and C. D. Polychronopoulos. An efficient approach for self-scheduling parallel loops on multiprogrammed parallel computers. In *Proceedings of the 18th International Workshop on Languages and Compilers for Parallel Computing*, Hawthorne, NY, October 2005.
15. Y. Zhang, M. Burcea, V. Cheng, R. Ho, and M. Voss. An adaptive OpenMP loop scheduler for hyperthreaded SMPs. In *Proceedings of the 17th International Conference for Parallel and Distributed Computing Systems*, San Francisco, CA, 2004.
16. Bigben: Pittsburgh Supercomputing Center. http://www.psc.edu/machines/cray/xt3/bigben.html.
17. SGI Altix: Columbia Supercomputer. http://www.nas.nasa.gov/Resources/Systems/columbia.html.
18. J. C. Browne, K. M. Chandy, J. Hogarth, and C. Lee. The effect on throughput in multi-processing in a multi-programming environment. *IEEE Transactions on Computers*, C-22(8):728–735, August 1973.
19. C. H. Sauer and K. M. Chandy. The impact of distributions and disciplines on multiple processor systems. *Communications of the ACM*, 22(1):25–34, 1979.

20. J. Ousterhout. Scheduling techniques for concurrent systems. In *Proceedings of the Conference on Distributed Computing Systems*, pages 22–30, 1982.
21. C. G. Rommel, D. Towsley, and J. A. Stankovic. Analysis of fork-join jobs using processor-sharing. Technical Report UM-CS-1987-052, University of Massachusetts, 1987.
22. M. R. Leuze, L. W. Dowdy, and K. H. Park. Multiprogramming a distributed-memory multiprocessor. *Concurrency: Practice and Experience*, 1(1):19–33, 1989.
23. S. K. Setia, M. S. Squillante, and S. K. Tripathi. Processor scheduling on multiprogrammed, distributed memory parallel computers. *SIGMETRICS Performance Evaluation Review*, 21(1):158–170, 1993.
24. C. McCann, R. Vaswani, and J. Zahorjan. A dynamic processor aladdress policy for multiprogrammed shared-memory multiprocessors. *ACM Transactions on Computer Systems*, 11(2):146–178, May 1993.
25. K. C. Sevcik. Application scheduling and processor aladdress in multiprogrammed parallel processing systems. *Performance Evaluation*, 19(2-3):107–140, 1994.
26. A. R. Miller. *Nonpreemptive run-time scheduling issues on a multitasked, multiprogrammed multiprocessor with dependencies, bidimensional tasks, folding and dynamic graphs*. PhD thesis, Department of Computer Science, University of Illinois at Urbana-Champaign, 1987.
27. S. Majumdar, D. L. Eager, and R. B. Bunt. Scheduling in multiprogrammed parallel systems. In *Proceedings of the 1988 ACM SIGMETRICS conference on Measurement and modeling of computer systems*, pages 104–113, Santa Fe, NM, 1988.
28. S. T. Leutenegger and M. K. Vernon. The performance of multiprogrammed multiprocessor scheduling algorithms. In *Proceedings of the 1990 ACM SIGMETRICS conference on Measurement and modeling of computer systems*, pages 226–236, Boulder, CO, 1990.
29. K. M. Chandy and P. F. Reynolds. Scheduling partially ordered tasks with probabilistic execution times. In *Proceedings of the Fifth Symposium on Operating Systems Principles*, pages 169–177, Austin, TX, 1975.
30. J. Bruno and P. Downey. Probabilistic bounds on the performance of list scheduling. *SIAM Journal of Computing*, 15(2):409–417, 1986.
31. S. Tongsima, C. Chantrapornchai, E. H.-M Sha, and N. L. Passos. Scheduling with confidence for probabilistic data-flow graphs. In *Proceedings of 7th Great Lakes Symposium on VLSI*, pages 150–155, Urbana, IL, 1997.
32. T. K. Som and R. G. Sargent. A probabilistic event scheduling policy for optimistic parallel discrete event simulation. In *Proceedings of 12th Workshop on Parallel and Distributed Simulation*, pages 56–63, Banff, Alberta, Canada, May 1998.
33. A. Burns, S. Punnekkat, L. Strigini, and D. R. Wright. Probabilistic scheduling guarantees for fault-tolerant real-time systems. In *Proceedings of Seventh IFIP International Working Conference on Dependable Computing for Critical Applications*, pages 361–378, San Jose, CA, January 1999.
34. S. Fujita and H. Zhou. Multiprocessor scheduling problem with probabilistic execution costs. In *Proceedings International Symposium on Parallel Architectures, Algorithms and Networks*, pages 121–126, Dallas/Richardson, TX, December 2000.
35. K. Li and Y. Pan. Probabilistic analysis of scheduling precedence constrained parallel tasks on multicomputers with contiguous processor allocation. *IEEE Transactions on Computers*, 49(10):1021–1030, 2000.

36. H. Moulin. Split-proof probabilistic scheduling. In *New Trends in Co-operative Game Theory*, January 2005.
37. H. Özsoy. Coordinated splitting in probabilistic scheduling. In *Public Economic Theory*, Marseille, France, 2005.
38. T. Glatard, , J. Montagnat, and X. Pennec. Probabilistic and dynamic optimization of job partitioning on a grid infrastructure. In *Proceedings of the 14th Euromicro Conference on Parallel, Distributed and Network-based Processing*, Montbilard-Sochaux, France, February 2006.

Data Sharing Conscious Scheduling for Multi-threaded Applications on SMP Machines

Shlomit S. Pinter and Marcel Zalmanovici

IBM Haifa Research Lab, Haifa University, Mount Carmel,
31905 Haifa, Israel
{shlomit, marcel}@il.ibm.com

Abstract. Extensive use of multi-threaded applications that run on SMP machines, justifies modifications in thread scheduling algorithms to consider threads' characteristics in order to improve performance. Current schedulers (e.g. in Linux, AIX) avoid migrating tasks between CPUs unless absolutely necessary. Unwarranted data cache misses occur when tasks that share data run on different CPUs, or are far apart time-wise on the same CPU. This work presents an extension to the Linux scheduler that exploits inter-task data relations to reduce data cache misses in multi-threaded applications running on SMP platforms, thus improving runtime, memory throughput, and energy consumption. Our approach schedules the tasks to the CPU that holds the relevant data rather than to the one with highest affinity. We observed improvements in CPU time and throughput on several benchmarks. For the Chat benchmark, the improvement in CPU time and cache misses is over 30% on average.

1 Introduction

The demand for greater computing capacity has lead to an increased use of multi-processor machines. Symmetric multi-processing (SMP) is a specific implementation of multiprocessing in which multiple CPUs are physically connected via a common high-speed bus and share resources such as memory, peripherals, and OS. With the rise in the number of parallel multi-threaded applications, the popularity of SMP has increased as well because it provides a way to utilize the application level parallelism for performance gain.

Schedulers in many operating systems, such as Linux, UNIX and AIX, implement variations on processor affinity thread (task) scheduling. The observation behind this choice is the desire to reuse data (and instructions) remaining in the processor's cache from a previous dispatch of the thread. We observed that in SMP machines unnecessary cache misses occur when tasks that share data run on different CPUs.

Several studies have examined the affinity of a task to a processor based on how fast a task can run on a processor in heterogeneous processor environments [4] [7] [15]. Those studies provide a variety of algorithms for different timing metrics and conditions for scheduling tasks to CPUs. Another type of processor affinity looks at the resources bound to processors [8] [15]. These studies attempt to optimize a certain metric under constraints, e.g. the task must execute on a processor that has access to the required resource. Another type, which has received far less attention, is based on tasks' affinity to cache contents, i.e., data affinity. The difficulty in utilizing such affinity is in its dynamic nature. Our work suggests and measures new methods that produce scheduling based on data affinity information.

The potential for performance improvements from exploiting data affinity is disputed. Squillante's theoretical work and simulations [14] have exhibited promising potential. Conversely, Gupta [6] in his simulations and Vaswani [12] in his measurements and tests have concluded that exploiting data affinity has a negligible effect for multi-threaded applications. These results should be rethought in view of architectural advances and the ever-growing use of multi-threading in today's applications.

We found that most applications can benefit from data affinity, regardless of the pessimistic claims mentioned above. Applications consisting of long-living, frequently synchronizing, and memory-intensive threads benefit the most. Moreover, the steady growth in cache sizes implies that a large portion of a task's data will reside in processor's cache, allowing optimizations to ignore the exact data access patterns of threads, thus simplifying data affinity based optimizations. Furthermore, the increase in the relative cost of cache misses [1] makes optimizations that reduce them, such as ours, attractive. In addition, the likelihood that the instructions and data remain in the cache between consecutive dispatching decreases as the number of threads in the system grows; this is a problem which can be alleviated by improving data affinity. Our scheme, as opposed to CPU affinity, maintains cache hotness within an époque by batching together threads that share data. However, data affinity may introduce additional thread preemption or migration that should be carefully traded-off with the extra cache misses contributed by CPU affinity.

We propose an algorithm that endeavors reduction of data cache misses by applying a paradigm whose essence is 'run the task on the processor holding the currently required data', as opposed to CPU affinity. This paradigm ignores the processor on which the task previously ran and focuses on what data it is about to work on and its location. Furthermore, our approach batches together tasks that use the same data and runs them in succession to maximize cache hotness utilization. The scheduler maintains information on data fragments (DF) shared by multiple tasks. A newly implemented syscall provides DF hints at strategic locations. Hints are generated by the compiler, the user, or potentially by the scheduler. A DF can be a set of variables, parts of arrays, etc. Based on available hints, the scheduler dynamically batches together ready-to-run tasks according to their current DF. Each batch is assigned a CPU and its tasks run in succession to minimize data cache misses. When a task accesses a different DF it is migrated to the appropriate batch. The load balancer attempts to preserve batches during migration.

We formally defined our optimization problem and implemented our scheduling algorithm in the Linux kernel version 2.6. Experiments were conducted on a few benchmarks. The results are very encouraging; cache misses were reduced by up to an order of magnitude on several tests, throughput in benchmarks such as the Chat benchmark [16] doubled in some cases and the total application runtime and cache misses were reduced on most tests.

2 Model

The scheduler of an OS handles the lists of running, waiting, and blocked threads. Its responsibilities include scheduling threads onto CPUs, determining their execution order and load balancing the system. In this section we present a system model and use it to describe how threads are dynamically mapped for execution, thereby allowing us to identify sources of overhead incurred by threads contending on the data cache.

2.1 Hardware Model

Our model of an SMP machine consists of processors and caches.

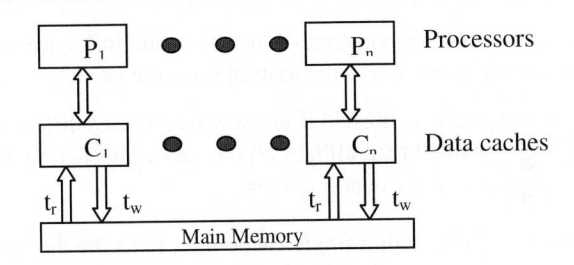

```
    ┌────┐         ┌────┐
    │ P₁ │  ● ● ●  │ Pₙ │     Processors
    └─⇕──┘         └─⇕──┘
    ┌────┐         ┌────┐
    │ C₁ │  ● ● ●  │ Cₙ │     Data caches
    └↑↓──┘         └↑↓──┘
     tr tw          tr tw
    ┌─────────────────────┐
    │    Main Memory      │
    └─────────────────────┘
```

Fig. 1. SMP Architecture

For simplicity, the following assumptions are made on the model:

- Each processor has a single cache and all caches are of equal size.
- The hardware's "snoopy protocol" is 'write-invalidate'.
- The cache can contain all the data for the present run of a thread.

The SMP hardware utilizes some variation of a "snoopy protocol" in order to keep the data in caches coherent. Common policies are: *Write-invalidate* protocols that allow multiple readers, but only one writer at a time. Every write to a shared cache line (block) must be preceded by the invalidation of all other copies of the same line. Local writes to exclusive lines are cheap. *Write-update/(broadcast)* schemes follow a quite an opposite approach. The word to be written to a shared line is distributed to all others, and caches containing that block can update it, thus preventing the stale state.

2.2 System Model

Threads use the machine resources on a need basis. The threads' running order and run time on each époque depend on their resource usage. Once a thread is selected for execution on some CPU, it can use all its resources (e.g. memory, caches and bus). The scheduling process controls when and on which CPU the thread will execute.

The following notations are used:

- Each cache is divided into k lines, which can be individually filled, whereas the write back to memory is done for the whole cache by the flush operation.

- Time to fill a cache line is t_r and t_w to write it. T_r and T_w are for *whole* cache.
- The execution time of a single thread v is $ex(v)$.
- Using a cache takes one of the following four forms:

Read	Write	Description
$t_r = 0$	$t_w = 0$	No writes were done by the previous thread thus cache is not dirty, and the needed data is already in cache.
$l \cdot t_r$	$t_w = 0$	l needed data lines are not present in memory; reading the l lines takes $l \cdot t_r$ time units; no writes are performed.
$t_r = 0$	$j \cdot t_w$	j data cache lines are dirty and flushing them back to memory takes $j \cdot t_w$ time units. Needed data is in cache.
$l \cdot t_r$	$j \cdot t_w$	The thread needs to fill the cache with l lines of data ($l \cdot t_r$), and flush changed data requiring write ($j \cdot t_w$).

The following example demonstrates how the cache influences scheduling results. To simplify the example, the following assumptions are used:

- The whole data cache is flushed if the new thread uses different data fragment
- During execution the thread utilizes all the cache attached to the CPU.
- Each thread uses a *single* data fragment.

Example. Assume a cache with a single cache line. Let t_0 be the time in which thread v is mapped to processor P with cache C_P. The execution of v can start at $t_1 = t_0 + T_r$ when the required data is read to cache, and end at $t_2 = t_1 + ex(v) + T_w$. In our example:

- Threads set, TH = $\{v_0, v_1, v_2, v_3, v_4\}$, $ex(v_n) = 1, 0 \leq n \leq 4$, $T_r = T_w = 4$
- Number of processors (P) = number of caches (C) = 2

The following constraints are given for the whole run of the application:

- v_2, v_3 and v_0 use the same data fragment – DF1.
- v_1 and v_4 use the same data fragment - DF2.

Threads' story: The program starts with thread v_0 running. It spawns four additional threads $v_1 - v_4$ and finishes. In a CPU affinity based scheduler, by default $v_1 - v_4$ are scheduled to run on the same CPU as their parent, v_0. Assume v_0 ran on P_1. Since having all four remaining threads run on P_1 causes imbalance, the load balancer will move two threads to P_2. Two possible scheduling scenarios are presented in Figures 2 and 3.

As can be seen from the figures below, not having to read the data for threads v_2, v_3 and v_4 decreases the total run time from 22 time units, to *only* 14! Next section presents a scheduling optimization problem that integrates the parameters in our model.

	0	1	2	3	4	5	6	7	8	9	10	11	12	13	14	15	16	17	18	19	20	21	22
P_1	$T_r(v_0)$				$ex(v_0)$	$T_r(v_1)$				$ex(v_1)$	$T_w(v_1)$				$T_r(v_2)$				$ex(v_2)$	$T_w(v_2)$			
P_2						$T_r(v_3)$				$ex(v_3)$	$T_w(v_3)$				$T_r(v_4)$				$ex(v_4)$	$T_w(v_4)$			

Fig. 2. Pure CPU affinity based scheduling

	0	1	2	3	4	5	6	7	8	9	10	11	12	13	14	15	16	17	18	19	20
P_1	$T_r(v_0)$				$ex(v_0)$	$ex(v_2)$	$ex(v_3)$			$T_w(v_3)$											
P_2							$T_r(v_1)$			$ex(v_1)$	$ex(v_4)$		$T_w(v_4)$								

Fig. 3. Data affinity based scheduling following the constraints on data sharing

3 Scheduling to Reduce Cache Miss Penalty

Assume a set $V = \{v_1, v_2, ..., v_n\}$ of currently running tasks to schedule. The number of CPUs available to the scheduler is denoted by m. Since, in our model, every CPU has its own cache, m is also the number of caches. $DF(v_i)$ is the data fragment used by v_i. $f(v_i, v_j)$ is an asymmetric penalty function for the context switch from v_i to v_j. Its characteristics, in our model, are defined below. l is the number of cache lines read by v_j, ϕ denotes an empty CPU and the times t_r and T_w are as defined in Section 2.2.

$$f(v_i, v_j) = \begin{cases} 0 & DF(v_i) = DF(v_j), v_i \text{ did not invalidate data.} \\ l \cdot t_r & v_i = \phi \text{ or } DF(v_i) \neq DF(v_j), v_i \text{ did not write, } v_j \text{ read } l \text{ lines.} \\ T_w & DF(v_i) = DF(v_j), \text{ flush of } v_i\text{'s data required.} \\ l \cdot t_r + T_w & DF(v_i) \neq DF(v_j), v_i \text{ invalidated data, } v_j \text{ read } l \text{ lines.} \end{cases} \quad (1)$$

To improve system throughput we seek to minimize the penalty induced by the context switches between the tasks of V on m CPUs, based on the data fragment they use. We define X_W to be the set of all permutations over a subset $W \subseteq V$ and let $w \in X_W$. $F(w)$, the penalty of permutation w is defined by: $F(w) = \sum_{i=1}^{|w|-1} f(v_i, v_{i+1})$. The minimal penalty F^* over W is $F^*(W) = \min_{w \in X_W} F(w)$. The collection of all partitions of V into m equal-sized subsets is denoted by $\Pi_{(V)}$, where each partition in this collection is of size $\left\lceil \frac{n}{m} \right\rceil = k$ (switching to padding task costs 0). Given a partition $S_m \in \Pi_{(V)}$, the minimal penalty for scheduling the partition is: $\tilde{F}(S_m) = \max_{X \in S_m} F^*(X)$ and our optimization problem is to find T such that $T = \min_{Z \in \Pi_{(V)}} \tilde{F}(Z)$.

Solving equation T when the system's state changes, is impractical. Next we consider a special simplified case where during a thread switch all the cache contents are replaced if the new thread uses a different data fragment. The assumptions previously made still hold; mainly that each task uses exactly one DF and that fragment fits precisely into a processor's cache. In the simplified penalty function g we pay a price only when switching between tasks that use different data fragments.

$$g(v_i, v_j) = \begin{cases} 0 & DF(v_i) = DF(v_j) \\ T_r + T_w & v_i = \phi \text{ or } DF(v_i) \neq DF(v_j) \end{cases} \quad (2)$$

An optimal algorithm for a single CPU would be: Go over all ready threads and put each thread in the bin (batch) corresponding to the DF that it uses (d bins), and then schedule the bins in arbitrary order. Changing currently running bin is done **only** after all tasks in the bin finished their quota. The runtime of the algorithm on a set of n tasks is $O(n)$. Adding and removing a task to the ready list can be performed in $O(1)$.

For multiple CPUs an additional step for distributing the bins between CPUs is added after partitioning the threads into bins. This general partitioning problem is NP-complete. However, based on the dynamic programming pseudo-polynomial partitioning by Cieliebak et al. [17] for $m=O(1)$ CPUs our algorithm will run in $O\left(\dfrac{d \cdot n^m}{m^{m-1}}\right)$.

4 Data Affinity Based Algorithm

Our scheduler implementation is an enhancement of the Linux 2.6.x scheduler. The Linux scheduler allocates a time slice to each thread/task in the system. An époque, i.e. the time it takes for all tasks to get a chance to run, may vary due to the many heuristics utilized. Tasks that spend much of their time submitting and waiting on I/O requests (I/O bound) have their time slice enlarged. Tasks that tend to run until preempted, spending most time executing code (CPU bound) receive time slice penalties.

The scheduler keeps the tasks in run-queues. A run-queue is a list of runnable tasks which may run in arbitrary order. There exists one list per processor and each task can be on exactly one. It contains two priority arrays; an active and an expired array. Each array contains one queue of runnable tasks per priority level. The 2.6.x scheduler can locate the next highest priority task and pull it off the priority list in constant time.

The scheduler is called explicitly by kernel code that is about to yield CPU and also whenever a task is to be preempted. The scheduler performs the following steps: it recalculates the time slice of the tasks that ended theirs and moves them to the expired array, determines the next highest priority task in the active array and switches to it.

The load balancer complements the scheduler. It ensures that the run-queues are balanced by moving tasks from the busiest run-queue to the relatively under utilized one that invoked it. It is invoked when a run-queue is idle and also via timer interrupt. An imbalanced run-queue contains 25% more tasks than the one on whose behalf the load balancer runs. From the tasks allowed to migrate, the load balancer prefers *expired* ones, since these are probably cache cold. It also favors high priority tasks because of their importance. Tasks are moved as long as an imbalance still exists.

4.1 The Implemented Algorithm

The basic idea is to schedule the current set of tasks in a way that will minimize cache misses, thus resulting in an overall reduction of execution time. This is done by supplying additional information to the scheduler. The algorithm's goal is as follows:

- Tasks that use the same data fragments (DFs) at some point in time are mapped to the same CPU. If a task accesses multiple DFs, it may be reassigned to a different CPU each time it changes DFs; this is done before actually accessing the data (i.e. yield and reschedule).

The mapping of a task to a processor (run-queue) occurs in the following cases: when it is first created (parent's processor is the default), whenever the load balancer is called, and when it returns from a wait queue. We would also like a mapping to occur when the task changes DFs. From the initial mapping onward until (if at all) it is migrated, the task's time slice is calculated according to the existing Linux policies.

The theoretic formulation in the previous section provides optimal scheduling when all information is known a priori. Unfortunately, real-life systems are dynamic in nature; therefore, there is a need to devise heuristics that use dynamically available info.

The information about which DF each task is using at a given time is passed from user space to the scheduler via a syscall. The syscall is called with the application ID, current (if any) and next DF IDs when a task is created and whenever it changes DFs.

Our scheduler works *online*, holding a list of application descriptors, one for each application that provides DF information. Each descriptor contains a map between DFs and <CPU, priority group> pairs. At any point in time all the tasks in a priority group use the same DF and run in succession on the same CPU. The load balancer is used to rectify 'mistakes' made when in the initial state.

The algorithm consists of several procedures that decide what to do in the following situations: a new DF is encountered, a new task enters the system, a task switches DFs, a task returns from a wait queue, a task dies, and when the system is imbalanced. We next describe those procedures as applied to a single application task.

- When a task arrives with a new DF, the load is verified for all CPUs, in terms of number of currently running tasks. Since the tasks' time-slices are similar in length, this constitutes a good measure for selecting the least loaded CPU.
- Upon arrival of a task with known DF, the scheduler locates its entry in the application map, moves it to the DF's CPU (if it is not there already), and places it in the priority group for that DF.
- When a task changes DFs, the scheduler removes it from its old priority group, preempting it if necessary (this may actually be cheaper than the cache misses), then treats it as if a new thread has arrived with a new or known DF.
- Upon return from a wait queue, a task's current DF entry is looked up in the application map. From there on, as before, it is handled based on whether or not its current DF is known.
- When a task dies (and when changing DFs), a counter in the DF's structure is updated. When it reaches 0, the priority group is flagged as 'may be removed'.
- When the load balancer is called, it checks if the run-queues are imbalanced. If tasks need to be moved, the scheduler tries to identify whether the source of the imbalance is in the managed or unmanaged tasks. When managed threads cause the imbalance, an attempt is made to move an entire priority group. If this is impossible, for example because a group's task is running, the load balancer reverts to unmanaged threads. In rare cases, when the problem is sustained, separating a priority group is allowed.

5 Experimental Results

To assess the performance of our scheduler we tested a custom made benchmark and a few known multi-threaded benchmarks that utilize shared data; achieving significant improvement on the well known Chat benchmark, and a more modest gain on the Hack benchmark, which uses processes. The custom benchmark tests scenarios that are not exploited by the other benchmarks. Syscalls were added manually to the benchmarks. We discuss the results and provide reasons for the large improvements.

All benchmarks were run on an Intel Xeon, dual processor at 3.2GHz, hyper threaded with L1 cache of 8KB, L2 of 256KB, L3 of 2MB and main memory of 1GB. The hyper-threading ability was turned off for most tests thus creating a 2-way machine. All tests were run using the 2.6.4 Linux distribution of the Linux kernel.

5.1 Chat Benchmark

The Chat benchmark (http://lbs.sourceforge.net) simulates chat rooms with multiple users exchanging messages through TCP sockets. It is based on the Volano Java benchmark that was used in prior papers to show limitations of the 2.4 scheduler [9].

A room consists of 20 users each sending 100 byte messages to the server, which broadcasts them to every other user in the room. Four threads are created per user (80 per room) two on the client side and two on the server side. 100 messages sent by a user translate to 20*100*(1+19)=40,000 transmitted messages per room. At the end of a run, the client side reports the total time and the throughput in messages per second. A lower run-time and higher throughput indicate a more efficient kernel scheduler.

The Chat benchmark was tested with all pairs of parameters from these sets: rooms = {10, 20, and 30} and messages per room = {500, 1000, and 1500}. The results are displayed in Table 1 and represent the average over five runs for each pair.

Table 1 demonstrates that the total gain increases as the number of messages grows. The results for some combinations are less than *half the original scheduler time* with *twice the throughput*! Another statistic worth mentioning is that the standard deviation over the five runs of each combination is considerably smaller for the new algorithm.

We further investigated the pairs that exhibited the largest improvement using Oprofile to count L2 cache misses. As can be seen from Table 2, there is a strong correlation between the number of cache misses and the runtime/throughput.

Table 1. Chat benchmark results. Number of messages ranges from 2 million for the 10 rooms with 500 messages combination to 18 million for 30 rooms and 1500 messages per room

Room Number		10			20			30		
Message Number		500	1000	1500	500	1000	1500	500	1000	1500
Avg. Time	Vanilla	5.794	15.303	26.86	11.01	30.7	55.01	16.83	38.93	72.0
	New	4.954	10.935	15.85	7.89	16.85	**24.6**	10.81	22.4	35.2
Avg. Throughput	Vanilla	35228	26362	67594	370625	121874	64336	103133	87981	11559
	New	420382	373323	108010	509062	221189	**139266**	158516	152722	23413

Table 2. Chat benchmark Oprofile results averaged over 5 runs; L2 misses divided by 3000

Room \ Message		10, 1500	20, 1500	30, 1500
Client side	Vanilla	49	88	133
	New	16	34	45
Server side	Vanilla	100	125	206
	New	43	58	68

5.2 Custom Benchmark

Our benchmark consists of a small, highly configurable, application whose parameters include: number of threads (using pthread), number of distinct data fragments (DF), size of those DF, and amount of work done by the threads on the common data. Threads are started in a loop and never sleep voluntarily. We used the Oprofile sampling tool to count the L2 and L3 cache misses (L1 ignored because of its size). L3 may be larger than L2 due to unused pre-fetch into L3.

Table 3. The effect of the number of threads on runtime and cache misses. All other parameters are unchanged; iteration number = 10×10^6; DF number = 4; Oprofile numbers divided by 3000.

Thread Number		8			32			128		
DF size		1000	8000	20000	1000	8000	20000	1000	8000	20000
Avg. Run-time	Vanilla	30.63	31.71	32.18	123.7	128.7	129.5	511	510	509
	New	27.84	27.56	27.44	113.5	110.5	110	439	442	439
Avg. L2 Miss	Vanilla	11236	6844	8509	72467	50232	48096	334290	247784	219611
	New	1610	305	95	9934	2445	899	49347	9630	3491
Avg. L3 Miss	Vanilla	11247	11648	11015	78941	63422	48697	367062	272297	241738
	New	1611	334	70	12467	2389	821	48952	9394	3125

Table 4. The effect of the number of iterations on runtime and cache misses. All other parameters are unchanged; DF number = 4; DF size = 8000; Oprofile numbers divided by 3000.

Iterations Number		1×10^6			10×10^6			100×10^6	
Thread Number		8	32	128	8	32	128	8	32
Avg. Runtime	Vanilla	2.94	12.5	51.11	31.71	128.7	510	312	1283
	New	2.75	11.1	44.2	27.56	110.5	442	277	1101
Avg. L2 Miss	Vanilla	630	5572	25781	6844	50232	247784	2537530	501624
	New	31	246	965	305	2445	9630	96430	24230
Avg. L3 Miss	Vanilla	771	5555	27132	11648	63422	272297	2732690	634042
	New	32	240	955	333	2389	9394	94145	23910

The results in Table 3 emphasize the fact that if a system is not conscious of data affinity, unnecessary CPU cycles are lost in moving data from one CPU cache to another or during cache replacement. Many scheduling cycles are saved in our method. The results in Table 4 are obvious. The longer the threads run repeatedly accessing the same DF access, even a tiny gain gradually increases to noticeable size.

6 Related Work and Future Enhancements

Processor affinity scheduling has been extensively studied. Squillante [14] and Gupta [6] showed its potential through simulations on several affinity-scheduling algorithms and measuring metrics. Vaswani [12] focused on quantifying the effect of processor reallocation on performance. Devarakonda [5] revealed a number of problems related to exploiting cache affinity in Unix-like systems.

Affinity based on how fast a task can run on a processor in a heterogeneous processor environment has been studied in [4] [7] [15]. Affinity that looks at the resources that are bound to a processor has been studied in [8] [15]. Affinity based on cache contents, closest to our work, was studied by Torrellas et al [3].

Linux has been widely used for scheduler experiments, especially in version 2.4 trying to deal with the queue lock contention bottleneck. For example, [10] proposed the multi-queue scheduler to enhance scalability on large scale SMP machines. Molloy et al. [11] proposed the ELSC scheduler. There was also Priority Level Scheduler (PLS). Yamamura et al [13] tackled the cache miss problem occurring in kernel code during the walk over the task structures held in a CPU's run-queue.

For practical usage, automatic insertion of the syscall by the compiler is necessary. Further study of the tradeoff between task preemption and the saved cache misses, and the tradeoff between DF sizes vs. the number of data fragments are needed.

References

1. N. P. Jouppi, D. W. Wall - Available instruction-level parallelism for superscalar and superpipelined machines – Proceeding of the 3rd ASPLOS conference, Apr 1989, pp. 272-282.
2. Y. Etsion, D. Tsafrir, D. Feitelson – Effects of Clock Resolution on the Scheduling of Interactive and Soft Real Time Processes – ACM SIGMETRICS, pp. 172-183, Jun 2003.
3. J. Torrellas, A. Tucker, A. Gupta - Evaluating the Performance of Cache Affinity Scheduling in Shared-Memory Multiprocessors, JPDC, Vol. 24, pp. 135-151, 1995.
4. E. Horowitz, S. Sahni – Exact and Approximate Algorithms for Scheduling Nonidentical Processors – J. ACM, vol. 23, no. 2, pp. 317-327, 1976.
5. M. Devarakonda, A. Mukherjee - Issues in Implementation of Cache-Affinity Scheduling – USENIX Technical Conference and Exhibition, pp. 345-357, 1992.
6. A. Gupta et al - The impact of operating system scheduling policies and synchronization methods of performance of parallel applications - ACM SIGMETRICS, Vol. 19, May 1991.
7. E. L. Lawler, C. U. Martel - Scheduling periodically occurring tasks on multiple processors - Information Processing Letters, vol. 7, pp. 9-12, Feb. 1981.
8. D. H. Craft - Resource management in a decentralized system - ACM SIGOPS Operating Systems Review, Vol. 17, Issue 5, Oct. 1983.
9. R. Bryant, B. Hartner – Java Technology, Threads and Scheduling in Linux, *Java Technology Update,* Volume IV, Issue 1 Jan 2000.
10. M. Kravetz et al - Enhancing Linux Scheduler Scalability - 5th ALS, Nov 2001.
11. S. Molloy, P. Honeyman – Scalable Linux Scheduling – CITI Technical Report, May 2001.

12. R. Vaswani et al - The implications of cache affinity on processor scheduling for multiprogrammed, shared memory multiprocessors – 13th ACM SOSP, pp. 26-40, Oct. 1991.
13. S. Yamamura et al -Speeding Up Kernel Scheduler by Reducing Cache Misses - Proceedings of the FREENIX Track 2002 USENIX Annual Technical Conference, pp. 275-285.
14. M. S. Squillante, E. D. Lazowska – Using Processor-Cache Affinity Information in Shared-Memory Multiprocessor Scheduling – IEEE TPDS archive, Volume 4(2), pp. 131-143, 1993.
15. R.W. Conway, W.L. Maxwell, L.Miller - Theory of Scheduling - Addison-Wesley, 1967.
16. Linux Benchmark Suite Home page - http://lbs.sourceforge.net/
17. G. J. Woeginger, Z. L. Yu - On the equal-subset-sum problem - Information Processing Letters, 42(6), pp. 299-302, 1992.

Topic 4: Compilers for High Performance

William Jalby, Oscar Plata, Barbara Chapman, and Paul Kelly

Topic Chairs

Welcome to Euro-Par's Topic 4, which provides a forum for the presentation of the latest research results and practical experience in Compilers for High Performance. Topic 4 deals with compilation for high performance architectures. Papers were invited targeting both general-purpose platforms and specialised hardware designs such as graphic coprocessors or low-power embedded systems. We were also keen to solicit papers that explore more general issues, such as program analysis, languages, run-time systems, and feedback-directed optimisation.

We thank the many helpful referees, who provided at least four reports on each of the nine papers submitted. After vigorous, good-natured and pleasurable debate among the program committee members, five were accepted.

The quality of submissions was uniformly high, and without exception, the papers we were unable to accept this time represent sound work which we would encourage the authors to submit next year in more mature form.

Euro-Par's tight page limit makes it a forum for work which is focussed on interesting new ideas, rather than extensive experimental evaluation of more established material. We believe this makes for a lively programme of presentations, and we are confident there will be plenty of interesting questions and discussion.

Compiler Technology for Blue Gene Systems

Stefan Kral, Markus Triska, and Christoph W. Ueberhuber

Institute for Analysis and Scientific Computing,
Vienna University of Technology,
Wiedner Hauptstrasse 8-10, A-1040 Wien, Austria
mailto:skral@complang.tuwien.ac.at
http://www.math.tuwien.ac.at/ascot/

Abstract. Standard compilers are incapable of fully harnessing the enormous performance potential of Blue Gene systems. To reach the leading position in the Top500 supercomputing list, IBM had to put considerable effort into coding and tuning a limited range of low-level numerical kernel routines by hand. In this paper the Vienna MAP compiler is presented, which particularly targets signal transform codes ubiquitous in compute-intensive scientific applications. Compiling FFTW code, MAP reaches as much as 80% of the optimum performance of Blue Gene systems. In an application code MAP enabled a sustained performance of 60 Tflop/s to be reached on BlueGene/L.

1 Introduction

Blue Gene Servers. Top-performing supercomputers are usually based on the fastest processors available. In their latest hardware development, IBM went a radically different way, building Blue Gene servers [3] on an embedded-systems processor with low-power consumption, the IBM PowerPC 440.

To support scientific computing applications efficiently, IBM added a functional unit for double-precision scalar and 2-way SIMD floating-point arithmetic, extending the existing processor design by an auxiliary processor unit, yielding the PowerPC 440 FP2.

One node of a Blue Gene server comprises two PowerPC 440 FP2 processors (one dedicated to communication, the other one to computation), shared memory, and high-speed network interconnect hardware. The biggest installation built to date—BlueGene/L—is made up of the unprecedented number of 65,536 nodes integrated into a single distributed memory system. As of November 2005, Blue Gene servers take three out of the ten top positions on the Top500 supercomputing list, including the number one and two.

Automatic Performance Tuning Software. State-of-the-art numerical libraries in the field of linear algebra and signal processing are not based upon predetermined and fixed algorithms for performing the requested calculation, but utilize automatic performance tuning [4] to search the space of different algorithms and implementations for members of this set showing optimal runtime behavior. Rather than relying on formal performance models (covering the utilization of the memory hierarchy, arithmetic operation count, instruction count, calling overhead of a procedure, and other relevant properties of the target architecture), they take actual runtime measurements obtained in numerical experiments to guide the process of automatic self-adaptation.

Automatic performance tuning systems often use automatically generated kernels that are long sequences of straight line code. For achieving high performance, these libraries heavily rely on the quality of the C compiler.

Experiments have uncovered a number of shortcomings of general purpose compilers when applying them to long, automatically generated straight line code, which opens up a performance gap between code generated by general purpose compilers and assembly code written by a skilled hand-coder.

The MAP Tool Chain. The Vienna MAP compiler tool chain aims at closing this performance gap, addressing domain-specific straight line codes produced by special-purpose program generators like `genfft` [8].

This paper describes a version of the MAP compiler targeting IBM's Blue Gene systems. The MAP compiler comprises a set of generic components arranged in the form of an open tool chain, communicating through a very narrow human-readable interface, which allows for (i) easy conservation of high-level information by means of annotation, (ii) introspection and injection of code, and (iii) easy experimentation with different arrangements of compilation stages.

Synopsis. Section 2 presents and discusses important properties of the target processor. Section 3 describes a 2-way single-instruction multiple-data (SIMD) vectorizer extracting parallelism out of basic blocks, Section 4 a versatile peephole optimizer for utilizing fused multiply-add (FMA) instructions, and Section 5 a Blue Gene specific backend that optimizes effective address calculations.

New contributions presented in this paper are improvements of (i) the vectorization method and of (ii) address-generation in the backend.

Section 6 demonstrates the impressive effects of the presented components and techniques on the performance of FFTW [9], the de-facto standard for the computation of discrete Fourier transforms (DFTs), running on Blue Gene servers.

2   The Blue Gene Processor

IBM's Blue Gene processor, the PowerPC 440 FP2, is a low-frequency (700 MHz) 32 bit processor with 32 integer registers, 32 SIMD floating-point registers, a short (7-stage) pipeline, large split L1 caches (32 kB for instructions, 32 kB for data), a fast non-pipelined multiplier, and support for 2-way super-scalar out-of-order execution. Integer registers are 32 bit, SIMD registers 128 bit wide.

Although the processor is a dual-issue design, not all conceivable pairs of instructions may be executed in parallel. Scalar arithmetic, SIMD data-reordering, and SIMD arithmetic use the same functional unit, and cannot be executed in parallel. At most one instruction per cycle may access naturally aligned memory.

The PowerPC 440 FP2 supports scalar [17] and 2-way SIMD [2] floating-point arithmetic, both operating on the same 2-way SIMD register file, with scalar instructions working on the lower half of SIMD registers. Floating-point addition, subtraction, and multiplication are all fully pipelined, and available in double precision only. Support for single-precision floating-point data is offered for data loads/stores and by explicit rounding operations. Both for the scalar and the SIMD case, FMAs are available, which doubles the peak performance and improves the accuracy of the results by avoiding intermediate rounding.

FP2 offers a huge collection of vertical (inter-operand style) SIMD FMAs, including instructions that (i) perform different operations on different parts of the SIMD registers (e. g., addsub), (ii) use one part of a register as input for both operations, and (iii) combine a swap with an arithmetic operation.

Native support for horizontal (intra-operand style) SIMD is, however, completely missing in FP2. Emulating horizontal SIMD operations with a sequence of vertical and data reordering operations is considerably more expensive than on other SIMD ISAs (Tables 1 and 2).

Table 1. Instruction Count for Horizontal (H) and Vertical (V) Addition and Subtraction Operations. Uniform instructions perform two additions or two subtractions, while mixed instructions perform an addition and a subtraction.

op	3DNow!	Ext. 3DNow!	SSE2	SSE3	IA64	**FP2**
H / uniform	1	1	3	1	3	5
H / mixed	2	1	4	2	3	5
V / uniform	1	1	1	1	1	1
V / mixed	2	2	2	1	1	1

Table 2. Instruction Count for Data Reordering Operations. Uniform unpacks (unpackXX) combine the lower parts of two registers, while mixed unpacks (unpackXY) combine the lower part of one register with the upper part of another.

op	3DNow!	Ext. 3DNow!	SSE2	SSE3	IA64	**FP2**
unpackXX	1	1	1	1	1	2
unpackXY	2	2	2	2	1	2

Scalar computation can only be done in the lower half of the registers and some data may need to be moved. Mixing scalar and SIMD code is possible, but not at uniform cost.

The application binary interface (ABI) used in the Blue Gene environment [12] defines approximately half the registers as callee-saved, which can be a considerable disadvantage for small leaf procedures.

The PowerPC 440 FP2 lacks two important features for calculating effective addresses efficiently. First, the PowerPC ISA does not offer a combined *shift by a constant and add* instruction. Second, FP2 does not support *register+immediate* forms [5] for SIMD loads/stores.

3 The MAP Vectorizer for Blue Gene

Unlike SIMD-style vector computers, SIMD floating-point ISA extensions on general purpose processors operate on very short vectors. As this allows expressing parallelism on a very low level, not only loop-based vectorization techniques [19], but also more fine-grained ones, that extract the parallelism already present within a basic block, can be utilized.

To get the highest possible performance, a basic block vectorizer tries to maximally cover a scalar DAG with SIMD instructions natively supported by the target machine.

While our approach has some similarity to existing work like [6, 13, 14], our work is biased towards different assumptions about the class of input codes and about the target hardware. (i) As linear transform codes are highly structured, any divide-and-conquer based vectorization approach incurs high costs when connecting vectorized sub-graphs. (ii) Unlike SIMD ISAs on some DSPs, SIMD ISAs present on general purpose microprocessors do not allow scalar and SIMD to be mixed efficiently. (iii) As numerical kernels used in automatic performance tuning systems can be very large, finding a compromise between vectorization runtime and code quality is a key issue. (iv) Accesses to interleaved complex numbers naturally translate to 2-way SIMD memory instructions, which massively prunes the search space. However, for some kernels that do not have this kind of access, e.g., real FFTs, the vectorizer has to consider all combinations of all possible pairs of DAG inputs and DAG outputs.

Vectorization consists of two major steps, that are alternated until either the scalar DAG is covered with SIMD instructions or failure is discovered.

First, the vectorizer combines pairs of scalar variables to SIMD variables, ensuring that no scalar variable occurs in two SIMD variables and that the producers of the respective variables may be joined into a (pseudo) SIMD instruction.

Second, as the vectorizer combines two scalar instructions to one SIMD instruction, it propagates the layout requirements of the inputs and outputs of the newly extracted SIMD instruction, triggering the creation of new pairs.

Non-deterministic choice in this search process is handled by using depth-first search with chronological backtracking.

In an attempt to prune the search tree, the vectorization engine tries to detect failure branches early, allowing to traverse a much smaller part of the search space without missing any relevant part.

To further restrict the search space, pairs of scalar variables that can not occur as part of any solution are filtered out before vectorization is started.

The scalar DAG traversal order can have a profound impact both on the solution order and on the vectorization runtime. Earlier versions of the vectorizer [7, 15] always started at the outputs of the DAG, i.e., store instructions, traversing the DAG in a bottom-up fashion.

To improve on this, we added a top-down traversal style and borrowed the concept of domain variables [18] from constraint programming (CP). Domain variables allow the vectorizer to keep track of all pairs of scalar variables that may be formed in the future. When traversing the scalar graph, the scalar variable that occurs in the smallest number of pairs, is picked as the next node to be visited (first-fail principle).

The combination of these traversal methods allows finding the optimal vectorization even for relatively large codes that use exclusively real arithmetic—a class of codes notoriously hard to vectorize.

Vectorization may yield more than one solution. A branch-and-bound based method is used to gradually find better and better solutions, until either optimality is proven or a time limit is reached. Generally, finding an optimal solution takes much less time than proving its optimality.

While all previous prototypes of the vectorizer have been specifically adapted to exactly one target architecture, the new version uniformly supports all target architectures taking target-specific data (as presented in Table 1) as input.

4 The MAP Optimizer for Blue Gene

The MAP optimizer only focuses on improving local structures (peepholes), rewriting sequences of instructions logically connected by data dependencies. Because of the locality of the approach, the global structure of the code, determined by the vectorizer, remains largely unchanged.

Implemented as a committed-choice term rewriting system, the optimizer is based on one or more sets of rewriting rules, each with a different priority. Out of all applicable rules, the rewriting engine picks one with the highest priority, and uses it to substitute instructions within a peephole with a semantically equivalent sequence of instructions. If no rule is applicable, a fixed-point is reached and the optimization terminates.

The optimizer uses two kinds of rules working in synergy. (i) *Improving rules* aim at an immediate improvement in code quality. Examples include rules for fusing two neighboring instructions into one, or rules handling horizontal SIMD instructions with neighboring SIMD swaps. (ii) *Assisting rules* do not immediate improve the code, but rather adapt the DAG such that improving rules may be applied. Examples include rules moving SIMD swaps or multiplications by constants within the DAG.

Apart from commonplace compiler optimizations [16], the optimizer tries to (i) shorten path lengths within peepholes, (ii) reduce the number of source operands by identifying domain-specific code patterns (e. g., the butterfly-ish code patterns typically occuring in FFT codes), and (iii) reduce the total number of instructions, both by eliminating superfluous instructions and by hiding some instructions in other ones, in particular by utilizing variants of SIMD FMAs.

5 The MAP Backend for Blue Gene

Unlike the two previously presented components, the MAP backend consists of a relatively large number of parts.

5.1 Effective Address Generation

All integer instructions in the code produced by the MAP compiler are devoted to either fulfilling the ABI calling convention or to calculating effective addresses.

While the code for fulfilling the calling convention has a constant size (regardless of the actual size of the procedure to be compiled) for procedure prolog and epilog, the code for the calculation of effective addresses may grow linearly with the number of memory accesses in the procedure to be compiled.

Experiments have shown that the portion of the code needed for the calculation of effective addresses often has a significant negative performance impact in case of algorithms having a high ratio of the memory access count compared to the number of arithmetic operations. All fast algorithms for linear signal transforms possess this property.

The IBM PowerPC 440 FP2 processor has DSP-like addressing mode limitations for SIMD loads/stores, minimizing the number of integer auxiliary instructions in of particular importance.

Basic Ideas. The calculation of effective addresses of elements of variably strided arrays (the actual stride is not known at compile time) can be done straightforwardly by using integer multiplication instructions. However, these instructions are expensive (low throughput, high latency) on all general purpose processors, including the IBM PowerPC 440 FP2.

The common approach to addressing this problem is *strength reduction*, which replaces complex instructions with sequences of simpler (high throughput, low latency) instructions like integer additions, subtractions, and shifts.

Implemented Solution. Doing strength reduction in a hard-coded fashion implies making instruction selection decisions without properly considering the temporal context, thereby missing opportunities to (i) reuse already calculated factors still residing in the register file and (ii) pick factors that could be beneficial for some proximate address calculation to be carried out in the near future.

To produce high-quality code, the MAP backend interleaves integer instruction selection and integer register allocation, thus removing a classical compiler optimization barrier.

Premature commitment to one particular factorization or reduction is avoided by utilizing a blended mixture of well-established search methods, depth-first iterative deepening (DFID) and dynamic programming (DP).

As exhaustive search for an optimal solution may not be possible for all but the smallest codes, the backend (i) looks at reasonably sized sub-problems, (ii) solves these sub-problems optimally, and (iii) combines the respective optima to one solution of the original problem. The quality of this solution depends on the amount of overlap of the sub-problems considered and on the size of these sub-problems.

To control the amount of search performed, the backend offers a set of parameters to directly control the speed and quality of the search, allowing to trade compilation time for code quality, by specifying the size and the amount of overlap of the sub-problems.

5.2 Register Allocation

The MAP backend performs register allocation for all register files holding non-integer data in one pass, using the farthest-first policy [1, 10].

5.3 Scheduling

The MAP backend implements a set of various schedulers, covering a wide range from domain-specific high level scheduling to target-processor specific low level code reordering.

High Level. Two high level schedulers are part of the MAP backend. Both of them aim at a minimization of the register pressure.

The first high level scheduler implements an FFT specific topological sort of the computation DAG, attempting to enhance locality by minimizing variable life-span. This scheduler is directly derived from the scheduler of `genfft`, the program generator of FFTW.

The second high level scheduler performs local code reordering, trying to further reduce the register pressure for codes exhibiting a non-regular structure, e. g., SIMD-vectorized FFT codes.

Medium Level. The medium level scheduler reorders instructions taking latencies into account, thereby increasing the register pressure. By avoiding all dispensable movement, it preserves the original instruction order—obtained by high level scheduling—as much as possible.

Low Level. The low level scheduler specifically addresses execution properties of the target processor, implementing a list-scheduling algorithm that provides a runtime estimation of a given basic block. This scheduler is based on an in-order, super-scalar execution model of the target processor and handles both pipelined and non-pipelined instructions (like integer multiplication) well.

Execution models incorporate information about (i) instruction latencies, (ii) instruction throughput, (iii) issuing and decoding constraints, (iv) the mapping of instructions to functional units, and (v) register forwarding features.

6 Performance Results

To assess the performance impact of the presented techniques on Blue Gene systems, we compiled the compute-intensive numerical kernels of FFTW 2.1.5 with the following setups. *xlc_scalar* uses the XL C compiler without automatic vectorization. *xlc_vect* uses XL C with automatic vectorization. *xlc_mapvect* uses the MAP vectorizer and optimizer, producing C code with SIMD intrinsics compiled by XL C. *map_vect* uses the MAP vectorizer, optimizer, and backend.

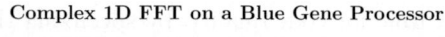

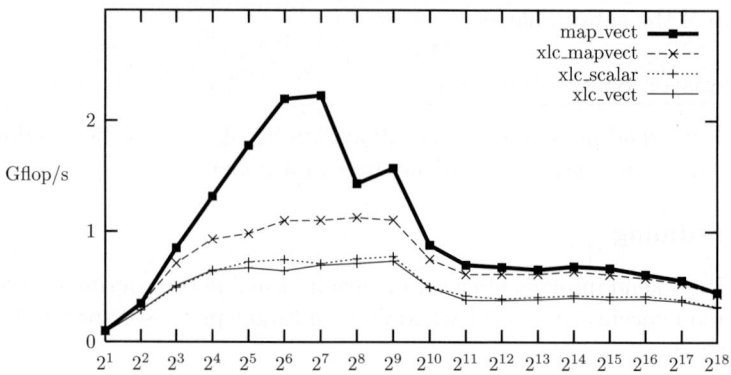

Fig. 1. Performance of Power-of-two 1D FFTs on the IBM PowerPC 440 FP2

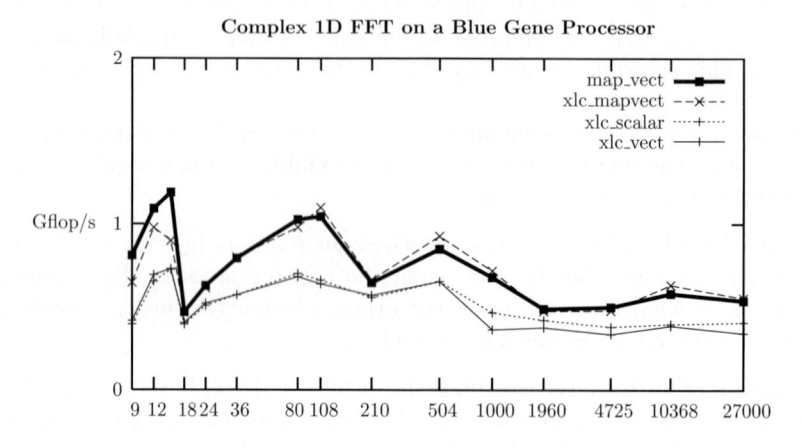

Fig. 2. Performance of Non-power-of-two 1D FFTs on the IBM PowerPC 440 FP2

Figs. 1 and 2 show the single-processor FFT performance achieved on Blue Gene systems by using various compilers and settings. All performance data are displayed in pseudo-Gflop/s, i.e., $5N \log N/T$.

Performance of Power-of-two Sizes. With very short vectors, calling the FFTW framework dominates the total cost. For medium sizes (2^3 to 2^9), all data fits into L1 cache, and the performance peaks—the MAP generated code for length 2^7 has 2230 pseudo Mflop/s, as opposed to 709 pseudo Mflop/s of the XL C compiled code. For transform lengths bigger than 2^{10}, data no longer fits into L1 cache, and the performance falls sharply.

Performance of Non-power-of-two Sizes. The performance shown in Fig. 2 is much more uneven than in the power-of-two case, because the chosen vector lengths have a larger number of different factors, leading to the use of many relatively small routines.

Effect of the Backend. We have examined the performance attributed to the compiler backend used (*xlc_mapvect* vs. *map_vect*), finding that the MAP backend produces much better code for compilation units consisting of one large basic block, while XL C profits from being able to perform its optimizations on units larger than one basic block, e. g., by loop unrolling.

It is noticeable that the backend does not give a significant performance gain in the non-power-of-two case (Fig. 2). This is due to the fact that FFTW normally does not include large kernels for non-power-of-two sizes as base cases. A comparable performance level as in the power-of-two case could be obtained if large non-power-of-two kernels were included into the library.

Instruction Count. For all codes investigated, the MAP vectorizer and optimizer for Blue Gene significantly reduced the instruction count by utilizing FP2 SIMD instructions. While the biggest part of the gain can be attributed to vectorization, the optimizer also has its share in code quality, by utilizing FP2 specific instructions, eliminating many SIMD swaps and multiplications.

For SIMD codes, the address generation part of the backend improves the code quality, by minimizing the number of integer instructions.

As FFTW kernels can be very large, minimizing the instruction count helps avoid hitting L1 instruction cache capacity limits.

Superior Performance Level. In the best cases, code produced by the MAP compiler runs at 80% of the performance that the best algorithm known in the literature could theoretically achieve on the target hardware.

MAP-compiled FFTW codelets enabled the material science code Qbox [11] to run with a sustained performance of 60 Tflop/s on BlueGene/L, thus reaching the second highest performance ever achieved by an application code.

7 Conclusion

The MAP compiler tool chain covers *all* stages of compilation that are important for achieving high performance in numerical software for linear signal processing transforms.

First, the code produced by a special purpose program generator, like FFTW's `genfft`, is vectorized, seeking an optimal utilization of the 2-way SIMD floating-point unit of IBM's PowerPC 440 FP2 processors.

Next, the MAP optimizer tries to minimize SIMD data reordering overhead and maximize utilization of FMAs and other FP2 specific idioms.

Finally, the code is compiled down to assembly, using (i) an optimal algorithm for register allocation for basic blocks, (ii) several levels of scheduling, and (iii) a clever instruction selection method for dealing with effective address generation on a processor with DSP-like addressing mode restrictions.

Performance data gathered in experiments with FFTW by itself and in the context of large application codes demonstrate the impressive performance—up to 60 Tflop/s—to be obtained by using the MAP compiler tool chain.

References

1. L. A. Belady. A study of replacement algorithms for virtual storage computers. IBM Systems Journal, 5(2):78101, July 1966.
2. K. Dockser. Oedipus Architecture: Extensions to PowerPC BookE for Hummer2. Technical report, IBM, August 2001.
3. J. E. Moreira et al. Blue Gene/L Programming and Operating Environment. IBM Journal for Research and Development, 49(2/3), 2005.
4. M. Puschel et al. SPIRAL: Code Generation for DSP Transforms. Proceedings of the IEEE, 93(2):232275, 2005.
5. S. Chatterjee et al. Design and exploitation of a high-performance SIMD floating-point unit for Blue Gene/L. IBM Journal for Research and Development, 49(2/3), 2005.
6. R. J. Fisher and H. G. Dietz. Compiling for SIMD Within A Register. In Proceedings of the 11th Workshop on Languages and Compilers for Parallel Computing (LCPC), pages 290304, 1998.
7. F. Franchetti, S. Kral, J. Lorenz, and C. W. Ueberhuber. Efficient Utilization of SIMD Extensions. IEEE Special Issue on Program Generation, Optimization, and Platform Adaptation, 93(2), 2005.
8. M. Frigo. A Fast Fourier Transform Compiler. Proceedings of the ACM SIGPLAN Conference on Programming Languages Design and Implementation (PLDI), 34(5):169180, May 1999.
9. M. Frigo and S. G. Johnson. FFTW: An adaptive software architecture for the FFT. In Proceedings of the IEEE Intl. Conference on Acoustics, Speech, and Signal Processing, volume 3, pages 13811384. IEEE, 1998.
10. J. Guo, M. Garzaran, and D. Padua. The power of Beladys algorithm in register allocation for long basic blocks. In LNCS on Languages and Compilers for Parallel Computing, volume 2958, pages 374390. Springer-Verlag, 2004.
11. F. Gygi, E. Draeger, B. R. de Supinski, R. K. Yates, F. Franchetti, S. Kral, J. Lorenz, C. W. Ueberhuber, J. Gunnels, and J. Sexton. Large-Scale First- Principles Molecular Dynamics Simulations on the BlueGene/L Platform using the Qbox Code. In Proceedings of the ACM/IEEE Conference on Supercomputing, 2005. Gordon Bell Prize runner-up.
12. S. Hoxey, F. Karim, B. Hay, and H. Warren (editors). The PowerPC Compiler Writers Guide. Warthman Associates, 1996.
13. S. Larsen and S. Amarasinghe. Exploiting superword level parallelism with multimedia instruction sets. ACM SIGPLAN Notices, 35(5):145156, 2000.
14. R. Leupers and S. Bashford. Graph-based code selection techniques for embedded processors. ACM Trans. Design Autom. Electron. Syst., 5(4):794814, 2000.
15. J. Lorenz, S. Kral, F. Franchetti, and C. W. Ueberhuber. Vectorization techniques for the Blue Gene/L double FPU. IBM Journal for Research and Development, 49(2/3), 2005.
16. S. S. Muchnick. Advanced Compiler Design and Implementation. Morgan Kaufmann, 1997.
17. E. Sikha and R. Simpson. The PowerPC Architecture: A Specification for a New Family of RISC Processors. Morgan Kaufmann, 2nd edition, 1995.
18. P. van Hentenryck. Constraint Satisfaction in Logic Programming. MIT Press, 1989.
19. H. Zima and B. Chapman. Supercompilers for Parallel and Vector Computers. ACM Press, New York, 1991.

SCAN: A Heuristic for Near-Optimal Software Pipelining

F. Blachot[1,2], Benoît Dupont de Dinechin[2], and Guillaume Huard[1]

[1] ID Laboratory, Grenoble, France*
[2] STMicroelectronics, Grenoble, France

Abstract. Software pipelining is a classic compiler optimization that improves the performances of inner loops on instruction-level parallel processors. In the context of embedded computing, applications are compiled prior to manufacturing the system, so it is possible to invest large amounts of time for compiler optimizations.

Traditionally, software pipelining is performed by heuristics such as iterative modulo scheduling. Optimal software pipelining can be formulated as integer linear programs, however these formulations can take exponential time to solve. As a result, the size of loops that can be optimally software pipelined is quite limited.

In this article, we present the SCAN heuristic, which enables to benefit from the integer linear programming formulations of software pipelining even on loops of significant size. The principle of the SCAN heuristic is to iteratively constrain the software pipelining problem until the integer linear programming formulation is solvable in reasonable time.

We applied the SCAN heuristic to a multimedia benchmark for the ST200 VLIW processor. We show that it almost always compute an optimal solution for loops that are intractable by classic integer linear programming approaches. This improves performances by up to 33.3% over the heuristic modulo scheduling of the production ST200 compiler.

1 Introduction

In scientific and multimedia applications, most of the execution time is spent in loops. In case of instruction-level parallel processors such as superscalar and VLIW, instruction scheduling of inner loops can significantly increase performances, in particular with software pipelining [1]. The mainstream software pipelining technique is called modulo scheduling [2,3,4].

Modulo scheduling solves a 1-periodic cyclic scheduling problem with the objective of minimizing the period or *initiation interval* (II). This is achieved by computing first a lower bound $MinII$ on the II. Then cyclic scheduling is attempted for increasing II values starting at $MinII$, until a solution is found. In general it is NP-hard to know what is the minimum II of a modulo scheduling problem, but modulo schedules whose II equals $MinII$ are clearly optimal.

The modulo scheduling problem at a given II can be formulated as an integer programming problem [5,6]. The formulation of Eichenberger et al. uses

---

* This work has been partially funded by the MOAIS INRIA project.

a mix of $\{0, 1\}$ and of integer variables, while the formulation of Dupont-de-Dinechin only uses $\{0, 1\}$ variables (but in a larger number). We implemented the formulation of Dupont-de-Dinechin as it is adapted from the integer linear programming formulation of resource-constrained project scheduling [7]. Solving such formulations enables to modulo schedule at the minimum II, an appealing possibility in cases the compilation time is not severely constrained. However, the time required to solve these formulations grows exponentially with the size of the modulo scheduling problem, making classic integer linear programming approaches intractable beyond a few tenths of instructions.

In this article, we propose a new approach for modulo scheduling that enables to benefit from integer linear programming formulations even on problems with a significant number of instructions. The main idea is to restrict the solution space to the areas where the formulation can be solved in reasonable time. Here solved means either computing a solution or proving that no solution exists. For the integer linear programming formulations discussed above, some constants and the number of variables depend on a parameter called the time horizon of the loop. Although a theoretical bound exists for the time horizon, it is usually much higher than required by an optimal solution.

The principle of our approach is to heuristically reduce the time horizon in order to solve the formulation. The issue with this reduction is that it might transform a feasible modulo scheduling problem into an infeasible one. Thus, we have to explore the solution space along two parameters: performance, represented by the II; the time horizon, that must be kept small enough so the formulation can be solved. Our approach takes advantage of an empirical knowledge on the general shape of the solution space, which we deduced from a large set of experiments. We called this approach the SCAN heuristic.

Although reducing the time horizon may eliminate all optimal solutions, our results show that SCAN reaches the optimal performance (II equals $MinII$) for most of the loops of our benchmark, including those that appeared intractable with the original integer linear programming formulations. For the remaining loops, we have no way to know if modulo scheduling with $II = MinII$ is feasible, but the II results of the SCAN heuristic are consistently close to $MinII$ and better than those of the modulo scheduling heuristic of the ST200 production compiler. Overall, our approach results in improvements of up to 33.3% for the most difficult loops as demonstrated by our experiments.

This article is organized as follows. In section 2, we present the modulo scheduling problem and we review the integer linear programming (ILP) formulation currently used by the SCAN heuristic. In section 3, we develop our findings about the time horizon and its relations with the ILP formulation. We characterize the search space of the initiation interval and the time horizon values. Then we present how the SCAN heuristic searches for the best tractable initiation interval by adjusting the time horizon accordingly. Finally, section 4 reports the experimental results of the SCAN heuristic on a multimedia benchmark.

2 The Cyclic Scheduling Problem

Cyclic scheduling, also known as software pipelining in the case of instruction scheduling, is a widely studied problem [1]. Modulo scheduling [4] is a class of software pipelining techniques that build 1-periodic schedules. This section introduces our notations for modulo scheduling and presents the integer linear programming formulation of modulo scheduling we use.

2.1 Cyclic Scheduling Problem Formulation

Consider the problem of scheduling a loop with a possibly large number of iterations. The loop can be represented by a finite, directed multigraph $G = (I, E_{dep}, \theta, \omega)$. The vertex set I contains the instructions of the loop body and each instruction $I_i \in I$ generates a set of instruction instances $\{I_i^k | k \in \mathbb{N}\}$, one for each iteration of the loop.

The directed edges E_{dep} model the dependence constraints between instructions: each edge $I_i \longrightarrow I_j \in E_{dep}$ is labeled with a pair $(\theta_i^j, \omega_i^j) \in \mathbb{N} \times \mathbb{N}$ where θ_i^j is the dependence latency and ω_i^j the dependence distance. Such dependence expresses the fact that the execution of I_i^k (instance of I_i at iteration k) must start θ_i^j cycles before the execution of $I_j^{k+\omega_i^j}$ (instance of I_j at iteration $k+\omega_i^j$).

Each instruction I_i is also associated with a execution time p_i and a resource requirements vector $\vec{b_i}$ of size r. The total availability of the processor resources is also given by a vector $\vec{B}$. Typical resources are issue width, functional units and memory ports. Execution time of fully pipelined instructions is $p_i = 1$.

The cyclic scheduling problem is to determine a schedule $\sigma : I \times \mathbb{N} \to \mathbb{N}$ for the instruction instances I_i^k that respects the dependence constraints:

$$\forall I_i \longrightarrow I_j \in E_{dep}, \forall k \geq 0 : \sigma_i^k + \theta_i^j \leq \sigma_j^{k+\omega_i^j} \qquad (1)$$

and the resource constraints: at any clock cycle, the sum of resources used cannot be greater than the available resources $\vec{B}$.

Among all cyclic schedules, the 1-periodic schedules are especially interesting in instruction scheduling as they enable simple code generation. Such schedules, also known as modulo schedules, are defined by:

$$\exists II \in \mathbb{N}, \forall I_i \in I, \forall k \in \mathbb{N} : \sigma_i^k = \sigma_i^0 + k \times II \qquad (2)$$

The period of the schedule, usually called the initiation interval in the literature, is denoted II. This is the performance metric of modulo schedules: the lower the initiation interval, the greater the execution throughput.

The initiation interval of any modulo schedule is limited by a lower bound $MinII$ defined as $\max(MIIRec, MIIRes)$ [4], where $MIIRec$ (recurrence minimum initiation interval) is related to dependence circuits and $MIIRes$ (resource minimum initiation interval) is related to resource uses:

$$MIIRec \stackrel{def}{=} \max_{C \text{ circuit in } G} \left\lceil \frac{\sum_{I_i \to I_j \in C} \theta_i^j}{\sum_{I_i \to I_j \in C} \omega_i^j} \right\rceil \quad (3)$$

$$MIIRes \stackrel{def}{=} \max_r \left\lceil \frac{\sum_{I_i \in I} p_i b_i^r}{B^r} \right\rceil \quad (4)$$

The *time horizon* is defined as the maximum number of cycles between the execution of two instruction instances of the same iteration:

$$H = \max_{I_i, I_j \in I} (\sigma(I_i, k) - \sigma(I_j, k)) \forall k \in \mathbb{N} \quad (5)$$

2.2 Modulo Scheduling by Integer Linear Programming

Integer linear programming (ILP) is a well known technique to formulate and solve combinatorial problems such as scheduling and routing. Several ILP formulations have been proposed for the modulo scheduling problem [5,8]. Based on the efficient ILP formulations used in resource-constrained project scheduling, B. Dupont-de-Dinechin recently introduced another formulation for modulo scheduling [6], which we use for the SCAN heuristic. Compared to the latest formulation of Eichenberger et al. [8], this new formulation only uses $\{0, 1\}$ variables and has stronger linear programming relaxations [7].

This new ILP formulation can be summarized as follows. For the sake of simplicity, we removed the objective function, the equations related to registers pressure and we assume that instructions have unit execution time, which is the case for our target processor. The complete formulation is available in [6].

$$\sum_{t=0}^{H-1} x_i^t = 1 \quad \forall i \in [1, n] \quad (6)$$

$$\sum_{s=t}^{H-1} x_i^s + \sum_{s=0}^{t+\theta_i^j - II\omega_i^j - 1} x_j^s \leq 1 \quad \forall t \in [0, H-1], \forall (i, j) \in E_{dep} \quad (7)$$

$$\sum_{i=1}^{n} \sum_{k=0}^{\lfloor \frac{H-1}{II} \rfloor} x_i^{t+k \times II} \vec{b_i} \leq \vec{B} \quad \forall t \in [0, II-1] \quad (8)$$

$$x_i^t \in \{0, 1\} \quad \forall i \in [1, n], \forall t \in [0, H-1] \quad (9)$$

Let n denote number of instructions. Each x_i^t is a $\{0, 1\}$ variable that is 1 if the instruction I_i is scheduled at time t, else it is 0. In this formulation, the equations correspond to: unique scheduling dates (6), dependence constraints (7) and resource constraints (8).

Any solution of this ILP formulation yields a valid modulo schedule at initiation interval II whose time horizon is at most H. By searching iteratively for the minimum II and with a large enough H, we eventually find an optimal solution of the modulo scheduling problem. Unfortunately, the resolution time of such integer linear program grows exponentially with the number of instructions.

3 The SCAN Heuristic

The ILP formulation introduced in section 2.2 depends on two parameters: the initiation interval II, which we want to minimize; the time horizon H, which bounds the span of the schedule of a given loop iteration. In this section, we describe how the SCAN heuristic drastically reduces the time to solve the the integer linear programs, based on a characterization of the solution space on the parameters H and II. It relies on the observation that, usually, a modulo schedule exists at a given II with a small time horizon.

3.1 The Search for the Time Horizon

The time horizon H of a solution to a modulo scheduling problem instance at a given II is not known in advance. A trivial lower bound is deduced from the longest path in the dependence graph. An upper bound is given in [9], which roughly equals the number of instructions times the sum of the initiation interval and the maximal dependence distance: $O(n \times (II + max(\omega_i^j)))$.

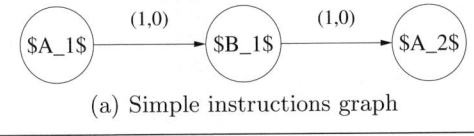

$A\_1\$ \xrightarrow{(1,0)} \$B\_1\$ \xrightarrow{(1,0)} \$A\_2\$$

(a) Simple instructions graph

resources/cycle	1	2	3	4	5	6	7	8	9	10
A	$A_{1,1}$		$A_{1,2}$	$A_{2,1}$	$A_{1,3}$	$A_{2,2}$	$A_{1,4}$	$A_{2,3}$	$A_{1,5}$	$A_{2,4}$
B		$B_{1,1}$		$B_{1,2}$		$B_{1,3}$		$B_{1,4}$		$B_{1,5}$

(b) cyclic schedule of initiation interval of 2 and time horizon of 4

resources/cycle	1	2	3	4	5	6	7	8	9	10
A	$A_{1,1}$		$A_{1,2}$		$A_{1,3}$	$A_{2,1}$	$A_{1,4}$	$A_{2,2}$	$A_{1,5}$	$A_{2,3}$
B		$B_{1,1}$		$B_{1,2}$		$B_{1,3}$		$B_{1,4}$		$B_{1,5}$

(c) cyclic schedule of initiation interval of 2 and time horizon of 6

Fig. 1. A simple instance of software pipelining. Graph of three instructions (1(a)) and two cyclic schedules (1(b)) and (1(c)) with time horizon 4 and 6.

A main issue is that the number of variables of the ILP formulation of section 2.2 directly depends on the value of the time horizon. Given that in most cases the time horizon of the optimal solution is close to its lower bound [9], the idea of using the ILP formulation with a small time horizon is natural. The difficult part is to determine which value of H is sufficient to keep the problem solvable at a given II. Unfortunately, this value is highly dependent on the interference between dependence and resource constraints.

Possible candidate values are difficult to guess as illustrated in figure 1: in this example, three instructions form a dependence chain and require two type of resources (type A for instructions A_1 and A_2 and type B for instruction B). Each resource type is limited to one instruction at a time. Because of this limitation and of the dependences, an optimal initiation interval of 2 (the resources lower bound) is achievable with a time horizon of 4 and 6 but not 5.

3.2 Characterization of the Search Space

An inappropriate choice for the value of H makes the integer linear program either infeasible or intractable. As appropriate choices of H for a given II are difficult to guess, we conducted experiments on all our benchmark loops to determine the shape of this search space. We tried all the possible (H, II) values with II ranging from the lower bound of the problem to the II found by a modulo scheduling heuristic and H ranging from the lower bound to the upper bound. For each possible couple of values, we reported if the problem was infeasible, feasible or reached a given timeout (in which case we consider it as intractable).

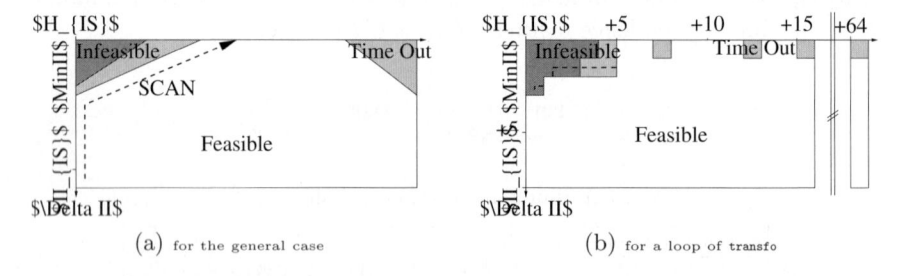

Fig. 2. Characterization of the search space and the SCAN heuristic

For all the loops, the observed search space has the same general shape: a timeout area for large H values, an infeasible area for small values of both II and H and a timeout area between the infeasible area and the rest which is feasible. This characterization is depicted in figure 2. For some loops, the infeasibility area reduces to an empty part (along with the timeout border) or the slope of the separating line might change (at worst this is an horizontal line: at some point we are not able to find a lower II whatever the H value). This leads to:

- infeasibility usually results from a too small value for H, in this case the solution is simply to increase it.
- intractability is more difficult to handle. It might result from a choice of H which is either too small, in the timeout border on the side of the unfeasible area, or too large, in the large timeout area.

Because of the non-predictable location of the timeout points, it is not possible to use a dichotomy for a given II value to find the appropriate H value. As it will be shown in the section 4, the value chosen for the timeout does not change the shape of the solution space. By choosing a smaller timeout value, the feasible area is just smaller and included in the area found with a larger value.

3.3 The SCAN Heuristic

The main idea of the SCAN heuristic is to change the values for II and H in order to progress along the line that separates the unfeasible area from the feasible one. This enables in most cases to reach the part with the lowest initiation interval while remaining within the feasible area. The algorithm is illustrated by figure 2 and can be described as follows:

1. start from II and H found by a classical modulo scheduling heuristic
2. solve the linear program, there are two cases:
    (a) if the program is unfeasible or stopped by the timeout, $H = H + 1$
    (b) if the program is feasible, $II = II - 1$
3. repeat step 2 until a global timeout or reaches $MinII$
4. return the best solution (lowest H in the set of lowest II feasible solutions)

4 Experimental Results

We performed experiments using our implementation of the SCAN heuristic integrated in the production compiler developed by STMicroelectronics for the ST200 processor. The ST200 is a VLIW processor that executes up to 4 instructions by cycle and has clean pipelines (instructions can be viewed as unit execution time with a latency of either 1 or 3 towards dependent instructions). The optimal scheduler used by the SCAN heuristic is linked with the CPLEX9.0 solver from ILOG for the ILP resolution. The compilation is performed on a cross-compiler running on a Pentium 4 1.8 GHz system with 1 Gb of RAM.

4.1 Space Characterization

The test suite is the multimedia benchmark used internally by STMicroelectronics for its compiler performance validation process. It contains 169 inner loops taken from speech coding, audio and video applications. These loops vary at the structural level (from sequential to highly parallel) as well as in the number of instructions of their body (from 12 to 114 instructions).

We performed an exhaustive search on all the loops of our benchmark to characterize the (H, II) space described in section 3.2. For this search, we used a timeout value of 3000 seconds for each point. Overall, the computation of these results ran for almost two weeks but validated our characterization.

Figure 3 shows the solution space for a difficult loop dbuffer with three different timeout values. We notice on this example that the slope of the timeout

(a) Solution space with a timeout of 3000s

(b) Solution space with a timeout of 100s

(c) Solution space with a timeout of 50s

(d) Behavior of Scan with a timeout of 50s

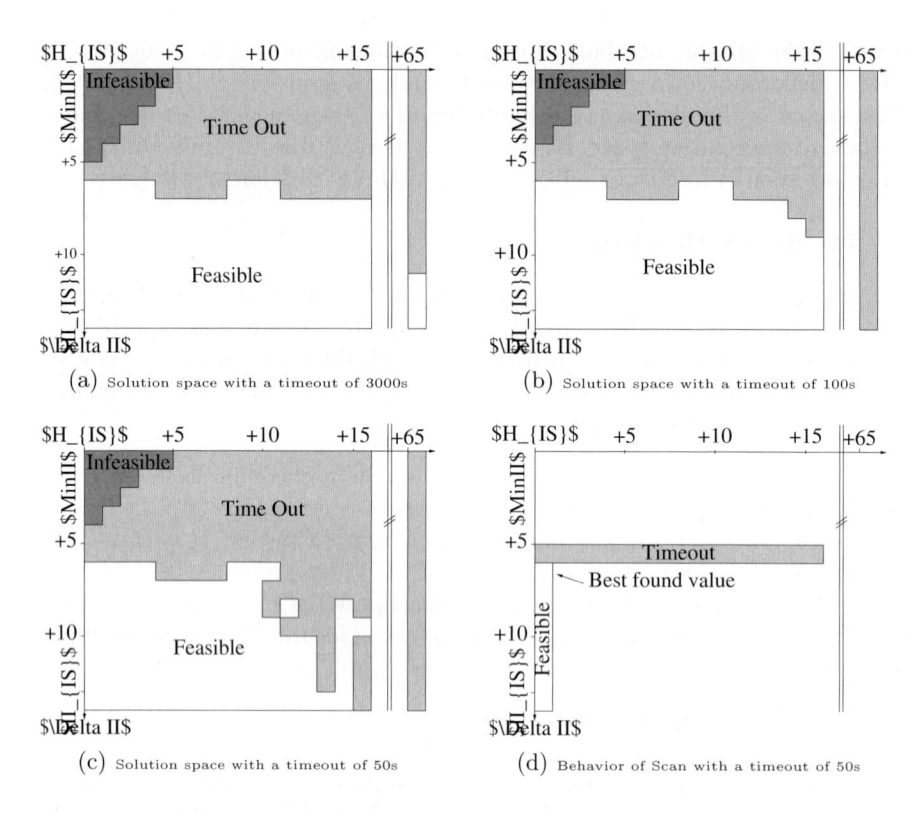

Fig. 3. Different solution spaces for a loop of *dbuffer* and behavior of SCAN

area frontier is almost horizontal and that the two timeout areas are connected. But it still conforms to our characterization whatever the timeout value.

4.2 Performance of the SCAN Heuristic

The ST200 production compiler integrates a heuristic modulo scheduler that schedules 65.4% of the loops at II equal to $MinII$. Precisely, among the 169 inner loops of our benchmark, 108 are scheduled optimally by the heuristic modulo scheduler (106 at the lower bound and 2 proved unfeasible at a lower initiation interval using an exact resolution). Thus 61 loops could be possibly improved after heuristic modulo scheduling. For this 61 loops, an exhaustive search on the initiation interval and horizon was done with a timeout of 3000s. A better solution was found for 47 loops, the other being unsolvable in 3000s. With a timeout of 75s by point, the SCAN heuristic is able to find all these better solutions; thus the SCAN heuristic finds the better known solutions of our benchmark.

The improvements of the SCAN heuristic over the heuristic modulo scheduler can be significant: for 8 loops the gain is close to 20% and maximum is 33.3%. Among the noticeable results, the main loop of the `fft32x32s` 32-bit fractional

Table 1. Improvements of the SCAN heuristic for different timeout values

timeout	SCAN vs. HMS	MinII vs. SCAN	time for scan	MinII vs. ILP	time for ILP
5s	4.10%	1.39%	3.36s	2.86%	5.48s
10s	4.21%	1.28%	5.95s	2.78%	8.86s
25s	4.28%	1.20%	9.19s	2.66%	19.75s
75s	4.29%	1.19%	22.10s	2.21%	52.57s
500s	4.29%	1.19%	98.34s	1.59%	277.40s

radix-4 Fourier transform, which appeared intractable using the ILP formulation because of its 83 instructions, has been improved by 21.4%.

Table 1 illustrates how the SCAN heuristic improves on average the 169 loops of our benchmark for the different values of the timeout listed in the first column. The second column contains the average II improvements of the SCAN heuristic over the production heuristic modulo scheduler (HMS). The third column contains the average II increase of the SCAN heuristic over $MinII$ and the average time spent per loop in the SCAN heuristic. The fourth column contains the average II increase of the ILP modulo scheduler over $MinII$ and the average time spent per loop in the ILP modulo scheduler. From these figures, the SCAN heuristic appears quite effective even at low timeout values.

5 Conclusions

We presented a heuristic that takes advantage of integer linear programming formulations of modulo scheduling. Such formulations when solved yield optimal software pipelines, but resolution times are worst case exponential. In practice, only loops that comprise less than a few tenths of instructions can benefit from integer linear programming formulations of modulo scheduling.

The SCAN heuristic we propose makes integer linear programming formulations of modulo scheduling applicable to significantly larger loops, by walking on the boundaries of the practically solvable solution space. The solution space we consider is bi-dimensional, one dimension being the software pipeline period II and the other a heuristic restriction on the schedule time horizon H. The SCAN heuristic takes advantage of an empirical characterization of the search space and evolves these parameters towards close to optimal solutions.

We implemented the SCAN heuristic and an integer linear programming formulations of modulo scheduling in the STMicroelectronics production compiler for the ST200 VLIW processor. The experiments we conducted show that our space characterization holds for all of the considered loops. Furthermore, the use of the SCAN heuristic on the difficult loops of a multimedia benchmark produced results up to 33.3% better than the heuristic modulo scheduler of the production compiler. The performance and flexibility of the SCAN heuristic make it perfectly suitable for production use in embedded code compilation.

References

1. Allan, V.H., Jones, R.B., Lee, R.M., Allan, S.J.: Software pipelining. ACM Comput. Surv. **27**(3) (1995) 367–432
2. Rau, B.R., Glaeser, C.D.: Some scheduling techniques and an easily schedulable horizontal architecture for high performance scientific computing. In: MICRO 14: Proceedings of the 14th annual workshop on Microprogramming, IEEE Press (1981) 183–198
3. Lam, M.: Software pipelining: an effective scheduling technique for vliw machines. In: PLDI '88: Proceedings of the ACM SIGPLAN 1988 conference on Programming Language design and Implementation, ACM Press (1988) 318–328
4. Rau, B.R.: Iterative modulo scheduling: an algorithm for software pipelining loops. In: Proceedings of the 27th annual international symposium on Microarchitecture, ACM Press (1994) 63–74
5. Eichenberger, A.E., Davidson, E.S., Abraham, S.G.: Optimum modulo schedules for minimum register requirements. In: ICS '95: Proceedings of the 9th international conference on Supercomputing, ACM Press (1995) 31–40
6. de Dinechin, B.D.: From machine scheduling to vliw instruction scheduling. ST Journal of Research **1**(2) (2005)
7. Azem, S., Dupont de Dinechin, B., Artigues, C.: Résolution d'un problème d'ordonnancement modulo sur une architecture vliw par la programmation linéaire en nombre entiers. In: ROADEF'2006: actes du 7ème congrès de la Société Francaise de Recherche Opérationnelle et d'Aide à la Décision, Lille (2006)
8. Eichenberger, A.E., Davidson, E.S.: Efficient formulation for optimal modulo schedulers. In: PLDI '97: Proceedings of the ACM SIGPLAN 1997 conference on Programming language design and implementation, ACM Press (1997) 194–205
9. Eisenbeis, C., Sawaya, A.: Optimal loop parallelization under register constraints. Technical Report RR-2781, INRIA (1996)

Code Generation for STA Architecture

J. Guo, T. Limberg, E. Matus, B. Mennenga, R. Klemm, and G. Fettweis

Vodafone Chair Mobile Communication Systems, TU Dresden, Germany
{guojie, limberg, matus, mennenga, klemm, fettweis}@ifn.et.tu-dresden.de

Abstract. This paper presents a novel compiler backend which generates assembly code for Synchronous Transfer Architecture (STA). STA is a Very Long Instruction Word (VLIW) architecture and in addition it uses a non-orthogonal Instruction Set Architecture (ISA). Generating efficient code for this architecture needs highly optimizing techniques. The compiler backend presented in this paper is based on Integer Linear Programming (ILP). Experimental results show that the generated assembly code consumes much less execution time than the code generated by traditional ways, and the code generation can be accomplished in acceptable time.

1 Introduction

Code generation has been the focus of many research works. In order to generate efficient code for irregular architectures, Integer Linear Programming (ILP) modelling for code generation has been explored extensively in the recent past.

Kent Wilken et al. [1] developed an instruction scheduling model and efficient basic block partition. Timothy Kong et al. [4] developed a register allocation model for regular and irregular architectures. Because the interdependencies between the phases in code generation may lead to a significant decrease of code quality, Daniel Kästner et al. [5] built two sets of ILP formulations for phase-coupled code generation. Besides using some ideas of Wilken and Kästner, our ILP model is built in order to be aware of the STA features.

This paper is organized as follows: Section 2 gives a brief introduction to the STA features. Section 3 and 4 explain the compiler backend in detail. The results of our experimental evaluation are summarized in Section 5. Section 6 concludes and gives an outlook.

2 Synchronous Transfer Architecture (STA)

Figure 1 gives an overview over the STA concept. STA [6] processors are built up from modules, each with a set of input and output ports. The output ports are buffered. The buffer at the output holds the result of the last operation, until the next operation of the belonging module is executed. The data at an input port is selected from a set of connected output ports by a multiplexer. Thus, data can be obtained directly from an output register of connected FU, which lowers the requirement for additional register strongly. This kind of data

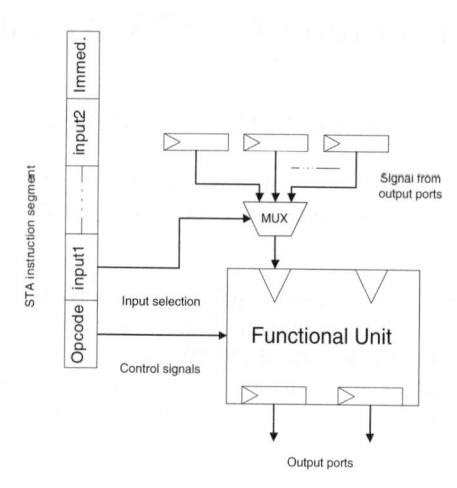

Fig. 1. STA modules

transfer will be called *direct data routing* (DDR) in the rest of this paper. Direct data routing can dramatically reduce the amount of required registers in a register file.

For the sake of generality and simplicity, register files and memory read or write ports are also implemented as STA modules. Each module behavior is fully qualified by an opcode and multiplexer control lines. The opcode controls the operation on the module. Multiplexer control lines select the input operands. The opcodes and multiplexer control lines for all modules are aligned in a VLIW.

Different from common architectures, minimizing register access in STA can result in much better results. The operands can't be made available one single clock cycle after storing them. When a FU uses a value in hardware register, the value is transferred to the read port of register file in the first clock cycle. In the next cycle, the FU can select this read port as its input and does operation. Due to the limited number of register read ports, additional wait cycles may be introduced, if more operands need to be read than ports are available.

3 Analysis for Integer Linear Programming (ILP)

Our ILP-based compiler backend can be divided into several stages: Firstly, an ILP model is built, which includes a set of formulations for all possible states. In a second step the useful data (exact value of the variables in formulations) is created from Medium-Level Intermediate Representation (MIR). Each kind of data is used by one or more different formulations in the model. Then the solvable formulations are generated and the results will be found by using CPLEX (a software for solving different kinds of optimization problems). In a final step, a program transforms the results of CPLEX into assembly code. The developed ILP model is restricted to basic block [7]. The proposed process of the data file generation consists of the following six phases:

3.1 Memory Spill Code Generation

Firstly, we made a global data flow analysis [7] to find the IN and OUT data for each basic block. Then all the registers are divided into two groups: group I is used to handle IN and OUT data and group II is used for register operations, which manipulate the temporary results within one basic block.

Our simulations with many different basic blocks have shown that *data direct routing* is mostly used in the results of CPLEX, so the number of register operations used for saving temporary results, which appear only in one basic block, is very small. Thus in our current implementation, we assigned 80% of all available registers to group I and the remaining 20% to group II. If the amount of registers in group I is not enough, memory spill code (load and store) will be generated in MIR. Otherwise, the remaining registers in addition to the registers of group II can be used in order to determine an optimal tradeoff between *direct data routing* or register access within each basic block.

3.2 MIR Analysis

Suppose instruction i_1 and i_2 are in the same basic block of MIR, and i_1 appears before i_2 and they may have one of the following dependence:

1. i_1 writes a location that i_2 reads (RAW)
2. i_1 reads a location that i_2 writes (WAR)
3. i_1 writes a location that i_2 writes (WAW)
4. i_1 reads a location that i_2 reads (RAR)

For guaranteeing the correctness of the assembly code, the first two kinds of dependencies are considered in the scheduling. The third dependence WAW is avoided by location renaming [7] in our MIR. Assume $\mathbb{I}$ is the set of all the instructions in the MIR, then these two kinds of dependence can be observed as:

1. Set $\mathbb{DDR\_U} \subset \mathbb{I} \times \mathbb{I}$ contains for all read after write instruction pairs (possible to be used for data direct routing of STA)
2. Set $\mathbb{U} \subset \mathbb{I} \times \mathbb{I}$ contains for all write after read instruction pairs (necessary to be considered in the scheduling)

In the rest of this paper we refer to these two coupled instructions in a pair of the first category as a *DDR pair*.

3.3 Inter-blocks Read and Write Analysis

Since register allocation is done for IN and OUT data in the first step, the following data sets are defined:

1. IN (the input data in registers come from other basic blocks or from itself in the last iteration)
2. OUT (the data in registers which will be as the input data in the other basic blocks or itself in the next iteration)
3. $\mathbb{U}_{\text{IN}} \subset \mathbb{RI} \times \mathbb{I}$ (dependence between inter-block register read instructions and MIR instructions)
4. $\mathbb{U}_{\text{OUT}} \subset \mathbb{I} \times \mathbb{WI}$ (dependence between MIR instructions and inter-block register write instructions)

3.4 Inner-blocks Read and Write Analysis

We assume there are always a pair of virtual read instruction $r_{u,v}$ and write instruction $w_{u,v}$ between DDR pair (i_u, i_v). There are two possibilities: only one instruction uses the result of i_u (Case 1) and more than one instructions use the result of i_u (Case 2).

In case 2, if we consider these *DDR pairs* independently, more than one write instructions may take place. Actually, it is enough to have only one write instruction, because the data from i_u exists already in the register after the first write instruction. In this case, only one register write instruction is generated in our compiler. If a true register write exists, a new read instruction is generated directly before each instruction which uses i_u's result in our compiler. Read instructions can be only executed on the read port. If the data is stored in the register's read port until last use, this read port cannot perform other read instructions for several clocks. It may decrease the instruction level parallelism.

3.5 Machine Resource Analysis

One of the key concepts in our model is the *execution time* of the instructions. One type of FUs have same execution time. Different operations which can be assigned to the same type of FU have the same execution time. The following parameters are predefined:

- E_u: the execution time of instruction i_u, $i_u \in \mathbb{I} \cup \mathbb{RI} \cup \mathbb{WI}$.
- $E_{u,v}^w / E_{u,v}^r$: the execution time of the virtual write/ read in a *DDR* pair (i_u, i_v).

The machine resource is divided into three classes in our model. There are Functional Units (FUs), the read and write ports of register, and registers. Each instruction can be explicitly matched into one type of FU in our architecture. According to the type of FUs, we divided instructions in $\mathbb{I}$ into some FUi sets.

4 Integer Linear Programming(ILP) Model

Let C be the number of clock cycles needed for the execution of all the very long instruction words in a basic block, then our optimization model is explained as follows:

- **Objective:** minimizing C
- **Self Constraint** This constraint ensures each instruction is scheduled exactly once in the basic block scheduler.
- **Data Routing Constraint** It describes the state of each FU with the consideration of RAW dependence.
- **Machine Constraint** This constraint guarantees that the scheduling is performed without exceeding machine resources in each cycle.
- **Dependence Constraint** It describes the WAR dependence.

4.1 Self Constraint

For an instruction $i_u \in \mathbb{I}$, we define a series of binary variables $x(u,j), j \in \mathbb{J}$, where $\mathbb{J}$ is the set of all clock cycles: $1..C$. The value of $x(u,j)$ is 1 when i_u is scheduled to start in cycle j, otherwise 0.

Equation (1) guarantees that each instruction i_u must be started and can be only started once.

$$\sum_{j=1}^{C} x(u,j) = 1, \forall i_u \in \mathbb{I} \tag{1}$$

In the same way, we define the corresponding binary variables for inner-block read and write instructions: $r(u,v,j), w(u,v,j), (i_u, i_v) \in \mathbb{DDR\_U}, j \in \mathbb{J}$. Between each *DDR pair*, only one read and one write instruction may start.

4.2 Data Routing Constraint

Observing a *DDR pair* $(i_u, i_v) \in \mathbb{DDR\_U}$ as shown in Figure 2 and 3. At first, i_v uses the result of i_u, so i_v should start after i_u is finished. Because the data transfer in our STA architecture is synchronous, i_v can also use the result of i_u in the same cycle i_u is finished. The equation (2) describes this constraint.

$$\sum_{j=1}^{C} x(u,j) * j + E_u \leq \sum_{j=1}^{C} x(v,j) * j, \qquad (i_u, i_v) \in \mathbb{DDR\_U} \tag{2}$$

Secondly, the write instruction $w_{u,v}$ must start in the cycle when it can get the result of instruction i_u. The equation (3) describes this constraint.

$$\sum_{j=1}^{C} x(u,j) * j + E_u \leq \sum_{j=1}^{C} w(u,v,j) * j, \qquad (i_u, i_v) \in \mathbb{DDR\_U} \tag{3}$$

Thirdly, as explained in Section 3.3, the read instruction $r_{u,v}$ is placed directly before instruction i_v. Equation (4) is used to describe this constraint.

$$\sum_{j=1}^{C} r(u,v,j) * j + E^r_{u,v} = \sum_{j=1}^{C} x(v,j) * j, \qquad (i_u, i_v) \in \mathbb{DDR\_U} \tag{4}$$

With the above three constraints, there are only five possible cases according to the sequence of virtual write and read instructions:

1. $w_{u,v}$ is at least $E_{u,v}^w$ cycles before $r_{u,v}$ (Fig2.a)
2. $w_{u,v}$ is in the same cycle as i_v (Fig2.b)
3. $w_{u,v}$ appears after i_v (Fig3.a)
4. $w_{u,v}$ is in one of the cycle from the same cycle of $r_{u,v}$ to one cycle before i_v (Fig3.b)
5. $w_{u,v}$ is before $r_{u,v}$, but less than $E_{u,v}^w$ cycles (Fig3.c)

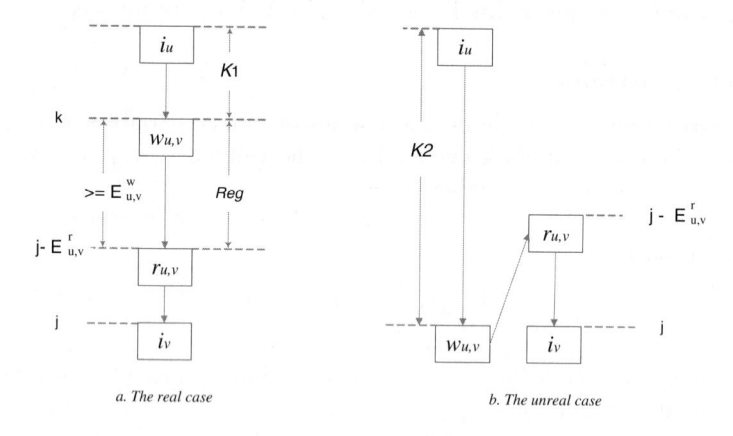

Fig. 2. The real and unreal case in our model

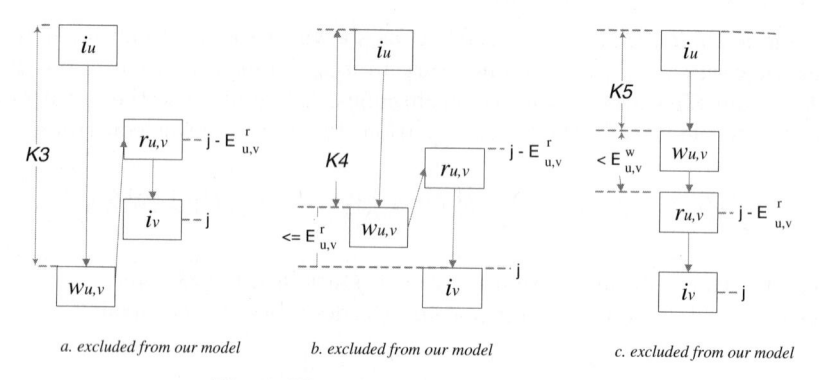

Fig. 3. The excluded cases in our model

Because a feasible write operation must save the data in the register before any read operation can access it, case 1 (Fig2.a) is considered as *real* and register is used to transfer the result of i_u to i_v. Other cases are all defined as *unreal*, but only case 2 (Fig2.b) is allowed and *direct data routing* is used to transfer the result. Case 3, 4 and 5 (Fig3) are excluded for conciseness of the modelling.

For excluding case 3 (Fig3.a), we further define that the read instruction can appear at most $E_{u,v}^r$ cycles before the corresponding write instruction.

Equation (5) is used to represent this constraint. Equation (4) has defined that $r_{u,v}$ is exactly $E_{u,v}^r$ cycles before i_v. So $w_{u,v}$ can not appear after i_v.

$$\sum_{j=1}^{C} w(u,v,j) * j - E_{u,v}^r \leq \sum_{j=1}^{C} r(u,v,j) * j, \qquad (i_u, i_v) \in \mathbb{DDR\_U} \quad (5)$$

Equation (6) and (7) are used to exclude case 4 (Fig3.b) and case 5 (Fig3.c) respectively.

$$w(u,v,j) + r(u,v,j+l) \leq 1, \quad (i_u, i_v) \in \mathbb{DDR\_U}, \forall j \in \mathbb{J}, l \in (0..E_{u,v}^r - 1) \quad (6)$$

$$w(u,v,j+l) + r(u,v,j) \leq 1, \quad (i_u, i_v) \in \mathbb{DDR\_U}, \forall j \in \mathbb{J}, l \in (0..E_{u,v}^w - 1) \quad (7)$$

Thus, the occupation of FU can be uniformly expressed with the remaining two cases by relative concise ILP equations. In $K1$ cycles (Fig 2.a), the corresponding FU is *occupied* until a register write instruction for saving the result of i_u appears. If $w_{u,v}$ appears in the same cycle as i_v (Fig 2.b), we define that the result of i_u is transferred to i_v by *direct data routing*, and the FU is occupied until i_v appears. Detailed analysis and other machine resource constraints are discussed in the Section 4.3. The formulations for the case that more than one instructions use the result of i_u are the extension of the above equations. They will also be explained together with the machine resource in the Section 4.3.

4.3 Machine Resource Constraint

The Constraints for Functional Units

There are two cases that one of the FU is occupied: an instruction is executed on this FU or the output register of this FU should be kept. We define a new set of binary variables $D_{link}(u,j)$, $j \in \mathbb{J}$ to represent the occupation of a FU. $D_{link}(u,j)$ takes the value of 1 if the FU is occupied in cycle j, otherwise 0. Because the coming instruction on the same FU has the same execution time of i_u and STA has pipeline structure, so this FU can be available for this instruction $E_u - 1$ cycles before write instruction takes place.

$$D_{link}(u,j) \geq \sum_{k=1}^{j} x(u,k) - \sum_{k=1}^{j+E_u-1} w(u,v,k) \quad \forall (i_u, i_v) \in \mathbb{DDR\_U}, \forall j \in \mathbb{J} \quad (8)$$

Equation (8) can be considered as a stair function with j as X-axis and the lower bound of $D_{link}(u,j)$ as Y-axis. $D_{link}(u,j)$ must take the value 1 from the cycle of i_u to $E_u - 1$ cycles before $w_{u,v}$ takes place, while not restricted in other cycles. In every cycle, the number of occupied functional units is bounded by the total available number N_{FUi} in the architecture. Thus, the equation of resource constraint for each type of FU can be derived:

$$\sum_{i_u \in FUi} D_{link}(u,j) \leq N_{FUi}, \forall j \in \mathbb{J} \quad (9)$$

The Constraints for Register Read and Write Ports

The access of register in each cycle is limited by the read and write ports. In Section 4.2, we have defined the *real* and *unreal* cases for the read and write instructions. Then, two sets of binary variables, $w_{real}(u,v,j)$ and $r_{real}(u,v,j), \forall j \in \mathbb{J}$, are defined to describe the usage of read and write ports in these two cases. The value 1 represents a real read/write operation in cycle j, and 0 means no register operation.

The equation (10) checks if a read instruction takes place before the corresponding write instruction. If so, $w_{real} \geq 0$, there is no restriction for this binary variable. The constraint $w_{real} \geq 1$ can only occur where $w(u,v,j) = 1$ and the read instruction appears after write.

$$w_{real}(u,v,j) \geq w(u,v,j) - \sum_{k=1}^{j-E_{u,v}^r} r(u,v,k), \forall j \in \mathbb{J}, \forall (i_u, i_v) \in \mathbb{DDR\_U} \quad (10)$$

In the same way, the equation (11) checks if a write instruction appears in the previous cycle of where the read instruction is.

$$r_{real}(u,v,j) \geq r(u,v,j) - \sum_{k=j+E_{u,v}^r}^{C} w(u,v,k), \forall j \in \mathbb{J}, \forall (i_u, i_v) \in \mathbb{DDR\_U} \quad (11)$$

If more than one instructions use the result of i_u, a set of binary variables $w_b(u,j), \forall j \in \mathbb{J}$ is introduced to trace the first write instruction (equation (12)), and the write instruction should still be placed after the instruction i_u (equation (13)).

$$\sum_{j=1}^{C} w_b(u,j) * j \leq \sum_{j=1}^{C} w_{real}(u,v,j) * j, \forall (i_u, i_v) \in \mathbb{DDR\_UB} \quad (12)$$

$$\sum_{j=1}^{C} x(u,j) * j + E_u \leq \sum_{j=1}^{C} w_b(u,j) * j, \forall i_u \in \mathbb{B} \quad (13)$$

In the above equations, $\mathbb{DDR\_UB}$ is defined as all DDR pairs in case 2 defined in Section 3.4 and $\mathbb{B}$ as the set of all instructions whose result is used by more than one instructions. We further define $\mathbb{DDR\_US}$ as the set of DDR pairs in case 1 in Section 3.4. Suppose N_{write} and N_{read} as the number of available register ports, then the register port constraints can be formulated with the following equations:

$$\sum_{(i_u, i_v) \in \mathbb{DDR\_US}} w_{real}(u,v,j) + \sum_{u \in \mathbb{B}} w_b(u,j) \leq N_{write}, \forall j \in \mathbb{J} \quad (14)$$

$$\sum_{(i_u, i_v) \in \mathbb{DDR\_U}} r_{real}(u,v,j) \leq N_{read}, \forall j \in \mathbb{J} \quad (15)$$

The Constraints for Registers

The following formula expresses a stair function expressed by the binary variables $reg(u,v,j)$, whose lower bound jumps from 0 to 1 in the cycle of the write instruction, and return to 0 in the cycle of read instruction.

$$reg(u,v,j) \geq \sum_{k=1}^{j} w(u,v,k) - \sum_{k=1}^{j} r(u,v,k), \forall j \in \mathbb{J}, \forall (i_u, i_v) \in \mathbb{DDR\_U} \quad (16)$$

$$reg(u,v,j) \geq 0, \forall j \in \mathbb{J}, \forall (i_u, i_v) \in \mathbb{DDR\_U} \quad (17)$$

The meaning is clear in the *real* case. For the *unreal* case, because the read instruction appears before write instruction, the right side of the formula (16) could be -1 in some cycles. But the $reg(u,v,j)$ is constrained to take non-negative values. Thus, it can represent the buffer occupation for both *real* and *unreal* cases.

The case 2 defined in Section 3.4 should also be considered here. We have introduced that after first write instruction is performed, the data exists in a register. It can be read again and again, and the hardware register can store other data as soon as the last read instruction takes place. A set of binary variables $r_b(u,j)$ is defined to trace the last read instruction of the branch:

$$\sum_{j=1}^{C} r_b(u,j) * j \geq \sum_{j=1}^{C} r_{real}(u,v,j) * j, \forall (i_u, i_v) \in \mathbb{DDR\_UB} \quad (18)$$

Thus, with another set of binary variables $reg_b(u,j)$, the register occupation in the case 2 can be calculated:

$$reg_b(u,j) \geq \sum_{k=1}^{j} w_b(u,v,k) - \sum_{k=1}^{j} r_b(u,v,k), \forall j \in \mathbb{J}, \forall (i_u, i_v) \in \mathbb{DDR\_UB} \quad (19)$$

$$reg_b(u,j) \geq 0, \forall j \in \mathbb{J}, \forall i_u \in \mathbb{B} \quad (20)$$

Assume $N_{register}$ is the available number of hardware registers (group II in Section 3.1). The constraint for registers is expressed in the following equation:

$$\sum_{(i_u, i_v) \in \mathbb{DDR\_US}} reg(u,v,j) + \sum_{i_u \in \mathbb{B}} reg_b(u,j) \leq N_{register}, \forall j \in \mathbb{J} \quad (21)$$

4.4 Dependence Constraint

We introduce this constraint in two categories: First, write after read dependence; Second, instructions in MIR and inter-block read and write instructions.

For the first category, assume instruction i_m reads a location before instruction i_n writes it. Equation (22) represents this dependence.

$$\sum_{j=1}^{C} x(m,j) * j + E_n - 1 \leq \sum_{j=1}^{C} x(n,j) * j, \quad (i_m, i_n) \in \mathbb{U} \quad (22)$$

For the second category, assume inter-block read instruction y_p reads a data from a register location, then instruction i_m uses this data, $(y_p, i_m) \in \mathbb{U}_{\text{IN}}$. For the instruction y_p, we also define a set of binary variables $y(p,j), y_p \in \mathbb{RI}, j \in \mathbb{J}$. The equation (23) expresses the modelling that y_p are placed directly before i_m. The reason is explained in Section 3.4.

$$\sum_{j=1}^{C} y(p,j) * j + E_p = \sum_{j=1}^{C} x(m,j) * j, \qquad (y_p, i_m) \in \mathbb{U}_{\text{IN}} \qquad (23)$$

Assume inter-block write instruction z_p writes the result of i_m to a register. Equation (24) represents their dependency.

$$\sum_{j=1}^{C} x(m,j) * j + E_m \leq \sum_{j=1}^{C} z(p,j) * j, \qquad (i_m, z_p) \in \mathbb{U}_{\text{OUT}} \qquad (24)$$

These inter block read and write instructions also occupy read/write ports, so the constraint equations (14) and (15) must be extended to accommodate the inter block instructions. Equation (25) and (26) describe the constraints. N_{read} and N_{write} the numbers of register read port and write port.

$$\sum_{(i_u.i_v) \in \text{DDR\_US}} w_{real}(u,v,j) + \sum_{i_u \in \text{B}} w_b(u,j) + \sum_{i_q \in \text{WI}} z(q,j) \leq N_{write}, \forall j \in \mathbb{J} \quad (25)$$

$$\sum_{(i_u.i_v) \in \text{DDR\_U}} r_{real}(u,v,j) + \sum_{i_p \in \text{RI}} y(p,j) \leq N_{read}, \forall j \in \mathbb{J} \qquad (26)$$

5 Experimental Results

We implemented all the formulations in our compiler by using one user licence of ILOG CPLEX (9.1 version) on one CPU. Our DSP has SIMD and VLIW architectural features [9]. The compiler center part performs vectorization for each application in MIR [10], then compiler backend generates corresponding assembly code with VLIW instruction set. Table 1 lists all the machine resource in our experiments. The execution time of decoder is 0; sfpu, sld, sst, vfpu, vldst are 2 clock cycles; seq is 3 clock cycles; the rest FUs are 1 clock cycle.

Table 2 shows the performance of our code generator for some signal processing benchmarks. According to the different partitions of basic blocks, we did three groups of tests. Each basic block in three groups contains 10, 15 and 20 instructions respectively. E[cycles] shows the execution time of the ILP generated assembly code in clock cycles. F[%] is the ratio (ILP_cycles/trad_cycles*100%). T[s] is the solving time of the CPLEX in seconds. As the length of the basic block becomes larger, the solving time of the CPLEX has exponential increment.

The execution time of the assembly code can be reduced about 50% in our experiments. Furthermore, these assembly code use much less memory than the former code. ILP-based code generator uses output buffers of the FUs very efficiently, that leads to less registers' requirement and much less memory spill code.

Table 1. Machine Resource Table

Scalar Unit		Vector Unit		Scalar MEM		Vector MEM		Other Unit	
name	amount	name	amount	name	amount	name	amount	name	amount
Salu	4	Valu	1	Sld	1	Vldst	1	Icu	1
Smul	1	Vshift	1	Sst	1			Preg	16
Sshift	1	Vif	1					Seq	1
Sif	1	Vfpu	1					Decoder	1
Slogic	1	Vreg	8						
Sfpu	1								
Sreg	32								

Table 2. Performance comparison between ILP and traditional code generator

benchmarks	group one			group two			group three		
name	E[cycles]	F[%]	T[s]	E[cycles]	F[%]	T[s]	E[cycles]	F[%]	T[s]
firparallel	114	65.1%	1.44	90	51.4%	2.9	82	46.9%	87.8
iirparallel	133	67.9%	1.73	105	53.6%	2.4	89	45.4%	19
firserial	75	64.1%	0.94	60	51.3%	2.5	55	47%	3.7
iirserial	62	54.9%	0.66	49	43.4%	5.1	43	38.1%	5.9
lmsparallel	1079	89.6%	19.7	904	75.1%	63.5	779	64.7%	842.1
lmsserial	149	50.5%	2.1	129	43.7%	7.9	103	34.9%	45
fft648	922	72.7%	29.7	723	56.9%	330.5	681	53.7%	1558.8
fft1288	1125	78.2%	35.1	890	61.8%	481.8	828	57.5%	1890.6
fft2568	1292	72.5%	44.3	1022	57.4%	555.8	951	53.4%	2092.1
dct2d88	767	59.2%	86.5	667	51.5%	249.8	607	46.9%	708.5

6 Conclusion and Future Work

The advantage of this work is, that it can improve the instruction level parallelism with block partition in acceptable time. Furthermore, the code generation is very flexible: the number of machine resource and the optimality of the solutions can be easily adjusted by the users. The limitation of such ILP-based code generator is its scalability. Within acceptable time, it can only optimize the code in the small basic blocks, and the performance of such code generator is also related to the block partition. As the future work, we will make better block partition for our code generator, then some global heuristic algorithms will also be implemented.

References

1. Kent Wilken, Jack Liu, and Mark Heffernan. "Optimal Instruction Scheduling Using Integer Programming." Proceedings of the ACM SIGPLAN 2000 conference on Programming language design and implementation, p121-133.

2. Krishnan Kailas, Kemal Ebcioglu and Ashok Agrawala. "CARS: A New Code Generation Framework for Clustered ILP Processors." Proceedings of the 7th International Symposium on High-Performance Computer Architecture, p133.
3. J.Hennessy and D.Patterson. "Computer Architecture, a Quantitative Approach." Morgan Kaufmann Publishers, Inc., San Francisco, CA, 2nd edition, 1996.
4. Timothy Kong and Kent Wilken. "Precise Register Allocation for Irregular Architectures." Proceedings of the 31st annual ACM/IEEE international symposium on Microarchitecture.
5. Daniel Kästner. "Retargetable Postpass Optimisation by Integer Linear Programming", PHD Thesis, Universität des Saarlandes, Germany, Oktober 2000.
6. Gordon Cichon, P. Robelly, H. Seidel, E. Matus, M. Bronzel and Gerhard Fettweis. "Synchronous Transfer Architecture (STA)." SAMOS-04, p126-130, June, 2004.
7. Steven S. Muchnick. "Advanced Compiler Design Implementation." Morgan Kaufmann Publishers, 1997.
8. A.V.Aho, R.Sethi, and J.D.Ullman. "Compilers, Principles, Techniques and Tools." Addison-Wesley, Redding, MA, 1985.
9. Gordon Cichon, P. Robelly, H. Seidel, M. Bronzel and Gerhard Fettweis. "SAMIRA: A SIMD-DSP Architecture targeted to the Matlab source language." GSPx'04, USA, 27. -30. July 2004.
10. P. Robelly, G. Cichon, H. Seidel and Gerhard Fettweis. "Automatic code generation for SIMD DSP architectures: An algebraic approach." PARELEC'04, Dresden, Germany, 07. - 10. September 2004.

Multi-dimensional Kernel Generation for Loop Nest Software Pipelining*

Alban Douillet[1], Hongbo Rong[2], and Guang R. Gao[1]

[1] University of Delaware Newark, DE 19716, USA
[2] Microsoft Corporation Redmond, WA 98052, USA

Abstract. Single-dimension Software Pipelining (SSP) has been proposed as an effective software pipelining technique for multi-dimensional loops [16]. This paper introduces for the first time the scheduling methods that actually produce the kernel code. Because of the multi-dimensional nature of the problem, the scheduling problem is more complex and challenging than with traditional modulo scheduling. The scheduler must handle multiple subkernels and initiation rates under specific scheduling constraints, while producing a solution that minimizes the execution time of the final schedule.

In this paper three approaches are proposed: the *level-by-level* method, which schedules operations in loop level order, starting from the innermost, and does not let other operations interfere with the already scheduled levels, the *flat* method, which schedules operations from different loop levels with the same priority, and the *hybrid* method, which uses the level-by-level mechanism for the innermost level and the flat solution for the other levels. The methods subsume Huff's modulo scheduling [8] for single loops as a special case. We also break a scheduling constraint introduced in earlier publications and allow for a more compact kernel. The proposed approaches were implemented in the Open64/ORC compiler, and evaluated on loop nests from the Livermore, SPEC200 and NAS benchmarks.

1 Introduction

Software pipelining (SWP) is an important loop scheduling technique that overlaps the execution of consecutive iterations of a loop to explore instruction-level parallelism [9,8,13,1,7,10]. Traditionally, it is applied to the innermost loop of a loop nest. The schedule can be extended to outer loops by hierarchical reduction [9,11,17]. Loop transformations can be performed to the innermost loop before SWP [2,18,12].

Single-dimension Software Pipelining (SSP) [16] is a unique resource-constrained framework for software pipelining a loop nest. The scheduling technique overlaps the iterations of any loop in a loop nest that satisfies the dependence constraints. The compilation framework is shown in Fig. 1. First, the loop level deemed the most profitable is selected and the *multi-dimensional data dependence graph (n-D DDG)* is simplified into a *one-dimensional DDG (1-D DDG)* and sent as input to the scheduler [16]. The kernel is then computed. If the register pressure is reasonable [5], registers are allocated [14] and the final code is generated [15].

---
* This work was supported in part by the DOD, by DARPA contract No.NBCH30904, by NSF grants No.0103723 and No.0429781, and by DOE grant No.DE-FC02-OIER25503.

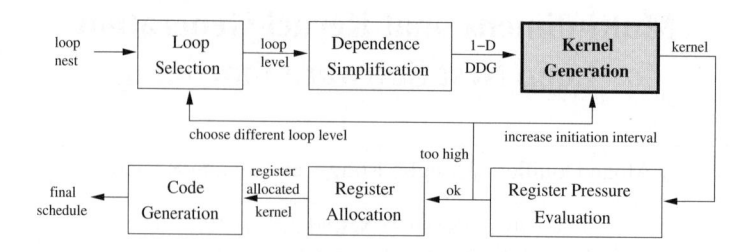

Fig. 1. SSP Compilation Framework

In this paper, we present for the first time algorithms for the *kernel generation* step. The computed kernel must minimize the execution time of the final multi-dimensional schedule. The problem is complex - as it involves the overlapping of operations from several loop levels (dimensions) of a loop nest, a challenge not encountered in traditional modulo scheduling. The kernel is partitioned into *subkernels*, one per loop level and each with its own initiation interval. Those subkernels interact with each other and optimizing one subkernel could have a negative impact on the others. Moreover, when the scheduler fails and the initiation interval must be increased, which subkernel should be chosen?

Three approaches are proposed and studied. The *level-by-level* approach generates the subkernels in order, starting from the innermost. Once a subkernel has been computed, it cannot be altered. The *flat* approach does not lock a subkernel once fully scheduled. Operations from any loop level may be considered and undo previous decisions made in a different subkernel. A larger solution space can therefore be explored. Finally, the *hybrid* approach schedules the innermost subkernel first and locks it. The other operations are then scheduled using the flat method. It allows for a shorter compilation time than the flat method while exploring a large solution space and focusing resources on the innermost loop. The three approaches subsumes Huff's scheduler [8] as a special case when the loop nest is a single loop.

We also break an SSP limitation that forced operations from different loop levels to be scheduled in distinct stages and that may artificially bloat the size of the kernel. We prove that, with minor modifications to the code generator and without code size increase, operations other than innermost can actually be scheduled in the same stage than operations from a different level.

The proposed approaches and heuristics associated with them have been implemented in the Open64/ORC compiler and analyzed on loop nests from the Livermore and NAS benchmarks. Experimental results show that the hybrid approach avoids the pitfalls of the two other approaches and produces schedules on average twice faster than modulo-scheduling schedules. Because of its large search space, the flat approach may not reach a good solution fast enough and showed poor results.

The rest of the paper is organized as follows. First, the SSP technique is reviewed. In section 2, the kernel generation problem for SSP is presented. Section 3 explains how to schedule operations from different levels into the same stage. The next section presents the scheduling methods in details. The last three sections are devoted toward experiments, related work, and conclusion, respectively.

2 The SSP Kernel Generation Problem

2.1 Single-Dimension Software Pipelining

Single-dimension Software Pipelining (SSP) [16,15,14,5] is a resource-constrained software pipelining method for both perfect and imperfect loop nests with a rectangular iteration space. Unlike other approaches [9,6,17,11], SSP does not necessarily software pipeline the innermost loop of a loop nest, but directly software pipelines the loop level estimated to be the most profitable. From the SSP point of view, the loop levels enclosing the selected loop are ignored. Therefore, the selected loop becomes the outermost loop. Within an iteration of the outermost loop, inner loops run sequentially.

Beside being able to software pipeline any loop level and overlap the execution of the prolog and epilog of the inner loops, the advantage of SSP over modulo-scheduling (MS) is that instruction-level parallelism or data cache reuse properties present in the outer loops are now accessible. Without prior iteration space transformations, a faster schedule with better cache performance can be found. If the innermost loop level is chosen, SSP is equivalent to classical modulo scheduling. SSP retains the simplicity of modulo scheduling, and yet may achieve significantly higher performance [16].

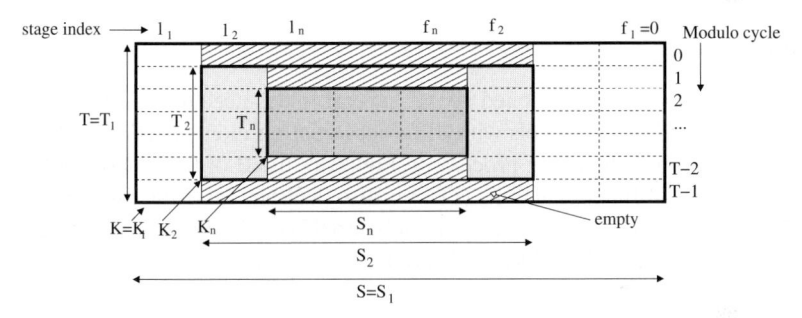

Fig. 2. Generic SSP Kernel

The final SSP schedule is derived from the kernel. Unlike the MS kernel, the SSP kernel has multiple initiation intervals and is composed of one subkernel K_i per loop level i in the loop nest. Each subkernel has its own number of stages S_i and initiation interval T_i. We note f_i and l_i the index of the first and last stages of K_i in the full kernel. Some slots are empty because of the kernel nesting constraints presented next (Fig. 2).

2.2 Problem Statement

The operations in the kernel must obey the scheduling constraints. A possible conservative definition of those constraints is given below. The modulo property and the resource constraints are identical to those used in MS. However, the dependence constraints now include the number of unused cycles term (uc, defined in Sec. 4), corresponding to the empty stages. The other constraints only exist in SSP. Let $\sigma(op)$ be the schedule time of operation op in the kernel. The constraints are:

- **Modulo Property:** operations are issued every T cycles. T is the initiation interval of the kernel.
- **Resource Constraints:** at any given cycle of the kernel, a hardware resource is not allocated to more than one operation.
- **Dependence Constraints:** $\sigma(op_1) + \delta \leq \sigma(op_2) + k * T - uc(op_1, op_2, k)$ for all the dependences from the 1-D DDG from op_1 to op_2 where δ is the latency of the dependence and k the distance.
- **Sequential Constraints:** $\sigma(op) + \delta \leq S_p * T_n$ for every positive dependence $\vec{d} = <d_1, ..., d_n>$ originating from op in the original multi-dimensional DDG and where d_p is the first non-null element in the subvector $< d_2, ..., d_n >$.
- **Kernel Nesting Constraints:** operations from different loop levels cannot be scheduled in the same stage and a stage cannot be enclosed between stages of deeper loop levels.

The SSP kernel generation problem can then be formulated as follows: given a set of loop nest operations and the associated 1-D DDG, schedule the operations so that the scheduling constraints are honored and the initiation interval of each subkernel is minimized. Even when the loop nest is a single loop, the problem is NP-hard [19].

2.3 Issues

To satisfy the constraints mentioned above, several issues need to be solved. First, the kernel is composed of subkernels with different initiation intervals (II) which must be respected during the scheduling process. In Fig. 3(a), the II of the innermost kernel is 2 and the number of functional units (FUs) is also 2. When inserting op_4, op_3 must be ejected to maintain the current II. Also, if a subkernel is rescheduled to a different cycle, one must make sure that the subkernel is not truncated as shown in Fig. 3(b)

(a) Strict Initiation Rate of Subkernels

(b) Truncation of Subkernels

(c) Useless II Increment Decision

(d) Non-Optimal II Increment Decision

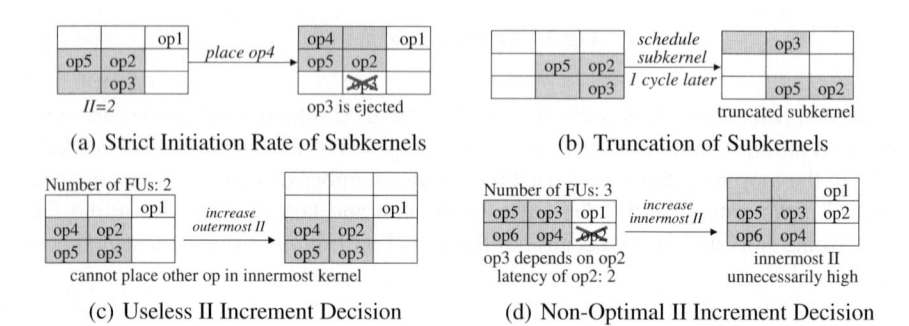

Fig. 3. Kernel Generation Issues

Then, the multiple II feature raises issues of its own. When the scheduler cannot find a solution and the II of one subkernel must be incremented, which subkernel should be chosen? Fig. 3(c) and 3(d) shows examples of inefficient schedules because of poor II increment decisions.

3 Breaking the Kernel Nesting Constraints

The kernel nesting constraints were originally introduced for implementation reasons. We now show that those constraints are unnecessary and can be removed. The advantage is two-fold. First, it gives more freedom to the scheduler which may be able to find a more compact kernel as shown in Fig. 4: $op1$, an operation from the outermost loop can now be scheduled in the same stage as operations from the innermost level. Second, because the number of stages may decrease, so may the register pressure.

To produce a correct final schedule from such a kernel, it is sufficient to conditionally emit the operations of the kernel. During the emission of stages for the execution of loop level i only operations from level i and deeper are emitted. If operations from other loop levels are present in the stage, they are simply ignored.

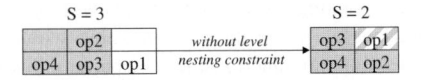

Fig. 4. Removal of the Kernel Nesting Constraints

Since the innermost loop is most frequently executed, it is not desirable to put operations from other levels into the innermost subkernel, in case they artificially increase its II. Also, the conditional emission of operations in the innermost stages requires code duplication to be used. Therefore, we will instead enforce a weaker limitation, called the *innermost level separation limitation*, that forbids outer loop operations to be scheduled into the innermost loop stages.

4 Solution

The algorithm framework, shared by the three approaches, is derived from Huff's algorithm [8,13,1] and shown in Fig. 5. Starting with the minimum legal II [16] for each loop level, the scheduler proceeds as follows. The minimum legal scheduling distance ($mindist$) between any two dependent operations is computed. Using that information, the earliest and latest start time, $estart$ and $lstart$ respectively, of each operation is computed. The difference $lstart - estart$, called $slack$, is representative of the scheduling freedom of an operation. The operations are then scheduled in the kernel in a heuristic-based order. If the scheduling of the current operation does not cause any resource conflict, the choice is validated. Otherwise, the conflicting operations are ejected. In both cases, the $estart$ and $lstart$ values of the ejected or not-scheduled operations are updated accordingly. The process is repeated until all the operations are scheduled. After too many iterations without success (max_op_try attempts), the II of one subkernel is incremented and the scheduler starts over. After max_II_try II increments, the scheduler gives up. If a solution is found, the scheduler enforces the sequential constraints and returns successfully. The different steps are detailed in the next subsections.

The proposed approaches are correct. As shown in the next subsections, the generated kernel respects all the scheduling constraints. Because the algorithm is based

SSP_SCHEDULER(*approach*, *priority*, *II_increment*, *max_II_try*, *max_op_try*):
  for each loop level i **do**
    set T_i to the minimum legal II for that level
  end for
  for *max_II_try* attempts **do**
    initialize *mindist* table and modulo resource table
    compute *slack* of operations
    for *max_op_try* attempts **do**
      choose next operation *op* according to *approach* and *priority*
      if no operation left **then**
        enforce sequential constraints
        return *success*
      end if
      schedule operation *op*
      eject operations violating resource constraints with *op*
      eject operations violating dependence constraints with *op*
      eject operations violating innermost level separation limitation with *op*
      update slack and MRT
    end for
    choose level i to increase II according to *II_increment*
    increment T_i by 1
  end for
  return *failure*

Fig. 5. Scheduling Framework

on modulo scheduling, the resource constraints are also honored. Moreover, when applied to a single loop, the method is Huff's algorithm and therefore subsumes modulo scheduling as a special case.

4.1 Scheduling Approaches

Three different scheduling approaches are proposed. With *level-by-level* scheduling, the operations are scheduled in the order of their loop levels, starting from the innermost. Once all the operations of one level are scheduled, the entire schedule becomes a virtual operation from the point of view of the enclosing level. The virtual operation acts as a white box both for dependences and resource usage. Operations within the virtual operation cannot be rescheduled. The method is simple and fast. However, the early scheduling decisions made in the inner loops might prevent the scheduler from reaching more beneficial solutions later in the scheduling process. Fig. 6 shows an example where we assume 2 functional units and a dependence between op_1 and op_2 with a latency of 2 cycles. The level-by-level scheduler must increase T_1 to 3 in order to schedule $op1$ whereas the flat scheduler can reschedule inner operations to obtain a kernel with $T_1 = 2$.

Flat scheduling considers operations from all loop levels as potential candidates. When backtracking, conflicting operations from all levels can be ejected from the schedule. The main advantage of this approach is its flexibility. Early decisions can always be

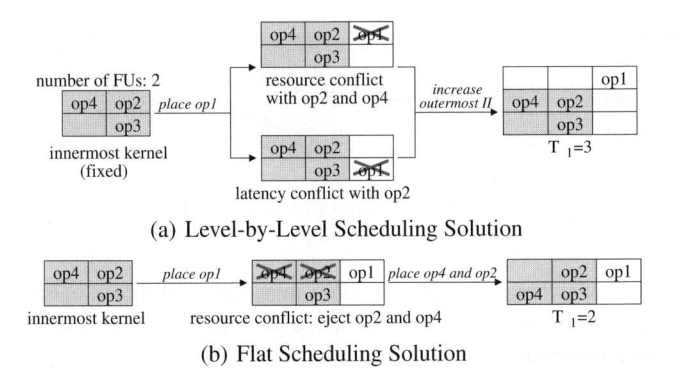

Fig. 6. Advantage of the Flat Approach over the Level-by-Level Approach

undone. Such flexibility leads to a larger solution space, and potentially better schedules. On the down side, the search space might become too large and slow down the scheduler.

The *hybrid* approach embeds the flat scheduling into a level-by-level framework. The innermost level is scheduled first. Its kernel becomes a virtual operation and the flat scheduling method is used for the other loop levels. The hybrid approach is intuitively a good compromise between level-by-level and flat scheduling, as confirmed by the experimental results. It can find better solutions than the level-by-level method without the cost in compile time of the flat method.

4.2 Enforcement of the SSP Scheduling Constraints

The dependence constraints, $\sigma(op_2) - \sigma(op_1) \geq \delta - k * T + uc(op_1, op_2, k)$, is enforced through the $mindist$ table. Because the table is generated before scheduling the operations, uc must be expressed independently of the yet unknown schedule time of op_1 and op_2. The following tight upper bound is proposed:

$$\sigma(op_2) - \sigma(op_1) \geq \frac{T}{T_n} * (\delta + 2 * T - (k+2) * T_n)$$

The right-hand side of the equation is the $mindist(op_1, op_2)$. By construction [4], the dependence constraints are always enforced.

The sequential constraints are not enforced during the scheduling process, but as a posteriori transformation once a schedule that satisfies the other constraints has been found. At that time, empty stages are inserted in the schedule until the sequential constraints are verified. The need for extra stages occurs rarely enough to justify such a technique.

To enforce the innermost level separation limitation without any extra computation cost, the schedule is conceptually split into three *scheduling blocks*: $before, innermost$ and $after$. Operations that lexically appear before (after, respectively) the innermost loop are scheduled independently into the 'before' ('after') scheduling block (Fig. 7). Innermost operations are scheduled into the 'innermost' scheduling block. Within each

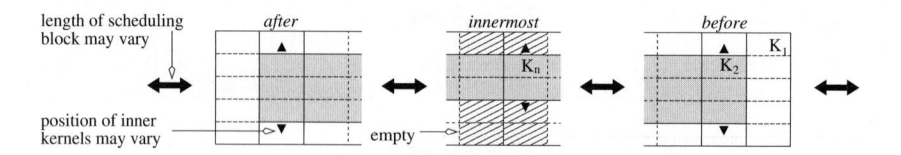

Fig. 7. Scheduling Blocks Example

scheduling block, the length of the schedule may vary without breaking the separation limitation and final length of the full schedule is only known at the very end. The modulo resource reservation table is shared between the three blocks.

When an operation is scheduled or when an operation is ejected, the slack of dependent operations must be recomputed. Usually, such an update is incremental. However, a dummy START and a dummy STOP operations are used to mark the boundaries of each scheduling block. As the slack is computed relatively to the distance between the START and STOP operations, if a dummy operation of one block is ejected and rescheduled, the slack of every operation within this block has to be recomputed.

4.3 Kernel Integrity

In the level-by-level approach, the subkernels are computed separately and therefore the initiation intervals are always respected. Truncation is avoided by forbidding the subkernels from being scheduled in cycles that would cause it.

In the flat approach, the initiation intervals are enforced by scheduling an operation first within the current boundaries of its subkernel. If impossible, the operation is scheduled at some other cycle. The subkernel is then correspondingly moved. All the operations that are then not within the boundaries of the kernel anymore are ejected. Therefore, subkernels cannot be truncated.

4.4 Operation Selection

The operation selection order is determined by a two-level priority mechanism. The primary priority is based on the loop level of the operation. In *innermost first* order, the operations are scheduled in depth order, starting from the innermost. In *lexical order*, the operations are scheduled in the order they appear in the original source code. In *block lexical* order, the operations are scheduled in the order of scheduling blocks: before-innermost-after. In *unsorted* order, the primary priority is bypassed. Then, 3 secondary priorities are used to break ties. With *slack* priority, the operations with a smaller slack are scheduled first. Critical operations, i.e. operations that use resource used at 90% or more in the schedule, have their slack divided by two to increase their priority. With *smaller lstart* priority, the operations with a smaller latest start time are scheduled first. The priority can be seen as a top-down scheduling approach. With *larger estart* priority, the operations with a larger earliest start time are scheduled first. It is a bottom-up scheduling approach.

4.5 Operation Scheduling

The legal range of schedule cycles for an operation selected for scheduling is defined by $[estart, lstart]$. If the operation is scheduled for the first time, $estart$ is chosen. Otherwise, the next value in the legal range since the last scheduling attempt is chosen. If there is none, the other scheduled operations, the availability of resources, and the II of the level of the operation are ignored and $estart$ is chosen. Conflicts created by the decision will be solved by later ejecting the scheduled operations involved in the conflicts.

4.6 Initiation Interval Increment

Several heuristics are proposed to decide which subkernel should have its II incremented when the scheduler times out. With *lowest slack first*, the average slack of the operations of each level is computed. The loop level with the lowest average slack is selected. With *innermost first*, the first level (from the innermost to the outermost) in which not all the operations have been already scheduled is selected. The heuristic is to be used only with the innermost first scheduling priority. With *lexical*, the first loop level in lexical order in which not all the operations have been already scheduled is selected. The heuristics is to be used only with the lexical scheduling priority.

5 Experiments

The proposed solution was implemented in the Open64/ORC2.1 compiler. 19 loop nests of depth 2 or 3, extracted from the NAS, SPEC2000, and Livermore benchmark suites, were software-pipelined at the outermost level and run on an Itanium2 workstation.

5.1 Comparison of the Scheduling Approaches

The best execution time for each approach was measured (Fig. 8). On average, hybrid and level-by-level schedules are twice faster than MS schedules. In several occasions, the flat solution is slower. Even when given as much as 10 times more attempts to find a solution, the flat scheduler fails and had to increment the initiation intervals, resulting in a slower final schedule. In one case (*liv-5*), the flat schedule was able to perform better than the level-by-level approach. As expected, the hybrid approach combined the advantages of the two other methods and, for all benchmarks but *liv-3*, produces a kernel with best execution time. Therefore, the hybrid approach should be the method of choice to generate SSP kernels.

The register pressure was also measured. On average, the register pressure in SSP schedules is 3.5 times higher than with MS schedules, in line with results from previous publications. The hybrid and level-by-level approaches have comparable register pressures, whereas the pressure is lower for the flat approach as the initiation intervals are higher. For *hydro*, the register pressure was too high with the level-by-level approach. It was observed that the register pressure is directly related to the speedup results. The higher the initiation intervals, the lower the register pressure and the execution time of the schedules.

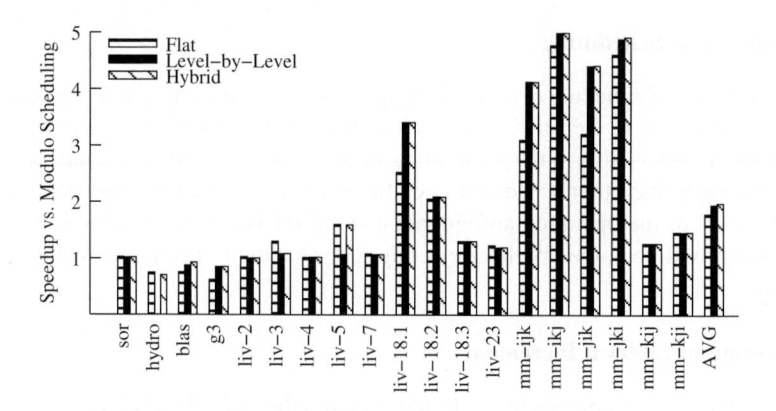

Fig. 8. Execution Time Speedup vs. Modulo Scheduling

5.2 Comparison of the Heuristics

Fig. 9 compares the results of the different operation selection heuristics for each scheduling approach. The minimum execution time and register pressures were recorded and the relative difference of each heuristic to the minimum was computed for each test case. The average is shown in the figure. The first letter U, L, I, or B stand for the primary selection method: Unsorted, Lexical, Innermost first or Block lexical respectively. The second letter S, E, or L for the secondary method: Slack, largest Estart or smallest Lstart. Level-by-Level scheduling was only tested for the unsorted primary method because all methods are equivalent when a single loop level is scheduled at a time.

For the flat scheduler, the best heuristic is highly dependent on the benchmark being evaluated. On average, each heuristic produces schedules 7.5% slower than the best schedule. Those high variations are explained by the size of the search space. For the two other methods, the choice of the heuristics have little influence on the execution time of the final schedule.the quality of the computed solution.

The II increment heuristics were also compared for the two approaches that use them: flat and hybrid. For the flat scheduler, the slack and lexical order produce the fastest schedules and are on average below 8% of the best schedule. The innermost order can produce schedules 30% slower. Because the slack order is not dependent on the operation selection heuristic used, it is to be preferred. For the hybrid scheduler, the impact of

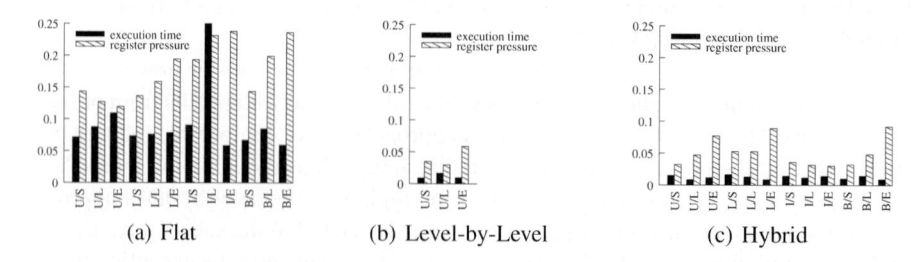

Fig. 9. Comparison of the Operation Selection Heuristics

the II increment heuristics is limited. Indeed, the innermost level, which contains most of the operations, is treated as a special case. Therefore there is not much scheduling pressure left for the other levels (2 to 3 maximum).

6 Related Work

SSP is not the only method to software pipeline loop nests. But, it is the first that has performed a complete and systematic study on each of the subjects: scheduling, register allocation, and code generation [16,14,15].

Modulo-scheduling techniques were extended to handle loop nests through hierarchical reduction [9,17,11], in order to overlap the prolog and the epilog of the inner loops of successive outer loop iterations. Although seemingly similar in idea to the level-by-level approach proposed here, hierarchical reduction software pipelines every loop level of the loop nest starting from the innermost, dependencies and resource usage permitting. The dependence graph needs to be reconstructed each time before scheduling each level, and cache effects are not considered. SSP only tries to software pipeline a single level and to execute its inner loops sequentially. MS has also been combined with prior loop transformations [2,18,12].

Finally, there exists other theoretical loop nest software pipelining techniques such as hyperplane scheduling [3]. Such method not consider fine-grain resources such as function units and registers.

7 Conclusion

This paper proposed for the first time kernel generation methods and heuristics for the Single-dimension Software Pipelining framework and break the kernel nesting constraints introduced in earlier publications [15]. We proved that each technique enforces all the SSP scheduling constraints. Experiments demonstrated that, although the level-by-level and hybrid approaches show comparable schedules in terms of execution and register pressure, the hybrid method is to be preferred because it outperforms the level-by-level approach in some cases. The flat method was victim of its own large search space and could not find good solutions in a reasonable amount of time and had to settle for kernels with larger initiation intervals. The choice of the heuristics have little influence on the final schedules for the hybrid and level-by-level approach.

References

1. Allan, V.H., Jones, R.B., Lee, R.M., Allan, S.J.: Software pipelining. ACM Comput. Surv. **27**(3) (1995) 367–432
2. Carr, S., Ding, C., Sweany, P.: Improving software pipelining with unroll-and-jam. In: Proc. of HICSS'96, IEEE Computer Society (1996) 183–192
3. Darte, A., Schreiber, R., Rau, B.R., Vivien, F.: Constructing and exploiting linear schedules with prescribed parallelism. ACM Trans. Des. Autom. Electron. Syst. **7**(1) (2002) 159–172
4. Douillet, A.: A Compiler Framework for Loop Nest Software-Pipelining. PhD thesis, University of Delaware, Newark, Delaware, USA (2006)

5. Douillet, A., Gao, G.R.: Register pressure in software-pipelined loop nests: Fast computation and impact on architecture design. In: Proc. of LCPC'05, Springer-Verlag (2005)
6. Gao, G.R., Ning, Q., Dongen, V.: Extending software pipelining techniques for scheduling nested loops. In: Proc. of LCPC'94. (1994) 340–357
7. Govindarajan, R., Altman, E.R., Gao, G.R.: A framework for resource-constrained rate-optimal software pipelining. IEEE Trans. Parallel Distrib. Syst. **7**(11) (1996) 1133–1149
8. Huff, R.A.: Lifetime-sensitive modulo scheduling. In: Proc. of PLDI'93, ACM Press (1993) 258–267
9. Lam, M.: Software pipelining: an effective scheduling technique for vliw machines. In: Proc. of PLDI '88, ACM Press (1988) 318–328
10. Llosa, J.: Swing modulo scheduling: A lifetime-sensitive approach. In: Proc. of PACT'96, IEEE Computer Society (1996) 80
11. Muthukumar, K., Doshi, G.: Software pipelining of nested loops. In: Proc. of CC'01, Springer-Verlag (2001) 165–181
12. Petkov, D., Harr, R., Amarasinghe, S.: Efficient pipelining of nested loops: unroll-and-squash. In: Proc. of IPDPS'02, IEEE (2002)
13. Rau, B.R.: Iterative modulo scheduling: an algorithm for software pipelining loops. In: Proc. of MICRO 27, ACM Press (1994) 63–74
14. Rong, H., Douillet, A., Gao, G.R.: Register allocation for software pipelined multi-dimensional loops. In: Proc. of PLDI'05. (2005) 154–167
15. Rong, H., Douillet, A., Govindarajan, R., Gao, G.R.: Code generation for single-dimension software pipelining of multi-dimensional loops. In: Proc. of CGO'04. (2004) 175–186
16. Rong, H., Tang, Z., Govindarajan, R., Douillet, A., Gao, G.R.: Single-dimension software pipelining for multi-dimensional loops. In: Proc. of CGO'04. (2004) 163–174
17. Wang, J., Gao, G.R.: Pipelining-dovetailing: A transformation to enhance software pipelining for nested loops. In: Proc. of CC '96, Springer-Verlag (1996) 1–17
18. Wolf, M.E., Maydan, D.E., Chen, D.K.: Combining loop transformations considering caches and scheduling. Int. J. Parallel Program. **26**(4) (1998) 479–503
19. Wood, G.: Global optimization of microprograms through modular control constructs. In: Proc. of MICRO 12, IEEE (1979) 1–6

Towards a Versatile Pointer Analysis Framework*

R. Castillo, A. Tineo, F. Corbera, A. Navarro, R. Asenjo, and E.L. Zapata

Dpt. of Computer Architecture, University of Málaga,
Complejo Tecnologico, Campus de Teatinos, E-29071. Málaga, Spain
{rosa, tineo, corbera, angeles, asenjo, ezapata}@ac.uma.es

Abstract. Current pointer analysis techniques fail to find parallelism in heap accesses. However, some of them are still capable of obtaining valuable information about the way dynamic memory is used in pointer-based programs. It would be desirable to have a unified framework with a broadened perspective that can take the best out of available techniques and compensate for their weaknesses. We present an early view of such a framework, featuring a graph-based shape analysis technique. We describe some early experiments that obtain detailed information about how dynamic memory arranges in the heap. Furthermore, we document how def-use information can be used to greatly optimize shape analysis.

1 Introduction

Pointer analysis is a field of study that has drawn a great deal of attention over the past few years. The problem of calculating pointer-induced aliases must be solved so that compilers can safely disambiguate memory references. Static knowledge of pointer-aliasing is key to perform optimizations related to parallelism and locality. While stack-pointer and array aliases allow for successful techniques to be applied, heap-directed pointers render such techniques ineffective. Therefore, new approaches must be taken.

We present in this work a pointer analysis framework that can accommodate several pointer analysis techniques, both existent and new. It is designed as an extensible framework based in Java. High-level program transformations are favored with the use of a near-source IR obtained with *Cetus* [1], a parsing tool aimed towards source-to-source translations. A key part of the framework is a newly designed graph-based shape analysis algorithm [2], that can obtain very detailed information about the arrangement of recursive data structures in the heap. Section 2 introduces our shape analysis technique in the context of the overall framework.

To better understand the shape analyzer capabilities, we have conducted some preliminary tests with typical heap-directed structures. These tests prove that memory configurations are accurately captured. We even discovered that the

---

* This work was supported in part by the Ministry of Education of Spain under contract TIC2003-06623.

analysis times can be greatly reduced by driving the analysis with def-use information. Section 3 documents our experiments with the shape analyzer.

On its own, the shape analysis is a great tool for programmer support, as it can be used by developers to check how dynamic structures are really used in their programs. Better still, higher-level client analysis modules can be built over the shape analyzer. In particular, we focus in dependence analysis in the context of loops that traverse dynamic recursive data structures. Such dependence analysis is needed for automatic parallelization of pointer-based programs and for locality exploitation, which are the final goals of our research. As a prerequisite for the dependence test, we need to automatically detect induction pointers. Section 4 covers the dependence test as a client analysis and how we tackle the automatic detection of induction pointers.

Finally, Section 5 comments some related work and Section 6 concludes with the main contributions and ideas for future work.

2 Shape Analysis Within the General Framework

Fig. 1 gives an overview of the general layout for our pointer analysis framework. First, we take an input program and parse it with the Cetus tool. Cetus is a compiler infrastructure specially aimed towards the development of compilation passes of high-level nature. It is written in Java and its source code is publicly available under a non-restrictive license. Cetus can parse C, C++ and soon Java, to a unique *intermediate representation* or IR, where transformations can be performed. Cetus IR is regarded to be close the source code, which is suitable for transformations related to pointer analysis.

Within Cetus, we can design compilation passes that are required by the pointer analysis techniques that follow. Such passes would perform preconditioning transformations, like expression simplification, statement reordering, etc., or would extract information, like data types, CFG, etc., as needed by the subsequent analysis. The results of the analysis can then be used to perform optimizations related to parallelism or locality, modifying the original program to obtain an optimized version.

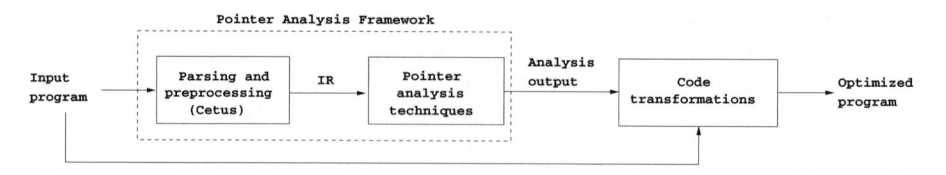

Fig. 1. General layout for the pointer analysis framework

Currently, we are focusing on the preprocessing and analysis phases. They conform the pointer analysis framework. Later, we can concentrate on using the results of the different pointer analysis techniques implemented to generate threaded versions of the programs.

The shape analyzer tool [2] is a cornerstone of our pointer analysis framework. Due to space limitations, only the main features and design principles will be described. It provides detailed information about the arrangement of memory locations in the heap for pointer-based programs. That information can be used for several purposes like: (i) data dependence analysis, by determining if two accesses may reach the same memory location; (ii) locality exploitation, by capturing the way memory locations are traversed to determine when are they likely to be contiguous in memory; and (iii) programmer support, to help detecting incorrect pointer usage or documenting complex data structures.

Our shape analyzer works as an iterative data-flow algorithm. It is flow-sensitive, context-sensitive and field-sensitive, although it lacks proper interprocedural support at the current state (we plan to add complete interprocedural support in the near future) and thus functions bodies must be inlined. The algorithm works by performing *abstract interpretation* over the pointer statements in the program until a *fixed-point* is reached. As result of the analysis, shape graphs are generated. Such graphs capture memory configurations arising in the heap in a conservative way. Fig. 2 shows an outline of the algorithm operation in the presence of (a) loops and (b) pointer statements, such as `ptr = ptr2` or `ptr = ptr2->sel`.

(a) **Loop statement class**

```
fun run(ShapeGraph sg)
    ShapeGraph oldSummary = EMPTYGRAPH;
    ShapeGraph newSummary = sg.copy();
    while(newSummary != oldSummary)
        Statement nextStmt = StatementList.next();
        while(nextStmt != NULL)
            sg = nextStmt.run(sg);
            nextStmt = StatementList.next();
        oldSummary = newSummary;
        newSummary.join(sg);
    return newSummary;
    //Return overall effect of loop
```

(b) **Pointer statement class**

```
fun run(ShapeGraph sg)
    ShapeGraphSet sgs = sg.splitBySel();
    //Breaks into possible graphs
    foreach(sg in sgs)
        sg.materializeNode();
        //Focus over currently accessed node
        sg.abstractSemantics();
        //Apply semantics of pointer statement
        sg.normalize();
        //Summarize compatible nodes
    foreach(sg in sgs)
        sgOut.join(sg);
    return sgOut;
```

Fig. 2. Outline of shape analysis algorithm regarding loops and pointer statements

Shape graphs are formed by nodes, links and CLSs (Coexistent Links Sets), which codify possibilities of connectivity between memory locations in the program. Graphs change according to the *abstract semantics* of the pointer statements present in the program. Fig. 3 sketches how graph change when analysing the first five statements in the creation of a singly-linked list. Dynamically allocated memory pieces are represented by nodes, and joined together with links. The last graph is also accompanied by its CLSs description, showing the combination of links that are possible for each node.

At compile time, the size and connectivity of recursive data structures is usually unknown. However, our representation of such structures must be finite, i.e., we must provide mechanisms to capture all possible memory configurations arising in the program in a finite number of bounded-size graphs. Graphs are

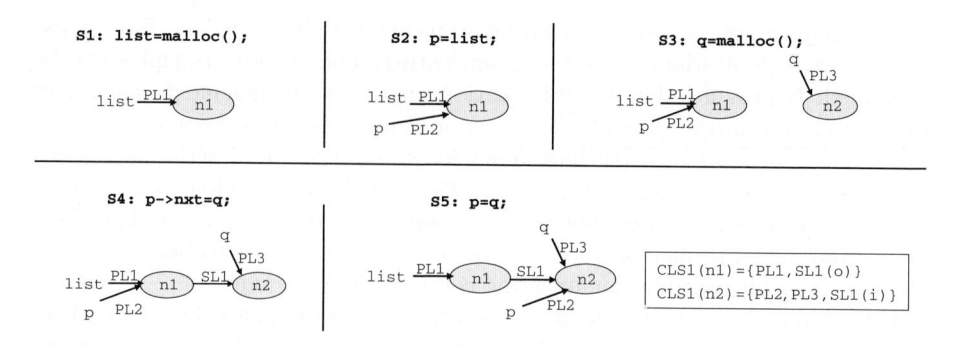

Fig. 3. Graphs are modified according to the abstract semantics of each statement

assured to be bounded by the *summarization* process: whenever nodes are regarded as *similar enough*, they are merged in to a so-called summary nodes. Similarity is determined by pointer alias relationships and adjustable *properties*. In fact, properties are a key instrument to fine-tune summarization decisions and therefore control how precisely graphs capture the features of the memory configuration.

Summarizing implies loosing information in favor of a bounded representation. We provide as well a dual operation to focus over previously summarized nodes: *materialization*. This operation can regain precision where pointer accesses are occurring because it performs *strong update* [3] [4], discarding unnecessary links in most situations. However, highly connected and summarized graphs can make impossible for the materialization operation to recover exactly the intended links, leaving some conservative ones.

Our analysis computes all possible memory configurations for every statement in the program. At any point during the analysis, there can be several graphs per statement to reflect all possible memory configurations that can reach the statement from different control flow paths. Different graphs represent mutually exclusive pointer arrangements over memory. Since the number of stack-declared pointer variables is fixed and known at compile time, the number of graphs per statement is limited by the different and mutually exclusive combinations of pointer over nodes and their properties.

The shape analyzer tool has been written in Java, taking in all new features of the latest Java 1.5 release. A big effort has been spent in making this tool as robust as it can be, so the object-oriented approach seemed a natural choice. Developing in such a manner facilitates writing extensions and performing maintenance tasks. Besides, a Java design makes it easier to blend with the extended version of Cetus that serves as front-end for the pointer analysis framework.

Fig. 4 is a simplified view of how elements interact within the pointer analysis framework: first, the input program is parsed by Cetus, this way we achieve an IR where we can easily operate; second, our specific preprocessing pass is run over Cetus IR to translate the program to the format required by the shape analyzer; third, the shape analyzer outputs the graphs for the program, which

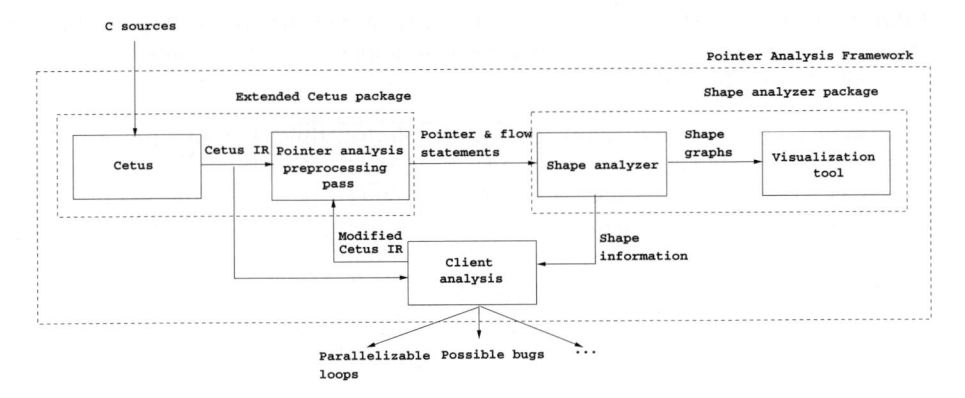

Fig. 4. Different modules working together within the pointer analysis framework

can be viewed in the companion visualization tool. Finally, client analysis techniques can be added to produce output results based on shape information, like parallelizable loops, possible bugs, etc. These techniques can drive the analysis to make it more effective as we will see later.

3 Experimental Results

We present now some early experimental results regarding the shape analyzer. For these tests we have considered six programs. The first four are typical kernels of applications that deal with recursive data structures. For the last two tests, we consider the product of a sparse matrix by a sparse vector, first based on singly-linked lists, then based on doubly-linked lists. Sparse structures are usually built with pointers to avoid wasting storage capacity with many empty values. Table 1 describes the structures tested and displays some metrics for the analysis performed.

The first column identifies each test, while the second column holds the number of analyzed statements. All available pointer and flow statements are considered. The tests that consider linked lists (singly-linked and doubly-linked) first create the lists, then traverse them. The tests working with trees (n-ary and binary) perform structure traversing during the trees creation, as each new element is added as a leave starting from the root. The sparse matrix is created as a header list (rows), whose elements point to other lists (columns), while the sparse vectors are created as lists. In the fifth test, the structures are based in simply-linked lists (s), while on the sixth test, they are based on doubly-linked lists (d). Regarding the product algorithm, first the input matrix and vector are created, then the output vector is built as the matrix and input vector are traversed. The output for each test is a graph that captures the structures created and traversed. The complete codes and resulting graphs are available through our website[1].

---
[1] http://www.ac.uma.es/~asenjo/research/codes.html

Table 1. Structures tested in the shape analyzer, number of analyzed statements, time spent on the analysis, total number of generated graphs, and nodes, links and CLSs per graph, in average (and maximum) values

Data structure	# stmts	Time	# graphs	Nodes, links & CLSs per graph
Singly-linked list	17	0.47 sec	62	2.51 (4) / 3.64 (7) / 4.75 (13)
Doubly-linked list	19	0.52 sec	74	2.59 (4) / 6.90 (13) / 4.55 (13)
N-ary tree	17	0.62 sec	372	2.61 (4) / 6.39 (12) / 9.38 (22)
Binary tree	25	2.02 sec	435	2.73 (4) / 10.58 (20) / 23.84 (65)
Matrix-vector(s)	83	1.14 min	2477	7.56 (12) / 26.10 (40) / 29.34 (50)
Matrix-vector(d)	97	1.55 min	2931	7.60 (12) / 30.95 (48) / 30.37 (50)

The third column shows times for the tests. Only the time for the actual shape analysis is shown (no parsing or preprocessing), as measured in a Pentium IV 2.4 GHz with 1 GB RAM. We think that times are very reasonable for such a detailed analysis. Within the first four examples of synthetic codes, the highest time is that of the binary tree analysis, probably due to its more complex CFG. It should be noted that more possible flow paths make the analysis more costly, as it has to consider all possibilities conservatively. On the other hand, the first three examples run in less than a second. The matrix by vector product takes longer, clocking at more than 1 minute, which is only reasonable considering there are quite some more statements to analyze than in previous tests.

The fourth column indicates the total number of graphs generated for each test. This metric gives an idea about the internal cost of analyzing different structures and traversals. The numbers range from a few dozens to a few thousands. Next columns show the total number of nodes, links and CLSs per graph, as average values with the maximum in brackets. The number of nodes per graph is essentially constant in the first four tests, as it depends mostly on the number of simultaneously live pointers, which is usually one for the structure handle and two for navigating it. The matrix by vector test has three times more nodes because there are three different structures, instead of one. The number of links depends on the amount of different links that each element has. Typically each element in a recursive data structure does not have more than two links.

Finally, CLSs are the elements where most of the complexity reside: they describe how nodes and links can combine to create all possible memory configurations arising in the program. The highest maximum is for the binary tree among all tests, but the maximum average is attained in the matrix by vector program based on doubly-linked lists.

To sum up, we can say that the shape analyzer can effectively analyze common data structures for pointer-based codes. Generated graphs accurately capture heap structures. Furthermore, we think that such graphs can be obtained in manageable times, specially for such a complex technique. Let us not forget that we are performing fixed-point abstract interpretation of pointer and flow statements to create and modify very detailed graphs. Despite this encouraging results, it is clear that this is a costly technique which is not likely to succeed if used for whole program analysis. Instead it would be better used within a client analysis module that would focus on *local analysis*.

In this regard, we discovered that def-use information can be used to identify the statements directly involved in the creation of the recursive data structures that are traversed in the segment of code under analysis. A def-use chain establishes a relationship between the definition point where a value is created and points where it is used. With that information we can automatically determine what are the statements that actually define the shape of dynamic memory and discard all other statements. With this approach we avoid to analyze irrelevant statements that could slow down the shape analysis.

We have tried this approach on the matrix by vector examples. Let us revisit them now, having *pruned* all traversal statements that are not involved in the output vector creation. The new values for the tests are shown in table 2, where the original values for the unprocessed versions are also displayed for reference.

Table 2. The matrix by vector product analyzed in original (o) and pruned (p) forms, based in singly-linked (s) or doubly-linked (d) lists

Data structure	# stmts	Time	# graphs	Nodes, links & CLSs per graph
Matrix-vector(o,s)	83	1.14 min	2477	7.56 (12) / 26.10 (40) / 29.34 (50)
Matrix-vector(p,s)	66	7.52 sec	772	5.69 (10) / 19.28 (36) / 19.91 (48)
Matrix-vector(o,d)	97	1.55 min	2931	7.60 (12) / 30.95 (48) / 30.37 (50)
Matrix-vector(p,d)	77	9.22 sec	823	5.45 (10) / 21.29 (42) / 19.68 (48)

The results prove that def-use driven shape analysis works best, as the analysis time has been reduced dramatically. Pruned tests produce the same output graphs than their original counterparts, thus capturing memory configuration without any loss in precision. This example motivates us to tightly integrate shape analysis within client analysis that focus on the statements of interest.

When trying to include def-use chains generation within the framework, we realized that their computation is easier in the SSA form of the program. This led us to implement SSA support as a Cetus extension. The cost of providing SSA support within the framework is not only justified by its use to drive shape analysis. It is also a required module for other pointer analysis and optimizations techniques. In our approach to SSA, we obtain the *dominator tree* in the first place. Then a slightly modified version of Cytron's algorithm [5] is used for constructing the SSA form. We modified the algorithm to remove unnecessary ϕ-functions that could hinder client analysis performance. We also made it more efficient by renaming each ϕ-function just once, instead of twice.

4 Dependence Test as a Client Analysis

The shape analysis algorithm is a basic element in our pointer analysis framework. However, we are aware that it is not sufficient as a stand-alone analysis technique. In order to take full advantage of its power it must be coupled with a higher-level client analysis that can determine regions to be analyzed for a given purpose. One of such client analysis is data dependence analysis for loops that traverse dynamic recursive data structures. Ultimately, this kind of analysis can determine what loops can be safely parallelized in an automated basis.

Let us focus now on a common situation. Usually, data structures are created at the initialization phase of programs and later, they are traversed to perform certain calculations. Often, most of the execution time of the program occurs at such traversals, where the structure change no more, but their values do. In such scenario, a client analysis could identify the statements that create the structure, call the shape analyzer over those statements to obtain shape information, and then use that information to look for data dependencies in the loops of interest.

In fact, the dependence detection can also be performed with the help of the shape analyzer, by marking or *touching* traversed nodes with accessing information. More precisely, nodes in the graph can be marked as having been read or written. This is achieved by using the *touch property* in the context of a loop-carried dependence test, similarly to [6]. As a prerequisite of such dependence analysis, induction pointers must be identified for loops that traverse recursive data structures.

Induction pointers, also called *navigator pointers*, are used in loops to traverse recursive structures, establishing their traversal pattern. That pattern, along with the shape of the structure, allows to detect dependencies between accesses. Of course, induction pointers already introduce an inherent dependence between different iterations of a loop, something known as the *pointer-chasing problem*. However, there are techniques to overcome it, provided that no other dependencies exist. In the next example, p is the *induction pointer*.

```
while (p!= NULL){
    p -> x = p -> y * 5;
    p = p -> next;
}
```

Being able to automatically detect induction pointers is a must for our compiler analysis framework, because they are needed to identify loops that traverse recursive data structures, and thus are candidate for parallelization in our approach. We have chosen Hwang and Saltz's method [7] for identifying induction pointers in a program, based on the calculation of def-use chains of statements that construct and traverse recursive data structures. As commented above, we have already added def-use chains generation support within our framework, so including this method comes as a straightforward addition.

5 Related Work

In the past few years pointer analysis has attracted a great deal of attention. A lot of studies have focused on stack-pointer analysis, like [8] and [9], while others, more related to our work, have focused on heap-pointer analysis, like [10] and

[11]. Both fields require different techniques of analysis. Unfortunately, heap-pointer techniques have failed to achieve aggressive optimizations. We think this is partly caused by techniques being isolated from other complementary pointer analysis techniques.

We are particularly fond of the work by Sagiv et al. [4], [12]. Their use of abstract interpretation/abstract semantics, along with materialization, have been adapted for the development of our framework. It is worth noting though, that their analysis is much more costly, meaning they can only analyze simple operations over singly-linked lists. Otherwise, analysis times and memory use become prohibitive. Also, their technique is only able to correctly analyze simple structures because they lack the support to handle structures like doubly-linked lists or heterogeneous trees. In our approach we have strived and achieved to obtain suitable graph abstractions for this kind of data structures. Besides, we think the analysis run at manageable times for such a complex technique. Finally, it should be noted that our technique is able to analyze structures based in pointer arrays, which is unheard of for a shape analysis technique, as far as we know.

We have been inspired for this work by existing research compiler frameworks: Polaris [13], which permitted the development of some noteworthy optimizations in array-based Fortran programs; ORC [14], which covers the whole compilation process and targets Itanium processors; SUIF [15], used by many researchers to implement their compiler techniques; or Soot [16], that features different modules for bytecode optimizations in Java programs. Polaris and SUIF ended their life cycle, ORC and Soot seem to concentrate on low level optimizations, and none of them focuses primarily on pointer analysis. Our plan is not to outdo these long established frameworks, but to swerve in a more specific direction where there is still plenty of room for optimizations related to parallelism and locality.

6 Conclusions and Future Work

As main contribution, we have introduced how a detailed shape analysis technique can be a valuable tool within a pointer analysis framework. Such a framework can combine different techniques towards better exploitation of parallelism. We have presented some early experiments that prove that shape analysis can be greatly improved when combined with information derived from other pointer analysis techniques, namely def-use chains.

We have also added support for the SSA form and def-use chains in Cetus. This support is useful in three ways: first, it helps to identify the statements that must be analyzed for correct shape analysis; second, it allows for automatic induction pointer recognition in the context of pointer-chasing loops, a key instrument for finding parallelism in recursive data structures; third, it allows for easy implementation of many pointer techniques that require SSA and/or def-use chains, enhancing the possibilities of the framework.

Only an early view of the pointer analysis framework has been presented. Still much work is needed to implement more pointer analysis techniques and make them work together towards finding unexploited parallelism in pointer-based programs. Also, we plan to conduct more experiments with benchmarks programs to fully test the capabilities of the techniques implemented.

References

1. Johnson, T.A., Lee, S.I., Fei, L., Basumallik, A., Upadhyaya, G., Eigenmann, R., Midkiff, S.P.: Experiences in using Cetus for source-to-source transformations. In: The 17th International Workshop on Languages and Compilers for Parallel Computing (LCPC '04), West Lafayette, Indiana, USA (2004)
2. Tineo, A., Corbera, F., Navarro, A., Asenjo, R., Zapata, E.: Shape analysis for dynamic data structures based on coexistent links sets. In: 12th Workshop on Compilers for Parallel Computers, CPC 2006, A Coruña, Spain (2006)
3. Plevyak, J., Chien, A., Karamcheti, V.: Analysis of dynamic structures for efficient parallel execution. In: Int'l Workshop on Languages and Compilers for Parallel Computing (LCPC'93). (1993)
4. Sagiv, M., Reps, T., Wilhelm, R.: Solving shape-analysis problems in languages with destructive updating. ACM Transactions on Programming Languages and Systems **20(1)** (1998) 1–50
5. Cytron, R., Ferrante, J., Rosen, B.K., Wegman, M.N., Zadeck, F.K.: Efficiently computing static single assignment form and the control dependence graph. In: ACM Transactions on Programming Languages and Systems (ACM'91). (1991) 13(4): 451–490
6. Tineo, A., Corbera, F., Navarro, A., Asenjo, R., Zapata, E.: A novel approach for detecting heap-based loop-carried dependences. In: The 2005 International Conference on Parallel Processing (ICPP'05), Oslo, Norway (2005)
7. Hwang, Y.S., Saltz, J.: Identifying DEF/USE information of statements that construct and traverse dynamic recursive data structures. In: Lecture Notes in Computer Science. Volume 1366 of Languages and Compilers for Parallel Computing (LCPC'97 Issue). (1998) 131–145
8. Hind, M., Pioli, A.: Which pointer analysis should I use? In: Int. Symp. on Software Testing and Analysis (ISSTA '00). (2000)
9. Wilson, R., Lam, M.: Efficient context-sensitive pointer analysis for C programs. In: ACM SIGPLAN'95 Conference on Programming Language Design and Implementation, La Jolla, CA (1995)
10. Ghiya, R., Hendren, L.J.: Is it a tree, a DAG, or a cyclic graph? A shape analysis for heap-directed pointers in C. In: 23rd ACM SIGPLAN-SIGACT Symposium on Principles of Programming Languages, St. Petersburg, Florida (1996)
11. Chase, D., Wegman, M., Zadek, F.: Analysis of pointers and structures. In SIGPLAN Conference on Programming Languages Design and Implementation (1990) 296–310
12. Sagiv, M., Reps, T., Wilhelm, R.: Parametric shape analysis via 3-valued logic. ACM Transactions on Programming Languages and Systems (2002)

13. Blume, W., Eigenmann, R., Faigin, K., Grout, J., Hoeflinger, J., Padua, D., Petersen, P., Pottenger, W., Rauchwerger, L., Tu, P., Weatherford, S.: Parallel programming with Polaris. IEEE Computer **29(12)** (1996) 78–82
14. Wu, C., Lian, R., Zhang, J., Ju, R., Chan, S., Liu, L., Feng, X., Zhang, Z.: An overview of the Open Research Compiler. In: The 17th International Workshop on Languages and Compilers for Parallel Computing (LCPC '04). (2005) 17–31
15. Wilson, R.P., French, R.S., Wilson, C.S., Amarasinghe, S.P., Anderson, J.M., Tjiang, S.W.K., Liao, S.W., Tseng, C.W., Hall, M.W., Lam, M.S., Hennessy, J.L.: Suif: An infrastructure for research on parallelizing and optimizing compilers. SIGPLAN Notices **29(12)** (1994) 31–37
16. Vallée-Rai, R., Hendren, L., Sundaresan, V., Lam, P., Gagnon, E., Co, P.: Soot - a Java optimization framework. In: Proceedings of CASCON 1999. (1999) 125–135

Topic 5: Parallel and Distributed Databases, Data Mining and Knowledge Discovery

Patrick Valduriez, Wolfgang Lehner, Domenico Talia, and Paul Watson

Topic Chairs

Managing and efficiently analysing the vast amounts of data produced by a huge variety of data sources is one of the big challenges in computer science. The development and implementation of algorithms and applications that can extract information diamonds from these ultra-large, and often distributed, databases is a key challenge for the design of future data management infrastructures. Today's data-intensive applications often suffer from performance problems and an inability to scale to high numbers of distributed data sources. Therefore, distributed and parallel databases have a key part to play in overcoming resource bottlenecks, achieving guaranteed quality of service and providing system scalability. The increased availability of distributed architectures, clusters, Grids and P2P systems, supported by high performance networks and intelligent middleware provides parallel and distributed databases and digital repositories with a great opportunity to cost-effectively support key everyday applications. Further, there is the prospect of data mining and knowledge discovery tools adding value to these vast new data resources by automatically extracting useful information from them.

We solicited submissions in either Experience and Application or System and Research in distributed and parallel data management. We received 15 paper submissions. We thank all authors for their submissions. All papers were reviewed by 4 reviewers. We selected the 4 following papers to be presented at EuroPar 2006, in one session: "Dynamic and Distributed Reconciliation in P2P-DHT Networks", "HyParSVM - A New Hybrid Parallel Software for Support Vector Machine Learning on SMP Clusters", "Supporting a Real-Time Distributed Intrusion Detection Application on GATES", and "A Semantic Web Service Based Approach to Supporting Augmented Provenance on the Grid".

Dynamic and Distributed Reconciliation in P2P-DHT Networks*

Vidal Martins[1,2] and Esther Pacitti[1]

[1] ATLAS Group, INRIA and LINA, University of Nantes, France
[2] PPGIA/PUCPR - Pontifical Catholic University of Paraná, Brazil
Firstname.Lastname@univ-nantes.fr

Abstract. Optimistic replication can provide high data availability for collaborative applications in large scale distributed systems (grid, P2P, and mobile systems). However, if data reconciliation is performed by a single node, data availability remains an important issue since the reconciler node can fail. Thus, reconciliation should also be distributed and reconciliation data should be replicated. We have previously proposed the *DSR-cluster* algorithm, a distributed version of the IceCube semantic reconciliation engine designed for cluster networks. However *DSR-cluster* is not suitable for P2P networks, which are usually built on top of the Internet. In this case, network costs must be considered. The main contribution of this paper is the *DSR-P2P* algorithm, a distributed reconciliation algorithm designed for P2P networks. We first propose a P2P-DHT cost model for computing communication costs in a DHT overlay network. Second, taking into account this model, we propose a cost model for computing the cost of each reconciliation step. Third, we propose an algorithm that dynamically selects the best nodes for each reconciliation step. Our algorithm yields high data availability with acceptable performance and limited overhead.

1 Introduction

Large-scale distributed collaborative applications are getting common as a result of rapid progress in distributed technologies (grid, P2P, and mobile computing). Consider a professional community whose members wish to elaborate, improve and maintain an on-line virtual document, e.g. notes on classical literature or common bibliography, supported by a P2P system. They should be able to read and write application data. In addition, user nodes may join and leave the network whenever they wish, thus hurting data availability.

Optimistic replication is largely used as a solution to provide data availability for these applications. It allows asynchronous updating of replicas such that applications can progress even though some nodes are disconnected or have failed. This enables asynchronous collaboration among users. However, concurrent updates may cause replica divergence and conflicts, which should be reconciled.

---

* Work partially funded by the ARA Massive Data Project.

In most existing solutions [11,13] reconciliation is typically performed by a single node (reconciler node) which may introduce bottlenecks. In addition, if the reconciler node fails, the entire replication system may become unavailable.

In [9], we proposed the *DSR-cluster* algorithm (Distributed Semantic Reconciliation for cluster), a distributed version of the semantic reconciliation engine of IceCube [6,11] for cluster networks. Tentative *actions*, stored at action logs, are reconciled using *constraints*. Other reconciliation objects, such as *clusters*, are also necessary to produce the global schedule. *DSR-cluster* avoids bottlenecks, speeds up large scale reconciliation, and provides high data availability in case of node failures during reconciliation for cluster networks. In addition, *DSR-cluster* employs a distributed approach for storing reconciliation objects (*actions*, *clusters*, *constraints*, etc.) using a distributed hash table (DHT) [12,14] in order to provide high data availability.

DSR-cluster proceeds in 5 distributed reconciliation steps. However, it does not take into account network costs during these steps. A fundamental assumption behind *DSR-cluster* is that the communication costs among cluster nodes are negligible. This assumption is not appropriate for P2P systems, which are usually built on top of the Internet. In this case, network costs may vary significantly from node to node and have a strong impact on the performance of reconciliation. Thus, network costs should be considered to perform reconciliation efficiently and to avoid network overload due to the communication with far distant nodes.

In this paper, we propose the *DSR-P2P* algorithm, a distributed reconciliation algorithm designed for P2P networks. The main contributions of this paper are: (1) a DHT cost model for computing communication costs of a P2P network using a DHT overlay network; (2) the DSR-P2P cost model for computing the cost of each reconciliation step based on DHT cost model; (3) the *DSR-P2P* algorithm for selecting the best reconciler nodes based on the DSR-P2P cost model (4); and experimental results that show that our cost-based approach yields high data availability with acceptable performance and limited overhead.

The rest of this paper is organized as follows. Section 2 describes the basis of the *DSR-P2P* semantic reconciliation solution for P2P networks. Section 3 introduces the DHT cost model. Section 4 describes the DSR-P2P cost model and the dynamic allocation algorithm for selecting the best reconciler nodes. Section 5 shows implementation and experimental results. Section 6 compares our work with the most relevant related works. Finally, Section 7 concludes this paper.

2 P2P Distributed Semantic Reconciliation

In this section, we describe the main terms and assumptions we consider for DSR-P2P followed by the main DSR-P2P algorithm itself.

We assume that DSR-P2P is used in the context of a virtual community which requires a high level of collaboration and relies on a reasonable number of nodes (typically hundreds or even thousands of interacting users) [15]. The P2P network we consider consists of a set of nodes which are organized as a distributed hash table (DHT) [12,14]. A DHT provides a hash table abstraction over multiple computer nodes. Data placement in the DHT is determined by a hash function which maps data identifiers into nodes.

In our solution, a *replica R* is a copy of a collection of objects (e.g. copy of a relational table, or an XML document). A *replica item* is an object belonging to a replica (e.g. a tuple in a relational table, or an element in an XML document). We assume *multi-master* replication, i.e. a replica R is stored in several nodes and all nodes may read or write R. Conflicting updates are expected, but with low frequency.

In order to update replicas, nodes produce *tentative* actions (henceforth actions) that are executed only if they conform to the application semantics. An *action* is defined by the application programmer and represents an application-specific operation (e.g. a write operation on a file or document, or a database transaction). The application semantics is described by means of constraints between actions. A *constraint* is the formal representation of an application invariant (e.g. an update cannot follow a delete).

On the one hand, users and applications can create constraints between actions to make their intents explicit (they are called *user-defined constraints*). On the other hand, the reconciler node identifies conflicting actions, and asks the application if these actions may be executed together in any order (*commutative* actions) or if they are mutually dependent. New constraints are created to represent semantic dependencies between conflicting actions (they are called *system-defined constraints*).

A *cluster* is a set of actions related by constraints, and a *schedule* is a list of ordered actions that do not violate constraints.

With DSR-P2P, data replication proceeds basically as follows. First, nodes execute local actions to update replicas while respecting user-defined constraints. Then, these actions (with the associated constraints) are stored in the DHT using the replica identifier as key. Finally, reconciler nodes retrieve actions and constraints from the DHT and produce a global schedule, by performing conflict resolution in 6 distributed steps based on the application semantics. This schedule is locally executed at every node, thereby assuring eventual consistency [11]. The replicated data is eventually consistent if, when all nodes stop the production of new actions, all nodes will eventually reach the same value in their local replicas.

In order to avoid communication overhead and due to dynamic connections and disconnections, we distinguish *replica nodes*, which are the nodes that hold replicas, from *reconciler nodes*, which is a subset of the replica nodes that participate in distributed reconciliation.

We now present DSR-P2P in more details. First, we introduce the reconciliation objects necessary to DSR-P2P. Then, we present the six steps of the DSR-P2P algorithm.

2.1 Reconciliation Objects

Data managed by DSR-P2P during reconciliation are held by *reconciliation objects* that are stored in the DHT giving the object identifier. To enable the storage and retrieval of reconciliation objects, each reconciliation object has a unique identifier. DSR-P2P uses six reconciliation objects:

- **Communication costs (noted CC):** it stores the communication costs to execute each DSR-P2P step, estimated by every replica node, and used to choose reconcilers before starting reconciliation. These costs are computed in terms of latency times.
- **Action log R (noted L_R):** it holds all actions that try to update the replica R.
- **Action groups of R (noted G_R):** actions that manage a common replica item are put together into the same action group in order to enable the parallel checking of semantic conflicts among actions (each action group can be checked independently of the others); every replica R may have a set of action groups, which are stored in the *action groups of R* reconciliation object.
- **Clusters set (noted CS):** all clusters produced during reconciliation are included in the *clusters set* reconciliation object; a cluster is not associated with a replica.
- **Action summary (noted AS):** it comprises constraints and action memberships (an action is a *member* of one or more clusters).
- **Schedule (noted S):** it is a list of ordered actions.

The node that holds a reconciliation object is called the *provider node* for that object (e.g. *cost provider* is the node that currently holds CC). Provider data are guaranteed to be available using known DHT replication solutions [7]. DSR-P2P's liveness relies on the DHT liveness.

2.2 DSR-P2P Algorithm

DSR-P2P executes reconciliation in 6 distributed steps as showed in Figure 1.

- **Step 1 node allocation:** a subset of connected replica nodes is selected to proceed as reconciler nodes.
- **Step 2 actions grouping:** for each replica R, reconcilers put actions that try to update common replica items of R into the same group, thereby producing G_R.
- **Step 3 clusters creation:** reconcilers split action groups into clusters of semantically dependent conflicting actions (actions that the application judge safe to execute together, in any order, are semantically independent, even if they update a common replica item); clusters produced in this step are stored in the clusters set, and the associated action memberships are included in the action summary.
- **Step 4 clusters extension:** user-defined constraints are not taken into account in clusters creation; thus, in this step, reconcilers extend clusters by adding to them new conflicting actions, according to user-defined constraints; the associated action memberships are also included in the action summary.

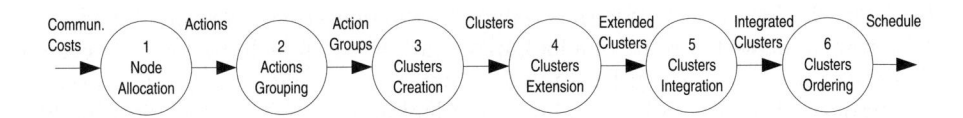

```
Commun.        Actions          Action       Clusters    Extended       Integrated    Schedule
 Costs  ┌───┐          ┌───┐   Groups ┌───┐          ┌───┐  Clusters ┌───┐  Clusters ┌───┐
        │ 1 │          │ 2 │          │ 3 │          │ 4 │          │ 5 │          │ 6 │
───────▶│Node│────────▶│Actions│─────▶│Clusters│────▶│Clusters│────▶│Clusters│────▶│Clusters│──────▶
        │Allocation│   │Grouping│     │Creation│     │Extension│    │Integration│  │Ordering│
        └───┘          └───┘          └───┘          └───┘          └───┘          └───┘
```

Fig. 1. DSR-P2P Steps

- **Step 5 clusters integration:** clusters extensions lead to clusters overlappings (an overlap occurs when different clusters have common actions, and this is identified by analyzing action memberships); in this step, reconcilers bring together overlapping clusters, thereby producing integrated clusters.
- **Step 6 clusters ordering:** in this step, reconcilers produce the global schedule by ordering actions of integrated clusters; all replica nodes execute this schedule.

At every step, the DSR-P2P algorithm takes advantage of data parallelism, i.e. several nodes perform simultaneously independent activities on a distinct subset of actions (e.g. ordering of different clusters). No centralized criterion is applied to partition actions. In fact, whenever a set of reconciler nodes request data to a provider, the provider node naively supplies reconcilers with about the same amount of data (the provider node knows the maximal number of reconcilers because it receives this information from the node that launches reconciliation).

3   DHT Cost Model

In this section, we propose a basic cost model for computing communication costs in DHTs. On top of it, we can build customized cost models (e.g. in the next section we elaborate a customized cost model for selecting DSR-P2P reconciler nodes).

In our model, we define communication costs (henceforth costs) in terms of latency times. We assume links with variable latencies and constant bandwidths. We intend to consider variable bandwidths in a future work.

Most DHT data access operations consist of a lookup, for finding the address of the node n that holds the requested information, followed by direct communication with n [5]. In the lookup step, several hops may be performed according to nodes' neighborhoods. Therefore, our DHT cost model relies on two metrics: lookup cost and direct cost. The *lookup cost*, noted $lc(n, id)$, is the latency time spent in a lookup operation launched by node n to find the data item identified by id. Similarly, *direct cost*, noted $dc(n_i, n_j)$, is the latency time spent by node n_i to directly access n_j.

Node n could easily compute the **lookup cost** $lc(n, id)$ by executing the lookup operation and measuring the associated time. However, this approach overloads the node that replies the lookup operation as it receives a lot of lookup messages. Furthermore, the network is overloaded. To avoid these problems, we propose that each node computes its lookup costs by taking advantage of cost information held by its neighbors. We illustrate this solution with an example.

In Figure 2a, let n_4 be a node that replies lookup operations searching for $id=x$; let arrows indicate the route of a lookup operation (e.g. if n_2 looks for x it makes this route: $n_2 \to n_3 \to n_4$); let a number over an arrow be the latency between the associated nodes. In this example, the lookup cost $lc(n_2, x)$ is 100 (i.e. 40 + 60), and $lc(n_1, x)$ is 150 (i.e. 50 + 40 + 60). Instead of executing the lookup operation to compute $lc(n_1, x)$, n_1 can ask n_2 for $lc(n_2, x)$ and add to this cost the latency between n_1 and n_2 (i.e. $lc(n_1, x) = lc(n_2, x) + 50$). The advantage of this incremental approach is locality and to avoid network overload.

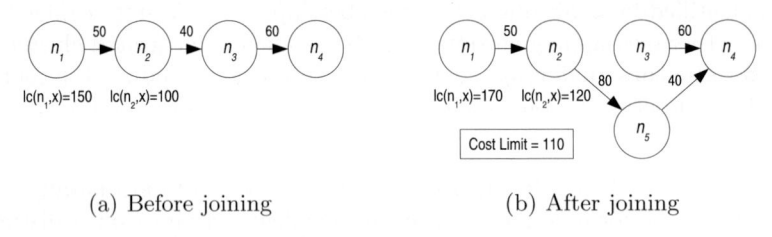

(a) Before joining (b) After joining

Fig. 2. Computing Lookup Costs

Joins and leaves change the neighborhoods of nodes and, accordingly, the routes of lookup messages. As a result, lookup costs must be refreshed. However, we should avoid the refreshment at distant nodes to avoid network overload. To cope with this problem, we introduce two definitions: *cost limit* and *relevant joins and leaves*. *Cost limit* is the maximal acceptable cost for looking up an identifier (it can be a parameter or an adaptively computed value). A join or leave is *relevant* for a node n if it changes the cost for looking up an identifier in which n is interested, such that the old or the new lookup cost does not overtake *cost limit*. Thus, we propose that nodes refresh their lookup costs only in the presence of relevant joins and leaves. We illustrate this approach with an example. In Figure 2b, let cost limit be 110; and consider that n_5 joins the DHT of Figure 2a taking the place of n_3 in the route towards $id=x$. The join of n_5 is relevant only to n_2 as n_2 updates $lc(n_2, x)$ from 100 (a value that does not overtake cost limit) to 120. In contrast, the join of n_5 is not relevant to n_3 and n_4 since the associated lookup costs remain unchanged. This join is not relevant to n_1 either, because both, the old lookup cost (i.e. 150) and the new one (i.e. 170), overtake cost limit. Thus, n_1, n_3 and n_4 do not participate in the refresh operation.

We now present how we compute **direct cost**. Node n could easily compute the direct cost between n and the provider node for id (henceforth *home(id)*) by measuring the latency between n and *home(id)*. However, this approach may overload *home(id)*. To avoid this problem, we propose that nodes locally estimate direct costs. Two equivalent approaches may be used for this estimation: (1) for DHTs that do not rely on nodes' physical location for choosing nodes' neighbors, the latency between a node n and any other node can be estimated based on the latencies between n and its neighbors in the DHT; (2) for location-aware DHTs, where n's neighbors are supposed to be closer to n than other nodes, the same estimation can be made based on the latencies between n and some other nodes randomly selected from a bootstrap list (list of nodes that are likely connected). The advantage of the estimated approach is locality, and its drawback is lack of accuracy. In the performance evaluation we compare the estimated and exact approaches.

The *home(id)* may change due to joins and leaves. Thus, direct costs must also be refreshed. In our solution, *dc(n, home(id))* is refreshed at node n whenever *home(id)* changes and the associated lookup cost (i.e. *lc(n, id)*) is smaller than cost limit. To compute the refreshed value, we use the same strategy employed for computing the initial value. The principle of this approach is to avoid the execution of refreshment operations at far distant nodes, and its advantage is to avoid network overload.

4 DSR-P2P Node Allocation Algorithm

In this section, we present a dynamic distributed algorithm for allocating nodes to DSR-P2P steps using the DHT cost model. We first present the DSR-P2P cost model for each reconciliation step. Next, we describe how the cost provider node selects reconcilers based on DSR-P2P cost model. Finally, we present our approach for managing the dynamic behavior of DSR-P2P costs.

4.1 DSR-P2P Cost Model

The DSR-P2P cost model takes into account each reconciliation step defining a new metric: node step cost. A *node step cost*, noted *cost(i, n)*, is the sum of lookup and direct costs estimated by node n for executing step i of DSR-P2P algorithm. By analyzing the DSR-P2P behavior in terms of lookup and direct access operations at every step, we produced a cost formula for each step of DSR-P2P, which are showed in Table 1. There is no formula associated with step 1 because it is not performed by reconciler nodes.

Table 1. DSR-P2P Cost Model

i	$Cost(i,n)$
2	$lc(n, L_R) + 2dc(n, n_{L_R}) + lc(n, G_R) + dc(n, n_{G_R})$
3	$lc(n, G_R) + 3dc(n, n_{G_R}) + lc(n, CS) + 2dc(n, n_{CS}) + lc(n, AS) + dc(n, n_{AS})$
4	$2lc(n, AS) + 3dc(n, n_{AS}) + lc(n, CS) + 3dc(n, n_{CS})$
5	$lc(n, AS) + 3dc(n, n_{AS}) + lc(n, CS) + dc(n, n_{CS})$
6	$lc(n, CS) + 3dc(n, n_{CS}) + lc(n, AS) + 2dc(n, n_{AS}) + lc(n, S) + dc(n, n_S)$

As an example, let us explain *cost(2, n)*. In the second step of DSR-P2P ($i=2$), node n takes actions from the action log R (L_R) and produces the action groups of R (G_R). Thus, the first term in the associated formula (*lc(n,L_R)*) represents the lookup cost for finding L_R provider. The second term ($2dc(n,n_{L_R})$) corresponds to the direct costs for taking actions from L_R provider (request and reply). The third term (*lc(n,G_R)*) represents the lookup cost for finding G_R

provider, and the last term ($dc(n,n_{G_R})$) corresponds to the direct cost for storing groups in G_R provider (only request). Similarly, all formulas can be explained.

4.2 Allocating Nodes

Node allocation is the first step of DSR-P2P algorithm. It aims to select for every succeeding step a set of reconciler nodes that can perform reconciliation with good performance. In this subsection, we describe how reconciler nodes are chosen and we illustrate that with an example.

The cost provider, i.e. the node that currently holds the communication costs reconciliation object, is the node responsible for allocating reconcilers. The allocation works as follows. Replica nodes locally estimate the costs for executing every DSR-P2P step, according to the DSR-P2P cost model, and provide this information to cost provider. The node that starts reconciliation computes the maximal number of reconcilers per step (noted $maxRec$), as described in [10], and asks cost provider for allocating at most $maxRec$ reconciler nodes per DSR-P2P step. As a result, the cost provider selects the best nodes for each step, and notifies these nodes about DSR-P2P steps they should execute.

In our solution, the cost management is parallel and independent of reconciliation. Moreover, it is network optimized since replica nodes do not send messages to cost provider, informing their estimated costs, if the node step costs overtake the *cost limit*. For these reasons, the cost provider does not become a bottleneck.

We now illustrate the allocation algorithm using an example. Table 2 shows the lookup and direct costs of our example, which were computed using a Chord DHT [14] with 4 connected nodes (i.e. n_0, n_1, n_4, and n_6). In a DHT, a node that is close to a reconciliation object (e.g. n_0 is close to AS ($id=1$)) may be far distant of others (e.g. n_0 is far distant of L_R ($id=5$)). As a result, a node that is suitable for a DSR-P2P step may not be worth in other steps. For this reason, every DSR-P2P step has its own set of reconcilers.

Table 2. Lookup and direct costs based on the DHT cost model. Each column has the identifier of a reconciliation object (id) and the node that holds this object ($home(id)$). Reconciliation object identifiers are: CS-0, AS-1, L_R-5, G_R-6, S-7. Each cell provides a specific lookup or direct cost, e.g. the cell in the 1st line and 3rd column indicates that n_0 spends 148.8ms to lookup L_R ($id=5$) stored in n_6 whereas the cell in the 2nd line and 3rd column indicates that a direct access between n_0 and n_6 costs 81.8ms.

Node	Cost Metric	Reconciliation Objects ($id \rightarrow home(id)$)				
		$0 \rightarrow n_0$	$1 \rightarrow n_1$	$5 \rightarrow n_6$	$6 \rightarrow n_6$	$7 \rightarrow n_0$
n_0	Lookup id	0	0	148.8	148.8	0
	Access $home(id)$	0	37.8	81.8	81.8	0
n_1	Lookup id	132.0	0	116.8	116.8	132.0
	Access $home(id)$	37.8	0	66.0	66.0	37.8
n_4	Lookup id	35.4	148.8	0	0	35.4
	Access $home(id)$	74.4	58.4	17.7	17.7	74.4
n_6	Lookup id	0	163.6	0	0	0
	Access $home(id)$	81.8	66.0	0	0	81.8

Table 3. Node step costs associated with the DHT considered in Table 2

Node	DSR-P2P steps (i)				
	2	3	4	5	6
n_0	543.0	**432.0**	113.4	**113.4**	**75.6**
n_1	431.6	522.4	**245.4**	**169.8**	**415.2**
n_4	**53.1**	444.5	**731.4**	433.8	634.0
n_6	**0**	**393.2**	770.6	443.4	622.8

Table 3 shows the estimated costs that the cost provider receives from the replica nodes. These costs are computed by applying on the DSR-P2P cost model (Table 1) the lookup and direct costs of the DHT cost model (Table 2). We show in bold the two less expensive costs associated with each DSR-P2P step. Thus, in our example, if the maximal number of reconcilers is 2, the cost provider selects as reconcilers for each DSR-P2P step the nodes of Table 3 whose costs are in bold (i.e. $Step_2 = \{n_4, n_6\}, Step_3 = \{n_0, n_6\}, Step_4 = \{n_0, n_1\}, Step_5 = \{n_0, n_1\}, Step_6 = \{n_0, n_1\}$), and notifies its decision to these nodes.

4.3 Managing the Dynamic Behavior of DSR-P2P Costs

The costs estimated by replica nodes for executing DSR-P2P steps change as a result of disconnections and reconnections. To cope with this dynamic behavior and assure reliable cost estimations, a replica node n_i works as follows:

- **Initialization:** whenever n_i joins the system, n_i estimates its costs for executing every DSR-P2P step. If these costs do not overtake the cost limit, n_i supplies the cost provider with this information.
- **Refreshment:** while n_i is connected, the join or leave of another node n_j may invalidate n_i's estimated costs due to routing changes. Thus, if the join or leave of n_j is relevant to n_i, n_i recomputes its DSR-P2P estimated costs and refreshes them at the cost provider.
- **Termination:** when n_i leaves the system, if its DSR-P2P estimated costs are smaller than cost limit (i.e. the cost provider holds n_i's estimated costs), n_i notifies its departure to the cost provider.

5 Validation and Performance Evaluation

To validate and study the performance behavior of DSR-P2P, we implemented it and simulated the overlay P2P network based on Chord (we used SimJava [4] for simulations). In this section, we present our performance model and the experimental results.

The performance model takes into account the strategy for selecting reconciler nodes (noted *Allocation*), the action log size (i.e. the number of actions to be reconciled, noted *Nb-Actions*) based on IceCube setup, and the network topology based on BRITE [2]. We define three strategies for selecting reconcilers: random selection (RDM); cost-based selection using precise costs for direct

Table 4. Performance Parameters

Parameter	Definition	Values
Allocation	Strategy for selecting reconciler nodes	CB/P; CB/E; RDM
Nb-Actions	Number of actions to be reconciled	106 - 10000
Nb-Nodes	Number of connected nodes	1024; 20000
Bandwidth	Network bandwidth	1Mbps; 10Mbps
Avg-Latency	Average latency among nodes	51ms - 263ms
Sd-Latency	Standard deviation of network latency	15ms - 96ms

communication (CB/P); and cost-based selection using estimated costs for direct communication (CB/E). A network topology is defined by its bandwidth (noted *Bandwidth*), the number of connected nodes (noted *Nb-Nodes*), the average latency among these nodes (noted *Avg-Latency*), and the associated standard deviation (noted *Sd-Latency*). Latency values follow a uniform distribution. We produced 3 network instances for every network topology definition. We also produced 3 action logs for each action log size. By combining action logs with network instances, we generate several distinct reconciliation scenarios that avoid over fitted results. Table 4 describes the parameters of the performance model.

The first experiment (Figure 3a) studies the reconciliation performance with locally estimated direct costs (recall that this approach reduces network load and avoids the overload of provider nodes, but it is not precise). For this experiment, we defined 4 network topologies and produced 12 network instances that are different only wrt. latency parameters (all topologies have $Bandwidth = 1$Mbps and $Nb\text{-}Nodes = 1024$). We used 3 action logs with $Nb\text{-}Actions = 1005$. Figure 3a shows the reconciliation performance using precise costs (CB/P), estimated costs (CB/E), and random allocation (RDM). In 3 topologies, the cost-based approaches (i.e. CB/P and CB/E) are equivalent and more efficient than the random approach. In the best case, which corresponds to a real P2P network, the CB/P reduces the reconciliation time of RDM in 37% whereas CB/E provides a performance improvement of 30%. Due to the small difference between CB/P and CB/E (i.e. 7%), we consider the estimated approach worth to avoid overload problems. Notice that the experimental conditions (i.e. constant bandwidth and uniform distribution of latencies) are strongly promising for random selection. We can improve the performance of cost-based approaches by changing these conditions (i.e. by providing variable bandwidths and distributing latencies in a way that some nodes are very close to each other making up clusters of nodes).

Due to the lack of space, we describe three additional experiments in a single graph, which corresponds to Figure 3b. The goal of these experiments is to show that the reconciliation time is improved because cost-based selection is used, and for faster network we have the best improvements compared with RDM. For instance, for a network of 10 Mbps and 1024 connected nodes using cost-based selection (CB/P-10-1024) we improved the random approach (RDM-1-1024) by a factor of 4. Notice that in this case both network bandwidths are different. For equal network bandwidths, the cost-based approach

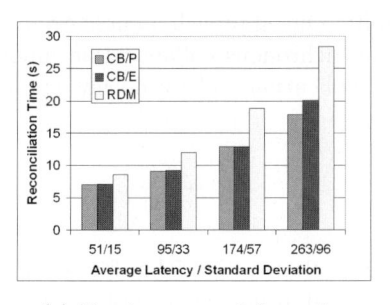

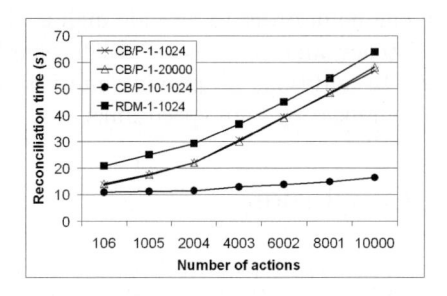

(a) Varying network latencies

(b) Varying actions, nodes, band.

Fig. 3. DSR-P2P Reconciliation Time

(CB/P-1-1024) still outperforms the random approach. Finally, increasing the number of connected nodes up to 20000 (CB/P-1-20000) does not degrade the DSR-P2P performance because it relies on a DHT and due to our allocation algorithm.

Liveness is an important issue in dynamic systems. DSR-P2P provides a greater degree of availability, scalability and fault-tolerance than the centralized solution. In contrast, since DSR-P2P depends on network communication, its reconciliation time (e.g. 57s for 10000 actions in a 1Mbps network with average latency of 229ms) is worse than the centralized counterpart (e.g. about 3s for 10000 actions). However, 57s remains an acceptable time for reconciling 10000 actions in a P2P network. The centralized solution, although more efficient than DSR-P2P, is unsuitable for P2P networks due to its low availability in dynamic environments.

6 Related Work

In the context of P2P networks, there has been little work on managing data replication in the presence of updates. Most of data sharing P2P networks consider the data they provide to be very static or even read-only. Freenet [3] partially addresses updates which are propagated from the updating peer downward to close peers that are connected. However, peers that are disconnected do not get updated. P-Grid [1] is a structured P2P network that exploits epidemic algorithms to address updates. It assumes that conflicts are rare and their resolution is not necessary in general. In addition, P-Grid assumes that probabilistic guarantees instead of strict consistency are sufficient. Moreover, it only considers updates at the file level in a single master-mode. In OceanStore [8] every update creates a new version of the data object. Consistency is achieved by a two-tiered architecture: a client sends an update to the object's *inner ring* (primary copies) and some secondary replicas in parallel. Once the update is committed, the *inner ring* multicasts the result of the update down the dissemination tree. OceanStore assumes an infrastructure comprised of servers that are connected by high-speed links. Different from the previous works, we propose to distribute the reconciliation engine in order to provide high availability. Our approach assures eventual consistency among replicas, which enables asynchronous collaboration among users. In addition, we provide multi-master replication and we do not assume servers linked by high-speed links.

7 Conclusion

In this paper, we proposed the DSR-P2P, a distributed algorithm for semantic reconciliation in P2P networks. Our main contributions are a cost model for computing communication costs in DHTs and an algorithm that takes into account these costs and the DSR-P2P steps to select the best reconciler nodes. For computing communication costs, we use local information and we deal with the dynamic behavior of nodes. In addition, we limit the scope of event propagation (e.g. joins or leaves) in order to avoid network overload.

We validated DSR-P2P through implementation and simulation. The experimental results showed that our cost-based reconciliation outperforms the random approach by a factor of 30% over scenarios that are favorable for the random approach (constant bandwidth and uniform distribution of latencies). In addition, the number of connected nodes is not important to determine the reconciliation performance due to the DHT scalability and the fact that reconcilers are as close as possible to the reconciliation objects. Compared with the centralized solution, which is more efficient but lowly available, our algorithm yields high data availability with acceptable performance and limited overhead. As future work, we plan to include variable bandwidths in our cost model.

References

1. K. Aberer, P. Cudré-Mauroux, A. Datta, Z. Despotovic, M. Hauswirth, M. Punceva, and R. Schmidt. P-grid: a self-organizing structured p2p system. *SIGMOD Rec.*, 32(3), 2003.
2. BRITE. http://www.cs.bu.edu/brite/.
3. I. Clarke, T.W. Hong, S.G. Miller, O. Sandberg, and B. Wiley. Protecting free expression online with freenet. *IEEE Internet Computing*, 6(1), 2002.
4. F. Howell and R. McNab. Simjava: a discrete event simulation library for java. In *Web-based Modeling and Simulation*, 1998.
5. R. Huebsch, J.M. Hellerstein, N. Lanham, I. Stoica, B.T. Loo, and S. Shenker. Querying the internet with pier. In *Proc. of VLDB Conference*, 2003.
6. A. Kermarrec, A. Rowstron, M. Shapiro, and P. Druschel. The icecube approach to the reconciliation of divergent replicas. In *Proc. of ACM PODC*, 2001.
7. P. Knezevic, A. Wombacher, and T. Risse. Enabling high data availability in a dht. In *Proc. of DEXA Workshops*, 2005.
8. J. Kubiatowicz, D. Bindel, Y. Chen, P. Eaton, D. Geels, R. Gummadi, S. Rhea, H. Weatherspoon, W. Weimer, C. Wells, and B. Zhao. Oceanstore: An architecture for global-scale persistent storage. In *Proc. of ACM ASPLOS*, 2000.
9. V. Martins, E. Pacitti, and P. Valduriez. Distributed semantic reconciliation of replicated data. In *Proc. of CDUR*. IEEE France and ACM SIGOPS France, 2005.

10. V. Martins, E. Pacitti, and P. Valduriez. A dynamic distributed algorithm for semantic reconciliation. In *Distributed Data and Structures 7 (WDAS)*, 2006.
11. N. Preguiça, M. Shapiro, and C. Matheson. Semantics-based reconciliation for collaborative and mobile environments. In *Proc. of IFCIS CoopIS*, 2003.
12. S. Ratnasamy, P. Francis, M. Handley, R. Karp, and S. Schenker. A scalable content-addressable network. In *Proc. of ACM SIGCOMM*, 2001.
13. Y. Saito and M. Shapiro. Optimistic replication. *ACM Comput. Surv.*, 37(1), 2005.
14. I. Stoica, R. Morris, D. Karger, M.F. Kaashoek, and H. Balakrishnan. Chord: A scalable peer-to-peer lookup service for internet applications. In *Proc. of ACM SIGCOMM*, 2001.
15. S. Whittaker, E. Isaacs, and V. O'Day. Widening the net: workshop report on the theory and practice of physical and network communities. *ACM SIGCHI Bulletin*, 29(3), 1997.

HyParSVM – A New Hybrid Parallel Software for Support Vector Machine Learning on SMP Clusters

Tatjana Eitrich[1], Wolfgang Frings[1], and Bruno Lang[2]

[1] Central Institute for Applied Mathematics, Research Centre Jülich, Germany
t.eitrich@fz-juelich.de
[2] Applied Computer Science and Scientific Computing Group, Department of Mathematics, University of Wuppertal, Germany

Abstract. In this paper we describe a new hybrid distributed/shared memory parallel software for support vector machine learning on large data sets. The support vector machine (SVM) method is a well-known and reliable machine learning technique for classification and regression tasks. Based on a recently developed shared memory decomposition algorithm for support vector machine classifier design we increased the level of parallelism by implementing a cross validation routine based on message passing. With this extention we obtained a flexible parallel SVM software that can be used on high-end machines with SMP architectures to process the large data sets that arise more and more in bioinformatics and other fields of research.

1 Introduction

Support vector machines are well-known data mining methods for classification and regression problems [1]. Their popularity is mainly due to their applicability in various fields of data mining, such as text mining [2], biomedical research [3], and many more. Their accuracy is excellent and in many cases they outperform other machine learning methods such as neural networks. SVMs have their roots in the field of statistical learning which provides the reliable generalization theory [4]. Several properties that make this learning method successful are well-known, e.g. the kernel trick [5] for nonlinear classification and the sparse structure of the final classification function. In addition, SVMs have an intuitive geometrical interpretation, and a global minimum can be located during the SVM training phase. In comparison to genetic algorithms or neural networks, less experience is required for using them, which helps researchers to get started with SVM software quite fast. The main drawback of current SVM models is their high computational complexity for large data sets [6]. This can in fact restrict the applicability of SVMs since the amount of data for classification modeling increases dramatically. Therefore the development of highly scalable parallel SVM algorithms is a new important topic of current SVM research. Some algorithms for parallel SVM learning already do exist, but most of them are limited to heuristics for distributed training on reduced data sets. These are not useful as stand-alone systems for high quality learning on large data.

In this paper we propose an efficient parallel support vector machine software well suited for multi-processor shared memory (SMP) clusters that become more and more available. Our algorithm can be used in serial and parallel mode. The parallel implementation provides pure MPI and OpenMP modes as well as a hybrid mode which combines fine and coarse grained parallelization aspects to a well scalable SVM learning method.

The remainder of this paper is organized as follows. In Sect. 2 we briefly review the basic concepts of support vector machine learning and describe the SVM parameter optimization problem, which leads to the computational challenges we address in this paper. We limit the discussion to the issues that are essential for understanding the following sections. Since the field of parallel SVM methods is quite new and implementations are rare, we give a detailed review of existing approaches for parallel data mining and support vector machine learning in Sect. 3. One aim of this paper is therefore to present the current state-of-the-art in parallel support vector machine design. In Sect. 4 we explain the structure of our new parallel SVM software *HyParSVM*. In Sect. 5 we present first experimental results on the IBM p690 cluster JUMP at Research Centre Jülich. Finally, Sect. 6 contains a summary and shows directions for future work.

2 Theoretical Background

In this paper we consider the well-known supervised binary classification problem [7]. Given a training set (reference data) of the form

$$\{(\boldsymbol{x}^i, y_i) \in \mathbb{R}^n \times \{-1, 1\}, \quad i = 1, \ldots, l\},$$

where $l \in \mathbb{N}$ is the number of given instances and $n \in \mathbb{N}$ the number of attributes in the data set, the task of support vector machine learning is to find a hypothesis function $h : \mathbb{R}^n \to \mathbb{R}$ that can be used to classify unseen data. The hypothesis function, the sign of which is used to classify a point $\boldsymbol{x}$, is of the form

$$h(\boldsymbol{x}) = \sum_{i:\alpha_i > 0} y_i \alpha_i K(\boldsymbol{x}^i, \boldsymbol{x}) + b^*.$$

It is mainly controlled by the so-called Lagrange multipliers α_i ($i = 1, \ldots, l$). They can be determined via the solution of the quadratic programming (qp) problem

$$\left.\begin{aligned}&\min_{\boldsymbol{\alpha} \in \mathbb{R}^l} \quad \frac{1}{2} \sum_{i,j=1}^{l} y_i y_j \alpha_i \alpha_j K(\boldsymbol{x}^i, \boldsymbol{x}^j) - \sum_{i=1}^{l} \alpha_i \\ &\text{s.t.} \quad \sum_{i=1}^{l} y_i \alpha_i = 0, \quad 0 \leq \alpha_i \leq C \quad (1 \leq i \leq l).\end{aligned}\right\} \quad (1)$$

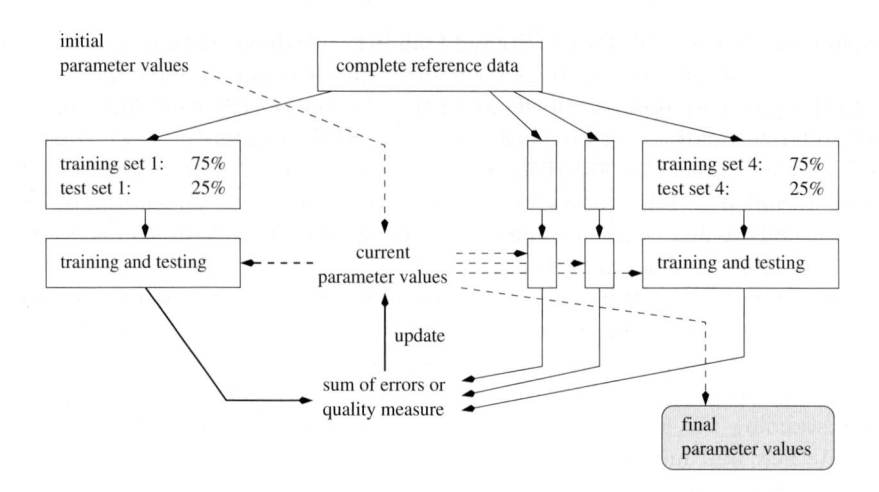

Fig. 1. Structure of parameter tuning with a 4-fold cross validation method

The function $K : \mathbb{R}^n \times \mathbb{R}^n \to \mathbb{R}$ is known as the kernel [1] and measures similarity between input vectors. $C \in \mathbb{R}_+$ is an SVM internal error penalization parameter which controls the trade-off between a large margin and the corresponding training errors. We refer to [1] for a detailed description of the SVM learning problem. Usually, for SVM learning either the L_1-norm or the L_2-norm approach is used. In this paper we work with the L_1-norm approach (1) and avoid the discussion about SVM internal algorithmics. Our software is able to handle both methods. All details to our flexible serial implementation are given in [8] where we presented a comparison between these methods and observed a superiority of the L_1-norm model for unbalanced classification problems.

One of the main challenges when using SVM-based methods is parameter selection. Several data dependent parameter values need to be adjusted [9]. Different methods for tuning the parameters have been proposed [10]. One of them is a search procedure that iteratively creates new parameter values using quality results from k-fold cross validation. In Fig. 1 we explain this method for $k = 4$. A k-fold cross validation includes k SVM training and test stages as well as a final combination of the results to obtain a quality measure value [9]. We are working with our implementation of the decomposition method which includes the fast projection method proposed in [11]. However, a single SVM training is expensive for large data. Thus, a complete validation takes a very long time. Our work is aimed at speeding up the SVM parameter optimization time. Please note that parameter tuning usually means to perform a large number of validation stages. Efficient and fast methods are of great interest since they allow for an extensive scan of the parameter space and usage of additional parameters, e.g. for sensitive classification of highly unbalanced data [9].

3 History of Parallel Support Vector Machines

Most sequential data mining algorithms have large runtimes, but the volume of data available for analysis is growing rapidly, i.e. the number of attributes as well as the number of instances both increase. In addition to improvements of the serial algorithms the development of parallel techniques may help to avoid computational bottlenecks. This section gives an overview of activities concerning large scale data mining, particularly the problem of classification using machine learning techniques like SVMs.

Parallel Data Mining

The first parallel data mining algorithms have emerged a decade ago. In [12] the general differences between parallel data mining and other numerical parallel algorithms are explained. The design of scalable data mining algorithms requires meeting several challenges, e.g., the enormous memory requirements have to be supported by the computing system. Various algorithms, especially for supervised learning methods, have been parallelized.

- A parallel algorithm for data mining of association rules was presented in [13]. It has been designed for work on shared memory multiprocessors.
- The *ScalParC* software [14], designed in 1998, was one of the first methods for parallel decision tree classification. Parallel decision tree applications are still of interest, mainly in the important field of Grid computing [15].
- Clustering is useful in various fields, i.e., pattern recognition and learning theory. The runtime complexity of a serial k-means clustering algorithm is high for problems of large size. Therefore parallel clustering methods have been developed. We refer to [16] for a master-slave approach.
- K-nearest neighbor methods have received a great deal of attention since they are applied frequently in bioinformatics, but performance is a serious problem for many implementations. In [17] a parallel algorithm was introduced to overcome the problem of runtime.
- Artificial neural networks (ANNs) are well-known data mining methods with high learning cost when the models are large. An approach for speeding up their implementation by using parallel environments is given in [18].
- Bayesian networks for unsupervised classification tasks include time consuming steps which can be parallelized. A description is given in [19].
- Boosting is a method for improving the accuracy of any given learning algorithm [20] and is often used within the context of supervised learning. A framework for distributed boosting is presented in [21]. The method requires less memory and computational time than serial boosting packages.

Parallel Support Vector Machine Approaches

Efficient and parallel support vector machine learning is a young and emerging field of research, but the number of truly parallel implementations is small.

Most approaches just try to increase the efficiency of the serial algorithms and to overcome the problem of large scale applications by dividing the data into subsets.

- Different approaches for splitting a large data set into small subsets have been implemented [22]. Usually results of the individual training stages are merged to finally obtain a single SVM model. The individual optimization steps can be run in parallel.
- A fast SVM algorithm, which uses caching, digest and shrinking policies is given in [23].
- The clustering-based SVM [24] is a learning method that scans the data set before training the SVM. It selects the data which are supposed to maximize the benefit of learning and is useful for very large problems when a limited amount of computing resources is available. So far it is only applicable for linear problems.

In addition, various projects exist where a simple parallelization scheme is used to speed up the learning process.

- In [25] a parallel optimization step is proposed. It approximates the kernel matrix by block diagonal matrices and splits the original problem into subproblems which can be solved independently from each other with standard algorithms. This step is used to remove non-support vectors before SVM training.
- Parallel training of several binary SVMs for solving multiclass problems is described in [26].
- Parallel cross validation methods do exist for the *WEKA* machine learning package [27].
- Parallel parameter optimization techniques such as grid search or pattern search have been studied for SVM parameter fitting [28].

These approaches can be interpreted as coarse grained parallelization techniques for SVM methods at a high level which is independent from the inner solver for the problem (1). However, the computational bottleneck of a single SVM training on a large data set can be avoided only by implementing a fine grained parallel support vector machine training. The following methods have been proposed.

- Parallel computation of the kernel matrix for high dimensional data spaces is implemented in [29]. The speedup is limited because of high communication costs. Therefore an approximation method that reduces the kernel matrix was implemented, too. The method is applicable only for commonly used kernels which are inner product-based and requires changes in the algorithm for each kernel.
- A distributed SVM algorithm for row-wise and column-wise data distribution is described in [26], which so far can be used for linear SVMs only.
- A promising parallel MPI-based decomposition solver for training support vector machines has been implemented recently [30].
- A parallel support vector machine for multi-processor shared memory (SMP) clusters has been introduced in [31].

4 A New Hybrid Parallel SVM Software

In [31] we have discussed a mixed library/loop-based shared memory parallelization for a single SVM training. We have continued to optimize the parallel code, i.e., in addition to the mixed parallelization we implemented two versions of the parallel SVM training that perform library- or loop-based parallelization exclusively (except for the distributed kernel computations). The first one is based on calls to the shared memory parallel version of the *ESSL* (*Engineering Scientific Subroutine Library*) [32], whereas the second one implements OpenMP loop level parallelism. This scheme was realized for the outer decomposition loop, as well as the projection method and the inner solver. The settings may be chosen independently for each routine by using C preprocessor macro names. The code is written in *Fortran90*, and the *IBM XL Fortran compiler* is used. We observed satisfactory speedups for moderate numbers of processors on the IBM supercomputer JUMP (Juelich Multi Processor) at Research Centre Jülich [33]. For a larger number of processors the speedup values tended to stagnate or even decreased. The training routine comprises some sequential parts that cannot be parallelized, e.g., the iterative working set selection scheme. These parts consume approximately 5% of the training time for data sets with more than 10000 points. In addition, the working set size, an important parameter for the decomposition loop that determines the size of the qp problem (1), which is solved within the parallel OpenMP mode, is limited by the available memory. Therefore the *ESSLsmp* routines have limited scalability for increasing numbers of threads. All in all, for the data we have analyzed, the attainable speedup was limited to values between 5 and 10. For a large number of threads (> 12) the speedups started to decrease. In this paper we present a parallel software which speeds up the SVM learning process to a greater extent by exploiting an additional level of parallelism.

So far, the parallel shared memory SVM training had been embedded into the serial validation loop as it is shown in Fig. 2. At this higher level we added a new parallelization scheme. A pure extension of the shared memory approach was not reasonable since usage of more than 32 processors on the JUMP supercomputer would mean to assign the validation tasks to different nodes which do not share the same memory and can communicate with MPI-based functions only. Therefore we implemented a hybrid parallel support vector machine with an MPI-based cross validation routine. Using a coarse grained parallelization scheme the k validation steps for a k-fold cross validation are distributed to p processes, each of which performs a training-and-testing step for k/p data sets. Each training may in turn be executed by multiple threads, as shown in Fig. 2. Since I/O is necessary only at the beginning of the program, we could use a simple data distribution scheme. A single ("master") process reads the complete training data, preprocesses it and then calls MPI collective broadcast operations to distribute the validation matrix to the other processes. Inside the validation loop each process uses the matrix k/p times to extract private training and test data. Each process accumulates results of the local validation tests during execution of the program. At the end of the validation loop, MPI

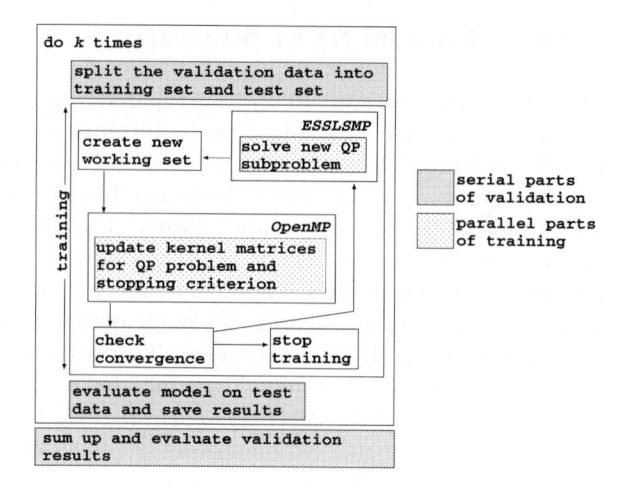

Fig. 2. Shared memory parallel SVM training as part of the validation loop to be parallelized

collective reduction operations compute the overall results, and the master process calculates the overall quality measure. Each validation step consists of a single SVM training on a data matrix with n features and $l \cdot (1 - 1/k)$ instances. It is known that training time is quadratic in the number of instances and linear in the number of features and does not heavily depend on other parameters except the outer SVM parameters which do not change during a single validation process. Due to this relatively balanced load and the fact that variances in time are data dependent and unpredictable, the assignment of validation jobs to processes was implemented in a straight forward way. As it can be seen in Fig. 2 each step of the cross validation method previously comprised some non-parallel parts (dark grey), which we have parallelized now with a distributed memory approach to increase the efficiency of the overall scheme. The additional speedup obtained by the hybrid parallelization is particularly useful in the context of parameter search, since a large number of validation steps may be necessary here. Sophisticated parameter search is usually performed iteratively and new paths in the parameter space are defined based on former results. For simple tuning approaches like grid search, where the validation runs are independent and can be processed in parallel, the MPI-parallelelism of our hybrid software may be turned off.

5 Experimental Results

We performed our tests on the Juelich Multi Processor. JUMP is a distributed shared memory parallel computer consisting of 41 frames (nodes). Each node contains 32 IBM Power4+ processors running at 1.7 GHz, and 128 GB shared main memory. The 1312 processors have an aggregate peak performance of 8.9 TFlop/s. For our tests we have used a QSAR data set from pharmaceutical

Table 1. Comparison of (running time in seconds : speedup : efficiency) for 8-fold cross validation using the L_1-norm approach with a Gaussian kernel. A data set with 40000 instances and 50 features was tested.

		# processes			
		1	2	4	8
# threads	1	6105 : 1.0 : 1.00	3074 : 2.0 : 1.00	1566 : 3.9 : 0.98	834 : 7.3 : 0.91
	2	3157 : 1.9 : 0.95	1599 : 3.8 : 0.95	815 : 7.5 : 0.94	453 : 13.5 : 0.84
	3	2168 : 2.8 : 0.93	1109 : 5.5 : 0.92	577 : 10.6 : 0.88	348 : 17.5 : 0.73
	4	1641 : 3.7 : 0.93	847 : 7.2 : 0.90	444 : 13.7 : 0.86	284 : 21.5 : 0.67
	5	1362 : 4.5 : 0.90	703 : 8.7 : 0.87	366 : 16.7 : 0.84	187 : 32.7 : 0.82
	6	1172 : 5.2 : 0.87	609 : 10.0 : 0.83	326 : 18.7 : 0.78	165 : 37.0 : 0.77
	7	1054 : 5.8 : 0.83	549 : 11.1 : 0.79	299 : 20.4 : 0.73	155 : 39.4 : 0.70
	8	978 : 6.2 : 0.78	518 : 11.9 : 0.74	290 : 21.9 : 0.68	158 : 42.9 : 0.67

industry with 40000 instances and 50 features. We show results for an SVM with the Gaussian kernel. However, with our flexible implementation any other kernel function is applicable, since the kernel function itself is not parallelized. The user may integrate his own kernel function into the software. We believe that this concept of a non-parallel kernel function is crucial for a flexible usage of the parallel SVM software as it allows for the classification of data sets with widely differing characteristics. Due to the fact that we focus on a parallelization scheme, no accuracy results for the data in this paper are given. In our tests we observedm that parallel computataion of (1) does not change the global solution. Concerning verification and improvement of SVM quality we refer to our work [8,9,28].

In the following we present the results for an 8-fold cross validation task using the hybrid software with the *ESSLsmp*-based inner parallelization. The working set size of the decomposition method was set to the largest possible value of $40000 \cdot 7/8 = 35000$, which is the size of the qp problems to be solved in the validation loop. For the allocation of matrices and vectors during computation each process needed 12 GB of memory, which was then used by the threads assigned to each process. In cases where only a smaller amount of memory is available the working set size may be reduced. This will cause the decomposition method to optimize the vector α iteratively. As we mentioned in the last chapter, each validation step is expected to consume approximately the same amount of time. For our data set the timings were between 751 and 778 seconds with a mean value of 763. These results were obtained with one thread and a single process on JUMP. Thus, the time differences between the steps are negligible and the assignment of steps to the available processes may indeed be implemented without a special mapping method. In Table 5 we show speedup and efficiency values for various combinations of processes and threads. The additional level of parallelism successfully increased the achievable speedup. Most interesting it the last column. The efficiency decreases from 0.91 down to 0.67 for 32 processors. If additional 8 processors are added, the efficiency increases to 0.82 and decreases again for further more processors. For tests with more than 32 processors two nodes of JUMP are used; all other tests were run on a single node. With using two nodes, memory bandwidth limitations become visible. However, our speedup values are promising – for 64 processors the SVM validation time decreased with a factor of 43 by using 8 processes with 8 threads each.

6 Summary and Future Work

In this paper we presented the new *HyParSVM* software for parallel SVM learning. This software, which is under development at the Research Centre Jülich, helps speeding up the data mining pipeline in various fields of classification applications. The hybrid implementation is very flexible and shows promising results on the JUMP supercomputer. In addition to the hybrid SVM software the user may increase the level of parallelism even more by using a parallel parameter tuning method which calls the *HyParSVM* cross validation routine, e.g. on different nodes of a SMP cluster.

Our future work will be aimed at further improvement of the *HyParSVM* software. The shared memory parallelization of the training routine will be enhanced and tested for larger data sets. We will analyze which parallel scheme – *ESSLsmp* or OpenMP-based constructs – gives the best speedups. The influence of the working set size onto the scalability will be investigated.

References

1. Cristianini, N., Shawe-Taylor, J.: An introduction to support vector machines and other kernel-based learning methods. Cambridge University Press (2000)
2. Joachims, T.: Text categorization with support vector machines: learning with many relevant features. In: Proceedings of ECML-98, 10th European Conference on Machine Learning, Chemnitz, Germany, Springer Verlag (1998) 137–142
3. Yu, H., Yang, J., Wang, W., Han, J.: Discovering compact and highly discriminative features or feature combinations of drug activities using support vector machines. In: 2nd IEEE Computer Society Bioinformatics Conference (CSB 2003), Stanford, CA, USA, IEEE Computer Society (2003) 220–228
4. Vapnik, V.N.: Statistical learning theory. John Wiley & Sons, New York (1998)
5. Schölkopf, B.: The kernel trick for distances. In: NIPS. (2000) 301–307
6. Chen, N., Lu, W., Yang, J., Li, G.: Support vector machine in chemistry. World Scientific Pub Co Inc (2004)
7. Thrun, S., Mitchell, T.M.: Learning one more thing. In: IJCAI. (1995) 1217–1225
8. Eitrich, T., Lang, B.: On the advantages of weighted L_1-norm support vector learning for unbalanced binary classification problems (2006) to appear.
9. Eitrich, T., Lang, B.: Efficient optimization of support vector machine learning parameters for unbalanced datasets. JCAM (2005) in press.
10. Chapelle, O., Vapnik, V.N., Bousquet, O., Mukherjee, S.: Choosing multiple parameters for support vector machines. Machine Learning **46**(1) (2002) 131–159
11. Serafini, T., Zanghirati, G., Zanni, L.: Gradient projection methods for quadratic programs and applications in training support vector machines. Optimization Methods and Software **20**(2-3) (2005) 353–378
12. Skillicorn, D.: Strategies for parallelizing data mining. In: Proceedings of the Workshop on High-Performance Data Mining at IPPS/SPDP. (1998)

13. Parthasarathy, S., Zaki, M., Ogihara, M., Li, W.: Parallel data mining for association rules on shared-memory systems. Knowledge and Information Systems **3**(1) (2001) 1–29
14. Joshi, M.V., Karypis, G., Kumar, V.: ScalparC: a new scalable and efficient parallel classification algorithm for mining large datasets. In: IPPS: 11th International Parallel Processing Symposium, IEEE Computer Society Press (1998)
15. Hofer, J.: Distributed induction of decision tree classifier within the grid data mining framework: Gridminer-core. AURORA Technical Report 2004-04, Institute for Software Science, University of Vienna, Vienna (2004)
16. Kantabutra, S., Couch, A.L.: Parallel k-means clustering algorithm on NOWs. NECTEC Technical Journal **1** (2000) 243–248
17. Callahan, P.B.: Optimal parallel all-nearest-neighbors using the well-separated pair decomposition. In: 34th Symp. Found. of Comp. Science, IEEE. (1993) 332–340
18. Misra, M.: Parallel environments for implementing neural networks. Neural Computing Surveys **1** (1997) 48–60
19. Jin, R., Yang, G., Agrawal, G.: Shared memory parallelization of data mining algorithms: techniques, programming interface, and performance. IEEE Transactions on Knowledge and Data Engineering **17**(1) (2005) 71–89
20. Schapire, R.E.: A brief introduction to boosting. In: Proceedings of the Sixteenth International Joint Conference on Artificial Intelligence. (1999)
21. Lazarevic, A., Obradovic, Z.: The distributed boosting algorithm. In: KDD '01: Proceedings of the seventh ACM SIGKDD international conference on knowledge discovery and data mining, New York, NY, USA, ACM Press (2001) 311–316
22. Graf, H.P., Cosatto, E., Bottou, L., Dourdanovic, I., Vapnik, V.: Parallel support vector machines: the cascade SVM. In: Advances in Neural Information Processing Systems 17. MIT Press, Cambridge, MA (2005) 521–528
23. Dong, J.X., Suen, C.Y.: A fast SVM training algorithm. International Journal of Pattern Recognition **17**(3) (2003) 367–384
24. Yu, H., Yang, J., Han, J.: Classifying large data sets using SVMs with hierarchical clusters. In: ACM SIGKDD. (2003) 306–315
25. Dong, J.X., Krzyzak, A., Suen, C.Y.: A fast parallel optimization for training support vector machines. In Perner, P., Rosenfeld, A., eds.: Proceedings of 3rd International Conference on Machine Learning and Data Mining. (2003) 96–105
26. Poulet, F.: Multi-way distributed SVM algorithms. In: Proc. of ECML/PKDD 2003 Int. Workshop on Parallel and Distributed Algorithms for Data Mining. (2003)
27. Celis, S., Musicant, D.R.: Weka-parallel: machine learning in parallel. Computer Science Technical Report 2002b, Carleton College (2002)
28. Eitrich, T., Lang, B.: Parallel tuning of support vector machine learning parameters for large and unbalanced data sets. In: Computational Life Sciences (CompLife 2005). Volume 3695 of LNCS., Springer (2005) 253–264
29. Qiu, S., Lane, T.: Parallel computation of RBF kernels for support vector classifiers. In: SDM. (2005)
30. Serafini, T., Zanghirati, G., Zanni, L.: Parallel decomposition approaches for training support vector machines. In: ParCo. (2003) 259–266
31. Eitrich, T., Lang, B.: Shared memory parallel support vector machine learning. Technical Report FZJ-ZAM-IB-2005-11, Research Centre Jülich (2005)
32. IBM: ESSL - engineering and scientific subroutine library for aix version 4.1 (2003)
33. Detert, U.: Introduction to the JUMP architecture. (2004)

Supporting a Real-Time Distributed Intrusion Detection Application on GATES

Qian Zhu, Liang Chen, and Gagan Agrawal

Department of Computer Science and Engineering
The Ohio State University, Columbus OH 43210
{zhuq, chenlia, agrawal}@cse.ohio-state.edu

Abstract. Increasingly, a number of applications across computer sciences and other science and engineering disciplines rely on, or can potentially benefit from, analysis and monitoring of *data streams*. We view the problem of flexible and adaptive processing of distributed data streams as a grid computing problem. In our recent work, we have been developing a middleware, GATES (Grid-based AdapTive Execution on Streams), for enabling grid-based processing of distributed data streams.

This paper reports an application study using the GATES middleware system. We focus on the problem of intrusion detection. We have created a distributed and self-adaptive real-time implementation of the algorithm proposed by Eskin using our middleware. The main observations from our experiments are as follows. First, our distributed implementation can achieve detection rates which are very close to the detection rate by a centralized algorithm. Second, our implementation is able to effectively adjust the adaptation parameters.

1 Introduction

Increasingly, a number of applications across computer sciences and other science and engineering disciplines rely on, or can potentially benefit from, analysis and monitoring of *data streams*. In the stream model of processing, data arrives continuously and needs to be processed in *real-time*, i.e., the processing rate must match the arrival rate. There are several trends contributing to the emergence of this model. First, scientific simulations and increasing numbers of high precision data collection instruments (e.g. sensors attached to satellites and medical imaging modalities) are generating data continuously, and at a high rate. The second is the rapid improvements in the technologies for Wide Area Networking (WAN). As a result, often the data can be transmitted faster than it can be stored or accessed from disks within a cluster.

The important characteristics that apply across a number of stream-based applications are: 1) the data arrives continuously, 24 hours a day and 7 days a week, 2) the volume of data is enormous, typically tens or hundreds of gigabytes a day, and the desired analysis could also require large computations, 3) often, this data arrives at a distributed set of locations, and all data cannot be communicated to a single site, 4) it is often not feasible to store all data for processing at a later time, thereby, requiring analysis in *real-time*.

We view the problem of flexible and adaptive processing of distributed data streams as a grid computing problem. We believe that a distributed and networked collection of computing resources can be used for analysis or processing of these data streams. Computing resources close to the source of a data stream can be used for initial processing of the data stream, thereby reducing the volume of data that needs to be communicated. Other computing resources can be used for more expensive and/or centralized processing of data from all sources.

In our recent work, we have been developing a middleware for enabling grid-based processing of distributed data streams [6,5]. Our system is referred to as GATES (Grid-based AdapTive Execution on Streams). One of the important characteristic of this middleware is that it can enable an application to achieve the best accuracy, while maintaining the *real-time* constraint. For this, the middleware allows the application developers to expose one or more *adaptation* parameters. An adaptation parameter is a tunable parameter whose value can be modified to increase the processing rate, and in most cases, reduce the accuracy of the processing. Examples of such adaptation parameters are, rate of sampling, i.e., what fraction of data-items are actually processed, and size of summary structure at an intermediate stage, which means how much information is retained after a processing stage. The middleware automatically adjusts the values of these parameters to meet the real-time constraint on processing, through a *self-adaptation* algorithm. Self-adaptation algorithms currently implemented in the middleware are described in our earlier papers [6,5].

This paper reports an application study using the GATES middleware system. We focus on the problem of intrusion detection, which a widely studied problem in computer security and data mining [1]. We have created a distributed and self-adaptive real-time implementation of the algorithm proposed by Eskin [3]. This implementation generates local models using data received at each node, and then combines these local models to create a global model. We use the functionality of GATES in two different ways. First, as network records typically arrive at multiple locations, a flexible distributed implementation can avoid high communication costs associated with a centralized implementation. Second, as data arrival rates can vary significantly, it is important for an intrusion detection implementation to choose the right trade-off between accuracy and processing rate, to continue to meet real-time constraints.

We have carried-out a number of experiments to evaluate our distributed implementation. The main observations from our experiments are as follows. First, our distributed implementation can achieve detection rates which are very close to the detection rate by a centralized algorithm. Second, our implementation is able to adjust the adaptation parameters. When the rate of data arrival is low, it chooses a small value of the adaptation parameter, EM convergence threshold, resulting in the best detection rate. On the other hand, when the data arrival rate is very high, it chooses a larger value of this parameter, resulting in somewhat lower accuracy, but still maintaining the same rate of processing.

Input: k, # of EM clusters, $D = \{d_1, d_2, \ldots, d_n\}$, set of n $10-dimentional$ points, λ, probability for the set of intrusions, c, anomaly detection threshold.
Output: intrusion detection result.
var
    M_t = probability distribution for normal elements at time t
    A_t = probability distribution for anomalous elements at time t
    C_t = number of intrusions detected at time t
begin
    M_t = GMM generated by Expectation Maximization (EM) algorithm on D
    A_t = a uniform distribution
    Logistic Regression (LR) on D using 3 categorical attributes
    for t = 1 to n
        $LL_t(D) = |M_t|log(1-\lambda) + \sum_{x_i \in M_t} log(P_{Mt}(x_i)) + |A_t|log(\lambda) + \sum_{x_i \in A_t} log(P_{At}(x_i))$
        $M_t = M_{t-1} - x_t$
        $A_t = A_{t-1} \bigcup x_t$
        if $(LL_t - LL_{t-1}) > c$
        then
            $C_t = C_t + 1$
        else
            $M_t = M_{t-1}$
            $A_t = A_{t-1}$
        if (the result says 0 but LR says 1)
        then d_t is intrusion
        else if (the result says 1 but LR says 0)
        then d_t is normal
        else d_t remains the same from the result
    endfor
end

Fig. 1. Pseudo-code for the Anomaly Detection Algorithm

2 Anomaly Detection Algorithm

Intrusion detection problem has been extensively studied in recent years. There are many different approaches for modeling normal and anomalous data, based on which the detections are carried out. A survey and comparison of anomaly detection techniques can be found in[1].

Our goal in this paper is to demonstrate that distributed and adaptive versions of anomaly detection can be implemented using our middleware, GATES. For this purpose, we have chosen an existing algorithm by Eskin [3]. One reason for choosing this algorithm is that many other anomaly detection approaches require training models over clean data. This can lead to problems since online data is not clean and once the anomalies hidden in the data have been detected as normal, further detections will also fail. Eskin's algorithm, in comparison, has the advantage that it can identify anomalies without clean data.

We now briefly describe this algorithm. This method identifies anomalies buried within the dataset. An assumption is made that the number of normal elements in the data set is significantly larger than the number of anomalous elements. The pseudo-code for the algorithm is shown in Figure 1. We use Gaussian Mixture Model (GMM) to represent the distribution of the normal elements in the dataset. This is because it has the property of being able to represent any distribution as long as the number of Gaussians in the mixture is large enough[7]. Further details of the use of EM algorithm to generate GMM can be found in [4]. Once we have the model, anomaly detection begins with first assuming every element is normal. Motivated by the model of anomalies, we use GMM to test each element to determine whether it is an intrusion or not. This is based on the difference of the loglikelihood by treating it as a normal element and as an intrusion. As compared to Eskin's original algorithm, we also use logistic regression[2] to further improve the performance of the algorithm.

3 GATES Middleware and Distributed Anomaly Detection Implementation

In this section, we initially describe the GATES middleware system, and then describe our distributed anomaly detection implementation.

3.1 Overview of the GATES System

GATES (Grid-based AdapTive Execution on Streams) is a middleware that supports the flexible and adaptive analysis of distributed data streams. A key goal is to able to allow the most accurate analysis while still meet the real-time constraint. For this purpose, GATES applies *self-adaptation* algorithm. In summary, GATES has the following features:

- It is designed to use the existing grid standards and tools to the extent possible. Specifically, GATES is built on the Open Grid Services Architecture (OGSA) model and uses the initial version of Globus Toolkits (GT) 3.0's API functions. Therefore, all components of GATES, including applications, exist in the form of Grid services.
- It supports distributed processing of one or more data streams, by facilitating applications that comprise a set of *stages*. For analyzing more than one data stream, at least two stages are required. Each stage accepts data from one or more input streams and outputs zero or more streams. The first stage is applied near sources of individual streams, and the second stage is used for computing the final results. However, based upon the number and types of streams and the available resources, applications can also take more than two steps. GATES's APIs are designed to facilitate specification of such stages.
- It flexibly achieves the best accuracy that is possible while maintaining the real-time constraint on the analysis. To do this, the system monitors the arrival rate at each source, the available computing resources and memory, as well as the available network bandwidth. Then it automatically adjusts

the accuracy of the analysis by tuning the parameter within a certain range specified by the user.

The *self-adaptation* algorithm used in GATES has been evaluated using a number of stream-based data mining applications, including counting samples and finding frequent itemsets in distributed data streams, using data stream processing for computational steering, and clustering evolving data streams[6,5]. Results from the evaluation show that GATES is able to self-adapt effectively, and achieve the highest accuracy possible while maintaining the real-time processing constraint, regardless of the resource availability, network bandwidth, or processing power.

3.2 Real-Time Distributed Intrusion Detection on GATES

The use of GATES middleware can provide two advantages in implementing intrusion detection. First, it can allow a distributed implementation. Many scenarios for intrusion detection involve data arriving at multiple locations. One possible solution for handling such cases is to forward all data to a single node, however, this can result in high communication and computation overheads. Second, it can allow for an *adaptive* implementation, which can trade-off rate of processing and accuracy. Therefore, it can allow the implementation to meet real-time constraints, and allow best accuracy for the given data arrival rate and available computational resources.

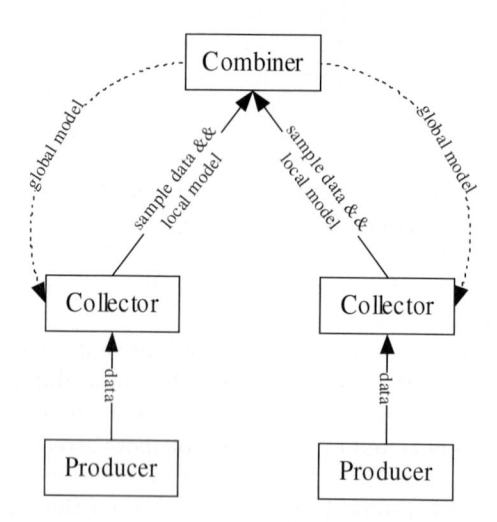

Fig. 2. Communication Topology for the Distributed Intrusion Detection Application

Figure 2 shows the hierarchical structure of the real-time distributed intrusion detection application built on GATES. Our implementation is composed of three stages. The entire procedure of the intrusion detection is divided into a *data preprocessing* stage, a stage which performs *local model generation* and *detects intrusions*, and a *global model generation* stage. We now describe each of these stages in more details.

Producer is simply the data source, where some initial preprocessing can be performed. We chose 10 out of 41 attributes (7 continuous and 3 categorical) for each network data record. The attributes with the most significant variance are included as they most likely to contain more information for distinguishing intrusions from normal data. The filtered data is sent to the second stage.

Collector first collects the data from the *Producer* and applies the EM algorithm to generate local GMMs. These are then used in the anomaly detection algorithm we described earlier, for detecting local intrusions. Another function of this stage is to forward *samples*, i.e. normal data points which have been detected as local intrusions, to the final stage. The goal is to allow the global model to capture their distribution. Finally, once the global model is generated, it is sent back to this stage. Here, the anomaly detection algorithm is applied again to get the global intrusions.

Note that the global model is improved iteratively, i.e., we approach closer to the true probability distribution after each iteration, and have greater accuracy in intrusion detection.

Combiner generates the global model based on the local model parameters and samples. For GMM, the local model parameters used are the mean vector, covariance matrix for each Gaussian model, and their weights contributing to the mixture. We use Kullback-Leibler(KL)-divergence [8] as the measurement to decide how similar two distributions are. KL-divergence is the most natural comparison measure since it is linearly related to the average loglikelihood of the data generated by one model with respect to the other. It is also a well-behaved differentiable function of the model parameters, unlike the other measures. Hence, we combine two local models if the KL-divergence between them is below a user-defined threshold, in which case, a modification has to be made to the global model. Otherwise, new models would be added in. The Combiner sends back the global model to the Collector to end the processing associated with each iteration.

As we stated earlier, GATES uses programmer declared *adaptation parameters* to achieve self-adaptation. In our implementation, we have used two different adaptation parameters. The first parameter is the rate of sampling between the Collector and the Combiner. If the Collector sends a very large number of samples to the Combiner, it increases the communication as well as the computations at the Combiner stage. On the other hand, a very small number of samples will result in a less accurate global model. Similarly, the convergence threshold for the EM algorithm also impacts the accuracy and the processing rate. The smaller the value of this parameter, the more accurate local model we can get, which also leads to a better global model. However, a small value of this parameter also results in more computations for the EM algorithm to converge, making the Collector overloaded.

Note that the self-adaptation algorithm in GATES can only adjust one parameter at a time. Thus, in our experiments, we keep one of them fixed, while allowing the middleware to adjust the other.

4 Experimental Evaluation

This section presents results from the real-time distributed intrusion detection application built on GATES. We had several goals in our experiments. First, we wanted to demonstrate that distributed implementations can achieve high accuracy. Second, we wanted to evaluate the middleware's ability to adjust the two adaptation parameters, under different conditions. Finally, we also evaluate the improvements in accuracy achieved through logistic regression, which is an area where we had improved Eskin's algorithm.

The dataset we used is KDD-CUP'99 Network Intrusion Detection data, which contains a wide variety of intrusions simulated in a military network environment. It consists of approximately 4,900,000 network connection records with more than 80% as intrusions. Each connection record has 41 attributes, including categorical and continuous ones. According the requirement from the anomaly detection algorithm[3], the majority distribution should have at least 90% of the entire dataset. Therefore, we randomly duplicate the normal data and choose part of the intrusion data, which result in a data set with 335,892 records in total and only 9.04% are intrusions. Each type of intrusion is evenly distributed and comes in a burst.

We conducted our experiments in a Linux cluster. Each node has a Pentium III 933MHz CPU with 512MB of main memory and 300GB local disk space. The interconnection network is a switched 100Mb/s Ethernet. In all our experiments, the number of mixtures used in the centralized anomaly detection algorithm is 3 and the distributed version also results in the same number of mixtures.

4.1 Experiment 1: Adjustable EM Threshold vs. Fixed Sampling Rate

In this experiment, we fix the sampling rates at certain values and let GATES adjust the EM threshold. The data production (arrival) rates vary from 100kb/s, 80kb/s, 50kb/s, 30kb/s to 10kb/s. Figure 3 shows how the EM threshold parameter converges to the ideal values with different data production rates. As expected, when the production rate is small, EM threshold converges to a smaller value since it can have enough time to perform the EM algorithm without being overloaded from its upstream. Also notice that the smallest EM threshold GATES converges to is 0.000012, which is very close to the EM threshold used in the centralized algorithm, 0.00001.

The detailed results, including processing time, detection rate, and false positive rate, are shown in Table 1. A centralized version, with no time constraints, takes 923 seconds, and is able to detect 97.63% of the intrusions. It also has a 8.08% rate of false positives. The best accuracy from the distributed version

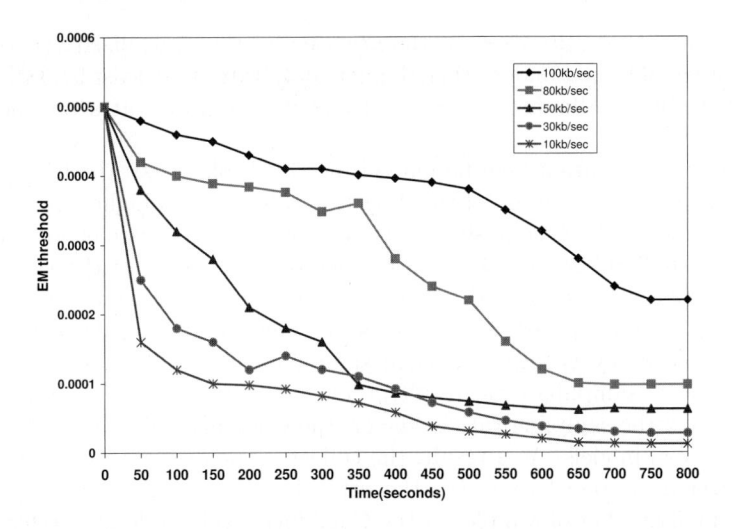

Fig. 3. Adaptation of EM Threshold with Different Data Production Rates

Table 1. Intrusion Detection Results

		Exe. time(sec)	Detection rate	False positive
Centralized		923.0	**97.63%**	**8.08%**
Distributed	Sample rate=40%	667.8	82.79%	5.95%
	Sample rate=20%	637.2	91.38%	7.39%
	Sample rate=16%	618.7	86.48%	6.35%
Producing rate	Sample rate=13%	609.8	83.71%	6.04%
= 100kb/s	Sample rate=10%	602.1	82.57%	5.83%
Distributed	Sample rate=40%	710.1	84.72%	6.22%
	Sample rate=20%	698.9	92.09%	7.57%
	Sample rate=16%	674.2	88.69%	6.83%
Producing rate	Sample rate=13%	653.3	86.21%	6.32%
= 80kb/s	Sample rate=10%	642.9	84.58%	6.17%
Distributed	Sample rate=40%	766.2	88.07%	6.88%
	Sample rate=20%	738.6	92.44%	7.63%
	Sample rate=16%	708.1	90.71%	7.19%
Producing rate	Sample rate=13%	692.6	89.00%	6.95%
= 50kb/s	Sample rate=10%	680.4	87.85%	6.%
Distributed	Sample rate=40%	795.6	90.20%	7.22%
	Sample rate=20%	766.7	94.63%	7.78%
	Sample rate=16%	732.1	93.13%	7.67%
Producing rate	Sample rate=13%	719.8	91.38%	7.55%
= 30kb/s	Sample rate=10%	706.9	90.10%	7.10%
Distributed	Sample rate=40%	862.1	93.74%	7.72%
	Sample rate=20%	798.9	**95.36%**	**7.90%**
	Sample rate=16%	762.4	95.16%	7.87%
Producing rate	Sample rate=13%	741.8	94.29%	7.82%
= 10kb/s	Sample rate=10%	728.3	93.50%	7.66%

is 95.36%, which is quite close to the accuracy of the centralized version. The best accuracy obtained under other (higher) data rates is at least 91.38%, which shows that the middleware is able to tradeoff processing rates and accuracy effectively.

Two other observations can be made from this table. First, the false positive rate is always a fixed fraction of the detection rate, i.e., the higher the detection rate, the higher is the false positive rate. The false positive rate is not impacted by whether the implementation is centralized or distributed, or the value of the adaptation parameters. Second, the trends between the choice of the sampling rate and detection rate are quite interesting. Across different data production rates, best accuracy is achieved when sampling rate is 20%. Both lower and higher values of sampling rates result in lower detection rates. The reason is as follows. When the sampling rate is higher, the Combiner takes a longer time to compute global models. As a result, the collector operates with an older model for a longer duration of time. On the other hand, when the sampling rate is lower, a small number of samples at the Combiner results in lower quality global models.

Table 2. Improvements Through Logistic Regression

	without L.R.		with L.R.	
	Detection Rate	False Positive	Detection Rate	False Positive
Centralized	92.36%	8.35%	97.63%	8.08%
Producing rate=100kb/s	86.45%	7.64%	91.38%	7.09%
Producing rate=80kb/s	87.12%	7.82%	92.09%	7.57%
Producing rate=50kb/s	87.45%	7.88%	92.44%	7.63%
Producing rate=30kb/s	89.52%	8.04%	94.63%	7.78%
Producing rate=10kb/s	90.21%	8.12%	95.36%	7.90%

We also now evaluate the benefits of using logistic regression. We have implemented logistic regression using three categorical attributes. The results are shown in Table 2. The detection rate increases from 92.36% to 97.63% with the false positive dropping from 8.35% to 8.08% for the centralized version. Comparing the best results from the real-time distributed intrusion detection implementation, we can get 95.36% as the detection rate and 7.90% as the false positive rate, compared with 90.21% and 8.12% without using the logistic regression, respectively. The overall observation is that once a data record fails anomaly test, i.e. either normal data is detected as intrusion or an intrusion is detected as being normal, the categorical attributes, namely, the protocol type, the service information, and the flag, can correct the detection results.

Two other observations from our implementation are shown through Figures 4 and 5. Figure 4 shows that as the processing proceeds, we are having smaller KL-divergence comparing to the true model, namely, the global model generated from our algorithm is closer to the true model, as expected. As we can see from the Figure 5, we are getting better detection rates for each processing round.

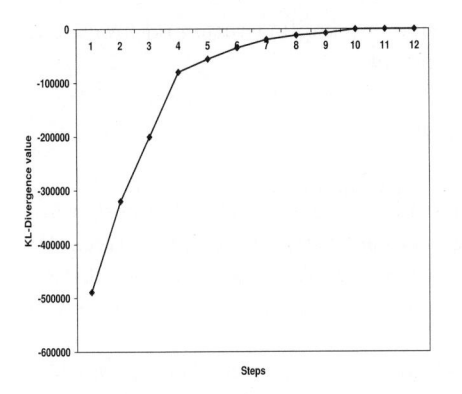

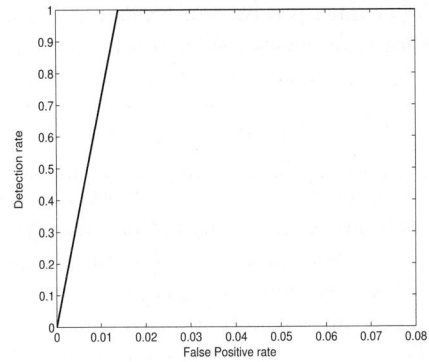

Fig. 4. KL-divergence of the Global Model Compared to the True Model

Fig. 5. ROC for the Intrusion Detection Application

We have used a ROC curve in this Figure, which is a graphical representation of the false positive rate versus the detection rate. The reason for our observation is that we have more data to generate the global model.

4.2 Experiment 2: Adjustable Sampling Rate vs. Fixed EM Threshold

The other experiment we carried out involved a fixed EM threshold, and sampling rate as the adaptation parameter. Again, under different data production rates, we observed how the middleware is able to converge to a stable value of the adaptation parameter. The results are shown in Figure 6. As expected,

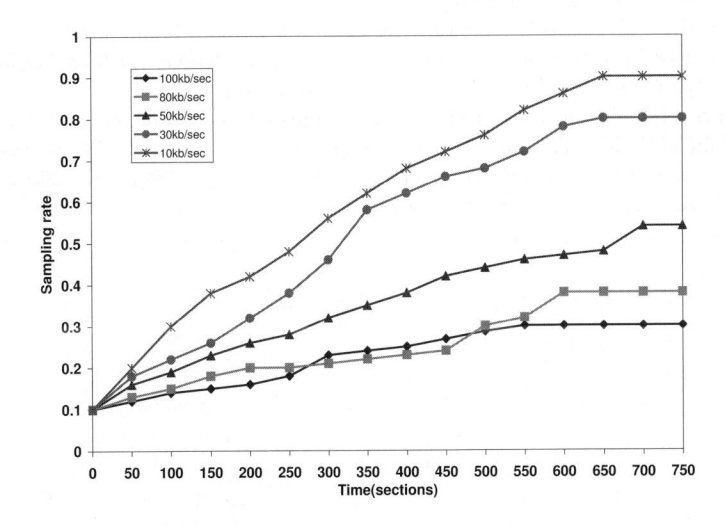

Fig. 6. Adaptation of Sampling Rate with Different Production Rates

higher data production levels result in a smaller sampling rate, and lower data production levels result in a higher sampling rate.

5 Conclusion

This paper has reported an application study using the GATES middleware, which has been developed for supporting grid-based streaming applications. We have focused on the problem of intrusion detection. We have created a distributed and self-adaptive real-time implementation of the algorithm proposed by Eskin using our middleware. The main observations from our experiments are as follows. First, our distributed implementation can achieve detection rates which are very close to the detection rate by a centralized algorithm. Second, our implementation is able to effectively adjust the adaptation parameters.

References

1. C.Warrender, S. Forrest, and B.Pearlmutter. Detecting intrusions using system calls: alternative data models. In *Proceedings of the 1999 IEEE Symposium on Security and Privacy*, pages 133–145, 1999.
2. D.Hosmer and S.Lemeshow. *Applied Logistic Regression*. Wiley, New York, 1989.
3. E.Eskin. Anomaly detection over noisy data using learned probability distributions. In *Proceedings of the Seventeenth International Conference on Machine Learning*, pages 255–262, June 2000.
4. J.A.Belmis. A gentle tutorial of the em algorithm and its application to parameter estimation for gaussion mixture and hidden markov models. In *Technical report*, UC Berkeley, USA, April 1998.
5. L.Chen and G.Agrawal. Supporting self-adaptation in streaming data mining applications. In *Proceedings of IEEE International Parallel & Distributed Processing Symposium(IPDPS)*, April 2006.
6. L.Chen, K.Reddya, and G.Agrawal. Gates: A grid-based middleware for distributed processing of data streams. In *Proceedings of IEEE Conference on High Performance Distributed Computing(HPDC)*, pages 192–201, June 2004.
7. N.Gilardi, T.Melluish, and M.Maignan. Conditional gaussian mixture models for environmental risk mapping. In *Proceedings of the 2002 12th IEEE Workshop on Neural Networks for Signal Processing (NNSP)*, pages 777–786, Sept. 2002.
8. S.Kullback. *Information Theory and Statistics*. Wiley, New York, 1959.

On the Use of Semantic Annotations for Supporting Provenance in Grids

Liming Chen[1], Zhuoan Jiao[2], and Simon J. Cox[2]

[1] School of Computing and Mathematics
University of Ulster, Newtownabbey, Co. Antrim, BT37 0QB, UK
`l.chen@ulster.ac.uk`
[2] School of Engineering Sciences
University of Southampton, Southampton SO17 1BJ, UK
`{z.jiao, s.j.cox}@soton.ac.uk`

Abstract. There has seen a strong demand for provenance in grid applications, which enables users to trace how a particular result has been arrived at by identifying the resources, configurations and execution settings. In this paper we analyses the requirements of provenance support and discusses the nature and characteristics of provenance data on the Grid. We define a new conception called augmented provenance that enhances conventional provenance data with extensive metadata and semantics. A hybrid approach is proposed for the creation and management of augmented provenance in which semantic annotation is used to generate semantic provenance data and the database management system is used for execution data management. The approach has been applied to a real world application, and tools and GUIs are developed to facilitate provenance management and exploitation.

1 Introduction

The essence of Grid computing is the sharing and reuse of distributed, heterogeneous resources for coordinated problem solving in dynamic, multi-institutional virtual organizations (VO). In service-oriented grid infrastructures such as OGSA [1] and WSRF [2], grid resources are regarded as services, and problem solving amounts to the discovery and composition of the required services into a workflow, plus the enactment of the workflow. Problem solving on the Grid is dynamic, collaborative and distributed, e.g. VOs are formed or disbanded on-demand, and services may be published and withdrawn by different stakeholders. In such dynamic environments, it is vital to record the problem solving process for later use such as in interpreting results, verifying that the correct process took place or tracing where data came from.

There has seen an increasing demand for provenance in grid applications [3], which enables users to trace how a particular result has been obtained by identifying the resources, configurations and execution settings. However, current grid architectures lack approaches, mechanisms, and tools to deal with this issue. In this paper we analyse the requirements of provenance support and discuss the nature and characteristics of provenance data on the Grid. We define a new conception called *augmented provenance* that enhances conventional provenance data with extensive metadata and semantics. We propose a hybrid approach for the creation and management of augmented provenance by exploiting the emerging Semantic Web technologies and the latest database technologies. The cornerstone of the approach is the use of ontologies for metadata modeling, and semantic annotations for provenance data population. Special emphasis is placed on semantics, i.e. the ontological relationships among the diversity of provenance data, which enables deep use of provenance data by reasoning.

The paper is organized as follows: Section 2 introduces the concept of augmented provenance. Section 3 describes a hybrid approach for recording and managing augmented provenance. We give an application example in Section 4, and discuss related work and our experience in Section 5. Section 6 concludes the paper and points out some future work.

2 Augmented Provenance

Provenance is defined, in the Oxford English Dictionary, as (i) the fact of coming from some particular source, origin, derivation; (ii) the history or pedigree of a work of art, manuscript, rare book, etc. This definition regards provenance as the derivation from a particular source to a specific state of an item, which particularly refers to physical objects. For example, in museum and archive management a collection is required to have archival history regarding its acquisition, ownership and custody.

In the context of Grid computing, we focus on electronic data produced by computer systems, and we define the provenance of a piece of data as the process that led to that piece of data [4]. A process in the service-oriented grid architecture refers to the execution of a workflow, which is a specification of a service composition. Therefore, the provenance of a piece of data is, in essence, the description of the process that resulted in that data item.

Grids have the characteristics of dynamic provisioning and across-institutional sharing. In such environments a workflow consists of services from multiple organizations in a dynamic VO. The success of workflow execution depends on domain knowledge for service selection and configuration, and mutual understanding of service providers and consumers on service functionalities and execution. The complexity of problem solving process requires not only the execution data of a workflow (e.g. the inputs and outputs of services, the configuration of service control parameters), but also rich metadata data about the services themselves (e.g. their usages, the runtime environment setting, etc.), in order to validate, repeat and further investigate the problem solving process at a later stage. A number of requirements for provenance data are identified and described below.

Firstly, provenance should include metadata at multiple levels of abstraction, i.e. process level, service level and data level. For example, an instantiated workflow instance is a provenance record for the data derived/generated from it, but the workflow instance itself also needs provenance information, e.g. the workflow specification it was instantiated from, the reason a particular set of input values were chosen, etc. Similar provenance requirement applies to services and data.

Secondly, provenance should include metadata from multiple categories including data, knowledge, decision, conclusion, etc. Each category of provenance has its roles and uses, and different applications have different emphases and requirements for provenance. For instance, in biology attention is paid on the transformation process of data; in engineering the focus is on the process creation; and in medical information system the emphasis is on the underlying decision-making process and results that may be more relevant to annotation. As provenance is not only used to validate, repeat and analyze previous executions but more importantly to further advance investigation and exploration based on present results, we are particularly interested in the knowledge and decision provenance, e.g. how a decision was arrived at.

Thirdly, provenance data should be interoperable, accessible and machine processable for sharing among distributed users. This requires provenance data and rich relationships among them be formally modeled and represented. Relations can be regarded as a kind of knowledge model and be used to encode domain knowledge. Appropriate organization of metadata help data retrieval and more importantly, discovery of new knowledge or pattern based on reasoning.

To meet the aforementioned requirements, we face two challenges: the first is how to capture all provenance data. While it is desirable to collect provenance data automatically, it becomes clear that not all provenance data can be captured automatically, especially regarding the rich metadata about services, workflows, knowledge and decisions. The second challenge is how to make provenance data interoperable, sharable and understandable for both humans and machines on the Grid.

Based on the above analysis and inspired by the Semantic Web technologies, we argue that ontologies and semantic annotation should be used for the acquisition, modeling, representation and reuse of provenance data. The reasons are (1) ontologies can model both provenance data and their contexts in an unambiguous way; (2) provenance data generated via semantic annotation are accessible, shareable and machine processable on the Grid; and (3) the Semantic Web technologies and infrastructure can be exploited to facilitate provenance data acquisition, representation, storage and reasoning. For example, it is straightforward to adopt Semantic Web Services for capturing the semantic metadata.

To differentiate from traditional provenance understanding, we introduce the concept of *augmented provenance*, defined as: the augment provenance of a piece of data is the process that leads to the data and its related semantic metadata.

3 A Hybrid Approach to Augmented Provenance

Augmented provenance contains execution data, e.g. the values of inputs and outputs of services; as well as semantic metadata, e.g. the descriptive information about the workflows, services and parameters. The different nature of these two types of data is reflected in the way they are captured, modeled, represented and stored. To support the heterogeneity of provenance data on the Grid a hybrid approach is proposed, which combines the emerging Semantic Web technologies with the database technologies to handle a workflow's semantic metadata and execution data respectively. The overall architecture is illustrated in Figure 1.

3.1 Managing Semantic Metadata

Managing semantic metadata for augmented provenance involves the metadata creation, semantic enrichment, representation and storage. By using the Semantic Web technologies, our idea is to formally model the semantic metadata in ontologies, thus their creation and enrichment can be accomplished in one process through semantic annotations. The generated metadata can be represented in semantic web languages such as RDF or OWL[1], and stored in semantic repositories such as 3Store [5] or Instance Store [6].

The above idea is realized in the architecture by a number of components, namely the Services, Ontologies, Semantic Metadata Repositories, Workflow Construction Environment and Query Tools. Central to the architecture is the Ontologies component containing various domain-related ontologies that specify ontological concepts, their relationships and constraints.

The Services component consists of distributed, internet-accessible services. Such services are generally described in WSDL[1] published in UDDI[2] and invoked by SOAP[1]. However these technologies do not provide formal support for service metadata and semantics. Our approach is to generate the service-level semantic metadata by semantically annotating services using ontologies, and store them in the Semantic Metadata Repositories. Composing services into a workflow is performed in the Workflow Construction Environment component. Service semantic metadata are linked to the workflow and the overall semantic metadata about the workflow are created through semantic annotations and stored in the Semantic Metadata Repositories as well.

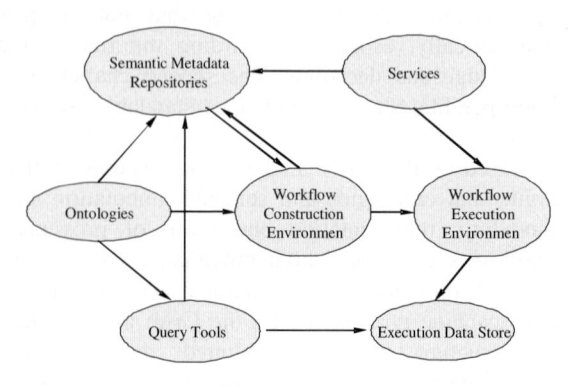

Fig. 1. The architecture for augmented provenance

The Query Tools component is for finding the required semantic metadata and execution data of the augmented provenance, as discussed later.

3.2 Managing Execution Data

Execution data include the input/output values of services, values of services control parameters, and data produced by the workflow. They have the nature of few metadata and semantics attached, but large in volume. For example, the simulation result of an aero-engine design could reach multi-gigabytes in size. Therefore, we leverage database technologies in the Execution Data Store component to facilitate the execution data storage and retrieval.

---

[1] RDF, OWL, WSDL and SOAP are W3C standards. Please refer to www.w3.org
[2] UDDI: www.uddi.org

The Workflow Execution Environment component is responsible for extracting the execution data from the workflow before executing it. It analyses a workflow script to collect initial default or user-defined input values. During the runtime it interprets the workflow script and binds individual constituent services with corresponding inputs and invokes the service. Intermediate results may be returned to the environment and used as inputs to the successive services. The collected and generated data are archived in the Execution Data Store.

3.3 Querying Augmented Provenance Data

Augmented provenance consists of semantic metadata and execution data, and they are represented and managed using different mechanisms. However, semantic metadata and execution data are closely linked and can be cross-referenced. When a workflow template is built with attached semantic metadata in the workflow construction environment, it is stored in the Semantic Metadata Repositories, together with a specifically generated unique ID (UUID, Universally Unique IDentifier [7]) as a handle for later reference. An instantiated workflow template creates a workflow instance which is executed in the Workflow Execution Environment. The executable workflow instance is stored, under its own unique ID, together with associated input/output data and possibly some simple metadata (e.g. the instance creation time, name of its creator, etc) in the database. The one-to-many relationships between the workflow template ID and the workflow instance IDs are also stored in the database, so that users can reference the semantic metadata of the workflow instances through the workflow template ID.

We have implemented the Query Tools component to provide dual query mechanisms for flexible and efficient provenance data search and retrieval. Semantic queries on workflows can be framed using ontologies and are answered through semantic matching. Once a workflow template ID becomes available, its executable instances can be found easily based on the ID by launching a database query.

The separation of semantic metadata and execution data has the following advantages: Firstly, semantic metadata can be formally modeled using ontologies and represented in expressive web ontology languages. This helps capture domain knowledge and enhance interoperability. Secondly, workflow execution usually produces large volume of data that have little added value for reasoning, but storing them in the database made the data searchable and easy to share. Finally, the hybrid query mechanism provides flexibility and alternatives – users can perform semantics based query or direct database query or a combination of the two to meet application requirements.

4 GEODISE: A Case Study of Augmented Provenance

Engineering Design Search and Optimisation (EDSO) is a computationally and data intensive process whereby existing engineering modeling and analysis capabilities are exploited to yield improved designs. An EDSO process usually comprises many different tasks. For example, the design optimization of an aero-engine or wing may involve: specify the wing geometry in a parametric form; generate a mesh for the design; decide which analysis code to use and carry out the analysis; decide the optimisation schedule; and finally execute the optimisation run coupled to the analysis code. Apparently a problem solving process in EDSO is a process of constructing and executing a workflow.

The Grid Enabled Optimisation and Design Search in Engineering (GEODISE) project [8] aims to aid engineers in the EDSO process by providing a range of Grid services comprising a suite of design optimization and search tools, computation packages, data management tools, analysis and knowledge resources. Additionally, GEODISE also intends to manage design provenance so that previous designs can be validated, repeated and further explored to lead to better designs.

We have applied the proposed hybrid approach for augmented provenance in GEODISE to help engineers answer provenance-related questions in the design process. Figure 2 shows the provenance management system.

To formally model EDSO metadata, we have developed GEODISE domain ontology and service ontology. We regard a workflow as a composite service, therefore, the service ontology can be used for modelling both service and workflow metadata. The GEODISE service ontology is based on OWL-S [9] upper service ontology which is an OWL-based Web Service ontology. It further extends OWL-S by incorporating EDSO specific metadata such as *algorithmUsed, reviousService, followingService, derivedFrom, leadTo*, etc., as shown in Figure 3. The left column displays the main concepts while the right column lists concept properties.

Semantic metadata annotation API is developed for capturing augmented provenance data [10] [11]. A front-end GUI is provided to help users enrich the automatically extracted service metadata using EDSO domain and service ontologies. The annotation API is also used to capture and annotate workflow metadata during workflow construction. The generated semantic metadata for both services and workflows are represented in OWL and stored in the Semantic Metadata Repositories, implemented using the Instance Store technology [6].

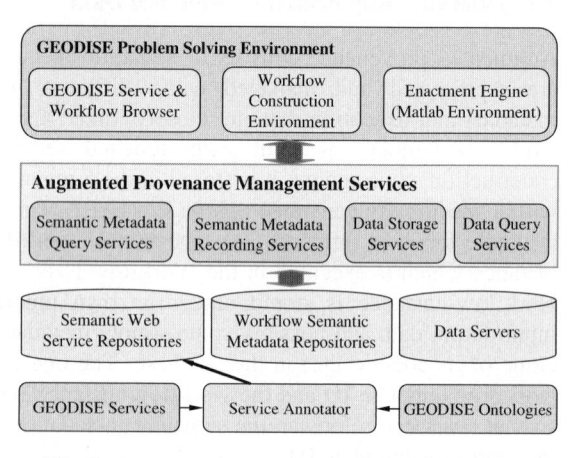

Fig. 2. Augmented provenance management system

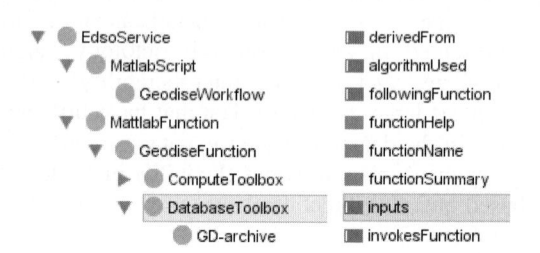

Fig. 3. An example of GEODISE service ontology

The execution data are managed by the GEODISE database toolbox [13]. The database toolbox exposes its data management capabilities to the client applications through Java API, as well as a set of Matlab functions. The Java API has been used by the workflow construction environment to archive, query, and retrieve the workflow instances for reuse and sharing. As Matlab provides the workflow enactment engine in GEODISE, the toolbox's Matlab function interfaces enable data to be archived, queried and retrieved on the fly at the workflow execution time. Data related to a workflow instance are logically grouped together using the *datagroup* mechanism supported by the database toolbox.

Querying augmented provenance in GEODISE is supported through semantic and database query tools, as shown in Figure 4. The semantic query GUI utilises the description logic based reasoning engine Racer [12] to reason over semantic metadata, and the construction of query expressions are supported by the service ontology. Here are two examples of using the dual query mechanism:

- Find the data derivation pathway for a given design result. Actions: querying the database to find the workflow instance that is responsible for the result. Additional semantic metadata about the work-flow instance can be obtained using the Semantic query GUI based on the workflow template ID. The retrieved workflow script can be enacted in the enactment engine (Matlab) for a re-run if necessary.

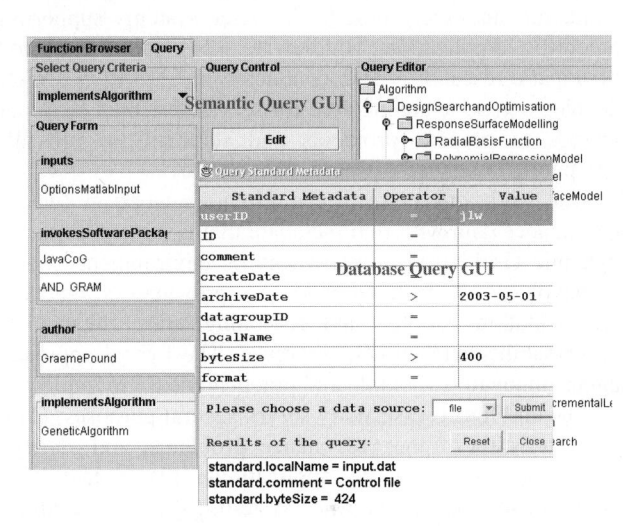

Fig. 4. Query GUIs for augmented provenance

- Find information about the optimisation service used in the workflow that generates the given result. Actions: based on the above search, the workflow template ID is available and can be used in the Semantic query GUI to find the information about the optimisation service used in the workflow.

We have also wrapped the semantic query functionalities as web services, thus making the provenance management system easy to be integrated into service-oriented grid applications.

5 Related Work and Discussion

Provenance has traditionally been used and explored in museum, library and archival management systems where it is mainly referred to the acquisition and creation information, and the history of the ownership and custody of a resource. Research on provenance of computer-generated data has been conducted under different banners, including audit trail, lineage, dataset dependence and execution trace. Such research is mainly undertaken in domain specific applications such as geographic information system [15] and satellite image processing [16]. The common features of Chimera Virtual Data System [17], CCLRC metadata manager [18] and systems developed in [19][20] are that they try to trace the movement of data between data sources and obtain information on the *where* and *why* of a data item of interest as a result of a database operation.

Recently research on the provenance of service-based problem solving processes has attracted more attention with the prevalence of service-oriented computing paradigm. An initial attempt has been made in myGrid project [21] in which derivation provenance (log files) has been annotated and recorded for experiment validation and recreation [22]. Other systems supporting provenance include the Scientific Application Middleware [23] and the e-notebook [24]. An on-going systematic research is also conducted in EU PROVENANCE project which aims to develop a generic architecture for service-oriented provenance system [4] [25]. It also intends to propose protocols and standards to formally standardize provenance computing in service-oriented architecture.

Our work differs from the previous work in two aspects: Firstly we extend provenance data with rich metadata that is particularly useful in open, distributed and dynamic Grid-based problem solving environments. Secondly, we utilize the latest Semantic Web technologies for provenance metadata acquisition, modeling, representation, storage and reasoning, thus enhancing interoperability, machine processability and knowledge reuse. The hybrid approach of managing provenance data is innovative, flexible and practically easy to implement and to use.

The GEODISE case study serves several purposes: (1) it helps identify the generic characteristics of the provenance problems, and clarify user requirements in the context of service-based applications; (2) it helps to pin down the software requirements for a provenance system; (3) the successful design/implementation and operation of the provenance system have demonstrated and proved our conception of provenance, its design approaches and implementation rationale. Through the case study we have learnt two important lessons with regards to the use of provenance system, namely, tools should be provided for end users in their familiar working environments; and easy-to-use tools should hide as much technical details as possible that are not relevant to the end users.

6 Conclusions

The complexity of dynamic problem solving in service-oriented grid infrastructure requires rich semantic metadata in order to verify and further investigate previous results. This gives rise to the conception of augmented provenance, which denotes both semantic metadata and execution data. We argue that the Semantic Web technologies, i.e. ontologies, semantic annotation, representation and storage, can be exploited for augmented provenance management. To this end, a hybrid approach is proposed together with an architecture that defines the core components and functionalities for realizing augmented provenance systems. We have developed a suite of generic APIs and front end GUIs in the context of GEODISE to implement the augmented provenance system. The approach is applicable for broader grid application domains.

The design and implementation of GEODISE provenance system is pioneering in many aspects. Firstly, the research provides a proof of concept for augmented provenance and provenance systems. Secondly, it provides guidelines towards the construction of a basic provenance system. Finally, it demonstrates a possible design and implementation pattern for provenance-enabled applications. In the future we shall focus on the seamless integration and interaction between provenance systems and domain-specific application systems, and in particular the design of a straightforward, easy-to-use query interface. We shall also futher investigate the security and scalability issues.

Acknowledgement

This work is supported by the UK EPSRC GEODISE e-Science pilot project (GR/R67705/01). The authors gratefully acknowledge the contributions from and discussion with EPSRC projects myGrid (GR/R67743/01) and AKT (GR/N15764/01(P)).

References

1. Foster, I., Kesselman, C., Nick, J., Tuecke, S.: Grid Services for Distributed System Integration, Computer, 35(6), 37-46, (2002)
2. WSRF: www.globus.org/wsrf/
3. Fox, G. and Walker, D.: e-Science gap analysis, technical report, http://www.nesc.ac.uk/technical_papers/UKeS-2003-01/GapAnalysis30June03.pdf, (2003)
4. Moreau, L., Chen, L., Groth, P., Ibbotson, J., Luck, M., Miles, M., Rana, O., Tan, V., Willmott, S. and Xu, F.: Logical architecture strawman for provenance systems, Technical report, University of Southampton. (2005)
5. Harris, S., Gibbins, N.: 3Store: Efficient Bulk RDF Storage. Proceedings of 1st International Workshop on Practical and Scalable Semantic Systems, Florida, USA, 1-15. (2003)
6. Horrocks, I., Li, L., Turi, D., Bechhofer, S.: The instance store: DL reasoning with large numbers of individuals, Proceedings of the 2004 Description Logic Workshop, BC, Canada, 31-40, (2004)
7. Mealling, M., Leach, P.J., Salz, R.: A UUID URN Namespace, IETF, October 2002. http://www.ietf.org/rfc/rfc4122.txt
8. GEODISE Project: http://www.geodise.org
9. OWL-S: www.daml.org/services/owl-s
10. Chen, L., Cox, S.J., Tao, F., Shadbolt, N.R., Goble, C., Puleston, C.: Empowering Resource Providers to Build the Semantic Grid. In Proceedings of the IEEE/WIC/ACM International Conference on Web Intelligence (WI'04), 271-278, (2004)
11. Chen, L., Shadbolt, N.R., Tao F., Goble, C.: Managing Semantic Metadata for Grid Services, International Journal of Web Service Research, (In Press), (2006)

12. Haarslev, V., Möller, R.: Racer: A Core Inference Engine for the Semantic Web, Proceedings of the 2nd International Workshop on Evaluation of Ontology-based Tools (EON2003), Florida, USA, 27-36, (2003)
13. Jiao, Z., Wason, J.L., Song, W., Xu, F., Eres, H., Keane, A.J., Cox, S.J.: Databases, Workflows and the Grid in a Service Oriented Environment, Euro-Par 2004, Parallel Processing, Lecture Notes in Computer Science, No.3149, 972-979.
14. Xu, F., Eres, M.H., Baker, D.J, Cox, S.J.: Tools and Support for Deploying Applications on the Grid, Proceedings of the IEEE International Conference on Services Computing (SCC 2004).
15. Lanter, D.P.: Design of a lineage-based meta-data base for GIS. Cartography and Geographic Information Systems, 18(4):255–261, (1991)
16. Frew, J., Bose, R.: Earth science workbench: A data management infrastructure for earth science products, Proceedings of the 13th International Conference on Scientific and Statistical Database Management. (2001)
17. Foster, I., Vockler, J., Wilde, M., Zhao, Y.: Chimera: A virtual data system for representing, querying, and automating data derivation, Proceedings of the 14th International Conference on Scientific and Statistical Database Management, 37-46, (2002)
18. CCLRC Data Management Group: http://www.e-science.clrc.ac.uk/web/groups/Data-Management
19. Boss, R.: A conceptual framework for composing and managing scientific data lineage, Proceedings of the 14th International Conference on Scientific and Statistical Database Management, 47-55, (2002)
20. Buneman, P., Khanna, S., Tan, W.-C.: Why and where: A characterisation of data provenance, Proceedings of the Int. Conf. on Databases Theory (ICDT), (2001)
21. myGrid project: www.mygrid.org.uk
22. Zhao, J., Wroe, C., Goble, C., Stevens, R., Quan, D., Greenwood, M.: Using Semantic Web Technologies for Representing e-Science Provenance, Lecturer Notes in Computer Science, No.3298, 92-106, (2004)
23. Myers, J.D., Chappell, A.R., Elder, M., Geist, A., Schwidder, J.: Reintegrating the research record, IEEE Computing in Science & Engineering, 44–50, (2003)
24. Ruth, P., Xu, D., Bhargava, B.K., Regnier, F.: E-notebook middleware for accountability and reputation based trust in distributed data sharing communities. In Proc. of 2nd Int. Conf. on Trust Management, LNCS2995, 161-175, (2004)
25. Szomszor, M., Moreau, L.: Recording and reasoning over data provenance in web and grid services, Proceedings of the International Conference on Ontologies, Databases and Applications of Semantics (ODBASE 03), (2003)

Topic 6: Grid and Cluster Computing: Models, Middleware and Architectures

Domenico Laforenza, Alexander Reinefeld,
Dieter Kranzlmüller, and Luc Moreau

Topic Chairs

Grid computing is a major research area with strong involvement from both academia and the computing industry. The common vision is that Grid computing represents the culmination of truly general distributed computing across various resources in a ubiquitous, open-ended infrastructure to support a wide range of different application areas. Recently the CoreGrid (http://www.coregrid.net) Executive Committee reached an agreement on the following definition: a Grid is ?a fully distributed, dynamically reconfigurable, scalable and autonomous infrastructure to provide location independent, pervasive, reliable, secure and efficient access to a coordinated set of services encapsulating and virtualizing resources (computing power, storage, instruments, data, etc.) in order to generate knowledge?. Although significant progress has been made in the design and deployment of Grids, many challenges still remain before the goal of a user-friendly, efficient, and reliable grid can be realized. Grid research issues cover many areas of computer science to address the fundamental capabilities and services that are required in a heterogeneous environment, such as adaptability, scalability, reliability and security, and to support applications as diverse as ubiquitous local services, enterprise-scale virtual organizations, and Internet-scale distributed supercomputing.

Therefore, Grid research will greatly benefit from interactions with the many related areas of computer science, making Euro-Par 2006 an excellent venue to present results and discuss issues. This years conference will feature 7 papers, selected from 35 original submissions by using the high quality review process of Euro-Par. For this endeavor, the chairs assembled a team of 45 experts in this domain to perform a minimum of 3 reviews per paper. The selected papers represent work in the area of execution of MPI on the Grid, instant Grids and virtual private Grids, protocols for distributed shared memory, I/O support, storage services, resource brokering in Grids, as well as self-healing aspects.

We would like to cordially thank our colleagues, which helped in the review process, and we invite you to study the papers in this topic on the following pages.

Supporting Efficient Execution of MPI Applications Across Multiple Sites*

Enol Fernández, Elisa Heymann, and Miquel Àngel Senar

Departament d'Arquitectura de Computadors i Sistemes Operatius
Universitat Autònoma de Barcelona, Barcelona, Spain
enol@aomail.uab.es, {elisa.heymann, miquelangel.senar}@uab.es

Abstract. One of the main goals of the CrossGrid Project [1] is to provide explicit support to parallel and interactive compute- and data-intensive applications. The CrossBroker job manager provides services as part of the CrossGrid middleware and allows execution of parallel MPI applications on Grid resources in a transparent and automatic way. This document describes the design and implementation of the key components responsible for an efficient and reliable execution of MPI jobs splitted over multiple Grid sites, executed either in an on-line or batch manner. We also provide details on the overheads introduced by our system, as well as an experimental study showing that our system is well-suited for embarrassingly parallel applications.

1 Introduction

Large-scale Grid computing requires job-management services that address new concerns arising in Grid environments. This 'job management' involves all aspects of the process of locating various types of resources, arranging these for use, utilizing them and monitoring their state. In these environments, job-management services have to deal with a heterogeneous multi-site computing environment that, in general, exhibits different hardware architectures, loss of centralized control, and as a result, inevitable differences in policies. Additionally, due to the distributed nature of the Grid, computers, networks and storage devices can fail in various ways.

Most systems described in the literature follow a similar pattern of execution when scheduling a job over a Grid. There are typically three main phases, as described in [2]:

– Resource discovery, which generates a list of potential resources to be used.
– Information gathering on those resources and the selection of a best set.
– Job execution, which includes file staging and cleanup.

---

* This work was made in the frame of the "int-eu.grid" project (sponsored by the European Union), and supported by the MEyC-Spain under contract TIN 2004-03388, and partially supported by the NATO under contract EST.EAP.CLG 981032.

Many Grid initiatives follow these scheduling phases by providing the middleware infrastructure to develop applications on computational grids and to manage resources. The job management system that we have developed in the CrossGrid project follows the same approach in scheduling jobs. However, our system, known as CrossBroker, is targeted to the kinds of applications that have received very little attention to date. Most existing systems have focussed on the execution of sequential jobs, the Grid being a large multi-site environment where jobs run in a batch-like way. Crossgrid jobs are computationally intensive applications mostly written with the MPICH library using the Globus2 device [3], taking advantage of being executed on multiple Grid sites.

From the scheduling point of view, support for parallel applications introduces the need for co-allocation. There are studies [4][5] that evaluate different co-allocation strategies, although the kind of jobs and grid environment these use are not applicable in CrossGrid and are focused on simulation.

To the best of our knowledge, only a basic support for running MPICH-G2 jobs is included in the Globus Toolkit by using the *globusrun* command and the DUROC services [6]. However this command requires a manual intervention of the user to discover and select resources, and to stage all necessary files to the remote sites and it does not support a reliable co-allocation mechanism to synchronize the start-up of all subjobs. GCM [7] deals with the execution of multi-site jobs using PACX-MPI [8], but it does not include a mechanism for the automatic selection of sites. Our job-management service supports MPICH-G2 job execution by performing the three main scheduling phases in an automatic and reliable way.

The rest of this paper is organized as follows: Section 2 briefly outlines the overall architecture of our resource-management services, Section 3 describes the particular services that support submission of MPI applications on a Grid environment. Section 4 describes some experimental evaluation of our system, and Section 5 summarizes the main conclusions to this work.

2 General Architecture of the CrossBroker

This section briefly describes the global architecture of our scheduling approach. A more detailed explanation can be found in [9]. The scenario that we are targeting consists of a user who has a parallel application and wishes to execute this on grid resources. The user can submit the job in either an on-line or batch manner. On-line submission is made when the application must start *immediately*, i.e. in a period of time very close to the time of submission. This kind of submission is suitable for interactive applications. It is worth observing that batch submissions do not require an immediate application start.

When users submit their application, our scheduling services are responsible for optimizing scheduling and node allocation decisions on a user basis. Specifically, they carry out three main functions:

1. Select the "best" resources that a submitted application can use. This selection will take into account the application requirements needed for its

execution. The most important requirement for on-line jobs is the availability of free machines at submission time; therefore, if there are no free machines, the job will be cancelled. In the case of batch submission, the application can wait for a free slot in the Grid sites and also for resources where other specified requirements are satisfied.
2. Perform a reliable submission of the application onto the selected resources. This involves the proper co-allocation of resources when the application is distributed among multiple sites.
3. Monitor the application execution and report on job termination.

Figure 1 presents the main components that constitute our resource management services. A user submits a job to a Scheduling Agent (SA) through a User Interface, command line or Migrating Desktop. The job is described by a JobAd (Job Advertisement) using the EU-Datagrid Job Description Language (JDL) [10], which has been conveniently extended with additional attributes to reflect the requirements of parallel applications.

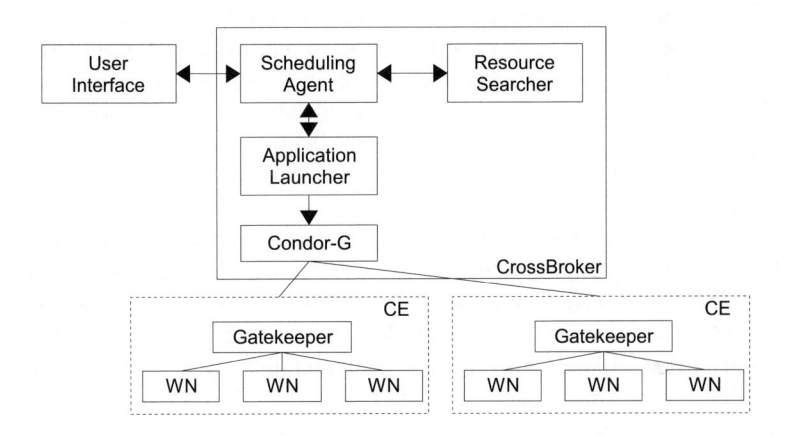

Fig. 1. CrossBroker Resource-Manager Architecture

Once the job has reached the SA, the Resource Searcher (RS) is asked for resources to run the application. The main duty of the RS is to perform matchmaking between job needs and available resources. Using the job description as input, the RS returns as output a list of possible resources within which to execute the job. The matchmaking process is based on the Condor ClassAd library [11], which has been extended with a set matchmaking capability, as described in [9].

The SA then selects the best resource (or group of resources) from the list returned by the RS taking into account its current state and the job requirements. The computing resources (or group of resources), also referred to as Computing Element (CE) in CrossGrid terminology, are passed to the Application Launcher, which is responsible for the co-allocation and the actual submission of the job. Due to the dynamic nature of the Grid, the job submission may fail on that particular site. Therefore, the Scheduling Agent will try other sites from the returned list until the job submission either succeeds or fails.

The Application Launcher is also in charge of the reliable submission of parallel applications on the Grid. Currently, two different launchers are used for MPI applications, one allowing execution on one site, described in detail in [9], and one allowing execution on multiple sites, described in the following section.

3 MPICH-G2 Job Management

An MPI application for grid execution has to be compiled with MPICH-G2 [3], a device which allows the submission to multiple grid sites, thus using the set matchmaking capability of our Resource Searcher for the automatic search of resources.

As we have already mentioned, an MPICH-G2 application can be executed on multiple sites using the *globusrun* command. The *globusrun* call performs subjob synchronization through a barrier mechanism. But when executing jobs with *globusrun*, it falls to the users to decide which sites to use, and it is these same users who should be aware of the need to ask for the status of their own application, resubmitting the application again if something is amiss, and so on. Any failure or delay in the startup of a subjob may block permanently the application given that the remaining subjobs will stay within the synchronization barrier. As a consequence, resources will be occupied but no progress will be achieved in application execution.

The lack of reliability exhibited by the *globusrun* command has been overcome by Condor-G [12], which constitutes a dependable submission system for the Grid. Unfortunately, Condor-G only supports sequential applications. We have modified the submission of MPICH-G2 jobs in such a way that the whole application is decomposed into a set of independent tasks - submitted to the Grid - which are submitted to Condor-G (and treated as sequential tasks). Additionally, we have included the necessary synchronization actions within each task so as to generate a reliable co-allocation of all tasks. We can thus react to the synchronization-related problems experienced by *globusrun* and avoid any blocking situation during the launching phase of the job.

Ideally, MPI applications should always run soon after submission. However, there may be situations in which not all the remote resources involved in an execution are available, causing the ready resources to stay idle until all subjobs start. Our job-management service features a special mechanism to deal with these situations for batch MPI jobs. Whenever a batch MPI-G2 application is submitted, an agent (rather than the actual application) is submitted to the remote sites. This agent, based on Condor Glide-In [12], is used to gain control of remote machines independently of the local-site job scheduler. Each machine acquired by the agent, is configured as two virtual machines, in order to create a separate group of dedicated resources for two types of applications: batch MPI on the one hand, and sequential, on the other hand. From a logical point of view, MPI batch jobs will then run on one virtual machine and sequential jobs will run on the other one. MPICH-G2 subjobs are submitted to the batch virtual machine and will wait until all subjobs are ready for execution. Meanwhile, the sequential virtual machine is used to execute other jobs using backfilling scheduling, hence attaining better utilization of resources. In the case of on-line applications, the agent submitted does not create two virtual machines, but rather immediately starts the application to ensure a faster start-up time.

In order to ensure the co-allocation of the different subjobs that make up one application, the Scheduler Agent launches an MPICH-G2 application launcher (MPI-AL), through Condor-G. This MPI-AL follows a two-step commit protocol:

- In the first step, all the subjobs (with their agents) are submitted to the remote sites.
- A second step guarantees that all subjobs have a machine for their execution, and that they have executed the MPI Init call. Synchronization is achieved through a barrier released by the MPI-AL. After such synchronization, the subjobs will then be allowed to run.

In order to avoid blocking situations, the MPI-AL will wait for several minutes for on-line jobs to execute their MPI Init call. If this call is not performed before the time is exhausted, the whole job will then be aborted. In the case of batch jobs, time-out will occur when the site's local scheduler removes the job.

Figure 2 depicts how execution over the multiple sites of a batch job is performed. In this example scenario, we have N subjobs constituting an MPICH-G2 application. These subjobs will be executed on different sites. For the sake of simplicity, Fig. 2 only shows 2 sites. The A arrows show subjobs submission to the remote machines. These subjobs will stage agent executable and will start it to gain control of the node. Once the virtual machines are available, the actual application is submitted to the batch virtual machine. This is shown by the B arrows. Once the subjobs are executing on the remote machines, the MPI-AL releases the barrier and starts monitoring their execution and writes an application global-log file, providing complete information of the jobs execution. This monitoring is shown by the C arrows in Fig. 2, and constitutes the key point for providing both reliable application execution and robustness.

In the event of the application ending correctly or of there being a problem with the execution of any subjob, the MPI-AL records this in a log file that will be checked by the SA, which will then take the correct action, in accordance with that information. This provides a reliable once-only execution of the application without user intervention.

4 Experimental Results

In this section we present an experimental evaluation of our system. First, we measure the overhead introduced by our software when running MPICH-G2 jobs and following this, we evaluate the performance of a real application executed on the Grid.

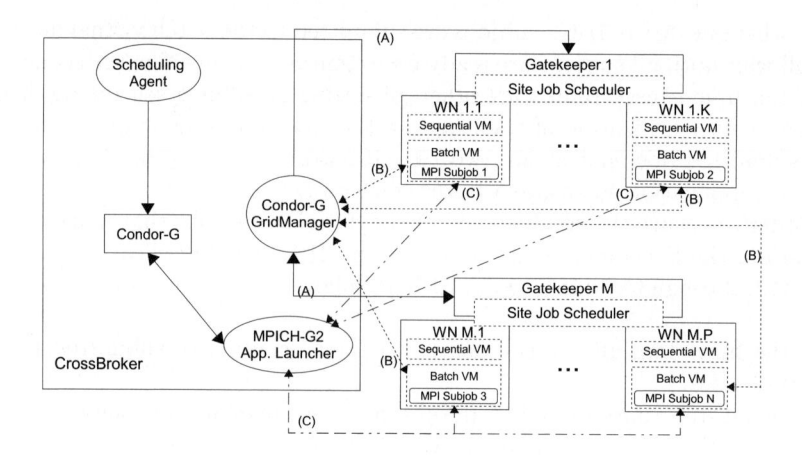

Fig. 2. MPI execution on multiple sites

4.1 MPICH-G2 Overhead

Submitting jobs to the Grid by using the CrossBroker incurs an initial overhead due to the different actions taken before the real job execution. This overhead depends on whether the MPICH-G2 job will run in an on-line or batch manner. Maximum overhead is incurred in the second case because it requires the following steps:

1. Submission of the different subjobs to the remote sites. This involves contacting the Globus gatekeeper, which in turns contacts the site job scheduler to create a basic job that starts our agent in a particular node.
2. Once the basic job has started in the remote site, the agent files are downloaded from the CrossBroker to the node that will execute the application.
3. Virtual Machine set-up: the virtual machines are created. The CrossBroker is notified, which in turn will submit the user application to the agent.
4. Subjob start-up: user application files are downloaded and the job is started in the remote node.

We have submitted a synthetic MPICH-G2 application in order to measure the impact of the different steps. The sites used were the following:

– UAB: cluster in Barcelona with 6 heterogeneous CPUs. The CrossBroker used for these tests is also located in Barcelona.
– FZK: remote cluster with 16 CPUs (4 nodes with 4 CPUs each) located in Karlsruhe (Germany).
– IFCA: remote cluster with 6 CPUs in 2 dual nodes. This cluster is located in Santander (Spain).

Figure 3 shows the time (in seconds) from job submission until the application starts running using different combinations of the above sites. In addition to the one-site submission, submissions using CPUs from two and three sites are also shown. When more than one site was used, CPUs have were distributed equally among all sites. The time obtained is the sum of the four steps mentioned above. The figure shows that, in general, overhead mostly depends on the sites used. With the increase in the number of sites involved in an execution, this overhead also increases. It should be observed that these measures have been obtained for a worst-case scenario, in which all the steps would be taken.

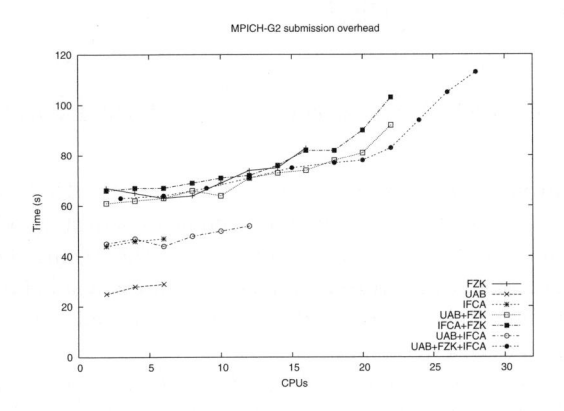

Fig. 3. MPICH-G2 submission overhead for multiple sites

In order to show the real influence of the different steps involved in application submission, Fig. 4 depicts the first step (submission to the remote sites) and agent download (second step) for the same scenario. Globus submission depends on the queue status of the site's local scheduler. In these tests, submission was made to sites with empty local queues (PBS, Condor), hence the subjobs start as soon as possible. As can be seen in Fig. 4, site submission remains almost constant, despite the number of CPUs used, and depends on the sites used: UAB and FZK jobs start earlier than those at IFCA. This is the minimum delay for application execution in our environment, and is similar to that obtained using *globusrun* directly or when using the on-line scheduling in CrossBroker (the second and third step ares not executed in such a case).

Figure 4 also shows how agent download is the most time-consuming step taken in submitting MPICH-G2 applications. This time depends on the limited bandwidth between the CrossBroker machine and each of the nodes in which the jobs is to be executed. As the number of CPUs increases, bandwidth is shared among all these nodes and the downloading process takes longer. The download takes longer in sites located at greater distance (FZK), i.e. having less bandwidth between the site and the CrossBroker. Downloading time is limited to less than a minute, which is negligible for batch applications intended to run for a much longer period of time.

The third step, virtual machine set up, is the time elapsed from the agent file download to the creation of the virtual machines on the remote machines. This

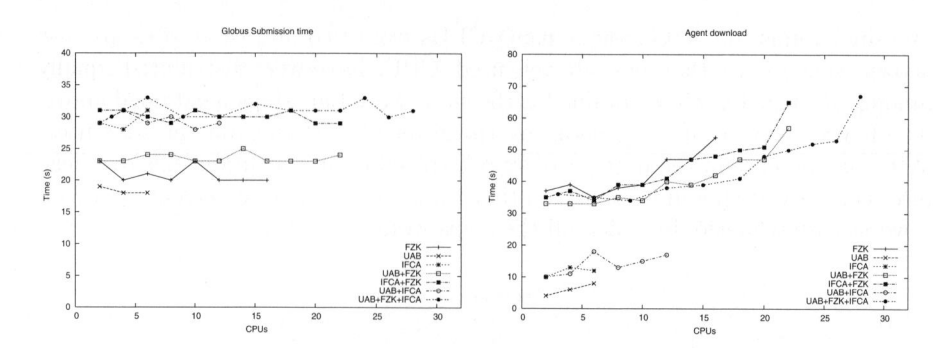

Fig. 4. Left: Globus submission time. Right: Agent download time

time is usually around one second, immediately after the CrossBroker submits the actual application to the batch virtual machine. The last step involves downloading the application binaries. As in the case of the agent, the downloading process depends on the sites involved. This step can be avoided if the user specifies a pre-staged binary in the job description or else makes use of the storage facilities available in the remote sites.

Although the overall overhead is not significant for batch applications, download times for the agent could be avoided by permanently installing the needed files in the remote nodes. In such a case, the CrossBroker would therefore only need to submit a simple job that initiates the agent without any previous download.

4.2 MPICH-G2 Application Execution

MPICH-G2 allows the execution of any MPI application using different-cluster nodes. Applications making heavy use of collective operations that are fairly sensitive to high-latency links are not suitable for this kind of environments. However, there are many applications that exhibit a computation/communication ratio that make them attractive for executiong over multiple sites. Many embarrassingly parallel applications are suited to such applications.

As an example, we have used a Master-Worker application developed in the CrossGrid Project to measure the impact of multiple-site execution. This application trains a neural network to find Higgs Boson [13]. The master node assigns a list of files with input data to each of the workers, and the training is repeated until the obtained error reaches a certain bound. The needed files are downloaded from each of the workers using replica management tools [14].

We have executed the application on the same sites used in section 4.1. In Fig. 5 the execution time is shown for the same CPU combination shown in the previous subsection. Depicted time (in seconds) includes the overhead for on-line scheduling, so implying there is therefore no creation of virtual machines in the remote nodes. This overhead is around 20 to 30 seconds, depending on the used sites.

The first measure uses two nodes, one master and one worker; for the remaining measures, the number of workers has been increased. Application scalability

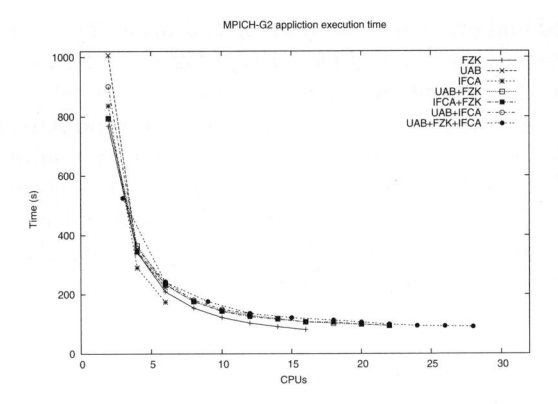

Fig. 5. Neural-net application execution time

is good for a small number of CPUs (less than 10), but does not scale so well for greater CPU numbers. However, application behaviour is not greatly affected when using multiple sites. In general, the use of more sites introduces a somewahat larger bigger overhead (around 10 seconds), although it also allows the use of faster CPUs not available in one-site executon. These results show that it is possible to exploit such a Grid environment as a large cluster for executing similar applications, without being limited to the number of CPUs available on a single site.

5 Conclusions

We have described the main components of the resource-management system that we have developed in order to provide automatic and reliable support for MPI jobs over grid environments. The system consists of three main components: a Scheduling Agent, a Resource Searcher and an Application Launcher.

The Scheduling Agent is the central element that records the job queue submitted by the user and carries out subsequent actions to run the application effectively on the suitable resources. The Resource Searcher has the responsibility of providing groups of machines for any MPI job, taking the application requirements into account. Finally, the Application Launcher is the module that, in the final stage, is responsible for ensuring reliable application execution and co-allocation on the selected resources.

The job-management service provides a reliable on-line and batch MPICH-G2 submission to a Grid. It uses agents to take control of remote machines, allowing the implementation of backfilling scheduling policies for sequential jobs, while all the MPICH-G2 application subjobs are waiting for the proper co-allocation of resources. The Application Launcher guarantees execution without blocking machines, and takes the appropriate decisions in order to guarantee resubmission of failed parallel jobs (due to crashes or failures with the network connection, resource manager or remote resources) and exactly-once execution.

We have tested and evaluated our system, measuring the overhead introduced by the CrossBroker when submitting MPICH-G2 jobs. This overhead is introduced in the case of batch submission and is less than a minute - a short duration for the kind of applications that the batch submission is targeted at,- which usually take much longer to execute. We have also tested system utility by the execution of a master-worker application. This application does not make a heavy use of communications, showing similar scalability both in Grid and in one-site-only execution. Many embarrassingly parallel applications should behave in a similar way, and therefore are also suitable for this environment.

References

1. EU-CrossGrid: http://www.eu-crossgrid.org (2004)
2. Schopt, J.M.: Ten Actions When Grid Scheduling. In: Grid Resource Management - State of the Art and Future Trends. Kluwer Academic Publishers (2003)
3. Karonis, N.T., et alters: Mpich-g2: A grid-enabled implementation of the message passing interface. J. Parallel Distrib. Comput. **63**(5) (2003) 551–563
4. Bucur, A., Epema, D.: The performance of processor co-allocation in multicluster. In: 11th Int. Symp on High Perf. Distr. Comp. (2002)
5. Wang, L., et alters: Resource co-allocation for parallel tasks in computational grids. In: Int. Workshop on Challenges of Large Apps. in Dist. Env. (2003)
6. Czajkowski, K., Foster, I., Kesselman, C.: Resource co-allocation in computational grids. In: Proceedings of the HPDC-8. (1999) 219–228
7. Lindner, P., et alters: Gcm: a grid configuration manager for heterogeneous grid enviromnents. Int. J. Grid and Utility Computing **1**(1) (2005) 4–12
8. Gabriel, E., et alters: Distributed computing in a heterogenous computing environment. In: EuroPVMMPI'98. (1998)
9. Heymann, E., et alters: Managing mpi applications in grid environments. In: European Across Grids Conference. (2004) 42–50
10. Pazini, F.: Jdl attibutes. Technical report, European Datagrid Project (2001)
11. Raman, R., et alters: Matchmaking: Distributed resource management for high throughput computing. In: HPDC-7, Chicago, IL (1998)
12. Thain, D., et alters: Condor and the grid. In: Grid Computing: Making the Global Infrastructure a Reality. John Wiley & Sons Inc. (2003)
13. Gutiérrez, A., et alters: Parallelization of a neural net training program in a grid environment. In: PDP 2004. (2004) 258–265
14. Cameron, D., et alters: Replica management services in the european datagrid project. In: UK e-Science All Hands Conference. (2004)

Private Virtual Cluster:
Infrastructure and Protocol for Instant Grids

Ala Rezmerita, Tangui Morlier, Vincent Neri, and Franck Cappello

INRIA/LRI, University Paris-Sud, Orsay, France
{rezmerit, tmorlier, neri, fci}@lri.fr

Abstract. Given current complexity of Grid technologies, the lack of security of P2P systems and the rigidity of VPN technologies make sharing resources belonging to different institutions still technically difficult. We propose a new approach called "Instant Grid" (IG), which combines various Grid, P2P and VPN approaches, allowing simple deployment of applications over different administration domains. Three main requirements should be fulfilled to make Instant Grids realistic: 1) simple networking configuration (Firewall and NAT), 2) no degradation of resource security and 3) no need to re-implement existing distributed applications. In this paper, we present Private Virtual Cluster, a low-level middleware that meets these three requirements. To demonstrate its properties, we have connected with PVC a set of firewall-protected PCs and conducted experiments to evaluate the networking performance and the capability to execute unmodified MPI applications.

1 Introduction

Sharing resources in a secure way, over the Internet, is attractive for a broad range of users and communities. Audio and video over IP, file sharing, file storage and distributed computing are examples of applications concerning many communities of users. However, despite the continuous progress in Grid, P2P and VPN technologies, sharing resources over different administration domains still raises technical difficulties. Grid technologies allow sharing resources between the participants of virtual organizations [1]. Compared to previously existing technologies, Grid middleware provides tools for inter-domain security and resource management, assuming pre-existing local software and policies in every Grid site. The current trend towards the use of Services [3, 2] responds to the complexity of managing heterogeneous resources and sharing policies by providing a standard interface between the user and the resources.

However, installing Grid middleware is still complex and requires the skills of networking, security and OS experts. Moreover, providing a standard but novel interface to the users imposes, in many cases, to re-implement or to adapt the applications. P2P systems allow simple resource sharing between large communities of users. However, they exhibit two major limitations: 1) the security is very limited and generally not considered in these systems and 2) they run dedicated applications. Albeit Jxta [4] provides a communication layer to deploy and run P2P applications, it has a major limitation by exposing only a Java interface to the application. Installing and using a VPN (Virtual Private Network), using technologies like VTun [5] or IPsec [6], allows users registered in the VPN to share their resources as if they were in a LAN. However, VPN's have their own limits: 1) installation and maintenance require OS and networking experts, administrator authorization and 2) they are static.

In fact, existing technologies restrict resource sharing to Grid and VPN experts or users of unsecured and dedicated P2P systems. This situation motivates the research presented in this paper towards a more spontaneous and dynamic Grid approach called "Instant Grid" (IG), in reference to popular "Instant Messaging" environments. Three mains requirements should be addressed in an IG environment: 1) Connectivity. Firewall and NAT settings may preclude the deployment of cross-domain applications. Moreover, the user may have no technical knowledge on how to setup correctly firewalls and NATs. Thus, an IG environment should use a set of firewall and NAT configuration and/or traversing techniques, transparent to the user and acceptable by domain administrators. 2) Security. Sharing resources across administrative boundaries should not lower the security level of the hosting sites and the shared resources, 3) Compatibility. Sharing resources should not imply specific application or runtime developments.

In this paper, we propose PVC (Private Virtual Cluster), an environment for Instant Grids. PVC design considers the following context: 1) resource sharing is established when required and 2) security is based on classical OS mechanisms (currently access rights and sandbox or virtual machines in the near future), used commonly in LAN's and clusters. PVC turns dynamically a set of resources belonging to different administration domains into a cluster where existing cluster runtime environments and applications can be run.

The next section presents the related work concerning the three issues. Section 3 describes the general architecture of PVC and gives details on the protocol implementation. The evaluation of PVC is presented in Section 4.

2 Related Works

In this section, we present the existing projects and technologies related to the three main issues of Instant Grids: connectivity, security and compatibility.

2.1 Connectivity, Security

One of the most popular projects providing connectivity among peers in different administration domains is JXTA [4]. Based on proxy technologies, JXTA proposes two communication approaches: a rendezvous and a pipe binding protocol. The two methods use a relay to forward messages between peers which results in significant communication overhead. To provide secure communication between peers, JXTA uses a virtual transport layer based on TLS (Transport Layer Security). The difficulty of installing and configuring the proxy limits the usage of JXTA in the Instant Grid context.

Ibis [7] is another project providing NAT and Firewalls traversing techniques to connect resources in different administration domains. Several approaches are successively proposed to bypass Firewall/NAT: a direct connection, a simultaneous TCP SYN connection and a proxy connection. For user identification and secure communication a standard SSL/TLS infrastructure, performing data encryption and peer authentication over a socket connection.

CODO [8] provides end-to-end connectivity for distributed applications over firewalls/NAT protected domains in a secure way. It consists in firewall agents (FAs), placed on the firewall machines and client libraries (CLs) linked with the application. The FA communicates with CL to dynamically add and delete rules needed to establish direct connections. Authentication and security are based on X.509 certificates. The major limitation of CODO is that it currently supports only firewalls based on Netfilter and it assumes the installation on firewalls of the FAs that is not always possible.

Another simple and practical NAT traversal technique is UDP/TCP hole punching [9]. This technique enables two clients behind NAT, to set up a direct peer-to-peer UDP/TCP session with the help of a rendezvous server. Following the statistics given in the article describing this technique, about 82% of the NATs support hole punching for UDP, and about 64% for TCP streams.

With the popularity of DSL network, the use of NAT increases dramatically. Unfortunately, NAT imposes another limitation for the direct connection of peers. Two projects propose NAT discovery and bypassing techniques. The first one is STUN RFC [11]. This standard describes the techniques to discover the NAT type and an UDP protocol to traverse it. The standard classifies NATs in four classes and the traversing technique works for three of them.

The second one is UPnP [10]. The UPnP Forum proposes an API for communications with the NAT device allowing opening firewall ports for direct connection. This technique is particularly suitable for the objectives and constraints of Instant Grids. Obviously, this method does not work with the NAT devices that are not UPnP compatible or if the administrator does not enable it.

2.2 Compatibility, Virtualization

To the best of our knowledge, only JXTA provides compatibility and a virtualization layer in addition to connectivity and security. Its approach is based on unique IDs, by which the network resources can be addressed independently of their physical address.

Several projects allow the creation of a virtual cluster from independently administered domains through machine and network virtualization. A VioCluster [12] logically moves machines between virtual domains, allowing a cluster to dynamically grow and shrink based on resource demand. Network virtualization in VioCluster is made by a hybrid version of VIOLIN [13] which gives to a machine the ability to connect to the private network.

Cluster-On-Demand(COD) [14] shares the same objective. COD was inspired by Oceano [15]. Its main difference is its dynamic resource management between multiple clusters by reinstalling the base OS on resources. The VNET [16] is a virtual private network tool implementing a virtual local area network over a wide area, for virtual machines in Grids. VNET is a simple proxy scheme that works entirely at user level and uses the Layer Two Tunneling Protocol (L2TP).

3   General Principles of Private Virtual Cluster

The objective of Private Virtual Cluster (PVC) is to provide, in a transparent way, an execution environment for existing cluster applications over nodes distributed on the Internet. The main difference between PVC and VPN is its capability to dynamically connect firewall protected nodes, without any intervention of domain administrators and without breaking the security rules of the domains hosting the nodes. Compared to other projects presented in the related work section, PVC provides a fully integrated environment.

PVC itself is a distributed system working as a) a daemon process (*peer*) running on each participating host and b) a brokering service. The role of each local peer daemon is to establish a secure direct connection between the local peer and the other participating peers, subsequently leaving the connection control to the application. The role of the brokering service is to help establishing these connections by 1) collecting and advertising the peer connection requests and 2) tunneling some communications between peers when direct connections are not established, 3) translating network addresses from virtual to real and 4) transporting security negotiation messages. Typically, the brokering service may also help with failure detection, although this feature is not yet implemented.

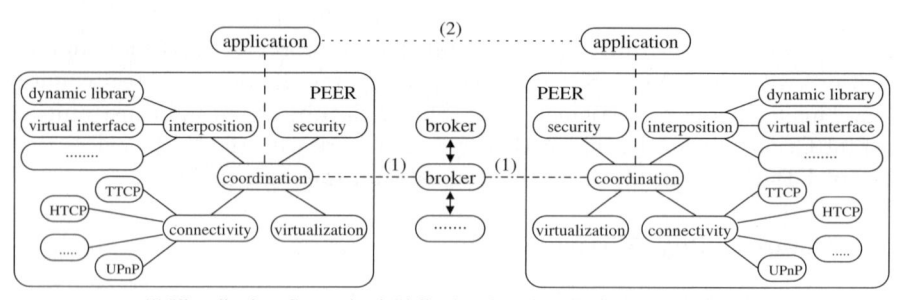

(1) Virtualization, Connection initialization, Security (2) Direct communication

Fig. 1. PVC architecture

Figure 1 presents the modular architecture of PVC. The peer daemon encapsulates five modules for: 1) operation coordination, 2) communication interposition, 3) network virtualization for the application, 4) security checking and 5) peer-to-peer direct connection establishment. The modular architecture offers the possibility to extend and adapt each module to fit with the target environment. The brokering service is implemented as a set of replicated nodes, connected to the Internet and accept inbound communications from PVC peers.

All daemon modules are coordinated locally by the coordinator, which also participates in the global coordination of a PVC deployment. The coordinator runs a workflow through the four other modules to establish the direct connection between the local peer and distant ones. The coordinator also exchanges messages with the brokering service to implement the global coordination.

The interposition module intercepts the application connection requests and transfers them to the coordination module. It may be implemented in various ways (network calls overloading, virtual network interface) offering high adaptability to the system configuration. The intercepted requests are routed on a virtual network simulated by the PVC virtualization module, which features its own IP range and domain name service.

In this virtual network, the PVC security module checks the respect of pre-existing security policies and authenticates the virtual cluster participants. Different security standard and specific methods may be adapted to the PVC architecture (SSL certificates, standard security challenges, etc.). The connectivity module transparently helps the cluster application to establish direct connections between virtual cluster nodes (peers). Like all the other modules, a variety of techniques can be used depending on participants host configuration as well as its local network environment (firewall, NAT). Standard (UPnP) and original mechanisms (Traversing-TCP, TCP Hole Punching) may be used to establish direct connection between peers. In the following parts, we will focus three key modules: virtualization, security and connectivity.

3.1 Domain Virtualization

One of the PVC objectives is to allow the execution of cluster applications without any modification. Cluster applications generally use the socket model as interface with the communication network. Following this constraint we have chosen to use a domain virtualization at IP level. The virtualization layer establishes an IP domain over resources belonging to different administration domains having public or private (possibly conflicting) addresses. Like in a VPN, an overlay network featuring virtual IP addresses is built on top of the actual network.

To avoid the conflict between real and virtual networks used by the resource, we use a specific IP class defined by a RFC [17] for experimental purposes (class E ranging between 240.0.0.1 and 255.255.255.254). The use of these IP addresses guaranties that no real machine uses them (such addresses are actually not routed on the Internet). A virtual DNS, configurable by the PVC members, is associated with this experimental IP class.

3.2 Security Policy in PVC

The main objective of the security mechanism is to fulfill the security policy of every local domain and enforce a cross-domain security policy. Two security levels are implemented: 1) local to the administration domain and 2) between domains. The intra and inter-domain security policies could be configured by local system administrator who could also define the global policy. When a connection is requested by the application, the local peer first checks that it can accept inbound and outbound communications with other peers outside the administration domain, according to the local policy. Then, it checks that IP addresses of external peers and the ports to be used are granted by the global policy.

Without a strong access control mechanism, someone may take advantage of the brokering service and pretend to take part of the virtual cluster. To avoid this, every virtual cluster has a master peer (a peer managed by the virtual cluster administrator) implementing the global security policy. Only the master peer can dynamically register new hosts. Before opening the connection, every peer checks that the other peer belongs to the same virtual cluster. The cross authentication is performed using master information and crossing the brokering service. A key point in the design is that the security protocol does not need to trust the brokering service. This protocol ensures that: 1) only the participating hosts of a cluster can be connected to each other and 2) only trusted connections are returned to the cluster application.

In the current implementation, each host connected to PVC has its own private and public key. Every participant to a virtual cluster knows the public key of its master before connecting to PVC infrastructure. The master peer registers the participation of a new peer, asking the brokering service to store its public key previously encoded with the master's private key. During the establishment of the connection, both peers obtain the other side's public key from the brokering service and decode the received message with the master's public key. This mechanism ensures that only the master registers other participants on the brokering service.

The peer's mutual authentication consist in a classical security challenge-response: the client generates a cryptographically random string M, encrypts it with server public key and sends it to the server; the server decrypts the message with its private key, encrypts the obtained value using its private key and returns the result, $Es(M)$, to the client; the client decrypts $Es(M)$ using the server's public key, obtaining $Ds(Es(M))$; if that value is equal to the original M, the client is satisfied of the server's identity. Similarly, the server picks a random string L, encrypts it and sends it to the client, which returns $Ec(L)$ to the server. The server checks that $Dc(Ec(L))$ equals L and it thereby satisfied with the client's identity. The security is implemented using OpenSSL Crypto library [23]. The experimental results are presented in Section 4, where we discuss the overhead of PVC in secure connection establishment.

3.3 Inter-domain Connectivity Techniques

As the major objective of PVC is to establish direct connections between distributed peers, the connectivity module can host several connection protocols. In the current implementation, we have integrated three techniques in PVC. In this section, we present the integration of these techniques.

Integration of a Firewall configuration protocol. In the last two years, UPnP project became very popular. The principal vendors of domestic network devices incorporated UPnP in their routers. Using UPnP, a PVC peer can communicate with the router and can open the ports for direct connections.

To guarantee the safety of the local area network, the port forwarding rules must be erased from the router when they are no longer needed. If during the connection initialization, the authentication fails, or if the peer detects the end of the connection, it erases immediately the target rule from the router. If the PVC peer fails before the end of the connection, the application that starts PVC, running on the same peer re-launches it. A security issue may occur if the host running PVC fails before the end of the connection, and another host takes its private address. In this case, all the rules related to the host should be removed from the router. We use a distributed architecture to detect node failure and handle firewall rule deletion. Every node of a domain runs a monitoring daemon. These daemons periodically check the rules present on the firewall and ping the corresponding host so that they may delete the rules related to a faulty machine.

TCP hole punching. Widely used for applications such as online gaming and voice over IP, TCP Hole Punching allows connection establishment between two hosts behind NATs in different administration domains.

Both clients establish a connection with the broker that observes the public addresses (given by NAT) and private addresses of the clients and shares this information between the peers. After this exchange, the clients try to connect to each other's NAT devices directly on the translated ports. If NAT devices use the previously created translation states then a direct connection is possible.

The advantage of using this method is that it does not require special privileges or specific network topology information. However, this technique does not work with all type of NATs as their behavior is not standardized.

A novel technique: Traversing-TCP. Traversing TCP (TTCP) is derived from the TCP protocol and it works with firewalls that are not running stateful packet inspection.

It essentially consists in 1) transporting, using the PVC Broker, the initiating TCP packet (SYN) blocked by the firewalls or NAT on the server side and 2) injecting the packet in the server IP stack.

Figure 2 presents in details the TTCP technique. Plain lines show the packets corresponding to the TCP standard. Dashed lines correspond to specific Traversing-TCP messages. A TTCP connection behaves as follows:

Definitions: Server node: S, Broker: B, Client node: C, initializing packet: SYN

1. The peer on S connects to B and waits for new connection demand;

2. C sends SYN to S. It opens the Firewall of C but it is stopped by the Firewall of S;

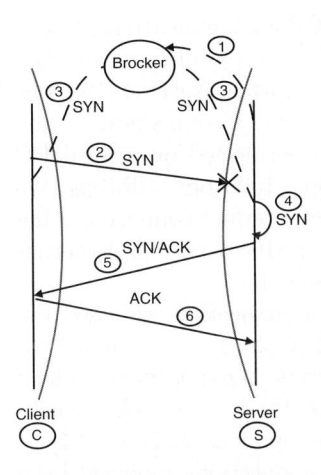

Fig. 2. Traversing-TCP

3. The peer on C sends the SYN packet information to B. B forwards it to the peer on S;

4. The peer on S injects the SYN packet to the IP stack on S;
5. To this SYN packet S replies with a SYN/ACK packet. The SYN/ACK packet opens Firewall on S and is accepted by the Firewall on C (previously opened);
6. The initialization of the TCP connection ends with an ACK packet from C to S: the TCP connection is established.

TTCP works under the following device configurations: 1) The firewall must authorize the outgoing packets and must accept all packets from established connections; 2) Following the [11] classification, TTCP should work with all NAT's, except symmetric NAT (which maps a port to a quadruplet: the internal host-port and external host-port).

Note that RST packets sent as rejection notification are also captured by a PVC client peer and not forwarded to the client IP stack. Following our experience with the DSL-Lab platform (cf. the evaluation section) and related work [18], these requirements fit many professional and domestic configurations.

After connection establishment, the communication can continue following classical TCP operations. The communication between the two peers is direct, bypassing the broker and ensuring high communication performance.

4 Performance Evaluation

In this paper, we focus on the performance evaluation of the whole workflow for establishing a virtual cluster. We measure the overhead of PVC and demonstrate its capabilities by running unmodified MPI applications deployed over a set of firewall protected PCs, connected to the Internet by ADSL connections.

4.1 Experimental Protocol

PVC was designed to have a minimal overhead for TCP communications. In our first experiments, we demonstrate this property with two types of tests: the first one compares network performance with/without PVC, using NetPerf [19]. The second one evaluates the overhead of PVC for establishing a connection.

The evaluation test for network performance was performed on a local PC cluster with three different Ethernet networks: 1Gbps, 100Mbps, 10Mbps. We have used standard PCs with BroadCom TG3 Ethernet interface connected using Netgear EN106 10Mbps Ethernet switch, a Netgear FS105 100Mbps Ethernet switch and a D-LINK D65-1216T Gigabit Ethernet switch.

To evaluate the system overhead of establishing a connection, we used the DSL-Lab [22] platform: a set of resources connected to the Internet by a DSL network. The same platform was used for the second type of experiments, which consisted of the execution of real cluster applications. We ran the NAS benchmarks [20], the MPIPOV program and the scientific application DOT [21] to evaluate the capability of PVC to establish all the connections required by a complex distributed environment like the MPICH runtime environment.

4.2 Evaluation Results

Bandwidth Overhead. Figure 3 presents the bandwidth (in Mbps) of PVC (with/without firewall) and the reference (without PVC), using NetPerf on 10Base-T, 100Base-T and 1000Base-T Ethernet networks. To simulate the firewall on both sides we used Linux Netfilter Iptables.

	1Gbps	100Mbps	10Mbps
Reference (without PVC)	715 (5)	94.0 (0.4)	6.9 (0.1)
PVC without firewall	720 (5)	94.0 (0.4)	6.9 (0.1)
PVC with firewall	717 (3)	94.8 (0.5)	6.9 (0.1)

Fig. 3. Bandwidth of PVC and reference, as measured by NetPerf, on three ethernet networks, in Mbps. Values in parenthesis are standard deviations.

Figure 3 demonstrates that the network rates computed by NetPerf are statistically similar. The difference between the two series of measurements is lower than the standard deviation for all the tests. We can conclude that PVC does not reduce the available bandwidth: once the connection is established, PVC does not interact with the application any more, leaving the network rate unaltered.

Connection overhead. For the establishment of direct connections between the peers, PVC uses the TCP-Traversing, TCP hole punching techniques or the firewall configuration protocol UPnP. To compare these three methods, we evaluated their overhead using a specific test suite. In our test suite, a client makes 1000 consecutive connections to a server and then tears them down.

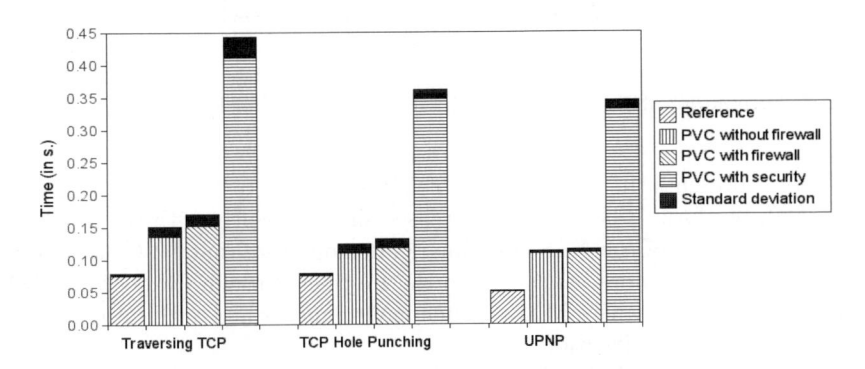

Fig. 4. Overhead of Private Virtual Cluster

Figure 4 shows the mean costs for the TCP connection establishment. In the presence of firewall the overhead of PVC using Traversing-TCP resp. (TCP Hole punching) technique is 76ms (42ms) and 60ms (40ms) without firewall. The overhead observed for PVC with UPnP is the same in both cases and is about 60ms. This overhead encompasses the costs of interception of application connection attempt and communication with the broker.

Security overhead. In the current version of PVC, security is implemented using OpenSSL Crypto library [23]. The first step of the verification (membership to the same virtual cluster) is coupled with the resolution of the virtual name, avoiding the substantial increase of overhead. The second step of the security protocol, which is done after the connection establishment, increases significantly the overhead. Figure 4 shows the mean costs of authentication in PVC during connection establishment.

Since PVC intervenes only at the beginning of the end-to-end communication, the observed overhead remains reasonable in the context of the distributed applications.

Running MPI applications. We successfully ran several of the NAS benchmarks class A (EP, FT, CG and BT) on the PVC architecture. However, due to the network performance between the nodes of the DSL platform, FT, CG and BT do not scale with the number of nodes. Only EP with its low communication to computation ratio is scalable.

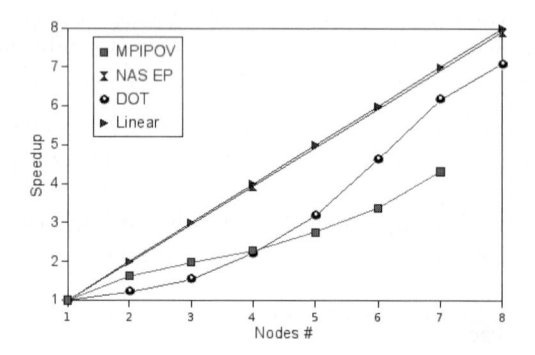

Fig. 5. Speedup of MPI applications with PVC over a set of DSL connected nodes

Figure 5 presents the performance of EP according to the number of nodes in the DSL platform. The speedup increases almost linearly with the number of nodes. The results of the NAS benchmarks demonstrate that PVC successfully transforms a set of nodes connected to the Internet through Firewall and NAT into a virtual cluster where MPI runtime environments and applications can be executed without modification.

The purpose of the DOT [21] program is to compute electrostatic potential energy between charged molecules. It operates in a master/slave mode. During the computation, the amount of data communication is low, but at the beginning and at the end, some large arrays must be communicated. Figure 5 presents the speedup for the computation of the example provided with the DOT distribution, using from one to eight workers. The master is running on the first node.

The MPIPOV test measures the execution time for the computation of a graphical rendering application parallelized with MPI. MPIPOV uses a master-worker algorithm. Compared to the NAS EP, MPIPOV requires more communications of image rendering parameters and results. We perform the test by splitting the image in 32 sub-images. Figure 5 presents the speedup for the computation of the same image using from one to seven workers. The master is running on a separated node.

Over all applications, the speedup evolves in a non-linear way. Obviously, the scalability of the MPI application performance on DSL networks depends on the communication to computation ratio of the application and the individual performance of the heterogeneous platform components. We did not tune the application in order to improve the performance since the purpose of the experiments is only to demonstrate the capability of PVC to run unmodified MPI runtime environments and unmodified, non-trivial MPI applications. However, even without tuning, the test demonstrates that the ADSL platform can provide significant speedups for some non-trivial MPI applications.

5 Conclusion

In this paper, we have presented and evaluated the performance of a lightweight middleware called PVC (Private Virtual Cluster) designed to dynamically establish virtual clusters over resources connected by the Internet and protected by firewalls and NAT. PVC derives from a mix of Grid, P2P and VPN concepts. It features three main properties required for "Instant Grids": 1) a security model that does not reduce the security level of the domains and resources to connect, 2) the capability to run cluster applications and runtime environments without modification, and 3) negligible communication overhead.

PVC is itself a distributed system running as coordinated peer daemons executed on volunteer participants. Its architecture is modular and uses a set of adaptable modules for its main functions: coordination, security, connectivity, virtualization, interposition. Modules can be extended or modified to fit with the target environment (e.g. interposition by virtual network interface or by shared libraries, firewall-traversing protocols based on UPnP, TCP hole punching or other.). We have detailed two important mechanisms: the security model and an original traversing technique called "Traversing TCP". TTCP allows establishing direct connections between resources protected in different administration domains, except if the communicating resources are protected by a firewall running state-full packet inspection.

Our performance evaluation demonstrates a moderate overhead (60 ms) for the connection establishment, and a negligible bandwidth and latency reduction compared to standard TCP communication. By establishing a virtual cluster, at the IP level, with negligible communication overhead, PVC can be used to deploy and run unmodified cluster applications and runtime environments. We demonstrate this capability by running the MPI version of the NAS benchmarks and the POV-Ray program on a set of PCs connected to the Internet by protected DSL connections. Altogether, PVC features a set of characteristics allowing non-OS and network specialists to deploy and run existing cluster applications over multiple administration domains, with minimal performance overhead.

References

[1] Ian Foster and Carl Kesselman. *The Grid 2: Blueprint for a New Computing Infrastructure*. Morgan Kaufmann Publishers Inc., San Francisco, CA, USA, 2003.
[2] M. Humphrey et all. An early evaluation of WSRF and WS-notification via WSRF.net. In *Proceedings of the 5th IEEE/ACMInternational Workshop on Grid Computing*, pages 172-181, Washington, DC, USA, 2004. IEEE Computer Society.
[3] S. Tuecke, K. Czajkowski, and I. Foster. Open Grid Services Infrastructure (OGSI) version 1.0. *Global Grid Forum*, 2003.
[4] Li Gong. JXTA: A network programming environment.*IEEE IC*, 5(3):88-95, 2001.
[5] M. Krasnyansky. Virtual tunnels over tcp/ip networks.*http://vtun.sourceforge.net/*.
[6] R. Thayer, N. Doraswamy, and R. Glenn. Rfc 2411 - ip security document roadmap. USA, 1998. RFC Editor.
[7] A. Denis et all. Wide-area communication for grids: An integrated solution to connectivity, performance and security problems. In *Proceedings of the 13th IEEE International Symposium on High Performance Distributed Computing*, pages 97-106, Washington, DC, USA, 2004. IEEE Computer Society.
[8] S. Son, B. Allcock, and M. Livny. CODO: Firewall traversal by cooperative on-demand opening. In *Proceedings of the 14th IEEE International Symposium on High Performance Distributed Computing*, Washington, DC, USA, 2005. IEEE CS.
[9] P. Srisuresh, B. Ford and D. Kegel. Peer-to-peer communication across NATs. *USENIX Annual Technical Conference*, 2005.
[10] http://www.upnp.org/standardizeddcps/.
[11] C. Huitema J. Rosenberg, J.Weinberger and R.Mahy. Rfc 3489 - STUN - simple traversal of UDP through NATs. USA, March 2003. RFC Editor.
[12] P. Ruth, P. McGachey, X. Jiang, and D. Xu. VioCluster: Virtualization for dynamic computational domains. *IEEE IC on Cluster Computing (Cluster 2005)*, 2005.
[13] X. Jiang and D. Xu. Violin: Virtual internetworking on overlay infrastructure. Technical report, Purdue University, 2003.
[14] J. Chase et all. Dynamic virtual clusters in a grid site manager. *The 12th International Symposium on High Performance Distributed Computing)*, 2003.
[15] K. Appleby et all. Oceano-SLA based management of a computing utility. *In Proc. 7th IFIP/IEEE International Symposium on Integrated Network Management*, 2001.
[16] W. Townsley, A. Valencia, A. Rubens, G. Pall, G. Zorn, and B. Palter. Layer two tunneling protocol l2tp. USA, August 1999. RFC Editor.
[17] J. Reynolds and J. Postel. Rfc 1340 - assigned numbers. USA, 1992. RFC Editor.
[18] A. Wool. A quantitative study of firewall configuration errors. *IEEE Computer*, volume 37, number 6, pages 62-67, 2004.
[19] R. Jones. Netperf: http://netperf.org/, 1999.
[20] D. H. Bailey et all. The NAS parallel benchmarks: summary and preliminary results. In *Proceedings of the 1991 ACM/IEEE conference on Supercomputing*, pages 158-165, NY, USA, 1991. ACM Press.
[21] LF Ten Eyck, J Mandell, VA Roberts, and ME Pique. Surveying molecular interactions with DOT. *Proc. ACM/IEEE SC 1995 Conference*, 1995.
[22] http://www.lri.fr/ rezmerit/dsllab/
[23] http://www.openssl.org/

Reducing Communication Overhead and Page Faults in SDSM Platforms

Artemis A. Christopoulou[1] and Eleftherios D. Polychronopoulos[1]

High Performance Information Systems Laboratory
Computer Engineering & Informatics Department
University of Patras, 26500 Rio, Greece
{aac, edp}@hpclab.ceid.upatras.gr

Abstract. In this paper we present a new dynamic, cache coherence protocol for Software Distributed Shared Memory (SDSM) systems that adopt the scope-consistency model[7]. We initially outline our basic protocol, called Reduced Message Protocol (RMP), and then propose two enhancements: the Multiple Home RMP (RMP-MH) and the Lock Migration RMP (RMP-LM). The experimentation we conducted with the proposed protocols, exhibits significant improvements by reducing two of the major latency factors in SDSM platforms: the total communication messages and the overall number of page faults. To demonstrate the efficiency and the effectiveness of the RMP protocols, we used SPLASH as well as synthetic application benchmarks.

Keywords: Cache Coherence Protocols, Memory Consistency Models, Software DSM Systems, Clusters, Grids.

1 Introduction

The advances of the last two decades in software environments for distributed and cluster computing, along with the improvement in the networking technology, have brought clusters in the proscenium of today's massive multiprocessor systems. When it comes to most enterprize and IT applications, cluster computing of today dominates over tightly coupled multiprocessor systems or the proprietary supercomputer designs of the previous decade. On the other hand, programming clusters of computational nodes, in order to take advantage of parallel execution, is more complex than programming for shared memory systems (the default model supported by multiprocessors). The communication cost of message passing implementations has fallen drastically over the years, but remains far higher than shared memory models, especially in the case of fine grain parallel execution.

Our research builds upon previous approaches which combine the convenience and low cost of clusters with the programming simplicity of shared memory systems, hiding away the distributed architecture via efficient communication libraries. These libraries provide for the transparent integration of message passing in a shared memory model as seen by the applications or the programmer. SDSM models, have been the subject of significant research in the past two decades and constitute the underlying framework for our research work.

In this paper we present a new dynamic Reduced Message Protocol (RMP) for Software Distributed Shared Memory systems (SDSM) which adopts the scope-consistency model[7]. Our main objective is the improvement of the SDSMs' cache coherence protocol, enabling them to function in Wide Area Networks as well as improving their performance in small to medium-sized clusters. Our ultimate goal is to incorporate the proposed cache coherence protocol in wide area clusters as well as define a computational platform in the Grid for parallel processing, based on new SDSM platforms with advanced features in communication and computation mechanisms[13]. Recently there is active research interest in this area, namely, using SDSMs platforms for investigating and testing new methods for Grids[11][12].

Software DSMs are typically categorized into write-invalidate and write-update, on the basis of the cache coherence protocol used to inform the processors for memory page modifications. In write-update protocols the modifications of a page are sent to the processors and the page copies are updated, while in write-invalidate protocols only write-notices for a modified page are sent and the page copies are invalidated. Several protocols have been proposed which adapt between write-invalidate and write-update. [4], [5] and [6] are some of them. In this paper, we propose a new adaptive cache coherence protocol, which was implemented in the Software DSM JiaJia[1]. Our protocol exploits the characteristics of the Scope consistency model[7] used by Jiajia, in order to improve the system's performance.

The rest of the paper is organized as follows. Section 2 describes the new protocol, section 3 presents the experimental evaluation, section 4 describes related work. Finally, the conclusions of the paper is drawn in Section 5.

2 Cache Coherence Protocol

The proposed protocol is based on the JiaJia protocol and specifically in it's write-vector version [1]. The main functions of JiaJia and our protocol have been analyzed in [14]. The protocol's objective is the reduction of page faults inside critical sections and the reduction of the total sent messages. This objective is achieved by piggybacking to the acq-grant message the modifications of the pages, expected to be used by the acquirer. Since the information is sent in existing messages, the number of messages is greatly reduced, although the total amount of transferred bytes remains the same. This provides for a significant benefit, since a great part of communication burden is due to message initialization, a cost greatly reduced when the same bytes are sent in fewer messages. The pages which are piggybacked in the lock grant message are only the pages which have the same home as the lock, since only these pages are available to the processor that sends the lock-grant message. Thus, when a processor acquires a lock, it receives within the acq-grant message updates of modified pages instead of simply write-notices. Consequently, during the next critical section

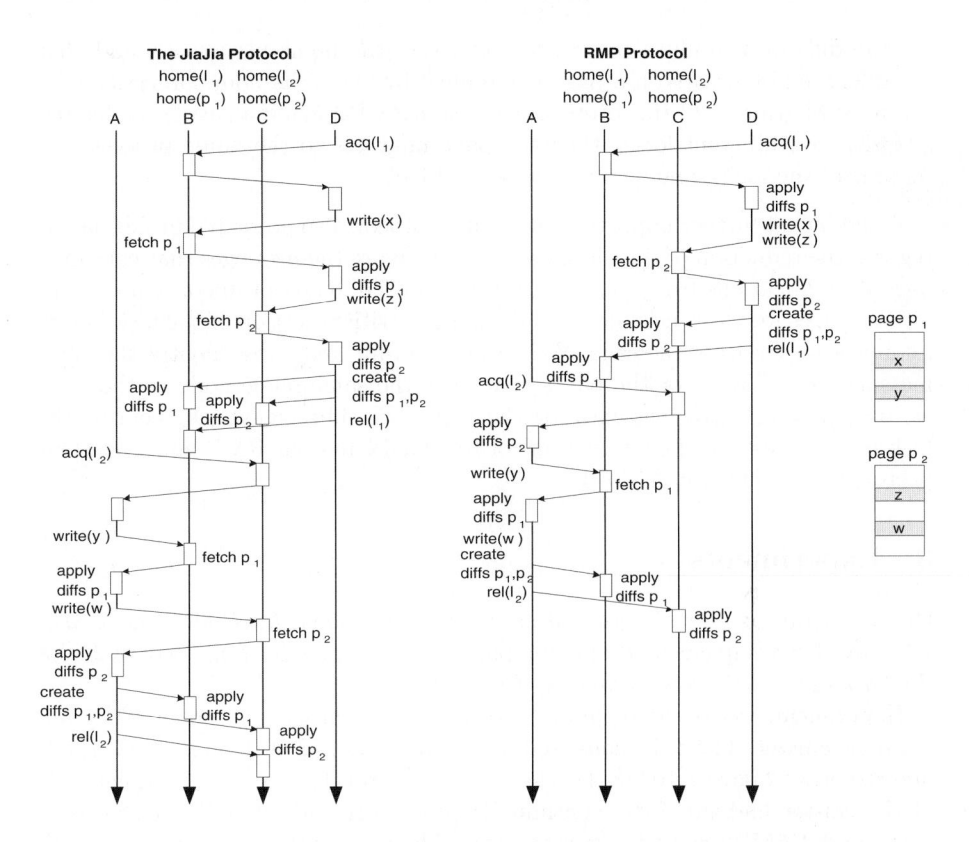

Fig. 1. JiaJia - RMP Protocol

requests for the corresponding pages will be facilitated by the local copies instead of resulting in page faults, minimizing the exchange of messages with the page owner. In order to piggyback to a lock grant message as many pages as possible, we perform page migrations as follows. When a locks home processor receives write notices for the lock, it records the pages that were modified. The pages that are frequently modified during critical sections of one lock migrate to the locks home processor.

There is a small chance that a page contains two different variables which are protected by different locks with different homes. In this case the question that comes up is to which processor the page should migrate. Our protocol follows a greedy approach according to which the page migrates to the first processor that asks for it and cannot migrate to any other processor after that. In addition, we have implemented two alternate approaches to the above problem: lock-migration and multi-home pages.

– Multi-home Pages: In this protocol variation(RMP-MH), a page is allowed to have more than one home, so that it can migrate to the homes of multiple locks associated with it. The additional actions taken in this case are that

the diffs of a modified page are sent to all the home processors, and that diffs are also created if a page is modified by one of its home processors.
– Lock Migration: In this protocol variation(RMP-LM), if a page is associated with two different locks, then the page migrates to the home processor of one of the locks, and so does the second lock.

A significant further improvement of our protocol, also presented in this paper, regards the reduction of the diff messages, which are the messages that contain a page's modifications from some node. In the initial JiaJia protocol, a node sent diff messages for the pages modified during the critical section of the lock before issuing a lock release message. We reduce the diff and subsequently the total messages as follows. For the pages that have the same home node as the lock, we do not send extra messages for the page diffs, but piggyback them to the lock release message. The operations of the JiaJia and our RMP and variations protocols, are shown in figure 1.

3 Experiments

Our main protocols *RMP* as well as its two variations, *RMP-MH* and *RMP-LM* have been implemented and compared against the initial protocol of JiaJia *JiaJia* and it's write vector version *JiaJiaWV*.

Experiments were carried out on two different systems, a 4-processor SMP and a 4-node cluster. The SMP consisted of four processors 2 of them having 512KB and the other 2 having 1024KB cache and 512MB total main memory. Each node of the cluster had two Intel Pentium III processors with 256KB per-processor cache and 256MB per-node main memory. The nodes were interconnected with a 1000Mbps Ethernet network. In all systems, the operating system used was Linux and the application binaries were created with the gcc compiler.

Our protocols were evaluated using four applications, Water and Raytrace from the SPLASH suite, the TSP problem from the JIAJIA SDSM distribution and one of our synthetic benchmarks.

Water simulates forces between different molecules. It uses an array of data structures, each corresponding to a molecule. The array is statically divided into equal parts, each of which is assigned to a processor. Processors use locks to protect the update of force values relating to the molecules. Barriers are used to ensure that all processes perform calculations corresponding to the same time step, as well as to guarantee global memory consistency at the beginning of each step.

Raytrace renders a three-dimensional scene using ray-tracing. A hierarchical uniform grid(similar to an octree) is used to represent the scene, and early ray termination and antialiasing are implemented. A ray is traced through each pixel in the image plane, and reflects in unpredictable ways of the objects it strikes. Each contact generates multiple rays, and the recursion results in a ray tree per pixel. The image plain is partitioned among processors in contiguous blocks of pixel groups, and distributed task queues, and the primitives that describe the scene. In this application the data access patterns are highly unpredictable.

Table 1. Protocols' communication variables

Application	Variable	JiaJia	JiaJiaWV	RMP	RMP-LM	RMP-MH
Synthetic	Total Messages	109458	109458	5474	5474	5474
	Messages in bytes	399210264	21139224	20552784	20552784	20552784
	Getp Requests	48120	48120	700	700	700
	Diff Messages	4800	4800	42	42	42
TSP	Total Messages	11399	11341	4212	2527	3921
	Messages in bytes	28525052	5422928	5956708	5904940	5771676
	Getp Requests	3394	3355	730	312	273
	Diff Messages	1503	1510	513	97	838
Water	Total Messages	1986	2000	1916	1921	2058
	Messages in bytes	4186660	2402076	2453024	2460712	2613048
	Getp Requests	426	433	380	388	233
	Diff Messages	189	189	193	201	406
Raytrace	Total Messages	22567	22414	12329	12382	12280
	Messages in bytes	33803804	12008220	11785268	11787940	11782188
	Getp Requests	3993	4003	1199	1199	1200
	Diff Messages	3827	3784	1529	1536	1523

TSP solves the travelling salesman problem using a branch and bound algorithm. The major shared data structures of TSP include a pool of partially evaluated tours, a priority queue containing pointers to tours in the pool, a stack of pointers to unused tour elements in the pool, and the current shortest path. Processors evaluate the partial paths successively and alternately until the shortest path is found. Locks are used to ensure exclusive accesses to shared objects.

The last application is a synthetic application which we used to stress the proposed protocol and evaluate its maximum performance. In this application there are four locks each of which protects forty variables in forty different memory pages. For each lock, each processor modifies the forty variables belonging to the lock and this procedure is repeated one hundred times.

Table 2. Execution Time in SMP

Programming Model	Synthetic	TSP	Water	Raytrace
Posix Threads	0,04	14,96	2,92	16
JiaJia	18,8	12,26	5,56	63

In the SMP each application was programmed and executed twice, once using Posix Threads and once using the SDSM JiaJia. These experiments were made in order to show the overhead of an SDSM and the communication cost among its nodes. In table 2 we see the total execution time for each application for each case.

The results show that all the applications besides TSP take more than double time to execute using JiaJia compared to Posix threads. This clearly indicates that the overhead of an SDSM is significant, and therefore there is great need to make it as efficient as possible.

The main way to improve the SDSM's efficiency was by reducing the total amount of messages sent among the SDSM nodes. Our experiments prove that the cost in time of sending a message is given by the following type:

$$t(size) = t_{init} + t_{send} * size, \qquad (1)$$

which means that the cost of sending a message is analogous to its size plus an initialization cost. In the SMP system it was measured that t_{init} is 228,8007 and t_{send} is 0,0235, while in the cluster t_{init} is 121,1614 and t_{send} is 0,0337. All time values are measured in microseconds.

Since the initialization cost of sending a message is that big, it is expected that if we send the same amount of bytes in a smaller number of messages, we can achieve a performance improvement. Our new protocol was designed to achieve this goal.

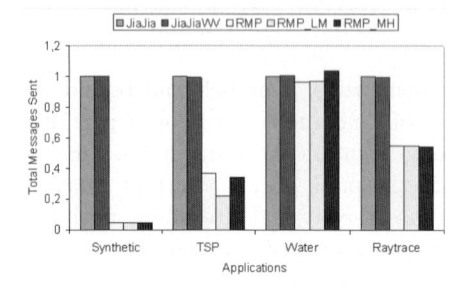

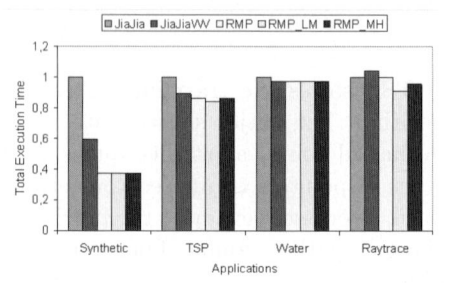

Fig. 2. Messages sent

In Figure 2 we can see for each application and for each protocol the total number of messages and it's total execution time, while in table 1 we can more details about see the messages' reduction.

All the results shown have been normalized with the *JiaJia* results.

Checking the number of sent messages, we see a great reduction in our synthetic application. In this application forty pages are modified during a critical section and these pages must be sent to the other nodes before they are accessed by them. In our protocol modifications of these pages are piggybacked to the lock release and lock grant messages and as a consequence the total number of sent messages is greatly reduced. Tsp uses locks as its synchronization method. Every key used protects a lot of pages and since modifications of the pages are sent with the lock release and lock grant messages, the total number of sent messages of the application is significantly reduced. In Water both locks and barriers are used for the synchronization, but the most page modifications occur in critical sections enclosed by barriers. Consequently the number of total sent messages is little affected by our protocol. Contrary to Water, Raytrace uses only locks for the synchronization which produces a reduction in the sent messages as in the other two applications.

The reduction of sent messages in our protocols lead also to a reduction of the execution time. In our synthetic application the reduction reaches the 63% compared to the initial protocol of JiaJia and the 37% compared to JiaJia's write vector version. In Raytrace the reduction is 9% and 13%, in TSP 16% and 6% while Water shows a little reduction of the total execution time.

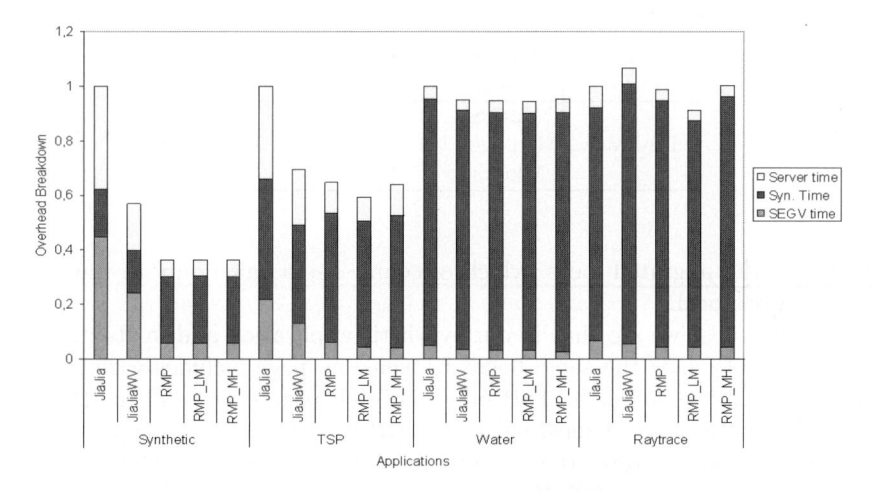

Fig. 3. Overhead breakdown

Another performance metric for the comparison of the protocols is their overhead, which consists of the synchronization time, the SEGV time and the server time. The synchronization time is the time spent for barriers, locks and unlocks, the SEGV time is the time spent due to page faults and the server time is the time a processor spends to serve other processors requests(i.e. lock or page requests). The protocol overhead breakdown is shown figure 3.

In general, we can see that the SEGV as well as the server time is reduced, while the synchronization time is increased. This happens because there are less page faults which naturally reduces the SEGV time. Subsequently a processor has less get page requests to serve and the server time is also reduced. On the other hand, at a lock request, a processor receives larger messages, since in the lock grant messages, modifications of some pages are piggybacked and as a result the synchronization time is increased.

If we compare the three new protocols, we see very little variations. Actually, the cases in which one of the variations needs to be taken are few. In table 3 we can see in detail for each application how many pages and how many locks migrate. MIpages is the number of pages that migrated to a node, while MOpages is the number of pages that migrated from a node. These numbers are not equal in RMP-MH, and for this reason they are given separately. Last, Mlocks is the

Table 3. Protocols' migration variables

Applications	Variable	RMP	RMP-LM	RMP-MH
Synthetic	MIpages	120	120	120
	MOpages	120	120	120
	Mlocks	-	0	-
TSP	MIpages	35	35	36
	MOpages	35	35	35
	Mlocks	-	1	-
Water	MIpages	7	6	12
	MOpages	7	6	6
	Mlocks	-	5	-
Raytrace	MIpages	0	0	0
	MOpages	0	0	0
	Mlocks	-	0	-

number of the migrated locks, which of course has a value only in the case of RMP-LM protocol.

By the results, we see that in our synthetic application and in Raytrace the three protocols show the same behavior, since no page includes variables protected by different locks. In TSP only one such page exist, and in RMP-MH this pages obtains two homes, while in RMP-LM one of the locks migrates to the page's home. In Water there are quite a few pages with variables protected by different locks. In RMP-MH six pages obtain two homes, while in RMP-LM, five locks to migrate. In this application we conclude that the best of the three variations is the RMP-LM. It manages to include more pages in the lock operations without any extra burden as in RMP-MH. For this reason, RMP-LM has a larger speedup compared to the JiaJia protocol. Although the benefits of the variations are not quite clear, we believe that their benefits will be greater when applications are run in a different system when nodes are interconnected with a slower network, as in Grids.

4 RelatedWork

Since the introduction of Ivy [10], the first Software DSM, many techniques have been proposed to improve SDSM performance. Here we will focus on adaptive techniques between write invalidate, write update, and prefetching techniques.

In [3], the proposed protocol tries to predict in various ways for each lock it's next acquirer(s). At a lock release, diffs of pages modified during the last critical section are sent to the processors that belong to the set of the locks next acquirers.

A dynamic adaptation between write invalidation and write update is described in [6]. Initially, for one page, the protocol switches from write invalidate to write update if the most collaborating processes are in need of that page and the page faults exceeds an experimental threshold. In [4] the pages are categorized in migratory, producer/consumer and falsely-shared. Adaptation is based on the category in which each page belongs. Migratory and producer/consumer pages are managed in a single-writer mode and may be updated, while falsely-shared pages are managed in multiple-writer mode and under invalidated protocol.

Three adaptive techniques are proposed in [5]: adaptation between single and multiple writer, dynamic page aggregation and adaptation between write invalidation and write update. The adaptive protocol between the write invalidate and write update, updates the pages that the processor is expected to access and invalidates the others but there is a limit so that no more than eight pages can be updated. For barrier based applications, each processor p records for a particular page from which processors it has received page requests and sends updates to these processor and invalidates to the others. For lock based applications, the pages that are protected by one lock are recorded and updates are sent for these pages while invalidates for the others.

Finally, in [9] a prefetching technique is proposed, in which the data is invalidated after a repetitive synchronization pattern and is prefetched at proper times.

5 Conclusion

In this paper we introduced the RMP cache coherence protocol for SDSM systems. This protocol in some cases adopts write-invalidate and in some cases write-update method. The difference with previous adaptive protocols is that in the new protocol the updates of the pages is done without sending any extra messages, but rather by piggybacking the information in the existing lock grant messages. Since the updates are successful, we achieve sending in general the same amount of bytes but in significantly less number of messages. We must also note that the performance improvements are even bigger because the messages are sent in the beginning of a critical section, while in other previous protocols, such as in JiaJia, there would be many more page faults and consequently, messages sent during critical sections. Apart from page requests, we further reduce messages owed to page diffs, which are piggybacked to lock release messages.

In order to send as many pages as possible in grant lock messages, we associate a lock with the pages modified during a critical section of the lock and if a page is associated with only one lock, it migrates to the lock's home. This happens in all three proposed protocols. However, in order to include pages that contain variables protected by different locks, we implemented two additional variations of RMP. In the first variation, we allowed a page to have more than one homes, and in the second, we permit, if necessary, a lock to migrate. The second variation achieves better results than the first since it has no extra burden like sending modifications of pages to it's multiple homes. On the other hand, it sends updates of more pages and achieves even less page faults. The significant reduction of messages in all three proposed protocols, results respectively in major reduction in the applications total execution time.

References

1. W. Shi, PhD thesis, Institute of Computing Technology, Chinese Academy of Sciences, 1999.
2. W. Hu, W. Shi and Z. Tang, *Optimizing Home-based Software DSM Protocols*, Cluster Computing: The Journal of Networks, Software and Applications, Baltzer Science Publishers, 2001.
3. C. B. Seidel, R. Bianchini, and C. L. Amorim, *The Affinity Entry Consistency Protocol*, Proceedings of the 1997 International Conference on Parallel Processing, August 1997.
4. L. Whately, R. Pinto, M. Rangarajan, L. Iftode, R. Bianchini, and C. L. Amorim, *Adaptive Techniques for Home-Based Software DSMs*, Proceedings of the 13th Symposium on Computer Architecture and High-Performance Computing, September 2001.
5. C. Amza, A.L. Cox, S. Dwarkadas, K. Rajamani, and W. Zwaenepoel, *Adaptive Protocols for Software Distributed Shared Memory*, Proceedings of the IEEE, Special Issue on Distributed Shared Memory, vol.87, no.3, pages 467-475, March 1999.
6. M. Ng and W. Wong, *Adaptive Schemes for Home-based DSM Systems*, Proceedings of the 1st Workshop on Software Distributed Shared Memory, pages. 13-20. June 1999
7. L. Iftode, J. P. Singh, and K. Li, *Scope consistency: A bridge between release consistency and entry consistency*, Proceedings of the 8th ACM Annual Symp. on Parallel Algorithms and Architectures (SPAA'96), pages 277-287, June 1996.
8. H. C. Yun, S. K. Lee, J. Lee, and S. Maeng, *An Efficient Lock Protocol for Home-based Lazy Release Consistency*, Proceedings of Cluster Computing and the Grid, 2001.
9. S. K. Lee, H. C. Yun, J. Lee, and S. Maeng, *Adaptive Prefetching Technique for Shared Virtual Memory*, Proceedings of 3rd International Workshop on Software Distributed Shared Memory System, Brisbane Australia, May 2001.
10. K. Li, *A shared virtual memory system for parallel computing*, Proceedings of the 1988 International Conference on Parallel Processing (ICPP88), pages 94101, 1988.
11. Louis Rilling and Christine Morin, *A Practical Transparent Data Sharing Service for the Grid*, Proceedings Fifth International Workshop on Distributed Shared Memory (DSM 2005), Cardiff, UK, May 2005.
12. G. Antoniu, L. Bouge, and M. Jan, *JuxMem: Weaving together the P2P and DSM paradigms to enable a Grid Datasharing Service*, Kluwer Journal of Supercomputing, 2004.
13. G. Tournabitis, E. Polychronopoulos, *Multithreaded Home-based Lazy Release Consistency for Clusters of SMPs*, Technical Report, HPCLAB-TR-250206, February 2006.
14. A. Christopoulou, E. Polychronopoulos, *A Dynamic Lock Protocol for Scope-Consistency in Software DSM Systems*, Technical Report, HPCLAB-TR-100106, January 2006.

Flexible I/O Support for Reconfigurable Grid Environments

Marc-André Hermanns[1], Rudolf Berrendorf[2], Marcel Birkner[2], and Jan Seidel[2]

[1] Central Institute for Applied Mathematics
Research Centre Jülich, 52425 Jülich, Germany
[2] Department of Computer Science
University of Applied Sciences Bonn-Rhein-Sieg, 53754 St. Augustin, Germany

Abstract. With growing computational power of current supercomputers, scientific computing applications can work on larger problems. The corresponding increase in dataset size is often correlated to an increase in needed storage for the results. Current storage area networks (SANs) balance I/O load on multiple disks using high speed networks, but are integrated on the operating system level, demanding administrative intervention if the usage topology changes. While this is practical for single sites or fairly static grid environments, it is hard to extend to a user defined per-job basis. Reconfigurable grid environments, where computing and storage resources are coupled on a per-job basis, need a more flexible approach for parallel I/O on remote locations.

This paper gives a detailed overview of the abilities of the transparent remote access provided by TUNNELFS, a part of the VIOLA parallel I/O project. We show how TUNNELFS manages flexible and transparent access to remote I/O resources in a reconfigurable grid environment, supporting the definition of the amount and location of persistent storage services on a per-job basis.

1 Introduction

With the enormous increase in computational power of today's high performance computers, scientific applications can exceed old boundaries of problem size and complexity. With the increasing rate at which data can be produced, the question arises as to where to store the data in an efficient way. Result data will normally be too large to be kept in memory, while the application continues processing different data sets. Also, memory-intensive applications often need to swap data to disk, and reread it at a later point in time. With increasing size of the data sets to be loaded into memory and stored on persistent storage devices, I/O can easily become the bottleneck of modern scientific applications. To develop efficient and portable applications, it is therefore essential for the middleware to provide efficient I/O mechanisms for scientific application developers.

Almost 10 years after the specification of MPI-2 [1] in 1996, most currently available MPI implementations support the API for MPI-IO. ROMIO [2] is a publicly available MPI-IO implementation that easily integrates into the MPICH MPI implementation [3]. ROMIO and MPICH are being developed at the Mathematics and Computer Science Department of Argonne National Laboratory. The use of the MPI API for I/O introduces a new layer of abstraction to a process's I/O accesses, increasing portability of the application. ROMIO uses the device abstraction ADIO [4] (Abstract Device Interface for I/O) to define multiple special purpose filesystem devices. In this way the MPI layer can call a generic ADIO function to handle a specific kind of I/O operation, with the ADIO layer choosing the correct device to use.

When using the MPI API for parallel I/O, it is assumed that all processes participating in each I/O call see the same underlying filesystem. In grid environments, where processes are interacting across inter-cluster boundaries with other processes, this is rarely the case and a common filesystem can introduce a high administrative overhead, like common user bases and common security policies. In static environments, where the status of available nodes in the grid is constant over a longer period of time, this is still feasible, as shown by the DEISA project [5], where a distributed GPFS filesystem is deployed over several supercomputing sites. In reconfigurable grids, where computing resources are added to and revoked from the resource pool much more frequently, this approach is a major administrative challenge.

In the VIOLA project [6] several sites with local clusters are connected by a 10 GBit/s dedicated network, while internally using either Gigabit Ethernet or Myrinet networks [6]. The MPI middleware for connecting the cluster sites is MP-MPICH [7]. MP-MPICH is an enhanced version of MPICH, providing a process environment for meta computing. The major advantage of MP-MPICH is the transparent use for different communication devices for inter- and intra-cluster communication. Thus clusters can still use the usually faster special purpose interconnect for intra cluster communication, instead of having to choose the least common denominator of all of the available communication devices. While earlier versions of MP-MPICH realize the inter-cluster connection with transparent router processes, providing a store-and-forward routing for messages between processes of different clusters, recent prototypes provide a secondary device to each process to handle intra- and inter-cluster communication by separate devices [8].

Within the parallel I/O subproject of VIOLA, we develop two ADIO devices for MP-MPICH to support the special needs for efficient parallel I/O in grid environments [9]. In Sect. 2 we introduce the overall design of the TUNNELFS client/server architecture. Section 3 presents possible I/O distribution strategies, used with TUNNELFS. In Sect. 4 we then state some of our results with the early version of the prototype using the VIOLA network. Sect. 5 places our current efforts into the context with other work, involving remote I/O for MPI. Section 6 concludes with a brief summary and future prospects in our work.

2 Design

To support special aspects of I/O access with different filesystem, ROMIO uses the ADIO layer [4] to decouple the implementation of MPI function calls and filesystem specifics. The MPI-IO functions work on driver functions of the ADIO layer that select the correct ADIO device to be used for a specific filesystem (e.g., UFS, NFS, PVFS2). To integrate transparent I/O for MPI applications, we defined two new ADIO devices, TUNNELFS and MEMFS [9]. TUNNELFS provides the service for transparent remote access (see Fig. 1, and MEMFS creates a multi-node shared virtual filesystem accessible with MPI-IO calls. Together they provide easy and efficient access to I/O on distributed memory resources.

All client/server communication regarding remote I/O is handled transparently by the ADIO device for TUNNELFS. The user therefore does not need to deal with explicit communication calls to the server. In MPI, file namespaces are implementation depended, as it may be required to provide additional information about the file, such as the type of the underlying filesystem. ROMIO uses prefixes separated by a colon, to explicitly identify a filesystem type (e.g., `nfs:file.dat`). TUNNELFS and MEMFS use the same prefix scheme for selection of the corresponding ADIO device (e.g., `tunnelfs:file.dat` or `memfs:file.dat`). For TUNNELFS the location of a file is not pinpointed to a specific server but rather to a class of servers belonging to the same *filesystem domain*, additionally coded in the filename. Filesystem domains and their usage are discussed in detail in Sect. 2.4.

2.1 Client/Server Architecture

TUNNELFS uses a parallel client/server architecture. As shown in Fig. 1, the user process, being the I/O client, issues an I/O call, descending through the MPI layers until the ADIO device layer is reached. Instead of calling system functions for disk I/O on the client's node, the TUNNELFS ADIO device transmits the request parameters and possible I/O data to the server, using MPI point-to-point communication. The I/O server receives the request of the client and acts upon it, using local MPI-IO calls. If the request is a read request, the server will issue the call locally and transfer the data to the client. If the request is a write request it will wait for the data to be sent by the client in a second message and then issue a local write call. Communication between clients and servers as well as between multiple servers is handled with point-to-point communication to avoid deadlocks and excessive blocking of server processes waiting for messages of other servers.

To maintain portability, flexibility and efficiency, the TUNNELFS I/O server uses only MPI function calls for communication and file I/O. Request header and data

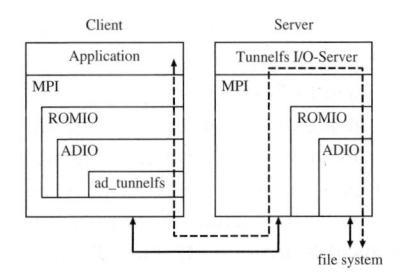

Fig. 1. Layers of the TUNNELFS device and I/O server

buffers are transferred in separate messages, to avoid repacking of buffers before and after transfer. The client/server protocol defines a single master I/O server that is derived from the middleware configuration files, common to all processes started. All client/server and server/server communication is issued upon an implicitly defined communicator, containing all user processes as well as all I/O server processes. As the communication involved in I/O requests is hidden from the user program, the user has no knowledge which server is appropriate for a specific file. Thus, initial requests are always sent to the global master I/O server, which either processes the request locally or delegates it to a different server. In case of request delegation the client is informed of the I/O server now responsible for this file.

2.2 I/O-Transfer-Staging

Though I/O calls usually handle large amounts of data, allowing arbitrary sized blocks for TUNNELFS is not feasible. The TUNNELFS protocol defines a two step data transfer for I/O calls. In a write call the buffer is transferred to the server process first, and then written to disk. Keeping large buffer sizes will result in a serialization of those two operations, maximizing the time needed for the complete operation. As MPI also defines that buffers involved in ongoing MPI operations must not be accessed until the operation involving the buffer has completed, transfers have to be broken down into smaller parts. Therefore the TUNNELFS servers stage data transfers to and from the client with its local I/O calls, to maximize overlapping of those operations.

For write requests, the client sends a request header, followed by a number of messages containing the I/O data. The number of I/O data messages is coded in the request header. After receiving the first I/O data package, the server can start writing the data to storage, while the rest of the I/O buffer is still in transfer. The write call is acknowledged by the server through a reply message. For read requests, the client sends the request header and the server replies directly with the parameters of the issued call, in particular the number of following I/O data packages. After the client processes the server reply, it will start receiving the messages containing the requested I/O data.

2.3 Multiple Distributed Server Support

The first prototype of the TUNNELFS and MEMFS devices supported any number of clients but only a single I/O server to be used with the application. This was partly due to the router concept of earlier versions of MP-MPICH, where avoidance of bottlenecks would have meant introducing a router for every I/O server defined. This would have implied an enormous overhead of additional processes. With the modification of MP-MPICH to support direct process to process inter-cluster communication without interference from use of router processes [8], multiple I/O server support becomes feasible for the MP-MPICH environment. With multiple I/O servers present in the application's MPI environment, I/O load sharing and I/O request delegation, as well as transparent file access becomes possible. The

TUNNELFS client/server protocol was extended to support the desired flexibility and transparency with the use of multiple I/O servers.

2.4 Filesystem Domains

As TUNNELFS is not a filesystem itself, but an interface to different remote filesystems, it has to provide a means for distinguishing those filesystems at runtime. The TUNNELFS I/O servers need information in addition to the filename, what the location of that file in the grid is. As nodes are assigned to the user application by the scheduling system, it is possible but not always practical, to request specific resources of a cluster for execution. With a uniform node architecture, i.e., no special purpose nodes available for reservation, it is not necessary to run the I/O server on one specific node, but rather on any node which has access to the desired filesystem. Within the grid, processes can then be assigned to specific classes, where each class has access to the same filesystem (e.g., the local filesystem on one cluster of the grid). These classes are called *filesystem domains*.

MP-MPICH defines so-called *metahosts* to classify all processes of a single cluster sharing a common internal interconnect. These metahosts comprise the nodes of the global meta computer. As nodes within one cluster usually share a global filesystem, filesystem domains contain all processes of one or more metahosts. The identifiers of a filesystem domain are implicitly defined by the user, as they are derived from the grid configuration defined by the user for this specific job. If the user defines a metahost with name "`metahostA`", clients have remote access to that filesystem via the prefix "`tunnelfs:metahostA:`".

To prevent the clients from governing too much information on data location, the logic for deciding which server is used for opening a file resides on the global master I/O server. Thus the client itself does not process the filename any further than the first separating colon. The TUNNELFS prefix is truncated from the filename and the rest of the filename is sent to the master I/O server for further processing. The master I/O server then decides whether the file is to be opened locally or the open request has to be delegated to a server of a different filesystem domain.

2.5 Distribution Schemes

During job execution user processes and I/O server processes are distributed in the grid. The configuration and placement of user processes can have direct influence on efficient placement of I/O server processes. Additionally I/O server processes have to be placed according to the required filesystem access of the application. Each filesystem domain that needs to be accessed during execution has to define at least one I/O server. During runtime of the application, the server placement should then be taken into account when clients are assigned to a specific server. As a part of the transparent access scheme, the distribution is completely handled by the TUNNELFS servers, with only minimal intervention by the user. The user can define hints in an `MPI_Info` object, referenced on special I/O calls like opening or setting a file view. The server will then use the given information for optimal server assignment.

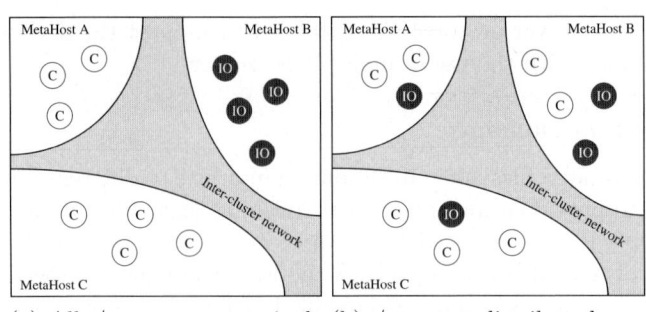

(a) All I/O servers on a single metahost (b) I/O servers distributed over metahosts

Fig. 2. Different placement scenarios for TUNNELFS

The servers are assigned to the clients by several different distribution schemes. The most simple distribution scheme is *single server*, where all clients use one server for a specific file. This is very close to the behavior of our first I/O server prototype, where only a single server was allowed in a job specification. One difference to having only one server in the system is that even though a single file is handled by one server exclusively, other files opened can be handled by other servers, providing a fairly balanced distribution of files over all available I/O servers. Another simple distribution scheme is *balanced global distribution*, where all defined servers are equally involved in sharing the I/O load. This scheme is not available for all types of files yet, as it is relying on a global filesystem that can handle multiple file handles on a single file without interfering with each other. This distribution scheme is used mostly for the MEMFS virtual filesystem, which can handle this efficiently. Figure 2(a) depicts a possible distribution of I/O servers, where all special I/O processes are started on a single metahost which is the only one providing I/O services to the compute grid.

A third distribution that might not result in a balanced distribution is *filesystem domain distribution*, as shown in Fig. 2(b). This is a specialization of the *balanced global distribution*, where clients are only assigned to servers of their filesystem domain. The goal of this distribution is to have an I/O server available via the faster local interconnect of the local cluster. Especially write operations can then be handled more efficiently, as the data can be transferred via local interconnect to a server, where it can be processed without further waiting time for the client. The user can influence the default mapping of clients to servers by providing additional information in an `MPI_Info` object, restricting the mapping to servers of a specific filesystem domain.

3 Distributed I/O Strategies

The main objective in development of the TUNNELFS ADIO device and server is efficient support for remote filesystems. The supported filesystems are primarily dependent on the other ADIO devices present in the ROMIO library. The MEMFS virtual filesystem is designed to be global over all TUNNELFS servers. Additionally, it supports multiple independent file handles on the same file. Colliding accesses are handled within MEMFS itself, so TUNNELFS can have multiple distributed handles open at the same time, without the need of collective operations on the file among all servers. Filesystems on the operating system level usually do not provide this service to an application, therefore the TUNNELFS servers have to use different strategies to support I/O load sharing among the servers when working on persistent filesystems.

As stated in Sect. 2.4, the TUNNELFS servers assume that every process within a filesystem domain has equal access to the filesystems of that metahost. Clients as well as servers are classified into these filesystem domains. Upon first registration during the initialization of MPI, the clients provide information about their filesystem domain to the main server. The main server can then use this information to compute an optimal distribution of clients to servers. For example, clients can be assigned to the nearest server. Having a server in reach of intra-cluster communication can speed up data transfer, as it is used in routed and cached I/O with intermediary I/O servers. With this filesystem domain distribution, the TUNNELFS servers support two sorts of I/O behavior, in regard to the configured views on the file.

3.1 Routed I/O

Routed I/O is a transfer of the router concept of MP-MPICH [7] from routing processes to the I/O server processes. Clients are assigned to I/O servers in a distributed fashion in a way that they are local to their metahost and therefore reachable via the local, typically faster internal network. After completion of the data transfer, the intermediary servers reply to the clients and then send the data to the responsible I/O server for persistent storing of the data. The clients can then continue with computation much earlier than having to wait for explicit storage of the data on the remote I/O server. The intermediary server uses the same file view the client would have used to write to the file, therefore no additional offset and file view calculation has to be done, and the server simply reuses the information. While write access is being accelerated by this strategy, read access is degraded, because of the store-and-forward data transfer and the additional hop of the intermediary server. Therefore this I/O strategy is not the best for all kinds of file accesses and is currently only used if `MPI_MODE_WRONLY` is specified upon opening the file. The clear benefit is the usability for arbitrary file views in comparison to cached I/O discussed in the next section. Figure 3(a) shows I/O message interchange between clients, intermediary servers and the responsible file server. This I/O strategy is still reasonable for scientific applications, as often the major I/O load is created by writing intermediary and result data.

3.2 Cached I/O

Cached I/O goes further and speeds up the read access, by keeping a local cache file of the clients view. This will imply coherence problems, if several clients have

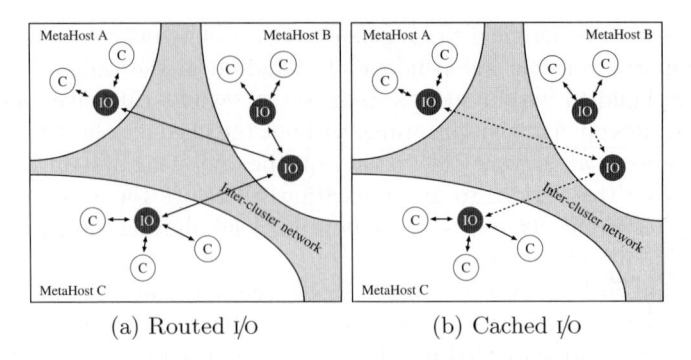

Fig. 3. Different TUNNELFS I/O strategies

access to the same region in a file. To avoid complicated handling of replicated data, cached I/O is restricted to file views with disjoint access patterns.

When a message passing library like MPI is used, the application problem domain is often already restricted to algorithms that do not need random access on global data, as this can be a big challenge to provide efficiently without hardware support of some kind. Thus disjoint access to global data is an access pattern very common in scientific applications using MPI. In Fig. 3(b) the message exchange for cached I/O is shown. It is very similar to the routed I/O scenario, but now the intermediary servers only synchronize with the responsible file server on special I/O requests, such as `MPI_File_close`, `MPI_File_set_view` and `MPI_File_sync`. This postponed synchronization, in regard to routed I/O, is illustrated by the dotted lines between the server processes.

4 Status

As the implementation of distributed TUNNELFS I/O servers has been completed recently, we cannot state solid performance data for it yet. With multiple servers balancing I/O load we expect to significantly exceed the already promising performance data of the single server TUNNELFS environment [9] presented in Fig 4. In this single server test, we are able to deliver I/O rates that are restricted by the network card of the server (1 GigE) rather than the processing on the

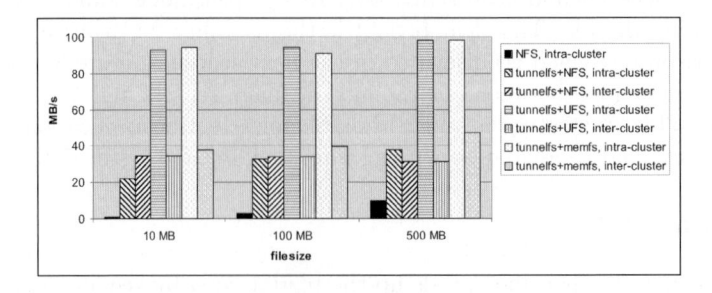

Fig. 4. Bandwidth with 12 I/O client processes and 1 I/O server on a shared file

server. The data was obtained using an MPI benchmark program that measured 8 different MPI I/O operations (individual/shared fp, explicit offset, collective/non-collective). Results are given as average bandwidth numbers over these operations. Two clusters were used, connected with a dedicated 100 km distance 10 GigE fibre network and 1 GigE network cards in each cluster node connected to a cluster switch. All cluster nodes running TUNNELFS I/O server processes are 4-way SMP nodes with 10.000 rpm SCSI disks accessed locally on a node by a UFS device, and an NFS-mounted RAID-5 based shared filesystem hosted by a cluster file server.

5 Related Work

Data access to remote locations from within MPI applications has been previously addressed with the RIO device, where the communication for I/O is directly bound to TCP/IP socket communication. The work was continued with RFS [10], focusing on client side caching to improve I/O performance. To the best of our knowledge, efficient remote data distribution using MPI datatypes for optimal placement on multiple servers has not been extensively addressed yet.

6 Conclusion and Future Work

We defined and implemented a client/server architecture for flexible, transparent and efficient I/O in grid environments. With our approach to distributed I/O server support, we have designed a very flexible infrastructure that can be adapted to applications' I/O needs on a per-job level. Users can define the number of I/O processes supporting their application as well as their placement in the grid. This enables substantial influence of the I/O infrastructure presented to applications by external configuration prior to runtime. The lifetime of TUNNELFS I/O server processes is limited to the scope of an applications execution time and are integrated into the application's MPI environment by the middleware. We introduce the term of filesystem domains to assist the user in a flexible specification for data location. TUNNELFS uses a client/server architecture, using MPI communication and I/O calls to maintain a maximum of portability and flexibility, while allowing the use of special purpose MPI devices for inter process communication.

With completion of TUNNELFS and MEMFS prototypes, further efforts will be aimed at an improved performance and robustness of the implementation, while using it with simulation applications in the VIOLA testbed.

Acknowledgments

This work was supported within the VIOLA project by the German Ministry of Education and Research under contract number FKZ 01AK605L. Thanks to Martin Pöppe, Boris Bierbaum and Carsten Clauss of the University of Technology Aachen for their very close support on MP-MPICH.

References

1. Message Passing Interface Forum (MPIF): MPI-2: Extensions to the Message-Passing Interface. University of Tennessee, Knoxville (1996) http://www.mpi-forum.org/docs/mpi-20-html/mpi2-report.html.
2. Thakur, R., Ross, R., Lusk, E., Gropp, W.: Users Guide for ROMIO: A High-Performance, Portable MPI-IO Implementation. Technical Report ANL/MCS-TM-234, Mathematics and Computer Science Division, Argonne National Laboratory (2004) http://www-unix.mcs.anl.gov/romio/.
3. Gropp, W., Lusk, E., Doss, N., Skjellum, A.: A High-Performance, Portable Implementation of the MPI Message Passing Interface Standard. Technical report, Mathematics and Computer Science Division - Argonne National Laboratory (1996) http://www-unix.mcs.anl.gov/mpi/mpich/.
4. Thakur, R., Gropp, W., Lusk, E.: An Abstract-Device Interface for Implementing Portable Parallel-I/O Interfaces. In: Proceedings of the 6th Symposium on the Frontiers of Massively Parallel Computation. (1996) 180–187
5. The DEISA Project Group: Distributed European Infrastructure for Supercomputing Applications (2005) http://www.deisa.org/.
6. The VIOLA Project Group: Vertically Integrated Optical testbed for Large scale Applications (2005) http://www.viola-testbed.de/.
7. Pöppe, M., Schuch, S., Bemmerl, T.: A Message Passing Interface Library for Inhomogeneous Coupled Clusters. In: Proceedings of the IEEE International Parallel and Distributed Processing Symposium (IPDPS 2003), Workshop on Communication Architecture for Clusters (CAC 2003), Nice, France (2003)
8. Bierbaum, B.: Implementation of a Multi Device Architecture for MetaMPICH. Master's thesis, Chair for Operating Systems, RWTH Aachen (2005) http://www.lfbs.rwth-aachen.de/~boris/diplomarbeit.pdf (In German).
9. Berrendorf, R., Hermanns, M.A., Seidel, J.: Remote Parallel I/O in Grid Environments. In: Proc. of the Sixth Conference on Parallel Processing and Applied Mathematics (PPAM). Lecture Notes in Computer Science, Poznan, Poland, Springer (2005) to appear.
10. Lee, J., Ma, X., Ross, R., Thakur, R., Winslett, M.: RFS: Efficient and Flexible Remote File Access for MPI-IO. In: Proceedings of the International Conference on Cluster Computing. (2004)

Storage Exchange: A Global Trading Platform for Storage Services

Martin Placek and Rajkumar Buyya

Grid Computing and Distributed Systems Laboratory and
NICTA Victoria Laboratory
Department of Computer Science and Software Engineering
The University of Melbourne, Australia
{mplac, raj}@csse.unimelb.edu.au

Abstract. The Storage Exchange (SX) is a new platform allowing storage to be treated as a tradeable resource. Organisations with varying storage requirements can use the SX platform to trade and exchange storage services. Organisations have the ability to federate their storage, be-it dedicated or scavenged and advertise it to a global storage market. In this paper we discuss the high level architecture employed by our platform and investigate a sealed Double Auction market model. We implement and experiment the following clearing algorithms: maximise surplus, optimise utilisation and an efficient combination of both.

1 Introduction

The Internet has proven to be a source of many exciting wide-area distributed computing applications, enabling its users to share and exchange resources across geographic boundaries. It is in this context we introduce the Storage Exchange (SX). Consumers and providers are able to submit their storage requirements and services along with budgetary constraints to the SX, which in turn employs a market model to determine successful trades. The motivation and long term goal behind our research and development of the SX platform has been to achieve Autonomic [1] management of storage. We envisage *Consumers* and *Providers* will employ brokers which may purchase or sell storage in an autonomic manner based on the organisations requirements.

The SX platform can be used in a collaborative manner, where participants use the model to exchange services for credits, or alternatively in an open marketplace where enterprises trade storage services. Whether in a collaborative or enterprise environment the incentives for an organisation to use our SX platform include: (i) *monetary gain:* Institutions providing storage services (*Providers*) are able to better utilise existing storage infrastructure in exchange for monetary gain. Institutions consuming these storage services (*Consumers*) have the ability to negotiate for storage services as they require them, without needing to incur the costs associated with purchasing and maintaining storage hardware. (ii) *common objectives:* There may be organisations which may wish to exchange storage services as they may have a mutual goal such as preservation of information [2]. (iii) *Spikes in Storage Requirements:* Research organisations may require temporarily access to mass storage [3] (e.g. temporarily store data generated from experiments) and in exchange may provide access to their storage services. (iv) *donate:* Institutions may wish to donate storage services, particularly if these services are going to a noble cause.

There are many considerations which need to be made when building a global scale platform such as the SX: security, high-availability, fault tolerance, reputation, monetary issues, consistency, operating environment, just to name a few. We have chosen to focus our efforts on the core components by proposing and developing a platform upon which we are able to develop a market model and begin the work necessary to realise the Storage Exchange.

2 Related Work

Applying economic models to manage computational resources has been the focus of much recent research [4,5,6]. These papers discuss the application of economic principles to manage the scheduling of jobs in a large scale environment such as the Grid [7]. Examples of different economic models and systems which use them include: (i) Commodity Market model (Mungi [8] and NimrodG [9]), (ii) Posted price model (NimrodG [9]), (iii) Auction model (Spawn [10] and Popcorn [11]) (iv) Barter model (Stanford Archival Repository Project [2], and MojoNation [12]).

FreeLoader [3] aggregates unused desktop storage to provide low-cost solution to storing massive datasets. Its specifically designed for research institutions which need to store large scientific datasets. This scenario is particularly useful for scientists engaged in high performance computing, where handling large datasets is common. FreeLoader aims to handle large immutable files (write-once-read-many). Farsite [13] is another system which demonstrates the resource potential to be gained from scavenging unused storage. Farsite operates within the boundaries of an institution providing a storage service logically similar to a file-server found in corporate environments.

Cooper et al [2] propose a bartering storage system for preserving information. Institutions which have common requirements and storage infrastructure can use the framework to barter with each other for storage services. The bartering model relies on their to be a *double coincidence of wants* [14]. OceanStore [15] is a globally scalable storage utility, providing paying users with a durable, highly available storage service by utilising untrusted infrastructure. Mungi [8] is Single-address-space operating system which employs economic principles to manage storage quota. MojoNation [12] uses digital currency (*Mojo*) to encourage users to share resources on its network, users which contribute are rewarded with *Mojo* which can be redeemed for services.

Freeloader [3] and Farsite [13] both demonstrate the storage potential that exists by scavenging storage from workstations. The following works [8,9,10,12] apply economic principles to effectively manage and foster the trade and exchange of services. The Storage Exchange aims to combine storage scavenging and economic principles to create a global platform allowing institutions to federate, trade, exchange and manage storage services.

3 System Overview

There are four main components which make up the SX platform, the Storage Client, Storage Broker, Storage Provider and the Storage Exchange itself (Figure 1). The SX platform has been designed to operate on global network such as the Internet, allowing organisations across geographic boundaries to trade and utilise storage services. Organisations have the ability to trade storage based on their current requirements, if a Storage Broker detects an organisation is running low on storage it may purchase storage, alternatively if it finds that there is an abundance of storage it has the ability to lease out the excess storage. The rest of this section discusses each of the components:

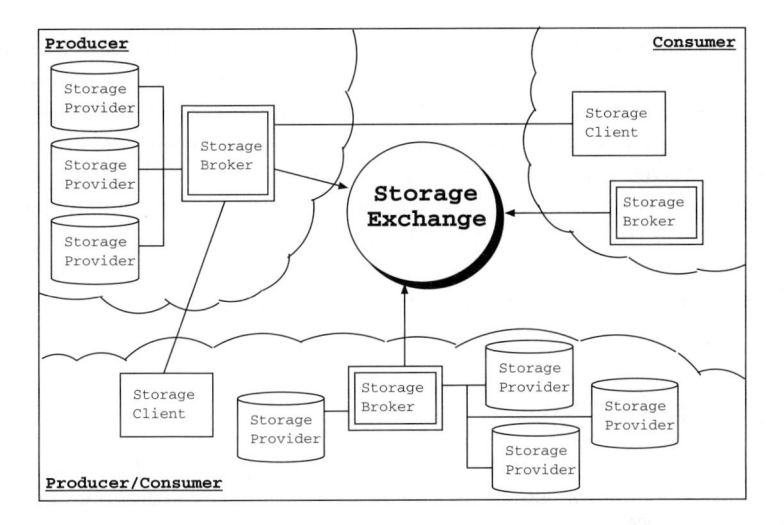

Fig. 1. Storage Exchange: Platform Architecture

Storage Provider: The Storage Provider is deployed on hosts within an organisation chosen to contribute their available storage. Whilst we envision the Storage Provider to be used to scavenge available storage from workstations, there is no reason why it can not be installed on servers or dedicated hosts. The Storage Provider is responsible for keeping the organisations broker up to date with various usage statistics and service incoming storage requests from Storage Clients.

Storage Client: An organisation wishing to utilise a negotiated storage contract will need to use a Storage Client. A user will configure the Storage Client with the storage contract details. The Storage Client then uses these details to authenticate itself with the provider's Storage Broker and upon successful authentication the Storage Client requests a mount for the volume. The provider's Storage Broker then looks up the Storage Providers responsible for servicing the storage contract and instructs them to connect to the Storage Client. Upon receiving a successful connection from the Storage Provider, the Storage Client provides an interface to the user (e.g. local mount point).

Storage Broker: For an organisation to be able to participate in the SX platform they will need to use a Storage Broker. The Storage Broker enables the organisation to trade and utilise storage services from other organisations. The Storage Broker needs to be configured to reflect how it should best serve the organisations interests. From a consumer's perspective the Storage Broker will need to know the organisations storage requirements and the budget it is allowed to spend in the process of acquiring them. From the Provider's perspective the Storage Broker needs to be aware of the available storage and the financial goals it is required to reach. Upon configuration, a Storage Broker will contact the Storage Exchange (SX) with its requirements.

Storage Exchange (SX): The Storage Exchange component provides a platform for Storage Brokers to advertise their storage services and requirements. The SX is a trusted entity responsible for executing a market model and determining how storage services are traded. When requests for storage are allocated to available storage services the Storage Exchange generates a storage contract. The storage contract contains a configuration of the storage policy forming a contract binding the provider to fulfill the service at the determined price. In a situation where either the provider or consumer breaches a storage contract, the SX will keep a record of reputation for each organisation which can be used to influence future trade allocations.

4 Trading Storage

This section covers topics key to making trading storage possible and begins by covering storage policies; which provide a way to quantify storage being traded. Followed by a discussion on the Double Auction (DA) market model and clearing algorithms we have investigated.

Storage Policy: Storage policies provide a way to quantify a storage service, this is essential regardless of chosen market model. Systems such as the one proposed in [16] use Storage Policies as a way to specify high-level Quality of Service (QoS) attributes, effectively abstracting away error prone low-level configurables from the administrator. Our use of Storage Policies allow Storage Brokers to quantify the service which they wish to lease out or acquire. When a trade is determined the storage policy will form the basis for a storage contract containing details of SLA (Service Level Agreement). The attributes which make up a storage policy are as follows:

1. **Storage Service Attributes:**
   (a) **Capacity**(C): Storage Capacity (GB) of volume.
   (b) **Upload Rate** (U): Rate (kb/sec) of transfer to the volume.
   (c) **Download Rate** (D): Rate (kb/sec) of transfer from the volume.
2. **Duration:**
   (a) **Time Frame** (T): Lifetime (sec) of storage policy.

Market Model: Decades of research and experiments [17,18,19,20] show that Double Auctions (DA) are effective and efficient market model. DAs have been shown to quickly converge towards a Competitive Equilibrium (CE). The CE is the intersection point of true demand and supply curves, yielding allocations which are near 100% efficient. From an economic stand point DAs are a sound and efficient market model. In a Double Auction (DA) [18] both buyers (Consumers) and sellers (Providers) may submit offers to buy and sell respectively. Providers and Consumers submit asks and bids simultaneously and hence participate in a *Double*-sided auction. The process of clearing determines the way in which trades are allocated amongst the asks and bids. There are two ways in which clearing may take place, continuously or periodically. Double Auctions cleared continuously are refered to as Continuous Double Auctions (CDA) and compatible bids and asks are cleared instantaneously. The New York Stock Exchange (NYSE) and Chicago Commodities market both employ a CDA market model. Double Auctions may also be cleared periodically, these are referred to as Clearinghouse (CH) or Call Markets. Bids and Asks are submitted sealed to a clearinghouse, which periodically processes the queued up bids and asks to determine a market clearing price. Call Markets are used to determine opening prices in continuous markets such as the NYSE.

As well as being economically sound there are two attractive features of Double auctions which come to our attention, (i) many trades can be cleared in an instant and using a sealed model (ii) the need to continuously broadcast the current market status to all participants is removed. Studies comparing Double Auctions [21,22] with other auction protocols (Dutch, English, First Price Sealed bid) found that Double Auctions possess least communication overhead. These remarkable properties have motivated our research and subsequent application of a DA market model in our SX platform. The Storage Exchange is responsible for executing a Clearinghouse variation of the DA model, which involves accepting sealed offers from provider and consumer brokers and periodically allocating trades amongst the queued up offers using a clearing algorithm. Consumers submitting bids do so in the form of Storage Request Bids (SRB). A SRB consists of a Storage Policy detailing the storage service and a bid price $SRB = (C, U, D, T, \$)$. A Provider submits a Storage Service Ask (SSA) representing the storage service they wish to lease out. An SSA consists of a Storage Policy representing the storage service they are selling, along with a cost function $SSA = (C, U, D, T, CostFunction(C, U, D, T))$. The cost function represents the Providers responsible and determines a cost based on Storage Policy attributes. The Storage Exchange uses the cost function to determine how much a Consumers would need to pay based on their Storage Policy. To achieve this the Storage Exchange substitutes consumers Storage Policy attributes into the Providers cost function to determine a price.

Clearing Algorithms: Periodically the Storage Exchange allocates trades amongst queued up SRBs with SSAs, the manner in which it does so is determined by the clearing algorithm it employs. We propose and investigate the following clearing algorithms in the context of our SX platform:

1. *First fit:* SRBs are allocated to SSAs on a first fit basis. An SSA is deemed to fit if it has the storage resources required by the SSA and the cost function returns a price within the SSA bid amount. SRBs are processed in the order which they have been queued up.
2. *Maximise Surplus:* This clearing algorithm aims to maximise the profit of the auction. An SRB is allocated to an SSA which results the maximum difference between Consumers bid price and result of Providers cost function.
3. *Optimise Utilisation:* This algorithm focuses on achieving better utilisation by trying to minimize the *left overs* that remain after an SRB is allocated to an SSA. A measure of fit is calculated (Algorithm 1) between an SRB and each SSA. A large measure of fit indicates that the remaining ratios have a large spread amongst each of the *Storage Service Attributes* and therefore would result in an SSA with potentially more waste, whereas a small population variance would indicate that the remaining *Storage Service Attributes* within the SSA would have less waste. Upon calculating a measure of fit between the considering SRB and each SSA, we allocate it to the SSA which returned the smallest measure of fit. SRBs are processed in the order which they have been queued up.

Algorithm 1. MeasureOfFit(S,A)

1: **Input**: Storage Request Bid S, Storage Service Ask A
2: **Output**: Measure of Fit F
3: $A = \{a_1, a_2, ..., a_n\}$ //*Storage Service Attributes*
4: //belonging to Available Storage Policy
5: $S = \{s_1, s_2, ..., s_n\}$ //*Storage Service Attributes* belonging to Storage Request
6: // calculate a remaining ratio for each of Storage Service Attributes
7: $R = \{r_1 = \frac{a_1 - s_1}{a_1}, r_2 = \frac{a_2 - s_2}{a_2}, ..., r_n = \frac{a_n - s_n}{a_n}\}$
8: // calculate the population variance amongst the remaining ratios
9: $F = \frac{1}{n}\sum_{i=1}^{n}(r_i - u_R)^2, where\ u_R = \frac{1}{n}\sum_{i=1}^{n}r_i$

---

4. *Max-Surplus/Optimise Utilisation:* This clearing algorithm (Algorithm 2) incorporates the last two allocation strategies and aims to draw a balance between the two. Parameter (k) serves to bias the balance, $(0.5 < k <= 1)$ means that importance will be given to utilisation, whereas a $k(0 <= k < 0.5)$ will give importance to achieving a better surplus. Algorithm 2 is applied to every SRB, in the order which they have been queued up.

5 Performance and Evaluation

Implementation: The Storage Provider and Storage Client components have been written in C. The Storage Client utilises the FUSE library [23] to provide a local mount point of the storage volume in user space. The Storage Broker and Storage Exchange have both been written in Java. Interactions between the Broker, Provider and Client have been implemented and tested. We have been able to successfully mount a replicated storage volume utilising scavenged

Algorithm 2. Max-Surplus/Optimise Utilisation Algorithm

1: **Input**: Storage Request Bid S, Storage Service Asks A, Balance k
2: **Output**: Selected Storage Policy P
3: $F \leftarrow \{\emptyset\}$ // a set to store MeasureOfFit values
4: $M \leftarrow \{\emptyset\}$ // a set to store Surplus calculations
5: **for all** $availableStoragePolicy \in A$ **do**
6:    **if** $availableStoragePolicy$ has greater resource attributes than S and S bid price is greater than $availableStoragePolicy$ reserve **then**
7:       $F \leftarrow F \cup \text{MeasureOfFit}(S, availableStoragePolicy)$
8:       $M \leftarrow M \cup \text{surplus}(S, availableStoragePolicy)$
9:    **end if**
10: **end for**
11: $minSurplus = \min(M)$, $worseFit = \max(F)$
12: $deltaMeasureFit = worseFit - \min(F)$, $deltaSurplus = \max(M) - minSurplus$
13: $currentHighScore = $ Large Negative Number
14: **for all** $availStorePl \in A$ **do**
15:    $ratioBetterFit = (worseFit - \text{MeasureOfFit}(S, availStorePl))/deltaMeasureFit$
16:    $ratioBetterSurplus = (\text{surplus}(S, availStorePl) - minSurplus)/deltaSurplus$
17:    $score = k * ratioBetterFit + (1 - k) * ratioBetterSurplus$
18:    **if** $score > currentHighScore$ **then**
19:       $currentHighScore = score$
20:       $P \leftarrow \{availStorePl\}$ // assign Storage Policy with max score
21:    **end if**
22: **end for**

storage made available by Providers. Communication between components is carried out via TCP socket communication. The Storage Exchange accepts offers from Storage Brokers and employs a clearing algorithm to allocate trades. Our performance evaluation focuses on the Storage Exchange and comparing the different clearing algorithms it employs.

Experiment Setup: We randomly generate a series of bids (SRB) and asks (SSA) which comply to the posting protocol used by Consumers and Providers. The cost functions in the SSAs are linear. The parameters we have used to generate our random set of offers are outlined in Table 1. Each experiment executed represents a single clearing period, that is assume the set of bids and asks generated were queued up over some period of time by the Storage Exchange, our experiment focuses on the sole process of clearing at the end of that period. With every experiment the same set of orders are loaded in the same order in the Storage Exchange to ensure each clearing algorithm is executed in exactly the same manner. Parameters with ranges are assigned with a randomly generated numbers within the specified range. Whilst our scenario has many more bids (600) than asks (50), the asks contain much larger storage service attributes, which would imply that Providers have a large quantity of storage they wish to sell to many consumers.

Results: Our experiment results have been broken down into four plots. The first two plots (Figure 2 and 3) focus on budget aspects while the second set of

Table 1. Experiment Parameters

Parameter	Description	Values
SRB	Number of Storage Request Bids	600
SRC_{range}	Storage Request Capacity range (GB)	5 - 50
SRU_{range}	Storage Request Up Rate range (kb/sec)	5 - 50
SRD_{range}	Storage Request Down Rate range (kb/sec)	5 - 50
$SRDU$	Storage Request Duration (sec)	20000
SSA	Number of Storage Service Asks	50
SAC_{range}	Storage Ask Capacity range (GB)	50 - 500
SAU_{range}	Storage Ask Up Rate range (kb/sec)	100 - 1000
SAD_{range}	Storage Ask Down Rate range (kb/sec)	100 - 1000
$SADU$	Storage Ask Duration (sec)	20000

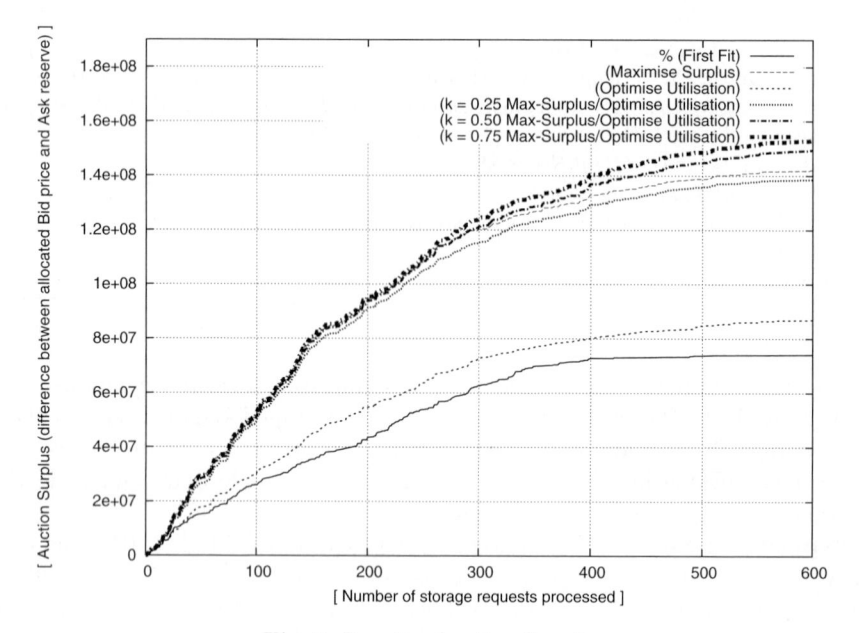

Fig. 2. Results: Auction Surplus

plots (Figure 4 and 5) focus on utilisation achieved. The horizontal axis in all the plots represents the number of bids that have been processed. We can see from the Auction Surplus plot that the *Maximise Surplus* algorithm achieves far better surplus than either *first fit* or *Optimise Utilisation*, but performs poorly in utilisation plots (Figure 4 and 5) which in turn has a bad impact on *Ask budget met*. The *Optimise Utilisation* algorithm achieves a far better utilisation (Figure 4 and 5) than *Maximise Surplus*, so much so it achieves the best in percentage of Ask budget met. Even though it performs well in utilisation it achieves a poor result in auction surplus. Finally when we apply *Max-Surplus/Optimise Utilisation* clearing algorithm we are able to achieve best Auction Surplus ($k = 0.75$) whilst achieving better utilisation than *Maximise Surplus*.

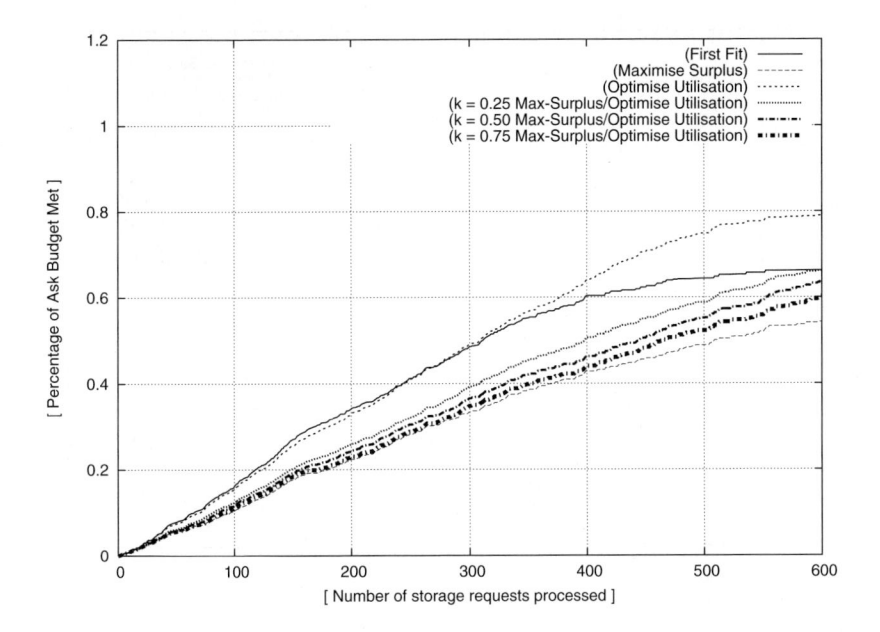

Fig. 3. Results: Percentage of Ask Budget met

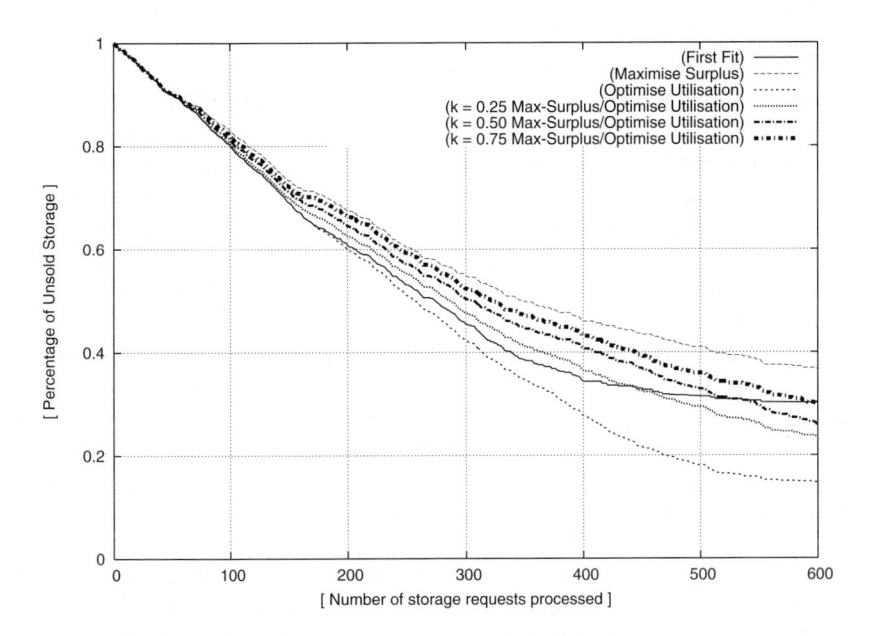

Fig. 4. Results: Percentage of Unsold Storage

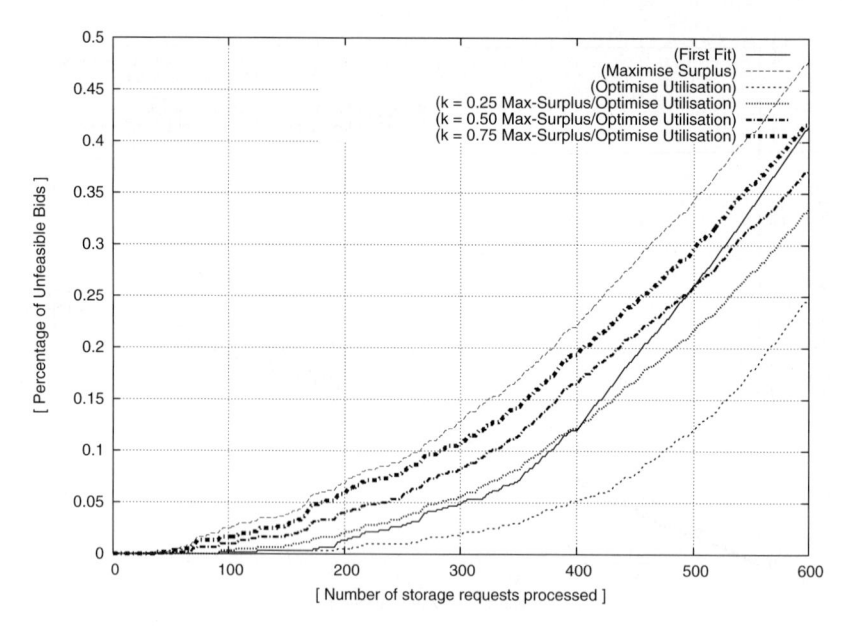

Fig. 5. Results: Percentage of Unfeasible Bids

6 Conclusion

The SX platform provides organisations with various storage services and requirements the capability to trade and exchange these services. Our platform aims to federate storage services, allowing organisations to find storage services which better meet their requirements whilst better utilising their available infrastructure. Organisations are able to scavenge storage services across their network of workstations and with the use of the SX platform lease it out globally. The Storage Exchange serves as a foundation for further research and development into utilising economic principles to achieve Autonomic management [24] of storage services.

In this paper we discuss our SX platform and apply a sealed Double Auction market model and evaluate various clearing algorithms which aim to maximise surplus, optimise utilisation and finally combining the previous two. Our results show that combining maximise surplus and optimise utilisation algorithms achieves better utilisation and consequently the best auction surplus. A couple of areas which require further research include:

1. *Determining a clearing price:* Whilst determining a clearing price in a double auction which deals with goods that are homogeneous and divisible abstract entities such as money and shares is found by looking where supply intersects demand, this is not applicable when dealing with heterogeneous goods [25] such as storage policies.

2. *Extending Experiment:* Conduct a more detailed assessment of our clearing aglorithms by varying parameters described in Section 5. Also determining a theoretically optimal clearing result would allow us to compare and guage the efficiency of our clearing algorithms.

References

1. Kephart, J.O., Chess, D.M.: The vision of autonomic computing. Computer **36**(1) (2003) 41–50
2. Cooper, B.F., Garcia-Molina, H.: Peer-to-peer data trading to preserve information. ACM Transactions on Information Systems **20**(2) (2002) 133–170
3. Vazhkudai, S.S., Ma, X., Freeh, V.W., Tammineedi, J.W.S.N., , Scott., S.L.: Freeloader: Scavenging desktop storage resources for scientific data. In: IEEE/ACM Supercomputing 2005 (SC—05), Seattle, WA, IEEE Computer Society (2005)
4. Wolski, R., Plank, J.S., Brevik, J., Bryan, T.: G-commerce: Market formulations controlling resource allocation on the computational grid. In: International Parallel and Distributed Processing Symposium (IPDPS), San Francisco, IEEEE (2001)
5. Buyya, R., Abramson, D., Giddy, J., Stockinger, H.: Economic models for resource management and scheduling in grid computing. The Journal of Concurrency and Computation: Practice and Experience (CCPE) (2002)
6. Weglarz, J., Nabrzyski, J., Schopf, J., eds.: Grid resource management: state of the art and future trends. Kluwer Academic Publishers, Norwell, MA, USA (2004)
7. Foster, I.T.: The anatomy of the grid: Enabling scalable virtual organizations. In: Euro-Par '01: Proceedings of the 7th International Euro-Par Conference Manchester on Parallel Processing, London, UK, Springer-Verlag (2001) 1–4
8. Heiser, G., Elphinstone, K., Vochteloo, J., Russell, S., Liedtke, J.: The Mungi single-address-space operating system. Software Practice and Experience **28**(9) (1998) 901–928
9. Buyya, R., Abramson, D., Giddy, J.: NimrodG: An Architecture of a Resource Management and Scheduling System in a Global Computational Grid. In: Proceedings of the 4th International Conference on High Performance Computing in Asia-Pacific Region. (2000)
10. Waldspurger, C.A., Hogg, T., Huberman, B.A., Kephart, J.O., Stornetta, W.S.: Spawn: A distributed computational economy. IEEE Transactions on Software Engineering **18**(2) (1992) 103–117
11. Regev, O., Nisan, N.: The popcorn market an online market for computational resources. In: ICE '98: Proceedings of the first international conference on Information and computation economies, New York, NY, USA, ACM Press (1998) 148–157
12. Wilcox-O'Hearn, B.: Experiences deploying a large-scale emergent network. In: Revised Papers from the First International Workshop on Peer-to-Peer Systems, Springer-Verlag (2002) 104–110
13. Adya, A., Bolosky, W.J., Castro, M., Cermak, G., Chaiken, R., Douceur, J.R., Howell, J., Lorch, J.R., Theimer, M., Wattenhofer, R.P.: Farsite: federated, available, and reliable storage for an incompletely trusted environment. SIGOPS Operating Systems Review **36**(SI) (2002) 1–14
14. Richard G. Lipsey and K.Alec Chrystal: Principles of Economics 9th Edition. Oxford University Press (1999)

15. Kubiatowicz, J., Bindel, D., Chen, Y., Eaton, P., Geels, D., Gummadi, R., Rhea, S., Weatherspoon, H., Weimer, W., Wells, C., Zhao, B.: Oceanstore: An architecture for global-scale persistent storage. In: Proceedings of ACM ASPLOS, ACM (2000)
16. Devarakonda, M.V., Chess, D.M., Whalley, I., Segal, A., Goyal, P., Sachedina, A., Romanufa, K., Lassettre, E., Tetzlaff, W., Arnold, B.: Policy-based autonomic storage allocation. In Brunner, M., Keller, A., eds.: DSOM. Volume 2867 of Lecture Notes in Computer Science., Springer (2003) 143–154
17. Smith, V.L.: An experimental study of competitive market behavior. The Journal of Political Economy **70**(2) (1962) 111–137
18. Friedman, D., Rust, J.: The Double Auction Market: Institutions, Theories and Evidence. Addison-Wesley Publishing (1993)
19. Gjerstad, S., Dickhaut, J.: Price formation in double auctions. In: E-Commerce Agents, Marketplace Solutions, Security Issues, and Supply and Demand, London, UK, Springer-Verlag (2001) 106–134
20. Rustichini, A., Satterthwaite, M.A., Williams, S.R.: Convergence to efficiency in a simple market with incomplete information. Econometrica **62**(5) (1994) 1041–63
21. Assuncao, M., Buyya, R.: An evaluation of communication demand of auction protocols in grid environments. In: Proceedings of the 3rd International Workshop on Grid Economics and Business (GECON 2006), World Scientific Press (2006)
22. Morali, A., Varela, L., Varela, C.: An electronic marketplace: Agent-based coordination models for online auctions. In: XXXI Conferencia Latinoamericana de Informática, Cali, Colombia (2005)
23. FUSE: http://sourceforge.net/projects/fuse/ (2000)
24. Pattnaik, P., Ekanadham, K., Jann, J.: Autonomic Computing and GRID. In: Grid Computing: Making the Global Infrastructure a Reality. Wiley Press, New York, NY, USA (2003)
25. Kalagnanam, J.R., Davenport, A.J., Lee, H.S.: Computational aspects of clearing continuous call double auctions with assignment constraints and indivisible demand. Electronic Commerce Research **1**(3) (2001) 221–238

Vigne: Towards a Self-healing Grid Operating System

Louis Rilling

IRISA/Université de Rennes 1/ENS Cachan, Brittany site - PARIS research group
Louis.Rilling@irisa.fr

Abstract. We consider building a Grid Operating System in order to relieve users and programmers from the burden of dealing with the highly distributed and volatile resources of computational grids. To tolerate the volatility of the nodes, the system should be self-healing, that is continuously adapt to additions, removals, and failures of nodes. We present the self-healing architecture of the Vigne Grid Operating System through three of its services: system membership, application management, and volatile data management. The experimental results obtained show that our approach is feasible.

1 Introduction

Grids gather large sets of services over a large set of resources provided by many independent organizations. The nodes of such distributed systems are in essence volatile: organizations may unilaterally decide to add or remove nodes at any time, and the failure rate increases with the number of nodes.

We consider building a Grid Operating System (GOS) in order to relieve users and programmers from the burden of dealing with such highly distributed and volatile resources. To achieve this goal, a GOS should provide users and programmers with simple abstractions of physically highly distributed resources, and transparently handle additions, removals, and failures of nodes.

We consider building *self-healing* systems. The self-healing property is a variant of fault-tolerance in which the system proactively maintains its degree of fault-tolerance. The mechanisms of the system must continuously adapt to additions, removals, and failures of nodes. This is an important property since assuming that human interventions quickly restore failed resources can not scale to large numbers of nodes. Moreover, no service of the system should depend on the stability of any set of nodes during the whole system's lifetime. However, current approaches like Globus [1] still rely on static hierarchies, defined by system administrators, and that prevent the system from being self-healing.

In this paper we present the self-healing architecture of the Vigne GOS, through the design of three of its services. One of the main contributions of this architecture is the application management service which decentralizes application control and provides applications with generic and transparent fault-tolerance policies. We implemented most of these three services and present in this paper experimental results obtained by simulations, which show the feasibility of our approach.

The paper is organized as follows. We precise our model of distributed system in Sect. 2 and give an overview of Vigne in Sect. 3. Then we present three self-healing services of Vigne, namely system membership in Sect. 4, application management in Sect. 5, and volatile data management in Sect. 6. We present an experimental evaluation of the self-healing properties of volatile data management in Sect. 7, and discuss related work in Sect. 8. Finally, Sect. 9 concludes.

2 System Model

The nodes of the system belong to many independent organizations. For this reason we consider a (large scale) distributed system composed of nodes which can fail, recover (or be added), and be gracefully removed (the last two events are called reconfigurations in the paper). Nodes fail in a fail-stop manner. Failures can be detected using (unreliable) failure detectors. We do not consider byzantine failures, as they are relevant to security. We focus our work on the scalability and self-healing aspects, and expect security issues to be tackled in future work.

Users run distributed as well as sequential applications on this system. Many users may run many applications simultaneously, using the system as a computational power provider.

3 Overview of Vigne

We consider building a Grid Operating System (GOS). As any operating system, a GOS virtualizes the physical resources to provide users and programmers with simple abstractions. A set of services is depicted in Fig. 1. In this paper we focus on the self-healing aspect of a GOS.

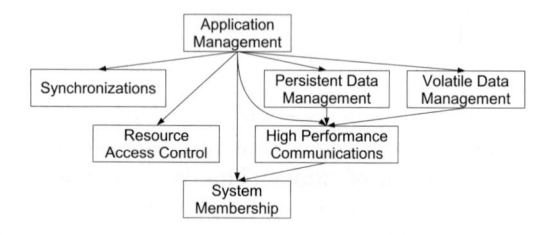

Fig. 1. Services of a Grid Operating System

The application management service (AMS) is the top-level service of the system. This service controls applications executions, and is the main service with which users interact. The AMS runs each application under the control of a dedicated self-healing agent called *application manager*. An application manager acts on behalf of the user to run efficiently the application and to ensure that it terminates correctly, despite node removals and failures.

The persistent data management service stores data in logical files that have location-independent names. The lifetime of these files is independent from the lifetime of applications. Conversely, the volatile data management service (VDMS) manages volatile data that is private to applications. The VDMS offers abstractions of shared data to build distributed applications communicating through the shared-memory paradigm. Since data managed by the VDMS is volatile and private to a single application, the VDMS can apply fault-tolerance mechanisms that are less costly in resources and performance than the mechanisms needed for persistent data. The VDMS is based on fault-tolerant *consistency protocols* allowing to replicate shared data to improve performance [2].

The system membership service (SMS) is the basis on which all other services are implemented. The SMS connects the nodes of the system in a scalable, decentralized, and self-healing manner. The SMS of Vigne is based on a *structured overlay network* designed in recent research in the peer-to-peer field [3].

The other services are not discussed in this paper. We briefly describe them. The synchronization service provides applications with synchronization primitives comprising distributed semaphores and barriers. The high performance communication service provides applications as well as higher level services with communication primitives that adapt to parallel communication links and to the various security policies used on the nodes. The resource access control service enforces the resource sharing policy defined by organizations for their nodes.

The main principle driving our approach is to simplify as much as possible the job of users and programmers without restricting the field of applications. In particular the GOS should relieve users and programmers from dealing with failures. The next sections describe the self-healing properties of Vigne's membership, application management, and volatile data management services.

4 System Membership

The system membership service of Vigne must achieve two goals despite continuous reconfigurations and failures: maintain the nodes of the system in connection, and deliver accurate membership information to higher level services. The system membership service is based on a structured overlay network built using the structure and routing algorithms of Pastry [3], and the maintenance algorithms of Bamboo [4]. The basic mechanism implemented by the overlay network is key-based routing, which allows to build self-healing distributed hash tables (DHT). The keys of such DHTs can be used as location-independent names, as the overlay network routes a message to a key without needing that the sender knows which node hosts the key. DHTs are a sound basis to build self-healing higher-level services. This is illustrated in the next sections for the application management service and the volatile data management service.

Structured Overlay Network. Pastry connects the nodes of the system in a logical ring. Nodes have numerical names, called ID and represented in hexadecimal, and are placed in clockwise order on the ring. Pastry maps a key (also represented in hexadecimal) to the node having the numerically closest ID (see Fig. 2, left part). Pastry routes messages to keys following the logical ring and shortcut links that allow to limit the average number of routing hops to $\log_{16} N$ with only $O(\log N)$ links per node, where N is the number of nodes connected to the system. The overlay network is made self-healing using redundant links to the neighbors in the ring, and gossiping protocols to refill the routing tables [4].

Distributed Hash Tables. On top of this structured overlay network we have implemented a distributed hash table (DHT) service that provides generic management of self-healing DHTs. This service allows higher level services to define any number of DHTs. DHTs are generically made self-healing by automatically moving and replicating keys using a per-DHT defined replication degree. The nodes hosting the replicas of a key are the numerically closest neighbors (clockwise and counter clockwise) of the node to which the overlay networks maps the key (see Fig. 2, right part). Automatic replication management of the keys can be customized by higher level services (for instance application management).

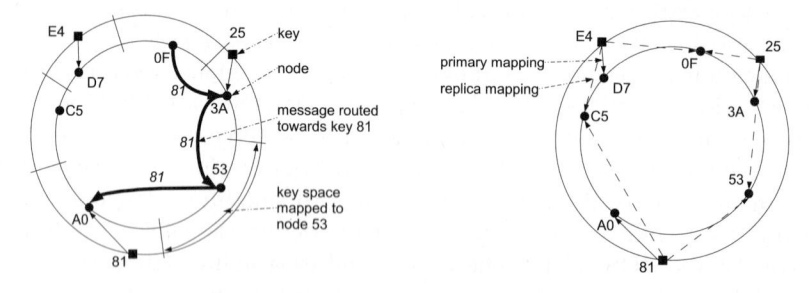

Fig. 2. Basic topology of a Pastry-based structured overlay network and mapping from keys to nodes (left part). Replication of keys in a self-healing DHT (right part).

5 Application Management

The application management service is the main interface of the system for users. This service controls the execution of all applications in order to minimize the execution time and to reliably execute each application, that is to ensure that the application correctly terminates despite failures. A discussion on minimizing execution time is out of the scope of this paper. To reliably execute applications, the application management service of Vigne controls the execution of each application through a dedicated self-healing agent called application manager. In this section we present the design of these application managers.

Application managers have three main features: control applications execution in a decentralized manner, transparently handle failures and reconfigurations, and allow to flexibly define fault-tolerance policies for each application.

Decentralized Control of Applications. Decentralization is achieved by placing application managers as keys in the application manager DHT (which is implemented using the system membership service, see Sect. 4). The keys are distributed over the whole system using secure hash functions, like SHA-1, to define the key's numerical name. Decentralization not only allows to avoid contention on certain nodes because of the load generated by application control, but also allows to limit the cost of a node removal or failure to the reconfiguration of few application managers. The cost of a node removal or failure is limited thanks to the locality properties of DHTs: only the application managers having a replica located on the removed (or failed) node are affected.

Transparent Handling of Failures and Reconfigurations. To relieve users from dealing with failures and reconfigurations, application managers transparently handle failures and reconfigurations from the point of view of users. Indeed, from the point of view of a user, a running application is represented by its application manager. To achieve transparency, an application manager is reachable through a location-independent name (its key in the application manager DHT), is self-healing by replicating itself using a group communication system and the DHT service, and, to handle all removals or failures of nodes that host components of the application, applies a fault-tolerance policy.

A group communication system is used to actively replicate application managers. The messages sent to an application manager are atomically multicast to the group of replicas of the application manager. Provided that an application manager can be defined as a deterministic input / output state machine, this ensures that all replicas output the same sequence of messages. To ensure this determinism, the failure detection mechanisms used by an application manager to monitor an application interact with the application manager through the group communication system (see Fig. 3).

The nodes hosting the replicas are automatically chosen using the DHT service, which makes application managers self-replicating and self-healing. However, to keep the replicas synchronized, creating replicas must be done under the control of the group communication system. To this end, the DHT service only informs application managers of the nodes on which they should replicate. For this reason, we chose to build a group communication system based on the architecture defined in [5], which offers the required flexibility.

Application Fault-Tolerance Policies. To relieve users from dealing with failures, an application manager applies a fault-tolerance policy defined for the application. Thanks to this feature, application managers are a powerful and flexible mechanism to provide applications with generic fault-tolerance with minimal efforts from users and programmers. The fault-tolerance policy can be a generic predefined policy, for instance based on checkpointing and restart, or a policy specifically designed for the application, for instance make the application rebuild lost data using data from a previous computing step [6]. In each case, the application manager enforces a fault-tolerance policy by reacting to suspicions of nodes hosting components of the application (see Fig. 3, right part).

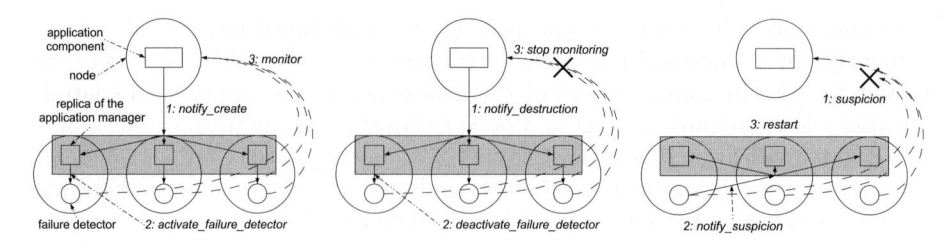

Fig. 3. Communications between application components, failure detectors, and the replicas of an application manager, when creating (left) or destroying (center) components, or suspecting nodes (right)

6 Volatile Data Management

The volatile data management service (VDMS) helps programmers to build distributed applications that use the shared memory paradigm to communicate, offering to these programmers abstractions of shared objects to which processes can access using location-independent names. To achieve this the VDMS of Vigne includes consistency protocols to provide the programmer with consistency models on the values of the copies of a shared object.

We have studied two protocols ensuring atomic consistency. These protocols are based on the write-invalidate scheme, in order to obtain performance (see [2] for a discussion). Before granting write access rights to a copy, the protocols ensure that all other copies are invalid. In our protocols, at each time one (and only one) copy is distinguished as the master copy. Other copies become valid by retrieving the value of the master copy.

These protocols are based on protocols designed by K. Li [7]. K. Li's protocols were designed for a static system having reliable FIFO communication channels. We improved K. Li's protocols to handle multiple and simultaneous reconfigurations and failures, and to tolerate unreliable communication channels. To handle reconfigurations and failures, we leverage the application management and system membership services. We tolerate unreliable communication channels in order to improve the scalability of memory consumption. We have proved in [8] that in our approach the amount of memory consumed per node does not depend on the number of nodes in the system, whereas this cost is linear in the number of nodes if communication channels are made reliable in the communication layer.

Both protocols eventually rely on the application manager to ensure fault-tolerance. However, we also handle reconfigurations and failures in the consistency protocols in order to limit the cost of the fault-tolerance mechanisms used by the application manager. For instance, in both protocols a copy may become useless when the node hosting it does not run processes of the application anymore. The removal or the failure of such nodes should not force the application manager to react (for instance by restarting the application). However, K. Li's protocols and their variants in the literature have to ensure that such copies are invalid before granting write access rights to another copy, and for this reason they block if a copy, even useless, can not send acknowledgments. Using these protocols without adapting them to dynamic reconfigurations and failures would force the application manager to perform costly fault-tolerance operations (for instance, restart the application) when no process of the application is lost.

In the first protocol, called STAT, object managers handle access requests from copies and redirect them to the master copy. In order to avoid contention, these object managers are distributed over a DHT. Compared to similar protocols, this allows the protocol to handle reconfigurations simply since object managers have location-independent names and remain reachable thanks to the self-healing management of the structured overlay network and of the DHT.

In the second protocol, called DYN, the copies organize themselves in chains of references towards the master copy. Compared to the STAT protocol, this avoids paying the latency of routing a message through the overlay network for each access request from a copy. However, reconfigurations break these chains. Therefore we added backup object managers, which are located in a DHT, and to which the master copy periodically publishes its location. A copy sends an access request to the backup object manager only when it suspects that its chain towards the master copy is broken.

7 Experimental Evaluation

We have implemented the membership, application management, and volatile data management services (VDMS) of Vigne, except advanced fault-tolerance policies and application manager replication which will be implemented and evaluated later. Based on this implementation, we present an evaluation of the self-healing property of the VDMS. An evaluation of the system membership service figures in [9].

We show the failure resilience of the consistency protocols of the volatile data management service, using a discrete event simulator coupled to the running Vigne prototype. To do this, we simulated the execution of a single writer multiple readers application on a set of 2000 volatile nodes. Node additions and failures were injected using traces collected in the Gnutella peer-to-peer file sharing application on the Internet [10], which represents an extreme case of volatility compared to an industrial grid environment (see the right part of Fig. 4 for the cumulative number of failures injected during the experiment).

Each component of the application runs 1000 loops composed of two phases. In the first phase, the writer writes a value to a shared object, and in the second phase all other components (the readers) read the value of the shared object. With this access pattern, each access from a component to its copy generates an access request. We ran experiments for 20 to 400 readers.

Upon a failure of a node hosting a component of the application, the application was immediately restarted. Coordinated checkpoints were taken before each iteration. In the simulator, coordinated checkpoints and restarts are done in null time, which allows us to observe the impact of node volatility on the other fault-tolerance mechanisms used in the protocols.

Figure 4 (left part) shows the progression of the application for both protocols and sample numbers of readers. The performance of the DYN protocol is much better than the performance of the STAT protocol. The progression of the application with the DYN protocol is almost linear even with a high number of readers (see Fig. 4, center part). In contrast, the performance of the STAT protocol degrades quickly when the number of readers increases. The nodes routing tables of the overlay network are continuously damaged and repairing which critically increases the latencies of the routed messages used in each access request. We also observe this effect with the DYN protocol each time the application is restarted, since messages are routed to reset the protocol (see Fig. 4).

These results suggest that the DYN protocol tolerates well frequent failures. Moreover, these results suggest that DHTs should be used with care in execution paths which are critical for application performance. These results also suggest that the STAT protocol is useless, but we showed in [8] that the STAT protocol exhibits better performance than the DYN protocol for applications having access patterns in which many write accesses are concurrent with other accesses.

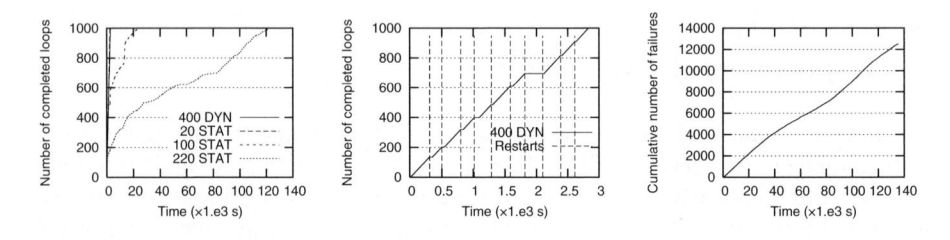

Fig. 4. Performance of STAT and DYN in a highly dynamic configuration

8 Related Work

Grid Operating Systems. Many grid infrastructures, including Globus [1], Legion [11], GridOS [12], and 9grid [13], provide operating system-like services. Globus, Legion, and GridOS are designed as middleware to ease the portability on heterogeneous operating systems, whereas 9grid is an integrated grid operating system which design is simple partly because it enforces that all applications run on top of the services. Between these two extreme approaches, Vigne adapts existing operating systems to, on the one hand, keep legacy interfaces and run legacy applications, and on the other hand, enforce resource sharing policies and provide generic application management.

In all these infrastructures, fault-tolerance is addressed for the services, but only to a limited extent because the systems rely on static hierarchies defined by system administrators. In contrast, Vigne's services are designed to be fully self-healing in order to continuously tolerate failures of any node in the system, without needing any action from system administrators.

In current grid infrastructures, fault-tolerance is not addressed for applications, the main assumption being that this task should be left to application-specific services. In contrast, as a true grid operating system should do, Vigne provides generic application fault-tolerance services that should meet the needs of most of the use cases, and helps applications to define custom mechanisms for the other use cases.

Membership. We based our system membership service on a Pastry-like structured overlay network. Other works, including JXTA [14] and NaradaBrokering [15], aim at providing infrastructures to build high level peer-to-peer services. JXTA is built on an hybrid structured peer-to-peer network and provides a loosely consistent DHT, which model differs from the DHTs we are using. JXTA's DHT only stores advertisements for resources bound to peers. In particular, this DHT does not manage the location and the replication of the objects for which it stores advertisements.

NaradaBrokering provides a communication infrastructure including scalable event-delivery and publish-subscribe to build high level services. NaradaBrokering's features are complementary to the features of our system membership service. However, the brokering infrastructure's design assumes that a set of nodes remains relatively stable, and the volatility of the nodes is mostly considered for the clients of the brokering services.

Application Management. Few projects include generic application management services to execute applications reliably. Chameleon [16] and XtremWeb [17] provide fault-tolerance mechanisms for a variety of programming models. In particular, Chameleon provides users with generic mechanisms for various fault-tolerance policies. In both systems, application management relies on a centralized entity (the main fault-tolerance manager in Chameleon, or the coordinator in XtremWeb), which is itself made reliable by replication on a static set of nodes. As a major contribution, our application managers decentralize application management, which is better to avoid contention and to resist to massive failures. Moreover application managers are replicated on dynamic sets of nodes, which allows them to adapt to any reconfiguration in the system.

Shared Data Management. Several projects, like JuxMem [18] or Pastis [19], provide mutable shared data management in large scale distributed systems composed of volatile nodes. Compared to our volatile data management service, these systems consider persistent data, which prevents them from providing programmers with an integrated fault-tolerance solution taking programs and data into account. In JuxMem, applications have to adapt to the fault-tolerance mechanisms used for the data, which makes fault-tolerance not fully transparent to the programmers. In Pastis, the replication degree of the data is maintained to keep the data available, but no mechanism allows application fault-tolerance mechanisms to synchronize with consistent versions of the data. Fault-tolerant volatile data management has only been studied in the context Distributed Shared Memory systems [20], which only consider clusters of workstations composed of at most a few hundreds of nodes that rarely undergo reconfigurations or failures.

9 Conclusion

In this paper we have presented the design of three self-healing services of the Vigne Grid Operating System. The self-healing property is important to ensure the availability of the system and to relieve users and programmers from dealing with reconfigurations and failures. This paper brings two contributions. As the the volatile data management service illustrates, the system membership service and the application management service that we presented constitute a sound basis for a grid operating system. In particular, with application managers our application management service decentralizes application control and provides applications with generic and transparent fault-tolerance. Thanks to application managers, the consistency protocols of our volatile data management service are the first ones based on the write-invalidate scheme for performance and tolerating multiple simultaneous failures. The experimental results on highly dynamic configurations suggest that the chosen self-healing architecture is feasible.

Future work includes enhancing the volatile data management service to enable programmers to choose between various consistency models, still without having to handle failures. We will also implement the group communication system described in [5]. This will complete the self-healing architecture of Vigne and will allow us to evaluate the application management service. In particular, we will be able to evaluate various fault-tolerance policies.

Acknowledgments

The author would like to thank Emmanuel Jeanvoine for his participation in designing Vigne, and Christine Morin for her valuable advices and comments.

References

1. Foster, I., Kesselman, C., eds.: The Grid: Blueprint for a New Computing Infrastructure. Morgan Kaufmann, San Francisco, CA, USA (1999)
2. Rilling, L., Morin, C.: A practical transparent data sharing service for the grid. In: Proc. Fifth International Workshop on Distributed Shared Memory (DSM 2005), Cardiff, UK (2005) Held in conjunction with CCGrid 2005.
3. Rowstron, A., Druschel, P.: Pastry: Scalable, decentralized object location, and routing for large-scale peer-to-peer systems. In: Proceedings of Middleware 2001. Volume 2218 of Lecture Notes in Computer Science., Springer (2001) 329–350
4. Rhea, S., Geels, D., Roscoe, T., Kubiatowicz, J.: Handling churn in a DHT. In: Proceedings of the USENIX Annual Technical Conference. (2004) 127–140
5. Mena, S., Schiper, A., Wojciechowski, P.: A step towards a new generation of group communication systems. In: Proceedings of Middleware 2003. Volume 2672 of Lecture Notes in Computer Science., Springer (2003) 414–432
6. Garbey, M., Ltaief, H.: Fault tolerant domain decomposition for parabolic problems. In: 16th International Conference on Domain Decomposition Methods. Lecture Notes in Computational Science and Engineering, Springer (2005) To appear.
7. Li, K., Hudak, P.: Memory coherence in shared virtual memory systems. ACM Transactions on Computer Systems **7**(4) (1989) 321–359

8. Rilling, L.: Système d'exploitation à image unique pour une grille de composition dynamique : conception et mise en œuvre de services fiables pour exécuter les applications distribuées partageant des données. PhD thesis, Université de Rennes 1, IRISA, Rennes, France (2005) In French.
9. Jeanvoine, E., Rilling, L., Morin, C., Leprince, D.: Using overlay networks to build operating system services for large scale grids. In: Proceedings of the fifth International Symposium on Parallel and Distributed Computing (ISPDC 2006), Timisoara, Romania (2006) To appear.
10. Saroiu, S., Gummadi, P.K., Gribble, S.D.: A measurement study of peer-to-peer file sharing systems. In: Proceedings of Multimedia Computing and Networking (MMCN) 2002, San Jose, CA, USA (2002)
11. Grimshaw, A.S., Wulf, W.A., Team, C.T.L.: The legion vision of a worldwide virtual computer. Communications of the ACM **40**(1) (1997) 39–45
12. Krauter, K., Maheswaran, M.: Architecture for a grid operating system. In: Proceedings of the First IEEE/ACM International Workshop on Grid Computing. Volume 1971 of Lecture Notes In Computer Science., Bangalore, India, Springer-Verlag (2000) 65–76
13. Mirtchovski, A., Simmonds, R., Minnich, R.: Plan 9 – an integrated approach to grid computing. In: 18th International Parallel and Distributed Processing Symposium (IPDPS'04) - Workshop on High-Performance Grid Computing, Santa Fe, New Mexico, USA, IEEE, CS Press (2004) 273a
14. Traversat, B., Abdelaziz, M., Pouyoul, E.: Project JXTA: A Loosely-Consistent DHT Rendezvous Walker. http://www.jxta.org/docs/jxta-dht.pdf (2003)
15. Pallickara, S., Fox, G.: NaradaBrokering: A middleware framework and architecture for enabling durable peer-to-peer grids. In: Proceedings of Middleware 2003. Volume 2672 of Lecture Notes in Computer Science., Springer (2003) 41–61
16. Kalbarczyk, Z.T., Iyer, R.K., Bagchi, S., Whisnant, K.: Chameleon: A software infrastructure for adaptive fault tolerance. IEEE Transactions on Parallel and Distributed Systems **10**(6) (1999) 560–579
17. Cappello, F., Djilali, S., Fedak, G., Herault, T., Magniette, F., Néri, V., Lodygensky, O.: Computing on large-scale distributed systems: XtremWeb architecture, programming models, security, tests and convergence with grid. Future Generation Computer Systems **21**(3) (2005) 417–437
18. Antoniu, G., Deverge, J.F., Monnet, S.: How to bring together fault tolerance and data consistency to enable grid data sharing. Concurrency and Computation: Practice and Experience (2006) To appear.
19. Busca, J.M., Picconi, F., Sens, P.: Pastis: A highly-scalable multi-user peer-to-peer file system. In: Proceedings of Euro-Par 2005. Volume 3648 of Lecture Notes in Computer Science., Springer (2005) 1173–1182
20. Shafi, H., Speight, E., Bennett, J.K.: Raptor: Integrating checkpoints and thread migration for cluster management. In: Proceedings of the 22nd International Symposium on Reliable Distributed Systems (SRDS'03), IEEE (2003) 141–152

Problems for Resource Brokering in Large and Dynamic Grid Environments[*]

Catalin L. Dumitrescu

CoreGRID Institute on Resource Management and Scheduling
Electrical Eng., Math. and Computer Science, Delft University of Technology
Mekelweg 4, Delft, 2628 CD Delft, The Netherlands
`c.dumitrescu@ewi.tudelft.nl`

Abstract. Running workloads in a Grid environment may become a challenging problem when no appropriate means are available for resource brokering. Many times resources are provided under various administrative policies and agreements that must be known in order to perform adequate scheduling decisions. Thus, providing suitable solutions for resource management is important if we want to cope with the increased scale and complexity of such distributed system. In this paper we explore the key requirements a brokering infrastructure must meet in large and dynamic Grid environments and illustrate how these requirements are addressed by a specialized infrastructure, DI-GRUBER - a distributed usage service level agreement (uSLA) brokering service. The accuracy function of the brokering infrastructure connectivity and the performance gains when a client scheduling policy is employed are analyzed in high detail. In addition, a performance comparison with a P2P-based distributed lookup service is performed to illustrate the performance differences between two different technologies that address similar problems (Grids that focus on federated resource sharing scenarios and P2Ps that focus on self-organizing distributed resource sharing systems, in which most of the communication is symmetric).

1 Introduction

The motivating scenarios of our work are large grid environments in which virtual organizations (VOs) and agreements appear and vanish with a high frequency (every day or week). Such VOs might be companies requiring outsourcing services over short time intervals or scientific communities that want to participate temporarily in different collaborations with access to other types of resources. In these environments, we distinguish between two types of entities participating: resource *providers* and resource *consumers*. They may be nested: a *provider* may function as a middleman, providing access to resources to which the provider has itself been granted access by some other *provider*. While sharing policies issues can arise at multiple levels in such scenarios, the dynamicity of such an environment is also a problem. *Providers* want to

---

[*] This research work is carried out for the CoreGRID IST project n[0]004265, funded by the European Commission.

express (and enforce) various sharing policies (what we call *usage service level agreements* or *uSLAs*) under which resources are made available to *consumers*. Consumers want to access and interpret uSLA statements published by *providers*, in order to monitor their agreements and guide their activities. Starting from this environment and interactions model, our main focus is the identification of requirements and the provisioning of the design ingredients for building a scalable distributed resource brokering service that supports uSLA expression, publication, discovery, interpretation, enforcement, and verification in large dynamic Grid environments. We build on much previous work concerning the specification and enforcement of resource uSLAs [1-5], information lookup, scheduling and brokering services [6-8], the GRUBER broker [9], and the DI-GRUBER version [10].

The main contributions of this paper are on three dimensions. First, we identify several requirements a brokering infrastructure has to meet when deployed in large and dynamic Grid environments. We base our judgment on our past experience with the GRUBER framework in the Grid3 [11] context and the enhanced version, DI-GRUBER. Second, we present several novel DI-GRUBER performance measurements, namely the brokering accuracy function of infrastructure components' connectivity, and the gains in performance when using automated decision point scheduling for the clients. Third, we realize a performance comparison with a P2P-based system for file management. The paper also introduces two major technical enhancements to the DI-GRUBER two layer brokering infrastructure: WS-Index Service-based infrastructure discovery [6] and a specific solution for handling infrastructures decision points' scheduling in order to meet the outlined requirements [10]. The first enhancement takes advantage of the WS-Index Service functionalities that acts as a lookup service. Each GRUBER decision point registers itself with a predefined list of WS-Index Services at startup and it is automatically deleted when it no longer provides brokering services. The second enhancement also takes advantage of the WS-Index Service to discover the most appropriate decision point (DP).

2 Brokering Key Requirements for Large and Dynamic Grids

This work targets Grids (and any large distributed systems in general) that may comprise hundreds of institutions and thousands of individual investigators where the participants often join or leave the environment [11]. Moreover, each individual investigator and institution may participate in, and contribute resources to multiple collaborative projects that can vary widely in scale, lifetime, and formality [10, 12]. Such globally distributed systems provide several key benefits over large centralized solutions, in particular: maintenance costs and upgrade operations are more easily handled and there are no single points of failure. Two main environment examples of this class are introduced next.

2.1 Grid Environment Examples

Open Science Grid (previously known as Grid3 [11]) is a multi-virtual organization (multi-VO) environment that sustains production level services required by various physics experiments. The Grid3 infrastructure had comprised more than 30 sites and

4500 CPUs, over 1300 simultaneous jobs and more than 2 TB/day aggregate data traffic. The participating sites were (are) the main resource providers under various conditions. Thus, we consider that OSG/Grid is a good example of the kind of environments we envisage for the work in this paper. However, with times we believe that this infrastructure can grow. For example, the number of sites can increase by means of new joins; the rate of jobs can jump when new scientific communities will want to solve high computer power consummative applications. Thus, the resource management infrastructure we envisage in this paper targets Grid environments ten to hundred times bigger than today OSG.

The other Grid testbed example is the LHC Computing Project (LCG). LCG targets to build and to maintain a data storage and analysis infrastructure for the entire high energy physics community that will use the LHC (Large Hadron Collider) [13]. The data from the LHC experiments will be distributed around the globe, according to a four-tiered model. Two of the goals of the LCG project include developing and deploying computing services based on a distributed Grid model, and managing acquisition, installation, and capacity planning for the large number of commodity hardware components. The expected size of the entire community is around 5000 scientists in 500 research institutes and universities worldwide. The analysis of the data, including simulations, requires around 100,000 CPUs. Such a distributed system presents a number of significant challenges; the most important one from our point of view is the provisioning of controlled resource sharing mechanisms so that different groups have fair access, based on their needs and contributions, to the infrastructure.

2.2 Resource Brokering Key Requirements

The resource brokering (and scheduling) problem in such Grid environments encompasses intertwined requirements, while the most important three ones in our vision are: *support for brokering of numerous resources, an adequate level of accuracy of the brokering infrastructure* and *fault-tolerant brokering*.

➢ *Support for Brokering of Dynamic and Numerous Resources (scalability):* dynamicity implies in our view that various communities, providers or VOs might join a Grid environment for short (days to weeks) time intervals in order to solve fast various problems. This dynamicity imposes certain technical requirements, such as rapid propagation of information about available resources in the brokering infrastructure and of the new administrative policies under which these resources are made available. When the environment is large (composed of hundreds to thousands simultaneous providers and more than thousands of consumers), the brokering solution must be scalable enough to handle such an infrastructure.

➢ *Adequate Level of Brokering Accuracy Independent of the Infrastructure:* regarding management information, an important problem is the accuracy of information provided by a brokering service in order to perform adequate scheduling decisions. Even more, for a distributed infrastructure, several operations have to be considered, such as propagation, reconciliation and removal. These operations may occur whenever new decisions are performed and new resources join or leave the environment. The entire brokering infrastructure must become aware of these changes in a timely fashion manner.

➤ *Fault-tolerant Resource Brokering:* fault tolerance is important from a client point of view. Even when a client cannot contact a brokering decision point, it still expects to perform scheduling operations over the Grid with a lower but acceptable execution performance. Also, when many clients perform queries, the brokering infrastructure must be able to cope with this request load. Even the reader might think about the P2P networks and their properties to re-organize, we pursue the path of scheduling the brokering decision points as any other resources. Thus, the employment of an adequate strategy becomes important in this approach.

3 Illustrating the Key Requirements in a Concrete Case

We now introduce the main concepts and tools used in this paper. We start with the WS-Index Service (monitoring and discovery service [6]) used as a supporting tool and introduce afterwards our brokering infrastructure (DI-GRUBER [10]) used as a vehicle for proving our assumptions.

3.1 WS-Index Service

WS-Index Service [6] is a standard component of the Globus Toolkit (one of the Grid technologies largely used in science and industry [14]). It provides specialized functions for resource and service monitoring and discovery, and it is used as the central rendezvous point by our brokering infrastructure. While someone might consider the WS-Index Service a bottle-neck, our previous experiments proved that its scalability is well beyond our needs. Thus, WS-Index Service's main function in our infrastructure is to act as a specialized directory of all DI-GRUBER decision points for all clients and the decision points themselves, and for infrastructure management.

3.2 DI-GRUBER (A Distributed Grid Resource uSLA-Based Broker)

GRUBER [9] is a prototype Grid V-PEP and S-PEP infrastructure that implements the brokering functionalities required for steering workloads in a distributed environment based on uSLAs. It is able to perform job scheduling based on notions such as sites, VOs, VO groups, and uSLAs at various levels [4]. Currently, GRUBER is implemented as a Grid Web Service using the Globus Toolkit (GT4) technologies [14]. As an additional clarification, GRUBER does not perform job submission by itself, but can be used in conjunction with various grid job submission infrastructures. So far, we have interfaced GRUBER for job execution with the Euryale and Pegasus planners, largely used on Grid3 [11].

However, managing uSLAs within environments that integrate participants and resources spanning many physical institutions is a challenging problem when a centralized infrastructure is employed. A single unified uSLA management decision point providing brokering decisions over hundreds to thousands of jobs and sites can easily become a bottleneck in terms of reliability as well as performance. DI-GRUBER, an extension to the GRUBER prototype, was developed as a distributed uSLA-based resource broker that allows multiple decision points to coexist and cooperate in real-time. DI-GRUBER targets to provide a scalable management service with the same functionalities as GRUBER but in a distributed approach [10]. It is a two layer resource brokering service, capable of working over large Grids, extending GRUBER with support for multiple brokering decision points that cooperate by periodically exchanging status information.

3.2 DI-GRUBER Enhancements to Meet Previous Requirements

DI-GRUBER was developed as a distributed uSLA-based grid resource broker that allows multiple decision points to coexist and cooperate in real-time. The problem is that without support for dynamic discovery of the brokering infrastructure, some of the advantages offered by this infrastructure become impractical. Here we outline how the interfacing and integration with the WS-Index Service practically fulfills the requirements enumerated in Section 2.

➤ *Transparent Decision Point Bootstrapping:* As already described, the ability to bring up a decision point is important in a large and dynamic Grid. Our proposed solution uses the functionalities offered by the WS-Index Service for various clients by employing the notion of rendezvous point. In our implementation, each DI-GRUBER decision point registers with a predefined WS-Index Service at startup, while it is automatically deleted when it vanishes.

➤ *Transparent Client Scheduling:* Further, all decision points and clients can use this registry to find information about the existing infrastructure and select the most appropriate point of contact. When we use the term *most appropriate*, we refer to metrics such as load and number of clients already connected. The scheduling policy employed by each client in selecting a decision point was the *least-used* (*LU*) strategy. Also, whenever a decision point stops responding, its clients query automatically the registry and select a new different decision point to communicate.

➤ *Failure Handling:* While dynamic DI-GRUBER decision point bootstrapping might be difficult to automate in a generic environment, the solution we have devised is simple. Every time a client fails to communicate or to connect with a decision point, it registers with the WS-Index Service a request fault. Such faults can be consumed by a specialized entity that based on various policies starts dynamically new decision points by means of the WS-GRAM service.

➤ *Brokering Infrastructure Accuracy Identification:* An important aspect of the work in this paper is to identify the accuracy of our brokering infrastructure function of the connectivity of each decision point to the rest of the network. This analysis falls into the same class of scenarios where a decision point has only partial knowledge, and is on the same path as the analysis of dealing with stalled information, measured and analyzed somewhere else [10].

4 DI-GRUBER Infrastructure Performance Results

Here we report on our latest results [10] while also considering some of our previous results. We used one to ten DI-GRUBER decision points deployed on the PlanetLab nodes [15]. Each decision point maintains a local view of the environment configuration and via periodic exchanges (in the experiments that follow every three minutes) with other decision points acquires the necessary knowledge about recent job dispatch operations or other changes in the system (new resources, new uSLAs).

The three metrics employed for analysis are **Throughput, Response** (or Average Response Time) and **Accuracy**. **Throughput** is defined as the number of requests completed successfully by the service per time unit. **Response** is defined by the following formula (with RT_i being the individual job time response and N being the number of jobs processed during the execution period): ***Response*** $= \Sigma_{i=1..N} RT_i / N$. Finally, we define the scheduling accuracy for a specific job (SA_i) as the ratio of free resources at the selected site to the total free resources over the entire grid. **Accuracy** is then the aggregated value of all scheduling accuracies measured for each individual job: ***Accuracy*** $= \Sigma_{i=1..N} (SA_i) / N$.

For all the experiments, we used synthetic workloads with a constant arrival rate of 1 job/s for each client or as soon as the previous scheduling decision was served that overlaid work for 60 VOs and 10 groups per VO. The experiment duration was one hour in all cases. Each of the 120 submission hosts ("clients") maintained a connection with one decision point; selected either under the *random* or the *least used scheduling* policy. The emulated environment was composed of 300 sites representing 40,000 nodes. The entire configuration was based on Grid3's landscape in terms of number of CPUs, disk space, network connectivity, etc., but ten times larger [10].

4.1 Decision Accuracy with Brokering Network Mesh Connectivity

First, we measure **Accuracy** of the brokering infrastructure function of the decision points' average connectivity. We consider practically three cases: *full connectivity* (DPs see each other), *half connectivity* (each DP collects information only from half of all the others), and *one-fourth connectivity* (each DP collects information only from a quarter of all the others). The results were achieved by means of the DI-GRUBER infrastructure in all three above configurations and are captured in Table 1.

Table 1. DI-GRUBER Accuracy Performance with Mesh Connectivity

	Connectivity	Util	Accuracy
Requests Handled by GRUBER	All	35%	75%
	One half	27%	62%
	One fourth	20%	55%
Total Request	All	41%	68%
	One half	30%	60%
	One fourth	21%	50%

We can observe that the performance of the brokering infrastructure drops substantially with connectivity degree of each individual decision point. As an additional note, the **Util** parameter is low because jobs do not start all in the beginning, but are scheduled every second during the entire execution period. In a nutshell, **Accuracy** drops almost linearly with clients' connectivity degree, intuitively.

4.2 Decision Point Scheduling and Performance Gains

Second, we focus on capturing the gains a client can achieve in term of performance when a *least-used* service selection policy is employed vs. the *random* scheduling policy we employed before. The results for the *random* scheduling policy are captured in Fig. 1 and Fig. 2, while the results for *least-used* scheduling policy are captured in Fig. 3 and Fig. 4. As can be observed in the first two figures, the distributed service provides a symmetrical behavior with the number of concurrent machines that is independent of the state of the Grid (lightly or heavily loaded). Also, with three decision points, **Throughput** increases slowly to about 4 job scheduling requests per second when all testing machines are accessing the service. With 10 decision points, the average **Response** time decreased even further to about 13 seconds, and the achieved **Throughput** reached about 7.5 queries per second [10].

Next two figures report the experiments performed when using the WS-Index Service and *LU* scheduling policy was employed by each client. We must mention that we also used this time a final GT4 release based implementation. As can be easily observed, the results show improvement in terms of both **Response** and **Throughput**. Clients achieved a more stable response time compared with the one in the previous set of tests. The **Response** metric's value is always less than 30 seconds for 3 decision points, and less than 10 seconds for 10 decision points. The **Throughput** metric's value shows even higher improvements, reaching a constant value of 5 queries per seconds for 3 decision points, while going us up as 16 queries per second for 10 decision points.

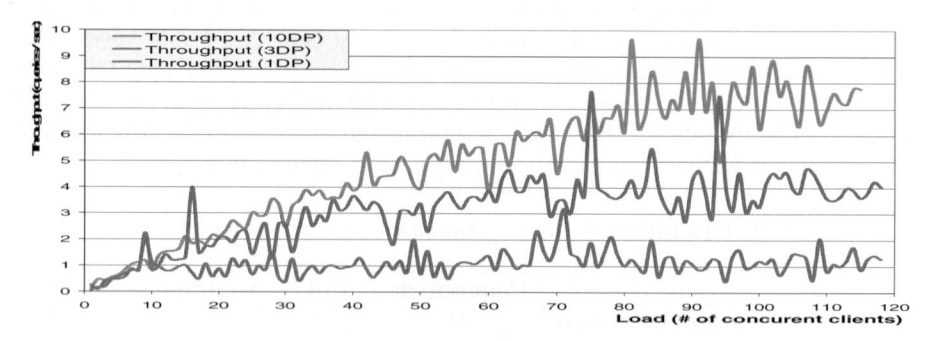

Fig. 1. DI-GRUBER Throughput (1, 3 and 10 Decision Points)

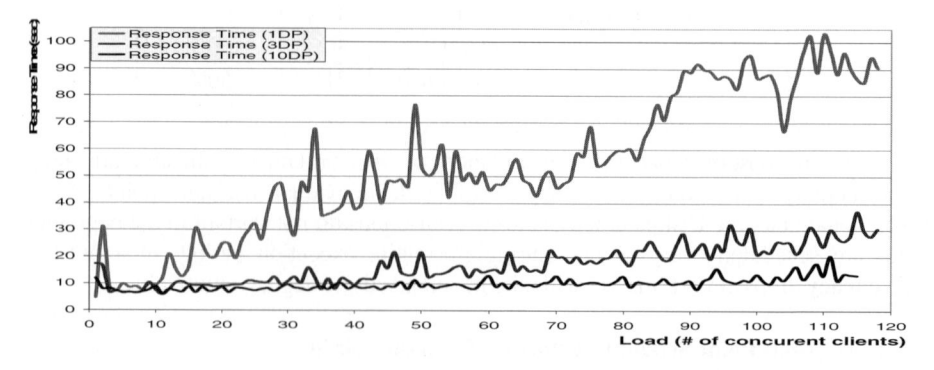

Fig. 2. DI-GRUBER Response (1, 3 and 10 Decision Points)

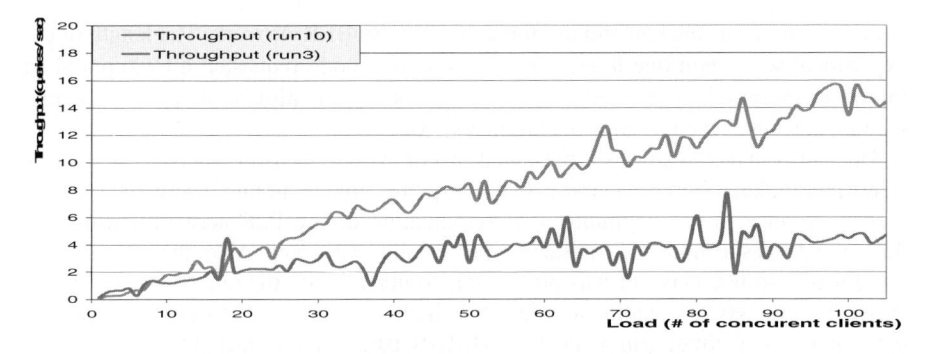

Fig. 3. DI-GRUBER Throughput (3 and 10 Decision Points)

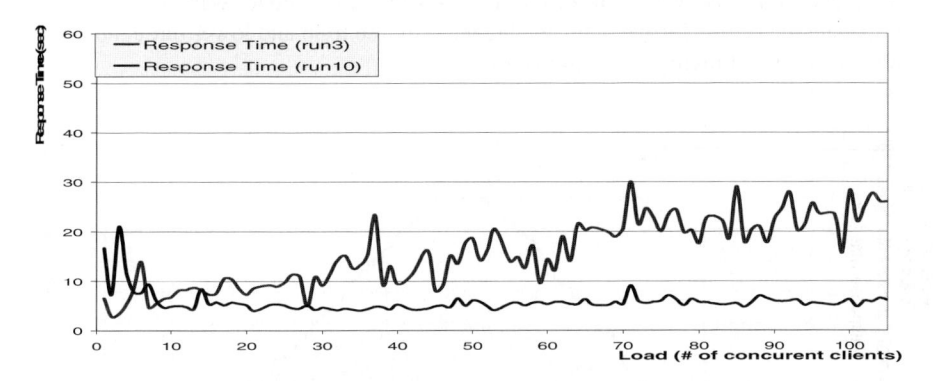

Fig. 4. DI-GRUBER Scalability Response (3 and 10 Decision Points)

However, on average, we find modest improvements for 3 decision points (19% higher throughput and 8% lower response time) and significant improvements for 10 decision points (68% higher throughput and 70% lower response times). From this observation we conclude that, practically, the request load was better balanced among the decision points and the infrastructure was able to achieve higher **Throughput** and lower **Response**.

4.3 Comparison with a FreePastry-Based Lookup Service (PAST)

For convincing the reader that even though DI-GRUBER's transaction throughput seems low compared to *'other transaction processing systems'*, we have performed further performance studies by means of DiPerF [16] on PlanetLab for a pretty well know distributed lookup service. The service chosen for testing was the PAST application [7], built on top of the PASTRY substrate.

The chosen setup was very similar to the one used for DI-GRUBER: the same PlanetLab nodes (around 120). This time we used five machines for running permanent PAST nodes, while the rest ones were brought up dynamically, joining and leaving the network in a controlled manner. Again, we used only one of the five nodes as the main contact point (a node situated at the University of Chicago). The rest ones were maintained as backup and to mimic the DI-GRUBER network. The length of the experiment was again one hour, while each joining node requested a lookup and an insert operation every second (or, if the previous operation took more than one second, at soon as the previous operation ended).

Our performance results are presented in Fig. 5. The measurements show that for insert and lookup operations, the PAST's response time is around 2.5 seconds with a higher variance in the beginning (the stabilization of the P2P network), while the throughput goes as up as 27 transaction per second in average. Also, the message lost rate for this ad-hoc network was pretty high compared with the one of DI-GRUBER. However, the network stabilization delay is higher for the P2P system (first 18% of the experimental time) compared with DI-GRUBER clients' instantaneous network join operation. Our last note is that all operations were performed and measured on the local nodes (insertion followed by lookup); each node was responsible to propagate the results further (thus the higher response time and lower throughput than in the case of employing the continuation).

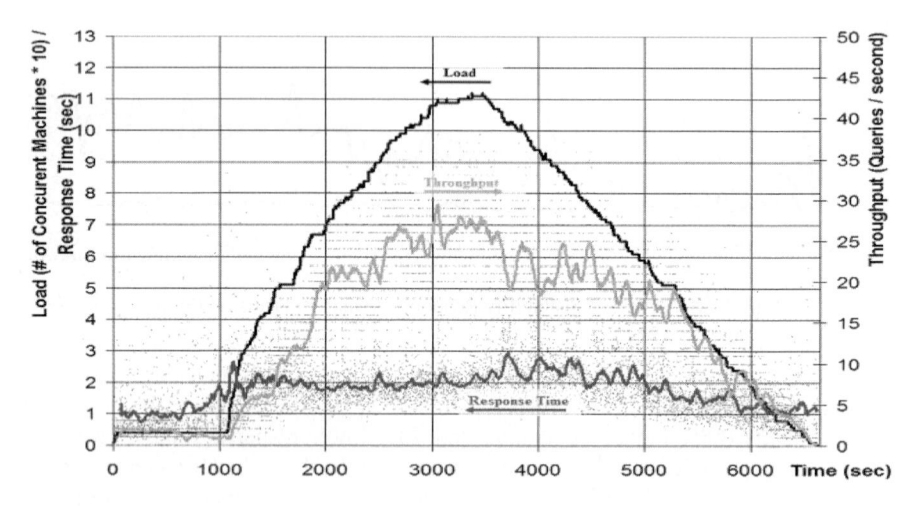

Fig. 5. PAST Network Response Time (left axis) and Throughput (right axis) for a variable Load (left axis * 10) on 120 PlanetLab Nodes

5 Conclusions

Resource management within large VOs that integrate participants and resources spanning multiple physical institutions is a challenging problem. The main question this paper addresses is *"what are the key requirements an already existing management infrastructure should meet in order to support large and dynamic Grid environments?"*. The contributions of this paper are represented by results we achieved on three dimensions: we have identified three key requirements for extending a resource management service for large and dynamic Grid environments (and any other distributed systems in general), analyzed these requirements by means of a real infrastructure in a real case scenario, and also compared the performance results of the considered infrastructure with the ones of a P2P-based service.

Our experimental results showed how the brokering accuracy decreases with the loss of connectivity for a single decision point instance, while the performance of the system almost doubles in the 10 decision points' case due to the better repartition of the clients with the DI-GRUBER's nodes. The last set of experiments, the comparison performance tests, convinced us that even though DI-GRUBER's performance may seem low compared with a cluster resource manager, its performance is comparable in a similar environment in terms of response time and throughput with a distributed P2P system that, however, employs less functionality than the Grid counterpart technology.

Acknowledgments. I would like to thank Ian Foster, Michael Wilde, Jens-S. Vöckler, Yong Zhao, Ioan Raicu and Luiz Meyer for their support and discussions during the development of the (DI-)GRUBER infrastructure.

Bibliography

1. Lupu, E., *A Role-based Framework for Distributed Systems Management*, in *Department of Computing*. 1998, University of London: London.
2. Dan, A., et al. *Web Services on Demand: WSLA-driven automated management*, S. Journal, Editor. 2004, IBM. p. 136.
3. Dumitrescu, C. and I. Foster. *Usage Policy-based CPU Sharing in Virtual Organizations.* in *5th International Workshop in Grid Computing*. 2004. Pittsburgh.
4. Dumitrescu, C., M. Wilde, and I. Foster. *A Model for Usage Policy-based Resource Allocation in Grids.* in *6th IEEE International Workshop on Policies for Distributed Systems and Networks (POLICY 2005)*. 2005. Stockholm, Sweden.
5. Pearlman, L., et al. *A Community Authorization Service for Group Collaboration.* in *IEEE 3rd International Workshop on Policies for Distributed Systems and Networks*. 2002.
6. Czajkowski, K., et al. *Grid Information Services for Distributed Resource Sharing.* in *10th IEEE International Symposium on High Performance Distributed Computing*. 2001: IEEE Computer Society Press.
7. Rowstron, A.I.T. and P. Druschel. *Storage Management and Caching in PAST, a Large-Scale, Persistent P2P Storage Utility.* in *Symposium on Operating Systems Principles*. 2001.
8. Ludwig, H., et al., *A Service Level Agreement Language for Dynamic Electronic Services*. IBM Research Report RC22316 (W0201-112), January 24, 2002.
9. Dumitrescu, C. and I. Foster. *GRUBER: A Grid Resource SLA Broker*. in *Euro-Par*. 2005. Portugal.
10. Dumitrescu, C., I. Raicu, and I. Foster. *DI-GRUBER: A Distributed Approach for Grid Resource Brokering*. in *Super Computing (SC'05)*. 2005. Seattle.
11. Foster, I. et al. *The Grid2003 Production Grid: Principles and Practice*. in *IEEE International Symposium on High Performance Distributed Computing*. 2004: IEEE Computer Science Press.
12. Dumitrescu, C., I. Raicu, and I. Foster. *Experiences in Running Workloads over Grid3*. in *Grid and Cooperative Computing* (GCC'05). Beijing. China. 2005.
13. *LHC Computing Project*. 2004.

14. M. Humphrey, G.W., K. Jackson, J. Boverhof, M. Rodriguez, Joe Bester, J. Gawor, S. Lang, I. Foster, S. Meder, S. Pickles, and M. McKeown. *State and Events for Web Services: A Comparison of Five WS-Resource Framework and WS-Notification Implementations.* in *4th IEEE International Symposium on High Performance Distributed Computing (HPDC-14).* 24-27 July 2005. NC.
15. Chun B., D.C., T. Roscoe, A. Bavier, L. Peterson, M. Wawrzoniak, and M. Bowman, *PlanetLab: An Overlay Testbed for Broad-Coverage Services.* ACM Computer Communications Review, 3, July 2003. 33(3).
16. Dumitrescu, C., et al. *DiPerF: Automated DIstributed PERformance testing Framework.* in *5th International Workshop in Grid Computing.* 2004. Pittsburgh.

Topic 7: Parallel Computer Architecture and Instruction Level Parallelism

Eduard Ayguadé, Wolfgang Karl,
Koen De Bosschere, and Jean-Francois Collard

Topic Chairs

We welcome you to the two Parallel Computer Architecture and Instruction Level Parallelism sessions of Euro-Par 2006 conference being held in Dresden, Germany. The call for papers for this Euro-Par topic area sought papers on all hardware/software aspects of parallel computer architecture, processor architecture and microarchitecture. This year 12 papers were submitted to this topic area. Among the submissions, 5 papers were accepted as full papers for the conference (41% acceptance rate).

Three of the accepted papers cover the hardware aspects of this Euro-Par topic. Ro and Gaudiot present and evaluate the design of hierarchically distributed dispatch queues, as an alternative to the traditional centralized dispatch queue. Authors show how their proposal can be designed with small-sized, distributed dispatch queues which consequently can be implemented with low hardware complexity and lead to high clock rates. Rui, Zhang and Hu present and describe the necessary hardware infrastructure on chip multiprocessors to support a hybrid strategy for prefetching that includes dynamic prefetching threads, automatically constructed, triggered, spawn and managed by hardware, and static prefetching threads, statically constructed by a binary-level optimization tool with the guide of profiling information. Finally, De Dios, Sahelices, Ibez, Vials and Llabera attack in their paper one of the major performance bottlenecks in parallel programs: synchronization. Authors present and show an inexpensive implementation of a novel hardware mechanism, named Request Bypass, to speed-up lock-based synchronizations in DSM multiprocessors.

The two other papers are related with code generation and architecture simulation. Bednarski and Kessler evaluate and compare two methods for optimal integrated VLIW code generation that fully integrate all steps of code generation (instruction selection, register allocation and instruction scheduling). The techniques are based on integer linear programming and dynamic programming, both previously proposed by the same authors. Colmenar, Garnica, Lanchares, Hidalgo and Miana present an architectural simulator able to model asynchronous superscalar architectures, with the aim of studying different architectural proposals for asynchronous processors. The novelty resides in the use of distribution functions to describe the probability of delays.

We are grateful to our referees for lending us their expertise and providing rigorous reviews. We hope that this collection of papers will prove to be interesting and useful to readers, and that the issues raised will stimulate many of them to further research in topics related to this topic. Enjoy!

Optimal Integrated VLIW Code Generation with Integer Linear Programming

Andrzej Bednarski and Christoph Kessler

PELAB, Department of Computer and Information Science,
Linköpings universitet, S-58183 Linköping, Sweden
andbe@ida.liu.se, chrke@ida.liu.se

Abstract. We give an Integer Linear Programming (ILP) solution that fully integrates all steps of code generation, *i.e.* instruction selection, register allocation and instruction scheduling, on the basic block level for VLIW processors.

In earlier work, we contributed a dynamic programming (DP) based method for optimal integrated code generation, implemented in our retargetable code generator OPTIMIST. In this paper we give first results to evaluate and compare our ILP formulation with our DP method on a VLIW processor. We also demonstrate how to precondition the ILP model by a heuristic relaxation of the DP method to improve ILP optimization time.

1 Introduction

We consider the problem of optimal integrated code generation for instruction-level parallel architectures such as VLIW processors. Integrated code generation solves simultaneously, in a single optimization pass, the tasks of instruction selection, instruction scheduling including resource allocation and code compaction, and register allocation.

In previous work [8], we developed a dynamic programming approach and implemented it in our retargetable framework called OPTIMIST [9]. However, there may be further general problem solving strategies that could likewise be applied to the integrated code generation problem. In this paper, we consider the most promising of these, *integer linear programming* (ILP).

ILP is a general-purpose optimization method that gained much popularity in the past 15 years due to the arrival of efficient commercial solvers and effective modeling tools. In the domain of compiler back ends, it has been used successfully for various tasks in code generation, most notably for instruction scheduling.

Wilken et al. [12] use ILP for instruction scheduling of basic blocks which allows, after preprocessing the basic block's data flow graph, to derive optimal solutions for basic blocks with up to 1000 instructions within reasonable time.

ILP formulations integrating instruction scheduling and resource allocation are either *time-based* or *order-based*. In time-based formulations the main decision variables indicate the time slot when an operation is to be started. In order-based formulations the decision variables represent the flow of the hardware resources among operations.

Gebotys *et al.* [5] give a time-based formulation that integrates instruction scheduling and resource allocation and computes time optimal schedules. Leupers and Marwedel [10] provide a time-based ILP formulation for code compaction of a given instruction sequence with alternative instruction encodings.

Zhang [16], Chang *et al.* [2] and Kästner [7] provide order-based and/or time-based ILP formulations for the combination of instruction scheduling with register allocation. Winkel [15] formulates an ILP model for post-pass optimization that can be solved efficiently for global instruction scheduling, including code motion and predication.

We know of only one ILP formulation in the literature that addressed all three tasks simultaneously, which was proposed by Wilson *et al.* [14, 13]. However, their formulation is for single-issue architectures only. Furthermore, their proposed model assumes that the alternatives for pattern matching in instruction selection be exposed explicitly for each node and edge of the basic block's data flow graph (DFG), which would require a preprocessing of the DFG before the ILP problem instance can be generated.

We provide an ILP formulation that fully integrates all three phases of code generation and extends the machine model used by Wilson *et al.* by including VLIW architectures with homogeneous register file. Moreover, our formulation does no longer need preprocessing of the DFG.

The remainder of this paper is organized as follows: After introducing some notation, we provide in Section 3 the ILP formulation for fully integrated code generation for VLIW processors. For a description of the DP approach of OPTIMIST, we refer to a recent article [8]. Section 4 evaluates the DP approach against the ILP approach, and draws some conclusions. Section 5 discusses further directions of ILP approach and Section 6 concludes the article.

2 Notation

We use uppercase letters to denote model parameters and constants provided to the ILP formulation. Lowercase letters denote solution variables and indexes.

Indexes i and j denote nodes of the DFG. We reserve indexes k and l for instances of nodes composing a given pattern. t is used for time index. We use the common notation $|X|$ to denote the cardinality of a set (or pattern) X.

As usual, instruction selection is modeled as a general pattern matching problem, covering the DFG with instances of patterns that correspond to instructions of the target processor. The set of patterns B is subdivided into patterns that consist of a single node, called *singletons* (B''), and patterns consisting of more than one node, with or without edges (B'). That is, $B = B' \cup B''$ such that $\forall p \in B', |p| > 0$ and $\forall p \in B'', |p| = 1$.

In the ILP formulation that follows, we provide several instances of each non-singleton pattern. For example, if there are two locations in the DFG where a multiply-accumulate pattern (MAC) is matched, these will be associated with two different instances of the MAC pattern, one for each possible location. We require that each pattern instance be matched at most once in the final solution. As a consequence, the model requires to specify a sufficient number of pattern instances to cover the DFG. For singleton patterns, we only need a single instance. This will become clearer once we have introduced the coverage equations where the edges of a pattern must correspond to some DFG edges.

2.1 Solution Variables

The ILP formulation uses the following solution variables:

- $c_{i,p,k,t}$ a binary variable that is equal to 1, if a DAG node i is covered by instance node k of pattern p at time t. Otherwise the variable is 0.
- $w_{i,j,p,k,l}$ a binary variable that is equal to 1 if DFG edge (i,j) is covered by a pattern edge (k,l) of pattern $p \in B'$ (see Figure 1).
- $s_{p,t}$ a binary variable that is set to 1 if a pattern $p \in B'$ is selected and the corresponding instruction issued at time t, and to 0 otherwise.
- $r_{i,t}$ a binary variable that is set to 1 if DFG node i must reside in some register at time t, and 0 otherwise.
- τ an integer variable that represents the execution time of the final schedule.

In the equations that follow, we use the abbreviation $c_{i,p,k}$ for the following expression $\sum_{\forall t \in 0..T_{max}} c_{i,p,k,t}$, and s_p for $\sum_{\forall t \in 0..T_{max}} s_{p,t}$.

2.2 Parameters to the ILP Model

The model we provide is sufficiently generic to be used for various instruction-level parallel processor architectures. Our ILP model requires the following parameters:

Data flow graph:
- G index set of DFG nodes
- E_G index set of DFG edges
- OP_i operation identifier of node i, representing a given DFG operation.
- OUT_i indicates the out-degree of DFG node i.

Patterns and instruction set:
- B' index set of instances of non-singleton patterns
- B'' index set of singletons (instances)
- E_p set of edges for pattern $p \in B'$
- $OP_{p,k}$ operator for an instance node k of pattern instance p. This relates to the operation identifier of the DFG nodes.
- $OUT_{p,k}$ is the out-degree of a node k of pattern instance p.
- L_p is an integer value representing the latency for a given pattern p. In our notation, each pattern is mapped to a unique target instruction, resulting in unique latency value for that pattern.

Resources:
- F is an index set of functional unit types.
- M_f represents the amount of functional units of type f, where $f \in F$.
- $U_{p,f}$ is a binary value representing the connection between the target instruction corresponding to a pattern (instance) p and a functional unit f that this instruction uses. It is 1 if p requires f, otherwise 0.
- W, is a positive integer representing the issue width of the target processor, i.e., the maximum number of instructions that can be issued per clock cycle.
- R denotes the number of available registers.
- T_{max} is a parameter that represents the maximum execution time budget for a basic block. The value of T_{max} is only required for limiting the search space, and has no impact on the final result. Observe that T_{max} must be greater (or equal) than the time required for an optimal solution, otherwise the ILP problem instance has no solution.

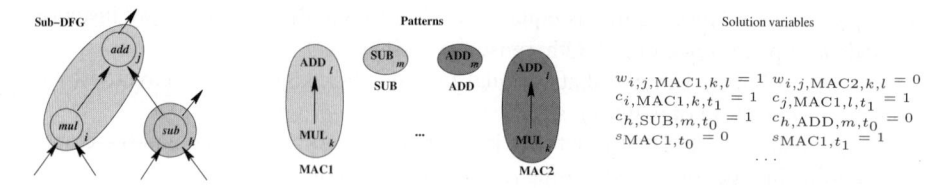

Fig. 1. Example of pattern matching

3 ILP Formulation

To provide the ILP model for fully integrated code generation for VLIW architectures, we first give equations for covering the DFG G with a set of patterns, *i.e.* the instruction selection. Secondly, we specify the set of equations for register allocation. Here we address regular architectures with general purpose registers, and thus only check that the register need does not exceed the amount of physical registers at any time. Next, we address scheduling issues. Since we are working on the basic block level, only flow dependences are considered. We assure that the schedule never exceeds available resources, and that instructions issued simultaneously fit into a long instruction word.

3.1 Instruction Selection

Our instruction selection model is suitable for tree-based and directed acyclic graph (DAG) data flow graphs. Also, it handles patterns in the form of tree, forest, and DAG patterns. The goal of instruction selection is to cover all nodes of DFG G with a set of patterns. For each DFG node i there must be exactly one matching node k in a pattern instance p. Equation (1) enforces this full-coverage property. Solution variable $c_{i,p,k,t}$ records for each node i which pattern instance node covers it, and at what time. Beside full coverage, Equation (1) also assures a requirement for scheduling, namely that for each DFG node i, the instruction corresponding to the pattern instance p covering it is scheduled (issued) at some time slot t.

$$\forall i \in G, \sum_{p \in B} \sum_{k \in p} c_{i,p,k} = 1 \qquad (1)$$

Equation (2) records the set of pattern instances being selected for DFG coverage. If a pattern instance p is selected, all its nodes should be mapped to distinct nodes of G. Additionally, the solution variable $s_{p,t}$ carries the information at what time t a selected pattern instance p is issued.

$$\forall p \in B', \forall t \in 0..T_{max}, \sum_{i \in G} \sum_{k \in p} c_{i,p,k,t} = |p| s_{p,t} \qquad (2)$$

If a pattern instance p is selected, each pattern instance node k maps to exactly one DFG node i. Equation (3) considers this unique mapping only for selected patterns, as recorded by the solution variables s.

$$\forall p \in B', \forall k \in p, \sum_{i \in G} c_{i,p,k} = s_p \qquad (3)$$

Equation (4) implies that all edges composing a pattern must coincide with exactly the same amount of edges in G. Thus, if a pattern instance p is selected, it should cover exactly $|E_p|$ edges of G. Unselected pattern instances do not cover any edge of G. Remark that in our model each pattern instance is distinct, and that we further assume that there are enough pattern instances available to fully cover a particular DFG.

$$\forall p \in B', \sum_{(i,j) \in E_G} \sum_{(k,l) \in E_p} w_{i,j,p,k,l} = |E_p| s_p \quad (4)$$

Equation (5) assures that a pair of nodes constituting a DFG edge covered by a pattern instance p corresponds to a pair of pattern instance nodes. If we have a match ($w_{i,j,p,k,l} = 1$) then we must map DFG node i to pattern instance node k and node j to pattern instance node l of pattern instance p.

$$\forall (i,j) \in E_G, \forall p \in B', \forall (k,l) \in E_p, \; 2w_{i,j,p,k,l} \leq c_{i,p,k} + c_{j,p,l} \quad (5)$$

Equation (6) imposes that instructions corresponding to a non-singleton pattern (instance) p are issued at most once at some time t (namely, if p was selected), or not at all (if p was not selected).

$$\forall p \in B', \; s_p \leq 1 \quad (6)$$

Equation (7) checks that the IR operators of DFG (OP_i) corresponds to the operator $OP_{p,k}$ of node k in the matched pattern instance p.

$$\forall i \in G, \forall p \in B, \forall k \in p, \forall t \in 0..T_{max}, \; c_{i,p,k,t}(OP_i - OP_{p,k}) = 0 \quad (7)$$

Equation (8) simply checks if the out-degree $OUT_{p,k}$ of node k of a pattern instance p equals the out-degree OUT_i of the covered DFG node i. As nodes in singleton patterns are always pattern root nodes, we only need to consider non-singleton patterns, *i.e.* the set B'.

$$\forall p \in B', \forall (i,j) \in E_G, \forall (k,l) \in p, \; w_{i,j,p,k,l}(OUT_i - OUT_{p,k}) = 0 \quad (8)$$

3.2 Register Allocation

Currently we address (regular) architectures with general-purpose register set. We leave modeling of clustered architectures for future work. Thus, a value carried by an edge not entirely covered by a pattern (active edge), requires a register to store that value. Equation (9) forces a node i to be in a register if at least one of its outgoing edge is active, where N is a large number considered to be infinity.

$$\forall t \in 0..T_{max}, \forall i \in G, \sum_{t_t=0}^{t} \sum_{(i,j) \in E_G} \sum_{p \in B} \left(\sum_{k \in p} c_{i,p,k,t_t} - \sum_{l \in p} c_{j,p,l,t_t} \right) \leq Nr_{i,t} \quad (9)$$

If all outgoing edges from a node i are covered by a pattern instance p, there is no need to store the value represented by i in a register. Equation (10) requires solution variable $r_{i,t}$ to be set to 0 if all outgoing edges from i are inactive at time t.

$$\forall t \in 0..T_{max}, \forall i \in G, \sum_{t_t=0}^{t} \sum_{(i,j) \in E_G} \sum_{p \in B} \left(\sum_{k \in p} c_{i,p,k,t_t} - \sum_{l \in p} c_{j,p,l,t_t} \right) \geq r_{i,t} \quad (10)$$

Finally, Equation (11) checks that register pressure does not exceed the number R of available registers at any time.

$$\forall t \in 0..T_{max}, \sum_{i \in G} r_{i,t} \leq R \qquad (11)$$

3.3 Instruction Scheduling

The scheduling is complete when each node has been allocated to a time slot in the schedule such that there is no violation of precedence constraints and resources are not oversubscribed. Since we are working on the basic block level, we only need to model the true data dependences, represented by DFG edges. Data dependences can only be verified once pattern instances have been selected, covering the whole DFG. The knowledge of the covered nodes with their respective covering pattern (i.e., the corresponding target instruction) provides the necessary latency information for scheduling.

Besides full coverage, Equation (1) constrains each node to be scheduled at some time t in the final solution. We need additionally to check that all precedence constraints (data flow dependences) are satisfied. There are two cases: First, if an edge is entirely covered by a pattern p (inactive edge), the latency of that edge must be 0, which means that for all inactive edges (i, j), DFG nodes i and j are "issued" at the same time. Secondly, edges (i, j) between DFG nodes matched by different pattern instances (active edges) should carry the latency L_p of the instruction whose pattern instance p covers i. Equations (12) and (13) guarantee the flow data dependences of the final schedule. We distinguish between edges leaving nodes matched by a multi-node pattern, Equation (12), and the case of edges outgoing from singletons, Equation (13). Active edges leaving a node covered by a singleton pattern p carry always the latency L_p of p.

$$\forall p \in B', \forall (i,j) \in E_G, \forall t \in 0..T_{max} - L_p + 1,$$

$$\sum_{k \in p} c_{i,p,k,t} + \sum_{\substack{q \in P \\ q \neq p}} \sum_{t_t=0}^{t+L_p-1} \sum_{k \in q} c_{j,q,k,t_t} \leq 1 \qquad (12)$$

$$\forall p \in B'', \forall (i,j) \in E_G, \forall t \in 0..T_{max} - L_p + 1,$$

$$\sum_{k \in p} c_{i,p,k,t} + \sum_{q \in B} \sum_{t_t=0}^{t+L_p-1} \sum_{k \in q} c_{j,q,k,t_t} \leq 1 \qquad (13)$$

3.4 Resource Allocation

A schedule is valid if it respects data dependences and its resource usage does not exceed the available resources (functional units, registers) at any time. Equation (14) verifies that there are no more resources required by the final solution than available on the target architecture. In this paper we assume fully pipelined functional units with an occupation time of one for each unit, i.e. a new instruction can be issued to a unit every new clock cycle. The first summation counts the number of resources of type f required by instructions corresponding to selected multi-node pattern instances p at time t. The second part records resource instances of type f required for singletons (scheduled at time t).

$$\forall t \in 0..T_{max}, \forall f \in F, \sum_{\substack{p \in B' \\ U_{p,f}=1}} s_{p,t} + \sum_{\substack{p \in B'' \\ U_{p,f}=1}} \sum_{i \in G} \sum_{k \in p} c_{i,p,k,t} \leq M_f \quad (14)$$

Finally Equation (15) assures that the issue width W is not exceeded. For each issue time slot t, the first summation of the equation counts for multi-node pattern instances the number of instructions composing the long instruction word issued at t, and the second summation for the singletons. The total amount of instructions should not exceed the issue width W, *i.e.*, the number of available slots in a VLIW instruction word.

$$\forall t \in 0..T_{max}, \sum_{p \in B'} s_{p,t} + \sum_{p \in B''} \sum_{i \in G} \sum_{k \in p} c_{i,p,k,t} \leq W \quad (15)$$

3.5 Optimization Goal

In this paper we are looking for a time-optimal schedule for a given basic block. The formulation however allows us not only to optimize for time but can be easily adapted for other objective functions. For instance, we might look for the minimum register usage or code length.

In the case of time optimization goal, the total execution time of a valid schedule is derived from the solution variables c as illustrated in Equation (16).

$$\forall i \in G, \forall p \in P, \forall k \in p, \forall t \in 0..T_{max}, c_{i,p,k,t} * (t + L_p) \leq \tau \quad (16)$$

The total execution time is less or equal to the solution variable τ. Looking for a time optimal schedule, our objective function is to minimize τ.

4 Evaluation

First, we provide two theoretical VLIW architectures for which we generate target code. Secondly we describe the experimental setup that we used to evaluate our ILP formulation against our previous DP approach and summarize the results.

4.1 Target Architectures

In order to compare OPTIMIST's DP technique to the ILP formulation of Section 3, we use two theoretical VLIW target platforms (Case I and Case II) with the following characteristics.

Case I: The issue width is a maximum of two instructions per clock cycle. The architecture has an arithmetic-logical unit (ALU). Most ALU operations require a single clock cycle to compute (occupation time and latency are one). Multiplication and division operations have a latency of two clock cycles. Besides the ALU, the architecture has a multiply-and-accumulate unit (MAC) that takes two clock cycles to perform a multiply-and-accumulate operation. There are eight general purpose registers accessible from

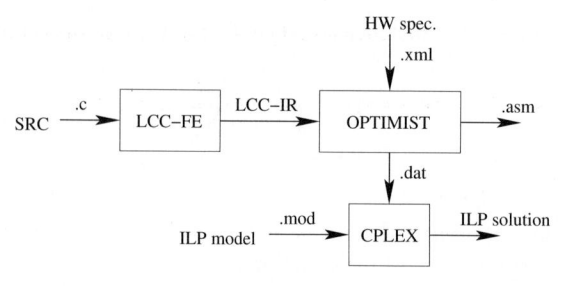

```
                        HW spec.
                         ↓ .xml
        .c        LCC-IR
SRC →  LCC-FE  →         OPTIMIST        .asm
                                         →
                         ↓ .dat
                   .mod              ILP solution
ILP model    →     CPLEX        →
```

Fig. 2. Experimental setup

any unit. We assume a single memory bank with unlimited size. A load/store unit (LS) stores and loads data in four clock cycles.

Case II: The issue width is of maximum four instructions per clock cycle. The architecture has twice as many resources as in Case I, *i.e.* two arithmetic-logical units, two multiply-and-accumulate units, and two load/store units with the same characteristics.

4.2 Experimental Setup

We implemented the ILP data generation module within the OPTIMIST framework. Currently our ILP model addresses VLIW architectures with regular pipeline, *i.e.* functional units are pipelined, but no pipeline stall occurs. We adapted hardware specifications in xADML [1, Chap. 8] such that they fit current limitations of the ILP model. In fact, the OPTIMIST framework accepts more complex resource usage patterns and pipeline descriptions expressible in xADML, which uses the general mechanism of reservation tables [3]. As assumed in Section 3, we use for the ILP formulation the simpler model with unit occupation time and a latency for each instruction. An extension of the ILP formulation to use general reservation tables is left to future work.

Figure 2 shows our experimental platform. We provide a plain C code sequence as input to OPTIMIST. We use LCC [4] (within OPTIMIST) as C front-end. Besides the source code we provide the description of the target architecture in xADML language For each basic block, OPTIMIST outputs the assembly code as result. If specified, the framework also outputs the data file for the ILP model of Section 3. The data file contains architecture specifications, such as the issue width of the processor, the set of functional units, patterns, *etc.* that are extracted from the architecture description document. It generates all parameters introduced in Section 2.2. Finally we use the CPLEX solver [6] to solve the set of equations.

Observe that for the ILP data we need to provide the upper bound for the maximum execution time (T_{max}). For that, we first run a heuristic variant of DP that still considers full integration of code generation phases, and provide its execution time (computed in a fraction of a second) as the T_{max} parameter to the ILP data.

4.3 Results

We generated code for basic blocks taken from various digital signal processing benchmark programs. We run the evaluation of the DP approach on a Linux (kernel 2.6.13)

PC with Athlon 1.6GHz CPU and 1.5GB RAM. The ILP solver runs on a Linux (kernel 2.6.12) PC with Athlon 2.4GHz CPU, 512MB RAM using CPLEX 9.

We should mention a factor that contributes in favor of the ILP formulation. In the OPTIMIST framework we use LCC [4] as C front-end. Within our framework we enhanced the intermediate representation with extended basic blocks [11] (which is not standard in LCC). As consequence, we introduced data dependence edges for resolving memory write/read precedence constraints. In the current ILP formulation we consider only data flow dependences. Thus, we instrumented OPTIMIST to remove edges introduced by building extended basic blocks. Removing dependence edges results in DAGs with larger base, *i.e.* with larger number of leaves, and in general a lower height. We are aware that the DP approach suffers from DAGs with a large number of leaves, as OPTIMIST early generates a large number of partial solutions. Further, removing those edges builds DFGs that may no longer be equivalent to the original C source code. However, it is still valid to compare the ILP and DP techniques, since both formulations operate on the same intermediate representation.

Table 1 reports our results for the Case I architecture. The first column indicates the name of the basic block. The second column reports the number of nodes in the DAG for that basic block. The third and fourth columns give the height of the DAG and the number of edges, respectively. Observe that the height corresponds to the longest path of the DAG in terms of number of DAG nodes, and not to its critical path length,

Table 1. Evaluation of ILP and DP fully integrated code generation approaches for the Case I architecture

Basic block		$|G|$	Height	$|E_G|$	DP		ILP	
					τ (cc)	t (sec)	τ (cc)	t (sec)
1)	iir filter bb9	10	4	10	10	0.3	10	0.9
2)	vec_max bb8	12	4	12	11	0.6	11	1.3
3)	dijkstra bb19	16	7	15	14	6.6	14	5.6
4)	fir filter bb9	16	3	14	15	61.3	15	7.8
5)	cubic bb16	17	6	16	14	15.0	14	5.7
6)	fir_vselp bb10	17	9	17	16	3.4	16	8.2
7)	matrix_sum loop bb4	17	8	17	16	4.0	16	8.8
8)	scalarprod bb2	17	8	18	17	1.2	17	15.8
9)	vec_sum bb3	17	8	18	16	1.4	16	11.8
10)	matrix_copy bb4	18	7	19	16	4.3	16	12.5
11)	cubic bb4	21	8	23	17	69.8	17	277.7
12)	iir filter bb4	21	6	17	20	3696.4	20	46.5
13)	fir filter bb11	22	6	27	19	89.7	CPLEX	
14)	codebk_srch bb20	23	7	22	17	548.8	17	63.1
15)	fir_vselp bb6	23	9	25	19	40.6	CPLEX	
16)	summatrix_un1 bb4	24	10	28	20	25.4	CPLEX	
17)	scalarprod_un1 bb2	25	10	30	19	14.9	CPLEX	
18)	matrixmult bb6	30	9	35	23	2037.7	AMPL	
19)	vec_sum unrolled bb2	32	10	40	24	810.9	AMPL	
20)	scalarprod_un2 bb2	33	12	42	23	703.1	AMPL	

whose calculation is unfeasible since the instruction selection is not yet known. The fifth column reports the amount of clock cycles required for the basic block, and in the sixth column we display the computation time (in seconds) for finding a DP solution. Columns seven and eight report the results for ILP. The computation time for the ILP formulation does not include the time for CPLEX-presolve that optimizes the equations.

In the tables we use three additional notations: CPLEX indicates that the ILP solver ran out of memory and did not compute a result. AMPL means that CPLEX-presolve failed to generate an equation system, because it ran out of memory. Where the DP ran out of memory we indicate the entry as MEM.

For all cases that we could check both techniques report the same execution time (τ). It was unexpected to see that the ILP formulation performs quite well and in several cases with an order of magnitude faster than DP. For cases 4), 12) and 14) in Table 1 the DP takes almost eight times, eighty times and nine times respectively longer than the ILP solver to compute an optimal solution. Since we removed the memory data dependence edges (as mentioned earlier) the resulting test cases present two, four and two unrelated DAGs for case 4), 12) and 14) respectively. We know that DP suffers from DAGs with a large number of leaves because a large number of selection nodes is generated already at the first step. For the rest of the test cases, DP outperforms the ILP formulation or has similar computation times. Observe that we reported for cases 3) and 5) that ILP takes shorter time to compute an optimal solution. But if we include the time of CPLEX-presolve, which runs for 7.1s in case 3) and 8.3s in case 5), the ILP times are worse or equivalent. For problems larger than 22 nodes, the ILP formulation fails to compute a solution. For problem instances over 30 nodes, the CPLEX-presolve does not generate equations because it runs out of memory.

Table 2 shows the results for the Case II architecture. The notations are the same as for Case I. We added an additional column in the ILP part, denoted t′, that reports the

Table 2. Evaluation of ILP and DP fully integrated code generation approaches for the Case II architecture

Basic block	$|G|$	Height	$|E_G|$	DP τ (cc)	DP t (sec)	ILP τ (cc)	ILP t (sec)	t′ (sec)
1) iir filter bb9	10	4	10	9	0.6	9	1.5	0.4
2) vec_max bb8	12	4	12	10	2.4	10	1.6	0.7
3) dijkstra bb19	16	7	15	14	73.5	14	10.7	4.2
4) fir filter bb9	16	3	14	9	2738.9	9	9.1	2.5
5) cubic bb16	17	6	16	12	1143.3	12	CPLEX	3.8
6) fir_vselp bb10	17	9	17	14	62.1	14	CPLEX	4.9
7) matrix_sum loop bb4	17	8	17	15	90.2	15	CPLEX	10.2
8) scalarprod bb2	17	8	18	15	10.0	—	CPLEX	CPLEX
9) vec_sum bb3	17	8	18	13	11.4	13	CPLEX	4.6
10) matrix_copy bb4	18	7	19	14	89.4	14	AMPL	4.1
11) cubic bb4	21	8	23	16	8568.7	—	AMPL	CPLEX
12) iir filter bb4	21	6	17	—	MEM	12	AMPL	7.4
13) fir filter bb11	22	6	27	—	MEM	—	AMPL	CPLEX
14) codebk_srch bb20	23	7	22	—	MEM	—	AMPL	CPLEX
15) fir_vselp bb6	23	9	25	16	7193.9	—	AMPL	AMPL

ILP computation time when the upper bound T_{max} is derived from a run of a heuristically pruned DP algorithm [1, Chap. 4] (this decreases the number of generated equations by providing a value of T_{max} closer to an optimal solution). The time for this DP run for preconditioning the ILP (within a fraction of a second) is not included in t′.

For the cases 4) and 12) in Table 2, DP performs worse than ILP. For the case 12) DP runs out of memory, whereas the ILP could compute a solution within 7.4s if T_{max} is close enough to the optimum. The case II results show that it is beneficial to spend time on minimizing T_{max}. We could gain four additional nodes in ILP problem size. For Case II, if the ILP computes a solution it outperforms the DP.

5 Future Work

The current ILP formulation lacks several features of the OPTIMIST framework. In this paper we considered target architectures that suit the ILP model. We plan to extend the formulation to handle clustered VLIW architectures, such as Veloci-TI DSP variants. For that, we will need to model operand residences (*i.e.*, in which cluster or register set a value is located). This will certainly increase the amount of generated variables and equations and affect ILP performance.

Also, we need to formulate the insertion of spill code. The current ILP formulation assumes a sufficient number of registers, which is not generally the case.

We also mentioned that the current ILP formulation is based on a simpler resource usage model that is limited to unit occupation times per functional unit and a variable latency per target instruction. It would be of interest to have a more general model using reservation tables for specifying arbitrary resource usage patterns and complex pipelines, which is already implemented in OPTIMIST's DP framework.

Finally, we will extend the scope of the optimization beyond the basic block level, in particular to integrated software pipelining of loops.

6 Conclusions

In this paper we provided an integer linear programming formulation for fully integrated code generation for VLIW architectures that includes instruction selection, instruction scheduling and register allocation. We extended the formulation by Wilson *et al.* [14] for VLIW architectures. In contrast to their formulation, we do no longer need to preprocess the DFG to expose instruction selection alternatives. Moreover, we have a working implementation where ILP instances are generated automatically from the OPTIMIST intermediate representation and a formal architecture description in xADML.

We compared the ILP formulation with our research framework for integrated code generation, OPTIMIST, which uses dynamic programming. We evaluated both methods on theoretical architectures that fit the ILP model restrictions. Where the ILP solver terminates successfully, the ILP-based optimizer mostly works faster than the dynamic programming approach; on the other hand, it fails for several larger examples where dynamic programming still provides a solution. Hence, the two approaches complement each other. Moreover, the ILP approach profits from preconditioning by a heuristic variant of DP.

Currently, our ILP formulation lacks support for memory dependences and for irregular architecture characteristics, such as clustered register files, complex pipelines, *etc.* We intend to complete the formulation as part of future work. Further we need to address insertion of spill code.

Acknowledgments. We thank Petru Eles and Alexandru Andrei from ESLAB of Linköpings universitet for letting us using their CPLEX installation. This research was partially funded by the Ceniit program of Linköpings universitet and by SSF RISE.

References

1. A. Bednarski. *Integrated Optimal Code Generation for Digital Signal Processors*. PhD thesis, Linköpings universitet, Linköping, Sweden, June 2006.
2. C.-M. Chang, C.-M. Chen, and C.-T. King. Using integer linear programming for instruction scheduling and register allocation in multi-issue processors. *Computers Mathematics and Applications*, 34(9):1–14, 1997.
3. E. S. Davidson, L. E. Shar, A. T. Thomas, and J. H. Patel. Effective control for pipelined computers. In *Proc. Spring COMPCON75 Digest of Papers*, pages 181–184. IEEE Computer Society Press, Feb. 1975.
4. C. W. Fraser and D. R. Hanson. *A Retargetable C Compiler: Design and Implementation*. Addison-Welsey Publishing Company, 1995.
5. C. H. Gebotys and M. I. Elmasry. Simultaneous scheduling and allocation for cost constrained optimal architectural synthesis. In *DAC '91: Proceedings of the 28th conference on ACM/IEEE design automation*, pages 2–7, New York, NY, USA, 1991. ACM Press.
6. I. Inc. CPLEX homepage. http://www.ilog.com/products/cplex/, 2005.
7. D. Kästner. *Retargetable Postpass Optimisations by Integer Linear Programming*. PhD thesis, Universität des Saarlandes, Saarbrücken, Germany, 2000.
8. C. Kessler and A. Bednarski. Optimal integrated code generation for VLIW architectures. To appear in *Concurrency and Computation: Practice and Experience*, 2006.
9. C. Kessler and A. Bednarski. OPTIMIST. www.ida.liu.se/~chrke/optimist, 2005.
10. R. Leupers and P. Marwedel. Time-constrained code compaction for DSPs. *IEEE Transactions on VLSI Systems*, 5(1):112–122, 1997.
11. S. S. Muchnick. *Advanced Compiler Design and Implementation*. Morgan Kaufmann Publishers, 1997.
12. K. Wilken, J. Liu, and M. Heffernan. Optimal instruction scheduling using integer programming. In *Proc. ACM SIGPLAN Conf. Programming Language Design and Implementation*, pages 121–133, 2000.
13. T. Wilson, G. Grewal, B. Halley, and D. Banerji. An integrated approach to retargetable code generation. In *Proc. 7th international symposium on High-level synthesis (ISSS'94)*, pages 70–75. IEEE Computer Society Press, 1994.
14. T. C. Wilson, N. Mukherjee, M. Garg, and D. K. Banerji. An integrated and accelerated ILP solution for scheduling, module allocation, and binding in datapath synthesis. In *The Sixth Int. Conference on VLSI Design*, pages 192–197, Jan. 1993.
15. S. Winkel. *Optimal Global Instruction Scheduling for the Itanium® Processor Architecture*. PhD thesis, Universität des Saarlandes, Saarbrücken, Germany, Sept. 2004.
16. L. Zhang. *SILP. Scheduling and Allocating with Integer Linear Programming*. PhD thesis, Technische Fakultät der Universität des Saarlandes, Saarbrücken (Germany), 1996.

Speeding-Up Synchronizations in DSM Multiprocessors[*]

A. de Dios[1], B. Sahelices[1], P. Ibáñez[2], V. Viñals[2], and J.M. Llabería[3]

[1] Dpto. de Informática. Univ. de Valladolid
{agustin, benja}@infor.uva.es
[2] Dpto. de Informática e Ing. de Sistemas, I3A and HiPEAC. Univ. de Zaragoza
{imarin, victor}@unizar.es
[3] Dpto. de Arquitectura de Computadores. Univ. Polit. de Cataluña
llaberia@ac.upc.es

Abstract. Synchronization in parallel programs is a major performance bottleneck. Shared data is protected by locks and a lot of time is spent in the competition arising at the lock hand-off. In this period of time, a large amount of traffic is targeted to the line holding the lock variable. In order to be serialized, the requests to the same cache line can either be bounced (NACKed) or buffered in the coherence controller. In this paper we focus on systems whose coherence controllers buffer requests.

During lock hand-off only the requests from the winning processor contribute to the computation progress, because the winning processor is the only one that will advance the work. This key observation leads us to propose a hardware mechanism named Request Bypass, which allows requests from the winning processor to bypass the requests buffered in the home coherence controller keeping the lock line. The mechanism does not require compiler or programmer support nor ISA or coherence protocol changes.

By simulating a 32 processor system we show that Request Bypass reduces execution time and lock stall time up to 35% and 75%, respectively. The programs limited by synchronization benefit the most from Request Bypass.

1 Introduction

The scalability of shared-memory programs is often limited by highly-contended critical sections guarded by mutual exclusion locks [1,2], where a large amount of traffic is generated during the lock hand-off. This traffic increases the time that the parallel program spends in serial mode, which reduces the benefits of parallel execution. Thus, optimizing lock transfer among processors is essential to achieve high performance in applications having highly-contended critical sections [3,4,5,6,7].

The processor architecture provides specific instructions to perform an atomic read-modify-write operation on a memory location. A lock is acquired by using

---

[*] This work was partly funded by grants TIN2004-07739-C02-01/02 (Spanish Ministry of Education/Science and European RDF) and the Diputación General de Aragón.

these instructions and it is released by performing a regular write. A Highly contended lock hand-off generates a burst of traffic aimed at the same memory location. This situation may be alleviated by queue-based software [4,8,9,1] or hardware locks [5,7,10,11]. Software mechanisms have a large overhead per synchronization access, even in the absence of contention. The proposed hardware mechanisms require modifications in software and/or in the coherence protocol, they need to handle queue breakdowns, and some of them require a hardware predictor to identify synchronization operations [11]. Request Bypass, the mechanism proposed in our work, does not add a significative overhead, does not use a predictor, and it does not require changes to cache or directory protocol.

In Distributed Shared-Memory (DSM) multiprocessors a coherence request is handled by the coherence controller of the node owning the corresponding line (home node). Moreover, the coherence controller is in charge of serializing all requests targeted to the same memory address. So, requests coming to a busy directory entry cannot be attended until the directory entry becomes free. A directory entry is busy whenever the coherence controller has started a coherence operation on such entry involving a third node whose reply has not been received yet. Requests to busy entries are handled in three ways in commercial DSM multiprocessors or in the literature: either bounced [12,13], forwarded to third nodes [14,15,16] or queued within the coherence controller [17]. Our base system uses request queuing because it has the potential to reduce network traffic, contention and coherence controller occupancy as it is shown by Chaudhuri and Heinrich in [17].

In this paper we are concerned with the lock hand-off in a DSM multiprocessor that queue requests to busy lines within the coherence controller. In order to speed up lock hand-off we propose to change the order in which the coherence controller selects the request to be processed once a line leaves the busy state. Instead of always selecting an already queued request we suggest processing first the request in the input port, if it exists, a technique we call Request Bypass. At the acquire phase of the lock hand-off, Request Bypass allows the request of the winning processor (that which is going to acquire the lock) to bypass the requests to the same line pending in the queue. A similar bypassing situation can arise when accessing shared variables inside critical sections and when releasing a critical section. The implementation we propose of Request Bypass does not require compiler or programmer support nor ISA or coherence protocol changes.

In Section 2 we use an example to describe a lock hand-off for a highly contended critical section in a baseline system, and in Section 3 we analyze the same example under Request Bypass. In Section 4 we present simulation results using Splash-2 benchmarks for 32 processors. We include a comparison of our proposal with Read Combining [17]. In Section 5 we discuss related work and we conclude in Section 6.

2 Lock-Transfer Contention

We first describe the baseline coherence controller. Next, we elaborate on an example case of a lock transfer among several processors. This example allows us to identify inefficiency sources and motivates the main idea of the paper.

2.1 Coherence Controller Model

The baseline model is based on a CC-NUMA multiprocessor with a MESI cache coherence protocol similar to the SGI Origin 2000 system [13]. Every memory line is allocated to one directory entry within a coherence controller which stores the line state and processes its requests.

Figure 1.a shows a logical view of part of the coherence controller structure. The coherence controller receives cache requests from the nodes. Requests can be of three different types: READ_SH, READ_OWN and UPGRADE. READ_SH and READ_OWN are used to request a line in Shared or Exclusive state, respectively, and UPGRADE is used to change the line state from Shared to Exclusive. Once the coherence requests are processed, the controller sends three types of replies: REPLY_SH and REPLY_EXCL supply a line in Shared or Exclusive state, respectively, and REPLY_UPGRADE acknowledges the change from Shared to Exclusive. When needed, the coherence controller sends line invalidation (INV) or cache-cache transfer (COPYBACK) requests.

Request processing is based on two structures handled by the coherence controller: a Busy State Queue (BSQ) and a Pending Request Queue (PRQ); each PRQ entry is in turn another queue. The incoming coherence requests are taken from the input port and processed. If a request requires some third-node reply, the involved line is flagged as busy and stored in BSQ. Any request targeted to such a busy line appearing in the meantime is serialized by enqueuing its identity (originating processor, request type, etc.) into the PRQ entry associated to the corresponding BSQ entry. Otherwise, a request targeted to a non-busy lines is processed.

After receiving the last reply a busy line is waiting for, the busy state in BSQ is cleared and the coherence controller begins processing the list of pending requests to such a line in PRQ. As before, if during such processing a request requires a third-node reply, the line is tagged as busy and the coherence controller stops processing the list. When there are no pending requests in PRQ that can be processed the controller listens to the input port. Whenever the protocol runs out of BSQ or PRQ entries the coherence controller resorts to sending NACKs.

2.2 Lock Hand-Off Example

Figure 1.b shows a typical critical section and the code used to acquire and release a lock variable. If the lock variable is already closed the code spins on a regular load instruction. Once the lock variable is released, the atomic test&set instruction tries to acquire it. Releasing is done by a regular store instruction.

Next we make a detailed study of the lock hand-off in a highly contended (n competing processors) critical section controlled by the lock variable B. Assume the lock is initially owned by a processor we call Owner, while the other $n-1$ processors are spinning on a local copy of B in Shared state. Consider that Owner is going to execute *Release(B)* and leave the critical section. After the lock hand-off is accomplished, one among the $n-1$ contending processors, we call Winner, will enter the critical section. The example in Figure 2.2.a shows such a scenario

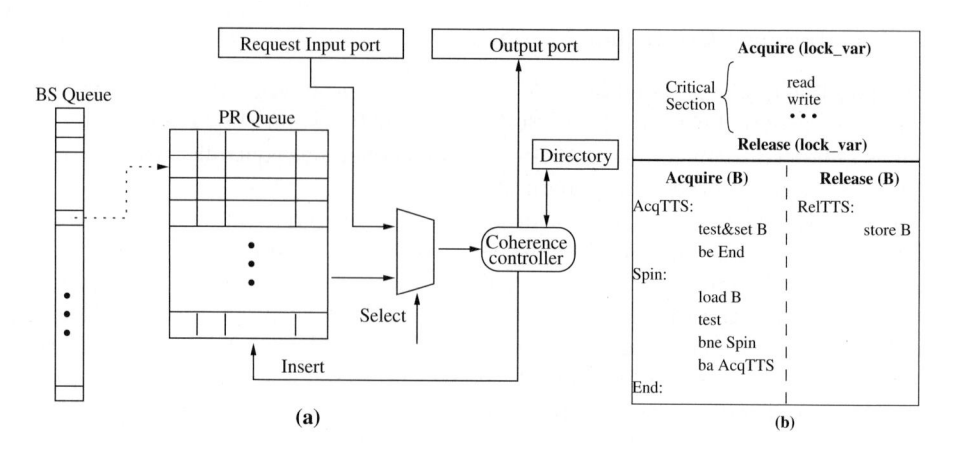

Fig. 1. (a) Logical view of part of the Coherence Controller. (b) Lock-based critical section skeleton (up), and code to acquire and release a lock variable (down).

of contention, where the Owner, the Winner, the Home node and the remaining $n-2$ contending processors are plotted from left to right, respectively.

When Owner executes *Release(B)* it sends out an UPGRADE request to the Home node, which in turn sends INVALIDATION requests to the $n-1$ processors having the line in Shared state. The $n-1$ processors invalidate the line and send INVALIDATION replies to Home, which collects all replies and sends the UPGRADE reply to Owner. This UPGRADE reply is the first activity ploted in Figure 2.2.a. The example continues as follows:

- The $n-1$ processors miss loading variable B and send READ_SH to Home.
- When Home receives the first READ_SH request, it sends a cache-cache transfer request to Owner (see (1) in Figure 2.2.a), puts the line in busy state, and buffers the remaining $n-2$ READ_SH into PRQ.
- The Owner replies (ACK) to Home and (REPLY_SH) to Winner.
- The Winner executes *test&set* instruction and sends an UPGRADE request to Home (see (2) in Figure 2.2.a). The UPGRADE request of the Winner may be delayed in the input port while PRQ is emptied of READ_SH requests.
- The Home process the $n-2$ READ_SH sending REPLY_SH (now the lock is open) to every processor.
- The $n-2$ contenders receive the lock open and execute test&set. All they send UPGRADE requests to Home (not shown in figure).
- The Home process the UPGRADE request of Winner and sends INVALIDATION requests to the Owner and the $n-2$ contenders (see (3) in Figure 2.2.a).
- The Home waits for all INVALIDATION ACKs and then sends an UPGRADE reply to Winner.
- The lock hand-off has been accomplished and the Winner can execute the critical section.

When processing the UPGRADE request of the remaining processors, the Home invalidates the previous copies of the line and sends it to the requesting processor,

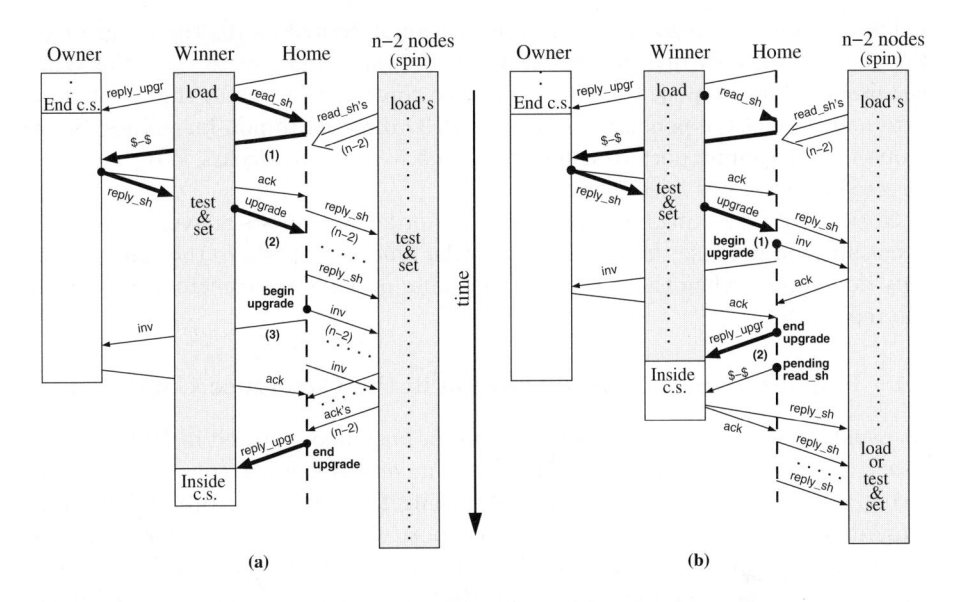

Fig. 2. Example of how a processor (Owner) frees a critical section and $n-1$ processors are contending for it. **(a)** Without bypassing. **(b)** With bypassing.

which sees the lock closed and resumes spinning. This line will be invalidated when processing the next UPGRADE request, so the processor will generate a new READ_SH that will be kept in PRQ to be serialized. While the contenders are generating the described traffic, the Winner processor is inside the critical section accessing the shared variables. Therefore, if the shared data and the lock variable are allocated to the same coherence controller, the accesses to shared variables may be delayed. This happens as long as the coherence controller is processing PRQ requests to the lock variable that do not put the lock line in Busy state.

This is because the coherence controller can be processing PRQ requests to the lock variable that does not put the lock line in Busy state. Later on, the winner releases the lock (store B) and generates a READ_OWN request. Again, it is delayed by all requests in PRQ that still contend for the line holding B.

Putting it all together, it can be expected that all of this message overhead will significantly increase the execution time of small, highly-contended, critical sections.

3 Bypassing PRQ Requests

From the above example we can stablish the following: whenever several processors compete to enter into a critical section, the request traffic originated by the loosing contenders can delay all the Winner execution phases (lock acquiring, shared data accesses and lock release). In this situation, processing PRQ requests before listen to the input port does not contribute to the progress of the Winner, the only one that will advance the work. In order to favor the Winner progress, we propose that the coherence controller listens to the input port before attending the pending requests of PRQ directed to non-busy lines. So, a request in the input port to a non-busy line is going to bypass PRQ requests directed to non-busy lines.

Notice that by issuing replies in a different order as requests arrive, correctness is not affected because the serialization order among requests to the same line is only determined when the coherence controller updates the directory and sends the reply.

3.1 Request Bypass Implementation in the Coherence Controller

As usual, an input port request targeted to a busy line is stored in PRQ, otherwise it is processed immediately. However, under a Request Bypass policy after receiving the last reply a busy line is waiting for, instead of processing PRQ requests associated to that line, the input port will be attended. Moreover, if a request appears in the input port while processing a PRQ entry, such a request will bypass all the outstanding work in PRQ. The Request Bypass policy can be easily implemented by adding a new state to each BSQ entry: the Ready state, which indicates the existence of outstanding work in PRQ.

Anyway, a Ready BSQ entry can become Busy if a request (coming either from the input port or from PRQ) require a third-node communication.

3.2 Lock Hand-Off Example with Request Bypass

Figure 2.2.b shows the previous example under PRQ bypassing. We suppose that, at the time the Winner's UPGRADE reaches Home, the coherence controller is processing the first READ_SH request of the remaining contenders (the losers). When such a request is completed, the coherence controller visits the input port and processes the UPGRADE request, bypassing the n-3 READ_SH requests kept in PRQ. In our example, processing the UPGRADE request requires only two invalidations to be sent out (see (1) in Figure 2.2.b), one to the owner processor and another one to the single contending processor having a copy of the lock line ($n-1$ invalidations required without bypassing).

Once the Winner's UPGRADE completes (see (2) in Figure 2.2.b), all the $n-3$ remaining READ_SH requests that were bypassed will be processed. However, in contrast with the previous situation, the losers receive the READ_SH reply with the lock closed and therefore remain spinning locally, not executing the test&set instruction nor generating any request (UPGRADE or READ OWN).

While the coherence controller is servicing READ_SH requests from PRQ, the Winner is inside the critical section, sending requests (may be some of them to the same controller) to access the shared variables, and sending a final request to release the lock. However, such requests are not delayed because they bypass PRQ.

3.3 Forward Progress Warranty

Bypassing PRQ requests can delay execution endlessly. As an example let us suppose that the code to acquire a critical section spins on a test&set instruction. In a highly contended critical section the coherence controller is receiving READ_OWN requests continuously. If the owner of the critical section is trying to release it by sending a READ_OWN request, and this request is queued in PRQ, then the owner stays indefinitely in the critical section. Bypassing READ_SH has a similar problem.

In order to warrant forward progress, we suggest limiting the maximum number of consecutive bypasses. We can implement this idea by incrementing a counter each time an input port request bypasses PRQ and decreasing the counter each time a request is processed in PRQ. If the counter has a value between 0 and $Max - 1$ then the controller works in Request Bypass mode. Otherwise, when the counter gets its maximum value the controller switches to default mode and remains in it until the counter decreases. Our experiments show good results with a 5-bit counter.

4 Experimental Results

Our simulations have been conducted with RSIM [18,19]. It is an execution-driven simulator performing a detailed cycle-by-cycle simulation of an out-of-order processor, a memory hierarchy, and an interconnection network. The processor implements a sequential consistency model using speculative load execution [20]. Coherence is based on a MESI protocol similar to the SGI Origin 2000 system [13]. The network is a wormhole-routed two-dimensional mesh network. Port contention, switches and links are accurately modeled. Table 1.a lists the processor, cache and memory system parameters.

As a workload we have chosen a SPLASH-II subset [21] having a significant amount of synchronization, see Table 1.b. In *Ocean* we use the optimization suggested in [22]. The applications have been compiled with a *test and test&set*-based synchronization library (Figure 1.b) implemented with the RMW instruction. Barriers are implemented with a simple binary tree.

Our results show execution time broken down into four categories: *lock*, *barrier*, *memory* and *compute*. The algorithm used to add a cycle into a given category works as follows: if the maximum allowed number of instructions can be committed from the ROB, the cycle is added to *compute*. Otherwise, the cycle is added to the stall category to which belongs the oldest instruction that can not be committed, as suggested in [19].

4.1 Results

In this section we present results for a baseline system without bypassing, and for a system enhanced with Request Bypass. We also consider a third system, by enhancing the baseline with Read Combining [17]. Chaudhuri and Heinrich propose Read Combining in the context of queuing coherence controllers, in

Table 1. Simulated system parameters and applications

Processor	1 Ghz		L2/Memory Bus	Split. 32–bits 3–cycle+1 arbit.
ROB	64–entry, 32–entry LS queue			
Issue	out–of–order issue/commit 4–ops/cyc.		Memory	4–way interleaved, 50–cycle DRAM
Branch	512–entry branch predictor buffer		Directory	SGI Origin–2000 based MESI
Cache			Cycle	16–cycle (without memory)
L1 inst.	Perfect		Interleaving	4 controllers per node
L1 data	128–Kbyte, direct mapped, write–back		BSQ/PRQ size	64/16–entries
	2 ports, 1–cycle, 16 outstanding misses		**Network**	Pipelined point–to–point
L2	1–Mbyte, 4–way associative, write–back		Network width	8–bytes/flit
	10–cycle access, 16 outstanding misses		Switch buffer size	64 flits
L1/L2 bus	Runs at processor clock		Switch latency	4–cycles/flit + 4 arbit.
Line size	64 bytes			

(a) Simulated system parameters

Code	Ocean	Barnes	Volrend	Water–Nsq	Water–Spt	FMM
Input	130x130	4K particles	head–scaleddown2	512 molecules	512 molec.	2K partic.

(b) Applications

order to speed up multiple read requests to the same line. In order to achieve this, Read Combining dictates that once the controller gets the line, it is stored in a fast data buffer which is repeatedly used to send out all the read requests replies. They show that Read Combining also benefits lock transfer by enabling a faster distribution of the cache line storing the lock variable. Finally, we evaluate the performance of merging Request Bypass with Read Combining.

Figure 3 shows the parallel execution time break into the former categories and normalized to the baseline system. From left to right we show, for each application, the baseline system (Baseline), the baseline system enhanced with the Read Combining (RC) [17], the baseline system enhanced with Request Bypass (Byp) and two ways of merging Request Bypass with Read Combining (RC+Byp and RC+BypS, see details below). We only show data for 32 processors because Splash-2 applications have small Acquire-related times with 16 processors [23].

By applying Request Bypass to the Baseline system, the Lock time becomes greatly reduced, from 12% to 75%. Reductions in the Barrier time can also be observed, to a greater or lesser extent, for all applications, from 1% to 48%. This reduction in the Barrier stall time can be explained as follows: when a critical section executes before a nearby barrier, reducing Lock-related stalls also reduces Barrier-related stalls, because the Barrier stall time of a given processor starts from when it reaches the barrier until the slowest processor exits the critical section and crosses the barrier the last. Summarizing, the overall time reduction obtained with Request Bypass varies from 1% to 35%.

The Lock time reduction achieved by Request Bypass is greater than Read Combining for most applications. Read Combining itself does not alleviate the delay experienced by the Winner processor because it does not acquire the critical

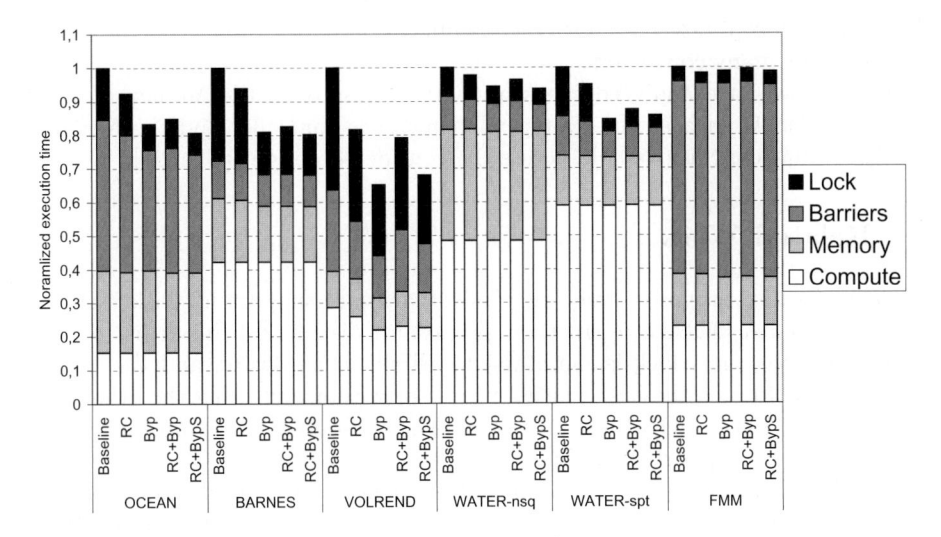

Fig. 3. Normalized execution time with 32 processors for the Baseline, Read Combining (RC), and Request Bypass (Byp) systems. RC+Byp (Blind merging) and RC+BypS (Selective merging) are the merged systems.

section until the contending processor requests kept in the buffer have been replied. Moreover, Read Combining exposes subsequent delays when accessing protected data and releasing the lock. Until the contending processors start busy-waiting on a local copy of the line, their requests will delay the progress of the Winner processor, firstly by delaying the requests made to the same coherence controller within the critical section, and secondly by delaying the request to release the lock.

Next we analyze the interaction between Request Bypass and Read Combining when applied simultaneously. Figure 3 presents data for two experiments merging Request Bypass and Read Combining namely Blind (RC+Byp) and Selective (RC+BypS). Blind merging implements both techniques simultaneously as they have been defined, resulting in a Lock time increase for all applications. This is because Read Combining speeds-up the READ_SH replies to contenders when a lock is released, and as a consequence, the UPDATE request in a Blind merging system bypasses less READ_SH requests than with Request Bypass working alone. So, more processors receive the lock variable opened, execute the test&set instruction, and have to be invalidated. Moreover, when the Winner wants to release the lock, its UPGRADE (or READ_OWN) request cannot bypass PRQ because the lock line is busy most of the time (the coherence controller is servicing the READ_OWN requests of the test&set instructions).

A Selective merging of Request Bypass and Read Combining tries to overcome the above problem by applying Read Combining only to lines which do not contain a lock variable. The execution time of Selective merging is similar to that of Request Bypass alone in our benchmarks. However, we can expect a better behavior in programs with a communication pattern where one processor produces for many consumers.

Finally, we have analyzed the sensitivity of results to the latency of some key components such as the router, the coherence controller and the memory, verifying that conclusions hold across the considered design space [23].

5 Related Work

Goodman et al. propose a very aggressive hardware support for locks (QLB - originally called QOSB) [5]. In their proposal a distributed linked list of processors waiting on a lock is maintained entirely in hardware, and the release transfers the lock to the first waiting processor without affecting the other contending processors. QOLB has proven to offer substantial speed up, but at the cost of software support, ISA changes and protocol complexity [7].

The DASH project provided a concept of queue locks in hardware for directory-based multiprocessors [24]. On a release, the lock is sent to the directory which randomly selects a waiting processor to acquire the lock.

Rajwar et al. propose to predict synchronization operations in each processor by building a speculative hardware-based queuing mechanism (IQOLB) for snoop-based and directory systems [10,11]. They use the notion of buffering external requests, applying it to cache lines supposed to contain a synchronization variable. The mechanism does not require any change to existing software or ISA, but requires changes in the cache or in the directory protocol in order to make the intelligent choices needed to implement the mechanism and some additional bits in directory entries.

The above described hardware queue-based mechanisms need to handle queue breakdowns (due to line eviction or multiprogramming). The mechanism proposed in our work does not use a predictor and does not require changes to cache or directory protocol.

The *combining pending read request* technique as proposed by Chaudhuri et al. [17], was initially intended to eliminate NACKs, but significantly accelerates lock acquiring in lock-intensive applications. It is based on buffering pending requests, so our work requires the same hardware support but uses a different selection heuristic.

6 Concluding Remarks

In this paper we introduce Request Bypass, a technique to speed-up the lock hand-off in DSM multiprocessors which use queuing at the coherence controller in order to serialize requests to busy lines. Under Request Bypass, the requests in the input port of the coherence controller are attended before the requests queued in the coherence controller which are directed to non-busy lines. The mechanism does not require compiler or programmer support nor ISA or coherence protocol changes.

When accessing a highly contended critical section, Request Bypass allows the Winner processor requests to bypass the queued requests of the contending processors, speeding-up the Winner execution of all critical section phases, namely lock acquiring, shared data accessing, and lock releasing.

Simulations performed in 32-processor systems show that Request Bypass reduces the overall execution time to some extent in all our tested benchmarks. The reduction is noticeable in programs with a large synchronization overhead, reaching 35% of execution time reduction and 75% of Lock time reduction. Read Combining also reduces both execution time and synchronization overhead, but to a lesser extent.

We have also merged naively Request Bypass and Read Combining. In this merged mode, when Read Combining operates on lock lines, it eliminates some of the benefits obtained with Request Bypass. This negative effect disappears when both techniques are applied selectively.

References

1. Mellor-Crummey, J., Scott, M.: Algorithms for scalable synchronization on shared memory multiprocessors. ACM Trans. on Computer Systems **9**(1) (1991) 21–65
2. Michael, M., Scott, M.: Implementation of atomic primitives on distributed shared memory multiprocessors. In: Proc. 1st HPCA. (1995) 221–231
3. Anderson, T.: The performance implications of spin-waiting alternatives for shared-memory multiprocessors. In: Proc. ICPP, volume II. (1989) 170–174
4. Anderson, T.: The performance of spin lock alternatives for shared-memory multiprocessors. IEEE Trans. on Parallel and Distributed Systems **1**(1) (1990) 6–16
5. Goodman, J., Vernon, M., Woest, P.: Efficient synchronization primitives for large-scale cache-coherent shared-memory multiprocessors. In: Proc. 3th ASPLOS. (1989) 64–75
6. Kagi, A.: Mechanisms for Efficient Shared-Memory, Lock-Based Synchronization. PhD thesis, University of Wisconsin. Madison (1999)
7. Kagi, A., Burger, D., Goodman, J.: Efficient synchronization: let them eat QOLB. In: Proc. 24th ISCA. (1997) 170–180
8. Graunke, G., Thakkar, S.: Synchronization algorithms for shared memory multiprocessors. IEEE Computer **23**(6) (1990) 60–69
9. Magnusson, P., Landin, A., Hagersten, E.: Queue locks on cache coherent multiprocessors. In: Proc. 8th ISPP. (1994) 165–171
10. Rajwar, R., Kagi, A., Goodman, J.: Improving the throughput of synchronization by insertion of delays. In: Proc. 6th HPCA. (2000)
11. Rajwar, R., Kagi, A., Goodman, J.: Inferential queueing and speculative push for reducing critical communication latencies. In: Proc. 17th ICS. (2003) 273–284
12. Kuskin, J., et al.: The stanford FLASH multiprocessor. In: Proc. 21th ISCA. (1994) 302–313
13. Laudon, J., Lenoski, D.: The SGI Origin: A CC-NUMA highly scalable server. In: Proc. 24th ISCA. (1997)
14. Barroso, L., et al.: Piranha: A scalable architecture based on single-chip multiprocessing. In: Proc. 27th ISCA. (2000) 282–293
15. Gharachorloo, K., et al.: Architecture and design of ALPHASERVER GS320. In: Proc. 9th ASPLOS. (2000) 13–24
16. James, D., Laundrie, A., Gjessing, S., Sohni, G.: Distributed directory scheme: Scalable coherence interface. IEEE Computer **23**(6) (1990)

17. Chaudhuri, M., Heinrich, M.: The impact of negative acknowledgments in shared memory scientific applications. IEEE Trans. on Parallel and Distributed Systems **15**(2) (2004) 134–152
18. Pai, V., Ranganathan, P., Adve, S.: RSIM: An execution-driven simulator for ILP-based shared-memory multiprocessors and uniprocessors. In: WCAE-3. (1997)
19. Pai, V., Ranganathan, P., Adve, S.: RSIM reference manual version 1.0. Technical report 9705, Dept. of Electrical and Computer Engineering, Rice University (1997)
20. Gharachorloo, K., Gupta, A., Hennessy, J.: Two techniques to enhance the performance of memory consistency models. In: Proc. ICPP. (1991) 355–364
21. Woo, S., et al.: The SPLASH-2 programs: Characterization and methodological considerations. In: Proc. 22th ISCA. (1995) 24–36
22. Heinrich, M., Chaudhuri, M.: Ocean warning: Avoid drowing. Computer Architecture News **31**(3) (2003) 30–32
23. de Dios, A., Sahelices, B., Ibáñez, P., Viñals, V., Llabería, J.M.: Speeding-up synchronizations in DSM multiprocessors. Tech. rep. DIIS RR-06-07, University of Zaragoza. Spain. (2006)
24. Lenoski, D., et al.: The stanford DASH multiprocessor. IEEE Computer **25**(3) (1992) 63–79

Design and Effectiveness of Small-Sized Decoupled Dispatch Queues[*]

Won W. Ro[1] and Jean-Luc Gaudiot[2]

[1] Department of Electrical and Computer Engineering
California State University, Northridge
wro@csun.edu
[2] Department of Electrical Engineering and Computer Science
University of California, Irvine
gaudiot@uci.edu

Abstract. Continuing demands for high degrees of Instruction Level Parallelism (ILP) require large dispatch queues in modern superscalar microprocessors. However, such large queues are inevitably accompanied by high circuit complexity which correspondingly limits the pipeline clock rates. This is due to the fact that most of today's designs are based upon a centralized dispatch queue which depends on globally broadcasting operations to wake up and select the ready instructions. As an alternative to this conventional design, we propose the design of hierarchically distributed dispatch queues, based on the access/execute decoupled architecture model. Simulation results based on 14 data intensive benchmarks show that our DDQ (Decoupled Dispatch Queues) design achieves performance comparable to a superscalar machine with a large dispatch queue. We also show that our DDQ can be designed with small-sized, distributed dispatch queues which consequently can be implemented with low hardware complexity and high clock rates.

1 Introduction

Reaching high degrees of Instruction Level Parallelism (ILP) through multiple-instruction issue and out-of-order execution has been an essential part of modern microprocessor design. During the last decade, superscalar architectures have dominated the commercial market by adopting a hardwired scheduling logic that enables dynamic instruction scheduling. However, conventional dynamic scheduling possesses an inherent scaling problem as far as the size of the *dispatch queue* is concerned since the wake up and select logic requires a one-cycle operation and cannot be pipelined [1].

Another important issue is how to solve the dramatically growing speed gap between processor and main memory. This performance gap causes long access latencies at cache misses and forces the cache miss instructions to be stalled. It

---

[*] This paper is based upon work supported in part by NSF grant CCF-0541403. Any opinions, findings, and conclusions or recommendations are those of the authors and do not necessarily reflect the views of NSF.

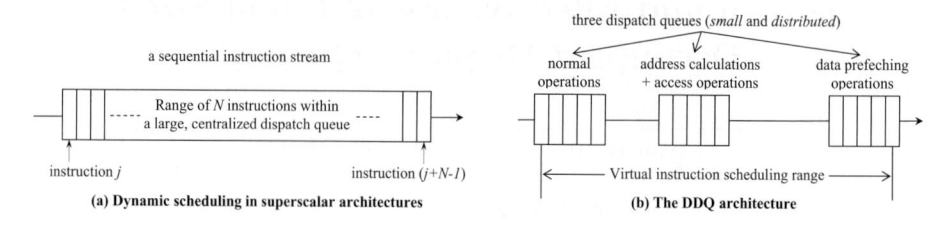

Fig. 1. Distributed instruction scheduling on the decoupled dispatch queues

consequently means all the instructions that depend on the cache miss instructions should stay inside the dispatch queue. In fact, those instructions would occupy the slots for considerable amounts of time, which would result in a reduction of the number of available entries in the dispatch queue. Therefore, the long memory latency also implies the need for a large dispatch queue. However, as described earlier, a large queue will eventually cause a scaling problem.

As an alternative to a large dispatch queue, we propose Decoupled Dispatch Queues (DDQ) which can be implemented with a three small-sized dispatch queues. It aims at reducing the critical path delay of a large queue. The basic motivation is to mask the long memory access latencies without increasing the size of a single dispatch queue. The DDQ enables asynchronous scheduling of three instruction groups which are separated according to the memory access role of the instructions (computation instructions, memory access instructions, and prefetching instructions); this means there is a dedicated dispatch queue for each of the three instruction groups. Three dispatch queues are, at any given moment, asynchronously dealing with different points of a sequential instruction stream. However, it is virtually operating as if we had a large queue (Fig. 1).

Performance evaluation is based on a cycle-time simulator which is developed from SimpleScalar 3.0 [2]. Compared to a superscalar architecture with a 256-entry dispatch queue, our DDQ achieves a similar performance (*98.5%*) with three 128-entry dispatch queues. When the dispatch queue is reduced by as much as one fourth (64 entries), the DDQ still performs at *91.3%* of the baseline performance. With 32-entry dispatch queues, the performance still remains as high as *86.7%*. Moreover, reduction in the queue size will eventually contribute to the higher clock rate.

The rest of the paper is organized as follows. In Section 2, we describe background research and previous work related to the complexity-effective dispatch queue design. Section 3 presents the detailed description of the proposed DDQ architecture. Section 4 includes experimental results and performance analysis. Conclusions and future work are included in Section 5.

2   Background Research

Access/execute decoupled architecture concepts are not new and we now describe them in some detail, while several related research projects are surveyed.

2.1 Access/Execute Decoupled Architectures

Access/execute decoupled architectures have been developed to tolerate long memory access latencies [3,4,5,6,7,8,9,10]. Latency tolerance is achieved by separating the original, single instruction stream into two streams: the *access stream* and the *execute stream*. By definition, the access stream includes memory access operations: load/store instructions and address calculation instructions. Other remaining instructions (commonly referred to as computation instructions) are included in the execute stream. Timely data prefetching can be achieved by running the access stream ahead of the execute stream; any processor stalling due to data delivery can be eliminated by the early execution of the access stream. The time difference between the access instruction produces a data element and the execute instruction needs the data is called the *slip distance*. The two independent instruction streams processed by each processing unit exploit instruction-level parallelism while providing memory latency tolerance. In general, the communications between the two streams are achieved via a set of FIFO queues.

2.2 Related Work

There have been several research projects which have sought to solve the complexity problem of a large dispatch queue by splitting it into multiple queues. Palacharla et al. have performed an initial analysis of the potential complexity of large window superscalar architectures [1]. They have proposed a dependence-based instruction queue design, in which the instructions are sent to separate FIFO queues based on the data dependencies. At the issue stage level, only the head instructions of each FIFO queue are considered for issuing. Their initial analysis demonstrates the advantage of a small-sized queue and has motivated further research on the clustered microprocessor design. The clustering is essentially related to the partitioning of a dispatch queue and functional units [1,11,12,13]. Also, clustered architectures separate the instructions based on register dependencies. Furthermore, the *speculative multithreading* technique has been developed [14,15] with the idea in mind of focusing on software separation (thread selection and scheduling) and speculative thread spawning on each separated processing unit.

Although the distributed queue design has been proposed in many prior research projects, none of them separate the instruction stream based on the memory access functionality as originally proposed in the early decoupled architectures. Actually, the access/execute decoupled architecture model can even be considered one type of clustered architectures. However, the difference lies in the separation of the instruction streams; the task separation is done according to the memory operations in the decoupled architectures and our DDQ.

Several previous projects in decoupled architectures also attempted to solve the complexity problem of superscalars [3,10]. However, none of them addressed the problem of the cache misses on the access processor. To the best of our knowledge, DDQ is the first work which proposes an implementation of data prefetching on the access/execute decoupled architectures (except our previous work in [16]).

3 DDQ: Hierarchically Decoupled Dispatch Queues

This section first describes the problems of traditional decoupled architectures and presents the idea behind our development of the proposed DDQ architecture. It also includes the hardware and software descriptions of the design.

3.1 Problems of the Access/Execute Decoupling

Our initial motivation is to solve the complexity problem of a monolithic dispatch queue in superscalar machines by using access/execute decoupled architectural concepts. As described earlier, the advantage of decoupled architectures can be exploited only if the slip distance is larger than the memory access latency. However, several factors in the current access/execute decoupled architecture designs prevent the access stream from running far ahead of the execute stream.

First of all, frequent synchronization between the two streams prohibits early execution of the access stream. In fact, the access stream also requires data from the execute stream; some control operations as well as data operations need data from the computation results of the execute stream. Therefore, synchronization between the two streams can happen at a certain point of the execution. We call this phenomenon a *loss of decoupling* event [17].

Secondly, frequent cache misses in the access stream prevent early execution of the access stream running on the access processor (AP). If a cache miss on the AP has a sufficient time until any instruction in the EP requires the data, the latency can be tolerated. However, frequent cache misses may cause the access processor to lag further behind. For example, two or more consecutive cache misses on the AP will slow down the execution of the access stream. From the above observations, we find that the cache misses in the access processor should be reduced.

3.2 Description of the DDQ Architecture

The DDQ architecture includes one additional processing unit to achieve data prefetching on the access processor. Consequently, our architecture requires one more stream separation in addition to the access steam and the execute stream. An additional stream named the *data prefetching stream* is intended to run ahead of the access stream, achieving another hierarchy of the prefetching from the memory to the L1 data cache. Fig. 2 shows the proposed DDQ architecture. It has a single fetch unit and separates three streams at the pre-decoding stage. The stream separation information (which indicates the stream to which the instruction belongs) is already annotated with each instruction at compile time (it is described in the following subsection).

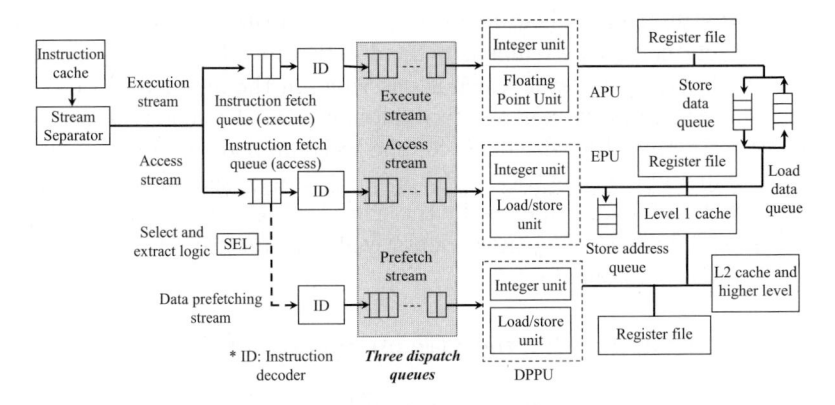

Fig. 2. The DDQ architecture

Three dedicated processing units for each of the execute stream, the access stream, and the data-prefetching stream are loosely combined; they are respectively the EPU (Execute Processing Unit), the APU (Access Processing Unit), and the DPPU (Data-Prefetching Processing Unit). The operations of the EPU and the APU are very similar to that in conventional access/execute decoupled architectures. The load data queue (LDQ) and the store data queue (SDQ) facilitate communications between the EPU and the APU. To guarantee the correctness of the communication order between the two processors, we use the indexed data queue concept which is first introduced in the DS (Decoupled Superscalar) architecture [10]. The indexed data queues are implemented to declare the FIFO order which is assigned at decoding time. However, the queue entries can be accessed out-of-order.

The basic idea behind the DPPU is similar to the speculative pre-execution concept [18,19], which extracts the future probable cache miss slices from the original code and executes them as an additional prefetching thread. The data-prefetching stream of the DDQ is equivalent to the p-thread in speculative pre-execution [18]. It contains the future probable cache miss instructions (target loads) and their backward slice (backward slice includes every instruction upon which the target loads have data dependencies). Access profiling is used to detect probable cache miss instructions at compiler time. The DPPU operation is very loosely coupled with the processor above it since data communications occur only through the L1 data cache.

The execution of the DPPU is triggered at runtime. When the stream separator detects the target load instructions, it triggers the execution of the DPPU. For that purpose, Select and Extract Logic (SEL) is implemented. When the triggering is initiated by the stream separator, the SEL is enabled and looks into the instruction fetch queue to select the instructions which are tagged as data-prefetching stream (those instructions have been pre-detected and tagged by the stream separator beforehand). After that, SEL extracts and sends those instructions to the instruction decoder which is dedicated to the data-prefetching dispatch queue. The extraction operation is a copy operation of the instruction bits, since the AP still needs to hold and execute those instructions. The separation information for the three streams is defined and embedded on each instruction by the DDQ binary translator which is described in the next subsection.

The main target of this design is to reduce the size of a single dispatch queue so that we can reduce the wire delay for the wake up and select logic. Although the total number of queue entries in the entire processor should be multiplied by three, the clock rate is only affected by the size of a single, largest dispatch queue. There are no data bypassing networks or wake up and select logic connected between two different queues.

3.3 Software Support for the Stream Separation

The DDQ binary code is produced by the DDQ binary translator which directly works on the SimpleScalar binary code. The tool analyzes the SimpleScalar binary code and separates it into three streams based on the instruction functionality. After that, the annotation field of each instruction of SimpleScalar binary is used to convey the each stream information (including information on the target load instructions) down to the hardware.

The separation of the access stream and the execute stream is very similar to that in conventional decoupled architectures. At the beginning, each load/store instruction is defined as the access stream. After that, the backward slice of the load/store instruction is included in the access stream. The remaining instructions of the code are separated as part of the execute stream. In our design, additional separation for the data-prefetching stream must be identified for the DPPU operations. Basically, the data-prefetching stream, which includes the probable cache miss instructions and their backward slice, is a subset of the access stream. If an instruction has been detected as a frequently miss-causing instruction by the access profiling, it is identified as a target load instruction for prefetching operation and defined as a part of the data prefetching stream. Finally, its backward slice is chased and included as the data prefetching stream. More detailed description can be found in our previous work in [16].

4 Experimental Results and Analysis

This section presents the experimental results and the performance analysis of the DDQ architecture.

4.1 Simulation Environment

The DDQ simulator has been designed based on the sim-outorder simulator of the SimpleScalar 3.0 tool set [4]. The baseline superscalar architecture for performance comparison has a 256-entry dispatch queue with 8-way issue and commit. In the DDQ model, each dispatch queue size is tested from 32, 64, to 128. The issue and commit width is also reduced to 4. The EPU is implemented with all the functional units except for the load/store units. The APU and DPPU only have integer units and load/store units. In addition, we assume 12 CPU cycles for L2 cache access latency and 120 cycles for memory access latency.

The set of benchmarks we have selected include 14 applications: six applications chosen from the Atlantic Aerospace Stressmark suite (pointer, update, field, neighborhood, transitive closure, and matrix), three benchmarks from the Atlantic Aerospace Data-Intensive Systems Benchmarks suite (data management, ray tracing, and fast Fourier transform), and five selected from the SPEC2000 suite (gzip, vortex, bzip2, art, and equake). The SPEC benchmarks have been compiled at peak optimization level and tested with the reference input set.

We have performed simulations with the above 14 benchmarks for the three different machine models: superscalar (*sus*), access/execute decoupled architecture (*aed*), and our model (*ddq*). For all simulation results, the performance is measured in terms of IPC (instructions per cycle) and normalized to the baseline superscalar models (*sus.256.8*). Note that the first term specifies the architecture model while the second and the third numbers correspond to the dispatch queue size and issue width. For example, *ddq.32.4* indicates a processor model for a DDQ configuration with 32-entry dispatch queues and 4-way issue width.

4.2 Performance Results and Analysis

We have simulated three different configurations of the DDQ by using three dispatch queue sizes: 32, 64, and 128 entries. They are respectively called *ddq.32.4*, *ddq.64.4*, and *ddq.128.4*. The performance results for the three configurations are shown in Fig. 3. The performance of each model is measured in terms of IPC and normalized to that of *sus.256.8*. Although we cannot quantify the expected clock rates of our design at this point, we know that the smaller dispatch queues in our design would ultimately contribute to higher clock rates. Indeed, as previous research indicates [1], the critical path delay shows a quadratic dependency on the dispatch queue size and issue width.

As the results indicate, the *ddq.128.4* configuration yields a performance comparable to the baseline superscalar model in most benchmarks. However, four benchmarks (*field*, *tr*, *fft*, and *art*) show a weak performance compared to the other benchmarks. *Field* does not encounter many cache misses with the superscalar model and did not benefit from the data prefetching. Also, *tr* suffers from a low branch hit-ratio which prevents a successful speculative prefetching. As for *fft*, the DPPU has too many instructions in the prefetching stream. It causes

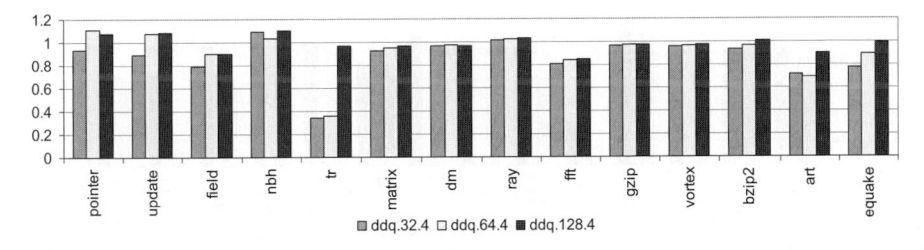

Fig. 3. Performance results for three DDQ models (normalized IPC to sus.256.8)

cache pollution that correspondingly degrades the performance. In addition, *art* does not provide good performance since it works too well with the baseline model which has a 256-entry instruction window. The wide range scheduling is very beneficial to *art* and diminishes the advantages of the DDQ approach. The other 10 benchmarks show very close or even better performance compared to the baseline architecture.

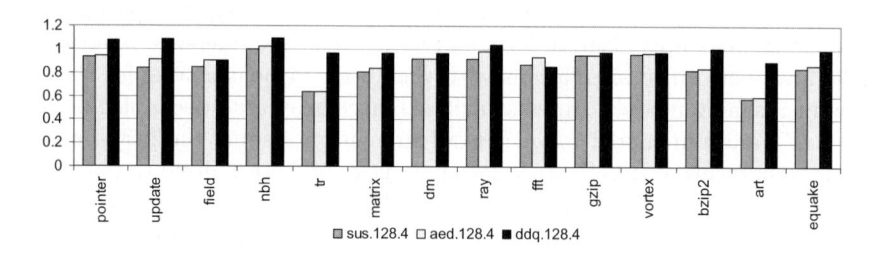

Fig. 4. Performance results with 128-entry queues (normalized IPC to sus.256.8)

For a further detailed analysis, Fig. 4 shows how *ddq.128.4* improves the performance over a traditional access/decoupled architecture (*aed.128.4*) and a superscalar (*sus.128.4*). The results demonstrate that *ddq.128.4* performs even better than the baseline model (*sus.256.8*) in 6 benchmarks in spite of having half-sized dispatch queues and half-sized issue width. However, *sus.128.4* and *aed.128.4* do not show good performance results in most benchmarks. In particular, *tr* and *art* shows noticeably low performance in the *sus* and *aed* configurations. We also performed benchmark simulations with 64-entry dispatch queues; the results show very similar tendency and characteristic. To avoid including too many redundant figures, only the average performance for the 64-entry configurations is presented later in this section.

Fig. 5 illustrates the performance of three architecture models with 32-entry dispatch queues and 4-way issue width; again, all results are normalized to *sus.256.8*. In this result, the dispatch queues in DDQ are as small as one eighths of the baseline model. However, the DDQ still reaches better than 80% of the baseline performance in 11 benchmarks. In contrast, more than half of the benchmarks (8 out of 14) cannot achieve 80% of the performance with *sus* and *aed*.

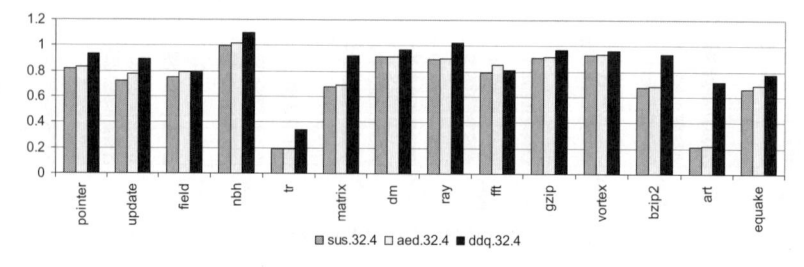

Fig. 5. Performance results with 32-entry queues (normalized IPC to sus.256.8)

More specifically, it should be noted that *tr* and *art* lose about 80% of the performance in those two configurations. Both models are affected much by the restriction of the queue size since neither configuration is assisted by prefetching.

The average performance over the 14 benchmarks is shown in Fig. 6. On average, *ddq.128.4* reaches up to 98.5% of the baseline performance with half-sized dispatch queues and half-sized issue width. However, the *aed.128.4* model experiences a 12.2% performance degradation. These results clearly demonstrate the advantage of the data prefetching operations of the DDQ. With *ddq.64.4*, the average performance still remains above 91%. More over, the *ddq* configuration remains in the range of over 86.7% of the baseline performance even for the smaller configurations such as 32-entry queues. However, *sus* and *aed* experience severe performance degradation when the dispatch queue is small.

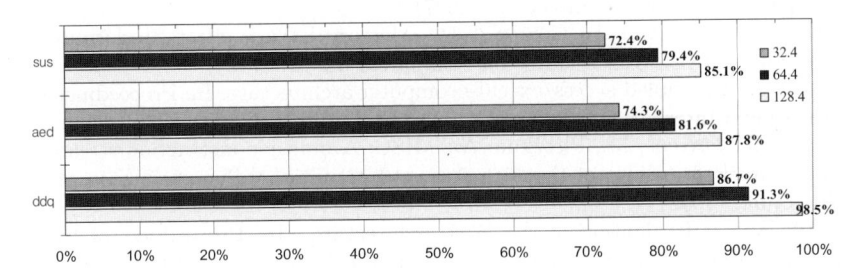

Fig. 6. Average performance with the small dispatch queue models

5 Conclusions and Future Work

The DDQ is based on the simple observation that the partitioning of a dispatch queue can reduce the complexity of a centralized design as well as the size of each component. Each processing unit is decoupled and works fairly independently of the others. The performance results show that the proposed architecture achieves performance comparable to that of the baseline superscalar architecture which has a large dispatch queue. In addition, our DDQ can be implemented with a faster clock since each processing unit has a smaller dispatch queue.

The main feature of the DDQ is having small dispatch queues, so that we can reduce the complexity and wire delay of the instruction scheduling logic. This eventually contributes to achieving higher clock rates. Even though the total number of queue entries over the DDQ grows with the each queue size times three, the clock rate is only affected by the size of a single dispatch queue. The three distributed dispatch queues do not require any data bypassing from the different functional units, nor share any instruction scheduling logic. Considering the clock rate improvement afforded by the size of a dispatch queue, these performance results are encouraging.

References

1. Palacharla, S., Jouppi, N.P., Smith, J.E.: Complexity-effective superscalar processors. In: Proceedings of the 24th Annual International Symposium on Computer Architecture. (1997)
2. Burger, D., Austin, T.: The simplescalar tool set. Technical Report CS-TR-97-1342, University of Wisconsin-Madison (1996)
3. Farrens, M., Nico, P., Ng, P.: A comparison of superscalar and decoupled access/execute architectures. In: Proceedings of the 26th Annual International Symposium on Microarchitecture. (1993)
4. Goodman, J.R., Hsieh, J.T., Liou, K., Pleszkun, A.R., Schechter, P.B., Young, H.C.: PIPE: A vlsi decoupled architecture. In: Proceedings of the 12th Annual International Symposium on Computer Architecture. (1985)
5. Jones, G.P., Topham, N.P.: A comparison of data prefetching on an access decoupled and superscalar machine. In: Proceedings of the 30th Annual International Symposium on Microarchitecture. (1997)
6. Kurian, L., Hulina, P.T., Coraor, L.D.: Memory latency effects in decoupled architectures. IEEE Transactions on Computers 43(10) (1994)
7. Smith, J.: Decoupled access/execute computer architecture. In: Proceedings of the 9th Annual International Symposium on Computer Architecture. (1982)
8. Tyson, G., Farrens, M., Pleszkun, A.: MISC: A multiple instruction stream computer. In: Proceedings of the 25th Annual International Symposium on Microarchitecture. (1992)
9. Wulf, W.A.: Evaluation of the WM architecture. In: Proceedings of the 19th Annual International Symposium on Computer Architecture. (1992)
10. Zhang, Y., Adams III, G.B.: Performance modeling and code partitioning for the DS architecture. In: Proceedings of the 25th Annual International Symposium on Computer Architecture. (1998)
11. Farkas, K.I., Chow, P., Jouppi, N.P., Vranesic, Z.: The multicluster architecture: Reducing cycle time through partitioning. In: Proceedings of the 30th Annual International Symposium on Microarchitecture. (1997)
12. Canal, R., Parcerisa, J.M., González, A.: Speculative data-driven multithreading. In: Proceedings of the 6th International Symposium on High Performance Computer Architecture. (2000)
13. Kemp, G.A., Franklin, M.: PEWs: A decentralized dynamic scheduler for ILP processing. In: Proceedings of the ICPP. (1996)
14. Krishnan, V., Torrellas, J.: A chip-multiprocessor architecture with speculative multithreading. IEEE Transactions on Computers 48(9) (1999)
15. Marcuello, P., González, A.: Clustered speculative multithreaded processors. In: Proceedings of the 13th International Conference on Supercomputing. (1999)
16. Ro, W.W., Gaudiot, J.L., Crago, S.P., Despain, A.M.: HiDISC: A decoupled architecture for data-intensive applications. In: Proceedings of the 17th IPDPS. (2003)
17. Bird, P., Rawsthorne, A., Topham, N.: The effectiveness of decoupling. In: Proceedings of the 7th International Conference on Supercomputing. (1993)
18. Collins, J.D., Wang, H., Tullsen, D.M., Hughes, C., Lee, Y.F., Lavery, D., Shen, J.P.: Speculative precomputation: Long-range prefetching of delinquent loads. In: Proceedings of the 28th Annual International Symposium on Computer Architecture. (2001)
19. Roth, A., Sohi, G.S.: Speculative data-driven multithreading. In: Proceedings of the 7th International Symposium on High Performance Computer Architecture. (2001)

Sim-async: An Architectural Simulator for Asynchronous Processor Modeling Using Distribution Functions

J.M. Colmenar[1], O. Garnica[2], J. Lanchares[2], J.I. Hidalgo[2],
G. Miñana[2], and S. Lopez[2]

[1] C. E. S. Felipe II, Complutense U. of Madrid
jmcolmenar@cesfelipesegundo.com
[2] Dept. of Computer Arch. and System Engineering, Complutense U. of Madrid
{ogarnica, julandan, hidalgo}@dacya.ucm.es,
guamiro@fdi.ucm.es, slopezal@dacya.ucm.es

Abstract. In this paper we present *sim-async*, an architectural simulator able to model a 64-bit asynchronous superscalar microarchitecture. The aim of this tool is to serve the designers on the study of different architectural proposals for asynchronous processors. *Sim-async* models the data-dependant timing of the processor modules by using distribution functions that describe the probability of a given delay to be spent on a computation. This idea of characterizing the timing of the modules at the architectural level of abstraction using distribution functions is introduced for the first time with this work. In addition, *sim-async* models the delays of all the relevant hardware involved in the asynchronous communication between stages.

To tackle the development of *sim-async* we have modified the source code of SimpleScalar by substituting the simulator's core with our own execution engine, which provides the functionality of a parameterizable microarchitecture adapted to the Alpha ISA. The correctness of *sim-async* was checked by comparing the outputs of the SPEC2000 benchmarks with SimpleScalar executions, and the asynchronous behavior was successfully tested in relation to a synchronous configuration of *sim-async*.

1 Introduction

Due to the current integration level and clock frequencies in microprocessor architectures, synchronization with a single clock source and negligible skew is an extremely difficult task. Fully asynchronous designs built using self-timed circuits replace the clock signal by local synchronization protocols. Then, these systems have no problems associated with the clock signal, and the global circuit performance corresponds to the performance of the average case because a new computation starts immediately after the previous has finished [1].

In the field of fully asynchronous systems, designers usually develop general purpose processors (like those presented in [2,3,4,5]) using high-level description languages like Occam, Tangram, Balsa or VHDL++. In addition, some works like [6,7] have proposed simulators of asynchronous processors, but they are slightly parameterizables and they do not model the asynchronous behavior at the architectural level of design. Albeit, these simulators are not able to run standard benchmarks.

As occurs in the synchronous paradigm, asynchronous systems designers need infrastructures for computer system modeling that abstract the implementation of hardware models. These infrastructures must be capable of model the data-dependant delay of a fully asynchronous system at the architectural level of abstraction, and also they have to be able to run complete applications. The main example of such a configurable, flexible and wide-spread toolset in the synchronous world is SimpleScalar [8]. SimpleScalar allows to modify cache, branch predictor or any other architectural parameter, and is able to run standard benchmarks in order to get comparable measures for any kind of data related to performance and also to custom statistics. Up to our best knowledge, such flexible infrastructures for simulation and architectural modeling of high-performance fully asynchronous processors have not been reported in literature.

Once argued the necessity of a modeling infrastructure, one of the key questions is how the tool will model the data-dependant computation delays of the modules that form an asynchronous processor. Since asynchronous circuits take distinct amounts of time when computing different values, it is possible to collect a large set of delays for a given circuit by running low-level simulations using a representative number of inputs. From that set of delays one may obtain the distribution function which characterizes the behavior of the circuit. *Sim-async* applies this idea inside out, that is, the simulator uses distribution functions (included as parameters) to dynamically select the delay for each computation of each one of the modules of the processor. This solution is introduced in this paper as a novelty related to the architectural asynchronous processor simulation.

Therefore, in this paper we present *sim-async*, an architectural simulator for asynchronous superscalar processor modeling. *Sim-async* is able to model, at the architectural level of abstraction, the data-dependant behavior of the modules of the processor by using distribution functions. In addition, *sim-async* is able to execute any test program compiled for the Alpha ISA, as SimpleScalar does.

The rest of the paper is organized as follows: Section 2 is devoted to describe the simulated processor microarchitecture and the functionality of its stages. In Section 3 we define the synchronization domains and detail the delays that model them, the implementation of those delays as input parameters of the simulator, and the communication protocol between the domains. In Section 4 we show the validation of the simulator by running the SPEC2000 benchmarks under both asynchronous and synchronous configurations. Finally, in Section 5 we explain the conclusions and the future work.

2 Description of the Processor Microarchitecture

Sim-async models the microarchitecture of a 64-bit fully asynchronous superscalar processor with out-of-order and speculative execution of instructions, and

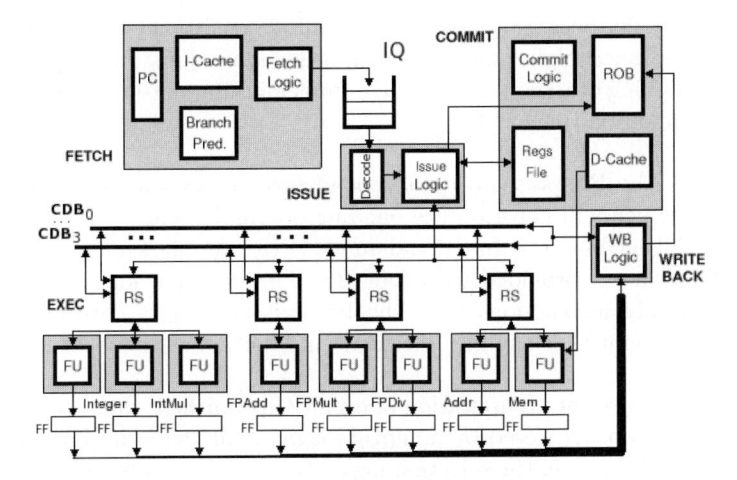

Fig. 1. Schema of the modeled microarchitecture. The logic involved in the communication between modules is not included within this schema.

this section is devoted to its introduction. The processor consists on five stages: *fetch*, *issue*, *exec*, *write-back* and *commit*. In Figure 1 we show the schema of the microarchitecture[1], where we have illustrated the Exec Unit with higher detail[2].

The implementation of the asynchronous processor is identical to the synchronous one, but substituting the clock network by a set of components that allows the communication of results between modules. For the sake of clarity, we briefly describe the functionality of each stage in this section.

Fetch. A parameterizable number of instructions is read from the I-cache taking into account the branch prediction. The instructions are moved to the instruction queue (IQ), where they wait for the issue stage. If one of the instructions in the middle of the fetch group is a taken conditional branch or an unconditional branch, then the subsequent instructions in the fetch group will be discarded.

Issue. As it is well-known, the design of the issue stage is crucial to obtain high performance on a superscalar processor. We have chosen the implementation called *instruction shelving* with reorder buffer (ROB) [9] for the issue stage because it decouples the instruction issue and the dependency checking. With shelving, the only fact that will provoke the block of the issue of instructions is the lack of free entries in the reservation stations (RS, or *shelving buffers*) or ROB, not the data dependencies, which are more frequently to appear.

This stage decodes and in-order issues a parameterizable number of instructions from the IQ to their corresponding RS and to the ROB. The issue is

---

[1] The twelve shadowed areas of the figure represent the different synchronization domains we have defined in the processor, but we postpone its explanation until the following section.
[2] This level of details of the execution unit will be useful in the following sections.

performed in-order because preserving sequential consistency for out-of-order issue requires a much higher effort than in-order issue does. In addition, due to the rarely blocking of issue with shelving, implementing out-of-order issue would only have a marginal benefit [9].

Execution. The RS preserve data dependencies maintaining the tags of the instructions which will generate the pending operands, and hold values waiting for the execution into the functional units (FU). As shown in Figure 1, the microarchitecture is provided with four RS. The dispatch logic decides which one of the ready instructions from the RS is issued to its corresponding FU taking into account that as older the ready instruction as sooner it is issued.

Write-back. Once the computation of each FU is finished, the result is held on its output flip-flop, triggered by a capture signal, till the write-back stage was completed. In this stage, the selection logic chooses the results to be distributed to the RS and the ROB through the number of instances (parameterizable) of the common data bus (CDB), also sending the tag of the instructions which generated each result. The wake-up logic of each RS compares the incoming tags with the tags of the pending instructions performing the update of values wherever a tag matches.

Commit. Each instruction at the ROB holds the result to be written to the register file or to the memory and the destination register or memory address. A parameterizable number of instructions is retired from the ROB maintaining program order, and branch prediction is checked each time this stage executes. Precise interruptions are also checked and the pipeline is flushed when a mispredicted branch is processed.

3 Modeling the Asynchrony

A synchronization domain consists in all the flip-flops triggered by the same signal and the combinational logic within their fan-in. In this paper we have defined twelve synchronization domains (see shadowed areas in Figure 1), where the communication between them is performed using a four-phase handshake protocol. The following subsections are devoted to present the temporal modeling of an individual domain and the assumed communication protocol.

3.1 Temporal Modeling of a Synchronization Domain

We have followed a mixed approach in the asynchronous paradigm to describe the temporal behavior of the domains. We have used the computation completion mechanism described in [10] to detect the end of the computation, and we have employed a bounded delay approach to model the behavior of the control logic.

In the asynchronous systems the delay spent on computing a data, detecting the computation completion and communicating the result to the receiver module through the synchronization protocol takes a different and unpredictable value for each input data. That delay comes from the combination of several other delays that appear during the operation of a module. Let's examine these delays, indicated with dotted lines in the scheme of Figure 2 (a).

The *computation delay*, t_c, is the delay spent by the module on computing the input data and generating the results. It is a variable delay because asynchronous circuits present a data-dependant behavior. In our simulator, each stage and FU of the microarchitecture receives its own t_c as a distribution function. Then, whenever the module makes a computation, the simulator randomly selects a computation delay taking into account the shape of the distribution. Thus, the actual delay for computing these data is not obtained, but the data-dependant behavior of the module is maintained.

The *completion detection delay*, t_{compl}, corresponds to the time spent by the completion detection logic (CD) on detecting a valid output and asserting the *compl* signal. This delay is included as a constant input parameter on *sim-async*.

We use a delay insensitive codification and a completion detection logic due to the variability of t_c. Therefore, as Martin showed in [11], the modules alternate a neutral or synchronization value (S) which does not mean any Boolean value, and the encoding of a valid output. The generation of that synchronization value takes t_{sync} time units (t.u.) and, after that, the module is ready to receive new incoming data. This delay is also an input parameter of *sim-async*. The logic that orders the generation of the synchronization value is omitted in the mentioned figure for the sake of clarity.

The modeling of the handshake protocol is divided on two delays: *request delay*, t_{req}, which is the time spent from the assertion of the *compl* signal to the assertion of the request signal, req_i; and *capture delay*, t_{cap}, which is the time spent from the falling edge of the acknowledge signal from the receiver module, ack_{i+1}, and the assertion of the *capture* signal. The time spent during the handshake is an uncertain delay that can be accurately obtained only by simulation because it mainly depends on the occupation or availability of the structures of the receiver module at each moment. Both t_{req} and t_{cap} are included as constant input parameters on *sim-async*.

Once the protocol is completed, the *capture* signal is asserted as a pulse. This assertion does not violate any timing assumptions because we consider t_{compl} to be longer than the setup delay of the destination register. In addition, the width of the pulse of *capture*, denoted as t_{cap-up}, must be higher than the hold delay of the register triggered by the *capture* signal because the generation of the synchronization value is ordered by the falling edge of that pulse. The t_{cap-up} delay is included as another constant input parameter of our simulator.

The delay spent from the fall of the *capture* signal and the assertion of the ack_i signal is denoted as t_{ack}, also included as an input parameter of *sim-async*.

3.2 Communication Protocol

The communication between domains is performed through channels implementing a four-phase handshake protocol like the one described in [12]. Figure 2 (b), shows the chronogram of an example of communication between the domain i

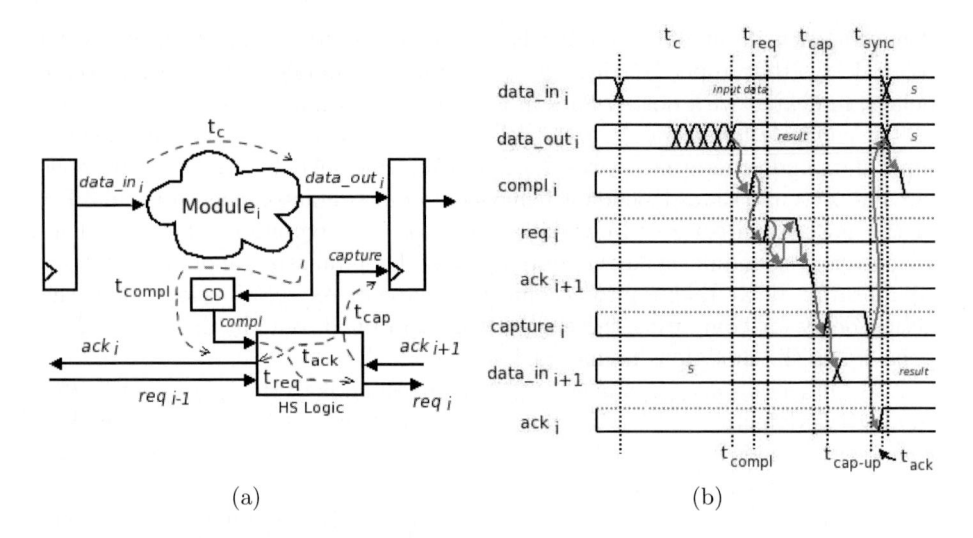

Fig. 2. (a) General scheme of a synchronization domain. Dotted lines are referred to delays. (b) Chronogram showing the delays of the logic involved on one computation of the module i and the communication of results to the next domain.

and its neighbor. We next explain this communication using the delays defined in the previous subsection.

The moment when the module starts to compute is the instant in which the $data\_in_i$ signal propagates the input data. Then, the module processes these data and, after a data-dependant delay, t_c, the result is propagated through the $data\_out_i$ signal. The $compl$ signal is asserted after t_{compl} t.u. and then the handshake logic activates the req_i signal in order to start the communication protocol. The receiver module is ready to process new data because ack_{i+1} is asserted. At that point, the handshake logic deasserts the request signal and waits for the fall of ack_{i+1}. The receiver module unsets the acknowledge signal and the communication protocol ends. After that, the handshake logic generates a pulse in the $capture$ signal. On the raising edge of $capture$ the destination register latches the results of the module and, on the falling edge of $capture$, the logic of the module return to the synchronization value before the next computation. In addition, the falling edge of $capture$ also provokes the assertion of ack_i, which indicates that the module i is ready to receive new input data.

4 Experimental Results

In order to validate *sym-async* we have run the SPEC2000 benchmarks on different timing configurations of the simulator. Then, we have compared the results of these executions with those obtained from the original SimpleScalar (*sim-safe* flavor) under the same cache and branch predictor configuration. The tests were run parameterizing *fetch, issue, write-back* and *commit* stages to process up to

Table 1. (a) Architectural configuration of the microarchitecture in the simulations. (b) Worst case delay of stages and FU operations in the asynchronous simulations.

Branch Predictor:	2-level PAg
Level 1	1024 entr, his 10
Level 2	1024 entr
BTB	4096 sets, 2-way
Instructions queue (IQ) size	100 entries
Integer RS queue size	6 entries
FP Addition RS queue size	3 entries
FP Mul, FP Div/Sqrt RS queue size	2 entries
Memory RS queue size	5 entries
Integer / FP Register File	32 / 32
ROB size	100 entries

(a)

Stage / FU Operation	T. U.
Fetch, Issue, Int/Logic, WB, Commit	1000
IntMul	7000
MemLoad, FPAdd, FPMul	4000
FPDiv/Sqrt	30000

(b)

four instructions each time they execute. Table 1 (a) shows the architectural configuration of the microarchitecture.

The first timing configuration tested was the fully asynchronous one. In this asynchronous configuration we used two distribution functions to characterize the computation delays of the modules: slow case (SC) and medium case (MC) functions. These functions were selected from the set of back-annotated gate-level simulations of related asynchronous circuits, and were normalized to the same upper bound (the worst case) of 1000 t.u..

The slow case (SC) function, shown in Figure 3 (a), whose average delay is near the worst delay, represents a slow behavior because the most of the data take a high delay. We had not made any assumptions about the implementation of the functional units, so they were individually characterized through the SC function. However, we considered the use of long-latency non-pipelined FU for FP operations and integer multiplications, so the normalization of the function was conveniently corrected to a higher upper bound for these slow non-pipelined FU, according with the Table 1 (b).

The medium case (MC) function, presented in Figure 3 (b), describes an asynchronous behavior where the average delay is close to the half of the worst delay. We have use this function to characterize the rest of the stages: *fetch*, *issue*, *write-back* and *commit*.

It is important to remark that the aim of these functions is not to be actual patterns of the modules of the modeled processor. We present these functions as typical examples of asynchronous circuit behaviors obtained from previous low-level simulations.

In order to establish the delays of the control logic, we have considered the work of Cheng in [10]. In that paper Cheng implemented a circuit for completion detection and synchronization (reset completion-detection) of data lines using a four-phase handshake protocol and dual-rail codification. He obtained an average delay of 0.28 ns for the completion detection circuit and 0.71 ns for the synchronization (reset). Considering that digital IC performance has tracked

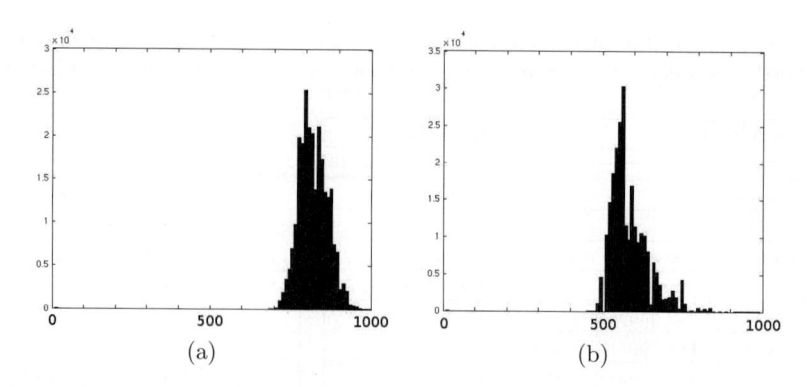

Fig. 3. (a) Slow case (SC) and (b) Medium case (MC) distribution functions

Moore's Law and improved by 30% annually, the delays of that circuit using current technology could be about 16 ps and 41 ps respectively.

Supposing that the critical path of the modules of the processor will be under 1.25 ns (that means a maximum frequency of 800 MHz in a synchronous version), which we have normalized to 1000 t.u., the normalized values for t_{compl} and t_{sync} taking into account the scaling are 12.8 t.u. and 32 t.u. respectively, which correspond to average delays. In our simulations we have conservatively doubled that delays to 26 t.u. and 64 t.u. for all the modules. The rest of the control logic delays, t_{req}, t_{cap} and t_{cap-up} were fixed to 5, 5 and 10 t.u. respectively, and t_{ack} was considered equal to t_{sync}.

We have checked that the outputs obtained by *sim-async* running the SPEC2000 benchmarks under the asynchronous configuration are identical to those generated by SimpleScalar for all the benchmarks. That is, *bzip* generates the same compressed file, *gcc* returns the same compilation statistics, and so on. In addition, we have compared the number of instructions committed on both simulators for the execution of those benchmarks and they only differ in a negligible range between 0.21% and -0.012% (attributed to the slightly different implementation of the system calls), as shown in Table 2 (a). Therefore, *sim-async* performs correct simulations and successfully executes the Alpha ISA.

With the aim of test that *sim-async* not only executes the Alpha ISA correctly, but it also correctly models the asynchronous behavior, we have made the comparison between the former simulations and those resulting from *sim-async* parameterized in order to model a synchronous processor. This is possible because the synchronous behavior is a particular case of the asynchronous one. That is, in a synchronous processor all the modules spend the same time on computing a data (the worst case of the slowest stage) and the communication protocol spends a delay of zero t.u. due to the clock signal.

Then, we set the parameters t_{compl}, t_{req}, t_{cap}, t_{cap-up}, t_{sync} and t_{ack} to zero t.u., and t_c was fixed to a distribution where all the delays were 1000 t. u. long, the worst case of the asynchronous simulations, but considering the slowest FU (IntMul, FPAdd and FPMul/Div) as fully-pipelined units. The *capture* signal

Table 2. (a) Number of instructions committed for several SPEC2000 benchmarks on SimpleScalar (*sim-safe*) and for *sim-async* under the asynchronous configuration. (b) Average differences between the instructions executed and the use of modules of *sim-async* on synchronous and asynchronous configurations running the SPEC2000.

SPEC	SimpleScalar	Async *Sim-async*	Diff (%)
ammp	45812883	45810845	-0.004
apsi	197579651	197612776	0.017
bzip	1819780172	1819780267	0.000
crafty	94419973	94420229	0.000
galgel	139306245	139310055	0.003
gap	82873902	82874407	0.001
gcc	2016139124	2016204817	0.003
gzip	601857009	601857104	0.000
lucas	19239488	19242782	0.017
mesa	1608605448	1608410610	-0.012
parser	268979662	269006191	0.010
perlbmk	205853718	205914747	0.030
sixtrack	11699655	11724227	0.210
swim	23557475	23562358	0.021
vortex	453666	454534	0.191

Async vs. Synch	% Avg Diff
# Insn Exec	0.132
Use of Fetch	-42.076
Use of Issue	-61.993
Use of Int	-66.062
Use of IntMul	-99.894
Use of FPAdd	-95.665
Use of FPMul	-97.733
Use of FPDiv	-99.921
Use of Addr	-79.314
Use of Mem	-81.116
Use of WB	-52.324
Use of Commit	-73.852

(a) (b)

(see Section 3) is only asserted if the receiving module is ready to accept new input data.

The synchronous simulations were run under the same architectural configuration described for the asynchronous simulations, and we obtained identical outputs and also identical number of committed instructions. In addition, we took some statistics in order to measure the asynchronous behavior. As shown in Table 2 (b), the number of instructions executed (including those speculative) is, on average, 0.132 % higher in the asynchronous configuration. This occurs because the average delays of the asynchronous stages are shorter than the synchronous worst case. Then, the asynchronous microarchitecture is able to advance on the execution of instructions faster than the synchronous one.

Albeit, the number of executions of the asynchronous modules is reduced in relation to the synchronous simulations. The average reduction ranges from the 42.076 % of the fetch stage to the 99.921% of the FPDiv functional unit, which remains idle almost all the time (see Table 2 (b)). This behavior corresponds to the one expected for an asynchronous circuit because the modules only compute when useful work has to be performed.

As an additional statistic, the speedup reached by the asynchronous configuration in relation to the synchronous one is, on average, 1.135 for the SPEC2000.

Thus, this comparison between both asynchronous and synchronous simulations verifies the correct modeling of the asynchronous behavior that *sim-async* performs by using distribution functions to characterize the computation delay of the modules of the microarchitecture.

5 Conclusions and Future Work

In this paper we have presented *sim-async*, an architectural simulator able to correctly model the behavior of a 64-bit asynchronous superscalar microarchitecture at the architectural level of abstraction. To tackle this goal, we have modified the source code of SimpleScalar by substituting the simulator's core with our own execution engine which provides the functionality of a parameterizable superscalar architecture adapted to the Alpha ISA.

In order to provide flexibility, we have defined twelve synchronization domains, and the delays involved on their computation, including them as parameters of *sim-async*. Albeit, due to the necessity of modeling a data-dependant behavior of the modules which form the simulated microarchitecture, we have introduced the idea of modeling the data-dependant computation delay of the modules by using distribution functions.

We have verified the correctness of *sim-async* by comparing the outputs of the SPEC2000 benchmarks run on the original SimpleScalar with those generated by *sim-async*. In addition, we have run simulations of *sim-async* where the delays were defining a synchronous microarchitecture. The number of instructions executed (including those speculative) was, on average, 0.132 % higher in the asynchronous configuration. This occurs because the average delays of the asynchronous stages are shorter than the synchronous worst case. In addition, the number of executions of the asynchronous modules suffered an important reduction in relation to the synchronous simulations. This behavior corresponds to the one expected for an asynchronous circuit because the modules only compute when useful work has to be performed. Then, the comparison between the asynchronous and the synchronous simulations shows that the modeling of the asynchronous behavior is correct. In addition, the asynchronous configuration of the processor presented an average speedup of 1.132 in relation to its synchronous counterpart.

Currently we are working on two ways: on one hand, we are tuning *sim-async* with the aim of reducing its execution time, which is still high (about thirty six hours each set of benchmarks). On the other hand, we are working on the implementation of the asynchronous modules of the microarchitecture in order to reach higher performance.

Acknowledgments

This research has been supported by Spanish Government Grant number TIN2005-05619.

References

1. D. Kearney, "Theoretical Limits on the Data Dependent Performance on Asynchronous Circuits," *Proc. of Intl. Symposium on Advanced Research in Asynchronous Circuits and Systems*, pp. 201–207, 1999.

2. A. J. Martin, A. Lines, R. Manohar, M. Nystroem, P. Penzes, R. Southworth, and U. Cummings, "The Design of an Asynchronous MIPS R3000 Microprocessor," in *Adv. Research in VLSI*, pp. 164–181, 1997.
3. D. K. Arvind and R. D. Mullins, "A Fully Asynchronous Superscalar Architecture," in *Proc. of the 1999 Intl. Conf. on Parallel Architectures and Compilation Techniques* (I. C. S. Press, ed.), pp. 17–22, 1999.
4. J. D. Garside, W. J. Bainbridge, A. Bardsley, D. M. Clark, D. A. Edwards, S. B. Furber, J. Liu, D. W. Lloyd, S. Mohammadi, J. S. Pepper, O. Petlin, S. Temple, and J. V. Woods, "AMULET3i - An Asynchronous System-on-Chip," in *Proc. of the 6th Intl. Symposium on Advanced Research in Asynchronous Circuits and Systems* (I. C. S. Press, ed.), pp. 162–175, April 2000.
5. Q. Zhang and G. Theodoropoulos, "Modelling SAMIPS: a Synthesisable Asynchronous MIPS Processor," in *Proc. of the 37th Annual Simulation Symposium*, pp. 205–212, April 2004.
6. C. Chien, M. A. Franklin, T. Pan, and P. Prabhu, "ARAS: Asynchronous RISC Architecture Simulator," *Proc. of the 2nd Working Conference on Asynchronous Design Methodologies (ASYNC'95)*, 1995.
7. V. Rebello, *On the Distribution of Control in Asynchronous Processor Architectures*. PhD thesis, 1997.
8. T. M. Austin, E. Larson, and D. Ernst, "SimpleScalar: An Infrastructure for Computer System Modeling," *IEEE Computer Journal*, vol. 35, 2, February 2002.
9. D. Sima, "Superscalar Instruction Issue," *IEEE Micro*, vol. 17, pp. 28–39, Sep.-Oct. 1997.
10. F. Cheng, "Practical Design and Performance Evaluation of Completion Detection Circuits," in *Proc. of the Intl. Conf. on Computer Design* (I. C. S. Press, ed.), pp. 354–359, 1998.
11. A. J. Martin, "Asynchronous Datapaths and the Design of an Asynchornous Adder," *Formal Methods in System Design*, vol. 1, pp. 119–137, July 1992.
12. T. H.-Y. Meng, R. W. Brodersen, and D. G. Messerschmitt, "Asynchronous Design for Programmable Digital Signal Processors," *IEEE Trans. on Signal Processing*, vol. 39(4), pp. 939–952, 1991.

A Hybrid Hardware/Software Generated Prefetching Thread Mechanism on Chip Multiprocessors

Hou Rui, Longbing Zhang, and Weiwu Hu

Key Laboratory of Computer System and Architecture,
Institute of Computing Technology, Chinese Academy of Sciences.
100080 Beijing, China
{hourui, lbzhang, hww}@ict.ac.cn

Abstract. This paper proposes a hybrid hardware/software generated prefetching thread mechanism on Chip Multiprocessors(CMP). Two kinds of prefetching threads appear in our hybrid mechanism. Most threads belong to Dynamic Prefetching Thread, which are automatically generated, triggered, spawn and managed by hardware; The others are of Static Prefetching Thread, targeting at the *critical delinquent loads* which can not be accurately or timely predicted by Dynamic Prefetching Thread. Static Prefetching Threads are statically generated by binary-level optimization tool with the guide of profiling information. Also, some aggressive thread construction policies are proposed. Furthermore, the necessary hardware infrastructure for CMP supporting this hybrid mechanism are described. For a set of memory limited benchmarks with complicated access patterns, an average speedup of 3.1% is achieved on dual-core CMP when constructing basic hardware-generated prefetching thread, and this gain grows to 31% when adopting our hybrid mechanism.

1 Introduction

Advances in integrated circuit technology afford great opportunities for Chip Multiprocessors(CMP). It is really a challenge to utilize multi-cores in CMP to accelerate sequential programs. Thread-based prefetching technique is a promising approach to achieve this purpose. It typically uses additional execution pipelines or idle thread contexts in a multithreaded processor(CMP or SMT) to execute helper threads that perform dynamic prefetching for the main thread. Pure hardware-generated prefetching thread mechanisms[1,3,5,7,8,12,16] are transparent to compiler. However, such mechanisms might be inaccurate or suffer from higher memory bandwidth because it is difficult for hardware to observe and analyze the large range runtime execution. Traditional software-generated prefetching thread techniques[2,4,10,11] are typically accurate due to the better understandability on program semantics and data structures, but might incur additional instruction overhead and can not observe runtime behaviors.

It is necessary to adopt the advantages of both traditional hardware and software methods. To the best of our knowledge, this paper firstly proposes a novel hybrid hardware/software generated prefetching thread mechanism on Chip Multiprocessors.

The main contributions of this work are: (1) A hybrid hardware/software generated prefetching thread mechanism on Chip Multiprocessors is proposed; (2) Two aggressive thread construction policies, known as "Self-Loop" and "Fork-on-Recursive-Call", are presented for Dynamic Prefetching Thread; (3) "Thread Merging" policy is proposed for Static Prefetching Thread, which also adopts "Multi-Chain" policy; (4) The necessary hardware infrastructure for CMP supporting this hybrid mechanism is designed.

The rest of this paper is organized as follows: Section 2 introduces Dynamic Prefetching Thread. Section 3 describes the challenges to Dynamic Prefetching Thread. A hybrid hardware/software generated prefetching thread mechanism is proposed in Section 4. And Section 5 is performance evaluation. Section 6 is conclusion.

2 Dynamic Prefetching Thread

Many researchers found that a small number of static loads, known as *delinquent loads*, are responsible for the vast majority of memory stall cycles. Furthermore, not all the instructions contribute to the address computation of the future delinquent load[2,3,7]. Motivated by these observations, we try to extract these sequence of instructions as prefetching thread from the executed instruction trace by means of hardware, and utilize idle cores to execute such threads that perform dynamic prefetching for the main thread. Such threads are called *Dynamic Prefetching Thread(DPT)*, which are automatically generated, triggered, spawned and managed by hardware. It should exit when meeting exceptions or interrupts. The operating system should make no response to these exceptions and interrupts except for TLB exception.

2.1 The Hardware Infrastructure Supporting Dynamic Prefetching Thread

Figure 1(a) illustrates the typical CMP architecture with DPT support. The black blocks are the necessary hardware infrastructure supporting DPT. The "DPT Generator" is in charge of extracting DPT, located off the pipeline critical path. It has no effects on the pipeline frequency due to its back-end work mode. The "shadow register" is used for quickly initializing the context of the new spawned thread.

The organization of DPT Generator is shown in Figure 1(b). The committed load instructions in original thread and their corresponding execution information(such as L2 hit/miss flag) are sent to the back-end DPT Generator. These load instructions will first probe the trigger pointer selector, "Spawn Table". Once a trigger pointer is identified,

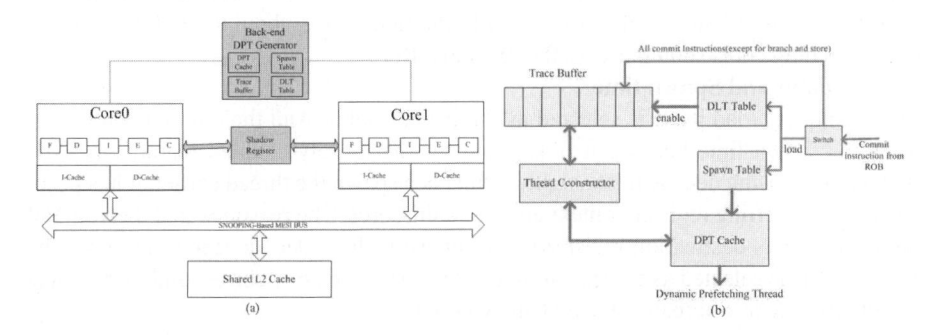

Fig. 1. The architecture of CMP with Dynamic Prefetching Thread support

the corresponding prefetching thread stored in DPT Cache is dispatched on idle core and run in parallel with original thread to perform dynamic prefetching for the targeted delinquent loads; otherwise it will query and update the Delinquent Load Table(DLT Table), which is in charge of identifying the delinquent load.

When any delinquent load is identified, DPT Generator begins to collect the committed instructions from the main core running original program. This collection does not stop untill the same delinquent load comes again or the Trace Buffer is full(If Trace Buffer is full, this mechanism is abort). After this collection, Thread Constructor performs a reverse walk of the trace to extract relevant instructions which contribute to the address computation of the targeted delinquent load. Then it produces a sequence containing these instructions in program order, oldest (lead) to youngest (candidate load). For simplicity, we only focus on the register dependence but ignore both memory and control-flow dependence during this reverse analysis. This policy is similar to Slice Processor[7], and we adopt it as our *basic policy*. Meanwhile, the trigger point is chosen for each Dynamic Prefetching Thread. These maps are recorded in DPT Cache.

The current CMP memory hierarchy is utilized to store prefetching results. No modifications are needed for memory hierarchy in this work.

Identify the delinquent load
The delinquent loads are identified at runtime via DLT Table. It is a PC-indexed table with 128 entries and each has 5-bit counters. One out-chip cache load miss(L2 Miss in our simulation) increases the corresponding counter by 4, otherwise decreases it by 1. A delinquent load is selected once the counter value exceeds 31. Predictor entry is allocated only when an L2 load miss occurs.

"Shadow Register" mechanism
The main core running original thread is to initialize the registers of the idle core when a DPT is dispatched. "Shadow Register" is for such quick initialization mechanism. It keeps the same data content with the main core. Some modifications are needed in pipeline to support this mechanism. The value and logical index of the destination register are attached with each issued instruction and reserved in ROB entries. Thus this information can be sent to the "Shadow Register" at commit time. The main core has the write privilege whereas the other cores running prefetching threads are only be allowed to read it. During the thread extraction phase, the live-in registers should be analyzed and used for marking some flags in renaming table of new core so as to differentiate the "Shadow Register" and local registers. Only the first access about the live-in registers on prefetching cores should access the "Shadow Register".

Trigger point and Spawn time
The delinquent load itself is selected as the trigger point. And the commit time is selected as spawn time because it is suitable for the loosely-coupled feature of CMP. Although choosing decode time as spawn time can spawn the thread earlier, it has problems in transporting register context among multi-cores. The reason is that the value of the instruction's destination register is still unavailable at decode time. Therefore the commit time is selected as the spawn time. It just needs to copy corresponding registers to initialize the new thread context at spawn time.

2.2 Aggressive Thread Construction Policies

(1) "Self-Loop" Policy

In basic policy, one Dynamic Prefetching Thread only prefetches one future instance of the static delinquent load. In "Self-Loop" policy, *the future N instances of the same delinquent load instruction are prefetched in the same Dynamic Prefetching Thread at one trigger point*(N=10 in our simulation). We accomplish this purpose via adding loop structure on basic-policy constructed thread code. The framework of new added loop structure is so stable that hardware implementation has high feasibility. This policy enlarges the prefetching range and helps the thread speculatively prefetch farther delinquent loads that are not visible in current pipeline. And it can also decrease the cost of thread initialization by merging multi-threads into one. Furthermore, "Self-Loop" policy need not copy register values between consecutive prefetching threads, since such threads are run on one core in our policy. This policy needs less prefetching cores(usually 1-4 cores are enough), thus releasing the access contention for "Shadow Register".

(2) "Fork-on-Recursive-Call" Policy

Most nodes in tree or graph structures connect two or more sub-nodes. This inherent memory parallelism can be exploited for prefetching. When the main program accesses one sub-tree or sub-graph, other idle cores can be utilized to speculatively access the other sub-tree or sub-graph. What's more, the recursive function is one of the primary methods used to access such structures. When any recursive call instruction is executed, a new prefetching thread is dispatched on one idle core starting from the next instruction address. Then the idle core begins to speculatively execute the following instructions. By means of this approach, idle cores are utilized to speculatively access the other sub-tree or sub-graph for prefetching. This is the "Fork-on-Recursive-Call" policy.

A hardware stack and Recursive Call Table are used for identifying the recursive call and recording the recursive entries for each recursive call. They work in back-end and are placed in DPT Generator. Any function call instruction(e.g, jal, jalr in MIPS ISA) at the top of ROB will trigger the following step:

(a) Looking up the Recursive Call Table to find whether this call is recursive. If some entry is found, then goto (b), else goto (c).

(b) The following PC of the current call instruction is sent to idle core to be speculatively executed. And exits here.

(c) The instruction's PC enters the hardware stack. It will look up the previous stack entries before entering the stack. If some entry matches, a recursive call is identified, and the PC is recorded in Recursive Call Table. Otherwise, it is just stored in the stack. The stack should be emptied if it is full.

Any return instruction(e.g, jr in MIPS ISA) should update the stack at commit time. If the stack is empty, nothing is done; otherwise the top stack entry is popped.

The store instructions are considered as nop operation since the speculatively executed thread is only used for prefetching and should not modify the architecture state. A counter is used to control the execution distance of prefetching thread. The prefetching thread also looks up the Recursive Call Table when any call instruction is executed. If one recursive call is identified, the counter begins to work and increase one for each instruction. In this work, the prefetching thread will not stop untill the counter exceeds 200 or some exception occurs.

3 The Challenges to Dynamic Prefetching Thread

The following three cases are great challenges for Dynamic Prefetching Thread.

(1) The loops with two or more delinquent loads. When there are two or more delinquent loads in the same loop structure, usually some of them are not timely prefetched by Dynamic Prefetching Threads. The reason is that each such threads usually targets at only one static delinquent load. If the number of processor core is small, several Dynamic Prefetching Threads separately targeting at different loads compete for the scarce idle cores. Thus some of prefetching threads have no chance to be dispatched.

(2) The loops with two or more levels. Larger prefetching range can be expected at the outer-level loop. Yet it is hard for the hardware to identify and collect the whole execution trace of the outer loop iterations. Therefore the prefetching timeliness and range are limited.

(3) The hot regions with complicated control flow. The instruction traces are unstable in this case. It is hard for hardware to analyze and conclude all the conditions at runtime. The prefetching accuracy might be quite low.

4 The Hybrid Hardware/Software Prefetching Thread Mechanism

Although software-generated prefetching thread might incur additional instruction overhead and can not observe runtime behaviors, it can overcome the challenges to Dynamic Prefetching Thread. We proposes a hybrid hardware/software generated prefetching thread mechanism on Chip Multiprocessors. Two kinds of prefetching threads appear in our hybrid mechanism. Most threads belong to Dynamic Prefetching Thread, which are automatically generated, triggered, spawn and managed by hardware; The others are of *Static Prefetching Thread(SPT)*, targeting at the *critical delinquent loads* identified by profiling information. SPT is statically generated by binary-level optimization tool. The software tool can understand the program semantics better, thus higher prefetching accuracy and larger prefetching range are anticipated for SPT.Furthermore, benefiting from the concentration on critical delinquent loads, SPT incurs little additional instruction overhead.

This hybrid mechanism is effectively composed of DPT and SPT where DPT is predominant. These two kinds of threads are efficiently combined by the identification of critical delinquent loads. An enhanced compilation flow and the corresponding profiling mechanism are proposed to support the identification of critical delinquent loads and the SPT construction. By the way, SPT has higher execution priority than DPT. All such threads are transparent to operating system.

4.1 Compilation in Hybrid Mechanism

The enhanced compilation supporting the hybrid mechanism is illustrated as the following steps:

(1) The program is compiled by general source-code compiler(e.g, gcc);
(2) The binary is run directly on CMP *without* Dynamic Prefetching Thread support. The instruction addresses of the TOP N most frequent load misses are collected via performance counter. Regarding these instructions, we call the set, which is composed of *(instruction address, the number of cache misses)* pairs, as **Miss_Set0**;
(3) The binary is run directly on CMP *with* Dynamic Prefetching Thread support. The instruction addresses of the TOP N most frequent load misses are collected via performance counter. Regarding these instructions, we call the set, which is composed of *(instruction address, the number of cache misses)* pairs, as **Miss_Set1**;
(4) Then the set of *critical delinquent loads*, which can not be accurately or timely prefetched by Dynamic Prefetching Thread, are identified according to the following formula:

$$Critical\_Set = \{x \mid \exists x, \exists y0, \exists y1,$$
$$(x, y0) \in Miss\_Set0,$$
$$(x, y1) \in Miss\_Set1,$$
$$and\ (y0 - y1)/y0 < \delta\}$$

In this formula, x is instruction address, y0 and y1 are the numbers of cache misses, and the δ is the assumed threshold for identifying critical instructions.

(5) Targeting at these critical delinquent loads, the binary-level SPT tool can extract more effective prefetching threads from original binary, attach them in a special program text segment, and regenerate the final version SPT-enhanced binary.

4.2 The Binary-Level SPT Tool

Firstly, the binary is loaded and disassembled. Guided by the relocation information in binary head section(e.g, ELF head), all basic blocks and their relationships(functions and branches) are identified. Then the control flow graph(CFG) is constructed. Secondly, the loop structures or functions containing *critical delinquent loads* are located, and the tool makes analysis on such zones based on several specific thread construction policies. All the instructions, which contribute to the address computation of the critical delinquent loads, are extracted. Such extracted instructions are Static Prefetching Thread, placed in a special program text segment at the bottom of original binary. During these analysis, the register live-ins of Static Prefetching Thread are also attained, which is helpful to choose a spawn point and insert a spawn instruction in original binary. Finally, some adjustments are necessary since original binary is modified, and then we get the SPT-enhanced binary by the SPT tool.

4.3 Thread Construction Policies for SPT

(1) "Thread Merging" Policy
"Thread Merging" policy is proposed to overcome the case where there are several delinquent loads in the same loop. In this policy, all the static delinquent loads in the same loop are prefetched by one prefetching thread.

According to profiling, SPT tool can observe that more than one critical delinquent loads appear in the same loop structure. Through analyzing the register and control dependence from the loop header to bottom(still ignoring memory dependence), all instructions contributing to the computation of these delinquent loads' addresses are extracted. The loop header is selected as the spawn point before which spawn instruction is inserted. Of course, "Self-Loop" can also be merged with "Thread Merging" policy.

(2) Multi-Chain Policy

Multi-Chain policy is described in [9]. We apply it to deal with the case where there are delinquent loads in loop structure with two or more levels. Such case is common in pointer-chasing applications, which tend to traverse composed data structures consisting of multiple independent pointer chains. Multi-Chain policy exploits this inter-chain memory parallelism. When the original thread accesses one pointer chain, Static Prefetching Threads simultaneously perform their speculative traversal of other possible future chains on idle cores. Consequently, the serialized memory latency can be tolerated by overlapping cache misses across independent pointer-chain traversals.

When SPT tool observes that there are little overlap work between the sequent critical delinquent loads in the loop with two or more levels, multi-chain policy is adopted to construct SPT. First, it analyze the inner loop, and extract all instructions contributing to the address computation of these delinquent load, including the loop induction variables and corresponding instructions. These extracted instructions are called as sub-thread. Meanwhile, the register live-ins of sub-thread are attained. Then the outer loop is analyzed, all instructions related are also extracted. These instructions are located before the sub-threads. Then the whole SPT is constructed. The spawn instruction is inserted directly before the entry of the inner loop.

4.4 Hardware Support for Hybrid Mechanism

(1) Extensions to the Instructions Set Architecture

Two additional instructions are needed. One is the *spawn* instruction. Its format is "spawn start-of-SPT". This instruction explicitly indicates one SPT dispatch and register context initialization. The other is the *stop* instruction, indicating that the prefetching thread is to be finished. It has no operator and is also used for DPT.

(2) Profiling mechanism for the TOP N out-chip load instructions

The critical delinquent loads identification needs to collect the top N most frequent out-chip loads for the execution of whole program. A hardware/software cooperative profiling mechanism is designed for such purposed.

A new Performance Counter(PC) is provided to record recent the top N most frequent out-chip load instruction, which is similar to Cache Miss Lookaside Buffer[13] aiming at releasing the access pressure on L2 cache. The new Performance Counter consists of *(process ID, instruction address, counter)* tuples with process ID and instruction address as index. It is implemented as content-indexed array(CAM). The Process ID is used for distinguishing the instructions from original or prefetching thread, and only the former is concerned. To improve the accuracy, the tuples are broken into two segments: HOT and LRU region. Once an out-chip load commits, a lookup in the PC(both the LRU and HOT segment) is performed. If it doesn't match, the least-used entry in LRU segment is replaced by the instruction and the counter is initialized as one; Otherwise, the corresponding counter is increased. Furthermore, if it matches the LRU segment and the counter is larger than the minimum in HOT segment, these two entries are exchanged. The size of LRU and HOT segments are important for the profiling accuracy. We find 32 is suitable in our simulation.

However, the Performance Counter can only record recent out-chip load instruction. In order to record the top N most frequent out-chip loads for the execution of whole program, software are needed to record and accumulate the performance counter at intervals. Such function is implemented in the timer interrupt entry of operating system. Since PC only works for profiling, this mechanism does not decrease the performance and has no additional power dissipation.

Table 1. Simulated CMP Processor Parameters

Processor core		Memory Hierarchy	
Number of cores / Frequency	2core/2GHz	Cache sizes	32KB IL1, 32KB DL1, 512KB L2
Fetch / Issue / Commit Width	4 / 4 / 4	Cache associativity	4-way L1, 8-way L2
I-window / ROB / LSQ size	64 / 128 / 64	Cache Hit/Miss latencies	L1:2/3 cycles, L2: 9/11 cycles
Int/FP registers	184	Cache line sizes/ports	L1:32B,2ports, L2:32B,4ports
LdSt/Int/FP units	2 /4 / 2	L1-L2, L2 cache Store policy	write-back
Execution latencies	similar to MIPS R10000	MSHRs	L1:64 , L2:128
Branch predictor	16K-entry gshare hybrid	Memory Bus	split transaction, 2words/cycle
RAS entries	16	Main memory latency	minimum 200 cycles
Hardware Supporting Hybrid Prefetching Thread			
Trace Buffer Size		256 entries	
DPT Cache size / associativity		32kB / 2 way	
Thread construction time		200 cycles	
Thread initiation time		6 cycles	
Shadow Register size / port		64*32B / 4w4r ports	
δ for Critical_Set		0.5	

5 Experiments

5.1 Simulation Methodology

The evaluation is performed by a detailed CMP architecture simulator based on SESC [17] implementing MIPS ISA, which is a cycle-accurate execution-driven simulator. The CMP cores are out-of-order superscalar processors. Table 1 lists the parameters in details. To demonstrate the performance potential of our architecture, we just use the dual core configuration for simplicity.

The memory limited benchmarks are selected from the Olden pointer intensive programs[6], and SPEC CPU2000. A large number of cache misses in these benchmarks are due to relatively irregular access patterns involving pointers, hash tables, tree/graph, indirect or complicated array references, or a mix of them, which are typically difficult for prefetching. The train sets are used for SPEC benchmarks to achieve reasonable simulation times. In addition, all benchmarks are compiled with gcc -O3 and simulated for one billion committed instructions after fast-forwarding the initialization with cache warmup.

The full mechanisms of Dynamic Prefetching Thread are simulated in details. Yet Static Prefetching Threads are constructed manually. SPT tool now can read and modify the MIPS binaries, while the implementation of SPT thread construction policies is still in development. These hand-generated Static Prefetching Threads demonstrate the performance potential of our hybrid mechanism.

5.2 Performance Evaluation

The performance speedup of our hybrid mechanism is illustrated in Figure 2. Since DPT is predominant in our hybrid mechanism, the speedup of DPT adopting basic and aggressive polices is also presented to make comparisons, and the speedup of pure SPT mechanism is ignored in Figure 2. With regards to the DPT mechanism, it can be observed that significant improvements are achieved with aggressive policies. The aggressive policies achieve 21.5% speedup on average while the basic policy only achieves 3.1% speedup. Furthermore, the performance can be further improved by the hybrid mechanism. SPT can overcome the challenges to DPT(especially for swim, mgrid, equake, em3d and mst). The performance speedup is increased to 31% on average when adopting the hybrid mechanism.

To understand the performance speedup, the prefetching coverage and timeliness information is provided to have a deep insight at the prefetching activity in Figure 3. Each bar is broken into eight segments according to the fractions of the miss latency hidden by prefetching, e.g, less than 10 cycles, between 10 and 50 cycles and so on.

For swim, mgrid,art, equake and mcf, most of the speedup benefits from the larger coverage and better timeliness achieved by DPT with "Self-Loop" policy. Through enlarging the prefetching range and number per prefetching thread, "Self-Loop" policy makes the thread generate more timely and more farther prefetching requests illustrated in Figure 3. For instance, the coverage of swim is about 2% in basic policy, and it increases to 21% in aggressive policies. And the performance speedup for swim also increases from 0 to 32% with the improvements of prefetching coverage and timeliness.

The "Fork-on-Recursive-Call" policy stimulates the performance improvements for treeadd, perimeter and tsp, since these benchmarks access tree-like structures via

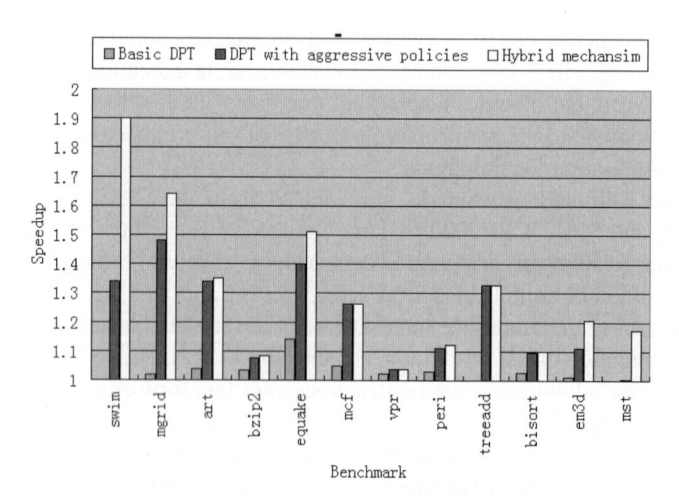

Fig. 2. The performance speedup of several prefetching thread mechanisms

recursive calls. This policy effectively exploits the memory parallelism indicated by the the kernel data structures and then improves the prefetching coverage and timeliness, especially for treeadd(31% performance improvement).

Although most benchmarks have been accelerated significantly by Dynamic Prefetching Thread, there are still considerable performance potential to be exploited by our hybrid mechanism. For swim, mgrid and equake, the "hot" loops always have several delinquent loads. Static Prefetching Threads constructed by "Thread Merging" policy prefetches these delinquent loads using one thread, leading to the significantly improved prefetching coverage demonstrated in Figure 3. For mcf and vpr, there are usually one delinquent loads in hot loop or thread contentions are scarce, Static Prefetching Threads almost have no effects. For mst and em3d, most pointer accesses have little overlap work, so Dynamic Prefetching Threads almost have no effects on them. Fortunately, it is observed that the kernel data structures are accessed by loops with two or more levels, "Multi-Chain" policy can effectively accelerate these benchmarks via higher-level prefetching(the performance improvement for em3d is 20%, mst is 18%). These phenomenons are demonstrated in Figure 3.

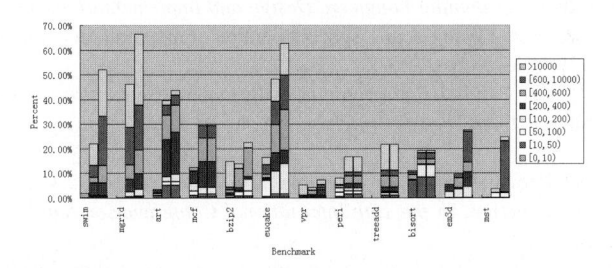

Fig. 3. The prefetching coverage and timeliness analysis. (For each group, the left bar: DPT with basic policy, the middle: DPT with aggressive policies, the right: hybrid mechanism.)

6 Conclusion

This paper firstly proposes a hybrid hardware/software generated prefetching thread mechanism on Chip Multiprocessors.This hybrid mechanism is effectively composed of Dynamic Prefetching Thread and Static Prefetching Thread. The former is predominant and dynamically generated by hardware, and the latter is complementary and statically generated by software. These two kinds of threads are efficiently combined by an enhanced compilation flow and the corresponding profiling mechanism.

For a set of memory limited benchmarks, an average speedup of 3.1% is achieved on dual-core CMP when constructing DPT with basic policy, and this gain grows to 21.5% when adopting aggressive policies. Although significant improvements can be achieved by DPT, the performance can still be further improved by the hybrid mechanism. SPT is an effective complement to DPT. The performance speedup is increased to 31% on average when adopting the hybrid mechanism.

Acknowledgements

We would appreciate the anonymous reviewers for their advices. This work is supported by the National Science Foundation for Distinguished Youth Scholar(60325205), the Basic Research Foundation of the ICT, CAS under Grant No.20056020; the 863 Hi-Tech Research and Development Program of China (2005AA1100102005AA119020); the National Grand Fundamental Research 973 Program of China, National Basic Research Program of China under No 2005CB321600.

References

1. A. Roth , G. Sohi. Speculative data-driven multithreading. In *7th HPCA*, pages 37-48, 2001.
2. J. Collins, H. Wang, etc. Speculative precomputation: Long-range prefetching of delinquent loads. In *the 28th ISCA*, pages 14-25, July 2001.
3. J. D. Collins, D. M. Tullsen, H. Wang, etc. Dynamic speculative precomputation. In *the 34th annual ACM/IEEE International Symposium on Microarchitecture*, pages 306-317, 2001.
4. S. Liao, P. Wang, etc. Post-Pass Binary Adaptation for Software-Based Speculative Precomputation. In *ACM Programming Language Design and Implementation*, June 2002.
5. Jeffery A. Brown, Hong Wang et al., Speculative Precomputation on Chip Multiprocessors. In *the 6th MTEAC*, November, 2002.
6. M. Carlisle. Olden: Parallelizing programs with dynamic data structures on distributed-memory machines. *PhD Thesis, Princeton University Department of Computer Science*, 1996.
7. A. Moshovos, D. Pnevmatikatos, and A. Baniasadi. Slice processors: An implementation of operation-based prediction. In *the 15th International Conference on Supercomputing*, pages 321-334, June 2001.
8. H. Zhou. Dual-core execution: building a highly scalable single-thread instruction window. In *the 14th PACT*, 2005.
9. N. Kohout, S. Choi and D. Yeung. Multi-chain prefetching: Exploiting memory parallelism in pointer-chasing codes. In *ISCA Workshop on Solving the Memory Wall Problem*, 2000.
10. T. Mowry and A. Gupta. Tolerating latency through software controlled prefetching in shared-memory multiprocessors. In *Journal of Parallel and Distributed Computing*, pages 87-106, June 1991.
11. C. Luk. Tolerating memory latency through softwarecontrolled pre-execution in simultaneous multithreading processors. In *the 28th ISCA*, pages 40-51, July 2001.
12. Ilya Ganusov and Martin Burtscher. Future Execution: A Hardware Prefetching Technique for Chip Multiprocessors. In *PACT 2005*, pages 350–360, 2005.
13. Brian N. Bershad, Dennis Lee et al. Avoiding Conflict Misses Dynamically in Large Direct-Mapped Caches. In *the 6th ASPLOS*. Pages: 158–170. 1994.
14. J. Huh, D. Burger, S. Keckler. Exploring the design space of future CMPs. In *the 10th PACT*, pages 199–210, September 2001.
15. Doug Burger, James R. Goodman. Billion-transistor architectures: there and back again. *Computer*, Page(s):22-28. Mar 2004.
16. O. Mutlu, J. Stark, C. Wilkerson, and Y. N. Patt. Runahead execution: an alternative to very large instruction windows for out-of-order processors. In *the 9th HPCA*, 2003.
17. Jose Renau, Basilio Fraguela, James Tuck et al., *http://sesc.sourceforge.net*. January, 2005.

Topic 8: Distributed Systems and Algorithms

Andrzej Goscinski, Gudula Rünger, Edgar Gabriel, and Christine Morin

Topic Chairs

Parallel computing is strongly influenced by the challenges of distributed systems, such as a need for a Single System Image, resource sharing and allocation, failures and a need for fault tolerance, long latencies, network partition, disconnected operation, demands of users wishing to solve more computationally and communication demanding problems, and opportunities created by grids and Web services. Distributed computing is the computing mainstream now; it is based on different forms of distributed systems: clusters, grids, peer-to-peer systems, web services, service oriented architectures. This topic provides a forum for research and practice, of interest to both academia and industry, about distributed computing and distributed algorithms. Submissions were encouraged in all areas of distributed systems and algorithms relevant to parallel computing, with emphasis on design and practice of distributed algorithms, analysis of the behaviour of distributed systems and algorithms, distributed fault-tolerance, distributed operating systems and databases, scalability, concurrency and performance in distributed systems, resource sharing and load balancing in distributed systems, distributed algorithms in telecommunications, distributed mobile computing, resource and service discovery, security in distributed systems, and standards and middleware for the distribution of parallel computations. Twenty papers were submitted in this topic. The subjects were varied, but a common theme of many of them is recovery, resource allocation, mutual exclusion, garbage collection and coordination. Other themes include load balancing, scheduling and consensus algorithms. Eight papers have been accepted.

Distributed Approximation Allocation Resources Algorithm for Connecting Groups

Fabien Baille[1], Lelia Blin[2,*], and Christian Laforest[2]

[1] LIAFA, Univ. Denis Diderot- Case 7014 2,
place Jussieu, F-75251 Paris Cedex 05
[2] Tour Evry II, IBISC, Univ. d'Evry, 523 place des terrasses,
91000 EVRY, France
lelia.blin@lami.univ-evry.fr

Abstract. This paper presents a distributed algorithm to allocate resources (links of a network) for interconnecting machines (forming a group) spread in a network. This is what we call a *connection structure* for this *group* of machines. An important innovative feature of our construction method is that we *prove* (not just simulate on particular and restricted cases) the fact that this structure has good properties in terms of, *simultaneously*, induced distances (for latency considerations) and cost (for cost considerations). Hence, we propose a *distributed multicriteria approximation* algorithm.

In applications like video-conferences or net-meetings, *members* of a *group* spread in a network, have to communicate with high QoS requirements. A possibility for a *provider* selling this service on his network is to allocate/rent resources (links) for the exclusive use of the members. We call this a *connection structure* for the group. We propose here a distributed protocol to construct connection structure with *high guaranty* of quality.

Minimizing induced distances in the structure. To optimize QoS requirements on latency between members, we want to construct a structure in which distances are minimized. However, these distances cannot be smaller than those of the underlying network. Hence, we want to design a structure in which the induced distances between members are as close as possible to those of the original graph. To capture this desired property, we focus in this paper on the minimization of two criteria, namely the *maximum* and *average* distances between members.

The cost of the structure. Steiner tree. As a connection structure is a definitive allocation of links, exclusively reserved for the group, these resources are not available for others applications during the existence of the group. Hence, the provider allocating the structure must *minimize* the total number of links

---

* Corresponding author.

used in the structure in order to minimize its exploitation costs and to keep the maximum number of available links for others services. In this paper, the number of edges in a graph or in a structure is called its *cost*. The minimum cost (weight) tree spanning a given set M of vertices is called the *Steiner tree* problem and is NP-complete.

Conflicting parameters. Simultaneous approximation. In the situation described above, the provider must minimize the distance parameters (to satisfy the customers) and, *simultaneously*, the cost of the structure. It was shown in [1] that it is impossible; the criteria cost and diameter are conflicting. To solve this conflict we do not strictly minimize the parameters but we *approximate* them.

A distributed algorithm. In spite of its interest in the guarantees that can be offered, this kind of approach generally suffers from a weakness: The methods to construct the structure are almost all centralized. To exploit them, the provider needs a global view of the state of its system before applying the algorithms. In general this is not possible in practice. To help him, we propose in this paper a *distributed* algorithm. The connection structure is then constructed by exchanging messages between members, with no centralized node doing all the job. Members just know local information to process. In this paper, we make reasonable hypothesis; we suppose that a routing table and an allocating mechanism is available in the system (see more details on the distributed model in section 1).

Another interesting parameter to investigate for our asynchronous protocol is then the *number of exchanged messages* (or messages complexity) during its execution to avoid to overload the network during the construction.

Representation by a graph. We model the network by a graph $G = (V, E)$ where V is the set of *vertices* representing the nodes of the network, E is the set of *edges* modeling its bidirectional physical links. Graphs considered here are unweighted, undirected and connected.

Let $M \subseteq V$ be the *group* of $m = |M|$ members that must be connected by allocating/renting links. The set of these links form what we call a *connection structure*; As the links are reserved for intra-group communications there is no external traffic disturbing the communications between members. Note that in terms of graphs a connection structure $S = (V_S, E_S)$ in a connected subgraph of $G = (V, E)$ spanning M ($M \subseteq V_S \subseteq V$, $E_S \subseteq E$).

Technically, the latency is represented by distances. For each pair u, v of vertices of V, we denote by $d_G(u, v)$ the *distance* between u and v. This is the minimum number of edges to cross to go from u to v in G.

Notation 1. *Let $G = (V, E)$ an unweighted, undirected graph and $M \subseteq V$ be the group.*

- *The **diameter** of M in G is:* $D_G(M) = \max\{d_G(u, v) : u, v \in M\}$
- *The **sum of distances** of M in G is:* $C_G(M) = \sum_{u,v \in M} d_G(u, v).$

– The **cost** of any graph (or structure) $G = (V, E)$ is its number of edges: $W(G) = |E|$.

We focus here on sum of distance since it is easy to get average distance from that. The minimum cost (weight) spanning structure of a group M is the well known *Steiner tree* denoted by $T^*(M)$.

We want to construct a structure S, spanning members of M such that its diameter $D_S(M)$ (resp. its sum of distance $C_S(M)$, resp. its cost $W(S)$) is no more than $\rho_D D_G(M)$ (resp. $\rho_a C_G(M)$ resp. $\rho_W W(T^*(M))$). The parameter ρ_D (resp. ρ_a resp. ρ_W) is the *approximation ratio* for the diameter (resp. sum of distances, resp. weight).

Known results. Due to space limitation we focus here on the main related works to ours. We underline the difference with our own contribution.

In [2] for example, reader can find treatment of approximation algorithms and of the NP-hard Steiner tree problem. However, classical approximation methods just deal with the optimization of *one* criterion: Only the weight is treated for the Steiner tree for example; the induced distances in such trees are not considered.

At the opposite, works on *spanners* investigate the problem to construct a structure of minimum weight, spanning *all the vertices* and inducing distances between *each pair* of vertices at most a given multiplicative factor of the one in the underlying graph. This approach is very interesting but unfortunately there are many non approximability results (see [3]). A variant for groups has been investigated in [4] and was also shown to be hard. Moreover all the existing methods are centralized.

We can see that the main constraint in spanner is to give guarantees for the distances between *each pair* of vertices. In other works, authors relax this constraints and investigate the construction of structures in which the maximum and/or average distance are minimized. For example, one can find in [5,1,6] approximation algorithms to construct trees spanning a given group with the objective to approximate simultaneously these parameters. In particular, [6] investigates exactly the three parameters considered in the present paper. However the algorithm given in [6] is not distributed. In the present paper we propose an alternative construction to [6] that allows us to obtain a distributed protocol. Indeed this version exhibits original local properties that are exploited. We show in this paper that this new approach leads to an efficient algorithm (in terms of message complexity) and good approximation ratios. For completeness, we present here the whole construction and all the proofs.

To finish we can cite [7] surveying several recent works on approximation distributed algorithms. However, there is no reference on our own subject, works mentioned focus on the optimization of only one criterion (we are multicriteria) and the distributed systems under consideration are synchronous (we deal here with asynchronous systems).

Outline of the paper. In Section 1, we describe our distributed algorithm. We prove its approximation ratios and number of exchanged messages in Section 2.

1 Our Tricriteria Approximation Distributed Algorithm

Our tricriteria algorithm proceeds in two phases: In the first phase, we construct the tree T_p, a ρ-approximation of a Steiner tree $T^*(M)$ of the group M. In a second phase, we reduce the induced distances between members in this tree; this is done by adding appropriated shortest paths without increasing significantly the weight of the initial tree.

Distributed model: A shortest path *routing function* is available as in many current networks. Each node u can use its local copy of the routing table to determine the distance $d_G(u,v)$ between itself and any node v. This routing function can also be used by node u to send messages by a shortest path of G to any node v. It can also be used to *allocate* a shortest path between u and v (reserve the links of this path). We suppose that these two operations induces a number of messages equal to the number of links crossed, that is $d_G(u,v)$. Moreover, we consider that each member is awake and knows the m distinct identities of the members.

1.1 Phase 1: Construction of an Approximated Steiner Tree

Phase 1 constructs a tree T_P satisfying $W(T_P) = O(\log(m)W(T^*(M)))$. Moreover, the first phase finds a member with particular properties, called the *Median* r of the group M. This particular member is essential for beginning the second phase.

As the identifiers of members of M are known by each node of M, each node can create an order. W.l.o.g. we suppose that the members are numbered $M = \{u_1, u_2, \ldots, u_m\}$ and that each member knows this order.

The construction of tree T_p is the following: Each member u_i connects itself to the *nearest* member in the set $\{u_1, \ldots, u_{i-1}\}$ (u_i can select this nearest member using its local routing table). These $m-1$ connections can be done in any order, even in parallel. At the end, the (intermediate) obtained structure is composed of $m-1$ shortest paths and is connected. It is not necessarily a tree (it may contain cycles), an appropriated DFS is done twice from node u_1 to cut potential cycles and to prune the structure to obtain the desired *tree* T_P. After this operation, each node $u \in M$ is in the tree T_P.

We give now the definition of the median and the way to construct it.

Definition 1 (Median of a group). *Let $G = (V, E)$ and $M \subseteq V$. A vertex $r \in M$ is a median of M (in M) if it satisfies:*

$$\sum_{u \in M} d_G(u, r) = \min\left\{\sum_{u \in M} d_G(u, v) : v \in M\right\}$$

To do that, u_1 starts the process by sending to u_2 the pair (S_1, u_1) where $S_1 = \sum_{v \in M} d_G(u_1, v)$. When it receives S_1, node u_2 can compute its own value

$S_2 = \sum_{v \in M} d_G(u_2, v)$. If $S_1 < S_2$ then u_2 sends (S_1, u_1) to u_3, otherwise it sends (S_2, u_2). This process can be continued from node to node, following the order $\{u_1, \ldots, u_m\}$. The last node u_m sends its result to u_1 that now knows the median r (since the minimum sum of distances has been filtered in successive transfers). It then broadcasts the result to the whole group.

1.2 Phase 2: Distributed Median-Control

The general idea of this phase 2 is to make a DFS from the *Median* r in T_P (constructed in previous phase). Each time the DFS reaches a member, it makes a particular test on distances: If the test is positive, it roughly means that the member is "too far" from the median in T_p *compared* to its distance in G (read in the routing table). In this case, we say that this member v_i is added in set S. This member connects to the *Median* by allocating a shortest path of the graph (using the routing mechanism). Note that this set S is *not* transported in the message. We only need its cardinality in the protocol; However, in the proof we will use this implicit set S. At the beginning, when $S = \emptyset$, we set $v_i = r$.

Message: For the protocol, we need to transport several information in the messages of type: $< DFS, Dist_{T_P}, Dist_G, cardS >$

- DFS is the name of the message.
- $Dist_{T_P}$ is the distance in tree T_P between the last node v_i inserted in set S and the node sending the DFS message. This parameter must be updated at each node (member or not).
- $Dist_G$ is the distance in G between v_i and the *Median* node r. This distance is read by v_i in its local routing table. Note that $Dist_G$ ($= d_G(r, v_i)$) does not change during the construction while v_{i+1} is not detected.
- Counter $cardS$ is the number of nodes in the current set S.

Local variables: For each node u in T_P, the main variables are:

- N_{T_P}: set of neighbors identifiers of u in tree T_P.
- TBS: (To Be Sent) set of neighbors of u in T_p to which node u has to forward the DFS message.
- $Visited$: boolean with value $True$ iff node u has received a DFS message. Init. at $False$.
- $Parent$: node identifier by which node u receives for the first time message DFS. Init. at NIL.
- $Last\_Card\_S$: integer counter containing the knowledge node u has of the cardinality of the current set S. Init. at 0.

We suppose in the following that a parameter α is already known by all the members of M.

Pseudo-code of the algorithm

> PROCEDURE: Executed only by Node $Median$ (node r), at the beginning
> $Last\_Card\_S := 0$; $Dist_{T_P} := 0$; $Dist_G := 0$;
> $NumConnections := 0$; $TBS := N_{T_P}$;
> Do
>   {Choose any $v \in TBS$; $TBS := TBS - \{v\}$;
>   Send $< DFS, Dist_{T_P}, Dist_G, cardS >$ to v;
>   Wait for a message $< DFS, Dist_{T_P}, Dist_G, cardS >$ from v;
> $Last\_Card\_S := cardS$;
>   If $cardS = 0$ then $Dist_{T_P} := 0$; else $Dist_{T_P} := Dist_{T_P} + 1$;}
> While $TBS \neq \emptyset$
>
> /* When $Last\_Card\_S = NumConnections$, the algorithm is finished and the final connection structure G_f is allocated. */

> PROCEDURE: Each time Node $Median$ r receives a connection from a node v
> $NumConnections := NumConnections + 1$;

> PROCEDURE: When a node $u \neq Median$ receives
> $< DFS, Dist_{T_P}, Dist_G, cardS >$ from node $v \in N_{T_P}$
> If $Visited = False$ then { /* First visit of the DFS */
> $Visited := True$; $Parent := v$; $TBS := N_{T_P} - \{Parent\}$;
> $Dist_{T_P} := Dist_{T_P} + 1$; $Last\_Card\_S := CardS$;
>   If $u \in M$ and $Dist_{T_P} + Dist_G > \alpha d_G(r,u)$ Then {
>   /* A new v_i is detected and added in set S */
> $Dist_{T_P} := 0$; $Dist_G := d_G(r,u)$; $cardS := cardS + 1$;
> $Last\_Card\_S := CardS$;
>   Connect to $Median$ r by a shortest path;}
>   }
> Else {/* $Visited = True$ */
>   If $Last\_Card\_S < CardS$ then {
> /* New v_i's have been discover in the exploration of the subtree of u */
> $Dist_{T_P} := Dist_{T_P} + 1$; $Last\_Card\_S = CardS$;}
>   else $Dist_{T_P} := Dist_{T_P} - 1$;
>   If $TBS = \emptyset$ then $a = Parent$; /* It is time to backtrack */
>   else {Choose any $a \in TBS$; $TBS := TBS - \{a\}$;}
>   Send $< DFS, Dist_{T_P}, Dist_G, cardS >$ to a;
> }

2 Proofs and Correctness

We express and prove in Section 2.1 a generic result giving the three approximation ratios obtained by our algorithm. In Section 2.2 we use this result to give

2.1 Three Simultaneous Approximation Ratios

Theorem 1. *Let $G = (V, E)$, $M \subseteq V$ ($m = |M|$), G_f be the connect structure of M returned (in the second phase* MEDIAN-CONTROL*) with T_P a ρ-approximation of the Steiner tree $T^*(M)$ of M, constructed in phase 1. The three simultaneous properties of our algorithm are the following:*

1. *A 2α-approximation for the sum of distances: $C_{G_f}(M) \leq 2\alpha C_G(M)$.*
2. *A 2α-approximation for the diameter of M: $D_{G_f}(M) \leq 2\alpha D_G(M)$.*
3. *A $\rho \left(1 + \frac{2}{\alpha-1}\right)$-approximation for the weight:*

$$W(G_f) \leq \rho \left(1 + \frac{2}{\alpha - 1}\right) W(T^*(M)).$$

For the clearness of the proof we will use Lemmas.

Lemma 1. *Let $G = (V, E)$, $M \subseteq V$ ($|M| = m$) and r be any vertex of V. Let A be any subgraph of G, spanning M and r, satisfying $d_A(u, r) \leq \alpha d_G(u, r)$ for all $u \in M$ for some $\alpha \geq 1$. We have:*

$$C_A(M) \leq 2\alpha m \sum_{u \in M} d_G(u, r)$$

Lemma 1 is an extension of a result of [8].

Proof. We upper bound $C_A(M)$ by using triangular inequality.

$$C_A(M) = \sum_{u,v \in M} d_A(u, v) \leq \sum_{u,v \in M} (d_A(u, r) + d_A(r, v))$$

$$= \sum_{u \in M} \sum_{v \in M} (d_A(u, r) + d_A(r, v))$$

$$= \sum_{u \in M} \left(m d_A(u, r) + \sum_{v \in M} d_A(r, v)\right)$$

$$= 2m \sum_{u \in M} d_A(u, r) \leq 2\alpha m \sum_{u \in M} d_G(u, r)$$

The last inequality follows from the property of distance in A. □

Lemma 2. *Let $G = (V, E)$, $M \subseteq V$ and $r \in M$ a median of M. We have:*

$$C_G(M) \geq m \sum_{u \in M} d_G(u, r)$$

Proof. $C_G(M) = \sum_{v \in M} \left(\sum_{u \in M} d_G(u, v)\right) \geq m \sum_{u \in M} d_G(u, r)$ □

We suppose here that a tree T_P spanning M is given, constructed in Phase 1. This is an approximation of the Steiner tree $T^*(M)$. Graph G_f in the following is the final connection structure constructed by our algorithm. The techniques used here and related proofs are adaptations and modifications of the ones of [5,6] for the distributed context.

Lemma 3. *For all $u \in M$ we have:* $d_{G_f}(r, u) \leq \alpha d_G(r, u)$

Proof. Let us consider the step where the DFS reaches vertex $u \in M$ for the first time. At this particular moment, let v_i be either the last vertex put in set S or vertex r (if set S is still empty). Two cases must be examined, depending on the test treating u in the algorithm. Note that $d_{T_P}(v_i, u)$ (resp. $d_G(r, v_i)$) is transported by the incoming DFS message in parameter $Dist_{T_P}$ (resp. $Dist_G$).

1. If $d_G(r, v_i) + d_{T_P}(v_i, u) > \alpha d_G(r, u)$ then u is put in set S and after the DFS a shortest path between r and u is added in G_f and in this case $d_{G_f}(r, u) = d_G(r, u)$.
2. Otherwise, $d_G(r, v_i) + d_{T_P}(v_i, u) \leq \alpha d_G(r, u)$. As G_f contains a shortest path between v_i and r in G and also contains the whole tree T_P, by using the triangular inequality we have:

$$d_{G_f}(r, u) \leq d_{G_f}(r, v_i) + d_{G_f}(v_i, u) \leq d_G(r, v_i) + d_{T_P}(v_i, u) \leq \alpha d_G(r, u) \quad \square$$

Lemma 4. $W(G_f) \leq \left(1 + \dfrac{2}{\alpha - 1}\right) W(T_P)$

Proof. Let $S = \{v_1, \ldots, v_k\}$ be the k vertices added in set S during the algorithm. Let $v_0 = r$ (median of M). A new vertex v_i is added in S by the algorithm when: $d_G(r, v_{i-1}) + d_{T_P}(v_{i-1}, v_i) > \alpha d_G(r, v_i)$. Making the sum on i, we obtain: $\alpha \sum_{i=1}^{k} d_G(r, v_i) < \sum_{i=1}^{k} d_G(r, v_{i-1}) + d_{T_P}(v_{i-1}, v_i)$. But, as $v_0 = r$ we get $d_G(r, v_0) = 0$ and: $\sum_{i=1}^{k} d_G(r, v_{i-1}) \leq \sum_{i=1}^{k} d_G(r, v_i)$. By combining we have: $(\alpha - 1) \sum_{i=1}^{k} d_G(r, v_i) < \sum_{i=1}^{k} d_{T_P}(v_{i-1}, v_i)$. As $v_1, \ldots, v_k$ is a prefix order of a subset of M visited in a DFS visiting exactly twice each edge of tree T_P we have: $\sum_{i=1}^{k} d_{T_P}(v_{i-1}, v_i) \leq 2W(T_P)$. Hence, $\sum_{i=1}^{k} d_G(r, v_i) < \dfrac{2}{(\alpha - 1)} W(T_P)$. But, $W(G_f) \leq W(T_P) + \sum_{i=1}^{k} d_G(r, v_i) \leq \left(1 + \dfrac{2}{(\alpha - 1)}\right) W(T_P)$. $\quad \square$

Proof. (**Theorem 1**)

1. **Result for the sum of distances:** Lemma 2 shows: $m \sum_{u \in M} d_G(u,r) \leq C_G(M)$ with $r \in M$ a median of M. With Lemma 1, and Lemma 3 we obtain: $C_{G_f}(M) \leq 2\alpha m \sum_{u \in M} d_G(u,r)$. Hence: $C_{G_f}(M) \leq 2\alpha C_G(M)$.

2. **Result for the diameter.** With the triangle inequality, with Lemma 3 and with the fact that $r \in M$, for all u and v in M we have:

$$d_{G_f}(u,v) \leq d_{G_f}(u,r) + d_{G_f}(r,v) \leq \alpha(d_G(u,r) + d_G(r,v)) \leq 2\alpha D_G(M).$$

Hence $D_{G_f}(M) \leq 2\alpha D_G(M)$.

3. **Result for the weight.** Lemma 4 gives the result since $W(T_p) \leq \rho W(T^*(M))$. □

2.2 Total Number of Exchanged Messages and Induced Approximation Ratios

In this Section we use all the previous analysis to compute the performance of our algorithm in terms of approximation ratios (Theorem 2) and number of exchanged messages (Theorem 3).

Theorem 2. *If α is a constant then our algorithm is a distributed, constant approximation for diameter and sum of distances parameters and a $O(\log(m))$ approximation for cost.*

Theorem 3. *If α is a constant and $m = |M|$ then our algorithm uses at most $O(m \log(m) D_G(M))$ messages.*

To prove these final results we need some Lemmas.

Lemma 5. $W(T_P) = O(\log(m) W(T^*(M)))$

Proof. The process of construction of our tree T_P can be viewed as another equivalent version of the *online* algorithm proposed in [9]. The authors show that if a Steiner tree is constructed step by step by connecting a new member to the nearest member already connected, then, the final tree is a $O(\log(m))$-approximation of the Steiner tree. This is in fact what we do here, by "simulating" this process. □

Proof. (**Theorem 2**) Apply Lemma 5 and Theorem 1 with $\rho = O(\log(m))$ □

Lemma 6. *Phase 1 uses at most $O(m \log(m) D_G(M))$ messages ($m = |M|$).*

Proof. The construction of the initial substructure requires at most $(m-1)D_G(M)$ messages. The message complexity of the pruning process of this structure to obtain T_P is linear in the number of edges in T_P and, from Lemma 5, it is at most $O(\log(m) W(T^*(M)))$. As $W(T^*(M)) \leq (m-1)D_G(M)$, the construction of T_P requires at most $O(m \log(m) D_G(M))$ messages. Determining the median of M just requires at most $O(m D_G(M))$ messages by the presented process. □

Lemma 7. *Phase 2 uses at most* $2\rho\left(1 + \dfrac{2}{\alpha - 1}\right) W(T^*(M))$ *messages.*

Proof. During phase 2, a DFS of tree T_P is performed and each edge of T_P is crossed exaclty twice. Moreover, some direct connections (by the routing functions) are made between some elements of M and the *Median* r. All the edges crossed by messages is exactly the set of edges of the final structure G_f. Hence, at most $2|E(G_f)|$ messages are exchanged during phase 2. As $|E(G_f)| = W(G_f)$, approximation ratio on the cost of Theorem 1 shows the desired result. □

Proof. (**Theorem 3**) Lemma 6 shows that phase 1 uses $O(m\log(m) D_G(M))$ messages. Lemma 5 shows that $\rho = O(\log(m))$; combined with Lemma 7 and the fact that $W(T^*(M)) \leq m D_G(M)$ we get the result. □

Theorem 3 just gives an estimation, an upper bound on the number of messages. In particular, each time a member sends a message to another member, in the worst case the number of messages is equal of the diameter $D_G(M)$. In practice many pairs of members are closer; this reduces the expected complexity. Each message transports three data: two distances and one cardinal. Each value is smaller than the numbers of vertices of G, thus can be represented on $O(\log n)$ bits.

3 Conclusion

In this paper we have proposed a tricriteria distributed algorithm for the construction of a connection structure for interconnecting a group of machines spread in a network modelled by a graph. This connection structure is allocated for the group and is optimized in terms of induced delays and total cost.

We *proved* the quality of the structure by approximation ratios and we proposed and proved an upper bound on the number of exchanged messages. This last parameter was evaluated by a worst case analysis. It could be refined by simulations for example. Moreover, our construction and proofs are parametric: We used two separated phases and parameters α and ρ; Hence, if a better distributed algorithm [1] to construct the initial approximation Steiner tree exists then it can directly be "plugged" in our method and its impact can easily be evaluated by using our analysis.

References

1. Irlande, A., König, J.C., Laforest, C.: Construction of low-cost and low-diameter steiner trees for multipoint groups. In M. Flammini, E. Nardelli, G.P., Spirakis, P., eds.: Procedings of Sirocco 2000, Carleton Scientific (2000) 197–210
2. Ausiello, G., Crescenzi, P., Gambosi, G., Kann, V., Marchetti-Spaccamela, A., Protasi, M.: Complexity and approximation. Springer (1999)
3. Elkin, M., Peleg, D.: The hardness of approximating spanner problems. In Reichel, H., S.Tison, eds.: STACS 2000. Number LNCS 1770, Springer Verlag (2000) 370–381

---

[1] Better in terms of approximation ratio **and** number of exchanged messages.

4. Laforest, C.: Construction of efficient communication sub-structures: Non-approximability results and polynomial sub-cases. In Springer, ed.: EUROPAR. Volume 2790 of LNCS. (2003) 903–910
5. Khuller, S., Raghavachari, B., Young, N.: Balancing minimum spanning trees and shortest-path trees. Algorithmica (14) (1995) 305–321
6. Laforest, C.: A good balance between weigth and distances for multipoint tre es. In: 6th International Conference On Principles Of DIstributed Systems (OPODIS). (2002)
7. Elkin, M.: Distributed approximation: a survey. SIGACT News **35**(4) (2004) 40–57
8. Wu, B., Chao, K., Tang, C.: Approximation algorithms for the shortest total path length spanning tree problem. Discrete applied mathematics (105) (2000) 273–239
9. Imase, M., Waxman, B.: Dynamic steiner tree problem. SIAM Journal on Discrete Mathematics **4**(3) (1991) 369–384

Rollback-Recovery Protocol Guarantying MR Session Guarantee in Distributed Systems with Mobile Clients*

Jerzy Brzeziński, Anna Kobusińska, and Michał Szychowiak

Institute of Computing Science
Poznań University of Technology, Poland
{Jerzy.Brzezinski, Anna.Kobusinska, Michal.Szychowiak}@cs.put.poznan.pl

Abstract. This paper presents rVsMR rollback-recovery protocol for distributed mobile systems, guarantying Monotonic Reads consistency model, even in case of server's failures. The proposed protocol employs known rollback-recovery techniques, however, while applying them, the semantics of session guarantees is taken into account. Consequently, rVsMR protocol is optimized with respect to session guarantees requirements. The paper includes the proof of safety property of the presented protocol.

Keywords: rollback-recovery, safety, mobile systems, Monotonic Reads session guarantee.

1 Introduction

Applications in mobile domain usually tend to be structured as client-server interactions. In such applications, clients accessing the data are not bound to particular servers, but they can switch from one server to another. This switching adds a new dimension of complexity to the problem of consistency and makes the management of data consistency from client's perspective very attractive. Therefore, in [TDP+94] a new class of consistency models, called *session guarantees* (or *client-centric* consistency models), has been proposed to define properties of the system, observed from client's point of view. Client-centric consistency models define four session guarantees: *Read Your Writes* (RYW), *Monotonic Writes* (MW), *Monotonic Reads* (MR) and *Writes Follow Reads* (WFR). RYW expresses the user expectation not to miss his own modifications performed in the past, MW ensures that order of writes issued by a single client is preserved, MR ensures that the client's observations of the data storage are monotonic and finally, WFR keeps the track of causal dependencies resulting from operations issued by a client.

In this paper we focus our attention on MR session guarantee. Below we give a couple of examples that demonstrate the usefulness of MR. First, let us imagine a mailbox of a traveling user, who opens the mailbox at one location, reads

---

* This work was supported in part by the State Committee for Scientific Research (KBN), Poland, under grant KBN 3 T11C 073 28.

emails, and afterwards opens the same mailbox at different location. The user should see at least all the messages he has read previously, which is impossible without MR. Further, imagine that user's appointment calendar is stored on-line in replicated database, and can be updated by both: the user and automatic meeting scheduler. The calendar program periodically refreshes its display by reading appointments from the database. The recently added (or deleted) meetings can not appear to come and go, which is ensured, when copies of the database held by servers are consistent with respect to MR [TDP+94]. Finally, consider a Web page replicated at two different stores S_1 and S_2. If a client first reads the page from S_1 and later again from S_2, then the second copy should be the same, or newer as the one read from S_1.

MR session guarantee is provided by appropriate consistency protocols [TDP+94, BSW05b]. In order to construct effective solutions, adjusted to real application requirements, these protocols should provide MR also in situations, when servers holding replicated data brake down. Unfortunately, as far as we know, none of the proposed consistency protocols preserving session guarantees, considers such a possibility; they generally assume non-faulty environments. Such assumption might be considered not plausible and too strong for certain mobile distributed systems, where in practice failures do happen. Therefore, this paper addresses a problem of providing MR session guarantee in case of server's failures.

We introduce the rollback-recovery protocol rVsMR for distributed mobile systems, which combines fault–tolerant techniques: logging and checkpointing with coherence operations of a formerly proposed VsSG consistency protocol [BSW05b]. As a result, the rVsMR protocol offers the ability to overcome the servers' failures, at the same time preserving MR session guarantee. Because of client's orientation, in rVsMR protocol run-time faults are corrected with any intervention from the user. The main contribution of this paper is a presentation of rollback-recovery protocol rVsMR of MR session guarantee and formal proof of its safety.

2 Related Work

Session guarantees have been introduced in the context of Bayou replicated storage system [TDP+94] to allow mobile clients to implicitly define sets of writes that must be performed by servers. Since in Bayou each server's state is maintained in the database, adding a persistent and crash resisting log is enough to provide fault–tolerance in case of server's failure. CASCADE — a caching service for distributed CORBA objects [CDFV00], is another system using consistency conditions based on session guarantees. In CASCADE it is assumed that processes do not crash during the execution and all communication links are eventually operational. The Globe system [KKST98] follows the approach similar to CASCADE, by providing a flexible framework for associating various replication coherence models with distributed objects. Among the coherence models supported by Globe are also client-based models, although they are combined with object-based consistency models in a single framework. Finally, Pastis —

a highly scable, multi-user, peer-to-peer file system [PBS05] implements a consistency model based on RYW session guarantee. In Pastis it is assumed that at least one replica is not faulty and all users allowed to write to a given file trust one another regarding the update of that file.

3 System Model, Basic Definitions and Notations

Throughout this paper, a replicated distributed storage system is considered. The system consists of a number of unreliable *servers* holding a full copy of a *shared objects* and *clients* running applications that access these objects. Clients are mobile, i.e. they can switch from one server to another during application execution. To access the shared object, clients select a single server and send a direct request to this server. Operations are issued by clients sequentially, i.e. a new operation may be issued after the results of the previous one have been obtained. In this paper we focus on failures of servers, and assume the *crash-recovery* failure model, i.e. servers may crash and recover after crashing a finite number of times [GR04]. Servers can fail at arbitrary moments and we require any such failure to be eventually detected, for example by failure detectors [SDS99].

The storage replicated by servers does not imply any particular data model or organization. Operations performed on shared objects are basically divided into *reads* and *writes*. The server, which first obtains the write from a client, is responsible for assigning it a globally unique identifier. Clients can concurrently submit conflicting writes at different servers, e.g. writes that modify the overlapping parts of data storage. Operations on shared objects issued by client C_i are ordered by a relation $\stackrel{C_i}{\hookrightarrow}$ called *client issue order*. A server S_j performs operations in an order represented by a relation $\stackrel{S_j}{\hookrightarrow}$. Operations on objects are denoted by w, r or o, depending on the operation type (write, read or these whose type is irrelevant). Every server maintains the set CR_{S_j} of indexes of clients from which it has directly received write requests and table RW_{S_j}, where the number of writes performed by S_j before read from C_i was obtained, is kept in position i. Relevant writes $RW(r)$ of a read operation r is a set of writes that has influenced the current state of objects observed by the read r. Formally, MW session guarantee is defined as follows [BSW05b]:

Definition 1. *Monotonic Reads (MR) session guarantee is a property meaning that:*

$$\forall C_i \forall S_j \left[r_1 \stackrel{C_i}{\hookrightarrow} r_2 |_{S_j} \implies \forall w_k \in RW(r_1) : w_k \stackrel{S_j}{\hookrightarrow} r_2 \right]$$

In the paper, it is assumed, that data consistency is managed by the VsSG *consistency protocol* [BSW05b]. The formerly proposed protocol VsSG [BSW05b] uses a concept of server-based version vectors for efficient representation of sets of writes required by clients. Server-based version vectors have the following form: $V_{S_j} = [v_1 \; v_2 \; ... \; v_{N_S}]$, where N_S is a total number of servers in the system and single position v_i is the number of writes performed by server S_j. Every write w in the VsSG protocol is labeled with a *vector timestamp*, denoted by $T(w)$ ($T : \mathcal{O} \mapsto V$) and set to the current value of the vector clock V_{S_j} of server S_j performing w for the first time. During writes, performed by server S_j, its version vector V_{S_j} is incremented in position j and a timestamped operation is recorded in history H_{S_j}. $\mathcal{O}_{S_j}$ is a set of all writes performed by the server in the past. The writes that belong to $\mathcal{O}_{S_j}$ come from direct requests received by S_j from clients or are incorporated from other servers during the synchronization procedure. The VsSG protocol eventually propagates all writes to all servers. At the client's side, vector R_{C_i} representing writes relevant to reads issued by the client C_i is maintained. The linearly ordered set $\left(\mathcal{O}_{S_j}, \stackrel{S_j}{\rightarrowtail} \right)$ of past writes is denoted by H_{S_j} and called *history* [BSW05b]. During synchronization of servers, their histories are *concatenated*. The concatenation of histories H_{S_j} and H_{S_k}, denoted by $H_{S_j} \oplus H_{S_k}$, consists in adding new operations from H_{S_k} at the end of H_{S_j}, preserving at the same time the appropriate relations [BSW05b].

Below, we propose formal definitions of fault-tolerance mechanisms used by the rVsMR protocol:

Definition 2. *A log Log_{S_j} is a set of triples:*

$$\{ \langle i_1, o_1, T(o_1) \rangle \; \langle i_2, o_2, T(o_2) \rangle \; ... \; \langle i_n, o_n, T(o_n) \rangle \},$$

where i_n represents the identifier of the client issuing a write operation $o_n \in \mathcal{O}_{S_j}$ and $T(o_n)$ is timestamp of o_n.

Definition 3. *Checkpoint $Ckpt_{S_j}$ is a couple $\langle V_{S_j}, H_{S_j} \rangle$, of version vector V_{S_j} and history H_{S_j} maintained by server S_j at the time t, where t is a moment of taking a checkpoint.*

In this paper we assume, that log and checkpoint are saved by the server in a *stable storage*, able to survive all failures [EEL+02]. Additionally, we assume that the newly taken checkpoint replaces the previous one, so just one checkpoint for each server is kept in the stable storage.

4 The rVsMR Protocol

For every client C_i that requires MR session guarantee when executing read r, results of all writes, which have influenced the read issued by a client before r cannot be lost. Unfortunately, at the moment of performing the operation, the server does not possess the knowledge, whether in the future the client will be interested in reading results of its writes or not. So, to preserve MR, the recovery protocol should ensure that outcomes of all writes performed by the server are not lost in the case of its failure.

In the proposed rVsMR protocol, we introduce a novel optimization that reduces the number of saved operations: we propose that every server S_j saves

Upon sending a request $\langle o \rangle$
to server S_j **at client** C_i

1: $W \leftarrow \mathbf{0}$
2: **if** (**not** iswrite(o)) **then**
3:     $W \leftarrow \max(W, R_{C_i})$
4: **end if**
5: send $\langle o, W \rangle$ to S_j

Upon receiving a request $\langle o, W \rangle$
from client C_i **at server** S_j

6: **while** $\left(V_{S_j} \not\geq W \right)$ **do**
7:     wait()
8: **end while**
9: **if** iswrite(o) **then**
10:     $CW_{S_j} \leftarrow CW_{S_j} \cup i$
11:     $V_{S_j}[j] \leftarrow V_{S_j}[j] + 1$
12:     timestamp o with V_{S_j}
13:     $Log_{S_j} \leftarrow Log_{S_j} \cup \langle i, o, T(o) \rangle$
14:     perform o and store results in res
15:     $H_{S_j} \leftarrow H_{S_j} \oplus \{o\}$
16:     $nWrites \leftarrow nWrites + 1$
17: **end if**
18: **if not** iswrite(o) **then**
19:     **if** $i \in CR_{S_j}$ **then**
20:       $secondRead \leftarrow TRUE$
21:     **else**
22:       $CR_{S_j} \leftarrow CR_{S_j} \cup i$
23:       $RW_{S_j}[i] \leftarrow nWrites$
24:     **end if**
25:     **if** $(RW_{S_j}[i] > 0)$ **and** $secondRead$ **then**
26:       $Ckpt_{S_j} \leftarrow \langle V_{S_j}, H_{S_j} \rangle$
27:       $Log_{S_j} \leftarrow \emptyset$
28:       $CR_{S_j} \leftarrow \emptyset$
29:       $secondRead \leftarrow FALSE$
30:       $nWrites \leftarrow 0$
31:       $RW_{S_j} \leftarrow \mathbf{0}$
32:     **end if**
33:     perform o and store results in res
34: **end if**
35: send $\langle o, res, V_{S_j} \rangle$ to C_i

Upon receiving a reply $\langle o, res, W \rangle$
from server S_j **at client** C_i

36: **if** iswrite(o) **then**
37:     $R_{C_i} \leftarrow \max(R_{C_i}, W)$
38: **end if**
39: deliver $\langle res \rangle$

Every Δt **at server** S_j
40: **foreach** $S_k \neq S_j$ **do**
41:     send $\langle S_j, H_{S_j} \rangle$ to S_k
42: **end for**

Upon receiving an update $\langle S_k, H \rangle$
at server S_j

43: **foreach** $w_i \in H$ **do**
44:     **if** $V_{S_j} \not\geq T(w_i)$ **then**
45:       perform w_i
46:       $V_{S_j} \leftarrow \max(V_{S_j}, T(w_i))$
47:       $H_{S_j} \leftarrow H_{S_j} \oplus \{w_i\}$
48:     **end if**
49: **end for**
50: signal()

On rollback-recovery
51: $\langle V_{S_j}, H_{S_j} \rangle \leftarrow Ckpt_{S_j}$
52: $CR_{S_j} \leftarrow \emptyset$
53: $secondRead \leftarrow FALSE$
54: $nWrites \leftarrow 0$
55: $RW_{S_j} \leftarrow \mathbf{0}$
56: $Log'_{S_j} \leftarrow Log_{S_j}$
57: $vrecover \leftarrow \mathbf{0}$
58: **while** $\{o'_j : T(o'_j) > vrecover\} \neq \emptyset$ **do**
59:     0choose $\langle i', o'_i, T(o'_i) \rangle$ **with minimal** $T(o'_j)$ **from** Log'_{S_j} **where** $T(o'_j) > V_{S_j}$
60:     $V_{S_j}[j] \leftarrow V_{S_j}[j] + 1$
61:     perform o'_j
62:     $H_{S_j} \leftarrow H_{S_j} \oplus \{o'_j\}$
63:     $CW_{S_j} \leftarrow CW_{S_j} \cup i'$
64:     $vrecover \leftarrow T(o'_i)$
65:     $nWrites \leftarrow nWrites + 1$
66: **end while**

Fig. 1. Checkpointing and rollback-recovery rVsMR protocol

only operations obtained directly from clients. Although only some of operations performed by S_j are saved, we prove that MR is fulfilled in case of S_j failure.

The server that obtains the write request directly from client C_i, logs the request to stable storage (Figure 1, l. 13), and only afterwards performs it (l. 14). The moment of taking a checkpoint is determined by obtaining a read request r_2, which follows another read r_1 issued by the same client. The server, which obtains operation r_2 from a client, checks first, whether such a read can be performed (by comparing the values of vectors V_{S_j} and W - l. 6). When performing read r_2 is possible, the server checks if it has already performed, since the latest checkpoint, any write operation that influenced the state of objects observed by the read r_1 (l. 25). When at least one such write has been performed, the server checkpoints its state (l. 26), performs the read operation (l. 33) and sends a reply to the client (l. 35). Otherwise, the new checkpoint need not be taken. After the checkpoint is taken, server logs are cleared (l. 27). Saving the state of server earlier would be unnecessary, as when write request is not followed by a read one, it does not violate MR. Essential is the fact, that first the checkpoint is taken, and only afterwards the content of log Log_{S_j} is cleared. (l. 27). After the failure occurrence, the failed server restarts from the latest checkpoint (l. 51) and replays operations from the log (l. 58-65) according to their timestamps, from the earliest to the latest one. Writes received from other servers during update procedure, and missing from the local history of S_j, are performed, but not logged (l. 45-47). Thus, such writes are lost after the failure occurrence. However, those writes are saved in the log or in the checkpoint of servers, which received them directly from clients. Hence, lost writes will be eventually obtained again in consecutive synchronizations.

5 Safety of rVsMR Protocol

Lemma 1. *Every write operation w issued by client C_i and performed by server S_j that received w directly from client C_i, is kept in checkpoint $Ckpt_{S_j}$ or in log Log_{S_j}.*

Proof. Let us consider write operation w issued by client C_i and obtained by server S_j.

1. From the algorithm, server S_j before performing the request w, saves it in the stable storage by adding it to log Log_{S_j} (l. 13). Because logging of w takes place before performing it (l. 14), then even in the case of failure operation w is not lost, but remains in the log.
2. Log Log_{S_j} is cleared after performing by S_j the second read request issued by the same client. However, according to the algorithm, read operations cause storing the information on writes by checkpointing the server's version vector V_{S_j} and history H_{S_j} in $Ckpt_{S_j}$ (l. 26). The checkpoint is taken before clearing log Log_{S_j} (l. 27). Therefore, the server failure, which occurs after clearing the log, does not affect safety of the algorithm because writes from the log are already stored in the checkpoint.

Lemma 2. *The rollback-recovery procedure recovers all write operations issued by clients and performed by server S_j that were logged in log Log_{S_j} in the moment of server S_j failure.*

Proof. Let us assume that server S_j fails. The rollback-recovery procedure recovers operations remembered in the log (l. 58), after recovering V_{S_j} and H_{S_j} from a checkpoint (l. 51). The recovered operation updates version vector V_{S_j} (l. 60), is performed by S_j (l. 61) and added to the server's S_j history H_{S_j} (l. 62).

Assume now, that failures occur during the rollback-recovery procedure. Due to such failures the results of operations that have already been recovered are lost again. However, since log Log_{S_j} is cleared only after the checkpoint is taken (line 27) and it is not modified during the rollback-recovery procedure (l. 56), the log's content is not changed. Hence, the recovery procedure can be started from the beginning without loss of any operation issued by clients and performed by server S_j after the moment of taking checkpoint.

Lemma 3. *Operations obtained and performed in the result of synchronization procedure and required by MR, are performed again after the failure of S_j, before processing a new read from a client.*

Proof. By contradiction, let us assume that server S_j has performed a new read operation r obtained from client C_i before performing again operation w, received during a former synchronization and lost because of S_j failure. According to VsSG protocol, before executing r the condition $V_{S_j} \geq R_{C_i}$ is fulfilled (l. 6) .

Further assume, that w issued by C_i before r, has been performed by server S_k. According to the protocol, after the reply from S_k is received by C_i, vector R_{C_i} is modified: $R_{C_i} \leftarrow \max(W, R_{C_i})$. This means that vector R_{C_i} is updated at least at position k: $R_{C_i}[k] \leftarrow k+1$. (l. 37). Server S_j, during synchronization procedure with S_k, performs w and updates its version vector: $V_{S_j} \leftarrow \max(V_{S_j}, T(w))$, which means that V_{S_j} has been modified at least in the position k (l. 46). However, if failure of S_j happens, the state of S_j is recovered accordingly to values stored in the checkpoint $Ckpt_{S_j}$ (l. 51) and in the log Log_{S_j} (l. 58-65). From the algorithm, while recovering operations from the log, the vector V_{S_j} is updated only at position j. Thus, if operation w_1 performed by S_j in the result of synchronization with server S_k is lost because of S_j failure, the value of $V_{S_j}[k]$ does not reflect the information on w. Hence, until the next update message is obtained, $V_{S_j}[k] < R_{C_i}[k]$, which contradicts the assumption.

Lemma 4. *The recovered server performs new read operation issued by a client only after all writes performed before the failure and required by MR are restored.*

Proof. By contradiction, let us assume that there is a write operation w performed by server S_j before the failure occurred, that has not been recovered yet, and that the server has performed a new read operation issued by client C_i. According to original VsSG protocol [BSW05b], managing only consistency not reliability issues, for reliable server S_j that performs new read operation, the condition $V_{S_j} \geq R_{C_i}$ is fulfilled (l. 6-7).

Let us consider which actions are taken when a write operation is issued by client C_i and performed by server S_j. On the server side, the receipt of the write operation causes the update of vector V_{S_j} in the following way: $V_{s_j}[j] \leftarrow V_{S_j}[j] + 1$ and results in timestamping w with the unique identifier (l. 12). The server that has performed the write sends a reply containing the modified vector V_{S_j} to the client. At the client side, after the reply is received, vector R_{C_i} is modified: $R_{C_i} \leftarrow \max(W, R_{C_i})$ (l. 37). This means that vector R_{C_i} is updated at least at position j: $R_{C_i}[j] \leftarrow \max[j] + 1$. If there is a write operation w performed by server S_j before the failure that has not been recovered yet, then $V_{S_j}[j] < R_{C_i}[j]$, which follows from the ordering of recovered operations (l. 59). This is a contradiction with $V_{S_j} \geq R_{C_i}$. Hence, the new read operation cannot be performed until all previous writes are recovered.

Theorem 1. *MR session guarantee is preserved by rVsMR protocol for clients requesting it, even in the presence of server failures.*

Proof. It has been proven in [BSW05b] that VsSG protocol preserves MR session guarantee, when none of servers fails. According to Lemma 1, every write operation performed by server S_j is saved in the checkpoint or in the log. After the server's failure, all operations from the checkpoint are recovered. Further, all operations performed before the failure occurred, but after the checkpoint was taken, are also recovered (according to Lemma 2). According to Lemma 4, all recovered write operations are applied before new reads are performed. Moreover, operations obtained by S_j during synchronization procedure and lost because of S_j failure, are also performed once again before new reads from C_i (from Lemma 3). Hence, for any client C_i and any server S_j, MR session guarantee is preserved.

Full versions of the theorems and proofs can be found in [BKS05].

6 Conclusions

Although our implementation of rollback-recovery protocol is based on the known techniques of operation logging and checkpointing of server's state, it is nevertheless unique in exploiting properties of Monotonic Reads session guarantee while applying these techniques. This results in checkpointing only the results of write operations, which are essential to provide MR. Furthermore, we have designed novel optimisations that reduce the number of saved operations. We believe that rVsMR protocol can be applied to other systems (Section 2), where it is required to maintain consistency for mobile clients.

Our future work encompasses the development of rollback-recovery protocols, which are integrated with other consistency protocols. Moreover, appropriate simulation experiments to quantitatively evaluate overhead of rVsMR protocol are being carried out.

References

[TDP+94] Terry, D.B., Demers, A.J., Petersen, K., Spreitzer, M., Theimer, M., Welch, B.W.: Session guarantees for weakly consistent replicated data. In: Proc. of the Third Int. Conf. on Parallel and Distributed Information Systems (PDIS 94), Austin, USA, IEEE Computer Society (1994) 140–149

[BSW05b] Brzeziński, J., Sobaniec, C., Wawrzyniak, D.: Safety of a server-based version vector protocol implementing session guarantees. In: Proc. of Int. Conf. on Computational Science (ICCS2005), LNCS 3516, Atlanta, USA (2005) 423–430

[CDFV00] Chockler, G., Dolev, D., Friedman, R., Vitenberg, R.: Implementing a caching service for distributed CORBA objects. In: Proc. of Middleware 2000: IFIP/ACM Int. Conf. on Distributed Systems Platforms. (2000) 1–23

[KKST98] Kermarrec, A.M., Kuz, I., van Steen, M., Tanenbaum, A.S.: A framework for consistent, replicated Web objects. In: Proc. of the 18th Int. Conf. on Distributed Computing Systems (ICDCS). (1998)

[PBS05] Picconi, F., Busca, J.M., Sens, P.: Pastis: a highly-scalable multi-user peer-to-peer file system. EuroPar 2005 (2005) 1173–1182

[GR04] Guerraoui, R., Rodrigues, L.: Introduction to distributed algorithms. Springer-Verlag (2004)

[SDS99] Sergent, N., Défago, X., Schiper, A.: Failure detectors: Implementation issues and impact on consensus performance. Technical Report SSC/1999/019, École Polytechnique Fédérale de Lausanne, Switzerland (1999)

[EEL+02] Elmootazbellah, N., Elnozahy, Lorenzo, A., Wang, Y.M., Johnson, D.: A survey of rollback-recovery protocols in message-passing systems. ACM Computing Surveys **34**(3) (2002) 375–408

[BKS05] Brzeziński, J., Kobusińska, A., Szychowiak, M.: Mechansim of rollback-recovery in mobile systems with mr. Technical Report RA-012/05, Institute of Computing Science, Poznań University of Technology (2005)

A Practical Single-Register Wait-Free Mutual Exclusion Algorithm on Asynchronous Networks

Hyungsoo Jung and Heon Y. Yeom

School of Computer Science and Engineering
Seoul National University
Seoul 151-742, Korea
{jhs, yeom}@dcslab.snu.ac.kr

Abstract. This paper is motivated by a need of practical asynchronous network systems, i.e., a wait-free distributed mutual exclusion algorithm (*WDME*). The *WDME* algorithm is very appealing when a process runs on asynchronous network systems and its timing constraint is so restricted that the process cannot perform a local-spin in a wait-queue, which forces it to abort whenever it cannot access the critical region immediately. The *WDME* algorithm proposed in this paper is devised to eliminate the need for processes to send messages to determine whether the critical region has been entered by another process, an unfavorable drawback of a naive transformation of the shared-memory mutual exclusion algorithm to an asynchronous network model. This drawback leads to an unbounded message explosion, and it is very critical in real network systems. Design of the *WDME* algorithm is simple, and the algorithm is practical enough to be used in current distributed systems. The algorithm has $O(1)$ message complexity which is suboptimal between two consecutive runs of critical section.

1 Introduction

The mutual exclusion (**ME**) problem is a classical problem in distributed computing, and it is crucial in the design of distributed systems, especially concurrent systems. Although a large number of researchers have delved into devising efficient shared-memory mutual exclusion (*SME*) algorithms, true masterpieces on distributed mutual exclusion (*DME*) algorithms on practical asynchronous networks are few and far between due to the more complex features of real asynchronous networks, such as communication delay. Therefore, this paper is concerned with a distributed mutual exclusion algorithm that is efficient enough to be used in real asynchronous networks. The proposed algorithm works in a wait-free manner, but it does not incur unnecessary communication messages.

Spin-wait algorithm. The mutual exclusion problem, which originally takes into consideration a spin-wait algorithm, has been studied for many years. Numerous solutions have been proposed, including a notable one proposed by Peterson in [12]. In Peterson's paper, a two-process solution is presented and then generalized to be applicable to an arbitrary number of processes. A refinement of Peterson's algorithm in which only single-writer variables are used was later presented by Kessels in [9]. Although Kessels' algorithm is more fine-grained than Peterson's, it still employs multi-reader shared variables.

The first local spin-wait algorithms were queue-lock algorithms in which read-modify-write primitives are used to enqueue blocked processes onto the end of a *spin queue* [1,6,11]. The most recent work that showed good results in the shared memory environment was made by Yang [13], which is refined further by Anderson in [2,3,4]. Through their works, they presented several results for the upper bound of remote memory reference (**RMR**) time complexity under the assumption of various memory access operations.

Wait-free algorithm. The wait-free *DME* algorithm, unlike the spin-wait versions, does not have to consider the time spent waiting in the contention area, which most traditional algorithms have been concerned about, because contending processes that fail to access the critical section will just escape from the contention area and try again later to gain access privilege. A closely related work is presented in Jayanti [8]. Jayanti considers nondeterministic timeouts to occur in real systems in a practical way. His algorithm, a shared memory mutual exclusion algorithm, achieves an appealing function to meet the demands of practical fields. A pure wait-free access behavior is comparable to the situation using Jayanti's algorithm where all processes except the one in the critical region abort.

However, the abortive algorithm, if it is deployed in asynchronous networks, generates unbounded remote memory references (or explicit messages) when a process keeps aborting because of its strict timing constraint on a waiting period, and this leads us to conclude that the abortive algorithm is not adequate for asynchronous network systems which have a definite message delay and a strict deadline on waiting time. In fact, the message delay affects the algorithm's efficiency significantly and makes the time spent waiting to enter a critical section a considerable factor in the algorithm's performance.

In this paper, we are mostly focused on designing a practical *WDME* algorithm which (1) guarantees mutual exclusion in asynchronous networks, (2) solves the problem arising from unbounded message complexity of unnecessary access messages under the wait-free access behavior, and (3) is simple and efficient enough to be deployed in real distributed systems. The *WDME* algorithm has some useful applications. An implementation of a wait-free distributed shared-data structure has an important advantage compared to the wait-version counterpart: a process does not need to perform *spin-wait*, which leads to a cycle consuming activity in local machines and unnecessary remote messages in asynchronous networks. If a distributed shared data structure can be constructed in a wait-free manner, it can improve the overall system efficiency in many kinds of practical distributed systems. Fraser [5] attempted to validate the practical usability of wait-free algorithms by implementing a wait-free shared data structure using his algorithm.

2 Preliminaries

The full definition of the problem can be found in other literatures, so only a general description is provided here. One is presented with n (n > 2) processes that communicate with each other through explicit messages, not shared variables.

The difference we insist on is the wait-free manner in accessing the critical section. We assume that shared variables are accessed using an asynchronous network. This causes another critical issue due to the potential *data race problem* that can be ignored in SMP environments in which processes can read and write shared variables in an atomic manner. The wait-free way of accessing the critical section causes a crucial problem due to unbounded message complexity, which has been optimized quite successfully in wait algorithms, but not in wait-free algorithms.

The program code of each process is largely divided into two parts: a *critical section* and a *non-critical section*. All that is known is that after entering the critical section, a process will eventually leave it and return to the noncritical section within a finite amount of time. Processes start execution at a specified location in the non-critical part with all of the variables set to initial values. If we describe the state of the program in a fine-grained form, during the computation, each process P_i is in one of the following four states: the *try* state, in which it attempts to enter the critical section, the *critical* state, in which it runs in the critical section, the *exit* state, in which it leaves the critical section, and the *normal* state, in which it does other local computations.

The standard properties of mutual exclusion algorithm are as follows.

Mutual exclusion. At any time t there is at most one process in the critical section.

Deadlock freeness. If the critical section is open and there is a process trying to enter the critical section, then some process eventually enters the critical section.

Progress. For each time t, if there is a process P_i not in the *normal* state at time t, there exists a time t' ($t' > t$) in which P_i makes progress to the next stage.

In addition to these standard conditions, we have to define other important condition that has immense importance in wait-free algorithms. A new condition is based on the criterion of message complexity, which has been proven to be the most crucial factor in determining an algorithm's performance on asynchronous networks, especially, the traffic it generates on the process-to-process interconnect. According to the RMR time complexity measure, a new condition is stated by the following question: *How can a process reduce or eliminate the remote memory reference to figure out whether the current winner has exited the critical section?*. The question is closely related to the RMR time complexity and is of utmost importance because the phenomenon entails excessive message traffic under the wait-free behavior. Within the framework of this condition, we mainly concentrate our effort upon devising a simple and novel algorithmic solution.

3 Algorithm

The **WDME** algorithm, proposed in this paper, is mainly based on the asynchronous shared memory model. This base algorithm is then partly transformed into an asynchronous network model.

3.1 Architecture

This section presents the architecture of the system in which the *WDME* algorithm works, and we give an explanation for each automaton we adopt. In Lynch [10], a shared memory model can be transformed into a network model by adding proper automata. This enables many asynchronous shared memory algorithms to be adapted to be run in asynchronous networks. We use this property to make our algorithm efficient by transforming the part of our algorithm originally designed to run on asynchronous shared memory into an asynchronous network algorithm. The premise underlying this strategy is that the asynchronous shared memory model is easier to design and faster to run than the asynchronous network model.

Figure 1 shows the architecture of the *WDME* algorithm for the special case of n processes and a single variable which supports a compare-and-swap (*CAS*) operation on the register. Since most of today's modern processor architectures support *CAS* operations, our assumption is very feasible. The architecture is

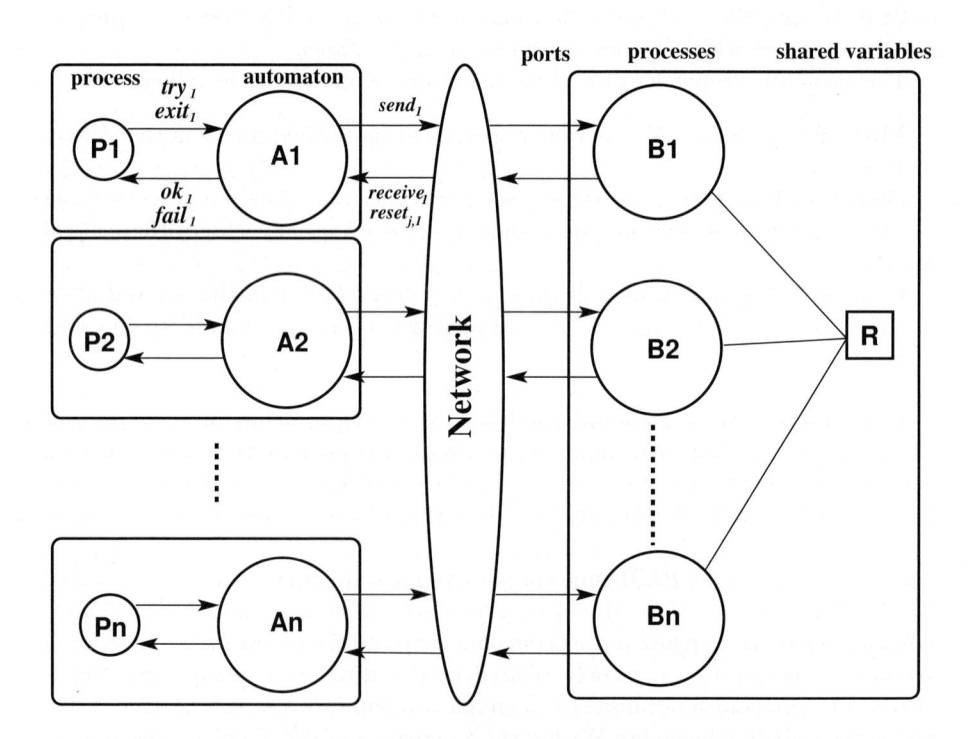

Fig. 1. Architecture of *WDME* Algorithm on Asynchronous Networks

largely composed of two models: an asynchronous network model on the left side and a shared memory model on the right side. Since we assume that all processes run on distributed systems, the asynchronous network model is required to represent our problem domain. In addition to the asynchronous network model, the shared memory model is also used in our architecture because of the favorable features it provides, such as design straightforwardness and algorithmic efficiency. Therefore, we embed the shared memory model into the asynchronous network model to achieve our goal. This simple embedded asynchronous network model resolves the drawback of naive transformations easily.

Automaton 3.1. Automaton A_i

Signature:
Input:
   try_i, a *try* invocation of process P_i
   $exit_i$, an *exit* invocation of process P_i
   $receive(v)_i$, a response from automaton B_i
   $reset(v)_{j,i}$, an invocation of a lock release from B_j

Output:
   ok_i, a response for try_i
   $fail_i$, a response for try_i
   $send(v)_i$, an invocation of automaton B_i

States:
$state \in \{yes, no\}$, a state of automaton A_i
$turn \in \{yes, no\}$, an asynchronous release of turn

Transitions:

```
 1: try_i                                 22:        ok_i
 2:    Effect:                            23:    else
 3:       if state = yes or               24:       if turn = yes
 4:          turn = yes then              25:          state := yes
 5:             send("try")_i             26:       else
 6:       else                            27:          state := no
 7:          fail_i                       28:          fail_i
 8:                                       29:
 9: exit_i                                30: reset(v)_{j,i}
10:    Effect:                            31:    Precondition: none
11:       state := no                     32:    Effect:
12:       send("exit")_i                  33:       turn := yes
13:                                       34:       state := yes
14: send(v)_i                             35:
15:    Precondition: none                 36: ok_i
16:    Effect: none                       37:    Precondition: none
17:                                       38:    Effect: none
18: receive(v)_i                          39:
19:    Effect:                            40: fail_i
20:       if v = "ok" then                41:    Precondition: none
21:          state := yes                 42:    Effect: none
```

Tasks:
none

We have three automata in our embedded asynchronous network model, including process P_i in the asynchronous network model and automaton B_i in the shared memory model. Two automata P_i and A_i work in the asynchronous network model since they should simulate distributed processes. Automaton B_i, which mainly simulates the non-blocking mutual exclusion algorithm, is described in the shared memory model with other automata B_k, ($k \in \{1, 2, .., n\}$).

The $WDME$ algorithm designed in this paper uses a single register to synchronize competing processes, and a single register is powerful enough to be suitable for practical purposes. For that purpose, we allow the register to support the CAS operation, which has an infinite consensus number as presented in Herlihy [7]. Then, we transform the algorithm into our model, which is message-suboptimal.

The reason we design the $WDME$ algorithm as a hybrid model, in other words, adopt a shared memory model in a network model, is simply for algorithmic efficiency. If we design the architecture purely in a network model, then every access to shared variables must entail corresponding communication messages and message delay. This degrades the algorithm's performance severely. So, we transform a pure mutual exclusion algorithm, which requires heavy accesses to shared variables, into a shared memory algorithm. The remaining part of the $WDME$ algorithm is concerned with the issues we mentioned previously, and the detailed explanation is presented in a later section.

3.2 Automaton

We can express that each process represented as P_i is the composition of an I/O automaton A_i, which is responsible for simulating process i of A and handling the front-end portion of the $WDME$ algorithm, and an I/O automaton B_i, which is responsible for simulating the shared memory mutual exclusion algorithm, the back-end portion of $WDME$. Various input and ouput interactions are described in Figure 1. The code for every automaton is expressed in I/O automaton format[10].

Automaton A_i. First, we present the code for automaton A_i. The code for A_i is shown in Automaton 3.1. Automaton A_i has inputs try_i, $exit_i$, $reset(v)_{j,i}$, and $receive(v)_i$, and outputs ok_i, $fail_i$, and $send(v)_i$. Automaton A_i is mainly responsible for handling the front-end part of the $WDME$ algorithm. It uses inputs try_i and $exit_i$, and outputs ok_i and $fail_i$ to interact with process P_i.

A_i needs a couple of state variables to work correctly. Each variable has obvious meanings. The *state* variable indicates states of A_i, *yes* or *no*. When it is in the *yes* state, automaton A_i, upon the arrival of a new *try* request from P_i, can send a trial message to automaton B_i (line 3-5). But, if it is in the *no* state, automaton A_i responds with a $fail_i$ whenever it receives *try* from automaton P_i (line 6-7). This immediate reply does not incur any messages, and this local, fast response continues until *state* is released by an asynchronous input invocation of $reset(v)_{j,i}$, which is invoked by the winner process' automaton B_j upon the "$exit_j$" invocation. Because our assumed model is asynchronous networks, this invocation can precede the response from automaton B_i. In this case, whenever a "fail" response is received from automaton B_i, we first check if *turn* has already been set by some fast process. If *turn* is *yes*, we simply set *state* to *yes*, otherwise, leave it. This is why we set both *turn* and *state* as *yes*.

This enables process P_i to send only a single message between the periods of each trial, and it eliminates the unbounded number of checking messages. Therefore we can assert that each waiting process has a $O(1)$ message suboptimal bound in each period of trial. This suboptimal message complexity can be achieved only by using our hybrid architecture.

Automaton 3.2. Automaton B_i

Signature :
Input:
    $receive(v)_i$, an invocation of A_i

Output:
    $send(v)_i$, a response
    $reset(v)_{i,j}$, an invocation of B_j by B_i

Shared variable:
$turn \in \{0,1,2,..,n\}$

States:
$state \in \{wait, done\}$
$result \in \{success, fail\}$

Transition

```
 1: receive("try")_i                         14:     turn := 0
 2:     Effect:                              15:     for every j, j ∈ {1,2,..,n}
 3:         if turn = 0 then                 16:         reset(v)_{i,j}
 4:             if CAS(0,&turn,i) then       17:     send("ok")_i
 5:                 result := ok             18:
 6:                 send("ok")               19: reset(v)_{i,j}
 7:             else goto fail               20:     Precondition: none
 8:         else                             21:     Effect: none
 9: fail:                                    22:
10:             send("fail")                 23: send(v)_i
11:                                          24:     Precondition: none
12: receive("exit")_i                        25:     Effect: none
13:     Effect:
```

Tasks:
none

Automaton B_i. The code for automaton B_i is shown in Automaton 3.2. Automaton B_i has input $receive(v)_i$ and outputs $send(v)_i$ and $reset(v)_{i,j}$. Automaton B_i is responsible for handling the back-end part of the *WDME* algorithm, which checks $turn$, a local shared register, and performs a *CAS* operation on the $turn$ register. If it succeeds in *CAS*, it replys *ok* to automaton A_i (line 4-6). Otherwise, it sends a *fail* message (line 8-10). As we mentioned previously, the *CAS* operation has an infinite consensus number, so the *WDME* algorithm works very efficiently even with a large number of competing processes.

Automaton B_i needs a single locally shared register $turn$ and a couple of state variables to work properly. The code is quite straightforward, so we do not give much detail about the code. The key feature we note in automaton B_i is the input $receive("exit")_i$. If B_i receives an *exit* message from automaton A_i, it flips the $turn$ register value, before propagating its *exit* state to all processes in a non-blocking manner through the invocation of $reset(v)_{i,j}$, which leads each automaton A_k, $k \in \{1,2,..,n\}$, to set its $turn$ and $state$ variables to *yes* (line 14-17). Finally, every waiting process is informed that the $turn$ register has been released.

4 Proof of Correctness

In this section, we show that the WDME algorithm preserves an important property which must be satisfied to be considered a mutual exclusion algorithm.

Lemma 1. *Automaton B_i preserves mutual exclusion.*

Proof. We will prove the lemma by contradiction. Assume that two automata B_i and B_j, $i \neq j$, are simultaneously allowed to execute in a critical section, which means that both automata consider the *turn* has been set by its process identifier. Let's assume there exists an execution that leads to this state. According to the code in Automaton 3.2, both automata perform CAS before setting the *turn* register to their identifiers. However, the CAS operation, which is an atomic *read-modify-write* with an infinite consensus number, allows only a single automata to set *turn* to its identifier. Therefore, there can be no legal execution which allows concurrent entrances into the critical section. □

Next, we prove the message complexity of each process in our algorithm to be less than one message between any two consecutive runs of the critical section, i.e., $O(1)$ message suboptimal algorithm.

Theorem 1. *The WDME algorithm has at least one process which executes in a critical section between two consecutive "try_i" invocations of B_i by process P_i ($i \in \{1,2,...,n\}$). The algorithm has $O(1)$, i.e., less than one, message suboptimal complexity between two consecutive runs of the critical section.*

Proof. The theorem above consists of two sentences, but both have the same meaning. Therefore, we need to prove just one of the two statements. We prove by contradicting the first sentence. Suppose the first part of theorem is false and consider an execution in which no process executes in the critical section between two consecutive "try_i" invocations of B_i by any process P_i. First, we can easily note that if the same automaton A_i sends a "try_i" message to automaton B_i consecutively, then the first "try_i" is failed by some other process P_j, $j \neq i$, and the *state* variable of A_i is released later by the process P_j. Otherwise, A_i can not send a second "try_i" message to B_i. Second, even though automaton A_i of process P_i is released by process P_j, we can not know when P_i will attempt to access the critical section again. The only thing we can guarantee is that there are more than zero trials by other processes until process P_i sends the second "try_i" message to B_i. Therefore, we conclude that there is at least one process which executes in the critical section because P_i can not send a second "try_i" message until it has been released through a $reset()_{k,i}$ invocation by the winner process P_k. This is a contradiction. Since the first statement is true, we can easily note that any waiting process P_i's "try_i" message cannot be sent to automaton B_i more than two consecutive times. Otherwise, this contradicts the first sentence, which proved that there exists at least one process in the critical section between two consecutive "try_i" messages. □

5 Conclusion

In this paper, we present a wait-free distributed mutual exclusion algorithm. The *WDME* algorithm preserves important properties that every correct mutual exclusion algorithms should obey. It successfully overcomes the drawback of naive transformations of shared memory algorithms into asynchronous network models, that is, unbounded message complexity of checking messages. The main features of the *WDME* algorithm are that (1) it eliminates unnecessary remote memory references that would have been generated by an asynchronous network algorithm and (2) it provides $O(1)$ message complexity between two consecutive trials to access the critical section.

We are working on developing a practical distributed system to work on distributed shared data structures. The *WDME* algorithm plays a crucial role in the prototype distributed system. We hope it will work very efficiently.

Acknowledgements

We thank the anonymous referees for thier very helpful suggestions on this paper. This improved the quality of this paper.

References

1. T. Anderson, "The performance of spin lock alternatives for shared-memory multiprocessors", *IEEE Transactions on Parallel and Distributed Systems*, 1(1):6.16, January 1990.
2. J. Anderson and Y.-J. Kim, "Fast and scalable mutual exclusion", *In Proceedings of the 13th International Symposium on Distributed Computing*, pages 180.194, September 1999.
3. J. Anderson and Y.-J. Kim, "Adaptive mutual exclusion with local spinning", *In Proceedings of the 14th International Symposium on Distributed Computing*, pages 29.43. Lecture Notes in Computer Science 1914, Springer-Verlag, October 2000.
4. J. Anderson and Y.-J. Kim, "A new fast-path mechanism for mutual exclusion", *Distributed Computing*, 14(1):17.29, January 2001.
5. Keir Fraser, "Practical lock-freedom", *Ph.D. thesis*, King's College, University of Cambridge, 2003.
6. G. Graunke and S. Thakkar, "Synchronization algorithms for shared-memory multiprocessors", *IEEE Computer*, 23:60.69, June 1990.
7. Maurice Herlihy, "Wait-free synchronization", *ACM Transactions on Programming Languages and Systems*, 13(1) pages 124.149, ACM, 1990.
8. Prasad Jayanti, "Adaptive and efficient abortable mutual exclusion", *In Proceedings of the twenty-second annual symposium on Principles of distributed computing*, pages 295.304, ACM, 2003.
9. J. Kessels, "Arbitration without common modifiable variables", *Acta informatica*, 17:135-141, 1982.

10. Nancy A. Lynch, *Distributed Algorithms*, Morgan Kaufmann. 1994.
11. J. Mellor-Crummey and M. Scott, "Algorithms for scalable synchronization on shared-memory multiprocessors", *In Proceedings of the Third ACM Symposium on Principles and Practice of Parallel Programming*, pages 106.113. ACM, April 1991.
12. G. Peterson, "Myth about the mutual exclusion problem", *Information Processing Letter*, 12(3):115-116, 1981.
13. J.-H. Yang, "A fast, scalable mutual exclusion algorithm", *Distributed Computing*, 9(1):51.60, August 1995.

Optimal and Practical WAB-Based Consensus Algorithms[*]

Lasaro Camargos[1,2], Edmundo R.M. Madeira[1], and Fernando Pedone[2]

[1] State University of Campinas, Brazil
{lasaro, edmundo}@ic.unicamp.br
[2] University of Lugano, Switzerland
fernando.pedone@unisi.ch

Abstract. In this paper we introduce two new WAB-based consensus algorithms for the crash-recovery model. The first one, B*-Consensus, is resilient to up to $f < n/2$ permanent faults, and can solve consensus in three communication steps. R*-Consensus, our second algorithm, is $f < n/3$ resilient, and can solve consensus in two communication steps. These algorithms are optimal with respect to the time complexity versus resilience tradeoff. We compare our algorithms to other consensus algorithms in the crash-recovery model.

Keywords: Consensus, crash-recovery, weak ordering oracles, Paxos.

1 Introduction

The consensus problem is a fundamental building block in fault-tolerant distributed systems. In a seminal paper, Fischer, Lynch, and Patterson have shown that consensus cannot be deterministically solved in a completely asynchronous distributed system subject to process failures [1]. This result implies that any consensus algorithm requires extensions to the pure asynchronous model if at least one process may crash during the execution.

Motivated by this theoretical bound, several approaches have been proposed to solve consensus by strengthening the asynchronous. Dolev et al. [2] and Dwork et al. [3] studied the minimum synchronization requirements needed by consensus. In [4], Chandra and Toueg introduced the concept of unreliable failure detectors, oracles that provide possibly incorrect information about process failures. Unreliable failure detectors encapsulate the synchronous assumptions needed to solve consensus and provide abstract properties to processes. The authors classified failure detectors in eight classes and showed that $\Diamond \mathcal{W}$ encapsulates the minimal assumptions needed to solve consensus [5]. Some proposals have also considered solving consensus using a leader election oracle Ω [6,7]. Intuitively, a leader election oracle ensures that nonfaulty processes eventually agree on the identity of some nonfaulty process, the leader.

Another way to circumvent the consensus impossibility result is to use randomization. The algorithms presented in [8,9] use a random number generator to

---

[*] The work presented in this paper has been partially funded by CNPq, Brazil (grant 141749/2003-2), and SNSF, Switzerland (project #200021-103556).

guarantee that with probability one processes will reach a decision. Algorithms similar to those in [8,9] were presented by Pedone et al. [10]. Instead of relying on randomization, however, progress is ensured using *weak ordering oracles*. Such oracles provide message ordering guarantees but, as unreliable failure detectors and Ω, they can make mistakes. More specifically, the algorithms in [10] use the *weak atomic broadcast (WAB)* oracle. WAB ensures that if processes keep exchanging broadcast messages, then some of these messages will be delivered in the same order by all nonfaulty processes. Weak ordering oracles are motivated by Ethernet broadcast, present in many clustered architectures.

Lower bounds on what consensus algorithms can achieve have been also considered in the literature. Lamport summarizes previous results (e.g., [11,12]) and presents new ones in [13]. These bounds show a tradeoff between resilience and time complexity (i.e., the number of communication steps needed to solve consensus). Briefly, the following results are stated: (a) To ensure progress, at least a majority of processes needs to be nonfaulty. (b) To allow a decision to be reached in two communication steps when more than one process is allowed to propose, more than two-thirds of the processes should be nonfaulty.

Despite the great interest that consensus has attracted and the multitude of algorithms that have been proposed to solve it, most works have considered system models which are of more theoretical than practical interest. This is mainly reflected in two aspects: the failure behavior of processes and the reliability of communication links. From a practical perspective, processes should be capable of re-integrating the system after a crash. Moreover, algorithms capable of tolerating message losses can make better use of highly-efficient communication means (e.g., UDP messages). We call such algorithms *practical*.

In this paper we introduce practical WAB-based consensus algorithms. Differently from those in [10], our protocols assume that processes can recover after failures and messages can be lost. The first one, B*-Consensus, is resilient to up to $f < n/2$ permanent failures; it solves consensus in three communication steps when the WAB oracle works. The second algorithm, R*-Consensus, is $f < n/3$ resilient and can solve consensus in two communication steps. Therefore, besides practical, our algorithms are also optimal regarding the time complexity versus resilience tradeoff.

The rest of the paper is organized as follows. We introduce our computational model and some definitions in Section 2. In Section 3 we present the B*-Consensus and the R*-Consensus algorithms. Correctness proofs for both algorithms can be found in [14]. In Section 4 we compare B*-Consensus and R*-Consensus to other practical consensus algorithms, and relate them to other works. Section 5 concludes the paper.

2 Model and Definitions

2.1 Processes, Communication, and Failures

We consider an asynchronous system composed of a set $\Pi = \{p_1, \ldots, p_n\}$ of processes, $n \geq 3$. Processes communicate by message passing. Messages can be lost or duplicated but not corrupted. Processes can crash and recover an unlimited number of times but do not behave maliciously (i.e., no Byzantine failures). To ensure liveness we assume that eventually a subset of processes remains up forever. Such processes are called *stable*.

In the following sections we provide a definition of the consensus problem and then augment the asynchronous model with further assumptions to consensus solvable.

2.2 The Consensus Problem

Processes executing consensus can propose a value, interact to accept a single value, and learn the decision. Similarly to [13], we consider that these roles can be played independently by each process. Characterizing processes as *proposers*, *acceptors*, and *learners* allows us to simplify the algorithm's presentation. It also better models some real systems, e.g., it adequately matches a system where clients propose values to servers and then, without participating in the decision protocol themselves, learn the value accepted.

Using the decomposition of roles, consensus is defined as follows:

Nontriviality: only a proposed value may be learned.
Consistency: any two values that are learned must be equal.
Progress: for any proposer p and learner l, if p, l and $n - f$ acceptors are *stable*, and p proposes a value, then l must learn a value.

2.3 Weak Ordering Oracles

Weak ordering oracles provide message ordering guarantees [10]. A WAB is a weak ordering oracle defined by the primitives w-broadcast(k, m) and w-deliver(k, m), where $k \in \mathbb{N}$ defines a w-broadcast instance, and m is a message. The invocation of w-broadcast(k, m) broadcasts message m in instance k; w-deliver(k, m) w-delivers a message m w-broadcast in instance k. WAB satisfies the following property:

– If w-broadcast$(k,-)$ is invoked in an infinite number of instances k, then (*Fairness*) for every instance k there is an instance $k' \geq k$ in which every stable process w-delivers a message and (*Spontaneous Order*) the first w-delivered message in instance k' is the same for every process that w-delivers a message in k'.

For example, consider an instance k in which processes p and q w-broadcast messages m_p and m_q respectively. If all non crashed processes in the system execute w-deliver$(k, \_)$, and their first invocation of w-deliver returns the same message $m \in \{m_p, m_q\}$, then the property is satisfied in k. If, otherwise, some non crashed process does not execute w-deliver, or if one process w-delivers m_p while another w-delivers m_q, the property is not satisfied in i.

WABs are motivated by empirical observation of the behavior of IP-multicast in some local-area networks (e.g., Ethernet). In such environments, IP-multicast ensures that most broadcast messages are delivered in the same order to all network nodes.

3 B*-Consensus and R*-Consensus

In this section we introduce two WAB-based consensus algorithms for the crash-recovery model: B*-Consensus and R*-Consensus. These protocols were inspired by those in [10] but, differently from them, tolerate an unbounded number of failures and message losses without losing consistency. For example, any process can crash and recover an unbounded number of times. To ensure progress, however, a certain number of processes is required to be stable: B*-Consensus requires $f < n/2$ stable processes, and R*-Consensus requires $f < n/3$. By requiring more stable processes, R*-Consensus may reach a decision in two communication steps, while B*-Consensus needs at least three steps. Therefore, they have optimal time complexity [13]. To the best of our knowledge, these are the first WAB-based algorithms to consider crash-recovery failures and message loses.

As the two algorithms share some behavior, in the following sections we initially describe their commonalities and then describe their particularities.

3.1 General Overview

R*-Consensus and B*-Consensus execute a sequence of rounds. In each round, proposers can propose a value, acceptors try to choose a value proposed in the round, and learners try to identify whether a decision has been made in the current round or if a new round must be started.

In a deciding round r, (i) a proposer w-broadcasts a value v, that is, it executes w-broadcast(r, v); (ii) acceptors w-deliver some proposed value, possibly interact to accept a value, and notify the learners; and (iii) the learners, after gathering enough acceptance messages, learn that a value was decided and tell the application. The main difference between the algorithms lies in the meaning of *enough*; in order to decide with fewer messages, increasing the resilience from $f < n/3$ to $f < n/2$, acceptors in B*-Consensus must execute an extra communication round before accepting a value.

The algorithms are divided in blocks of statements, each one executed until completion and one at time. The *Initialization* block runs when the algorithm is started. If the process is recovering from a crash, the *Recovery* block is run, instead. The other blocks have clauses triggered by message arrivals (receive and w-deliver), and only run after *Initialization* or *Recovery* have run.

In both algorithms, every process p keeps a variable r_p with the highest-numbered round in which p took part, and a variable $prop_p$ that either has the proposal for round r_p or $\bot$, meaning that any value can be proposed. Variables $prop_p$ and r_p are always logged together (see Algorithms 1 and 2), ensuring that processes replay rounds consistently, after recovering from a crash.

Skipping rounds. When a process p in round r_p sees a message sent in round $r_q > r_p$, p immediately jumps to r_q without performing rounds $r_p+1..r_q-1$. This allows processes that were down for a long time to rapidly catch up with the most advanced processes. Not every value is a valid proposal for every round: after deciding rounds, for example, only the decided value can be proposed. As only processes that finished round $r_q - 1$ initially know which values can be proposed in r_q, processes in earlier rounds must learn, maybe indirectly, which values are valid in r_q from processes in later rounds. This is accomplished by having each process' proposal attached to every message it sends (the last field of each message in the algorithm). The *Round Skipping Task*, presented in Algorithms 1 and 2, lines 8-16, runs on every message received/w-delivered before other clauses handle them. The algorithms in [15] can also skip rounds, but the procedure is more complicated than the one we present.

Proposers. Proposers are given a value by the application and try to pass it as the instance's decision. Due to message losses and process crashes, a consensus instance may not terminate in the first attempt, and may have to be retried. At any time, proposers can retry a consensus instance if they believe that the previous attempt has failed; consistency is kept even if the previous attempt is actually still running. To be able to learn that a round of the algorithm has

---

Algorithm 1. The B*-Consensus Algorithm

1: Initialization:
2: $r_p \leftarrow 0$
3: $prop_p \leftarrow est1_p \leftarrow est2_p \leftarrow \bot$
4: $Cset \leftarrow Sset \leftarrow \emptyset$

5: Recovery:
6: retrieve($r_p, prop_p, est1_p, est2_p$)
7: $Cset \leftarrow Sset \leftarrow \emptyset$

8: Round Skipping Task:
9: before executing receive($\_, r_q, \ldots$)
      or w-deliver(m, r_q)
10:   **if** $r_p > r_q$
11:     send (SKIP,$r_p,prop_p$) to q
12:   **if** $r_p < r_q$
13:     $r_p \leftarrow r_q$
14:     $prop_p \leftarrow prop_q$
15:     $est1_p \leftarrow est2_p \leftarrow \bot$
16:     $Cset \leftarrow Sset \leftarrow \emptyset$

17: To propose value v_p do as follows:
18: **if** $prop_p = \bot$
19:   $prop_p \leftarrow v_p$
20: w-broadcast (FIRST,$r_p, prop_p$)
      to acceptors
21: Acceptors execute as follows:
22: **upon** w-deliver (FIRST, $r_p, prop_q$)
23:   **if** $est1_p = \bot$
24:     $est1_p \leftarrow prop_q$
25:   log ($est1_p, r_p, prop_p$)
26:   send (CHECK, $r_p, est1_p, prop_q$)
      to acceptors

27: **upon** receive (CHECK, $r_p, est1_q, prop_q$)
28:   $Cset \leftarrow Cset \cup \{(\text{CHECK}, r_p, est1_q, prop_q)\}$
29:   **if** $|Cset| = \lceil (n+1)/2 \rceil$
30:     **if** $\forall (\text{CHECK}, r_p, est1_q, -) \in Cset$:
             $est1_q = v$
31:       $est2_p \leftarrow v$
32:     **else**
33:       $est2_p \leftarrow \top$
34:     log ($est2_p, r_p, prop_p$)
35:     send (SECOND, $r_p, est2_p, prop_p$)
        to learners

36: Learners execute as follows:
37: **upon** receive (SECOND, $r_p, est2_q, v_q$)
38:   $Sset \leftarrow Sset \cup \{(\text{SECOND}, r_p, est2_q, v_q)\}$
39:   **if** $|Sset| = \lceil (n+1)/2 \rceil$
40:     **if** $\forall (\text{SECOND}, r_p, est2_q, -) \in Sset$:
             $est2_q = v \neq \top$
41:       decide v
42:     **if** $\exists (\text{SECOND}, r_p, est2_q, -) \in Sset$:
             $est2_q = v \neq \top$
43:       $prop_p \leftarrow v$
44:     $r_p \leftarrow r_p + 1$
45:     $est1_p \leftarrow est2_p \leftarrow \bot$
46:     $Cset \leftarrow Sset \leftarrow \emptyset$

terminated we assume that each proposer is also a learner. So, if a proposer does not learn the decision of the consensus it has initiated after some time, it re-starts its execution by proposing in its current round. Proposers execute lines 17-20 of the algorithms.

3.2 The B*-Consensus Algorithm

Algorithm 1 presents the B*-Consensus algorithm.

Acceptors. In the B*-Consensus algorithm, every acceptor p will accept the first proposal w-delivered. That is, p takes this proposal as its first estimative ($est1_p$), and logs it together with the current round number (r_p) and a valid proposal. Then, p exchanges its estimative with other acceptors using CHECK messages, collecting $\lceil(n+1)/2\rceil$ estimatives. p uses them as its second estimative ($est2_p$) if they are all equal, or $\top$, otherwise. p then logs $est2_p$ and r_p and sends them both to the learners in SECOND messages.

Learners. Once a learner has received $\lceil(n+1)/2\rceil$ SECOND messages, it checks whether all carry the same estimative v. If that is the case, a decision has been reached and v is delivered to the application. Otherwise, p looks for at least one $v \neq \top$ in SECOND messages to be used as a proposition for the next round, so that any future rounds will only be able to decide v.

Algorithm 2. The R*-Consensus Algorithm

1: Initialization:
2: $r_p \leftarrow 0$
3: $prop_p \leftarrow , est1_p \leftarrow \bot$
4: $Sset \leftarrow \emptyset$

5: Recovery:
6: retrieve($r_p, prop_p, est1_p$)
7: $Sset \leftarrow \emptyset$

8: Round Skipping Task:
9: before executing receive($\_, r_q, \ldots$)
   or w-deliver(m, r_q)
10:   if $r_p > r_q$
11:     send (SKIP,$r_p, prop_p$) to q
12:   if $r_p < r_q$
13:     $r_p \leftarrow r_q$
14:     $prop_p \leftarrow prop_q$
15:     $est1_p \leftarrow \bot$
16:     $Sset \leftarrow \emptyset$

17: To propose value v_p do as follows:
18: if $prop_p = \bot$
19:   $prop_p \leftarrow v_p$
20: w-broadcast (FIRST,$prop_p$)
   to acceptors

21: Acceptors execute as follows:
22: **upon** deliver (FIRST,$r_p, prop_q$)
23:   if $est1_p = \bot$
24:     $est1_p \leftarrow prop_q$
25:     log ($est1_p, r_p, prop_p$)
26:     send (SECOND, $r_p, est1_p, prop_p$)
   to learners

27: Learners execute as follows:
28: **upon** deliver (SECOND,$r_p, est1_q, v_q$)
29:   $Sset \leftarrow Sset \cup \{(\text{SECOND}, r_p, est1_q, v_q)\}$
30:   if $|Sset| = \lceil(2n+1)/3\rceil$
31:     if $\forall(\text{SECOND}, r_p, est1_q, \_) \in Sset$:
   $est1_q = v$
32:       decide v
33:     if $\exists v_{maj}$, for $\lceil(n+1)/2\rceil$ (SECOND, $r_p, v, \_$)
   $\in Sset : v = v_{maj}$
34:       $prop_p \leftarrow v_{maj}$
35:     else
36:       $prop_p \leftarrow \bot$
37:     $r_p \leftarrow r_p + 1$
38:     $est1_p \leftarrow \bot$
39:     $Sset \leftarrow \emptyset$

3.3 The R*-Consensus Algorithm

Algorithm 2 presents the R*-Consensus algorithm.

Acceptors. In the R*-Consensus algorithm, as in B*-Consensus, every acceptor p accepts the first proposal it w-delivers, and logs it together with r_p and $prop_p$, so that these values will not be forgotten in case of crash. p then sends these values to all learners in SECOND messages.

Learners. Learners gather $\lceil (2n+1)/3 \rceil$ SECOND messages and check whether they contain the same estimative v. If that is the case, v is decided and delivered to the application. Otherwise, learners check if at least a majority of the estimatives are equal to v' in which case $prop_p$ is set to v', locking the value for future decisions. In any case, r_p is incremented.

4 Related Work

WAB-based consensus algorithms were introduced in [10]. This work assumed crash-stop failures and reliable channels. Here we presented WAB-based consensus algorithms that allow processes to recover and messages to be lost.

The problem of consensus in the crash-recovery model was previously studied in [7,15,16,17,18,19]. These approaches considered either a leader-election oracle or unreliable failure detectors (UFD) as extensions to the asynchronous model. In [15], Aguilera et al. showed that if the number of processes that never crash ("always-up processes") is bigger than the number of processes that eventually remain crashed or that crash and recovery infinitely many times, then consensus is solvable without stable storage; without this assumption stable storage is needed. As we do not bound the number of processes that are allowed to crash, our algorithms must use stable storage, although this is done sparingly. Differently from our approach, the algorithms in [16,19] keep all their variables in stable storage and cannot be considered practical.

In [13] some lower bounds on how fast, in terms of communication steps, a consensus algorithm can be are given. Roughly, if any value proposed by two or more proposers can be decided within two communication steps, then no more than $f < n/3$ processes can be unstable; to be able to decide in three communication steps, no more than $f < n/2$ processes can be unstable. Some algorithms found in the literature may decide in two communication steps and still be $f < n/2$ resilient. In these algorithms, however, only the value proposed by the coordinator[1] can be decided in two steps; deciding on a value proposed by other processes requires at least one message step more. As this extra step is important in several practical situations, e.g., when using consensus to implement atomic broadcast, in the following analysis we consider this extra step whenever it applies. Since WAB-based algorithms do not have the role of a coordinator, they do not suffer from this shortcoming.

The Paxos algorithm [7,18] relies on an Ω leader election oracle to solve consensus. In its normal form, Paxos needs at least four (plus one) communica-

---

[1] Leader and initiator are also names commonly used.

tion steps to decide on a value. By omitting the first phase of the algorithm, a simple optimization for good runs, two communication steps can be saved. Another variation of Paxos, Fast Paxos [20], eliminates the extra step by having any proposer proposing on behalf of the leader. In good runs, a decision can be reached in two communication steps. Since Paxos is $f < n/2$ resilient and Fast Paxos is $f < n/3$, they are optimal. Just like Paxos and Fast Paxos, the WAB-based algorithms we presented here are also optimal.

Hurfin et al. [17] presented an algorithm that has the same message pattern as Paxos in optimized mode, i.e., two (plus one) communication steps. Because it uses the rotating coordinator paradigm, the decision may be delayed when coordinators, elected deterministically, crash.

The algorithm relying on stable storage in [15] is $f < n/2$ resilient. In best-case runs, processes access stable storage twice in a round and reach decision within three (plus one) communication steps. B*-Consensus writes in disk twice in a round, while R*-Consensus writes only once. In Paxos, disk writes happen once per round in the optimized mode, and twice in the normal mode, that is, in each mode it has the same cost as one of our algorithms. Fast Paxos writes once per round, as does the algorithm in [17].

Table 1 summarizes consensus algorithms in terms of communication steps (i.e., expected latency), number of messages, their resilience, number of disk writes, and the oracle needed for termination. δ denotes the expected network delay assumed for the analysis of the algorithms. We consider both point-to-point and multicast communication, and assume that either one or the other is used at each configuration, but not both at the same time. Notice that we consider messages sent from a process to itself, as these messages also impose some processing cost at each machine.

Table 1. Consensus algorithms in the crash-recovery model

Protocol	Expected Latency	Number of Messages		Resilience	Oracle	Disk Writes
		Unicast	Broadcast			
B*-Consensus	3δ	$2n^2 + n$	$2n + 1$	$f < n/2$	WAB	2
R*-Consensus	2δ	$n^2 + n$	$n + 1$	$f < n/3$	WAB	1
Fast Paxos	2δ	$n^2 + n$	$n + 1$	$f < n/3$	Ω	1
Paxos (optimized)	3δ	$n^2 + n + 1$	$n + 2$	$f < n/2$	Ω	1
Paxos (normal)	5δ	$n^2 + 3n + 1$	$2n + 3$	$f < n/2$	Ω	2
Aguilera et. al	4δ	$3n + 1$	$n + 3$	$f < n/2$	UFD	2
Hurfin et. al	3δ	$n^2 + n + 1$	$n + 2$	$f < n/2$	UFD	1

From Table 1, the optimized version of Paxos takes the same number of communication steps as B*-Consensus but, due to its centralized nature, needs nearly half the messages. Two more communication steps are required when Paxos runs the first phase. Although the number of messages is half of B*-Consensus with point-to-point communication, it becomes almost the same when broadcast is available. Moreover, if the proposer is the current leader, then one communication step and one message can be saved in Paxos. When compared to

R*-Consensus, the optimized version of Paxos uses the same number of messages, and trades one communication step for better resilience: $f < n/2$ instead of $n < n/3$. Finally, notice that Paxos always degenerate to the normal case after the first try to achieve consensus fails using the optmized version of the protocol. Fast Paxos equals R*-Consensus in all criteria but the oracle. Aguilera et al.'s algorithm has the same resilience, latency and number of disk writes as B*-Consensus, but is more efficient in terms of messages. Hurfin et al.'s algorithm is just as efficient as R*-Consensus, but has better resilience. If it is important for more than one proposer to be able to have its proposal decided, then the WAB-based consensus algorithms become one communication step more efficient. Moreover, in our analysis we do not count the messages needed to implement the Ω and UFD abstractions.

5 Conclusions

In this paper we introduced B*-Consensus and R*-Consensus, two WAB-based algorithms that assume the crash-recovery model and tolerate message losses. Both algorithms can cope with any number of process failures without violating safety. B*-Consensus takes three communication steps to reach a decision and requires a majority of stable processes to ensure progress. R*-Consensus can decide in two communication steps, but requires more than two thirds of stable processes for progress. Both algorithms are optimal in terms of communication steps for the resilience they provide.

We compared our algorithms to other well-known consensus algorithms in the crash-recovery model. Due to their decentralized fashion, when using these protocols, any proposer may have its value decided within the minimal latency. This comes at the cost of resilience or extra messages. In the case of Fast Paxos, the only difference is the oracle used, Ω, and the number messages needed to implement it.

References

1. Fischer, M.J., Lynch, N.A., Paterson, M.S.: Impossibility of distributed consensus with one faulty process. Journal of the ACM (JACM) **32**(2) (1985) 374–382
2. Dolev, D., Dwork, C., Stockmeyer, L.: On the minimal synchronism needed for distributed consensus. Journal of the ACM (JACM) **34**(1) (1987) 77–97
3. Dwork, C., Lynch, N., Stockmeyer, L.: Consensus in the presence of partial synchrony. Journal of the ACM (JACM) **35**(2) (1988) 288–323
4. Chandra, T.D., Toueg, S.: Unreliable failure detectors for reliable distributed systems. Journal of the ACM (JACM) **43**(2) (1996) 225–267
5. Chandra, T.D., Hadzilacos, V., Toueg, S.: The weakest failure detector for solving consensus. Journal of the ACM (JACM) **43**(4) (1996) 685–722
6. Dutta, P., Guerraoui, R.: Fast indulgent consensus with zero degradation. Lecture Notes in Computer Science **2485** (2002)
7. Lamport, L.: The part-time parliament. ACM Transactions on Computer Systems (TOCS) **16**(2) (1998) 133–169

8. Ben-Or, M.: Another advantage of free choice (extended abstract): Completely asynchronous agreement protocols. In: Proceedings of the second annual ACM symposium on Principles of distributed computing, ACM Press (1983) 27–30
9. Rabin, M.O.: Randomized byzantine generals. In: Proc. of the 24th Annu. IEEE Symp. on Foundations of Computer Science. (1983) 403–409
10. Pedone, F., Schiper, A., Urban, P., Cavin, D.: Solving agreement problems with weak ordering oracles. In: 4th European Dependable Computing Conference (EDCC-4), Toulouse, France (2002)
11. Charron-Bost, B., Schiper, A.: Uniform consensus is harder than consensus. J. Algorithms **51**(1) (2004) 15–37
12. Keidar, I., Rajsbaum, S.: On the cost of fault-tolerant consensus when there are no faults: preliminary version. SIGACT News **32**(2) (2001) 45–63
13. Lamport, L.: Lower bounds for asynchronous consensus. Technical Report MSR-TR-2004-72, Microsoft Research (2004)
14. Camargos, L., Pedone, F., Madeira, E.: Optimal and practical wab-based consensus algorithms. Technical Report IC-05-07, Institute of Computing, State University of Campinas, Campinas, Brazil (2005)
15. Aguilera, M.K., Chen, W., Toueg, S.: Failure detection and consensus in the crash-recovery model. In: Proc. of the 12th International Symposium on Distributed Computing. (1998)
16. Dolev, D., Friedman, R., Keidar, I., Malkhi, D.: Failure detectors in omission failure environments. In: Symposium on Principles of Distributed Computing. (1997) 286
17. Hurfin, M., Mostefaoui, A., Raynal, M.: Consensus in asynchronous systems where processes can crash and recover. In: Proceedings Seventeenth IEEE Symposium on Reliable Distributed Systems, IEEE Comput., Soc, Los Alamitos, CA (1998) 280–286
18. Oki, B.M., Liskov, B.H.: Viewstamped replication: A new primary copy method to support highlyavailable distributed systems. In: PODC '88: Proceedings of the seventh annual ACM Symposium on Principles of distributed computing, New York, NY, USA, ACM Press (1988) 8–17
19. Oliveira, R., Guerraoui, R., Schiper, A.: Consensus in the crash-recover model. Technical Report TR-97/239, EPFL – Départment d'Informatique, Lausanne, Switzerland (1997)
20. Lamport, L.: Lower bounds for asynchronous consensus. Technical Report MSR-TR-2005-112, Microsoft Research (2005)

Self-stabilizing Deadlock Detection Under the OR Requirement Model

Christian F. Orellana[1,2], Cristian Ruz[1], and Yadran Eterovic S.[2]

[1] Escuela de Ingeniería Informática, Universidad Diego Portales
cristian.ruz@udp.cl
[2] Depto. de Ciencia de la Computación, Pontificia Universidad Católica de Chile
{cforella, yadran}@ing.puc.cl

Abstract. This article introduces a self-stabilizing deadlock-detection algorithm for the OR model. The algorithm is complete, because it detects all deadlocks, and it is correct, because it does not detect false deadlocks. Because of the self-stabilization property, the algorithm supports dynamic changes in the wait-for graph on which it works, and transient faults; also, it can be started in an arbitrary state. Previous deadlock-detection algorithms for the OR model are not guaranteed to recover from transient faults, nor can they be started in an arbitrary state. Once the algorithm terminates, each process knows if it is or not deadlocked; moreover, deadlocked processes know whether they cause or only suffer from deadlock.

1 Introduction

One of the main motivations to build distributed systems is the possibility of sharing resources among several processes. A process can acquire and release resources in a sequence that is unknown beforehand. The deadlock problem arises in this setting; being able to detect deadlocks is the first step to take actions and resolve them. A set of processes is said to be deadlocked when each process in the set is blocked, waiting for resources assigned to other processes in the same set. The presence of a deadlock is a stable property of a system; once a set of processes becomes deadlocked, it will remain in that state unless a resolution action is taken.

Knapp classified the deadlock-detection problem in six models, according to the type of requirements a process can make [1]; for most models, deadlock-detection algorithms have been proposed. Under the single-outstanding-request model, a process can request only one resource at a time [2]. Under the AND model, a process can request multiple resources simultaneously; requirements are satisfied when all the requested resources are assigned [2,3]. Under the OR model, a process also can request multiple resources simultaneously, but requirements are satisfied when any of the requested resources is assigned [4,5,6,7]. Under the AND/OR model, a process can request any number of resources in an arbitrary combination of AND and OR requirements [8]. Under the n-out-of-k model, a requirement for n resources is satisfied when k of them are assigned [9]. Under the unrestricted model, no assumption is made about the way in which a process makes its requirements.

Since Dijkstra introduced the concept of self-stabilization in 1974 [10], several self-stabilizing algorithms have been proposed, to solve many problems in distributed systems. Mutual exclusion and leader election are among the classical problems solved with this approach. Schneider wrote an early survey on the subject [11]. In general, a system is said to be self-stabilizing if, regardless of its initial global state, it reaches a legitimate global state in a finite number of steps [10]. The global state of a distributed system is the cartesian product of the local states of every process in the system. The definition of legitimate and illegitimate global states depends on the context of the problem being solved. The ability of regaining a legitimate global state that these systems present, makes them able to support transient faults. A transient fault is one that occurs once, and ceases to occur. Furthermore, self-stabilizing systems can be started in an arbitrary global state, even illegitimate ones, since they will reach a legitimate state nonetheless.

The dynamic nature of resource competition, in which processes are involved in a distributed system, makes the deadlock-detection problem suitable to be treated from a self-stabilizing perspective. In addition, transient-fault tolerance is a desirable property for a distributed deadlock-detection algorithm.

2 The OR Model

This article presents a self-stabilizing deadlock-detection algorithm for the OR requirement model. A process can make an OR request, for example, in a replicated distributed database system, where a read request for a replicated element is satisfied when any copy is read [1]. Also, in a store-and-forward communications network, packets can be forwarded whenever any buffer at the destination node is free [5]. In a similar way, in a message-routing system based on wormhole routing, a router can forward a received message to a neighbor router through one of several channels [12]; a requirement for an output channel is satisfied when any of them becomes available.

A useful way to represent resource requirements is by means of a directed graph, known as *Wait-For Graph* (WFG). In a WFG, each node represents a process in the system. Nodes with outgoing edges represent blocked processes, waiting for resources. On the contrary, nodes without outgoing edges represent active processes. An edge from node i to node j means that process i is waiting for a resource assigned to process j. In general, the deadlock-detection problem can be reduced to that of detecting cyclic structures on this graph. For example, the presence of a directed cycle in the WFG is a necessary and sufficient condition for the existence of deadlock under the AND model [1]. In Fig. 1(a), processes 1, 2, and 3 form a cycle, and are deadlocked.

Under the OR requirement model, the presence of a cycle in the WFG is a necessary — but not sufficient — condition for a deadlock to exist. If the edges

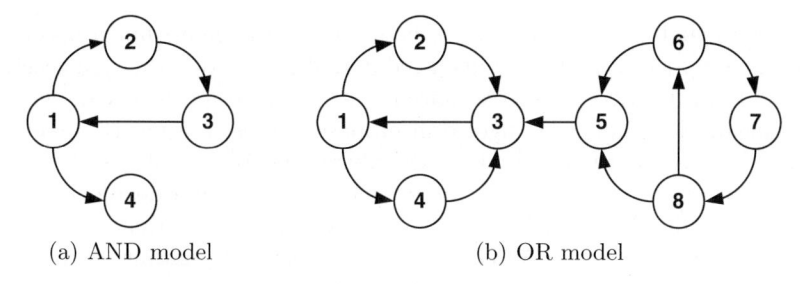

Fig. 1. Examples of deadlock. (a) Processes 1, 2, and 3 form a cycle, and are deadlocked under the AND model. (b) Processes 1, 2, 3, and 4 form a knot, and are deadlocked under the OR model. Processes 5, 6, 7, and 8, only suffer from deadlock.

represent OR requirements, there is no deadlock in Fig. 1(a), in spite of the cycle, because process 1 is waiting for the resource assigned to process 2 *or* the resource assigned to process 4.

Under the OR requirement model, a process is *blocked* if it has a pending OR requirement. A set of processes is *deadlocked*, if they form a *tie* in the WFG, and all of them are blocked. A tie in a graph is a set of nodes with no directed edges going to nodes outside the set. Another important notion is that of a *knot* in the WFG [13]. A node v is in a knot, if all nodes that are reachable from v by a directed path, can reach node v by a directed path; in that case, the knot is the set of nodes that are reachable from v. That is, a knot is a strongly connected component; moreover, a knot is a tie of blocked processes of which any subset is not a tie. Also, any tie of blocked processes contains at least one knot [5]; there is a path from every node in that tie to at least one knot.

Under the OR requirement model, deadlocked processes can be sorted into two groups. A process *suffers* from deadlock if it is in a tie. A process *causes* a deadlock if it is in a knot. According to these definitions, a process that causes a deadlock also suffers from deadlock. For example, in Fig. 1(b), all processes form a tie of blocked processes. Processes 1, 2, 3, and 4 form a knot, they are deadlocked, and they all cause deadlock. Processes 5, 6, 7, and 8, on the other hand, are not in a knot; they do not cause deadlock, but they are deadlocked nonetheless; they *only suffer* from deadlock. The distinction is important when trying to resolve the deadlock. In order to resolve all deadlocks, a process from each knot must be terminated; it would not help to kill processes that only suffer from deadlock.

Chandy, Misra, and Haas have proposed an algorithm to detect deadlocks under the OR model, based on the technique known as *diffusing computations* [4]. In their proposal, a process starts the algorithm when a request is not granted. Upon termination, a process is guaranteed to know that it is deadlocked only if it was deadlocked when the algorithm started. Nonetheless, in a set of deadlocked processes, at least one of them is able to report it. Cidon, Jaffe, and Sidi [5] proposed an algorithm based on detecting cycles of connected components, which they call *clusters*, and merging them into bigger clusters until a knot is found.

All processes that cause deadlock are detected. In the algorithm proposed by Lee and Lee [6], the *initiator* builds a reduced WFG locally, through receiving the paths from its successors. The initiator uses this graph to decide whether there is deadlock or not. The algorithm proposed by Natarajan [7] is based on the same principle as the one by Chandy, Misra, and Haas, but uses a periodic protocol that allows the choice of exactly one process from a deadlocked set of processes to report the deadlock. Some of these algorithms are dynamic, because they support changes in the WFG; however, they are not guaranteed to recover from transient faults, nor can they be started in an arbitrary state.

In the algorithm proposed in this paper, processes gather enough information about their successors to detect deadlocks. A process that is not deadlocked when the algorithm starts, but becomes deadlocked later, is able to report the deadlock. Thus, in a set of deadlocked processes, every process is able to report it. Additionally, each deadlocked process can decide if it is deadlocked because it is part of a knot, or because it only suffers from deadlock. The algorithm supports dynamic changes in the WFG; furthermore, it supports transient faults and can be started in an arbitrary state.

3 Self-stabilization

In a distributed system, processes are connected to each other according to some underlying network topology, which may be defined by virtual connections on top of a transport protocol. Each process has its own set of local variables, and can communicate with any other process through those connections. The local variables define the state of a process. A process might decide to change its local state depending on its current state and the state of some other processes. In a distributed system, a process can learn the state of other processes through message passing. The ability to change state is called a *privilege*; a process that has a privilege is called a *privileged* process. In a *step*, a privileged process changes its local state.

The cartesian product of the local states of every process defines the global state of the system. Global states can be sorted into two sets: legitimate and illegitimate. A self-stabilizing system converges in a finite number of steps to a legitimate global state, regardless of whether its initial global state is legitimate or not. It is because of this property that self-stabilizing systems can support transient faults. A transient fault is one that changes the local state but not the behavior of a process, and does not continue to occur. Even if a transient fault puts the system in an illegitimate global state, the system will eventually regain a legitimate global state. In addition, it is not necessary to define an initial global state, that is, local variables can be initialized arbitrarily.

In the system defined by the algorithm proposed in this paper, legitimate global states are characterized by the absence of privileges, and by the fact that a process decides that it is deadlocked if and only if it is really deadlocked. In a legitimate global state, every resource request that is not granted and every release of a resource pushes the system into an illegitimate global state, because new privileges appear in the system every time the WFG changes.

Flatebo and Datta [2], and Karaata and Line [3], have proposed self-stabilizing algorithms to solve the deadlock-detection problem under the AND model. Deadlocks are detected by finding cycles in the WFG; each node propagates the information about its predecessor nodes to its successors [3], or the information about its successors to its predecessors [2]. If a predecessor of a node is also a successor of the node, or viceversa, then there is a cycle in the WFG and a deadlock in the system. In both proposals, a global state is legitimate when a process knows that it is deadlocked if and only if it is deadlocked. In both proposals, a change in the WFG puts the system in an illegitimate state; the algorithms support changes produced by the processes that share resources. Moreover, the algorithm proposed by Karaata and Line [3] supports transient faults and arbitrary initialization; on the other hand, the algorithm proposed by Flatebo and Datta [2] does not.

Schneider provided a formalism to prove that a system is self-stabilizing with respect to a predicate over the global states of the system [11]. This state predicate identifies the correct operation of the system, by defining legitimate states. Every state that satisfies the predicate is legitimate, and states that do not satisfy the predicate are illegitimate. According to Schneider, a system is self-stabilizing with respect to a state predicate P, if it satisfies two properties: *closure* and *convergence*. The closure property says that, once the system reaches a state satisfying P, it cannot reach an illegitimate state through execution of the program. The convergence property says that, starting from an arbitrary global state, the system will reach a state satisfying P in a finite number of steps. In this paper, this formalism is used to prove the property of self-stabilization.

4 Self-stabilizing Deadlock Detection

The proposed algorithm is shown in Fig. 2.

Processes make requests for a resource to a distributed component called *resource allocator*. Whenever a resource allocator receives a request, the resource is assigned locally if it is available. In the other case, the request can not be satisfied.

The algorithm starts at a process, when a request is not granted. The requesting process blocks, and control is transferred to a thread that runs the detection algorithm. These threads maintain exact, up-to-date information about their neighbors in the WFG. The set of neighbors of a node v changes when one of them releases a resource, which is then reallocated to some waiting node. If it is reallocated to v, v is no longer blocked; otherwise, it has a different set of neighbors. The resource allocator can inform the detection-algorithm thread of these changes through atomic updates of the local variables *Succ* and *Pred*. No other event can change the set of neighbors, since the process is blocked.

4.1 Variables

Each process mantains eight local variables when executing the algorithm: *Succ*, *Pred*, *Succ**, *Pred**, *Deadlocked**, *Knot*, *Tie*, and *deadlocked*, which it can read

For node i:

(0.1) **if** $Succ = \emptyset \land (Succ^* \neq \emptyset \lor Deadlocked^* \neq \emptyset \lor Knot \neq \emptyset \lor Tie \neq \emptyset)$
    then $Succ^* := \emptyset; Deadlocked^* := \emptyset; Knot := \emptyset; Tie := \emptyset$
(1.1) **if** $Succ^* \neq (\cup_{j \in Succ} Succ_j^*) \cup Succ$
    then $Succ^* := (\cup_{j \in Succ} Succ_j^*) \cup Succ$
(1.2) **if** $Pred^* \neq (\cup_{j \in Pred} Pred_j^*) \cup Pred$
    then $Pred^* := (\cup_{j \in Pred} Pred_j^*) \cup Pred$
(2.1) **if** $Succ^* \neq \emptyset \land Knot \neq \{i\} \land Succ^* \subseteq Pred^*$
    then $Knot := \{i\}$
(2.2) **if** $Knot \neq \emptyset \land Succ^* \nsubseteq Pred^*$
    then $Knot := \emptyset$
(2.3) **if** $Succ^* \neq \emptyset \land Tie \neq \{i\} \land Succ^* \subseteq (Deadlocked^* \cup Pred^*)$
    then $Tie := \{i\}$
(2.4) **if** $Tie \neq \emptyset \land Succ^* \nsubseteq (Deadlocked^* \cup Pred^*)$
    then $Tie := \emptyset$
(2.5) **if** $Succ^* \neq \emptyset \land Deadlocked^* \neq ((\cup_{j \in Succ} Deadlocked_j^*) - \{i\}) \cup Knot \cup Tie$
    then $Deadlocked^* := ((\cup_{j \in Succ} Deadlocked_j^*) - \{i\}) \cup Knot \cup Tie$
(3.1) **if** $Succ^* \neq \emptyset \land (Succ^* \subseteq Deadlocked^*) \neq deadlocked$
    then $deadlocked := (Succ^* \subseteq Deadlocked^*)$

Fig. 2. The deadlock-detection algorithm

and write. Also, it is assumed that each process has read-only access to the local variables $Succ^*$, $Pred^*$, and $Deadlocked^*$ of its neighbors. Since the algorithm is self-stabilizing, there is no need to set specific initial values for the variables.

Variable $Succ$ represents the set of successors of the node i that is executing the algorithm, while variable $Pred$ represents the set of its predecessors. Variable $Succ^*$ represents the set of nodes that are reachable from the node i that is executing the algorithm, while variable $Pred^*$ represents the set of nodes that reach node i. Variable $Deadlocked^*$ represent the set of reachable nodes that are probably deadlocked. Variables $Knot$ and Tie are sets, and by execution of the algorithm can get two values: empty, or the identifier of the node that is executing the algorithm. Boolean variable $deadlocked$ indicate whether the process that is executing the algorithm is deadlocked or not.

Transient faults can change the value of any variable but $Succ$ and $Pred$, which are kept up to date by the resource allocator, and represent the view that a node has of the local connections on the WFG.

4.2 Notation

Each step of the algorithm is written as a guarded command. The guard is a predicate over the variables that the process can read: its own local variables and the ones from its neighbors. If the predicate is true, then there is a privilege in the system, and it is possible to execute the associated action. Actions are executed atomically until there are no more true guards at the node, with the non-local variables being read once, before evaluating the guards. When there are more than one true guard at a node, the action executed is always the one with minor number.

In Fig. 2 variable i represents the identifier of the process that is executing the algorithm. The local variables of neighbor j are represented as $Succ_j^*$, $Pred_j^*$, and $Deadlocked_j^*$.

4.3 The Algorithm

The algorithm begins at a node i when the process blocks waiting for resources and, therefore, it acquires a set of successors. Step (1.1) locally computes the set $Succ^*$, using the information available in variable $Succ$ and the information in variable $Succ^*$ of every successor. Because of this step, any change in the set $Succ$ is reflected in the local variable $Succ^*$, and propagated to predecessors nodes. In a symilar manner, step (1.2) computes the set $Pred^*$ using the information available in variable $Pred$ and the information in variable $Pred^*$ of every successor.

Step (2.1) sets the local variable $Knot$ to a set with i as its only element, when all successors of i are also its predecessors. Step (2.2) sets the local variable $Knot$ to empty when i has at least one successor node that is not a predecessor at the same time.

Step (2.3) sets the local variable Tie to a set with i as its only element, when all successors of i are deadlocked, or can reach i back. If that is not the case, step (2.4) sets the local variable Tie to empty.

Step (2.5) includes in local variable $Deadlocked^*$ the information in variables $Knot$ and Tie, and the information in variables $Deadlocked^*$ of every successor, and propagates this information to predecessor nodes.

Step (3.1) allows a node to decide whether it is deadlocked or not, setting boolean variable $deadlocked$ accordingly.

When a blocked process becomes active, step (0.1) reset all variables that depend on $Succ$ back to empty.

5 Properties of the Algorithm

This section proves the main theorem of this paper, which states that the proposed algorithm is complete, correct, and self stabilizing.

Lemma 1. *Once there are no privileges in the system, the following three statements are equivalent:*

1. *There is a path from node i to node j in the WFG*
2. $j \in Succ_i^*$
3. $i \in Pred_j^*$

Proof. (1⇒2) Assume there are no privileges in the system, and there is a path $x_0, x_1, \ldots, x_{n-1}, x_n$ in the WFG, with $x_0 = i$ and $x_n = j$. The local resource allocator ensures that $j \in Succ_{x_{n-1}}$. If $j \notin Succ_{x_{n-1}}^*$ then node x_{n-1} would be privileged; the guard from step (1.1) would be true. Since there are no privileges in the system, $j \in Succ^*_{x_{n-1}}$. Following the same reasoning, $j \in Succ^*_{x_{n-2}}$, or else x_{n-2} would be privileged. The same is true for all nodes in the path, including i.

(1⇒3) The proof is similar to the one given for (1⇒2). The local resource allocator ensures that $i \in Pred_{x_1}$, so $i \in Pred^*_{x_1}$ or else x_1 would be privileged. The same is true for all nodes in the path, including j.

(2⇒1) Let $j \in Succ^*_i$. Then, $j \in Succ_i$ or $j \in Succ^*_k$ for some $k \in Succ_i$, or else i would be privileged. If $j \in Succ_i$ then there is a path of length 1 from i to j in the WFG. Otherwise, if $j \in Succ^*_k$ then $j \in Succ_k$ or $j \in Succ^*_{k'}$, for some $k' \in Succ_k$. If $j \in Succ_k$ then there is a path of length 1 from k to j, and a path of length 2 from i to j. When there is a node m such that $j \in Succ_m$, it is possible to find a path from i to j in the WFG. Note that there is always a node m such that $j \in Succ_m$, or else j would never be included in a variable $Succ^*$.

(3⇒1) The proof is similar to the one given for (2⇒1). □

Lemma 2. *Once there are no privileges in the system, if node i causes deadlock then $deadlocked_i = true$.*

Proof. If node i causes deadlock, then it is in a knot. All nodes that are reachable from i by a directed path in the WFG are in variable $Succ^*_i$ (by Lemma 1). All nodes that reach i by a directed path in the WFG are in variable $Pred^*_i$ (by Lemma 1). Since i is in a knot, all reachable nodes from i can reach i back. Then $Succ^*_i \subseteq Pred^*_i$ and, after one execution of step (2.1), $Knot_i = \{i\}$. This is also true for all nodes in the knot, that is, $Knot_j = \{j\}$ for all j in $Succ^*_i$.

Because of step (2.5), $j \in Deadlocked^*_j$ for all nodes j such $Knot_j = \{j\}$. The information that each node keeps in variable $Deadlocked^*$ is propagated backwards in the graph in step (2.5), just like the information in variable $Succ^*$ in step (1.1).

Once there are no privileges in the system, all nodes in $Succ^*_i$ are also in $Deadlocked^*_i$. Therefore, $Succ^*_i \subseteq Deadlocked^*_i$ and, after one execution of step (3.1), variable $deadlocked_i = true$. □

Theorem 1 (Completeness). *Once there are no privileges in the system, if node i suffers from deadlock then $deadlocked_i = true$.*

Proof. If node i suffers from deadlock, then it is in a tie of blocked processes in the WFG. Let d_{ik} be the length of the longest simple path from i to a reachable knot k that does not include edges in k. There is at least one reachable knot. Let d_i be the maximum d_{ik} over all k. If $d_i = 0$ then i belongs to a knot and $deadlocked_i = true$ by Lemma 2. If $d_i = n > 0$ then for all successors v of i, $d_v < n$ or there is a path from v to i. For if $d_v \geq n$ and there is no path from v to i, there would be a longer path from i to a knot through v, and d_i would be strictly larger than n. Inductively, if $d_v < n$ then $deadlocked_v = true$ and $v \in Deadlocked^*_v$. Because of step (2.5), the information in variable $Deadlocked^*$ is propagated backwards in the WFG so, in time, $v \in Deadlocked^*_i$. If $d_v \geq n$, then $v \in Pred^*_i$ by Lemma 1. Hence, at some time, every successor v of i belonged to $Deadlocked_i^*$ or $v \in Pred_i^*$. Thus, the guard of step (2.3) had to be true and after the execution of the step, $Tie_i = \{i\}$. This is true for all nodes in the tie.

Since all nodes j in $Succ_i^*$ are in the tie, $Tie_j = \{j\}$. Because of step (2.5), $j \in Deadlocked_j^*$ and, in time, $j \in Deadlocked_i^*$. Therefore, $Succ_i^* \subseteq Deadlocked_i^*$ and, after one execution of step (3.1), variable $deadlocked_i = true$. □

Theorem 2 (Correctness). *Once there are no privileges in the system, if $deadlocked_i = true$ then node i suffers from deadlock.*

Proof. Let $deadlocked_i = true$ and suppose that i does not suffer from deadlock. Since i does not suffer from deadlock, then it reaches a node j that is not blocked. Thus $j \in Succ_i^*$ (by Lemma 1). Since j is not blocked, $Succ_j = \emptyset$. Because of step (0.1), $Deadlocked_j^* = \emptyset$, $Knot_j = \emptyset$ and $Tie_j = \emptyset$. Because of step (2.5), no other node appart from j can include j in its own variable $Deadlocked^*$, and j can not execute step (2.5) because $Succ_j^* = \emptyset$. Thus, j can not be included in variable $Deadlocked^*$ at any node, in particular i. Then, $j \notin Deadlocked_i^*$ and $j \in Succ_i^*$, that is $Succ_i^* \not\subseteq Deadlocked_i^*$. Because of step (3.1), variable $deadlocked$ can not be true, or i would be privileged. Since there are no privileges, $deadlocked_i$ must be false, leading to a contradiction. □

Lemma 3. *A privileged node looses its privilege in a finite number of steps.*

Proof. A privileged node has at least one true guard. After the execution of one step, the guard associated to that step becomes false.

If the execution of a step could make true guards associated to later steps then, in the worst case, each step will be executed once and, eventually, the privilege will be lost.

In the proposed algorithm all steps can only make true guards associated to later steps. The only exception is step (2.5) which could also make true the guard of step (2.3) or the guard of step (2.4). Note that the guards of steps (2.3) and (2.4) can not be both true at the same time.

If after one execution of step (2.5) the guard of step (2.3) becomes true, then $Succ^* \subseteq Deadlocked^* \cup Pred^*$. After the execution of step (2.3), $Tie = \{i\}$ and the guard of step (2.5) could become true again. If step (2.5) is executed again, the value of variable Tie does not change, and variable $Deadlocked^*$ now includes i. The guard of step (2.3) can not become true again, because variable Tie has not changed. The guard of step (2.4) can not become true because $Succ \subseteq Deadlocked^* \cup Pred^*$ still holds.

On the other hand, if after one execution of step (2.5) the guard of step (2.4) becomes true, then $Succ^* \not\subseteq Deadlocked^* \cup Pred^*$. After the execution of step (2.4), $Tie = \emptyset$ and the guard of step (2.5) could become true again. If step (2.5) is executed again, the value of variable Tie does not change, and variable $Deadlocked^*$ now does not includes i. The guard of step (2.4) can not become true again, because variable Tie has not changed. The guard of step (2.3) can not become true because $Succ \not\subseteq Deadlocked^* \cup Pred^*$ still holds.

Since the execution of one step can only make true a finite number of guards, then a privileged node looses its privilege after a finite number of steps. □

When a privileged node changes its state and produces privileges in neighbour nodes, it is called a *propagation of privilege* in this paper.

A node can propagate privileges only when it changes its local variables $Deadlocked^*$, $Succ^*$, and $Pred^*$. These variables can change initially when the WFG is modified.

Changes in variables $Deadlocked^*$ and $Succ^*$ can only propagate privileges to predecessor nodes; changes in variable $Pred^*$ can only propagate privileges to successor nodes. These privileges can not be propagated indefinitely, because once a node has received the privilege and updated its local variables, it will not receive the privilege because of the same change again.

A transient fault can generate privileges at a node. These privileges will be used locally (by Lemma 3) and will not be propagated. Thus, the effects of transient faults are always corrected locally by the algorithm.

This observation along with Lemma 3 conclude the following theorem.

Theorem 3 (Extinction of privileges). *Privileges produced in the system are eventually lost.*

An algorithm is said to be self-stabilizing with respect to a state predicate P if it satisfies the properties of closure and convergence, as defined by Schneider [11]. In the system defined by the proposed algorithm, legitimate states are defined by the following predicate:

> P: There are no privileges in the system and, for every node i, $deadlocked_i = true$ if and only if i forms part of a tie in the WFG.

The following two lemmas show that P satisfies both the closure and convergence properties.

Lemma 4 (Closure). *Once P is established, it is not falsified by execution of the algorithm.*

Proof. When P becomes true, there are no privileges in the system; therefore, no actions are executed and the state remains the same. Hence P is not falsified. □

Lemma 5 (Convergence). *Starting from an arbitrary initial state, once transient faults cease to occur, the system reaches a global state satisfying P within a finite number of steps.*

Proof. By Theorem 3, privileges eventually disappear from the system. Therefore, the first part of P is satisfied. Once there are no privileges in the system, by Theorems 1 and 2, the second part of P is satisfied. Hence P holds after a finite number of steps, once transient faults cease to occur. □

Lemmas 4 and 5 prove the following theorem.

Theorem 4 (Self-stabilization). *The proposed algorithm is a self-stabilizing deadlock-detection algorithm under the OR requirement model.*

6 Deadlock Resolution

In order to resolve a deadlock, at least one of the deadlocked processes must be terminated. Therefore, once a deadlock is detected, it becomes necessary to choose a victim to terminate. Terminating just any process does not necessarily resolve the deadlock. In Fig. 1(b), if process 5 were terminated, there would still be deadlock, because the knot in the WFG remains. To resolve a deadlock, it is not enough to kill a process that only suffers from deadlock; it is necessary to terminate one process from each knot.

Once there are no privileges in the system, each process knows whether it is deadlocked or not. And, in addition, deadlocked processes also know whether they are part of a knot or not. Variables *Knot* and *Tie* compute precisely that information. Processes that are part of a knot can start an algorithm to choose a victim such that, when terminated, the knot disappears.

Processes that are part of a knot have the same set of successors, formed by all processes in the knot. Thus, if all the nodes in the knot apply a rule —like $victim = \min(Succ^*)$— it is possible to choose exactly one victim to terminate from each knot.

No special actions need to be taken once a deadlock has been resolved. The immediate predecessors of the terminated processes would see a change in their variable *Succ*, and the detection algorithm would recompute for the new WFG.

7 Concluding Remarks

This article presents a self-stabilizing deadlock-detection algorithm for the OR requirement model. The algorithm is self-stabilizing, that is, it supports changes to the WFG, transient faults, and arbitrary initialization; previous algorithms for the OR model are not guaranteed to recover from transient faults or arbitrary initialization. The algorithm is complete and correct since it detects all deadlocks and it detects no false deadlocks, respectively. Hence, every process knows whether it is deadlocked or not and, moreover, deadlocked processes know whether they cause or only suffer from deadlock. In addition, the algorithm provides enough local information to implement actions and resolve the deadlocks detected.

References

1. Knapp, E.: Deadlock detection in distributed databases. ACM Computing Surveys **19**(4) (1987) 303–328
2. Flatebo, M., Datta, A.K.: Self-stabilizing deadlock detection algorithms. In: Proceedings of the 1992 ACM Annual Conference on Communications, Kansas City, Missouri (1992) 117–122
3. Karaata, M.H., Line, J.C.: Self-stabilizing algorithms for deadlock detection and identification in distributed systems. In: Proceedings of the ISCA Thirteenth International Conference on Parallel and Distributed Computing, Las Vegas, Nevada (2000) 320–325

4. Chandy, K.M., Misra, J., Haas, L.M.: Distributed deadlock detection. ACM Transactions on Computer Systems **1**(2) (1983) 144–156
5. Cidon, I., Jaffe, J.M., Sidi, M.: Local distributed deadlock detection by knot detection. In: Proceedings of the ACM SIGCOMM Conference on Communications Architecture & Protocols, ACM Press (1986) 377–384
6. Lee, S., Lee, Y.: A distributed algorithm for deadlock detection under OR-request model. In: Proceedings of the 18th IEEE Symposium on Reliable Distributed Systems, IEEE Press (1999) 298–299 Work in Progress, Fast Abstracts.
7. Natarajan, N.: A distributed scheme for detecting communication deadlocks. IEEE Transactions on Software Engineering **SE-12**(4) (1986) 531–537
8. Herman, T., Chandy, K.: A distributed procedure to detect AND/OR deadlock. Technical Report TR LCS-8301, Department of Computer Science, University of Texas, Austin, Texas (1983)
9. Bracha, G., Toueg, S.: A distributed algorithm for generalized deadlock detection. In: Symposium on Principles of Distributed Computing, Vancouver, British Columbia, Canada (1984) 285–301
10. Dijkstra, E.: Self-stabilizing systems in spite of distributed control. Communications of the ACM **17**(11) (1974) 643–644
11. Schneider, M.: Self-stabilization. ACM Computing Surveys **25**(1) (1993) 45–67
12. Schwiebert, L.: Deadlock-free oblivious wormhole routing with cyclic dependencies. IEEE Transactions on Computers **50**(9) (2001) 865–876
13. Holt, R.C.: Some deadlock properties of computer systems. ACM Computing Surveys **4**(3) (1972) 179–196

Incremental Distributed Garbage Collection Using Reverse Reference Tracking

M. Schoettner, R. Goeckelmann, S. Frenz, M. Fakler, and P. Schulthess

University of Ulm, Computer Science Faculty, 89069 Ulm Germany
michael.schoettner@uni-ulm.de

Abstract. Most modern middleware systems like Java Beans and .NET provide automatic garbage collection (GC). In spite of the many distributed solutions proposed in literature collection is typically limited to a single node and simple leasing techniques are used for remote references. In this paper we present a new incremental multistage GC. It has been implemented in the Plurix operating system but might easily be applied to other platforms. The scheme works incrementally and avoids blocking remote nodes. The reverse reference tracking scheme efficiently detects acyclic garbage and is also used for finding cyclic garbage without precomputing a global root set. To minimize network communication cycle detection splits into a local and a global detection part. Keeping the object markers in a separate stack avoids invalidation of replicated objects. Performance measurements show that the proposed distributed GC scheme scales very nicely.

1 Introduction

Garbage Collection (GC) relieves the programmer of explicit memory management and avoids memory leaks and dangling pointers. This is important on a single node system and almost indispensable in a distributed and persistent environment. As a consequence most modern middleware systems such as Java Beans and .NET provide automatic GC. These commercial GCs are typically based on scanning algorithms (mark and sweep) for a single node and fall back to a leasing scheme for remote references in distributed programs. In the literature numerous more sophisticated distributed GCs have been proposed [6].

Efficient GC for a distributed environment is more of a challenge than for a single machine. Basic scanning algorithms can not detect concurrent manipulation of pointers during the execution of the GC task and require suspending all other execution. Unfortunately in a distributed environment this means stopping all processing in the cluster. Incremental GC algorithms solve this problem, but often require read or write barriers and introduce programmed synchronization between the nodes in the cluster. Furthermore, in a distributed system all changes to objects including those introduced by the GC (e.g. temporary markers) must be propagated to remote object replicas. Hence small changes made on a single node may affect the entire cluster and decrease overall cluster performance.

In this paper we propose a reverse reference tracking scheme to collect incrementally all types of garbage – local or remote, cyclic or acyclic. Objects which are no longer referenced are called acyclic garbage. Garbage cycles consist of at least two objects referencing each other but neither of these objects is referenced from the root set.

The acyclic GC phase is a simple reference counting scheme with local and global parts of the computation. Unlike other scanning algorithms our reverse reference tracking avoids an atomic precomputation of the global root set and scales smoothly to larger clusters. The second phase collects cyclic garbage in an incremental fashion. Invalidation of remote replicates is avoided by storing the temporary marks separate from the candidate objects in small tables.

The remainder of the paper is organized as follows. In section 2 we briefly present relevant parts of the Plurix architecture followed by a discussion of related work in section 3. In section 4 we present our GC scheme which uses reverse reference tracking. Subsequently, we present the measurement results indicating the scalability of the proposed approach. The conclusions and an outlook on future work is given in the last section 6.

2 Plurix Architecture Aspects

Plurix is a native cluster operating system (OS) which simplifies distributed and parallel programming [3]. The entire OS is written in Java (with some minor language extensions for device drivers) and works in a type safe and object-oriented language framework continuing the OS development which was convincingly demonstrated by the Oberon system [2].

Distributed Shared Memory (DSM) in Plurix offers an elegant solution for distributing and sharing data in a cluster of loosely coupled PCs [8]. Applications running on top of the DSM are unaware of the physical location of objects. Remote objects are automatically transferred to an accessing node by the runtime system. Plurix implements a distributed heap (DHS) on top of the DSM which hosts language objects, kernel objects, code segments and device drivers.

Tracking references to objects is a requirement both for the GC scheme and for the object relocation facility. The latter is needed to compact the heap, to resolve false sharing (page thrashing) situations, and to support type evolution. We have developed the so-called *backpack* scheme to track all references to an object. The basic idea is that each object can track up to three references in its own header accommodating the majority of all reference situations (in-line backlinks). The reverse tracking links are called *backlinks*. If more than three references are tracked *backpacks* are created on demand. These are separate hash tables containing additional backlinks. A detailed description of backlinks and backpacks can be found in [1].

Any heap object may be registered and then looked up in the directories and subdirectories of a cluster-wide name service. This corresponds to the directory structure of traditional file systems but the functionality of the name service is extended to store symbol tables, configuration information, and to cover all naming issues occurring in the OS. Any heap object reachable from the name service root is not garbage and thus persistent.

The Plurix DHS detects memory access using the Memory Management Unit (MMU) of the CPU thus implementing a page-based DSM. Since individual pages and the allocated objects get replicated a distributed consistency protocol is necessary. Plurix uses a strong consistency model, called transactional consistency [3]. All actions in Plurix regarding the DHS are encapsulated in restartable transactions (TA) combined with an optimistic synchronization scheme. Before a page is modified by a TA the OS creates a shadow image. During the commit phase the addresses of all modified pages are multicast and the receiving nodes will invalidate these pages. Those nodes that detect a collision, abort themselves voluntarily.

In case of an abort all modified pages in a TA are discarded. Shadow images are used to reset the DHS is to the state before this conflicting TA. A token mechanism guarantees that only one node at a time can enter the commit phase. Currently, the token is passed according to a first wins strategy, but improved fairness strategies are currently being investigated. For a more detailed discussion about consistency management, fault tolerance, and persistence see [3].

3 Related Work

In this section we briefly discuss GC algorithms which were designed for distributed environments or whose ideas inspired our implementation. An excellent summary of basic GC algorithms is found in [6].

Copying Algorithms
These schemes copy all live objects (reachable from the root object set) from one part of the address space to another and the garbage objects are left in the source portion. After the "copy" action heap fragmentation is eliminated but copying many small objects (even if only logically) may be time consuming and expensive invalidations of live remote objects are unavoidable.

LeSergent and Berthomieu [5] have developed an copying algorithm for a distributed GC. Each process in the system has a uniform view of the DSM. The memory is divided into parts with equal size, e.g. physical pages. A single page may be dynamically assigned to one or more processes at a time. If a process tries to access a page which is not present the page is fetched across the network and locally assigned. For this algorithm it is necessary to lock pages if a process needs write access to it. As a consequence nodes may be blocked during the GC cycle.

Mark-and-Sweep Algorithms
These algorithms mark each object reachable from the root set. Unmarked objects are garbage. Setting marks within an object may lead to many invalidations of remote objects. It is preferable to store marks outside of the objects, e.g. in bitmap- or hash-tables. Hash-tables consume less physical memory than the bitmap approach but are still expensive in a scenario with many small objects (e.g. 32-64 byte) that are common in object-oriented languages.

A mark-and-sweep algorithm for a distributed system was developed by Yu and Cox [10] in 1996. They designed a GC scheme for the Treadmarks DSM system [4]. Here the heap is divided into blocks in which each process can allocate its own objects. After allocation, the process gains ownership of the object. Objects can be either "local" meaning that the process is owner of this object or "remote". "local" objects which are used by other nodes are marked as "exported". Remotely owned objects are "imported". Both kinds of objects are tracked using import-/export tables. References to "remote" objects are handled using weighted reference counters without using indirection objects i.e. the weight of an object may be less than the weight of all references to it. The GC itself is divided into a local and a global part. The local part is a mark-and-sweep scheme examining entries in the export table, but it is unable to detect distributed cyclic garbage. The global GC part will stop the cluster. All objects reachable from a local root are marked; references to other nodes are recorded and afterwards sent to the associated node, which continuous marking. These steps are repeated until no more references to other nodes exist.

Reference Counting Algorithms
These GC algorithms depend on a counter for each object, recording the number of existing references. The placement of the reference counter raises a problem similar to the placing of the marks of a mark-and-sweep algorithm. Although the GC is simple and does not block the cluster it cannot detect cyclic garbage without special provisions. Detecting cyclic garbage mostly depends on marking algorithms or removing internal counts (i.e. the reference counter is decremented for each pointer which potentially references another object from the same garbage cycle) [11]. This modifies all checked objects and hereby causes unwarranted invalidations.

Traditionally, the reference counter is included in the object and this forces a modification of the object each time a reference to it is created or destroyed. Invalidation of an object during the creation of a reference can be avoided by using weighted reference counting. But objects can not always be identified as garbage and are modified when a reference is deleted.

Reference counting GC faces additional problems if a node crashes. In this case references to an object are lost but the reference counter is not decremented. Now the reference counter never reaches 0 and the object will not be collected.

David Bacon [11] has presented a GC strategy which is based on reference counting but also collects cyclic garbage. In a separate structure (a so called RootBuffer) the algorithm remembers all objects which could potentially be cyclic garbage. Separate from the traditional reference counting mechanism, the GC scheme contains a second phase in which cyclic garbage is detected traversing all reachable objects starting with the objects included in the RootBuffer. During this computation the reference counter of reached objects is decremented to remove internal reference counts (references which points from one potentially cyclic garbage object to another), and the objects are marked. The algorithm is able to collect cyclic garbage in linear time but it needs to modify the traversed objects. Objects are cyclic garbage candidates if their reference counter is decremented but does not reach zero and is not incremented before the cyclic detection part of the GC is started. This condition may be true for many live objects leading to a large number of invalidations of replicated objects.

Algorithms Basing on an Inverse Reference Graph
The first GC depending on the inverse reference graph was made in 1991 by Piquer [13]. The algorithm uses Indirect Reference Counting based on a diffusion tree which eliminates the need for increment and decrement messages to adjust the reference counter of an object. This avoids race conditions which can lead to incorrect behavior

of distributed reference counters. Shapiro [9] extended this approach. Scion-Stub Pointer Chains uses parent pointers to track where references to an object are located. These pointers build the inverse reference tree from an object to its accessors. The GC works similar to traditional reference counting and is not able to collect cyclic garbage.

A similar approach was presented by Birrell [7]. The algorithm eliminates the reference chains by maintaining a set of identifiers for processes with references to an object. To determine this ID-set the transfers of object references to another process are handled by a remote procedure call. Premature collection of objects is prevented by forcing the sender of a reference to keep its copy until receipt is verified.

Another GC strategy depending on the inverse reference graph was presented by Matthew Fuchs [12]. The described algorithm solves the problem of discovering the distributed root set for a mark- and-sweep GC by starting with any object and traversing inverse pointers. The algorithm uses a three color marking to determine whether an object is garbage. An object is live, if the inverse pointer graph contains at least one root object. Root objects are separately marked so that they can be identified. The algorithm is interesting as it can collect garbage without knowledge of the current cluster state and because it is not necessary to know each root object, but it sets marks in shared objects and thus invalidates replicated objects.

4 Garbage Collection Using Reverse Reference Tracking

Plurix is designed for both distributed and parallel computing but also for cooperative working. Hence its GC must be capable to collect all types of garbage and run concurrently with other applications and without significantly degrading cluster performance. To achieve this goal, the GC should neither utilize excessive network capacity nor block the cluster during execution. The objective of keeping network traffic low requires that write access to objects must be kept to a minimum within the GC, as this would lead to invalidations of replicated objects or of 4 KB pages that could store dozens of objects within Plurix.

Non-cyclic Garbage Collection Using Reference Counting
Reference Counting is conceptually simple but in a distributed environment it is important to avoid frequent modification of objects. Piquer [13] has shown that instead of a reference counter backward references can be used, too.

In Plurix the bookkeeping of references is primarily used for relocation of objects but it can also used for GC at little additional cost. We merely count the number of references stored in backlinks within the object itself and in associated backpacks. An object is garbage if all backlink entries from the object are empty. Special root objects which are never garbage are marked by a special non-garbage flag by the OS.

The bookkeeping of references modifies objects only when the "in-line" backlinks are changed. The respective backpack table-object is deleted if the last object reference is removed. The memory management makes sure that backpacks do not co-reside with normal objects on a page, i.e. aborts of other TAs may only occur if both TAs try to modify a reference to an object.

Reference counting schemes also need to consider stack references and CPU registers. Because of the transactional processing in Plurix this can be done elegantly. The

GC runs as a separate TA thus seeing only committed and valid state of objects. Most TAs (e.g. processing an event) commit with an empty stack and with empty registers. Some TAs (e.g. for parallel computing) may commit with a non-empty stack that is consolidated during commit time including CPU registers - all references on the stack are recorded in backpacks during commit time. Postponing stack reference tracking is recommended because not all applications need this feature. Often the stack shrinks before commit and only the references from a small residual stack need to be tracked and only once during commit.

Plurix will find all objects in the heap by stepping from one object to the next. The reference counting algorithm can run concurrently on several nodes. The GC only has to check objects which are present locally, as each object must be present on at least one node. The backlinks of each such object are checked, and it is collected in case of garbage.

Acyclic GC can be run without causing additional network traffic during detection of garbage objects, as only local objects are inspected and the internal backlinks contain sufficient information about the state of an object. Network traffic and collisions only occur if a garbage object and all its references to remote objects are deleted.

Cyclic Garbage Collection Using Inverse References
The major challenge for a GC in a distributed system is to detect and collect cyclic garbage. After collecting non-cyclic garbage the remaining objects are either alive or part of cyclic garbage. Cyclic GC is used to break the cyclic structure of garbage objects so that these objects can be collected during the next execution of the non-cyclic GC. We have developed an incremental variation of mark-and-sweep to detect cyclic garbage. The marks are kept outside the objects to avoid invalidations. Backpacks provide all information for inverse reference tracking.

In traditional systems the set of root objects must be determined by obtaining the root subset from each node or running the GC simultaneously on each node. More easily our algorithm starts at an arbitrary object which is locally present and traverses the inverse reference graph searching for a root object. If none is found, the object is part of cyclic garbage and should be deleted. Thereby all references included in this object are removed. Other objects which were traversed during this GC scan are not yet collected because the remaining members of this cycle will be detected by the non-cyclic GC if the cycle is broken at an appropriate place. Otherwise the cyclic GC will identify the next candidate and so forth.

It is necessary to mark each traversed object to avoid endless loops during the execution of the GC. These marks must not be located inside the objects to reduce invalidations. Unlike traditional mark-and-sweep algorithms not all objects in the cluster need to be marked, therefore it is possible to place the marks in a separate "marking table" (MT, hashed or otherwise). As all objects in Plurix are located on 4 Byte borders, the least significant 2 Bits of each address or backlink are 0. These bits in the MT are conveniently used to remember whether an object has already been checked. In addition to the MT there is another table (i.e. an integer array) used during the GC which is organized as a stack. All encountered backlinks which are not already checked are placed on top of this handle stack (HS).

At the start of a GC TA the MT and HS are created unless older tables can be reused. Both tables are not shared so that TAs on other nodes are not affected by modifications of these tables. Additionally, the memory management allocates the HT and MS on a 4kB border and with a size of a multiple of 4 KB to avoid false sharing so that modifications do not cause invalidations of unconcerned objects. The size of both tables is limited by a configurable value, given to the GC TA at start time. This defines the maximum depth of cycles which might be detected by this TA but does not reduce the capability of the algorithm. If the GC is terminated due to an exhausted MT or MS, the GC can be restarted with a larger one. In contrast to the tables needed for general mark-and-sweep, the tables for cyclic garbage detection are very small. The GC has successfully detected a cyclic structure if the HS is empty and no root object has been found. In this case at least the object at which the GC has been started should be de-allocated. References from this object to another one are deleted. Depending on the remaining time, other objects in the MT may be deleted since they are not reachable from the root set.

The steps of the algorithm are described below and an example is shown in fig. 1:

1. The flag field of the object is checked. If it is marked as non-garbage the GC terminates because the object is a root object. The chosen object is reachable and not garbage.
2. The address of this object is inserted into MT. If the MT is exhausted go to step 7.
3. All backlinks of the object are inserted into the MT and pushed onto the HS; duplicates are ignored. If the MT or HS is exhausted go to step 7.
4. The MT entry for the object is marked. This object is now completely handled.
5. If the HS is not empty get next address of an object from the HS and go to step1.
6. If the HS is empty, the chosen object is part of cyclic garbage and can be deleted. MT and HS can be cleared and the algorithm will terminate.
7. The GC terminates without being able to detect a root object. The chosen object is treated to be non-garbage.

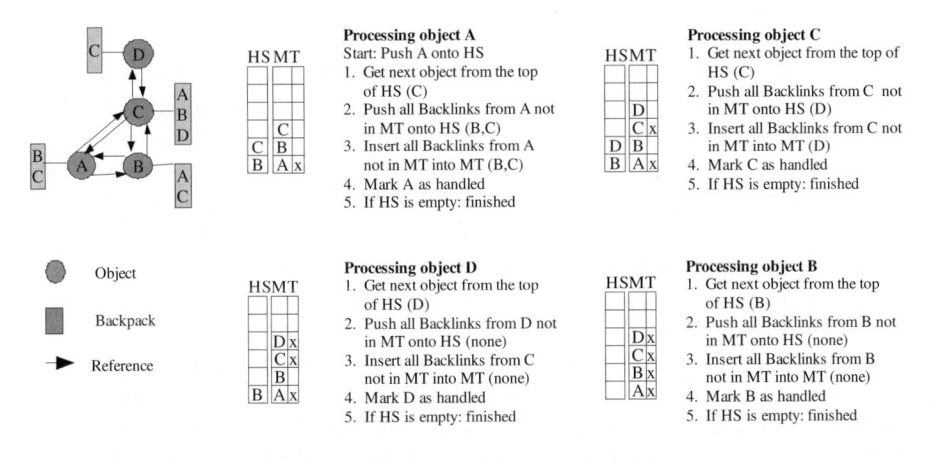

Fig. 1. Cyclic Garbage Detection Example

Because cyclic garbage detection can be a time consuming operation depending on the size of the cycle and the distribution of the affected objects, the cycle GC comes in two variants: local and global cyclic GC. Both variants may be aborted at any time without affecting the cluster state. Which variant and which parameters of the cyclic GC are started is configurable reflecting CPU load of the node, network load, and low memory, etc.

The local part of the cyclic garbage detection checks only those objects which are locally present. This can be determined by the flags (set by the MMU) in the page tables. For each candidate object the backpack or respectively the backlinks are inspected and the inverse reference tree is built. As soon as a backlink references a remote object the cyclic garbage detection stops and the object is regarded to be live. The GC will choose the next candidate object until the configured time slot if any expires. Since even in a distributed environment many objects are locally used the local phase is useful – effectively reducing network traffic.

The second part of the cycle detection GC works on the entire cluster. Again objects which are locally present are used as a start for cycle detection but all backlinks are checked. To reduce network traffic, the cluster wide cycle detection algorithm tries to detect a local root object before remote objects are transferred to the node. Remote pages are not requested until all local references are checked and no root object was yet found, hence the GC does not cause network traffic for objects which are reachable from the local root subset. To distinguish between local and remote the inverse reference stack of the GC is duplicated. One stack is used for objects locally present and the other for remote ones. When the local stack is exhausted and no root object was found, the backlinks preserved in the remote stack are inspected. If such a remote object contains backlinks to locally present objects these are checked before other remote backlinks are observed. This ensures, that remote pages are only requested if it is inevitable. In most cases a local root object is found, if the chosen object is not part of cyclic garbage, before all remote backlinks have been checked.

5 Measurements

Measurements were made on a cluster of 16 nodes (AthlonXP 2500+ with 512 MB RAM) using a switched FastEthernet. We compare our GC with a traditional blocking mark-and-sweep GC (BMSGC). Since blocking GCs are faster than the corresponding incremental solutions the execution time of BMSGC can be viewed as the lower bound. In the first part the measurements only use a single node. We allocated 13'800 objects whereof 1'600 were acyclic and 1'200 cyclic garbage. The cyclic garbage was spread over 36 cycles each containing between two and eight objects. Table 1 shows the execution times of different steps of the Plurix GC (PGC) and for the BMSGC. Times shown are an average of 10 independent runs.

The measurements show that the detection of acyclic garbage is much faster in PGC than in BMSGC whereas the situation is reversed for cyclic garbage on a single node. But BMSGC requires marking all objects each time the GC is called to determine if an object is garbage or not - and of course it represents a lower execution bound if we can afford to block all nodes.

Table 1. Execution times of PGC and BMSGC on a single node

Action	exec. time (ms)	#objects processed
PGC: acyclic garbage	6,03	2'653'000
PGC: cyclic garbage (detection only)	54,00	252'000
PGC: cycle detection & removal	55,00	251'000
BMSGC: remove marks	3,27	4'220'000
BMSGC: mark phase	18,07	763'600
BMSGC: delete objects	4,12	3'349'0004,12

In the second part we evaluated the performance of PGC in cluster operation. For these measurements we have allocated 61'600 objects whereof 12'000 are acyclic and 9'600 cyclic garbage. The latter included 4'000 objects having references to remote nodes. As the acyclic GC stage is able to check individual objects it can be executed concurrently on all nodes. Inter-node communication is necessary only if an object with a reference to a remote object is deleted because this requires deleting the associated backlink on the remote node. In this case the object deletion is increased by 784 μs - reflecting network latency. Of course this will be less significant for faster networks. In the best case there are no remote references and the acyclic GC will scale almost linearly with the number of nodes.

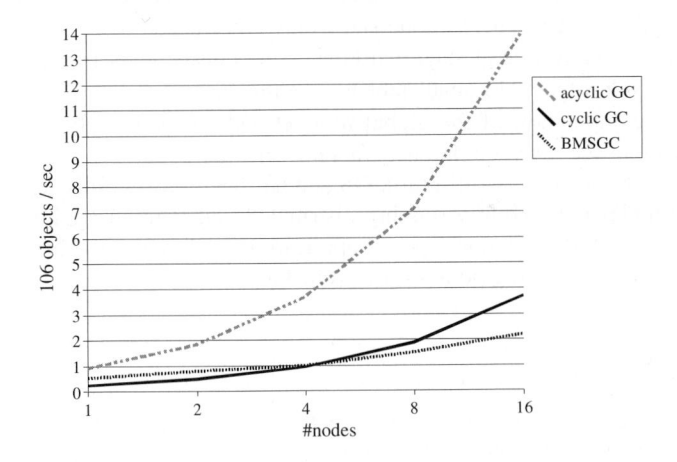

Fig. 2. Performance in cluster operation PGC & BMSGC

The performance of the cyclic garbage detection in cluster mode depends on the number of objects that need to be checked and the number of page requests to remote objects to be performed. If all local objects are referenced by some object that is part of the root set, no network communication is necessary. In that best case the scalability of the GC depends only on the distribution of objects in the cluster. For the applications we use (distributed and parallel ones) this is true for approximately 90% of all objects. This is the reason why PGC outperforms BMSGC in the cycle detection in cluster operation. Fig. 2 shows the scalability for the acyclic and cyclic stages of PGC and BMSGC. The throughput given in 10^6 objects per second has been computed using measurements with the example with 61'600 objects described above. Although for this example BMSGC offers a better performance on a single node PGC outperforms the BMSGC for four and more nodes and scales quite well. As we assume concurrent execution of PGC on all nodes periodically this is a very nice result.

Further measurements might be beneficial and we do not claim that there won't be a special case where the one or other sophisticated GC will be faster than the one we propose. But generally speaking we find that our approach scales well in a distributed DSM environment and that it is an interesting option for other distributed scenarios as well.

6 Conclusion

In this paper we have proposed an incremental multistage GC built on reverse reference tracking and keeping the reverse references in so-called backpack/backlinks. The proposed GC approach is easily be adapted to other distributed systems and does not limit the GC to a special environment. The first stage of our GC detects non-cyclic garbage and is basically a reference counting GC evolving directly from the backpack concept and scaling very nicely.

The cyclic phases deal with cyclic garbage and can be executed concurrently without blocking the cluster. There is one stage only working on local objects (avoiding network traffic) and a second stage working at the cluster level if necessary. Marks are stored outside objects in small tables avoiding invalidations of remote replicas. Furthermore, computation of the global root set and contacting all nodes is not required because of the reverse reference tracking scheme.

The GC algorithm is used in our Plurix OS and has been successfully tested in a cluster with 16 nodes concurrently running distributed and parallel applications. In the future we plan to study different types of applications and to develop heuristics to find good candidates as a starting point of the cyclic GC.

References

1. R. Goeckelmann, S. Frenz, M. Schoettner, P. Schulthess, "Compiler Support for Reference Tracking in a type-safe DSM", in: Proc. of the Joint Modular Languages Conf., Klagenfurt, Austria, 2003.
2. N. Wirt and J. Gutknecht, „Project Oberon", Addison-Wesley, 1992.
3. The Plurix project: www.plurix.de.
4. Amza C., Cox A.L., Drwarkadas S. and Keleher P., „TreadMarks: Shared Memory Computing on Networks of Workstations", in: Proc. of the Winter 94 Usenix Conference, 1994.
5. T. Le Sergent and B. Berthomieu, "Incremental multi-threaded garbage collection on virtually shared memory architectures", in: Proc. Int. Workshop on Memory Management, number 637 in Lecture Notes in Computer Science, pages 179-199, Utrecht (NL), 1992.
6. Richard Jones, "Garbage Collection: Algorithms for Automatic Dynamic Memory Management", JohnWiley and Sons, July 1996. With a chapter on Distributed Garbage Collection by Rafael Lins. Reprinted 1997 (twice), 1999, 2000.

7. A.Birrell et al. , "Distributed garbage collection for network objects", in Technical Report 116, DEC Systems Research Center, 1993.
8. K. Li, "IVY: A Shared Virtual Memory System for Parallel Computing", In Proceedings of the International Conference on Parallel Processing, 1988.
9. M. Shapiro, D. Plainfossé, P. Ferreira, L. Amsaleg, " Some Key Issues in the Design of Distributed Garbage Collection and References", in seminar on "Unifying Theory and Practice in Distributed Systems," Dagstuhl Int. Conf. and Res. Center for Comp. Sc., 1994.
10. W. M. Yu and A. L. Cox, "Conservative garbage collection on distributed shared memory systems", in: Proc. of the Int'l Conf. on Distributed Computing Systems (ICDCS-16), 1996.
11. D. F. Bacon and V. T. Rajan, "Concurrent Cycle Collection in Reference Counting Systems", Proc. European Conference on Object-Oriented Programming, June 2001, volume 2072 of Lecture Notes in Computer Science, Springer Verlag.
12. M. Fuchs, "Garbage Collection on an Open Network", in volume 986 of Lecture Notes in Computer Science, 1995.
13. J.M. Piquer, "Indirect Reference Counting, a distributed garbage collection algorithm" , in: PARLE'91- Parallel Architectures and Languages Europe, volume 505 of Lecture Notes in Computer Science, page 150-165, Eindhoven (NL), June 1991, Springer-Verlag
14. J. M. Piquer, "Indirect Mark and Sweep", in Baker HG (Ed.), Memory Management, Proc IWMM95 LNCS 986, Springer-Verlag, 268-282.

Run-Time Switching Between Total Order Algorithms*

José Mocito and Luís Rodrigues

University of Lisbon
{jmocito, ler}@di.fc.ul.pt

Abstract. Total order broadcast protocols are a fundamental building block in the construction of many fault-tolerant distributed applications. Unfortunately, total order is an intrinsically expensive operation. Moreover, there are certain algorithms that perform better in specific scenarios and given network properties. This paper proposes and evaluates an adaptive protocol that is able to dynamically switch between different total order algorithms. The protocol allows to achieve the best possible performance, by selecting, in each moment, the algorithm that is most appropriate to the present network conditions. Experimental results show that, using our protocol, adaptation can be achieved with negligible interference with the data flow.

1 Introduction

A total order broadcast protocol is a fundamental building block in the construction of many distributed fault-tolerant applications [1]. Informally, the purpose of such a protocol is to provide a communication primitive that allows processes to agree on the set of messages they deliver and, also, on their delivery order. Uniform total order broadcast is particularly useful to implement fault-tolerant services by using software-based replication [2].

Unfortunately, the implementation of such a primitive can be expensive both in terms of communication steps and number of messages exchanged. This problem is exacerbated in large-scale systems, where the performance of the algorithm may be limited by the presence of high-latency links. Several total order protocols have been proposed that use different strategies to offer good performance [3]. There is no protocol that outperforms all others in all scenarios: each protocol offers best results under different load profiles and/or network conditions.

In this paper we describe and evaluate a total order protocol that combines different algorithms and adapts itself to the running environment. The protocol allows a fluid transition between algorithms, never stopping the flow of application messages. Such feature can be very useful in fault-tolerant safety- and mission-critical systems, like air traffic or nuclear plant control, where stoppages and/or significant delays imposed by adaptive mechanisms may be unacceptable.

---

* This work was partially supported by the IST project GORDA (FP6-IST2-004758).

Table 1. Regular total order properties

TO1 - Total order: Let m_1 and m_2 be two messages that are *TO-broadcast*. Let p_i and p_j be any two correct processes that *TO-deliver*(m_1) and *TO-deliver*(m_2). If p_i *TO-delivers*(m_1) before *TO-delivers*(m_2), then p_j *TO-delivers*(m_1) before *TO-delivers*(m_2), and we note $m_1 < m_2$.

TO2 - Agreement: If a correct process in Ω has *TO-delivered*(m), then every correct process in Ω eventually *TO-delivers*(m).

TO3 - Termination: If a correct process *TO-broadcasts*(m), then every correct process in Ω eventually *TO-delivers*(m).

TO4 - Integrity: For any message m, every correct process *TO-delivers*(m) at most once, and only if m was previously *TO-broadcast* by some process $p \in \Omega$.

We evaluate the performance of an implementation of the protocol and show how it can be optimized to induce a low overhead in resource consumption. Finally, we discuss how the protocol can be configured to operate using different classes of failure detectors.

The rest of the paper is structured as follows. Section 2 clarifies the properties of total order broadcast services. Section 3 describes the adaptive protocol. Performance evaluation results are presented in Section 4. Section 5 discusses optimizations that have been applied in the protocol implementation. Failure detection issues are addressed in Section 6. Section 7 concludes the paper.

2 Total Order Broadcast

Informally, total order broadcast is a group communication primitive that ensures that messages sent to a set of processes are delivered by all those processes in the same order. Such a primitive is useful, for example, in the implementation of fault-tolerant services [1], for instance, using the state machine approach (active replication) [4].

Total order broadcast is defined on a set of processes Ω by the primitives (1) *TO-broadcast*(m) which issues message m to Ω, and (2) *TO-deliver(m)* which is the corresponding delivery of m. When a process p_i executes *TO-broadcast*(m) (resp *TO-deliver(m)*), we say that p_i "*TO-broadcasts* m" (resp "*TO-delivers* m"). The total order primitive characterized by the properties listed in Table 1 is known as regular total order. A stronger version, called uniform total order [3], can also be defined. The difference among these definitions is not relevant for understanding our adaptive protocol, thus we will not delve further in this topic.

Many algorithms exist to implement total order. To give the reader an insight on the possible alternatives, we briefly introduce two of the most used ones, namely the *sequencer-site* [5] and the *symmetric* [6,7] approach. Both methods have advantages and disadvantages.

In the sequencer-site approach one site is responsible for ordering messages on behalf of the other processes in the system. Sequencer-based protocols are appealing because they are relatively simple and provide good performance when message transit delays are small (they are particularly well suited for local area networks). However, in these protocols, a message sent by a process that is not the sequencer experiences a delivery latency close to $2D$, where D is the message transit delay between two system processes (i.e., the time to disseminate the message plus the time to obtain an order number from the sequencer process). Thus, sequencer-based approaches are inefficient in face of large network delays. Note that it is possible to design solutions where the sequencer role is rotated among processes [8].

In the symmetric approach, ordering is established by all processes in a decentralized way, using information about message stability. This approach usually relies on *logical clocks* [9] or *vector clocks* [10,6]: messages are delivered according to their partial order and concurrent messages are totally ordered using some deterministic algorithm. Symmetric protocols have the potential for providing low latency in message delivery when all processes are producing messages. In fact, symmetric protocols can exhibit a latency close to $D + t$, where t is the largest inter-message transmission time [11]. Unfortunately, this also means that all (or at least a majority [7]) processes must send messages at a high rate to achieve low protocol latency.

Several other alternatives exist. For a comprehensive survey, the reader is referred to [3]. However, from the two examples above, it should be clear that it is interesting to have a protocol that can dynamically adapt to changes in the operation envelope by switching, in run-time, from one algorithm to another.

3 An Adaptive Protocol

We now present a protocol that is able to switch from a total order algorithm to another total order algorithm in response to changes in the operation envelope (such as changes in the workload, network conditions, number of participants, etc). In this paper we do not focus on the conditions that trigger adaptation, as these are highly application dependent (for a concrete scenario, see [12]). Instead, we are interested in finding a generic switching procedure that can switch from one algorithm to the other with minimum interference in the data flow.

Such protocol can be built from scratch using a monolithic approach where all the functionality of every total order algorithm is embedded in a single unity. A more modular (and generic) way of reaching the same goal is to (re-)utilize independent implementations of total order algorithms and build the adaptive behavior on top of them. In steady-state, the adaptive protocol would simply receive *TO-broadcast/TO-deliver* requests/indications and forward them to the most appropriate algorithms.

To our knowledge, there is little work in the literature on how to efficiently perform this transition. Previous works on dynamic adaptation require messages to be buffered during the reconfiguration [13,14], the message flow to be stopped

```
 1: Initialization:                          18: upon TO-broadcast(msg) do
 2:    deliv ← ∅                             19:    TO-broadcast(curAlg,msg)
 3:    undeliv ← ∅                           20:    if switching = true then
 4:    curAlg ← TO-A {current algorithm}     21:        TO-broadcast(newAlg,msg)
 5:    newAlg ← ∅ {next alg.}
 6:    switching ← false                     22: upon TO-deliver(alg,msg) do
 7:    check[1..n] ← false                   23:    if alg = curAlg ∧ msg ∉ deliv then
                                             24:        deliver(msg)
 8: upon changeAlgorithm(newTO) do           25:        deliv ← deliv ∪ {msg}
 9:    rBroadcast(switch,newTO)              26:    else if msg ∉ deliv then
                                             27:        undeliv ← undeliv ∪ {msg}
10: upon rDeliver(switch,newTO) do
11:    newAlg ← newTO                        28: procedure endSwitch()
12:    switching ← true                      29:    for all msg ∈ undeliv ∧ msg ∉ deliv do
13:    TO-broadcast(curAlg,(flag,null,myself))  30:       deliver(msg)
                                             31:        deliv ← deliv ∪ {msg}
14: upon TO-deliver(curAlg,(flag,null,sender))  32:    undeliv ← ∅
    do                                       33:    check[i..n] ← false
15:    check[sender] ← true                  34:    curAlg ← newAlg
                                             35:    switching ← false
16: upon check[1..n] = true do
17:    endSwitch()
```

Fig. 1. Adaptive Total Order algorithm

in the current protocol [15], or some communication delay to be imposed during the transition between protocols [16]. Here we describe a generic transition protocol that does not require the traffic to be stopped, allowing a smooth adaptation to changes in the underlying network.

To be able to effectively transition from one algorithm to the other, all nodes need to agree on the point in the message flow where they switch. Also, both algorithms must provide FIFO ordering of messages (which is the most common case). The rational behind our proposal is to start broadcasting messages using both total order algorithms, during the switching phase, until a safe point is reached in every process. By using both algorithms simultaneously, no stoppage in the message flow is necessary. The protocol is listed in Figure 1.

Let us assume that the adaptation protocol is using algorithm TO-A to order messages and wants to switch to algorithm TO-B. The transition protocol works as follows. A control message is broadcast to all processes to initiate the reconfiguration (lines 8–9). When a node receives this message (line 10) it starts broadcasting messages using both total order algorithms. Also, the first message it broadcasts using algorithm TO-A is flagged. If no message is to be sent, then a flagged special null message is broadcast using TO-A, to allow faster protocol termination (flagged first message is not represented in the algorithm to preserve clarity). When a process starts receiving messages from both TO algorithms it performs the following steps (lines 22–27): messages received from TO-A are delivered as normally; messages received from TO-B are buffered in order. As soon as a flagged message is received from each and every node (line 15) the transition is concluded using the following "sanity" procedure (lines 28–35). Firstly, all messages received from TO-B that have not yet been delivered by TO-A are delivered in order. Finally, from this point on, all messages received from TO-A are simply discarded and no further message is sent using TO-A (until a new reconfiguration is needed). The TO-B algorithm is then used to broadcast and receive all the messages to be delivered.

Note that, after the transition is concluded, messages received from TO-B are delivered only if they have not been already received and delivered from TO-A (line 23). This is a necessary safeguard as the two total order algorithms do not necessarily deliver messages in the same order, nor at the same time. So there is a possibility that a message that has already been delivered from TO-A is received after the termination of the reconfiguration procedure from TO-B.

Also, the protocol presented does not allow concurrent adaptations. For one adaptation to happen, the previous (if any) should always have concluded.

4 Performance Evaluation

We evaluate the performance of our adaptive protocol from two different perspectives. First, we evaluate the overhead of the switching procedure. Then, we provide a comparative analysis on how different switching strategies interfere with the traffic flow during the reconfiguration.

4.1 Switching Overhead

To evaluate the switching overhead of our adaptive protocol we compare the performance of a system that always uses the same total order algorithm, with that of a system that is periodically switching between two algorithms. To make the comparison as fair as possible, we made our protocol switch between two instances of the same total order algorithm, which is also used as the non-adaptive protocol. Also, the network topology and working conditions did not change during the tests. In this way, we can isolate the cost of the switching procedure given that all the remaining factors remain unchanged.

The adaptive protocol was implemented in Java using the *Appia* [17] protocol composition and execution framework. The experiments were conducted in the SSFNet [18] network simulator and the scenario consists of a five node cluster, where all nodes are connected to each other by 100Mbps bi-directional links.

Two runs of the same experiment were performed: (A) one using a single total order protocol (non-adaptive), (B) and another using the proposed adaptive total order protocol, which is forced to switch periodically. Each run consists of every node broadcasting 5000 messages of 5KB in total order. The experiment ends when all nodes receive all the broadcast messages. The values presented are averages of the measurements conducted in each node.

Figure 2 presents the overall throughput results when the send rate is made variable. As depicted, both total order algorithms perform the same until they reach approximately 400 msg/s. After this point, the throughput of the non-adaptive protocol continues to grow while its value stabilizes for the adaptive protocol. This behavior is explained by the overhead introduced by the switching

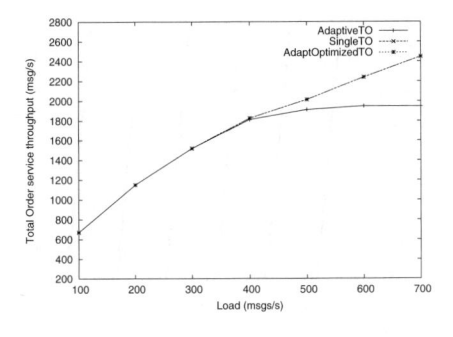

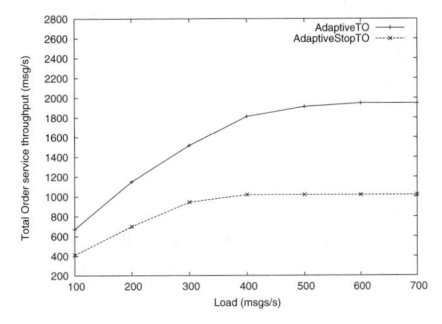

Fig. 2. TO throughput in non-adaptive, adaptive and optimized algorithms

Fig. 3. TO throughput in adaptive and stop algorithms

phase in the adaptive protocol. During this phase, the same set of messages is being broadcast by two total order algorithms at the same time, leading to an increase (approximately double) in the bandwidth usage. If the send rate is too high, the available bandwidth can be exhausted, leading to the stagnation observed in the throughput.

Thus, we can conclude that our switching protocol offers negligible overhead as long as there is enough network bandwidth to support the transmission of data in parallel during the reconfiguration. When the protocol operates close to the available bandwidth, the switching procedure introduces an overhead. This limitation can be addressed at the implementation level, by sending the payload of the messages using just one of the two algorithms. This optimization is described in Section 5 and its switching overhead is also depicted in Figure 2.

4.2 Comparative Analysis

As we noted in Section 1, most switching protocols require the message flow to be stopped in order to terminate the reconfiguration process. By not imposing a gap in the message flow, our protocol provides smooth transitions between algorithms, thus allowing applications that rely in its services to normally execute, even during the switching phase. Therefore, it should offer better overall throughput, as long as enough bandwidth is available to cope with the demand imposed by the transmission of messages using two algorithms at the same time. The same experiment described in 4.1 was conducted using a protocol that stops the message flow. This protocol operates by sending a stop request to all nodes and awaiting for a confirmation from each of these nodes. After confirming the stop request a node does not send further messages until the switch is complete. The performance of such protocol when compared to our proposal can be observed in Figure 3, which clearly shows that our approach always performs better.

Other protocols that try to minimize the cost of switching between algorithms have also been proposed. A previous work [16], proposes a solution that has some

Fig. 4. Latency in Adaptive TO

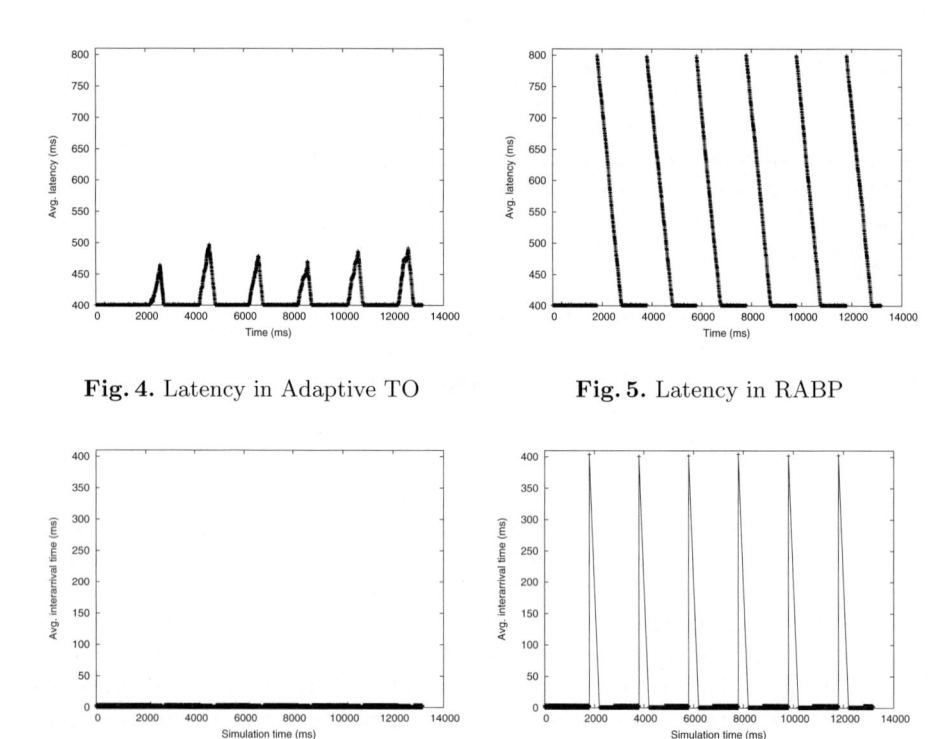

Fig. 5. Latency in RABP

Fig. 6. Inter-arrival time in Adaptive TO

Fig. 7. Inter-arrival time in RABP

similarities with our protocol, but differs from it by not requiring every node to wait for a "special" (in our algorithm the term is "flagged") message from every other node, and also for not making any assumptions about the failure model where it is executing (see Section 6). In [16], a special reconfiguration message is broadcast in total order. When a node receives such message, it stops the flow in the current algorithm, and re-issues all his undelivered messages in the next algorithm. It then starts using it to broadcast messages in total order. We will refer to this protocol by RABP (*Replacement of the Atomic Broadcast Protocol*).

The RABP strategy has the advantage of requiring less bandwidth during the switching phase. However, some delay is imposed to the message flow during the retransmission of the undelivered messages. To observe this side effect, the experiment was now conducted using our protocol and the RABP protocol. In Figures 4 and 5 we can observe how both compare in terms of latency. The spikes depicted correspond to the switching phases, in the time-line of the experiment. The inter-arrival time of messages was also measured and its evolution is shown in Figures 6 and 7. Finally, the number of messages delivered by a fixed period of time (10 ms) was also observed and the comparative results are depicted in Figures 8 and 9.

This experiment clearly showed that our proposal is able to keep a sustained delivery rate during the switching phase and performs similarly to RABP during

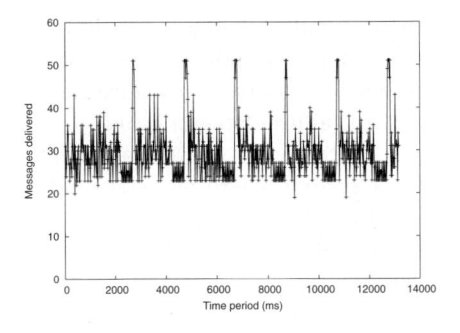

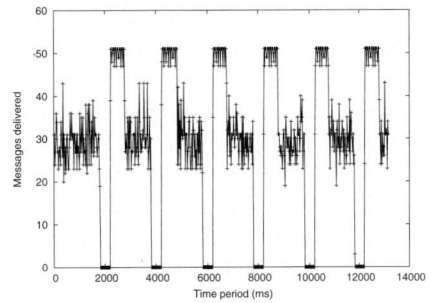

Fig. 8. Delivery rate in Adaptive TO

Fig. 9. Delivery rate in RABP

the remaining time. By not significantly delaying the message flow, our protocol can best suit environments where application stoppage, due to significant communication delays, is not desirable.

5 Implementation Optimization

When enough bandwidth is available, the (non-optimized) version of our protocol already implements the switching procedure with negligible overhead in the message flow. However, the experimental results provided in Section 4 showed that during the switching phase, when both protocols are being used to broadcast the same set of messages, the available bandwidth can be exhausted when the send rate and/or message payload is too high.

To overcome this problem we now describe an optimization to reduce the amount of data being transmitted by the adaptive protocol during this phase. The optimization consists of broadcasting using the first (and current) algorithm only the identifiers of the messages being transmitted. The messages payload is only transmitted using the second algorithm. In this manner, the amount of redundant information transmitted over the wire is reduced substantially. This optimization has a minor drawback: the protocol cannot deliver a message to the application until it is received by both total order algorithms. However, since both algorithms are executed in parallel, the impact of this feature is negligible.

Figure 2 shows that the optimization allows the protocol to continue increasing its throughput after the point where the non-optimized version stabilizes (approximately 400 msg/s), showing a behavior similar to the non-adaptive protocol (note that the lines for the optimized and the non-adaptive algorithms overlap in the figure).

6 On Failure Detection

To simplify the description of our protocol, in Section 3 we have not addressed the issue of failure detection during the switching protocol. Namely, we have stated that the protocol moves to the sanity step when it receives a flag from *every* participant (Figure 1, line 16). Without further changes, the protocol would simply block in the presence of a single failure. We now discuss how our protocol can be adapted to operate in the presence of faults. Our algorithm can operate in asynchronous systems augmented with failure detectors [19].

We start by discussing the operation of the protocol in a system augmented with a *Perfect Failure Detector* ($\mathcal{P}$) [19], i.e., a system where processes fail by crashing and crashes can be accurately detected by all correct processes. In this model, the transition condition should be set to "a flag is received by all *correct* processes". This model is actually used in all of our implementations, where the failure detection is encapsulated by a view-synchronous interface [20].

The protocol can also be modified to operate in an asynchronous system augmented with an unreliable failure detector (such as the $\diamond S$ failure detector proposed in [19]) as long as a majority of processes do not fail (naturally, in this case, the underlying total order algorithms, must also be designed for such a model). In this model, the transition condition should be set to "a flag is received by a majority of processes". However, in this configuration, correct processes that do not belong to the majority may be required to retransmit some messages. It is interesting to observe that the strategy proposed before for the $\mathcal{P}$ detector (perform the switch when a flag is received from *all* correct processes) and the strategy proposed in [16] (perform the switch when the *first* flag is received) can be seen as extreme point of a spectrum. Between these extreme cases, there is a range of alternative switching points, from which the "majority of processes" is the one that ensures less disruption in $\diamond S$ model.

7 Conclusions and Future Work

Several total order protocols exist that use quite different strategies. Such strategies may perform better in specific environments and/or working conditions. We presented an adaptive total order protocol that is able to switch in run-time between different total order algorithms. When the environment is dynamic, this allows the system to use the ordering strategy that is most favorable.

If one is not careful, the procedure to switch between algorithms can disrupt the message flow. Our work tackles this issue by proposing a novel switching strategy that performs the reconfiguration with negligible impact on the observed delivery rate. Evaluation results of an implementation of the protocol showed performance improvements in regard to competing approaches.

Planned future work on this subject will aim at embedding the resulting protocol in a database replication service based on the state machine approach [4].

References

1. Powell, D., ed.: Special Issue on Group Communication. Number 4 in 39. In: Communications of the ACM. ACM (1996) 50–97
2. Guerraoui, R., Schiper, A.: Software-based replication for fault tolerance. IEEE Computer **30**(4) (1997) 68–74

3. Défago, X., Schiper, A., Urbán, P.: Total order broadcast and multicast algorithms: Taxonomy and survey. ACM Computing Surveys **36**(4) (2004) 372–421
4. Schneider, F.B.: Implementing fault-tolerant services using the state machine approach: a tutorial. ACM Computing Surveys **22**(4) (1990) 299–319
5. Kaashoek, M., Tanenbaum, A.: Group communication in the Amoeba distributed operating system. In: Proceedings of the 11th International Conference on Distributed Computing Systems, IEEE (1991) 222–230
6. Peterson, L., Buchholz, N., Schlichting, R.: Preserving and using context information in interprocess communication. ACM Transactions on Computer Systems **7**(3) (1989) 217–146
7. Dolev, D., Kramer, S., Malki, D.: Early delivery totally ordered multicast in asynchronous environments. In: Digest of Papers, The 23th International Symposium on Fault-Tolerant Computing, IEEE (1993) 544–553
8. Chang, J., Maxemchuck, N.: Reliable broadcast protocols. ACM, Transactions on Computer Systems **2**(3) (1984)
9. Lamport, L.: Time, clocks and the ordering of events in a distributed system. Communications of the ACM **21**(7) (1978) 558–565
10. Birman, K., Joseph, T.: Reliable communication in the presence of failures. ACM, Transactions on Computer Systems **5**(1) (1987)
11. Rodrigues, L., Fonseca, H., Veríssimo, P.: Totally ordered multicast in large-scale systems. In: Proceedings of the 16th International Conference on Distributed Computing Systems, Hong Kong, IEEE (1996) 503–510
12. Rodrigues, L., Mocito, J., Carvalho, N.: From spontaneous total order to uniform total order: different degrees of optimistic delivery. In: Proceedings of the 21st ACM symposium on Applied computing (SAC'06) *(to appear)*, ACM Press (2006)
13. Liu, X., van Renesse, R.: Fast protocol transition in a distributed environment. In: Proceedings of the 19th ACM Conference on Principles of Distributed Computing (PODC 2000), Portland, OR (2000) 341
14. Chen, W.K., Hiltunen, M., Schlichting, R.: Constructing adaptive software in distributed systems. In: ICDCS '01: Proceedings of the The 21st International Conference on Distributed Computing Systems, Washington, DC, USA, IEEE Computer Society (2001) 635
15. van Renesse, R., Birman, K., Hayden, M., Vaysburd, A., Karr, D.: Building adaptive systems using Ensemble. Software: Practice and Experience **28**(9) (1998) 963–979
16. Rutti, O., Wojciechowski, P., Schiper, A.: Structural and algorithmic issues of dynamic protocol update. In: Proceedings of the 20th IEEE International Parallel and Distributed Processing Symposium (IPDPS'06), IEEE (2006)
17. Miranda, H., Pinto, A., Rodrigues, L.: Appia, a flexible protocol kernel supporting multiple coordinated channels. In: Proceedings of the 21st International Conference on Distributed Computing Systems, Phoenix, Arizona, IEEE (2001) 707–710
18. Nicol, D., Liu, J., Liljenstam, M., Yan, G.: Simulation of large-scale networks using SSF. In: Proceedings of the 2003 Winter Simulation Conference. (2003)
19. Chandra, T., Toueg, S.: Unreliable failure detectors for reliable distributed systems. Journal of the ACM **43**(2) (1996) 225–267
20. Birman, K., Joseph, T.: Exploiting virtual synchrony in distributed systems. Technical Report 87-811, Department of Computer Science, Cornell University, Ithaca, New York (1987)

On Greedy Graph Coloring in the Distributed Model

Adrian Kosowski and Łukasz Kuszner*

Department of Algorithms and System Modeling,
Gdańsk University of Technology, Poland

Abstract. In the paper we consider distributed algorithms for greedy graph coloring. For the largest-first (**LF**) approach, we propose a new distributed algorithm which is shown to color a graph in an expected time of $O(\Delta \log n \log \Delta)$ rounds, and we prove that any distributed **LF**-coloring algorithm requires at least $\Omega(\Delta)$ rounds. We discuss the quality of obtained colorings in the general case and for particular graph classes. Finally, we show that other greedy graph coloring approaches, such as smallest-last (**SL**) or dynamic-saturation (**SLF**), are not suitable for application in distributed computing, requiring $\Omega(n)$ rounds.

1 Introduction

Problem definition. We discuss the vertex coloring problem in a *distributed network*. Such a network consists of a set V of processors and a set E of bidirectional communication links between pairs of processors. It can be modeled by an undirected graph $G = (V, E)$. We denote $n = |V|$, $m = |E|$ and for each vertex v define its *open neighborhood* $N(v) = \{u : \{u,v\} \in E\}$ and *vertex degree* $\deg_G v = |N(v)|$. In order to distinguish neighbours of higher degree, we will use the symbols $N_>(v) = \{u \in N(v) : \deg(u) > \deg(v)\}$ and similarly $N_\geq(v) = \{u \in N(v) : \deg(u) \geq \deg(v)\}$.

To *color* the vertices of G means to give each vertex a positive integer color value in such a way that no two adjacent vertices get the same color. If at most k colors are used, the result is called a *k-coloring*. In many practical considerations, such as code assignment in wireless networks [1], it is desirable to minimise the number of used colors. The smallest possible positive integer k for which there exists a k-coloring of G is called the *chromatic number* $\chi(G)$. This value is bounded from above by $\Delta + 1$, where Δ denotes the maximum vertex degree of the graph.

Model of computation. We assume the common model used widely in previous research on the subject [3,11,16]. Moreover, we assume neither any global parameters known *a priori* for any vertex in a graph, nor unique identifiers. We allow each vertex of the graph to know only its own local state and local states

---

* Research supported by the State Committee for Scientific Research (Poland) Grant No. 4 T11C 047 25.

of neighboring vertices. To measure time complexity we use the number of *synchronized rounds* as such a measure is used in most cited material, even though the algorithms discussed here may be adapted for the asynchronous model.

When evaluating the performance of a random distributed coloring algorithm A on a graph G there are at least two random variables of interest: $C_A(G)$, the number of colors used by the algorithm to color the graph G, and $T_A(G)$, the number of rounds used to color G.

A good distributed algorithm is one where $C_A(G)$ is close to $\chi(G)$, the chromatic number of G, and where $T_A(G)$ is small relative to the number of vertices in G. The difference $C_A(G) - \chi(G)$ can be viewed as a measure of the effectiveness of the algorithm. It has to be remembered though, that in the general case approximating $\chi(G)$ within a factor of $n^{1/7-\varepsilon}$ is an NP-hard problem, for any $\varepsilon > 0$ [2].

1.1 Preliminaries: Greedy Graph Coloring in a Distributed Context

For a given graph G and the sequence of vertices $K = (v_1, v_2, \ldots, v_n)$, we will use the term *greedy coloring* to describe the following procedure of color assignment:

algorithm Greedy-Color(G, K):
  for $v := v_1$ **to** v_n **do**
    give vertex v the smallest possible color;

Definition 1. *A sequential coloring algorithm is an algorithm which determines a sequence K of vertices of G, and then colors G using the procedure* Greedy-Color(G, K).

Below we briefly recall the basic principles of the most common sequential algorithms; for a more detailed analysis of sequential coloring see [8,9].

- S algorithm: no assumptions are made concerning sequence K.
- LF algorithm: sequence K is formed by arranging the vertices of graph G in non-ascending order of degrees,
- SL algorithm: sequence K is formed by iteratively removing a vertex of minimal degree from the graph and placing it at the end of K.
- SLF (DSATUR) algorithm: sequence K is formed by dynamically arranging the vertices of graph G in non-ascending order of saturation degrees, where the saturation degree is the number of neighboring vertices which have already been colored (ties are broken by choosing the vertex of greater degree).

For some sequential algorithm A, we will call a coloring of a graph an A-*coloring* if it may be obtained by coloring the graph greedily using a sequence of vertices K legal for algorithm A. In particular, it is easy to observe that an S-coloring of graph G is equivalent to a *Grundy coloring* [5] of G, i.e. such a coloring, that no single vertex may have its color value decreased without affecting the color of some other vertex. All other sequential algorithms also produce Grundy colorings and may also have other, stronger properties (see [9,14,17] for an extensive characterisation).

Definition 2. *A distributed graph coloring algorithm* DA *is said to be a distributed implementation of sequential algorithm* A *(or simply: a distributed A-coloring algorithm) if all possible results of algorithm* DA *are correct* A*-colorings.*

1.2 Summary of Main Results

State-of-the-art results. For the general graph coloring problem some extremely fast algorithms have been described. Linial in [11] gave an algorithm working in $O(\log^* n)$ time but using $O(\Delta^2)$ colors. This result was improved later on by De Marco and Pelc [13]. The algorithm given in the paper uses $O(\Delta)$ colors and $O(\log^*(n/\Delta))$ rounds, but local computations are not even polynomially bounded. On the other hand, a very simple algorithm for coloring arbitrary graphs with $(\Delta + 1)$ colors was given by Johansson [7]. It was proved to run in $O(\log n)$ time, but the number of colors used by the algorithm is close to Δ even if the graph is bipartite. This is not surprising, since Johansson's algorithm has no mechanism for economizing on the number of colors. Further improvements were proposed in [4]. In that paper a very similar technique was used to compute a coloring of triangle-free graphs using $O(\Delta/\log \Delta)$ colors, but the algorithm can fail for some instances of the problem.

To the best of our knowledge, the first greedy distributed approach to graph coloring was studied by Panconesi and Rizzi [15]. The authors used a forest decomposition technique to achieve $O(\Delta^2 \log^* n)$ time. Recently an algorithm motivated by sequential LF-coloring was described in [6]. Analysis shows that it runs in $O(\Delta^2 \log n)$ time. However, it sometimes leads to colorings which are not LF-colorings, so the described algorithm is not a distributed implementation of the LF algorithm according to the Definition 2.

Our contribution. In Section 2 we describe a new approach to distributed LF-coloring, showing an algorithm based on iterated maximal independent set construction with $O(\Delta \log n \log \Delta)$ expected runtime, which is shown to be nearly optimal and improves earlier results from [6,15]. In Subsection 2.3 we briefly discuss the quality of colorings obtained using the proposed algorithm and compare it to its sequential counterpart. In Section 3 we show that every distributed implementation of the LF algorithm requires $\Omega(\Delta)$ time, whereas the SL and SLF approaches may in fact for some graphs require $\Omega(n)$ rounds to color.

2 A Distributed Implementation of the LF Algorithm

Before discussing the details of the distributed implementation of the LF algorithm, we present an equivalent characterization of a correct LF-coloring. For a given coloring of G, let $IS_{(d,c)} \in V$ denote the independent set of vertices of G of degree d and colored with color c.

Lemma 1. *An assignment of colors to G is a correct* **LF***-coloring iff for all $1 \leq d, c \leq \Delta + 1$ the set $IS_{(d,c)}$ is a maximal independent set in the subgraph $H_{(d,c)}$ of G induced by the set of vertices $\left(\bigcup_{c_i \geq c} IS_{(d,c_i)}\right) \setminus N\left(\bigcup_{d_i > d} IS_{(d_i,c)}\right)$.*

Proof. ($\Rightarrow$) Consider a coloring obtained by an arbitrary sequence of the **LF** algorithm. Clearly, $IS_{(d,c)}$ is an independent set. By contradiction, suppose that the set $IS_{(d,c)}$ is not maximal and can be extended by some vertex v of $H_{(d,c)}$. This implies that v is of degree $\deg_G v = d$ and has some color $c(v) > c$, which means that it is adjacent in G to a vertex u previously colored by the **LF** algorithm with color c, i.e. such that $v \in N(u)$ and $\deg_G u \geq d$. Hence we either have $v \in N(IS_{(d,c)})$ or $v \in N\left(\bigcup_{d_i > d} IS_{(d_i,c)}\right)$, a contradiction.

($\Leftarrow$) It suffices to observe that if a coloring of G fulfills the right hand side of the lemma, then it may be obtained by using the **LF** algorithm with the sequence of vertices: $K = (IS_{(\Delta,1)}, IS_{(\Delta,2)}, \ldots, IS_{(\Delta,\Delta+1)}, IS_{(\Delta-1,1)}, IS_{(\Delta-1,2)}, \ldots, IS_{(\Delta-1,\Delta)},$ $\ldots, IS_{(1,1)}, IS_{(1,2)})$, where the elements of each independent set may be enumerated in arbitrary order. □

We now propose a distributed algorithm in which each vertex v is characterised by three principal local state variables: $c(v)$ which will store the color of vertex v at the end of the coloring, $d(v)$ which constantly stores the degree of v in G, and a binary flag $f(v)$ which specifies whether v has already reached its final color. Using the terminology from Lemma 1, at any point of the execution of the algorithm we classify the vertices of G into three categories, depending on the information currently available to the vertex from its own local state and the local states of its neighbours:

- v is *correctly colored*, if v can determine that it will belong to $IS_{(d(v),c(v))}$ at the end of the coloring,
- v is *actively uncolored*, if v can determine that it will not belong to any of the sets $IS_{(d(v),1)}, \ldots, IS_{(d(v),c(v)-1)}, N\left(\bigcup_{d_i > d} IS_{(d_i,c)}\right)$, but cannot infer whether it will belong to set $IS_{(d(v),c(v))}$ at the end of the coloring.
- v is *passively uncolored*, if v can determine that it will not belong to any of the sets $IS_{(d(v),1)}, \ldots, IS_{(d(v),c(v)-1)}$, but cannot infer whether it will belong to the set $N\left(\bigcup_{d_i > d} IS_{(d_i,c)}\right)$ at the end of the coloring.

Theorem 1. *There exists a distributed graph coloring algorithm using local state variables $c(v), d(v), f(v)$, such that at any stage of execution each vertex belongs to exactly one of three categories: correctly colored, actively uncolored and passively uncolored. Moreover, a correctly colored vertex will remain correctly colored throughout the rest of the execution.*

Proof (sketch). Let us assume the interpretation of the state variables as in the earlier description. Initially, let $c(v) := 1$ and $f(v) := false$. The value $f(v)$ will be set to *true* when vertex v becomes correctly colored. Let us assume that throughout the algorithm the value $c(v)$ will never decrease and may only increase when $f(v) = false$ and it is certain that v will not belong to set $IS_{(d(v),c(v))}$.

We will show that under these assumptions it is possible to construct an algorithm such that a vertex v with $f(v) = false$ is actively uncolored if there does not exist a vertex $u \in N_>(v)$ such that $c(u) \leq c(v)$ and $f(u) = \mathit{false}$, or passively uncolored in the opposite case. Indeed, it suffices that the algorithm repeatedly performs the following two actions in successive rounds:

1. For each vertex v, if there exists a correctly colored vertex $u \in N_\geq(v)$ such that $c(u) = c(v)$, increase $c(v)$ by 1 and repeat the step if necessary.
2. For each actively uncolored vertex v, attempt to include v in the independent set $IS_{(d(v),c(v))}$. If successful, mark v as correctly colored ($f(v) := \mathit{true}$).

The inclusion of v in the independent set $IS_{(d(v),c(v))}$ performed in Action 2 may only fail if two neighbouring vertices attempt to join the same set simultaneously, thus implying the need for a separate tie-breaking mechanism. It is easy to see that Action 1 is performed only for actively uncolored vertices directly after losing a tie in action 2 and for passively uncolored vertices. Simple inductive reasoning shows that the earlier assumed condition for identifying passively and actively uncolored vertices is indeed correct, which completes the proof. □

Assuming that the actions of the algorithm presented in the proof of Theorem 1 are understood as rounds in the distributed model, we observe that during action 2 all actively uncolored vertices with the same value of variables d and c attempt to join the same independent set $IS_{(d,c)}$. For instance, directly after the initialisation of the algorithm the set of actively uncolored vertices is equal to the set of vertices of degree Δ, all of which attempt to join independent set $IS_{(\Delta,1)}$. The number of such vertices may be arbitrarily large (even equal to n in the case of Δ-regular graphs), thus necessitating an efficient approach to the distributed independent set problem, described in detail in Subsection 2.1.

2.1 Tie-Breaking in the Distributed Independent Set Problem

The problem of constructing a maximal independent set IS by adding subsets of candidate vertices in successive rounds, encountered in the proof of Theorem 1, has no deterministic solution in the distributed model. The first efficient probabilistic approach was proposed by Luby [12] for a parallel processing system, but due to its nature the algorithm may also be applied in a distributed setting. For a given graph G, let $S \subseteq V$ denote an independent set of vertices and let $P_i \subseteq V \setminus (S \cup N(S))$ be the set of candidates for inclusion into S in the i-th stage of the algorithm. The algorithm divides set P_i into three disjoint subsets, $P_i = W_i \cup N(W_i) \cup P_{i+1}$, where W_i denotes the independent set of vertices merged with S at the end of the stage (known as *winners*, $S := S \cup W_i$), $N(W_i)$ is the neighborhood of W_i which will never enter S (known as *losers*), and P_{i+1} is the set of candidates remaining for later consideration. The process continues until for some k we have $P_k = \emptyset$, then $IS := S$ is a maximal independent set, with respect to the set of all candidate vertices.

The details of the i-th stage of the algorithm may be written as follows. Let H_i denote the subgraph of G induced by set P_i and let E_i be its edge set. First, each vertex $v \in P_i$ is either assigned local state value $r(v) = 0$ and transferred to P_{i+1} with probability $1 - 1/(2\deg_{H_i} v)$, or *contends* for a place in W_i with probability $1/(2\deg_{H_i} v)$. Next, each of the contending vertices v draws a random number with uniform distribution. A contending vertex v becomes a winner if $r(v) > \max_{u \in N(v)} r(u)$, and becomes a loser in the opposite case. Luby showed that the described algorithm fulfills the following property.

Theorem 2 ([12]). *Let D_i denote the random variable given as the ratio $D_i = E_{i+1}/E_i$. Then the mean value of D_i is bounded by $\mathrm{E}[D_i] \leq 7/8$.*

Observe that for obvious reasons the value of variable D_i lies within the range $D_i \in [0, 1]$. Consequently, from the above theorem we obtain the following conclusion.

Corollary 1. *In each stage of the algorithm, the number of edges in the candidate set decreases by not less than $1/16$-th part with probability at least $1/15$, $\Pr[D_i \geq 1/16] \geq 1/15$.*

Let us now consider the random variable T describing the number of rounds performed by the maximum independent set algorithm before its completion. Let random variable L_i be defined as $L_i = \log_{16/15} E_i$. Initially, $L_0 = \log_{16/15} E_0 \in O(\log m) = O(\log n)$. In each stage of the algorithm (which may easily be implemented in the form of three rounds), the value $L_i - L_{i+1}$ is always non-negative and, by Corollary 1, not less than 1 with probability at least $1/15$. Hence we obtain the following statement.

Corollary 2. *The number of rounds T performed before the candidate set is empty has a probability distribution with mean value $\mathrm{E}[T] \in O(\log n)$ and probability mass function $f_T(x)$, bounded from above for $x \in \Omega(\log n)$ by that of the negative binomial distribution with a probability parameter of $1/15$.*

In further analysis it is important to remember that the probability distribution of variable T is understood in terms of randomly drawn local variables, and is independent of the structure of sets S and P_i.

2.2 An Algorithm for LF-Coloring in $O(\Delta \log n \log \Delta)$ Rounds

A formal description of the proposed LF-coloring algorithm is obtained by combining the approach from the proof of Theorem 1 with the results of considerations from Subsection 2.1. For the values $1 \leq d, c \leq \Delta + 1$ we will in parallel be constructing the maximal independent sets $IS_{(d,c)}$. At a given stage of construction of set $IS_{(d,c)}$, the set $S_{(d,c)}$ of known elements will consist of vertices v having $c(v) = c, d(v) = d, f(v) = true$, while the set $P_{(d,c)}$ of candidates will consist of actively uncolored vertices having $c(v) = c, d(v) = d, f(v) = false$. The complete pseudocode of distributed LF-coloring algorithm DLF is given below. Implementation details of independent set tie-breaking follow Subsection 2.1, and we assume that function $\mathrm{rnd}[a, b]$ returns an integer with uniform distribution from the range $[a, b]$. Note that when implementing the process of contention for independent set $IS_{(d,c)}$, at most d neighbours contend for one place in the independent set, thus it is sufficient to assume $[0, d^4]$ as the range from which random values $r(v)$ are drawn, without escalating the number of ties.

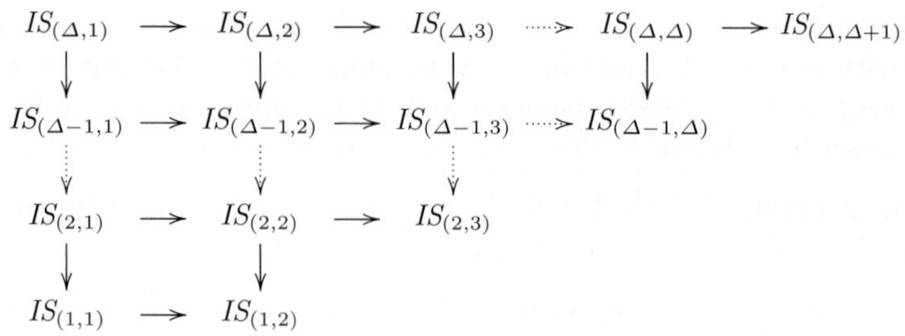

$$IS_{(\Delta,1)} \rightarrow IS_{(\Delta,2)} \rightarrow IS_{(\Delta,3)} \dashrightarrow IS_{(\Delta,\Delta)} \rightarrow IS_{(\Delta,\Delta+1)}$$
$$\downarrow \qquad \downarrow \qquad \downarrow \qquad \downarrow$$
$$IS_{(\Delta-1,1)} \rightarrow IS_{(\Delta-1,2)} \rightarrow IS_{(\Delta-1,3)} \dashrightarrow IS_{(\Delta-1,\Delta)}$$
$$\vdots \qquad \vdots \qquad \vdots$$
$$IS_{(2,1)} \rightarrow IS_{(2,2)} \rightarrow IS_{(2,3)}$$
$$\downarrow \qquad \downarrow$$
$$IS_{(1,1)} \rightarrow IS_{(1,2)}$$

Fig. 1. Illustration of worst-case time ordering of independent set construction

algorithm DLF(G):
Round 0:
  $f(v) := \textit{false}$; $d(v) = \deg_G(v)$;
Round $3k + 1$:
  if $f(v) = \textit{false}$
    then while $\exists_{u \in N_{\geq}(v)}(c(u) = c(v) \wedge f(u) = \textit{true})$
      do $c(v) := c(v) + 1$;
Round $3k + 2$:
  $r(v) := 0$;
  if $f(v) = \textit{false} \wedge c(v) < \min_{u \in N_{>}(v)}\{c(u) : f(u) = \textit{false}\}$
    then if $1 = \text{rnd}[1, 2 \cdot |\{u \in N(v) : d(u) = d(v) \wedge c(u) = c(v)\}|]$
      then $r(v) := \text{rnd}[0, d(v)^4]$;
Round $3k + 3$:
  if $r(v) > \max_{\{u \in N(v) : d(u) = d(v) \wedge c(u) = c(v)\}} r(u)$
    then $f(v) := \textit{true}$;

Theorem 3. *Algorithm* DLF *determines an* LF*-coloring of G in $O(\Delta \log n \log \Delta)$ rounds.*

Proof. The proof of correctness is complete when we observe that by Lemma 1 finding a correct LF-coloring of G is equivalent to determining maximal independent sets $IS_{(\Delta,1)}, IS_{(\Delta,2)}, \ldots, IS_{(\Delta,\Delta+1)}, IS_{(\Delta-1,1)}, IS_{(\Delta-1,2)}, \ldots, IS_{(\Delta-1,\Delta)}, \ldots, IS_{(1,1)}, IS_{(1,2)}$. It is a direct conclusion from Theorem 1 that the DLF algorithm does indeed determine these independent sets through the local variables $(d(v), c(v))$ of the vertices.

Careful analysis of algorithm DLF shows that for any fixed d and c, the process of construction of set $IS_{(d,c)}$ is dependant only on the construction of sets $IS_{(d_i,c_i)}$, for $d_i \geq d, c_i \leq c$. Moreover, vertices once added to an independent set are never removed from it. Without any time gain we may therefore assume that the construction of independent set $IS_{(d,c)}$ starts directly after the construction of sets $IS_{(d+1,c)}$ and $IS_{(d,c-1)}$ is complete. This would result in a time dependency diagram as shown in Figure 1, where a pointer denotes flow of control after completion of the preceding action. Let $T_{(d,c)}$ denote the random variable describing the number of rounds used for the construction of set $IS_{(d,c)}$. From Figure 1 it is easy to observe that the anticipated completion time T_{DLF} of the DLF algorithm is bounded by the expression:

$$T_{\mathsf{DLF}} \leq (\Delta + 1) \cdot \mathrm{E}\left[\max_{1 \leq d, c \leq \Delta+1} T_{(d,c)}\right]$$

However, a characterisation of the mass function of the distribution of $T_{(d,c)}$ is given by Corollary 2. Moreover, this distribution remains the same regardless of the nature of the constructed independent set, thus making the family of variables $T_{(d,c)}$ pairwise independent. By bounding the negative binomial distribution from above by the exponential distribution and performing a number of technical transformations (which we leave out), we obtain the following result:

$$\mathrm{E}\left[\max_{1 \leq d, c \leq \Delta+1} T_{(d,c)}\right] \in O(\log n \log \Delta)$$

Thus, we may finally write $T_{\mathsf{DLF}} \in O(\Delta \log n \log \Delta)$, which completes the proof. □

2.3 Quality Characteristics of Distributed LF-Colorings

As a natural consequence of Lemma 1, algorithm DLF produces worst-case colorings which never use more colors than the worst-case colorings given by a sequential implementation of LF. It is for instance known that any LF-coloring is optimal or near optimal for numerous graph classes, e.g. complete k-partite graphs, caterpillars, crowns, bipartite wheels [9]; as a result, algorithm DLF also performs well for all these graph classes. As a matter of fact, it is easy to observe that the worst case performance of DLF is exactly the same as that of LF, by the following fact (which we leave without proof).

Corollary 3. *A coloring of graph G is an* LF*-coloring iff it is a* DLF*-coloring.*

However, the sequential LF and distributed DLF algorithms may have a different probability of achieving a given coloring, thus affecting their average-case performance. Here we confine ourselves to an experimental comparison of the average number of colors used by LF and DLF for random graphs of different order, edge density and average vertex degree, the results of which are presented in Table 1. It can be clearly seen that the number of colors used by both algorithms is nearly identical, though as a rule marginally smaller for the sequential algorithm. Both LF and DLF clearly outperform all non-greedy algorithms based on the assignment of random colors from the range $[1, \Delta + 1]$.

3 Complexity Bounds on Distributed Greedy Coloring

We will now show that the most popular sequential coloring algorithms impose strong lower bounds on the expected computational time in a distributed setting. First, consider the SL and SLF algorithms. Both of them exactly color paths

Table 1. An experimental average-case comparison of the sequential and distributed LF-coloring algorithms. The tests were conducted for a sample of 100 uniform edge probability random graphs of fixed order n and edge density $\varphi = m/\binom{n}{2}$ (left table) or mean vertex degree $\varrho = 2m/n$ (right table), each of which was colored 10 times.

n	φ	Δ	C_{LF}	C_{DLF}	T_{DLF}	n	ϱ	Δ	C_{LF}	C_{DLF}	T_{DLF}
100	0.001	1.21	2.00	2.00	9.67	1000	4	11.79	4.63	4.68	22.87
	0.005	2.81	2.09	2.11	12.14		20	35.83	10.06	10.08	44.47
	0.025	7.00	3.42	3.48	15.68		100	132.38	29.49	29.61	117.57
500	0.001	3.63	2.44	2.52	15.57	5000	4	13.23	4.99	5.00	26.71
	0.005	8.43	3.95	3.97	19.21		20	38.53	10.25	10.30	48.88
	0.025	24.24	7.67	7.71	33.73		100	138.22	29.28	29.30	127.54
2500	0.001	9.61	4.00	4.00	22.59	25000	4	14.44	5.00	5.00	29.92
	0.005	26.51	7.99	8.00	37.67		20	41.51	10.73	10.88	51.40
	0.025	91.73	21.07	21.02	91.21		100	142.71	29.82	29.82	134.81

and rings, hence all distributed implementations of SL and SLF have the same property. Linial [11] proved that the exact coloring of a ring requires $\Omega(n)$ rounds, so we obtain the following conclusion.

Corollary 4. *Any distributed implementation of* SL *or* SLF *requires* $\Omega(n)$ *time.*

Now, let us consider the LF algorithm. We will show that any distributed implementation of LF algorithm requires $\Omega(\Delta)$ rounds. To achieve this, we construct a family of graphs G_d with $\text{diam}(G_d) = d$ and $\Delta(G) = 2d$. A representative of such a family is depicted in Figure 2. Vertices $v_0, v_1, \ldots, v_d$ induce a path of length d. Some additional components are connected to particular vertices to ensure that vertex v_i obtains a color $d - i + 1$ in each LF-coloring of G_d, that is component K_r depicts a complete graph with r vertices and a bold line between such a component and v_i illustrate that each of the vertices of K_r is connected to v_i. Similarly, when two components K_{r_1} and K_{r_2} are connected, each vertex from K_{r_1} is connected to each vertex from K_{r_2}, thus forming a clique $K_{r_1+r_2}$. We have $\deg v_i = d + i$ and $|N_>(v_i)| = d - i$, hence it is easy to observe that the only possible color for v_i in any LF-coloring is $d - i + 1$. However, if vertex v_d were to be removed from the graph, the colors of all other vertices of the path would decrease by 1. Thus we have shown that color of v_0 depends on the value of the length of the path, $d \in \Omega(\Delta)$. As information in our model can propagate only at the speed of one vertex per round we have the following.

Corollary 5. *Any distributed implementation of* LF *requires* $\Omega(\Delta)$ *time.*

Final conclusions. Taking into account the above corollaries, the proposed $O(\Delta \log n \log \Delta)$ implementation of LF presented in Section 2 may be considered not far from optimal among distributed LF-coloring algorithms, and to have lower complexity than any possible implementation of an SL-coloring or SLF-coloring algorithm. For graphs of bounded degree, the proposed distributed LF-coloring algorithm is, to the best of our knowledge, the first of the well known graph coloring heuristics running in $O(\log n)$ rounds.

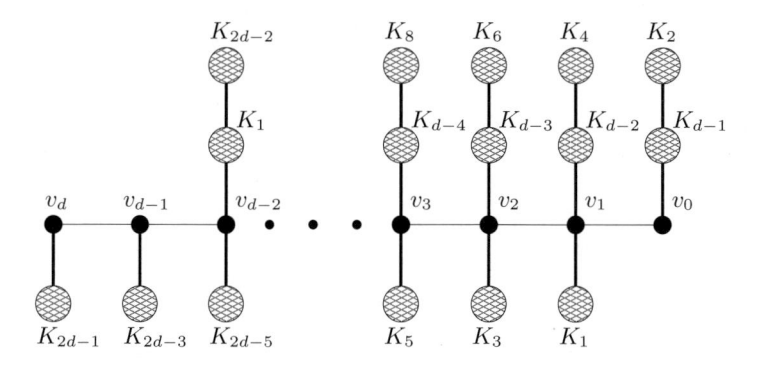

Fig. 2. A graph which requires $\Omega(\Delta)$ time to LF-color in the distributed model

References

1. Battiti, R., Bertossi, A. A., and Bonuccelli, M. A.: Assigning codes in wireless networks. Wireless Networks **5** (1999) 195–209.
2. Bellare, M., Goldreich, O., and Sudan, M., Free bits, PCPs and non-approximability — towards tight results, SIAM J. Comp. **27** (1998), 804–915.
3. Chaudhuri, P.: Algorithms for some graph problems on a distributed computational model. Information Sciences **43** (1987), 205–228.
4. Grable, D. A., and Panconesi, A.: Fast distributed algorithms for Brooks-Vizing colorings. J. Algorithms **37** (2000) 85–120.
5. Grundy, P. M.: Mathematics and games. Eureka **2** (1939), 6–8.
6. Hansen, J., Kubale, M., Kuszner, Ł. and Nadolski, A.: Distributed largest-first algorithm for graph coloring. Proc. Euro-Par, LNCS **3149** (2004), 804–811.
7. Johansson, Ö.: Simple distributed $(\Delta + 1)$-coloring of graphs. Inf. Process. Lett. **70** (1999) 229–232.
8. Kosowski, A. and Manuszewski, M.: Classical Coloring of Graphs. In: *Graph Colorings*, AMS Contemporary Math. 352 (2004), Providence, USA, 1–20.
9. Kubale, M.: Introduction to Computational Complexity and Algorithmic Graph Coloring. GTN (1998), Gdańsk, Poland.
10. Kubale, M. and Kuszner, Ł.: A better practical algorithm for distributed graph coloring. Proc. PARELEC (2002) 72–75.
11. Linial, N.: Locality in distributed graph algorithms. SIAM J. Comput. **21** (1992) 193–201.
12. Luby, M.: A simple parallel algorithm for the maximal independent set problem. SIAM J. Comput. **15** (1986) 1036–1053.
13. De Marco, G. and Pelc, A.: Fast distributed graph coloring with $O(\Delta)$ colors. Proc. SODA (2001) 630–635.
14. Olariu, S. and Randall, J.: Welsh-Powell opposition graphs. Inf. Process. Lett. **31** (1989) 43–46.
15. Panconesi, A. and Rizzi, R.: Some simple distributed algorithms for sparse networks. Distributed Computing **14** (2001), 97–100.
16. Panconesi, A. and Srinivasan, A.: Improved distributed algorithms for coloring and network decomposition problems. Proc. STOC (1992) 581–592.
17. Turner, J. S.: Almost all k-colorable graphs are easy to color. J. Algorithms **9** (1988) 63–82.

Topic 9: Parallel Programming: Models, Methods and Languages

José C. Cunha, Sergei Gorlatch, Daniel Quinlan, and Peter H. Welch

Topic Chairs

This topic provides a forum for the presentation of research results and practical experience in the development of parallel programs. Advances in algorithmic and programming models, design methods, languages, and interfaces are needed to produce correct, portable parallel software with predictable performance on different parallel and distributed architectures.

The topic emphasizes results that improve the process of developing high-performance programs, including high-integrity programs that are scalable with both problem size and complexity. Of particular interest are novel techniques by which parallel software can be assembled from reusable parallel components without compromising efficiency. Related to this is the need for parallel software to adapt, both to available resources and to the problem being solved.

This year, 13 papers were submitted to this topic. Each paper was reviewed by four reviewers and, finally, we were able to select 7 papers. Globally, the accepted papers discuss methods and programming language constructs to promote the development of correct and efficient parallel programs.

The approaches based on higher-order skeletons are discussed in two papers, for computations on two-dimensional arrays, and for dynamic task farming. Data parallel programming is discussed in another paper concerning the automatic parallelisation of "for-each" loops for grid and tree algorithms. Shared memory parallel programming is discussed in two papers that propose extensions to OpenMP, for handling irregular parallel algorithms, and for improved control and synchronisation of multiple threads. Improved support for multithreading models is also discussed in another paper that proposes a methodology towards more efficient memory management for threading libraries that are based on non-preemptive models. Distributed termination detection is discussed in a paper that proposes the concept of "partial quiescence" as a construct of a distributed programming language.

We would like to thank all the authors who submitted papers to this topic, and the external referees, for their contribution to the success of this conference.

Surrounding Theorem: Developing Parallel Programs for Matrix-Convolutions

Kento Emoto, Kiminori Matsuzaki, Zhenjiang Hu, and Masato Takeichi

Department of Mathematical Informatics,
University of Tokyo
{emoto, kmatsu, hu, takeichi}@ipl.t.u-tokyo.ac.jp

Abstract. Computations on two-dimensional arrays such as matrices and images are one of the most fundamental and ubiquitous things in computational science and its vast application areas, but development of efficient parallel programs on two-dimensional arrays is known to be hard. To solve this problem, we have proposed a skeletal framework on two-dimensional arrays based on the theory of constructive algorithmics. It supports users, even with little knowledge about parallel machines, to develop systematically both correct and efficient parallel programs on two-dimensional arrays. In this paper, we apply our framework to the matrix-convolutions often used in image filters and difference methods. We show the efficacy of the framework by giving a general parallel program for the matrix-convolutions described with the skeletons, and a theorem that optimizes the general program into an application-specific one.

1 Introduction

Computations on two-dimensional arrays, such as matrix computations, image processing, and difference methods, are both fundamental and ubiquitous in scientific computations and other application areas [7, 15, 11]. However, development of efficient parallel programs on two-dimensional arrays is known to be a hard task due to the necessity of considering data allocation, synchronization and communication between processors. *Skeletal parallel programming* is one promising solution to the situation [5, 16]. In this model, users build parallel programs by composing ready-made components (called *skeletons*) implemented efficiently in parallel for various parallel architectures. Since low-level parallelism is concealed in the skeletons, users can obtain a comparatively efficient parallel program without needing technical details of parallel computers or being conscious of parallelism explicitly.

We have proposed a skeletal framework on two-dimensional arrays [9], based on the theory of constructive algorithmics (also known as *Bird-Meertens Formalism*) [2, 4]. Our framework provides users, even with little knowledge about parallel machines, with a concise way to describe safe and efficient parallel computations over two-dimensional arrays, and theorems for deriving and optimizing parallel programs. The main features of our framework are: (1) *a novel use of the abide-tree representation* [2] in developing parallel programs for manipulating two-dimensional arrays; (2) *a strong support* for systematic development of both efficient and correct parallel programs in a highly abstract way; (3) *an efficient implementation* of basic skeletons in C++ and MPI on PC clusters, guaranteeing that programs composed with these parallel skeletons can run efficiently in parallel. To develop parallel programs in our framework, users construct a simple and general program that covers a class of problems, derive its efficient version using general techniques such as fusion, tupling and generalization, and then instantiate the general program to solve concrete problems. Usually, this derivation is summarized as a theorem (tool).

In this paper, we give a domain-specific tool and show the efficacy of the framework. We focus on computations known as matrix-convolutions [12], in which each element in the resulting array depends on its surrounding elements. This set of computations includes important and fundamental problems such as image filters, difference methods and the N-body problem (although this last problem seems more difficult than the others, it merely refers to not only the nearest neighbors but all the surrounding elements). The most general form $mconv$ is described with three components:

$$mconv\ f\ shrink = \mathsf{map}\ f \circ \mathsf{map}\ shrink \circ surrounds\ .$$

Here, $surrounds$ gathers all the surrounding elements for each element, $shrink$ picks the necessary parts up from those gathered elements, and f calculates the resulting element from them. This general form is parameterized by the two functions $shrink$ and f, and users can solve many problems by specifying suitable ones. For example, users can develop a sharpen-filter by choosing the function $shrink$ that reduces the surroundings into a 3×3 matrix, and the function f that calculates the weighted sum of the nine values. We can further optimize instances of the general program to application-specific ones with the *surrounding theorem*. The main contributions of this paper are as follows.

- We show the general parallel program for the matrix-convolutions described with parallel skeletons. Users can solve their problems as its instance.
- We give the *surrounding theorem* that enables users to get an efficient program easily. The experimental results show that the derived program can be executed efficiently in parallel.

Technical details of this paper are available in the master's thesis [8].

2 Notations

Notation in this paper follows that of Haskell [3], a pure functional language that can describe both algorithms and algorithmic transformation concisely.

Function application is denoted by a space and the argument may be written without brackets. Thus, $f\ a$ means $f(a)$ in ordinary notation. Functions are curried, i.e. functions take one argument and return a function or a value, and the function application associates to the left. Thus, $f\ a\ b$ means $(f\ a)\ b$. The function application binds more strongly than any other operator, so $f\ a \otimes b$ means $(f\ a) \otimes b$, but not $f\ (a \otimes b)$. Function composition is denoted by $\circ$, so $(f \circ g)\ x = f\ (g\ x)$ from its definition. Binary operators can be used as functions by sectioning as follows: $a \oplus b = (a\oplus)\ b = (\oplus b)\ a = (\oplus)\ a\ b$. Two binary operators $\ll$ and $\gg$ are defined by $a \ll b = a$, $a \gg b = b$. Pairs are Cartesian products of plural data, written like (x, y). A function that applies functions f and g respectively to the elements of a pair (x, y) is denoted by $(f \times g)$. Thus, $(f \times g)\ (x, y) = (f\ x, g\ y)$.

3 Skeletal Framework on Two-Dimensional Arrays

In this section, we introduce our parallel skeletal framework on two-dimensional arrays [9] based on the theory of constructive algorithmics [2,4].

3.1 Abide-Trees for Two-Dimensional Arrays

To represent two-dimensional arrays, we define the abide-trees, which are built up by three constructors $|\cdot|$ (singleton), $\ominus$ (above) and ϕ (beside) following the idea in [2].

$$\begin{aligned}\textbf{data } \textit{AbideTree } \alpha = & \ |\cdot|\ \alpha \\ | & \ (\textit{AbideTree } \alpha) \ominus (\textit{AbideTree } \alpha) \\ | & \ (\textit{AbideTree } \alpha) \phi (\textit{AbideTree } \alpha)\end{aligned}$$

Here, $|\cdot|\ a$, or abbreviated as $|a|$, means a singleton array of a, i.e. a two-dimensional array of a single element a. For two-dimensional arrays x and y of the same width, $x \ominus y$ means that x is located above y. Similarly, for arrays x and y of the same height, $x \phi y$ means that x is located on the left of y. Moreover, $\ominus$ and ϕ are associative operators and satisfy the following *abide* (a coined term from ab̲o̲v̲e̲ and bes̲i̲d̲e̲) property.

Definition 1 (Abide Property). *Two binary operators $\oplus$ and $\otimes$ are said to satisfy the abide property or to be abiding, if the following equation is satisfied:*

$$(x \otimes u) \oplus (y \otimes v) = (x \oplus y) \otimes (u \oplus v).$$

In the rest of the paper, we will assume that x has the same width as y when $x \ominus y$ appears, and that u has the same height as v for $u \phi v$.

Note that one two-dimensional array may be represented by many abide-trees, but these abide-trees are equivalent because of the abide property of $\ominus$ and ϕ. For example, we can express the following 2×2 two-dimensional array by two equivalent abide-trees.

$$\begin{pmatrix} 1 & 2 \\ 3 & 4 \end{pmatrix} \Rightarrow \begin{cases} (|1| \phi |2|) \ominus (|3| \phi |4|) \\ (|1| \ominus |3|) \phi (|2| \ominus |4|) \end{cases}$$

This is in sharp contrast to the quadtree representation of matrices [10], which does not allow such freedom.

From the theory of constructive algorithmics [4], it follows that each constructively built-up data structure (i.e., algebraic data structure) is equipped with a powerful computation pattern called homomorphism.

Definition 2 ((Abide-tree) Homomorphism). *A function h is said to be an abide-tree homomorphism, if it is defined as follows for a function f and binary operators $\oplus, \otimes$.*

$$\begin{aligned} h\ |a| &= f\ a \\ h\ (x \ominus y) &= h\ x \oplus h\ y \\ h\ (x \phi y) &= h\ x \otimes h\ y \end{aligned}$$

For notational convenience, we write $(\!|f, \oplus, \otimes|\!)$ to denote h. When it is clear from the context, we just call $(\!|f, \oplus, \otimes|\!)$ homomorphism. Note that $\oplus$ and $\otimes$ in $(\!|f, \oplus, \otimes|\!)$ should be associative and satisfy the abide property, inheriting the properties of $\ominus$ and ϕ.

Intuitively, a homomorphism $(\!|f, \oplus, \otimes|\!)$ is a function to replace the constructors $|\cdot|$, $\ominus$ and ϕ in an input abide-tree by f, $\oplus$ and $\otimes$ respectively.

Fig. 1. Intuitive Definition of Parallel Skeletons on Two-Dimensional Arrays

$$\text{map } f \begin{pmatrix} x_{11} & \cdots & x_{1n} \\ \vdots & \ddots & \vdots \\ x_{m1} & \cdots & x_{mn} \end{pmatrix} = \begin{pmatrix} f\, x_{11} & \cdots & f\, x_{1n} \\ \vdots & \ddots & \vdots \\ f\, x_{m1} & \cdots & f\, x_{mn} \end{pmatrix}$$

$$\text{reduce}(\oplus, \otimes) \begin{pmatrix} x_{11} & \cdots & x_{1n} \\ \vdots & \ddots & \vdots \\ x_{m1} & \cdots & x_{mn} \end{pmatrix} = \begin{matrix} (x_{11} \otimes \cdots \otimes x_{1n}) \oplus \\ \ddots \\ (x_{m1} \otimes \cdots \otimes x_{mn}) \end{matrix}$$

$$\text{zipwith } f \begin{pmatrix} x_{11} & \cdots & x_{1n} \\ \vdots & \ddots & \vdots \\ x_{m1} & \cdots & x_{mn} \end{pmatrix} \begin{pmatrix} y_{11} & \cdots & y_{1n} \\ \vdots & \ddots & \vdots \\ y_{m1} & \cdots & y_{mn} \end{pmatrix} = \begin{pmatrix} f\, x_{11}\, y_{11} & \cdots & f\, x_{1n}\, y_{1n} \\ \vdots & \ddots & \vdots \\ f\, x_{m1}\, y_{m1} & \cdots & f\, x_{mn}\, y_{mn} \end{pmatrix}$$

$$\text{scan}(\oplus, \otimes) \begin{pmatrix} x_{11} & \cdots & x_{1n} \\ \vdots & \ddots & \vdots \\ x_{m1} & \cdots & x_{mn} \end{pmatrix} = \begin{pmatrix} y_{11} & \cdots & y_{1n} \\ \vdots & \ddots & \vdots \\ y_{m1} & \cdots & y_{mn} \end{pmatrix} \text{ where } y_{ij} = \begin{matrix} (x_{11} \otimes \cdots \otimes x_{1j}) \oplus \\ \ddots \\ (x_{i1} \otimes \cdots \otimes x_{ij}) \end{matrix}$$

$$\text{scanr}(\oplus, \otimes) \begin{pmatrix} x_{11} & \cdots & x_{1n} \\ \vdots & \ddots & \vdots \\ x_{m1} & \cdots & x_{mn} \end{pmatrix} = \begin{pmatrix} z_{11} & \cdots & z_{1n} \\ \vdots & \ddots & \vdots \\ z_{m1} & \cdots & z_{mn} \end{pmatrix} \text{ where } z_{ij} = \begin{matrix} (x_{ij} \otimes \cdots \otimes x_{in}) \oplus \\ \ddots \\ (x_{mj} \otimes \cdots \otimes x_{mn}) \end{matrix}$$

Table 1. Parallel Complexity of the Skeletons for a Two-Dimensional Array of $n \times n$

	P processors	n^2 processors
map, zipwith	$O(n^2/P)$	$O(1)$
reduce	$O(n^2/P + \log P)$	$O(\log n)$
scan, scanr	$O(n^2/P + \sqrt{n^2/P}\log P)$	$O(\log n)$

3.2 Parallel Skeletons on Two-Dimensional Arrays

We introduce the parallel skeletons map, reduce, zipwith, scan and scanr for manipulating two-dimensional arrays. In the theory of constructive algorithmics [2, 4], these functions are known to be the most fundamental computation components for manipulating algebraic data structures and for being glued together to express complicated computations. Intuitive definitions of the skeletons are shown in Fig. 1. All the skeletons are implemented efficiently in parallel and their costs are shown in Table 1.

The skeletons map and reduce are two special cases of homomorphism. The skeleton map applies a function f to each element of a two-dimensional array while keeping the shape of the structure. The skeleton reduce collapses a two-dimensional array to a value using two abiding binary operators $\oplus$ and $\otimes$. They are defined formally as map $f = (\!|\cdot|\circ f, \ominus, \phi|\!)$, and reduce$(\oplus, \otimes) = (\!|id, \oplus, \otimes|\!)$.

The skeleton zipwith, an extension of map, takes two arrays of the same shape, applies a function f to corresponding elements of the arrays and returns a new array of the same shape. The skeletons scan and scanr, extensions of reduce, hold all values generated in reducing an array by reduce. The scan generates the result of reducing

```
int sharpen_filter(int **b, int **a, int n, int m){
  for(int i = 0; i < m; i++)
    for(int j = 0; j < n; j++)
      b[i][j] = f(a[i][j], a[i-1][j], a[i+1][j], a[i][j+1], a[i][j-1],
                  a[i-1][j+1], a[i-1][j-1], a[i+1][j+1], a[i+1][j-1]);
}
int f(int c, int n, int s, int e, int w, int ne, int nw, int se, int sw){
  return 5*c + (-1)*n + (-1)*s + (-1)*e + (-1)*w + 0*ne + 0*nw + 0*se + 0*sw;}
```

Fig. 2. C++ Code of the Sharpen Filter (Sequential Program)

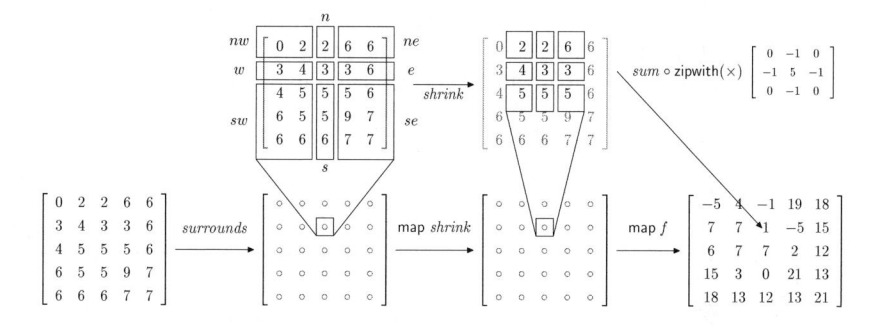

Fig. 3. An Image of the Sharpen Filter in the General Program

the upper-left subarray, while the scanr generates that of the lower-right subarray. We omit the formal definition of zipwith, scan and scanr for the space limitation.

4 Developing Parallel Programs for Matrix-Convolutions

In this section, focusing on the matrix-convolutions such as image filters and difference methods, we give the general form described with parallel skeletons, and then give the theorem to get optimized program from the general form.

The matrix-convolution is computation in which each element of the resulting array depends on the surrounding elements. For example, the sharpen-filter that sharpens the input image is one instance of the matrix-convolution. A pixel of the resulting image is the weighted sum of the surrounding pixels of the input image. Similarly, the difference method is another instance of matrix-convolution since it calculates the new value of each point from the old values of the surrounding points. We show a code in C++ for the sharpen-filter in Fig. 2, to give a concrete image of the problems dealt with here.

4.1 A General Form Described with Parallel Skeletons

As argued in the introduction, the most general form of this kind of computation is thought to consist of three components: gathering all the surrounding elements of each element to it, shrinking those to the necessary amount, and applying a function to get a new element from them. Thus, the program is described as follows:

$$mconv\ f\ shrink = \mathsf{map}\ f \circ \mathsf{map}\ shrink \circ surrounds\ .$$

The idea of our general from is illustrated in Fig. 3 that shows an image of execution of the sharpen-filter: (1) *surrounds* gathers all the surrounding elements for each element, (2) *shrink* picks the necessary parts up from those gathered elements, and (3) f calculates the resulting element from them. This general form has clear correspondences to the code in Fig. 2. The function f corresponds to f of the code, *shrink* corresponds to which elements are the arguments passed to f, and *surrounds* corresponds to for-loops. Thus, users can easily write their programs using the general form.

This general form is parameterized by the two functions *shrink* and f, and users can solve many problems by specifying application-specific ones, as shown below. The function *surrounds*, which is commonly used in those problems, has two-phase calculation as follows: (1) calculation of the parts of the northwest (i.e. c, n, w and nw) by scan, and (2) that of the other parts by scanr. Its definition is as follows.

$$surrounds = \mathsf{scanr}(\oplus_r, \otimes_r) \circ \mathsf{map}\ f_r \circ \mathsf{scan}(\oplus_f, \otimes_f) \circ \mathsf{map}\ f_f$$

where

$f_f\ a = (a, Nil, Nil, Nil)$

$(c_a, n_a, w_a, nw_a) \oplus_f (c_b, n_b, w_b, nw_b) = (\underbrace{c_b}_{c}, \underbrace{n_a \ominus |c_a| \ominus n_b}_{n}, \underbrace{w_b}_{w}, \underbrace{nw_a \ominus w_a \ominus nw_b}_{nw})$

$(c_a, n_a, w_a, nw_a) \otimes_f (c_b, n_b, w_b, nw_b) = (\underbrace{c_b}_{c}, \underbrace{n_b}_{n}, \underbrace{w_a \Diamond |c_a| \Diamond w_b}_{w}, \underbrace{nw_a \Diamond n_a \Diamond nw_b}_{nw})$

$f_r\ (c, n, w, nw) = (c, n, Nil, Nil, w, Nil, nw, Nil, Nil)$

$(c_a, n_a, s_a, e_a, w_a, ne_a, nw_a, se_a, sw_a) \oplus_r (c_b, n_b, s_b, e_b, w_b, ne_b, nw_b, se_b, sw_b)$
$= (\underbrace{c_a}_{c}, \underbrace{n_a}_{n}, \underbrace{s_a \ominus |c_b| \ominus s_b}_{s}, \underbrace{e_a}_{e}, \underbrace{w_a}_{w}, \underbrace{ne_a}_{ne}, \underbrace{nw_a}_{nw}, \underbrace{se_a \ominus e_b \ominus se_b}_{se}, \underbrace{sw_a \ominus w_b \ominus sw_b}_{sw})$

$(c_a, n_a, s_a, e_a, w_a, ne_a, nw_a, se_a, sw_a) \otimes_r (c_b, n_b, s_b, e_b, w_b, ne_b, nw_b, se_b, sw_b)$
$= (\underbrace{c_a}_{c}, \underbrace{n_a}_{n}, \underbrace{s_a}_{s}, \underbrace{e_a \Diamond |c_b| \Diamond e_b}_{e}, \underbrace{w_a}_{w}, \underbrace{ne_a \Diamond n_b \Diamond ne_b}_{ne}, \underbrace{nw_a}_{nw}, \underbrace{se_a \Diamond s_b \Diamond se_b}_{se}, \underbrace{sw_a}_{sw})$

Here, Nil is a special value to indicate that there is no value, and we treat it as an identity of $\ominus$ and $\Diamond$ for simplification of the notation. Thus, $Nil \ominus x = x$, $x \ominus Nil = x$, $Nil \Diamond x = x$, and $x \Diamond Nil = x$. Each element of the resulting array is a tuple of nine elements. The meaning of each element of the tuple is as follows: c is the center element; s is an array of the elements on the south of the element; similarly n, e and w are arrays of the elements on the north, east and west respectively; ne, nw, se and sw are arrays of the elements on the northeast, northwest, southeast and southwest. Note that this *surrounds* needs $O(n^4)$ memory space for a matrix of $n \times n$.

We show some examples written with the general form.

imagefilter ker = *mconv* (*conv ker*) $shrink_1$
FDM n ker = *iter n* (*mconv* (*conv ker*) $shrink_1$)
  where
  $shrink_1 = id \times B \times T \times L \times R \times BL \times BR \times TL \times TR$
  $B = (\!|\ \cdot\ |, \gg, \Diamond\ |\!)$, $T = (\!|\ \cdot\ |, \ll, \Diamond\ |\!)$, $L = (\!|\ \cdot\ |, \ominus, \ll\ |\!)$, $R = (\!|\ \cdot\ |, \ominus, \gg\ |\!)$,
  $BL = (\!|\ \cdot\ |, \gg, \ll\ |\!)$, $BR = (\!|\ \cdot\ |, \gg, \gg\ |\!)$, $TL = (\!|\ \cdot\ |, \ll, \ll\ |\!)$, $TR = (\!|\ \cdot\ |, \ll, \gg\ |\!)$

The function *imagefilter ker* is an image filter with the coefficient matrix *ker*, which is used to compute weighted sum of the surrounding pixels. The $shrink_1$ reduces each part of the gathered surrounding elements to the element closest to the center, and the function *conv ker* calculates the weighted sum of them. The functions B and T take the bottom row and the top row of the input array respectively. Similarly, each of L, R, BL, BR, TL and TR takes corresponding part of the input array. Figure 3 shows an image of execution of the sharpen-filter by the above general program. The function $FDM\ n\ ker$ performs the finite difference method, where *iter* is an iteration function and each iteration step is the same as image filters with specific coefficients.

The following example calculates the array of which element at (i, j) is the maximum in the i-th row and the j-th column, i.e. the maximum in the cross. The $shrink_{max}$ reduces each part of the gathered surrounding elements to the biggest element in the part, where the binary operator $\uparrow$ takes the bigger element. The function max_5 takes the maximum of the column and the row including the center element.

$$crossmax = mconv\ max_5\ shrink_{max}$$
$$\textbf{where}\ shrink_{max} = max \times \cdots \times max$$
$$max = (\!|id, \uparrow, \uparrow|\!)$$
$$max_5\ (c, n, s, e, w, \_, \_, \_, \_) = c \uparrow n \uparrow s \uparrow e \uparrow w$$

As shown in this example, *shrink* is allowed not only to shrink the shape of the surroundings but to perform some calculation.

4.2 Surrounding Theorem

In this section, we give the theorem to optimize the general form by fusing *shrink* to *surrounds*.

Image filters and difference methods usually have the *shrink* of the fixed size window that takes the fixed-size rectangle region (window) of the surrounding elements. The function that takes a fixed number of columns (rows) can be written as a homomorphism. For example, the function $right = (\!|\,|\cdot|, \ominus, \gg |\!)$ takes the right-most column, which is used in the examples in the previous section. Thus, we here consider the general *shrink* that consists of homomorphisms. It is defined as follows.

$$shrink = g_c \times h_n \times h_s \times h_e \times h_w \times h_{ne} \times h_{nw} \times h_{se} \times h_{sw}$$
$$\textbf{where}$$
$$h_n = (\!|g_n, \oplus_n, \otimes_n|\!),\quad h_s = (\!|g_s, \oplus_s, \otimes_s|\!),\quad h_e = (\!|g_e, \oplus_e, \otimes_e|\!)$$
$$h_w = (\!|g_w, \oplus_w, \otimes_w|\!),\quad h_{ne} = (\!|g_{ne}, \oplus_{ne}, \otimes_{ne}|\!),\quad h_{nw} = (\!|g_{nw}, \oplus_{nw}, \otimes_{nw}|\!)$$
$$h_{se} = (\!|g_{se}, \oplus_{se}, \otimes_{se}|\!),\quad h_{sw} = (\!|g_{sw}, \oplus_{sw}, \otimes_{sw}|\!)$$

Here, $\oplus_X$ and $\otimes_X$ are extended to satisfy the following equations: $Nil \oplus_X x = x$, $x \oplus_X Nil = x$, $Nil \otimes_X x = x$, and $x \otimes_X Nil = x$. The general form using this *shrink* uses $O(n^4)$ operations for a two-dimensional array of $n \times n$.

Then, we give the result of the optimization by fusing *shrink* to *surrounds*.

Theorem 1 (Surrounding). *Let the function shrink be defined by homomorphisms as above. Then, there exist a projection function proj and operators $\oplus'_f$, $\otimes'_f$, $\oplus'_r$ and $\otimes'_r$, whose complexity is bounded by the largest of $\oplus_X$ and $\otimes_X$, and the program*

$$mconv\ f\ shrink$$

is optimized to the following program.

$$\text{map } (f \circ proj) \circ \text{scanr}(\oplus'_r, \otimes'_r) \circ \text{map } {f_r}' \circ \text{scan}(\oplus'_f, \otimes'_f) \circ \text{map } {f_f}'$$

Proof. The theorem is proved by the promotion of map *shrink* with extending the tuples. See the master's thesis [8] for details.

The resulting program uses $O(n^2)$ operations for a two-dimensional array of $n \times n$, while the original general form uses $O(n^4)$ operations. The parallel complexity of the resulting program is $O((n^2/P + \sqrt{n^2/P} \log P) T_{(\oplus_X, \otimes_X)})$ for P processors, provided that the calculational complexity of $\oplus_X$ and $\otimes_X$ in the homomorphisms are $T_{(\oplus_X, \otimes_X)}$.

All the examples shown in the previous section have the *shrink* functions described with homomorphisms. Thus, we can apply this theorem to all of them, and they are executed in $O(n^2/P + \sqrt{n^2/P} \log P)$ complexity using the skeletons.

As mentioned above, the function that takes a fixed number of columns (rows) can be written as a homomorphism. Thus, this theorem holds for the *shrink* of the fixed size window that shrinks the surrounding elements to a fixed size, which is often seen in image filters and difference methods.

Corollary 1 (Fixed Size Window). *Let the function shrink be the fixed size window. Then, the program mconv f shrink is optimized to that of $O(n^2)$ operations.*

Note that the homomorphism taking $h \times w$ subarray of a two-dimensional array has the operators of $O(wh)$ complexity. Thus, the total complexity of the program of fixed size window is $O(n^2 wh)$.

Finally, we note that we may perform more optimizations by using the shifting of the edges instead of butterfly computations for the global computations of scan and scanr, provided that the operators influence only a fixed number of elements [8]. This leads to the parallel complexity of $O((n^2/P + \sqrt{n^2/P}) T_{(\oplus_X, \otimes_X)})$ for P processors.

5 Experimental Results

We implemented the program[1] using our parallel skeleton library [14] and did our experiment on a cluster (distributed memory). Each of the nodes connected with Gigabit Ethernet has a CPU of Intel® Xeon®2.80GHz and 2GB memory, with Linux 2.4.21 for the OS, gcc 2.96 for the compiler, and mpich 1.2.7 for the MPI.

Figures 4 and 5 show the speedups and the calculation times of the sharpen-filter. The program is an optimized one from the general form (an equivalent of the program in Fig. 2). The inputs are images of 1000×1000 and 2000×2000. The computation times of the program on one processor are 0.70s and 3.85s respectively.

The result shows programs described with skeletons can be executed efficiently in parallel, and proves the success of our framework. The program achieves almost linear speedups, and the total computational complexity of the optimized program is $O(n^2)$ (thus, its parallel complexity is $O(n^2/P)$ for small P). However, the serial performance

---

[1] The source code of the test program as well as the skeleton library are available at the web page http://www.ipl.t.u-tokyo.ac.jp/sketo/.

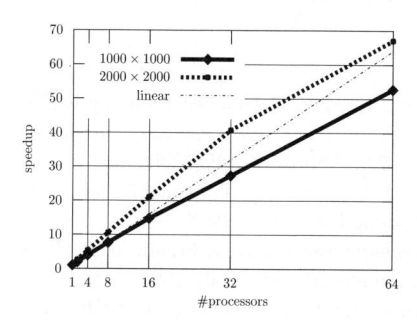

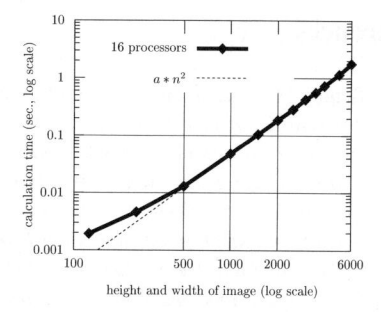

Fig. 4. Speedup of Image Filter

Fig. 5. Calculation Time vs. Size of Image

is rather poor due to the overhead of using general skeletons (i.e. scan and scanr). We think this problem can be solved by replacing the general skeletons with those specialized for this domain, and it can be automatically done by compilers (future work).

6 Related Work

SKiPPER [17] is a skeleton-based parallel programming environment for real-time image processing. It has skeletons specialized for image processing, while we use general skeletons on two-dimensional arrays. Thus, a program developed with SKiPPER may be faster than that written with our skeletons, but, the latter program can be easily composed with other programs and be optimized by fusion due to generality and solid foundation of our skeletons.

There are several other skeletal parallel approaches (libraries), such as eSkel [1], Muesli [13] and P3L [6]. Their formalizations of skeletons on two-dimensional arrays are not enough (e.g. they have no *scan* skeletons, and the reduction takes only one operator) to deal with matrix-convolutions suitably. Our skeletons have a solid foundation, so that we can easily deal with matrix-convolutions and perform optimizations.

7 Conclusion

In this paper, we proposed a general theorem, called *surrounding theorem*, for optimization of a general skeleton program into an efficient application-specific program. It can deal with a wide class of matrix-convolution problems including image filters and difference methods. The experimental results show that the optimized program can be executed efficiently in parallel. We are now working on making an automatic mechanism for translating the sequential code to our general form with skeletons, and further an optimization mechanism for the application-specific program with respect to its global communication and sequential performance.

Acknowledgment

This work is partially supported by the Grant-in-Aid for Scientific Research (B), No. 17300005, Japan Society for the Promotion of Science. We are grateful to the referees for their detailed and helpful comments.

References

1. A. Benoit, M. Cole, J. Hillston, and S. Gilmore. Flexible skeletal programming with eskel. In *Proceedings of 11th International Euro-Par Conference (Euro-Par'05)*, volume 3648 of *Lecture Notes in Computer Science*, pages 761–770. Springer-Verlag, 2005.
2. R. S. Bird. Lectures on Constructive Functional Programming. Technical Report Technical Monograph PRG-69, Oxford University Computing Laboratory, 1988.
3. R. S. Bird. *Introduction to Functional Programming using Haskell*. Prentice Hall, 1998.
4. R. S. Bird and O. de Moor. *Algebras of Programming*. Prentice Hall, 1996.
5. M. Cole. *Algorithmic Skeletons : Structured Management of Parallel Computation*. Research Monographs in Parallel and Distributed Computing, Pitman, London, 1989.
6. M. Danelutto, F. Pasqualetti, and S. Pelagatti. Skeletons for data parallelism in P3L. In *Proceedings of 3rd International Euro-Par Conference (Euro-Par'97)*, volume 1300 of *Lecture Notes in Computer Science*, pages 619–628. Springer-Verlag, 1997.
7. E. Elmroth, F. Gustavson, I. Jonsson, and B. Kagstroom. Recursive Blocked Algorithms and Hybrid Data Structures for Dense Matrix Library Software. *SIAM Review*, 46(1):3–45, 2004.
8. K. Emoto. A Compositional Framework for Parallel Programming on Two-Dimensional Arrays. Master's thesis, Graduate School of Information Science and Technology, the University of Tokyo, 2006. Available at http://www.ipl.t.u-tokyo.ac.jp/~emoto/master_thesis.pdf.
9. K. Emoto, Z. Hu, K. Kakehi, and M. Takeichi. A Compositional Framework for Developing Parallel Programs on Two Dimensional Arrays. Technical Report METR2005-09, Department of Mathematical Informatics, University of Tokyo, 2005.
10. J. D. Frens and D. S. Wise. QR Factorization with Morton-Ordered Quadtree Matrices for Memory Re-use and Parallelism. In *Proceedings of 4th ACM SIGPLAN Symposium on Principles and Practice of Parallel Programming (PPoPP'03)*, pages 144–154, 2003.
11. G. Hains. Programming with Array Structures. In A. Kent and J. G. Williams, editors, *Encyclopedia of Computer Science and Technology*, volume 14, pages 105–119. M. Dekker inc, New-York, 1994. Appears also in *Encyclopedia of Microcomputers*.
12. A. K. Jain. *Fundamentals of Digital Image Processing*. Prentice Hall, 1989.
13. H. Kuchen. A Skeleton Library. In *Proceedings of 8th International Euro-Par Conference (Euro-Par'02)*, volume 2400 of *Lecture Notes in Computer Science*, pages 620–629. Springer-Verlag, 2002.
14. K. Matsuzaki, K. Emoto, H. Iwasaki, and Z. Hu. A library of constructive skeletons for sequential style of parallel programming (invited paper). In *Proceedings of the First International Conference on Scalable Information Systems (INFOSCALE 2006)*. IEEE Press, 2006. To appear.
15. L. Mullin, editor. *Arrays, Functional Languages, and Parallel Systems*. Kluwer Academic Publishers, 1991.
16. F. A. Rabhi and S. Gorlatch, editors. *Patterns and Skeletons for Parallel and Distributed Computing*. Springer-Verlag, 2002.
17. J. Serot and D. Ginhac. Skeletons for Parallel Image Processing: an Overview of the SKIPPER Project. *Parallel Computing*, 28(12):1685–1708, 2002.

Dynamic Task Generation and Transformation Within a Nestable Workpool Skeleton*

S. Priebe

Philipps-Universität Marburg, Fachbereich Mathematik und Informatik
Hans-Meerwein-Straße, D-35032 Marburg, Germany
priebe@mathematik.uni-marburg.de

Abstract. Within a classical workpool skeleton a master process employs a set of worker processes to solve tasks contained in a task pool. In contrast to the usual statically fixed task set some applications generate tasks dynamically. Additionally often the need for dynamic task pool transformation arises, for example to combine newly generated partial tasks to form full tasks. We present an *extended workpool skeleton* for the parallel Haskell dialect *Eden* which provides both features and employs careful stream-processing and a termination detection mechanism. We also show how to nest the skeleton to alleviate the bottleneck a single master presents. Furthermore we demonstrate its efficiency by its fruitful use for the parallelisation of a DNA sequence alignment algorithm.

1 Introduction

Uneven task sizes arise naturally from many problems and are often an obstacle for parallelisation. Within parallel dialects of *Haskell* [1] the classical static distribution schemes (like parallel *map*) can hardly establish load-balance given unevenly sized tasks; therefore dynamic task distribution schemes are used. The well-known *workpool* scheme (also known as farm, master-worker, or client-server) [2] is mostly used to compensate for such irregularly sized tasks: A master administrates a statically fixed task pool out of which tasks are gradually assigned to currently idle workers, leading to a balanced workload. Such a scheme is often expressed as a high-level code template, known as a *skeleton* [3,4].

Some applications however expose their full task set only successively as the computation proceeds and need therefore a more general workpool skeleton whose worker processes are allowed to generate new tasks dynamically. Then a task will not only produce a result, but possibly also a set of new tasks for the global task pool. This introduces the problem of termination detection, which was not a problem before since a statically fixed task number makes it easy to determine termination. Now special care has to be taken to account for a dynamically growing and shrinking task pool. Emptiness of the the task pool does no longer mean that there is no more work to do, since new work may still be created by active workers.

---

* Supported by *Evangelisches Studienwerk Villigst e.V.*

To make things even more complicated, dynamically created tasks may be incomplete (e. g. due to limited local data) and need to be combined with other incomplete tasks before submission to a worker. Therefore means have to be provided for traversing and transforming the task pool on the fly. But since the task pool is often modelled as a lazy list and woven into a network of interdependent streams, one has to be extra careful during a transformation. When combining partial tasks, deadlocks can easily occur by searching for not yet existent partial partner tasks; additionally, all usual techniques (like delayed pattern matching and incremental functions) when dealing with lazy lists have to be considered.

All this means extra work for the master process, which worsens the bottleneck it already presents. One way to alleviate this is exchanging the single master process by a tree of master and submaster processes distributing the administration load. This corresponds to a nested workpool, in which a workpool is given other workpools as worker processes.

Contributions
- We present a new workpool skeleton for the parallel Haskell dialect *Eden* [5] which firstly enables worker processes to dynamically generate additional tasks for the task pool (together with the needed termination detection) and secondly permits dynamic transformation of the task pool (Sect. 2.2). A function aiding safe task pool transformation is also described.
- Via folding the workpool skeleton is then nested to reduce the administrative load of the master process (Sect. 2.3).
- We show the usefulness of the new workpool by applying it to the parallel alignment of DNA sequences (Sect. 3).

2 The Workpool Skeleton for Eden

2.1 A Short Glance at Eden

Eden extends Haskell with means for relocating the evaluation of a function application to other network nodes, enabling the evaluation of multiple expressions in parallel. A function embedded in a *process abstraction* by applying

```
process :: (Trans a, Trans b) => (a -> b) -> Process a b
```

can be run in parallel to the continuing evaluation of its parent expression on another processor by applying its arguments to a special application operator

```
(#)      :: (Trans a, Trans b) => Process a b -> a -> b
```

New processes are placed round-robin on available nodes. There is no shared memory, all data exchanges happen via communication based on PVM [6] message passing within Eden's implementation. The `Trans` context ensures that only those values can be communicated for which corresponding low-level communication routines exist. Not only finite values but also infinite lists (known as *streams*) can be transmitted. These are sent piecewise with each stream element being demanded strictly. For merging a set of streams into a single stream a nondeterministic `merge :: [[a]]->[a]` function is predefined. Stream communication plays a vital part in the following workpool skeleton.

2.2 The Workpool Skeleton

Now we present a workpool skeleton which provides dynamic task generation and task pool transformation. The basic scheme works as follows: A master process keeps a pool of tasks which are distributed to a set of worker processes on request. When a worker receives a task, it solves it and sends back the result together with a request for new work. This way each worker is busy most of the time and load balance is kept as tasks are assigned depending on the current work distribution.

We extend the basic scheme by two new features, dynamic task generation and task pool transformation: *Firstly*, when a worker processes a task new tasks may arise. These will be sent back to the master and appended to the global task pool, preserving task order. *Secondly*, sometimes it is helpful to be able to process and transform the task pool. A given transformation function `tt` will be applied to the task pool to combine incomplete tasks and replace them by complete ones, possibly changing task order. The resulting basic interaction scheme is shown in Fig. 1. All connections shown are stream connections; the thick pointers touching the worker processes Ⓦ are interprocess connections while all others reside within the master process. The full code for the extended workpool is shown in Fig. 2. Parameters are: The number of processors available, the number of advance requests for each worker, the worker function, the transformation function for the task pool, and finally a set of initial tasks. At first, the workpool demands the first cons of the list of worker processes via `touch` to trigger their creation using the predefined parallel zip function `eagerInstList`, then the `results` are given back. The main body is divided into two parts:

The *stream part* defines the parallel stream network according to Fig. 1. A set of `workerProcs` is created, which apply the worker function `f` to their input and attach their id number to the result as a request for new work. Their input `toWorkers` is a list of streams, each of which contains `tasks` for the corresponding worker according to its `requests`. The `initialRequests` are built based on an interleaved sequence (each of size `prefetch`) of worker numbers and provides an initial supply of tasks for each worker. The worker's outputs are merged to

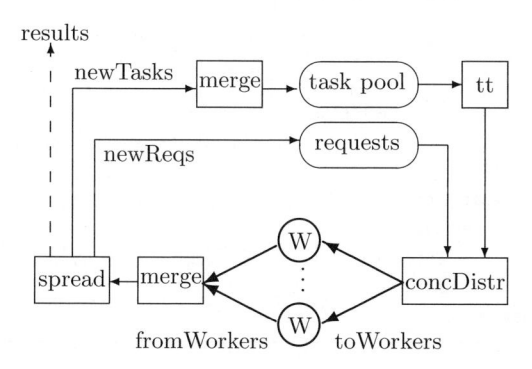

Fig. 1. Stream interconnections of workpool (seen from master process)

```
wpool :: (Trans t, Trans r) =>
         Int -> Int -> ([t] -> [(r, [t])]) ->
         (([t],[t],[t],Int) -> ([t],[t],[t],Int)) ->
         [t] -> [r]
wpool np prefetch f tt initialTasks =
  (touch fromWorkers) 'seq' results
  where touch []    = ()         -- Demand first constructor to
        touch (_:_) = ()         -- initiate worker creation

        -- 1) Stream ---------------------------------------------
        fromWorkers = eagerInstList workerProcs toWorkers
        workerProcs = [process (zip [n,n..] . f) | n<-[1..np]]

        toWorkers   = concDistr requests [1..np] tasks
        requests    = initialReqs ++ newReqs
        initialReqs = concat (replicate prefetch [1..np])

        taskpool                 = initialTasks ++ (merge newTasks)
        (_, _, tasks, _)         = tt (taskpool, [], [], 0)
        workerstream             = merge fromWorkers
        (newReqs, (x, newTasks)) = spread workerstream

        -- 2) State ----------------------------------------------
        ([], _, results, _) = terminate
                                ([], length initialTasks, [], 0)
                                workerstream

        terminate (is,t,rs,r) ( (_,(res,ntasks)) :ws)
          | t' >  r' = terminate (is', t', res:rs, r') ws
          | t' == r' = (is', t', reverse (res:rs), r')
          | t' <  r' = error ("Will never happen.")
          where ([], is', _, n) = tt (is++ntasks, [], [], 0)
                t'                = t + n
                r'                = r + 1
        terminate _            []
          = error "Workerstream empty!"
concDistr :: Eq a => [a] -> [a] -> [b] -> [[b]]
concDistr unsortedKeys allKeys vals =
  where vals' = zip unsortedKeys vals
        result = [ [v | (uk,v) <- vals', uk == k] | k <- allKeys]

-- TH splice creates: spread :: [(a,(b,[c]))] -> ([a], ([b], [[c]]))
$(do let empty     = ListE []
     let structure = TupE [empty, TupE [empty, empty]]
     let spread_fct = mkSpread structure
     return spread_fct)
```

Fig. 2. Workpool with dynamic task generation and task pool transformation

```
-- Call in Fig.2       spread []              = ([], ([], []))
-- creates two         spread ((a,(b,c)):rest) = (a:as,(b:bs,c:cs))
-- clauses:              where (as,(bs,cs)) = spread rest
mkSpread :: Exp -> [Dec]
mkSpread struct = [FunD "spread" clauses] where
  clauses  = [Clause pat1 body1 [], Clause pat2 body2 [ValD pat b []]]
  pat1     = [ListP []]
  body1    = NormalB (buildE struct (repeat (ListE [])))
  pat2     = [ConP "GHC.Base::" [pat2', VarP "rest"]]
  pat2'    = buildP struct [VarP [c] | c <- ['a','b'..]]
  body2    = NormalB (buildE struct lists)
  lists    = [AppE (AppE (ConE "GHC.Base::") (VarE [c]))
                    (VarE [c,'s']) | c <- ['a','b'..]]
  pat      = buildP struct [VarP [c,'s'] | c <- ['a','b'..]]
  b        = NormalB (AppE (VarE "spread") (VarE "rest"))
buildE :: Exp -> [Exp] -> Exp;   buildP :: Exp -> [Pat] -> Pat
buildE (TupE vs) ls = TupE (fst (traverse vs ls)) where
  traverse ((ListE []):rest) (l:ls) = (l : r, ls')
    where (r, ls')   = traverse rest ls
  traverse ((TupE ws):rest) ls      = ((TupE rec):r,ls'')
    where (rec, ls') = traverse ws ls
          (r, ls'')  = traverse rest ls'
  traverse [] ls                    = ([], ls)
```

Fig. 3. Template Haskell generation of spread for any tuple nesting (buildP omitted)

a single workerstream which is spread to yield a tuple of streams instead of a stream of tuples. Fig. 3 shows how Template Haskell [7] is used to flexibly create the needed version of spread. New requests are appended to the list of pending requests while new tasks are added to the task pool which gets transformed by tt. One could in principle also extract the results out of spread via x; but this would result in non-termination since after processing all tasks the master would wait forever for further worker messages containing new tasks.

Therefore the *state part* has been introduced to care for termination detection and result accumulation. The terminate function carries a state consisting of a set of incomplete tasks, the number of complete tasks in the task pool, the accumulated results, and the number of accumulated results. In addition to the continuous evaluations in the stream part, terminate traverses workerstream a second time in a stepwise fashion. For every answer from a worker process, terminate will run tt on the incomplete tasks extended by the received new tasks and update its t counter accordingly. The result counter r is incremented by 1, as every answer delivers exactly one result. If then the new counters t' and r' are equal, which means that for every complete task issued to the task pool a result has been received, the workpool terminates giving back the reversed list of results. If, on the other hand, t' > r', then the remaining incomplete tasks together with the new counters and the result list will be used for a tail-recursive call to `terminate`. The remaining case `t' < r'` can never happen since every step will yield only one result.

When constructing a proper task pool transformation function `tt` for the workpool one has to be careful because:

- In the stream part `tt` is applied once to a stream of tasks while in the state part it is applied many times to a finite task pool. It has to behave correctly in both situations.
- As interdependent task and result streams are used it is necessary to produce as much output as possible with as few inputs as possible. Therefore delayed matching (via the lazy matching operator ~ or selection functions `head` and `tail`) and the earliest possible production of results should be used.
- Transformation often means combination or comparison which implies searching the task pool. As the task pool is potentially infinite one runs the risk of searching for (and then blocking on) not yet existent tasks.

As `tt` will often in some way have to combine incomplete tasks to complete ones, we present in Fig. 4 a predefined function `ttransform` for doing this while taking some care of the aforementioned dangers. One has to provide only two arguments to `ttransform` to get a full version of `tt`: Firstly, a predicate `cp`, which checks whether a given task is *complete* or not. Secondly, a function `co` which takes a set of mixed complete and incomplete tasks and tries to *combine* as many incomplete tasks as possible. Its results are the already complete tasks together

```
ttransform,ttransform2 :: (t -> Bool) ->                    -- complete, cp
                          ([t] -> ([t],[t],Int)) ->         -- combine,  co
                          ([t], [t], [t], Int) -> ([t], [t], [t], Int)
ttransform cp co old@(tasks, incomplete, complete, n)       -- Step 1
  = if (not (null incomplete))
      then let (ct,it,d) = co incomplete
           in if (not (null ct))
                then ttransform  cp co (tasks,it,ct++complete,n+d)
                else ttransform2 cp co old
      else ttransform2 cp co old

ttransform2 cp co (t:ts, incomplete, complete, n)           -- Step 2
  | cp t      = (tts1,iis1, t:ccs1, d1+1)
  | otherwise = (tts2,iis2,ct++ccs2, d2+d)
  where (tts1,iis1,ccs1,d1) = ttransform cp co
                                  (ts, incomplete, complete, n)
        (ct,it,d)           = co (t:incomplete)
        (tts2,iis2,ccs2,d2) = ttransform cp co (ts, it, complete, n)
ttransform2 cp co ([],incomplete, complete, n) = ([], it, ct, n+d)
  where (ct,it,d) = co incomplete
```

Fig. 4. Higher-order function `ttransform` for task pool transformation

with the newly completed tasks, the currently not combinable incomplete tasks, and the number of newly generated complete tasks.

To avoid the above mentioned danger of blocking when trying to find partners for incomplete tasks we will make only a single traversal over the task list and use an accumulator to carry not yet combined tasks with us. For that purpose `ttransform` carries a state argument consisting of the remaining task stream, the accumulator, a stream of complete tasks (its result), and the number of new complete tasks (needed to correct termination detection counters). `ttransform` is divided in two steps: The first step postpones any matching on the input task stream and tries to combine incomplete tasks inside the accumulator as long as possible. Only if that fails, it matches the first task of the task stream and acts depending on its completeness. Complete tasks are immediately passed to the output stream while incomplete ones are tried to be completed. By considering data dependencies the user has to make sure that enough complete or completable tasks are generated in the right order by his application.

2.3 The Nested Workpool

A growing number of workers or tasks induces heavy traffic at the master process which then apparently quickly becomes a bottleneck for the whole workpool scheme. This can be avoided by having more independent workers which manage a buffer of tasks for themselves. In other words: We will replace each worker by another workpool for local task distribution. Fig. 5 shows the code for such a nested workpool with an even arbitrary nesting depth ≥ 1. For depth 1 the previously defined workpool is returned. The depth is controlled by the (equal) length of the first three argument lists which contain the number of workers (or submasters respectively), the prefetch, and the task transformation function for each level of the workpool tree. The nesting itself works by folding the zipped arguments for each level with the `wpool` function. The worker function `f` is used to close the workpool tree with a set of worker leafs. Note the use of `repeat`: No submaster will migrate tasks to masters above him, therefore newly created tasks will only be sent by the worker leafs to their respective master processes. Fig. 6 shows an example call of the nested workpool together with the resulting process tree; additionally for each argument it is shown to which level it applies. The termination detection of `wpool` fits smoothly into this nesting.

```
wpN :: (Trans t, Trans r) =>
  [Int] -> [Int] ->                                  -- #workers, prefetches
  [(([t],[t],[t],Int) -> ([t],[t],[t],Int))] -> -- transformations
  ([t] -> [(r,[t])]) ->                              -- worker function
  [t] -> [r]                                         -- tasks, results
wpN ns pfs tts f initTasks = results where
  (results,_)    = unzip ((foldr fld f (zip3 ns pfs tts)) initTasks)
  fld (n,pf,tt) wf = \ts -> zip (wpool n pf wf tt ts) (repeat [])
```

Fig. 5. Nested workpool

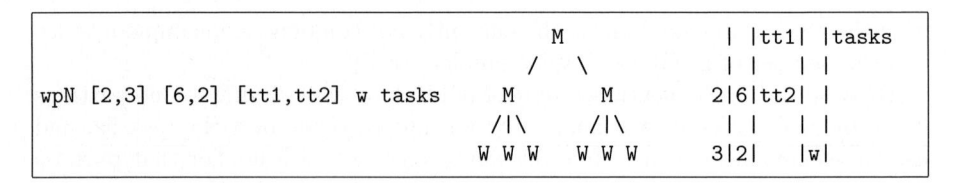

```
                                     M                  | |tt1| |tasks
                                    / \                 | |   | |
wpN [2,3] [6,2] [tt1,tt2] w tasks  M   M              2|6|tt2| |
                                  /|\ /|\               | |   | |
                                  W W W W W W         3|2|   |w|
```

Fig. 6. Two-level example call of wpN with process tree and argument distribution

3 Case Study: Parallel Sequence Alignment

We have used the extended workpool of Sect. 2.2 to parallelise the alignment of DNA sequences via the linear Needleman-Wunsch [8] algorithm. Although not being very efficient, the algorithm serves as a good example for wavefront parallelism [9]: Within a matrix structure the algorithm exhibits diagonal wavefront dependencies (see Fig. 7) which can be expressed as tasks for execution via the extended workpool. More specifically: Each block depends on its two left and upper neighbours in the matrix. Therefore each result produces incomplete tasks for its right and lower (not yet computed) neighbours. Two of these incomplete tasks will then be combined in the task pool to form a new complete task. Only elements of the first row and the first column can be computed given only one of their respective neighbours.

Fig. 8 shows on the left the relative speedup of the parallel sequence alignment algorithm using the extended workpool. All measurements were taken on a cluster of nine Linux PCs connected via 100 Mbit ethernet. The PCs are not completely identical, but this is compensated by the dynamic task distribution of the workpool. Sequences of length 10.000 have been tested with a varying block partitioning. The figure shows that a medium task granularity (block size 500) has paid off the most in our experiments. Larger tasks result in task shortage, while smaller tasks induce too much administrative overhead due to their large number. The nested workpool cannot be used to reduce that overhead, since tasks of different subworkpools would have to be combined. We are aware that our unoptimised implementation of a suboptimal algorithm is slower than modern imperative alignment solutions; it nevertheless serves as a good example of wavefront parallelism for our workpool.

Fig. 8 shows on the right an activity diagram for the execution of the parallel sequence alignment on nine processors (length 10.000, block size 500). Each row represents the activity of one processor during execution, starting on the left

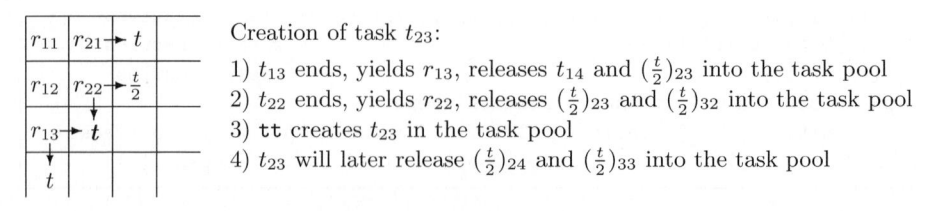

Creation of task t_{23}:

1) t_{13} ends, yields r_{13}, releases t_{14} and $(\frac{t}{2})_{23}$ into the task pool
2) t_{22} ends, yields r_{22}, releases $(\frac{t}{2})_{23}$ and $(\frac{t}{2})_{32}$ into the task pool
3) tt creates t_{23} in the task pool
4) t_{23} will later release $(\frac{t}{2})_{24}$ and $(\frac{t}{2})_{33}$ into the task pool

Fig. 7. Task creation for parallel sequence alignment ($\frac{t}{2}$ represents an incomplete task)

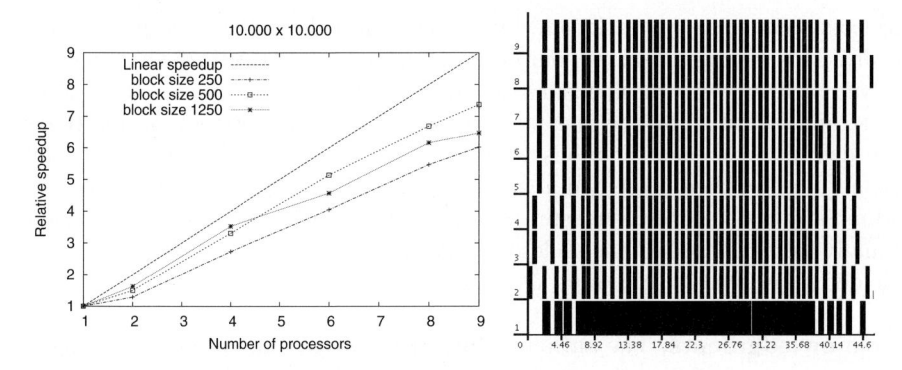

Fig. 8. Relative speedups and activity diagram (sequences of length 10.000, 9 nodes)

and ending on the right at around 45 seconds. White areas represent phases of inactivity or blocking on not yet available data (combined for better visibility), while black areas represent active computation or communication. The lowest row (processor 1) contains the master process which shows constant activity in distributing and combining tasks. The remaining rows show the activity of the worker processes. These are evenly loaded with tasks. Both start and end phase of the computation show clearly the growing and shrinking task availability induced by the diagonal wavefront traversal of the matrix described in Fig. 7.

4 Related Work

An older survey by Stephens [10] describes approaches to stream programming in general. Kahn et al. already described in [11] a model of functional processes communicating via streams. Also the big complex of Dataflow languages [12] has to be mentioned in the context of stream programming. In [2] we have already shown a basic workpool skeleton for Eden which we have extended by dynamic task creation, task pool transformation, and termination detection in this work. Martínez and Peña describe in [13] another workpool scheme for including dynamic task creation. Their approach also introduces state for the master and worker processes aiming at branch-and-bound algorithms which is not covered by our approach. However, only the master is allowed to create new tasks; furthermore they implement the scheme via continuous state updates and do not offer task pool transformation. Regarding our application we have not been able to find another application of parallel functional languages to DNA sequence alignment. A non-parallel application of Haskell, however, has been described in [14].

5 Conclusion

We have developed a new generalised workpool skeleton for the parallel Haskell dialect Eden by adding two features for dynamic task handling: Firstly, worker processes can generate tasks and insert them into the global task pool dynamically. This requires a more complicated termination detection which we have solved by a counting mechanism. Secondly, the master process is enabled to traverse and to transform the task pool to cope e. g. with incomplete tasks. To ease the definition of such functions we have given a function for task pool transformation. We then presented a way to nest the workpool skeleton to lower the administrative load of the master process by introducing additional submasters. Finally we have applied the skeleton to parallel DNA sequence alignment which yielded good relative speedups.

Acknowledgments. The author thanks Rita Loogen for carefully reading the paper and Hans-Philipp Annen, Simon Göbel, and Simon Wiesler for their work on the parallel sequence alignment.

References

1. Peyton Jones, S., et al.: Haskell 98: A Non-strict, Purely Functional Language (2003) See: http://www.haskell.org/onlinereport.
2. Klusik, U., Loogen, R., Priebe, S., Rubio, F.: Implementation Skeletons in Eden: Low-Effort Parallel Programming. In: IFL 2000, Aachen, LNCS 2011. (2001)
3. Cole, M.: Algorithmic Skeletons: Structured Management of Parallel Computation. MIT Press (1989)
4. Rabhi, F.A., Gorlatch, S.: Patterns and Skeletons for Parallel and Distributed Computing. Springer-Verlag (2003)
5. Loogen, R., Ortega-Mallén, Y., Peña, R.: Parallel Functional Programming in Eden. Journal of Functional Programming, Special Issue on Functional Approaches to High-Performance Parallel Programming (2004)
6. Oak Ridge National Laboratory: Parallel Virtual Machine (2003) See: http://www.csm.ornl.gov/pvm/pvm_home.html.
7. Sheard, T., Peyton Jones, S.: Template Meta-programming for Haskell. In: Haskell Workshop 2002, ACM Press (2002)
8. Needleman, S.B., Wunsch, C.D.: A general method applicable to the search for similarities in the amino acid sequence of two proteins. J. Mol. Biol. (48) (1970)
9. Anvik, J., et al.: Generating Parallel Programs from the Wavefront Design Pattern. In: Proceedings of the 7th International Workshop on High-Level Parallel Programming Models and Supportive Environments (HIPS'02). (2002)
10. Stephens, R.: A survey of stream processing. Acta Informatica **34**(7) (1997) 491–541
11. Kahn, G., MacQueen, D.: Coroutines and Networks of Parallel Processes. In: IFIP 77, North Holland (1977)
12. Hanna, J.R.P., Johnston, W.M., Millar, R.J.: Advances in dataflow programming languages. ACM Computing Surveys **36**(1) (2004) 1–34
13. Martínez, R., Peña, R.: Building an Interface Between Eden and Maple: A Way of Parallelizing Computer Algebra Algorithms. In: IFL 2003, Edinburgh. (2004)
14. Giegerich, R., Kurtz, S., Weiller, G.: An Algebraic Dynamic Programming Approach to the Analysis of Recombinant DNA Sequences. In: Proceedings of Workshop on Algorithmic Ascpects of Advanced Programming Languages, Paris. (1999)

Data Parallel Iterators for Hierarchical Grid and Tree Algorithms

Gerhard Zumbusch

Friedrich-Schiller-Universität Jena, Institut für Angewandte Mathematik,
Ernst-Abbe-Platz 2, 07743 Jena, Germany
zumbusch@mathe.uni-jena.de
http://cse.mathe.uni-jena.de

Abstract. The data parallel programming language construct of a "for-each" loop is proposed in the context of hierarchically nested arrays and unbalanced k-ary trees used in high performance applications. In order perform an initial evaluation, an implementation of an automatic parallelization system for C++ programs is introduced, which consists of a preprocessor and a matching library for distributed memory, shared memory and mixed model parallelism. For a full compile time dependence analysis and a tight distributed memory parallelization, some additional application knowledge about alignment of arrays or indirect data access can be put into the application's code data declarations. Results for a multigrid and a fast multipole benchmark code illustrate the concept.

1 Introduction

High performance computing should be about a single application of large scale such that both memory size and computing time of a parallel computer limit the precision of the solution computable. A single algorithm operates on a large data set, distributed over the local memories of the parallel processors. Data structures may be uniform or unstructured grids, cells or trees, that is large containers of relatively small, numerical data. It is usually not a good idea to move a substantial amount of data to another local memory or even to redistribute data during computation. Hence a data parallel programming style seems to be natural with operations performed on all elements of the large container. Further, the operations have to operate almost on a local neighborhood only, to be efficiently parallelizable. The "owner computes" paradigm guarantees local memory store operations, such that non-local load operations are the main source of inter process communication.

Parallelization of a sequential code can be done in several steps. First, local and global data dependence analysis can be applied. However, currently they fall short for more complicated data structures and require additional specification [1]. The second step of parallelization is a scheduling and mapping step. Independent operations have to be combined to larger blocks which are mapped to processes or threads. The mapping problem can be far more serious, because again global dependence analysis of the code is required. Once the data structures are instantiated, scheduling and mapping can be written as a large graph problems. However, at compile time the graph is unknown and solutions are available only in very simple cases such as arrays and uniform grids with regular access patterns. This is implemented in numerous data parallel array constructs.

Global dependence analysis of imperative sequential codes is unlikely to solve the scheduling and mapping problem at compile time in general. However, there often is application knowledge, such as geometric properties within a grid or tree, sufficient to enable an efficient parallelization by hand. It is not very economic, however popular to create a full featured parallel programming language for each application area and to incorporate this knowledge. On the other hand, standard library design for common languages is not able to forward this knowledge to an optimizing/parallelizing compiler. Hence there is a current trend to combine library and compiler or some kind of optimizing preprocessor in order to allow for application specific knowledge for parallelization in an abstract programming environment. Such effort include the use of expression templates and extensible source-to-source compilers/optimizers and tools like Rose [2]. A more general concept of application specific code optimization are telescoping languages [3].

The goal of the article is to discuss a preprocessor/library system for parallelization of array and more complex tree data structures common in high performance computing. The sequential programming language is extended by a single data parallel "foreach" construct together with data iterators defined by the library. Data structure dependent parallelization knowledge is confined to the library, application specific parallelization knowledge such as alignment or non-local references can be specified in the application code.

A model implementation uses C++ class libraries and a set of perl scripts, the m4 macro processor and the Gnu g++ compiler to do the local dependence analysis and the source-to-source transformations. Targets currently are distributed memory computers with MPI message passing, shared memory computers with pthreads and hybrid systems with MPI on processes and pthreads to spawn several threads per process. Hierarchies of grids in a multigrid code and unbalanced k-ary trees in a fast multipole code are used to demonstrate the concept. The emphasis of this paper is on the parallel programming style, especially for parallel *tree* algorithms, rather than its model implementation, which can easily be improved. We do acknowledge a large number of alternative solutions for parallel array style programming, which again is not the main subject here. We do *not* advocate the use of scripts and preprocessors for parallel computing, but would like to foster the development of more general and easier ways to incorporate domain specific knowledge into the parallelization of codes. However, even more sophisticated solutions will never be able to perform automatic parallelization of all possible codes.

2 Data Parallel Programming Paradigms

We consider different target architectures. The current preprocessor scans standard C++ code augmented by the "foreach" construct and emits multi-threaded code for shared memory computers with pthreads, message passing code for distributed memory computers with library calls based on MPI 1, and a combination of message passing between processes and multi-threading within. The message passing calls and parts of the data iterators are encapsulated in a C++ run-time library.

The strategy for distributed memory computing is based on the following considerations:

- Distribute large data structures. Each element is mapped to one process (owner), which is the only process to modify it: "owner computes". The mapping is implemented in the library and is application specific.
- Replicate small data structures and small numbers of operations thereon for each process instead of sending data.
- Use as few send and receive operations as possible. transfers.
- Transfer only data necessary. Perform a dependence analysis at compile-time to determine which data needs to be sent.

Basic message passing operations needed are matching point-to-point send/receive and global reduction operations. The overall performance of the parallel code relies on the pre-computed minimal communication pattern compared to dynamic distributed-shared-memory and related techniques.

Shared memory versions with global address space are easier to implement. The parallel iterator on large data structures with static mapping first cuts the data into several pieces of similar size, and then starts a thread which executes the iterator on each piece and finally waits for the threads to finish. Data is partitioned according to memory layout. Each thread is allowed to modify its own piece of data only, with the exception of global reduction of scalars. Such reductions are done locally for each thread, with a final reduction over all threads.

3 Array Operations

For illustration purposes only we begin with well known for-loops and distributed arrays. We use a block distribution and restrict ourselves to the important part of nearest neighbor communication. This occurs for Finite Difference discretizations on cartesian grids. Each element of an array is associated to a grid point which itself represents a geometrical location. The resulting compact difference stencils represent a finite geometric interaction distance between grid points. Hence it is a good idea to decompose the geometric domain for parallelization, which is done by an array block distribution. Of course, a parallel compiler is not able to figure this out without global code analysis. Hence, the application programmer provides the geometric interpretation implicitly for parallelization through the specification of a block distribution.

3.1 Sequential Semantic

We begin with a code snippet in C++ creating a one dimensional grid, an iterator for all interior grid points and two arrays on the grid. The grid is defined by an index set, the interval $(0, n+1($. Arrays are allocated according to this index set. Further we introduce an iterator on a subset of the grid, here $(1, n($, in order to treat boundary indices separately.

```
int n = 64;
Grid1 *g = new Grid1(0, n+1);
Grid1IteratorSub it(1, n, g);
DistArray1<double> x(g), y(g);
double e = 0.;
ForEach(int i, it,  x(i) += ( y(i+1) + y(i-1) )*.5;  e += sqr( x(i) ); )
```

The "foreach" loop based on the index set of the iterator expands to the sequential code

```
for(int i=1; i<n; i++) {x(i) += (y(i+1) + y(i-1))*.5;  e += sqr(x(i));}
```

but provides the semantic of independent operations, and consists of arbitrary (reentrant) C++ code including nested function calls. The result is guaranteed to be independent of the sequence. For the reduction of the variable e this is only true up to floating point rounding. A two dimensional grid example including different iterators inside and on the boundary reads like this. The grid is represented by a set of index tuples, here $(0, n+1(\times(0, n+1($. Iterator ita visits all tuples in $(1, n(\times(1, n($ and iterator ita all tuples except for $(1, n(\times(1, n($.

```
Grid2 *g2 = new Grid2(0, n+1, 0, n+1);
DistArray2<double> z(g2), a(g2);
Grid2IteratorSub ita(1, n, 1, n, g2);
ForEach('int i, int j', ita, 'z(i,j) = ( a(i-1,j-1) + a(i+1,j+1) +
                    a(i-1,j) + a(i+1,j) + a(i,j-1) + a(i,j+1) )/6.;')
Grid2IteratorOutside itb(1, n, 1, n, g2);
ForEach('int i, int j', itb, 'z(i,j) = 0.;')
```

Basically, the nesting of the i and j loops and the execution order is not specified. The "foreach" syntax including comma separator and ' ' quotation marks are due to an m4 preprocessor step and could be changed to semicolon and {} brackets for more C style. Of course there are many different ways to express this including array operations.

3.2 Code Analysis

In the current implementation a sequence of preprocessing steps identifies the variables and types used in the "foreach" loop, checks for data and loop dependence (including inter procedure analysis) and issues warnings if the code does not seem to be parallel, emits communication operations such as send/receive and reduce, transforms loop code and finally creates C++ source code. The code can be compiled with a run-time library which provides the implementations of grids, arrays and the remaining parts of the iterator. We briefly comment on some rationales.

Replicated data, i.e. scalars and small data structures allocated on each process can either be read-only (store is disallowed) in a loop or a reduction variable (simple load is disallowed). In the case of a reduction, special code is created for thread and/or message passing environments.

Distributed data container and iterators ought to match. Assume that the iterator and the arrays involved share the same index space and distribution. Then it is easy to detect the disallowed cases of non-local store (violates the owner-computes-rule), references to elements more distant than direct process neighbors (violates nearest neighbor communication) and loop carried dependence (non-local load together with local store). Only output-dependence for global objects (e.g. file descriptor cout) is allowed with non deterministic output order. Non-local load operations trigger appropriate send/receive message passing code, which is executed prior to the "foreach" loop.

The model can be generalized to non-neighbor communication, which raises the questions of appropriate data distributions. For indirect addressing see the following chapter on trees.

3.3 Arrays of Different Shapes

The computational model for arrays so far can be extended relaxed to arrays and iterators based on different grids. We are aiming at the multigrid application with a set of nested grids to be discussed later. A notation of grid alignment is used which is slightly different than the HPF guarantees a relationship between the distributions of the arrays. In order to use a fixed communication scheme, a finer grid is created aligned to a coarser one using a mapping function. Further, to be able to compute the communication patterns at compile time, the mapping is also passed to the "foreach" loop as a C++ template type.

To be more precise, assume two one dimensional arrays, a base array with integer index space $[n_0, n_1) \subset \mathbb{Z}$ and another, possibly larger array with index space $[m_0, m_1)$. We define a (truncated) affine monotone mapping $\pi : \mathbb{Z} \to \mathbb{Z}$, which can be written as $\pi(i) = \lfloor (i-k)/m \rfloor$ with constant $m \in \mathbb{N}$, $k \in \mathbb{Z}$. Each index $i \in [n_0, n_1)$ of the base array is mapped uniquely to process $p(i) \in \mathbb{N}_0$. A grid $[m_0, m_1)$ is said to be aligned, iff index $j \in [m_0, m_1)$ is mapped to process $p(\pi(j))$. The mapping and an example code, simplified versions of the following application multigrid code's restriction and prolongation operations, looks like this:

```
class fine { public: int map(int i) {return i / 2;} } f;
Grid1 *gf = new Grid1(0, 2*i+1, g, &f);
DistArray1map<double, fine> z(gf);
ForEach(int i, it, x(i)     = z(2*i)*.5 + ( z(2*i-1) + z(2*i+1) )*.25; )
ForEach(int i, it, z(2*i)   = x(i);
                   z(2*i+1) = ( x(i) + x(i+1) )*.5; )
```

The first "foreach" loop triggers a left process fetch for the array z on the finer grid, while the second one triggers a right fetch for array x, while still being a local store operation under transformation π.

4 Tree Operations

Now we consider the main target of the paper, namely algorithms on tree data structures with the same "foreach" loop syntax. For alternative ways to write tree iterators see [4]. With a fast multipole summation of particle-particle interactions in mind, a k-ary tree represents a hierarchical decomposition of the computational domain with particles at the leafs of the tree according to their geometrical location. The tree can be written as a directed acyclic graph starting from a root node. A useful data partition in distributed memory starts with a coarse sub-tree from the root node which is replicated on each process. The remaining nodes form a forest of trees, with each tree mapped to one process. The mapping may combine trees geometrically or by some graph partitioning scheme, see [5].

4.1 Communicationless Tree Traversal

The following code snippet shows part of the tree declaration, but hides the library's tree implementation.

```
class tree : public KAryTree<class tree, 2> {
public:     // generic binary tree provides  tree* child(int);
  complex<double> m, l, f, x;
... };
tree *root = new tree;
```

Tree creation proceeds by (parallel) insertion of particles or (parallel) sorting according to some partitioning scheme, where algorithms of different types are involved. Geometric domain decomposition, graph partitioning, space-filling curves and other techniques [5] are available to partition the data in an initial step or after a number of (time-) steps e.g. in a particle simulation. We consider iterators for tree traversal only.

```
TopDownIterator<tree> down(root);
ForEach(tree *b, down, b->f = b->l; )
ForEach(tree *b, down, '
  for (int i=0; i<2; i++)
     if (b->child(i))  b->child(i)->l += b->l; ')
```

The order of execution is no longer arbitrary, but partially ordered, in this case top down from root to the leaves, such that lots of parallelism is exposed. Operations on the replicated coarse tree are executed on all processes and operations on the remaining trees are executed by the respective owners. The first "foreach" loop shows a local assignment completely independent of the execution order. The second one performs a local store at the child nodes, such that a strict parent before child order has to be enforced. Different orders of load and store operations which lead to loop carried dependence trigger warning messages of the preprocessor using path matrix dependence analysis [1]. Except for possible global reduction operations no parallel communication is needed.

4.2 Bottom-Up Communication

A bottom up, leaf to root execution order is shown in the next example.

```
BottomUpIterator<tree> up(root);
ForEach(tree *b, up, '
  for (int i=0; i<2; i++)
    if (b->child(i))  b->m += b->child(i)->m; ')
```

Now, the processes first execute the operations on their own sub trees in a children before parent order. A communication step at the replicated coarse tree's leaf nodes is necessary to update them, which currently uses a global message passing gather operation. Afterwards the operations can be executed on all processes on the coarse tree. The preprocessor determines variables to be transferred (here: m). The library does the packing and un-packing. Additional communication would be needed for global reduction operations. Non-local store or different load operations leading to loop dependence would again trigger warning messages. The shared memory implementation performs a synchronization step instead of the communication, with coarse tree operations done by a single thread.

4.3 Communication Within a Geometrical Neighborhood

Besides parent to child and children to parent data flow, fast summation techniques also rely on neighborhood data on all tree levels. However, the nodes actually needed are a small fraction of the full tree and are often determined geometrically. Assume that the operation on node i requires data of 'neighbor' node j, which we denote by relation $i \wedge j$. Each fine tree node i is mapped to process $p(i)$. In a communication step, data of nodes $\bigcup \{j \mid i \wedge j,\ p(i) = p_1,\ p(j) = p_2\}$ has be send from p_2 to p_1. In order to do this efficiently, a hierarchical hull relation $i \triangleleft j$ is needed with $i \wedge j \Rightarrow i \triangleleft j$ and $i \triangleleft j \Rightarrow \mathrm{parent}(i) \triangleleft j$ and $i \triangleleft \mathrm{parent}(j)$. Using relation $\triangleleft$ on the coarse tree representation of the data partition is sufficient and each process is able to compute a superset of nodes to be transferred. Such relations are available for many tree codes and one might try to construct them based on more abstract specifications [6].

The following statements within class tree declaration define the relation $\triangleleft$ fetch, which guards all accesses to nodes pointed to by elements of the interaction list inter.

```
  Require( list<tree*> inter,  fetch );
  double x0, x1;
  int fetch(tree *b) { return (x0==b->x1) || (x1==b->x0); }
```

Statement 'require' both declares the variable inter and attaches the relation fetch to it. The following code shows a tree iterator using indirect addressing.

```
ForEach(tree *b, down, '
  for (list<tree*>::const_iterator i = b->inter.begin();
       i != b->inter.end(); i++)
    b->l += log(abs(b->x - (*i)->x)) * (*i)->m;  ')
```

Each process collects all nodes defined by ◁ on the coarse tree's leafs and may be needed by another process and forwards these. The preprocessor determines the variables actually needed (here: x and m) and looks for loop dependencies. The implementation proceeds with the tree traversal. During the creation of the interaction list for example, no data except for the tree structure is needed.

5 Applications

Basically we want to demonstrate the feasibility of the proposed way of expressing parallelism in numerical array and tree codes. It is essential to see some test cases can be written and translated to lower-level parallel code this way. The communication patterns, the number and volume of messages and the placement of thread synchronization points are identical to hand written code based on the same parallelization strategy. Since such a parallelization is known to be efficient for the given applications, we do not explore in detail the scalability for large numbers of processors or different hardware platforms. A direct comparison to a hand written parallelization is not expected to give new insight at this stage of development.

First we consider a test example for hierarchical arrays. The NAS multigrid benchmark code Fapin [7] implements a geometric multigrid $V_{0,1}$-cycle with one post-smoothing step for a Poisson equation on a set of nested three-dimensional cartesian grids with constant coefficients. The Fortran77 code was ported C++ using the distributed array classes and run for larger data sets (fine grid 513^3) than originally conceived. All message passing and multi-threaded timings are reported for an eight-processor (4 dual-core) AMD64 at 1.8GHz with Scientific Linux 4.1 and Gnu g++ compiler 3.4.3 in 64bit address mode binaries, optimization 'O3'. Compared are timings for Mpich (shmem device) and the native pthread library, see Table 1 left. The mapping of MPI processes and shared memory threads onto the four physical processors and their two processing cores is done dynamically by the operating system.

The parallel speedup on eight processor cores indicate that both different strategies, message-passing (shown vertically) and threads (shown horizontally) work almost equally well with slight advantages for the local address space message-passing. This advantage is probably due to a strong processor to MPI-process binding compared to an arbitrary mapping of processors at each synchronization point of the multi-threaded implementation with effects on access to and caching of local memory banks. This seems to outweigh message passing overhead, which is limited to the transfer of a small fraction of all data. Parallel efficiencies are well above 70% and seem to be limited by the memory bandwidth rather than less efficient coarse grid computations. The measured times for two and four processors involved showed larger variations in different runs due to operating system scheduling, with minimum times shown in the table. Without idle processors the effects vanish. Additional slackness due to more jobs than processors does not improve the numbers significantly, with slight improvements for additional message-passing processors.

Table 1. Parallel speedups of the message-passing (> 1 processes), multi-threaded (> 1 threads), and hybrid (> 1 processes and > 1 threads) program versions on 8 processor cores. Multigrid test case on a nested set of 3D arrays (left) and 2D adaptive fast multipole test case (right).

threads per process		1	2	4	8		1	2	4	8
no. of processes	1	1	1.50	2.44	5.27	1	1	1.85	3.59	6.71
	2	1.92	2.51	5.58	5.70	2	1.94	3.63	6.80	2.87
	4	2.38	5.77	5.57	5.67	4	3.91	7.64	6.79	4.28
	8	5.91	5.63	5.76	5.80	8	7.79	7.76	7.78	7.73

The second test case covers a hierarchical tree algorithm. Based on parts of the two-dimensional adaptive fast multipole C code FMM of the shared memory Splash-2 benchmarks [8], a C++ implementation using the distributed quad-tree was developed. For reasons of simplicity we consider only at most one particle per leaf cell, using a multipole Laurent series and a local polynomial with 20 complex coefficients each. The tree is partitioned on balanced coarse trees both for message-passing and for multi-threading, although the tree populated with $2 \cdot 10^6$ particles is unbalanced and the tree implementation works for arbitrary partitions. Measured are the times of one field evaluation by the fast multipole method. Parallel load-balancing and tree creation like in [5] could be inserted here additionally, at an initial step and after a couple of multipole evaluations, but involve algorithms of different types (e.g. parallel sorting or embarrassingly parallel) not discussed here. A parallel Barnes-Hut algorithm could be implemented similar to the fast multipole method, but is of higher complexity. The timings were made on the eight-processor system like before. The results are in Table 1 right.

We see again efficient parallelization both for message-passing and for multi-threading with larger advantages for local address space message-passing. Extremely high 97% parallel efficiency are obtained for 8 processes and any number of threads per process. Due to irregular memory access patterns, the shared memory version is slightly slower. Again we obtain large variations in measured times for the partially loaded computer with less than 8 jobs. For the 8 thread case in message passing (2 and 4 processes) we obtain some reproducible timing anomalies. The overall parallel efficiency of the tree code is extremely good as to be expected for large trees of this type.

6 Outlook

We have demonstrated that automatic parallelization does work even for hierarchical algorithms and data structures in high performance computing with a parallelization strategy similar to the ones used for parallelization by hand. However, domain or application specific language extensions were necessary, which in this case were provided by a combination of a source-to-source preprocessor and a dedicated library.

The parallelism detected so far has been used for coarse grain parallelism. Within a job it is currently not used further. A possible extension would be to exploit the dependence analysis also for code optimization of memory access patterns for hierarchical memory, instruction level parallelism, software-pipelining and for techniques like hyper-threading.

The "foreach" loops and data parallel iterators may also be implemented by a fully fledged C++ source-to-source translator instead of the current scripting solution, which would certainly be an improvement. The concept of telescoping languages would include to have the parallel iterator programming style interoperable with other (parallel) libraries and language extensions. Furthermore, we would like to see easier ways to exploit application specific knowledge for the parallelization in the future.

We would like to thank the anonymous referees for their helpful comments.

References

1. Hendren, L.J., Hummel, J., Nicolau, A.: Abstractions for recursive pointer data structures: Improving the analysis and transformation of imperative programs. In: Proc. ACM SIGPLAN 1992 conf. Programming language design and implementation, ACM (1992) 249–260
2. Quinlan, D., Schordan, M., Yi, Q., de Supinski, B.R.: Semantic-driven parallelization of loops operating on user-defined containers. In: 16th int. workshop LCPC 2003. Volume 2958 of LNCS., Springer (2004) 524–538
3. Kennedy, K., Broom, B., Chauhan, A., Fowler, R., Garvin, J., Koelbel, C., McCosh, C., Mellor-Crummey, J.: Telescoping languages: A system for automatic generation of domain languages. Proc. of the IEEE **93**(3) (2005) 387–408
4. Ananiev, A.: Algorithm alley: A generic iterator for tree traversal. Dr. Dobb's J. **25**(11) (2000) 149–154
5. Zumbusch, G.: Parallel Multilevel Methods. Adaptive Mesh Refinement and Loadbalancing. Teubner (2003)
6. Birken, K.: Semi-automatic parallelisation of dynamic, graph-based applications. In: Proc. Conf. ParCo'97, Elsevier (1998) 269–276
7. Bailey, D.H., Barszcz, E., Barton, J.T., Browning, D.S., Carter, R.L., Dagum, L., Fatoohi, R.A., Frederickson, P.O., Lasinski, T.A., Schreiber, R.S., Simon, H.D., Venkatakrishnam, V., Weeratunga, S.K.: The NAS parallel benchmarks. Inter. J. Supercomp. Appl. **5**(3) (1991) 63–73
8. Woo, S.C., Ohara, M., Torrie, E., Singh, J.P., Gupta, A.: The SPLASH-2 programs: Characterization and methodological considerations. In: Proc. 22nd annual int. symp. computer architecture, ACM (1995) 24–36

Implementing Irregular Parallel Algorithms with OpenMP
What's Missing and How to Solve It

Michael Süß and Claudia Leopold

University of Kassel, Research Group Programming Languages / Methodologies,
Wilhelmshöher Allee 73, D-34121 Kassel, Germany
{msuess, leopold}@uni-kassel.de

Abstract. Writing irregular parallel algorithms with OpenMP has been rarely practised in the past. Yet it is possible, and in this paper we will use a simple breadth–first search application as an example to show a possible stepping stone and deficiency of the OpenMP specification: It is very difficult to cancel the threads in a parallel region. A way to work around the issue within the existing OpenMP specification is sketched, while our main contribution is a proposal for an extension of OpenMP to solve the issue in an easier way.

1 Introduction

OpenMP[1] is a parallel programming system that aims to be powerful and easy to use, while at the same time allowing the programmer to write high performance programs. Its initial focus was on numerical applications involving loops written in Fortran or C/C++, but it includes the necessary constructs to deal with more kinds of parallel algorithms.

Irregular parallel algorithms involve subcomputations whose amount of work is not known in advance, and hence the work can only be distributed at runtime. Important subclasses include algorithms using taskpools, as well as speculative algorithms. We are concentrating on the first type, although the problem and solutions we present apply to other types as well. Examples for irregular algorithms are search and sorting algorithms, graph algorithms, and more involved applications like volume rendering.

According to Mattson [2], one of the initial designers of the OpenMP specification, OpenMP was never meant for irregular applications (where an irregular application in this context is one containing irregular algorithms as sketched above). Other people have tried to use OpenMP for this kind of applications, though, and have gotten mixed results [3,4,5]. This paper explores an important issue in developing irregular parallel algorithms with OpenMP, which is the missing ability to cancel threads in a parallel region. While a (not completely functional) workaround for the issue is suggested in Sect. 2, the main contribution of this paper is a proposal for new functionality to solve the problem in a convenient and easy to use way on the language level (in Sect. 3). The suggested additions to OpenMP are previewed in Fig. 1. A working implementation can

# **pragma omp cancelregion**:	request cancellation of parallel region
# **pragma omp exitregion**:	take current thread to end of parallel region
int omp_get_cancelled (void):	has the current region been cancelled ?
# pragma omp barrier **oncancel**:	execute scope if thread is cancelled in barrier
# **pragma omp onbarriercancel**:	execute scope if thread is cancelled in implicit barrier

Fig. 1. Thread cancellation in a nutshell

be found in a special version of the OMPi compiler[6], which is available from the authors on request.

As a running example, we use breadth–first search on a labyrinth. The algorithm and its implementation are explained shortly in Sect. 2, where we will also explain why thread cancellation is a problem. Furthermore, a workaround for the issue is presented in this section, while a more advanced solution on the language level is described in Sect. 3. At the end of the paper, Sect. 4 summarizes our findings and shows some prospects for future work.

2 Problem Description

In labyrinth search, the objective is to find the shortest path through a labyrinth, from a given entry to a single exit. This problem is not merely a theoretical one, but has practical relevance e. g. for mapping electrical circuits on a chip. We consider a breadth-first search algorithm, which is not necessarily the fastest choice, but is simple enough to serve as an example here and to still include all the problems we want to illustrate. A very broad sketch of the algorithm is presented in pseudocode in Fig. 2.

The algorithm starts by putting the entry position of the labyrinth into the taskpool (not shown in the pseudocode). Afterwards, it spawns a parallel region (line 1). Then, one of the threads takes a position out of the taskpool (line 4), marks it on a map as processed (line 5), evaluates all neighbours by checking the four possible directions for walls (line 6), and checks if an exit is found on any of them. If no exit was found and the neighbour-positions have not been evaluated before (this check is not shown in our pseudocode), the neighbours are put into the taskpool to be processed in the next step (line 7), possibly by a different thread. If an exit is found, a flag is set that indicates this fact (line 9). We need to be careful with the different positions in the taskpool, since only positions with the same distance to the start should be evaluated together, or else the breadth-first search will degenerate. Therefore, only positions with the same distance to the entry are kept in the taskpool, while the neighbours are put into a different one (called next_taskpool). As soon as the taskpool is empty,

```
 1  #pragma omp parallel
 2  {
 3    while (!exit_found) {
 4      while ((task = pop (taskpool)) != NULL) && (!exit_found)) {
 5        mark (labyrinth, task);
 6        if (!inspect_field_for_exit (task))   // inspect all neighbours
 7          push (neighbours (task), next_taskpool);  // no exit was found
 8        else
 9          exit_found = true;   // an exit was found
10        #pragma omp flush (exit_found)
11      }
12      #pragma omp barrier
13      #pragma omp single
14      {
15        taskpool = next_taskpool; // switch the taskpools
16        next_taskpool = NULL;
17      } // implicit barrier (includes flush)
18    }
19  } // end of parallel region with implicit barrier
```

Fig. 2. Parallel breadth first search, using a flag for thread cancellation

both taskpools are switched by a single thread, and the computation proceeds with the former next_taskpool (lines 15–16). When the algorithm depicted in Fig. 2 is done, a single thread follows the marks set in the labyrinth (line 5) from the exit point back to the entry point and identifies the shortest way.

In the figure, a flag is used to indicate when the threads in the parallel region should finish their work, because an exit was found (indicated by exit_found == true). We know of no other way in OpenMP to indicate that the threads should end their work in a parallel region. In Sect. 2.1, we will point to problems with this approach. Section 3 will present an extension of OpenMP that leads to an easier solution, which we will discuss in Sect. 3.5.

2.1 The Problem with Flags

When using flags to indicate that the parallel region should be aborted, great care has to be taken with checking these flags by the programmer. In our example, it might happen that one thread enters the while loop (line 3), finds an exit, sets the appropriate flag, and afterwards hangs in the barrier (line 12), because another thread does not enter the next iteration of the while loop at all, as the flag is indicating now that an exit was found! The program will exhibit undefined behaviour in this case (most likely a deadlock), because in OpenMP the sequence of barrier constructs encountered must be the same for every thread in the team. Thus, the code in Fig. 2 is not correct, and it is not safe to use without further adjustments that would make it even harder to read and explain!

Flags that indicate when a parallel region is to be cancelled give rise to yet another problem: Due to the OpenMP memory model, the flags have to be updated with a flush directive before their values are guaranteed to be up to date. This step is frequently missed by inexperienced OpenMP programmers [7]. The consequence is similar as sketched above: the program will potentially deadlock, because the thread which set the cancel flag has got its current correct value and will exit the loop, whereas other threads might still use the old value and continue with it[1].

Let us summarize the problems we have identified so far with thread cancellation in OpenMP:

- there is no easy way to branch out of a parallel region, the only possible workaround is to use flags
- it is difficult to work with flags indicating that a region should end, at least as soon as barriers come into play
- if one forgets to flush a flag, a deadlock may arise

While we have presented a workaround for the main problem (flags manually set and checked by the programmer), it is still cumbersome and error-prone. Therefore we will present another possible solution in Sect. 3, based on a proposal to add thread cancellation to OpenMP. The proposal is also useful for the following common scenarios, which could benefit from thread cancellation:

- a cancel button from a user interface was pressed
- a solution has been found in a speculative algorithm

3 Thread Cancellation

This section shows a possible way to extend OpenMP with thread cancellation support. Sect. 3.1 shortly introduces a few basic terms often used when talking and writing about thread cancellation. An actual specification of the new functionality is given in Sect. 3.2, followed by the rationale for some of our design decisions in Sect. 3.3 and a short discussion on implementation and performance issues in Sect. 3.4. Sect. 3.5 puts the specification in perspective, by applying it to the labyrinth example.

3.1 Terms

We speak of *forceful cancellation* when a thread has the ability to cancel another thread from the outside. The cancelled thread may get the opportunity to clean up after itself, yet it does not have the power to decide when to be cancelled, nor to prevent cancellation at all. Asynchronous cancellation in POSIX Threads is an example of forceful cancellation. *Deferred cancellation* is an important subcase of asynchronous cancellation, in which the cancelled thread is not terminated immediately, but only at certain predefined cancellation points. Deferred cancellation is supported in POSIX Threads as well. With *cooperative cancellation*, in contrast, a thread can only ask for the cancellation of another thread. The cancelled thread has the opportunity to honor this request and cancel itself, to process the request at a later time, or even to ignore it altogether. Java threads support cooperative cancellation.

---

[1] This is not an issue in our example, as there is a flush included in many OpenMP directives (e.g. in the implicit barrier on line 17). Nevertheless, when the code is only slightly altered, the problem may surface.

3.2 Specification

The following directives to support cooperative thread cancellation in OpenMP are proposed:

#pragma omp cancelregion

This directive asks all threads in the team to stop their work and go to the end of the parallel region, where only the master thread will continue execution as usual. The emphasis here is on *asks*. The threads in the team are not cancelled immediately, but merely an internal cancel flag is set. The threads are not interrupted in any way and have to poll the flag using one of the directives described below. An exception is the thread that called the directive: it is cancelled immediately by an implicit call of the exitregion directive (explained below). Invoking the cancelregion directive on an already cancelled region has no effect except for the implicit call to exitregion. It is the task of the programmer to check if the cancel flag has been set, using a new OpenMP runtime library function:

int omp_get_cancelled (void)

This function returns 1 (true) if the cancellation of the enclosing parallel region was requested, and 0 (false) otherwise.

#pragma omp exitregion

This directive is not only useful for thread cancellation, but can be invoked at any point in a parallel region to immediately end the execution of the calling thread. This is accomplished by jumping to the end of the present parallel region, right into its closing implicit barrier (which is of course honored).

There is a problem with the proposal so far: barriers. If a region containing barriers is cancelled, at least one thread (the one calling the cancelregion directive) will never reach that barrier. Without further adjustment, one or more of the other threads in the region could hang in the barrier and never recover, since the barrier is not completed.

#pragma omp barrier oncancel

A solution to this problem is proposed in the form of the **oncancel** clause for the barrier directive. A new scope is optionally added to the barrier directive by specifying the oncancel clause. The commands in this scope are carried out only if the present parallel region has been or is being cancelled while the thread is waiting on the barrier. This can be seen on line 12 of Fig. 3.

It is now possible to use barriers in combination with thread cancellation. It remains the task of the programmer to do the right thing when a thread waiting on a barrier is cancelled, although most of the time he will just free the resources associated with the thread and exit the parallel region afterwards (using the newly proposed exitregion directive). Note that if the thread is not finalized with exitregion, it will hang in the barrier again (or phrased differently: there is an implicit barrier at the end of the oncancel clause). The reasons for this design decision are given in Sect. 3.3. The oncancel code is carried out at most once per barrier and thread. Furthermore, if the region is already cancelled when a thread enters the barrier, it will immediately proceed with the oncancel code.

For implicit barriers (at the end of worksharing constructs), a similar construct is proposed:

#**pragma omp onbarriercancel**

The usage of this directive is similar to the oncancel clause suggested above, except that onbarriercancel is a standalone construct and must be specified immediately after the implicit barrier it references. This is shown on line 21 of Fig. 3.

If the directive is present, all commands in its scope are carried out if the region is cancelled before or while the thread is waiting on the barrier. A nowait clause on the referenced worksharing construct and the onbarriercancel directive cannot be specified together. The directive also cannot be specified after a combined parallel worksharing construct (e. g. #pragma omp parallel for), the reasons for this design decision are also given in Sect. 3.3.

OpenMP allows for nested parallelism, i.e., when a member of a team inside a parallel region encounters a new parallel construct, a new subteam is formed. Our proposed extensions apply to nested parallelism as follows: Cancellation requests from inside the subteam only cause members of the subteam to have their cancellation flag set. If another member of the original team requests cancellation however, the cancellation flags for all members of all subteams are set as well, although technically they are not in the same team.

3.3 Rationale

Some of the suggested changes could be emulated manually by the experienced OpenMP programmer (such as keeping track of the cancel state of each thread). As has been explained in Sect. 2, this is, however, an unnecessary burden and gets difficult when barriers are involved at the latest. Therefore, our proposal introduces the new functionality on a language level.

The exitregion directive can be seen as a convenient shortcut, but even without thread cancellation, it is useful as soon as one gets into deeply nested functions inside parallel regions. It allows the programmer to jump to the end of the parallel region immediately, thereby potentially saving many lines of code of conditional statements. If barriers are involved in the parallel region, care has to be taken with exitregion for the reasons described in Sect. 3.2, or else the program might deadlock.

We have decided against forceful cancellation as in POSIX Threads. On one hand, asynchronous cancellation makes resource deallocation practically impossible. Since one never knows when a thread is cancelled, there is no place to put cleanup code. POSIX Threads solves this problem by utilizing cleanup stacks, but these are difficult to handle and keep track of. The concept of having cancellation points and deferred cancellation in OpenMP, on the other hand, seemed like overkill, as the amount of functions which are cancellation points is difficult to handle for programmers. Therefore, this proposal suggests cooperative cancellation, which can be found in a similar way e.g. in Java threads. Other good arguments for the use of cooperative cancellation can be found in the Java documentation [8].

A major problem with cooperative cancellation are the barrier constructs. The suggested solution (oncancel clause, onbarriercancel directive) may seem like a lot of overhead to cope with barriers, but the proposal is still easier and more natural than the possible alternatives (such as disallowing barriers with thread cancellation, putting the burden on the programmer to carefully work around them with flags, cancelling barriers forcefully).

We have also decided against automatically including an exitregion directive at the end of an oncancel or onbarriercancel scope. The main reason for this is consistency, as automatically including the directive would cancel the threads waiting on barriers forcefully. This would be inconsistent with the rest of the proposal, where cooperative cancellation is employed. Another reason is nested parallelism. We have specified in Sect. 3.2 that cancelling a parallel region will cancel all subregions as well. But as a subregion might be presently doing uninterruptible work and may contain barriers, the decision not to cancel on barriers automatically allows these subregions to complete their work when interrupted from threads in the upper parallel region, while properly shutting down when cancelled from inside their subregion.

The reason for not allowing the onbarriercancel directive after combined parallel worksharing constructs is that the two main reasons for applying the directive are not valid after a combined directive. There is no need to take care of left over threads hanging in the implicit barrier at the end of the combined construct, as these threads are exactly where they would be if an exitregion clause was specified. There is also no need to clean up any resources, as the programmer must have already done this before the end of the parallel region.

During our internal discussions on the topic of thread cancellation, we have worked out a checklist that each and every proposal we came up with had to pass. This checklist and some explanations of why our proposal passes it are spelled out here to make our design decisions yet more clear:

1. **Backwards Source Compatibility**
   Old code must run unchanged, when translated with a compiler that understands thread cancellation. This is the case, as the behaviour of existing OpenMP–constructs is not changed, but only new clauses or directives are added.
2. **Nested Parallelism**
   Each proposal must clearly state how thread cancellation and nested parallelism play together. Our proposal does so, by declaring that when a parallel region is cancelled, all parallel regions that were created by a thread from the cancelled region have their cancel flag set as well.
3. **Barriers**
   Each proposal must cope with the case that a region is cancelled while one or more threads are waiting on a barrier (including implicit barriers), without producing deadlocks. Our proposal does so with the introduction of the oncancel clause and the onbarriercancel directive.

4. **No Resource Leaks**
   The programmer must have the option to free any resources before a thread is cancelled. Our proposal takes care of this by advocating cooperative cancellation, where the programmer checks if a cancellation request has been put up and can therefore deallocate / free all of his resources before exiting from a thread. Even resource deallocation while waiting on barriers is allowed with the introduction of the new oncancel clause and onbarriercancel directive.
5. **C / C++ / Fortran Compatibility**
   Each proposal must apply to all three supported languages of the OpenMP specification. Although our proposal only spells out the C syntax of the proposed changes, we believe that these are adaptable to C++ and Fortran as well.
6. **Simplicity**
   Each proposal must be as simple and easy to understand as possible, staying in line with the original OpenMP philosophy. Especially the barrier constructs made this a difficult task, but we think to have met that goal with the introduction of only three new directives, one new runtime library function and one new clause.

3.4 Implementation and Performance Issues

We have used the Ompi compiler [6] as a testing ground for our implementation. One of the benefits of employing cooperative cancellation is ease of implementation, and most of our changes were straightforward:

- adaptation of the compiler frontend to the new directives
- addition of new runtime library functions for exitregion, cancelregion, onbarriercancel and omp_get_cancelled
- a few more minor and locally restricted changes in the runtime library

The most difficult part was the implementation of exitregion, which must be able to jump out of deeply nested functions to the end of the parallel region. This was solved using setjmp / longjmp. The second difficulty was adapting the barriers to the oncancel clause. A total rewrite of the runtime support function for barriers was required.

Great care was taken not to impact performance with our changes. Our choice of cooperative cancellation enabled us to implement thread cancellation without any measurable impact on performance. None of our test applications showed any notable slowdown. Neither did the OpenMP Microbenchmarks [9], which we used to measure performance of our adapted barrier implementation.

3.5 Application

In this section, we apply the thread cancellation functionality to our labyrinth search example from Sect. 2. We had isolated three main problems there:

- there is no easy way to branch out of a parallel region, the only possible workaround is to use flags
- it is difficult to work with flags indicating that a region should end, at least as soon as barriers come into play
- if one forgets to flush a flag, a deadlock may arise

All these issues have been solved, as can be seen in Fig. 3. Firstly, it is easy now to branch out of a parallel region, as the cancelregion directive is a natural fit for the problem (see line 9). Just one directive, and the code will branch to the end of the parallel region on line 26. If barriers are involved like in our case, oncancel clauses have to be added (line 12), as well as an onbarriercancel clause at the end of the single worksharing construct (line 21). The second problem is also solved, as there is no need to work with programmer-managed flags to indicate that a parallel region should be finished. Last but not least, the third issue has been made obsolete: there is no need anymore to flush any cancel flags, as they are managed automatically by the OpenMP runtime system. We believe that this change alone will make errors less common in irregular parallel applications.

```
1   #pragma omp parallel
2   {
3     while (!omp_get_cancelled ()) {
4       while ((task = pop (taskpool)) != NULL) && !omp_get_cancelled ()) {
5         mark (labyrinth , task);
6         if (!inspect_field_for_exit (task))   // inspect all neighbours
7           push (neighbours (task), next_taskpool);   // no exit was found
8         else {
9           #pragma omp cancelregion   // an exit was found
10        }
11      }
12      #pragma omp barrier oncancel
13      {
14        #pragma omp exitregion
15      }
16      #pragma omp single
17      {
18        taskpool = next_taskpool;   // switch the taskpools
19        next_taskpool = NULL;
20      } // implicit barrier
21      #pragma omp onbarriercancel
22      {
23        #pragma omp exitregion
24      }
25    }
26  } // end of parallel region with implicit barrier
```

Fig. 3. Parallel breadth first search in a labyrinth, using new language constructs for thread cancellation

4 Concluding Remarks and Perspectives

In this paper, we have discussed a major problem with parallelizing irregular applications in OpenMP: lacking support for thread cancellation. A workaround and an extension to OpenMP have been suggested, whose main part is the cancelregion directive that enables cooperative cancellation.

A reference implementation of the extended OpenMP functionality can be found in a special release of the OMPi Compiler [6] that is available from the authors on request. In the future, we plan to explore more applications with OpenMP, trying to find ways to improve the specification in the process. Our progress will be visible in the UKOMP project [10]. The project will serve as our testing ground for new functionality we discover to be useful, and also enables other developers to give feedback on how they like our changes. Additionally, the proposals are being sent to the OpenMP ARB, for consideration of inclusion into the official OpenMP specification.

Acknowledgments

We are grateful to Björn Knafla for proofreading the paper and for his insightful comments. We thank Vassilios V. Dimakopoulos and his group at the University of Ioannina for providing the OMPi compiler. We thank the University Computing Centers at the RWTH Aachen, TU Darmstadt and University of Kassel for providing the computing facilities used to test our compiler and applications.

References

1. OpenMP Architecture Review Board: OpenMP specifications. http://www.openmp.org/specs (2005)
2. Mattson, T.G.: How good is OpenMP. Scientific Proramming **11** (2003) 81–93
3. Hisley, D., Agrawal, G., Satyanarayana, P., Pollock, L.: Porting and performance evaluation of irregular codes using OpenMP. Concurrency: Practice and Experience (2000)
4. Dedu, E., Vialle, S., Timsit, C.: Comparison of OpenMP and classical multi-threading parallelization for regular and irregular algorithms. In Fouchal, H., Lee, R.Y., eds.: Software Engineering Applied to Networking Parallel/Distributed Computing (SNPD), Association for Computer and Information Science (2000) 53–60
5. Nikolopoulos, D.S., Polychronopoulos, C.D., Ayguade, E.: Scaling irregular parallel codes with minimal programming effort. In: Supercomputing '01: Proceedings of the 2001 ACM/IEEE Conference on Supercomputing (CDROM), ACM Press (2001)
6. Dimakopoulos, V.V., Georgopoulos, A., Leontiadis, E., Tzoumas, G.: OMPi compiler homepage. http://www.cs.uoi.gr/~ompi/ (2003)
7. Süß, M., Leopold, C.: Common mistakes in OpenMP and how to avoid them. In: Proceedings of the International Workshop on OpenMP - IWOMP'06. (2006)
8. Sun Microsystems: Why Are Thread.stop, Thread.suspend, Thread.resume and Runtime.runFinalizersOnExit Deprecated. http://java.sun.com/j2se/1.4.2/docs/guide/misc/ threadPrimitiveDeprecation.html (1999)
9. Bull, J.M., O'Neill, D.: A microbenchmark suite for OpenMP 2.0. SIGARCH Comput. Archit. News **29**(5) (2001) 41–48
10. Süß, M.: University of Kassel OpenMP – UKOMP homepage. http://www.plm.eecs.uni-kassel.de/plm/index.php?id=ukomp (2005)

Toward Enhancing OpenMP's Work-Sharing Directives

Barbara M. Chapman[1], Lei Huang[1], Haoqiang Jin[2],
Gabriele Jost[3], and Bronis R. de Supinski[4]

[1] University of Houston, Houston TX 77004, USA
[2] NASA Ames Research Center, USA
[3] Sun Microsystems, Inc., USA
[4] Lawrence Livermore National Laboratory, USA
{chapman, leihuang}@cs.uh.edu, hjin@nas.nasa.gov,
gabriele.jost@sun.com, bronis@llnl.gov

Abstract. OpenMP provides a portable programming interface for shared memory parallel computers (SMPs). Although this interface has proven successful for small SMPs, it requies greater flexibility in light of the steadily growing size of individual SMPs and the recent advent of multithreaded chips. In this paper, we describe two application development experiences that exposed these expressivity problems in the current OpenMP specification. We then propose mechanisms to overcome these limitations, including thread subteams and thread topologies. Thus, we identify language features that improve OpenMP application performance on emerging and large-scale platforms while preserving ease of programming.

1 Introduction

OpenMP supports portable, high-level shared memory parallel programming and has been successfully deployed on small-to-medium shared memory systems (SMPs) and large-scale distributed shared memory platforms (DSMs). Its current version 2.5 [14] merges C/C++ and Fortran bindings and clarifies some concepts, especially with regard to the memory model. OpenMP 3.0 is expected to follow, and to consider a variety of new features. Among the many open issues are some tough challenges including extending OpenMP to SMP clusters and supporting other new architectures.

Several architectural trends to which we collectively call Chip MultiThreading (CMT) provide support for the simultaneous execution of two or more threads within one chip. It may be implemented through several physical processor cores in a chip (Chip MultiProcessor, CMP) [13], a single core with replication of features to maintain the state of multiple threads simultaneously (Simultaneous multithreading, SMT) [17] or their combination [9,10]. A hierarchical multithreading architecture results from using several of these chips in a single SMP. OpenMP was not designed for such hierarchical parallelism, nor to enable a programmer to assign different workloads to sibling threads in order to avoid resource contention. Traditionally, OpenMP targets computationally intensive, loop-based applications. CMT will probably dramatically increase the usage of OpenMP. Programmers will need language mechanisms that facilitate scalable parallel programming for these hierarchical systems, including flexibility in the assignment of work to threads.

In this paper, we describe two application development experiences from different domains that exposed problems with the expressivity of the current OpenMP specification. The first example involved porting an industrial seismic data processing application to OpenMP in order to create an easy-to-maintain version that exploited SMPs with hyperthreading. The language extensions we designed based on this effort turned out to have a much wider applicability. The second example comes from experiences gained while building scalable scientific applications on a large distributed shared-memory platform. Here too, the extensions facilitated an appropriate mapping of work to threads and led to a scalable parallel code. In each case, our inability to assign work to subsets of threads in the current thread team, and to orchestrate the work of different threads, in OpenMP 2.5 artificially limited performance. To overcome this, we propose a new clause for worksharing constructs that assigns the work to a subteam of the existing threads. Further, we introduce the notion of a topology, which gives a subteam a shape, and library routines to support these concepts. Finally, we also propose new constructs for improved work coordination between threads. We outline these applications and our proposed OpenMP extensions that facilitate programming them in the next two sections. Then, we discuss related work briefly before summarizing our findings.

2 Thread Subteams

Our experiences with commercial seismic data processing software initially motivated our thread subteam concept. Kingdom Suite from Seismic Micro-Technology, Inc. is an integrated geosciences interpretation software package for Windows systems used by the energy industry in the search for oil. OpenMP was applied to TracePak, an I/O-intensive module of Kingdom Suite to analyze and to process two-dimensional (2-D) and three-dimensional (3-D) post-stack seismic data [16]. Our goal was a parallel version for Windows-based SMPs with hyperthreading enabled. This version must be as close as possible to the original sequential code to simplify its maintenance, a common industrial requirement. Although our example could be programmed in a low-level style using thread IDs explicitly, this would require significant changes in the source code. In contrast, the suggested directives require only a minimal, localized modification of the source code and maintain the ease of programming that makes OpenMP a desired programming model. The subteam concept proposed here has been implemented in the OpenUH compiler [15]. It is comparatively straightforward, requiring less implementation effort than nested parallelism. WE are currently implementing our other proposals. Due to space limitations, implementation details will be addressed in a separate paper.

2.1 Seismic Data Processing on an SMT Platform

Fig. 1 shows the structure of the sequential program. This code iteratively reads data from an input file, processes it using different transform functions in a specified order, and then writes the results to an output file. The amount of seismic data typically handled in a job is quite large, ranging from 100MB to 100GB, and reading and writing consume considerable time.

Since OpenMP does not support parallel I/O, we decided that the best strategy to parallelize the code of Fig. 1 is to overlap the sequential I/O operations (lines 2 and 7)

```
1. for (i=0; i<N; i++) {
2.     ReadFromFile(i,...);
3.     for (j=0; j<ProcessingNum; j++)
4.         for (k=0; k<M; k++) {
5.             ProcessData();  //processing involves several
                                //different seismic functions
6.         }
7.     WriteResultsToFile(i);
8. }
```

Fig. 1. A sequential pseudo-code fragment for seismic data processing

with the parallelized computation (line 5), as illustrated by the timeline view in Fig. 2. A simple way to parallelize the computation is to enclose the innermost loop (k-loop) between threads in an "omp parallel for" directive. This approach, however, does not overlap the computation and I/O, and moreover, frequently entering and leaving parallel regions degrades performance. A dependence between the seismic data processing functions prevents parallelization of the outer loop (j-loop). In order to overcome these deficiencies, we enclose the entire loop nest in a paralel region as shown in Fig. 3. This version preloads the data needed for the first iteration of the i-loop (line 6). Then, we use "omp single nowait" and "omp for schedule(dynamic)" to enclose and to overlap the I/O operations and computation. One thread reads the data for the next iteration and another thread writes the results to an output file. The remaining threads share the work of the j loop (line 11 of Fig. 3). The dynamic schedule enables the threads performing I/O to subsequently join the computation.

The innermost, work-shared loop includes an implicit barrier at its end. Unfortunately, we cannot simply remove it since the data processing functions must follow a specific sequential order: each iteration uses results from the previous one. Thus although plenty of computation remains, the computing threads must wait at the implicit barrier until the I/O has completed, as shown in Fig. 4. Thus I/O operations and computation are not fully overlapped. Unfortunately, exchanging the order of the loops in the nest would, if possible, require a complete rewrite. However, a parallelization strategy that requires major code reorganization is unacceptable, as previously discussed.

2.2 Performance Improvement

In a normal run, the ratio of I/O and computation is about 1.2:1, where the I/O takes slightly longer than the computation. Thus, including the I/O threads in the barrier limits the overlap of I/O with computation. To determine how much removing this limitation

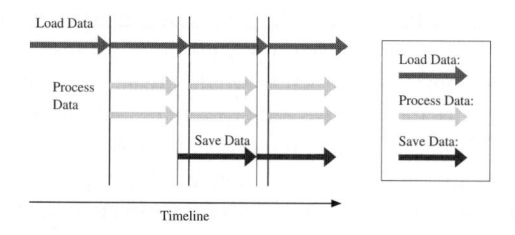

Fig. 2. Overlapping I/O with computation in the parallel seismic program

```
1.  #pragma omp parallel
2.  {   #pragma omp single
3.        {  //preload data to be used in the first iteration of the i-loop in line 6
4.           ReadFromFile(0,...);
5.        }
6.        for (i=0; i<N; i++) {
7.           #pragma omp single nowait
8.           {  //preload the data for next iteration of the i-loop
9.              ReadFromFile(i+1...);
10.          }
11.          for (j=0; j< ProcessingNum; j++)
12.             #pragma omp for schedule(dynamic)
13.             for(k=0; k<M; k++) {
14.                ProcessData();  //user configurable data processing functions
15.             } //here is the barrier
16.          #pragma omp single nowait
17.          {
18.             WriteResultsToFile(i);
19.          }
20.       }
21. }
```

Fig. 3. The OpenMP code for seismic data processing kernel

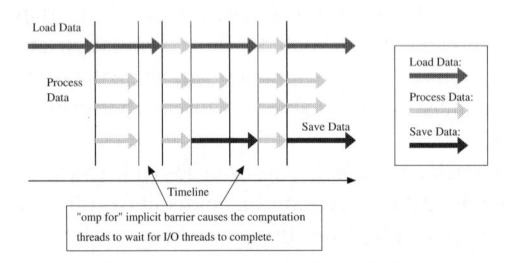

Fig. 4. Execution behavior of OpenMP seismic code

Fig. 5. Performance comparison: OpenMP vs. hybrid OpenMP and Windows API codes

would improve performance, we combined OpenMP with Windows threads for reading and writing files and achieved much greater overlap than with pure OpenMP. Fig. 5 shows results on an HP XW8200 with dual Xeon 3.4 GHz CPUs, 1MB L2 cache, 3GB memory, Intel extended memory 64, and hyperthreading technology. The compiler used was Microsoft Visual C++ in Visual Studio 2005 with OpenMP support. The hybrid version was 25% faster than standard OpenMP on four threads.

To achieve similar results with pure OpenMP, we require mechanisms to separate the computational threads from the data handling threads, and to synchronize their activities in the desired manner. We can achieve this with three parallel sections: read, write, and computation. The computation section would create a nested parallel region and share the work among its threads. We either prefetch data in the previous iteration, as in the code of Fig. 3, or use critical regions and arrays of variables. Unfortunately, each iteration of the outer i-loop requires a new parallel region if we are to retain the sequential program structure and the overheads for these are potentially high.

2.3 Thread Subteam as a Solution

Nested parallelism can dynamically create, exploit and terminate teams of threads and is well-suited to codes with needs that change over time. Our code structure is static. The relative amount of data and computation does not vary, and we expect the number of participating threads and their roles to remain the same. Nested parallelism is more powerful than we require. Thus, we propose a simpler mechanism that allows us to bind the execution of a worksharing or barrier construct to a subset of threads in the current team. Only the threads in the specified subteam participate in its work, including any barrier operations encountered. To synchronize the actions of multiple subteams, we may use existing OpenMP constructs and take advantage of the shared memory.

To realize this idea, we define an "onthreads" clause for worksharing and barrier directives. In contrast to nested parallelism, it refers only to existing threads. This clause permits us to specify that a worksharing directive is applied to a *subteam* of threads: participation in the associated work is restricted to the specified members. In particular, implicit and explicit barriers within the code it encloses do not block threads that are not part of the subteam. This clause would require minimal change to the current specification. In addition we can define an "onthreads" directive that could enclose arbitrary structured block of code within a parallel region. Work in the block would be carried out by the specified subteam of threads.

Using the thread subteam notation, we can rewrite the example code in Fig. 3 to that in Fig. 6. Line 5 and line 14 use the "onthreads" clause to limit the I/O to individual threads, while line 7 defines a subteam of threads to process the data. The integer expressions in parentheses use OpenMP's thread-ids and array section notation to specify the desired subset of threads. The implicit barrier at line 12 applies only to the threads defined in the subteam from line 7.

Additional syntax could enable the programmer to name these subsets. New run-time library routines would be provided to get the number of threads in a (named) subteam and a subteam-internal consecutive thread number. A programmer might also want to permute the order of threads in a subteam to specify schedules that enforce a certain work distribution, e.g. to support data reuse. Although none of these (except possibly the

```
1.  #pragma omp parallel
2.  {   #pragma omp single
3.          ReadFromFile(0,...);  //preloads data for first iteration of i-loop
4.      for (i=0; i<N; i++) {
5.          #pragma omp single onthreads(0)
6.              ReadFromFile(i+1...);  //preload data for next iter. of i-loop
7.          #pragma omp onthreads ( 2:omp_get_num_threads()-1 )
8.              for (j=0; j< ProcessingNum; j++)
9.                  #pragma omp for schedule(dynamic)
10.                 for (k=0; k<M; k++) {
11.                     ProcessData();  //user configurable data processing functions
12.             }  //here is the group-internal barrier
13.         #pragma omp barrier  //this ensures we are ready for next iter.
14.         #pragma omp single onthreads(1)
15.             WriteResultsToFile(i);
16.     }
17. }
```

Fig. 6. OpenMP seismic data processing kernel with the "onthreads" directive

library routines) are essential, they would greatly increase the expressive power of this construct. Interactions between subteams could be made explicit by providing notation for communication between subteams. This might help a programmer reason about the structure of this communication and avoid programming errors such as deadlock. The same construct might also enable point-wise synchronization between threads in a single subteam to avoid barriers. In the code fragment of Fig. 7, a post-wait notation does this succinctly and we have named the thread team, whose order is a permutation of the original thread numbers (used here only to illustrate the concept).

```
#pragma omp parallel
{
   #pragma omp team ComptreadsReordered = threads(omp_get_num_threads()-1:2:-1)
   for (i = 0; i < N; i++) { //executed by all threads
       #pragma omp single onthreads(0)
       {   ReadFromFile(i);
           #pragma omp post (dataready[i]) //signals reading is complete
       } //thread(0) independently does this reading and posting
       ........
       #pragma omp on ComptreadsReordered
       {   //subteam starts to work
           #pragma omp wait (dataready[i])  //after data is ready
```

Fig. 7. Excerpt from OpenMP code with named subteam and post/wait

The ability to divide work among subteams of threads, and thus to have different subteams working concurrently and independently, seems to be a fairly natural extension to the current API and it has a variety of potential uses. It would likely simplify the use of OpenMP within third party libraries. It also enables the specification of multidisciplinary code ensembles and permits components written in traditional programming languages to interact without the need to provide external file-based interactions. It supports the simpler case of multilevel parallelism with a fixed team of threads without the extra overheads and burden of nested parallelism.

3 Worksharing and Synchronization Across Loop Nests

Scientific and engineering computations must exploit large numbers of threads, not only in emerging, very large shared-memory systems, but also in smaller SMPs with CMPs. Writing scalable code requires special care. Two of the authors previously proposed a set of language features to enable the parallelization of multiple levels of loop nests [8]. These features specify an appropriate execution schedule and assign threads to loop levels, as well as additional synchronization that enables a pipelined execution scheme in the LU benchmark from the NAS Parallel Benchmarks [2]. They addressed scalability limitations in several applications despite the presence of sufficient inherent parallelism.

3.1 The LU Example

The LU application benchmark uses the symmetric successive over-relaxation (SSOR) method to solve a seven band block-diagonal system. Figure 8 illustrates its lower triangular phase. References to values of elements of array v in line 4 create dependences between loop iterations that prevent straightforward parallelization. However, a wave-front or a pipelined technique can enable considerable levels of parallelism to be exploited, since the value of an element of v can be computed once the new values are available from the previous iteration in each of the three dimensions.

A wave-front restructuring of the code reveals parallelism that can be expressed with the existing OpenMP parallel directive to update points on a diagonal plane concurrently. However, this method suffers from poor cache utilization. A pipelined approach, in which data are partitioned as blocks in selected dimensions, usually gives better cache performance. We illustrate the differences between wave-front and pipelined parallelism in Fig. 9. Expression of the parallelism in two dimensions would reduce the cost of pipeline startup and shutdown, and support good cache performance for this kernel. However, OpenMP currently can only successfully exploit parallelism in one dimension. Parallelization in multiple dimensions requires nested parallelism, which results in multiple one-dimensional pipelines and incurs high overheads [7].

```
1. for (k = 1; k < nz; k++) {
2.   for (j = 1; j < ny; j++) {
3.     for (i = 1; i < nx; i++) {
4.       v[k][j][i] =
             v[k][j][i] +
             a*v[k][j][i-1] +
             b*v[k][j-1][i] +
             c*v[k-1][j][i];
5.     ...
6.   }
7. }
8. }
```

Fig. 8. The LU computational kernel

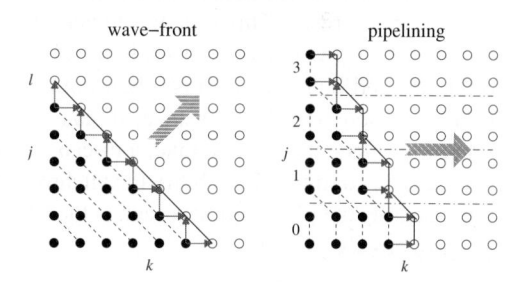

Fig. 9. Wave-front and pipelined algorithms. j,k are data dimensions. l in the left panel indicates a diagonal plane. Numbers in the right panel indicate data blocks mapped to different threads.

3.2 Thread Topology

We introduce the notion of a thread topology to support pipelined algorithms. A thread topology does not create new threads; instead, it *reshapes* the thread (sub)team and associates a new naming scheme with existing threads. We can use the topology to specify a variety of new schedules for worksharing directives. Our syntax requires the programmer to provide the number of dimensions in the topology and the coordinates in each dimension. We will also need a default strategy for mapping the linearly numbered threads to a Cartesian grid. The basic syntax of specifying a topology is:

`#pragma omp topology name(ndim,start,stop,stride,fixedorder)`

where `name` defines a name of the topology. The `ndim` argument specifies the number of dimensions. The arguments `start`, `stop`, and `stride` are arrays with one entry per dimension to specify the topological shape. `fixedorder` is a Boolean variable that tells the compiler whether or not the default strategy for associating these threads with the linear thread numbers must be applied. If not, the system can choose any mapping of threads to the topology. For example, if 16 threads exist, the directive can reshape threads into a $4 \times 2 \times 2$ grid with coordinates from start[]=(0,0,0) to stop[]=(3,1,1) and stride[]=(1,1,1) or any other numbering scheme we desire that has 16 threads. We can associate a defined topology with a worksharing construct using the "onthreads" clause. We use standard section notation to specify the target of the worksharing directive in each topological grid dimension. We use ":" to denote the entire dimension of an array. Dimensions not involved in the worksharing are marked via a dummy "*" and the computation is replicated in those dimensions. A runtime function "omp_get_coord(name,idim)" can obtain coordinates of a thread in the grid topology.

We illustrate the use of our topology notation in Fig. 10 for the LU computational kernel. We introduce a 2-D logical grid of threads with the same number of threads in each dimension. Our thread subteam clause maps the iterations of two different loops to threads using our grid topology through two worksharing constructs (this notation does not conform to current OpenMP rules). The 2-D topology is used to distribute the work in the i and j loop nests among threads.

Finally, we need a way to define synchronization between threads in a topology. We cannot use existing features of OpenMP, since the interaction required is not between iterations but threads. This is achieved here using post and wait directives with our 2-D thread-ids. In our example, each thread of the topology must wait for its neighbors to the left and below it to finish their computation except for where the thread does not have a neighbor. For instance, thread 0 does not have a neighbor and can start right away. Once its work is done, a thread signals its neighbors to the right and above that they can continue. The ability to synchronize between threads is very important for implementing the pipelined approach in the LU algorithm. In general, it enables loosely synchronous algorithms [12].

```
      mystart[0] = 0; mystart[1] = 0; ...  // assign values to mystart[:] and mystop[:]
      #pragma omp parallel {
      #pragma omp topology grid(2,mystart,mystop,mystride,1)
                                            // arrange threads logically into a square called grid
      iam1 = omp_get_coord(grid,1);
      iam2 = omp_get_coord(grid,2); // my coords in grid
1.    for (k = 1; k < nz; k++) {
          #pragma omp wait grid (iam1-1,iam2)  // wait for thread below to complete its portion
          #pragma omp wait grid (iam1,iam2-1)  // wait for thread on left to complete its portion
          #pragma omp for nowait onthreads(grid(:,*))  // share out to first dimension of grid
2.        for (j = 1; j < ny; j++) {
              #pragma omp for nowait onthreads(grid(*,:))  // share out to second dimension of grid
3.            for (i = 1; i < nx; i++) {
4.                v[k][j][i] = v[k][j][i] + a*v[k][j][i-1] +
                               b*v[k][j-1][i] + c*v[k-1][j][i];
5.            ...
6.            }
7.        }
          #pragma omp post grid(iam1,iam2+1)  // indicate to thread on right that it is ready
          #pragma omp post grid(iam1+1,iam2)  // indicate to thread above that it is ready
8.    }
      }
```

Fig. 10. The multilevel LU computational kernel using thread topology

4 Related Work

The NanosCompiler team has proposed groups of threads in association with parallel regions [5,6]. Their notation permits the user to specify the number of independent teams of threads that will be created. Since these thread groups are associated with the parallel region, additional notation is required to assign work to the individual groups. They also propose extensions to express the precedence relations in pipelined computations. These extensions are also valid in the scope of nested parallelism and are based on the ability to name worksharing constructs and to specify a predecessor-successor relationship between them to support synchronization. Our topology simplifies specifying the desired target sets and is more intuitive than the predecessor-successor relationship. Furthermore, it does not rely on nested parallelism and the associated overhead.

There have been a variety of proposals for multilevel loop parallelism. The SGI compiler for the Origin [11] provides the SGI NEST clause on the OMP DO directive. The NEST clause requires at least two variables as arguments to identify indices of subsequent DO-loops, which must be perfectly nested. It informs the compiler that the entire set of iterations across the identified loops can be executed in parallel. The compiler can then linearize the iteration space and divide it among the threads. Intel has proposed a new directive to enable wavefront execution schema. Although this might sometimes be appropriate, we expect that it will be hard to achieve good data locality in most cases. Our proposal explicitly enables control of work distribution and, thus, enables the expression of data locality.

New programming languages [1,3,4] are being proposed to facilitate high end application development in a multithreading environment. These languages address problems faced by levels of scaling that are far from those currently envisaged for hierarchical SMPs, and they provide a wealth of new ideas related to correctness, locality, efficiency of shared memory updates, and more. We will explore these ideas in the context of OpenMP.

5 Conclusions

OpenMP is a widely deployed shared memory programming API that offers the promise of performance and ease of use. It seems possible that the judicious addition of language features that increase the power of expressivity might also improve the achievable performance of a variety of OpenMP codes. In this paper, we introduced a unified notation for sharing work among subteams of threads and for flexibly executing multiple levels of loop nests in parallel. Table 1 lists the proposed new OpenMP constructs and clauses in the paper. This approach fits in well with existing features of the API. As our future work, we will conduct more detailed performance study of the proposed subteam concept implemented in the OpenUH compiler.

Table 1. Proposed new OpenMP Constructs and Clauses

Proposed OpenMP Directives/Clauses	Description
omp onthreads / onthreads (clause only)	Defines thread subteams for work sharing
omp topology name	Defines the thread topology
omp post / omp wait	Uses for point-wise synchronization

References

1. E. Allen, D. Chase, V. Luchangco, J-W. Maessen, S. Ryu, G.L. Steele Jr., S. Tobin-Hochstadt. "The Fortress Language Specification, Version 0.785." http://research.sun.com/projects/plrg/fortress0785.pdf
2. D. Bailey, T. Harris, W. Saphir, R. Van der Wijngaart, A. Woo, and M. Yarrow, "The NAS Parallel Benchmarks 2.0," RNR-95-020, NASA Ames Research Center, 1995.
3. Cray Inc., "Chapel Specification 0.4." http://chapel.cs.washington.edu/specification.pdf
4. P. Charles, C. Grothoff, V. Saraswat, C. Donawa, A. Kielstra, K. Ebcioglu, C. von Praun and V. Sarkar, "X10: an object-oriented approach to non-uniform cluster computing." in the proceedings of OOPSLA '05, pp. 519-538, 2005
   K. Ebcioglu, V. Saraswat and V. Sarkar. " X10: Programming for hierarchical parallelism and nonuniform data access (extended abstract)." OOPSLA 2004), October 2004.
5. M. Gonzalez, E. Ayguade, X. Martorell and J. Labarta. "Defining and Supporting Pipelined Executions in OpenMP." in the proceedings of WOMPAT 2001, July 2001.
6. M. Gonzalez, J. Oliver, X. Martorell, E. Ayguade, J. Labarta, and N. Navarro. "OpenMP Extensions for Thread Groups and Their Run-time Support." in the proceedings of LCPC'2000, New York (USA), pp. 317-331, August 2000.
7. H. Jin, G. Jost, J. Yan, E. Ayguade, M. Gonzalez, and X. Martorell, "Automatic Multilevel Parallelization Using OpenMP," Scientific Programming, Vol. 11, No. 2, pp. 177-190, 2003.
8. H. Jin and G. Jost. "Support of Multidimensional Parallelism in the OpenMP Programming Model," WOMPEI2003, Tokyo, Japan, October 2003, in the Proceedings of the International Symposium on High Performance Computing (ISHPC-V).
9. R. Kalla, B. Sinharoy, and J. Tendler. "IBM POWER5 chip: a dualcore multithreaded processor", in IEEE Micro, 24(2): 40-47, 2004.
10. P. Kongetira. "A 32-way Multithreaded SPARC Processor", in Hot Chips 16, http://www.hotchips.org/archives/hc16/.
11. MIPSPro 7 Fortran 90 Commands and Directives Reference Manual 007-3696-03. http://techpubs.sgi.com/.
12. Z. Liu, B. Chapman, Y. Wen, L. Huang and O. Hernandez. "Analyses and Optimizations for the Translation of OpenMP Codes into SPMD Style," Proc. WOMPAT 03, LNCS 2716, 26-41, Springer Verlag, 2003.
13. K. Olukotun, B. A. Nayfeh, L. Hammond, K. Wilson, and K. Chang, "The Case for a Single-Chip Multiprocessor", in Intl. Conf. on Architectural Support for Programming Languages and Operating Systems, 1996, pp. 2-11.
14. OpenMP Application Program Interface, Version 2.5, May 2005. http://www.openmp.org/drupal/mp-documents/spec25.pdf
15. "The OpenUH compiler project", http://www.cs.uh.edu/ openuh
16. Sesimc Micro-Technology, Inc., TracePak Module, http://www.seismicmicro.com/Prod_Geo.htm.
17. D. Tullsen, S. Eggers, and H. Levy, "Simultaneous Multithreading: Maximizing On-Chip Parallelism", Intl. Symp. on Computer Architecture, pp. 392-403, 1995.

Toward a Definition of and Linguistic Support for Partial Quiescence

Billy Yan-Kit Man[1], Hiu Ning (Angela) Chan[1], Andrew J. Gallagher[1], Appu S. Goundan[1], Aaron W. Keen[2], and Ronald A. Olsson[1]

[1] Department of Computer Science, University of California,
Davis, Davis, CA 95616 USA
{many, chanhn, gallagha, goundan, olsson}@cs.ucdavis.edu
[2] Computer Science Department,
California Polytechnic State University, San Luis Obispo, CA 93407 USA
akeen@csc.calpoly.edu

Abstract. The global quiescence of a distributed computation (or distributed termination detection) is an important problem. Some concurrent programming languages and systems provide global quiescence detection as a built-in feature so that programmers do not need to write special synchronization code to detect quiescence. This paper introduces *partial quiescence* (PQ), which generalizes quiescence detection to a specified part of a distributed computation. Partial quiescence is useful, for example, when two independent concurrent computations that both rely on global quiescence need to be combined into a single program. The paper describes how we have designed and implemented a PQ mechanism within an experimental version of the JR concurrent programming language. Our early results are promising qualitatively and quantitatively.

1 Introduction

In distributed programs, multiple processes cooperate to perform some task and communicate via messages to exchange information. One important, and well-studied, problem for such programs is to determine when the program's computation has completed, i.e., it has terminated normally or deadlocked. This *quiescence* problem is challenging because each process has only local information, but to solve the problem requires information about all processes (i.e., global state information). More formally, global quiescence (GQ) is defined as the state in which each process has terminated or deadlocked and there are no messages in the communication channels [14]. *Quiescence detection*, then, is the mechanism used to detect such a state in a distributed system.

Some programming languages and systems provide GQ detection as a built-in feature. That is, programmers do not need to write special synchronization code to detect quiescence. Instead, they can focus on writing application code. When quiescence is reached, the program can perform various actions such as simply terminating the program, outputting final results, gathering statistics from the overall computation, or initiating a new phase of the program, which might involve a new, corresponding phase of quiescence detection.

Although useful, GQ is limited to dealing with the state of *all* processes in a program. A more general, but more difficult to detect, property would define when a specified *part* of the program has become quiescent. For example, suppose we have two programs that use GQ and we want to combine them into a single program in which we want to perform different actions when each part of it becomes quiescent. This motivation led us to explore *partial quiescence* (PQ).

This paper proposes possible ways of defining PQ. It then discusses the particular definition selected and implemented in an experimental version of the JR concurrent programming language. (JR extends Java with a richer concurrency model [7,18,6].) The paper also shows how PQ leads to a different programming style for some problems. We compare the performance of using PQ detection and GQ detection. PQ might be a useful feature for other languages and libraries that define process or thread groups, as many do, and especially useful for those languages and libraries that already provide GQ detection.

Our work involves detecting quiescence (GQ or PQ) *dynamically*, i.e., during the actual execution of a concurrent program. An alternative approach involves *statically* determining various properties of concurrent programs, e.g., determining whether a concurrent program is deadlock-free. For example, [13] describes how to statically determine whether a given Ada program is deadlock-free and [9,16,17] describe how to statically verify that a program in notations such as process calculus with communication channels or timed automata with shared variables is *partial-deadlock* free. Partial deadlock means that a specified part of the program deadlocks. For example, [9] places restrictions on how processes communicate over channels and shows that the part of the program that uses "reliable" communication channels does not deadlock. The key difference between our approach and the static approaches, besides when the checking is performed, is that our approach treats quiescence as a normal part of program execution; the program itself is aware of its own quiescence and can react to quiescence as it desires.

The rest of this paper is organized as follows. Section 2 provides background on the general definition of GQ and how it has been incorporated as a built-in feature in some programming languages and systems; it describes how GQ is defined and implemented in JR and presents examples of programs that use GQ. Section 3 discusses the different ways of defining PQ. Section 4 discusses the definition of PQ we chose to provide in JR and gives examples of programs that use PQ. Section 5 presents an overview of our implementation and discusses its performance. Finally, Section 6 concludes. Further details appear in [12].

2 Background

2.1 Distributed Termination Detection (DTD)

As noted in Section 1, detecting the termination of a distributed computation is an important and challenging problem. A nice survey [14] describes DTD as follows. A distributed system consists of a collection of processes such that processes communicate with each other by sending *activation messages* via some communication channels. An activation message is used not only for communication purposes among processes, but also for creation of a new process. A process is *active* if it is working on some computation or processing activation messages addressed to it. A process is *passive* if it is waiting for an activation message or termination. All processes in the system behave based on the following rules:

1. Activation messages can be generated only by active processes.
2. An active process may change its state to passive at any time.
3. A passive process may change its state to active only if it receives an activation message.

The above rules ensure that no further activation messages can be created in a system where all processes are passive: messages cannot be generated spontaneously. When the system has reached such a state, i.e., all processes are passive and no activation messages are in transit, then the system is *quiescent*. Quiescence detection is defined as the mechanism used to detect the state in which there are no messages in transit and all processes are waiting [14]. This definition generalizes that of DTD to both detecting termination as well as deadlock: i.e., sensing when the system is in a state from which it can no longer continue. Two main categories of DTD algorithms, as classified in [14], are wave algorithms (e.g., [3,4,19]) and credit distribution and recovery algorithms (e.g., [15]).

2.2 Tools and Systems with Support for Termination Detection

In some languages and systems (e.g., Ada, Java, MPI, and Pthreads), programs that reach a deadlock state wait indefinitely for the user to terminate them manually. However, in some cases, tools can assist in such detection. For example, Umpire [20] and MPI-CHECK 2.0 [11] detect deadlocks for MPI programs.

Some other programming languages or systems provide GQ detection as a built-in feature. GARLIC [8] extends Ada 95 with distributed programming features; it detects termination based on the algorithm proposed in [5]. JR [7,18], SR [2,1], and Charm [19] allow a quiescent program to output final results, gather statistics from the overall computation, or simply terminate the program. In this regard, JR and Charm are similar: when a program quiesces, it can initiate new computation, for which the quiescence feature can be used again. SR's quiescence feature is not as powerful: it is intended only for the program to clean up and terminate, and programs cannot use quiescence repeatedly.

Implementation of a Quiescence Feature in JR and SR. The implementation uses an approach that differs from the general DTD algorithms described in Section 2.1 because of their particular model of computation. A distributed program consists of a group of "virtual machines" (VMs). Each VM represents an address space, or unit of program distribution, and contains several processes, which can share variables within that address space or send message to other processes on that VM or to processes on other VMs. Typically, the number of VMs is not very large, but it varies as the program executes. The implementation uses a centralized manager to record information about all VMs in one place so as to make it easy to implement various services, such as an explicit exit (stop) from the program code, which needs to shut down all VMs. The implementation of DTD involves the RTS (run-time system) on each VM and the centralized manager. When a VM can make no further progress (i.e., all of its processes have terminated or are waiting to receive a message), it sends an idle message to the manager. This message contains the number of messages this VM has sent to each other VM and the number of messages this VM has received from each other VM. If the manager has received an idle message from each VM, it checks that no messages are in transit, specifically: for each VM VM_a, the number of sends from VM_a to each other VM, VM_b, matches the number of receives from VM_a reported by VM_b, If so, then the system is globally quiescent.

Example JR Program Using GQ. The program in Figures 1 and 2 (from [18]) performs matrix multiplication. Its MMMain class reads in two N × N matrices, instantiates a MMMultiplier object, and registers the operation done as the quiescence operation.[1] Its MMMultiplier class contains the processes that perform the actual computation. These processes begin execution after MMMultiplier's constructor completes its execution. GQ is used to determine when these compute processes have finished their tasks. Once GQ has been detected, the registered operation done is invoked and its code outputs the resulting matrix. Without GQ detection, the programmer would need to write additional code to determine when the computation has terminated.

```
public class MMMain {
  private static MMMultiplier m;
  public static void main(String [] args) {
    double [][] A, B;  int N;  // A and B are NxN
    // read in NxN arrays A and B
    ...
    m = new MMMultiplier(A, B, N);
    // register done as the quiescence operation
    JR.registerQuiescenceAction(done);
  }
  private static op void done() { m.print(); }
}
```

Fig. 1. Matrix multiplication using GQ – MMMain class

```
public class MMMultiplier {
  double [][] A, B, C;  int N; // A, B, and C are NxN
  public MMMultiplier(double [][] A, double [][] B, int N) {
    this.A = A; this.B = B; this.N = N;   C = new double [N][N];
  }
  process compute ( (int r = 0; r < N; r++), (int c = 0; c < N; c++) ) {
    // compute the inner product for C[r,c]
    C[r][c] = 0.0;  for (int k = 0; k < N; k++) { C[r][c] += A[r][k] * B[k][c]; }
  }
  public void print() { /* output C */  ... }
}
```

Fig. 2. Matrix multiplication using GQ – MMMultiplier class

---

[1] Technically, the registration needs to be within a try/catch block.

If no GQ operation is registered, then the program simply terminates when it quiesces. The quiescence operation can initiate new activity and can re-register the GQ operation (either the same or different operation), which will be invoked when the newly initiated activity quiesces.

3 Definition of Partial Quiescence (PQ)

Although GQ is useful, it restricts the detection to determine the quiescent state of *all* processes in a given program. Some notion of PQ, which addresses the quiescence of *part* of the program, would be useful. We want, for example, to combine two programs (i.e., two independent concurrent computations) that use GQ into a single program in which we want to perform different actions when each part of it becomes quiescent.

The first step is to define what PQ means. A natural approach is to apply quiescence to a group of processes in a program. Modifying the definition of quiescence from Section 2.1 to apply to a specific group of processes yields:

> Quiescence of group A is defined as the state in which (1) there are no messages in the system in transit to group A and (2) all processes in group A have terminated or are waiting for a message.

Because PQ deals with the interactions of groups of processes, it is, in general, more difficult to detect. This definition fits well if the process group is "closed" [10], i.e., only processes in group A send messages to processes in group A. However, this definition is not realistic if the process group is "open" [10], i.e., a message for a process in group A can be generated by a process outside of the group; such a message appears, from within group A, to have been generated "spontaneously". More concretely, a detection mechanism could detect that all processes in group A are passive and no message in transit is destined for group A, and so it would decide that group A is partially quiescent. However, that decision could be followed by a process outside group A sending a message to a process in group A. (In contrast, such spontaneous message generation is not possible for GQ (Section 2.1).)

A definition of PQ can deal with this spontaneous generation problem in various ways. One way would be to alter the above definition with a third clause, e.g., "and (3) no process outside of group A can possibly send to a process in group A". However, such a definition might not be useful: just because a process outside of group A can send a message to a process in group A does not guarantee that it ever actually will. Moreover, in general, keeping track of such information in a system where communication paths between processes is determined dynamically would be costly.

Therefore, we choose a weaker definition of partial quiescence, namely one that modifies (1) from the earlier definition:

> Quiescence of group A is defined as the state in which (1) there are no messages in the system from group A in transit to group A and (2) all processes in group A have terminated or are waiting for a message.

4 JR Extended for Partial Quiescence

We have extended JR to support PQ. Now, JR programs can define groups of related processes and can register, for each process group, a *partial quiescence operation*. This section begins with examples to illustrate how PQ in the extended JR works and then discusses key aspects of the various mechanisms.

4.1 Expository Examples of PQ in JR

Multiple Matrix Multiplications. The main program in Figure 3 shows how to use PQ to perform two simultaneous matrix multiplications. A nice attribute of our PQ approach is that the same `MMMultiplier` class from Figure 2 works here. The main program creates two process groups, one for each matrix multiplication. It uses `JR.changeCreationGroup` to specify the group in which newly created processes will be placed for each new matrix computation. (There is one default process group.) The main program then registers the PQ operation for each process group. When either group quiesces, its PQ operation will be invoked and that code outputs the results.

In contrast, consider a variant of the original main program in Figure 1 that starts two matrix multiplications and that uses GQ. It would wait for *both* computations to finish before outputting the result from either.

In Figure 3, two process groups might quiesce at about the same time, in which case the outputs from their quiescence operation might be interleaved. Their outputs can be serialized by deleting the present code for `done1` and `done2` (but keeping their `op` declarations) and adding the code in Figure 4 to the end of the `main` method. This code uses JR's multi-way receive statement (`inni`) to

```
public class MMMain {
  private static MMMultiplier m1, m2;
  public static void main(String [] args) {
    double [][] A1, B1, A2, B2;  int N;   // A1, B1, A2, B2 are NxN
    // read in NxN arrays A1, B1, A2, B2
    ...
    ProcessGroup m_g1 = new ProcessGroup("Multiply Group1");
    ProcessGroup m_g2 = new ProcessGroup("Multiply Group2");
    JR.changeCreationGroup(m_g1); // processes within m1 will be in m_g1
    m1 = new MMMultiplier(A1, B1, N);
    JR.changeCreationGroup(m_g2); // processes within m2 will be in m_g2
    m2 = new MMMultiplier(A2, B2, N);
    // register partial quiescence operation for each process group
    JR.registerPartialQuiescenceAction(m_g1, done1);
    JR.registerPartialQuiescenceAction(m_g2, done2);
  }
  private static op void done1() { m1.print(); }
  private static op void done2() { m2.print(); }
}
```

Fig. 3. Multiple matrix multiplications using PQ – `MMMain` class

```
for (int i = 0; i < 2; i++) {
  inni void done1() {m1.print();}
  []   void done2() {m2.print();}
}
```

Fig. 4. Code to serialize output from the multiple matrix multiplications

wait for an invocation of either of the PQ operations; it services one at a time, thus serializing their outputs.

Barrier Synchronization. PQ, as noted earlier for GQ, allows JR programs to re-register a quiescence operation. Consider the program in Figure 5 (from [18]). It shows a group of worker processes synchronizing their iterations via a barrier, implemented with semaphores.[2] The program also contains a coordinator pro-

```
public class Barrier {
  private static final int N = 10; // number of workers
  private static sem done = 0;
  private static cap void () proceed[] = new cap void()[N];
  static { for (int i = 0; i < N ; i++) { proceed[i] = new sem; } }
  private static process worker( (int i = 0; i < N; i++) ) {
    while (...) { // iterations remain
      // code to implement one iteration of task i
      ...
      // barrier
      V(done);       // tell coordinator "I did iteration i"
      P(proceed[i]); // wait for coordinator to say "continue"
    }
  }
  private static process coordinator {
    while (...) { // iterations remain
      for (int w = 0; w < N; w++) { P(done); }
      for (int w = 0; w < N; w++) { V(proceed[w]); }
    }
  }
  public static void main(String [] args) {
  }
}
```

Fig. 5. Barrier synchronization using semaphores

cess that controls when workers begin their next iteration. The program uses an array of semaphores, `proceed` (one for each worker), rather than a single semaphore, to prevent a fast worker from "stealing" the message intended for a slow worker. With a single semaphore, a slow worker might be context switched after `V(done)` and before the `P(proceed)`, which would allow a fast worker to finish its iteration and get past the `P(proceed)`. (See [18] for details.)

This program can be rewritten using PQ and fewer semaphores, as shown in Figure 6. Worker processes no longer need to tell the coordinator that they are done (via the `done` semaphore); instead PQ will detect that. The role of the coordinator is no longer performed by a separate process. It is now the PQ operation that is invoked when all worker processes quiesce. Also, the `proceed` array of semaphores is now replaced with a single `proceed` semaphore: a fast

---

[2] In JR, the semaphore primitives P and V are just special cases of the message passing primitives **receive** and **send**.

```
public class Barrier {
  private static final int N = 10; // number of workers
  private static sem proceed;
  private static ProcessGroup WG;
  static {
    WG = new ProcessGroup("Worker Group");
    JR.changeCreationGroup(WG); // all worker processes will belong to process group WG
  }
  private static process worker( (int i = 0; i < N; i++) ) {
    while (...) { // iterations remain
      // code to implement one iteration of task i
      ...
      // barrier
      P(proceed); // wait for coordinator to say "continue"
    }
  }
  private static op void coordinator() { // no longer a process -- it's invoked on PQ.
    for (int w = 0; w < N; w++) { V(proceed); }
    if (...) // iterations remain
      JR.registerPartialQuiescenceAction(WG, coordinator); // re-register PQ op.
  }
  public static void main(String [] args) {
    JR.registerPartialQuiescenceAction(WG, coordinator); // register PQ op.
  }
}
```

Fig. 6. Barrier synchronization using partial quiescence

worker cannot overtake a slow worker since all workers must quiesce before the coordinator operation is invoked and tells any worker it may proceed.

4.2 Key Aspects of Partial Quiescence

As seen in the examples in the previous section, process groups allow the programmer to specify parts of the program for separate PQ detection. The names of process groups, specified by the string argument to the `ProcessGroup` constructor, are in a global namespace. For example, in a multi-VM program (Section 2.2), processes created in process group "A" on two different VMs are in the same process group. The programmer can also create a process group specific to a VM by using a per-VM unique identifier in the name.

PQ detection for a process group does not begin until the PQ operation has been registered. This avoids the following "startup problem". Suppose a process group has just been created, but no processes have yet been created within that group, for example, if the main program in Figure 3 registered its PQ operations before instantiating the `MMMultiplier` objects. Then, PQ detection would detect that the group has quiesced, which would not be too useful for the programmer. Just as in GQ, the PQ operation can start up new activity and can re-register another PQ operation.

The precise definition of PQ for JR differs slightly from that given in Section 3. The reason is that in JR a message is sent to an operation, which can be serviced by processes that might belong to different process groups. The PQ definition for JR, therefore, says "(1) there are no messages in the system *that are serviceable by a process in group* A *from group* A *in transit to group* A".

A program that uses PQ can be nondeterministic. For example, a message from outside a process group might be sent either before or after PQ is detected for that process group, thus affecting program behavior. However, such nondeterminism does not occur in the examples in this paper (or other practical examples we have written so far). It remains to be seen whether such nondeterminism is a problem in further practice.

PQ is an extension to, not a replacement for, GQ. A program is globally quiescent when all parts of it have become partially quiescent and the remaining processes not associated with any group have terminated or deadlocked, and no messages are in transit. Also, the extended JR has four additional PQ features for more complicated programming situations as illustrated in [12]. First, the program can disable or enable PQ detection features during execution. Second, an optional argument to the process group constructor can specify the number of processes expected in the group; quiescence of the group occurs when that number of processes have terminated or deadlocked. Third, process groups can be hierarchical. A parent group is defined to have become partially quiescent only when all of its child process groups have become quiescent. Fourth, a process can change its process group in the middle of execution.

5 Implementation and Performance

We have an initial implementation of PQ in an extension of JR version 1.00061 (based on Java 1.4). We are presently porting it to JR version 2.00001 (based on Java 1.5).

5.1 Implementation

The implementation of PQ adapts the centralized manager implementation of GQ described in Section 2.2. (The implementation of PQ for closed process groups (Section 3) could follow the GQ implementation rather directly, but with message counts specific to process groups.) When a process group is created on a VM, the RTS (run-time system) on the VM sends a message to the manager. The manager uses the process group name as the key into a hashtable; the hashtable entry contains the list of VMs on which the process group has been created and a capability for the PQ operation. When the PQ operation is registered, it is sent by the RTS to the manager. The manager then creates a thread to handle quiescent messages for this group (if such a thread has not already been created). The thread executes until the group becomes quiescent (as described in the following paragraph), at which point the thread invokes the PQ operation and terminates. If the PQ operation is re-registered, a new thread is created (Exactly when the thread is created is important so that the thread does not detect quiescence before the operation has been (re-)registered, i.e., to avoid the "startup problem" mentioned in Section 4.2.)

When the RTS on a VM detects that a process group on that VM becomes quiescent (i.e., all of its processes have terminated or are waiting to receive a message), it sends an idle message to the manager, where it is handled by the thread that is managing the process group. If the manager has received an idle message for the process group from each VM, it then sends a message to each

VM to confirm that the VM is indeed idle. If the manager receives such confirmation, then the process group is quiescent. Otherwise, it waits for idle messages from those VMs who reported they were not idle before it attempts confirmation again. This second, confirmation phase is necessary to account for one VM reporting that the process group is idle just after it sends a message to another process within the same process group on another VM that already reported that it was idle, i.e., to implement the modified PQ definition in Section 4.2.

5.2 Performance

Because PQ is a new language feature, we have no direct basis of comparison to assess the performance of our implementation. However, we have compared the performance of PQ in several programs with the performance of GQ in roughly comparable programs. Specifically, we compared the PQ matrix multiplication program (Section 4.1) with a variant of the original main program in Figure 1 that starts two matrix multiplications (GQ). The results show that over a range of different sized matrices PQ required 0%–1.5% additional time; the multi-VM versions of those programs required 0.4%–4.1% additional time. We also compared the (single VM) PQ and GQ versions of the barrier programs (Figure 6). The times for the two versions over a range of different numbers of workers were always within 3% of each other; the times for a multi-VM barrier program were always within 7% of each other. The PQ version of Figure 6 took 4-10% more time than the (GQ) program in Figure 5. We ran these tests on various PCs (1.4GHz and 2.0GHz uniprocessors; 2.4GHz and 2.8GHz dual-processors) running Linux; specific results, of course, varied according to platform. The actual code and execution times, and further results and explanations appear in [12].

6 Conclusion

This paper introduced the notion of partial quiescence and showed how it can be incorporated into a programming language. Having such a PQ mechanism can lead to a different style of programming, which in some cases is simpler as seen, for example, in the barrier example in Section 4.1. This paper also discussed the implementation of PQ and its performance, which differs only slightly from GQ's performance. Although our early results are promising, further experience is needed with using PQ mechanisms and measuring their costs. In particular, we plan to investigate further concurrent applications and see which might benefit from using PQ mechanisms. We plan to include PQ in the standard JR language release [6].

Acknowledgements

The referees provided thoughtful comments, which helped us to improve our presentation. One referee kindly provided additional references to related work.

References

1. G. R. Andrews and R. A. Olsson. *The SR Programming Language: Concurrency in Practice*. The Benjamin/Cummings Publishing Co., Redwood City, CA, 1993.
2. G. R. Andrews, R. A. Olsson, M. Coffin, I. Elshoff, K. Nilsen, T. Purdin, and G. Townsend. An overview of the SR language and implementation. *ACM Transactions on Programming Languages and Systems*, 10(1):51–86, January 1988.
3. E. W. Dijkstra and C. S. Scholten. Termination detection for disffusing computations. *Inform. Process. Lett.*, 11(1):1–4, 1980.
4. N. Francez. Distributed termination. *ACM Trans. Programming Languages and Systems*, 2(1):42–55, 1980.
5. J. Helary, C. Jard, N. Plouzeau, and M. Raynal. Detection of stable properties in distributed applications. In *PODC '87: Proceedings of the Sixth Annual ACM Symposium on Principles of Distributed Computing*, pages 125–136, 1987.
6. JR distribution. http://www.cs.ucdavis.edu/~olsson/research/jr/.
7. A. W. Keen, T. Ge, J. T. Maris, and R. A. Olsson. JR: Flexible distributed programming in an extended Java. *ACM Transactions on Programming Languages and Systems*, pages 578–608, May 2004.
8. Y. Kermarrec, L. Pautet, and S. Tardieu. GARLIC: generic Ada reusable library for interpartition communication. In *TRI-Ada '95: Proceedings of the Conference on TRI-Ada '95*, pages 263–269, New York, NY, USA, 1995. ACM Press.
9. N. Kobayashi. A partially deadlock-free typed process calculus. *ACM Trans. Program. Lang. Syst.*, 20(2):436–482, 1998.
10. L. Liang, S. T. Chanson, and G. W. Neufeld. Process groups and group communications: classifications and requirements. *IEEE Computer*, 23(2):56–66, 1990.
11. G. R. Luecke, Y. Zou, J. Coyle, J. Hoekstra, and M. Kraeva. Deadlock detection in MPI programs. *Concurrency and Computation: Practice and Experience*, 14:911–932, 2002.
12. Billy Yan-Kit Man. The design and implementation of partial quiescence in a concurrent programming language. Master's thesis, University of California, Davis, Department of Computer Science, March 2006. http://www.cs.ucdavis.edu/~olsson/students/.
13. S. P. Masticola and B. G. Ryder. Static infinite wait anomaly detection in polynomial time. In *Proceedings of 1990 International Conference on Parallel Processing*, pages II.78–II.87, University Park PA, 1990.
14. J. Matocha and T. Camp. A taxonomy of distributed termination detection algorithms. *The Journal of Systems and Software*, 43(3):pp 207–221, 1998.
15. Friedemann Mattern. Global quiescence detection based on credit distribution and recovery. *Inf. Process. Lett.*, 30(4):195–200, 1989.
16. U. Nestmann. What is a 'good' encoding of guarded choice? *Journal of Information and Computation*, 156:287–319, 2000.
17. K. Okano, S. Hattori, A. Yamamoto, T. Higashino, and K. Taniguchi. Specification of real-time systems using a timed automata model with shared variables and verification of partial-deadlock freeness. In *ICPP Workshop*, pages 576–581, 1999.
18. R. A. Olsson and A. W. Keen. *The JR Programming Language: Concurrent Programming in an Extended Java*. Kluwer Academic Publishers, Inc., 2004.
19. Amitabh B. Sinha, L. V. Kalé, and B. Ramkumar. A dynamic and adaptive quiescence detection algorithm. Technical Report 93-11, Department of Computer Science, University of Illinois, Urbana-Champaign, 1993.
20. J. S. Vetter and B. D. de Supinski. Dynamic software testing of MPI applications with Umpire. Technical report, Lawrence Livermore National Laboratory, 2000.

Tying Memory Management to Parallel Programming Models

Ioannis E. Venetis[*] and Theodore S. Papatheodorou

High Performance Information Systems Laboratory
Department of Computer Engineering and Informatics
University of Patras, Rion 26500, Greece
http://www.hpclab.ceid.upatras.gr

Abstract. Stand-alone threading libraries lack sophisticated memory management techniques. In this paper, we present a methodology that allows threading libraries that implement non-preemptive parallel programming models to reduce their memory requirements, based on the properties of those models. We applied the methodology to NthLib, which is an implementation of the Nano-Threads programming model, and evaluated it on an Intel based multiprocessor system with Hyper-Threading and on the SMTSIM simulator. Our results indicate that not only memory requirements drop drastically, but that execution time also improves, compared to the original implementation. This allows more fine-grained, but also larger numbers of parallel tasks to be created.

1 Introduction

Efficiency of parallel programming models has traditionally been measured in terms of two important metrics. On the one hand, execution time is used to indicate whether threading libraries and parallel applications have been implemented effectively. On the other hand, the amount of resources that are required to express and execute parallel tasks has also been of great importance. Especially usage of memory, which might be a scarce resource on some parallel systems, has been carefully analyzed in several cases. Work conducted towards this direction, mainly targets threading libraries that support multithreaded languages, i.e., parallel languages that support dynamic thread creation. This is due to the fact that the accompanying compilers are able to perform powerful analysis of memory requirements per function and propagate this information to the library.

In contrast, stand-alone threading libraries lack the knowledge about the memory requirements of an application and must therefore be pessimistic about them. Due to this fact, two different paths to implement such libraries have emerged. The basic requirement of the first one is to reduce execution time at all costs. Libraries implemented under this scheme, which we will refer to as *Descriptor on Stack (DOS)*, usually allocate a large region of memory during

---

[*] This work has been carried out while the first author was supported by a grant from the 'Alexander S. Onassis' Public Benefit Foundation and the European Commission through the 'POP' IST project (Grant No.: IST-2001-33071).

thread creation. This region is logically divided into two parts, one describing the parallel task (Descriptor) and the other being the stack of the task during execution. Although common sense suggests that this is a fast method to create threads, memory requirements are often excessive. Taking into account that contemporary stand-alone threading libraries set the default stack size somewhere between 1 and 4MB, makes it obvious that those libraries cannot support large numbers of threads. In order to overcome this problem, supporters of the second implementation strategy suggest that the first priority should be to minimize memory requirements, even if execution time suffers. In this case, during thread creation time, only a small descriptor is allocated and initialized, whereas a larger stack is assigned to the task when it is selected to run. We will refer to this strategy as *Lazy Stack Allocation (LSA)*. Effectively, stacks are traded in favour of smaller descriptors.

Current developments in computer architecture, such as *Simultaneous Multi-Threading (SMT)* [1], *HyperThreading* [2] and multicore processors, allow efficient execution of more fine-grained parallelism, in addition to a larger number of parallel tasks. This allows applications to express more of their inherent parallelism, creating a number of threads that might exceed the number of available execution contexts (ECs). Our view is to create an infrastructure that will allow stand-alone threading libraries to efficiently support the above execution scheme. In addition to these observations, taking into account that the above architectures usually have a lower memory per EC ratio, leads us to the conclusion that the LSA implementation strategy is more appropriate for such systems. However, those libraries are usually slower, thus invalidating the means provided by modern processors to efficiently execute parallel tasks. Hence, it becomes obvious that a new approach to tackle this problem is necessary, which will combine the benefits of both approaches.

In this paper we present a methodology that allows stand-alone threading libraries that implement non-preemptive parallel programming models to reduce their memory requirements. This is accomplished by taking into consideration the properties of this specific parallel programming paradigm. A prerequisite, however, is that threading libraries should be LSA enabled. Firstly, we present a method to convert a DOS into a LSA enabled library. In addition, this method takes into account an important factor that greatly affects speed of DOS enabled libraries, i.e., their self-identification mechanism. Based on this, we take a step further, compared to previous approaches, introducing two methods that reduce memory requirements even more. The first one allows us to compute *a priori* the total number of stacks that are required to run an application. The second one improves on LSA, by directly handing the stack of a terminating thread to the next one that should run on a processor. We will refer to this method as *Direct Stack Reuse (DSR)*. We demonstrate for the first time, to the best of our knowledge, that LSA and DSR enabled libraries can actually outperform DOS enabled libraries, in terms of both, execution time and memory requirements.

The rest of this paper is organized as follows: Section 2 presents related work, with respect to memory management techniques under several parallel programming models. Section 3 presents how to apply our methodology to convert DOS into LSA enabled libraries. In Section 4 and Section 5 we present how to further reduce memory requirements. In Section 6 we experimentally evaluate our approach. Finally, in Section 7 we conclude our paper.

2 Related Work

As already mentioned, much work has been done to reduce memory requirements in threading libraries that support multithreaded languages. For example, in the Lazy Task Creation [3] model, a thread is implemented as a serial call to a function, which allows it to run in the stack of the parent. If, however, the child suspends execution or more parallelism is required, the recorded return address of the parent is assigned to another processor and the corresponding stack frames are copied, in order for the parent to continue execution. The Lazy Threads [4] model employs several representations of parallelism and makes the compiler responsible for selecting the most efficient in each case. Accordingly, the compiler decides which is the best representation of a stack, after statically analyzing each function. For serial execution, a conventional stack is used, for threads with small and medium sized data a structure know as a *stacklet* is used, whereas for larger data sets a separate memory region is allocated. The usual case is to use a stacklet, which is a memory region that can hold more stack frames. However, initialization and release of a stacklet are quite expensive operations. The Capriccio [5] threading library also uses data from static analysis of the compiler. Similarly to the previous model, more stack frames are put in each memory region, with the associated management cost. A new region is allocated at the check points that the compiler inserts, according to the performed analysis. However, Capriccio implements an 1:N model, where more user-level threads are executed on top of only one kernel-level entity. Hence, true parallelism cannot be exploited and some of the optimizations are not valid in M:N models.

An interesting approach, that tries to simplify development of compilers for multithreaded languages, is the one proposed in StackThreads [6]. It provides basic functionality to compilers, in order to map the execution model of multithreaded languages to the execution model of the C programming language. More advanced management of stack frames can be built on top of this functionality. Memory is managed in the library, through the information that the compiler has to pass to it. Although generality is an important concern in StackThreads, that work targets a different set of threading libraries than our work.

With respect to stand-alone threading libraries, TiNy Threads [7] targets the Cyclops64 system, which has extremely limited memory. Only 4800KB are available for 150 ECs. It uses the DOS model and due to increased memory requirements has to limit the number of threads that can be created, actually invalidating the objective of the architecture. In threading libraries that are preemptive, such as POSIX threads, threads are usually created and immediately put into a ready queue for execution. In these cases, more threads than there are processors are usually active. This, in turn, implies that a large number of stacks must be available, in order to keep the state of threads that have run but are currently preempted. As a consequence, separation of descriptors and stacks can only be applied to efficiently recycle objects in such cases and not reduce usage of memory. In this paper, however, we target non-preemptive threading libraries, which have well defined entry and exit points for a thread. Exploiting this property, is what differentiates our work from previous approaches.

3 Retaining a Fast Self-identification Mechanism

DOS enabled libraries are thought to be fast for two reasons. Firstly, no stack has to be assigned to each thread before execution, because it has already been allocated at thread creation time. Secondly, the allocated memory region, that is split between the descriptor and the stack, is usually aligned at the region's size. This allows a thread to quickly perform self-identification, i.e., find the starting address of it's own descriptor and acquire important information about it's status. For example, if a $1MB(=2^{20}$ bytes) memory region is allocated, the starting address should have it's 20 last bits zero. Self-identification is performed in this case by reading the current value of the Stack Pointer and clearing the last 20 bits. The result is always the starting address of the memory region. Adding the size of the memory region and subtracting the size of the descriptor, returns the starting address of the latter. By dereferencing this value, all the information contained in the descriptor can be obtained. For a more detailed description of the mechanism, including figures, we refer the reader to [8].

Our first requirement, while switching to a LSA model, is to retain a fast self-identification mechanism. In order to achieve this goal, our methodology requires us to follow two steps. Firstly, stacks should be aligned as in the DOS model. We must point out that the stack under LSA is actually the same as the memory region in the DOS model, whereas descriptors are allocated separately. Secondly, after a thread has been selected to run and a stack has been assigned to it, a pointer to the descriptor (instead of the descriptor itself) should be put at the top of the stack. Thus, self-identification is performed almost in the same manner as in DOS enabled libraries. The difference lies in the last step, where the size of a pointer is subtracted from the computed value, instead of the size of a descriptor. This returns a pointer to the descriptor, which can be dereferenced, as in the DOS case. Hence, this mechanism is as fast as the original one.

4 Direct Stack Reuse

DOS requires a large memory region for each thread. If more threads than processors are created, this leads to unnecessarily high memory consumption. In contrast, LSA, in combination with the fact that a non-preemptive model guarantees that a thread will not be interrupted, allows the stack of a terminating thread to be inserted into a recycling queue and another stack to be assigned to the next thread. The same process is repeated for every thread that terminates, on each processor. Hence, LSA requires only two stacks per processor. If, however, only one thread is created for each processor, LSA will use the recycled stack when a new thread finally arrives, thus the number of stacks will be equal to the DOS case, in addition to a number of small descriptors.

LSA is a widely used method to reduce memory usage. However, with non-preemptive models more improvements can be achieved. Our second requirement, when switching to a LSA model, is to reduce the time required to assign a stack to a thread that is ready to run. When a context-switch occurs, under a non-preemptive model, a thread is actually terminating and it's stack is not needed anymore. Due to this observation, it is obvious that the stack of that thread can be directly reused by the new thread, without accessing queues or allocating a new one. This reduces time to find a stack for the new thread. Setting up the stack in this case, is as expensive as in the DOS case. The difference is that initialization just happens at a different point in the execution path.

Two important points have to be made clear, the first being that DSR can only be applied if LSA is also active. The second point is that DSR is a complementary mechanism to the recycling queues. Someone could conclude that recycling queues could be dropped from a library, since each thread directly uses the stack of the previous thread. However, there are cases where recycling queues are necessary. For example, a thread might block and voluntarily release the processor it is running on. In this case, the user-level scheduler selects a new thread for that processor. Obviously, the stack of the thread that blocked must be preserved, in order for it to be able to resume execution. Hence, the newly selected thread needs a new stack, which it will request from the queues.

5 Calculating the Number of Required Stacks

Although LSA already contributes to reduced memory requirements and DSR improves on that, eliminating one stack per processor for switching to a new thread, the non-preemptive nature of the programming models that are considered allow us to go even further. Specifically, it is possible to calculate a priori the number of required stacks that are necessary to run an application. To demonstrate this, we will use as an example NthLib [9], a threading library that implements the Nano-Threads programming model [10]. However, this specialization has an effect only in the initialization phase of a library. During context-switches, the following reasoning applies to every threading library.

Suppose that initially only LSA is enabled and that an application requests P processors. In this case, the library must create $P - 1$ Virtual Processors (VPs) more, since the first VP is the one that started the application. Currently, the requirements for the initialization phase are one stack for the first VP and two for each other, giving a total of $2 \cdot P - 1$. If DSR is also enabled, then there is no need for additional stacks. If not, then as soon as a VP, except of the first one, receives the first user-level thread for execution, one of the stacks is recycled, as it is thought to be the stack of the previous thread. Therefore, only one more stack is required per VP to perform context-switches, which sums up to $(2 \cdot P - 1) + (P - 1) = 3 \cdot P - 2$. Finally, one more stack is required for the main thread, because it voluntarily blocks and joins the other VPs in the parallel phase. Therefore, the final sums are $3 \cdot P - 1$ for LSA and $2 \cdot P$ for DSR.

This information can be used by threading libraries. During initialization, allocation of all stacks can be performed at once. In order to enforce the memory alignment requirements of the library, some additional pages of memory must be allocated. However, after the first stack has been correctly aligned, all subsequent stacks in that memory region will also be aligned. This is in contrast to DOS, where additional memory has to be allocated for each new thread, which is wasted. If we were to free that memory, execution time would suffer. Since the application is still in a serial phase, stacks can be put into the corresponding recycling queues without using any locking mechanism. This has two advantages. Firstly, the library requests only once memory for stacks. Memory allocators are usually slow when large memory areas are requested. Under DOS, this cost is paid for every thread. Under LSA and DSR, the cost is amortized among all threads that will run on the allocated stacks. Secondly, each processor is assigned the total number of stacks it requires during execution. This reduces contention on the recycling queues of the stacks to the minimum. Finally, it is a priori known that all allocated stacks will be used and no memory will be wasted.

Although the number of stacks can be predicted, the number of descriptors cannot. Despite that fact, we propose a similar pre-allocation technique for them. The first time a descriptor is requested, a larger area of memory is allocated. However, in contrast to stacks, all descriptors that fit into this area are not directly put into queues, due to the fact that mutual exclusion would be necessary. In addition, only the main thread of an application usually creates threads, hence spreading descriptors among all recycling queues would be inefficient. Therefore, each time a descriptor is needed, and none can be found in a recycling queue, we atomically increase the base address of the allocated area by the size of a descriptor and return the previous address. Since contention is very low, due to the fact that usually only one thread creates others, this atomic operation is very likely to complete very fast. Only when the allocated area is exhausted, does the library request more memory for descriptors. Currently, $2 \cdot P$ descriptors can fit into the area that is each time allocated in the new implementation of NthLib. In combination with the fact that a descriptor is only 512 bytes large, one can conclude that this is an effective method to reduce the number of expensive memory allocation requests, without actually sacrificing memory.

6 Experimental Evaluation

In order to evaluate the efficiency of our methodology, we applied it to NthLib. The original implementation of NthLib is DOS based. The new implementation has been developed so as to support all designs that were described, i.e., DOS, LSA and DSR. The one that is each time used is defined during compilation of the library. Supporting all designs was intentional, in order to make comparison among them easier. Having only one library, makes our results independent of other differences and details that two separate implementations would have.

The first system we used to evaluate our approach is a 4-processor, Hyper-Threading enabled system, running Linux 2.6.8. The second one is SMTSIM [1],

Table 1. Hardware configuration of the experimentation platform

	Intel processor based system	SMTSIM
Processors	4 Intel Xeon MP HTs, 2 GHz, 2 execution contexts/processor	1 Alpha based, 8 execution contexts
L1 Data Cache	8KB shared, 4-way assoc.	32KB, 2-way assoc., 10-cycle miss latency
L1 Inst. Cache	12KB shared execution trace	32KB, 2-way assoc., 10-cycle miss latency
L2 Cache	512KB shared, unified, 8-way assoc.	256KB, 2-way assoc., 15-cycle miss latency
L3 Cache	1MB shared, unified, 8-way assoc.	2MB, 2-way assoc., 125-cycle miss latency
D-TLB	64 entries	128 entries
I-TLB	2x64 entries	48 entries
DRAM	2GB	Depends on host system

a simulator that implements an Alpha processor with 8 ECs. More detailed hardware characteristics for both systems are summarized in Table 1. The compiler used is gcc 4.0.2 for both platforms, at the highest optimization level (-O3).

Due to space limitations, we present results for only one benchmark. We refer the reader to [8], for a more detailed description of applying our methodology to NthLib and a more thorough evaluation. The benchmark that we used, which we will refer to as *Empty*, follows the fork/join model. The master thread creates one million empty nano-threads, whereas the slave processors dispatch and execute them. The master thread blocks after it has created all threads, hence calling the user-level scheduler and joining the other processors to execute threads. This benchmark is appropriate for estimating the pure run-time overhead of thread management in NthLib. Moreover, it can be used to determine the number of stacks that an application requires and to estimate the minimum number of descriptors that must be allocated. This is due to the fact that nano-threads perform no computation in this benchmark. Therefore, they are consumed as fast as possible by the slave processors and are immediately recycled.

Fig. 1 summarizes the results for this benchmark. They are normalized with the time of the slowest benchmark, which is when LSA is enabled and the benchmark is run on one EC. The absolute execution times in this case were 2,56 seconds for the Intel based system and 643,5 million simulated clock cycles for SMTSIM. The stack size used for each nano-thread was set in all runs to the quite small size of 32KB, in order to allow the benchmark to successfully complete in most cases under the DOS scheme. If either LSA or DSR is enabled, pre-allocation of stacks and descriptors is also enabled. For SMTSIM, the horizontal axis represents the number of ECs used. For the Intel based system, the numbers of physical processors and ECs used on each one of them are mentioned. For example, (4, 1) means that 1 EC was used on each one of the 4 physical processors. A special case is the one denoted with (4, 1/2), where 2 ECs were used on 2 physical processors and 1 EC on the other 2 physical processors. Finally, notice that the benchmark could not complete when DOS was enabled and it

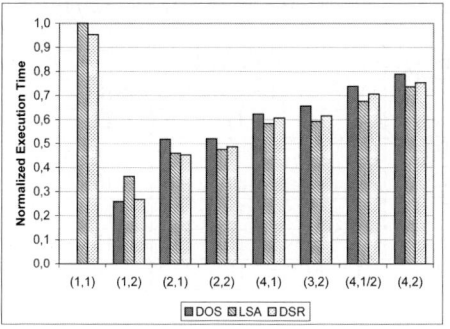

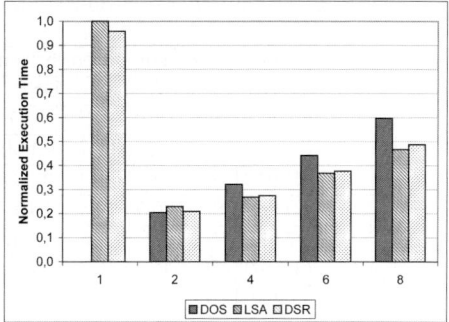

Fig. 1. Normalized execution time for *Empty* on the Intel based system and SMTSIM

was run on one processor, due to excessive memory requirements. The master thread must first create all threads in this case, before blocking to execute them.

The results indicate that LSA and DSR are from 2,65% (DSR, (4,1)) up to 12,54% (DSR, (2,1)) faster than DOS, on the Intel platform. For SMTSIM, the range is between 14,66% (DSR, 4) and 21,82% (LSA, 8). The exception occurs when two ECs are used on only one processor. In this case, DSR is 3,49% and LSA 40,41% slower than DOS, for the Intel platform, whereas for SMTSIM the difference is 2,48% and 12,38% respectively. More detailed measurements revealed that the cause of this inefficiency is the contention on the recycling queues of both, the stacks and the descriptors. Under LSA and DSR, nano-threads are created and also start executing faster after they have been selected to run. This means that both ECs try to acquire access to the queues in smaller time intervals. In combination with the fact that resources of the processor are shared between the ECs and one of them may stall, if a resource is not available, explains this odd behavior. DSR is faster than LSA in this case, due to the fact that stacks are not recycled but directly reused. This alleviates the queueing system significantly. Moreover, the differences for SMTSIM are smaller, compared to the Intel platform. Additionally, we observe that execution time rises, as more ECs are used, although with a smaller pace in the case of SMTSIM. The fact that SMTSIM exhibits better behaviour in all cases can be attributed to the following facts. Firstly, SMTSIM simulates 8 ECs on one processor. The Intel platform, in contrast, is a SMP system, where communication costs among processors are quite higher. Secondly, resources of the simulated processor in SMTSIM are dynamically shared among ECs. In the second system, however, resources of a processor are statically divided between both ECs. Therefore, even if a resource is available on one of them, the other cannot take advantage of it. Lastly, SMTSIM uses a very efficient hardware implementation of locks, based on the concept of a *lockbox* [1], which is exploited in our library and significantly reduces synchronization time among threads.

The other important factor that our implementation tries to minimize, apart from execution time, is usage of memory. Table 2 summarizes the memory requirements of our benchmark. Starting with the results for the Intel platform,

Table 2. The number of required Stacks (S) and Descriptors (D) for *Empty*

	Intel processor based system					SMTSIM					
	DOS	LSA		DSR			DOS	LSA		DSR	
	S	S	D	S	D		S	S	D	S	D
(1,1)	-	3	1000003	2	1000003	1	-	3	1000003	2	1000003
(1,2)	13012	6	284877	4	204662	2	8	5	31	3	15
(2,1)	9602	6	176701	4	107078	4	9	7	14	7	14
(2,2)	7342	10	104896	8	76378	6	11	11	16	11	15
(4,1)	10909	10	80893	8	73132	8	13	13	17	12	16
(3,2)	9993	16	73760	12	86505						
(4,1/2)	8558	16	70684	12	76668						
(4,2)	10496	22	65270	16	73812						

it becomes obvious that savings in memory are significant. The biggest difference between DOS and LSA appears when both ECs are used on each one of the physical processors and is 90,07%. The biggest difference between DOS and DSR appears when one EC is used on all physical processors and is 89,45%. The smallest difference appears in both cases when one EC is used on each of two physical processors and is 65,75% for LSA and 75,39% for DSR.

Different results are acquired for SMTSIM, where memory requirements for all cases are almost identical. This difference, compared to the Intel platform, can be explained, if we take into consideration that SMTSIM does not run an OS and delays that origin from it are not accounted for. As an example of the importance of this fact, we mention that on the Intel platform, a thread is created in about 30000 clock cycles, under the DOS scheme, whereas the time required to insert it into a ready-queue is only about 250. Almost all of the time to create the thread is spent in the OS, in order to allocate the required memory. In SMTSIM, however, this time is not measured and a thread is created in 60 cycles and inserted into a queue in 100. Therefore, we believe that for fine-grained benchmarks, that frequently interact with the OS, SMTSIM is not as accurate as required. Consequently, we believe that the results obtained on the Intel platform, with respect to memory requirements, reflect better reality. Furthermore, we believe that differences in execution time between DOS and both, LSA and DSR, would be higher for SMTSIM, if the time for memory allocation had been taken into account. However, SMTSIM gives a good estimation of execution times in all cases and can be used more reliably for applications where each thread has to perform more computations [8].

7 Conclusions

In this paper we presented a methodology that can be applied to non-preemptive parallel programming models, in order to reduce their memory requirements. We used a widely known methodology to convert a DOS into a LSA enabled library, demonstrating that it is possible to retain a fast self-identification mechanism. Furthermore, taking into account the fact that most contemporary high performance threading libraries implement non-preemptive parallel programming models, we introduced two more methods to reduce memory requirements. Those are based on the properties of non-preemptive programming models, which is what differentiates our work from previous approaches. Specifically, we introduced a mechanism that allows the stack of a terminating thread to be directly reused by the thread that is next to be run. In addition, we demonstrated how it is possible to calculate a priori the total number of stacks that an application requires. The latter can be exploited to reduce the amount of memory that would otherwise be wasted, due to the alignment requirements of memory regions and stacks. Finally, our performance evaluation proved that combining all of the above techniques, not only drastically reduces memory requirements to represent parallelism in threading libraries, but that it can also be faster than the traditional DOS approach, in contrast to general belief.

References

1. Tullsen, D., Eggers, S., Levy, H.: Simultaneous Multithreading: Maximizing On-Chip Parallelism. In: Proceedings of the 22nd Annual International Symposium on Computer Architecture, S. Margherita Ligure, Italy (1995) 392–403
2. Marr, D.T., Binns, F., Hill, D.L., Hinton, G., Koufaty, D.A., Miller, J.A., Upton, M.: Hyper-Threading Technology Architecture and Microarchitecture. Intel Technology Journal **Volume 6, Issue 1** (2002) 4–15
3. Mohr, E., Kranz, D.A., R. H. Halstead, J.: Lazy Task Creation: A Technique for Increasing the Granularity of Parallel Programs. IEEE Transactions on Parallel and Distributed Systems **Volume 2, Issue 3** (1991) 264–280
4. Goldstein, S.C., Schauser, K.E., Culler, D.E.: Lazy Threads: Implementing a Fast Parallel Call. Journal of Parallel and Distributed Computing **Volume 37, Issue 1** (1996) 5–20
5. von Behren, R., Condit, J., Zhou, F., Necula, G., Brewer, E.: Capriccio: Scalable Threads for Internet Services. In: Proceedings of the 19th Symposium on Operating System Principles, Bolton Landing, New York (2003) 268–281
6. Taura, K., Tabata, K., Yonezawa, A.: Stackthreads/MP : Integrating Futures into Calling Standards. Technical Report TR 99-01, University of Tokyo (1999)
7. del Cuvillo, J., Zhu, W., Hu, Z., Gao, G.R.: TiNy Threads: a Thread Virtual Machine for the Cyclops64 Cellular Architecture. In: Proceedings of the 5th Workshop on Massively Parallel Processing, Denver, Colorado (2005)
8. Venetis, I.E., Papatheodorou, T.S.: A Time and Memory Efficient Implementation of the Nano-Threads Programming Model. Technical Report HPCLAB-TR-210106, High Performance Information Systems Laboratory (2006)
9. Martorell, X., Labarta, J., Navarro, N., Ayguade, E.: A Library Implementation of the Nano-Threads Programming Model. In: Proceedings of the 2nd International EuroPar Conference, Lyon, France (1996) 644–649
10. Polychronopoulos, C., Bitar, N., Kleiman, S.: Nanothreads: A User-Level Threads Architecture. Technical Report 1297, CSRD, University of Illinois at Urbana-Champaign (1993)

Topic 10: Parallel Numerical Algorithms

Michel Cosnard, Hans-Joachim Bungartz,
Efstratios Gallopoulos, and Yousef Saad

Topic Chairs

Since the early days of supercomputing, numerical routines have caused the highest demand for computing power anywhere, making their efficient parallelization one of the core methodical tasks in high-performance computing. And still, many of today's fastest computers in the world are mostly used for the solution of huge systems of equations as they arise in the simulation of complex large scale problems in engineering and science.

Despite this long tradition, parallel numerical algorithms did not lose anything of their relevance. The efficient implementation of existing schemes on state-of-the-art parallel systems (such as clusters or hybrid systems), the challenges resulting from massively parallel systems, the design of easy-to-use portable software components, the recent endeavours to tackle optimization, control, and interactive steering scenarios, too – all this clearly shows that progress in computational science and engineering strongly depends on progress with parallel numerical algorithms. This crucial importance of parallel numerical algorithms certainly justifies to again having devoted a special workshop to this topic at Euro-Par, in addition to the discussion of special aspects of high-performance and grid computing in Topic 16.

Overall, fourteen papers were submitted to our Topic. With authors from Denmark, France, Germany, Greece, Lithuania, Spain, and Switzerland, Europe is the dominant male (as expected at Euro-Par), but three papers are authored by scientists from the United States, Australia, and China. Out of these fourteen submissions, eight were accepted as regular papers.

Both devising new parallel algorithms for numerical tasks and adapting existing ones to state-oft-the-art parallel systems are vigorously flourishing and still developing fields of research. Hence, it is no surprise that the eight research articles presented in this section cover a wide range of topics arising in the various subdomains of parallel numerical algorithms. At the conference, the presentations were arranged into three sessions on *PDE-Related Topics*, *ODE- or Particle-Related Topics*, and *Miscellaneous Topics*. This structure also reflects in the following part of the conference's proceedings.

In the *PDE-Related Topics* section, Raimondas Čiegis addresses a parallel locally one-dimensional scheme for the numerical solution of three-dimensional parabolic problems with nonlocal boundary conditions. Vincent Heuveline and Andrea Walther discuss online checkpointing strategies for parallel adjoint computations as they occur in the optimization or control of time-dependent flow problems. Finally, fault tolerance is the topic of Hatem Ltaief, Marc Garbey, and Edgar Gabriel. In their contribution, they study parallel fault tolerant algorithms for parabolic problems.

Ordinary differential equations and particle methods are the central topics of the second section. José M. Badía, Peter Benner, Rafael Mayo, and Enrique Quintana-Ortí deal with generalized algebraic Riccati equations, especially parallel schemes for large-scale and sparse ones. Load balancing strategies and their applicability to data-parallel embedded Runge-Kutta integrators are studied by Matthias Korch and Thomas Rauber. The third presentation, given by a team of eight scientists from ETH Zürich, focuses on a software framework for the portable parallelization of particle-mesh simulations, the combination of particle- and mesh-based methods.

The final third session gathers the two remaining papers of Topic 10. Rita Zrour, Pierre Chatelier, Fabien Feschet, and Rémy Malgouyres study the parallelization of a numerical and computationally intense problem from computer graphics: discrete radiosity methods for global illumination. Finally, an interdisciplinary team of five researchers from chemistry and computer science at TU München addresses the parallelization of matrix operations as a typical building block from numerical linear algebra, as they appear in optimal control-based quantum compilers.

Altogether, the contributions to Topic 10 at the 2006 Euro-Par in Dresden show once more the great variety of interesting, challenging, and important issues in the field of parallel numerical algorithms. Thus, we are already looking forward to the new results submitted to and presented at next year's Euro-Par conference.

Parallel LOD Scheme for 3D Parabolic Problem with Nonlocal Boundary Condition

Raimondas Čiegis

Vilnius Gediminas Technical University,
Saulėtekio av. 11, LT-10223 Vilnius, Lithuania
rc@fm.vtu.lt

Abstract. A parallel LOD algorithms for solving the 3D problem with nonlocal boundary condition is considered. The algorithm is implemented using the parallel array object tool *ParSol*, then a parallel algorithm follows semi-automatically from the serial one. Results of computational experiments are presented.

1 Problem Formulation

Boundary conditions are important part of any mathematical model. Recently new types of boundary conditions are proposed and investigated. Many physical and technological processes are described by mathematical models consisting of elliptic or parabolic problems with non-local boundary conditions. A review of such applications and mathematical results for analysis of one-dimensional problems is presented in the recent survey paper of Dehghan [9]. Numerical algorithms for solving linear and nonlinear parabolic problems with nonlocal boundary conditions are investigated in [5,7,8,11,12].

In this paper we consider parallel numerical algorithms for solving 3D parabolic problem with the additional integral boundary condition. Let $Q_T = \Omega \times [0, T]$, $\Omega = (0;1) \times (0;1) \times (0;1)$ be a domain with the boundary $\partial \Omega$. This boundary is split into two parts $\partial \Omega = \partial \Omega_1 \cup \partial \Omega_2$, $\partial \Omega_2 = \{X : (x_1, x_2, 0), 0 \leq x_j \leq 1, j = 1, 2\}$. In Q_T we consider a parabolic equation

$$\frac{\partial u}{\partial t} = \sum_{\alpha=1}^{3} \frac{\partial}{\partial x_\alpha}\left(k_\alpha(X,t)\frac{\partial u}{\partial x_\alpha}\right) - q(X,t)u + f(X,t), \tag{1}$$

subject to boundary conditions:

$$u(X,t) = \mu_1(X,t), \quad X \in \partial\Omega_1 \times (0,T],$$

$$u(X,t) = \mu_0(t)\mu_2(X), \quad X \in \partial\Omega_2 \times (0,T],$$

initial condition:

$$u(x_1, x_2, x_3, 0) = u_0(x_1, x_2, x_3), \quad X \in \Omega \cup \partial\Omega,$$

and the additional nonlocal condition:

$$\int_0^1 \int_0^1 \int_0^1 \rho(X,t)\, u(X,t)\, dx_3\, dx_2\, dx_1 = M(t). \tag{2}$$

Here $k_\alpha, q, d, \rho, f, u_0, M, \mu_j, j = 1, 2$ are given continuous functions, and the functions $u(X,t)$, $\mu_0(t)$ are unknown. Thus the initial-boundary problem (1)–(2) is over-specified, and the integral condition is used to identify the boundary condition function $\mu_0(t)$, i.e. we solve an inverse problem. When this boundary value is obtained, we can use any efficient method to solve a standard three-dimensional parabolic boundary value problem.

The existence and uniqueness of the solution of 2D problem is studied in [2]. The analysis of the forward Euler method and a modified Locally One Dimensional (LOD) scheme is presented in [10,16]. At each splitting step of the LOD scheme one-dimensional problems were approximated by the forward Euler method, thus the obtained LOD method was only conditionally stable.

The analysis of new finite difference schemes (including the LOD method) is presented in [3,4]. It is proved that integral (2) can be approximated by the trapezoidal rule, if the initial condition is approximated in consistent way.

High-performance computers with massive parallel processors are developing very fast and parallel numerical algorithms play an important role in large-scale scientific and engineering computations. Three groups of methods are widely used for solving multidimensional parabolic initial-boundary value problems: a) explicit algorithms, b) fully implicit approximations, c) splitting methods. In splitting methods the multidimensional problem is reduced to a sequence of one dimensional implicit difference systems with tridiagonal matrix. Special parallel versions of the serial factorization algorithm are used to implement LOD algorithms on multiprocessor computers. A reduction of communication costs is the second main problem in developing efficient parallel splitting algorithms for parallel computers with distributed memory.

In this paper we consider the LOD parallel algorithm for solving three dimensional problem (1)–(2) with the nonlocal boundary condition. The rest of the paper is organized as follows. In Section 2, we formulate the LOD finite-difference scheme. In Section 3 the parallel LOD algorithm is proposed. The parallel array object tool *ParSol* is used for its implementation. Then a parallel algorithm follows semi-automatically from the serial one. The complexity and scalability analysis of the parallel LOD algorithm is done. In Section 4 results of computational experiments are presented to test the accuracy and the efficiency of the parallel algorithm.

2 Locally One Dimensional Method

In Q_T we define a uniform grid $Q_{h\tau} = \omega_h \times \omega_\tau$:

$$\omega_h = \big\{(x_{1i}, x_{2j}, x_{3k}) : x_{\alpha,i} = ih,\ h = \frac{1}{J},\ 0 < i < J\big\},$$

$$\omega_\tau = \big\{t^n : t^n = n\tau,\ n = 1, 2, \ldots, N,\ N\tau = T\,\big\}.$$

Let γ_h be a boundary of ω_h, we split it into two parts $\gamma_h = \gamma_{1h} \cup \gamma_{2h}$. Let $U_{ijk}^n = U(x_{1i}, x_{2j}, x_{3k}, t^n)$ be a discrete approximation to the exact solution of differential problem (1)–(2).

We propose unconditionally stable LOD scheme, which approximates 3D parabolic problem and the integral condition:

$$\begin{cases} \dfrac{U^{n+j/3} - U^{n+(j-1)/3}}{\tau} = A_j U^{n+1/3} + \delta_{1j} f^{n+1}, \quad j = 1, 2, 3, \ X \in \omega_h, \\ U_{ijk}^{n+1/3} = (I - \tau A_2)(I - \tau A_3)\mu_1^{n+1}, \ X \in \gamma_{1h}(x_1 = 0) \cup \gamma_{1h}(x_1 = 1), \\ U_{ijk}^{n+2/3} = (I - \tau A_3)\mu_1^{n+1}, \ X \in \gamma_{1h}(x_2 = 0) \cup \gamma_{1h}(x_2 = 1), \\ U_{ijk}^{n+1} = \mu_1^{n+1}, \ X \in \gamma_{1h}, \\ U_{ijk}^{n+1} = \mu_0^{n+1} \mu_2, \ X \in \gamma_{2h}, \\ S_h U^{n+1} = M(t^{n+1}). \end{cases} \quad (3)$$

Boundary conditions are approximated consistently with the approximation of the differential equations [17]. Here we use the following difference operators:

$$A_\alpha U = \left(a_\alpha U_{\bar{x}_\alpha}\right)_{x_\alpha} - \frac{1}{3} q(X, t^n) U, \quad \alpha = 1, 2, 3,$$

$$a_{\alpha,\,ijk}^n = k_\alpha \left(x_{1i} - \frac{h}{2}\delta_{1\alpha},\ x_{2j} - \frac{h}{2}\delta_{2\alpha},\ x_{3k} - \frac{h}{2}\delta_{3\alpha}\right),$$

$$U_{x_1} = \frac{U_{i+1,jk} - U_{ijk}}{h}, \quad U_{\bar{x}_2} = \frac{U_{ijk} - U_{i,j-1,k}}{h}.$$

Integral condition (2) is approximated by the trapezoidal rule

$$S_h U^{n+1} := \sum_{i,j,k=0}^{J} c_i c_j c_k \rho_{ijk}^{n+1} U_{ijk} h^3 = M(t^{n+1}), \quad (4)$$

$$c_0 = \frac{1}{2}, \quad c_l = 1,\ l = 1, \ldots, J-1, \quad c_J = \frac{1}{2}.$$

We propose to change the simplest approximation of the initial condition

$$U^0 = u_0(X), \quad X \in \omega_h \cup \gamma_h$$

by the following one, which exactly satisfies the discrete nonlocal condition:

$$U^0 = \frac{M(t^0)\, u_0(X)}{S_h u_0}, \quad X \in \omega_h \cup \gamma_h. \quad (5)$$

Then the truncation error of the discrete initial condition is given by

$$|U^0 - u_0(X)| = \mathcal{O}(h^2),$$

but this error is not propagating in time due to the stability of the LOD method with respect to the initial condition. This new discretization of the initial condition is mass conservative, therefore the accuracy of approximation of the boundary condition μ^n is increased to the second order.

The LOD scheme is implemented as follows. The first two subproblems for $j = 1, 2$ are standard: we solve $(J-1)^2$ systems of linear equations, the matrix of each system is tridiagonal. Total costs of these two steps are $\mathcal{O}(J^3)$ floating point operations.

The serial implementation algorithm of the third step was proposed in [3]. By using the Dirichlet boundary condition at $\gamma_{1h}(x_1 = 1)$ and discrete 1D equations with operator A_3 we obtain the factorization coefficients $\tilde{\alpha}^{n+1}, \tilde{\beta}^{n+1}$ such that:

$$U_{ijk}^{n+1} = \tilde{\alpha}_{ijk}^{n+1} U_{i,j,k-1}^{n+1} + \tilde{\beta}_{ijk}^{n+1}, \quad 0 < i, j < J, \ k = J, \dots, 1.$$

Then the solution is expressed in the following form:

$$U_{ijk}^{n+1} = \alpha_{ijk}^{n+1} U_{ij0}^{n+1} + \beta_{ijk}^{n+1}, \quad i, j = 0, \dots, J, \tag{6}$$

$$\alpha_{ijk}^{n+1} = \tilde{\alpha}_{ijk}^{n+1} \alpha_{i,j,k-1}^{n+1}, \quad k = 1, \dots, J, \quad \alpha_{ij0}^{n+1} = 1,$$

$$\beta_{ijk}^{n+1} = \tilde{\alpha}_{ijk}^{n+1} \beta_{i,j,k-1}^{n+1} + \tilde{\beta}_{ijk}^{n+1}, \quad \beta_{ij0}^{n+1} = 0.$$

By using the discrete non-local condition we find the function:

$$\mu_0^{n+1} = \frac{M(t^{n+1}) - S_h \beta^{n+1}}{S_h(C)}, \quad C_{ijk} = \alpha_{ijk}^{n+1} \mu_{2ij}.$$

After determination of μ_0^{n+1} solution U^{n+1} is computed by using (6). The complexity of the third step of LOD scheme is equal to $\mathcal{O}(J^3)$.

3 Parallel Algorithm

Let us assume that we have p processors, which are connected by three dimensional mesh, i.e. $p = p_1 \times p_2 \times p_3$. The grid ω_h (a data set) is decomposed into a number of 3D subgrids by using a block distribution scheme. Then each subgrid ω_{hp} has

$$\frac{(J+1)}{p_1} \times \frac{(J+1)}{p_1} \times \frac{(J+1)}{p_1} = \frac{(J+1)^3}{p}$$

computational points of the grid ω_h and it is assigned to one processor, which is responsible for all computations of the local part of vector U.

Since the sub-domains are connected at their boundaries, processors dealing with neighbouring sub-domains have to exchange boundary information with each other at every time-step. More exactly, the update of vector U^{n+1} at grid points which lie beside cutting planes (i.e. boundary nodes of the local part of the vector U) needs a special attention, since information from the neighbouring processors is required to compute new values of U^{n+1}. Such information is obtained by exchanging data with neighbour processors in the specified topology of processors. The amount of exchanged data depends also on the grid stencil, which is used to discretize the PDE model. A star-stencil of seven points is used in (3), therefore local subgrids are enlarged by two *ghost* points in each dimension of the subgrid.

In the parallel algorithm the implementation of the third step of the LOD scheme is modified to the following one:

$$U^{n+1} = V^{n+1} + \gamma^{n+1} W^{n+1},$$

where V^{n+1} is a solution of the discrete boundary value problem

$$\begin{cases} \dfrac{V^{n+1} - U^{n+2/3}}{\tau} = A_3 V^{n+1}, & X \in \omega_h, \\ V^{n+1} = \mu_1(X, t^{n+1}), & X \in \gamma_{1h}, \\ V^{n+1} = \mu_0^n \mu_2(X, t^{n+1}), & X \in \gamma_{2h}. \end{cases} \quad (7)$$

Function W^{n+1} is a solution of the auxiliary problem

$$\begin{cases} W^{n+1}/\tau = A_3 W^{n+1}, & X \in \omega_h, \\ W^{n+1} = 0, & X \in \gamma_{1h}, \\ W^{n+1} = \mu_2(X, t^{n+1}), & X \in \gamma_{2h}. \end{cases} \quad (8)$$

Then we find μ_0^{n+1} by using the discrete nonlocal condition:

$$\gamma^{n+1} = \frac{M(t^{n+1}) - S_h V^{n+1}}{S_h W^{n+1}}.$$

Thus during implementation of the parallel LOD algorithm we solve $4(J-1)^2$ systems of linear equations with tridiagonal matrix.

The complexity of solving one tridiagonal system of J equations by the serial factorization algorithm is equal to $8J$ arithmetical operations.

For two processors the Gaussian elimination process is started simultaneously at the first and last equations and it goes in opposite directions. Processors exchange two factorization coefficients at the end of the first stage of the factorization algorithm. The total complexity of this modified algorithm is equal to $8J$ arithmetical operations.

For the case when a system is distributed between $p_1 > 2$ processors, we use the Wang parallel factorization algorithm [14]. It solves the tridiagonal system by using $17J$ arithmetical operations. The main idea is to reduce the given system to a new tridiagonal system of p_1 equations, where each processor has only one equation. Such small system is solved by using the serial factorization algorithm. The total costs of the parallel Wang algorithm in the worst case when

the simplest algorithm is used to broadcast data to the master process can be estimated as
$$T_p(J) = \frac{17J}{p_1} + 8p_1 + p_1(\alpha + \beta).$$

4 Complexity and Scalability Analysis

We will estimate the complexity of the LOD algorithm by counting basic operations. At each time step the following amount of work is done:

1. Coefficients of the LOD scheme (3) are computed. The complexity of this step is J^3.
2. $3(J-1)^2$ systems with tridiagonal matrix are solved. The complexity of this step is aJ^3.
3. Discrete approximations of integrals $S_h(V^{n+1})$, $S_h(W^{n+1})$ are computed. The complexity of this step is bJ^3.
4. The values of the solution on boundary γ_{2h} are updated with a known function μ_0^{n+1}. The complexity of this step is cJ^2.

As a result, the total complexity of the serial LOD algorithm can be expressed as
$$W = (1 + a + b)J^3 + cJ^2 = (1 + a + b)J^3 + \mathcal{O}(J^2). \tag{9}$$

The communication step is implemented before updating vectors $U^{n+j/3}$, $j = 1, 2, 3$ and only neighbouring processors are communicating with each other. Each processor exchanges with its six neighbours vector elements corresponding to boundary points of the local subdomain. A total amount of data, exchanged between two processors, is equal to $J^2/p^{2/3}$ elements. This can be done in
$$T_{1,p}(J) = \alpha + \beta \frac{J^2}{p^{2/3}}$$
time, by using the *odd–even* data exchange algorithm. Here α is the message startup time and β is the time required to send one element of data.

When the required information is exchanged, processors compute in parallel coefficients of local part of the matrix. The complexity of this step is given by
$$T_{2,p}(J) = \frac{J^3}{p}.$$

Parallel computation of integrals $S_h(V^{n+1})$ and $S_h(W^{n+1})$ requires global communication among all processors during summation of local parts of integrals. The complexity of reduce operation depends strongly on the architecture of the parallel computer (see [13]). We will estimate the time required to reduce local values of integrals between p processors by $B(p) = R(p)(\alpha_b + \beta_b)$, where $R(p)$ depends on the algorithm used to implement the MPI_ALLREDUCE operation and the architecture of the computer. For the simplest reduce algorithm, when every processor sends its result to the master processor, who finishes computation of the integral and broadcasts the global sum to all processors, $R(p) = p$. Thus the complexity of parallel computation of both integrals and updating boundary values on γ_{2h} is given by

$$T_{3,p}(J) = 2R(p)(\alpha_b + \beta_b) + b\frac{J^3}{p} + c\frac{J^2}{p^{2/3}}.$$

The time required to solve all $4(J-1)^2$ systems of linear equations is estimated as

$$T_{4,p}(J) = \frac{4}{3}\frac{17aJ^3}{8p} + 4J^2\left(8p^{1/3} + p^{1/3}(\alpha + \beta)\right).$$

Summing up all obtained estimates we compute the complexity of the parallel LOD algorithm

$$T_p(J) = \left(1 + \frac{17a}{6} + b\right)\frac{J^3}{p} + 6\left(\alpha + \beta\frac{J^2}{p^{2/3}}\right) + c\frac{J^2}{p^{2/3}} + 2R(p)(\alpha_b + \beta_b). \quad (10)$$

According to the definition of the isoefficiency function, we must find the rate at which the problem size W needs to grow with p for a fixed efficiency of the algorithm. Let $H(p, W) = pT_p - W$ be the total overhead of a parallel algorithm. Then the *isoefficiency* function $W = g(p, E)$ is defined by the implicit equation (see [14]):

$$W = \frac{E}{1-E}H(p, W).$$

The total overhead of the parallel LOD algorithm is given by

$$H(p, W) = \frac{11}{6}aJ^3 + 6\alpha p + (6\beta + c)\, p^{1/3}J^2 + 2pR(p)(\alpha_b + \beta_b)$$

$$= \frac{11a}{6(1+a+b)}W + 6\alpha p + \frac{(6\beta+c)p^{1/3}}{(1+a+b)^{2/3}}W^{2/3} + 2pR(p)(\alpha_b + \beta_b).$$

The first term defines a range of possible values of E. This term in $H(p, W)$ arises due to the fact the parallel algorithm does not coincide with the serial LOD algorithm. For simplicity of notation we take E such, that

$$\frac{11aE}{6(1+a+b)(1-E)} = \frac{1}{2}.$$

Since it is impossible to get the isoefficiency function in a closed form as a function of p, we will analyze the influence of each individual term. The component that requires the problem size to grow at the fastest rate determines the overall asymptotic isoefficiency function. After simple computations we get the following three isoefficiency functions

$$W = \mathcal{O}(p), \quad W = \mathcal{O}(p), \quad W = \mathcal{O}\bigl(pR(p)\bigr).$$

Thus the the overall asymptotic isoefficiency function is defined by the overheads of the global reduction operation. Let us assume that processors are connected by three dimensional mesh $p^{1/3} \times p^{1/3} \times p^{1/3}$. Then the global *reduce* and *broadcast* operations can be implemented with $R(p) = p^{1/3}$. Thus the problem size W has to grow as $\mathcal{O}(p^{4/3})$ to maintain a certain efficiency. For a hypercube mesh we have smaller costs of the global reduction operation $R(p) = \log p$, then isoefficiency function is close to linear $W = \mathcal{O}(p \log p)$.

We note, that in the case of a moderate number of processors $p = \mathcal{O}(J)$, the costs of global reduction operation can be ignored and the isoefficiency function $W = \mathcal{O}(p)$ depends linearly on p.

Parallel numerical objects. Special tools are developed to simplify parallelization of sequential algorithms, e.g. *Diffpack* tool [15] and *PETSc* toolkit [1]. We have developed new tool *ParSol* of parallel numerical arrays, which can be used for semi–automatic parallelization of data parallel algorithms, that are implemented in C++. Such algorithms are usually constructed for solving PDEs and systems of PDEs on logically regular rectangular grids. *ParSol* is a library of parallel array objects, a functionality of which is similar to *Distributed Arrays* in *PETSc*. We list the following main features of *ParSol* (see [6]): a) created for C++ programming language, b) based on HPF ideology, c) the library heavily uses such C++ features as OOP and template, d) MPI 1.1 standard is used to implement parallelization.

ParSol arrays have a number of advantages for programming mathematical algorithms, such as virtual indexing, built-in array operations, automated management of dynamically allocated memory, periodic boundary conditions. *ParSol* arrays simulate numerical objects of linear algebra and many useful basic vector operations are supported within the *ParSol* library, e.g. parallel computation of vector norms, the inner product of two vectors, scaling of vectors.

The LOD algorithm can not be described as a simple data parallel algorithm, but *ParSol* library is used to implement the algorithm (3) and only the Wang algorithm requires a special treatment.

5 Results of Computational Experiments

In this section we present some results of computational experiments. Computations were performed on IBM SP5 computer at CINECA, Bologna. We have solved problem (1)–(2) with the following coefficients and the exact solution:

$$k_\alpha(X,t) = 1 + (x_1^2 + x_2^2 + x_3^2)t, \quad q(X,t) = (x_1 + x_2 + x_3)t^2,$$
$$M(t) = e^t(A^3 + B^3), \quad A = 2(e^{0.5} - 1), \quad B = 2(2 - e^{0.5}),$$
$$\rho(X,t) = 1 + x_1 x_2 x_3, \quad u(X,t) = \exp\left(0.5(x_1 + x_2 + x_3) + t\right).$$

In order to scale the computation time for different space steps $h = 1/(J-1)$, a solution was computed in time intervals $[0, T(J)]$, where

$$T(40) = 0.4, \quad T(80) = 0.04, \quad T(120) = 0.005, \quad T(160) = 0.001.$$

In Table 1 we present the values of experimental speedup $S_p(J) = \dfrac{T_1(J)}{T_p(J)}$ and efficiency $E_p(J) = \dfrac{S_p(J)}{p}$ coefficients for different sizes of the discrete problem.

Table 1. The speedup and efficiency coefficients for the LOD method. CPU time of the sequential algorithm (in s): $T_1(40) = 64.8$, $T_1(80) = 105.2$, $T_1(120) = 94.98$, $T_1(160) = 131.0$.

p	$S_{p,40}$	$E_{p,40}$	$S_{p,80}$	$E_{p,80}$	$S_{p,120}$	$E_{p,120}$	$S_{p,160}$	$E_{p,160}$
2	1.979	0.990	2.001	1.000	2.115	1.058	1.955	0.978
4	3.880	0.970	4.062	1.016	4.236	1.059	4.662	1.166
8	7.043	0.880	7.684	0.961	8.284	1.036	9.388	1.173
16	11.54	0.721	14.30	0.894	15.23	0.952	18.10	1.131
32	18.72	0.585	26.73	0.835	29.67	0.927	34.77	1.087

It follows from results, presented in Table 1, that the parallel LOD algorithm scales well.

Remark 1. We see that a superlinear speedup of the parallel algorithm is obtained, when more processors are used. This effect is due to special properties of cash memory usage in SP5 processors. We implemented a simple test, where matrix operations $A := A + B$, $C := C - D$ were performed many times. The dimension of matrix is taken to be $160 \times 160 \times 160$. The following results were obtained:

$$T_1 = 35.3, \quad T_2 = 15.3, \quad T_4 = 7.18, \quad T_8 = 2.83, \quad T_{16} = 1.29, \quad T_{32} = 0.65.$$

Acknowledgment

The work has been performed under the Project HPC–EUROPA (RII3–CT-2003-506079), with the support of the European Community – Research Infrastructure Action under the FP6 "Structuring the European Research Area" Programme. I gratefully acknowledge the hospitality and excellent working conditions in CINECA, Bologna. In particular I thank Dr. Giovanni Erbacci for his help.

This work was also supported by the Lithuanian State Science and Studies Foundation within the framework of the Eureka Project EUREKA E!3691 OPTCABLES .

References

1. S. Balay, K. Buschelman, V. Eijkhout, W.D. Gropp, D. Kaushik, M.G. Knepley, L. Curfman McInnes, B.F. Smith, H. Zhang. PETSc Users Manual. ANL-95/11 - Revision 2.3.0, Argonne National Laboratory, 2005.

2. J.R. Cannon, Y. Lin, A. L. Matheson. Locally explicit schemes for three–dimensional diffusion with non–local boundary specification. *Appl. Anal.*, **50**, 1–19, 1993.
3. R. Čiegis. Economical difference schemes for the solution of a two dimensional parabolic problem with an integral condition, *Differential Equations*, **41**(7), 1025–1029, 2005.
4. R. Čiegis. Numerical Schemes for 3D Parabolic Problem with Non–Local Boundary Condition, In: *Proceedings of 17th IMACS World congress, Scientific Computation, Applied Mathematics and Simulation*, June 11-15, 2005, Paris, France, T2-T-00-0365, 2005.
5. R. Čiegis. Finite-Difference Schemes for Nonlinear Parabolic Problem with Nonlocal Boundary Conditions. In: M. Ahues, C. Constanda, A. Largillier (Eds.) *Integral Methods in Science and Engineering: Analytic and Numerical Techniques*, ISBN 0-8176-3228-X, Birkhauser, Boston, 47–52, 2004.
6. R. Čiegis, A. Jakušev, A. Krylovas and O. Suboč. Parallel algorithms for solution of nonlinear diffusion problems in image smoothing. *Math. Modelling and Analysis*, **10**(2), 155–172, 2005.
7. R. Čiegis, A. Štikonas, O. Štikonienė, O. Suboč. Monotone finite-difference scheme for parabolical problem with nonlocal boundary conditions. *Differential Equations*, **38**(7), 1027–1037, 2002.
8. R. Čiegis, A. Štikonas, O. Štikonienė, O. Suboč. Stationary problems with nonlocal boundary conditions. *Mathematical Modelling and Analysis*, **6**, 178–191, 2001.
9. M. Dehghan Efficient techniques for the second-order parabolic equation subject to nonlocal specifications *Applied Numer. Math.*, **52**, 39–62, 2005.
10. M. Dehghan Locally explicit schemes for three–dimensional diffusion with non–local boundary specification. *Applied Mathematics and Computation*, **135**, 399–412, 2002.
11. G. Ekolin. Finite difference methods for a nonlocal boundary value problem for the heat equation. *BIT*, **31**(2), 245–255, 1991.
12. G. Fairweather, J.C. Lopez-Marcos. Galerkin methods for a semilinear parabolic problem with nonlocal boundary conditions. *Adv. Comput. Math.*, **6**, 243–262, 1996.
13. R. Hockney. Performance parameters and benchmarking on supercomputers. *Parallel Computing*, **17**, 1111–1130, 1991.
14. V. Kumar, A. Grama, A. Gupta, G. Karypis. *Introduction to parallel computing: design and analysis of algorithms*. Benjamin/Cummings, Redwood City, 1994.
15. H.P. Langtangen and A. Tveito. *Advanced Topics in Computational Partial Differential Equations. Numerical Methods and Diffpack Programming*. Springer, Berlin, 2003.
16. B.J. Noye, M. Dehghan. New explicit finite difference schemes for two–dimensional diffusion subject specification of mass. *Numer. Meth. for PDE*, **15**, 521–534, 1999.
17. A.A. Samarskii. *The theory of difference schemes*. Marcel Dekker, Inc., New York–Basel, 2001.

Online Checkpointing for Parallel Adjoint Computation in PDEs: Application to Goal-Oriented Adaptivity and Flow Control

Vincent Heuveline[1] and Andrea Walther[2]

[1] Universität Karlsruhe, Institute for Applied Mathematics, Karlsruhe, Germany
vincent.heuveline@math.uni-karlsruhe.de
[2] Technische Universität Dresden, Institute of Scientific Computing, Dresden, Germany
Andrea.Walther@tu-dresden.de

Abstract. The computation of derivatives for the optimization of time-dependent flow problems is based on the integration of the adjoint differential equation. For this purpose, the knowledge of the complete forward solution is required. Similar information is needed for a posteriori error estimation with respect to a given functional. In the area of flow control, especially for three dimensional problems, it is usually impossible to store the full forward solution due to the lack of memory capacities. Additionally, adaptive time-stepping procedures are needed for efficient integration schemes in time. Therefore, standard optimal offline checkpointing strategies are usually not well-suited in that framework.

We present a new online procedure for determining the checkpoint distribution on the fly. Complexity estimates and consequences for storing and retrieving the checkpoints using parallel I/O are discussed. The resulting checkpointing approach is integrated in HiFlow, a multipurpose parallel finite-element package with a strong emphasis in computational fluid dynamic, reactive flows and related subjects. Using an adjoint-based error control for prototypical three dimensional flow problems, numerical experiments demonstrate the effectiveness of the proposed approach.

1 Introduction

In time-dependent flow control as well as in the framework of goal-oriented a posteriori error control, the calculation of adjoint information forms a basic ingredient to generate the required derivatives of the cost functional (see e.g. [8]). However, the corresponding computations may become extremely tedious if possible at all because of the sheer size of the resulting discretized problem as well as its nonlinear character, which requires keeping track of the complete forward solution to be able to integrate the corresponding adjoint differential equation backwards. This fact still forms a main bottleneck in the overall optimization process despite the ever-growing size of memory devices. For that reason, several checkpointing techniques have been developed. Here, only a few intermediate states are stored as checkpoints. Subsequently, the required forward information is recomputed piece by piece from the checkpoints according to the adjoint calculation. Hence, checkpointing methods seek for an acceptable compromise between memory requirements and run time increase due to re-computations that cannot be avoided.

If the number of time steps for integrating the differential equation describing the state is known a priori, one very popular checkpointing strategy is to distribute the checkpoints equidistantly over the time interval. However, it was shown in [18] that this approach is not optimal. One can compute optimal checkpointing schedules in advance to achieve for a given number of checkpoints an optimal, i.e. minimal, run time increase [7]. This procedure is referred to as offline checkpointing and implemented in the package revolve [7]. However, in the context of flow control, the partial differential equations to be solved are usually stiff, and the solution process relies therefore on some adaptive time stepping procedure. Hence, the number of time steps performed is known only after the complete integration. This fact makes an offline checkpointing intractable. Instead, one may apply a straightforward checkpointing by placing a checkpoint each time a certain number of time steps has been executed. This transforms the uncertainty in the number of time steps to a uncertainty in the number of checkpoints needed. This approach is used by CVODES [17]. However, when the amount of memory per checkpoint is very high one certainly wants to determine the number of checkpoints required a priori. For that purpose, we propose a new procedure for online checkpointing that distributes a given number of checkpoints during the integration procedure. This new approach yields a time-optimal adjoint computation for a given number of checkpoints. The present paper focus on practical aspects, i.e. the specific online checkpointing algorithm and the consequences for the computation of adjoint information on parallel computers. Furthermore, it describes the coupling of the presented online checkpointing software with the package HiFlow (see www.hiflow.de) for the parallel computation of adjoints. A companion paper [10] concentrates on the theoretical aspects of the goal-oriented adaptivity and the online checkpointing algorithm.

The outline of this paper is as follows. Section 2 is dedicated to the derivation of adjoint-based a posteriori error control for flow problems and its link to flow control. The new online checkpointing strategy is presented in Section 3. Here also complexity estimates for the resulting checkpointing strategy and the usage of the algorithm on parallel computers are addressed. First numerical experiments are presented in Section 4. Finally, conclusions are drawn in Section 5.

2 Adjoint Based Techniques for Error Estimation

2.1 Problem Formulation

Let u denote the state variables, g the control variables, $J(u,g)$ the objective functional, and $G(u,g) = 0$ the constraints. A standard formulation for the related optimization problem reads

Problem 1: Find controls g and states u such that $J(u,g)$ is minimized subject to $G(u,g) = 0$.

Our goal in this paper is to address the case where the constraints are defined by means of time-dependent partial differential equations. Even though the derived method is very general, we concentrate on the case of instationary, incompressible, viscous flows modeled by means of the Navier-Stokes equations, i.e., ignoring the controls g we have

$$\frac{\partial u}{\partial t} - \nu \Delta u + u \cdot \nabla u + \nabla p = f \quad \text{in } (0,T) \times \Omega, \tag{1}$$

$$\nabla \cdot u = 0 \quad \text{in } (0,T) \times \Omega, \quad u|_{t=0} = u_0, \tag{2}$$

where $u \in \mathbb{R}^d$ describes the velocity field and $p \in \mathbb{R}$ the pressure. We assume that the velocity field is subject to adequate boundary conditions. A standard approach to solve such problems is based on the solution of an adjoint system backward in time to compute the gradient of the functional $J(u,g)$ (see e.g. [8]). The state variables appear in the coefficients and right-hand sides of the adjoint equations and must be available as the solver marches backward in time. Generally for flow control problems the storage of the state variables for every time step results in a huge amount of data. Therefore, we propose a checkpointing technique that relies on the storage of a few selected time steps. One then recomputes the information required by the adjoint calculation time step per time step.

Similarly to Problem 1, the proposed framework for checkpointing can be used in the context of goal-oriented a posteriori error estimation for time-dependent problems. Again one considers a state equation defined by partial differential equations $F(u)$. We suppose that these equations are discretized by means of a Galerkin method (e.g. finite-element method) and that the corresponding discrete solution is denoted by u_h. The goal is to determine the discretization error with respect to some functional $J(\cdot)$, i.e. $J(u) - J(u_h)$. This problem can be formulated as a control problem similar to Problem 1. In the remainder of this paper we will consider this setup for the derivation of the proposed checkpointing strategy.

2.2 A General Paradigm for Dual-Based a Posteriori Error Estimation

In this section, we outline the concepts related to dual-based error estimation following the general paradigm introduced in Eriksson et al. [3]. We refer to Machiels et al. [13], Oden and Prudhomme [14], and Giles [6] for related approaches to goal-oriented error estimation.

Let $A(\cdot;\cdot)$ be a differentiable semi-linear form and $F(\cdot)$ a linear functional defined on some function space V. For $u \in V$, $A'(u;v)(\cdot)$ denotes the directional derivative of $A(u;\cdot)$ in the v direction. The second derivative of $A(u;\cdot)$ is refereed to by $A''(\cdot;\cdot)(\cdot,\cdot)$. We seek a solution $u \in V$ to the variational equation

$$A(u;\varphi) = F(\varphi) \quad \forall \varphi \in V. \tag{3}$$

This problem is approximated by a *Galerkin method* using a sequence of finite dimensional subspaces $V_h \subset V$ parameterized by a parameter h. The corresponding discrete problem seeks $u_h \in V_h$ satisfying

$$A(u_h; \varphi_h) = F(\varphi_h) \qquad \forall \varphi_h \in V_h. \tag{4}$$

We assume that equations (3) and (4) possess unique solutions. A key feature of the discrete problem (4) is the property of *Galerkin orthogonality*, which reads in the general nonlinear case

$$A(u; \varphi_h) - A(u_h; \varphi_h) = 0 \qquad \forall \varphi_h \in V_h. \tag{5}$$

Suppose that the quantity $J(u)$ has to be computed, where $J(\cdot)$ is a differentiable functional defined on V. To control the error with respect to the functional J we introduce the following dual problem

$$A'(\overline{uu_h}; \varphi)(\hat{z}) = J'(\overline{uu_h})(\varphi) \qquad \forall \varphi \in V, \quad \text{where} \tag{6}$$

$$A'(\overline{uu_h}; \varphi)(\psi) = \int_0^1 A'(su + (1-s)u_h; \varphi)(\psi)\, ds,$$

$$J'(\overline{uu_h})(\varphi) = \int_0^1 J'(su + (1-s)u_h)(\varphi)\, ds.$$

We assume that (6) possesses a solution. Based on the dual solution $\hat{z}$ and due to the Galerkin orthogonality (5), we obtain the following error representation

$$J(u) - J(u_h) = A'(\overline{uu_h}; e)(\hat{z}) = A(u; \hat{z}) - A(u_h; \hat{z})$$
$$= A(u; \hat{z} - \hat{z}_h) - A(u_h; \hat{z} - \hat{z}_h)$$
$$= F(\hat{z} - \hat{z}_h) - A(u_h; \hat{z} - \hat{z}_h) = \rho(u_h, \hat{z} - \hat{z}_h)$$

for any $\hat{z}_h \in V_h$, where $\rho(u_h, \cdot) = F(\cdot) - A(u_h; \cdot)$ describes the *primal* residual, and $e := u - u_h$. In practice, the previously derived error representation cannot be used directly since the adjoint problem (6) involves the unknown solution u. One alternative is to replace the exact solution u by its approximation u_h in the adjoint problem (6). The resulting adjoint problem reads

$$A'(u_h; \varphi)(z) = J'(u_h; \varphi) \qquad \forall \varphi \in V. \tag{7}$$

One can show that the following modified error representation holds

$$J(u) - J(u_h) = \rho(u_h, z - z_h) + R, \tag{8}$$

for any $z_h \in V_h$, where the remainder term R depends on the second order derivatives of $A(\cdot; \cdot)$ and $J(\cdot)$. The remainder term vanishes if $A(\cdot; \cdot)$ and $J(\cdot)$ are linear.

From now on, we consider procedures based on the error representation (8) for the a posteriori error control with respect to the functional J. The remainder term is neglected since, in our context, it involves higher order terms with respect to the discretization parameter h which can be omitted for h small enough.

The solution of the dual problem (7) needed for the error representation related to (8) corresponds in our context of time-dependent problems to the adjoint problem which has to be solved backward in time.

2.3 Galerkin Discretization in Time and Space

We consider a discretization of the problem (1)-(2) using a Galerkin finite element discretization simultaneously in space and in time. This setup allows us to rely on the error representation (8) for the error control. Following the lines of Eriksson and Johnson [4,5] we consider the dG(r)-method for the time discretization, i.e. we allow discontinuous functions in time. This discontinuity can be used to decouple the considered system on each subinterval $I_n = (t_{n-1}, t_n]$ of the time interval $(0, T]$, where $0 = t_0 < \cdots < t_n < \cdots < t_N = T$, $k_n = t_n - t_{n-1}$. For simplicity, we consider for each time step t_n a unique regular spatial mesh. Then, we can write the solution process as a standard time-stepping scheme. For $r = 0$, the corresponding dG(0)-method is equivalent to the backward-Euler scheme.

The Galerkin space discretization using conforming mixed finite elements with continuous pressure is based on a variational formulation of the Navier-Stokes equations (1)-(2). For this purpose, we employ standard Hood-Taylor finite elements [12] for the trial and test spaces (for a detailed description see, e.g., [1]). This choice for the trial and test functions guarantees a stable approximation of the pressure since the Babuska-Brezzi inf-sup stability condition is satisfied uniformly in h (see [2] and references therein). The advantage, when compared to equal order function spaces for the pressure and the velocity, is that no additional stabilization terms are needed.

Based on this space discretization, the arising nonlinear algebraic systems are then solved implicitly in a fully coupled manner by means of a damped Newton method. The linear subproblems are solved by the Generalized Minimal Residual Method (GMRES) (see [15]) preconditioned by means of a geometric multigrid iteration (see [19]). Two specific features characterize the scheme we consider: varying orders of the FEM ansatz on the mesh hierarchy and a Vanka-type smoother. This somewhat technical part is described in full detail in [9].

3 Online Checkpointing Algorithms

Having a fixed number of checkpoints to store intermediate states but an unknown number of time steps for which the adjoint has to be computed on the base of the forward trajectory, one has to decide on the fly, i.e., during the forward integration, where to place the checkpoints. Hence, without knowing how many time steps are left to perform, one has to analyze the current distribution of the checkpoints. Depending on the time steps performed so far, one may then discard the contents of one checkpoint to store the current available state. Obviously, one may think that this procedure could not be optimal since it may happen that one reaches the final time just after replacing a checkpoint, in which case another checkpoint distribution may be advantageous. A surprising efficient heuristic strategy to rearrange the checkpoints is implemented by the online procedure **arevolve** [11]. Here, a checkpoint distribution is judged by computing an approximation of the overall re-computation cost caused by the current distribution. This number is compared with an approximation of the re-computation cost if one resets a checkpoint to the currently available state. Despite the fact that significant simplifications are made for approximating the required re-computations, the resulting checkpointing schemes are comparatively cheap. Naturally, the optimal cost can be computed only afterwards when the number of time steps is known.

3.1 Optimal Online Checkpointing

However, a main drawback of **arevolve** is that it is not possible to prove an upper bound on the deviation from the optimal checkpointing schedule because a heuristic is used to judge the current checkpointing distributions. In this paper, we present online checkpointing strategies for an a priori unknown number l of time steps and a given number of checkpoints c under the assumption that

$$l \leq \binom{c+2}{c} = \frac{(c+2)(c+1)}{2} = \sum_{i=1}^{c+1} i \equiv b_c . \tag{9}$$

Hence, the upper bound b_c on the number of time steps is directly determined by the number of checkpoints c. Let $F_l(x)$ denote the execution of the lth time step corresponding to the discretized PDE. Using p as a pointer to the next state where a checkpoint is set and s as a flag if a checkpoint has to be set, the proposed online checkpointing procedure reads as follows:

Algorithm 1. Online Checkpointing Algorithm

    Start: Set $i = 0$, $o = c$, $p = c$, $s = 1$
    for $l = 0, 1, \ldots$
        1. Evaluate $x_{l+1} = F_l(x_l)$
        2. **If** termination criterion fulfilled **then** start reversal
            If $s = 1$ **then**
                Store state x_l in checkpoint i
                $i = i + 1$
                If $i > o$ **then** $i = 1$
        3. **If** $l + 1 = p$ **then** $s = 0$
        4. **If** $l = p$ **then**
            $p = p + o$, $o = o - 1$, $i = o$
            If $o > 0$ **then** $s = 1$ **else** $s = 0$
        5. **If** $l = p$ **and** $o = -1$ **then** error: $l > b_c$

For a given value of c, this algorithm stores the states $0, \ldots, c-1$ in the checkpoints $0, \ldots, c-1$. Subsequently, the state $c+1$ is copied to the checkpoint $c-1$. Then the states $c+2, \ldots, 2c-1$ are stored in the checkpoints $1, \ldots, c-2$ by overwriting the information already contained in these memory pads. This process continues until either the termination criterion is fulfilled or the number of time steps exceeds the upper bound b_c. If a reversal is started in step 2, the optimal offline checkpointing provided by **revolve** is applied. Analyzing the described online checkpointing in more detail, we can prove the following complexity result:

Table 1. Upper bound b_c

c	10	20	40	80	160	320
b_c	66	231	861	3321	13041	51681

Theorem 1 (Optimal Online Checkpointing). *Let the number of available checkpoints equal c. Then the online checkpointing procedure given by Algorithm 1 ensures a time-minimal adjoint computation storing no more than c checkpoints at any time for any number l of time steps if l satisfies the inequality $l \leq b_c$.*

Proof. See [10]. ∎

Hence, provided that the number of time steps does not exceed the upper bound b_c one can compute the adjoint of a time step sequence with an a priori unknown length using up to c checkpoints at any time with the optimal, i.e. minimal, run time. This minimal run time is given by the number of time step evaluations in addition to the evaluations of adjoint time steps. Since each adjoint time step has to be executed exactly once, only the number of time steps performed can vary for different checkpointing approaches. In [7], checkpoint strategies were studied for an a priori known number l of time steps the adjoint of which has to be calculated. It was shown that the minimal number of time step executions is given by an explicit formula in the following way: Let $t(c, l)$ denote the minimal number of time steps evaluated to compute the adjoint of l time steps storing up to c checkpoints at any time. Then $t(c, l)$ has the explicit form

$$t(c, l) = rl - \beta(c + 1, r - 1) + 1, \tag{10}$$

where r is the unique integer satisfying $\beta(c, r - 1) < l \leq \beta(c, r) \equiv \binom{c+r}{c}$. Surprisingly, the checkpoint algorithm proposed in this paper reaches this minimal number of time steps even for an unknown number l of time steps as long as l does not exceed the upper bound b_c. The constant b_c grows quadratically in the number of checkpoints as illustrated by Table 1. Therefore, already a moderate number of checkpoints ensures an optimal run time for a reasonable number of time steps to be reversed. For example, usually no more than 200 checkpoints are required for the problems considered in this paper.

3.2 Online Checkpointing on Parallel Computers

The optimal online checkpointing of Algorithm 1 has been implemented as an extension of the optimal offline checkpointing software **revolve** [7]. It is planed for a future version of **revolve** to incorporate the heuristics of **arevolve** in the case of online checkpointing and $l > b_c$. Here, one would perform the optimal online checkpointing as long as $l \leq b_c$. If l exceeds b_c the heuristics of **arevolve** will be applied to avoid a break down of the overall adjoint computation.

Applying the checkpointing routine **revolve** on a parallel computer, one faces two very different situations: The first possibility is that all checkpoints can be kept in main memory. Then the access time to all checkpoints is negligible as assumed in the theoretical analysis contained in [7]. However, the maximal number of checkpoints may be considerably limited due to this approach. Taking advantage of new features of parallel IO filesystems such as Lustre allows to extend the number of checkpoints by storing checkpoints also on disc. Then the access cost of the checkpoints is no longer negligible for all checkpoints because of the parallel I/O. Hence, one has to take the memory access costs into account resulting in a so-called multi-stage checkpointing. For this purpose, we present the following result:

Theorem 2 (Number of Checkpoint Writes). *Let $l > c+2$ be the number of time steps the adjoint of which is computed using c checkpoints and the online checkpointing Algorithm 1. If w_i denotes the number of times data is written onto the checkpoint i during the first integration to state x_l, then one has for*

$$l = \sum_{i=1}^{j}(c+2-i) + q \in \left\{ \sum_{i=1}^{j}(c+2-i), \ldots, \sum_{i=1}^{j+1}(c+2-i) - 1 \right\}$$

if $q \in \{0,1\}$: $w_0 = 1$, $w_i = j$ $0 < i \leq c-j$, $w_i = c-i+1$ $c-j < i < c$
if $q = 2$: $w_0 = 1$, $w_i = j$ $0 < i < c-j$, $w_i = c-i+1$ $c-j \leq i < c$
if $q > 2$: $w_0 = 1$, $w_i = j+1$ $0 < i < \min\{q-1, c-j\}$,
                      $w_i = j$ $q-2 < i < c-j$,
                      $w_i = c-i+1$ $c-j \leq i < c$.

Proof. For $l \leq c+2$, the checkpointing schedule is trivial. Therefore, we do not consider this case here. For $c + 2 < l \leq b_c$, one can divide the range $\{0, \ldots, b_c\}$ into the c ranges

$$R_j \equiv \{l_j, \ldots, u_j\} \equiv \left\{ \sum_{i=1}^{j}(c+2-i), \ldots, \sum_{i=1}^{j+1}(c+2-i) - 1 \right\} \quad 0 \leq j < c-1$$

$$R_{c-1} \equiv \{l_{c-1}, \ldots, u_{c-1}\} \equiv \{b_c - 3, \ldots, b_c\}.$$

This separation is based on the definition of b_c. Applying Algorithm 1, checkpoint 0 stores the initial state x_0 and is not overwritten afterwards. Furthermore, the states $1, \ldots, c-1$ are stored in the checkpoints $1, \ldots, c-1$ since $l > c+2$. Then, for each range R_j with $j > 0$ and $\tilde{l} < l$ for all $\tilde{l} \in R_j$, the checkpoint $c - j$ stores the state l_j and is not overwritten afterwards. Furthermore, the states $l_j + 1, \ldots, l_j + c - j - 1 = l_j + 1, \ldots, u_j - 1$ are stored in the checkpoints $1, \ldots, c - j - 1$. For j_u with $l \in R_{j_u}, l_{j_u} \leq l \leq u_{j_u}$, one has that $l = l_{j_u} + q$ with $q \leq c - j + 1$. Then, the checkpoint $c - j_u$ stores the state l_{j_u} if $q > 1$. If $q > 2$, additionally the states $l_{j_u} + 1, \ldots, l_{j_u} + q - 2$ are stored in the checkpoints $1, \ldots, q - 2$. Summarizing these observations proves the assertion. ∎

The checkpoint write counts proved in the last theorem form a first step to allow larger checkpoint numbers based for example on parallel IO filesystems such as

Lustre. The remaining part is an analysis of the checkpoint write and read counts for the reversal process initiated by revolve. This topic is currently investigated to allow an overall minimization of the access time to the checkpoints. From the results obtained so far in this direction, a suitable strategy seems to be that one assigns the more expensive checkpoints, i.e., the checkpoints distributed on the file system to the checkpoints with higher numbers and to assign the less expensive checkpoints, i.e., the checkpoints in main memory to the checkpoints with lower numbers.

4 Numerical Experiments

The HiFlow package is a multipurpose parallel finite-element package with a strong emphasis in computational fluid dynamic, reactive flows and related subjects. It is developed in C++, and its design takes great advantage of the object-oriented concepts and of the generic programming capabilities offered by this language. The overall design of this project is highly modular and allows an interplay of its different submodules. The computations presented in this paper rely especially on two submodules: *HiFlowOpti* and *HiFlowNavierStokes*. The *HiFlowOpti* submodule contains generic solvers for optimal control and parameter identification as well as experimental design. This module has been extended by means of the checkpointing strategy described in the previous section. The *HiFlowNavierStokes* module contains the solvers related to the resolution of the instationary Navier-Stokes equations. In both modules all methods are available for both sequential and parallel platforms. The numerical experiments presented in this paper have been performed on the high performance computer HP XC 6000 at the Computing Center of the University Karlsruhe. This parallel computer is based Itanium2 processors with a frequency of 1.5 GHz. On each node 8 GB RAM are available.

4.1 Three Dimensional Benchmark Channel Flow

In order to validate the proposed checkpointing strategy we consider the three dimensional benchmark configuration proposed by Schäfer et al. [16]. The proposed setup consists of a flow channel around a cylinder with squared crossed section. The height and width of the channel are $H = 0.41m$, and the diameter of the cylinder is $D = 0.1m$. The goal of this benchmark is to compute accurately the drag and lift forces acting on the cylinder, where the cost functional is the averaged value of the drag over the interval $I = [50, 100]$.

We stress that our aim in this section is to illustrate the capabilities of the proposed checkpointing strategy. The exact analysis of the impact of such a technique in relation with adjoint-based a posteriori error estimation is beyond the scope of this paper and is described in more detail in [10].

In Table 2, results of the proposed checkpointing scheme are presented. For a fixed amount of available memory, we consider three different levels of refinement

Table 2. Results for the proposed checkpointing scheme for various computational setups with l time steps. The amount of available memory capacity is equal for all configurations and results in c checkpoints.

	# Unknowns (space)	l	c	N_{new}	N_{old}
global refinement	$1.2\ 10^6$	912	600	310	1222
global refinement	$1.0\ 10^7$	702	36	664	1366
local refinement	$8.2\ 10^5$	854	715	137	991

in space. For these three configurations the discretization in space is so fine that the full storage of the forward solution in main memory would be impossible even on the considered parallel platform. For this application, the number of available checkpoints in main memory is a priori fixed due to the enormous amount of memory needed for each checkpoint but the number of time steps is a priori unknown. Therefore, one alternative checkpointing strategy would be to first perform a pure function evaluation without adjoint computations to determine the number of time steps to perform and then to apply revolve for the distribution of the checkpoints. The required number of additional time steps needed by this alternative is given by N_{old} in the last column of Table 2. Using the new optimal online checkpointing proposed in this paper, the number of additional forward steps can be reduced significantly, as shown by the column N_{new} in Table 2. As can be seen, the equation $N_{old} = N_{new} + l$ holds since the new checkpointing approach does not require an extra integration to determine the value of l. For the most memory consuming case of the 3D-channel with global refinement leading to 10^7 unknowns we impose the number of checkpoints to be equal to $c = 36$. The performances which are measured in Table 2 with respect to the number of extra forward steps clearly show the high efficiency of the proposed scheme.

5 Conclusion

We present a provable optimal, i.e., time-minimal, online checkpointing procedure. In the present paper, we focus on the practical aspects, that is the specific application of revolve and its coupling with the parallel finite-element package HiFlow for solving optimal control problems and goal-oriented error estimation on parallel machines. The main advantage of the presented online checkpointing is that it guarantees a time-minimal run time for an a priori unknown number of time steps as long as this number does not exceed a given upper bound. Due to the semi-implicit time stepping applied, this upper bound is only a very weak restriction. Additionally, we proved an explicit formula for the number of times data is written onto the checkpoints during the generation of the checkpoint distribution. This forms the first step to allow an improved checkpointing strategy if parallel IO filesystems such as Lustre are used.

References

1. S.C. Brenner and R.L. Scott. *The mathematical theory of finite element methods.* Springer, Berlin-Heidelberg-New-York, 1994.
2. F. Brezzi and R. Falk. Stability of higher-order Hood-Taylor methods. *SIAM J. Numer. Anal.*, 28(3):581–590, 1991.
3. K. Eriksson, D. Estep, P. Hansbo, and C. Johnson. Introduction to adaptive methods for differential equations. *Acta Numerica*, 4:105–158, 1995.
4. K. Eriksson and C. Johnson. Adaptive finite element methods for parabolic problems, I: A linear model problem. *SIAM J. Numer. Anal.*, 28:43–77, 1991.
5. K. Eriksson and C. Johnson. Adaptive finite element methods for parabolic problems, II, IV, V. *SIAM J. Numer. Anal.*, 32:706–740, 32:1729–1763, 1995.
6. M.B. Giles. On adjoint equations for error analysis and optimal grid adaptation. In D.A. Caughey and M.M. Hafez, editors, *In Frontiers of Computational Fluid Dynamics 1998*, pages 155–170. World Scientific, 1998.
7. A. Griewank and A. Walther. Revolve: An implementation of checkpointing for the reverse or adjoint mode of computational differentiation. *ACM Trans. Math. Software*, 26:19–45, 2000.
8. M.D. Gunzburger. *Perspectives in flow control and optimization.* Advances in Design and Control 5. Philadelphia, SIAM., 2003.
9. V. Heuveline. On higher-order mixed FEM for low Mach number flows: Application to a natural convection benchmark problem. *Int. J. Num. Meth. Fluids*, 41(12):1339–1356, 2003.
10. V. Heuveline and A. Walther. Towards the economical computation of adjoints in PDEs using optimal online checkpointing. In preparation, 2006.
11. M. Hinze and J. Sternberg. A-revolve: An adaptive memory- and run-time-reduced procedure for calculating adjoints; with an application to the instationary Navier-Stokes system. *Opti. Meth. Softw.*, 20:645–663, 2005.
12. P. Hood and C. Taylor. A numerical solution of the Navier-Stokes equations using the finite element techniques. *Comp. and Fluids*, 1:73–100, 1973.
13. L. Machiels, A.T. Patera, and J. Peraire. Output bound approximation for partial differential equations; application to the incompressible Navier-Stokes equations. In S. Biringen, editor, *Industrial and Environmental Applications of Direct and Large Eddy Numerical Simulation*. Springer, 1998.
14. J.T. Oden and S. Prudhomme. On goal-oriented error estimation for elliptic problems: Application to the control of pointwise errors. *Comput. Methods Appl. Mech. Eng.*, 176:313–331, 1999.
15. Y. Saad. *Iterative methods for sparse linear systems.* Computer Science/Numerical Methods. PWS Publishing Company, 1996.
16. M. Schäfer and S. Turek. Benchmark computations of laminar flow around cylinder. *Notes on numerical fluid mechanics*, 52:856–869, 1996.
17. R. Serban and A.C. Hindmarsh. CVODES: An ODE solver with sensitivity analysis capabilities. UCRL-JP-20039, LLNL, 2003.
18. A. Walther and A. Griewank. Advantages of binomial checkpointing for memory-reduced adjoint calculations. In M. Feistauer et al., editor, *Numerical mathematics and advanced applications*, pages 834–843. Springer, 2004.
19. P. Wesseling. *An introduction to multigrid methods.* Wiley, Chichester, 1992.

Parallel Fault Tolerant Algorithms for Parabolic Problems*

Hatem Ltaief, Marc Garbey, and Edgar Gabriel

Department of Computer Science, University of Houston
4800 Calhoun Road, Houston, TX 77204, USA
{ltaief, garbey, gabriel}@cs.uh.edu

Abstract. With increasing number of processors available on nowadays high performance computing systems, the mean time between failure of these machines is decreasing. The ability of hardware and software components to handle process failures is therefore getting increasingly important. The objective of this paper is to present a fault tolerant approach for the implicit forward time integration of parabolic problems using explicit formulas. This technique allows the application to recover from process failures and to reconstruct the lost data of the failed process(es) avoiding the roll-back operation required in most checkpoint-restart schemes. The benchmark used to highlight the new algorithms is the two dimensional heat equation solved with a first order implicit Euler scheme.

1 Introduction

Today's high performance computing (HPC) systems offer to scientists and engineers powerful resources for scientific simulations. At the same time, the reliability of the system becomes a paramount key: systems with tens of thousands of processors face inherently a larger number of hardware and software failures, since the mean time between a failure is related to the number of processors and network interface cards (NICs). This is not necessarily a problem for short running application utilizing a small/medium number of processors, since rerunning the application in case a failure occurs does not waste a large amount of resources. However, for long running simulations requiring many processors, aborting the entire simulation just because one processor has crashed is often not an option, either because of the significant amount of resources being involved in each run or because the application is critical within certain areas.

Nowadays, a single failing node or processor on a large HPC system does not imply, that the entire machine has to go down. Typically, the parallel application utilizing this node has to abort, all other applications on the machine are not affected by the hardware failure. The reason that the parallel application, which utilized the failed processor, has to abort is mainly because the most widespread parallel programming paradigm MPI [1], is not capable of handling process failures. Several approaches how to overcome this problem have been proposed, most

---

* Research reported here was partially supported by Award 0305405 from the National Science Foundation.

of them relying on some forms of checkpoint-restart [2,3]. While these solutions require few modifications of the application source code, checkpoint-restart has inherent performance and scalability limitations. Another approach suggest by Fagg et. all [4] defines extensions to the MPI specification giving the user the possibility to recognize, handle and recover from process failures. This approach does not have built-in performance problems, requires however certain changes in the source code, since it is the responsibility of the application to recover the data of the failed processes.

In the last couple of years, several solutions have been proposed how to extend numerical applications to handle process failures on the application level. Geist et al. suggest a new class of so-called naturally fault tolerant algorithms [6] based on mesh-less methods and chaotic relaxation. In-memory checkpointing techniques [7] avoid expensive disk I/O operations by storing regular checkpoints in the main memory of neighbor/spare processes. In case an error occurs, the data of the failed processes can be reconstructed by using these data items. However, the application has to roll-back to the last consistent distributed checkpoint, loosing all the subsequent work and adding a significant overhead for applications running on thousands of processors due to coordinated checkpoints. Further, while it is fairly easy to recover numerically from a failure with a relaxation scheme applied to an elliptic problem, the problem is far more difficult with the time integration of a parabolic problem. As a matter of fact the integration back in time is a very ill-posed problem. Further time integration of unsteady problem may run for very long time and are more subject to process failures.

In this paper, we concentrate on the heat equation problem that is a representative test case of the main difficulty and present a new explicit recovery technique which avoid the roll-back operation and is numerically efficient.

The paper is organized as follows: section 2 defines our test-system and describes two different fault tolerant algorithms. Section 3 discusses implementation issues with respect to the communication and checkpointing scheme applied in our algorithms. Section 4 presents some results for the recovery operation. Finally, section 5 summarizes the results of this paper and presents the ongoing work in this area.

2 Description of the Fault Tolerant Algorithms

The work presented in this paper is based on the Fault Tolerant MPI (FT-MPI) framework developed at the University of Tennessee. FT-MPI extends the MPI specification by giving applications the possibility to discover process failures. Furthermore, several options how to recover from a process failure are specified: the application can either continue execution without the failed processes (COMM_MODE_BLANK) or replace them (COMM_MODE_REBUILD). The current implementation of the specification is based on the HARNESS framework [5]. HARNESS provides an adaptive, reconfigurable runtime environment, which is the basis for the services required by FT-MPI. While FT-MPI is capable of surviving the simultaneous failing of $n-1$ processes in an n processes job, it remains up to the application developer to recover the user data, since FT-MPI does not perform any (transparent) checkpointing of user-level data items.

2.1 Definition of the Problem

The model problem used throughout the paper is the two dimensional heat equation as given by

$$\frac{\partial u}{\partial t} = \Delta u + F(x,y,t), \ (x,y,t) \in \Omega \times (0,T), \ u_{|\partial\Omega} = g(x,y), \ u(x,y,0) = u_0(x,y). \quad (1)$$

We suppose that the time integration is done by a first order implicit Euler scheme, $\frac{U^{n+1}-U^n}{dt} = \Delta U^{n+1} + F(x,y,t^{n+1})$, and that Ω is partitioned into N subdomains Ω_j, $j = 1..N$.

For the sake of simplicity, the explanations in the paper will be restricted to the one dimensional heat equation problem $\Omega = (0,1)$, discretized on a regular cartesian grid, which leads to

$$\frac{U_j^{n+1} - U_j^n}{dt} = \frac{U_{j+1}^{n+1} - 2U_j^{n+1} + U_{j-1}^{n+1}}{h^2} + F_j^{n+1}. \quad (2)$$

Furthermore, we assume that $dt \sim h$. In case process j fails, the most recent values for U_j are not available for continuing the computations, assuming that the runtime environment can survive process failures. On each *up and running* process the last computed solution is still available. The goal of the approach presented in the paper is therefore to design an algorithm which reconstructs the solution of the failed process(es) efficiently based on the checkpointed data.

The general fault tolerant approach is based on periodic checkpoints of the local data, e.g. to persistent disks or spare processes. Furthermore, processes are not coordinated for the checkpointing procedure for performance reasons, e.g. each process might save its local data at different time steps. As soon as a process failure occurs, the runtime environment will report it through a specific error code to the application. The application initiates the necessary operations to recover first the MPI environment and replace the failed process(es). In a second step, the application has to ensure, that the data on the replacement processes is consistent with the other processes. For this, the last checkpoint of the failed processes has to be retrieved. However, since the checkpoint of each process might have been taken at a different time step, this data does not yet provide a consistent state across all processes. Therefore, we discuss two mathematical methods based on time integration for constructing a consistent state from the available, inconsistent checkpoints. This difficulty is characteristic of a time dependent problem with no easy reversibility in time.

Figure 1 gives an example for the status of different process(es) after respawning a failed process and retrieving the last available checkpoint for this process. The thick lines represent the available data from which the recovery procedure will start. The circle lines correspond to the lost solution which we are trying to retrieve mathematically. The dashed lines are the boundary interfaces

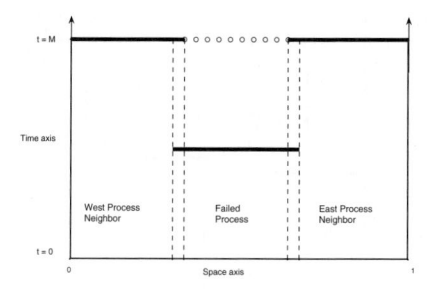

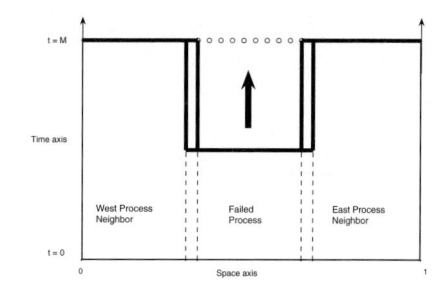

Fig. 1. Available data on main memory processors before starting the reconstruction algorithms

Fig. 2. Reconstruction procedure in one dimension using forward time integration

between subdomains. In the following, we will review two numerical methods to reconstruct a uniform approximation of U^M at a consistent time step M on the entire domain Ω.

2.2 The Forward Implicit Reconstruction

For the first approach, the application process j has to store every K time steps its current solution $U_j^{n(j)}$. Additionally, the artificial boundary conditions $I_j^m = \Omega_j \cap \Omega_{j+1}$ have to be stored for all time steps $m < M$ since the last checkpoint. The solution U_j^M can then be reconstructed with the forward time integration (2). Figure 2 demonstrates how the recovery works. The vertical thick lines represent the boundary data that need to be stored, and the intervals with circles are the unknowns of the reconstruction process.

The major advantage of this method is that it is using the same algorithm as in the standard domain decomposition method. The only difference is, that it is restricted to some specific subsets of the domain. Thus, the identical solution U_j^M as if the process had no failures can be reconstructed. The major disadvantage of this approach is the increased communication volume and frequency. While checkpointing the current solution $U_j^{n(j)}$ is done every K time steps, the boundary values have to be saved each time step for being able to correctly reconstruct the solution of the failed process(es).

2.3 The Backward Explicit Reconstruction

This method does not require the storage of the boundary conditions of each subdomain at each time step, but it allows to retrieve the interface data by computation instead. For this, the method requires the solution for each subdomain j at two different time steps, $n(j)$ and M, with $M - n(j) = K > 0$. The solution at time step M is already available on each *running* process after the failure occurred. The solution at time step $n(j)$ corresponds to the last solution on subdomain j saved to the spare memory before the failure happened. This is the starting point for the *local* reconstruction procedure. In this approach, only the replacement of the crashed process and its neighbors are involved (figure 3)

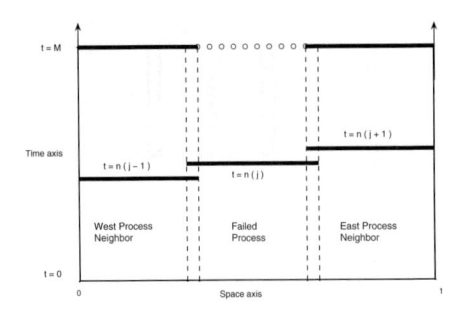

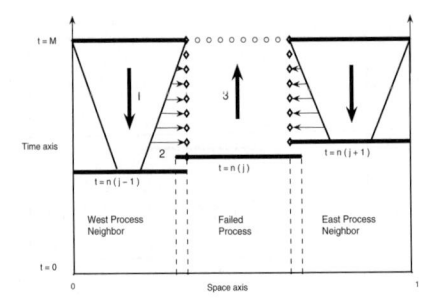

Fig. 3. Available data on main memory processors before starting the reconstruction algorithms

Fig. 4. Reconstruction procedure in one dimension using explicit backward time stepping

in the reconstruction. The reconstruction process is split in two explicit numerical schemes. The first one comes from the Forward Implicit scheme (2) which provides an explicit formula when going backward in time:

$$U_j^n = U_j^{n+1} - dt \frac{U_{j+1}^{n+1} - 2U_j^{n+1} + U_{j-1}^{n+1}}{h^2} - F_j^{n+1}. \quad (3)$$

The existence of the solution is granted by the forward integration in time. Two difficulties arise: first, the numerical procedure is instable, and second, one is restricted to the cone of dependence (Step 1 in figure 4). We have in Fourier modes $\hat{U}_k^n = \delta_k \hat{U}_k^{n+1}$, with $\delta_k \sim -\frac{2}{h}(\cos(k\ 2\ \pi\ h) - 1)$, $|k| \leq \frac{N}{2}$. The expected error is at most in the order $\frac{\nu}{h^L}$ where ν is the machine precision and L is the number of time steps which we can compute backwards. Therefore, the backward time integration is still accurate up to time step L with $\frac{\nu}{h^L} \sim h^2 \iff L \sim \frac{\log \nu}{\log h} - 2$. Then, the precision may deteriorate rapidly in time. Thus, to stabilize the scheme, one can use an hyperbolic regularization such as on the telegraph equation. Further details regarding this result can be found in [9,10].

To construct the solution outside the cone of dependencies and therefore to determine the solution at the subdomain interface, we used a standard procedure in inverse heat problem, the so-called space marching method [8] (Step 2 in figure 4). This method is second order but may require a regularization procedure of the solution obtained inside the cone using the product of convolution $\rho_\delta * u(x, t)$, where $\rho_\delta = \frac{1}{\delta\sqrt{\pi}} \exp(-\frac{t^2}{\delta^2})$. The space marching scheme is given by:

$$\frac{U_{j+1}^n - 2 U_j^n + U_{j-1}^n}{h^2} = \frac{U_j^{n+1} - U_j^{n-1}}{2\ dt} + F_j^n. \quad (4)$$

Equation 4 is unconditionally stable, given that $\delta \geq \sqrt{\frac{2\ dt}{\pi}}$. The neighbors of the failed processes apply these two methods successively. At the end of the procedure, these processes are able to provide to the replacement of the crashed process the artificial boundary conditions. Then, the respawned process can rebuild its lost data using the forward time integration as shown in section 2.2

(Step 3 in figure 4). The backward explicit time integration is well known to be an ill-posed problem and works only for few time steps. Indeed, for example if we assume $\nu = 10^{-12}$ and $h = 0.05$, the solution computed may blow up for $L > 7$. This would be equivalent to set at most the frequency of backup time step to $K = 9$. We refer to [9] for more details on the accuracy of our numerical scheme. Let us mention also that neither the backward explicit scheme nor the space marching scheme are limited to Cartesian grids.

In the following, we would like to compare the communication and checkpointing overhead imposed by the two methods described up to now.

2.4 Performance Comparison of the Checkpointing Operations

The two methods described in the sections 2.2 and 2.3 have different requirements with respect to what data has to be checkpointed by each process. While the backward explicit scheme requires only the storage of the domain of each process every K time steps, the forward implicit scheme requires additionally saving the boundary values in every time step.

For the performance comparison between both methods, two different codes have been evaluated, one based on a two dimensional domain decomposition and a code using a three dimensional domain decomposition. For the two-dimensional tests (1), we tested three different problem sizes per processor ($50*50$, $100*100$ and $200*200$) on four different processor configurations (9, 16, 25 and 36 processes). The processes are arranged in a regular two dimensional mesh. Each column of the processor-mesh has a separate checkpoint processor assigned to it. The data each checkpoint process receives is stored in their memory, avoiding therefore expensive disk I/O operations. More details to the checkpointing scheme are given in the section 3.

The cluster used for the 2-D testcase consisted of 154 Intel Itanium 2 processors with 4 GB of main memory and a Gigabit Ethernet interconnect. The results for the configuration described above are displayed in figures (5-8). The abscissa gives the checkpointing frequency in number of time steps between each checkpoint, while the ordinate shows the overhead compared to the same code and problem size without any checkpointing. As expected, the overhead is decreasing with increasing distance between two subsequent checkpoints. Furthermore, saving the boundary conditions each time step adds only a negligible overhead, especially for the largest test case with 36 processes and $200*200$ problem size per process. However, for higher dimension problem, saving at each time step the artificial boundary conditions slows the code execution down significantly. While interface conditions are one dimension lower than the solution itself, the additional message passing is interrupting the application work and data flow. Figure 9 and 10 shows the checkpointing time cost on the 3D version of the model problem defined in (1) for a small ($50*50*50$) and a larger problem size ($102*102*102$). For the large problem, saving the boundary conditions with a backup frequency of ten time steps slows the application down dramatically on

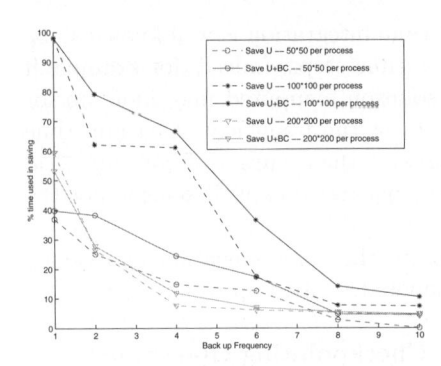

Fig. 5. Asynchronous checkpointing overhead with 9 processes

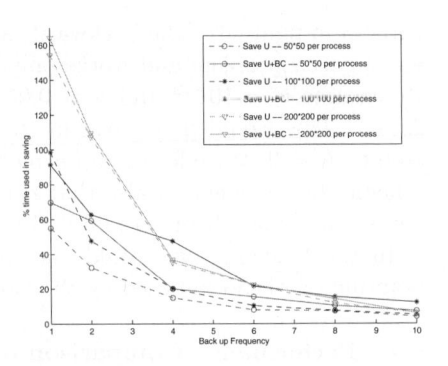

Fig. 6. Asynchronous checkpointing overhead with 16 processes

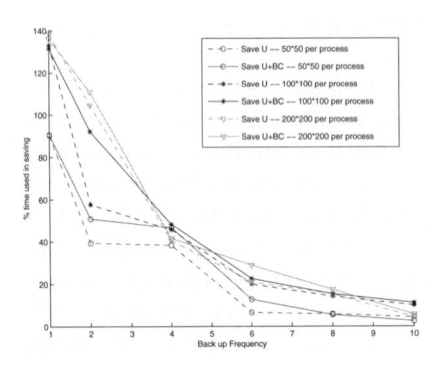

Fig. 7. Asynchronous checkpointing overhead with 25 processes

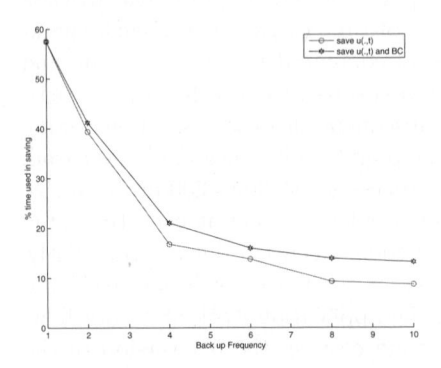

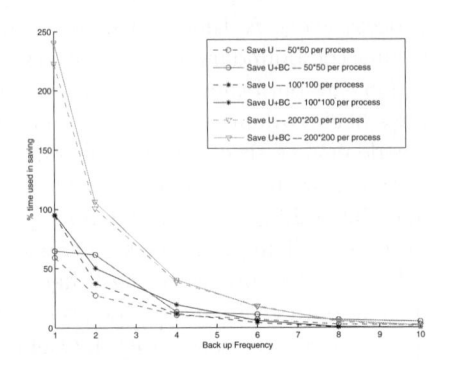

Fig. 8. Asynchronous checkpointing overhead with 36 processes

Fig. 9. Asynchronous checkpointing overhead for the small 3-D test case

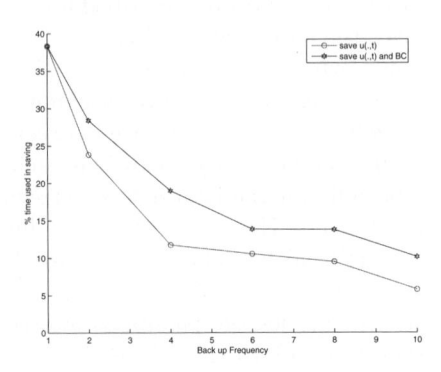

Fig. 10. Asynchronous checkpointing overhead for the large 3-D test case

an Intel EM64T cluster with a Gigabit Ethernet network. As shown in 10, the overhead compared to the method not requiring to store the boundary condition can double in the worst case. Figure (5-6-7-8) for the two dimensional problem and figure (9-10) for the three dimensional problem make us confident, that saving the local solution each 9 time steps brings a small overhead (between 5% and 15%) on the overall execution time. Moreover, such numerical methods are very cheap in term of computation and very fast. Therefore, the focus of the following sections is on the implementation aspects of the backward explicit scheme.

3 Implementation Details

As described in section 2.4, the checkpointing infrastructure utilized in the 2-D test case is implemented by using two groups of processes: a solver group composed by processes which will only solve the problem itself and a spare group of processes whose main function is to store the data from solver processes using local asynchronous checkpointing and non-blocking communications.

The communication between the solver processes and the checkpointing processes is handled by an inter-communicator. Since it does not make sense to have as many checkpointing processes as solver processes, the number of spare processes is equal to the number of solver processes in the x-direction. Thus, while the solver processes are arranged in the 2-D cartesian topology, the checkpoint processes are forming a 1-D cartesian topology.

Figure 11 gives a geometrical representation of the two groups with a local numbering. This figure shows furthermore, how the local checkpointing is applied for a configuration of 16 solver processes and 4 spare processes. Each spare

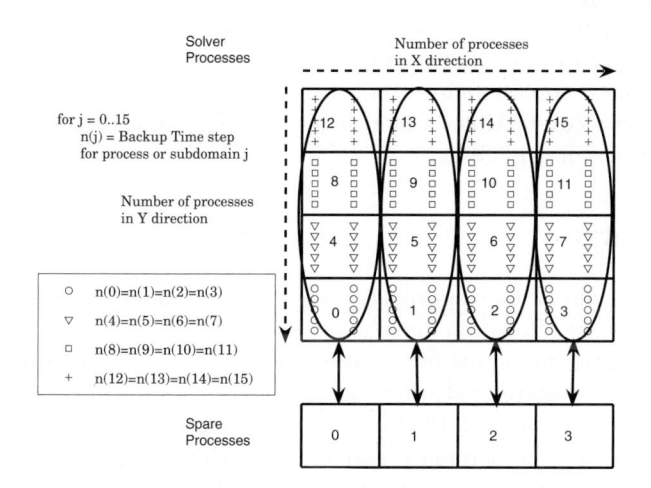

Fig. 11. Scheme of the local checkpointing

process is in charge of storing the data of a single subgroup, depicted by the ellipses in figure 11. To further improve the performance, asynchronous checkpointing has been used. Thus, each spare process stores the solution of only few solver processes of its subgroup at each time step. This approach further reduces the load on the network. As an example, suppose the backup time step is set to 10. The solver processes $j = [0-1-2-3]$ with $floor(j/4) = 0$ will send to the checkpoint process $[0-1-2-3]$ respectively their solution at the 1st time step and then at the 11th, at the 21st time step and so one. The solver processes $j = [8-9-10-11]$ with $floor(j/4) = 2$ will send it to the spare process $[0-1-2-3]$ respectively at the 3rd time step and then at the 13th, at the 23rd time step etc... Before starting the numerical reconstruction procedure discussed in section 2.3, the spare process(es) will send the last local solution(s) of the failed process(es) to their replacement process(es). Additionally, the last checkpoint of all neighbors of the failed process(es) will be distributed to them, since this data is required in the algorithm presented previously as well. From that on, the three step *local* reconstruction procedure can start and only the crashed process(es) and its neighbors will be involved.

4 Results

In the following, we would like to present the costs of a recovery operations in case a process failure occurs. Using the 2-D testcase described in section 2.4 we simulated a process failure and measured the execution time required for respawning the failed process and to reconstruct the data of this process using the backward explicit scheme. Two testcases using 9 and 16 solver processes have been analyzed. The recovery time for both cases was in the order of 2% of the overall execution time of the same simulation for the same number of time steps. For the 9 processor case, the average recovery time of the application was 0.16 seconds for the small problem size and 0.2 seconds for the largest one. The recovery time for the 16 processes test cases was in the same range, the recovery operation after a process failure took up to 0.34 seconds. These results show, that while the recovery time is increasing with the number processes used, its overall effect is still negligible.

5 Summary

This paper discusses two approaches on how to handle process failures for parabolic problems. Based on distributed and uncoordinated checkpointing, the numerical methods presented here can reconstruct a consistent state in the parallel application, despite of storing checkpoints of various processes at different time steps. The first method, the forward implicit scheme, requires for the reconstruction procedure the boundary variables of each time step to be stored along with the current solution. The second method, the backward explicit scheme, only requires checkpointing the solution of each process every K time steps. Performance results comparing both methods with respect to the checkpointing overhead have been presented. We presented the results for recovery time of a 2-D heat equation. Currently ongoing work is focusing on the implementation of these explicit methods for a 3D Reaction-Convection-Diffusion code simulating the Air Quality Model [11].

References

1. MPI Forum : *MPI: A Message-Passing Interface Standard.* Document for a Standard Message-Passing Interface, University of Tennessee, 1993, 1994.
2. Sriram Sankaran, Jeffrey M. Squyres, Brian Barrett, Andrew Lumsdaine, Jason Duell, Paul Hargrove, and Eric Roman, The LAM/MPI Checkpoint/Restart Framework: System-Initiated Checkpointing, in *International Journal of High Performance Computing Applications*, 2004.
3. G. Bosilca, A. Bouteiller, F. Cappello, S. Djilali, G. Fedak, C. Germain, T. Herault, P. Lemarinier, O. Lodygensky, F. Magniette, V. Neri, and A. Selikhov, MPICH-V: Toward a Scalable Fault Tolerant MPI for Volatile Nodes, in SC'2002 Conference CD, IEEE/ACM SIGARCH, Baltimore, MD, 2002.
4. Graham E. Fagg, Edgar Gabriel, Zizhong Chen, Thara Angskun, George Bosilca, Jelena Pjesivac-Grbovic, and Jack J. Dongarra, 'Process Fault-Tolerance: Semantics, Design and Applications for High Performance Computing', in 'International Journal of High Performance Computing Applications', Volume 19, No. 4, pp. 465-477, Sage Publications 2005.
5. Beck, Dongarra, Fagg, Geist, Gray, Kohl, Migliardi, K. Moore, T. Moore, Papadopoulous, Scott, and Sunderam, 'HARNESS: a next generation distributed virtual machine', Future Generation Computer Systems, 15, 1999.
6. C. Engelmann and G. A. Geist. "Super-Scalable Algorithms for Computing on 100,000 Processors". Proceedings of International Conference on Computational Science (ICCS) 2005, Atlanta, GA, USA, May 2005.
7. Zizhong Chen, Graham E. Fagg, Edgar Gabriel, Julien Langou, Thara Angskun, George Bosilca, Jack J. Dongarra, 'Fault Tolerant High Performance Computing by a coding approach', Proceedings of the 2005 ACM SIGPLAN Symposium on Principles and Practice of Parallel Programming (PPoPP '05), Chicago, IL, June 15-17, 2005, ACM Press.
8. D.A. Murio, The Mollification Method and the Numerical Solution of Ill-posed Problems, Wiley, New York (1993).
9. M. Garbey, H. Ltaief, *Fault Tolerant Domain Decomposition for Parabolic Problems*, Domain Decomposition 16, New York University, January 2005. To appear.
10. W. Eckhaus and M. Garbey, Asymptotic analysis on large time scales for singular perturbation problems of hyperbolic type, SIAM J. Math. Anal. Vol 21, No 4, pp867-883 (1990).
11. Dupros, M.Garbey and W.E.Fitzgibbon, A Filtering technique for System of Reaction Diffusion equations, Int. J. for Numerical Methods in Fluids, in press 2006.

Parallel Solution of Large-Scale and Sparse Generalized Algebraic Riccati Equations

José M. Badía[1], Peter Benner[2], Rafael Mayo[1], and Enrique S. Quintana-Ortí[1]

[1] Depto. de Ingeniería y Ciencia de Computadores, Universidad Jaume I,
12.071–Castellón, Spain
{badia, mayo, quintana}@icc.uji.es
[2] Fakultät für Mathematik, Technische Universität Chemnitz,
D-09107 Chemnitz, Germany
benner@mathematik.tu-chemnitz.de

Abstract. We discuss a parallel algorithm for the solution of large-scale generalized algebraic Riccati equations with dimension up to $\mathcal{O}(10^5)$. We survey the numerical algorithms underlying the implementation of the method, in particular, a Newton-type iterative solver for the generalized Riccati equation and an LR-ADI solver for the generalized Lyapunov equation. Experimental results on a cluster of Intel Xeon processors illustrate the benefits of our approach.

Keywords: generalized algebraic Riccati equation, Newton's method, generalized Lyapunov equation, LR-ADI iteration, parallel algorithms.

1 Introduction

Consider the (generalized) *algebraic Riccati equation* (ARE)

$$0 = A^T X E + E^T X A - E^T X B R^{-1} B^T X E + C^T Q C =: \mathcal{R}(X), \qquad (1)$$

where $A, E \in \mathbb{R}^{n \times n}$, $B \in \mathbb{R}^{n \times m}$, $C \in \mathbb{R}^{p \times n}$, $R \in \mathbb{R}^{m \times m}$ is symmetric positive definite, and $Q \in \mathbb{R}^{p \times p}$ is symmetric positive semidefinite. Under certain conditions, the ARE (1) has a unique symmetric positive semidefinite solution $X \in \mathbb{R}^{n \times n}$ [17,19]. This particular solution is usually required in applications.

Solving the ARE (1) is the key step in many computational methods for model reduction, filtering, and controller design of dynamical linear systems (see, among many, [10,13,14,15,19,20,24] and the references therein). In general, numerical methods for solving AREs have a computational cost of $\mathcal{O}(n^3)$ floating-point arithmetic operations (flops) and require storage for $\mathcal{O}(n^2)$ numbers [17,19,24]. While current desktop computers provide enough computational power to solve problems with state-space dimension n in the hundreds using libraries such as SLICOT (http://www.slicot.org) or the MATLAB control-related toolboxes, large-scale applications clearly require the use of advanced computing techniques.

Over the last few years we have developed a library of parallel algorithms for the solution of (dense) AREs on parallel architectures [6]. The library, PLiCOC, employs the kernels in LAPACK, BLAS, and ScaLAPACK [3,8], enabling the solution of equations with n in the order of a few thousands. However, this approach is still insufficient for very large-scale AREs arising in weather forecast, circuit simulation, VLSI chip design, and air quality simulation, among others (see, e.g., [4,9,11,12]). Dynamical systems leading to AREs with dimension n as high as $\mathcal{O}(10^4) - \mathcal{O}(10^5)$ and sparse matrix pairs (A, E) are common in these applications.

In this paper we consider a different method that exploits the sparse structure of the matrix pair (A, E) in (1), and thus allows the solution of much larger AREs. (Throughout the paper the term "sparse" will refer to both unstructured sparse matrices and structured ones, such as banded matrices.) The codes employ the parallel kernels in ScaLAPACK [8]. Depending on the specific structure of A and E, the (unstructured) sparse linear system solvers in MUMPS [2] or the banded linear system solver in ScaLAPACK are also employed.

The rest of the paper is structured as follows. In Section 2 we review a specialized formulation of Newton's iterative method for the solution of large-scale sparse AREs. The major computational task in this method is the solution of a large-scale sparse (generalized) Lyapunov equation. A variant of the iterative Lyapunov solver introduced in [18,22], based on a low-rank *alternating direction implicit* (LR-ADI) method, is then summarized in Section 3. Following the description of the numerical methods, in Section 4 we offer some details on the parallelization of the corresponding algorithms. In Section 5, experiments on a cluster of Intel Xeon processors report the potential of the ARE solver. Finally, we give some concluding remarks in Section 6.

2 Newton's Method for the ARE

In this section we review a variant of Newton's method, described in [23], which delivers a full-rank approximation of the solution of large-scale sparse AREs. Here we focus on the implementation details relevant to an efficient (parallel) implementation.

Starting from an initial solution X_0, Newton's method for the ARE [16] proceeds as follows:

Newton's method
1) Compute the Cholesky factorization $Q = \bar{Q}\bar{Q}^T$
2) Compute the Cholesky factorization $R = \bar{R}\bar{R}^T$
3) $\bar{C} := \bar{Q}^T C$
4) $\bar{B} := E^{-1}BR^{-1} = ((E^{-1}B)\bar{R}^{-T})\bar{R}^{-1}$
repeat with $j := 0, 1, 2, \ldots$
   5) $K_j := E^T X_j B R^{-1} = E^T X_j E \bar{B}$
   6) $\hat{C}_j := \begin{bmatrix} \bar{C} \\ \bar{R}^T K_j^T \end{bmatrix}$
   7) Solve for X_{j+1}:
   $0 = (A - BK_j^T)^T X_{j+1} E + E^T X_{j+1}(A - BK_j^T) + \hat{C}_j^T \hat{C}_j$
until $\|X_j - X_{j-1}\| < \tau \|X_j\|$

Provided $(A-BR^{-1}B^T X_0, E)$ is a stable matrix pair (i.e., all its eigenvalues lie on the open left half plane), this iteration converges quadratically to the desired symmetric positive semidefinite solution of the ARE [16], $X_\infty = \lim_{j \to \infty} X_j$. In practice, (A, E) is often a stable matrix pair, so that setting $X_0 := 0$ is enough to guarantee the convergence of the iteration. Thus, no globalization strategy is required to guarantee convergence. Note that a line search procedure in [7] can be used to accelerate initial convergence, though.

In real large-scale applications, $m, p \ll n$, and both A and E are sparse, but the solution matrix X is in general dense and, therefore, impossible to construct explicitly. However, X is often of low-numerical rank and thus can be approximated by a full-rank factor $\hat{R} \in \mathbb{R}^{n \times r}$, with $r \ll n$, such that $\hat{R}\hat{R}^T \approx X$. The method described next aims at computing this "narrow" factor $\hat{R}$ instead of the explicit solution.

2.1 Exploiting the Rank-Deficiency of the Solution

Let us review how to modify Newton's method in order to avoid explicit references to X_j. Note that all but one of the computations in Steps 1–4 of Newton's method involve matrices of small dimensions and therefore can be performed employing dense linear algebra kernels even if the matrices are sparse. In particular, the cost of the two Cholesky factorizations in Steps 1 and 2 is $m^3/3 + p^3/3$ flops, and obtaining $\bar{C} \in \mathbb{R}^{p \times n}$ from there in Step 3 requires $n^2 p$ additional flops. On the other hand, computing $\bar{B} \in \mathbb{R}^{n \times m}$ in Step 4 requires $2m^2 n$ flops plus the cost of solving the system $E^{-1}B$. The cost of this latter operation (via, e.g., a direct method) strongly depends on the sparsity degree and pattern of the coefficient matrix E, and the solver that is employed. For unstructured sparse matrices, this cost is difficult to determine *a priori*.

Assume for the moment that, at the beginning of the iteration, we maintain $\hat{R}_j \in \mathbb{R}^{n \times r_j}$ such that $E^{-T}\hat{R}_j\hat{R}_j^T E^{-1} = X_j$. Then, in the first step of the iteration, we can compute K_j as

$$K_j := E^T X_j E \bar{B} = \hat{R}_j(\hat{R}_j^T \bar{B}),$$

which initially requires a (dense) matrix product, $M := \hat{R}_j^T \bar{B}$, at a cost of $2r_j mn$ flops, and then a (dense) matrix product, $\hat{R}_j M$, with the same cost. Even for large-scale problems, as m is usually a small order constant, this represents at most a quadratic cost. In practice, r_j usually remains a small value during the iteration so that this cost becomes as low as linear.

The matrix product $\bar{R}^T K_j^T$ needed in the second step of the iteration for the construction of $\hat{C}_j$ presents only a moderate cost, $2m^2 n$ flops, and therefore does not require any special action.

The key of this approach lies in solving the Lyapunov equation in the third step for a full-rank factor $\hat{R}_{j+1}$, such that $E^{-T}\hat{R}_{j+1}\hat{R}_{j+1}^T E^{-1} = X_{j+1}$. We describe how to do so in the next section.

3 Low Rank Solution of Lyapunov Equations

In this section we introduce a generalization of the Lyapunov solver proposed in [18,22], based on the cyclic *low-rank alternating direction implicit* (LR-ADI) iteration.

Consider the Lyapunov equation to be solved at each iteration of Newton's method
$$0 = (A - BK^T)^T Y E + E^T Y (A - BK^T) + \hat{C}^T \hat{C}, \qquad (2)$$
where, for simplicity, we drop all subindices in the expression. Here, $A, E \in \mathbb{R}^{n \times n}$, $B, K \in \mathbb{R}^{n \times m}$, and $\hat{C} \in \mathbb{R}^{(p+m) \times n}$. Recall that we are interested in finding a full-rank factor $S \in \mathbb{R}^{n \times s}$, with $s \ll n$, such that $SS^T \approx Y$. Then, in the jth iteration of Newton's method, $\hat{R}_j := S$ and $r_j := s$.

A generalization of the LR-ADI iteration, tailored for equation (2), can be formulated as follows:

LR-ADI iteration
1) $V_0 := ((A - BK^T)^T + \sigma_1 E^T)^{-1} \hat{C}^T$
2) $S_0 := \sqrt{-2\,\alpha_1}\, V_0$
repeat with $l := 0, 1, 2, \ldots$
    3) $V_{l+1} := V_l - \delta_l ((A - BK^T)^T + \sigma_{l+1} E^T)^{-1} V_l$
    4) $S_{l+1} := [S_l, \; \gamma_l V_{l+1}]$
until $\|\gamma_l V_l\|_1 < \tau \|S_l\|_1$

In the iteration, $\{\sigma_1, \sigma_2, \ldots\}$, $\sigma_l = \alpha_l + \beta_l\,\jmath$, is a cyclic set of (possibly complex) shift parameters (that is, $\sigma_l = \sigma_{l+t}$ for a given period t), $\gamma_l = \sqrt{\alpha_{l+1}/\alpha_l}$, and $\delta_l = \sigma_{l+1} + \overline{\sigma_l}$, with $\overline{\sigma_l}$ the conjugate of σ_l. The convergence rate of the LR-ADI iteration strongly depends on the selection of the shift parameters and is super-linear at best [18,22,26].

At each iteration the column dimension of S_{l+1} is increased by $(p+m)$ columns with respect to that of S_l so that, after $\bar{l}$ iterations, $S_{\bar{l}} \in \mathbb{R}^{n \times \bar{l}(p+m)}$. For details on a practical criterion to stop the iteration, see [5,22]. Note that the LR-ADI iteration does not guarantee full colum rank of the S_l. This could be achieved using a column compression based on a rank-revealing LQ factorization. As the full-rank property is irrelevant for the approximation quality, we will not discuss this any further. Possible positive effects on the efficiency of the algorithm will be reported elsewhere.

From the computational view point, the iteration only requires the solution of linear systems of the form
$$((A - BK^T)^T + \sigma E^T)V = W \quad \Leftrightarrow \quad ((A + \sigma E) - BK^T)^T V = W, \qquad (3)$$
for V. Now, even if A and E are sparse (and therefore, so is $\bar{A} := A + \sigma E$), the coefficient matrix of this linear system is not necessarily sparse. Nevertheless, we can still exploit the sparsity of A, E by relying on the *Sherman-Morrison-Woodbury (SMW) formula*
$$(\bar{A} - BK^T)^{-1} = \bar{A}^{-1} + \bar{A}^{-1} B (I_m - K^T \bar{A}^{-1} B)^{-1} K^T \bar{A}^{-1}.$$

Specifically, the solution V of (3) can be obtained following the next five steps:

SMW formula
1) $V := A^{-T}W$
2) $T := \bar{A}^{-T}K$
3) $F := I_m - B^T T$
4) $T := TF^{-1}$
5) $V := V + T(B^T V)$

Steps 1 and 2 require the solution of two linear systems with sparse coefficient matrix $\bar{A}$. The use of direct solvers is recommended here as iterations l and $l + t$ of the LR-ADI method share the same coefficient matrices for the linear system. The remaining three steps operate with dense matrices of small-order; specifically, $F \in \mathbb{R}^{m \times m}$, $T \in \mathbb{R}^{n \times m}$ so that Steps 3, 4, and 5 only require $2m^2 n$, $2m^3/3 + m^2 n$, and $4mn(m + p)$ flops, respectively.

4 Parallel Implementation

The numerical algorithms described in the previous two sections for Newton's method and the LR-ADI iteration are composed of a few dense linear algebra operations (Cholesky factorizations, matrix products, and linear systems) involving dense matrices of relatively small order, and the solution of sparse linear systems (via direct methods) with large-scale coefficient matrices.

Our approach for dealing with these matrix operations is based on the use of existing parallel linear algebra and communication libraries. In Fig. 1 we illustrate the multilayered architecture of libraries employed by our ARE solver included in *SpaRed* (a parallel library for model reduction of large-scale sparse linear systems; http://www.pscom.uji.es/modred/SpaRedW3/SpaRed.html).

The solver employs the parallel kernels in ScaLAPACK. Depending on the structure of the state matrix pair (A, E) the banded linear system solver in ScaLAPACK or the sparse linear system solvers in MUMPS are also invoked. Table 1 lists the specific routines employed for each one of the major operations in the algorithm. In the table we do not include the operations required to compute the shift parameters for the LR-ADI iteration as that part of the algorithm was not described. The shifts are currently obtained using an Arnoldi-type iteration. The implementation employs PARPACK, and requires the solution of sparse linear systems and the sparse matrix-vector product. For the first operation, depending on the structure of the coefficient matrix we again use ScaLAPACK or MUMPS. The sparse matrix-vector product is parallelized as a sequence of dot products with the matrix cyclically distributed by rows.

5 Experimental Results

All the experiments presented in this section were performed on a cluster of $n_p = 16$ nodes using IEEE double-precision floating-point arithmetic ($\varepsilon \approx 2.2204 \times$

Table 1. Parallelization of the major matrix operations that appear in the ARE solver

		Newton's method		
Step	Operation	Structure of matrices	Parallel library	Routine
1	Factorize Q	All dense	ScaLAPACK	p_potrf
2	Factorize R	All dense	ScaLAPACK	p_potrf
3	Compute $\bar{Q}^T C$	All dense	PBLAS	p_gemm
4.1	Solve $E^{-1}B$	Sparse E/ dense (BR^{-1})	MUMPS or ScaLAPACK	_mumps_c or p_gbsv
4.2	Solve $((E^{-1}B)\bar{R}^{-T})\bar{R}^{-1}$	All dense	PBLAS	p_trsm×2
5	Compute $\hat{R}_j(\hat{R}_j^T \bar{B})$	All dense	PBLAS	p_gemm×2
6	Compute $\bar{R}^T K_j^T$	All dense	PBLAS	p_gemm
7	Solve Lyapunov eq.	Sparse E, A/ dense B, K_j	SpaRed	LR-ADI iter.
		LR-ADI iteration		
Step	Operation	Structure of matrices	Parallel library	Routine
1	Compute V_0	Sparse E, A/ dense $B, K, \hat{C}$	SpaRed	SMW formula
3	Compute V_{l+1}	Sparse E, A/ dense B, K, V_l	SpaRed	SMW formula
		SMW formula		
Step	Operation	Structure of matrices	Parallel library	Routine
1	Solve $\bar{A}^{-T}W$	Sparse $\bar{A}$/ dense W	MUMPS or ScaLAPACK	_mumps_c p_gbsv
2	Solve $\bar{A}^{-T}K$	Sparse $\bar{A}$/ dense K	MUMPS or ScaLAPACK	_mumps_c p_gbsv
3	Compute $I_m - B^T T$	All dense	PBLAS	p_gemm
4	Solve TF^{-1}	All dense	ScaLAPACK	p_gesv
5	Compute $V + T(B^T V)$	All dense	PBLAS	p_gemm×2

10^{-16}). Each node consists of an Intel Xeon processor@2.4 GHz with 1 GByte of RAM. We employ a BLAS library specially tuned for this processor that achieves around 3800 MFLOPs (millions of flops per second) for the matrix product (routine DGEMM from Goto BLAS, http://www.tacc.utexas.edu/resources/software/). The nodes are connected via a *Myrinet* multistage network and the MPI communication library is specially developed and tuned for this network. The performance of the interconnection network was measured by a simple loop-back message transfer resulting in a latency of 18 μsec. and a bandwidth of 1.4 Gbit/s.

In order to evaluate the performance of our ARE solvers we employ two different examples:

Example 1. This standard system is obtained from a finite difference discretization (with equidistant grid) of a 2-D heat equation. The dimension of the system is given by the number of grid points, n_0, in one direction,

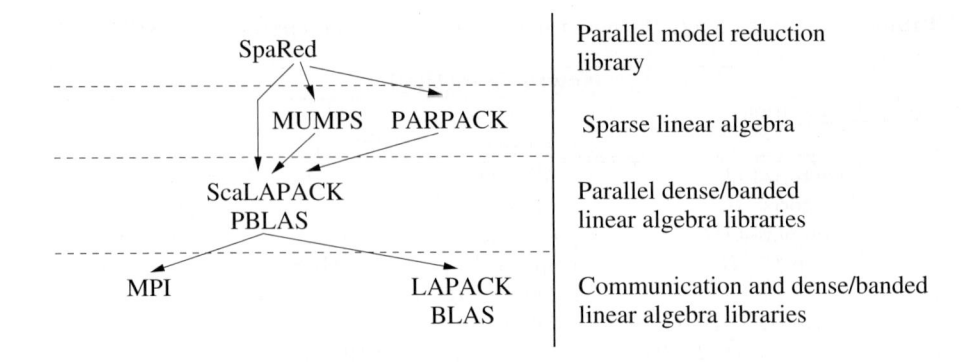

```
         SpaRed                      Parallel model reduction
                                     library
  -----------/-\---------------
            /   \
           /     \
          MUMPS  PARPACK             Sparse linear algebra
   -------|-/---------------
          |/
        ScaLAPACK                    Parallel dense/banded
        PBLAS                        linear algebra libraries
  ----------/------------------
           /
   MPI                 LAPACK        Communication and dense/banded
                       BLAS          linear algebra libraries
```

Fig. 1. Multilayered architecture of the ARE solver

so that $n = n_0^2$. The system is single-input, single-output (SISO); that is, $m = p = 1$.

Example 2. This model arises in a manufacturing method for steel profiles [21,25]. The goal is to design a control that achieves moderate temperature gradients when the rail is cooled down. The mathematical model corresponds to the boundary control for a 2-D heat equation. A finite element discretization, followed by adaptive refinement via bisection results in a generalized systems of order n=79841 with 7 inputs and 6 outputs.

In both examples we employ weight matrices $Q = I_p$ and $R = I_m$.

We next report the execution times of the sparse ARE solver in Table 2 using $n_p = 16$ processing nodes. For ScaLAPACK a logical 4×4 grid was selected with the distribution block size equal 32. Other logical topologies/block sizes did not offer significative differences. A modest number of shifts is used for the LR-ADI iteration due to the need of storing the LU factors in the current implementation of the algorithm. We could overcome this restriction by recomputing the factorization at each iteration. However, doing so would produce a notable raise in the execution time. A different approach would be that of maintaining the factors stored on disk. The LR-ADI iteration was stopped after 100 iterations for Example 2. Although more iterations were strictly necessary according to the convergence criterion employed for the LR-ADI iteration, stopping the iteration at this point for this particular example guaranteed Newton's method to converge to the desired stabilizing solution of the equation.

Table 2. Execution times of the ARE solver on n_p=16 nodes

	n	#iter. Newton	#shifts LR-ADI	Avg. #iter. LR-ADI	r	Ex. time
Example 1	160000	11	10	80	160	1h 25m 5s
Example 2	79841	5	20	100	1300	1h 4m 30s

The parallel performance of the ARE solver is highly dependent on the efficacy of the underlying parallel sparse linear system solver and the sparsity pattern of the matrix pair (A, E). Due to the problem dimensions and structure, the number of shifts selected for each example, the numerical tools that were employed (MUMPS for the solution of the sparse linear systems in both cases), and the specifications of the hardware resources (1 Gbyte of RAM per node), virtual memory was required to solve the problems using less than $n_p=16$ nodes; in those cases where the problem could still be solved using storage on disk, I/O resulted in much larger execution times.

6 Concluding Remarks

We have presented a solver for large-scale generalized algebraic Riccati equations with sparse matrix pair (A, E). The method involves the solution of large-scale linear systems, with sparse coefficient matrix, and well-known dense linear algebra operations on small-scale matrices. These operations are available in parallel linear algebra libraries such as ScaLAPACK, PLAPACK, MUMPS, SuperLU, etc. The experimental results on a cluster of moderate dimensions illustrates a parallel efficacy that enables the solution of equations with dimension up to $\mathcal{O}(10^5)$ in a relatively short time. Using these parallel algorithms, the solution of large-scale optimal control problems and model reduction of large-scale systems via relative error methods become thus feasible.

Acknowledgments

This research has been partially supported by the DAAD programme Acciones Integradas HA2005-0081. José M. Badía, Rafael Mayo, and E.S. Quintana-Ortí were supported by the CICYT project TIN2005-09037-C02-02 and FEDER, and project No. P1B-2004-6 of the *Fundación Caixa-Castellón/Bancaixa and UJI*.

References

1. J. Abels and P. Benner. CAREX – a collection of benchmark examples for continuous-time algebraic Riccati equations (version 2.0). SLICOT Working Note 1999-14, November 1999. Available from http://www.slicot.org.
2. P.R. Amestoy, I.S. Duff, J. Koster, and J.-Y. L'Excellent. MUMPS: a general purpose distributed memory sparse solver. In *Proc. PARA2000, 5th International Workshop on Applied Parallel Computing*, pages 122–131, 2000.
3. E. Anderson, Z. Bai, C. Bischof, J. Demmel, J. Dongarra, J. Du Croz, A. Greenbaum, S. Hammarling, A. McKenney, and D. Sorensen. *LAPACK Users' Guide*. SIAM, Philadelphia, PA, third edition, 1999.
4. A.C. Antoulas. *Approximation of Large-Scale Dynamical Systems*. SIAM Publications, Philadelphia, PA, 2005.
5. J.M. Badía, P. Benner, R. Mayo, and E.S. Quintana-Ortí. Balanced truncation model reduction of large and sparse descriptor linear systems. In preparation.

6. P. Benner, E.S. Quintana-Ortí, and G. Quintana-Ortí. Solving Linear-Quadratic Optimal Control Problems on Parallel Computers. Preprint SFB393/05-19, Sonderforschungsbereich 393 *Numerische Simulation auf massiv parallelen Rechnern*, TU Chemnitz, 09107 Chemnitz, FRG, March 2006. Available from http://www.mathematik.tu-chemnitz.de/preprint/2005/SFB393_19.html.
7. P. Benner and R. Byers. An exact line search method for solving generalized continuous-time algebraic Riccati equations. *IEEE Trans. Automat. Control*, 43(1):101–107, 1998.
8. L.S. Blackford, J. Choi, A. Cleary, E. D'Azevedo, J. Demmel, I. Dhillon, J. Dongarra, S. Hammarling, G. Henry, A. Petitet, K. Stanley, D. Walker, and R.C. Whaley. *ScaLAPACK Users' Guide*. SIAM, Philadelphia, PA, 1997.
9. C.-K. Cheng, J. Lillis, S. Lin, and N.H. Chang. *Interconnect Analysis and Synthesis*. John Wiley & Sons, Inc., New York, NY, 2000.
10. B.N. Datta. Linear and numerical linear algebra in control theory: Some research problems. *Linear Algebra Appl.*, 197/198:755–790, 1994.
11. R. Freund. Model reduction methods based on Krylov subspaces. *Acta Numerica*, 12:267–319, 2003.
12. R. W. Freund and P. Feldmann. Reduced-order modeling of large passive linear circuits by means of the SyPVL algorithm. In *Technical Digest of the 1996 IEEE/ACM International Conference on Computer-Aided Design*, pages 280–287. IEEE Computer Society Press, 1996.
13. M. Gawronski. *Balanced control of flexible structures*, volume 211 of *Lecture Notes in Control and Information Sciences*. Springer-Verlag, London, UK, 1996.
14. M. Gawronski. *Dynamics and control of structures: A modal approach*. Mechanical Engineering Series. Springer-Verlag, Berlin, FRG, 1998.
15. M. Green. Balanced stochastic realization. *Linear Algebra Appl.*, 98:211–247, 1988.
16. D.L. Kleinman. On an iterative technique for Riccati equation computations. *IEEE Trans. Automat. Control*, AC-13:114–115, 1968.
17. P. Lancaster and L. Rodman. *The Algebraic Riccati Equation*. Oxford University Press, Oxford, 1995.
18. J.-R. Li and J. White. Low rank solution of Lyapunov equations. *SIAM J. Matrix Anal. Appl.*, 24(1):260–280, 2002.
19. V. Mehrmann. *The Autonomous Linear Quadratic Control Problem, Theory and Numerical Solution*. Number 163 in Lecture Notes in Control and Information Sciences. Springer-Verlag, Heidelberg, July 1991.
20. R. Ober. Balanced parametrizations of classes of linear systems. *SIAM J. Cont. Optim.*, 29:1251–1287, 1991.
21. T. Penzl. Algorithms for model reduction of large dynamical systems. *Linear Algebra Appl.*, 415:322–343, 2006. (Reprint of Preprint SFB393/99-40, TU Chemnitz, 1999.)
22. T. Penzl. A cyclic low rank Smith method for large sparse Lyapunov equations. *SIAM J. Sci. Comput.*, 21(4):1401–1418, 2000.
23. T. Penzl. LYAPACK Users Guide. Preprint SFB393/00-33, Sonderforschungsbereich 393 *Numerische Simulation auf massiv parallelen Rechnern*, TU Chemnitz, 09107 Chemnitz, FRG, 2000. Available from http://www.tu-chemnitz.de/sfb393/sfb00pr.html.
24. V. Sima. *Algorithms for Linear-Quadratic Optimization*, volume 200 of *Pure and Applied Mathematics*. Marcel Dekker, Inc., New York, NY, 1996.

25. F. Tröltzsch and A. Unger. Fast solution of optimal control problems in the selective cooling of steel. Preprint SFB393/99-7, Sonderforschungsbereich 393 *Numerische Simulation auf massiv parallelen Rechnern*, TU Chemnitz, 09107 Chemnitz, FRG, 1999. Available from http://www.tu-chemnitz.de/sfb393/sfb99pr.html.
26. E.L. Wachspress. The ADI model problem, 1995. Available from the author.

Applicability of Load Balancing Strategies to Data-Parallel Embedded Runge-Kutta Integrators

Matthias Korch and Thomas Rauber

University of Bayreuth, Department of Computer Science
{matthias.korch, rauber}@uni-bayreuth.de

Abstract. Embedded Runge-Kutta methods are among the most popular methods for the solution of non-stiff initial value problems of ordinary differential equations (ODEs). We investigate the use of load balancing strategies in a data-parallel implementation of embedded Runge-Kutta integrators. Since the parallelism contained in the function evaluation of the ODE system is typically very fine-grained, our aim is to find out whether the employment of load balancing strategies can be profitable in spite of the additional overhead they involve.

1 Introduction

In this paper, we consider the parallel solution of initial value problems (IVPs) of ordinary differential equations (ODEs) defined by

$$\mathbf{y}'(t) = \mathbf{f}(t, \mathbf{y}(t)), \quad \mathbf{y}(t_0) = \mathbf{y}_0, \quad \mathbf{y} : \mathbb{R} \to \mathbb{R}^n, \quad \mathbf{f} : \mathbb{R} \times \mathbb{R}^n \to \mathbb{R}^n. \quad (1)$$

The numerical solution of large IVPs is a computationally intensive task. Therefore, efforts have been taken to find efficient parallel solution methods, e.g., extrapolation methods [1], waveform relaxation techniques [2], and iterated Runge-Kutta methods [3]. Most of these approaches develop new numerical algorithms with a larger potential for parallelism, but with different numerical properties.

Non-stiff ODE systems can be solved efficiently by embedded Runge-Kutta (ERK) methods with stepsize control. Popular methods are, for example, DOPRI5(4) and DOPRI8(7) [4]. An ERK method with s stages which uses the argument vectors $\mathbf{w}_1, \ldots, \mathbf{w}_s$ to compute the two new approximations $\eta_{\kappa+1}$ and $\hat{\eta}_{\kappa+1}$ from the two previous approximations η_κ and $\hat{\eta}_\kappa$ is represented by the computation scheme

$$\mathbf{w}_l = \eta_\kappa + h_\kappa \sum_{i=1}^{l-1} a_{li} \mathbf{f}(t_\kappa + c_i h_\kappa, \mathbf{w}_i), \quad l = 1, \ldots, s,$$

$$\eta_{\kappa+1} = \eta_\kappa + h_\kappa \sum_{l=1}^{s} b_l \mathbf{f}(t_\kappa + c_l h_\kappa, \mathbf{w}_l), \quad \hat{\eta}_{\kappa+1} = \eta_\kappa + h_\kappa \sum_{l=1}^{s} \hat{b}_l \mathbf{f}(t_\kappa + c_l h_\kappa, \mathbf{w}_l). \quad (2)$$

The coefficients a_{li}, c_i, b_l, and $\hat{b}_l$ are determined by the particular ERK method used.

Because, in general, all coefficients a_{ij} may be non-zero and an evaluation of the right hand side function $\mathbf{f}(t, \mathbf{w})$ may access all components of the argument vector $\mathbf{w}$, the stages $l = 1, \ldots, s$ have to be computed sequentially. However, ERK methods possess a large potential for data-parallelism across the ODE system, since the function evaluations of individual ODE components can be performed in parallel. Experiences from earlier experiments (e.g., [5,6], cf. [2]) suggest that this type of parallelism can be exploited efficiently only if the ODE system is sufficiently large and the communication network of the parallel computer system is fast in relation to the speed of the processors or the function evaluations of the ODE components are computationally intensive. In general, the obtainable performance strongly depends on the characteristics of the IVP (cf. Section 5). But if these conditions are fulfilled, general ERK solvers can work efficiently on small or medium-sized shared-memory multiprocessors (SMMs). However, on larger SMMs and on most modern distributed-memory multiprocessors (DMMs) the speedups obtainable with current implementations are not yet satisfactory [6].

Therefore, it is desirable to find new implementations that can deliver a higher efficiency. Two possible approaches take advantage of special properties of either the ERK method [7] or the ODE system [5,6]. In this paper, we follow a different approach which requires no assumptions about particular properties of the method or the ODE system. We investigate if an improvement in performance can be achieved by the application of dynamic load balancing strategies. We show that if the load balance can be achieved with only little overhead, a higher performance can be obtained if the right hand side function of the problem is irregular and also in other situations where a load imbalance limits scalability, while the performance for regular problems is still competitive with solvers with a static work distribution.

2 Motivation and Computational Structure

Knowing from previous experiments that the communication costs of general ERK solvers on DMMs are to high for most problems to achieve satisfactory speedups, we concentrate our initial investigations of load balancing strategies on SMMs, because on such machines load balancing strategies with little overhead can be realized.

The computational kernel of a data-parallel implementation of a general ERK solver with a static blockwise data distribution can be realized as shown in Fig. 1. Since the evaluation of the right hand side function f cannot start before the parallel computation of the corresponding argument vector has been completed, a barrier operation must be executed before each stage. Since no further synchronization operations are used, the scalability is mainly determined by the efficiency of the barriers, the waiting times due to memory operations and the waiting times of the processors at the barriers.

In practice, the processors may not arrive at the barriers simultaneously for several reasons: (1) The function f can be irregular, i.e., a different number of instructions is required to evaluate the individual components. Examples are shown in Section 3. (2) A parallel computer can be heterogeneous, i.e., the processors work at different speeds. (3) The operating system scheduler may temporarily suspend threads of the ERK solver to execute other processes. (4) On systems with non-uniform memory access times (NUMA), the latency of memory operations can vary between one and several thousands of cycles. (5) On systems with simultaneous multithreading (SMT) support, the threads of the ERK solver compete for the functional units of the processors.

Considering these facts, asynchronous techniques that adaptively assign work to the participating processors might be able to improve the performance. Therefore, based on

```
1:  me := my_thread_id;

2:  barrier();
3:  for (j := first_component[me]; j ≤ last_component[me]; j++)
4:      v := hf_j(t + c_1 h, η);
5:      for (i := 2; i ≤ s; i++) w_i[j] := η[j] + a_{i1} v;
6:      η_{κ+1}[j] := b_1 v; η̂_{κ+1}[j] := b̂_1 v;

7:  for (l := 2; l ≤ s; l++)
8:      barrier();
9:      for (j := first_component[me]; j ≤ last_component[me]; j++)
10:         v := hf_j(t + c_l h, w_l);
11:         for (i := l + 1; i ≤ s; i++) w_i[j] += a_{il} v;
12:         η_{κ+1}[j] += b_l v; η̂_{κ+1}[j] += b̂_l v;
```

Fig. 1. Computational kernel of a data-parallel ERK implementation for shared address space using a static blockwise data distribution

our experience with load balancing of task-based irregular applications [8,9] we have implemented different strategies that realize a dynamic work distribution and apply them to several test problems. Our aim is to investigate if and under which conditions a performance improvement can be achieved on modern SMMs, and which performance bottlenecks still remain to be resolved.

3 Test Problems

We consider four test problems which exhibit different characteristics and are therefore suitable for the investigation of different aspects of our load balancing strategies. Figure 2 shows the number of instructions and the number of cycles required to evaluate the individual components of these problems on a Pentium 4 processor.

- **EMEP** [10] is the chemistry part of the EMEP-MSC-W ozone chemistry model. The dimension of this problem is 66. This problem exhibits the most irregular structure of all problems in our testset because the equations that model the concentrations of the individual species have a widely varying complexity.
- **MEDAKZO** [10]. The medical Akzo Nobel problem has been derived from two 1D partial differential equations (PDEs) which describe the penetration of antibodies into a tissue that has been infected by a tumor. The dimension of this system $n = 2N$ depends on the discretization parameter N. The number of instructions required to evaluate the ODE components differs between odd and even components. But if a blockwise data distribution is used, the load is nearly evenly balanced.
- **STARS** [11,12] describes a 3D n-body problem. The original second order ODE system has been transformed into a first order system by substitution of the first derivative. The system dimension is $6N$, where N is the number of stars. We consider two orderings of the components: STARS-CON uses a consecutive ordering of the first and second derivative and leads to a very uneven load balance. STARS-MIX interleaves the two derivatives and thus balances the load more evenly.

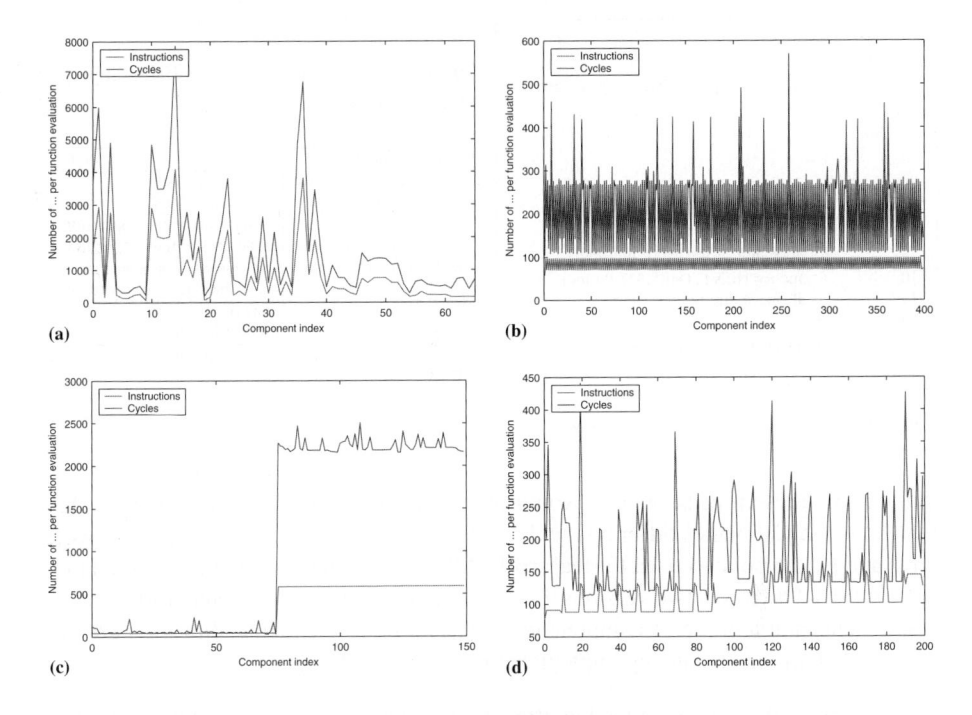

Fig. 2. Number of instructions and number of cycles required to evaluate the individual components of some test problems on a Pentium 4 processor. (a) EMEP, (b) MEDAKZO, $N = 200$, (c) STARS-CON, $N = 25$, (d) BRUSS2D-ROW, $N = 10$.

- **BRUSS2D** [13] results from a 2D PDE system that describes the chemical reaction of two substances. We consider two orderings: a row-oriented ordering, BRUSS2D-ROW, where the concentrations of the two substances are stored consecutively, and a mixed row-oriented ordering, BRUSS2D-MIX, where the components of the two substances are interleaved. The evaluation costs of the components of the two substances are slightly different. Elements at the boundary of the discretization grid require a special treatment. The balance of the decision tree used to identify boundary elements influences the regularity of the ODE system. The decision tree realized for BRUSS2D-MIX is more evenly balanced than that of BRUSS2D-ROW.

4 Load Balancing Strategies

The realization of a load balancing strategy for an ERK solver leads to the problem of scheduling the iterations of the irregular loops in lines 3 and 9 of Fig. 1 dynamically. The problem of loop scheduling has been considered previously by several authors, e.g., [14,15]. We have implemented three different load balancing strategies that pay regard to the special context of the loops within the ERK solver. All strategies start with the same blockwise work distribution as the static implementation. But after a processor has finished its own range of components, it 'steals' work from other processors and

```
 1:  me := my_thread_id;
 2:  for (l := 1; l ≤ s; l++)
 3:      next_work_unit_l[me] := first_work_unit[me];
 4:  for (l := 1; l ≤ s; l++)
 5:      barrier();
 6:      loop
 7:          work_unit := FETCH_AND_INC(
 8:              next_work_unit_l[me]);
 9:          if (work_unit > last_work_unit[me])
10:              me := NEXT_THREAD_ID(me);
11:              if (me = my_thread_id) break;
12:          else
13:              PROCESS_WORK_UNIT(l,
                    work_unit);
```

Fig. 3. Load balancing strategy 'Simple'

Table 1. Overview of the load balancing implementations

Name	Strategy	Granularity	Synchronization
SCIL	simple/increment	components	lock
SPIL	simple/increment	cache lines	lock
SCIA	simple/increment	components	atomic operations
SPIA	simple/increment	cache lines	atomic operations
SCRA	simple/random	components	atomic operations
SPRA	simple/random	cache lines	atomic operations
TC	task queue	components	lock
TP	task queue	cache lines	lock
IC	interval queue	components	lock
IP	interval queue	cache lines	lock

thus 'helps' these processors that would otherwise arrive late at the next barrier. The load balancing strategies differ by the data structures used to represent tasks, i.e., work units and by the policies used to steal work. All strategies have been implemented in C with POSIX Threads in two versions that support two different task granularities: single components and a group of components that fit into one cache line. Table 1 shows an overview of all load balancing implementations discussed in this article.

Strategy 'Simple'. As a simple but effective load balancing strategy with small overhead we have realized a strategy that was also used in the *volrend* application [16] included in the SPLASH-2 benchmark suite. Every thread provides a counter that points to the next work unit to be processed in the range initially assigned to the thread. During execution, each thread fetches and increments the counter corresponding to its thread ID. As long as the counter points to a work unit that lies within the range assigned to its thread ID, the thread processes the corresponding work unit and then fetches and increments the counter again. If the value of the counter leaves the range assigned to the thread ID, the thread changes its thread ID and continues with the corresponding counter. The pseudocode of this algorithm is displayed in Fig. 3.

We have implemented two strategies for changing the thread ID: 'Increment' computes the new ID as in [16] by $((\text{old\_id} + 1) \mod \#\text{threads})$. 'Random' reduces the probability that many threads work on the same counter simultaneously by changing the thread ID according to an initially generated random permutation.

Since the counters can be accessed by several threads simultaneously, we must ensure that the threads are not preempted when they read and increment the counters. We achieve this by either using locks or atomic Fetch & Inc. The lock based implementations use mutex variables of type `pthread_spinlock_t` if available, e.g., on Linux-based systems, or `pthread_mutex_t` on other systems. The implementations based on atomic Fetch & Inc currently support the following platforms:

- IA64: Fetch & Add is available as a machine instruction.
- x86: We emulate Fetch & Inc by a loop using Compare & Swap.
- AIX: The operating system kernel provides the function `fetch_and_add()`.

Strategy 'Task Queue'. This is a more sophisticated strategy than 'Simple' that reduces contention but requires a higher sequential overhead. It realizes one task queue per thread with FIFO (first in, first out) access order. The tasks are represented by integer values specifying the index of the associated work unit. Hence, the FIFO queues can be implemented by fixed size arrays of integers with head and tail pointers. Locks (pthread_mutex_t or pthread_spinlock_t if available) are used to avoid race conditions when the queues are accessed by several processors simultaneously.

The load balancing algorithm based on the 'Task Queue' strategy is shown in Fig. 4. At the beginning of each stage the queues are initialized according to the same block-wise work distribution as used in the static implementation. But the size of the queues is not revealed to the other threads before the barrier has been executed, so that no other thread that still works on the preceding stage will steal work from a queue during this initialization phase. After the initialization, the threads fetch tasks from the head of their local queues until their local queues get empty. When a thread finds no more tasks in its own queue, it tries to 'steal' work from another thread, i.e, it tries to move tasks from another thread's queue into its own queue. The stealing heuristics tries to steal half of the average queue size tasks from the tail of the queue with the largest size. The stolen tasks are appended at the tail of the target queue. But before the tasks are removed from the source queue, the size of the source queue is decreased by the number of tasks to be stolen, so that it appears less attractive to other threads searching for work. The new size of the target queue is hidden to the other threads, thus pretending it were still empty until all tasks have been transferred. Therefore, no thread will try to steal work from the target queue during the task transfer.

Strategy 'Interval queue'. Analyzing the behavior of the 'Task Queue' strategy we observe that the queues always store consecutive intervals of work units. A different data structure, called interval queue, can provide similar operations as the task queue but stores only the lowest and the highest index of the range of work units contained in the queue. It therefore leads to a significantly lower overhead. To fetch a work unit, the local thread only needs to increment the start index of the interval by 1; m tasks can be stolen at once in time $O(1)$ by decrementing the end index of the interval by m. Hence, the load balancing algorithm based on the interval queue can be realized similarly as in Fig. 4 by replacing lines 4–6 by

> REWIND(my_queue, first_work_unit[me], last_work_unit[me]);

and lines 21–28 by

> work_unit := STEAL(queue[index], my_queue, work_units_to_steal);
> UNLOCK(queue[index]);

where REWIND(Q, A, B) initializes the interval stored in queue Q to $[A, B]$ and STEAL(Q_S, Q_T, m) decreases the end of the interval of the source queue Q_S by m, initializes the target queue Q_T with the stolen interval of size m and returns the first element of this interval.

```
 1: me := my_thread_id;
 2: my_queue := queue[me];

 3: for (l := 1; l ≤ s; l++)
 4:     REWIND(my_queue);
 5:     for (j := first_work_unit[me]; j ≤ last_work_unit[me]; j++)
 6:         HIDDEN_APPEND(my_queue, j);
 7:     barrier();
 8:     REVEAL_SIZE(my_queue);
 9:     loop
10:         LOCK(my_queue);
11:         if (EMPTY(my_queue))
12:             UNLOCK(my_queue);
13:             loop
14:                 sum := ∑_{k=1}^{#threads} SIZE(queue[k]);
15:                 index := argmax_{1 ≤ k ≤ #threads} SIZE(queue[k]);
16:                 if (sum = 0) goto Stage_Complete;
17:                 if (TRYLOCK(queue[index]))
18:                     if (SIZE(queue[index]) > 0) break;
19:                     UNLOCK(queue[index]);
20:             work_units_to_steal := STEAL_HEURISTICS(SIZE(queue[index]), sum);
21:             PREPARE_STEALING(queue[index], work_units_to_steal);
22:             work_unit := STEAL(queue[index]);
23:             REWIND(my_queue);
24:             for (k := 1; k ≤ work_units_to_steal; k++)
25:                 HIDDEN_APPEND(my_queue, STEAL(queue[index]));
26:             FINISH_STEALING(queue[index], work_units_to_steal);
27:             UNLOCK(queue[index]);
28:             REVEAL_SIZE(my_queue);
29:         else
30:             work_unit := FETCH(my_queue);
31:             UNLOCK(my_queue);
32:         PROCESS_WORK_UNIT(l, work_unit);
33:     label Stage_Complete;
```

Fig. 4. Load balancing strategy 'Task Queue'

5 Runtime Experiments

Runtime experiments with the implemented load balancing strategies have been performed on three symmetric multiprocessors (SMPs): a 4-way 2.0 GHz Opteron 270 SMP, a 4-way 1.5 GHz Itanium 2 SMP, and an IBM p690 with 32 POWER4+ cores at 1.7 GHz. As a basis for the assessment of the implementations we use the average execution time per step measured by executing a limited number of time steps and dividing the resulting execution time by the number of steps executed. As a reference for the speedup calculation and the evaluation of the overhead we use a sequential implementation similar to Fig. 1, which contains no synchronization operations. All experiments presented in the following have been performed using the ERK method DOPRI5(4).

Sequential Overhead. First, we investigate the sequential overhead of the parallel implementations, that is the percentage of time they run slower than a sequential implementation when executed on one processor. As an example, Table 2 shows the sequential overhead measured for BRUSS2D-ROW with $N = 1000$. For this problem the highest overheads have been observed.

Table 2. Overhead of the parallel implementations for BRUSS2D-ROW with $N = 1000$ in %

Target system	static	IC	IP	SCIA	SCIL	SCRA	SPIA	SPIL	SPRA	TC	TP
Opteron 270 SMP	0.4	37.7	7.7	41.5	40.7	41.1	9.0	9.3	9.1	53.1	9.8
Itanium 2 SMP	4.7	29.6	6.2	10.9	29.8	10.9	5.0	6.1	4.9	35.0	6.8
IBM p690	5.1	274.4	37.4	131.6	266.8	124.7	10.8	21.5	6.4	299.2	42.2

Comparing the different target systems, the highest overheads have been measured on the IBM p690. On this system, pthread_spinlock_t is not available and we have to use pthread_mutex_t instead. Further, we used an AIX kernel function to realize atomic Fetch & Inc instead of inline assembler instructions as on the other two systems. Comparing the Itanium 2 and the Opteron SMP, we observe lower overheads on the Itanium 2 system. Hence, it appears that the spinlocks require less instructions on the Itanium 2 than on the Opteron. Also, the implementations that use atomic operations to realize Fetch & Inc run more efficiently on the Itanium 2 since on this machine only one machine instruction needs to be executed while on the Opteron more instructions are required to emulate Fetch & Inc by Compare & Swap. Only the overhead of our reference implementation based on a static work distribution is higher on the Itanium 2 than on the Opteron. This is due in part to cache effects caused by interferences between memory accesses to data and instructions.

In general, on all machines the overhead of the load balancing implementations is higher than that of the static implementation. The highest overhead was observed for the implementations that use single components as work units. The lowest overhead of the load balancing implementations is obtained by the implementations based on the 'Simple' strategy which use atomic operations to realize Fetch & Inc.

Scalability on the Itanium 2 SMP. In this section, we give a detailed discussion on the results measured on the Itanium 2 SMP. Because on this system the overhead of the load balancing implementations is lower than on the other two systems, we can observe relatively high improvements over the static work distribution. The speedup diagrams for the problems discussed in the following are shown in Fig. 5.

The most irregular test problem is EMEP. Due to its low dimension, only small speedups can be obtained. The best speedup of the static work distribution is 1.14 obtained on two processors. The implementations of 'Simple' that use cache lines as work units obtain a slightly better speedup of up to 1.17 on three processors. Using MEDAKZO with $N = 2400$ we obtain significantly better speedups. Except for TC and SCIL, all implementations obtain their maximum speedup on four processors. Using the static work distribution, a speedup of 2.74 is possible. But all load balancing implementations that use cache lines as work units obtain higher speedups. The best speedups between 3.02 and 3.06 are achieved by IP, SPRA, and SPIA. For STARS-CON and STARS-MIX we use $N = 1000$ stars. The load balancing implementations obtain nearly perfect speedups between 3.95 and 3.99 for STARS-CON and between 4.00 and 4.01 for STARS-MIX. With the static work distribution, only a speedup of 2.00 can be obtained for STARS-CON due to the severe load imbalance, but for STARS-MIX a speedup of 4.01 has been measured. For BRUSS2D with $N = 1000$ the speedups of the two orderings are similar. The static implementation obtains speedups of 3.32 and 3.30,

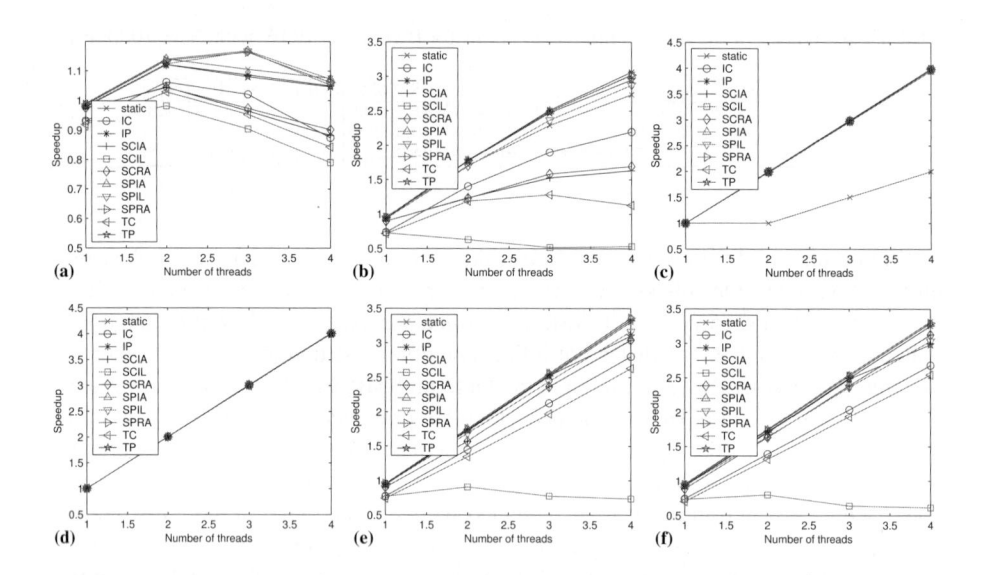

Fig. 5. Speedups measured on the Itanium 2 SMP. **(a)** EMEP, **(b)** MEDAKZO, $N = 2400$, **(c)** STARS-CON, $N = 1000$, **(d)** STARS-MIX, $N = 1000$, **(e)** BRUSS2D-ROW, $N = 1000$, **(f)** BRUSS2D-MIX, $N = 1000$.

respectively, for the two orderings on four processors. Since BRUSS2D-MIX is nearly evenly balanced, the best load balancing implementations, SPRA and SPIA, only obtain a slightly worse speedup of 3.29 for this ordering. But for BRUSS2D-ROW these two implementations obtain a better speedup than the static implementation of 3.36.

Scalability on Other Systems. The speedups measured on the Opteron 270 SMP and the IBM p690 are summarized in Table 3. On the Opteron system the load balancing implementations are similarly successful as on the Itanium 2 SMP. Thus, except for MEDAKZO, for every problem, at least one load balancing implementation obtains a higher speedup than the static work distribution. But on the IBM p690, due to a higher overhead, the load balancing implementations cannot obtain a higher performance than the static work distribution for most problems. Only for STARS-CON, which is characterized by a severe load imbalance, a significantly better speedup can be achieved.

Table 3. Summary of the speedups measured on the Opteron 270 SMP and on the IBM p690

		Opteron 270 SMP			IBM p690		
		Static	Load balancing		Static	Load balancing	
Problem	Parameter	Speedup	Speedup	Best implementation	Speedup	Speedup	Best implementation
EMEP		1.37	1.50	SPIA	1.35	1.35	TP
MEDAKZO	$N = 2400$	3.17	3.07	SPIA	3.23	2.88	SPRA
STARS-CON	$N = 1000$	1.97	3.98	IC, IP, TC, TP	15.65	30.13	SCRA
STARS-MIX	$N = 1000$	3.94	3.98	all except SCIL	30.60	30.24	SCIA
BRUSS2D-ROW	$N = 1000$	2.10	2.11	IC, IP	27.19	25.33	SPRA
BRUSS2D-MIX	$N = 1000$	2.26	2.28	IP	25.03	23.54	SPRA

6 Conclusions

Our results show that load balancing strategies can successfully be applied to data-parallel ERK solvers even though they require a larger sequential overhead than a static work distribution. They are particularly successful for ODE systems which lead to a severe load imbalance, but if special machine instructions are exploited to reduce the overhead, an improvement can be obtained even for some problems with a well-balanced right hand side function. However, our current load balancing implementations leave room for improvements, and a further investigation of load balancing strategies might lead to new insights.

References

1. Ehrig, R., Nowak, U., Deuflhard, P.: Massively parallel linearly-implicit extrapolation algorithms as a powerful tool in process simulation. In D'Hollander, E.H., et al., eds.: Parallel Computing: Fundamentals, Applications and New Directions. Elsevier (1998) 517–524
2. Burrage, K.: Parallel and Sequential Methods for Ordinary Differential Equations. Oxford Science Publications (1995)
3. van der Houwen, P.J., Sommeijer, B.P.: Parallel iteration of high-order Runge-Kutta methods with stepsize control. J. Comput. Appl. Math. **29** (1990) 111–127
4. Prince, P.J., Dormand, J.R.: High order embedded Runge-Kutta formulae. J. Comput. Appl. Math. **7**(1) (1981) 67–75
5. Korch, M., Rauber, T.: Scalable parallel RK solvers for ODEs derived by the method of lines. In: Euro-Par 2003. Parallel Processing. LNCS 2790, Springer (2003) 830–839
6. Korch, M., Rauber, T.: Optimizing locality and scalability of embedded Runge-Kutta solvers using block-based pipelining. J. Par. Distr. Comp. **6**(3) (2006) 444–468
7. Jackson, K.R., Nørsett, S.P.: The potential for parallelism in Runge-Kutta methods. Part 1: RK formulas in standard form. SIAM J. Numer. Anal. **32**(1) (1995) 49–82
8. Hoffmann, R., Korch, M., Rauber, T.: Performance evaluation of task pools based on hardware synchronization. In: SC '04: Proceedings of the 2004 ACM/IEEE conference on Supercomputing, Washington, DC, USA, IEEE Computer Society (2004) 44
9. Korch, M., Rauber, T.: A comparison of task pools for dynamic load balancing of irregular algorithms. Concurrency and Computation: Practice and Experience **16** (2004) 1–47
10. Lioen, W.M., de Swart, J.J.B.: Test Set for Initial Value Problem Solvers, Release 2.1. CWI, Amsterdam, The Netherlands. (1999)
11. Hussels, H.G.: Schrittweitensteuerung bei der Integration gewöhnlicher Differentialgleichungen mit Extrapolationsverfahren. Diploma thesis, University of Cologne, Cologne, Germany (1973)
12. Lecar, M.: Comparison of eleven numerical integrations of the same gravitational 25-body problem. Bulletin Astronomique **3** (1968) 91
13. Hairer, E., Nørsett, S.P., Wanner, G.: Solving Ordinary Differential Equations I: Nonstiff Problems. 2nd rev. edn. Springer, Berlin (2000)
14. Banicescu, I., Carino, R., Pabico, J., Balasubramaniam, M.: Design and implementation of a novel dynamic load balancing library for cluster computing. Parallel Computing **31**(7) (2005) 736–756
15. Tabirca, S., Tabirca, T., Yang, L.T., Freeman, L.: Evaluation of the feedback guided dynamic loop scheduling (FGDLS) algorithms. IEICE Trans. Inf. & Syst. **E87-D**(7) (2004) 1829–1833
16. Nieh, J., Levoy, M.: Volume rendering on scalable shared-memory MIMD architectures. In: Proceedings of the Boston Workshop on Volume Visualization, ACM Press (1992) 17–24

A Software Framework for the Portable Parallelization of Particle-Mesh Simulations

I.F. Sbalzarini[1], J.H. Walther[1,2], B. Polasek[1], P. Chatelain[1], M. Bergdorf[1],
S.E. Hieber[1], E.M. Kotsalis[1], and P. Koumoutsakos[1]

[1] Institute of Computational Science, ETH Zürich, CH-8092, Switzerland
[2] DTU, Mechanical Engineering, DK-2800 Lyngby, Denmark

Abstract. We present a software framework for the transparent and portable parallelization of simulations using particle-mesh methods. Particles are used to transport physical properties and a mesh is required in order to reinitialize the distorted particle locations, ensuring the convergence of the method. Field quantities are computed on the particles using fast multipole methods or by discretizing and solving the governing equations on the mesh. This combination of meshes and particles presents a challenging set of parallelization issues. The present library addresses these issues for a wide range of applications, and it enables orders of magnitude increase in the number of computational elements employed in particle methods. We demonstrate the performance and scalability of the library on several problems, including the first-ever billion particle simulation of diffusion in real biological cell geometries.

1 Introduction

A large number of problems in physics and engineering can be described by particles. Examples include Molecular Dynamics (MD) simulations of nano-devices, Smooth Particle Hydrodynamics (SPH) simulations of astrophysics, and Vortex Methods (VM) for fluid dynamics. The dynamics of particle methods are governed by the interactions of the N computational elements, resulting in an N-body problem with a nominal computational cost of $\mathcal{O}(N^2)$. For short-ranged particle interactions, as in simulations of diffusion [1], the computational cost scales linearly with the number of particles. In the case of long-range interaction potentials such as the Coulomb potential in MD or the Biot-Savart law in VM, Fast Multipole Methods (FMM) [2] reduce the computational cost to $\mathcal{O}(N)$. Alternatively, these long-range interactions can be described by the Poisson equation which, when solved on meshes, results in the hybrid Particle-Mesh (PM) algorithms as pioneered by Harlow [3,4]. The computational cost of hybrid methods scales as $\mathcal{O}(M)$, where M denotes the number of mesh points used for resolving the field equations. Due to the regularity of the mesh, PM methods are one to two orders of magnitude faster than FMM [5].

The parallelization of PM and FMM techniques is complicated by several factors:

- the simultaneous presence and spatial distribution of particles and meshes prohibits a single optimal way of parallelization,
- complex-shaped computational domains and strong particle inhomogeneities require spatially adaptive domain decompositions,
- particle motion may invalidate the existing domain decomposition, causing rising load imbalance and requiring additional communication in multi-stage ODE integration schemes,
- exploiting the symmetry of the particle interactions requires sending back of ghost contributions to the corresponding real particle,
- inter-particle relations constrain decompositions and data assignment,
- the global nature of the tree data structure hampers the parallel implementation of FMM methods.

Many of the available domain decomposition, load balancing, solver, interpolation, and data communication methods are applicable to a wide range of particle or PM algorithms, regardless of the specific physics that are being simulated [6]. In this paper we present the newly developed Parallel Particle Mesh library PPM [5] and its extension to FMM. The PPM library provides a generic, physics-independent infrastructure for simulating discrete and continuum systems, and it bridges the gap between general libraries and application-specific simulation codes.

The design goals of PPM include ease of use, flexibility, state-of-the-art parallel scaling, good vectorization, and platform portability. The library is portable through the use of standard languages (Fortran 90 and C) and libraries (MPI) and it was successfully compiled and used on distributed memory, shared memory, and vector architectures on 1 to 512 processors. Computational efficiency is achieved by dynamic load balancing, dynamic particle re-distribution, explicit message passing, and the use of simple data structures.

2 Particle Concepts and Particle-Mesh Techniques

The use of the PPM library requires that the simulated systems are formulated in the framework of PM algorithms [6]. The field equations are solved using structured or uniform Cartesian meshes. As a result, the physical and computational domains are rectangular or cuboidal in two or three dimensions. Complex geometries are handled by immersed boundaries, through the use of source terms in the corresponding field equations, or through boundary element techniques. Adaptive multi-resolution capabilities are possible using mapping concepts as adapted to particle methods [7].

The simultaneous presence of particles and meshes requires different concurrent domain decompositions. These decompositions divide the computational domain into a minimum number of *sub-domains* with sufficient granularity to provide adequate load balancing. The concurrent presence of different decompositions allows to perform each step of the computational algorithm in its optimal environment with respect to load balance and the computation-to-communication ratio. For the actual computations, the individual sub-domains are treated as independent problems and extended with *ghost mesh layers* and *ghost particles* to allow for communication between them. *Ghosts* are copies of true mesh points or particles that either reside on a neighboring processor or account for periodic boundary conditions.

The PPM library supports *connections* and *relations* between particles, such as particle pairs, triplets, quadruplets, etc. These relations may describe a physical interaction, such as chemical bonds in molecular systems, or a spatial coherence, such as a triangulation of an immersed boundary or an unstructured mesh.

2.1 Topologies

A *topology* is defined by the decomposition of space into sub-domains with the corresponding boundary conditions, and the assignment of these sub-domains onto processors. Multiple topologies may coexist and library routines are provided to *map* particle and field data between them as described below. Fields are defined on *meshes*, which in turn are associated with topologies. Every topology can hold several meshes.

In order to achieve good load balance, the *SAR heuristic* [8] is used in the PPM library to decide when problem re-decomposition is advised, i.e. when the cost of topology re-definition is amortized by the gain in load balance. Moreover, all topology definition routines can account for the true computational cost of each particle, for example defined by the actual number of its interactions, and the effective speeds of all processors.

The PPM library provides a number of different adaptive domain decomposition techniques for particles, meshes, and volumes, the latter defining geometric sub-domains with neither meshes nor particles present [5]. The sub-domains can be assigned to the processors in various ways. The PPM-internal method assigns contiguous blocks of sub-domains to processors until the accumulated cost for a processor is greater than the theoretical average cost under uniform load distribution. The average is weighted with the relative processor speeds to account for heterogeneous machine architectures. In addition, four different Metis-based [9] and a user-defined assignment are available.

At the external boundaries of the computational domain, Neumann, Dirichlet, free space, symmetric, and periodic boundary conditions are generally supported, but depend on the particular mesh-based solver that is being employed. More involved boundary conditions and complex boundary shapes are represented inside the computational domain by defining connections among the particles, or by using immersed interfaces.

2.2 Data Mapping

PPM topologies implicitly define a data-to-processor assignment. Mapping routines provide the functionality of sending particles and field blocks to the proper processor, i.e. to the one that "owns" the corresponding sub-domain(s) of the computational space. Three different mapping types are provided for both particles and field data:

- a *global mapping*, involving an all-to-all communication,
- a *local mapping* for neighborhood communication, and
- *ghost mappings* to update the ghost layers.

The global mapping is used to perform the initial data-to-processor assignment or to switch from one topology to another, whereas the local mapping is mainly used to account for particle motion during a simulation. Communication is scheduled by solving the minimal edge coloring problem using the efficient approximation algorithm by Vizing [10], and connections between particles are appropriately accounted for. Ghost mappings are provided to receive ghost particles or ghost mesh points, and to send ghost contributions back to the corresponding real element, for example after a symmetric particle-particle interaction or a particle-to-mesh interpolation.

All mapping types are organized as stacks. A mapping operation consists of four steps: (1) defining the mapping, (2) pushing data onto the send stack, (3) performing the actual send and receive operations, and (4) popping the data from the receive stack. This architecture allows data stored in different arrays to be sent together to minimize network latency, and mapping definitions to be reused by repeatedly calling the push/send/pop sequence for the same, persisting mapping definition.

3 Particle-Particle Interactions

The evaluation of Particle-Particle (PP) interactions is a key component of PM algorithms. The overall dynamics of the system may be governed by local particle interactions, sub-grid scale phenomena may require local particle-based corrections [11], or differential operators can be evaluated on irregular locations [12]. The PPM library implements symmetric and non-symmetric PP computations using a novel type of cell lists [5], Verlet lists [13], and the full $\mathcal{O}(N^2)$ direct method. The last is based on distributing the particles among processors in equal portions, irrespective of their location in space.

4 Particle-Mesh Interpolation

All hybrid PM methods involve interpolation of irregularly distributed particle quantities from particle locations onto a regular mesh, and interpolation of field quantities from the grid points onto particle locations. These interpolations are utilized for two purposes, namely: (1) the communication of the particle solver with the field solver, and (2) the reinitialization of distorted particle locations ("remeshing"). The PPM library provides routines that perform these operations. The interpolation of mesh values onto particles readily vectorizes: interpolation is performed by looping over the particles and receiving values from mesh points within the support of the interpolation kernel. Therefore, the values of individual particles can be computed independently. The interpolation of particle values onto the mesh, however, leads to data dependencies as the interpolation is still performed by looping over particles, but a mesh point may receive values from more than one particle. To circumvent this problem, the PPM library implements the following technique [14]: when new particles are created in the course of remeshing, we assign colors to the particles such that no two particles within the support of the interpolation kernel have the same color. Particle-to-mesh interpolation then visits the particles ordered by color to achieve data independence. In combination with appropriate directives, this coloring scheme enables the compiler to safely vectorize the loops, as confirmed by a test on a NEC SX-5 vector computer. The wall-clock time for interpolating 2 million particles onto a 128^3 mesh using the M_4' interpolation kernel [15] decreases from 30 s to 2.7 s when using the present coloring scheme, and the vector operation ratio increases from 0.4% to 99%.

5 Mesh-Based Solvers

In PPM, meshes can be used to solve the field equations associated with long-range particle interactions [4], or to discretize the differential operators in the governing equations of the simulated physical system. A large class of pair interaction potentials in particle methods can be described by the Poisson equation as it appears in MD of charged particles via electrostatics (Coulomb potential), fluid mechanics in stream-function/vorticity formulation (Biot-Savart potential), and cosmology (gravitational potential). The PPM library provides fast parallel Poisson solvers based on FFTs and geometric Multi-Grid (MG) in both two and three dimensions. The library architecture is however not limited to Poisson solvers.

6 ODE Solvers

Simulations using particle methods entail the solution of systems of ODEs [6]. The PPM library provides a set of explicit integration schemes to solve these ODEs. Parallelism is achieved by mapping the integrator stages of multi-stage schemes – i.e., the intermediate function evaluations – along with the other particle quantities. The set of available integrators currently includes forward Euler with and without super time stepping [16], 2-stage and 4-stage standard Runge-Kutta schemes, Williamson's low-storage third order Runge-Kutta scheme [17], and 2-stage and 3-stage TVD Runge-Kutta schemes [18].

7 Parallel File I/O

File I/O in distributed parallel environments exist in *distributed* and *centralized* modes. By distributed we denote the situation where each processor writes its part of the data to its local file system. Centralized I/O on the other hand produces a single file on one of the nodes, where the data contributions from all processors are consolidated. The PPM library provides a parallel I/O module which supports both binary and ASCII read and write operations in both modes.

Write operations in the centralized mode can concatenate or reduce (sum, replace) the data from individual processors; read operations can transparently split the data in equal chunks among processors, or send an identical copy to each one. To improve performance of the centralized mode, network communication and file I/O are overlapped in time using non-blocking message passing.

8 Parallel FMM

The PPM library uses FMM to evaluate Dirichlet and free-space boundary conditions for the computationally more efficient MG solver. The implementation of the parallel FMM module is based on the topology, tree, and mapping routines provided by the library core. Hereby, the FMM module creates multiple temporary topologies. The first topology comprises the sub-domains defined at the level of the tree that holds at least as many sub-domains as there are processors. Subsequent levels of the FMM tree structure are also declared as PPM topologies. By means of the user-defined assignment scheme, individual processors operate on disjoint sub-trees to minimize the amount of communication. The particles are then mapped according to the finest topology, which contains all the leaf boxes as sub-domains. The PPM tree directly provides index lists to the particles in each box. This allows straightforward computation of the expansion coefficients on the finest level, without requiring communication. The computed leaf coefficients are shifted to parent boxes by recursively traversing the tree toward its root, cf., e.g., [19]. Since the topologies are defined such that each processor holds a disjoint subtree, the expansions can be shifted without communication.

To evaluate the potential at the locations of a set of target particles, these particles are first mapped onto the finest-level PPM topology. A pre-traversal of the tree then decides which expansion coefficients and source particles are needed for each target point. This is done by traversing the tree from the root down to the leafs using a stack data structure. On each level, we check if the corresponding box is already far enough away from the target particle. This is done by comparing the distance (D) between the target particle and the center of the box to the diameter (d) of the box. If the ratio $D/d > \theta$, the expansion of that box is used and the traversal stops.

When evaluating the potential, expansion coefficients or particles from other processors may be needed. Before evaluating the potentials, the expansion coefficients from all processors are thus globally communicated. This can be done since the data volume of the coefficients is much smaller than the original particle data. Required particles are received on demand as additional ghosts using the regular PPM ghost mapping routines.

9 Benchmarks Results

Benchmark results are presented for the PPM FMM, for a remeshed SPH (rSPH) client application [20,5] for the simulation of compressible fluid dynamics, and for simulations of diffusion in the Endoplasmic Reticulum (ER) of live cells [21,22].

In addition, preliminary results with a simple PPM MD client have shown it to reach the same performance as the dedicated MD program FASTTUBE [23].

The benchmarks for the FMM are collected on a Linux cluster of 16 2.2 GHz AMD Opteron 248 processors, connected by standard gigabit Ethernet. This architecture is chosen since it provides the most stringent test for the communication-intense tree data structure of the FMM. To have access to larger numbers of processors, the subsequent application benchmarks are preformed on the IBM p690 computer of the Swiss National Supercomputing Centre (CSCS). The machine consists of 32 Regatta nodes with 8 1.3 GHz Power4 processors per node. The nodes are connected by a 3-way Colony switch system. Furthermore, code vectorization is assessed on the NEC SX-5 computer of CSCS.

9.1 The Fast Multipole Method

The test cases for the PPM FMM involve 10^5 source points with a uniformly random distribution in a cubic box. The potential induced by these points is computed at the locations of 10^5 target points, also uniformly randomly distributed in the same cube. Fig. 1(a) shows the wall-clock time as a function of the number of particles. The acceptance factor θ for the tree traversal is set to 1.5, and we vary the expansion order (l). The scaling of the FMM is compared to the $\mathcal{O}(N^2)$ scaling of the direct evaluation method. Fig. 1(b) shows the parallel speedup of the PPM FMM on up to 16 processors of the Linux cluster. The wall-clock time is 452 seconds on 1 processor and 36.8 seconds on 16. The observed loss in efficiency is mainly caused by the global communication of the expansion coefficients.

9.2 A Remeshed Smooth Particle Hydrodynamics Application

The first application benchmark considers an rSPH client for the simulation of three-dimensional compressible flows. For the parallel performance to be independent of the particular flow problem, we consider a computational domain

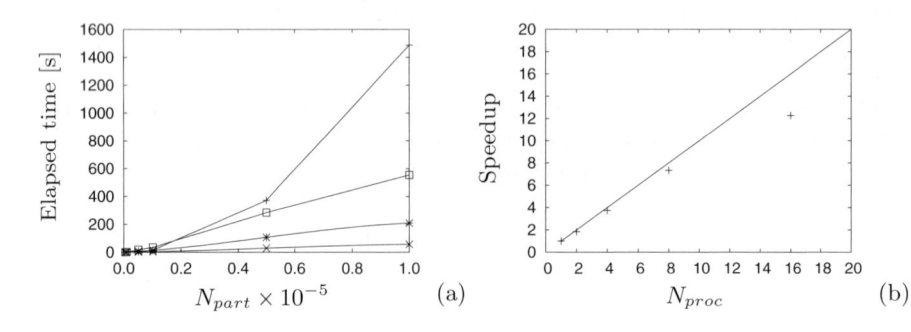

Fig. 1. Performance of the Fast Multipole Method (FMM) implementation in the PPM library. (a) Serial performance of the FMM as function of number of particles and the order (l) of the expansion: -+-: direct calculation; -□-: $l = 9$; -*-: $l = 5$; -×-: $l = 1$. (b) Parallel speedup using 10^5 particles and $l = 5$ on up to 16 nodes of the 2.2 GHz Opteron Linux cluster. —: linear scaling; +: measurement.

fully populated with particles. The speedup and parallel efficiency of the rSPH client are shown in Fig. 2. The maximum number of particles considered for this test case is 268 million with a parallel efficiency of 91% on 128 processors and also 91% on 32 processors. This compares well with the 85% efficiency of the GADGET SPH code by Springel et al. [24] on 32 processors of the same computer model (IBM p690). One time step of a fixed-size simulation using 16.8 million particles takes 196.9 seconds on 1 processor and 7.3 seconds on 128 processors.

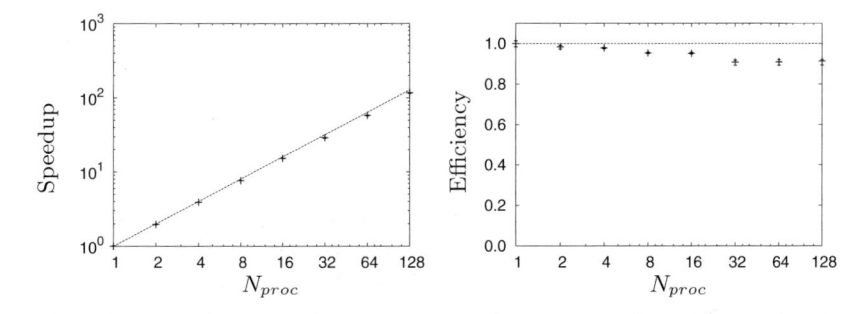

Fig. 2. Parallel speedup and efficiency of the PPM rSPH client for a scaled-size problem starting with 2 million particles on one processor. Each point is averaged from 5 samples, error bars indicate min-max span. Timings are performed on the IBM p690.

9.3 Diffusion in the ER

We present a client application for the simulation of three-dimensional diffusion in the ER, an organelle of live cells. This test demonstrates the capability of the

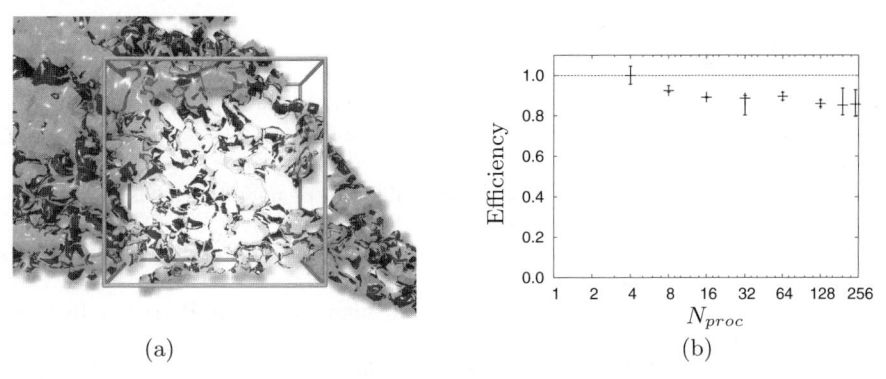

Fig. 3. (a) Visualization of the simulated concentration distribution inside a real ER as reconstructed from a live cell. The volume indicated by the cube is initially empty and we simulate the diffusive influx. The solution at time 0.25 is shown. The edge length of the box is 50, and the computational diffusion constant is 7.5. (b) Parallel efficiency of the PPM diffusion client for a fixed-size problem with 3.4 million particles distributed inside the ER. Each point is averaged from 5 samples, error bars indicate min-max span. Timings are performed on the IBM p690.

library in handling complex-shaped domains with irregular load distribution. The problem size is fixed at 3.4 million particles, distributed inside the ER. Using an adaptive recursive orthogonal bisection, the complex ER geometry is decomposed, and the sub-domains are distributed among 4 to 242 processors. Fig. 3(a) shows a visualization of the solution, and Fig. 3(b) shows the parallel efficiency of the present PPM client. The simulations sustain 20% of the peak performance of the IBM p690, reaching 250 GFlop/s on 242 processors at 84% efficiency. The load balance is 90 to 95% in all cases, and one time step takes 14 seconds on 4 processors. This client was used to perform simulations of diffusion using up to 1 billion particles on 64 processors, thus demonstrating the good scaling in memory of the PPM library.

10 Summary

The lack of efficiently parallelized and user friendly software libraries has hindered the wide-spread use of particle methods. We have initiated the development of a generic software framework for hybrid Particle-Mesh simulations. The PPM library described in this paper provides a complete infrastructure for parallel particle and hybrid Particle-Mesh simulations for both discrete and continuum systems. It includes adaptive domain decompositions, load balancing, optimized communication scheduling, parallel file I/O, interpolation, data communication, and a set of commonly used numerical solvers, including a parallel FMM.

We have demonstrated the library's parallel efficiency and versatility on a number of different problems on up to 242 processors. All applications showed parallel efficiencies reaching or exceeding the present state of the art, and favorable run-times on large systems.

References

1. Degond, P., Mas-Gallic, S.: The weighted particle method for convection-diffusion equations. Part 1: The case of an isotropic viscosity. Math. Comput. **53**(188) (1989) 485–507
2. Greengard, L., Rokhlin, V.: The rapid evaluation of potential fields in three dimensions. Lect. Notes Math. **1360** (1988) 121–141
3. Harlow, F.H.: Particle-in-cell computing method for fluid dynamics. Methods Comput. Phys. **3** (1964) 319–343
4. Hockney, R.W., Eastwood, J.W.: Computer Simulation using Particles. Institute of Physics Publishing (1988)
5. Sbalzarini, I.F., Walther, J.H., Bergdorf, M., Hieber, S.E., Kotsalis, E.M., Koumoutsakos, P.: PPM – a highly efficient parallel particle-mesh library for the simulation of continuum systems. J. Comput. Phys. **215**(2) (2006) 566–588
6. Koumoutsakos, P.: Multiscale flow simulations using particles. Annu. Rev. Fluid Mech. **37** (2005) 457–487
7. Bergdorf, M., Cottet, G.H., Koumoutsakos, P.: Multilevel adaptive particle methods for convection-diffusion equations. Multiscale Model. Simul. **4**(1) (2005) 328–357

8. Moon, B., Saltz, J.: Adaptive runtime support for direct simulation Monte Carlo methods on distributed memory architectures. In: Proceedings of the IEEE Scalable High-Performance Computing Conference, IEEE (1994) 176–183
9. Karypis, G., Kumar, V.: A fast and high quality multilevel scheme for partitioning irregular graphs. SIAM J. Sci. Comput. **20**(1) (1998) 359–392
10. Vizing, V.G.: On an estimate of the chromatic class of a p-graph. Diskret. Anal. **3** (1964) 25–30 in Russian.
11. Walther, J.H.: An influence matrix particle-particle particle-mesh algorithm with exact particle-particle correction. J. Comput. Phys. **184** (2003) 670–678
12. Eldredge, J.D., Leonard, A., Colonius, T.: A general deterministic treatment of derivatives in particle methods. J. Comput. Phys. **180** (2002) 686–709
13. Verlet, L.: Computer experiments on classical fluids. I. Thermodynamical properties of Lennard-Jones molecules. Phys. Rev. **159**(1) (1967) 98–103
14. Walther, J.H., Koumoutsakos, P.: Three-dimensional particle methods for particle laden flows with two-way coupling. J. Comput. Phys. **167** (2001) 39–71
15. Monaghan, J.J.: Extrapolating B splines for interpolation. J. Comput. Phys. **60** (1985) 253–262
16. Alexiades, V., Amiez, G., Gremaud, P.A.: Super-time-stepping acceleration of explicit schemes for parabolic problems. Comm. Numer. Meth. Eng. **12**(1) (1996) 31–42
17. Williamson, J.H.: Low-storage Runge-Kutta schemes. J. Comput. Phys. **35** (1980) 48–56
18. Shu, C.W., Osher, S.: Efficient implementation of essentially nonoscillatory shock-capturing schemes. J. Comput. Phys. **77**(2) (1988) 439–471
19. Cheng, H., Greengard, L., Rokhlin, V.: A fast adaptive multipole algorithm in three dimensions. J. Comput. Phys. **155** (1999) 468–498
20. Chaniotis, A.K., Poulikakos, D., Koumoutsakos, P.: Remeshed smoothed particle hydrodynamics for the simulation of viscous and heat conducting flows. J. Comput. Phys. **182**(1) (2002) 67–90
21. Sbalzarini, I.F., Mezzacasa, A., Helenius, A., Koumoutsakos, P.: Effects of organelle shape on fluorescence recovery after photobleaching. Biophys. J. **89**(3) (2005) 1482–1492
22. Sbalzarini, I.F., Hayer, A., Helenius, A., Koumoutsakos, P.: Simulations of (an)isotropic diffusion on curved biological surfaces. Biophys. J. **90**(3) (2006) 878–885
23. Werder, T., Walther, J.H., Jaffe, R.L., Halicioglu, T., P., K.: On the water–carbon interaction for use in molecular dynamics simulations of graphite and carbon nanotubes. J. Phys. Chem. B **107** (2003) 1345–1352
24. Springel, V., Yoshida, N., White, S.D.M.: GADGET: a code for collisionless and gasdynamical cosmological simulations. New Astron. **6** (2001) 79–117

Parallelization of a Discrete Radiosity Method

Rita Zrour, Pierre Chatelier, Fabien Feschet, and Rémy Malgouyres

LLAIC, IUT, 63172 Aubière Cedex
{zrour, chatelier, feschet, remy.malgouyres}@llaic3.u-clermont1.fr

Abstract. In this paper we present three different parallelizations of a discrete radiosity method achieved on a cluster of workstations. This radiosity method has lower complexity when compared with most of the radiosity algorithms and is based on the discretization of surfaces into voxels and not into patches. The first two parallelizations distribute the tasks. They present good performance in time but they did not distribute the data. The third parallelization distributes voxels and required the transmission of small part of the voxels between machines. It improved time while distributing data.

1 Introduction

The radiosity method has been widely used in different fields concerned with the exchange of energy. It is mainly used in computer graphics (3D rendering) and in image synthesis to simulate global illumination and lighting effects. Most radiosity methods require large memory and time computation. This leads researchers to parallelize radiosity algorithms trying to minimize both time and memory.

Many parallelizations of different radiosity algorithms have been proposed [10,14,3,2,13,9,4,11,7,12,8,16]. These algorithms are characterized by their memory system and their parallelization procedure that contribute together to the final parallelization. The memory system can be a shared memory system [8] or a distributed memory system. The distributed memory system can be divided into two branches: the cluster of workstations and the distributed shared memory. The cluster of workstations is composed of many machines where the exchanges between the machines is done via the message passing strategy [2,10,14]. The distributed shared memory system uses shared variables for communication [3,12,13,16]. As for the parallelization procedure, radiosity algorithms are not easy to parallelize because the radiosity equation contains form factor calculation or visibility information that put restrictions and dependencies when dividing the data among a distributed memory system. Many parallel methodologies tried to find some local characterizations that can minimize these interdependencies. In [3] the environment is split into sub-environments and a visibility mask is used for the transfer of light. In [4] an idea of group iterative approach was used to split the radiosity solving between a master and slaves.

The purpose of this paper is to parallelize the discrete radiosity method proposed by Chatelier and Malgouyres [1]. This method is different from most radiosity methods because it is based on discretization of surfaces into voxels and not into patches. In terms of complexity the method is quasi-linear in time and space and has a lower complexity than other methods for large scenes containing a lot of details. The parallelizations presented in this article have been tested on a distributed memory system, a cluster of workstations composed of bi-processors hyperthreaded Xeon Pentium processors.

The paper is organized as follows. Section 2 explains the sequential algorithm and its complexity. Section 3 presents the distribution of tasks which includes two parallelizations, the local parallelization and the distribution of the computation. Section 4 details the data distribution. Finally Section 5 states some conclusions and perspectives.

2 Sequential Algorithm

2.1 Discretization of the Radiosity Equation

Radiosity is defined as the total power of light leaving a point. The continuous radiosity equation has been discretized in [6]. The voxel-based radiosity equation is the following:

$$B(x) = E(x) + \rho_d(x) \sum_{\vec{\sigma} \in D} B(V(x, \vec{\sigma})) \frac{\cos \theta(x, V(x, \vec{\sigma}))}{\pi} \hat{A}(\vec{\sigma}) \quad (1)$$

This equation shows that the total power of light $B(x)$ leaving a voxel x depends on two terms, the first is the proper emittance of this voxel as a light source (the $E(x)$ term), and the second is some re-emission of the light it receives from its environment (the sum). The term $B(\cdot)$ that is present in both sides of the equality, reflects interdependence between a point and its environment ; it does not consider any outgoing direction, so it supposes that each point emits light uniformly in every direction (diffuse hypothesis). The factor $\rho_d(x)$ indicates that a point re-emits only a fraction of the light that it receives. D is a set of discrete directions in space. $V(x, \vec{\sigma})$ is a visibility function, returning the first point y seen from x in the direction of $\vec{\sigma}$. The term $\hat{A}(\vec{\sigma})$ is the fraction of a solid angle associated to the direction $\vec{\sigma}$ it quantifies how much of an object is seen from a point, we call it a *direction factor*. The $\cos \theta(x, V(x, \vec{\sigma}))$ expresses that incident light is more effective when it comes perpendicularly to the surface. Finally the π factor is a normalization term deriving from radiance considerations.

This equation is usually solved using Gauss-Seidel relaxation requiring thus many iterations to obtain a convergence to the correct radiosity solution [15].

2.2 Computing Visibility Using Discrete Lines

The space can be partitioned into parallel 3D discrete lines, along a given direction. It means that a voxel belongs to one and only one of these lines which can be easily computed. In [1] Chatelier and Malgouyres relied on this space partition concept, that can allow with some constraints, to reflect the visibility between the voxels. So, given a voxel $(x, y, z) \in Z^3$, this voxel belongs to the discrete line $L_{i,j}$ where (i,j) are two integers calculated from both the coordinates of the voxel and the directing vector of the direction. The set of all $L_{i,j}$'s represents the space partition into 3D discrete lines.

An important information in the radiosity equation is the visibility factor, reflecting the visibility between the voxels. Given a direction, a voxel belongs to one and only one line characterized by a couple of integers (i,j). Putting each voxel in its line (i,j) does not ensure voxels visibility ; to ensure the correct visibility between the voxels, a possible solution consists in doing a precomputation step that sorts the voxels according to different lexicographic orders ; usually eight lexicographic orders are needed but four are sufficient because of symmetries.

2.3 Algorithm and Complexity

The sequential algorithm of the radiosity comprises the following steps:

1. Discretize the scene into voxels.
2. Sort the voxels according to the four lexicographic orders.
3. Choose the set of directions.

For each iteration, traverse all the directions and for each direction the following steps are done:

1. Choose from the four pre-computed lexicographic orders the one that is suitable depending on the directing vector of the direction.
2. Traverse the voxels and put each one in the list (i,j) to which it belongs.
3. Propagate light between successive voxels of the same list.

The sequential algorithm requires 2 parameters, the number of iterations "I" and the number of directions "D". If N is the number of voxels of the scene, the time complexity is: $4 \times O(NlogN) + I \times D \times (O(N) + O(N))$. The term $4 \times O(NlogN)$ represents the four lexicographic sorting of the voxels. It is negligible, because the four sorting are pre-computed once. The term $I \times D \times (O(N)+O(N))$ reflects the dispatch of the voxels in their lists and the propagation of the light in the lists. Note that "I" is a small constant. "D" is a constant that is independent of the number of voxels ; it depends on the intensity of the light sources. It is increased to avoid aliasing effects when the sources are small but having high intensities. As for the space complexity, we just need to store the voxels in memory. More details about the complexity can be found in [1].

3 Distribution of the Tasks

3.1 Local Parallelization

The first parallelization is local and uses threads. Our radiosity method contains two major computations that are expensive for large scenes and that can be locally parallelized using threads. These computations are the dispatching of the voxels and the propagation of the light in the lists.

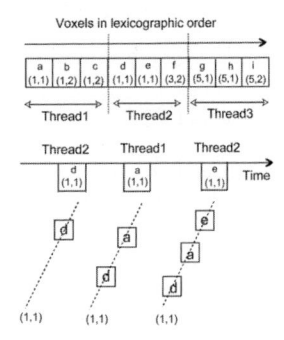

Fig. 1. Distributing the voxels in their lists using threads. There is a risk of loosing the correct lexicographic order.

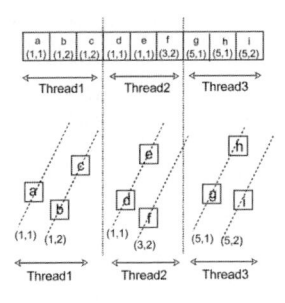

Fig. 2. Distributing the voxels in their lists using threads. Distinct data structures are created for each thread.

Multithreaded Voxels Dispatch. The distribution of the voxels in their lists consists in traversing the voxels in the lexicographic order and putting each voxel in the list (i, j) it belongs to. The partition of this work among the threads consists in giving each thread an equal consecutive part of the voxels ordered in the lexicographic order (see Fig. 1). The threads then start the distribution of the voxels in their lists. If one data structure is created, a problem of priority will appear when the threads begin to work on the same lists. The problem arises from the fact that the distribution should respect the lexicographic order, thus giving priority to one thread on the others. For example for Fig. 1 thread 1 has priority on thread 2 and thread 3. The voxels "a", "d" and "e" belonging to the same list but to different threads may not be in the correct lexicographic order if thread 2 puts its voxel "d" before thread 1 has put its voxels "a". A sorting of the lists in the lexicographic order after the threads finish their work is possible however this sorting has a complexity $O(N log N)$, N being the number of voxels. This sorting can be avoided by creating distinct data structures for each thread in which it will put its voxels in their lists (Fig. 2).

Multithreaded Light Propagate. The propagation of light in the lists consists in propagating light between consecutive voxels of the same list. The lists

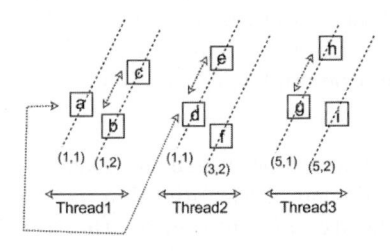

Fig. 3. Propagating light in the lists using threads

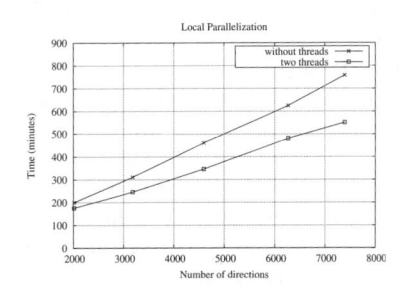

Fig. 4. Local parallelization

The effects of threads on Time

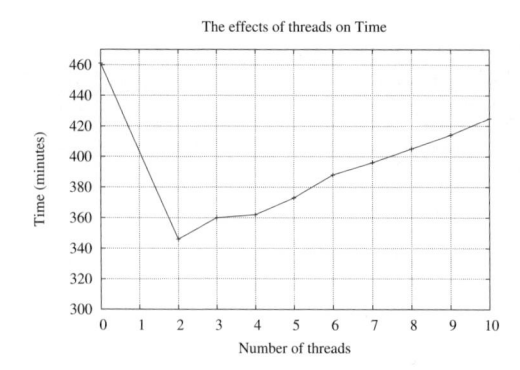

Fig. 5. The effects of increasing the number of threads on time (lower time obtained using two threads)

created by the threads in Sect. 3.1 are not complete lists since each thread creates its own lists. Gathering the partial lists to propagate light in them is an expensive step in time and it should be avoided. A complete list is the concatenation of partial lists, so each thread can propagate light inside its partial lists of voxels. Then an additional propagate is performed to propagate light between the partial lists (see Fig. 3).

Results. The local parallelization tests are done on a living room scene composed of 17×10^5 voxels. The results with 6 iterations and different number of directions without threads (sequential algorithm) and using two threads are shown on Fig. 4. The time improvement varies between 12 and 30 percent. It is important to note that tests are done with 2 threads because it was noticed that when the number of threads increases time improvement decreases. Fig. 5 tests the effects of increasing the number of threads for the living room scene using ≈ 5000 directions and 6 iterations, the optimal time is obtained with 2 threads. This is normal since the work is tested on bi-processors machines.

3.2 Distribution of the Computation

The discrete radiosity method is linear with respect to the number of directions, so it is possible to divide the summation done on all the directions in Equation (1), into several summations. Each machine of the cluster of workstations will take an equal set of directions. This parallelization requires the presence of all the voxels on the machines, so voxels should be duplicated on the machines. It offers two major advantages:

- It minimizes the communication between the machines via the message passing interface ; the exchanges are done at the end of each iteration when all the machines have accomplished the radiosity computation for the set of directions given to them. These exchanges consist in adding the radiosity value gained by each voxel duplicated on the machines.
- It offers the ability of using threads to increase the parallelization.

Figure 6 shows the results of this parallelization for the living room scene with 6 iterations and 2000 directions. The time has been almost divided by the number of machines when no threads are used and it decreases even more with the use of threads. The efficiency of this parallelization can reach 98 percent. This parallelization showed good execution times but its main limitation is that it cannot be used for large scenes that do not fit in the RAM.

4 Data Distribution

The local parallelization as well as the distribution of the computation improve the time, however they do not distribute the voxels. For complex scenes that do not fit into the main memory, it is important to exploit a parallelization capable of distributing the voxels together while minimizing the time. In this section, a complete parallelization is proposed.

4.1 Principle

The data distribution dispatches the lists of voxels. Distributing lists of voxels has not the same implication as distributing the voxels. It guarantees having complete lists on one machine i.e. a set of all the voxels included in the lists. Having complete lists on each machine offers the advantage of propagating light in the lists belonging to one machine without having to exchange radiosities between voxels of another machine.

The distribution of the lists among the machines is done by assigning to each machine a distinct set of (i,j). However the (i,j) of each voxel are not constants, they change with the direction. This may splits the complete lists into non-complete ones necessitating some transmission of voxels between the machines to rebuild complete lists. Figure 7 illustrates the method principle. When changing the direction, the lists present on the machines may not be complete: for example the list number $(6,7)$ is present on 2 different machines which requires the exchange of voxels to build one complete list.

The major steps of this parallelization are detailed in next sections and can be summarized as follows:

For each iteration, traverse all the directions and for each direction:

1. Find (iteration equal to 1) or load (iteration different than 1) from memory the suitable list distribution that guarantees the presence of complete lists on each machine and establishes as well a load balancing.
2. Depending on the (i,j) of each voxel decide if it will stay or leave the machine, if it will leave fill it in the corresponding data structure.
3. Exchange data i.e. send and receive implicated voxels.
4. Merge the received voxels with the voxels already present on the machine. The merge is necessary for keeping the voxels sorted in their lexicographic order (see Section 4.4)
5. Dispatch the voxels in their lists.
6. Propagate light in the lists.

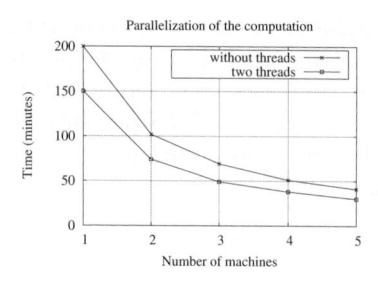

Fig. 6. Distribution of the computation

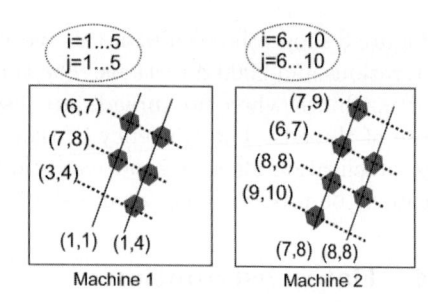

Fig. 7. The lists distribution on two different machines

4.2 Sorting the Directions

The Lexicographic Order. When changing the direction, we may need to change the lexicographic order that is used to sort the lists. In the sequential algorithm, a precomputation step sorting the voxels according to the four different lexicographic orders was done [1]. When parallelizing the lists, voxels are exchanged between the machines, so this precomputation step can not be done. To solve this problem, a sorting of the direction according to their directing vector (a, b, c) is done ; we can distinguish eight possibilities according to different signs of a, b, and c. For each of these eight sorting will correspond a set of directions satisfying the sign of the directing vector and leading thus to a particular lexicographic order. This will maintain the lexicographic order stable for many directions and will lead to just eight changes in the lexicographic order during an iteration $((x \uparrow, y \uparrow, z \uparrow), (x \uparrow, y \uparrow, z \downarrow), (x \uparrow, y \downarrow, z \uparrow), (x \uparrow, y \downarrow, z \downarrow), (x \downarrow, y \downarrow, z \downarrow), (x \downarrow, y \downarrow, z \uparrow), (x \downarrow, y \uparrow, z \downarrow), (x \downarrow, y \uparrow, z \uparrow))$.

Close Directions. Minimizing the exchanges of voxels between the machines is an interesting aspect that could minimize the rate of exchange as well as the time wasted for sending and gathering data to or from other machines. It has been noticed that close direction would generate small changes in the value of (i, j) which will insure that most of the voxels stays in the machine and only a small amount will leave to another machine. Sorting the directions in such a way they are close to each other can be done by a special traversing of the discrete sphere as shown on the Fig. 8.

Final Sorting. Two sorting for the directions were needed. The best way to achieve these two sorting can be done by traversing every 1/8 quadrant of the discrete sphere using the traversing mentioned in Fig. 8.

4.3 Filling Data and Load Balancing

Filling Data. At every direction, each machine should traverse all of its voxels and calculate the new value of (i, j) (space partition to discrete lines) of each of them to decide whether it should be kept or send. The voxels to be sent are stored in a data structure labelled with respect to the destination they should go to facilitate the filling and the sending steps. The lexicographic order of the voxels is maintained in order to avoid resorting at the reception.

Load Balancing. The exchanges of voxels between the machines at each direction may cause an unbalance in the number of voxels between the machines. To achieve good performance and avoid having one machine waiting for the others to finish their work, it was important to find a suitable lists distribution that can establish load balancing by giving to each machines the appropriate interval $((i_1..i_n), (j_1...j_n))$ of lists to be treated. A static distribution was not efficient since the distribution changes with the direction so a dynamic load balancing was needed. The following steps are done by each cluster machine to obtain the appropriate lists distribution:

- Count the number of voxels belonging to each list in one machines ; lists are not build yet however the number of voxels in each list can be calculated by calculating the (i,j) of each voxel.
- Exchange the voxels number and do the summation between the same (i,j) lists on different machines so that each machine will have a global (i,j) distribution of the whole scene.
- Find the (i,j) distribution that distribute lists among the machine giving each machine almost the same number of voxels ; for this the elastic algorithm [5] is done. This algorithm finds the suitable distribution depending on the number of processors. It is achieved by summing rows to find the suitable row distribution, then summing column to find the column distribution.

It should be stated that it is desirable to have a load balance at every direction however doing it is expensive in terms of time because counting the number of voxels in each list for large scenes is expensive. Exchanging these numbers between the machines to do the summation is also expensive. One interesting solution equivalent to doing the load balance at every direction consists in taking advantage of having the same directions in the same order for all the iterations. So it is possible to do the load balance for all the directions for the first iteration, memorize the distribution found for each direction then for the rest of iterations reload the correct distribution from the memory. Keeping the distribution in memory is not expensive because it maintains $(numberMachines)^{3/2}$ integers at every directions. Figure 10 shows the importance of increasing the frequency of doing the load balancing. Tests are done for the cabin room composed of 3 millions voxels, using four machines, $\approx$ 5000 directions and 6 iterations. It is clear that the best time is obtained at the value 100 i.e. when doing the load balance at every direction.

4.4 Exchanging and Merging Data

The exchange and the merge are two important operations that should handle the sent and received voxels according to the data structures of each machine. The exchange consists in sending and receiving the set of voxels between the machines just after the filling operation terminates. The Merge operation merges the voxels owned by the machine (ordered in their lexicographic order) together with each of the packs received (also ordered in their lexicographic order) from other machines ; it is expensive, however most of the time the exchanges are of small size (10 percent of voxels owned by the machines) and between limited number of machines (the direction are close, the couple (i,j) varies slowly), so it becomes less expensive.

4.5 Results

The data distribution tests were examined for a cabin room composed of 3 millions voxels. The tests are done for $\approx$ 5000 directions and 6 iterations. The speed-up are shown on Fig. 9. A speed-up of 2.7 is achieved with 8 machines. In the sequential algorithm we had just two important times, the dispatch and the propagate of light in the lists, however in the parallel version we still have the dispatch propagate time that is divided by the number of machines and in addition, we have four operations that are added by the parallelization (Filling, Load balancing, Exchange, Merge). The time study for these four operations is viewed in Fig. 11. It can be seen that the filling process is an expensive operation in time but it decreases with the number of machines and its time is divided by this number. The load balancing time does not vary a lot and it is almost constant with respect to the number of machines. As for the exchange

Fig. 8. Directions ordering on 1/8 of the sphere to minimize the exchange of voxels

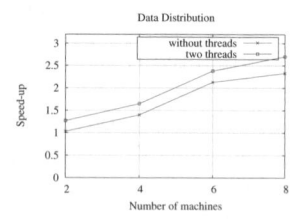

Fig. 9. Data distribution speed-up for different number of machines

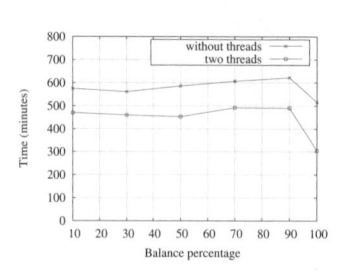

Fig. 10. Effect of increasing the frequency of doing load balance on time

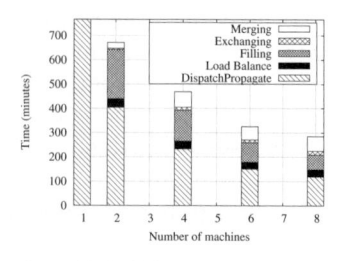

Fig. 11. Data distribution: Time distribution for each operation

time, it is negligible when compared to other times since only small percentage of voxels (10 percent) is usually exchanged between the machines. Finally the merging time depends on the quantity of exchanges between the machines and it does not vary too much.

5 Conclusion and Perspectives

In this paper we have examined three parallelizations. The first parallelization is just a local one that can help to minimize the time of the other parallelizations. The parallelization of the computation has shown good results, but it is limited to scenes that fit in the main memory. The data distribution has improved the sequential time ; it has a low speed-up because it demanded the construction of complete lists on each machine and necessitated though many additional operations that are added to the sequential operations. Despite its low speed-up, it has distributed the data, which is important when dealing with large scenes that do not fit in memory. As for the perspectives and further works, many tests are also needed to minimize the time of the data distribution. We also intend to apply the data distribution proposed in this paper with the computational distribution to exploit different levels of parallelizations ; this is expected to increase the speed-up and to minimize the time even further. Preliminary experiments have confirmed those hypotheses.

Acknowledgments. The authors wish to acknowledge the support of Conseil Regional d'Auvergne within the framework of the Auvergrid project.

References

1. Chatelier, P., Malgouyres, R.: A Low Complexity Discrete Radiosity Method. Discrete Geometry for Computer Imagery. **3429** (2005) 392–403
2. Arnaldi, B., Priol, T., Renambot, L., Pueyo, X.: Visibility Masks for solving Complex Radiosity Computations on Mutliprocessors. Parallel Computing **23(7)** (1988) 887–897
3. Renambot, L., Arnaldi, B., Priol, T., Pueyo, X.: Towards Efficient Parallel Radiosity for DSM-based Parallel Computers Using Virtual Interfaces. Proceedings of the IEEE symposium on Parallel rendering (1993) 76–86
4. Funkhouser, TA.: Coarse-grained parallelism for hierarchical radiosity using group iterative methods. Proceedings of the 23rd annual conference on Computer graphics and interactive techniques (1996) 343–352
5. Feschet, F., Miguet, S., Perroton, L.: Parlist: a parallel data structure for dynamic load balancing. Journal of Parallel and Distributed Computing **51** (1998) 114–135
6. Malgouyres, R.: A Discrete Radiosity Method. Discrete Geometry for Computer Imagery **2301** (2002) 428–438
7. Smith, B., Arvo, J., Donald, G.: A clustering algorithm for radiosity in complex environments. Proceedings of the 21st annual conference on Computer graphics and interactive techniques **28** (1994) 435–442
8. Podehl, A., Rauber, T., Runger, G.: A Shared-Memory Implementation of the Hierarchical Radiosity Method. Theoretical Computer Science **196** (1998) 215–240

9. Bouatouch, K., Mebard, D., Priol, T.: Parallel Radiosity Using a Shared Virtual Memory. Proceedings of Advanced Techniques in Animation, Rendering and Visualization (1993) 71–83
10. Sturzlinger, W., Wild, C.: Parallel Visibility Calculations for Radiosity. Proceedings of the Third Joint International Conference on Vector and Parallel Processing: Parallel Processing (1994) 405–413
11. Yu, Y., Oscar, H., Yang, T.: Parallel progressive radiosity with adaptive meshing. Journal of Parallel and Distributed Computing **42(1)** (1997) 30–41
12. BarLev, A., Itzkovitz, A., Raviv, A., Schuster, A.: Parallel Vertex-To-Vertex Radiosity on a Distributed Shared Memory System. Proceedings of the 5th International Symposium on Solving Irregularly Structured Problems in Parallel. J. Diff. Eq. **1457** (1998) 238–250
13. Sillion, F., Hasenfratz, J.M.: Efficient Parallel Refinement for Hierarchial Radiosity on a DSM Computer. Third Eurographics Workshop on Parallel Graphics and Visualisation (2000) 61–74
14. Guitton, P., Roman, J., Subrenat, G.: Implementation Results and Analysis of a Parallel Progressive Radiosity. Proceedings of the IEEE symposium on Parallel rendering (1995) 31–38
15. Cohen, M.F., Greenberg, D.F.: The hemi-cube: a radiosity solution for complex environments. SIGGRAPH '85: Proceedings of the 12th annual conference on Computer graphics and interactive techniques **19(3)** (1985) 31–40
16. Martnez, R., Sbert, M., Szirmay-Kalosv, L.: Parallel Multipath with Bundles of Lines. Third Eurographics Workshop on Parallel Graphics and Visualisation (2000)

Parallelising Matrix Operations on Clusters for an Optimal Control-Based Quantum Compiler

T. Gradl[1], A. Spörl[2], T. Huckle[1], S.J. Glaser[2], and T. Schulte-Herbrüggen[2]

[1] Department of Mathematics and Computer Science
[2] Department of Chemistry, Technical University Munich, Lichtenbergstrasse 4,
D-85747 Garching, Germany
huckle@in.tum.de and tosh@ch.tum.de
http://www5.in.tum.de/~huckle/

Abstract. Quantum control plays a key role in quantum technology, *e.g.* for steering quantum hardware systems, spectrometers or superconducting solid-state devices. In terms of computation, quantum systems provide a unique potential for coherent parallelisation that may exponentially speed up algorithms as in Shor's prime factorisation. Translating quantum software into a sequence of classical controls steering the quantum hardware, *viz.* the quantum compilation task, lends itself to be tackled by optimal control. It is computationally demanding since the classical resources needed grow exponentially with the size of the quantum system. Here we show concepts of parallelisation tailored to run on high-end computer clusters speeding up matrix multiplication, exponentials, and trace evaluations used in numerical quantum control. In systems of 10 spin qubits, the time gain is beyond a factor of 500 on a 128-CPU cluster as compared to standard techniques on a single CPU.

We are currently in the midst of a second quantum revolution. The first one gave us new rules that govern physical reality. The second one will take these rules and use them to develop new technologies.
                                                Dowling and Milburn, 2003 [1]

Scope

For exploiting the power of quantum systems, one has to steer them by classical controls such as voltage gates, radio-frequency pulses, or laser beams. Here the aim is to provide computational infrastructure for doing so in an optimal way, because the shapes of these controls critically determine the performance of the quantum system in terms of overlap of its actual final states with the desired target states. Standard engineering methods solve related problems for systems of classical physics. For quantum control, however, the calculations of optimal shapes become more complicated (on conventional classical computers): the quantum states have to be represented by matrices, the dimensions of which grow exponentially with system size.

Here we present the adaptation of a number of matrix operation routines to high-end parallel computer clusters while using the symmetry of quantum spin

```
┌─────────────────────────┐         ┌─────────────────────────────┐
│   Program, Module       │         │  Quantum Algorithm, Module  │
└─────────────────────────┘         └─────────────────────────────┘
  Compiler │           │              Precompiler │            │
           ↓           ↓                          ↓            ↓
┌──────────────────┐                 ┌──────────────────┐
│ Assembler Code   │   "direct"      │ Universal Gates  │   by optimal control
└──────────────────┘                 └──────────────────┘
  Assembler│           │              Assembler │            │
           ↓           ↓                        ↓            ↓
┌─────────────────────────┐         ┌─────────────────────────────────┐
│      Machine Code       │         │ Machine Code of Quantum Controls│
└─────────────────────────┘         └─────────────────────────────────┘
```

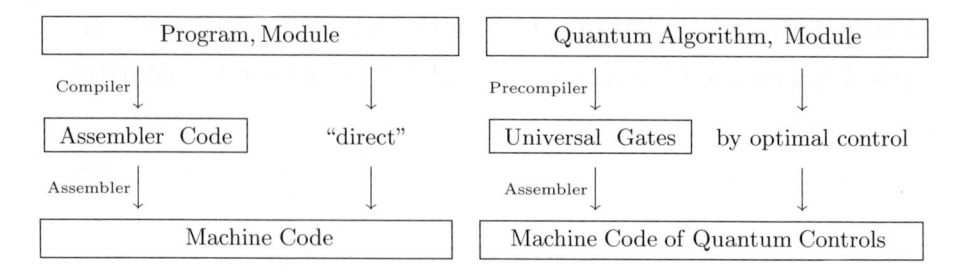

Fig. 1. Compilation in classical computation (left) and quantum computation (right). In the quantum scenario, the machine code has to be time-optimal or dissipation-protected, otherwise decoherence wipes out the coherent superpositions the quantum bits (qubits) build upon. By assembling universal quantum gates timeoptimality is in general not reached, while compilation by quantum control lends itself to minimise losses.

systems to minimise computational and communication effort as well as storage costs.

These methods of using classical computer clusters are tantamount to exploiting the power of present and future quantum resources.

1 Algorithms on Parallel Clusters for Quantum Control

1.1 Quantum Parallelism

Controlling quantum systems also offers a great potential for performing computational tasks or for simulating the behaviour of other quantum systems [2,3]. This is because the complexity of many problems [4] reduces upon going from classical to quantum hardware. It roots in Feynman's observation [2] that the resources required for simulating a quantum system on a classical computer increase exponentially with the system size. In turn, he concluded that using quantum hardware might therefore *exponentially decrease* the complexity of certain classical computation problems. Coherent superpositions of quantum states used as so-called 'qubits' can be viewed as a particularly powerful resource of quantum parallelism unparalleled by any classical system. Important applications are meanwhile known in quantum computation, quantum search and quantum simulation: most prominently, there is the exponential speed-up by Shor's quantum algorithm of prime factorisation [5,6], which relates to the general class of quantum algorithms [7,8] solving hidden subgroup problems in an efficient way [9].

1.2 Quantum Compilation as Control Problem

Moore's Law of increasing classical computing power concomitant to miniaturising microchips is often quoted to make a strong case for predicting that within the next decade computer hardware scales will reach sizes that inevitably have to include quantum effects. Among the generic tools needed for advances in quantum technology (see *e.g.* Ref. [1] for a survey), quantum control plays a major

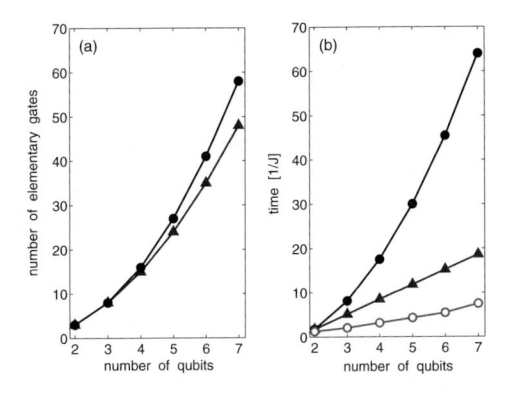

Fig. 2. (a) Gate complexity of the QFT in linear spin chains. Standard-gate decomposition (•) [13] and optimised scalable gate decomposition (▲) [14]. (b) Time complexity of the QFT in linear spin chains. Upper traces give analytical times associated with the decompositions of part (a): standard-gate decompositions (•) [13] and optimised scalable gate decompositions (▲) [14]. Lowest trace: speed-up by time-optimal control with shortest numerical realisations obtained (○) rounded to 0.01 J^{-1}. Details in Ref. [12].

role. As illustrated in Fig. 1, this may be exemplified by envisaging the process of compiling a quantum module into the machine language of a concrete quantum hardware device as an instance of quantum control. To this end, there are two different approaches: one may either use (i) universal elementary quantum gates [10] to synthesise a quantum computational module from prefabricated standard building blocks, the so-called universal quantum gates, or (ii) one may prefer to generate the quantum module directly from the experimentally available controls with the help of gradient-flow based numerical algorithms implementing tools of optimal control [11,12]. While decomposition into universal gates is inspired by discrete level permutations, direct compilation exploits the differential geometry of smooth manifolds for governing unitary quantum dynamics in an optimal way. Recently we have shown that one may thus obtain dramatic speed-ups, *e.g.* for the Quantum Fourier Transform (QFT) in spin systems (see Fig. 2 and [12]) or for realising multiply controlled gates in solid-state devices. Here the speed-up translates into a gain of some two orders of magnitude in approaching the error-correction threshold [15]. The approach is very general and holds for spin and pseudo-spin systems whose dynamics are Lie-algebraically closed.

1.3 Gradient Flow Algorithms for Quantum Control

Our algorithmic tools of optimal quantum control [11] for obtaining these results are based on gradient flows [16,17] tailored to the unitary group of Hamiltonian quantum evolution [18,19,20]. Let U_G denote the unitary representation of a quantum gate, *i.e.* the target matrix. On the other hand, define by $U(T) := e^{-it_M H_M} \cdots e^{-it_k H_k} \cdots e^{-it_1 H_1}$ the propagator brought about by a sequence of

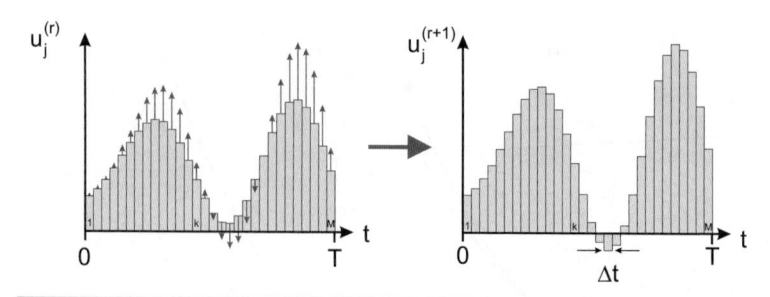

1. set initial controls $u_j^{(0)}(t_k)$ for all times t_k with $k = 1, 2, \ldots M$ at random or by guess;
2. starting from $U_0 = \mathbb{1}$, calculate the forward-propagation for all $t_1, t_2, \ldots t_k$ (for simplicity $\Delta t := t_{k+1} - t_k$ uniform)

$$U^{(r)}(t_k) = e^{-i\Delta t H_k^{(r)}} e^{-i\Delta t H_{k-1}^{(r)}} \ldots e^{-i\Delta t H_1^{(r)}} \quad ;$$

3. likewise, starting with $T = t_M$ and $\lambda(T) = \text{const} \cdot U_G$, compute the back-propagation for all $t_M, t_{M-1}, \ldots t_k$;

$$\lambda^{(r)}(t_k) = e^{i\Delta t H_k^{(r)}} e^{i\Delta t H_{k+1}^{(r)}} \ldots e^{i\Delta t H_M^{(r)}} \lambda(T) \quad ;$$

4. calculate $\frac{\partial h(U(t_k))}{\partial u_j} = \text{Re tr}\{\lambda^\dagger(t_k)(-iH_j)U(t_k)\}$;
5. with $u_j^{(r+1)}(t_k) = u_j^{(r)}(t_k) + \varepsilon \frac{\partial h}{\partial u_j}\big|_{t=t_k}$ update all the piece-wise constant Hamiltonians to $H_k^{(r+1)}$ and return to step 2.

Fig. 3. Top trace: updating the vector of control amplitudes u_j by the gradients (arrows) evaluated via the iterative scheme, the GRAPE-algorithm [11] in the box below. Gradients are calculated in step 4, while the controls are updated as in step 5.

evolutions of the quantum system under M piece-wise constant Hamiltonians H_k. Then the optimal control problem can be cast into the tasks:

$$\text{maximise} \quad f(U(T)) = \text{Re tr}\{U_G^\dagger U(T)\} \tag{1}$$
$$\text{subject to} \quad \dot U(t) = -iHU(t) \quad ,$$

where the Hamiltonian H comprises drift and control terms $H = H_{\text{drift}} + \sum_j u_j H_j$ with u_j as element of the (real) control-amplitude vector. As usual, the boundary condition may be included by a Lagrange parameter λ so that one may finally exploit the corner stone of control theory, Pontryagin's maximum principle, in a quantum setting (for details see, e.g, [21,11]) to require for the real-valued function h that $\frac{\partial h}{\partial u_j} = \text{Re tr}\{\lambda^\dagger(-iH_j U)\} \to 0$ at all times t_k. So the task can readily be solved by gradient flows iteratively improving the classical controls driving the quantum system into maximal overlap with the target.

1.4 Computational Tasks: Previous Performance and Goals

As sketched in Fig. 3, from a computational point of view, the GRAPE-algorithm makes heavy use of (1) matrix multiplication, (2) matrix exponentials, (3) trace

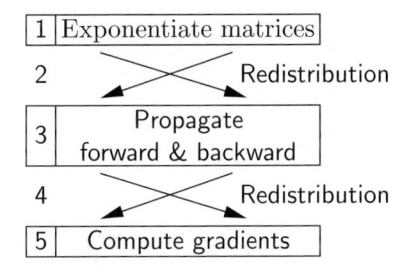

```
1 | Exponentiate matrices |
2         ╲╱         Redistribution
           ╱╲
3 | Propagate
    forward & backward |
4         ╲╱         Redistribution
           ╱╲
5 | Compute gradients |
```

Fig. 4. Redistribution of matrices is needed between steps of the GRAPE algorithm

evaluation, and (4) step-size optimisation in conjugate gradients. The standard C++ code with CBLAS matrix-matrix multiplication [22,23,24] could deal with quantum systems up to 7 spins on an AMD Athlon Processor with 2.13 GHz and 1 MB RAM, taking months of CPU time. With the CPU time roughly growing by a factor of 8 per additional spin qubit, 10-spin systems are clearly out of reach unless one could speed up the calculations some 500 times.

In the present work we set out to reach this benchmark on a high performance cluster. The employed system consists of 128 AMD *Opteron 850* CPU (2.4 GHz), four on each node; the nodes are connected with an *Infiniband* network[1]. For parallel programming the MPI standard was used.

In order to exploit the power [25] of high-end computer clusters, here we address features of distributed matrix multiplication, concepts of broadcasting no more than the necessary information to the nodes of processors, reducing communication effort between different processors as well as exploiting symmetries of the matrix representations of the pertinent quantum mechanical system Hamiltonians for speeding up matrix exponentials. Some of the symmetries are not coincidental: fully controllable quantum systems are Lie-algebraically closed [19]. They are thus largely confined to finite-dimensional spin- or pseudo-spin systems, whose representations in terms of Kronecker or tensor products of Pauli matrices often entail persymmetric matrices (*vide infra*).

2 Parallel Matrix Multiplication

This section compares the implementation of two algorithms for multiplying a series of matrices ("propagation"), as needed in steps 2 and 3 of the GRAPE algorithm. The algorithms differ in run-time and, as will turn out to be most decisive, in memory demand. For comparing performance in terms of run-time, just considering the time required in the propagation step does not suffice; as we will see, it is also important to understand how it is embedded into the whole of the GRAPE algorithm. To this end, consider Fig. 4: the propagation (step 3 in this figure) is preceded by the computation of the exponential matrices $e^{\pm i \Delta t H_k^{(r)}}$ (step 1). The exponentials of all the matrices ($k = 1 \ldots M$) are distributed over

---

[1] http://www.lrr.in.tum.de/Par/arch/infiniband/

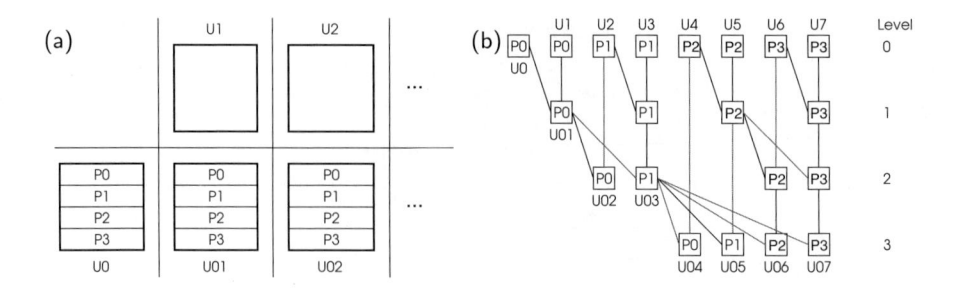

Fig. 5. (a) Slice-wise matrix multiplication provides a simple way of parallelisation. U_{0k} denotes the $(k+1)$-fold product $U_k U_{k-1} \cdots U_0$ according to step 2 in the GRAPE-algorithm (Fig. 3). The resulting complexity is $\mathcal{O}(M \cdot N^3/p)$. Communication between the processors P is needed solely for broadcasting the matrices U_k prior to propagation. (b) Scheme for tree-like propagation. In this example, propagation is carried out in three steps. Red lines indicate communication between processors P_0 through P_3.

all the processors. However, in step 3 it is not granted every processor exactly needs those matrices it computed in step 1. For this very reason an intermediate redistribution of matrices among processors is required not only in step 2 but also upon proceeding from step 3 to step 5 (gradient computation).

2.1 Slice-Wise Propagation

The matrix matrix multiplication AB can most easily be split into jobs distributed to different CPUs by taking say the rows a_ℓ of A separately as

$$AB = (a_1; a_2; \ldots a_N)B = (a_1 B; a_2 B; \ldots; a_N B) \quad . \tag{2}$$

This scheme is readily extendible to k out of the M matrices in step 2 and 3 of the GRAPE-algorithm above (see Fig. 5(a)). However, each processor then refers to $k-1$ matrices, which means that they have to be broadcasted in step 2 of figure 4. Also, the workspace required by each processor is of the order of $\mathcal{O}(M \cdot N^2)$. The time complexity in this straightforward scheme can easily be evaluated, because the total number of operations is evenly distributed among the available processors. So the order of operations is $\mathcal{O}(M \cdot N^3/p)$, where $N := 2^n$ denotes the dimension of the matrix and p is the number of processors.

Moreover, here there are no stability concerns, as unitary matrices are known to allow for numerically stable algorithms [26,27]: the computation of the product of two unitary matrices is well-conditioned and this clearly extends to multiple products in general. However, Section 2.2 will reveal this is not necessarily the case in all other schemes.

For further acceleration some simplifying features can be used: step 4 of the GRAPE-algorithm, for instance, takes the trace Re tr$\{\lambda^\dagger HU\}$, where λ and U are fully occupied, but H is a sparse matrix representation of the spin-control Hamiltonian, which acts by permutation and scalar multiplication on the rows

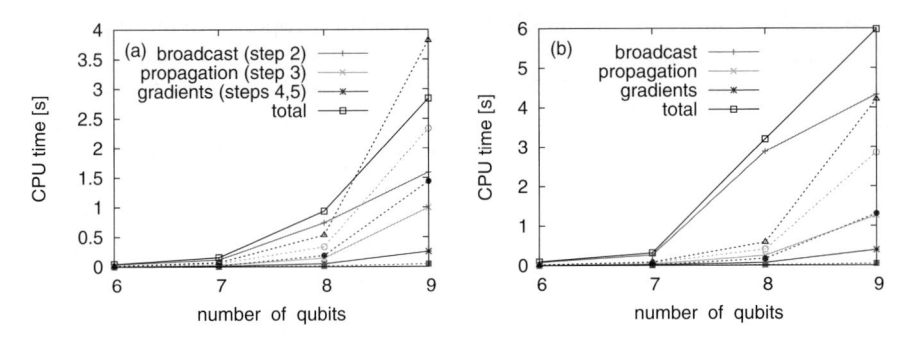

Fig. 6. Comparison of performance against system size with (a) 32 and (b) 64 parallel processors under slicewise (—) and tree-like (- -) propagation. Note the difference in broadcasting time (red lines). Steps 2-5 refer to the numberings as in Fig. 4.

of U. Thus, instead of two matrix-matrix multiplications $\lambda^\dagger H$ and $(\lambda^\dagger H)U$, a multiplication of $\lambda^\dagger$ with a row-transformed U suffices.

As shown in Fig. 6, all these considerations result in valuable speed-ups for quantum systems of up to 9 spin qubits. In larger systems, though, workspace becomes limiting, as every processor requires M matrices. A way of compensation would be to compute the exponential matrices on demand, *i.e.* at forward and backward propagation respectively. However, this would be a really bad solution because computing the exponential takes most of the computation time already. Therefore system sizes beyond 9 spins cannot be computed with the slice-wise propagation and require alternative methods as described next.

2.2 Tree-Like Propagation: The Parallel Prefix Algorithm

A different approach for computing the propagation is the parallel prefix algorithm [28] depicted in Fig. 5(b). In general, it is applicable to arbitrary combinations of number of processors p and digitisation M. Yet in this article we confine ourselves to $p = M/2$, which is also the maximum reasonable number for p. Our code performs forward and backward propagation simultaneously thus increasing the overall number of processes to M. Under the assumption that $p = M/2$, the computation time of the algorithm is $\mathcal{O}(\log_2 M \cdot N^3)$. In contrast to slice-wise propagation, parallel prefix requires communication during the propagation (red lines in Fig. 5(b)): they sum up to $\sum_{l=2}^{\log_2 M}[Broadcast(N^3, p = 2^{l-1}) + (l-1) \cdot Send(N^3)]$, provided the times for *Broadcast* and *Send* are not influenced by other ongoing communication. Recalling the computation time of $\mathcal{O}(M \cdot N^3/p)$ for the slice-wise propagation and assuming $p = M/2$, parallel prefix should never be faster (neglecting effects like memory prefetching). Fig. 6 shows this is indeed true. On the other hand, parallel prefix does not require all the matrices $U(t_k)$ in all processes, which eliminates the broadcast time prior to the propagation step (see Fig. 4). It is this advantage that is large enough to outweigh the slower propagation time.

Even more important is yet another gain from this property: reduced memory demand. In our current implementation the maximum number of matrices stored at a single process is $\mathcal{O}(\log_2 M)$ [P_0 produces one result in every level], which is already much less than the $\mathcal{O}(M)$ of the slice-wise propagation. In case this should be inacceptable, the number can be reduced to $\mathcal{O}(1)$ by implementing a slightly different communication pattern from the one used here.

The stability of parallel prefix-matrix multiplication deserves a closer look [29]: following Mathias [30], in the general case, this multiplication is numerically unstable. He derives the following first-order upper bound to the error

$$\left|\mathrm{rd}(U_1 \cdots U_M) - (U_1 \cdots U_M)\right| \leq 2(N-1)\,\varepsilon \sum_{k=1}^{M-1} |B_k| + \mathcal{O}(\varepsilon^2)\,, \qquad (3)$$

where $|B_j|$ stands for the product of absolute values of the largest matrix elements within the factors U_k. However, with all these matrices being unitary in our case, the same error estimate turns into the "well-behaved" form

$$\left|\mathrm{rd}(U_1 \cdots U_M) - (U_1 \cdots U_M)\right| \leq 2(N-1)M\varepsilon + \mathcal{O}(\varepsilon^2) \quad. \qquad (4)$$

In coincidence with the linearity in M, we observed $||U_{\text{tree}} - U_{\text{reg}}||_2$ increasing linearly with the propagation step, where at $k = M = 128$, a value of 1.5×10^{-13} was reached in a nine-qubit system (not shown). This underpins the computation is numerically stable for unitary matrices used in quantum dynamics.

3 Optimising Matrix Exponentials

Computing matrix exponentials numerically is a notoriously intricate problem [31,32]. Here we compare the standard Padé-approximation with the generic QR-approach and a symmetry-adapted QR-variant thus allowing optimised LAPACK routines [33,34] to be employed. With controllable qubit systems permitting pseudo-spin representations in terms of Pauli matrices, their Hamiltonian generators of the exponential map often show 'persymmetry' [26,35], i.e., a matrix representation that is symmetric with respect to the anti-diagonal and which may be induced by the Pauli matrices (vide infra). Defining J_N as the $N \times N$ reversal matrix (obtained by reversing the columns of the identity matrix), the persymmetry of a matrix A is equivalent to the condition $J_N A J_N = A^T$.

Lemma 1. *(1) A (finite) Kronecker or tensor product of persymmetric matrices is again persymmetric. (2) The same is true in the tensor product of an even number $2r$ of matrices that are themselves 'anti-persymmetric' due to the sign change $J_N A J_N = -A^T$.*

Proof. Assertion (1) simply follows from the fact that forming the tensor product and taking the transpose with respect to the anti-diagonal commute. Part (2) is due to the construction $J_N = J_2 \otimes J_2 \cdots J_2 \otimes J_2$, since one finds $J_N(A_1 \otimes \cdots \otimes A_k)J_N = (J_2 A_1 J_2 \otimes \cdots \otimes J_2 A_k J_2) = (-1)^{2r}(A_1^T \otimes \cdots \otimes A_k^T)$. ■

Remark. Let $\{\sigma_x, \sigma_y, \sigma_z\} = \{\begin{pmatrix} 0 & 1 \\ 1 & 0 \end{pmatrix}, \begin{pmatrix} 0 & -i \\ i & 0 \end{pmatrix}, \begin{pmatrix} 1 & 0 \\ 0 & -1 \end{pmatrix}\}$ be the Pauli matrices. By (2), the drift term comprises the persymmetric terms $\sigma_z \otimes \sigma_z + \alpha(\sigma_x \otimes \sigma_x + \sigma_y \otimes \sigma_y)$ for any real α, while the control terms are tensor products of the unit matrix with σ_x, σ_y, which in turn are persymmetric by (1).

Moreover, we can write the Hamiltonian H in the form $H = D + C$ with a real diagonal matrix D and a multilevel matrix C of the recursive form $C = C_1 \otimes \cdots \otimes C_n$ with 2×2 matrices

$$C_k := \begin{pmatrix} 0 & \gamma_k \\ \gamma_k^* & 0 \end{pmatrix}. \tag{5}$$

Each of the small matrices C_k can be transformed into a real circulant matrix by the unitary diagonal matrix $V_k := \operatorname{diag}(1, \gamma_k/|\gamma_k|)$. Therefore, by the Kronecker products of the small matrices V_k we can transform the entire matrix H to a real symmetric persymmetric matrix $\tilde{H}$ of the form

$$\tilde{H} = \begin{pmatrix} A_1 & r\mathbf{1} \\ r\mathbf{1} & A_2 \end{pmatrix} \tag{6}$$

with real r, symmetric A_1 and A_2, and the identity matrix $\mathbf{1}$. Now, the persymmetry leads to the relation $A_2 = JA_1J$ and thereby to the similarity transform

$$\begin{pmatrix} \mathbf{1} & \mathbf{1} \\ J & -J \end{pmatrix} \begin{pmatrix} A_1 & r\mathbf{1} \\ r\mathbf{1} & A_2 \end{pmatrix} \begin{pmatrix} \mathbf{1} & J \\ \mathbf{1} & -J \end{pmatrix} = \begin{pmatrix} A_1 + A_2 + 2r\mathbf{1} & 0 \\ 0 & A_1 + A_2 - 2r\mathbf{1} \end{pmatrix}. \tag{7}$$

Consequently, the computation of the eigenvalues of H for the matrix exponential $\exp(i\tau H)$ can be reduced to solving the same problem for two real matrices of half the size. For the exponentials we have been using two different methods:

1. classical scaling and squaring algorithm based on the Padé approximation;
2. finding the eigendecomposition of persymmetric $\tau H = UDU^{-1}$ for

$$\exp(i\tau H) = \exp(iUDU^{-1}) = U\exp(iD)U^{-1} = U \operatorname{diag}\left(\exp(i \cdot d_j)\right) U^{-1};$$

   with H being hermitian, this method is numerically stable [31,32]; moreover, using persymmetry, the eigendecomposition then reduces to real symmetric matrices of just half the size.

The major disadvantage of the classical scaling and squaring method with full-sized matrices can be circumvented by a series expansion instead of the Padé approximation, because then only sparse matrix matrix products arise. Expanding in terms of Chebychev polynomials is superior to a Taylor expansion [36,37]. Moreover, the error δ for approximating $\exp(i\tau H)$ by a Chebychev series with m terms can asymptotically be estimated by the Bessel function of the first kind $|J(m, ||\tau H||)| \approx \delta$ given the norm of the effective Hamiltonian $||\tau H||$. Thus one can predict the number of steps required for a given accuracy. Unfortunately, due to the control amplitudes sometimes $||\tau H|| \geq 100$; then the Chebychev expansion only pays if low accuracy suffices. Yet in future code this approximation will be included as an option for matrices of small norm or cases not demanding high accuracy.

Table 1. Contributions of Parallelised Matrix Operations to Overall Speed-up

Subroutine	Fraction of CPU Time		Weighted
	with 1 CPU	with 128 CPUs	Speed-up
maxStepSize	0.9	0.713	521
getGradient	0.091	0.287	52.6
expm	0.075	0.049	43.0
propagation	0.01	0.194	6.0
gradient	0.006	0.044	3.5
optimiseCG	1	1	576

4 Conclusions and Outlook

We have shown how using the potential parallelism inherent in coherent quantum superpositions relies on methods of classical control theory in order to steer the quantum systems. By the nature of quantum matrix mechanics, this requires powerful matrix calculations backed by architecture-adapted redistribution (Fig. 4). Here we demonstrated a speed-up by more than a factor of 500 for a 10 spin system by way of various matrix-operation techniques (see Tab. 1): slice-wise propagation is advantageous in systems up to 9 spin qubits, while tree-like propagation pays for systems from 10 qubits onwards. Moreover, by making use of the symmetry properties induced by the pseudo-spin structure of controllable spin-qubit systems, faster matrix exponentials are feasible. To further improve gradient-flow algorithms, sparse Hamiltonians will be exploited for matrix multiplications along the lines of Ref. [38] or by using Strassen's method.

By the current extensions, larger spin-qubit systems are in reach thus allowing a broader numerical basis for deducing quantum computational complexity measures. Related symmetry properties can be used in wider classes of quantum systems therefore making the presented tools broadly applicable in quantum technology and control. From a general point of view of numerics, the partial problems of computing the exponential $\exp(iH)$ for a hermitian matrix H and furthermore the task of evaluating a sequence of products of unitary matrices are of broader interest and important far beyond the GRAPE-algorithm. Here, the special case of unitary matrices may lead to improved stability and allow algorithms that are fast but usually known to be unstable or only weakly stable.

Acknowledgements

This work was supported in part by the integrated EU project QAP and by *Deutsche Forschungsgemeinschaft*, DFG, in the incentive SPP 1078 *Quanten-Informationsverarbeitung*, QIV. Helpful discussion with Michael Riss on the 'tree-like' matrix multiplication is gratefully acknowledged.

References

1. Dowling, J., Milburn, G.: Quantum technology: The second quantum revolution. Phil. Trans. R. Soc. Lond. A **361** (2003) 1655–1674
2. Feynman, R.P.: Simulating physics with computers. Int. J. Theo. Phys. **21** (1982) 467–488
3. Feynman, R.P.: Feynman Lectures on Computation. Perseus Books, Reading, MA. (1996)
4. Papadimitriou, C.H.: Computational Complexity. Addison Wesley, Reading, MA. (1995)
5. Shor, P.W.: Algorithms for Quantum Computation. In: Proceedings of the Symposium on the Foundations of Computer Science, 1994, Los Alamitos, California, IEEE Computer Society Press, New York (1994) 124–134
6. Shor, P.W.: Polynomial-Time Algorithms for Prime Factorisation and Discrete Logarithm on a Quantum Computer. SIAM J. Comput. **26** (1997) 1484–1509
7. Jozsa, R.: Quantum Algorithms and the Fourier Transform. Proc. R. Soc. A. **454** (1998) 323–337
8. Cleve, R., Ekert, A., Macchiavello, C., Mosca, M.: Quantum Algorithms Revisited. Proc. R. Soc. A. **454** (1998) 339–354
9. Ettinger, M., Høyer, P., Knill, E.: The Quantum Query Complexity of the Hidden Subgroup Problem is Polynomial. Inf. Process. Lett. **91** (2004) 43–48
10. Deutsch, D.: Quantum Theory, the Church-Turing Principle, and the Universal Quantum Computer. Proc. Royal Soc. London A **400** (1985) 97–117
11. Khaneja, N., Reiss, T., Kehlet, C., Schulte-Herbrüggen, T., Glaser, S.J.: Optimal Control of Coupled Spin Dynamics: Design of NMR Pulse Sequences by Gradient Ascent Algorithms. J. Magn. Reson. **172** (2005) 296–305
12. Schulte-Herbrüggen, T., Spörl, A.K., Khaneja, N., Glaser, S.J.: Optimal Control-Based Efficient Synthesis of Building Blocks of Quantum Algorithms: A Perspective from Network Complexity towards Time Complexity. Phys. Rev. A **72** (2005) 042331
13. Saito, A., Kioi, K., Akagi, Y., Hashizume, N., Ohta, K.: Actual Computational Time-Cost of the Quantum Fourier Transform in a Quantum Computer using Nuclear Spins . quant-ph/0001113 (2000)
14. Blais, A.: Quantum Network Optimisation. Phys. Rev. A **64** (2001) 022312
15. Spörl, A.K., Schulte-Herbrüggen, T., Glaser, S.J., Bergholm, V., Storcz, M.J., Ferber, J., Wilhelm, F.K.: Optimal Control of Coupled Josephson Qubits. quant-ph/0504202 (2005)
16. Brockett, R.W.: Dynamical systems that sort lists, diagonalise matrices, and solve linear programming problems. In: Proc. IEEE Decision Control, 1988, Austin, Texas. (1988) 779–803 see also: Lin. Alg. Appl., 146 (1991), 79–91.
17. Helmke, U., Moore, J.B.: Optimisation and Dynamical Systems. Springer, Berlin (1994)
18. Glaser, S.J., Schulte-Herbrüggen, T., Sieveking, M., Schedletzky, O., Nielsen, N.C., Sørensen, O.W., Griesinger, C.: Unitary control in quantum ensembles: Maximising signal intensity in coherent spectroscopy. Science **280** (1998) 421–424
19. Schulte-Herbrüggen, T.: Aspects and Prospects of High-Resolution NMR. PhD Thesis, Diss-ETH 12752, Zürich (1998)
20. U. Helmke, K. Hüper, J.B. Moore, T. Schulte-Herbrüggen.: Gradient Flows Computing the C-Numerical Range with Applications in NMR Spectroscopy. J. Global Optim. **23** (2002) 283–308

21. Butkovskiy, A.G., Samoilenko, Y.I.: Control of Quantum-Mechanical Processes and Systems. Kluwer, Dordrecht (1990)
22. Lawson, C.L., Hanson, R.J., Kincaid, D., Krogh, F.T.: Basic Linear Algebra Subprograms for FORTRAN usage. ACM Trans. Math. Soft. **5** (1979) 308–323
23. Dongarra, J.J., Croz, J.D., Hammarling, S., Hanson, R.J.: An Extended Set of FORTRAN Basic Linear Algebra Subprograms. ACM Trans. Math. Soft. **14** (1988) 1–17
24. Dongarra, J.J., Croz, J.D., Hammarling, S.: A Set of Level 3 Basic Linear Algebra Subprograms. ACM Trans. Math. Soft. **16** (1990) 1–17
25. Dongarra, J., Duff, I., Sørensen, D., van der Vorst, H.: Numerical Linear Algebra on High-Performance Computers. SIAM (1998)
26. Golub, G.H., van Loan, C.F.: Matrix Computations. The Johns Hopkins University Press, Baltimore (1989)
27. Higham, N.J.: Accuracy and Stability of Numerical Algorithms. SIAM (1996)
28. Ladner, R.E., Fischer, M.J.: Parallel Prefix Computation. J. ACM **27** (1980) 831–838
29. Demmel, J., Heath, M., van der Vorst, H.: Parallel Numerical Linear Algebra. Acta Numerica **2** (1993) 111–198
30. Mathias, R.: The Instability of Parallel Prefix Matrix Multiplication. SIAM J. Sci. Comput. **16** (1995) 956–973
31. Moler, C., van Loan, C.: Nineteen Dubious Ways to Compute the Exponential of a Matrix. SIAM Rev. **20** (1978) 801–836
32. Moler, C., van Loan, C.: Nineteen Dubious Ways to Compute the Exponential of a Matrix, Twenty-Five Years Later. SIAM Rev. **45** (2003) 3–49
33. Anderson, E., Bai, Z., Bischof, C., Blackford, L.S., Demmel, J., Dongarra, J., Croz, J.D., Greenbaum, A., Hammarling, S., McKenney, A., Sørensen, D.: LAPACK User's Guide, Third Edition. SIAM (1999)
34. Blackford, L.S., Choi, J., Cleary, A., D'Azevedo, E., Demmel, J., Dhillon, I., Dongarra, J., Hammarling, S., Henry, G., Petitet, A., Stanley, K., Walkerx, D., Whaley, R.C.: ScaLAPACK User's Guide. SIAM (1997)
35. Cantoni, A., Butler, P.: Eigenvalues and Eigenvectors of Symmetric Centrosymmetric Matrices. Lin. Alg. Appl. **13** (1976) 275–288
36. Rivlin, T.J.: The Chebychev Polynomials. Wiley, New York (1974)
37. Veshtort, M., Griffin, R.: SPINEVOLUTION: A Powerful Tool for the Simulation of Solid and Liquid State NMR Spectra. J. Magn. Reson. **178** (2006) 248–282
38. Grote, M., Huckle, T.: Parallel Preconditioning with Sparse Approximate Inverses. SIAM J. Sci. Comput. **18** (1997) 838–853

Topic 11: Distributed and High-Performance Multimedia

Geoff Coulson, Harald Kosch, Odej Kao, and Frank Seinstra

Topic Chairs

It is increasingly common for information - whether it be scientific, industrial, or otherwise - to be composed of multiple media items, e.g., video-based, image-based, linguistic, or auditory items. As digital video may produce over 100 Mbytes of data per second, and image sets routinely require Terabytes of storage space, traditional resource management techniques in both end-systems and networks are rapidly becoming bottlenecks in the handling of such information. Moreover, in emerging multimedia applications, the generation, processing, storage, indexing, querying, retrieval, delivery, shielding, and visualization of multimedia content are fundamentally intertwined processes, all taking place at the same time and - potentially - in different administrative domains.

As a result of these trends, a range of novel and challenging research questions arise, which can be answered only by applying techniques from the parallel, distributed, and Grid computing fields. The scope of this topic therefore embraces issues from high-performance processing, coding, indexing, and retrieval of multimedia data over parallel architectures for multimedia servers, databases and information systems, up to highly distributed architectures in heterogeneous, wired and wireless networks.

This year 7 papers were submitted to this topic area. We thank all the authors for their submissions. All the papers were reviewed by 4 referees, and 4 papers were ultimately selected (although unfortunately one paper was subsequently withdrawn, leaving 3 papers to be presented). The quality of the submissions was extremely high. In the resulting session you will find two particularly distinguished papers each of which was in the best 10 papers conference-wide. These are: "Supporting Reconfigurable Parallel Multimedia Applications", and "Providing VCR in a Distributed Client Collaborative Multicast Video Delivery Scheme". Our third paper, which was also highly rated, is entitled: "Linear Hashtable Predicted Hexagonal Motion Estimation Algorithm for Parallel Video Processing".

Supporting Reconfigurable Parallel Multimedia Applications

Maik Nijhuis, Herbert Bos, and Henri E. Bal

Department of Computer Science, Vrije Universiteit, Amsterdam
{maik, herbertb, bal}@cs.vu.nl

Abstract. Programming multimedia applications for System-on-Chip (SoC) architectures is difficult because streaming communication, user event handling, reconfiguration, and parallelism have to be dealt with. We present Hinch, a runtime system for multimedia applications, that efficiently exploits parallelism by running the application in a dataflow style. The application has to be implemented as components that communicate using streams. Reconfigurability is supported by a generic component interface. Measurements have been performed on a SpaceCake SoC architecture simulator. Hinch can easily be ported to other shared-memory architectures.

1 Introduction

The problem we address in this paper is the complexity of programming embedded System-on-Chip (SoC) architectures with multiple processing units operating in parallel. These architectures are becoming more and more popular in the field of consumer electronics, especially in the area of multimedia applications. This trend, which is already visible, is likely to become even more important in the future for several reasons. First, media applications increasingly require complex processing, for example new coding algorithms and dynamic picture-in-picture. Second, the processing is applied to increasing amounts of data per time-unit, such as multichannel HDTV. Third, media applications often exhibit much potential parallelism. Fourth, hardware vendors have already opted for multi-core processors as the key to speed, partly because it is hard to crank up clock speeds at the same rate as in the past (e.g., Cell[1], Network processors[2], Xeon, and Opteron). Unfortunately, programming such parallel hardware is challenging and very much an open research problem.

In this paper, we present a solution that facilitates application development for SoC architectures. While our target domain includes a broad range of applications, we focus our examples and discussion on TV sets, for ease of explanation. The SpaceCake SoC architecture[3], developed by Philips, is used as the experimentation platform. For testing purposes, the applications can also run natively on Linux.

By careful observation of several applications, we obtained a core set of properties that a multimedia run time system should cater for. We used these properties to derive requirements for our system. Our system, named Hinch, exploits the following properties of typical multimedia applications:

1. The application consists of several kernels that perform a specific operation, such as motion estimation. To manage large numbers of kernels, we should be able to group

them into *components*, which in turn can be grouped into higher-level components. The resulting application will be organized hierarchically like a tree with kernels at the leaf nodes.
2. The configuration of kernels changes as a result of asynchronous user inputs and other events. For example, by pressing a button a user may add a picture-in-picture to the television screen. Hinch supports events and allows the kernel configuration to change at run time. As we typically do not want to stop the complete application as reconfiguration occurs, we allow subtrees in the tree-based hierarchy to be reconfigured without interfering with user experience.
3. Individual nodes in the tree communicate either via streaming channels (e.g., a motion estimator kernel calculates motion vectors that are used by a motion compensated de-interlacer kernel), or via events (e.g., a component sends an event that a sub-program has started). Hinch supports both streaming and events.
4. Multimedia applications exhibit both task- and data-parallelism. Hinch is able to map the tree-based component hierarchy on hardware such that both forms of parallelism are exploited. For task parallelism this implies that different kernels are mapped on different functional units. For data parallelism, multiple instances of a kernel have to run concurrently.

While it is well known how to support the individual properties, to the best of our knowledge, Hinch is the first system that supports all, while greatly simplifying the SoC programmers' task. Moreover, measurements show Hinch incurs only little overhead and achieves a parallelization efficiency of about 95 % with 9 processors. The major difficulties we encounter are combining task- and data-parallelism[4], supporting dynamic reconfiguration and handling asynchronous events. Hinch can be used as a lower layer in a programming environment for building multimedia applications. We plan to combine Hinch with higher level layers such as SPC-XML[5].

The remainder of this paper is organized as follows. In Section 2 we will explain reconfigurability requirements. In Section 3 we will describe the design of Hinch. In Section 4 we will show the results of using Hinch on the SpaceCake architecture. Related work is discussed in Section 5. Finally, the paper is concluded in Section 6.

2 Reconfigurability

Many multimedia applications need support for reconfiguration. This can be due to user input (e.g., the user wants to add a picture-in-picture), or to available resources (e.g., scaling down quality when less bandwidth is available). Reconfiguration can be performed by adjusting parameters of application components (component reconfiguration) or by adding and removing components while the application is running (application reconfiguration). In this section, we give two examples of reconfigurable applications. Both applications are used for the experiments described in Section 4.

2.1 Add/Remove Components

In a dynamic Picture-In-Picture (PiP) application, the user can add or remove small subpictures (of different TV channels) on the screen. The structure of this application is shown in Fig. 1. The downscale components reduce the size of their inputs and the blend component merges the downscaled images into the main background image. When a picture is added, input and downscale components are created and connected to the blender. The blender is then notified that it has to blend one more picture into its output. When a picture is deleted, the blender is notified, the connection to the blender is removed and the picture-in-picture input and downscale components are destroyed. The notifications to the blender are an example of component reconfiguration.

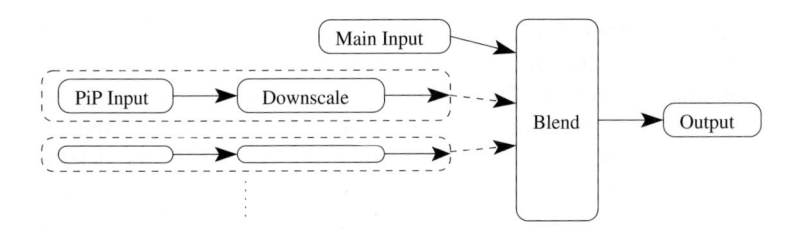

Fig. 1. Picture-in-Picture application

2.2 Replace Components

Temporal upscaling (increasing the frame rate) of a movie can be done at different quality levels that have different computation requirements. A trivial temporal upscaler simply copies existing images to create new images. An advanced temporal up scaler can perform motion estimation and use the motion vectors to compute the new images.

We have created an application ('Tups') that performs temporal upscaling of an image sequence by a given factor. Copy mode and motion estimation mode are both supported. The application layout for both modes is shown in Fig. 2. The application can dynamically switch between the two modes by replacing the middle component, for instance depending on available resources.

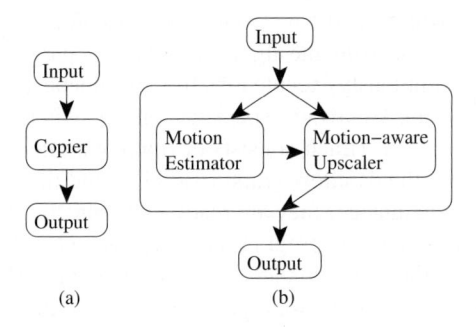

Fig. 2. Tups application: copy mode (a) and motion estimation mode (b)

3 Run Time System

In this section we explain the design of Hinch and the model for applications that use it. Hinch, as well as the applications that use it, are written in C. Hinch consists of several modules that provide functionality to the application and/or to other Hinch modules. These modules are also described below.

3.1 Application Layout

A Hinch application consists of components that are actors in a dataflow process network[6]. The application is run by executing *iterations* of the dataflow graph, in which each actor is fired one or more times. One firing corresponds to running one iteration of the component. A graph iteration begins by scheduling the initial component(s). The other components are scheduled as soon as their predecessors in the dataflow graph have finished. There is no restriction on the shape of the dataflow graph. For a video processing application, the components typically contain image processing kernels. One iteration then consists of processing one image frame from the video stream.

Dataflow graphs can be nested using special grouping components. Grouping components contain child components and connect these into a dataflow graph. When an iteration of a grouping component is run, an iteration of the dataflow graph inside the grouping component is run. When this 'inner' iteration has finished, the successors of the grouping component in the higher level dataflow graph are scheduled.

All components have a generic interface, which provides an abstraction of the component. This interface contains functions to create, configure and destroy instances of the component, to get its properties, and to run an iteration of the component. A grouping component has extra functions to add, remove, replace, and (dis)connect its contained children. All functions except the run function are used to configure the application, at the start of the application. An application can also be reconfigured at run time using these functions. The connections between children correspond to dataflow dependencies. Connections can also include a data stream, as will be explained in the next subsection.

A central job queue in shared memory is used to accomplish automatic load balancing. To enable usage of Hinch on distributed memory machines, we plan to scale the job queue to a distributed version using the algorithms of [7]. A parallel program consists of multiple threads that continuously execute jobs from the queue, and add new jobs to the queue that are ready to run. We avoid expensive context switches by running a single thread per processor. A job in Hinch consists of running an iteration of a component. When a job has finished, the dataflow graph is used to find the successors of this job. These are added to the job queue if they are ready to be run.

To build a parallel program, the programmer only has to build components, specify the connections between these components, and call some initialization routines. Using the job queue, Hinch makes sure that the program runs in parallel. We plan to add an XML layer on top of Hinch for specifying the component connections. This XML layer can then also be used for performance prediction using PAM-SoC[8].

The Hinch application model resembles that of Koala[9]. A Hinch component iteration corresponds to the execution of a Koala task. As in Koala, Hinch components can only be coupled if their interfaces match and components can be grouped recursively.

3.2 Streams

Components can communicate in a streaming fashion using a stream module. This modules provides similar streaming functions as the StreamIt[10] language and a communication abstraction similar to Space-Time Memory[11]. A component does not know with whom it is communicating. It merely has to read and write the appropriate streams, which are parameters to the `run` function. This way, a component can easily be reused in another part of the application or in a different application.

Streams are implemented using an efficient zero-copy protocol. The producer can write to the stream after allocating a write buffer. When it is finished writing, it commits the buffer. The consumer can then read the data from the same buffer. The zero-copy protocol is only possible at shared memory machines. For distributed memory machines, a different implementation must be used.

Multicast streams are also supported by Hinch. These streams are shared by multiple dataflow connections. There is also support for reading a fixed amount of old data, which has been used in previous iterations, from the stream. This is similar to the peek functionality in StreamIt[10].

3.3 Task Parallelism

Task parallelism is supported by Hinch in two ways, as shown in Fig. 3. The circles indicate components, and the arrows indicate connections between them. We assume one iteration of the application processes one image frame using multiple image processing components. The first type of task parallelism is running multiple iterations concurrently in a pipeline style. When component A has finished frame 1, A could be processing frame 2 while B is processing frame 1. The second type of task parallelism is running independent tasks concurrently. Components V and W have no dependency so they can be concurrently active in the same iteration. Both types of task parallelism can be combined. For example, if V and W both have finished frame 2 they can start processing frame 3 while X is processing frame 2.

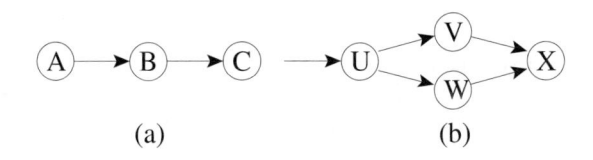

Fig. 3. Task parallelism. Pipeline-style (a) and independent tasks (b)

3.4 Data Parallelism

Normally, a component iteration processes the data it gets by reading its input streams once (say one image frame). However, often this processing can be done in parallel, in which case a component iteration consists of processing a *slice* (multiple lines) instead of a frame. Hinch has slicing helper functions for components, which contain common code needed to code slicing. These functions tell the component which slice of which frame is to be processed, do the appropriate stream reading and writing, and handle out-of-order execution of sliced iterations. By using these functions, exploiting data parallelism becomes easy.

One might argue that data parallelism can also be obtained by reducing the stream granularity from a frame to a slice. However, processing a slice usually requires the data in the previous and next slice for the pixels at the boundary, for example with convolution kernels. Programming a component is more difficult in this case because boundary conditions have to be dealt with. In our approach, this is not necessary since the whole frame is in a continuous memory area.

3.5 Reconfiguration and Event Handling

Reconfigurability is supported by the general component interface. Components can be dynamically created, destroyed, grouped, and connected at run time. To avoid race conditions, the application parts that are to be reconfigured are made idle before reconfigurating.

The replace function in the grouping component interface can replace a child component by a component that has the same I/O interface. Without the replace function, this has to be accomplished by removing all connections to the old child, removing the child, adding the new child, and creating the connections to the new child. With the replace function, removing and creating the connections is not necessary because the new child has the same interface.

Asynchronous events can easily be handled by buffering them in an event queue, as shown in Fig. 4. The event queue is periodically emptied by the manager component, which is run at the end of every iteration. It regulates the number of concurrent active iterations in the application using the control flow connection to the start of the application. The manager component can halt the program for reconfiguration by lowering the number of active iterations to zero.

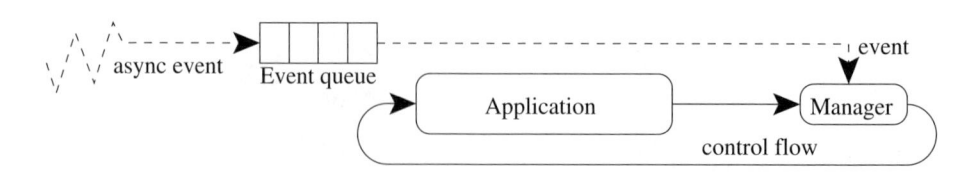

Fig. 4. Event queue (Application overview)

4 Experiments

To verify the usefulness of Hinch, we measured parallelization efficiency and reconfigurability overhead for the picture-in-picture (PiP) and temporal up scaler (Tups) applications. The temporal up scaler doubles the amount of image frames in its input stream by inserting a new image frame after each image. All measurements were done using 720x576 video files. I/O (reading the files and writing the final result) is not included in the measurements.

Since SpaceCake[3] hardware is not yet available, all experiments are run using simulation software, which simulates a tile with multiple TriMedia cores. A TriMedia is a VLIW processor aimed at multimedia applications. At a tile, each TriMedia has its own level 1 cache. The level 2 cache is shared between all TriMedias. The SpaceCake architecture allows multiple tiles to be combined. We plan to add support for multiple tiles in the future.

The TriMedia cycle counter is used for all measurements. For the experiments on a single processor, we turned off all inter-processor synchronization, e.g., locking of shared variables. All other measurements use a parallel version. The number of slices is set to 4 for the picture-in-picture applications and 9 for the temporal up scaling applications. These settings yielded the best results.

4.1 Parallelism

Figure 5 shows the speedup of processing 96 image frames with four variants of the PiP application. These variants (PiP-0, PiP-1, PiP-2 and PiP-3) process zero, one, two, and three pictures-in-picture, respectively. PiP-0 does not exhibit much speedup because it is a trivial application that merely copies its input to its output. However, the speedup stays constant when run at a larger number of nodes, which shows that the overhead of Hinch does not increase with the number of nodes. PiP-1 reaches maximum speedup at 8 nodes. There is no more parallelism to exploit when PiP-1 is run at 9 nodes.

Figure 6 shows the speedup of the Tups application in copy mode (tups-copy), motion estimation mode (tups-me), and in a reconfigurable mode (tups-reconf). Tups-copy creates the newly inserted images by copying the previous image. Tups-me generates the newly inserted images from the two adjacent images using motion estimation techniques. Tups-reconf is a mixture of tups-copy and tups-me and will be explained in the next subsection. These applications process 68 image frames. Again, the trivial application (tups-copy) does not exhibit a good speedup and its speedup stays constant when run at a larger number of nodes.

Both Fig. 5 and Fig. 6 show that Hinch provides the means to efficiently parallelize these multimedia applications. At nine nodes, the efficiency of PiP-2 and tups-me are 94,2 % and 95,8 %, respectively.

4.2 Reconfigurability

We have created four reconfigurable applications. Three of these are variants of the PiP application, the other is a variant of the Tups application. These applications process an equal amount of frames as their non-reconfigurable counterparts. The four variants are:

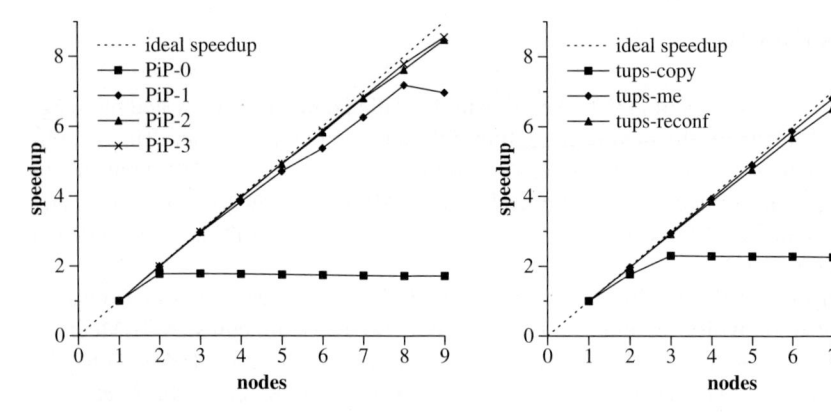

Fig. 5. PiP application speedup

Fig. 6. Tups application speedup

1. PiP-01. Half of the frames have no pictures-in-picture, the other half have one picture-in-picture.
2. PiP-12. Half of the frames has one picture-in-picture, the other half has two pictures-in-picture.
3. PiP-012. One third of the frames has no pictures-in-picture, one third has one picture-in-picture, and one third has two pictures-in-picture.
4. Tups-reconf. Half of the generated frames is generated in copy-mode. For the other half motion-estimation mode is used.

All reconfigurations are matched by their inverse to cancel out latency differences. Because we are interested in average reconfiguration latency, multiple reconfiguration pairs are performed at regular intervals. The PiP variants perform eight reconfigurations in total. Tups-reconf performs four reconfigurations.

Overhead. To measure reconfigurability overhead, we compared the run time of these applications to the run time of equivalent static applications. For example, the run time of PiP-01 is compared to the average run time of PiP-0 and PiP-1.

Figure 7 shows the overhead of the reconfigurable applications using one to eight processors. We have omitted measurements at 9 processors due to the result of PiP-1 at 9 processors. (PiP-1 does not scale beyond 8 processors, and all PiP overheads are (partly) based on the performance of PiP-1.) An overhead factor of 1.01 means the reconfigurable program is 1 percent slower than the corresponding static applications.

Although reconfiguration occurs very often (once every 12 frames for the PiP applications, once every 17 frame for Tups-reconf), the overhead is at most 8 %. When the application is stopped for reconfiguration, the amount of parallelism in the application drops until the application is run sequentially. Thus, on average there is less parallelism to exploit in the reconfigurable applications and the reconfigurable applications will perform relatively worse on larger numbers of nodes. This causes the reconfigurability overhead to increase with the number of nodes, which is is clearly visible in Fig. 7.

Latency. We have measured the latency of reconfigurations. This is the time between the occurrence of the asynchronous event and the completion of the reconfiguration. The latency includes waiting until the application is idle and performing the reconfiguration.

Figure 8 shows the average reconfiguration latency. Because anomalies occur at 1 and 2 nodes, we have not included measurements at 1 and 2 nodes. These anomalies occur when the event is generated just before or after many computations. For example, in a sequential application it can happen that all active iterations are about to start the manager (see Fig. 4). When an event occurs at this point, there are no computation jobs in the job queue (only small manager jobs) and the application quickly becomes idle. On the other hand, latency will be high if many computation jobs have been scheduled when the event occurs. When the applications are run at higher number of nodes this effect becomes less visible and the average latency goes to 40 ms which is a single image frame in a 25 Hz video stream.

The individual latency measurements show that the latency depends on the complexity of the program before reconfiguration. This is because the complexity determines the (average) number of outstanding computations and thereby the time before the application is idle. For example, in the Tups-reconf application, the latency of going from copy-mode to motion-estimation mode is 10 ms (at 3 or more nodes). The latency of going from motion-estimation mode to copy-mode varies between 56 and 85 ms at 3 or more nodes.

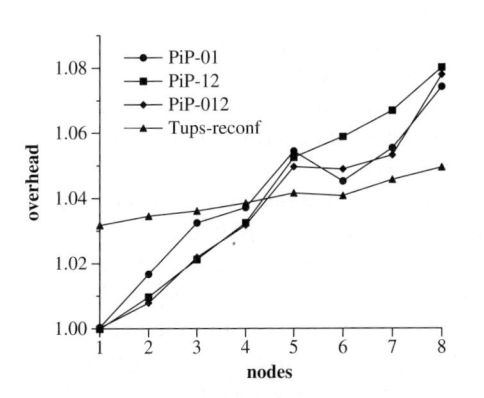

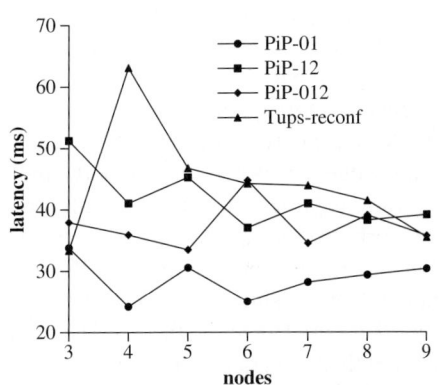

Fig. 7. Reconfigurability overhead

Fig. 8. Reconfigurability latency

5 Related Work

There are many other systems that simplify programming multimedia applications for embedded architectures by providing abstractions. Often these systems include hardware design. Some systems mainly focus on hardware design, like Cheops[12] and Imagine[13].

At Philips Research many of these systems have been developed, e.g., TTL[14], YAPI[15] and C-HEAP[16]. These systems model an application as a Kahn Process Network (KPN)[17], which is a number of independent tasks that communicate using

FIFO channels. These tasks and FIFO channels can be implemented in hardware or software, using shared or distributed memory. Task parallelism is supported this way, however, data parallelism is not. Load balancing is mostly done statically by mapping the tasks to fixed resources. C-HEAP has the most advanced reconfiguration support of these systems.

The SmartCam project[18] aims at simplifying building image processing applications for use in smart camera's. Skeletons[19] are used to provide an implementation abstraction and to exploit data parallelism. Tasks can be mapped on different hardware in different (heterogeneous) hardware architectures. Memory is distributed in these architectures. Unlike Hinch, SmartCam does not do reconfiguration or dynamic task scheduling. User events are absent in the SmartCam application domain.

The Model Integrated Real-Time Imagine Processing System[20] is a programming environment for building image processing applications, which are run on a network of DSPs. MIRTIS generates a parallel image processing application from sequential kernel code, data dependencies and the application graph. The generated application includes a run time system for a distributed computing platform. MIRTIS supports both data and task parallelism. The mapping of the application to the parallel hardware is done statically. Dynamic load balancing and runtime reconfiguration are therefore not supported. User events do not occur in the applications MIRTIS supports. MIRTIS deliberately does not have support for (designing) specialized hardware.

The Nizza framework[21] has a similar structure as Hinch. It also processes streaming multimedia applications in dataflow-style, allowing task parallelism. Data parallelism can be exploited using 'combinatorial' modules. The application can stop Nizza if it wants to perform reconfiguration and restart Nizza afterwards. Unlike Hinch, Nizza targets desktop applications instead of embedded applications. Therefore, Nizza does not have support for (designing) specialized hardware and distributed memory. To our knowledge, Nizza does not support handling user events.

Various projects are dealing with programming SoC architectures in the domain of network processing. Among these projects are NP-Click[22], NEPAL[23], Shangri-La[24], the system described in [25], and Netbind [26]. Like multimedia applications, network processing applications also process streams of data (network packets) by multiple kernels. However, these kernels are much smaller than multimedia kernels. To exploit this fine grained parallelism, the systems cooperate closely with the hardware.

Table 1 provides a summary of the features of the mentioned multimedia programming systems and Hinch. The first two columns indicate the presence of support for

Table 1. Comparison of related work. '+' = excellent support, 'o' = supported, '-' = bad / no support, 'N/A' = not applicable

Feature	Task par.	Data par.	Load balancing	Dist. mem	Reconfigurability	Events	Hardware
C-HEAP	+	-	o	o	+	o	o
SmartCam	+	+	o	o	-	N/A	o
MIRTIS	+	+	o	o	-	N/A	-
Nizza	+	o	+	-	o	-	-
Hinch	+	+	+	-	+	+	-

task- and data-parallelism, respectively. The load balancing column indicates the quality of the load balancing features of the system. The Dist. mem column indicates if the system targets distributed memory architectures. A '-' in this column means the system targets shared memory architectures. The events column indicates if the system has support for handling asynchronous user events. Finally, the hardware column shows if the system has support for specialized acceleration hardware in the target architecture. Table 1 shows that distributed memory and special hardware are currently not supported by Hinch. We plan to include support for this in the future.

6 Conclusion

We have presented Hinch, a runtime system for multimedia applications. Hinch has support for streaming, event handling, reconfiguration, data parallelism and task parallelism, amongst others. Hinch also provides automatic load balancing of the application when run on a shared-memory architecture, by running the application in a data-flow style. Experiments show that applications using Hinch can be efficiently parallelized and the overhead of reconfigurating a running program is low.

Future work includes building a layer on top of Hinch that provides a simple interface for specifying multimedia applications. This layer will automatically generate optimized applications that use Hinch. We plan to include performance prediction in this layer. Other future work tracks are adding support for specialized hardware and more complex memory architectures, such as multiple SoC tiles.

Acknowledgments

We would like to thank Philips Research, especially Paul Stravers, for their support. This work is performed in the context of STW/PROGRESS project DES.6397, which is supported by the Dutch government.

References

1. Kahle, J.A., Day, M.N., Hofstee, H.P., Johns, C.R., Maeurer, T.R., Shippy, D.: Introduction to the Cell multiprocessor. IBM Journal of Research and Development **49**(4/5) (2005) 589
2. Intel Corporation: Network processors. http://www.intel.com (2006)
3. Stravers, P., Hoogerbrugge, J.: Single chip multiprocessing for consumer electronics. In Bhattacharyya, ed.: Domain-Specific Processors. Marcel Dekker (2003)
4. Bal, H.E., Haines, M.: Approaches for integrating task and data parallelism. IEEE Concurrency **6**(3) (1998) 74–84
5. González-Escribano, A., van Gemund, A.J., noso Payo, V.C.: An XML structured representation for nested-parallel programming languages. In: Proc. CPC, Chiemsee (2004) 149–160
6. Lee, E.A., Parks, T.M.: Dataflow process networks. In: Proc. of the IEEE. (1995) 773–799
7. van Nieuwpoort, R.V., Kielmann, T., Bal, H.E.: Efficient load balancing for wide-area divide-and-conquer applications. In: Proc. PPoPP'01, Snowbird, UT (2001)
8. Varbanescu, A.L., Sips, H., van Gemund, A.: PAM-SoC: A toolchain for predicting MPSoC performance. In: Proc. EuroPAR '06, Dresden (2006)

9. van Ommering, R., van der Linden, F., Kramer, J., Magee, J.: The Koala component model for consumer electronics software. Computer **33**(3) (2000) 78–85
10. Thies, W., Karczmarek, M., Amarasinghe, S.P.: StreamIt: A language for streaming applications. In: Proc. CC 2002, Grenoble, France (2002) 179–196
11. Ramachandran, U., Nikhil, R.S., Harel, N., Rehg, J.M., Knobe, K.: Space-time memory: A parallel programming abstraction for interactive multimedia applications. In: Proc. PPoPP'99, Atlanta, Georgia (1999) 183–192
12. Bove, V., Watlington, J.: Cheops: A reconfigurable data-flow system for video processing. IEEE Trans. on Circuits and Systems for Video Technology **5**(2) (1995) 140–149
13. Serebrin, B., Owens, J.D., Chen, C.H., Crago, S.P., Kapasi, U.J., Khailany, B., Mattson, P., Namkoong, J., Rixner, S., Dally, W.D.: A stream processor development platform. In: Proc. 20th International Conference on Computer Design, Freiburg, Germany (2002)
14. van der Wolf, P., de Kock, E., Henriksson, T., Kruijtzer, W., Essink, G.: Design and programming of embedded multiprocessors: an interface-centric approach. In: Proc. CODES+ISSS. (2004) 206–217
15. de Kock, E.A., Smits, W.J.M., van der Wolf, P., Brunel, J.Y., Kruijtzer, W.M., Lieverse, P., Vissers, K.A., Essink, G.: Yapi: application modeling for signal processing systems. In: Proc. 37th Design Automation Conference, New York, NY, USA, ACM Press (2000) 402–405
16. Nieuwland, A., Kang, J., Gangwal, O.P., Sethuraman, R., Busá, N., Goossens, K., Llopis, R.P., Lippens, P.: C-HEAP: A heterogeneous multi-processor architecture template and scalable and flexible protocol for the design of embedded signal processing systems. Design Automation for Embedded Systems **7**(3) (2002) 233–270
17. Kahn, G.: The semantics of a simple language for parallel programming. In Rosenfeld, J.L., ed.: Information processing, Stockholm, Sweden, North Holland, Amsterdam (1974) 471–475
18. Caarls, W., Jonker, P., Corporaal, H.: Skeletons and asynchronous RPC for embedded data- and task parallel image processing. In: Proc. MVA2005, Tokyo (2005) 384–387
19. Cole, M.I.: Algorithmic Skeletons: Structured Management of Parallel Computation. MIT Press & Pitman (1989)
20. Moore, M.S., Sztipanovitz, J., Karsai, G., Nichols, J.: A model-integrated program synthesis environment for parallel/real-time image processing. In: Proc. Par. Dist. Methods for Image Processing. (1997) 31–45
21. Tanguay, D., Gelb, D., Baker, H.H.: Nizza: A framework for developing real-time streaming multimedia applications. Technical Report HPL-2004-132, HP Labs, Palo Alto (2004)
22. Shah, N., Plishker, W., Keutzer, K.: NP-Click: A programming model for the Intel IXP1200. In: Proc. NP-2 conjunction with HPCA-9, Anaheim, California (2003)
23. Memik, G., Mangione-Smith, W.: Nepal: A framework for efficiently structuring applications for network processors. In: Proc. NP-2 in conjunction with HPCA-9, Anaheim, California (2003)
24. Chen, M.K., Li, X.F., Lian, R., Lin, J.H., Liu, L., Liu, T., Ju, R.: Shangri-La: achieving high performance from compiled network applications while enabling ease of programming. In: Proc. ACM SIGPLAN PLDI, New York, NY, USA, ACM Press (2005) 224–236
25. Ramaswamy, R., Weng, N., Wolf, T.: Application analysis and resource mapping for heterogeneous network processor architectures. In: Proc. NP-3 in conjunction with HPCA-10, Madrid, Spain (2004) 103–119
26. Campbell, A.T., Chou, S.T., Kounavis, M.E., Stachtos, V.D., Vincente, J.: Netbind: A binding tool for constructing data paths in network processor-based routers. In: Proc. IEEE OPENARCH, New York, NY (2002)

Providing VCR in a Distributed Client Collaborative Multicast Video Delivery Scheme*

X.Y. Yang[1], P. Hernández[1], F. Cores[2], A. Ripoll[1],
R. Suppi[1], and E. Luque[1]

[1] Computer Science Department, ETSE, Universitat Autònoma de Barcelona,
08193-Bellaterra, Barcelona, Spain
[2] Computer Science & Industrial Engineering Department, EPS, Universitat de Lleida, 25001, Lleida, Spain

Abstract. In order to design a high scalable video delivery technology for VoD systems, two representative solutions have been developed: multicast and P2P. Each of them has limitations when it has to implement VCR interactions to offer true-VoD services. With multicast delivery schemes, part of system resources has to be exclusively allocated in order to implement VCR operations, therefore the initial VoD system performance is considerably reduced. The P2P technology is able to decentralize the video delivery process among all the clients. However, P2P solutions are for video streaming systems in Internet and do not implement VCR interactivity. Therefore, P2P solutions are not suitable for true-VoD systems. In this paper, we propose the design of VCR mechanisms for a P2P multicast delivery scheme. The new mechanisms coordinate all the clients to implement the VCR operations using multicast communications. We compared our design with previous schemes and the results show that our approach is able to reduce the resource requirements by up to 16%.

1 Introduction

Recent advances in high-performance network and video codification technology have made it feasible for the Video on Demand (VoD) server to implement the interaction capabilities of a classic Video Cassette Recorder (VCR), offering the true-VoD service. However, VCR interactions, such as pause or fast forward, increase the resource requirements in the delivery process and reduce the VoD system performance.

Certain researchers have proposed delivery policies that take advantage of the multicast feature. A multicast scheme allows clients to share delivery channels and decrease the server and network resource requirements. Patching multicast policy [3], for example, dynamically assigns clients to join on-going multicast channels and patches the missing portion of video with a unicast channel. The disadvantage of a multicast solution is the complexity of implementing interactive operations, because there is not a dedicated channel per client.

---
* This work was supported by the MCyT-Spain under contract TIC 2004-03388.

In [2], the authors introduce techniques to implement jump operations. The techniques are able to join the client with another on-going multicast channel (B-Channels), after a jump operation. The B-Channel has to be delivering the desired new playback point of the client. Emergency channels (I-Channels) are used in case no such B-Channel exists. *Split and Merge* (SAM) in [7], is a protocol to merge I-Channels with B-Channels, reducing the cost of a VCR operation. In SAM, the merge process is performed by synchronized buffers that are statically allocated in the central access nodes. The centralized buffer management of SAM is not scalable (the VCR request blocking probability grows linearly in accordance with the VCR frequency). In [1], the authors decentralize buffer among clients to reducing the merging time and the VCR operation resource requirements. In [4], the authors analyze the optimal number of I-Channels that a multicast VoD system has to allocate in order to implement VCR operations. Despite the fact that multicast policies are able to offer true-VoD with VCR, part of the server resource is, exclusively, allocated for VCR interactions. Consequently, VCR operations considerably reduce the system global performance.

Most recently, the peer-to-peer (P2P) paradigm has been proposed to decentralize the delivery process to all clients. Delivery schemes like Chaining, DirectStream o P2Cast [6] are the most representatives. In these schemes, clients cache the most recently received video information in the buffer and forward it to the next clients using unicast channels. Other P2P architectures such as CoopNet and PROMISE[5] assumes that a single sender does not have enough outbound-bandwidth to send one video and use n senders to aggregate the necessary bandwidth. All previous P2P schemes use unicast communications between clients producing a high network overhead. Furthermore, since a client just sends data to only one client, the unicast P2P schemes achieve poor client collaboration efficiency. In [9], we proposed a P2P delivery scheme, called DDCM. In DDCM, each client (one peer) is able to send video information to a set of m clients using only one multicast channel. Furthermore, the DDCM is able to synchronize a set of clients (n peers) to create one collaboration group to replace the server in order to send video information to m peers, providing a collaboration mechanism from n-peers to m-peers. In [10], the collaboration mechanisms of DDCM are incorporated in a distributed VoD architecture with multiple service nodes. Each service node is able to create an independent P2P system to extend the global scalability.

In this paper, we propose mechanisms to implement VCR operations in the DDCM scheme. In our design, any client actively collaborates with the server to implement the join-back process of VCR operations. The new mechanisms are able to create local channels to send different playback points of one video. The jump operations could be implemented by the local channels without requiring the server resource. In a pause operation, a client extends the service time and, consequently it also extends the collaboration time. The mechanisms are able take advantage of pause operations to increase the client collaboration efficiency and reduce the system resource requirement. The reduced resource requirement by pause operations is used implement fast-forward and reverse operations.

In our study, we evaluated DDCM performance with new VCR mechanisms, compared with Patching policy. With a normal VOD system workload and with new VCR mechanisms, the experimental results show our approach is able to reduce the requirement up to 16%.

The remainder of this paper is organized as follows: we dedicate the section 2 to show the key ideas behind our VCR operation mechanisms. Performance evaluation is shown in section 3. In section 4, we indicate the main conclusions of our results and future studies.

2   P2P VoD Architecture to Provide VCR Operations

Our VoD architecture is based on multicast communications and client collaborations that decentralize the video information delivery process. Our design is not a server-less P2P architecture; the server has every video in the service catalogue. The server is responsible for establishing every client collaboration process. The collaboration scheme is designed as two policies: Patch Collaboration Manager (PCM) and Multicast Channel Distributed Branching (MCDB). The objective of PCM is to create multicast channels to service groups of clients, and allows clients to collaborate with the server in the delivery of portions of video. The objective of MCDB, however, is to establish a group of clients to eliminate on-going multicast channels that have been created by PCM.

In the explanation, we assume that video is encoded with a Constant Bit-Rate (CBR). The video information is delivered with invariable size network packets and a video is composed by L *video blocks*. Furthermore, we assume that each client has local buffers to cache a limited number of *video blocks*.

Different VCR operations, such as slow motion, are considered in [1], but we concentrate our explanation on jump forward, jump backward, fast forward, fast reverse and pause which are the most typical VCR operations. We will first explain the jump operation and then we will comment on other VCR operations that could be implemented with jumps. The explanation of our new VCR mechanisms consists of 4 points. We dedicate first two points (Section 2.1 and 2.2) to overview the client collaboration mechanism, third point to presenting details about the jump operation and the last point to discuss remaining VCR operations.

2.1   New Client Admission by PCM

Fig.1a shows the main idea of PCM. When the server is sending video block 3 to the channel $Ch3$, the client (C4) sends a control message to the server, indicating the video of interest and the size of the local buffer. The PCM tries to find an on-going multicast channel that is sending the requested video to service the client. Only on-going multicast channels ($Ch3$) with offset ($O(C3h)$) smaller than client-buffer size could be used to service the client request. The $O(Ch3)$ of channel $Ch3$ is the number of the video block that the channel $Ch3$ is currently sending. The multicast channel is called Complete Stream. The client will not receive the first portion of the video from the Complete Stream, since these

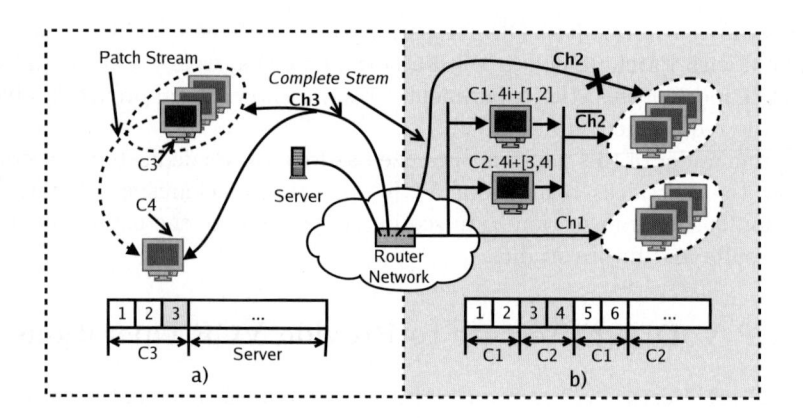

Fig. 1. a) PCM Collaboration Process. b)MCDB Collaboration Process.

video blocks (1, 2 and 3) have already been delivered by the server; therefore the server needs another channel to send the first portion of the video (the channel is called Patch Stream). In order to create the Patch Stream, the server finds a *collaborator* and requests the client's collaboration. The *collaborator*($C3$) has to have the desired video blocks and enough bandwidth to create the Patch Stream. If there is no any available *collaborator*, the server creates the Patch Stream using server resources. Finally, the client joins the Complete Stream, establishes communication with the *collaborator* and starts receiving video blocks.

In the establishment of client collaboration by PCM, the policy needs information about clients' availability for collaboration. In our design, the server uses a table structure (*Collaborator_Table*) to register the information about *video blocks* that are cached by each client in its buffer. Each client is responsible for announcing their availability to collaborate with server. The announcement message is sent only once after the client admission or after a VCR operation; the network overhead is therefore negligible.

2.2 Branching Process of MCDB Policy

Fig.1b shows the main idea behind the MCDB. The MCDB periodically checks the *Collaborator_Table* in order to find out if there are enough *collaborators* to replace some on-going multicast channels. Every channel having another channel that is sending the same video, but with a larger offset, is a candidate to be replaced by MCDB. The MCDB replaces an on-going multicast channel ($Ch2$) with a local multicast channel $\overline{Ch2}$. In order to create the local channel, a group of collaborative clients are synchronized to cache video blocks from another multicasting channel ($Ch1$). The cached blocks are delivered by the collaborative clients to generate the channel $\overline{Ch2}$. When $Ch2$ is replaced by $\overline{Ch2}$, we say that $Ch2$ is branched from $Ch1$ and $\overline{Ch2}$ is a *branch*-channel of $Ch1$.

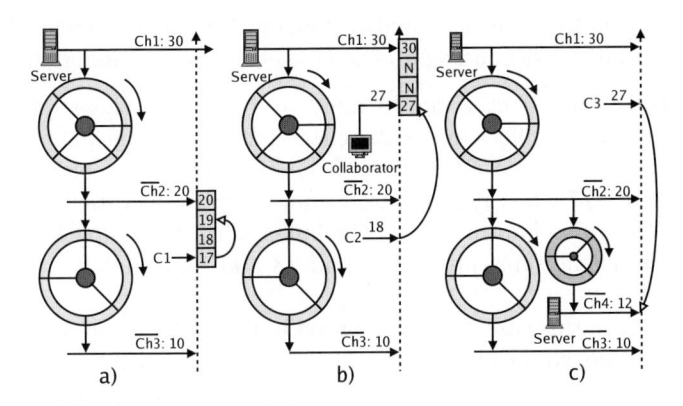

Fig. 2. Jump Operation: Serviced by a) the Client Buffer. b) by PCM. c) by MCDB.

In the example of Fig.1b, we have two channels ($Ch1$ and $Ch2$) that are separated by a gap of 2 video blocks[1]. In order to replace channel $Ch2$, MCDB selects clients C1 and C2 to create a group of collaborators. Clients C1 and C2 are both able to cache 2 blocks and a total of 4 video blocks can be cached by these two collaborators. The C1 caches every block of $Ch1$ whose block-number is $4i+1$ or $4i+2$, being $i = [0..L/4-1]$. For example, C1 has to cache blocks 1, 2, 5, 6 and so on. C2 caches every block of $Ch1$ whose block-number is $4i+3$ or $4i+4$ (3, 4, 7, 8 and so on). All the cached blocks have to be in the collaborator's buffer for a period of time. In this case, the period of time is the playback time of 2 video-blocks. After the period of time, the cached blocks are delivered to channel $\overline{Ch2}$ which is used to replace $Ch2$. It is not difficult to see that the MCDB is able to create multiple local multicast channels with different offsets to collaborate the delivery process of several points of a video. In the case of Fig.1b, the offset of the local multicast channels ($\overline{Ch2}$) is 2 video-blocks lower than $Ch1$'s.

In the branching process, two parameters are determined by the MCDB: 1) The client collaboration buffer size (B_{C_i}). It is the buffer size of client C_i used by MCDB. 2) Accumulated buffer size (BL) is the total size of the collaborative buffer ($\sum_{C_i \in CG} B_{C_i}$, being CG the group of clients). The value of these two parameters is determined by MCDB under 2 constraints: a) A client cannot use more buffer than it has. b) A client only uses one channel in the collaboration process. For more details, see [9].

2.3 Jump Forward/Backward Operation

The Fig.2a shows how the server delivers minute 30 of a video with a multicast channel ($Ch1$). The *branch*-channel ($\overline{Ch2}$) is created by MCDB and is delivering

---

[1] The separation or gap ($G(Ch2, Ch1)$) between two channels is calculated as $O(Ch1) - O(Ch2)$, being $O(Ch1) \geq O(Ch2)$.

minute 20 of the video. The local multicast channel ($\overline{Ch3}$) is branched from $\overline{Ch2}$. There are 3 situations depending on the new playback position after a jump.

Situation 1: The client wants to jump to a position whose information is already in the client buffer. In Fig.2a, client C1 is receiving information from $\overline{Ch2}$ and has cached video information from 17 to 20 in the buffer. Client C1 is playing minute 17 when it performs a jump forward operation to minute 19. In this case, the video information in the buffer is enough to perform the jump operation. The client skips minute 17 and 18 and immediately starts playing minute 19.

Situation 2: In this case, the information in the buffer is not able to perform the jump operation, so the client contacts with the server and the server uses the PCM policy to service the jump operation. The PCM policy finds an on-going multicast channel in which the offset is bigger than the new playback point, and finds a *collaborator* to send the Patch Stream. Fig.2b shows the delivery process after the jump forward operation of C2 to minute 27. In this example, the client will join multicast $Ch1$ to receive information from minute 30. A *collaborator* will send the video of minute 27, 28 and 29. If the server is not able to find a *collaborator*, the server creates a unicast channel to deliver the Patch Stream.

Situation 3: In this case, the PCM policy is not able to service the client with a new playback position. The server opens a new multicast channel to send the video information and triggers the MCDB policy. The MCDB policy forms a new collaborative group and the new multicast channel will finally be replaced by a local *branch*-channel. Fig.2c shows C3's jump backward action to minute 12. Notice that the PCM is not able to service the jump action because we assumed that C3 is not able to cache more than 5 minutes of video.

In three situations, a client could need the portion of buffer that is currently collaborating with the server. In such a case, the client notifies the server and stops the collaboration. Furthermore, if the buffer is not completely used in the VCR operation, the client starts caching the video information from the new playback point. Once the buffer is filled, the client sends a control message to the server to announce the new collaborative buffer capacity.

2.4 Pause and Fast Forward/Reverse Operations

The behaviour of the pause operation is quite similar to a jump. In this case, the new playback position is determined by the time that the client makes the pause. During a pause, the client continues buffering the video information. If the client is collaborating with the server, the client stops the collaboration and use all the buffer to cache more video information. However, the client-buffer could eventually overflow. In such a situation, the client temporarily stops the service. After the pause operation, the client consumes all the video blocks in the local buffer and then performs a jump action to the point where the video information is no longer in the local buffer. The jump action is then managed as a normal jump operation.

In order to implement fast forward and reverse, we assume that the server has a VCR-version for each video. The VCR-version of a video requires the same bit-rate as a normal video but with a shorter playback time and lower frame rate. In the case of fast reverse, the video is encoded in the reverse order. The advantage of this technique is its flexibility to implement any speed of fast playback and it does not need more client network incoming bandwidth. However, it has more storage requirements. When a client issues a fast forward, the client contacts with the server to start the delivery process of the VCR-version of the video. After the VCR operation, the new playback point of the normal video is calculated and the client issues a jump operation.

3 Performance Evaluation

In this section, we show the simulation results. We have designed and implemented an object-oriented VoD simulator. Patching and PCM+MCDB with new VCR mechanisms are incorporated in the simulator that is also used in [10]. The experimental study, we evaluated VCR operations' influence and calculated the resource requirements to offer a true-VoD service. We took into account different type as well as the frequency and the duration of VCR operations. The comparative evaluation is based on the *Server Stress* and *Client Stress*. They are defined as the average amount of server and client bandwidth (Mbps) used to service all client requests.

3.1 Workload and User Behaviour Model

We assumed that the inter-arrival time of client requests follows a Poisson arrival process with a mean of $\frac{1}{\lambda}$, where λ is the request rate. We used Zipf-like distribution to model video popularity. The probability of the ith most popular video being chosen is $\frac{1}{i^z \cdot \sum_{j=1}^{N} \frac{1}{j^z}}$ where N is the catalogue size and z is the *skew* factor that adjusts the probability function. For the whole study, the default *skew* factor and the video length were fixed at 0.729 (typical video shop distribution) and 90 minutes, respectively. We assumed that each video was encoded

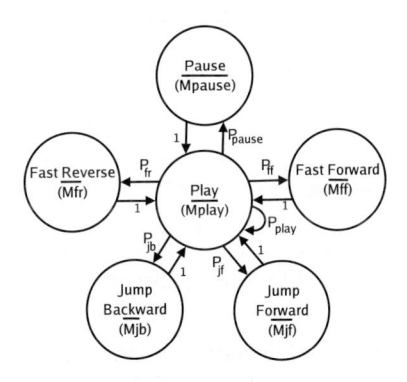

Fig. 3. Client Behaviour Model

in MPEG-2 and requires 1.5 Mbps and the service time is fixed at 24 hours. We assumed that each client is able to cache 5 minutes of the MPEG-2 video.

We assumed that the client interaction behaviour follows the model in Fig.3 which was used in [8]. The model does not try to reflect reality but includes the most common VCR operations and is able to capture two parameters: 1) VCR operation frequency. 2) the duration of each VCR operation. The VCR operation frequency is modelled by P_i's and the duration is evaluated by $\overline{M_i}$'s. The value of $\overline{M_{ff}}$ and $\overline{M_{fr}}$ indicates the average time, in seconds, that a client uses Fast Forward/Reverse. The amount of original video, in units of time, that is visualized in a Fast Forward/Reverse action depends on the fast visualization speed and the duration of the VCR operation. We assumed that the speed of any fast visualization is 2X. In the case of Jump Forward and Jump Backward, the value of $\overline{M}$ indicates the average length of video, in seconds, that will be skipped. Following this model, a client starts to visualize the video during a mean of $\overline{M_{play}}$ seconds, after this, the client issues a VCR operation with a probability of $1 - P_{play}$ or continues with playback. We assumed that the client always returns normal playback after a VCR action and that the duration of each VCR action is uniformly distributed in the interval $[\overline{M_i} \times 0.5 - \overline{M_i} \times 1.5]$.

3.2 VCR Interaction Effect

In this section, we evaluate PCM+MCDB performance with VCR operations. All the probabilities (P_i) of the clients behaviour model are fixed at 0.1 except P_{play}, which is 0.5. The time of each VCR interaction is set at 5 minutes, including playback. With these values of P_i and $\overline{M_i}$, each client could perform an average of 18 VCR operations during playback. We changed the client request rate from 1 to 40 requests per minute. The video catalogue is fixed at 200 videos.

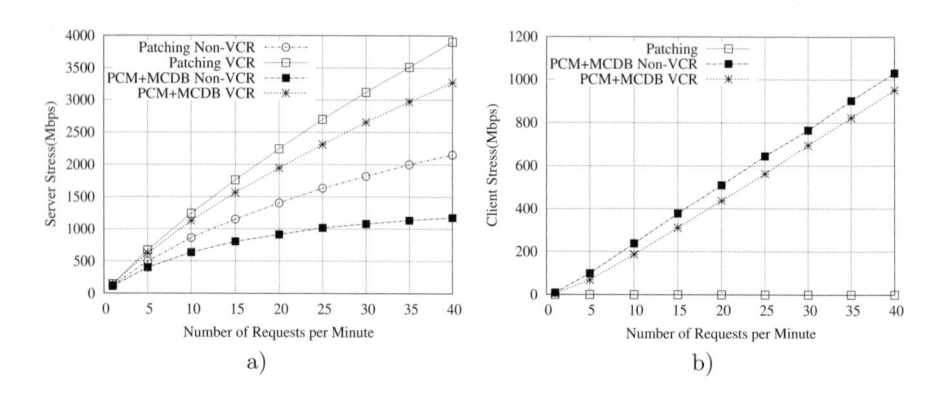

Fig. 4. VCR Interaction Effect on: a) Server Stress. b) Client Stress.

Fig.4a shows the server stress in servicing 200 videos. Without any VCR operation, the Patching policy demands 1818Mbps in order to serve 30 requests per minute. Compared with a Patching policy, PCM+MCDB reduces it by

29% (1282Mbps vs. 1818Mbps). With VCR operations and with Patching, the server stress increase up to 3117Mbps. With our mechanism, the server stress is reduced to 2654Mbps; 15% lower.

Fig.4b shows the client stress generated by PCM+MCDB policy. The Patching policy does not introduce any local overhead. With VCR operation, the client stress of our approach reduce about only 59Mbps(10%). These results indicate that the VCR operations do not affect the client collaboration capacity to decentralize the system load. The explanation for this result is:

1. The pause operations increase the service time of clients and, consequently, the time of the collaboration.
2. With jump forward operations, client service time is lower, so the client will reduce the collaboration time. However, part of the video information is also skipped, requiring less server resource.
3. In jump backward operations, the increase in server resource due to the replay of part of the video information could be rewarded with an increase in collaboration time.
4. The same explanation of jump forward/backward is applicable to fast forward/reverse. Even though, these VCR operations require extra server stress to send VCR-version of video.

3.3 VCR Interaction Frequency and Duration Effect

In this section, we evaluate the effect of VCR frequency and duration on server stress. We set the client request rate at 20 requests per minute.

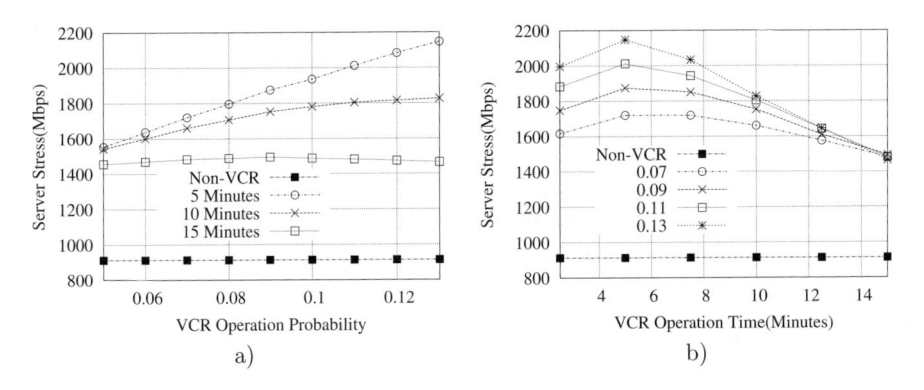

Fig. 5. a) VCR Interaction Probability Effect. b) VCR Interaction Duration Effect.

Fig.5a shows server stress in accordance with the P_is value. Different lines indicate the different durations of each VCR operation (5, 10 and 15 minutes). The value of P_i determines the number of fast forward/reverse operations that introduce extra server stress. As we can see, the server stress increases according with the VCR frequency. However, the increase also depends on the VCR durations. With $\overline{M_i} = 15Min$ the value of P_i does not affect the server stress.

These results suggest that when the duration of fast view operations is long, our approach is able to create client collaborations to decentralize these operations.

Fig.5b shows server stress in accordance with the $\overline{M_i}$s value. Different lines indicate different VCR interaction frequencies. The duration of VCR operations affects the server resource requirements for implementing VCR operations and client collaboration times. On the one hand, a longer duration of VCR operations means that the server has to send more VCR-version of video information to implement fast forward and reverse. On the other hand, a longer duration of a pause operation increases the client collaboration time, thus reducing server stress. As we can see in Fig.5b, there are two different tendencies. The first tendency is when the duration is longer than buffer size. In this case, a longer duration means less server stress and suggests that the VCR mechanism is able to efficiently use the increase in client collaboration time to reduce server stress. The second tendency happens with $\overline{M_i} < 5Min$. In this case, a longer duration increases server stress because the short jumps could be implemented with local buffer of the clients.

4 Conclusions and Future Works

We have proposed and evaluated distributed VCR mechanisms to provide true-VoD in a P2P architecture. Our mechanisms enable clients to efficiently collaborate with VoD servers to implement VCR operations.

Offering multiple videos, experimental results show that PCM+MCDB P2P delivery scheme achieves a reduction in server resource of up to 29%, compared with Patching. Several common VCR operations are analyzed in the experimental study. The experimental results show our mechanisms are very suitable to implement VCR operations with long durations because PCM+MCDB P2P delivery scheme is able to take advantage of the extra client collaboration time, introduced by VCR operations. Comparing with the Patching policy, our mechanisms are able to reduce server resource requirements up to 16%.

We have started several future research projects. First, we are developing a VoD system prototype with the P2P VCR mechanisms. Even though the partial experimental results in laboratory with the prototype have demonstrated the validity of the simulation results, we have to continue working on its implementation. Secondly, we are studying a mechanism to encourage clients to collaborate with the server even they are not playing any video.

References

1. E.L. Abram-Profeta and K.G. Shin. Providing Unrestricted VCR Functions in Multicast Video-on-Demand Servers. In *Proceedings of the IEEE International Conference on Multimedia Computing and Systems, ICMCS*, 66–75, 1998.
2. K.C. Almeroth and M.H. Ammar. The Use of Multicast Delivery to Provide a Scalable and Interactive Video-on-Demand Service, In *IEEE Journal of Selected Areas in Communications*, vol. 14, pages 1110–1122, 1996.

3. Y. Cai, W. Tavanapong and K. A. Hua. Enhancing patching performance through double patching. In *Proceeding of 9th Intl Conf. On distributed Multimedia Systems*, pages 72–77, September 24-26 2003.
4. N.L.S. Fonseca and H.K. Rubinsztejn. Channel allocation in true video-on-demand systems. In *Global Telecommunications Conference, 2001. GLOBECOM '01. IEEE*. Pages:1999-2004 vol.3, 2001
5. Mohamed Hefeeda, Ahsan Habib, Boyan Botev, Dongyan Xu, and Bharat Bhargava. Promise: peer-to-peer media streaming using collectcast. In *Proceedings of MULTIMEDIA '03*, 45–54, New York, USA, 2003.
6. K. A. Hua, M. Tantaoui, and W. Tavanapong. Video delivery technologies for large-scale deployment of multimedia applications. In *Proceedings of the IEEE*, volume 92, September 2004.
7. Wanjiun. Liao and Victor O.K. Li. The Split and Merge (SAM) Protocol for Interactive Video-on-Demand Systems In *INFOCOM '97*, pp 1349, 1997.
8. M.A. Tantaoui, K.A. Hua and S. Sheu A Scalable Technique for VCR-like Interactions in Video-on-Demand Applications. In *The 4th International Workshop on Multimedia Networks Systems and Applications (MNSA 2002) in conjunction ICDCS 2002 Vienna, Austria*, pages 246-251, July 2-4, 2002
9. X. Y. Yang, P. Hernández, F. Cores, A. Ripoll, R. Suppi and E. Luque. Dynamic distributed collaborative merging policy to optimize the multicasting delivery scheme. In *EuroPar'05*, Lisbon, Portugal, 879–889, August 2005.
10. X.Y. Yang, P. Hernández, F. Cores, L. Souza, A. Ripoll, R. Suppi, and E. Luque. $dvodp^2p$: Distributed p2p assisted multicast vod architecture. In *Proceedings of the IEEE International Parallel & Distributed Processing Symposium (IPDPS'06)*, Rhodes, Greece, April 2006.

Linear Hashtable Motion Estimation Algorithm for Distributed Video Processing

Yunsong Wu[1,2] and Graham Megson[2]

[1] Jiangxi Science & Technology Normal University
Nanchang, China
[2] School of Systems Engineering, Reading University
Reading, UK
{sir02yw, g.m.megson}@rdg.ac.uk

Abstract. This paper presents a parallel Linear Hashtable Motion Estimation Algorithm (LHMEA). Most parallel video compression algorithms focus on Group of Picture (GOP). Based on LHMEA we proposed earlier [1][2], we developed a parallel motion estimation algorithm focus inside of frame. We divide each reference frames into equally sized regions. These regions are going to be processed in parallel to increase the encoding speed significantly. The theory and practice speed up of parallel LHMEA according to the number of PCs in the cluster are compared and discussed. Motion Vectors (MV) are generated from the first-pass LHMEA and used as predictors for second-pass Hexagonal Search (HEXBS) motion estimation, which only searches a small number of Macroblocks (MBs). We evaluated distributed parallel implementation of LHMEA of TPA for real time video compression.

Keywords: Parallel Algorithm, Distributed Computing, Distributed Video Coding, Linear Hashtable, Motion Estimation.

1 Introduction

In this paper, a parallel Linear Hashtable Motion Estimation Algorithm (LHMEA) for the Two-Pass Algorithm (TPA) constituted by LHMEA and Hexagonal Search (HEXBS) to predict motion vectors for inter-coding [1] is proposed. The objective of the motion estimation scheme is to achieve good quality video with very low computational time and low transmission rate. It is hard to find software solutions that efficiently code high-quality video in real-time or faster. We propose and evaluate distributed parallel implementations of the LHMEA of TPA on clusters of workstations for real time video compression as test. It discusses how distributed video coding on load balanced multiprocessor systems can help, especially on motion estimation. The software platform used for these is the Parallel Virtual Machines (PVM) programming model and C respectively. The effect of load balancing for improved performance will also be discussed. This paper is only concerned with the Block Matching Algorithms (BMA), which is widely used in MPEG2, MPEG4, and H.263. In BMA, each block of the current video frame is compared to blocks in reference frame in the vicinity of its corresponding position. It is highly desired to speed up the process of compression without introducing serious distortion. The HEXBS is a widely accepted fast motion estimation algorithm [2]. The Linear

Algorithm and Hexagonal Search Based Two-Pass Algorithm (LAHSBTPA) previously proposed has an improvement over the HEXBS on compression rate, PSNR and compression time. In the last 20 years, many fast algorithms have been proposed to reduce the exhaustive checking of candidate Motion Vectors (MV). Such as Two Level Search (TS), Two Dimensional Logarithmic Search (DLS) and Subsample Search (SS) [3], the Three-Step Search (TSS), Four-Step Search (4SS) [4], Block-Based Gradient Descent Search (BBGDS) [5], and Diamond Search (DS) [6], [7] algorithms. A very interesting method called HEXBS has been proposed by Zhu, Lin, and Chau [8]. The fast BMA increases the search speed by taking the nature of most real-world sequences into account while also maintain a prediction quality comparable to Full Search. Most algorithms suffer from being easily trapped in a non-optimum solution. LHMEA based TPA sorts out this problem. Normally video encoders are very effective reducing the size of the video stream, but the processing cost is very high for high quality video sequences. Although there are hardware video encoders available, they have severe restrictions (resolution, coding options, etc). A more flexible choice is to use distributed parallel implementations. Processing video with high performance distributed computing has great potential and good future, but the studies in these fields mainly concentrated on Group of Pictures (GOP) separation.

To take advantage of the potential processing power of distributed computing, we use distributed programming techniques based on message passing. We have used PVM because there are free implementations available and it is a widely accepted standard.

Various image and video compression algorithms use parallel processing. Approaches used can largely be divided into four areas. The first is the use of special purpose architectures designed specially for image and video compression. An example of this is the use of an array of DSP chips to implement a version of MPEG. The second approach is the use of VLSI techniques. The third approach is algorithm driven, in which the structure of the compression algorithm describes the architecture, e.g. pyramid algorithms. The fourth approach is the implementation of algorithms on high performance parallel computers.

The TPA which we have proposed has achieved best result in all the algorithms in the survey. To further improve the result and speed, the most suitable and easiest way is using parallel algorithm to implement the algorithm on high performance parallel computers. In the first-pass coding of TPA, LHMEA is employed to search all Macroblocks (MB) in the picture. Because LHMEA is based on a linear algorithm, which fully utilizes optimized computer's structure based on addition, so it is easy to be paralleled. Meanwhile HEXBS is one of the best motion estimation methods to date. The new method proposed in this paper achieves the best results so far among all the algorithms investigated on compression rate, time and PSNR.

Contributions from this paper are:
1. The TPA achieves the best results among all investigated BMA algorithms.
2. Improved Hashtable is used in video encoding.
3. The parallel algorithm improves LHMEA of TPA. It implements and shows better compression speed, and fair compression rate and PSNR than original TPA.
4. Work load balancing algorithm is implemented in the hashtable image encoding process.

The rest of the paper is organized as follows. Section 2 continues with an introduction to improved LHMEA and TPA and gives experimental result showing TPA's advantage over other algorithms. The proposed parallel algorithm and its implementation for LHMEA are introduced in Section 3. Experimental results showing paralleled hashtable compared with the original are also included in Section3. The paper concludes in Section 4 with some remarks and discussions about the proposed scheme.

2 Sequential and Parallel Implementation of Linear Hashtable Motion Estimation Algorithm (LHMEA)

Our method attempts to predict the motion vectors using linear algorithm.[1][2] It uses hashtable method into video compression. After investigating of most traditional and on-the-edge motion estimation methods, we use latest optimization criterion and prediction search method. Spatially MBs' information is used to generate the best motion vectors[8]. We designed a vector hashtable lookup matching algorithm which is more efficient method to perform an exhaustive search: it considers every macroblock in the search window. This block-matching algorithm calculates each block to set up a hashtable. It is a dictionary in which keys are mapped to array positions by a hash function. We try to find as few variables as possible to represent the whole macroblock. Through some preprocessing steps, "integral projections" are calculated for each macroblock. These projections are different according to different algorithm. The aim of these algorithms is to find best projection function. The algorithms we present here has 2 projections. One of them is the massive projection, which is a scalar denoting the sum of all pixels in the macroblock. It is also DC coefficient of macroblock. The other is A of Y=Ax+B (y is luminance, x is location.) Each of these projections is mathematically related to the error metric. Under certain conditions, the value of the projection indicates whether or not the candidate macroblock will do better than best-so-far match.

2.1 Sequential Implementation of LHMEA

The followings are the pseudo code, theory time, practical time calculation of linear hashtable motion estimation algorithm. The algorithm is used in pre-computation part of in MPEG codec and implemented in both sequential and parallel ways. In the program, we try to use polynomial approximation to get such y=mx+c; y is luminance value of all pixels; x is the location of pixel in macroblocks. The way of scan y is from left to right, from top to button. Coefficients m and c are what we are looking for. As shown in the figure below.

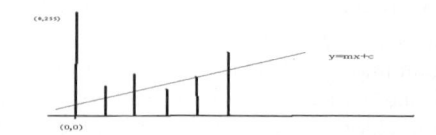

Fig. 1. Linear algorithm for discrete algorithm

In this function y=f(x), x will be from 0 to 255 in a 16*16 pixels macroblock, y=f(x)=mx+c.

$$m = \frac{N * \sum_{i=0}^{N}(x_i * y_i) - \sum_{i=0}^{N} x_i * \sum_{i=0}^{N} y_i}{N * \sum_{i=0}^{N} x_i^2 - \sum_{i=0}^{N} x_i * \sum_{i=0}^{N} x_i} \quad (1)$$

$$c = \frac{\sum_{i=0}^{N} y_i * \sum_{i=0}^{N} x_i^2 - \sum_{i=0}^{N} x_i * \sum_{i=0}^{N} x_i * y_i}{N * \sum_{i=0}^{N} x_i^2 - \sum_{i=0}^{N} x_i * \sum_{i=0}^{N} x_i} \quad (2)$$

Here we state the pseudo code to calculate the hashtable function: The function to implement the algorithm is encapsulated in MyMotionSearchPreComputation Mpeg-Frame *frame)

Sequential Code:
Step 1: if ((psearchAlg == VECTOR_HASH || psearchAlg == HEX_VECTOR_HASH || psearchAlg == HEX) && (frame->type == I_FRAME || frame->type == P_FRAME))
Step 2: EnterTimeCount(0)
Step 3, Paral: if (IsSetUpHashTablePVM) { call PVM Motion Search PreComputation;}
   else{
Step 4:   if(HashTableSearchType) InitMHashTable();
Step 5:   for (y = 0; y < Fsize_y - 16; y++) {
Step 6:   for (x = 0; x < Fsize_x - 16; x++)
Step 7:   { call different hashtable setup functions }
Step 8: if (use HashTable) { add M,C,X,Y into hashtable }}}}
Step 9:   LeaveTimeCount(0);

MB is transferred by hash function to hash coefficients, M,C,X,Y generated are added into hashtable.

In previopus research methods, when people try to find a block that best matches a predefined block in the current frame, matching was performed by SAD (calculating difference between current block and reference block). In Linear Hashtable Motion Estimation Algorithm (LHMEA), we only need to compare two coefficients of two blocks. In current existing methods, the MB moves inside a search window centered on the position of the current block in the current frame. In LHMEA, the coefficients move inside hashtable to find matched blocks. If coefficients are powerful enough to hold enough information of MB, motion estimators should be accurate. So LHMEA increases speed and accuracy to a large extent.

From the pseudo code above, we can get calculation time in theory:

The precomputation complexity is the function (3)

$$T_{seq}(n_{seq}, s_{seq}, \phi_{seq}) = n_{seq} \times s_{seq} \times \phi_{seq} \quad (3)$$

The variables inside the function are

1. n_{seq} : reference frame number, which is also number of I, P frames
2. s_{seq} : frame size, which in the program is

$$(width\_frame) \times (length\_frame) \qquad (4)$$

3. ϕ_{seq} : the complexity to calculate the hash function per macroblock, which will be explained later.

So the complexity of the linear hashtable motion estimation algorithm depends on the three variables.

To demonstrate the complexity of calculation, the following example is given:

The video sequence used in the experimentation is three YUV (352x240 pixels) test sequence, which is known as Flower Garden sequence. There are 150 frames in the original sequences, which sub sampled to the 4:1:1 format in the YUV color space. The video sequence was divided into several sections (GOPs), each of which contained 15 frames to be compressed and a reference frame. A frame pattern of IBBPBBIBBPBBPBB was used. The average time is defined as the overall execution time of the group, including the I/O time, the computation time and the communications time. The motion vector search algorithm used is the LHMEA based TPA and produces integer pixel motion vectors.

We calculate it in details here to demonstrate how it is working.

n_{seq} =50 out of 150 frames.

$s_{seq} = (width\_frame - MB\_size) \times (length\_frame - MB\_size)$ =75264.

According to the complexity of calculate Macroblock,

ϕ_{seq} depends on the hash function calculation method.

For the coefficients m and c we mentioned earlier:

$$m = \frac{N * \sum_{i=0}^{N}(x_i * y_i) - \sum_{i=0}^{N} x_i * \sum_{i=0}^{N} y_i}{N * \sum_{i=0}^{N} x_i^2 - \sum_{i=0}^{N} x_i * \sum_{i=0}^{N} x_i} \qquad (1)$$

$$c = \frac{\sum_{i=0}^{N} y_i * \sum_{i=0}^{N} x_i^2 - \sum_{i=0}^{N} x_i * \sum_{i=0}^{N} x_i * y_i}{N * \sum_{i=0}^{N} x_i^2 - \sum_{i=0}^{N} x_i * \sum_{i=0}^{N} x_i} \qquad (2)$$

In the C codec, We only calculate $\sum_{i=0}^{N}(x_i * y_i)$ and $\sum_{i=0}^{N} y_i$, because for a 16x16 macroblock, $\sum_{i=0}^{N} x_i$, $\sum_{i=0}^{N} x^2_i$, N, $N*\sum_{i=0}^{N} x_i^2 - \sum_{i=0}^{N} x_i * \sum_{i=0}^{N} x_i$ can be pre-calculated before calling the function.

In the codec, pseudo code decides the complexity of ϕ_{seq} is as following:

so $\phi_{seq} = 16*16*[\ 1\ (*)+5\ (+)\ ]+4\ (*)+2\ (+)$

```
for(iy=0;iy<MB_size;iy++){
  for(ix=0;ix<MB_size;ix++){
    temp1= frame->ref_y[y+iy][x+ix];
    sum_yi += temp1;
    sum_xiyi += count* temp1;
    count ++; }}
    (*pnowBuildTable)[y][x].B = 0.125*sum_yi;
    (*pnowBuildTable)[y][x].A
    =(12*sum_xiyi-6*(total_size+1)*sum_yi)>>10;
```

In this example total sequential time in theory is

$T_{seq}(n,s,\phi)$

$= n_{seq} \times s_{seq}(M^2_{frame\_dimension}) \times \phi_{seq}(N^2_{MB\_dimension} \times \gamma_{hashfunctioncomplexity})$ (5)

$= n_{seq} \times M^2_{frame\_dimension} \times N^2_{MB\_dimension} \times \gamma_{hashfunctioncomplexity}$ (6)

$=50*[(Fsize\_x-MB\_size)*(Fsize\_y-MB\_size)]*\{16*16*[1\ (*)+5\ (+)\ ]+4\ (*)+2\ (+)\}$
$=978432000\ (*)+4824422400\ (+)$

Practical sequential time counting: $T_{seq}(n,s,\phi) = 7.2763(s)$

2.2 Parallel Implementation of LHMEA

In the parallel implementation, to parallelize an encoder, we divide each reference frames (which can be I or P frame) into equally sized regions. Current frames are also divided into non-overlapped regions. These regions are going to be processed in parallel to increase the encoding speed significantly. Each region is divided into non-overlapping range blocks. Each region will be sent to corresponding slaves and generates its own hashtable table. The slave will be alive until encoding finishes. Slaves will generate its own hashtable and Motion Vectors table, sending MVs table back to the master. However, there is an upper limit on the number of PEs that can be used due to the limited spatial resolution of a video sequence. Also a massive spatial parallel algorithm usually needs to tolerate a relatively large communication over-head. In our approach of spatial parallelism, load balancing was implemented to ensure an equal distribution of the frame data among the processors.

Here we state the pseudo code to calculate the hashtable function in parallel. The function to implement the algorithm is encapsulated in PreComputation()

Parallel Code:
Input: part of reference frame from master
Output: part of hashtable

Step 1: rcode=pvm_upkint (FrameData,Fsize_X*(rows),1); /*Get Data from Master*/
Step 2: /*Give Data from buffer to Reference Frame, */
for(i=0;i< Fsize_X *(rows);i++)
{prevFrame.ref_y[tempy][tempx] = FrameData[i];}
Step 3: For (i=0;i< rows;i++)
Step 4: for (k=0;k< Fsize_x-16; k++){
Step 5: for(iy=0;iy<16;iy++){
Step 6: for(ix=0;ix<16;ix++)
{calculate sum_xi*yi and sum_yi for each Pixel;}}
Step 7: calculate M and C for each Pixel;}}}

The structure of the algorithm can be demonstrated in the figure 2.
The reference frame are divided into several parts,

$$rows = (width\_frame - width\_MB)/N\_PCs + searchwindows$$

are sent to clients.

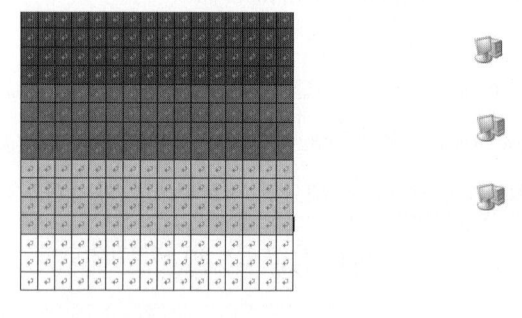

Fig. 2. The Parallel Structure of Hashtable

From the pseudo code above, we can get calculation time in theory:
The precomputation complexity is function

$$T_{paral}(n_{paral}, s_{paral}, \phi_{paral}) = n_{paral} \times s_{paral} \times \phi_{paral} \qquad (7)$$

The variables inside the function have similar meaning as in sequential function

1, $n_{paral} = n_{seq}$

2, s_{paral} : frame size, which is whole frame divided by Number_PCs

$$((width\_frame - width\_MB)/N\_PCs + searchwindows) * (length\_frame) \qquad (8)$$

3, per macroblock. pseudo code decides the complexity of $\phi_{paral} = \phi_{seq}$

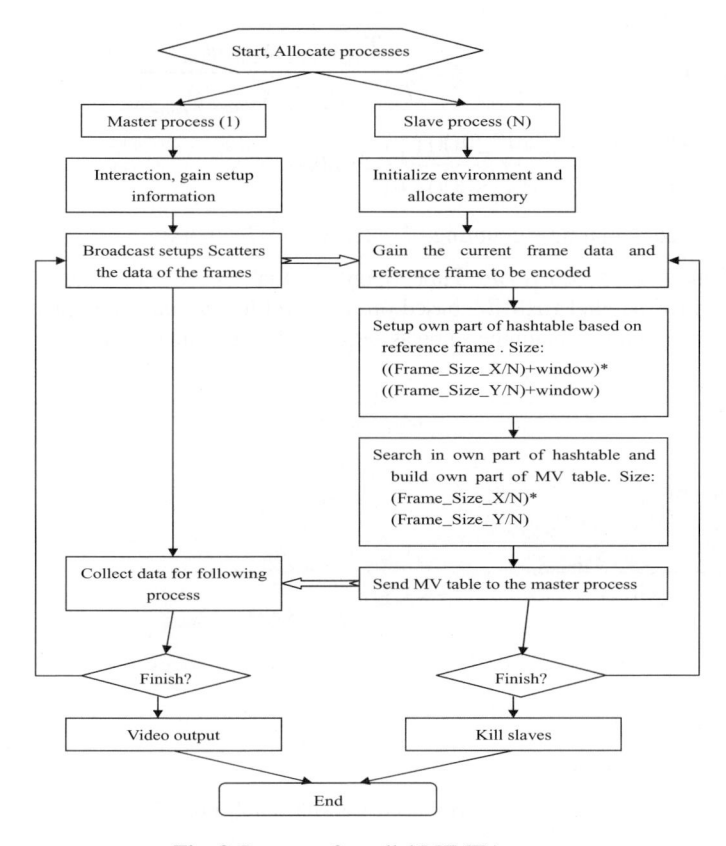

Fig. 3. Process of parallel LHMEA setup

Using the same example sequences of frames and number of slaves equal to 4, if we use 2 slaves, each slave will get *rows=352/2 + search window=216* for each. If we use 4 slaves, each slave will get *rows=352/4 + search window*, 108,128,128,108 for each. We use biggest one to calculate totally time for slaves.

In this case:

When the number of PCs=2:

$$T_{paral}(n_{paral}, s_{paral}, \phi_{paral}) = n_{paral} \times s_{paral} \times \phi_{paral}$$

$$= n_{paral} \times s_{paral}(M^2_{frame\_dimension}) \times \phi_{paral}(N^2_{MB\_dimension} \times \gamma_{hashfunctioncomplexity}) \quad (9)$$

$$= n_{paral} \times \frac{M^2_{frame\_dimension}}{N\_PCs} \times N^2_{MB\_dimension} \times \gamma_{hashfunctioncomplexity} \quad (10)$$

=50*[(rows-MB_size)*(Fsize_x-MB_size)]*{16*16[1(*)+5(+)]+4(*)+4(+)}
50*[124*336]*[260(*)+1284(+)]
= 541632000 (*)+2674828800 (+)

Speedup:
$$\tau = \frac{T_{seq}(n_{seq}, s_{seq}, \phi_{seq})}{T_{paral}(n_{paral}, s_{paral}, \phi_{paral})} = \frac{978432000\,(*) + 4824422400\,(+)}{541632000\,(*) + 2674828800\,(+)} = 1.8065$$

Practical sequential time counting: $T_{seq}(n, s, \phi) = 7.2763(s)$

The figure 4 & 5 below are Time Spent, Actual Speed Up, Theory Speed Up comparison for parallel LHMEA based on the 150 Flower Garden Sequences. PSNR and compression rate remain the same as sequential algorithm [1][2].

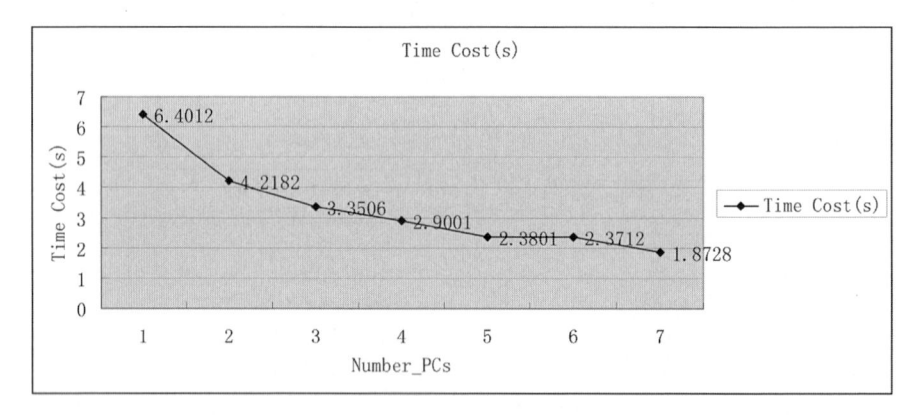

Fig. 4. Time cost decrease with Number of PCs

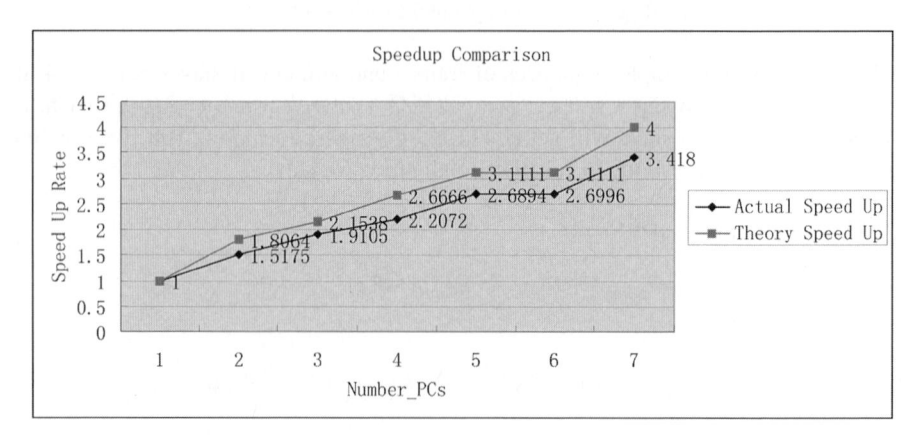

Fig. 5. Actual Speed Up, Theory Speed Up comparison

In theory, the speedup should be in linear increasing with number of PCs. The reason why it does not match a linear model is that we are not sending exact Frame/Number_PCs data to slaves, instead, we send Fsize_y/Number_PCs plus search windows size rows data to slaves. Also it is limited by resolution of images.

More data ($2 \times window\_size \times Frame\_width$) will be calculated than the original frame. In theory, the larger number of PCs, the more redundant data. The curve of speedup-Number PCs will have less descent when the PCs number increases.

Time cost also depends on the speed of PCs. We use a network of workstations comprises similar workstations linked together by a common network e.g. Ethernet. When CPU clock is counted, the faster the PC, the less time it takes.

3 Conclusion

In the paper, a parallel Linear Hashtable Motion Estimation Algorithm (LHMEA) and Hexagonal Search Based Two-Pass Algorithm (TPA) in video compression is proposed based on the LHMEA. The hashtable is used in video compression and implemented with parallel computing in motion estimation. The algorithm searches in the hashtable to find the motion estimator in-stead of by full search algorithm in whole frame. Then the LHMEA was implemented in parallel algorithm. The speedup of paralleled LHMEA is compared to the original sequential LHMEA. The parallel video coding is implemented inside frame rather than between frames. The key point in the method is to find suitable hash function to produce the hashtable.

References

1. Yunsong Wu, Graham Megson, "Linear Predicted Hexagonal Search Algorithm with Moments", ICIC 2005, Part I, Springer LNCS 3644, pp. 136 – 145, (2005).
2. Yunsong Wu, Megson G, "Two-pass hexagonal algorithm with improved hashtable structure for motion estimation Pro-ceedings." IEEE Conference on Advanced Video and Signal Based Surveillance, pp. 564 – 569, (2005).
3. Ce Zhu, Xiao Lin, Lappui Chau, and Lai-Man Po, "Enhanced Hexagonal Search for Fast Block Motion Estimation", IEEE Trans on Circuits and Systems for Video Technology, Vol. 14, No. 10, (Oct 2004)
4. Qiang Peng; Yulin Zhao, "Study on parallel approach in H.26L video encoder", PDCAT'2003. Proc of the Fourth International Conference, p:834 – 837 Aug. (2003)
5. K. Shen, L. A. Rowe, E. J. Delp. "A Parallel Implementation of an MPEGI Encoder: Faster Than Real-Time!". Proc of the SPIE - The International Society for Optical Engineering, ~01.2419p, p:407-418.
6. M. Ribeiro, 0. Sinnen, L. Sousa. "MPEG-4 Natural Video Parallel Implementation on a Cluster". 12th (RECPAD2002), Portugal, June (2002).
7. H. Ning, J. T. Li and S. X. Lin. "A Study of Parallelism in MPEG-4 Video Encoder", Journal of Computer Engineering and Applications, Vol 38, pp.9-12, July, (2002)
8. Alexis M. Tourapis, Oscar C. Au, Ming L. Liou, "Predictive Motion Vector Field Adaptive Search Technique (PMVFAST) Enhancing Block Based Motion Estimation", Proc Visual Communications and Image Processing, San Jose, CA, January (2001)

Topic 12: Theory and Algorithms for Parallel Computation

Danny Krizanc, Michael Kaufmann, Pierre Fraigniaud, and Christos Zaroliagis

Topic Chairs

Parallelism exists at all levels in computing systems from circuits to grids. Effective use of parallelism crucially relies on the availability of suitable models of computation for algorithm design and analysis, and on efficient strategies for the solution of key computational problems on prominent classes of platforms. The study of foundational and algorithmic issues has led to many important advances in parallel computing and has been well represented in the Euro-Par community over that past two decades. A distinctive feature of this topic is the variety of results it as reported over the years that address classical problems as well as the new challenges posed by emerging computing paradigms. This year was no different.

Thirteen papers were submitted to the topic of which five were accepted as full papers for the conference. The resulting papers run the gamut from low-level architectural issues to high-level algorithmic analysis. What they have in common is the same basic theoretical approach to problem-solving. The topics covered include: a hierarchical version of the Craig, Landin and Hagersten (CLH) queue lock which achieves locality while maintaining many of the desirable performance properties of CLH locks and overcoming the fairness issues of previous approaches; the first competitive analysis for the age or freshness of state returned by algorithms for maintaining wait-free data objects in multiprocessor and real-time systems; a new parallel algorithm for the two dimensional cutting stock problem; an on-line adaptive solution to the problem of performing parallel prefix operations on a set of processors running at different and possibly changing speeds; and an efficient algorithm in the Bulk Synchronous Parallel model of computing for the problem of finding all the maximal contiguous subsequences of a sequence of numbers.

A Hierarchical CLH Queue Lock

Victor Luchangco, Dan Nussbaum, and Nir Shavit

Sun Microsystems Laboratories

Abstract. Modern multiprocessor architectures such as CC-NUMA machines or CMPs have nonuniform communication architectures that render programs sensitive to memory access locality. A recent paper by Radović and Hagersten shows that performance gains can be obtained by developing general-purpose mutual-exclusion locks that encourage threads with high mutual memory locality to acquire the lock consecutively, thus reducing the overall cost due to cache misses. Radović and Hagersten present the first such *hierarchical* locks. Unfortunately, their locks are backoff locks, which are known to incur higher cache miss rates than queue-based locks, suffer from various fundamental fairness issues, and are hard to tune so as to maximize locality of lock accesses.

Extending queue-locking algorithms to be hierarchical requires that requests from threads with high mutual memory locality be consecutive in the queue. Until now, it was not clear that one could design such locks because collecting requests locally and moving them into a global queue seemingly requires a level of coordination whose cost would defeat the very purpose of hierarchical locking.

This paper presents a hierarchical version of the Craig, Landin, and Hagersten CLH queue lock, which we call the *HCLH queue lock*. In this algorithm, threads build implicit local queues of waiting threads, splicing them into a global queue at the cost of only a single CAS operation.

In a set of microbenchmarks run on a large scale multiprocessor machine and a state-of-the-art multi-threaded multi-core chip, the HLCH algorithm exhibits better performance and significantly better fairness than the hierarchical backoff locks of Radović and Hagersten.

1 Introduction

It is well accepted that on small scale multiprocessor machines, queue locks [1,2,3,4] minimize overall invalidation traffic by allowing threads to spin on separate memory locations while waiting until they are at the head of the queue. Their advantage over backoff locks [5] is not only in performance, but also in the high level of fairness they provide in accessing a lock.

Large scale modern multiprocessor architectures such as cache-coherent nonuniform memory-access (CC-NUMA) machines, have nonuniform communication architectures that render programs sensitive to memory-access locality. Such architectures include clusters of processors with shared local memory, communicating with each other via a slower communication medium. Access by a processor to the local memory of its cluster can be two or more times faster than access to the remote memory in another cluster [6]. Such machines also have large per-cluster caches, further reducing the cost of communication between processors on the same cluster. A recent paper by Radović and Hagersten [6] shows that performance gains can be obtained by developing *hierarchical locks*: general-purpose mutual-exclusion locks that encourage threads with high mutual memory locality to acquire the lock consecutively, thus reducing the overall level of cache misses when executing instructions in the critical section.

Radović and Hagersten's locks are simple backoff locks: *test-and-test-and-set* locks, augmented with a *backoff scheme* to reduce contention on the lock variable. Their hierarchical backoff mechanism allows the backoff delay to be tuned dynamically so that when a thread notices that another thread from its own local cluster owns the lock, it can reduce its delay and increase its chances of acquiring the lock. The dynamic shortening of backoff times in Radović and Hagersten's lock introduces significant fairness issues: it becomes likely that two or more threads from the same cluster will repeatedly acquire a lock while threads from other clusters starve. Moreover, because the locks are test-and-test-and-set locks, they incur invalidation traffic on every modification of the shared lock variable, which is especially costly on CC-NUMA machines, where the cost of updating remote caches is higher.

We therefore set out to design a hierarchical algorithm based on the more advantageous queue-locking paradigm. A queue lock uses a FIFO queue to reduce contention on the lock variable and provide fairness: if the lock is held by some thread when another thread attempts to acquire it, the second thread adds itself to the queue, and does not attempt to acquire the lock again until it is at the head of the queue (threads remove themselves from the queue when they execute their critical section). Thus, once a thread has added itself to the queue, another thread cannot acquire the lock twice before the first thread acquires it. Several researchers have devised queue locks [1,2,3,4] that minimize overall invalidation traffic by allowing threads to spin on separate memory locations while waiting to check whether they are at the head of the queue. However, making a queue-lock hierarchical implies that requests from threads with high mutual memory locality be consecutive in the queue. To do so one would have to somehow collect local requests within a cluster, integrating each cluster's requests into a global queue, a process which naïvely would require a high level of synchronization and coordination among remote clusters. The cost of this coordination would seemingly defeat the very purpose of hierarchical locking.

This paper presents a hierarchical version of what is considered the most efficient queue lock for cache-coherent machines: the *CLH queue lock* of Craig, Landin, and Hagersten [2,4]. Our new *hierarchical CLH queue lock* (HCLH) has many of the desirable performance properties of CLH locks and overcomes the fairness issues of backoff-based locks. Though it does not provide *global FIFO ordering* as in CLH locks—that is, FIFO among the requests of all threads—it does provide what we call *localized FIFO ordering*: lock-acquisitions of threads from any given cluster are FIFO ordered with respect to each other, but globally, there is a preference to letting threads from the same cluster follow one another (at the expense of global FIFO ordering) in order to enhance locality.

The key algorithmic breakthrough in our work is a novel way for threads to build implicit local queues of waiting threads, and splice them to form a global queue at the cost of only a single *compare-and-swap* (CAS) operation.

In a bit more detail, our algorithm maintains a *local queue* for each cluster, and a single *global queue*. A thread can enter its critical section when it is at the head of the global queue. When a thread wants to acquire the lock, it adds itself to the local queue of its cluster. Thus, threads are spinning on their local predecessors. At some point, the thread at the head of the local queue attempts to splice the entire local queue onto the global queue, so that several threads from the same cluster appear consecutively in the global queue, improving memory-access locality. This splicing into the global queue requires only a single CAS operation, and happens without the spliced threads knowing they have been added to the global queue (except, of course, for the thread doing the splicing): they continue spinning on their local predecessors. The structure of our lock maintains other desirable properties of the original CLH queue lock: It avoids extra pointer manipulations by maintaining only an implicit list; each thread points to its predecessor through a thread-local variable. It also uses a CLH-like recycling scheme that allows the reuse of lock records so that, as in the original CLH algorithm, L locks accessed by N threads require only $O(N + L)$ memory.

We evaluated the performance of our new HCLH algorithm on two nonuniform multiprocessors: a large-scale Sun Fire™ E25K[7] SMP (*E25K*) and a Sun Fire™ T2000, which contains a UltraSPARC® T1[8] 32-thread 8-core multithreaded multiprocessor (*T2000*). In a set of microbenchmarks, including one devised by Radović and Hagersten [6] to expose the effects of locality, the new HCLH algorithm shows various performance benefits: it has improved throughput, better locality, and significantly improved fairness.

In Section 2, we describe our algorithm in detail, in Section 3, we present and discuss the experimental results and we conclude and touch on future work in Section 4.

2 The HCLH Algorithm

In this section, we explain our new queue-lock algorithm in detail. We assume that the system is organized into clusters of processors, each of which has a large cache that is shared among the processors local to that cluster, so that intercluster communication is significantly more expensive than intracluster communication. We also assume that each cluster has a unique *cluster id*, and that every thread knows the cluster_id of the cluster on which it is running (threads do not migrate to different clusters). An HCLH lock consists of a collection of local_queues, one per cluster, and a single global_queue.

As in the original CLH queue lock [2,4], our algorithm represents a queue by an *implicit* linked list of elements of type qnode, as follows: A queue is represented by a pointer to a qnode, which is the tail of the queue (unless the queue is empty—see below for how to determine whether the queue is empty). Each thread has two local variables, my_qnode and my_pred, which are both pointers to qnodes. We say that a thread *owns* the qnode pointed to by its `my_qnode` variable, and we maintain the invariant that at any time, all but one qnode is owned by exactly one thread; we say that one qnode is owned by the lock. For any qnode in a queue (other than the head of the queue), its predecessor in the queue is the qnode pointed to by the `my_pred` variable of its owner. This is well-defined because we also maintain the invariant that the qnode owned by the lock is either not in any queue (while some thread is in the critical section) or is at the head of the global queue. A qnode consists of a single word containing three fields: the `cluster_id` of the processor on which its current owner (or most recent owner, if it is owned by the lock) is running, and two boolean fields, `successor_must_wait` and `tail_when_spliced`. The `successor_must_wait` field is the same as in the original CLH algorithm: it is set to `true` before being enqueued, and it is set to `false` by the qnode's owner upon exit from the critical section, signaling the successor (if any) that the lock is available. Thus, if a thread is waiting to acquire the lock, it may do so when the `successor_must_wait` field of the predecessor of its qnode is `false`. We explain the interpretation of `tail_when_spliced` below.

Threads call the procedure `acquire_HCLH_lock()` when they wish to acquire the lock. Briefly, this procedure first adds the thread's qnode to the local queue, and then waits until either the thread can enter the critical section or its qnode is at the head of the local queue. In the latter case, we say the thread is the *cluster master*, and it is responsible for splicing the local queue onto the global queue. We describe the algorithm in more detail below. Pseudocode appears in Figure 1.

A thread wishing to acquire the lock first initializes its qnode (i.e., the qnode it owns), setting `successor_must_wait` to `true`, `tail_when_spliced` to `false`, and the `cluster_id` field appropriately. The thread then adds its qnode to the end (tail) of its local cluster's queue by using CAS to change the tail to point to its qnode. Upon success, the thread sets its `my_pred` variable to point to the qnode it replaced as the tail. We call this qnode the *predecessor qnode*, or simply the *predecessor*.

Then `wait_for_grant_or_cluster_master()` (not shown) is called, which causes the thread to spin until one of the following conditions is true:

1. the predecessor is from the same cluster, the boolean flag `tail_when_spliced` is `false`, and the boolean flag `successor_must_wait` is `false`, or
2. the predecessor is not from the same cluster or the predecessor's boolean flag `tail_when_spliced` is `true`.

In the first case, the thread's qnode is at the head of the global queue, signifying that it owns the lock and can therfore enter the critical section. In the second case, as we argue below, the thread's qnode is at the head of the local queue, so the thread is the cluster master, making it responsible for splicing the local queue onto the global queue. (If there is no predecessor—that is, if the local queue's tail pointer is null—then the thread becomes the cluster master immediately.) This spinning is mostly in cache and hence incurs almost no communication cost. The procedure `wait_for_grant_or_cluster_master()` (not shown) returns a boolean indicating whether the running thread now owns the lock (if not, the thread is

```
qnode* acquire_HCLH_lock(local_q* lq, global_q* gq, qnode*
my_qnode) {
  // Splice my_qnode into local queue.
  do {
    my_pred = *lq;
  } while (!CAS(lq, my_pred, my_qnode));
  if (my_pred != NULL) {
    bool i_own_lock = wait_for_grant_or_cluster_master(my_pred);
    if (i_own_lock) {
      // I have the lock. Return qnode just released by previous owner.
      return my_pred;
    }
  }

  // At this point, I'm cluster master.  Give others time to show up.
  combining_delay();

  // Splice local queue into global queue.
  do {
    my_pred = *gq;
    local_tail = *lq;
  } while (!CAS(gq, my_pred, local_tail));

  // Inform successor that it is new master.
  local_tail->tail_when_spliced = true;

  // Wait for predecessor to release lock.
  while (my_pred->successor_must_wait);

  // I have the lock. Return qnode just released by previous owner.
  return my_pred;
}

void release_HCLH_lock(qnode* my_qnode) {
  my_qnode->successor_must_wait = false;
}
```

Fig. 1. Procedures for acquiring and releasing a hierarchical CLH lock. The acquire_HCLH_lock() procedure returns a qnode to be used for next lock acquisition attempt.

the cluster master). It is at this point that our algorithm departs from the original CLH algorithm, whose nodes do not have cluster_id or tail_when_spliced fields, in which only the first case is possible because there is only one queue. The chief difficulty in our algorithm is in moving qnodes from a local queue to the global queue in such a way that maintains the desirable properties of CLH queue locks. The key to achieving this is the tail_when_spliced flag, which is raised (i.e., set to true) by the cluster master on the last qnode it splices onto the global queue (i.e., the qnode that the cluster master sets the tail pointer of the global queue to point to).

If the thread's qnode is at the head of the global queue, then, as in the original CLH algorithm, the thread owns the lock and can enter the critical section. Upon exiting the critical section, the thread releases the lock by calling release_HCLH_lock(), which sets successor_must_wait to false, passing ownership of the lock to the next thread, allowing it to enter the critical section. The thread also swaps its qnode for its predecessor (which was owned by the lock) by setting its my_qnode variable.

Otherwise, either the predecessor's `cluster_id` is different from mine or the `tail_when_spliced` flag of the predecessor is raised (i.e., `true`). If the predecessor has a different `cluster_id`, then it cannot be in the local queue of this thread's cluster because every thread sets the `cluster_id` to that of its cluster before adding its qnode to the local queue. Thus, the predecessor must have already been moved to the global queue and recycled to a thread in a different cluster. On the other hand, if the `tail_when_spliced` flag of the predecessor is raised, then the predecessor was the last node moved to the global queue, and thus, the thread's qnode is now at the head of the local queue. It cannot have been moved to the global queue because only the cluster master, the thread whose qnode is at the head of the local queue, moves qnodes onto the global queue.

As cluster master, a thread's role is to splice the qnodes accumulated in the local queue onto the global queue. The threads in the local queue are all spinning, each on its predecessor's qnode. The cluster master reads the tail of the local queue and then uses a CAS operation to change the tail of the global queue to point to the qnode it saw at the tail of its local queue, and sets its `my_pred` variable to point to the tail of the global queue that it replaced. It then raises the `tail_when_spliced` flag of the last qnode it spliced onto the global queue, signaling to the (local) successor of that qnode that it is now the head of the local queue. This has the effect of inserting all the local nodes up to the one pointed to by the local tail into the CLH-style global queue in the same order they were in in the local queue.[1] To increase the length of the combined sequence of nodes that is moved into the global queue, the cluster master waits a certain amount of time for threads to show up in the local queue before splicing into the global queue. We call this time the *combining delay*. With no combining delay, we achieved little or no combining at all, since the time between becoming cluster master and successfully splicing the local queue into the global queue was generally so small. By adding a simple adaptive scheme (using exponential backoff) to adjust the combining delay to current conditions, we saw combining rise to the level we hoped for.

Once in the global queue, the cluster master acts as though it were in an ordinary CLH queue, entering the critical section when the `successor_must_wait` field of its (new) predecessor is `false`. The threads of the other qnodes that were spliced in do not know they moved to the global queue, so they continue spinning as before, and each will enter the critical section when the `successor_must_wait` field of its predecessor is `false`. And as in the case above, and in the original CLH algorithm, a thread simply sets its qnode's `successor_must_wait` field to `false` when it exits the critical section.

---

[1] Note that in the interval between setting the global tail pointer and raising the `tail_when_spliced` flag of the last spliced qnode, the qnodes spliced onto the global queue are in both local and global queues. This is okay because the cluster master will not enter the critical section until after it raises the `tail_when_spliced` flag of the last spliced qnode, and no other thread from that cluster can enter the critical section before the cluster master, since all other threads from that cluster are ordered after the cluster master's in the global queue.

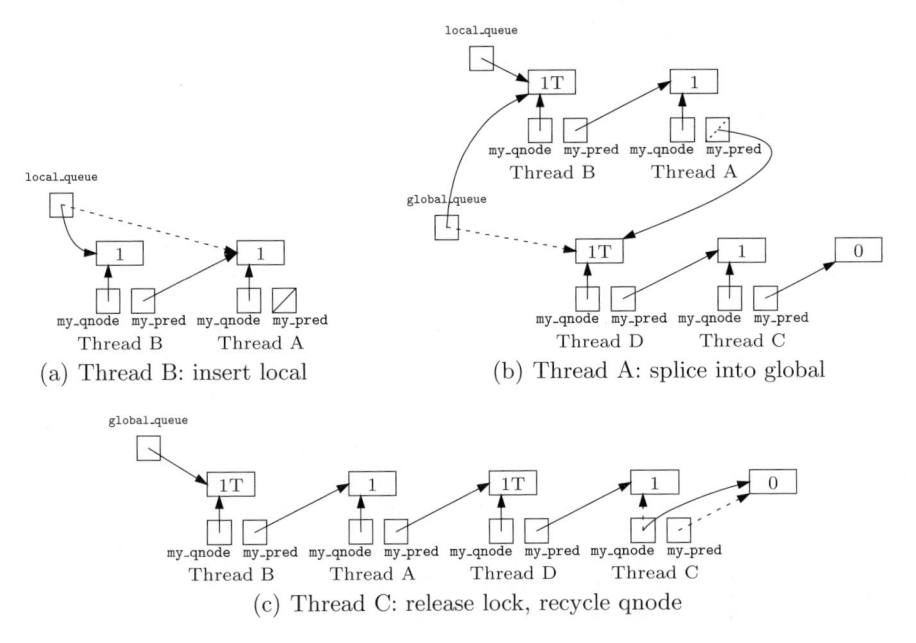

Fig. 2. Lock acquisition and release in a hierarchical CLH lock

Figure 2 illustrates a lock acquisition and release in a hierarchical CLH lock. The successor_must_wait flag is denoted by 0 (for false) or 1 (for true) and the raised tail_when_spliced flag by a T. We denote the thread's predecessor, local or global (they can be implemented using the same variable), as my_pred. The local queue already contains a qnode for a thread A that is the local cluster master since its my_pred is null. In part (a), thread B inserts its qnode into the local queue by performing a CAS operation on the local queue's tail pointer. In part (b), thread A splices the local queue consisting of the qnodes of threads A and B onto the global queue, which already contains the qnodes of threads C and D, spliced at an earlier time. It does so by reading the local queue pointer, and using CAS to change the global queue's tail pointer to the same qnode it read in the local queue's tail pointer, and then raising the tail_when_spliced flag of this qnode (marked by a T). Note that in the meantime other qnodes could have been added to the local queue but the first among them will simply be waiting until B's tail_when_spliced flag is raised (marked by T). In part (c), thread C releases the lock by lowering the successor_must_wait flag of its qnode, and then setting my_qnode to the predecessor qnode. Note that even though thread D's qnode has its tail_when_spliced flag raised, and it could be a node from the same cluster as A, A was already spliced into the global queue and is no longer checking this flag, only the successor_must_wait flag.

As can be seen, the structure of the HCLH algorithm favors sequences of local threads, one waiting for the other, within the waiting list in the global queue. As with the CLH lock, additional efficiency follows from the use of implicit

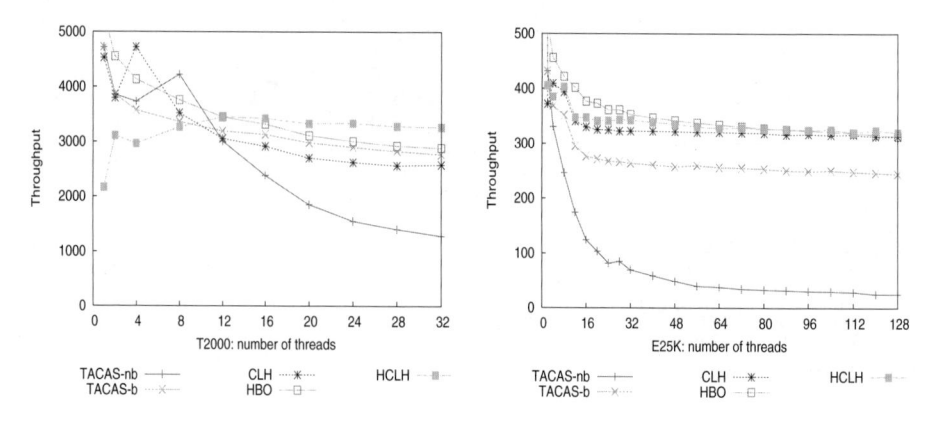

[Figure 3 chart]

Fig. 3. Traditional microbenchmark performance, measured for each lock type. Results for the T2000 are on the left; those for the E25K are on the right. Throughput is measured in thousands of lock acquire/release pairs per second.

pointers which minimizes cache misses, and from the fact that threads spin on local cached copies of their successor's qnode state.

3 Performance

In this section, we present throughput figures using the same two microbenchmarks suggested by Radović and Hagersten[6]. For lack of space, we present only a subset of the relevant results. A full version of the results can be found in http://research.sun.com/scalable/pubs/hclh-main.pdf. In particular, we omit locality data and fairness data (both of which show our algorithm in a good light), along with uncontested-performance data (in which area our algorithm suffers as compared to the others).

These experiments were conducted on two machines: a 144-processor Sun Fire^{TM} E25K[7] SMP (*E25K*) with 4 processor chips per cluster (two cores per chip), and a prototype Sun Fire^{TM} T2000 UltraSPARC® T1[8]-based single-chip multiprocessor (*T2000*) with 8 cores and 4 multiplexed threads per core. We compared the following locking primitives:

TACAS-nb: The traditional test-and-compare-and-swap lock, without backoff.
TACAS-b: The traditional test-and-compare-and-swap lock, with exponential backoff.
CLH: The queue-based lock of Craig, Landin, and Hagersten[2,4].
HBO: The hierarchical backoff lock of Radović and Hagersten[6]. The HBO backoff mechanism allows the backoff parameters to be tuned dynamically so that when a thread that notices that another thread from its own cluster owns the lock, it can reduce the delay between attempts to acquire the lock, thus increasing its chances of acquiring the lock.

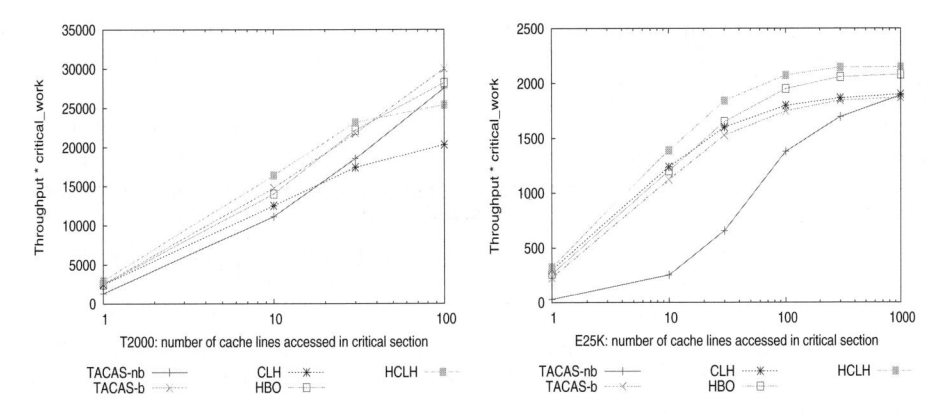

Fig. 4. New microbenchmark performance, measured for each lock type. Results for the T2000 are on the left; those for the E25K are on the right. The Y measures of work per unit time, as given by thousands of lock acquire/release pairs per second multiplied by `critical_work` performed in the critical section. `critical_work` varies along the x-axis. T2000 tests were run with 32 threads; E25K tests were run with 128 threads.

HCLH: Our hierarchical CLH lock. We choose cluster sizes of 8 for the E25K and 4 for the T2000, which make the most sense for the respective architectures.

Results omitted due to lack of space show that HBO's locality is considerably better than random, and HCLH's is considerably better than HBO's for large numbers of (hardware and software) threads. Performance results for the traditional microbenchmark are presented in Figure 3 as run on each platform. This is a variant of the simple loop of lock acquisition and release used by Radović and Hagersten [6] and by Scott and Scherer [9]. As one might expect given its locality advantage, HCLH outperforms all other candidates on both platforms. On the T2000, this superiority asserts itself on tests of 12 or more threads; on the E25K, the effect of improved locality doesn't really assert itself until around 80 threads, and even from there on up, the separation between HCLH, HBO and (somewhat surprisingly) CLH is minimal.

Performance results for our version of Radović and Hagersten's new microbenchmark are presented in Figure 4. In this microbenchmark, each software thread acquires the lock and modifies `critical_work` cache-line-sized blocks of shared data. After exiting from the critical section, each thread performs a random amount of noncritical work. On the E25K, HCLH outperforms the others along the entire range, with HBO close behind, and CLH and TACAS-b not far back. (In fact, CLH slightly outperforms HBO for small `critical_work` values.)

4 Conclusions

Hierarchical mutual-exclusion locks can encourage threads with high mutual memory locality to acquire the lock consecutively, thus reducing the overall level of cache misses when executing instructions in the critical section. We present the HCLH lock—a hierarchical version of Craig, Landin, and Hagersten's queue lock—with that goal in mind.

We model our work after Radović and Hagersten's hierarchical backoff lock, which was developed with the same ends in mind. We demonstrate that HCLH produces better locality and better overall performance on large machines than HBO does when running two simple microbenchmarks.

Compared with the other locks tested (including HBO), the HCLH lock's uncontested performance leaves something to be desired. We have achieved some preliminary success in investigating the possibility of bypassing the local queue in low-contention situations, thus cutting this cost to be near to that of CLH, which is only slightly worse than that of HBO and the others. This is a topic for future work.

Acknowledgements

We wish to thank Brian Whitney for getting us access to the E25K machine and Rick Hetherington for providing us architectural details of the T2000 machine.

References

1. Anderson, T.: The performance implications of spin lock alternatives for shared-memory multiprocessors. IEEE Trans. Parallel and Distributed Systems **1**(1) (1990) 6–16
2. Craig, T.: Building FIFO and priority-queueing spin locks from atomic swap. Technical Report TR 93-02-02, University of Washington, Dept of Computer Science (1993)
3. Mellor-Crummey, J., Scott, M.: Algorithms for scalable synchronization on shared-memory multiprocessors. ACM Trans. Computer Systems **9**(1) (1991) 21–65
4. Magnussen, P., Landin, A., Hagersten, E.: Queue locks on cache coherent multiprocessors. In: Proc. 8th International Symposium on Parallel Processing (IPPS). (1994) 165–171
5. Agarwal, A., Cherian, M.: Adaptive backoff synchronization techniques. In: Proc. 16th International Symposium on Computer Architecture. (1989) 396–406
6. Radović, Z., Hagersten, E.: Hierarchical Backoff Locks for Nonuniform Communication Architectures. In: HPCA-9, Anaheim, California, USA (2003) 241–252
7. Sun Microsystems: Sun Fire E25K/E20K Systems Overview. Technical Report 817-4136-12, Sun Microsystems (2005)
8. Kongetira, P., Aingaran, K., Olukotun, K.: Niagara: A 32-way multithreaded sparc processor. IEEE Micro **25**(2) (2005) 21–29
9. Scott, M., Scherer, W.: Scalable queue-based spin locks with timeout. In: Proc. 8th ACM SIGPLAN Symposium on Principles and Practices of Parallel Programming. (2001) 44–52

Competitive Freshness Algorithms for Wait-Free Data Objects

Peter Damaschke, Phuong Hoai Ha, and Philippas Tsigas

Department of Computer Science and Engineering, Chalmers University of Technology,
S-412 96 Gothenburg, Sweden
{ptr, phuong, tsigas}@cs.chalmers.se

Abstract. Wait-free concurrent data objects are widely used in multiprocessor systems and real-time systems. Their popularity results from the fact that they avoid locking and that concurrent operations on such data objects are guaranteed to finish in a bounded number of steps regardless of the other operations interference. The data objects allow high access parallelism and guarantee correctness of the concurrent access with respect to its semantics. In such a highly-concurrent environment, where many wait-free write-operations updating the object state can overlap a single read-operation, the age/freshness of the state returned by this read-operation is a significant measure of the object quality, especially for real-time systems.

In this paper, we first propose a freshness measure for wait-free concurrent data objects. Subsequently, we model the freshness problem as an online problem and present two algorithms for it. The first one is a deterministic algorithm with asymptotically optimal competitive ratio $\sqrt{\alpha}$, where α is a function of the execution-time upper-bound of wait-free operations. The second one is a competitive randomized algorithm with competitive ratio $\frac{\ln \alpha}{1+\ln 2 - \frac{2}{\sqrt{\alpha}}}$.

1 Introduction

Concurrent data objects play a significant role in multiprocessor systems, but also create challenges on consistency. In concurrent environments like multiprocessor systems, consistency of a shared data object is guaranteed mostly by mutual exclusion, a form of locking. However, mutual exclusion degrades the system's overall performance due to lock convoying, i.e. other concurrent operations cannot make any progress while the access to the shared object is blocked. Mutual exclusion also contains risks of deadlock and priority inversion. To address these problems, researchers have proposed *non-blocking algorithms* for shared data objects. Non-blocking methods do not involve mutual exclusion, and therefore do not suffer the problems that blocking can cause. Non-blocking algorithms are either lock-free or wait-free. *Lock-free* [11] algorithms guarantee that regardless of both the contention caused by concurrent operations and the interleaving of their sub-operations, always at least one operation will progress. However, there is a risk for starvation as progress of other operations could cause one specific operation to never finish. *Wait-free* [10] algorithms are lock-free and moreover they avoid starvation. In a wait-free algorithm every operation is guaranteed to finish in a limited number of steps, regardless of actions of other concurrent operations. Non-blocking algorithms have been shown to be of big practical importance [7,8,18], and recently NOBLE, which is a non-blocking inter-process communication library, has been introduced [23]. As a result, many aspects of concurrent data objects have been researched deeply such as consistency conditions [1,9,20], concurrency hierarchy [6] and fault-tolerance [17].

In this paper, we look at another aspect of concurrent data objects: the freshness of the object states returned by read-operations. Freshness is a significant property for shared data in general and has achieved great concerns in databases [3,12,19] as well as in caching systems [13,15,16]. Briefly, freshness is a yardstick to evaluate how fresh/new a value of a concurrent object returned by a read-operation is, when the object is updated and read concurrently. For concurrent data objects, although read-operations are allowed to return any value written by other concurrent operations, they are preferred to return the freshest/latest one of these valid values, especially in reactive/detective systems. For instance, monitoring sensors continuously concurrently input data via a concurrent object and the processing unit periodically reads the data to make the system react accordingly. In such systems, the freshness of data influences how fast the system reacts to environment changes.

However, there are few results on the freshness problem in the literature. Simpson [21,22] suggested a freshness specification for a single-writer-to-single-reader asynchronous communication mechanism, which is different from atomic register suggested by Lamport [14]. Simpson's communication model with a single writer and a single reader is not suitable for fully concurrent shared objects that many readers and many writers can concurrently access.

These issues motivate us to define and attack the freshness problem for wait-free shared objects. We model the problem as an online problem and then present two algorithms for it. The first one is a deterministic algorithm, which is a natural adaptation from an online search algorithm called *reservation price policy* [5]. The algorithm achieves a competitive ratio $\sqrt{\alpha}$, where α is a function of execution-time upper-bound of wait-free operations. Subsequently, we prove that the algorithm is optimal by proving that $\sqrt{\alpha}$ is the best competitive ratio for deterministic algorithms. The second is a new competitive randomized algorithm with competitive ratio $\frac{\ln \alpha}{1+\ln 2 - \frac{2}{\sqrt{\alpha}}}$. The randomized algorithm is nearly optimal since our results [4] from an elaboration on the EXPO search algorithm [5] showed that $O(\ln \alpha)$ is an asymptotically optimal competitive ratio for randomized freshness algorithms.

The paper is organized as follows. Section 2 describes the freshness problem and models it as an online problem. Section 3 presents the optimal deterministic algorithm. Section 4 presents the randomized algorithm. The competitive ratio in this case is the expected value against an oblivious adversary. (We presume that the reader is familiar with competitive analysis of online algorithms, cf. [2].)

2 Problem and Model

Linearizability [9] is the correctness condition for concurrent objects. It requires that operations on the objects appear to take effect atomically at a point of time in their execution interval. This allows a read operation to return any of values written by concurrent write operations, which is illustrated by Figure 1.

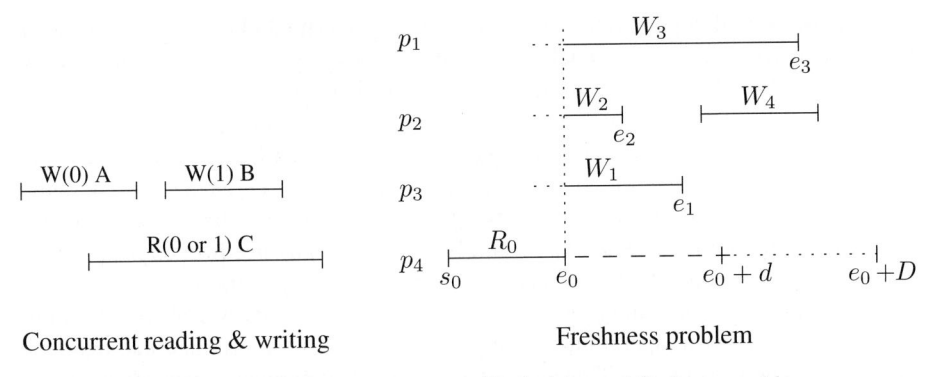

Fig. 1. Illustrations for concurrent reading/writing and freshness problem

We use "W(x) A" ("R(x) A") to stand for a write (read) operation of value x to (from) a shared register by process A. It is correct for C to return either 0 or 1 with respect to linearizability. However, from freshness point of view we prefer C to return 1, the newer/fresher value of the register. The freshness problem is to find a solution for read operations to obtain the freshest value from a shared object. Intuitively, if a read operation lengthens its execution interval by putting some delay between the invocation and the response, it can obtain a fresher value but it will respond more slowly from application point of view. Therefore, the freshness problem is to design read-operations that both respond fast and return fresh values.

The freshness problem is especially interesting in reactive systems, where monitoring sensors continuously and concurrently input data for a processing unit via a concurrent data object. The unit periodically reads the data from the object and subsequently makes the system react to environment changes accordingly. In order to react fast, the read-operation used by the unit must both respond fast and return a value as fresh as possible. If the read-operation responds immediately at time e_0 and an environment change occurs at time $e_0 + \epsilon$, the system must wait for a period T until the next read in order to observe the change. In this scenario, the system will react faster if the read-operation delays a bit to return the fresh value at $e_0 + \epsilon$. The system will subsequently react according to the change at time $e_0 + \epsilon$ instead of waiting until time $e_0 + T$ to be able to observe the change, where $\epsilon \ll T$ (Assume that processing time is negligible.).

The freshness problem is illustrated by Figure 1. In the illustration, a read operation R_0 runs concurrently to three write operations W_1, W_2 and W_3 on a concurrent shared object. In this paper, read/write operations imply operations on the same object. The actual execution interval of a operation i is defined from the time s_i the operation starts to the time e_i it takes effect (i.e. linearization point [9]). A time axis is from left to right. The value returned by R_0 becomes fresher if there are more end-points e_i appear in the interval $[s_0, e_0]$. In the illustration, if R_0 delays the time-point e_0 to $e'_0 = e_0 + d$, the execution interval $[s_0, e'_0]$ will include two more end-points e_1 and e_2 and thus the value returned is newer. However, the delay will also make the read-operation respond more slowly. This implies that R_0 needs to find the time delay d so as to maximize the freshness value $f_d = \frac{k(|we_d|)}{h(d)}$, where $|we_d|$ is the number of new write-endpoints earned by delaying R_0's read-endpoint an interval d and k, h are increasing functions that depend on real applications. The k and h functions should be increasing in order to model progressive systems. Each application may specify its own functions k and h according to the relation between the latency and freshness in the application.

Assume that the shared object supports a function for read operations to check how many write operations (with their timestamp) are ongoing at a time[1]. A write-timestamp wt shows the *start-point* of the corresponding write operation whereas a read-timestamp rt shows the *end-point* of the corresponding read operation. The timestamp objective is to help R_0 ignore W_4 due to $rt_0 < wt_4$. Note that R_0 only needs to consider write-endpoints of write operations that occur concurrently to R_0 in its original execution interval $[s_0, e_0]$, e.g. R_0 will ignore W_4. Therefore, in the freshness problem, the number of concurrent write operations that have not finished at the original read-endpoint e_0 is known and is called M. This number is also the total number of considered write-endpoints, i.e. $|we_d| \leq M$.

The most challenging issue in the freshness problem is that the end-points of concurrent write operations appear unpredictably. In order to analyze the problem, we consider it as an online game between a player and an oblivious adversary where the malicious adversary decides when to place the write-endpoints e_i on-the-fly and the player (the read operation) decides when she should stop and place her read-endpoint e'_0. The online game starts at the original read-endpoint e_0 and the player knows the total number of write-endpoints M that the adversary will use throughout the game. At a time t, the player knows how many of M end-points have been used by the adversary so far, i.e. $|we_t|$, (by comparing M with the number of ongoing write operations that ran concurrently with the original read operation) and computes the current freshness value $f_t = \frac{k(|we_t|)}{h(t)}$. For each f_t observed, without knowledge of how the value will vary in the future, the player must decide whether she accepts this value and stops or waits for a better one. In this online game, the player's goal is to minimize the competitive ratio $c = \frac{f_{max}}{f_{chosen}}$, where f_{chosen} is the freshness value chosen by the player and f_{max} is the best value in this game, which is chosen by the adversary. The duration of this game D is the upper bound of execution time of the wait-free read/write operations and is known to the player. This implies that all the M write-endpoints must appear at a time-point in the interval, i.e. $|we_D| = M$.

In summary, we define the freshness problem as follows. Let M be the number of ongoing wait-free write operations at the original read-endpoint e_0 of a wait-free read operation and D be the execution-time upper-bound of these wait-free read/write operations. The read operation needs to find a delay $d \leq D$ for its new end-point e'_0 so as to achieve an optimal freshness value $f_d = \frac{k(|we_d|)}{h(d)}$, where $|we_d|$ is the number of write-endpoints earned by the delay d and k, h are increasing functions that reflect the relation between latency and freshness in real applications. The read-operation is only allowed to read the object data and check the number of ongoing write-operations. The write-operation is only allowed to write data to the object. We assume the time is discrete, where a time unit is the period with which the read operation regularly checks the number of ongoing write operations on the shared object. The extended read operation is still wait-free with an execution-time upper-bound $2D$.

---

[1] The assumption is practical since this can be done by adding a list of timestamps of ongoing write operations to the shared object.

The rest of this paper presents two competitive online algorithms for the freshness problem. The first one is an optimal deterministic algorithm with competitive ratio $\sqrt{\alpha}$, where $\alpha = \frac{h(D)}{h(1)}$. The second one is a nearly-optimal randomized algorithm with competitive ratio $\frac{\ln \alpha}{1+\ln 2 - \frac{2}{\sqrt{\alpha}}}$. Note that the competitive ratios do not depend on k and M, which are related to the number of end-points.

3 Optimal Deterministic Algorithm

Modeling the freshness problem as an online game, we observe that the freshness problem is a variant of online search [5]: In that problem, a player searches for the maximum (minimum) price in a sequence of prices that unfolds daily. For each day i, the player observes a price p_i and must decide whether to accept this price or to wait for a better one. The game ends when the player accepts a price, which is also the result.

Inspired by an online search algorithm called *reservation price policy* [5], we suggest a competitive deterministic algorithm for the freshness problem. In addition to the fact that the player is searching for the best in a sequence of freshness values that unfolds sequentially in a foreknown range, there are more restrictions on the adversary. Freshness values f_t at time t must fulfill:

$$\frac{f_{t-1} * h(t-1)}{h(t)} = \frac{k(|we_{t-1}|)}{h(t)} \leq f_t = \frac{k(|we_t|)}{h(t)} \leq \frac{k(M)}{h(t)} \quad (1)$$

The restrictions come from the fact that the adversary cannot remove the end-points she has placed, i.e. $|we_{t-1}| \leq |we_t| \leq M$, where $|we_t|$ is the number of end-points that have appeared until a time t, and the freshness value at the time t is $f_t = \frac{k(|we_t|)}{h(t)}$, where k, h are increasing functions. The restrictions make the adversary in the freshness problem weaker than the adversary in the online search problem, and intuitively the player in the freshness problem should benefit from this. However, we will prove that this is not the case for deterministic algorithms (cf. Theorem 2).

Before presenting the deterministic freshness algorithm, we need to find upper/lower bounds on freshness values f_t. Since $1 \leq t \leq D$, from Equation (1) it follows $f_t \leq \frac{k(M)}{h(1)}$. On the other hand, since M ongoing write-operations must end at time-points in the interval D, the player is ensured a freshness value $f_{min} = \frac{k(M)}{h(D)}$ by just waiting until $t = D$. Therefore, the player considers to stop at a freshness value f_t only if $f_t \geq \frac{k(M)}{h(D)}$. We have $\frac{k(M)}{h(D)} \leq f_t \leq \frac{k(M)}{h(1)}$.

Deterministic Algorithm: The read operation accepts the first freshness value that is not smaller than $f^* = \frac{k(M)}{\sqrt{h(1)h(D)}}$.

Indeed, let f^* be the threshold for accepting a freshness value and f_{max} be the highest value chosen by the adversary. The player (the read operation) waits for a value $f_t \geq f^*$. If such a value appears in the interval D, the player accepts it and returns it as the result. Otherwise, when waiting until the time D, the player must accept the value $f_{min} = \frac{k(M)}{h(D)}$.

Case 1: If the player chooses a big value as f^*, the adversary will choose $f_{max} < f^*$, causing the player to wait until the time D and accept the value $f_{min} = \frac{k(M)}{h(D)}$. The competitive ratio in this case is $c_1 = \frac{f_{max}}{\frac{k(M)}{h(D)}} < \frac{f^*}{\frac{k(M)}{h(D)}}$.

Case 2: If the player chooses a small value as f^*, the adversary will place f^* at a time t, causing the player to accept the value and stop. Right after that, the adversary places all M end-points, achieving a value $f_{max} = \frac{k(M)}{h(t)} \leq \frac{k(M)}{h(1)}$ (equality occurs when the adversary chooses $t = 1$). The competitive ratio in this case is $c_2 = \frac{\frac{k(M)}{h(1)}}{f^*}$.

The player chooses f^* so as to make $c_1 = c_2$, which results in $f^* = \frac{k(M)}{\sqrt{h(1)h(D)}}$ and the competitive ratio $c = c_1 = c_2 = \sqrt{\frac{h(D)}{h(1)}}$. This leads to the following theorem.

Theorem 1. *The suggested deterministic algorithm is competitive with competitive ratio $c = \sqrt{\alpha}$, where $\alpha = \frac{h(D)}{h(1)}$.*

We now prove that no deterministic algorithm can do better.

We use a logarithmic vertical axis for freshness. Let LF denote the logarithm of freshness. More specifically, we normalize the LF axis so that freshness $\frac{k(M)}{h(D)}$ corresponds to point 0 and freshness $\frac{k(M)}{h(1)}$ corresponds to point $\ln \frac{h(D)}{h(1)} = \ln \alpha$. One unit on the LF axis multiplies the freshness by factor e (Euler's number).

We also introduce some parameters that characterize the status of a game. Let t be the time, initially $t = 1$. At any moment, let f be the maximum LF the adversary has already reached during the history of the game, and g the maximum LF the adversary can still achieve at a given time. LF value $g(t)$ at time t corresponds to freshness $k(M)/h(t)$, unless f is already larger, in which case we have $g = f$. However in the latter case the game is over, without loss of generality: The adversary cannot gain more and would therefore decrease the freshness as quickly as possible, in order to make the player's position as bad as possible, hence an optimal player would stop now. (The dotted polyline in Figure 2 illustrates the case $f = g(t)$ in which the player should stop at time t.)

The horizontal axis is for the logarithm of $h(t)$. We normalize it so that $h(1)$ corresponds to point 0 and $h(D)$ corresponds to point $\ln \frac{h(D)}{h(1)} = \ln \alpha$). Note that, in these logarithmic coordinates, g simply decreases at unit speed, starting at point $\ln \alpha$. Finally, let c denote the current LF. We remark that c can decrease at most at unit speed but can jump upwards arbitrarily as long as $c \leq g$.

Theorem 2. *The optimal deterministic competitive ratio is asymptotically (subject to lower-order terms) $\sqrt{\alpha}$, where $\alpha = \frac{h(D)}{h(1)}$.*

Proof. We only need to show an adversary strategy that enforces the claimed competitive ratio. Our logarithmic coordinates make the argument rather simple: The adversary starts with $c = \frac{\ln \alpha}{2} = \frac{\ln \frac{h(D)}{h(1)}}{2}$. Then she decreases c at unit speed until the player stops. Immediately after this moment, c jumps to g if $c > 0$ at the stop time (Case 1), otherwise c keeps on decreasing at unit speed (Case 2). Clearly, we have constantly $g - c = \frac{\ln \alpha}{2}$

until the stop time. Let p be the player's value of LF. In Case (1) we finally get $f = g$, hence $f - p = g - c = \frac{\ln \alpha}{2}$ (cf. the dashed polyline c_1 in Figure 2). In Case (2), f has still its initial value $\frac{\ln \alpha}{2}$ whereas $p \leq 0$, hence $f - p \geq \frac{\ln \alpha}{2}$ (cf. the line c_2 in Figure 2). Thus the competitive ratio is at least $e^{\frac{\ln \alpha}{2}} = \sqrt{\alpha}$. □

We have shown that a deterministic player cannot benefit from the constraints on the behaviour of freshness in time (compared to the unrestricted online search problem).

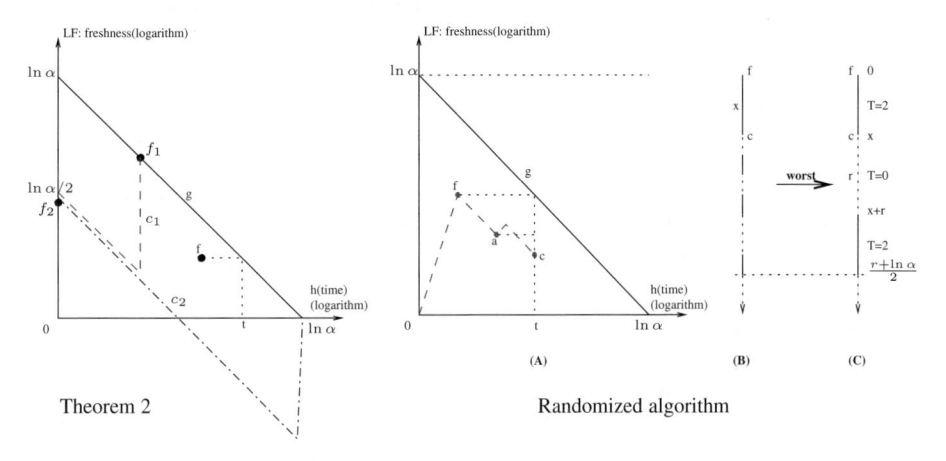

Fig. 2. Illustrations for Theorem 2 and the randomized algorithm

4 Competitive Randomized Algorithm

Next we present a randomized algorithm for the freshness problem, against the oblivious adversary [2]. It achieves a competitive ratio $c = \frac{\ln \alpha}{1 + \ln 2 - \frac{2}{\sqrt{\alpha}}}$, where $\alpha = \frac{h(D)}{h(1)}$.

As discussed in the previous section, our problem is a restricted case of online search. We model the problem by a game between an (online) player and an adversary. The adversary's profit is the highest freshness ever reached. The player's profit is the freshness value at the moment when she stops. Note that for a player running a randomized strategy, the profit is the expected freshness value, with respect to the distribution of stops resulting from the strategy and input. We shall make use of a known simple transformation of (randomized) online search to (deterministic) one-way trading [5]: The player has some budget of money she wants to exchange while the exchange rates may vary over time. Her goal is to maximize her gain. The transformation is given as follows: The budget corresponds to probability 1, and exchanging some fraction of money means to stop the game with exactly that probability. Note that a deterministic algorithm for online search has to exchange all money at *one* point in time. For the freshness problem, it is possible to apply a well-known competitive randomized algorithm EXPO [5]. Applying the EXPO algorithm on the freshness problem achieves a competitive ratio $\ell \frac{2^{\ell-1+1/\ln 2}}{2^{\ell-1+1/\ln 2} - \frac{1}{\ln 2}}$, where $\ell = \log_2 \alpha$. That means for the freshness problem our randomized algorithm is better than the EXPO algorithm by a constant factor $\frac{1+\ln 2}{\ln 2}$ when α becomes large.

Theorem 3. *There is a randomized algorithm for the freshness problem with expected competitive ratio $\frac{\ln \alpha}{1+\ln 2 - \frac{2}{\sqrt{\alpha}}}$ against an oblivious adversary, where $\alpha = \frac{h(D)}{h(1)}$.*

Proof. We start with some conventions. We imagine that the money, both exchanged and non-exchanged, is "distributed" on the LF axis. Formally, the allocation of money on the LF axis at any time is described by two non-negative real *density functions* S and T, where $S(x)$ is the density of not yet exchanged money in point x of the LF axis, $T(x)$ is similarly defined for the money that has been already exchanged. What functions S and T specifically are, and how they are modified by the opponents' actions, will be described below. Let the total amount of money be $\ln \alpha$ by convention. (Recall that scaling factors do not influence the competitive ratio.)

The *value* of every piece of *exchanged* money is the freshness value of its position on the LF axis. Note that the total value of exchanged money defined in this way, i.e. the integral over the value-by-density product, is the player's profit in the game. Moreover, the player can temporarily have some of the money in her *pocket*.

The idea of the strategy is to guarantee some concentration of exchanged money immediately below the final f, either some constant minimum density of T or, even better, a constant amount at one point not too far from f. We want to keep T simple in order to make the calculations simple. (The well-known δ_x symbol used below denotes the distribution with infinite density at a single point x but with integral 1 on any interval that contains x. We also use the same notations f, g, c as earlier.) Locating much money instantaneously is risky because c may jump upwards, and then this money has little value compared to the adversary's. On the other hand, since c decreases at most with unit speed, the player may completely abstain from exchanging money as long as c is increasing, and wait until c goes down again. These preliminary thoughts lead to the following strategy.

In the beginning, let the not-yet-exchanged money be located on the LF axis on interval $[0, \ln \alpha]$ with density 1, that is, we have $S = 1$ on this interval. Remember that g decreases at unit speed. The player puts the money above g in her pocket. Whenever f increases, she also puts the money below the new f in her pocket. Hence we always have $S = 1$ on $[f, g]$, and $S = 0$ outside. The player continuously locates exchanged money on the LF axis, observing the following rule: *If you have money in your pocket and c is positive and decreasing, and $T(c) < 2$ at the current c, then set $T(c) := 2$. If the game is over (because of $f = g$) and not all money is exchanged yet, put the rest r on the current c.* Note that the adversary must set the final c nonnegative.

Filling-up density T to 2 is always possible: The player uses the one unit of money from S that she gets per time unit from the region above the falling g, and the money from S that she got directly from the current points c when f went upwards.

Obviously, the player produces a density function T that is constantly 2 on certain intervals and 0 outside, plus some component $r\delta_c$. We make some crucial observations regarding the final situation: (1) T has density 2 on interval $(c, f]$, or we have $c = f$. (2) The *gaps* with $T = 0$ between the "$T = 2$ intervals" have total length at most r.

These claims follow easily from the strategy: (1) Either c begins decreasing, starting from the last f, and T is filled up to 2 all the time when $c > 0$, as we saw above, or the final c equals the final f. (2) Whenever f went upwards, the player has taken from S the money corresponding to the increase of f, and later she has transferred it to T

and located it at the same points again. Hence, only on intervals not "visited" again by c we have $T = 0$, and the money taken from S on these intervals is still in the player's pocket and thus contributes to r.

Figure 2-(A) illustrates the player's behavior. The dashed line represents a variation of c in a game; point c is the final value of c when the game ends, i.e. $f = g(t)$. For all values v on the LF axis between f and a and between a and c, the player sets $T(v) = 2$.

Using (1),(2) we now analyze the profit the player can guarantee herself. Remember that the value of exchanged money located on the LF axis decreases exponentially. Let $x = f - c$ (final values). Both r and x depend on the input, i.e., the behavior of c in time. The total amount of money is fixed, it equals $\ln \alpha$. For any fixed r, x, the worst case is now that the gaps in T sum up to the maximum length r and are as high as possible on the LF axis, that is, immediately below point c, because in this case all exchanged money outside $[c, f]$ has the least possible value. That is, T has only one gap, namely interval $[c - r, c]$.

Figure 2-(C) illustrates the worst case corresponding to an instance -(B), where solid lines represent ranges on the LF axis with $T = 2$. In the worst case, the adversary shifts all solid lines except for $[c, f]$ to the lowest possible position so as to minimize the player's profit.

Hence, a lower bound on the player's profit, divided by the value at f, is given by

$$\min_{r,x} \left(2 \int_0^x e^{-t} dt + r e^{-x} + 2 \int_{x+r}^{(r+\ln \alpha)/2} e^{-t} dt \right),$$

where we started integration (with $t = 0$) at point f and go down the LF axis (cf. Figure 2-(C)). Verify that, in fact, $\int T dt = \ln \alpha$. The above expression evaluates to

$$2 + (r - 2 + 2e^{-r})e^{-x} - 2e^{-(r+\ln \alpha)/2} > 2 + (r - 2 + 2e^{-r})e^{-x} - 2/\sqrt{\alpha}.$$

For any fixed x, this is minimized if $2e^{-r} = 1$, that is, $r = \ln 2$. Since now $r - 2 + 2e^{-r} = \ln 2 - 2 + 1 < 0$, the worst case is $x = 0$, which gives $1 + \ln 2 - 2/\sqrt{\alpha}$. The adversary earns $\ln \alpha$ times the value at f. □

References

1. Attiya, H., Welch, J.L.: Sequential consistency versus linearizability. ACM Trans. Comput. Syst. **12**(2) (1994) 91–122
2. Borodin, A., El-Yaniv, R.: Online computation and competitive analysis. Cambridge University Press (1998)
3. Cho, J., Garcia-Molina, H.: Synchronizing a database to improve freshness. In: Proc. of the ACM SIGMOD Intl. Conf. on Management of Data. (2000) 117–128
4. Damaschke, P., Ha, P.H., Tsigas, P.: One-way trading with time-varying exchange rate bounds. Technical report CS:2005-17, Chalmers University of Technology (2005)
5. El-Yaniv, R., Fiat, A., Karp, R.M., Turpin, G.: Optimal search and one-way trading online algorithms. Algorithmica **30**(1) (2001) 101–139
6. Gafni, E., Merritt, M., Taubenfeld, G.: The concurrency hierarchy, and algorithms for unbounded concurrency. In: Proc. of Symp. on Principles of Distributed Computing (PODC). (2001) 161–169

7. Harris, T.: A pragmatic implementation of non-blocking linked lists. In: Proc. of the Intl. Symp. on Distributed Computing (DISC). (2001) 300–314
8. Hendler, D., Shavit, N., Yerushalmi, L.: A scalable lock-free stack algorithm. In: Proc. of the ACM Symp. on Parallel Algorithms and Architectures (SPAA). (2004) 206–215
9. Herlihy, M., Wing, J.: Linearizability: a correctness condition for concurrent objects. ACM Trans. on Programming Languages and Systems **12**(3) (1990) 463–492
10. Herlihy, M.: Wait-free synchronization. ACM Trans. on Programming and Systems **11**(1) (1991) 124–149
11. Herlihy, M.: A methodology for implementing highly concurrent data objects. ACM Trans. on Programming Languages and Systems **15**(5) (1993) 745–770
12. Kang, K.D., Son, S.H., Stankovic, J.A.: Managing deadline miss ratio and sensor data freshness in real-time databases. IEEE Trans. on Knowledge and Data Engineering **16**(10) (2004) 1200–1216
13. Labrinidis, A., Roussopoulos, N.: Exploring the tradeoff between performance and data freshness in database-driven web servers. The VLDB Journal **13**(3) (2004) 240–255
14. Lamport, L.: On interprocess communication. part ii: Algorithms. Distributed Computing **1**(2) (1986) 86–101
15. Li, W.S., Po, O., Hsiung, W.P., Candan, K.S., Agrawal, D.: Engineering and hosting adaptive freshness-sensitive web applications on data centers. In: Proc. of the Intl. Conf. on World Wide Web. (2003) 587–598
16. Ling, Y., Chen, W.: Measuring cache freshness by additive age. SIGOPS Oper. Syst. Rev. **38**(3) (2004) 12–17
17. Malkhi, D., Merritt, M., Reiter, M.K., Taubenfeld, G.: Objects shared by byzantine processes. Distrib. Comput. **16**(1) (2003) 37–48
18. Michael, M.M., Scott, M.L.: Simple, fast, and practical non-blocking and blocking concurrent queue algorithms. In: Proc. of Symp. on Principles of Distributed Computing (PODC). (1996) 267–275
19. Pacitti, E., Simon, E.: Update propagation strategies to improve freshness in lazy master replicated databases. The VLDB Journal **8**(3-4) (2000) 305–318
20. Shao, C., Pierce, E., Welch, J.L.: Multi-writer consistency conditions for shared memory objects. In: Proc. of the Intl. Symp. on Distributed Computing (DISC). (2003) 106–120
21. Simpson, H.R.: Correctness analysis for class of asynchronous communication mechanisms. Computers and Digital Techniques, IEE Proc.- **139**(1) (1992) 35–49
22. Simpson, H.R.: Freshness specification for a class of asynchronous communication mechanisms. Computers and Digital Techniques, IEE Proc.- **151**(2) (2004) 110–118
23. Sundell, H., Tsigas, P.: NOBLE: A non-blocking inter-process communication library. In: Proc. of the Workshop on Languages, Compilers and Run-time Systems for Scalable Computers. LNCS, Springer Verlag (2002)

A Parallel Algorithm for the Two-Dimensional Cutting Stock Problem

Luis García, Coromoto León, Gara Miranda, and Casiano Rodríguez

Dpto. Estadística, I. O. y Computación
Universidad de La Laguna
E-38271 La Laguna, Tenerife, Spain
{lgforte, cleon, gmiranda, casiano}@ull.es

Abstract. Cutting Stock Problems (CSP) arise in many production industries where large stock sheets must be cut into smaller pieces. We present a parallel algorithm - based on Viswanathan and Bagchi algorithm (VB) - solving the Two-Dimensional Cutting Stock Problem (2DCSP). The algorithm guarantees the processing of best nodes first and does not introduce any redundant combinations - others than the already present in the sequential version. The improvement is orthogonal to any other sequential improvements. Computational results of an OpenMP implementation confirm the optimality of the algorithm. We also produce a new syntactic based reformulation of the 2DCSP problem which leads to a concise representation of the solutions. A highly efficient data structure to store subproblems is introduced.

1 Introduction

Cutting Stock Problems (CSP) arise in many production industries where large stock sheets (glass, textiles, pulp and paper, steel, etc.) must be cut into smaller pieces. CSP can be classified [1,2] attending to several characteristics: the number of dimensions (1D, 2D, 3D), the number of available surfaces and patterns, the shape of the patterns (regular or irregular), the orientation, etc.

The Constrained 2 Dimensional Cutting Stock Problem (2DCSP) is one of the most interesting variants of CSP and targets the cutting of a large rectangle S of dimensions $L \times W$ in a set of smaller rectangles using orthogonal guillotine cuts. That means that any cut must run from one side of the rectangle to the other end and be parallel to the other two edges. The produced rectangles must belong to one of a given set of rectangle types $\mathcal{D} = \{T_1 \ldots T_n\}$ where the i-th type T_i has dimensions $l_i \times w_i$. Associated with each type T_i there is a profit p_i and a demand constraint b_i. The goal is to find a feasible cutting pattern with x_i pieces of type T_i maximizing the total profit:

$$\max \sum_{i=1}^{n} x_i p_i \text{ subject to } x_i \leq b_i \text{ and } x_i \in \mathbb{N}$$

Though a large number of heuristics have been proposed [3,4,5,6], the number of exact algorithms is not so extensive. The optimal algorithms fall in two categories: depth-first searches [7] and best-first search methods [8,9,10]. To our knowledge, not many parallel exact algorithms have been devised [11,12].

Wang [13] was the first to make the observation that all guillotine cutting patterns can be obtained by means of horizontal and vertical builds of pieces (Figure 1). Her idea was exploited by Viswanathan and Bagchi [8] to propose a brilliant best-first search algorithm (VB) which uses Gilmore and Gomory [14] dynamic programming solution - for the unbounded version of the problem - to build an upper bound. The algorithm resembles A* algorithms and uses two lists OPEN and CLIST to yield the set of feasible solutions. At each step, the best build of pieces (or meta-rectangle) from OPEN is chosen and combined with the already found best meta-rectangles (elements in CLIST) to produce horizontal and vertical builds. Later, Hifi [9] and Cung, Hifi and Le-Cun [10] proposed a modified version of VB algorithm (called MVB) introducing an initial lower bound and rules to find in constant/time duplicated/dominated patterns. The efficiency of MVB is also a consequence of other two novelties: CLIST is represented by a bidimensional data structure and VB upper bound is reduced combining it with the solution of a One-Dimensional Knapsack Problem. The Knapsack Problem results from mapping the 2DCSP bidimensional constraints onto one dimensional area constraints (similar proposals were made by Tschöeke and Holthöfer in [11]).

Niklas et al. in [12] proposed a parallel version of Wang's approximation algorithm [13]. Unfortunately, Wang's method does not always yield optimal solutions in a single invocation and is slower than VB algorithm [8]. Tschöeke and Holthöfer parallel version [11] starts from the original VB algorithm and uses the Paderborn Parallel Branch-and-Bound Library (PPBB-LIB [15]). Due to the asynchronous nature provided by the PPBB-LIB skeleton, the algorithm does not guarantee the processing of best subproblems first. Another consequence is the generation of unwanted duplicates which aren't produced by the sequential version. In the worst case an exponential growth of elements may result. The authors proposed a stamp-based mechanism to hinder the generation of duplicates.

Though the next section is devoted to introduce VB algorithm, it contains some contributions. Namely, it emphasizes a new syntactic based reformulation of the problem and proposes a concise representation of the solutions. More important, a highly efficient data structure to store subproblems is introduced. In this section we also present an improvement to avoid unwanted repetitions of computations which are common to the two parallel loops. The parallel algorithm is presented in section 3. On each step the best subproblem from OPEN is all-to-all reduced and combined with each of the elements in CLIST. The design of the parallel algorithm was suggested by the bidimensional data structure proposed by Cung and others [10,11] to hold VB CLIST. The bidimensional structure leads to two traversing loops which can be parallelized. The performance gain is compatible with any other sequential improvements. Some computational results are shown in section 4. Finally, the conclusions and future works are given in section 5.

2 A Sequential Algorithm

Reformulating the 2DCSP. Given two meta-rectangles α and β of dimensions (α^l, α^w) and (β^l, β^w) the vertical build $\alpha|\beta$ is a meta-rectangle of dimensions $(\max\{\alpha^l, \beta^l\}, \alpha^w + \beta^w)$. The horizontal build $\alpha - \beta$ is a meta-rectangle of dimensions $(\alpha^l + \beta^l, \max\{\alpha^w, \beta^w\})$. Using this idea, a feasible solution can be represented by a formula like $(T_2|T_3)|(T_1 - T_2)$. Even better, we may use postfix expressions avoiding the need for parenthesis, i.e. $T_2T_3|T_1T_2 - |$, and leading to a compact representation of the syntax tree (see Figure 1).

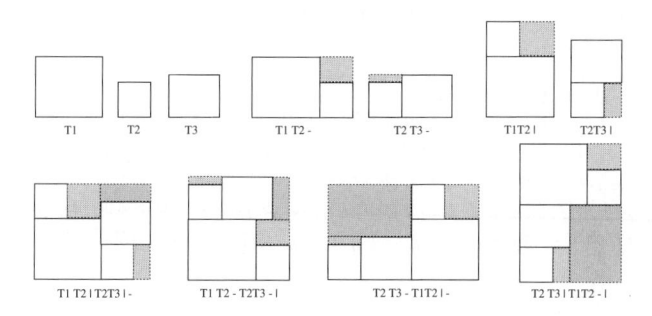

Fig. 1. Examples of vertical and horizontal builds, shaded areas represent waste

Figure 2 presents a context free grammar G and the associated semantic rules defining the attributes value (g), length (l), width (w) and number of used patterns (i) associated with meta-rectangles $\alpha \in L(G)$. Given the aforementioned syntax directed definition the 2DCSP can be reformulated as:

$$\max\{\alpha^g \text{ such that } \alpha \in L(G), \alpha^l \leq L, \alpha^w \leq W \text{ and } \alpha^i \leq b_i \text{ for any pattern } i\}$$

Being $L(G)$ the language generated by the grammar G. Observe how semantic geometrical properties can be embedded into the syntactic structure easing the expression of constant-time dominance rules as described in [10]. Moreover, this notation makes possible to easily build new patterns compositions and to manage and represent the problem (partial) solutions.

Syntax	Semantic Rules
$S \to S_1 S_2 |$	$S^g = S_1^g + S_2^g$ $S^l = \max\{S_1^l, S_2^l\}; S^w = S_1^w + S_2^w$ $S^i = S_1^i + S_2^i$
$S \to S_1 S_2 -$	$S^g = S_1^g + S_2^g$ $S^l = S_1^l + S_2^l; S^w = \max\{S_1^w, S_2^w\}$ $S^i = S_1^i + S_2^i$
$S \to T_i$ for each $T_i \in \mathcal{D}$	$S^g = c_i; S^l = l_i; S^w = w_i; S^i = S^i + 1$

Fig. 2. Syntax Directed Definition for the 2DCSTP. S^i is initialized to 0.

```
1   OPEN := {T_1, T_2, ..., T_n}; CLIST := ∅; f' := UpperBound();
2   BestSol := Heuristic(); B := BestSol^g;
3   repeat
4     choose α meta-rectangle from OPEN with higher f' value;
5     return(Bestsol) if B = f'(α);
6     insert α in CLIST at entry (α^l, α^w);
7     for x := 0 to L - α^l do {
8       forall β ∈ CLIST_x such that β^i ≤ b_i - α^i do {
9         γ = αβ-; γ^l = α^l + β^l; γ^w = max(α^w, β^w);  /* horizontal build */
10        γ^g = α^g + β^g; γ^i = α^i + β^i ∀i;
11        if (γ^g > B) then { free OPEN from B to γ^g; B = γ^g; BestSol = γ; }
12        if (f'(γ) > B) then { insert γ in OPEN at entry f'(γ); }
13      }
14    }
15    for y := 0 to W - α^w do {
16      forall β ∈ CLIST_y such that β^i ≤ b_i - α^i do {
17        γ = αβ|; γ^l = max(α^l, β^l); γ^w = α^w + β^w;  /* vertical build */
18        γ^g = α^g + β^g; γ^i = α^i + β^i ∀i;
19        if (γ^g > B) then { free OPEN from B to γ^g); B = γ^g; BestSol = γ; }
20        if (f'(γ) > B) then { insert γ in OPEN at entry f'(γ); }
21      }
22    }
23    return(Bestsol) if OPEN = ∅;
24  forever;
```

Fig. 3. Modified Version of Viswanathan and Bagchi Algorithm (MVB)

The Modified Viswanathan and Bagchi Algorithm. In VB original version the combination is achieved traversing the whole CLIST, discarding non feasible solutions. To alleviate this, Cun and others [10] introduced the skillful data structure depicted in Figure 4. This way using two loops (see Figure 3), one for the horizontal combinations (lines 7-14) and another for the vertical combinations (lines 15-22) only problems holding the geometry constraints are visited. There is one loss however. Observe that $(\alpha\beta-)^i = (\alpha\beta|)^i$ and $(\alpha\beta-)^g = (\alpha\beta|)^g$ for any α and β and any pattern i. When using the original list data structure these common values are computed only once. The decoupling on two loops implies the repetition of such calculus (lines 10 and 18). We can reduce this overhead as follows: During the horizontal loop (lines 7-14) we save a pointer to α inside the data structure representing the meta-rectangle β (for that we use an extra field, let us call it *current*). On a second field (call it *horizontal*) we store a pointer to $\alpha\beta-$. Now if during the vertical loop (lines 15-22) the meta-rectangle β (line 16) has its *current* field pointing to α we can recover the values $\alpha^i + \beta^i$ stored in $\alpha\beta-$ using the *horizontal* field of β.

Data Structure to Represent the Upper Bounds. On any best-first search algorithm, subproblems are sorted by the value of their upper bounds. Maintaining this usually very large set is often cause for performance degradation.

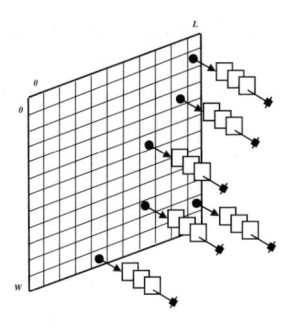

Fig. 4. Data Structure to store CLIST

Since along the execution of any branch-and-bound the lower bounds keep ascending and the upper bounds descending we can state that during the search all the upper bounds fall in the interval $[best_0, upper_0]$. We denote by $best_0$ the initial heuristic value and by $upper_0$ the upper bound of the initial problem. That suggest a natural solution: to have an array $[best_0 \ldots upper_0]$ of pointers to linked lists of subproblems. Subproblems with the same upper bound go to the same linked list. Insertion then can be done in constant time. Notice that insertion using the classical list approach [8,13] leads - for the VB algorithm but it is also the general case for branch-and-bound - to $\mathcal{O}(2^n)$ time since in the worst case the list grows exponentially with the number of patterns. The other main operation involved, choosing/extracting the subproblem with the largest upper bound consists now in descending the interval searching for a non void pointer. Full segments of memory can be freed any time the lower bound improves (line 19). When memory is an issue and there is no space to afford storing the whole interval $[best_0, upper_0]$ the data structure becomes a tree-of-intervals (Figure 5). The root node is now a smaller interval $[0, C]$ where $C = \frac{upper_0 - best_0}{d}$. Each item in $[0, C]$ is a pointer to an interval of size $\frac{C}{d} = \frac{upper_0 - best_0}{d^2}$ and so on.

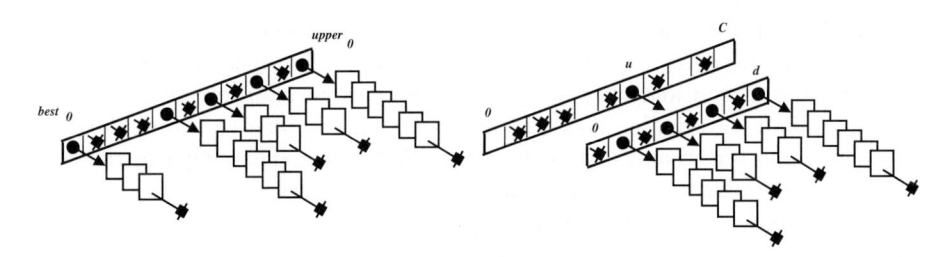

Fig. 5. Data Structure to store OPEN

The idea is extremely simple - but as far as we know it is the first time that it is proposed - and it can be applied to any best-first search branch-and-bound algorithm and therefore to algorithmic skeletons [15,16,17] supporting this technique. Taking advantage of this structure we have sorted the subproblems with same uppers by their lower values. This can change the search order and cause the exploration of more nodes but the best solution value will increase more quickly and we will be able to discard more subproblems.

3 The Parallel Algorithm

In order to improve the sequential scheme, we propose a parallel algorithm that introduces a parallel generation of subproblems (meta-rectangles or builds) from a certain best subproblem. The general operation of this parallel scheme follows the same structure than the sequential scheme presented before (Figure 3). The main difference appears in the subproblem generation loops. Each processor involved in the resolution of the problem works on a section of the bidimensional CLIST. It combines the current best subproblem with the subproblems contained in its matrix section. The work distribution will depend on the processor characteristics. It can be a dynamic or static (cyclic, block) distribution.

Each processor keeps a replicated copy of CLIST. Meanwhile, OPEN will be distributed and only contains the subproblems generated by its owner processor. These structures allow the processors to work independently in the generation of new subproblems. After every combination of the current subproblem with the previous best ones, each processor has its own best current subproblem. To determine which is the global best current subproblem, it is necessary to add an all-to-all reduction point. Once all processors have the new best subproblem, they can begin with the generation work. The same reduction point is used to update the best solution current value.

A brief description of the OpenMP implementation is given below:

1. Each thread initializes its OPEN and CLIST variables. The master thread creates the initial subproblems and inserts them into its OPEN list.
2. At every iteration of the search loop, the global best subproblem must be identified. Each thread makes public its best subproblem by updating the corresponding entry at a shared static structure. The master thread determines the thread identifier having the best current subproblem and writes it to a shared variable. Then the slave threads are able to access the best global subproblem and copy it to a private variable. The owner of the best global subproblem must remove it from its OPEN list.
3. If the subproblem is not the solution and is not dominated/duplicated, it is inserted in each local CLIST (notice that operations for the detection of dominated/duplicated builds are not included in the pseudocode of Figure 3). If the subproblem is discarded, go to the previous step.
4. The horizontal new builds are generated in a first loop and in the second loop are generated the vertical ones. These loops have a *parallel-for* pragma, so each thread will do combinations of the subproblem with certain sections of the CLIST matrix. The new subproblems are inserted into the corresponding thread OPEN list.

5. Once all the new subproblems have been created and inserted into the lists, each thread must find its current best subproblem and copy it to the static shared array. The same is done with the best solution.
6. These steps must be followed until the solution is found.

The main problem of the implementation done on shared memory deals with the use of dynamic linked lists. These lists have to be modified by all threads and there is no mechanism available to ensure the integrity of data. OpenMP compilers usually ensure the integrity of the static structures, that is, arrays or structs stored in the static segment or in the execution stack. But this is not the case when dealing with data structures allocated in the heap. By this reason, some additional operations are necessary to update the shared dynamic data structures used by our implementation.

The exposed parallel algorithm can be easily implemented on a distributed memory scheme. As in this case, each processor would have its own OPEN and CLIST variables. A barrier point would be necessary to do the reduction of the best subproblem and send it to every processor in the team.

4 Computational Results

For the computational study, we have selected some instances from the ones available at [18]. From the instances proposed at [19] we have selected problem 1 from category 1 (cat1_1) and problem 2 from category 3 (cat3_2). The algorithms have been also tested with the problem instances exposed in [9]. Tests have been run over La Laguna University cluster (tarja). The cluster provides a Bull NovaScale 6320 SMP server that consists of 32 Intel Itanium 2 processors at 1.5GHz. The compilers used are: gcc 3.3.3 and $Intel\ C/C++$ 8.1.20.

Table 1 presents the results for the sequential algorithms. Columns labelled "Time" show execution times in seconds and the labelled "Gen.", "Comp." and "Ins." show the number of average generated, computed and inserted nodes respectively. Notice that the generated nodes are the nodes that represent any build created during the search process. Nodes removed from OPEN to be combined with all the previous best subproblems are the computed nodes. Inserted nodes represent the non duplicated generated nodes that can be inserted into OPEN. The results grouped under the name "Initial Version" are for an initial implementation based on VB algorithm. Sequential times for the modified version described in section 2 are also shown in the table under the label "Improved Version". As we can see, the modified sequential implementation introduces great improvements over the original version. The differences are due to the new data structures. They make possible to easily sort elements in OPEN and find duplicated/dominated nodes. But, when the number of generated nodes increases, the insertion of subproblems into OPEN turns too heavy for the first implementation. By this reason, large problem instances are not approachable by this version.

Table 2 shows the results for the parallel implementation of the improved algorithm. The columns labelled "Ins." and "Gen." show the number of nodes,

Table 1. Sequential Results - Original and Improved Versions

Problem Instance	Original Version				Improved Version			
	Gen.	Comp.	Ins.	Time	Gen.	Comp.	Ins.	Time
cat1_1	71631356	80575	124683	329,054	71631356	42842	122307	74,031
cat3_2	5073790	10968	146468	437,434	5067565	10966	145802	29,698
CL_10_24_01	1142805	4533	24654	2,714	1136429	3116	25335	1,042
CL_10_24_03	3547161	9551	32554	6,146	3544239	6625	33535	2,908
CL_10_24_09	1699757	7051	31819	4,873	1685099	4908	32049	1,580
CL_10_51_01	825620	2359	30913	4,582	848216	1732	31703	1,068
1	652	48	198	0,001	825	42	266	0,001
1_	27543	974	1979	0,017	35694	578	2555	0,079
2	11014	240	5804	0,171	11142	161	6011	0,017
2_	4414	136	2571	0,021	4586	91	2745	0,008
3	29935	1120	1868	0,046	31467	628	2141	0,038
3_	282	44	182	0,002	290	32	194	0,002
A1	21757	688	4031	0,061	25026	440	4057	0,104
A2	184186	5813	11749	0,694	172684	2511	9642	0,238
A3	24331	379	4188	0,069	24538	262	4384	0,036
A4	64866	558	27373	3,303	68394	374	29002	0,174

Table 2. Improved Version - Parallel Results

	Thread 1		Thread 2		Thread 3		Thread 4		Thread 5		Thread 6		Thread 7		Thread 8			
	Ins.	Gen.	Ins.	Gen.	Ins.	Gen.	Ins.	Gen.	Ins.	Gen.	Ins.	Gen.	Ins.	Gen.	Ins.	Gen.	Comp.	Time
cat1_1																		
Th 1	122	71631															42	112.09
Th 2	72	35551	52	36079													50	73.49
Th 4	39	18896	28	19874	35	18093	26	17515									56	73.96
Th 8	20	11443	16	13898	14	8216	15	9427	18	7068	10	5657	20	9503	10	7894	57	68.99
cat3_2																		
Th 1	145	5067															10	32.08
Th 2	43	2621	44	2919													11	5.59
Th 4	32	175	31	179	23	148	23	110									3	0.77
Th 8	30	200	19	126	21	184	17	128	25	172	28	228	12	112	22	117	5	1.01

in thousands, inserted and generated by every thread. The number of computed nodes during the process is shown (also in thousands) in column "Comp.". Computational time, in seconds, invested in the search process is presented in column "Time". Parallel speedups strongly depend on the particular problem. That is a result of changing the search space exploration order when more than one thread is collaborating in the resolution. Even in worse cases (cat1_1) we can improve sequential times. A fair work load distribution between threads is difficult to obtain since it is not only needed to fairly distribute the subproblems to generate but also the ones to be inserted. Before doing a certain combination we are not able to know if a build will be valid or not (to be inserted).

5   Conclusions

An exact algorithm for the resolution of the Two-Dimensional Cutting Stock Problem has been presented. The implementation is based on VB and MVB algorithms. First of all, we have presented a new reformulation of the problem. A new syntax is introduced for the representation of the solutions. This notation helps in the detection of similar properties between different subproblems, making possible to efficiently detect duplicated combinations. By these representations we also are able to easily build the solution found. New data structures have been designed in order to efficiently manage insertions, combinations and dominance/duplication detections. The new data structure to manage subproblems in OPEN allows to do insertions in constant time independently of the number of nodes in the lists. The idea can be easily extend to any best-first search, branch-and-bound or algorithmic skeletons giving support to these techniques. In order to avoid the unnecessary recomputation of some subproblems, we have added a mechanism to store subproblems related to a particular element in CLIST. All these new features introduce an important improvement in the sequential exact algorithm. For being able to afford larger problem instances, we have presented a general parallel algorithm. The algorithm proposes a parallelization of the new build generation loop. On most space search algorithms, the slightest changes in the search order may cause dramatic consequences on the execution time. Super and sublinear speedups may occur since the parallel algorithm alters the sequential order. A first parallel implementation has been developed over shared memory. Porting an existing C application to OpenMP even if the algorithm is straightforwardly parallel can be sometimes a nightmare due to the lack of support to qualify dynamic memory variables as shared or private.

Future work targets improvement of both, the upper bound and the initial heuristic lower bound. This improvement in the bounds will allow to highly reduce the search space. In relation to the parallel algorithm we would like to develop a message passing implementation in order to compare with the one presented. A deep study of the work load distribution is also required.

Acknowledgements

This work has been supported by the EC (FEDER) and by the Spanish Ministry of Education inside the 'Plan Nacional de I+D+I' with contract number TIN2005-08818-C04-04. The work of G. Miranda has been developed under the grant FPU-AP2004-2290.

References

1. Sweeney, P.E., Paternoster, E.R.: Cutting and Packing Problems: A categorized, application-orientated research bibliography. Journal of the Operational Research Society **43**(7) (1992) 691–706
2. Dyckhoff, H.: A Typology of Cutting and Packing Problems. European Journal of Operational Research **44**(2) (1990) 145–159

3. Dowsland, K.A., Dowsland, W.B.: Packing Problems. European Journal of Operational Research **56**(1) (1992) 2–14
4. Burke, E., Kendall, G.: Applying Simulated Annealing and the No Fit Polygon to the Nesting Problem. In Arabnia, H.R., ed.: Proceedings of the International Conference on Artificial Intelligence (IC-AI'99). Volume 1., CSREA Press (1999) 51–57 ekb@cs.nott.ac.uk, gxk@cs.nott.ac.uk.
5. Maouche, S., Bounsaythip, C.: Optimizing Textile Shape Placement by Tree Genetic Annealing. In: Proceedings of the Society for Computer Simulation Conference (SCSC'96). (1996) Salah.Maouche@univ-lille1.fr.
6. Roussel, G., Maouche, S.: Automatic Lay Planning for Irregular Shapes on Plain Fabric. Search in Direct Graph and ε-Admissible Resolution. In: Proceedings of the XVI IFIP-TC7 Conference, Compiègne, France (1993)
7. Christofides, N., Whitlock, C.: An Algorithm for Two-Dimensional Cutting Problems. Operations Research **25**(1) (1977) 30–44
8. Viswanathan, K.V., Bagchi, A.: Best-First Search Methods for Constrained Two-Dimensional Cutting Stock Problems. Operations Research **41**(4) (1993) 768–776
9. Hifi, M.: An Improvement of Viswanathan and Bagchi's Exact Algorithm for Constrained Two-Dimensional Cutting Stock. Computer Operations Research **24**(8) (1997) 727–736
10. Cung, V.D., Hifi, M., Le-Cun, B.: Constrained Two-Dimensional Cutting Stock Problems: A Best-First Branch-and-Bound Algorithm. Technical Report 97/020, Laboratoire PRiSM - CNRS URA 1525. Université de Versailles, Saint Quentin en Yvelines. 78035 Versailles Cedex, FRANCE (1997)
11. Tschöke, S., Holthöfer, N.: A New Parallel Approach to the Constrained Two-Dimensional Cutting Stock Problem. In Ferreira, A., Rolim, J., eds.: Parallel Algorithms for Irregularly Structured Problems, Berlin, Germany, Springer-Verlag (1995) 285–300 sts@uni-paderborn.de, http://www.uni-paderborn.de/fachbereich/AG/monien/index.html.
12. Nicklas, L.D., Atkins, R.W., Setia, S.K., Wang, P.Y.: The Design and Implementation of a Parallel Solution to the Cutting Stock Problem. Concurrency - Practice and Experience **10**(10) (1998) 783–805
13. Wang, P.Y.: Two Algorithms for Constrained Two-Dimensional Cutting Stock Problems. Operations Research **31**(3) (1983) 573–586
14. Gilmore, P.C., Gomory, R.E.: The Theory and Computation of Knapsack Functions. Operations Research **14** (1966) 1045–1074
15. Tschöke, S., Polzer, T.: Portable parallel branch-and-bound library - PPBB-lib (1996)
16. Alba, E., et al: MaLLBa: A Library of Skeletons for Combinatorial Optimization. In: Proceedings of Euro-Par. Volume 2400 of Lecture Notes in Computer Science., Paderborn (GE), Springer-Verlag (2002) 927–932
17. Le-Cun, B., Roucairol, C.: BOB : a unified platform for implementing branch-and-bound like algorithms (1995)
18. Group, D.O.R.: Library of Instances (Two-Constraint Bin Packing Problem) http://www.or.deis.unibo.it/research_pages/ORinstances/2CBP.html.
19. Hopper, E., Turton, C.H.: An Empirical Investigation of Meta-heuristic and Heuristic Algorithms for a 2D Packing Problem (1999) http://people.brunel.ac.uk/~mastjjb/jeb/orlib/files/strip1.txt.

A BSP/CGM Algorithm for Finding All Maximal Contiguous Subsequences of a Sequence of Numbers[*]

Carlos Eduardo Rodrigues Alves[1], Edson Norberto Cáceres[2], and Siang Wun Song[3]

[1] Universidade São Judas Tadeu, Brazil
prof.carlos_r_alves@usjt.br
[2] Universidade Federal de Mato Grosso do Sul, Brazil
edson@dct.ufms.br
[3] Universidade de São Paulo, Brazil
song@ime.usp.br

Abstract. Given a sequence A of real numbers, we wish to find a list of all non-overlapping contiguous subsequences of A that are *maximal*. A *maximal* subsequence M of A has the property that no proper subsequence of M has a greater sum of values. Furthermore, M may not be contained properly within any subsequence of A with this property. This problem can be solved sequentially in linear time. We present a BSP/CGM algorithm that uses p processors and takes $O(|A|/p)$ time and $O(|A|/p)$ space per processor. The algorithm uses a constant number of communication rounds of size at most $O(|A|/p)$. Thus the algorithm achieves linear speed-up and is highly scalable.

1 Introduction

Given a sequence of real numbers, the *maximum subsequence problem* finds the contiguous subsequence with the maximum sum [1]. A more general problem is the *all maximal subsequences problem* [2] which finds a list of all non-overlapping contiguous subsequences with maximal sum. These two problems arise in several contexts in Computational Biology. Many applications are presented in [2], to identify transmembrane domains in proteins expressed as a sequence of amino acids and to discover CpG islands. Csuros [3] mentions other applications that require the computation of such subsequences, in the analysis of protein and DNA sequences, determination of isochores in DNA sequences, etc.

Linear time sequential algorithms are known to solve both problems [1,2]. Wen [4] presents a EREW PRAM algorithm that solves the maximum subsequence problem of n given numbers in $O(\log n)$ time using $O(n/\log n)$ processors. For this same problem, Alves, Cáceres and Song [5] present a BSP/CGM parallel

---

[*] Partially supported by CNPq 30.0317/02-6, 30.5218/03-4, 55.0094/05-9 and 62.0123/04-4, and FUNDECT-MS Proc. 41/100117/03. We also acknowlegde the comments of the anonymous referees.

algorithm on p processors that requires $O(n/p)$ computing time and constant number of communication rounds. Dai and Su [6] present a PRAM EREW work-optimal algorithm that solve the all maximal subsequences problem in $O(\log n)$ time with $O(n)$ operations.

In this paper we present a BSP/CGM algorithm to solve the all maximal subsequences problem. Given a sequence A of numbers, this algorithm uses p processors and finds all the maximal subsequences in $O(|A|/p)$ time, with $O(|A|/p)$ space per processor, and requires a constant number of communication rounds in which at most $O(|A|/p)$ data are transmitted. Unlike the parallel solution for the basic maximum subsequence problem, it is not at all intuitive that one can find a parallel algorithm for this problem that requires only a constant number of communication rounds. In this sense, this is also an important result in a theoretical viewpoint. Finding a BSP/CGM algorithm with constant number of communication rounds for a problem with linear sequential complexity is not always possible, as shown by the list ranking problem where the best known BSP/CGM algorithm requires $O(\log n)$ communication rounds [7].

2 Preliminary Definitions and Results

Given a sequence A of real numbers, denote its elements by a_i, $1 \leq i \leq |A|$. Subsequences of A are indicated as $A_i^j = (a_{i+1}, ..., a_j)$. The superscript indicates the rightmost position in the subsequence, and the subscript is one less than the leftmost position. If the subscript and the superscript are equal, the subsequence is empty. A particular subsequence of A can be denoted by some other upper-case letter, but all indices will refer to sequence A. To indicate the indices of the first (leftmost) and last (rightmost) positions of a sequence X we use $L(X)$ and $R(X)$. For coherence with the previous notation we have $X = A_{L(X)}^{R(X)} = (a_{L(X)+1}, ..., a_{R(X)})$. Notice that $L(X)$ indicates one position to the left of the actual beginning of X. The concatenation of sequences X_1, X_2, ... X_n will be denoted by $\langle X_1, X_2, ... X_n \rangle$. The sum of the values of a subsequence X will be denoted by $Score(X)$. If X is empty, then we define its score to be zero. As the sum of prefixes of A is very important in this paper, we use $PS(j)$ to denote $Score(A_0^j)$. We consider $PS(0) = 0$. Notice that $Score(A_i^j) = PS(j) - PS(i)$. For a subsequence $X = A_i^j$, the minimum and the maximum among all values of $PS(k)$, for $i \leq k \leq j$, will be denoted by $Min(X)$ and $Max(X)$, respectively.

We consider the BSP/CGM (Bulk Synchronous Parallel/Coarse-Grained Multicomputer) model [8,9], with p processors each with $O(n/p)$ local memory, where n is the input size of the problem. A BSP/CGM algorithm consists of alternating local computation and global communication rounds. In each communication round, each processor can send/receive messages with at most $O(n/p)$ data. A BSP/CGM algorithm attempts to minimize the number of communication rounds as well as the total local computation time.

A *maximum* scoring subsequence of X is one with the largest score among all scores of subsequences of X. When ties occur, we choose the subsequence of minimum length. If there is no positive number in X, we assume that there is no maximum scoring subsequence. It is easy to see that prefixes and suffixes of a maximum subsequence always have positive scores, because the deletion of a prefix or suffix with non-positive score would lead to a better subsequence. The problem of finding all maximal subsequences of A is more complicated. Ruzzo and Tompa [2] define the set of maximal subsequences recursively, as follows.

Definition 1. *Given a sequence A of real numbers, the set of maximal subsequences of A is empty if A has no positive values. Otherwise, let $\langle A_1, M, A_2 \rangle$ be a decomposition of A in three subsequences where M is the maximum scoring subsequence of A (A_1 and A_2 may be empty sequences). Then the set of maximal subsequences of A is the union of the set $\{M\}$, the set of maximal subsequences of A_1, and the set of maximal subsequences of A_2.*

i	1	2	3	4	5	6	7	8	9	10	11	12	13	14	15	16	17	18	19	20	21	22
a_i	5	-3	-1	5	-9	0	3	3	7	-9	3	-6	3	-1	0	3	-3	0	7	-4	0	-6

Fig. 1. Example sequence to be used throughout the text

Consider the sequence $A = (a_1, a_2, \ldots, a_{22})$ of Figure 1. The maximal subsequences are $A_0^4 = (5, -3, -1, 5)$, $A_6^9 = (3, 3, 7)$, $A_{10}^{11} = (3)$, and $A_{12}^{19} = (3, -1, 0, 3, -3, 0, 7)$, with respective scores of 6, 13, 3, and 9.

Ruzzo and Tompa also give two necessary and sufficient properties that a subsequence X must have to be maximal in sequence A. They are stated in the following theorem. For a proof, see [2].

Theorem 1. *A subsequence X is maximal in A iff it has both properties below:*

Property Pr1. *For any proper subsequence Y of X, $Score(Y) < Score(X)$.*
Property Pr2. *There is no proper supersequence of X that has Property Pr1.*

Notice that the score of a sequence with property Pr1 must be positive. Subsequences of A that have property Pr1 will be called *Pr1-subsequences*. We can restate the definition of a maximal subsequence in terms of these properties.

Definition 2. *Given a sequence A of real numbers, the list of maximal subsequences of A, denoted MList (A), is the list of all subsequences that have Properties Pr1 and Pr2, ordered with respect to $L(.)$. This list is indexed starting at 1 with the leftmost subsequence.*

Property Pr1 can also be stated in terms of prefix sums, by the following lemma. In this paper we omit all the proofs. They can be found in [10].

Lemma 1. *A subsequence A_i^j is Pr1-subsequence iff for all m, $i < m < j$, $PS(i) < PS(m) < PS(j)$.*

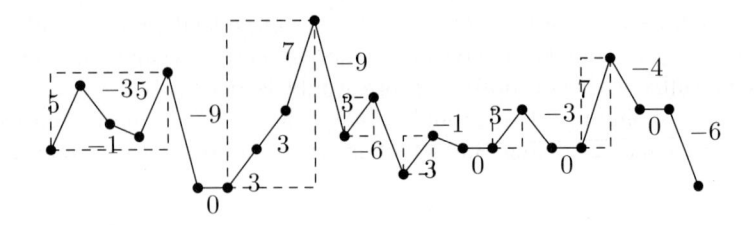

Fig. 2. Sequence $A = (5, -3, -1, \ldots, 0, -6)$ and some Pr1-subsequences

In Figure 2, we plot the function $PS(.)$, so that positive (negative) values in the sequence are represented by ascending (descending) line segments. A Pr1-subsequence X is indicated by a rectangular box with $(L(X), PS(L(X)))$ and $(R(X), PS(R(X)))$ as lower-left and upper-right corners, respectively. The plotted curve touches the box only in these corners. Notice that the first three Pr1-subsequences in Figure 2 are maximal subsequences of A, but the last three are not (they are subsequences of the same A-maximal, namely A_{12}^{19}).

We say that A_i^j, $i < j$, is a *Pr1-prefix* if $PS(i) < Min(A_{i+1}^j)$ and it is a *Pr1-suffix* if $Max(A_i^{j-1}) < PS(j)$. A Pr1-subsequence is both a Pr1-prefix and a Pr1-suffix.

Corollary 1. *If P is a Pr1-prefix and S is a Pr1-suffix, $\langle P, S \rangle$ is a Pr1-subsequence iff $Min(P) < Min(S)$ and $Max(P) < Max(S)$.*

One can observe [2] that (i) any Pr1-subsequence of a sequence A is contained in a maximal subsequence of A (maybe not properly), and (ii) given a sequence A, any two distinct maximal subsequences of A do not overlap or touch each other. The parallel algorithm is based on finding lists of maximal subsequences in segments of the original sequence A. Consider a subsequence X of A. We will say that a subsequence is an *X-maximal subsequence*, or just an *X-maximal*, if it is maximal in X, that is, it is a Pr1-subsequence and has no proper supersequence that is a Pr1-subsequence of X. (As an abuse of our notation we write the plural of X-maximal as X-maximals.) Thus we want to find the set of all A-maximals.

Lemma 2. *Let $Z = \langle X, Y \rangle$ for some non-empty X and Y. Then there is at most one Z-maximal M that overlaps both X and Y. If there is such M, it has an X-maximal as a prefix and a Y-maximal as a suffix. The X-maximals to the left of M and the Y-maximals to the right of M are also Z-maximals.*

This lemma shows that it is possible to build $MList(A)$ working incrementally. This is important for the proposed parallel algorithm, where the sequence A is divided into subsequences that are treated separately. Their maximal subsequences are used later to find the A-maximals. The parallel algorithm deals with the following subproblem: given a subsequence X of A and its list of maximal subsequences $MList(X)$, find, if possible, an X-maximal that is a prefix (or suffix) of a larger A-maximal. This clearly involves $MList(X)$ and the rest of sequence A. However, some X-maximals need not be considered as possible

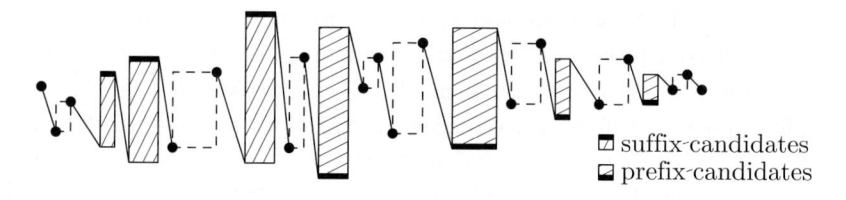

Fig. 3. A sequence X, $MList(X)$, $PList(X)$ and $SList(X)$. The first (last) maximal is not a suffix (prefix) candidate because of the first condition of the definition. The other maximals that are not candidates fall in the second condition - observe the bottom of the prefix candidates and the top of the suffix candidates. The descending lines represent sequences of non-positive numbers.

prefixes or suffixes of larger A-maximals, regardless of what is outside X. The efficiency of our algorithm is based on this important notion, so we formalize it in the following definitions and lemmas. We deal with prefix candidates first.

Definition 3. *Given a subsequence X of A, $PList(X)$ is the ordered list of all X-maximals, with the exception of those X-maximals M for which one of the two conditions are satisfied: (1) $Min(M) \geq PS(R(X))$, or (2) there is an X-maximal N to the right of M such that $Min(M) \geq Min(N)$. (The elements of $PList(X)$ are indexed starting at 1 with the leftmost subsequence.)*

Informally, $PList(X)$ gives us the list of all X-maximals that are potential candidates to be merged to the right to give larger maximals. Notice that we excluded from $PList(X)$ those X-maximals (satisfying conditions 1 and 2) that can never give larger maximals. Consider $X = A_0^{14}$ of the example sequence (see Figure 1 and Figure 2). There are four X-maximals, namely A_0^4, A_6^9, A_{10}^{11}, and A_{12}^{13} (indicated by the first four boxes of Figure 2). A_0^4 does not belong to $PList(X)$ because of condition 1. A_{10}^{11} does not belong to $PList(X)$ because of both conditions 1 and 2. Thus $PList(X) = (A_6^9, A_{12}^{13})$.

We have following properties [10]: (i) If X is a subsequence of A, $PList(X)$ contains all X-maximals that can be a proper prefix of an A-maximal, (ii) if M is a sequence in $PList(X)$ and $i \in \,]L(M), R(X)]$ then $Min(M) < PS(i)$, that is, $A_{L(M)}^{R(X)}$ is a Pr1-prefix, and (iii) if M is a sequence in $PList(X)$ and $i \in \,]R(M), R(X)]$ then $Max(M) \geq PS(i)$. A consequence of these properties is that $PList(X)$ is in a non-increasing order of $Max(.)$ and a strictly increasing order of $Min(.)$. Figure 3 illustrates $PList(X)$ (and $SList(X)$, defined shortly).

We also a need similar definition for possible suffixes of A-maximals. The associated properties are given below. Notice the exchanging roles of $Max(.)$ and $Min(.)$, "left" and "right", etc.

Definition 4. *Given a subsequence X of A, $SList(X)$ is an ordered list of all X-maximals, with the exception of those X-maximals N for which one of the two conditions below are satisfied: (1) $Max(N) \leq PS(L(X))$, or (2) there is a X-maximal M to the left of N such that $Max(N) \leq Max(M)$. (The elements of $SList(X)$ are indexed starting at 1 with the rightmost subsequence.)*

Properties: (i-a) If X is a subsequence of A, $SList(X)$ contains all X-maximals that can be a proper suffix of an A-maximal, (ii-a) if N is a sequence in $SList(X)$ and $i \in [L(X), R(N)[$ then $PS(i) < Max(N)$, that is, $A_{L(X)}^{R(N)}$ is a Pr1-suffix, and (iii-a) if N is a sequence in $SList(X)$ and $i \in [L(X), L(N)[$ then $PS(i) \geq Min(N)$. Notice that at most one X-maximal may belong to both $PList(X)$ and $SList(X)$, namely a maximum subsequence of X. Any other element of $SList(X)$ must be to the left of any element of $PList(X)$. See Figure 3 for an illustration of $PList(X)$ and $SList(X)$ (when these lists are disjoint).

3 The Parallel Algorithm

Consider p processors $P_1, P_2, \ldots, P_p$. Assume that the input sequence A is divided into p subsequences, each of size $l = \lceil |A|/p \rceil$ except the last one, which may be smaller. We call these subsequences $AP_i = A_{l(i-1)}^{li}$. At the beginning, each AP_i is already stored in the local memory of processor P_i. At the end, processor P_i will contain the information (position and score) of all A-maximals that start or end within AP_i. Lemma 3 shows how to find the local maximals.

Lemma 3. *In $O(|A|/p)$ time and space and with one communication round of size $O(p)$, each processor P_i may obtain: (i) its local lists of maximals ($MList(AP_i)$), prefix candidates ($PList(AP_i)$) and suffix candidates ($SList(AP_i)$), and (ii) $PS(L(AP_j))$, $Min(AP_j)$ and $Max(AP_j)$ for all $j \in [1, p]$.*

We now consider a basic procedure to join lists of maximals. We will see how $MList(Z)$ may be obtained from $MList(X)$, $MList(Y)$, $PList(X)$ and $SList(Y)$ when $Z = \langle X, Y \rangle$. The following lemma states the condition for two local maximal subsequences to be merged to form a larger one.

Lemma 4. *Given $M \in PList(X)$ and $N \in SList(Y)$, $A_{L(M)}^{R(N)}$ is a Pr1-subsequence iff $Min(M) < Min(N)$ and $Max(M) < Max(N)$.*

Properties (i) and (i-a) state that we may search for a Z-maximal that overlaps X and Y using only $PList(X)$ and $SList(Y)$. Algorithm 1 does this. We use $Pl = PList(X)$ and $Sl = SList(Y)$ for short, indexing them as stated in Definitions 3 and 4. The algorithm returns the indices of the chosen candidates for prefixes and suffixes of the new Z-maximal.

Algorithm 1. Joining Two Lists of Maximals
Require: Lists Pl and Sl, with $|Pl|$ and $|Sl|$ candidates, respectively.
Ensure: Flag f that indicates if a new maximal was found, indices i_p and i_s of
  the candidates that define this maximal.
1: $i_p \leftarrow 1$, $i_s \leftarrow 1$, $f \leftarrow false$
2: **while** $i_p \leq |Pl|$ and $i_s \leq |Sl|$ and not f **do**
3:   **if** $Max(Pl[i_p]) \geq Max(Sl[i_s])$ **then**
4:     $i_p \leftarrow i_p + 1$
5:   **else if** $Min(Pl[i_p]) \geq Min(Sl[i_s])$ **then**

```
 6:         i_s ← i_s + 1
 7:     else
 8:         f ← true
 9:     end if
10: end while
```

It can be shown that, given $Z = \langle X, Y \rangle$, $Pl = PList(X)$ and $Sl = SList(Y)$, Algorithm 1 finds the only Z-maximal that overlaps X and Y, if it exists, in $O(|Pl| + |Sl|)$ time and $O(1)$ additional space.

The parallel algorithm performs a single joining step, using a constant number of communication rounds, involving all the local maximals found in the local step. This step is based on the simple observation that a non-local maximal must start inside some AP_i and end in some AP_j with $1 \leq i < j \leq p$, so it must have some sequence in $PList(AP_i)$ as prefix and some sequence in $SList(AP_j)$ as suffix. The problem is to find a *relevant* set of Pr1-subsequences of A that cross processor boundaries. By *relevant* we mean all the A-maximals that cross processor boundaries must be contained in this set.

We say that a prefix candidate and a suffix candidate *match* if they define a Pr1-subsequence of A. The following definition states the conditions for a match.

Lemma 5. *For $M \in PList(AP_i)$ and $N \in SList(AP)_j$, $1 \leq i < j \leq p$, $A_{L(M)}^{R(N)}$ (the sequence that has M as prefix, N as suffix and contains AP_k, $i < k < j$) is a Pr1-subsequence iff $Min(M) < Min(N)$, $Max(M) < Max(N)$, $Min(M) < \min_{i<k<j} Min(AP_k)$ and $Max(N) > \max_{i<k<j} Max(AP_k)$.*

After the local step described in Lemma 3 the processors cannot determine which candidates match because they have access only to their own lists of candidates. However, given a particular prefix or suffix candidate, the extra conditions of Lemma 5 allow the determination of the processors where a match for this candidate may be found. So the first step in the global joining operation is to *tag* each candidate with the number of the processor(s) that may contain a match for it.

Lemma 6. *For $i \in [1, p]$ it is possible to tag all the elements of $PList(AP_i)$ and $SList(AP_i)$ based on the values of $Max(AP_j)$ and $Min(AP_j)$ for all $j \in [1, p]$. Each tag indicates which processor may contain a match for a particular candidate. Each candidate is tagged at most once, with two exceptions per processor at the most. The time required is $O(|A|/p)$ and the space required is $O(p)$.*

Algorithm 2 presents the tagging process, based on a case by case study [10]. This algorithm contains the tagging procedure for the prefix candidates of $PList(AP_i)$, called Pl for short. $PTagList(i)$ is called Tl for short. Figure 4 illustrates the tagging of prefix candidates.

Algorithm 2. Tagging a List of Prefix Candidates

Require: Lists Pl and Tl, with $|Pl|$ and $|Tl|$ elements, respectively.
Ensure: Tagging of the elements of Pl.

1: $i_p \leftarrow 1, i_t \leftarrow 1, f \leftarrow false$
2: **while** $i_p \leq |Pl|$ and $i_t \leq |Tl|$ and not f **do**
3:   **if** $Max(Pl[i_p]) \geq Max(Tl[i_t])$ **then**
4:     $i_p \leftarrow i_p + 1$
5:   **else if** $Min(Pl[i_p]) \geq Min*(Tl[i_t])$ **then**
6:     $i_t \leftarrow i_t + 1$
7:   **else**
8:     tag $Pl[i_p]$ with $tag(Tl[i_t])$
9:     **if** $Min(Pl[i_p]) < Min(Tl[i_t])$ **then**
10:       $f \leftarrow true$
11:     **end if**
12:   **end if**
13: **end while**

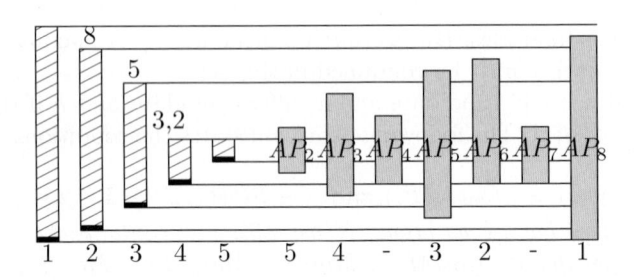

Fig. 4. We consider the tagging of elements of $PList\,(AP_1)$, represented as shaded bars on the left. The darkened bars in the right represent the data from other processors. The numbers below the bars represent the indices in $PList\,(AP_i)$ and $PTagList\,(1)$.

After the tagging procedure, each prefix/suffix candidate may be associated with two other processors: the one which contains it and the one specified in the tag. Some candidates have no tags and may be ignored. A few candidates have two tags and have to be duplicated for the next phase. The next phase involves checking the existence of cross-processors Pr1-subsequences of A, that is, Pr1-subsequences that start within AP_i and ends within AP_j for some pair (i, j), $1 \leq i < j \leq p$. This is done by checking the elements of $PList\,(AP_i)$ that are tagged with j and elements of $SList\,(AP_j)$ that are tagged with i. These elements must be in the local memory of one single processor for verification by Algorithm 1. The rule to choose which processor does the verification is simple: the one whose list of candidates is larger receives the data from the other one. In case both lists have the same size, a deterministic rule is used to break the tie. For example, if $i + j$ is even then P_i does the job, otherwise P_j does it.

The following lemma summarizes the complexity of the tagging process.

Lemma 7. *After tagging the prefix and suffix candidates, all cross-processors Pr1-subsequences that may be A-maximals can be found in $O(|A|/p)$ time and space and two communication rounds of sizes $O(p)$ and $O(|A|/p)$. The number of sequences is at most $2p$.*

It should be noticed that some of the new Pr1-subsequences may not have Property Pr2. The important thing here is that the procedure just described does not miss any possible A-maximal. The next step is to find the Pr1-subsequences that are really A-maximals. All processors broadcast the information about the new Pr1-subsequences found. Every processor then eliminates the Pr1-subsequence that are contained in another Pr1-subsequence. The presented procedure does not generate two Pr1-subsequences that overlap, unless one is contained in the other. That is because if two Pr1-subsequences overlap then their union is also a Pr1-subsequence. Each Pr1-subsequence is related to a different pair of processors. All that must be verified is which pairs generated new sequences, done by Algorithm 3. It takes $O(p)$ time and space.

Algorithm 3. Removing Pr1-subsequences that not A-maximals

Require: List L (with $|L|$ elements) of pairs of processors for which there is a cross-processor Pr1-subsequence.
Ensure: List N (with n elements) of pairs of processors for which there is a cross-processor A-maximal.
1: **for** $k \leftarrow 1$ to p **do**
2:     $V[k] \leftarrow k$
3: **end for**
4: **for** $k \leftarrow 1$ to $|L|$ **do**
5:     $i \leftarrow$ smallest component of $L[k]$
6:     $j \leftarrow$ largest component of $L[k]$
7:     **if** $j > V[i]$ **then**
8:        $V[i] \leftarrow j$
9:     **end if**
10: **end for**
11: $n \leftarrow 0$, $k \leftarrow 1$
12: **while** $k < p$ **do**
13:     **if** $V[k] > k$ **then**
14:        $n \leftarrow n + 1$
15:        $N[n] \leftarrow (k, V[k])$
16:        $k \leftarrow V[k]$
17:     **else**
18:        $k \leftarrow k + 1$
19:     **end if**
20: **end while**

A final step is done locally by each processor. By examining the list of new A-maximals, processor P_i verifies if there is an A-maximal that contains its entire local subsequence AP_i, which means that its own local set of maximals $MList(AP_i)$ should be discarded. This can be done in time $O(p)$. If there is a cross-processors A-maximal that starts or ends within AP_i, a final scan of $MList(AP_i)$ will eliminate the local maximals that are contained in a larger A-maximal. This final scan can be done in time $O\left(\log(|A|/p)\right)$.

Theorem 2. *Using a BSP/CGM with p processors, all maximal subsequences of a sequence A (already distributed in the p local memories) can be found in time $O(|A|/p)$, using $O(|A|/p)$ local space and $O(1)$ communication rounds.*

4 Conclusion

We have presented a parallel algorithm that finds all maximal subsequences of a sequence A with linear speed-up and high scalability. The size of the communication rounds is bounded by $O(|A|/p)$. communication rounds should be much lower than $|A|/p$. Experimenting with a sequence X of random numbers we conjecture that the average size of *PList* (X) is $O(\log(|X|))$. The running time of the whole algorithm is dominated by the time of the first step to find the local maximal subsequences. To derive this parallel $O(|A|/p)$ time and $O(|A|/p)$ space per processor algorithm, requiring a constant number of communication rounds, we explored the properties of those local maximals that are potential candidates to be merged together to form larger maximals, as well as an efficient merge process to join candidate local maximals.

References

1. Bentley, J.: Programming Pearls. Addison-Wesley (1986)
2. Ruzzo, W.L., Tompa, M.: A linear time algorithm for finding all maximal scoring subsequences. In: Proceedings of the 7th International Conference on Intelligent Systems for Molecular Biology, AAAI Press (1999) 234–241
3. Csuros, M.: Algorithms for finding maximal-scoring segment sets. In: Proceedings WABI2004 - 4th Workshop on Algorithms in Bioinformatics. Volume 3240 of Lecture Notes in Computer Science., Springer Verlag (2004) 62–73
4. Wen, Z.: Fast parallel algorithm for the maximum sum problem. Parallel Computing **21** (1995) 461–466
5. Alves, C.E.R., Cáceres, E.N., Song, S.W.: BSP/CGM algorithms for maximum subsequence and maximum subarray. In: Proceedings 11th European PVM/MPI Users' Group Meeting. Volume 3241 of Lecture Notes in Computer Science., Springer Verlag (2004) 139–146
6. Dai, H.K., Su, H.C.: A parallel algorithm for finding all successive minimal maximum subsequences. In: Latin American Theoretica Informatics. Volume 3887 of Lecture Notes in Computer Science., Springer Verlag (2006) 337–348
7. Dehne, F., Ferreira, A., Caceres, E., Song, S.W., Roncato, A.: Efficient parallel graph algorithms for coarse grained multicomputers and BSP. Algorithmica **33**(2) (2002) 183–200
8. Valiant, L.: A bridging model for parallel computation. Communication of the ACM **33**(8) (1990) 103–111
9. Dehne, F., Fabri, A., Rau-Chaplin, A.: Scalable parallel geometric algorithms for coarse grained multicomputers. In: Proc. ACM 9th Annual Computational Geometry. (1993) 298–307
10. Alves, C.E.R., Cáceres, E.N., Song, S.W.: A BSP/CGM algorithm for finding all maximal contiguous subsequences of a sequence of numbers. Technical report, Universidade de São Paulo (2005)

On-Line Adaptive Parallel Prefix Computation

Jean-Louis Roch, Daouda Traoré, and Julien Bernard*

Équipe MOAIS (CNRS-INRIA-INPG-UJF)
Laboratoire d'Informatique de Grenoble
38330 Montbonnot Saint Martin, France
{Julien.Bernard, Jean-Louis.Roch, Daouda.Traore}@imag.fr
http://moais.imag.fr

Abstract. We consider parallel prefix computation on processors of different and possibly changing speeds. Extending previous works on identical processors, we provide a lower bound for this problem. We introduce a new adaptive algorithm which is based on the on-line recursive coupling of an optimal sequential algorithm and a parallel one, non-optimal but recursive and fine-grain. The coupling relies on a work-stealing scheduling. Its theoretical performance is analysed on p processors of different and changing speeds. It is close to the lower bound both on identical processors and close to the lower bound for processors of changing speeds. Experiments performed on an eight-processor machine confirms this theoretical result.

1 Introduction

Given $x_0, x_1, \ldots, x_n$, the prefix problem is to compute the n products $\pi_k = x_0 \circ x_1 \circ \ldots \circ x_k$ for $1 \leq k \leq n$, where $\circ$ is an associative operation. Prefix computation is a common operation in many algorithms including the evaluation of polynomials and modular additions [1], packing problems, loop parallelization [2].

The iterative sequential prefix computation requires n operations $\circ$. However, any parallel prefix circuit of depth d contains at least $2n - d$ operations $\circ$ (see section 3). The minimal parallel time is $\Omega(\log n)$ on a machine without concurrent write. Ladner and Fischer [3] proposed a parallel algorithm which takes a time of $2 \log n$ and $2n$ operations. Fich [4] proved that any algorithm of time $\log n$ requires $4n$ operations. Then Ladner-Fisher's algorithm is fine grain and asymptotically optimal on $\frac{n}{\log n}$ processors. It can be scheduled on $p < \frac{n}{\log n}$ identical processors in time $\frac{2n}{p} + O(\log n)$. Since it carries out $2n$ operations, it is not optimal for a fixed p. Nicolau and Wang [2] showed that a strict lower bound for the parallel time on p identical processors is $\left\lceil \frac{2n}{p+1} \right\rceil$ for $n \geq \frac{p(p+1)}{2}$. They provided an algorithm, based on a cutting in $(p+1)$ blocks and a pipeline between blocks, which reaches this lower bound. Most implementations either on dedicated distributed architectures or circuits [1] are based on an off-line block partitioning, with a block size depending on p.

---
* This work is supported by the French government ANR ARA-SSIA BGPR/SafeScale and a France-Mali grant.

The drawback of such optimal algorithms for a fixed number p of processors is that the number of operations is at least $2\frac{p}{p+1}$ times greater than the number n of operations performed by an optimal sequential algorithm. Thus, although optimal on p processors, those algorithms are not efficient on a machine with processors of different and possibly changing speeds. This is the practical case for a multi-processor machine concurrently used by several users, since the load of the processors varies during the execution. In this case, the scheduling must be on-line.

To resolve this problem, we use an on-line work-stealing (see section 2) implemented in Kaapi [5,6]. Bender and Rabin [7] extended this work-stealing to processors of changing speeds: they analyze the time of an algorithm with respect to Π_{ave}, the average speed per processor. On this model, we provide in section 3 a lower bound $\frac{2n}{p \cdot \Pi_{ave} + \Pi_{max}}$ for computation time of parallel prefix, where Π_{max} is the maximal speed of a processor during execution. We prove this bound is tight on uniform processors: for $n \gg p$, a block algorithm – with an off-line partitioning based on the relative speeds of the processors – reaches it.

In order to suit to processors with changing speeds, section 4 presents our new adaptive algorithm that performs an on-line block partitioning without assumption on the processor speeds. It is based on the recursive coupling of a sequential optimal algorithm and a fine grain parallel one which is scheduled by work-stealing. Its execution time, including on-line scheduling overhead, is $T_p \leq \frac{2n}{(p+1) \cdot \Pi_{ave}} + O\left(\frac{\log n}{\Pi_{ave}}\right)$ which is close to the lower bound, and asymptotically optimal when processors are identical.

Finally, in section 6, we present experimental comparisons on a eight-processor machine between this algorithm and an optimal one with off-line static partitioning. Two cases are considered: dedicated processors and processors disturbed by additional processes. Even for small values of n (100) and of p (1 to 8), our adaptive algorithm has performances analogous to the optimal one when the machine is dedicated to the computation, and it is faster when the machine concurrently executes other processes (multiuser case), which is the practical case that motivates this work.

The coupling of two algorithms, a sequential one and a parallel one, is inspired by Daoudi *et al* [8] where it is applied to algorithms with the same number of operations on identical processors. However, its use for processors with different and possibly changing speeds, as well as the technique used to analyze its complexity are original. This technique is very general, we think that it can be applied to other problems.

2 Notations and On-Line Scheduling by Work-Stealing

Let W_∞ be the critical-path in number of operations for an execution on an unbounded number of processors; W is the total number of arithmetic operations performed (work) by a given execution of the parallel algorithm. Note that W does not include scheduling operations but may depend both on the number of processors and the scheduling used for the considered execution. Let T_p be the execution time of the algorithm when scheduled on p physical processors, including scheduling overhead;

Cilk [9] and Kaapi [5,6] are parallel programming interfaces that support recursive parallelism and implement an on-line work-stealing scheduling based on the work first principle. The principle of work-stealing is simple. Each processor serially executes the tasks it has locally created according to a depth-first order. When a processor P_v becomes idle, it steals the oldest ready task (breadth first order) on a non-idle processor P_w, randomly chosen. For any series-parallel program with critical path W_∞ and work W on p identical processors, a work-stealing schedule ensures with high probability that $T_p \leq \frac{W}{p} + O(W_\infty)$ [9,7].

Bender and Rabin [7] extended this theorem to heterogeneous processors of different and possibly changing speeds. The authors proposed a model that encompasses the practical use of a parallel architecture concurrently shared by various processes and users. Let $\Pi_i(t)$ be the instantaneous speed of processor i at time t, measured as the number of operations ○ per unit of time. For a computation with duration T, let Π_{ave} be the average speed by processor: $\Pi_{ave} = \frac{\sum_{t=1}^{T} \sum_{i=0}^{p-1} \Pi_i(t)}{p \cdot T}$. In [7], a high utilisation schedule of factor β is used which is defined by the following property: if there are $i < p$ idle processors, then the fastest idle processor is at most β times faster than the slowest busy processor. The parameter β can be tuned to optimize the performances of the system by reducing the number of migrations. Indeed, it is not even necessary to define a particular value of β [7] and in the sequel, we will assume $\beta = O(1)$.

To implement such a high utilisation schedule, the previous work-stealing is only modified in [7] when a processor P_v steals a work on an active processor P_w that has no ready work to be stolen in its local queue. Then, if P_w is slower than P_v by at least a β factor, then the work in progress on P_w is preempted and migrated on P_v. In the case when the processors speeds do not change too much, the following theorem bounds the execution time T_p.

Theorem 1. *(see theorem 6 and 8 in [7]) With high probability, the number of successful steal operations is $O(p \cdot W_\infty)$ and the execution time T_p is bounded by*

$$T_p \leq \frac{W}{p \cdot \Pi_{ave}} + O\left(\frac{W_\infty}{\Pi_{ave}}\right).$$

The next section stands a lower bound for parallel prefix on this model.

3 Lower Bound for Parallel Prefix on Processors with Varying Speeds

In this paragraph, a lower bound is first given for p processors with varying speeds. Then, an off-line algorithm is provided that proves this lower bound is tight on p processors with constant and known speeds Π_i. The next theorem stands the lower bound with respect to Π_{ave} and also to the maximal speed Π_{max} of all processors: $\Pi_{max} = \max_{i=0,\ldots p-1; t=1,\ldots T} \Pi_i(t)$.

Theorem 2. *A lower bound on the time T_p of any parallel prefix computation on p processors with average speed Π_{ave} and maximal speed Π_{max} is*

$$T_p \geq \frac{2n}{p \cdot \Pi_{ave} + \Pi_{max}}.$$

Proof. Let G be the computation DAG representing the execution of the parallel algorithm. G has $n+1$ leaves corresponding to the inputs $(x_i)_{i=0,n}$; each internal node matches an operation $\circ$ with two inputs. Let A be the predecessor graph of a node that computes the output π_n. In π_n, each input x_i is operand of exactly one operation $\circ$; then A is a binary tree with $n+1$ leaves. Thus A contains exactly $\#A = n$ operations $\circ$. Besides, let d be the depth of G and B be the complementary DAG of A in G. Since any prefix π_i is a successor of the leaf x_o and the depth of A is at most d, at most d prefixes are computed in A; thus B computes at least $n-d$ prefixes and then contains at least $\#B = n-d$ operations $\circ$. As a consequence, the number $\#G = \#A + \#B$ of nodes $\circ$ in G is at least $2n - d$ (note that this first part of the proof is similar to the one established in [4], theorem 2, for restricted prefix circuits of depth $d = \log n$).

During T_p, at most $p \cdot \Pi_{ave}$ operations $\circ$ are computed; then, $p \cdot \Pi_{ave} \cdot T_p \geq 2n - d$. Besides, since G has a critical path with d operations $\circ$, $\Pi_{max} . T_p \geq d$. Putting things together gives $(p \cdot \Pi_{ave} + \Pi_{max}) . T_p \geq 2n$. □

To prove that this lower bound is tight, we now introduce a parallel algorithm that reaches it in the restricted case where processors have uniform known speeds.. For the sake of simplicity, the algorithm is first explicited in the case of p identical processors, and after extended to processors with uniform speeds. On p identical processors $(P_i)_{i=0,\ldots,p-1}$, the algorithm is based on a partitioning of the $n+1$ entries $(x_i)_{i=0,\ldots,n}$ in $p+1$ blocks $B_0, \ldots, B_p$ of approximately the same size. To simplify, we suppose that each block B_i contains $K = \frac{n}{p+1}$ consecutive elements.

Step 1. In parallel for $i = 0, \ldots, p-1$, we compute on processor i the sequential prefix of the block B_i. Let α_i denote the last prefix of the block B_i. We notice that the prefixes $(\pi_j)_{j=1,\ldots,K}$ of the block B_0 are thus computed.
Step 2. We compute the $p-1$ prefixes $\beta_0 = \alpha_0$, $\beta_1 = \alpha_0 \circ \alpha_1, \ldots, \beta_{p-1} = \alpha_1 \circ \ldots \circ \alpha_{p-1}$ of values $\alpha_0, \ldots, \alpha_{p-1}$.
Step 3. On processor 0, we compute the product by β_{p-1} of each element of the block B_p to obtain the prefixes $\pi_{pK}, \ldots, \pi_n$. And, in parallel for $i = 1, \ldots, p-1$, we compute on processor i the product by β_{i-1} of each element of the block B_i. We notice that these products are independant, even if they are made sequentially. All the prefixes π_i thus are obtained.

The execution time of this algorithm is $2K + p - 1 \simeq \frac{2n}{p+1}$, thus asymptotically optimal. Its number of operations $2n - (2k + p - 1)$ is strictly optimal because it reaches the lower bound $2n - d$. Moreover, we notice that by taking $K = 2$ and by executing step 2 in a recursive way, it is the algorithm of Ladner and Fisher [3] which takes $W = 2n$ operations $\circ$ with a critical path $W_\infty = 2\log_2 N$.

We now extend the previous algorithm to the case of p processors with uniform speeds $\Pi_{max} = \Pi_0 \leq \Pi_1 \leq \ldots \leq \Pi_{p-1}$ by tuning the block sizes in the partitioning. Let $n_0 = \frac{n}{1 + p \cdot \Pi_{ave} \cdot \Pi_{max}^{-1}}$. Blocks B_0 and B_p, each of size n_0, are assigned to processor 0. For $1 \leq i \leq p-1$, the processor i is assigned a block B_i of size $n_0 \frac{\pi_i}{\Pi_{max}}$. We also have $2n_0 + \sum_{i=1}^{p-1} n_i = n$. Step 1 and 3 takes a time $\frac{n_0}{\Pi_{max}} = \frac{n_i}{\Pi_i} = \frac{n}{\Pi_{max} + p \cdot \Pi_{ave}}$. So the whole time is $\frac{2n}{\Pi_{max} + p \cdot \Pi_{ave}} + p$, then asymptotically equal to the lower bound of theorem 2.

However, this algorithm assumes that relative speeds of the processors are known. It is not suited to the case of processors with varying speeds. In the next section, we present an on-line parallel algorithm that adapts automatically to the speeds of the processors by work-stealing.

4 Parallel Adaptive Algorithm

Our parallel algorithm with adaptive grain is based on the coupling of two algorithms: a sequential process P_s which sequentially computes prefixes and minimizes the number of operations and a variant of the preceding parallel algorithm, but with fine grain and scheduled by work-stealing on the $p-1$ other processes. Initially, the process P_s starts the prefix computation of 1 to n. Let $a = \frac{n}{p+1}$ and $b = \frac{p}{p+1} n$, the prefixes of 1 to a and b to n will be computed by this process P_s. However, the interval of indices $[a, b]$ can be stolen and cut out recursively by processes P_v that become inactive. The algorithms for P_s and processes P_v are as follows:

Sequential algorithm on process P_s

1. P_s sequentially computes the prefixes starting from index 1 (i.e. π_1), until an index u_1 such that the interval $[u_1, u_2]$ of indices was stolen by a process P_v.
2. P_s preempts P_v and recovers the last index $k \leq u_2$ computed by P_v, which thus already computed $r_{u_1} = x_{u_1}, r_{u+1} = r_{u_1} \circ x_{u_1+1}, \ldots, r_k = r_{k-1} \circ x_k$. P_s sends the value π_{u_1-1} to P_v and starts again P_v (see below).
3. P_s computes $\pi_k = \pi_{u_1-1} \circ r_k$. Then it takes again the sequential computation of the prefix of $k+1$ to n starting from $k+1$ while returning at step 1. We speak about jump operation. For each jump operation, P_s makes an operation $\circ$.
4. P_s stops when it computed π_n (the prefixes of indices of b to n cannot be stolen). After having computed π_n, it becomes a thief process and executes the algorithm P_v.

Parallel algorithm on $p-1$ processes P_v

- When it is preempted by P_s (see algorithm of P_s), P_v already computed partial prefix locally $r_{u_1}, \ldots, r_{u_k}$ of interval $[u_1, u_k]$. It then receives the value of the last prefix $\beta = \pi_{u_1-1}$ computed by P_s. It then finalizes the interval $[u_1, u_k]$ by computing the products $\pi_i = \beta \circ r_i$ for $u_1 \leq i \leq u_k$.

These products are parallel. On inactivity of another process thief, a half of these computations remaining to be made on P_v in this interval can then be stolen.

- When it is inactive, process P_v chooses a processor until finding an active process P_w. It can be either P_s or another thief process. If the victim is P_s, the steal is possible only if P_s has a remaining interval of indices ranging between a and b.
  1. P_v cuts the stealable interval on P_w in two parts. P_v extracts the right part $[u_1, u_2]$ of the interval and steals it. The left part remains on P_w.
  2. P_v starts computation on the stolen interval $[u_1, u_2]$. It can be either a computation of a local prefix (i.e. $r_{u_1} = x_{u_1}, r_{u_1+1} = r_{u_1} \circ x_{u_1+1}, \ldots$) or the finalization of computations of prefixes starting from already computed values r_k (i.e. $\pi_{u_1+1} = \pi_{u_1} \circ r_{u_1+1}, \pi_{u_1+2} = \pi_{u_1} \circ r_{u_1+2}, \ldots$).

The program stops when all the processors are inactive. The main point of this algorithm is that a process that become slow will be preempted by the sequential process or will be stolen by a parallel process. The following section analyzes the complexity of this adaptive algorithm.

5 Asymptotic Optimality of the Adaptive Algorithm

We use the modified work-stealing schedule (theorem 1) to execute the adaptive algorithm for the computation of parallel prefixes on p processors of changing speed. As in [7], we assume that the speed of the processor vary within a constant factor: there is a constant $c \geq 1$ such that $\max_{i,t} \Pi_i(t) \leq c . \min_{i,t} \Pi_i(t)$.

Theorem 3. *With high probability,*

$$T_p \leq \frac{2n}{(p+1)\Pi_{ave}} + O\left(\frac{\log n}{\Pi_{ave}}\right) \sim_{n \to \infty} \frac{2n}{(p+1)\Pi_{ave}}.$$

Proof. For the analysis, we cut out the execution in two successive phases, ϕ_1 and ϕ_2. The phase ϕ_1 is until P_s has computed π_n. Then, the phase ϕ_2 starts when P_s becomes a work-stealer. Let n_{seq} (resp. j) be the number of prefixes (resp. jumps) computed by P_s during ϕ_1. Let x (resp. y) be the number of final prefix computed by the other processes (work-stealers) in ϕ_1 (resp. ϕ_2). Then $n = n_{seq} + j + x + y$ and $W = n_{seq} + 2j + 2x + 2y$. Let I_1 (resp. I_2) be the total number of operations performed by idle processors during ϕ_1 (resp. ϕ_2).

1. During the phase ϕ_1 of time $T_p(\phi_1)$, the sequential prefix algorithm is always executing on a processor and makes $n_{seq} + j$ operations ∘. We note Π_{seq} the average speed of this algorithm: $\Pi_{seq} = \frac{n_{seq}+j}{T_p(\phi_1)}$.

   At each unit of time, the $p-1$ others processors make the parallel part of the adaptative algorithm and make in total $2x + y + j$ operations ∘ and I_1 inactivity operations. During ϕ_1, the average speed per processor for this part of the algorithm is $\Pi_{ave}(\phi_1) = \frac{2x+y+j+I_1}{(p-1).T_p(\phi_1)}$.

   The total number of operations ∘ in the phase ϕ_1 is $W(\phi_1) = n_{seq} + 2j + 2x + y$.

2. During the phase ϕ_2, the sequential part of the adaptative algorithm is finished. The p processors finalize the y prefix computations that were anticipated in parallel and not finished in the phase ϕ_1. During the phase ϕ_2, the p processors make thus y operations $\circ$ and I_2 inactivity operations. The average speed per processor during ϕ_2 is $\Pi_{ave}(\phi_2) = \frac{y+I_2}{p \cdot T_p(\phi_2)}$.

For the sake of simplicity, we assume $\Pi_{seq} = \Pi_{ave}(\phi_1) = \Pi_{ave}(\phi_2) = \Pi_{ave}$ (without loss of generality, since they are within a constant factor c). During ϕ_2, the processors which don't execute the sequential algorithm (i.e. $p-1$ at each unit of time) make $2x + y + J$ operations $\circ$ with a critical path $W_\infty(\phi_1) \leq 2.\log_2 n$ related to recursive cutting. By applying the theorem 1, we obtain $T_p(\phi_1) \leq \frac{2x+y+j}{(p-1)\Pi_{ave}} + O(\frac{\log n}{\Pi_{ave}})$. Thus, $T_p(\phi_1) = \frac{n_{seq}+j}{\Pi_{seq}} = \frac{n_{seq}+j}{\Pi_{ave}}$. And then, $(p+1)T_p(\phi_1) = (p-1)T_p(\phi_1) + 2T_p(\phi_1) \leq \frac{2n_{seq}+2x+y+3j}{\Pi_{ave}} + O\left((p-1)\frac{\log n}{\Pi_{ave}}\right)$. As $n = n_{seq} + x + y + j$, We obtain: $(p+1)\Pi_{ave}T_p(\phi_1) \leq 2n - y + j + O((p-1)\log n)$. In addition, by applying theorem 1 to ϕ_2, we obtain $T_p(\phi_2) \leq \frac{y}{p\Pi_{ave}} + O\left(\frac{\log n}{\Pi_{ave}}\right)$ Thus, $(p+1)\Pi_{ave}T_p = (p+1)\Pi_{ave}T_p(\phi_1) + (p+1)\Pi_{ave}T_p(\phi_2) \leq 2n + j + \frac{y}{p} + O\left(p \log n\right)$.

The number j of jumps is lower than the number of successful stealing i.e. $O(\log n)$ since $W_\infty(\phi_1) \leq 2\log n$ (theorem 1). Moreover, by using $(p-1)(n_{seq} + j) = (2x+y+j+I_1)$ and $n_{seq} \geq \frac{2n}{p+1}$, we obtain $y \leq I_1 \leq (p-2)\log n$. Finally, we have: $(p+1)\Pi_{ave}T_p \leq 2n + O(p\log n)$. □

We can note that the proof and the theorem remain valid in the more general and realistic case where $\Pi_{seq} \geq \Pi_{ave}(\phi_1)$ (the sequential algorithm is always executed by a processor faster than the average of the processors) and $\Pi_{ave}(\phi_2) \geq \Pi_{ave}(\phi_1)$ (the sequential processor added in phase 2 is faster than the average of the other processors).

6 Experimental Results

We implemented the algorithms on a eight-processor SMP machine, with 31 GB of memory (Intel's Itanium-2 at 1.5 GHz) and in the multi-user context under the GNU/Linux 2.6.7 system. The adaptive and parallel algorithms are implemented with Kaapi [5,6].

The experiments consist in the computation of prefixes of 10000 elements (double) with a time of 1ms per operation $\circ$ while varying p the number of processors from 1 to 8. The optimal sequential time of reference is 10s.

Tables 1 and 2 give the execution times obtained by the two parallel algorithms (with fixed grain on p processors and adaptive grain). For each experiment, we made 10 measurements and we kept the times of the fastest and the slowest execution and the average time of the 10 executions.

Table 1 compares the execution times when there are no other computations in progress on the processors. We notice that measurements of time are stable (variation between minimum and maximum lower than 6% for the algorithm with fixed grain and lower than 8% for the algorithm than adaptive grain). We

Table 1. Comparison of the times of the three algorithms on p identical processors

	Sequential	Static				Adaptive			
		p=2	p=4	p=6	p=8	p=2	p=4	p=6	p=8
Lower bound $\frac{2n}{(p+1)\Pi_{ave}}$	10.00	6.67	4.00	2.86	2.22	6.67	4.00	2.86	2.22
Min	10.087	6.73	4.04	2.89	2.82	6.73	4.03	2.88	2.23
Avg	10.09	6.74	4.05	2.93	2.87	6.73	4.04	2.89	2.24
Max	10.09	6.75	4.06	3.00	2.99	6.73	4.04	2.89	2.25

Table 2. Comparison of the times of the algorithms on p perturbated processors. Each column reports, the minimal, average and maximal times of 10 executions. For each of those 10 executions, the adaptive algorithm is the fastest.

	Static				Adaptive			
	p=2	p=4	p=6	p=8	p=2	p=4	p=6	p=8
Lower bound $\frac{2n}{(p+1)\Pi_{ave}}$	7.49	4.50	3.22	2.50	7.49	4.50	3.22	2.50
Min	8.34	7.33	4.97	3.67	7.55	6.03	3.77	2.94
Avg	9.97	8.15	5.60	4.05	9.21	7.23	4.47	3.34
Max	10.41	8.57	5.77	4.31	10.28	8.12	5.23	3.86

check the optimality of the algorithm with fixed grain whose time is with less than 8% of the lower bound. Moreover, we check the optimality of the adaptive algorithm which is also less than 5% of the lower bound.

In table 2, additional processes of load are injected to disturb the load of the machine and to simulate the behavior of a real machine, disturbed by other users. In the aim of reproducibility, each experiment on $p \leq 8$ processors is disturbed by $9 - p$ artificial processes of duration larger than 10s. We can check in table 2 that the adaptive algorithm is at least 7% faster.

We note that the time are very changing but we observe that in the case of the minimum time, the adaptive algorithm is not so far from the lower theoretical bound ($\Pi_{ave} = \frac{8}{9}$). We think this is due to the scheduling of the system.

In conclusion, the adaptive algorithm brings a guaranteed performance when the machine is divided between several users, while adapting automatically to the available resources during the execution. Moreover its performance remains close to optimal even in the ideal case where the processors are all dedicated to the application. It thus appears to be more powerful than the sequential algorithm or than a fixed parallel algorithm.

This is confirmed by another experimentation where each elementary test corresponds to simultaneous launching in competition of the nine programs: adaptive algorithm on eight processors, sequential algorithm and the fixed parallel algorithm for the seven values $p = 2, \ldots, 8$ processors. Table 3 summarizes the results on a 10 test campaign. For 10 executions, the adaptive algorithm is always the fastest.

Table 3. Comparison of the times of the 9 algorithms simultaneously launched – On the 10 executions of each test, the adaptive algorithm was the fastest

	Sequential	Static p=2	p=4	p=6	p=7	p=8	Adaptive p=8
Min	20.53	18.70	16.41	13.93	13.54	12.25	10.79
Max	22.96	20.04	17.23	15.86	14.56	13.66	13.06
Avg	21.69	19.24	16.89	15.13	13,89	13.16	12.09
Median	22.00	19.26	16.96	15.12	13.76	13.12	12.18

Its average time of execution is on average 19% times shorter than that of the optimal fixed parallel algorithm on 8 processors, with variations to 40% on one of the tests.

7 Conclusion

Motivated by the use of multi-processor machines shared between several users, we introduced a new parallel algorithm for the prefix computation which adapts automatically and dynamically to the available processors. This algorithm performs an asymptotically optimal number of operations. It is equivalent to that of the sequential algorithm when only one processor is available and to that of an optimal parallel algorithm when p identical processors are available. In the case of p variable processors speeds, its time is equivalent to that of an optimal algorithm on p identical processors speed equal to the average speeds. These theoretical results are validated by the experiments made on a SMP machine with 8 processors. A first perpsective is to validate it on the national French heterogeneous grid GRID'5000 within the ANR BGPR-Safescale project.

More generally, our adaptive algorithm is based on the recursive and dynamic coupling of two algorithms, a sequential one, optimal in terms of number of operations, and a parallel one with a maximum degree of parallelism. Both the algorithm and its analysis are applied to the prefix computation, for which any parallel algorithm requires more operations than the sequential algorithm. However, we think that both this scheme and its theoretical analysis are more general and may apply to other problems, in particular for the resolution of exact linear systems.

References

1. Dimitrakopoulos, G., Nikolos, D.: High-speed parallel-prefix vlsi ling adders. IEEE Trans. Computers **54**(2) (2005) 225–231
2. Wang, H., Nicolau, A., Siu, K.Y.S.: The strict time lower bound and optimal schedules for parallel prefix with resource constraints. IEEE Trans. Comput. **45**(11) (1996) 1257–1271
3. Ladner, R., Fischer, M.: Parallel prefix computation. J. ACM **27**(4) (1980) 831–838

4. Fich, F.E.: New bounds for parallel prefix circuits. In: STOC '83: Proceedings of the 15th ACM symp. Theory of computing, New York, NY, USA, ACM Press (1983) 100–109
5. Jafar, S., Gautier, T., Krings, A.W., Roch, J.L.: A checkpoint/recovery model for heterogeneous dataflow computations using work-stealing. In Springer-Verlag, L., ed.: EUROPAR'2005, Lisboa, Portogal (2005)
6. MOAIS Project: KAAPI homepage. `http://gforge.inria.fr/projects/kaapi/` (since 2005)
7. Bender, M.A., Rabin, M.O.: Online scheduling of parallel programs on heterogeneous systems with applications to cilk. Theory Comput. Syst. **35**(3) (2002) 289–304
8. Daoudi, E.M., Gautier, T., Kerfali, A., Revire, R., Roch, J.L.: Algorithmes parallèles à grain adaptatif et applications. TSI **24** (2005) 1–20
9. Frigo, M., Leiserson, C.E., Randall, K.H.: The Implementation of the Cilk-5 Multithreaded Language. In: Proceedings of the ACM SIGPLAN Conference on Programming Language Design and Implementation (PLDI'98). (1998)

Topic 13: Routing and Communication in Interconnection Networks

Jose A. Gregorio, Bettina Schnor, Angelos Bilas, and Olav Lysne

Topic Chairs

Welcome to the Euro-Par 2006 Topic 13 on Routing and Communication in Interconnection Networks. Interconnection networks is a key area in the quest for performance in parallel and distributed computers, and this topic is dedicated to techniques that improve the state of the art in interconnecting parallel computers, workstations or clusters.

All aspects of communication in modern systems were considered by the programme committee of this topic. These included advances in the design, implementation, and evaluation of interconnection networks, network interfaces, system and storage area networks, on-chip interconnects, communication protocols, routing and communication algorithms, among others.

This year 22 papers were submitted to this topic. All papers were reviewed, at least, by three referees. Using the referees? reports as basis of discussion, the programme committee picked four of the submitted papers for publication and presentation at the conference.

We would like to thank all the authors for their submissions to this Euro-Par 2006 topic. We owe special thanks to the more than one hundred external referees who provided competent and timely review reports. Their effort ensured the high quality of this track and we expect you will find this topic to be highly stimulating and quite informative.

A Model for the Development of AS Fabric Management Protocols*

Antonio Robles-Gómez, Eva M. García, Aurelio Bermúdez,
Rafael Casado, and Francisco J. Quiles

Instituto de Investigación en Informática de Albacete (I^3A)
Universidad de Castilla-La Mancha
02071 Albacete, Spain
arobles@dsi.uclm.es

Abstract. Advanced Switching (AS) is a switching fabric architecture based on the PCI Express technology. In order to support high availability, AS includes important features, such as device hot addition and removal, redundant pathways, and fabric management failover. This work presents an AS model developed in OPNET. The contribution of this tool is that it can help researchers to design and evaluate management mechanisms for this new technology. It can also be used to analyze other key aspects of the architecture, such as routing, congestion, and quality of service.

Keywords: Advanced switching, modeling, network management, network availability.

1 Introduction

The Advanced Switching specification [1] has been developed by the Advanced Switching Interconnect Special Interest Group (ASI-SIG). It is a chip-to-chip and backplane interconnect switched fabric architecture. Unlike similar technologies, such as InfiniBand [5] and Quadrics [9], AS can be seen as the last step in the evolution of the traditional PCI bus [6]. In particular, AS inherits most of the physical and link characteristics of PCI Express [8]. However, it offers a bigger application space, including multiprocessing and peer-to-peer communications. The first commercial AS-compliant products have just started to appear in the marketplace [11].

To guarantee network availability, AS provides a fabric management mechanism, which basically configures and monitors the status of the network. Consider, for example, the occurrence of a failure in a network device. The management mechanism must detect that failure, discover the resulting topology, and finally obtain and distribute to the endpoints a new set of routes for packet delivery. All these tasks are performed by the fabric manager (FM), a software entity running on one or more AS endpoints.

---

* This work is supported by the following projects: TIC2003-08154-C06-02 (Ministerio de Ciencia y Tecnología), PBC05-007-1 (Junta de Comunidades de Castilla-La Mancha), and PCTC0622 (Universidad de Castilla-La Mancha).

The internal behavior of the management mechanism is currently an open issue for vendors and researchers. The AS specification only considers a set of configuration data structures –called *capabilities*– into each device, and the management packets – called *PI-4 packets*– used to exchange them among devices.

This paper presents a simulation model that provides the necessary support – capabilities and PI-4 packets– to develop management mechanisms. In order to be able to evaluate future proposals, our simulator allows measuring accurately control overhead and the time expended by the management process.

Our AS model is an evolution of a previous model [2] developed for the InfiniBand technology [5]. There are many differences between both technologies, such as source routing instead of distributing routing, and passive instead of active switches. These differences completely justify the development of a new tool to design specific management mechanisms for AS.

The AS model has been developed using the OPNET Modeler software [7]. This tool provides support to model and analyze communication networks and distributed systems. In OPNET, network devices are modeled through node models, which are built using basic modules. Fig. 1 shows an example. Each module can generate, send, receive, and consume packets from other modules. The behavior of a module is programmed via its process model. It consists of a finite state machine (see Fig. 4) containing blocks of C/C++ code and calls to the OPNET API.

The remainder of this paper is organized as follows. The next section describes the way we have modeled the AS network devices. Then, Section 3 introduces the modeling of the fabric management support. After that, we revise some tasks in the management mechanism that we plan to develop in the future. Finally, Section 5 gives some conclusions and future work.

2 Modeling the AS Architecture

Our model[1] is made up of AS x1 links, 16-port switches, and fabric endpoints. This section presents the way in which these network devices are modeled and it details some architectural issues closely related to the fabric management process, as flow control and the port state machine.

2.1 Network Components

We have defined AS links starting from the basic OPNET point-to-point bidirectional link model. The specified bandwidth for these links is 2.5 Gbps. However, bandwidth is reduced by 8b/10b encoding to 2.0 Gbps. So far, we have not considered transmission errors. To implement cut-through switching, we have programmed the link model in such a way that the receiver port can process a packet once the header has been received.

We have also modeled a multiplexed virtual cut-through switch [4]. Fig. 1 shows the modules implementing two switch ports –numbered as 7 and 8–, the switch arbitration unit, and the crossbar.

---

[1] The source code of our AS model will be freely available for the OPNET community, at the "Contributed Models" depot of the OPNET support center [7].

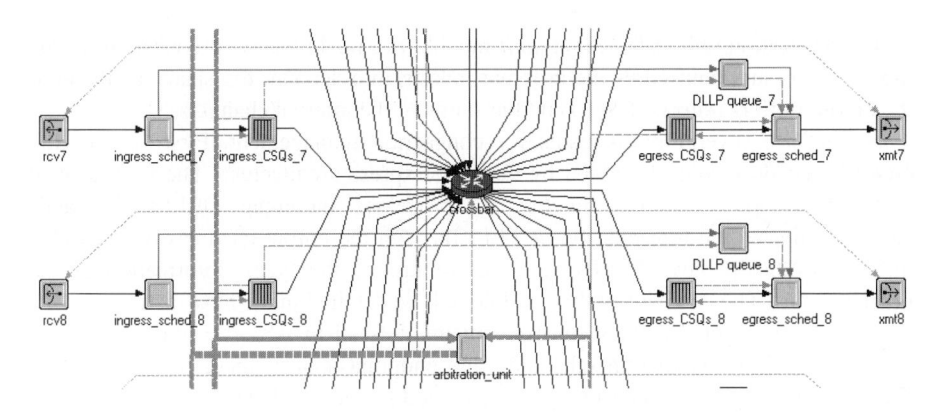

Fig. 1. A detail of the switch model

Each input channel contains a point-to-point receiver (*rcv* module in Fig. 1) connected to the link. A selector (*ingress_sched* module) delivers flow control packets (DLLP, data link layer packet) to the flow control unit. The rest of packets (TLP, transaction layer packet) are sent to the *ingress_CSQs* module.

AS defines three types of virtual channels: unicast bypassable VCs (BVC), unicast ordered VCs (OVC), and multicast VCs (MVC). Each BVC implements an ordered queue and a bypass one. Packets marked as "bypassable" (the *OO* field in Fig. 2 is unset) are delivered to the bypass queue if they cannot progress due to lack of credit. Packets at this queue can be "bypassed" by other packets at the ordered queue. On the other hand, OVCs and MVCs only support ordered queues. In our model, the number of virtual channels of each type and the size of the associated input and output buffers are defined as switch attributes.

A traffic class (TC) mechanism allows the grouping of traffic flows for similar treatment. The traffic class of a packet is defined at the source endpoint. When a packet reaches a port, the *Traffic Class* field at the header is used to obtain the corresponding VC, by using a set of fixed TC/VC mapping tables. The *ingress_CSQs* module in Fig. 1 performs this mapping, and stores the packet at the tail of the input buffer associated with the corresponding virtual channel.

In order to simplify the hardware, AS states that unicast packets use source routing. Endpoints include path information into the packets, by filling up the *Turn Pool*, *Turn Pointer*, and *D* (direction) fields in the packet routing header (shown in Fig. 2). These values are used at each intermediate switch to obtain the output port. In our model, unicast packets are routed when they reach the header of the input buffers. On the other hand, multicast packets require to look up into a specific forwarding table. These tables are stored at the switches and are defined by the management process.

31 30 29 28 27 26 25 24 23 22 21 20	19	18 17 16 15 14	13 12	11 10 9	8	7	6 5 4 3 2 1 0
Header CRC	F E C N	Credits Required	T O S O	Traffic Class	P C R C	P	PI
D	Turn Pool							

Fig. 2. AS routing header

The arbitration unit (*arbitration_unit* module in Fig. 1) receives requests from the input buffers and configures the crossbar, taking into account the space available at the output buffers (*egress_CSQs*) and the status of the internal channels.

AS defines several mechanisms for congestion management. First, it uses the credit-based flow control defined by the PCI Express architecture. The flow control unit (*DLLP queue* module in Fig. 1) processes incoming DLLPs and activates/deactivates the transmission of TLPs through the output channels. It must also inject periodically new DLLPs, in order to update the credit information at the neighbor port. The behavior of this module will be detailed in the next section.

Additional optional congestion mechanisms defined in AS are status-based flow control, minimum bandwidth scheduler, and endpoint source injection rate limiting. These mechanisms are not currently implemented in our simulator.

To conclude the description of the switch model, the output channel arbitration unit (*egress_sched* module in Fig. 1) receives requests from the port flow control unit and output buffers, and decides the packet that will be finally delivered to the physical link, through the transmitter module (*xmt*). Before sending a TLP, this module must consider the credit available at the corresponding neighbor input buffer, which is periodically notified by the flow control unit.

The endpoint model (not shown here) incorporates a communication port, including exactly the same modules as a switch port. There is also an *application* module which generates and consumes upper-level packets. Parameters for traffic generation, such as packet size and injection rate, are defined as simulation attributes.

2.2 Port Behavior and Flow Control Unit Model

Fig. 3 shows the set of possible states for a port, as defined in the AS specification. This behavior has been considered in our model. Once the device is powered-on, a port initialization phase starts. Each port tries to synchronize with a potential neighbor device. To do that, the port transits from *DL_Inactive* to *DL_Init*, and sends DLLPs through the link. If the port does not receive a response, it returns to the *DL_Inactive* state. After some time, it will try to synchronize again.

If the port receives a response from the neighbor, they must negotiate the number of virtual channels they are going to use in the communication. When the negotiation process finishes, the port transits to the *DL_Protected* state. In this state, the transmission of certain management packets (PI-0:0, for FM election, and PI-4, for device discovery and configuration) is allowed.

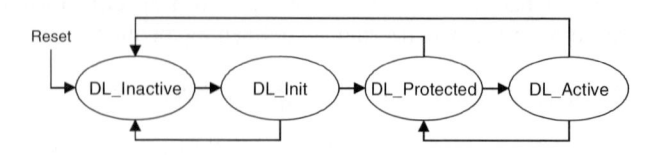

Fig. 3. Port state machine

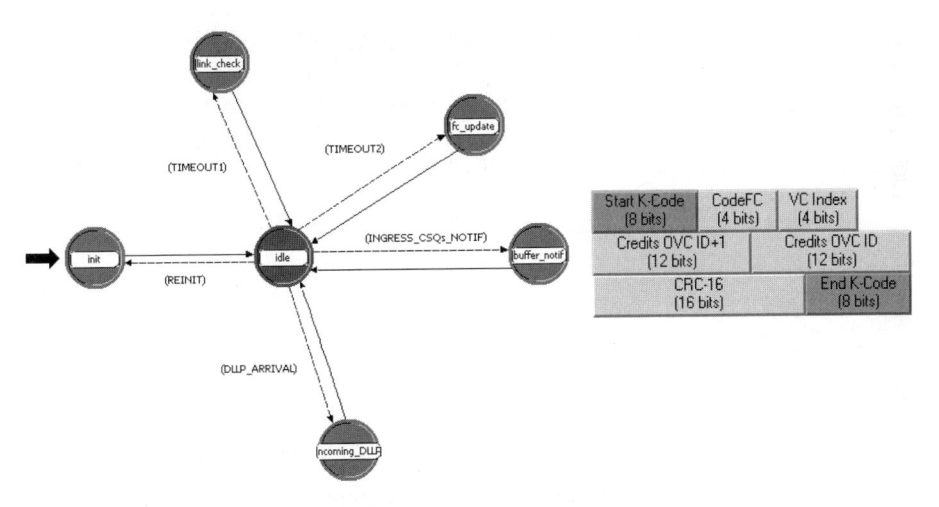

Fig. 4. (*left*) Flow control unit behavior and (*right*) a DLLP with credit information for two successive OVCs

The FM can order the port to transit to the *DL_Active* state by means of a PI-4 packet. In this state, the port is completely operational, allowing the transmission of all packet types. In the same way, the FM can order a transition from *DL_Active* to *DL_Protected*. Finally, the port will return to *DL_Inactive* if the link or the neighbor device is taken down, and DLLPs are not received during a period of time.

The flow control unit (*DLLP queue* module in Fig. 1) models the port behavior we have just described. Moreover, it implements the flow control tasks enumerated in the previous section. Fig. 4(*left*) shows the finite state machine in the corresponding OPNET process model.

The *init* state performs some initialization tasks. In the *idle* state, the process model is waiting for the occurrence of some simulation event. Periodically, the machine enters the *link_check* state and begins the port initialization phase.

In order to inform the neighbor port about the credit available at the local input buffers, the process enters the *fc_update* state periodically. In this state, the corresponding DLLPs are generated and injected. Fig. 4(*right*) shows the format of an *FC_Update* DLLP defined in OPNET.

The *buffer_notif* state is reached when the flow control unit receives a notification from the *ingress_CSQs* module, reporting about a variation in the occupation of an input buffer.

When a DLLP arrives to the flow control unit, it is processed at the *incoming_DLLP* state. According to its type, the flow control unit either continues the initialization process, or communicates the available neighbor credit to the *egress_sched* module.

Finally, the process model returns to the *init* state if for a certain time interval the flow control unit has not received DLLPs with information about the neighbor credit.

(a) Scenario used

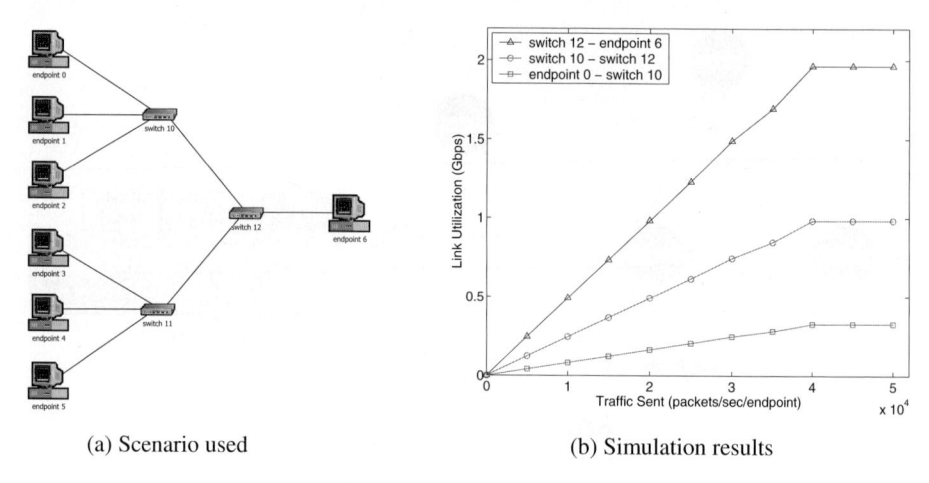

(b) Simulation results

Fig. 5. Example of validation

2.3 Flow Control Validation

Several tests have been conducted in order to validate the AS model. As an example of this process, the topology in Fig. 5(a) has been used to check the correct implementation of the credit-based flow control mechanism. In this scenario, the six endpoints on the left side inject packets to the endpoint located on the right, assuming the existence of only one OVC in each device.

We have run several simulations varying the packet injection rate at the source endpoints. Fig. 5(b) shows the link utilization at each level in the topology. We can see that the maximum link bandwidth (2 Gbps) is never exceeded on the link connecting switch *12* and endpoint *6*.

Additionally, results for the other two series show that the flow control is correctly working. Note that the utilization of the link connecting switch *10* and switch *12* is exactly half of the utilization at the next hop, and the utilization of the link connecting endpoint *0* and switch *10* is the third part of the previous one.

3 Fabric Management Model

We are interested in the development of fabric management mechanisms for the AS technology. This section describes the aspects of the specification that provide support for this purpose, and how they have been modeled into the simulator. These features are the configuration space in each device, and the protocol that allows the FM to access it.

3.1 Device Configuration Space

The device configuration space is a storage area that contains a set of fields to specify device characteristics as well as fields used to control the device. This information is presented in the form of structures called capabilities. Each capability structure defines a specific characteristic of the device.

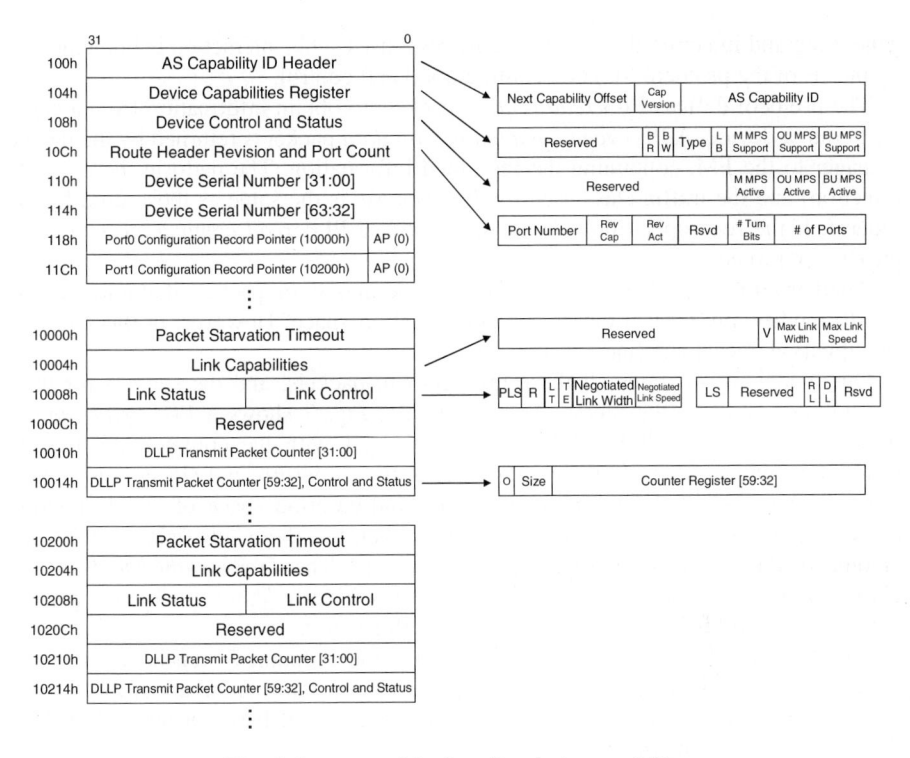

	31		0
100h	AS Capability ID Header		
104h	Device Capabilities Register		
108h	Device Control and Status		
10Ch	Route Header Revision and Port Count		
110h	Device Serial Number [31:00]		
114h	Device Serial Number [63:32]		
118h	Port0 Configuration Record Pointer (10000h)		AP (0)
11Ch	Port1 Configuration Record Pointer (10200h)		AP (0)

10000h	Packet Starvation Timeout	
10004h	Link Capabilities	
10008h	Link Status	Link Control
1000Ch	Reserved	
10010h	DLLP Transmit Packet Counter [31:00]	
10014h	DLLP Transmit Packet Counter [59:32], Control and Status	

10200h	Packet Starvation Timeout	
10204h	Link Capabilities	
10208h	Link Status	Link Control
1020Ch	Reserved	
10210h	DLLP Transmit Packet Counter [31:00]	
10214h	DLLP Transmit Packet Counter [59:32], Control and Status	

Fig. 6. Structure of the *baseline device* capability

The configuration space is made up of up to 16 blocks of 4 Gbytes of storage, called apertures. All the capability structures reside at the aperture 0. Additional data associated with the capabilities may be stored in any aperture.

Our model allows defining any AS capability. Currently, it includes the *baseline* and the *spanning tree* capabilities. The reason is that these capabilities are needed by several management processes, such as the topology discovery and the FM election.

In particular, the *baseline* capability includes device control and status information. Fig. 6 shows part of the contents of this capability. Each register is marked with the corresponding offset inside the aperture 0. The offsets could be different to the ones shown in this figure. The first six 32-bit blocks in the *baseline* capability contain general information for the device, such as its type –endpoint or switch– and serial number, the number of ports supported, and the maximum packet size. Next (from offset 118h in Fig. 6), we can find up to 256 32-bit blocks that point to the information about each particular port in the device. This information includes link speed and width, and current port state. In the Fig. 6 we only show the information corresponding to ports 0 and 1.

3.2 PI-4 – Node Configuration and Control Protocol

A *device_manager* module in the endpoint and switch models is defined. Its function consists of receiving requests from the FM, accessing to the capabilities in the device configuration space –by means of read and write operations–, and, if necessary, generating and injecting the corresponding responses. This interaction is implemented by means of the protocol for node configuration and control.

The protocol defines PI-4 *read request* packets to obtain information from a capability into a device. A PI-4 *read completion with data* packet is returned by the device manager to the FM, containing the requested information. The path (in the opposite direction) and the traffic class used by the response is the same as those used by the request. If the read operation was not successful, a PI-4 *read completion with error* packet is returned.

Apart from read packets, the PI-4 protocol defines *write* packets that allow to the FM to modify any data in the device configuration space. However, in this case the specification does not define a response packet.

Our model incorporates all these management packets, and the support for their transmission through the fabric. As an example, Fig. 7 shows a PI-4 *read request* packet defined in OPNET. PI-4 *read* packets start with the common AS routing header (shown in Fig. 2). In the request packet, the *Aperture* and *Offset* fields determine the position of the information in the configuration space of the destination device that the FM is requesting for. In this packet, the *Req Code* field specifies the amount of data to read (up to eight 32-bit blocks). The *Transaction Number* field allows the FM to match completions with requests. Finally, in the completion packet, the *Data Payload* field contains the requested information.

Fig. 8 shows an example. The FM –located at the endpoint *7*– repeatedly accesses the baseline capability in switch *12*, in order to obtain information about the activity of its ports. Fig. 9 shows some of the packets exchanged between the FM and the device manager during this process.

The FM sends a first PI-4 *read request* packet to the device manager in switch *12* to get general information about this device (located at *Offset*=100h in Fig. 6). The corresponding response indicates that the destination device is a switch implementing a total of 16 physical ports. Then, the FM injects two request packets (*Offset*=118h and *Offset*=138h respectively) to obtain the pointers to every port information. Each response packet will contain a block of 8 pointers.

After receiving the pointers, the FM generates sixteen new requests to access to the information about the corresponding ports. The *Link State* field in each response packet reports about the activity of the corresponding port. In this case, the state of

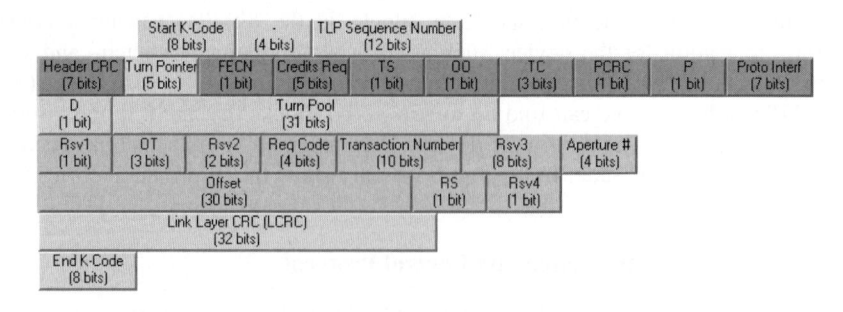

Fig. 7. PI-4 *read request* packet

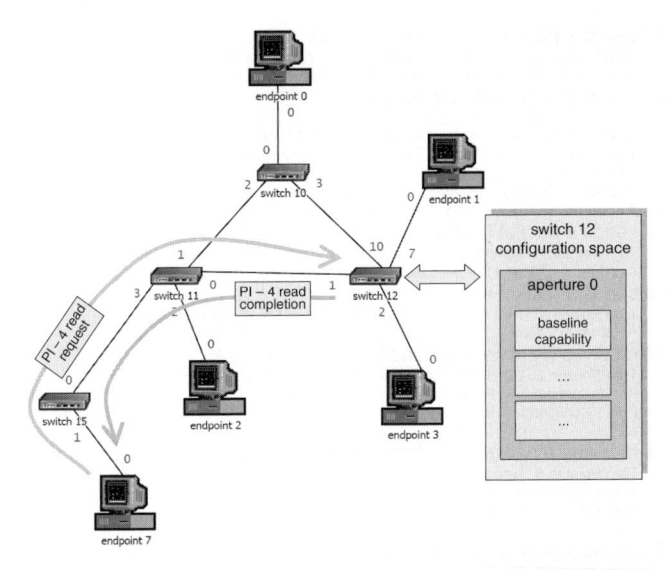

Fig. 8. Example of irregular fabric topology composed of 4 switches and 5 endpoints. Small numbers at link ends represent port numbers.

port *0* is *DL_Inactive* (we can see in Fig. 8 that it is unconnected) and the state of port *1* is *DL_Protected* (it is connected to switch *11*).

To sum it up, the FM has generated 19 PI-4 *read request* packets, and it has received 19 PI-4 *read completion with data* packets.

4 Fabric Management Tasks

The model described provides support to develop and evaluate management mechanisms for the AS technology. Network management involves a wide set of different tasks. Our work will be focused on those tasks related to network topology monitoring, computation of paths among devices, and their distribution to the source endpoints. In this section, we briefly describe the entire management process, and how it is modeled into our simulator.

As we have seen in Section 2.2, when a fabric device is powered-on, it enters to an initialization phase, exchanging credit information with potential neighbors and negotiating the available amount of virtual channels. When this negotiation concludes, it can transmit and receive management packets through its active links (*DL_Protected* state in Fig. 3).

If the device runs a FM driver, it triggers a FM election process. This process elects the primary and secondary fabric managers, the only endpoints that can configure the fabric. If the primary FM fails, the secondary one takes over. The FM election process is completely defined in the AS specification.

The first task of the (primary) FM is to discover the fabric topology. The discovery process is performed by using the PI-4 *read* packets described in the previous section. The particular implementation of this task is not detailed in the specification, and can be performed in either a centralized or distributed way [10].

```
FM sends a packet to discover a device
    Turn Pointer: 8  Turn Pool: ECh  Direction: 0  Transaction Number: 0
    Aperture: 0  Offset: 100h  Request Scale: 1  Request Code: 6
FM receives a packet from switch 12
    Turn Pointer: 8  Turn Pool: ECh  Direction: 1  Transaction Number: 0
    Next Capability Offset: 900h  CapVersion: 0  ID: F000h
    Type: SWITCH  Block Write: 1  Block Read: 1  Loopback: 1
    MVC MPS Support: 0  OVC MPS Support: 8  BVC MPS Support: 8
    BVC MPS Active: 8  OVC MPS Active: 8  MVC MPS Active: 0
    # of Ports: 16  # of Turn bits: 4  Rev Act: 0  Rev Cap: 0  Port Number: 1
FM sends a packet to obtain the pointer block 1
    Turn Pointer: 8  Turn Pool: ECh  Direction: 0  Transaction Number: 1
    Aperture: 0  Offset: 118h  Request Scale: 1  Request Code: 8
FM sends a packet to obtain the pointer block 2
    Turn Pointer: 8  Turn Pool: ECh  Direction: 0  Transaction Number: 2
    Aperture: 0  Offset: 138h  Request Scale: 1  Request Code: 8
FM receives a packet including the pointer block 1
    Turn Pointer: 8  Turn Pool: ECh  Direction: 1  Transaction Number: 1
    Port 0 Configuration Record Pointer: 10000h  AP: 0
    Port 1 Configuration Record Pointer: 10200h  AP: 0
    ...
FM receives a packet including the pointer block 2
    Turn Pointer: 8  Turn Pool: ECh  Direction: 1  Transaction Number: 2
    Port 8 Configuration Record Pointer: 11000h  AP: 0
    Port 9 Configuration Record Pointer: 11200h  AP: 0
    ...
FM sends a packet to obtain port 0 information
    Turn Pointer: 8  TurnPool: ECh  Direction: 0  Transaction Number: 3
    Aperture: 0  Offset: 10000h  Request Scale: 1  Request Code: 6
FM sends a packet to obtain port 1 information
    Turn Pointer: 8  Turn Pool: ECh  Direction: 0  Transaction Number: 4
    Aperture: 0  Offset: 10200h  Request Scale: 1  Request Code: 6
...
FM receives a packet including port 0 information
    Turn Pointer: 8  TurnPool: ECh  Direction: 1  Transaction Number: 3
    Timeout: -1  VLink: 0  MaxLink Width: 1  MaxLink Speed: 1
    Peer Link State: DL_Inactive  Training Prog: 0  TError: 0  Link Width: 1
    Link Speed: 1  Link State: DL_Inactive  Retrain Link: 0  Disable Link: 0
FM receives a packet including port 1 information
    Turn Pointer: 8  TurnPool: ECh  Direction: 1  Transaction Number: 4
    Timeout: -1  VLink: 0  MaxLink Width: 1  MaxLink Speed: 1
    Peer Link State: DL_Protected  Training Prog: 0  TError: 0  Link Width: 1
    Link Speed: 1  Link State: DL_Protected  Retrain Link: 0  Disable Link: 0
...
```

Fig. 9. Sequence of PI-4 packets to obtain topological information about switch *12* in Fig. 8

After discovery, the FM configures fabric devices. This task includes, for example, the distribution of paths to endpoints. Moreover, fabric ports must be activated in order to allow the reception and transmission of all packet types (*DL_Active* state in Fig. 3). In this case, PI-4 *write* packets are used.

Once the network has been configured and activated, the FM remains monitoring its state. The specification provides an event-reporting mechanism to notify topology changes. In particular, the device manager in a detecting device can report the FM about a change in the state of a local port, through a PI-5 packet. After detecting a change, the FM must update again the set of fabric routes.

In our model, we have defined a *FM* module at the endpoint model, which models the behavior of a centralized fabric manager. At this moment, the election process has not been modeled. We indicate the endpoint that hosts the primary FM by activating a particular node attribute. In the remaining endpoints, the *FM* module is inactive.

The *FM* module handles several data structures to store the fabric topology, the set of paths between endpoints, and other configuration information. At this moment, it can discover the configuration information about a particular device, and detect a topological change (i.e. the addition or removal of any fabric component) by means of the event-reporting mechanism. Currently, we are developing a discovery algorithm which can obtain the entire fabric topology. Later, we will focus on the path computation and dynamic distribution tasks, without stopping upper-level traffic.

5 Conclusions and Future Work

The model presented in this paper embodies key physical and link layer features of Advanced Switching. Unlike classical simulation tools, our model incorporates the fabric management entities defined in the specification and the packets that allow the fabric manager to access to the configuration information in fabric devices. It also includes the behavior of a port upon a change in its neighbor. At this moment, a basic fabric management mechanism is being developed. As future work, we plan to improve each management task, in order to optimize the performance of the entire process. In particular, we plan to reuse previous proposals [3], and to design specific protocols for this architecture.

References

1. Advanced Switching Interconnect Special Interest Group, Advanced Switching Core Architecture Specification Revision 1.0. December 2003, http://www.asi-sig.org
2. Bermúdez, A., Casado, R., Quiles F. J., Pinkston T. M., Duato, J.: Modeling InfiniBand with OPNET. In Proceedings of the 2nd Annual Workshop on Novel Uses of System Area Networks, February 2003
3. Bermúdez, A., Casado, R., Quiles F. J, Duato, J.: Handling topology changes in InfiniBand. IEEE Transactions on Parallel and Distributed Systems (accepted for publication)
4. Duato, J., Yalamanchili, S., Ni, L.: Interconnection Networks: An Engineering Approach. Morgan Kaufmann Publishers, 2003
5. InfiniBand Architecture Specification (1.2). November 2002, http://www.infinibandta.com/
6. Mayhew, D., Krishnan, V.: PCI Express and Advanced Switching: evolutionary path to building next generation interconnects. In Proceedings of the 11th Symposium on High Performance Interconnects (HOTI'03), 2003
7. OPNET Technologies, Inc., http://www.opnet.com/
8. PCI-SIG, PCI Express Base Specification Revision 1.0a. April 2003, http://www.pci-sig.org
9. Petrini, F., Frachtenberg, E., Hoisie, A., Coll, S.: Performance evaluation of the Quadrics interconnection network. Journal of Cluster Computing, 6(2): 125-142, April 2003
10. Rooholamini, M.: Advanced Switching: a new take on PCI Express. October 2004, http://www.asi-sig.org/press/Articles/
11. Stargen, http://www.stargen.com/

On the Influence of the Selection Function on the Performance of Fat-Trees*

F. Gilabert, M.E. Gómez, P. López, and J. Duato

Dept. of Computer Engineering
Universidad Politécnica de Valencia
`fragivil@gap.upv.es`

Abstract. Fat-tree topology has become very popular among switch manufacturers. Routing in fat-trees is composed of two phases, an adaptive upwards phase, and a deterministic downwards phase. The unique downwards path to the destination depends on the switch that has been reached in the upwards phase. As adaptive routing is used in the ascending phase, several output ports are possible at each switch and the final choice depends on the selection function. The impact of the selection function on performance has been previously studied for direct networks and has not resulted to be very important. In fat-trees, the decisions made in the upwards phase by the selection function can be critical, since it determines the switch reached in the upwards phase, and therefore the unique downwards path to the destination. In this paper, we analyze the effect of the selection function on fat-trees. Several selection functions are defined, compared and evaluated. The evaluation shows that selection function has a great impact on fat-trees.

Keywords: selection function, adaptive routing, fat-tree, interconnection networks.

1 Introduction

Clusters of PCs have grown in popularity in the last years due to their excellent cost-performance ratio. The interconnection network has a great impact in the performance of these systems. Several switch-based point-to-point commercial networks are currently available. As long as high degree switches are available, multistage networks (MINs) have become very popular. Among them, the fat-tree topology is the preferred choice (e.g.: Mellanox [13], Myricom [15], Quadrics [14]). Routing is one of the most important design issues of interconnection networks. The routing strategy determines the path that each packet follows between a source–destination pair. Routing is deterministic if only one path is provided for every source–destination pair, or adaptive, if several paths are available. Adaptive routing better balances network traffic, thus allowing the network to obtain a higher throughput. The routing algorithm is implemented by means of the routing and selection functions [3]. The routing function supplies a set of suitable routing options to reach the destination. A choice from this set is made by the selection function based on network status.

---
* This work was supported by the Spanish MCYT under Grant TIC2003-08154-C06-01.

Many works [1], [2], [3] has pointed out the great influence that the routing function has on network performance especially for direct networks. There are also some papers [10], [8], [6] that analyze the influence of selection function, showing that it has some small impact on performance. In these networks, the routing function is adaptive along the whole path. However, this is not the case of fat-trees topologies. Routing in fat-trees is performed in two phases, an upwards adaptive one and a downwards deterministic one. The unique path to follow in the downwards phase is determined by the selected upwards path. So, in fat-trees the selection made in the upwards path is responsible of balancing network traffic. Thus, we expect that the selection function will have a greater influence on interconnection network performance.

The rest of the paper is organized as follows. Section 2 presents some background on the fat-tree topology. Section 3 contains references to related work. In Section 4, we propose several selection functions for fat-trees, analyzing their performance in Section 5. Finally, some conclusions are drawn.

2 Fat-Trees

A multistage interconnection network (MIN) is a regular topology in which switches are identical and organized as a set of stages. Each stage is only connected to the previous and the next stage using regular connection patterns. Depending on the interconnection scheme employed between two adjacent stages, several MINs have been proposed. In this paper, we focus on the fat-tree topology.

A fat-tree topology is based on a complete tree. Unlike traditional trees, fat-trees get thicker near the root. A set of processors is located at the leaves and each edge of the tree corresponds to a bidirectional channel. However, the degree of the switches increases as we go nearer to the root, which makes the physical implementation unfeasible. Hence, some alternative implementations have been proposed in order to use switches with constant degree, as the k-ary n-trees. A k-ary n-tree is composed of $N = k^n$ processing nodes and nk^{n-1} switches, with k input and k outputs ports. In what follows, we will use either the term fat-tree or k-ary n-tree to refer to k-ary n-trees.

Each processing node is represented as a n-tuple $\{0, 1, ..., k-1\}^n$, and each switch is defined as a pair $\langle s, o \rangle$, where s is the stage where the switch is located at, that is $s \in \{0..(n-1)\}$, and o is a $(n-1)$-tuple $\{0, 1, ..., k-1\}^{n-1}$. Figure 1.(a) shows a 2-ary 3-tree, with 8 processing nodes and 12 switches. We consider stage 0 as the closest one to the processing nodes.

In a fat-tree, two switches $\langle s, o_{n-2}, ..., o_1, o_0 \rangle$ and $\langle s', o'_{n-2}, ..., o'_1, o'_0 \rangle$ are connected by an edge if $s' = s + 1$ and $o_i = o'_i$ for all $i \neq s$. On the other hand, there is an edge between the switch $\langle 0, o_{n-2}, ..., o_1, o_0 \rangle$ and the processing node $p_{n-1}, ..., p_1, p_0$ if $o_i = p_{i+1}$ for all $i \in \{n-2, ..., 1, 0\}$. Descending links will be labeled from 0 to k-1, and ascending links from k to $2k-1$ (see Figure 1.(b)).

In fat-trees, minimal routing from a source to a destination can be accomplished by sending packets forward to one of the nearest common ancestors of both source and destination and, from there, backward to destination. When crossing stages in the forward direction, several paths are possible, so adaptive routing is provided. Each switch can select any of its upward output ports. Then, the packet is turned around and sent

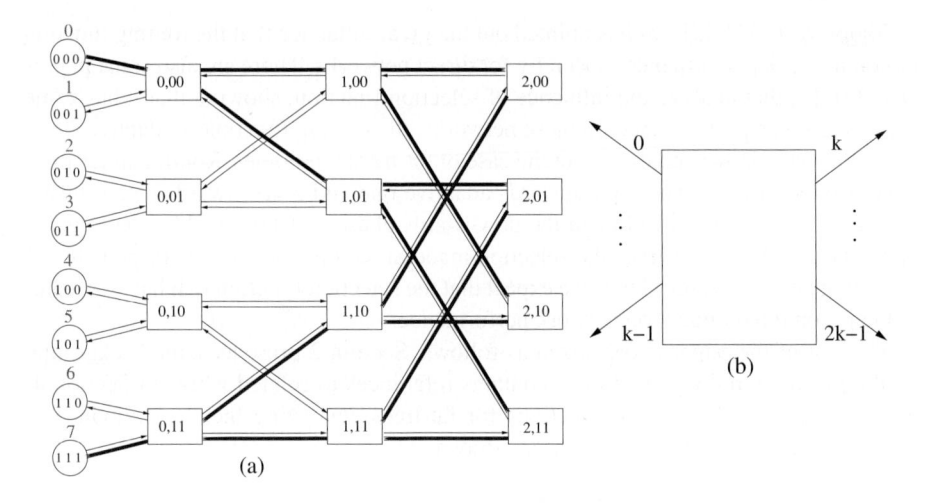

Fig. 1. a) The four possible paths from source 0 to destination 7 in a 2-ary 3-tree. (b) Link numbering in switches of a k-ary n-tree.

backwards to its destination. Once the turnaround is crossed, a single path is available up to the destination node. The stage up to which the packet must be forwarded up is obtained by comparing the source and destination components beginning from the $n-1$ (the most significant one). The fist pair of components that differs indicates the last stage to forward up the packet. For instance, in order to send a packet from the node $p_{n-1}, ..., p_1, p_0$ to the node $p'_{n-1}, ..., p'_1, p'_0$, the packet must be sent up to the stage i, if $p_j = p'_j$ for $j \in \{n-1..i+1\}$ and $p_i \neq p'_i$. Once at stage i, the descending path is deterministic. At each stage, the descending link to choose is indicated by the component corresponding to that stage in the destination n-tuple. In the example, from stage i, the packet must be forwarded through the p'_i link; at stage $i-1$ through link p'_{i-1}, and so on.

For instance in Figure 1.(a), a packet generated at node 0 whose destination is node 2 will be forwarded up to stage number one (through switch 0,00 and choosing either path to 1,00 or 1,01). From any of these switches, the remaining bits of the destination node (10 in our example) correctly forwards the packet.

3 Related Work

Although there is not any work about the impact of selection function in fat-trees, there are some previous works about this issue on other topologies. In [4], Duato proposed a time-dependent selection function for hypercubes, which prevents a message from using certain virtual channels until the time a message has been waiting exceeds some threshold value. In [1], Badr and Podar showed that the *zigzag* selection function is optimal for meshes, in the sense that it maximizes the probability of a message reaching the destination without delay. In [2], [5], and [6] the authors, analyzed the impact of the selection function in the routing algorithm performance in meshes and tori. In [9],

Koibuchi et al. evaluated a selection function that is not specific to any topology. Finally, in [10] Martínez et al. analyzed the impact of selection function in the context of irregular topologies.

4 Selection Functions

An adaptive routing algorithm is composed of the routing and selection functions. The routing function supplies a set of output channels based on the current and destination nodes. The selection function selects an output channel from the set of channels supplied by the routing function.

All the selection functions proposed in this paper take into account the state of the output physical link offered by the routing function and then applies some criteria to select one of them. Virtual cut-trough switching with credit-based flow control is assumed. Notice that, if virtual channel multiplexing is available, the selection function does not select the virtual channel of the physical link that will store the packet. The virtual channel will be selected when the packet is transferred through the link, and the first virtual channel with a free buffer will be selected.

The selection functions presented below provide a preferred ascending link (i). When this link is not free, it performs a linear rotative search starting at link $i+1$ until it finds a free link, if any.

We propose a possible hardware implementation for each selection function. Unless said otherwise, all the selection functions can be implemented in two steps: the first one will obtain the preferred link, and the second implements the linear rotative search by using a programmable priority encoder which takes the preferred link as an extra input. The encoder will give the highest priority to the input represented by this value. The first step changes according to the particular selection function to implement. Taking into account the set of physical links and the physical link that is preferred, it changes the order of the set of physical links in order to put the preferred one in the first position.

We have tested the following selection functions:

First Free (FF). The FF selection function selects the first physical link which has free space. It uses a lineal search, starting at the first ascending physical link (the kth, port according to our notation). FF can be implemented by using a plain priority encoder.

Static Switch Priority (SSP). In a given stage, the SSP selection function assigns the highest priority to a different ascending link at each switch. The idea is to create a disjoint high priority ascending path for each switch of the first stage. Hence, packets coming from different switches at the first stage will reach different switches at the last one, thus balancing the traffic. The high priority physical link for the switch $\langle s, o_{n-2}, ..., o_1, o_0 \rangle$ is the ascending link labeled $k + o_s$. SSP needs the hardware mentioned above, connecting the switch component $\langle o_s \rangle$ to the programmable priority encoder.

Static Destination Priority (SDP). The SDP selection function assigns priorities to physical links at each switch depending only on the packet destination. The preferred physical link is given by the least significant component of the packet destination, which represents the port that the destination is attached to in the first stage. That is,

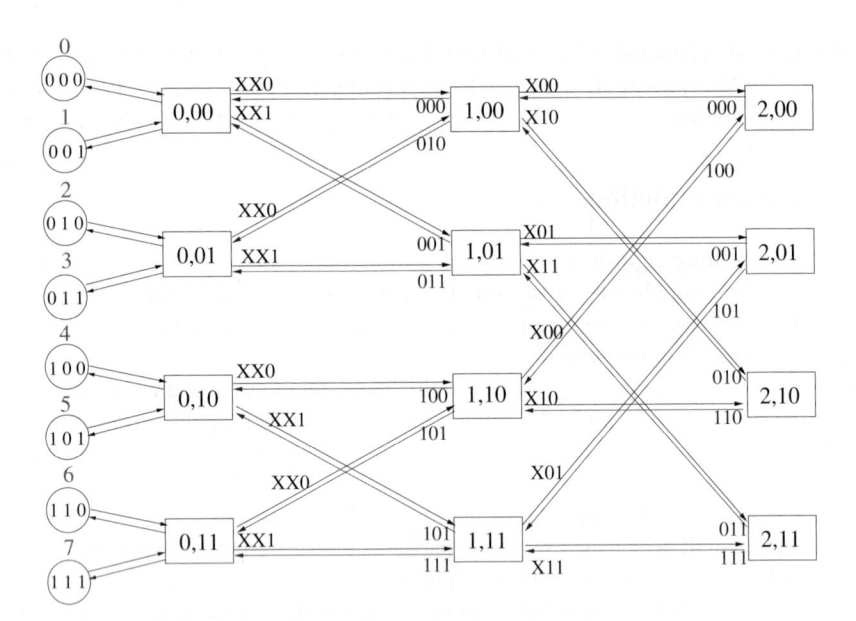

Fig. 2. Preferred links for each destination in SADP for a 2-ary-3-tree

a packet sent to processing node $p_{n-1}, ..., p_1, p_0$ has as the preferred link the $k + p_0$ link. Thus the ascending paths of two packets with destination nodes $p_{n-1}, ..., p_1, p_0$ and $p'_{n-1}, ..., p'_1, p'_0$ are not disjoint only if $p_0 = p'_0$. SDP uses the last component of the packet destination to control the programmable priority encoder.

Static Origin Priority (SOP). The SOP selection function assigns priorities to physical links depending only on the packet source. The preferred physical link is given by the least significant component of the packet source (which represents the port that the origin is attached to in the first stage). That is, for a packet sent from processing node $p_{n-1}, ..., p_1, p_0$, it is $k + p_0$. The hardware implementation of SOP is the same as SDP, but connecting the last component of the packet source to the programmable priority encoder.

Stage And Destination Priority (SADP). The SADP selection function takes into account both the stage at which the switch belongs to and the component of the packet destination corresponding to that stage, (i.e., a switch located at stage s considers the sth component of the destination address). That is, at the switch $\langle s, o_{n-2}, ..., o_1, o_0 \rangle$, the highest priority physical link for a packet with destination $p_{n-1}, ..., p_1, p_0$ will be $k + p_s$. As a consequence, each switch located at the top of the tree concentrates traffic destined to all processors whose id differ only by the most significant digit. Indeed, the paths to these destination are disjoint, as each one is reachable through different output ports of the switch. Figure 2 illustrates this selection function.

Packets are classified according to their destination considering all the components and not only the last one component as SDP does. The main difference between SADP and SDP is that in SDP packets destined to different nodes can have the same preferred links if their destination nodes have the same least significant component (p_0). On the other hand, in SADP, only those packets that share the first switch and are destined to the same node will share all the preferred links along the complete upwards phase.

To implement SADP, we need an additional multiplexer to select a different component of the packet destination at each stage. The output of this multiplexer is connected to the programmable priority encoder.

Cyclic Priority (CP). The CP selection function uses a round robin algorithm to choose a different physical link each time a packet is forwarded. The implementation of CP needs a counter that is incremented each time a packet is routed. The counter is connected to the programmable priority encoder.

More Credits (MC). Since we use credits to implement the flow control mechanism, the MC selection function selects the link which has the highest number of credits available. This number is determined by the sum of the credits available in the all the virtual channels of the physical link. The implementation of MC is more complex, as it needs several comparators to select the link with more available credits.

Random Priority (RP). It selects a random physical link each time a packet is transmitted. This function obtained similar performance results as CP. This is due to the fact that with a high number of packets to transmit, CP and RP selects each physical link the same number of times without considering the source or destination of the packets. The implementation of RP is complex, because it is difficult to obtain by hardware a truly random number. As it obtains similar results to CP, we finally decide to not to consider this selection function in this paper.

5 Performance Evaluation

5.1 Network Model

To evaluate the different selection functions proposed below, a detailed event-driven simulator has been implemented. The simulator models a k-ary n-tree with adaptive routing and virtual cut-through switching. Each router has a full crossbar with queues both at the input and output ports. We assumed that it takes 20 clock cycles to apply the routing algorithm and the selection function, and switch and link bandwidth has been assumed to be one flit per clock cycle and fly time through the link has been assumed to be 8 clock cycles. These values were used to model Myrinet networks [7]. Credits are used to implement the flow control mechanism. Each physical input port can be multiplexed into up to 3 virtual channels, with space to store two packets. Also, each output port link has a two-packet output buffer.

Packet size is 8 Kb and packet generation rate is constant and the same for all the processors in the network. We have evaluated two different traffic patterns: uniform and complement. In the uniform traffic pattern, message destination is randomly chosen among all the processors in the network, while in the complement traffic pattern each processor sends all its messages to the opposite node. Thus, in a network with N processors the processor i sends messages to the processor $N - i - 1$. The complement traffic patterns has two interesting properties in fat-tree networks. The first one is that all the packets have to reach the upper stage in order to arrive to their destination, hence, the selection function must be applied several times. The second one is that each processor node only sends messages to one destination. This proves useful because with a good selection function, the preferred ascending path of two packets should not cross each other.

5.2 Evaluation Results

We have evaluated a wide range of k-ary-n-tree topologies. We have evaluated from 2-ary-2-tree (4 nodes) to 2-ary-8-tree (256 nodes), from 4-ary-2-tree (16 nodes) to 4-ary-6-tree (4096 nodes), from 8-ary-2-tree (64 nodes) to 8-ary-4-tree (4096 nodes), from 16-ary-2-tree (256 nodes) to 16-ary-3-tree (4096 nodes) and for a 32-ary-2-tree (1024 nodes). Due to space limitations, we show here only a subset of the most representative simulations.

Figure 3.(a) shows results for a very small network (4-ary 2-tree, 16 nodes). The behavior of the selection functions is not very different, with the exception of FF. FF always returns the same preferred ascending link, therefore an ascending path needs to be saturated before another one is selected. Hence, there is a really unbalanced link

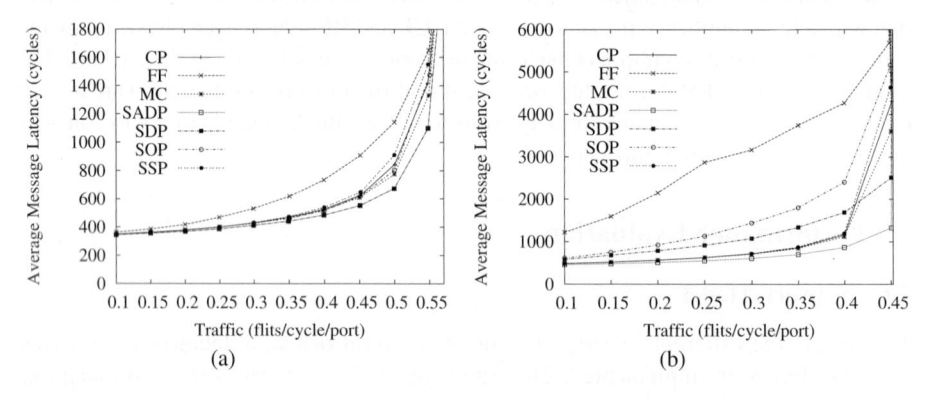

Fig. 3. Average message latency versus traffic with uniform traffic pattern. (a) 4-ary-2-tree. (b) 4-ary-4-tree.

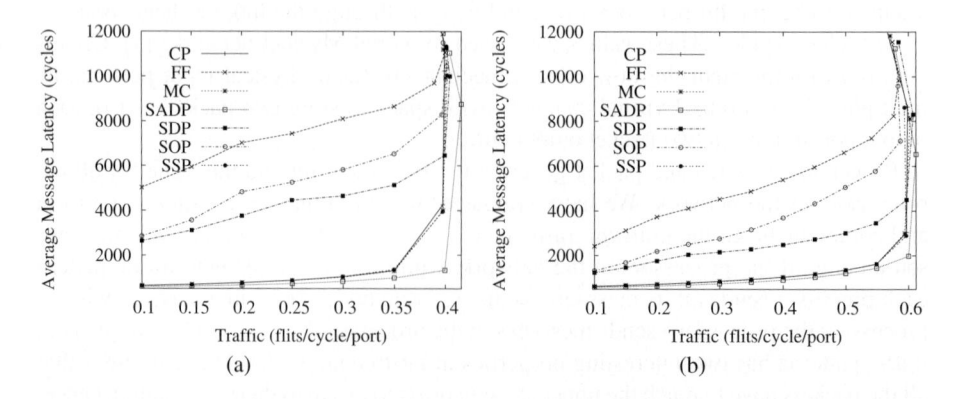

Fig. 4. Average message latency versus traffic with uniform traffic pattern. (a) 4-ary-6-tree with one virtual channel. (b) 4-ary-6-tree with three virtual channels.

utilization as those that belong to the preferred ascending paths always have a higher utilization than the others. On the other hand, due to the fact that there are only 2 stages, the rest of selection functions have almost the same performance, because with a low number of stages there is a low number of different paths that can be chosen to reach any destination.

Figure 3.(b) shows the results for a 4-ary 4-tree (256 nodes). As it can be seen, with 4 stages there are more differences in packet latency for the different selection functions. Despite SOP and SDP achieve a better performance than FF, they still have a high network latency. Their main drawback is that they select paths based only on the p_0 component of the packet origin (SOP) or destination (SDP). Therefore, the probability of obtaining disjoint paths and, thus, an even network utilization is quite low. On the other hand, CP, SSP and MC have almost the same performance, because they do a good job balancing the network utilization. However, in CP and MC there is not any mechanism to try avoiding that the ascending paths of packets cross each other. In SSP, each switch at each stage tries to send the ascending traffic to a different switch, but it does not take into account the destination of packets. Hence, the ascending paths of packets with different origin and destination processing nodes may cross. SADP achieves the best performance because, assuming that the preferred ascending path is free, only the ascending paths of packets sent to the same destination will cross each other.

Figure 4.(a) shows results for a larger network. In this case, the differences among selection functions are higher. This is due to the fact that with more stages, there are more different ascending paths to choose from, therefore more opportunities to balance traffic. On the other hand, the selection functions with a poor traffic balance achieve even worse results than in the previous examples.

Figure 4.(b) shows the effects of using virtual channel multiplexing. The use of virtual channels reduces the effect of the head-of-line blocking. By using three virtual channels, all the selection functions have a better performance, but it is also important to balance network traffic. Although all the selection functions benefits from the use of virtual channels, the ones that have a balanced use of the links give better results than the other ones.

We have also analyzed the impact of the arity (k) of the tree on performance (not shown). The difference among the evaluated selection functions keeps qualitatively the same. This is due to the fact that the impact of any selection function is greater when there is a high number on stages.

Figure 5 shows the utilization of network links for uniform traffic when injecting traffic at the saturation rate of each selection function. As can be expected, the selection functions that better balances traffic are the ones that obtains the best performance.

Figure 4 show the performance of the selection functions for a medium sized network 4-ary 4-tree (256 nodes). As expected, FF has the worst performance. SOP and SDP with complement traffic have worse performance than the one obtained with uniform traffic, because they concentrates in the same switch of the last stage all the packets with the same least significant component of the packet source or destination, and in the complement traffic pattern all the packets reach the last stage. SSP with complement

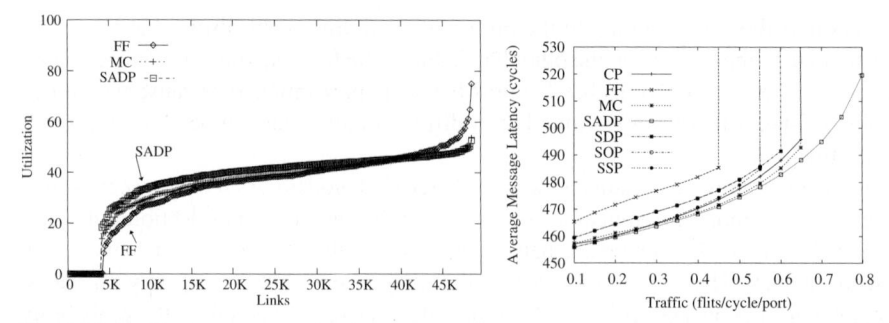

Fig. 5. Link utilization at saturation in a 4-ary-6-tree with uniform traffic for three selection functions

Fig. 6. Average message latency vs. accepted traffic with complement traffic for a 4-ary-4-tree

traffic also shows worse performance, as in SSP the chosen link does not depend on the source or destination node. Therefore, packets with different destinations crosses each other.

Both CP and MC achieve a good performance with complement traffic. MC has a better performance than CP because it takes into account the current number of credits of the link, and this allows MC to select the links that are less saturated. SADP shows the best performance of all. Remember that, with SADP, the preferred ascending path of two packets only can cross each other if the packets have the same destination and in the complement traffic pattern, every packet sent from a different node has a different destination. As a consequence, it achieves a good balance on link utilization.

We have also analyzed the hot spot traffic pattern. As we expected, all the selection functions proposed here obtain a very low performance, because we do not use any congestion control mechanism. On the other hand, other traffic patterns have been also analyzed (like perfect shuffle and bit reversal) and the overall results are qualitatively similar to the ones presented in this paper.

6 Conclusions

The selection function in fat-trees has a strong impact on network performance. As in the descending phase routing is deterministic, it is very important to choose the ascending path correctly. In this paper, we have proposed several selection functions for fat-trees, which use one of the two following strategies: the first one is to give always priority to one ascending path; the second one is to dynamically balance the link utilization without the use of preferred paths. FF, SDP, SOP, SSP and SADP use the first approach, while MC and CP use the second one. From the results of all the simulations we have performed, we can say that it is important to balance the utilization of the links, that is what CP and MC basically do. But an alternative way of achieving this good traffic balance is by correctly choosing ascending paths (SADP).

SADP provides the best results, because it chooses the preferred path in a manner so that the ascending paths of two packets only can cross each other if the packets have the same origin switch or destination node. This provides a balance of link utilization even better than the one provided by MC. For example, a 4-ary-6 with uniform traffic using SADP as selection function reaches, for a medium network load, a 24.53% lower latency than the same fat-tree using MC and the latter has a more complex implementation. If we compare SADP with the naivest selection function (FF), SADP decreases latency by a factor of 8.9.

References

1. S. Badr and P. Podar, An Optimal Shortest-Path Routing Policy for Network Computers with Regular Mesh-Connected Topologies, IEEE Transactions on Computers, Oct. 1989.
2. W. J. Dally and H. Aoki, Deadlock-Free Adaptive Routing in Multicomputer Networks Using Virtual Channels, IEEE Transactions on Parallel and Distributed Systems, Apr. 1993.
3. J. Duato, S. Yalamanchili and L. Ni, Interconnection Networks. An Engineering Approach. Morgan Kaufmann, 2004.
4. J. Duato, Improving the Efficiency of Virtual Channels with Time-Dependent Selection Functions, Proc. Parallel Architectures and Languages Europe '92, 1992.
5. C. Glass and L. M. Ni, The Turn Model for Adaptive Routing, Journal of the Association for Computing Machinery, Sep. 1994.
6. W. Feng and K. Shin, Impact of Selection Functions on Routing Algorithm Performance in Multicomputer Networks, Proc. ACM International Conference on Supercomputing, 1997.
7. J. FLich, M.P. Malumbres, P. Lopez, J. Duato, Improving Routing Performance in Myrinet Networks, Proc. 14th International Parallel and Distributed Processing Symp., 2000.
8. A. Funahashi, M. Koibuchi, A. Jouraku, H. Amano, The Impact of Output Selection Function on Adaptive Routing, Proc. International Conference on Computers And Their Applications, Mar. 2001.
9. M. Koibuchi, A. Jouraku, H. Amano, MMLRU Selection Function: A Simple and Efficient Output Selection Function in Adaptive Routing, IEICE Transactions, 2005.
10. J.C. Martínez, F. Silla, P. López and J. Duato, On the Influence of the Selection Function on the Performance of Networks of Workstation, Proc. High Performance Computing, 2000.
11. F. Petrini and M. Vanneshi, k-ary n-trees: High Performance Networks for Massively Parallel Architectures, Proc. International Parallel Processing Symp., 1997.
12. F. Petrini and M. Vanneschi, Performance analysis of wormhole routed K-ary N-trees, International Journal of Foundations of Computer Science, 1997.
13. Mellanox homepage http://www.mellanox.com/
14. Quadrics homepage http://www.quadrics.com/
15. Myricom homepage http://www.myri.com/

Scalable Ethernet Clos-Switches

Norbert Eicker[1] and Thomas Lippert[1,2]

[1] Central Institute for Applied Mathematics, John von Neumann Institute for Computing (NIC), Research Center Jülich, 52425 Jülich, Germany
{n.eicker, th.lippert}@fz-juelich.de
[2] Department C, Bergische Universität Wuppertal, 42097 Wuppertal, Germany

Abstract. Scalability of Cluster-Computers utilizing Gigabit-Ethernet as an interconnect is limited by the unavailability of scalable switches that provide full bisectional bandwidth. Clos' idea of connecting small crossbar-switches to a large, non-blocking crossbar – wide-spread in the field of high-performance networks – is not applicable in a straight-forward manner to Ethernet fabrics. This paper presents techniques necessary to implement such large crossbar-switches based on available Gigabit-Ethernet technology. We point out the ability to build Gigabit-Ethernet crossbar switches of up to 1152 ports providing full bisectional bandwidth. The cost of our configuration is at about €125 per port, with an observed latency of less than 10μsec. We were able to find a bi-directional point-to-point throughput of 210 MB/s using the ParaStation Cluster middle-ware[2].

1 Introduction

Sophisticated software accelerators enable Gigabit-Ethernet[1] to act as an alternative in the field of interconnects for Cluster-Computing[2]. Since small- and medium-sized switches are available economically priced, this technology is able to serve as an inexpensive network for Clusters with up to ~ 64 nodes – as long as the communication requirements of the applications allow to disobey high-end technologies like Myrinet, InfiniBand, InfiniPath or Quadrics. Nevertheless, in this role Gigabit Ethernet suffers from the unavailability of large, reasonably priced switches. Thus, for large Clusters one either has to purchase one expensive monolithic switch providing full bisectional bandwidth or is forced to accept the handicap imposed by cascaded, medium-sized switches. The latter configuration is afflicted with decreasing accumulated bandwidth from stage to stage.

In the early 50's Clos already proposed a way out of this dilemma[3]. Originally in the field of telephony networks he suggested to set up a special topology of cascaded crossbar-switches providing full bisectional bandwidth. This idea is widespread in the field of high-performance networks. Actually this scheme is used by *e.g.* Myrinet or InfiniBand.

In order to solve the problem discussed above at least in principle, it is possible to use a similar setup with Gigabit Ethernet switches, too. Unfortunately, some specific features of the Ethernet protocol inhibit to actually exploit the bandwidth provided by this topology to a large extent. This work will present a way out of this dilemma.

The abilities of the switch building-blocks play an essential role for the constructions of Ethernet Clos-switches. On the one hand, they have to support virtual LANs (VLAN)[9]. On the other hand, it is necessary to modify the switches routing tables on the level of MAC addresses. Switches fulfilling these conditions are usually called to be "level 2 manageable".

This paper is organized as follows: In the next section, Clos ideas are briefly reviewed. The following two sections discuss some essential features and extensions of the Ethernet protocol, namely spanning trees, VLANs and multiple spanning trees. Based on this fundament we will sketch the setup of cascaded Ethernet crossbar switches in section 5. This includes presenting our testbed and discussing the need for explicit routing tables. Section 6 displays the results produced using the testbed. We end with conclusions and give a brief outlook on further work done in the context of the ALiCEnext project in Wuppertal.

1.1 Related Work

The idea of extending the basic fat tree Ethernet topology is wide-spread. Nevertheless, all projects targeting the improvement of the accumulated bandwidth of a Cluster's Ethernet fabric do not implement full bisectional bandwidth. Instead, special network topologies are developed which are well suited for the communication pattern of a specific class of applications the corresponding Cluster is dedicated to. Good examples of this philosophy are the network of the McKenzie Cluster[12] dedicated to astrophysics or the flat neighborhood networks[13] of the KLAT2 and KASY0 machines used for computational fluid dynamics.

Another approach getting along without any switch is the ALiCEnext mesh network dedicated to lattice QCD[14]. This network only supports pure nearest neighbor communication.

All these concepts prove to be efficient only as long as communication patterns are used they were specifically tuned for. As soon as other applications with different communication patterns are involved, either the fabric has to be re-cabled or a significant performance penalty has to be accepted. In particular such concepts are not feasible for general purpose Clusters running various applications with diverse communication patterns, using different numbers of nodes, serving many users in parallel, etc.

The Viking project[11] pursues a concept similar to the one presented in this paper. I.e. it uses VLANs to create many independent spanning trees. Since the main target of Viking are metropolitan area networks, a static configuration like the one proposed in the present work is not feasible. Instead, they use so called node controller and manager instances in order to adapt the configuration of the Ethernet fabric dynamically to the constraints of the current hardware configuration and the needs of the actual communication load. Furthermore, since full bisectional bandwidth is not the main goal of the Viking project, the proposed network topology is not the one presented by Clos but a grid like.

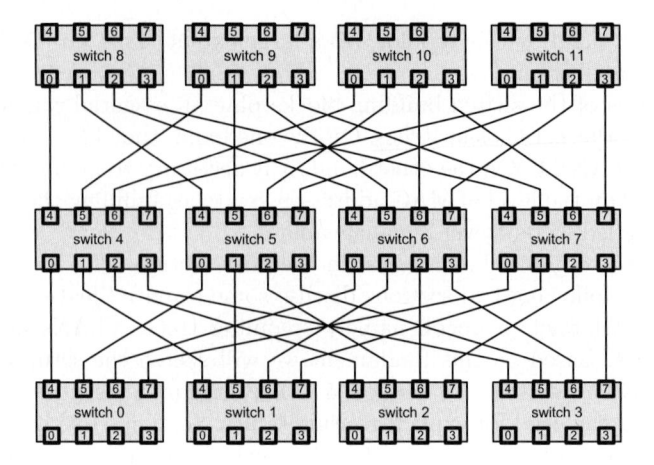

Fig. 1. Example of a full 3-stage Clos-network based on 8-port switches. The full hierarchical switch provides 8×4 ports with full bisectional bandwidth.

2 Clos-Switches

In 1953 Clos[3] introduced the concept of multiple cascaded switches interconnected in a mesh like topology. Originally having telephone networks in mind the idea behind this topology was twofold: Make the network more fault tolerant, *i.e.* more robust in the case of the loss of one or more switches and increase the scalability of the accumulated bandwidth of such systems substantially.

At last, Clos' idea paved the way for multi-stage crossbar networks providing full bisectional bandwidth. The maximum size of a fully connected network is no longer limited by the number of ports offered by the biggest switch available. Of course, with increasing number of ports more switching hierarchies are necessary, each adding to the switching latency.

Today, actually all switched high performance networks (*e.g.* Myrinet[4], Quadrics[5] or InfiniBand[6]) make use of Clos' idea in order to provide full connectivity to large fabrics. This is necessary since the atomic crossbars available for these technologies typically offer not more than $\mathcal{O}(32)$ ports.

The basic topology of a 3-stage Clos-network is sketched in figure 1. It is easy to prove that at any level the same number of connections are available and full bisectional bandwidth is provided in this sense. In order to maximize the usable connectivity, an appropriate routing strategy has to be introduced. The main result of this work is a technique devising a routing strategy for a hierarchy of Gigabit-Ethernet switches. Furthermore figure 1 serves as an introduction of some terminology used in the course of this work:

- Switches connected to nodes are called *level-1 (or L1) switches*. In the example of figure 1 these are the switches 0, 1, 2, 3, 8, 9, 10 and 11.
- Switches connecting level-1 switches are called *level-2 (or L2) switches*. Switches 4, 5, 6 and 7 are the example's L2 switches.

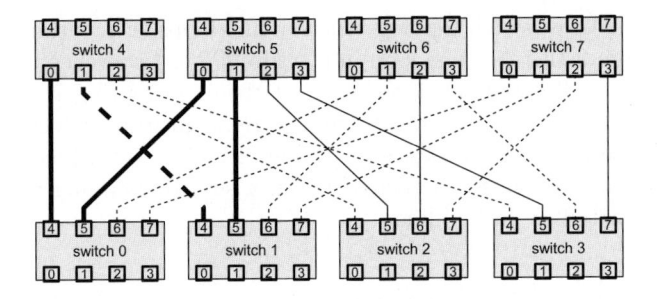

Fig. 2. Basic loop appearing in Clos-switch topologies marked by fat lines. In order to suppress such loops, STAs will switch off the dashed links.

3 Spanning Trees

Trying to construct Clos' topology based on Ethernet technology, a first limitation origins in the use of spanning trees. These are needed to avoid loops within the network fabric. While this feature is inevitable for an Ethernet fabric to work at all, it prevents from using the multiple paths between two switches in parallel. This leads to an accumulated bandwidth found to be identical to the one delivered by cascaded switches.

The very importance of the absence of loops within an Ethernet fabric lies in the fact of missing restricted lifetimes of packets on Ethernet level. This will enable packets to live forever, if the routing-information within the switches creates loops due to misconfiguration. Furthermore – even if the routing is set up correctly – the existence of broadcast packets within the Ethernet protocol provokes packet storms inside the fabric: Whenever a switch receives a broadcast packet on a given port, this packet will be forwarded to all other ports irrespective of available routing information. If there are at least two connections between two switches, a broadcast package sent from one switch to another via a first connection will be sent back to the originating one using the second connection. Once the first switch is reached again, the packet will be sent on its original way again and a loop is created.

Unfortunately, Ethernet broadcast packets play a prominent role within the Internet protocol family, since ARP messages at least on Ethernet hardware are implemented using this type of communication[7]. Thus, every time the MAC-address corresponding to a destination's IP address is unknown, broadcast messages are sent on Ethernet level. Consequently, it is almost impossible to prevent this kind of Ethernet packets in practice.

To beware an Ethernet fabric of this vulnerability, spanning trees were introduced [8]. The main idea here is the detection of loops within a given network fabric and the selective deactivation of such connections, which would eventually close loops. Unfortunately, this will happen on a quite fundamental level of the switch's functioning and thus prevent this link from carrying any data at all.

Investigating Clos-switch topology one can find loops even within the simplest example. Figure 2 sketches one loop in a setup of 2×4 switches[1]. In fact, even this simple setup hosts many loops each of them preventing it from working correctly. On the other hand, the spanning tree algorithm detects these loops and – at the same time – disables all the additionally introduced bandwidth. In figure 2 all connections deactivated by an assumed spanning tree algorithm (STA) are depicted as dashed lines.

4 Virtual LANs and Multiple Spanning Trees

Particularly important for our purposes is the concept of virtual local area networks (VLAN)[9]. This implements multiple virtually disjunct local area networks (LAN) on top of a common Ethernet hardware layer. In order to realize this feature a new level of indirection is introduced by explicitly tagging every native Ethernet packet – including broadcasts – as a member of a distinct VLAN.

This creates a twofold benefit: The topology of the network fabric can be rearranged remotely just by reconfiguration of the switches without physically touching any hardware at all. Additionally – if supported by the operating system – it is possible to assign a given computer to different VLANs at the same time without the need of extra communication hardware.

This technology is widely used in order to map a company's organization virtually onto an uniform physical network fabric in a very flexible way. Thus, it is not surprising that many so-called department switches support this feature.

In this context the idea of spanning trees has to be extended. The reason for providing each VLAN with its own spanning tree is threefold:

- For security reasons broadcast messages have to be restricted to the VLAN they were created in. Otherwise, depending on the high-level protocol[2] the possibility is given to spoof data between different VLANs.
- Within each VLAN there might be loops. These loops would compromise the functionality of the fabric as a whole if they are not eliminated.
- As long as a physical connection is available, the connectivity within each VLAN has to be guaranteed, even if the different VLANs as a whole would build loops. Configurations with physical connections inevitable for the correct functioning of one VLAN but closing a loop within another can easily be constructed. Such dichotomy can only be cured by spanning trees assigned to each VLAN separately.

In order to meet this constraints the STA discussed above was extended to the concept of multiple spanning trees (MST)[10].

---

[1] Actually, loops already appear in 2×2 setups. Since figure 2 also sketches the effects of STAs, the 2×4 setup was chosen.
[2] Here everything above Ethernet protocol level is seen as high-level.

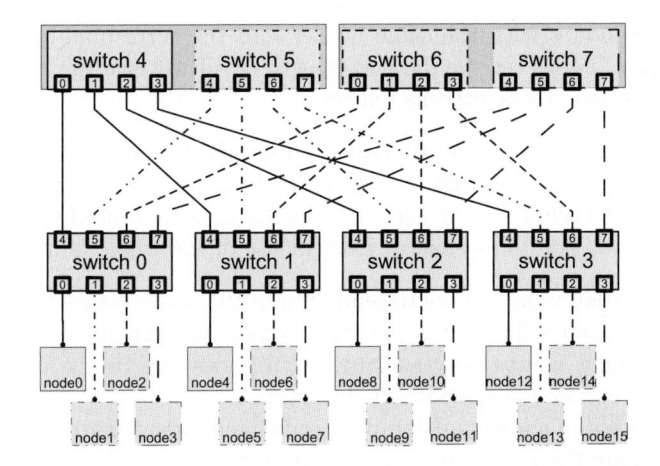

Fig. 3. Crossbar configuration with virtual switches. Each VLAN is depicted with a different line-mode. Lines connecting nodes and L1 switches are only in node → switch direction exclusively used by one VLAN; in switch → node direction each link is used by any VLAN.

5 Configuration Setup

Putting together the technologies described in the last sections the problem of loops can be eliminated without loosing the additional bandwidth of the Clos-topology at the same time. We proceed as follows:

- Setup various VLANs, each forming a spanning tree.
- As many VLANs as nodes attached to a single L1 switch are needed. This fixes the number of VLANs to half the number of ports a switch provides.
- Configure node-ports (*i.e.* ports with nodes attached) to use – depending on the port – a specific VLAN whenever receiving inbound traffic. This implements the required traffic shaping.
- Configure all the node-ports to send outbound traffic from every VLAN[3]. *I.e.* data from every VLAN (and thus from every node) can be sent to any other node, irrespective of the VLAN the sending node is mapped to.

It is essential that all traffic sent from switches directly to a node is not spoiled by any VLAN information[4]. Hence, from the nodes' point of view the complex network topology is completely transparent and no modification has to be done to the nodes' configuration. Even nodes not supporting the VLAN technology at all can be used within this setup.

Figure 3 sketches the typical setup of the crossbar configuration. Here VLANs are depicted by line styles, *i.e.* L2 switches with a distinct border only carry

---
[3] This "port overlapping" is a feature marked as optional in the VLAN standard[9].
[4] This might introduce additional overhead on the L1 switches; in practice this proved to be negligible.

traffic sent by nodes with the same border into the corresponding VLAN. On the other hand, nodes receive data irrespectively of the sending node's VLAN. Traffic shaping is implemented as follows:

- Traffic sent from one node to another connected to the same L1 switch does not touch any other switch. *E.g.* `node6` will talk directly to `node4`.
- The sending node's switch port chooses the L2 switch used to emit data to other L1 switches. This ensures efficient use of the whole fabric.

Assuming `node0` to `node3` try to send data to the nodes connected to `switch2` concurrently, `node0` will send via `switch4`, `node1` via `switch5`, *etc.* Hence, there are 4 independent routes between any two L1 switches in the example. This assures the bisectional bandwidth of the setup.

Figure 3 shows an additional detail. `switch4` and `switch5` are only logical and assumed to share the same hardware. Because of 4 L1 switches only 4 ports of a logical L2 switch are occupied. Thus, another logical switch can use the remaining 4 ports. Again, the configuration is realized via the VLAN mechanism. This guarantees the absence of data-exchange between ports of a physical L2 switch dedicated to different VLANs. Since both – logical and physical L2 switches – have to handle the VLANs anyhow no further effort is introduced.

5.1 The ALiCEnext Testbed

Our testbed used to implement this configuration consists of 144 dual-Opteron nodes of the ALiCEnext[15] Cluster located at Wuppertal University. They are connected via 10 SMC 8648T Gigabit-Ethernet switches[16]. Providing 48 port, each L1 switch serves 24 nodes leading to 24 VLANs. Thus, 6 L1 switches are needed. The other 24 ports are connected to the 4 remaining switches. Every L1 switch is connected via 6 lines to each of the 4 L2 switches. Thus, each physical L2 switch hosts 6 virtual L2 switches leading to a total of 24 logical ones as implied by the number of VLANs and the number of nodes connected to each L1 switch[5].

Unfortunately, at least the implementation of the MST algorithm the SMC 8648T provides is not robust enough to detect the – admittedly very special – setup of our Clos-switch topology correctly. In fact, the switches locked up and the network was unusable[6]. Consequently, automatic loop detection and elimination provided by the MST mechanism was switched off explicitly. This enforces special care when setting up VLANs and cabling. In particular, the default VLAN which includes all ports of the different switches and thus contains countless loops has to be eliminated from the fabric.

---

[5] In principle 3 L2 switches are sufficient to build a 144 port crossbar fabric. The extra ports in our setup were used to implement a connection to the outside world.
[6] Interestingly, plugging the 144 cables between L1 and L2 switches one after the other worked out. This leads to the assumption that switches cannot handle the flood of Hello BPDUs. Of course, plugging all cables whenever a switch restarts is no viable solution.

With this a first prove of concept was obtained by assuring that complete connectivity between the nodes is seen: pings were send from every node to any other node. Furthermore services like ARP[7] and multicasts work out of the box.

Nevertheless, a detailed investigation of the fabric unveiled a problem buried deeper in this setup. In fact, communication between nodes attached to the same VLAN, *i.e.* connected to the same port number of different switches, worked as expected. Communication from one VLAN to another worked in principle, too, but we observed significantly reduced performance.

5.2 Routing Tables

Investigating the dynamic routing tables of the L2 switches the problem was disclosed. These are created on the fly while listening to network traffic between the nodes. Since all inbound traffic is sent via specific VLANs, a L2 switch will never see traffic of nodes sending into different VLANs. In figure 3 *e.g.* switch5 will never see any traffic from node0. On receiving traffic addressed to yet unknown nodes, switches will start to broadcast to all ports. This introduces a plethora of useless traffic. Due to congestion this leads to packet loss and significantly reduced throughput.

To prevent congestion one has to harness the switches with static routing tables. They will shape the traffic addressed to a distinct node in a given VLAN to a specific port. One has to keep in mind that the size of such routing tables is proportional to both, the number of VLANs and the number of nodes connected to the fabric. Thus the tables needed for the testbed will have $24 \times 144 = 3456$ entries. Correspondingly, the routing tables of the entire ALiCEnext machine (528 ports) will contain 12672 entries. Switches providing such numbers of static entries are available; *e.g.* the SMC 8648T allows up to 16k entries[8].

The sheer number of routing entries entail that programming the switches cannot be done in a manual way. Instead, we wrote programs that collect the required data and create the corresponding configuration files. Furthermore, the collected data allow some automatic debugging of the cabling between the nodes and the L1 switches as well. The deployment of configuration files to the switches is automated, too.

6 Results

First we determined the basic parameters of the involved building-blocks. The measurements were carried out on two nodes of the ALiCEnext Cluster with their Gigabit-Ethernet ports directly connected, *i.e.* no switch in between. The very efficient ParaStation protocol was used in order to reduce the message

---

[7] Since ARP is based upon Ethernet broadcasts these work, too. Interestingly, request and response might use different VLANs and, thus, different spanning trees.

[8] This limits scalability for this building-blocks to actually ~ 680 nodes.

Table 1. Performance results

	Back-to-back	single switch	3-stage crossbar
Throughput / node [MB/s]	214.3	210.2	210.4
Latency [μsec]	18.6	21.5	28.0

latencies as much as possible[9]. As a high-level benchmark the Pallas MPI Benchmark suite (PMB)[17] was employed. We applied two tests, `pingpong` for latency determination and `sendrecv` for bandwidth measurements.

Performance numbers for directly connected ports are in the left column of table 1. Latency is half-round-trip time of 0 byte messages as determined by `pingpong`. The low latency found is due to the ParaStation protocol[10]. Throughput is for 512 kByte messages. Larger messages give slightly less throughput ($\sim$ 200MB/sec) due to cache effects when accessing the main memory. The message size for half-throughput was found to be 4096 Byte for all tests.

To determine the influence of a single switch stage the benchmark was repeated using two nodes connected to the same switch. The corresponding results are marked as "single switch" in table 1. Obviously, there is almost no influence of the switch on the throughput. Since the total latency rises from 18.6μsec to 21.5μsec, each switch stage is expected to introduce an additional latency of 2.9μsec. We anticipate a total latency of $\sim$ 27.5μsec when sending messages through all three stages of the testbed. This corresponds to a latency of $\sim$ 9μsec from the switch alone. Throughput is expected to be unaffected.

The above tests were done using a single pair of processes. In order to show bisectional bandwidth we have to concurrently employ as many pairs as possible. Furthermore, the processes have to be distributed in a way that communicating partners are connected to disjoint L1 switches, forcing all traffic to go via the L2 switches and stress the fabric to the hilt. At the time we ran our benchmarks 140 processors were accessible to us, leading to 70 pairs.

The numbers for the 3-stage crossbar presented in table 1 are worst case number. *I.e.* the result for the pair showing the least throughput is displayed there. Looking at the average value of all pairs, throughput is $\sim$ 5% bigger. The best performing pair even gives a result of $\sim$ 218MB/sec. The total throughput observed is larger than 15 GB/s.

Based on an observed latency of 28.0μsec the latency actually introduced by the crossbar-switch was found to be 9.4μsec, *i.e.* in the expected range. This is well below the numbers available for many big, monolithic Gigabit-Ethernet switches with full bisectional bandwidth – at a much lower price! We expect this number to be constant up to 1152 ports[11].

---

[9] ParaStation uses a fine-tuned high-performance protocol in order to reduce the overhead of general-purpose protocols like TCP.
[10] On the same hardware a fine-tuned TCP-setup will reach about 28μsec on MPI-level; out of the box the MPI latency over TCP is often in the range of 60 − 100μsec.
[11] Which is a theoretical limit since size of routing tables restricts us to $\sim$ 680 ports.

7 Conclusion and Outlook

We presented a new way to set up a scalable crossbar switch based on of-the-shelf Gigabit-Ethernet technology. The switch itself is completely transparent to the node-machines. Using the ALiCEnext Cluster at Wuppertal University we showed the concept to work as expected. Full bisectional bandwidth could be achieved at a price of less than €125 per port[12] even with more expensive level 2 manageable switches[13]. In this work we demonstrated our concept to actually work for 144 ports[14].

We submitted an international patent for our approach which is pending[18].

Acknowledgments. We thank the ALiCEnext team in Wuppertal for patience and kind support.

References

1. IEEE standard 802.3z, IEEE standard 802.3ab
2. ParaStation (http://www.par-tec.com).
3. Charles Clos, "A Study of Non-blocking Switching Networks", The Bell System Technical Journal, 1953, vol. 32, no. 2, pp. 406-424
4. http://www.myri.com
5. http://www.quadrics.com
6. http://www.infinibandta.org
7. David C. Plumme, "An Ethernet Address Resolution Protocol" RFC 826, November 1982
8. IEEE standard 802.1D
9. IEEE standard 802.1Q, IEEE standard 802.3ac
10. IEEE standard 802.1s
11. S. Sharma, K. Gopalan, S. Nanda, and T. Chiueh, "Viking: A Multi-Spanning-Tree Ethernet Architecture for Metropolitan Area and Cluster Networks" in IEEE INFOCOM, 2004.
12. J. Dubinski, R. Humble, U.-L. Pen, C. Loken, and P. Martin, "High Performance Commodity Networking in a 512-CPU Teraflops Beowulf Cluster for Computational Astrophysics", eprint arXiv:astro-ph/0305109, Paper submitted to the SC2003 conference.
13. http://aggregate.org/FNN/ and http://aggregate.org/SFNN/
14. Z. Fodor, S.D. Katz, G. Papp, "Better than $1/Mflops sustained: a scalable PC-based parallel computer for lattice QCD", Comput. Phys. Commun. 152 (2003) 121-134.
15. http://www.alicenext.uni-wuppertal.de
16. http://www.smc.com
17. Pallas MPI Benchmark now available from Intel as Intel MPI Benchmark (IMB) http://www.intel.com/software/products/cluster/mpi/mpi_benchmarks_lic.htm
18. Patent: "Data Communication System and Method", EP 05 012 567.3.

---

[12] Based on prices found at `http://www.pricewatch.com`.
[13] The use of more simple (and cheaper) switches lacking the possibility of defining routing tables is not feasible due to congestion.
[14] In the meantime the ALiCEnext crossbar was scaled up without problems. Now it uses 34 switches (22 L1 and 12 L2) to support 528 ports.

Towards a Cost-Effective Interconnection Network Architecture with QoS and Congestion Management Support*

A. Martínez[1], P.J. García[1], F.J. Alfaro[1], J.L. Sánchez[1], J. Flich[2], F.J. Quiles[1], and J. Duato[2]

[1] Departamento de Sistemas Informáticos, Escuela Politécnica Superior
Universidad de Castilla-La Mancha, 02071 - Albacete, Spain
{alejandro, pgarcia, falfaro, jsanchez, paco}@dsi.uclm.es

[2] Dept. de Informática de Sistemas y Computadores, Facultad de Informáica
Universidad Politécnica de Valencia, 46071 - Valencia, Spain
{jflich, jduato}@disca.upv.es

Abstract. Congestion management and quality of service (QoS) provision are two important issues in current network design. The most popular techniques proposed for both issues require the existence of specific resources in the interconnection network, usually a high number of separate queues at switch ports. Therefore, the implementation of these techniques is expensive or even infeasible. However, two novel, efficient, and cost-effective techniques for provision of QoS and for congestion management have been proposed recently. In this paper, we combine those techniques to build a single interconnection network architecture, providing an excellent performance while reducing the number of required resources.

1 Introduction

High-speed interconnection networks have become a major issue on the design of several computing and communication systems, including systems for parallel computing since they provide the low-latency and high-performance demanded by parallel applications. Unfortunately, the network is also becoming the most expensive and power consuming part of these systems.

On the other hand, networks have been traditionally overdimensioned in order to avoid high link utilization, but currently this is an expensive practice. Therefore, new and clever solutions for the problems related to high link utilization are needed.

One of these problems is network congestion. If not managed, congestion dramatically degrades network performance because it leads to blocked packets[1]

---

* This work was partly supported by the Spanish CICYT under grant TIC2003-08154-C06, by Junta de Comunidades de Castilla-La Mancha under grant PBC-05-005, by the Spanish State Secretariat of Education and Universities under FPU grant, and by UPV under Grant 20040937.

[1] We are considering lossless networks like InfiniBand, Quadrics, or Myrinet.

that prevent the advance of other packets stored in the same queue, even if they are requesting free resources further ahead. Moreover, a high utilization of the links may also degrade the performance observed by the users, which leads to the necessity of techniques to provide the traffic with quality of service (QoS). In this case, it is necessary to avoid interferences from best-effort traffic, which only demands a "deliver when possible" service, and guarantee that traffic with strict requirements is properly served.

Many techniques have been proposed both for provision of QoS and for congestion management. Unfortunately, most of them rely on the use of a considerable number of queues at the switches. For instance, the use of virtual output queues has been proposed for handling congestion, but it requires as many queues per switch port as end-points in the network. This increases switch cost due to the silicon area required for implementing such number of buffers.

Regarding QoS provision, the use of virtual channels (VCs) is a common solution, but, although current interconnect standards propose 16 or even more VCs, most commercial components do not offer so many VCs because it is too expensive in terms of silicon area. In fact, the trend followed nowadays by interconnect manufacturers in their new products is to increase the number of switch ports instead of increasing the number of VCs per port [1]. Note that for high-speed, single-chip switches, proposals requiring many queues could be considered if external DRAM is available for implementing the buffers. However, in this case, the low latencies demanded by QoS-requiring traffic could not be provided.

Recently, two novel, efficient, and cost-effective techniques both for provision of QoS and for congestion management have been proposed. The first one [2] consists in a full QoS support with only two VCs. The RECN [3,4] congestion management strategy is the second one. The implementation of both proposals requires a very small number of queues per port, while they offer the same effectiveness as other more silicon-requiring techniques.

In this paper, we study in detail how these two techniques can be combined to form a single architecture that uses a reduced number of queues per port. We also show the performance achieved by the resulting switch architecture in comparison with the performance reached by more expensive solutions.

The rest of this paper is structured as follows. In Section 2, we review the techniques proposed for congestion management and, in particular, the RECN mechanism. Next, in Section 3 we review the proposals for QoS support in interconnection networks and specially the proposal that uses only two VCs. Section 4 presents the proposed interconnection network architecture, whose performance evaluation is presented in Section 5. Finally, in Section 6 some conclusions are drawn.

2 Dealing with Congestion in Interconnection Networks

The risk of congestion in interconnection networks is a well-known problem, and many strategies have been proposed to deal with it. The simplest of those strategies are the network overdimensioning and the dropping of packets in congestion situations. However, none of them are suitable for modern interconnection networks: Overdimensioning the network implies a high cost and power consumption, while the dropping of packets implies packet retransmission that increases packet latency, so current interconnection networks are usually lossless.

Other more elaborated techniques have been specifically proposed for avoiding or eliminating congestion. For instance, proactive strategies are based on reserving network resources for each data transmission, requiring a traffic planification based on network status [5]. However, this status information is not always available, and the resource reservation procedure introduces significant overhead. On the other hand, reactive congestion management is based on notifying congestion to the sources contributing to its formation, in order to cease or reduce the traffic injection from those sources [6]. Unfortunately, these solutions are not quite efficient due to the delay between congestion detection and notification.

Other congestion management strategies focus on eliminating the main negative effect of congestion: The head-of-line (HOL) blocking. This phenomenon happens when a blocked packet at the head of a FIFO queue prevents the advance of other packets at the same queue, even if those packets require available resources. This effect may degrade network performance dramatically, since data flows not contributing to congestion may advance at the same speed than congested flows. In fact, an effective HOL blocking elimination would turn congestion harmless. In that sense, many HOL blocking elimination strategies have been proposed: virtual output queues (VOQs) [7], dynamically allocated multi-queues (DAMQs) [8], congestion buffers [9], etc. Most of these techniques rely on allocating different buffers for storing separately packets belonging to different flows.

In general, traditional HOL blocking elimination techniques are scalable or efficient, but not scalable and efficient at the same time. For instance, the use of VOQs at network level requires as many queues at each port as end-points in the network, being so an effective but not scalable technique. A variation of VOQ uses as many queues at each port as output ports in a switch [10]. So, this technique is scalable, but it does not eliminate completely HOL blocking, only the switch's internal HOL-blocking.

Recently, a new HOL blocking elimination technique has been proposed: RECN [3,4]. RECN eliminates HOL blocking in a scalable and efficient way.

2.1 RECN Description

RECN (Regional Explicit Congestion Notification)[3] is a congestion management strategy that focuses on eliminating HOL blocking. In order to achieve it, RECN detects congestion and dynamically allocates separate buffers for each congested flow, assuming that packets from non-congested flows can be mixed in the same buffer without producing significant HOL blocking. Therefore, maximum performance is achieved even in the presence of congestion.

RECN requires the use of some source deterministic routing in order to address a particular network point from any other point in the network. In fact, RECN has been designed for PCI Express Advanced Switching (AS) [11], a technology that uses source routing[2]. AS packet headers include a turnpool made up of 31 bits, which contains all the turns (offset from the input port to the output port) for every switch in a route. Thus, a switch, by inspecting the appropriate turnpool bits, can know in advance if a packet that is coming through one of its input ports will pass through a particular network point.

In order to separate congested and non-congested flows, RECN adds a set of additional queues (set aside queues, SAQs) to the standard queue at every input and output port of a switch. While standard queues will store non-congested packets, SAQs are dynamically allocated and used to store packets passing through a congested point. Every set of SAQs is controlled by means of a CAM (Content Addressable Memory). Every CAM line contains information required for identifying a congested point and for managing the associated SAQ.

Whenever an input or output standard queue receives a packet and fills over a given threshold, RECN detects congestion[3]. Then, a congestion notification is sent upstream to the packet sender port (an input port of the same switch or an output port of an upstream switch). These notifications include the turnpool required to reach the congested point from the notified port. Upon reception of a notification, a port allocates a new SAQ and fills the corresponding CAM line with the received turnpool. Since that moment, every packet received in this port will be stored in the allocated SAQ if it will pass through the associated congested point (this can be deduced from the packet turnpool). As non-congested packets are stored in different queues (the standard ones), congested packets cannot cause HOL blocking.

Furthermore, if any SAQ becomes congested, another notification will be sent upstream, and the receiving port should allocate a new SAQ. This procedure can be repeated until the notifications reach the sources. Therefore, there will be SAQs for storing congested packets at every point where otherwise these packets could produce HOL blocking. Moreover, congested packets cannot fill the port memory completely as RECN uses a SAQ-specific Xon/Xoff flow control.

RECN also detects congestion vanishment at any point, in such a way that the SAQs assigned to this point can be deallocated and later re-allocated for new congested points. This allows RECN to eliminate HOL blocking while using a reduced number of SAQs. Further details about RECN can be found in [3,4].

3   QoS Support in Interconnection Networks

During the last decade, several switch architectures with QoS support have been proposed. Among the most recent proposals are the industry standards Infini-

---

[2] However, note that RECN could be applied on any network technology if it allows the use of source deterministic routing.
[3] Actually, in order to detect at input ports which output port is congested, RECN divides each standard queue into several small detection queues [4].

Band and PCI Express Advanced Switching (AS). The InfiniBand standard [12] considers up to 16 VCs, while the AS specification [11] incorporates up to 20 VCs (16 unicast, 4 multicast). However, the implementation of such number of VCs would require a significant fraction of silicon area and it would make packet processing a time-consuming task. Consequently, as far as we know, no implementation of these standards includes the full number of proposed VCs.

Several proposals have been presented in order to reduce the hardware required for QoS provision, some of them using only two VCs. For instance, the Avici TSR [7] is able to segregate premium traffic from regular traffic. However, it is limited to this classification and cannot consider more categories. Also, the architecture proposed in [13] maps multiple priority levels onto two queues. However, this proposal is aimed at a single-stage router based on a single buffered crossbar with small buffers at the crosspoints that are split into two VCs.

In contrast, the technique proposed in [2] uses two VCs while being simpler and more generic, as it is shown in the next section.

3.1 Full QoS Support with 2 VCs

The key idea of the proposal explained in [2], is quite simple: Assuming that the links are not oversubscribed, all the traffic flows through the switches seamlessly. Therefore, it is possible to use only two VCs at the switch ports. One of these VCs is used for QoS packets and the other one for best-effort packets. In [2], a connection admission control (CAC) is used to guarantee that QoS traffic will not oversubscribe the links.

Another cornerstone of this proposal is to reuse at the switches the scheduling decisions taken at the network interfaces regarding the injection of traffic from the different classes. Specifically, it is assumed that packets are ordered at network interfaces according to a static priority criterion. In this way, every packet would be stamped with a priority (or service) level (typically, 8 or 16 levels). This is necessary because packets arriving at the switches come in the order specified by the interfaces, and the switch must merge these packet flows at the output ports. The ordering established at network interfaces does not need to be changed at any switch in the path because queuing delays for QoS traffic will be short.

Although it is assumed that QoS traffic does not oversubscribe any link, no assumption is made about best-effort traffic. However, network interfaces are still able to assign the available bandwidth (the fraction not consumed by QoS traffic) to best-effort traffic in the configured proportions. In this way, switches can still take into account the modest QoS requirements of this kind of traffic. Obviously, this is a coarse-grain QoS provision.

Note that this proposal does not aim at achieving a higher performance but, instead, at drastically reducing buffer requirements while reaching the same levels of performance and behavior as systems with many more VCs. In this way, an effective QoS support could be implemented at an affordable cost.

Note also that some aspects of this proposal could be simplified or improved if it is combined with the RECN strategy. For instance, instead of the CAC applied to QoS traffic in [2], the RECN mechanism could detect if some traffic flows start producing congestion, and immediately segregate these flows from the non-congested ones. Moreover, regarding best-effort traffic, RECN can make important contributions. Specifically, it can guarantee the maximum throughput for best-effort traffic, avoiding also that congested flows affect non-congested traffic.

As RECN also requires a reduced number of resources, the combination of both techniques would allow to provide effective QoS support and congestion management at a low cost in terms of silicon area. The architecture we propose for combining both techniques is explained in the following section.

4 Proposed Interconnection Network Architecture

Our proposal consists in a interconnection network architecture able to support QoS and to cope with congestion while requiring reduced resources. The strategies we propose affect both network interfaces and switches.

(a) Switch organization

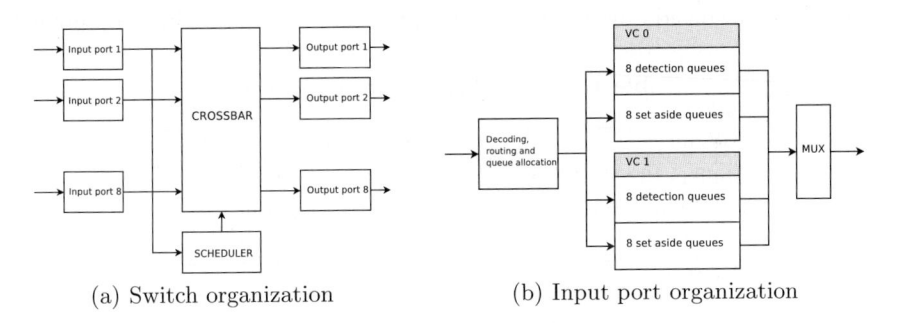

(b) Input port organization

Fig. 1. Proposed architecture

Figure 1 (a) shows a logical view of the switch organization, which consists in a combination of input and output buffering. Note that all the switch components are intended to be implemented in a single chip. This is necessary in order to offer the low cut-through latencies demanded by current parallel applications.

The innovations of our proposal are in the port design. The organization of an input port can be seen at Figure 1 (b). There are only two VCs: VC 0 is intended for QoS traffic, while VC 1 is intended for best-effort traffic. As can be seen, each VC is further divided into 16 queues. The first 8 queues of each VC are the detection queues, so each queue corresponds to each switch output port. The next 8 queues of each VC are the SAQs, where congested traffic is stored.

We include an additional field in the CAM lines used for managing the SAQs: The service level (SL). This new field will be used, in addition to the turnpool, for assigning a SAQ to a specific point and to a specific SL. Therefore:

- When a standard queue reaches the detection threshold, the corresponding congestion notification includes now, in addition to the turnpool, the SL of the packet responsible of the detection.
- Each allocated SAQ contains traffic of a single SL.
- A single turnpool may be replicated for several SAQs, with different SLs.

So, the congestion detection process is slightly different than in RECN without QoS support. As a detection queue may reserve several SAQs (each for a different SL), a bit mask is required to control which SLs have reserved a SAQ.

The output ports of the switch replicate the structure of the inputs, with two main differences. There is no need to decode the messages and detection queues refer to the outputs of the next switch. The network interface design would be very similar to this, although, in this case, a VC exists for each SL.

The scheduling at the switches goes as follows. There is a strict precedence of VC 0 (QoS traffic) over VC 1 (best-effort traffic): As long as there are ready packets of VC 0, no one from VC 1 is eligible. Among the queues inside each VC, a simple round-robin algorithm is applied. Note that RECN ensures that none of the SAQs will occupy all the buffer space and, therefore, this simple scheduling is sufficient.

The area requirements study for this design is based on the process detailed at [14]. We do not have yet a detailed Verilog switch design, but we can obtain good estimations by considering the area consumption of each individual component.

Table 1. Area consumption by components

Module	Area 0.18 μm	Area 0.13 μm
Buffers (32 × 16$Kbytes$)	64 mm^2	32 mm^2
Crossbar and datapath	10 mm^2	5 mm^2
Scheduler	5 mm^2	3 mm^2
Total	79 mm^2	40 mm^2

In Table 1, the aforementioned estimations can be found. The memory area consumptions are taken from memory datasheets [15]. The number of buffers in the switch comes from 8 ports × 2 VCs × input and output. Keep in mind that at the placement and routing phase of the design process, the wiring introduced could increase these figures. Therefore, this design would take 100-150 mm^2 using 180 nanometers technology.

5 Performance Evaluation

We have evaluated the proposed architecture by means of simulations. In this section, we will detail the simulated scenarios and we will offer results showing the behavior of our proposal in comparison with the one of traditional switches.

5.1 Simulated Architecture

We have supposed a workload of 8 SLs, with decreasing priority, such that SL 0 has the highest priority and SL 7 has the lowest. We have also assumed that SLs from 0 to 3 are QoS-requiring traffic, and share the same VC in the two-VC architecture. Moreover, we also suppose that SLs from 4 to 7 are best-effort traffic, and share the other VC in the two-VC scheme.

We have run simulations for three architectures. First, we have tested the performance of the ideal architecture, using VOQ at the network level combined with a VC per SL (*VOQ Net*). Also, we have tested a more realistic architecture, using VOQ at the switch level and as many VCs as SLs (*VOQ Switch*). Finally, our proposed architecture, combining the use of only 2 VCs at the switches with the use of RECN (*RECN 2 VCs*).

The network used in the tests is a folded (bidirectional) butterfly multi-stage interconnection network (MIN) with 128 ports (3 stages). We have chosen a MIN because it is a usual topology for computer clusters and IP routers. However, our proposal is valid for any other network topology, including direct networks. Other assumptions are based on the AS specifications [11]. For instance, maximum packet size is 2 Kbytes and link bandwidth is 8 Gb/s. Another assumption is the use of source routing, since it is needed by RECN.

The *VOQ Switch* and *RECN 2 VCs* architectures consider 8 ports and 32 Kbytes of buffer space per port. Note that the buffer space per VC in the *RECN 2 VCs* case is bigger than in the *VOQ Switch* case in a factor of 4. Also note that the scheduler considers 64 queues per port in the *VOQ Switch* design (8 VCs × 8 VOQs), while in the *RECN 2 VCs* case 32 queues (8+8 from each VC) are considered. For the sake of clarity, we do not consider in this study the saving in silicon area and the gain in scheduler speed due to these facts.

The *VOQ Net* architecture assumes a space of 2 maximum size packets per VC. This is the minimum required to assure a full throughput under a full load between two ports (one packet size plus a round-trip time, rounded to full packets). In a 128 end-points network, it requires 128 × 8 VCs × 4 Kbytes = 4 Mbytes per port of buffering. It is clearly too much for a single chip, and it should be implemented in external DRAM, but in that case the cut-through latency of the switch would be much higher. Nevertheless, in order to provide a reference of the ideal performance, we have not considered this additional delay in the *VOQ Net* simulations.

5.2 Traffic Model

In all the tests we have used self-similar traffic. This traffic is composed of bursts of packets heading to the same destination. The packets' sizes are governed by a Pareto distribution, as recommended in [16]. In this way, many small size packets are generated, with an occasional large size packet. The periods between bursts are modelled with a Poisson distribution and the distribution of the bursts destinations is uniform. If the burst size is long, there is a lot of temporal and spatial locality and should show worst-case behavior because at a given moment, many packets are grouped going to the same destination. Regarding burst length, we have used a long one of 30 Kbytes for the four best-effort classes and a shorter one of 5 Kbytes for the QoS classes.

5.3 Simulation Results

We have considered two traditional QoS metrics in the performance evaluation: Throughput and latency. Packet loss is not considered because no packets are dropped due to the use of credit-based flow control. However, note that inappropriate results of latency may lead to dropped packets at the application level. For this reason, we also consider maximum latency.

We first analyze the results for the best-effort traffic classes, which are more likely to suffer from congestion. Figure 2 shows the global throughput of the network for the unregulated traffic. It can be seen that our *RECN 2 VCs* proposal, from an input load of 80% on, offers a 25% improvement over the *VOQ Switch* architecture and only losses a 5% from the ideal, infeasible *VOQ Net* architecture.

Figure 3 depicts the detailed throughput results for each one of the best-effort classes (SLs 4 to 7). While our proposal offers results very close to those

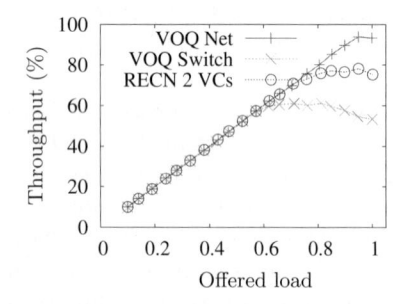

Fig. 2. Global throughput results for best-effort SLs

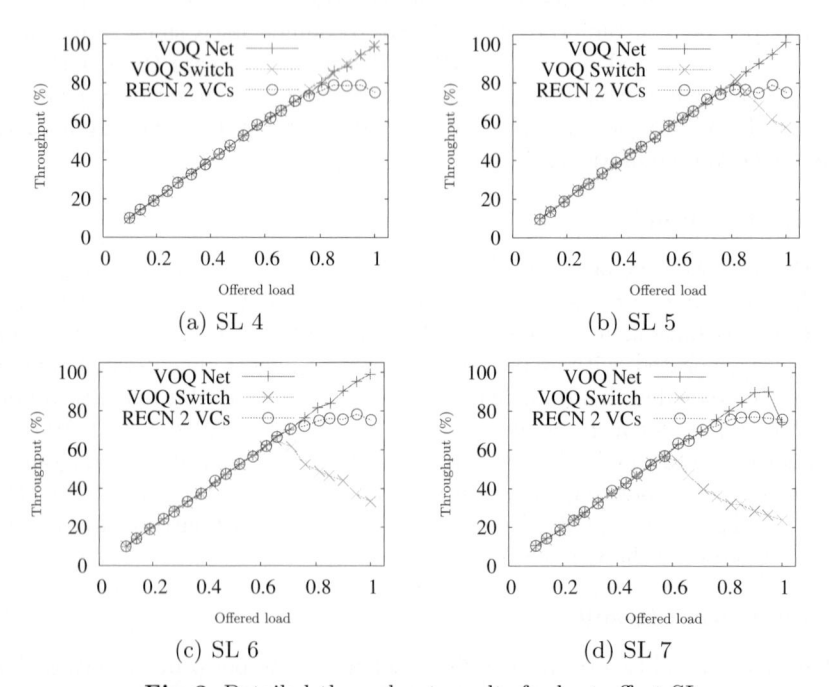

(a) SL 4

(b) SL 5

(c) SL 6

(d) SL 7

Fig. 3. Detailed throughput results for best-effort SLs

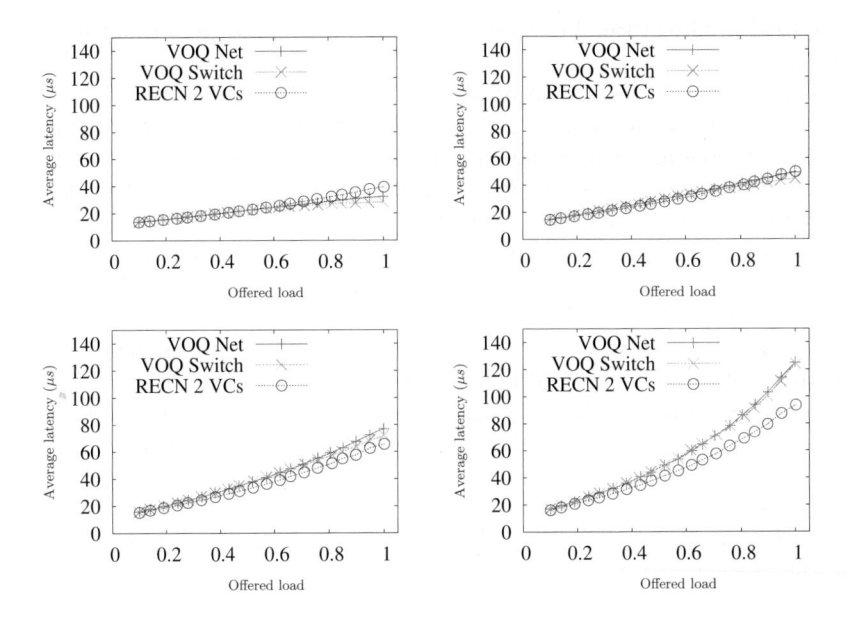

Fig. 4. Average latency results for QoS SLs

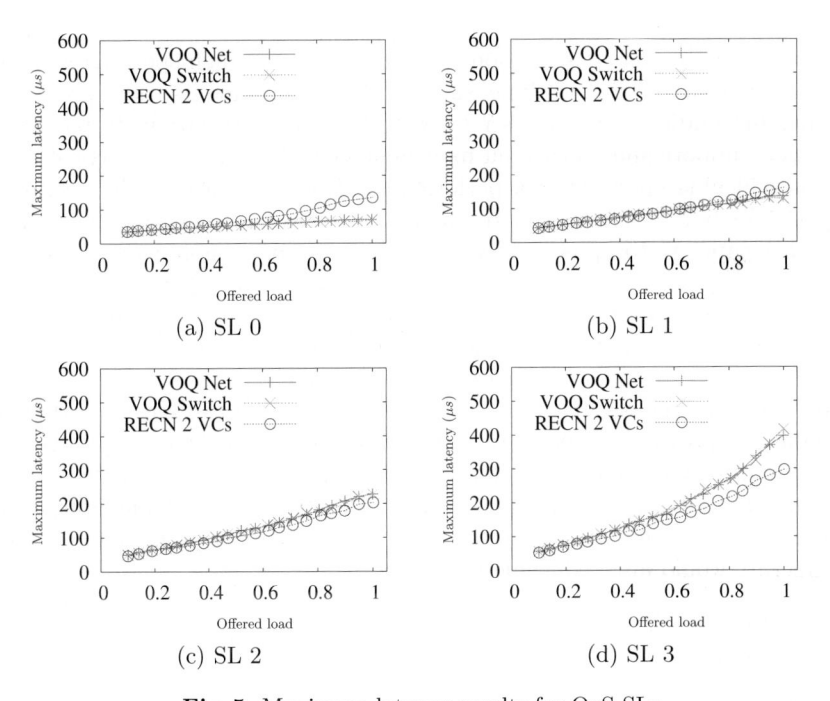

(a) SL 0

(b) SL 1

(c) SL 2

(d) SL 3

Fig. 5. Maximum latency results for QoS SLs

of the ideal architecture, *VOQ Switch* offers a very poor performance for SLs 6 and 7.

The next part of the evaluation deals with the QoS traffic. In this case, the three architectures offer 100% throughput for QoS traffic (not shown). Figure 4 shows the interesting average latency results for the QoS SLs. Note that these results are very similar for all the architectures. Although our proposal offers slightly worse results for the SL 0, the SL 3 benefits from the RECN technique. Almost identical results have been obtained for maximum latency (Figure 5). Due to space constraints maximum jitter results are not shown, but are also similar to those of latency.

These results show that QoS-requiring flows get the performance they need when using our architecture, although they are sharing a single VC. Therefore, our proposal is able to offer QoS support at the same level as a proposal that doubles the required number of queues (*VOQ Switch*) and even at the same level as an ideal and expensive architecture (*VOQ Net*). Moreover, in the case of heavy congestion in the QoS SLs, the *VOQ Switch* case will suffer strong degradation while the RECN mechanism included in our architecture would solve the problem.

6 Conclusions

Due to cost and power consumption constraints, current high-speed interconnection networks cannot be overdimensioned. Therefore, some solutions are needed in order to handle the problems related to high link utilization. In particular, both QoS support and congestion management techniques have become essential for achieving good network performance. However, most of the techniques proposed for both issues require too many resources for being implemented.

In this paper we propose a new network architecture able to face the challenges of congestion management and, at the same time, QoS provision, while being more cost-effective than other proposals. Our proposal is based on the combination of two novel techniques that provide congestion control and QoS support while requiring a reduced number of resources.

According to the results presented in this paper, we can conclude that our proposal can provide an adequate QoS while properly dealing with congestion. We provide advanced techniques for the buffer management, which allow a good performance under heavy and unbalanced load, while still providing appropriate QoS levels. Since all this is achieved with a reduced number of resources, this architecture would also reduce network cost.

References

1. Minkenberg, C., Abel, F., Gusat, M., Luijten, R.P., Denzel, W.: Current issues in packet switch design. In: ACM SIGCOMM Computer Communication Review. (2003)

2. Martínez, A., Alfaro, F.J., Sánchez, J.L., Duato, J.: Providing full QoS support in clusters using only two VCs at the switches. In: Proceedings of the 12th International Conference on High Performance Computing (HiPC). (2005) Available at http://www.i3a.uclm.es/documentos/2/congresos/Congreso2_134_HiPC05.pdf.
3. Duato, J., Johnson, I., Flich, J., Naven, F., García, P., Nachiondo, T.: A new scalable and cost-effective congestion management strategy for lossless multistage interconnection networks. In: Proceedings of the 11th Symposium on High Performance Computer Architecture (HPCA). (2005)
4. García, P., Flich, J., Duato, J., Johnson, I., Quiles, F., Naven, F.: Dynamic evolution of congestion trees: Analysis and impact on switch architecture. Lecture Notes in Computer Science (HiPEAC 2005) **3793** (2005) 266–285
5. Wang, M., Siegel, H.J., Nichols, M.A., Abraham, S.: Using a multipath network for reducing the effects of hot spots. IEEE Transactions on Parallel and Distributed Systems **6** (1995) 252–268
6. Thottetodi, M., Lebeck, A., Mukherjee, S.: Self-tuned congestion control for multiprocessor networks. In: Proc. of 7th. Int. Symp. on High Performance Computer Architecture. (2001)
7. Dally, W., Carvey, P., Dennison, L.: Architecture of the Avici terabit switch/router. In: Proceedings of the 6th Symposium on Hot Interconnects. (1998)
8. Tamir, Y., Frazier, G.: Dynamically-allocated multi-queue buffers for vlsi communication switches. IEEE Transactions on Computers **41** (1992)
9. Smai, A., Thorelli, L.: Global reactive congestion control in multicomputer networks. In: Proc. 5th Int. Conference on High Performance Computing. (1998)
10. Anderson, T., Owicki, S., Saxe, J., Thacker, C.: High-speed switch scheduling for local-area networks. ACM Transactions on Computer Systems **11** (1993) 319–352
11. Advanced Switching Interconnect Special Interest Group: Advanced Switching Core Architecture Specification. Revision 1.1. (2005)
12. InfiniBand Trade Association: InfiniBand architecture specification volume 1. Release 1.0. (2000)
13. Chrysos, N., Katevenis, M.: Multiple priorities in a two-lane buffered crossbar. In: Proceedings of the IEEE Globecom 2004 Conference. (2004)
14. Simos, D.: Design of a 32x32 variable-packet-size buffered crossbar switch chip. Technical Report FORTH-ICS/TR-339, Inst. of Computer Science, FORTH (2004) http://archvlsi.ics.forth.gr/bufxbar/.
15. Virtual Silicon Technology, Inc.: eSi-RAM/2P^{tm} Two-port register file SRAM. Data Sheet (2004)
16. Jain, R.: The art of computer system performance analysis: techniques for experimental design, measurement, simulation and modeling. John Wiley and Sons, Inc. (1991)

Topic 14: Mobile and Ubiquitous Computing

Alois Ferscha, Alexander Schill, Gianluigi Ferrari, and Valerie Issarny

Topic Chairs

Mobile networking, mobile systems and applications and ubiquitous computing infrastructures are of strongly growing importance in the IT sector in general, and particularly for the parallel and distributed computing community. Mobile internet access has become a commonplace service. Many conventional software products such as e-mail systems, databases, or enterprise resource planning have been adapted towards mobile usage patterns. The world of ubiquitous information processing will soon be revolutionizing our day-to-day routines. Major components are sensor networks, radio frequency identification technology, and whole new layers of data management assembling their low-level signals to high-level knowledge.

While there has been tremendous progress during the last decade in the networking sector and at the mobile computing software level, many challenging research and development issues remain. Examples include the optimization of mobile communication channels, the automated planning of scalable mobile networks or the inherent support of ad-hoc routing and computing. Moreover, highly adaptive, personalized, and context-aware mobile computing software will be one of the major challenges in distributed computing of the future.

For parallel and distributed processing, these trends also mean that new, more flexible techniques of accessing global grid infrastructures are becoming available. However, mobile systems are not only access technologies but can also build the core of loosely-coupled parallel processing scenarios, even in an ad-hoc fashion. Moreover, the whole area of ubiquitous computing will yield a tremendous increase of information, for example based on the output of thousands of sensors, challenging the capacity of parallel processing architectures at the higher layers.

The international research community has shown strong interest in these and other related aspects, also resulting in a very large number of submissions to this topic area of EuroPar 2006. The reviewers and their colleagues did a great job ensuring three quality reviews for each paper. Based on their judgements and an intensive interactive discussion, we finally selected 9 papers to be presented. These contributions equally cover the areas of mobile networking and associated protocols, of ubiquitous computing based on specialized sensor network solutions, and of mobile computing with new software solutions and architectures. I would like to thank the organizers of EuroPar 2006 to have made this topic possible, and I would especially like to thank Alois Ferscha, Gianluigi Ferrari, and Valerie Issarny for their review work and for the excellent cooperation.

Alexander Schill, Topic Chair

Multi-rated Packet Transmission Scheme for IEEE 802.11 WLAN Networks

Namgi Kim

Telecommunication R&D Center, Samsung Electronics,
416, Maetan-3, Youngtong, Suwon, Kyounggi, 442-600, Korea
namgi.kim@samsung.com

Abstract. In a multirate wireless network such as IEEE 802.11 WLAN, the connection having a good channel condition uses a high transmission rate and the connection having a poor channel condition uses a low transmission rate. However, this coexistence of different transmission rates degrades the total system performance of the network. In order to eliminate this performance abnormality and improve protocol capacity, we propose a new packet transmission algorithm, the RAT (Rate-Adapted Transmission) scheme. The RAT scheme distributes the wireless channel fairly based on the channel occupancy time. Moreover, it efficiently transmits packets even in a single station using rate-based queue management. Therefore, the RAT scheme obtains not only the inter-rate contention gain among stations but also the intra-rate contention gain among connections in a single station. By simulation, we show that the proposed RAT scheme is superior to the default IEEE 802.11 MAC DCF access method and the modified OAR (Opportunistic Auto Rate) scheme.

1 Introduction

Recently, WLAN (Wireless LAN) has achieved tremendous growth and has become the prevailing technology for wireless access for mobile devices. WLAN has been rapidly integrated with the wired Internet and has been deployed in offices, universities, and even public areas. Moreover, the IEEE 802.11 WLAN standard is considered as the most popular wireless access method for ad-hoc mobile communications.

The wireless channel condition varies over time and space due to the dynamic features of the wireless environments such as mobility, interference, and location. To cope with this channel variation, the physical specifications of the IEEE 802.11 WLAN provide multiple transmission rates by employing different channel modulation and coding schemes. The IEEE 802.11b standard [1] provides four different physical transmission rates from 1Mbps up to 11Mbps. The IEEE 802.11a [2] and 802.11g [3] standards provide eight different transmission rates from 6Mbps up to 54Mbps. This multiple rate capability enables a wireless station to dynamically choose a physical transmission rate depending on the channel condition.

The efficient selection of the physical transmission rate affects the WLAN performance significantly. To choose the best rate among multiple transmission rates at a given time, numerous rate control algorithms have been proposed [5,6,7,8,9,10,11,12]. These algorithms enhance the WLAN performance by dynamically changing transmission rates according to the variable channel conditions. However, problems still exist. Although these rate control algorithms efficiently utilize the wireless medium, there is consider-able performance degradation when some connections transmit data at lower physical rates than others. This performance degradation with mixed transmission rates is due to the CSMA/CA protocol, which is used in the IEEE 802.11 MAC DCF (Distributed Coordination Function) channel access method [4].

Therefore, in this paper, we investigate this performance abnormality. We then propose a novel packet transmission strategy for improving protocol capacity in multi-rate wireless networks.

2 Background and Motivation

2.1 Performance Abnormality with Multiple Transmission Rates

For investigating the performance abnormality, twelve mobile stations send data frames to wired stations over the WLAN AP (Access Point). The default transmission rate is set as 11Mbps. We started the experiments without low transmission rate nodes and thereafter increased the number of low rate nodes gradually.

Fig. 1(a) displays the result of the experiments with multiple transmission rates. When there is no low transmission rate node, the total system throughput is 5.253 Mbits/s. However, the throughputs are degraded drastically when the number of low rate nodes is increased. In particular, the throughput is degraded by almost half even when only one 1Mbps rate node is involved in the transmissions.

Fig. 1(b) shows he channel occupancy time when 11Mbps and 1Mbps rates are mixed in the same experiment. On average, about 77 % of the total time is used

(a) Throughput degradation

(b) Channel occupancy time

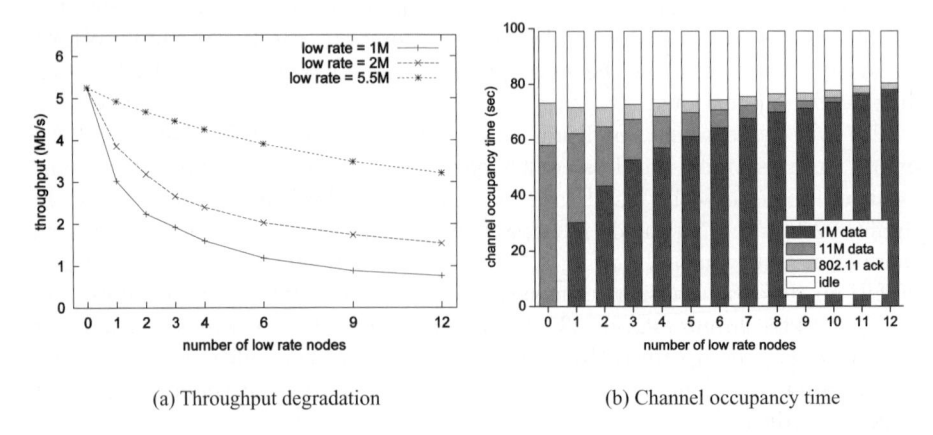

Fig. 1. Performance degradation with multiple transmission rates

for actual data transmission and the other time is idle. In the figure, the portion of 1Mbps data is increased as the number of 1Mbps rate nodes is increased. However, the time portion used by 1Mbps nodes increases exponentially corresponding to the number of 1Mbps nodes. This nonlinearity is because the IEEE 802.11 MAC DCF channel ac-cess method, which is founded on the CSMA/CA protocol, equally distributes the wireless channel based on access probability. The uniform channel access probability guarantees long term fairness when all connections use the same transmission rate. However, if the transmission rates of connections are different, the low rate connection requires more channel resources than the high rate connection to send the same amount of data. Consequently, if the channel access probability is equal, the low rate connection captures the channel much longer than the high rate connection and the fairness is broken. In this situation, the wireless medium is not fully utilized and the total system throughput is considerably degraded even when only few low rate connections are involved in transmissions [13].

2.2 Related Works

In efforts to improve the performance of the IEEE 802.11 WLAN, many researchers have studied the WLAN protocol and have proposed new algorithms. They analyzed system throughput of IEEE 802.11 DCF MAC protocol [13,14,15] and proposed numerous new algorithms to improve the WLAN performance from many points of view such as fairness, service differentiation, and system throughput [5,6,7,8,9,10,11,12].

As noted earlier, all IEEE 802.11 physical specifications support variable data transmission facility at multiple rates. A simple way to select the best physical transmission rate is to change the rates based on the history of successes or failures of previous packet transmissions. The representative algorithm based on this proactive approach is the ARF (Auto Rate Fallback) scheme [5]. The ARF scheme is simple and easy to implement. However, it does not prevent performance degradation when the transmission rates are mixed. Furthermore, the ARF scheme cannot react quickly when the wireless channel condition fluctuates.

The RBAR (Receiver-Based Auto Rate) scheme [6] is another rate adaptation algorithm for improving WLAN system performance. RBAR uses feedback information from the receiver to sense the wireless channel conditions. Due to the more accurate channel estimation, the RBAR scheme yields significant throughput gains compared to the ARF scheme. However, the RBAR scheme also fails to cope with the performance degradation with multiple transmission rates. It does not consider the throughput degradation arising from data transmission of low rate connections.

The OAR (Opportunistic Auto Rate) scheme [7] attempts to maximize the system performance by exploiting a good quality channel via burst packet transmissions. The OAR scheme opportunistically transmits multiple packets in a burst whenever the channel quality is good. Due to the opportunistic gain, the OAR scheme outperforms the RBAR and is enable to handle the performance degradation arising from multiple transmission rates. However, in the OAR, a burst packet transmission is only possible when the packets queued in the network interface have the same destination. There is little performance gain when most stations have packets destined for more than one station such as downstream traffic in an infrastructure topology and all traffic in an ad-hoc topology.

The RBAR and the OAR schemes require a modified RTS / CTS mechanism for channel estimation. Instead of the signal quality of the RTS frames, the SNR (Signal to Noise Ratio) scheme [8] estimates the channel quality using the received signal strength measured from the received frames. Accordingly, the SNR scheme does not require the RTS / CTS mechanism or any change in the current IEEE 802.11 WLAN standard. However, the SNR scheme is a link adaptation scheme only. It does not consider the performance degradation arising from multiple transmission rates.

A simple solution for solving the performance abnormality problem is to combine the OAR and the SNR schemes. By simply combining the OAR and SNR schemes, we can arrive at a new feasible solution for preventing the performance degradation with-out modification of the existing standard. We call this solution the MOAR (Modified OAR) scheme. However, the MOAR scheme still has limitations originated from the original weaknesses of the OAR scheme. Even though the MOAR scheme avoids the performance abnormality with multiple transmission rates, it is only effective when the station transmits packets to only one destination and the serialized packets in the queue head to the same station. This limitation restricts the performance gain considerably in dynamic wireless networks such as ad-hoc networks. Therefore, we propose a new packet transmission strategy for improving IEEE 802.11 WLAN protocol capacity in a multiple rate network.

G. Tan and J. Guttag have proposed TBR (Time Based Regulator) scheme [16] that removes the performance degradation in multiple destined packet environments. However, the TBR scheme does not work in ad-hoc networks. It only runs on the AP and matches just with infrastructure topology. It cannot improve performance when distributed nodes content each other to transmit packets. In addition, the TBR scheme requires the slight modification of the existing MAC standard in case of only existing upstream one-way traffic such as UDP. Therefore, we propose a new packet transmission strategy for improving IEEE 802.11 WLAN protocol capacity in a multiple rate network.

3 Rate-Adapted Transmission Scheme

When multiple stations contend for a channel in the IEEE 802.11 WLAN, the default DCF MAC access method probabilistically gives an equal chance for channel access to all stations. This channel distribution method, however, degrades the system performance severely when stations use multiple transmission rates together. In multiple rate transmissions, the packet transmission at a low rate occupies the channel too long in comparison to packet transmission at a high rate. Therefore, the RAT scheme at-tempts to share the channel based on the occupancy time rather than access opportunity. Accordingly, when multiple stations with different transmission rates contend to obtain the channel, the RAT scheme grants the stations constant channel occupancy time. Thus, the stations at a high rate multiply transmit as many packets as possible within the granted occupancy time. We call this capacity improvement the inter-rate contention gain.

Even in a single station, performance degradation occurs when multiple connections with different rates contend to transmit packets. In the IEEE 802.11 WLAN, a station transmits packets sequentially based on the arrival sequence order. However, when multiple packets at different rates are transmitting in a station, the packet at a low transmission rate occupies the channel for a long time and blocks fast packet transmissions at high transmission rates. As a result, the high rate connection is deprived of its share by the low rate connection and the total system performance of the WLAN is degraded. To avoid this performance degradation, the RAT scheme adopts rate-based queues. Through the rate-based queues, the RAT scheme sends packets adaptively depending on the physical transmission rates in a station. When a packet comes into the network interface, the RAT scheme classifies the packet according to the physical transmission rate of its connection and inserts the packet into one of the rate-based queues. Then, when the station obtains the channel, the RAT packet scheduler

```
01:   while (the data queue is non-empty) {
02:       // T_unit is allowed occupancy time per connection
03:       T_used := 0;      // used occupancy time
04:       rq_s := select_rate_queue();
05:       n_c := number_of_connections(rq_s);
06:       rate_tx := trasmission_rate(rq_s);   // transmission rate of rq_s

07:       do {
08:           f_d := dequeue(rq_s);
09:           T_used += (f_d.length / rate_tx);
10:           enque(que_tx, f_d);
11:           if (head_frame(rq_s).dst != f_d.dst) {
12:               while (que_tx is non-empty) {
13:                   f_tx := dequeue(que_tx);
14:                   do {
15:                       result := transmit(f_tx);
16:                   } while (result != success);
17:                   if (que_tx is non-empty) {
18:                       idle(t_SIFS);
19:                   }
20:                   else {
21:                       idle(t_DIFS);
22:                   }
23:               }   // end of while
24:           }   // end of if
25:       } while (T_used < T_unit * n_c);
26:   }
```

Fig. 2. The RAT scheduler algorithm

selects the appropriate rate queue and transmits multiple packets in the queue up to the channel occupancy time. Consequently, the high rate connection is not interfered with by the low rate connection. We call this capacity improvement the intra-rate contention gain. Fig. 2 describes in detail the RAT algorithm.

4 Simulation Experiments

Through simulations, we compared our RAT scheme with the default DCF MAC access method and the MOAR scheme in the simulations. We modified an NS simulator [17] to follow the IEEE 802.11b WLAN parameters.

4.1 Network Topologies

First, we evaluated the system performances according to the network topologies, an infrastructure topology and an ad-hoc topology, shown in Fig. 3. The infrastructure topology has one AP and many mobile nodes. The AP is located in the center of the mobile nodes and functions as a centralized controller. Thus, the downstream traffic is delivered from the AP to the mobile nodes and the upstream traffic is delivered from the mobile nodes to the AP.

In the infrastructure topology, we measured the total system throughputs with mixed physical transmission rates. For the experiment, the AP connects with 20 mobile nodes. The default rate of each connection is 11Mbps, the highest physical transmission rate. The rate for the low connection is 1Mbps, the lowest physical transmission rate. Each connection sends 300Kbits UDP data per second. The unit of channel occupancy time for the MOAR and RAT schemes is 8ms.

Fig. 4(a) depicts the system throughput for upstream traffic in the infrastructure topology. In the figure, the vertical axis represents the total system throughput and the horizontal axis represents the ratio of the number of low rate nodes to the number of total nodes. As the results indicate, the default DCF MAC access method degrades system throughputs severely in proportion to the number of low rate nodes. This drastically decreases the system performance even when only a few nodes send data at 1Mbps. However, the MOAR and

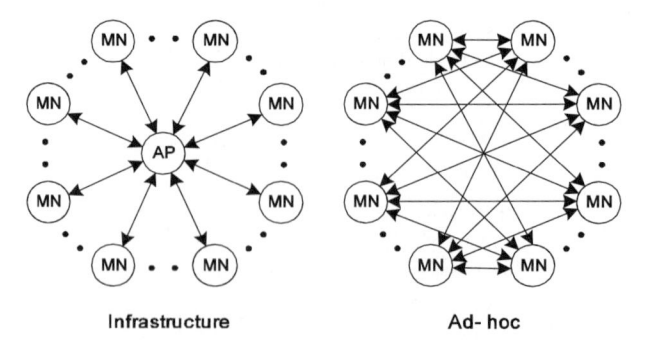

Fig. 3. Network topologies

(a) Upstream traffic

(b) Downstream traffic

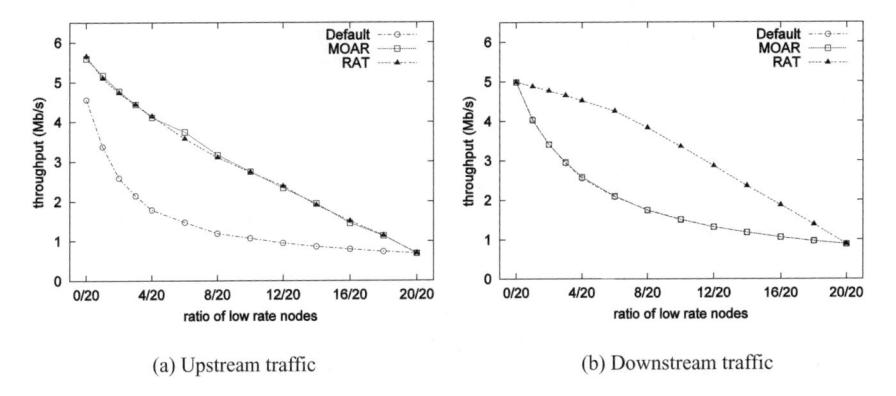

Fig. 4. System throughput in infrastructure topology

RAT scheme are not seriously affected by the low rate nodes. They yield smooth degradation of the system throughput depending on the number of 1Mbps rate nodes. This is because, when multiple stations with different rates contend for the channel, the MOAR and RAT schemes extract the inter-rate contention gain by distributing the channel based on the occupancy time.

Fig. 4(b) displays the system throughput for downstream traffic in the infrastructure topology. In this experiment, the AP transmits packets to all mobile nodes. Similar to the upstream experiment, the default DCF method does not have good performance in the downstream traffic. Moreover, the MOAR scheme shows poor performance corresponding with that of the default DCF method. However, the RAT scheme displays good performance in comparison to the other two schemes. As noted earlier, the MOAR scheme distributes the channel based on the occupancy time. Thus, it has good performance when faced with inter-rate contention. The MOAR scheme, however, does not consider the case where the multiple connections at different rates contend to send packets in a single node. It does not improve system performance at all when connections in a single station contend to send packets with multiple transmission rates. On the other hand, the RAT scheme adaptively transmits packets based on the physical transmission rates. It sends multiple packets for high rate connection using rate-based queue management even in a single station. Accordingly, the RAT scheme improves protocol capacity when faced with an intra-rate contention environment as well as an inter-rate contention environment.

For the next simulation, we evaluated the system performance in the ad-hoc net-work topology. In the ad-hoc topology, the mobile node transmits data to other mobile nodes in distributed manner without the centralized AP. Thus, every node simultaneously sends packets to multiple nodes at different rates. Fig. 5 shows the results of total system throughput in the fully connected ad-hoc network topology. As can be seen in the figure, the results are similar to those of the downstream traffic in the infra-structure topology. The default DCF method and the MOAR scheme show poor performance. However, the RAT scheme shows better performance than the other schemes even in the ad-hoc network topology.

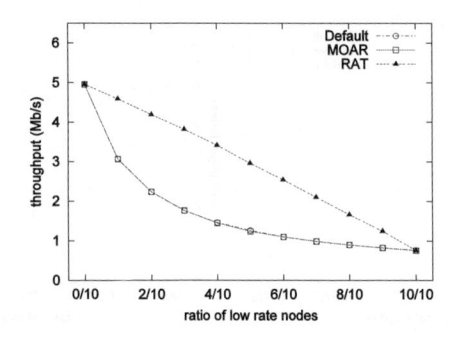

Fig. 5. System throughput in ad-hoc network topology

4.2 TCP Traffic

In this section, we evaluated the system performance with TCP traffic type using FTP. Fig. 6(a) shows the system throughputs for the upstream TCP traffic in the infrastructure network topology. Hence, the mobile nodes transmit TCP DATA packets to the AP in upstream and the AP transmits TCP ACK packets to the mobile nodes in down-stream. Similar to the UDP experiment, the throughputs of the default DCF method sink rapidly in proportion to the increase of low rate nodes. However, contrary to the upstream UDP experiment, the upstream TCP throughput of the MOAR scheme is poor and similar to that of the default DCF method. This is because TCP is a bidirectional protocol and the throughputs of both directions influence the total TCP performance mutually. In TCP, a TCP ACK packet is generated by successfully transmitted a TCP DATA packet, and the next TCP DATA packet is also generated by successfully transmitted a TCP ACK packet. Thus, in order for a mobile node to transmit multiple TCP DATA packets in a burst for upstream, the AP should transmit multiple TCP ACK packets at once to a mobile node in downstream. However, when we use the MOAR scheme, the AP generally transmits TCP ACK packets one by one to all mobile nodes. This is because the TCP ACK packets from the AP differently head to multiple destinations in downstream. Consequently, a mobile node in the MOAR scheme does not receive multiple TCP ACK packets simultaneously and does not send multiple TCP DATA packets using the burst packet transmission mechanism. As a result, the throughput of the MOAR scheme fails to exceed that of the default DCF method. However, the throughput of the RAT scheme is much better than that of the MOAR scheme and the default DCF method. This is because the RAT scheme makes it possible for the AP to transmit multiple TCP ACK packets even in multiple destined connections. Therefore, TCP DATA packets are also delivered efficiently using the burst packet transmission mechanism for an upstream TCP traffic environment.

Fig. 6(b) depicts the system throughputs for the downstream TCP traffic. Similar to the UDP experiment, the throughputs of the default and the MOAR scheme sink rap-idly in proportion to the increase of low rate nodes. However, the RAT scheme shows relatively good throughput in all cases. The total system

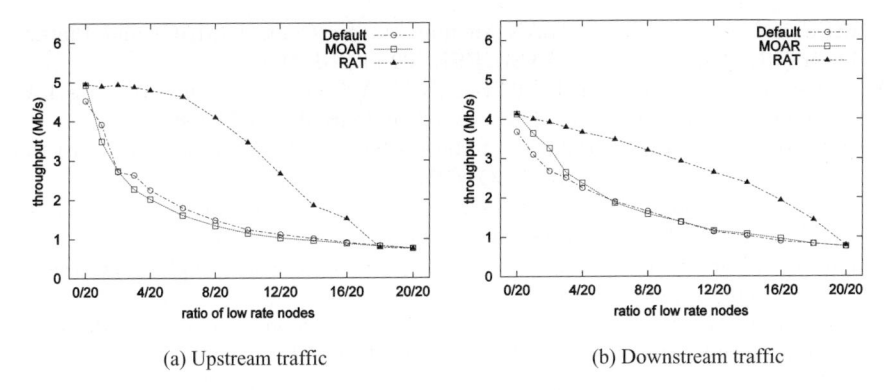

Fig. 6. System throughput for TCP traffic in infrastructure topology

(a) Upstream traffic (b) Downstream traffic

throughput is slightly de-graded in comparison to that of the UDP case. This is because the TCP ACK consumes a portion of the network resources.

5 Conclusion

In this paper, we proposed a new rate control algorithm, the RAT scheme. The RAT scheme distributes channel resources based on the channel occupancy time. It gives equal time shares to all stations. In addition, the RAT scheme guarantees the channel occupancy time even in a single station by adopting rate-based queue management. As a result, the RAT scheme improves protocol capacity in the face of intra-rate contentions as well as inter-rate contentions.

Through simulations we showed that the RAT scheme is superior to the default DCF method and the MOAR scheme. The RAT scheme displays good performance in all network topologies. Moreover, it uniquely enhances the TCP performance among the compared schemes. In addition, the RAT scheme is practical and easy to implement. This is because the RAT scheme does not require modification of the existing IEEE 802.11 specification. Consequently, the RAT scheme is suitable for multiple rate networks, especially ad-hoc networks where topologies dynamically change.

References

1. IEEE Std 802.11b-1999: Part 11: Wireless LAN Medium Access Control (MAC) and Physical Layer (PHY) specifications: High-speed Physical Layer Extension in the 2.4GHz Band. Supplement to ANSI/IEEE Std 802.11. (1999)
2. IEEE Std 802.11a-1999: Part 11: Wireless LAN Medium Access Control (MAC) and Physical Layer (PHY) specifications: High-speed Physical Layer Extension in the 5GHz Band. Supplement to ANSI/IEEE Std 802.11. (1999)
3. IEEE Std 802.11g-2003: Part 11: Wireless LAN Medium Access Control (MAC) and Physical Layer (PHY) specifications: Further Higher Data Rate Extension in the 2.4GHz Band. Amendment to IEEE 802.11 Std. (2003)

4. IEEE Std 802.11: Wireless LAN Medium Access Control (MAC) and Physical Layer (PHY) specifications. ANSI/IEEE 802.11 Std. (1999)
5. Kamerman, A., Monteban, L.: WaveLAN-II: A High-Performance Wireless LAN for the Unlicensed Band. Bell Labs Technical Journal. (1997) 118–133
6. Holland, G., Vaidya, N., Bahl, P.: A Rate-Adaptive MAC Protocol for Multi-Hop Wire-less Networks. ACM MOBICOM'01. (2001) 236–251
7. Sadeghi, B., Kanodia, V., Sabharwal, A., Knightly, E.: Opportunistic Media Access for Multirate Ad Hoc Networks. ACM MOBICOM'02. (2002) 24–35
8. Pavon, J., Choi, S.: Link Adaptation Strategy for IEEE 802.11 WLAN via Received Signal Strength Measurement. IEEE ICC'03. (2003)
9. Cantieni, G., Ni, Q., Barakat, C., Turletti, T.: Performance Analysis under Finite Load and Improvements for Multirate 802.11. Computer Communication Journal. (2005)
10. Haratcherev, I., Langendoen, K.: Hybrid Rate Control for IEEE 802.11. ACM MOBIWAC'04. (2004)
11. Heusse, M., Rousseau, F., Guillier, R., Duda, A.: Idle Sense: An Optimal Access Method for High Throughput and Fairness in Rate Diverse Wireless LANs. ACM SIGCOMM'05. (2005)
12. Hoffmann, C., Manshaei, M., Turletti, T.: CLARA: Closed-Loop Adaptive Rate Allo-cation for IEEE 802.11 Wireless LANs. IEEE WirelessCom'05. (2005)
13. Heusse, M., Rousseau, F., Berger-Sabbatel, G. Duda, A.: Performance Anomaly of 802.11b. IEEE INFOCOM'03. (2003)
14. Bianchi, G.: Performance Analysis of the IEEE 802.11 Distributed Coordination Function. IEEE Journal on Selected Areas in Communications. (2000)
15. Foh, C., Zukerman, M.: Performance Analysis of the IEEE 802.11 MAC Protocol. European Wireless 2002. (2002)
16. Tan, G., Guttag, J.: Time-based Fairness Improves Performance in Multi-rate Wireless LANs.The USENIX Annual Technical Conference. (2004)
17. http://www.isi.edu/nsnam/ns/

Comparison of Different Methods for Next Location Prediction

Jan Petzold, Faruk Bagci, Wolfgang Trumler, and Theo Ungerer

Institute of Computer Science, University of Augsburg,
Eichleitnerstr. 30, 86159 Augsburg, Germany
{petzold, bagci, trumler, ungerer}@informatik.uni-augsburg.de

Abstract. Next location prediction anticipates a person's movement based on the history of previous sojourns. It is useful for proactive actions taken to assist the person in an ubiquitous environment. This paper evaluates next location prediction methods: dynamic Bayesian network, multi-layer perceptron, Elman net, Markov predictor, and state predictor. For the Markov and state predictor we use additionally an optimization, the confidence counter. The criterions for the comparison are the prediction accuracy, the quantity of useful predictions, the stability, the learning, the relearning, the memory and computing costs, the modelling costs, the expandability, and the ability to predict the time of entering the next location. For evaluation we use the same benchmarks containing movement sequences of real persons within an office building.

1 Introduction

Can the movement of people working in an office building be predicted based on room sequences of previous movements? In our opinion people follow some habits, but interrupt their habits irregularly, and sometimes change their habits. Moreover, moving to another office fundamentally changes habits too. Thus location prediction methods need to exhibit some features: high prediction accuracy, a short training time, retention of prediction in case of irregular habitual interrupts, but an appropriate change of prediction in case of habitual changes.

Location predictions with such features could be used for a number of applications in ubiquitous and mobile environments.

- Smart doorplates that are able to direct visitors to the current location of an office owner based on a location-tracking system and predict if the office owner is soon coming back [14].
- Similarly, next location prediction within a smart building can be used to prepare the room which is presumably entered next by a habitant, e.g. by phone call forwarding.
- Outdoor movement patterns can be used to predict the next region a person will enter.
- Elevator prediction could anticipate at which floor an elevator will be needed next.

– Routing prediction for cellular phone systems may predict the next radio cell a cellular phone owner will enter based on his previous movement behaviour.

We considered the first application in more detail and used benchmarks with movement data of four persons over several months. The benchmarks are called Augsburg Indoor Location Tracking Benchmarks. They are publicly available [9], and are applied to evaluate several prediction techniques and to compare the efficiency of these techniques with exactly the same evaluation set-up and data.

Our aim is to investigate how far machine learning techniques can dynamically predict room sequences and time of room entry independent of additional knowledge. Of course the information could be combined with contextual knowledge as e.g. the office time table or personal schedule of a person, however, in this paper we focus on dynamic techniques without contextual knowledge.

Time of arrival at the predicted location depends on the sojourn time at the current location plus the rather constant time to move to the predicted location. The sojourn time was modelled into the presented Bayesian network. We tested also a time prediction which calculated the mean and the median of the previous sojourn times within a location. The best results were reached by the median. The time prediction is independent of the location prediction method and can easily be combined with any of the regarded methods. Therefore we restrict this comparison to location prediction only.

Several prediction techniques are proposed in literature — namely Bayesian networks, Markov models or Hidden Markov models, various neural network approaches, and the state predictor methods. The challenge is to transfer these algorithms to work with context information. In this paper we choose five approaches, a dynamic Bayesian network, a multi-layer perceptron, an Elman net, a Markov predictor, and a state predictor. In the case of the Markov predictor and the state predictor we use additionally a version which is optimized by confidence estimation.

There are a lot of methodological problems for a fair comparison of such diverse methods. The models are different and hard to compare. We chose the same set-up to model all methods and for each method the best model that we could find. Moreover we had the choice either to combine all persons within a single model thus potentially making improvements by detecting correlations between person movements or to model each person separately. We chose the latter simpler model because simulations with the combined model using the Augsburg Benchmarks showed no improvements.

The main criterion for comparison is the average prediction accuracy of the different methods. Another question concerns the model and the modelling costs of the technique. Which parameters exist and influence the model? What happens if one parameter is changing? We call this the stability of the techniques. Can the model simply be extended by more or other locations? The answer to this question allows to assess how well the model can be transferred to other applications.

Further interesting questions concern the efficiency of training of a predictor, before the first useful predictions can be performed, and of retraining, i.e. how long it takes until the predictor adapts to a habitual change and provides again useful predictions. Predictions are called useful if a prediction is accurate with a certain confidence level. Moreover, memory and performance requirements of a predictor are of interest in particular for mobile appliances with limited performance ability and power supply.

The next section states related work on context prediction. Section 3 introduces shortly the five approaches and the applied location models. For detailed information about the basic techniques use the stated references. Section 4 gives the evaluation results. The paper ends with the conclusions.

2 Related Work

The Adaptive House project [7] of the University of Colorado developed a smart house that observes the lifestyle and desires of the inhabitants and learned to anticipate and accommodate their needs. Occupants are tracked by motion detectors and a neural network approach is used to predict the next room the person will enter and the activities he will be engaged. Patterson et al. [8] presented a method of learning a Bayesian model of a traveller moving through an urban environment based on the current mode of transportation. The learned model was used to predict the outdoor location of the person into the future.

Markov chains are used by Kaowthumrong et al. [5] for active device selection. Ashbrook and Starner [1] used location context for the creation of a predictive model of user's future movements based on Markov models. They propose to deploy the model in a variety of applications in both single-user and multi-user scenarios. Their prediction of future location is currently time independent, only the next location is predicted. Bhattacharya and Das [2] investigate the mobility problem in a cellular environment. They deploy a Markov model to predict future cells of a user. An architecture for context prediction was proposed by Mayrhofer [6] combining context recognition and prediction. Active LeZi [4] was proposed as good candidate for context prediction.

There are several publication of our group which present next location prediction in an office building. In [10] we proposed the basic state predictor technique which is similar to the Markov predictor, but an automaton is used for the prediction. In [11] an enhancement by confidence estimation techniques is presented. Vintan et al. [15] applied a multi-layer perceptron and Petzold et al. [12] proposed a dynamic Bayesian network to predict indoor movements of several persons.

The contribution of this paper is the comparison of five different prediction methods including the new Elman net approach and the confidence estimation applied to the Markov predictor. According to our knowledge no comparative studies of different methods with the same evaluation setups and benchmarks exist.

3 Prediction Methods

Figure 1 shows the next location prediction principle which is used by each investigated model. The input consists only of the sequence of the last visited locations and the entry time of these locations. The output is the possible next location and the appropriate entry time.

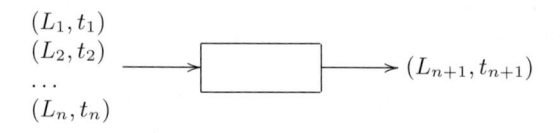

$$(L_1, t_1)$$
$$(L_2, t_2)$$
$$\ldots$$
$$(L_n, t_n)$$
$$\longrightarrow \boxed{} \longrightarrow (L_{n+1}, t_{n+1})$$

Fig. 1. Next location prediction

Dynamic Bayesian Network

In order to predict the next location of a person, a dynamic Bayesian network was chosen. Additionally the time is predicted when the person is probably entering the next location. In different simulations we looked for the best settings [12]. We detected that the prediction of next location is independent of the time parameter like the time of day and the weekday. Therefore we chose this proposed dynamic Bayesian network without these time dependencies for the comparison. As history we elected 2 for a better comparison based on similar memory costs.

Multi-Layer Perceptron

For next location prediction we chose the simplest multi-layer perceptron with one hidden layer and used a modified back-propagation algorithm for learning [15]. In principal each location would be represented by a single input and a single output neuron. However, we chose a binary encoding because it saves computing costs. This fact is interesting for mobile devices which must achieve some energy and real-time restrictions. The optimal parameter values for the network structure and the learning algorithm were determined by many simulation runs and are summarized in table 1.

Table 1. Optimal parameter values of the multi-layer perceptron

parameter	investigated values	optimal value
network structure		
history	[1;6]	2
number of hidden neurons	{5;7;...;15}	9
learning algorithm		
threshold	{0.1;0.3;...;0.9}	0.1
learning rate	[0.05;0.30]	0.10
number of backward steps	[1;5] and unlimited	1

Elman Net

The Elman net is another neural network method which expands the multi-layer perceptron by another hidden layer – the context layer. The context neurons provide storage for the activation states of the hidden neurons. This generates a dependency between two propagations within the net, since the hidden neurons get information from the input and the context neurons across the weighted connections to perform the next step. The number of the context neurons corresponds with the number of hidden neurons. To find the optimal parameters of the net many simulations were performed. Table 2 shows the investigated and the optimal values of the parameters separated in parameter of the network structure and the learning algorithm. Since the Elman net is a recurrent network, the information about previous locations is modelled in the context cells. Therefore the history consists only of the current location.

Table 2. Optimal parameter values of the Elman net

parameter	investigated values	optimal values
network structure		
encoding	binary, one to one	one to one
number of hidden neurons	[5;20]	5
history	[1;5]	1
learning algorithm		
initialization	random, fix	fix
activation function	$\tanh(x)$, $\frac{1}{1+\exp(-x)}$	$\tanh(x)$
learning cycles	[5;150]	31
learning rate η	[0.1;0.7]	0.1
momentum α	[0.00;0.05;...;0.95]	0.00

Markov Predictor

Markov models seem a good approach for the next location prediction based on location histories. A Markov model regards a pattern of the last visited locations of a user to predict the next location. The length of the regarded pattern is called the order. Thus a Markov model with order 3 uses the last three visited locations. For all patterns the model stores the probabilities of the next location which is calculated from the whole sequence of the visited locations by the user. A simple Markov model is the Markov predictor [3,13]. A Markov predictor stores for every pattern the frequencies of the next locations. For the comparison we chose an order of 2. Furthermore we will compare a Markov predictor which is optimized by confidence estimation [11].

State Predictor

A disadvantage of the Markov predictor is its bad relearning capability because of the frequency counter. After a habit change the new habit must be followed as often as the previous habit before the prediction is changed. The state predictors [10] prevent this problem. They use a finite automaton which is called two-state predictor for every pattern thus replacing the frequency counter of the Markov predictor. A state predictor with order 2 is used in the comparison.

The basic state predictor method can be significantly improved by some confidence estimation techniques [11]. One of the proposed methods, the confidence counter method, is independent of the used prediction algorithm. This method estimates the prediction accuracy with a saturation counter. Figure 2 shows a two-bit counter that consists of 4 states.

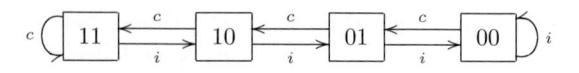

Fig. 2. Confidence counter

The initial state is state 10. Let s be the current state of the confidence counter. If a prediction result is proved as correct (c) the counter will be incremented, that means the state graph changes from state s into the state $s + 1$. If $s = 11$ the counter keeps the state s. Otherwise if the prediction is incorrect (i) the counter switches into the state $s - 1$. If $s = 00$ the counter keeps the state s. If the counter is in the state 11 or 10 the predictor is assumed as confident, otherwise the predictor is unconfident and the prediction result will not be supplied.

For the state predictor the prediction accuracy will be considered separately for every pattern. The confidence counter can also be applied with other techniques, in the evaluation a Markov predictor using the counter will be considered.

4 Evaluation

To evaluate the five techniques we chose the Augsburg Indoor Location Tracking Benchmarks taken from the Context Database of the University of Linz [9]. These benchmarks consist of two sets, the summer and the fall data. The used benchmarks contain the movements of four persons in an office building. The prediction accuracy is calculated for every person for all predictions from all rooms except the own office. For our comparison we tried to use models with similar memory costs. Therefore we didn't always choose the best setting for every technique. In fact we elected history 2 for the Markov predictor and the state predictor, which perform better with longer history but at the expense of large tables.

In the following we will compare all techniques on the basis of different criterions. The results are summarized in table 3.

Table 3. Comparison on the basis of the criterions

	Bayesian network	multi-layer perceptron	Elman net	Markov predictor	state predictor	Markov predictor with counter	state predictor with counter
accuracy (%)	78.82	76.45	79.68	76.53	70.89	81.14	81.88
(quantity (%))	(89.89)	($\approx$ 100)	(100)	(90.47)	(90.47)	(78.40)	(74.38)
stability (%)	29.67	32.59	71.57	24.67	29.97	24.95	23.99
learning	fast	slow	slow	fast	fast	fast	fast
relearning	slow	slow	slow	slow	fast	slow	fast
memory (bit)	6,500	3,880	7,215	36,960	2,730	37,380	3,150
computing costs	inefficient chain rule	training until $E < t$, otherwise one propagation	training over many learning cycles, otherwise one propagation	table look-up	table look-up	table look-up	table look-up
modelling costs	medium	high	high	low	low	low	low
expandability	yes	no	no	yes	yes	yes	yes
time prediction	integrated	parallel	parallel	parallel	parallel	parallel	parallel

Prediction accuracy. The prediction accuracy is calculated with the fall data; the summer data is used for the training. We assume that a prediction is needed after every location change. That means the number of requested predictions p is equal to the number of location changes. The Bayesian network, the Markov predictor and the state predictor cannot predict the next location if the current pattern occurs the first time. The number of predictions which cannot be delivered by these three techniques were denoted by p_n. In contrast the neural networks deliver a prediction if the code of the output vector corresponds to a location. Thus the Elman net predicts always a location ($p_n = 0$). Now we can determine the number of deliverable predictions $p_d = p - p_n$. Let c be the number of correct predictions then the prediction accuracy a is calculated as follows:

$$a = \frac{c}{p_d}$$

It isn't essential to make a prediction in our application, rather a prediction is an added value. Therefore we consider in the calculation of the accuracy only predictions which provide a result.

Table 3 shows the average prediction accuracy of the four persons for all techniques. If we consider the five techniques without confidence counter enhancement, the Elman net reaches the highest average prediction accuracy. If we consider the state predictor and Markov predictor using the confidence counter and compare them with all other methods, the state predictor with confidence counter delivers now the best accuracy.

Quantity. The number of deliverable predictions p_d is smaller than the number of requested predictions p. This gap can be determined by the quantity q:

$$q = \frac{p_d}{p}$$

The Elman net reaches a quantity of 100% because the net produces always an output vector. The quantity of the multi-layer perceptron is nearly 100% since not every code of the output vector corresponds with a location. The Bayesian network, the Markov predictor and the state predictor reach nearly the same quantity. With an optimization like the confidence estimation the quantity decreases.

Stability. The stability shows the impact of the change of a parameter. For the Bayesian network the history and the time parameters are a possibility to optimize the prediction accuracy. The multi-layer perceptron and the Elman net hold a multitude of parameters which can influence the prediction accuracy. Therefore the Elman net shows the worst stability. The Markov predictor and the state predictor give only the possibility to choose the order. Table 3 shows as stability the difference between the minimum and the maximum of the prediction accuracies reached with different parameters.

Learning. The learning phase of the neural networks takes a long time since the networks must be trained before they can be effectively used. The Bayesian network, the Markov predictor and the state predictor with and without confidence counter could make a prediction already after the second occurrence of a pattern.

Relearning. The neural networks, the Markov predictor without and with confidence counter need a long time for relearning. The Bayesian network relearns also slow. The state predictor with and without confidence counter relearns after two changes the new habit.

Memory costs. For the memory costs we calculated the minimal number of bits which will be needed to store the current state of the technique. For the evaluation all models were chosen to exhibit similar memory costs. In general, the memory costs of both neural networks are very low and independent from the number of location changes of a person. The Bayesian network needs only a small memory space, but the memory depends on the number of location changes. The state predictor requires the least memory. Against this the Markov predictor has the highest memory costs since the Markov predictor stores all frequencies. The costs of both predictors are dependent on the number of location changes. The confidence counter arises the costs insignificantly. Table 3 shows the memory costs for an upper limit of 500 location changes.

Computing costs. Because of the training process the computing costs of the neural networks are very high. The Elman net needs many learning cycles and the multi-layer perceptron will be trained until the error is less than a threshold. The computing costs during the use of both neural networks are low because both execute only one backward propagation. The Bayesian network calculates the probabilities by the chain rule resulting in relatively high computation costs. The computing costs of the Markov predictor and the state predictor with and without the confidence counter consist of one table look-up.

Modelling efforts. For modelling the Bayesian network possible dependencies of the used variables must be extracted from the available data. Both neural networks require a high effort for modelling generated by the search for the optimal parameters. The costs for modelling the Markov predictor and the state predictor are low. A decision will only be needed concerning the length of the order. If a confidence counter is used the number of the counter states and the barrier must be determined.

Expandability. The expandability means the possibility to use the model with more locations. In the used benchmarks there are 15 locations. If we use a scenario with locations like the cells in a mobile network, the neural network models cannot be reused. A new modelling process with search for the optimal parameter is necessary. That means the neural network models cannot be reused without additional costs in another application. The Bayesian network, the Markov predictor and the state predictor with and without confidence counter can be expanded for more context without additional costs.

5 Conclusion

The paper compared five prediction techniques on the basis of different criterions. The comparison of the different techniques showed that there isn't an ultimative prediction technique. The user must decide which is the most important criterion for the application.

The Elman net reached the highest prediction accuracy, since it is a recurrent neural network which is affected by previous inputs. But both neural networks require high modelling costs, additional costs to expand for more contexts, and show the lowest stability. If the time prediction is the most important criterion the Bayesian network must be chosen. The state predictor should be applied if the prediction accuracy or the memory costs are the main facts. Compared to the Markov predictor, the state predictor relearns faster and uses less memory. The use of the confidence counter improves the prediction accuracy of state and Markov predictors.

Next step could be the test of a hybrid predictor which uses different techniques in parallel. A selector within the hybrid predictor selects the estimated best prediction among these different predictors. With the hybrid predictor the advantages of the different methods can be joined. A further question is: can the confidence estimation also improve the prediction accuracy of the other approaches. In this paper we considered only next location prediction. A further investigation should be to expand the techniques to predict locations at a certain time in future.

References

1. Daniel Ashbrook and Thad Starner. Using GPS to learn significant locations and predict movement across multiple users. *Personal and Ubiquitous Computing*, 7(5):275–286, 2003.
2. Amiya Bhattacharya and Sajal K. Das. LeZi-Update: An Information-Theoretic Framework for Personal Mobility Tracking in PCS Networks. *Wireless Networks*, 8:121–135, 2002.
3. I-Cheng K. Chen, John T. Coffey, and Trevor N. Mudge. Analysis of Branch Prediction via Data Compression. In *ASPLOS VII*, pages 128–137, Cambridge, Massachusetts, USA, October 1996.
4. Karthik Gopalratnam and Diane J. Cook. Active LeZi: An Incremental Parsing Algorithm for Sequential Prediction. In *Sixteenth International Florida Artificial Intelligence Research Society Conference*, pages 38–42, St. Augustine, Florida, USA, May 2003.
5. Khomkrit Kaowthumrong, John Lebsack, and Richard Han. Automated Selection of the Active Device in Interactive Multi-Device Smart Spaces. In *Workshop at UbiComp'02: Supporting Spontaneous Interaction in Ubiquitous Computing Settings*, Gothenburg, Sweden, 2002.
6. Rene Mayrhofer. An Architecture for Context Prediction. In *Advances in Pervasive Computing*. Austrian Computer Society (OCG), April 2004.
7. Michael C. Mozer. The Neural Network House: An Environment that Adapts to its Inhabitants. In *AAAI Spring Symposium on Intelligent Environments*, pages 110–114, Menlo Park, CA, USA, 1998.
8. Donald J. Patterson, Lin Liao, Dieter Fox, and Henry Kautz. Inferring High-Level Behavior from Low-Level Sensors. In *5th International Conference on Ubiquitous Computing*, pages 73–89, Seattle, WA, USA, 2003.
9. Jan Petzold. Augsburg Indoor Location Tracking Benchmarks. Context Database, Institute of Pervasive Computing, University of Linz, Austria. http://www.soft.uni-linz.ac.at/Research/Context_Database/index.php, January 2005.
10. Jan Petzold, Faruk Bagci, Wolfgang Trumler, and Theo Ungerer. Global and Local Context Prediction. In *Artificial Intelligence in Mobile Systems 2003 (AIMS 2003)*, Seattle, WA, USA, October 2003.
11. Jan Petzold, Faruk Bagci, Wolfgang Trumler, and Theo Ungerer. Confidence Estimation of the State Predictor Method. In *2nd European Symposium on Ambient Intelligence*, pages 375–386, Eindhoven, The Netherlands, November 2004.
12. Jan Petzold, Andreas Pietzowski, Faruk Bagci, Wolfgang Trumler, and Theo Ungerer. Prediction of Indoor Movements Using Bayesian Networks. In *Location- and Context-Awareness (LoCA 2005)*, Oberpfaffenhofen, Germany, May 2005.
13. Sheldon M. Ross. *Introduction to Probability Models*. Academic Press, 1985.
14. Wolfgang Trumler, Faruk Bagci, Jan Petzold, and Theo Ungerer. Smart Doorplate. In *First International Conference on Appliance Design (1AD)*, Bristol, GB, May 2003. Reprinted in Pers Ubiquit Comput (2003) 7: 221-226.
15. Lucian Vintan, Arpad Gellert, Jan Petzold, and Theo Ungerer. Person Movement Prediction Using Neural Networks. In *First Workshop on Modeling and Retrieval of Context*, Ulm, Germany, September 2004.

SEER: Scalable Energy Efficient Relay Schemes in MANETs

Lin-Fei Sung, Cheng-Lin Wu, Yi-Kai Chiang, and Shyh-In Hwang

Department of Computer Science and Engineering
Yuan Ze University, Chung-Li 320, Taiwan
{lfsung, clwu, ykchiang, shihin}@mmlab.cse.yzu.edu.tw

Abstract. In Mobile Ad Hoc Networks (MANETs), *broadcasting* is widely used to support many applications. Several adaptive broadcast schemes have been proposed to reduce the number of rebroadcasting, and can consequently reduce the chance of contention and collision among neighboring nodes. In practice, broadcasting is power intensive especially in dense networks. Thus, a good energy-efficient relay scheme should be able to further maximize the system lifetime without sacrificing the reachability of broadcasting. In this paper, we propose two Scalable Energy Efficient Relay (SEER) schemes that use probabilistic approaches to achieve higher performance and to prolong the system lifetime. In the schemes, each node uses some energy-based heuristic method to independently determine an appropriate rebroadcast probability. Nodes with more residual energy are responsible for forwarding more broadcast messages. One important feature is that such heuristic knowledge is obtained by self-contained local operation. To further improve the effectiveness of broadcasting, we also study how to dynamically adjust the rebroadcast probability according to node mobility. The simulation results show that our proposed approach outperforms the related scheme when the number of broadcast messages, broadcast reachability, and system lifetime are taken into consideration altogether.

1 Introduction

A mobile ad hoc network (MANET) is defined as a collection of mobile nodes where each node is free to move around. In a MANET, broadcasting is an important communication operation for route discovery, address resolution, and many other network services. For instance, on-demand routing protocols such as AODV [9] and DSR [4] use the broadcast operation to disseminate control packets (e.g., the request of discovering a new route to a destination) for maintaining routing-related information at each node. The most straightforward way of broadcasting is by flooding. However, the radio signals are likely to overlap with each other in a geographical area. Broadcasting by blind flooding suffers from the increasing of serious redundancy, contention, and collision, which is known as a broadcast storm problem [8].

Some works [2], [8], [11] have investigated to improve the effectiveness of broadcasting in MANETs. Despite the optimization effort to reduce rebroadcast messages, the approaches mentioned above fail to take energy issues into consideration. The following requirements concerning how to consume energy in an efficient way are important in broadcast protocols. First, it should minimize the number of rebroadcast messages on one hand, while still maintaining good latency and reachability on the other hand. Then, energy consumption situation should be considered at each node when making decisions about whether to rebroadcast the received messages. A simple idea is that nodes with more battery power should be responsible for forwarding more data in behalf of its neighbors. This implies that nodes with lower residual energy can decide to sleep to save their precious energy.

In this paper, we address three important issues on designing an energy-efficient broadcast protocol based on probabilistic schemes. First, the knowledge of global network energy consumption should be available for reference at each node. Here, we use self-contained local operations to approximate the average network energy. Note that nodes should not need to know information about neighbors multiple hops away for our calculating process. Second, each node can compare its residual energy with such maintained energy-based knowledge to determine an appropriate rebroadcast probability based on the principle that nodes with more residual energy are responsible for forwarding more broadcast messages. Third, the rebroadcast probability at a node can be adjusted according to node distribution and node mobility to further improve the effectiveness of broadcasting.

The remainder of this paper is organized as follows: Section 2 gives a brief review of related work. Section 3 presents a detailed description of our SEER schemes. Section 4 provides simulation results to compare the performance of our methods with that of other existing scheme. Finally, conclusions are drawn in Section 5.

2 Related Work

The efficiency of broadcasting protocol can significantly affect the performance of many applications in MANETs. Some works [2], [8], [11] have investigated the inefficiency problem of broadcasting by blind flooding. When node density is high, blind flooding approach may cause (1) redundant transmissions, (2) higher collision rate, and (3) congestion of wireless medium that seriously impair the performance of the entire network. In this section, we briefly review some adaptive broadcast techniques that attempt to minimize the number of rebroadcast messages while maintaining good latency and reachability. These methods can be categorized into three groups: probabilistic, counter-based, and area-based methods.

In simple probabilistic method [8], a mobile node rebroadcasts received messages with a fixed probability P. Clearly, when $P = 1$, this method is equivalent to flooding. [2] follows from results in percolation theory [7] that probabilistic approaches exhibit a certain type of bimodal behavior in sufficiently large networks: in some executions, the broadcast message dies out quickly and hardly any node gets it; in the remaining executions, a substantial fraction of the nodes gets the message. It is also demonstrated that the optimal rebroadcast probability is around 0.65. [11] argues that this value is not likely to be globally optimal and attempt to dynamically adjust the rebroadcasting probability with the node distribution and node movement.

Besides probabilistic methods, Ni et al. [8] introduced a counter-based approach, in which a counter is used to record the number of receiving the same message. A mobile node inhibits the rebroadcast when the counter is larger than a given threshold. The more copies a node receives indicates the higher chance of its neighbors having already received the same message, and more likely it is a rebroadcast redundant. In their approach, a random assessment delay (RAD) is initiated for counting the number of received copies of the current message. It is obvious that this approach is not suitable for delay-sensitive applications. Ni et al. [8] also discussed area-based schemes, including distance-based and location-based approaches. In distance-based approach, a node may hear the same message several times. If the distance to the nearest node is smaller than some distance threshold D, the rebroadcast transmission is canceled. In location-based approach, GPS (Global Positioning System) receivers [5] is used to assist for calculating an additional area. This value is compared to a predefined coverage threshold $A(0 < A < 0.61)$ to determine whether the rebroadcast should be carried on or not.

Although many broadcast protocols have been proposed to reduce redundant rebroadcast messages, most of them do not take energy consumption into account. When several nodes drain of power due to unbalanced energy consumption, it may lead to network partition and shorten the network lifetime. In this paper, we address this problem by combining the probabilistic approaches and energy consumption balancing to maximize the system lifetime while maintaining a high reachability. Our energy-efficient relay schemes adopt the strategy that nodes with more residual energy are responsible for forwarding more broadcast messages. Besides, we also utilize neighbor connectivity information to dynamically determine an appropriate rebroadcasting probability for various network topologies.

3 SEER Design

One solution to maximize long-term network lifetime for frequent broadcast operation over entire network is to inhibit some nodes with lower residual energy from unnecessary rebroadcasting. We present two schemes to do so. In the first scheme, we accumulate a network-wide energy-related knowledge to assist its rebroadcast decision. And, the second scheme further exploits neighbor connectivity information to improve the overall broadcast throughput.

3.1 Network-Wide Energy-Related Heuristic

Intuitively, for energy conservation purpose, nodes with relatively higher residual energy should be responsible for forwarding more broadcast messages. This implies that nodes with relatively lower residual energy can decide to sleep to save their precious energy. However, in fact, it is hard for a node to accurately determine whether its residual energy is relatively higher or lower than most others. Hence, it is desirable if some network-wide energy-related heuristic can be maintained at each node to help independently distinguish its relative energy level from others. To satisfy the above requirement, we propose an energy-based diffusion algorithm in which each node uses a local operation to approximate the average energy of the entire network, which is called system energy approximation (SEA). A node is called a sub-critical node if its residual energy is less than the SEA value; otherwise, it is called a super-critical node. The algorithm shown below is executed in each node.

Algorithm 1. The Energy-based Diffusion Algorithm

```
Initially SEA := residual energy level and received_SEA_list is
empty

1:    for every periodic time interval t do
2:        if received_SEA_list is not empty then
3:            compute new SEA by averaging all SEA values from
              received_SEA_list and its residual energy level
4:        send <SEA> to all neighbors

5:    upon receiving <SEA_i> from a neighbor n_i
6:        if <SEA_i, n_i> is not in received_SEA_list then
7:            add <SEA_i, n_i> to received_SEA_list with an expiration
              time
8:        else
9:            replace it with new <SEA_i, n_i> and reset its
              expiration time

10:   when an entry <SEA_i, n_i> has expired
11:       remove this <SEA_i, n_i> from received_SEA_list
```

In Algorithm 1, SEA is initially equal to its own residual energy level and the *received_SEA_list* is set to empty. In lines 1 to 4, each node sends the $<SEA>$ message to all neighbors within every time interval t. If the *received_SEA_list* was not empty before sending $<SEA>$ message, the SEA value will be recomputed by averaging all SEA values from *received_SEA_list* and its residual energy. Upon receiving $<SEA>$ message from a neighbor n_i, the $<SEA, n_i>$ entry will be added to *received_SEA_list* with an expiration time if the $<SEA, n_i>$ entry has not been added yet. Otherwise, replace it with new $<SEA, n_i>$ entry and reset its expiration time (lines 5 to 9). We use the expiration time field to guarantee that the SEA value of this entry is fresh. If a node moves away and does not send its SEA value before a pre-determined expiration time, its SEA value is removed from *received_SEA_list*. To reduce protocol overhead, a node can periodically piggyback $<SEA>$ value on the data packet by forwarding.

Initial result about the average-based diffusion algorithm was provided in [6], which gave the convergence proof in mobile environment. The correctness of our energy-based diffusion algorithm follows in the same manner as in the average-based diffusion algorithm, since they have the same averaging operation to approximate a network-wide knowledge. Different from the average-based diffusion algorithm, we feedback the residual energy level to each averaging operation to guarantee the new SEA value can adjust according to the current energy consumption situation of entire network.

3.2 Original SEER Scheme

Our first scheme is based on the basic gossiping protocol proposed by [2]. Our scheme is different from the original gossiping in that only the super-critical nodes need to rebroadcast messages to its neighbors with probability p and discard the received messages without further forwarding with probability $1-p$. A super-critical node rebroadcasts a given message at most once. Hence, if the message has been received again, it is dropped. Note that the sub-critical nodes do not participate in the message forwarding to save the precious energy. This simple scheme is called SEER-1 (p).

Following the results in percolation theory [7], SEER-1(p) exhibits a certain type of bimodal behavior. We assume that all nodes have been initialized their residual energy in a uniform distribution with a given range and let the forwarding probability p of super-critical nodes be equal to 1. As mentioned before, the SEA value obtained at each node approaches to the actual average network energy. The rebroadcast probability of a node in SEER-1(1) is equal to the probability that its residual energy is greater than the SEA value, which is about 0.5.

One problem of SEER-1(p) scheme is how to set the rebroadcast probability p. In SEER-1(1), the rebroadcast probability of each node is around 0.5. Intuitively, this value is not likely to be the globally optimal. For instance, in a denser area, each node has more neighbors whose coverage areas overlap significantly. Rebroadcast messages from nodes in a dense neighborhood will reach the same nodes many times. To reduce such redundancy, the rebroadcast probability in these areas should be set lower. On the contrary, the rebroadcast probability should be set higher in sparse areas to achieve better reachability.

3.3 Adaptive SEER Scheme

As mentioned earlier, only selecting nodes with higher residual energy to participate message forwarding is our primary aim for SEER-1(p). However, using predefined fixed probability p falls in a dilemma between reachability, the number of rebroadcasting messages, and the system lifetime as node movement. It is desirable if the nodes, including both super-critical and sub-critical nodes, can dynamically adjust its rebroadcast probability on-the-fly. In the remainder of this section, we discuss how to optimize the SEER-1(p) scheme by taking connectivity with neighbors into account.

Neighborhood Detection. To dynamically adjust the rebroadcast probability as neighbor connectivity changes, we propose a packet-monitoring-based neighbor detection algorithm to estimate the number of neighbors on-the-fly. Different from the mechanism using periodical HELLO messages, there is no extra message overhead in our algorithm. The pseudocode is shown in Algorithm 2. In lines 1 to 5, each node continuously monitors the incoming broadcast packets and record the number of packets received. For every periodical time interval t at each node, if no broadcast packet p_i is received within t, it updates nbr_count with the counter of p_i and removes the entry of p_i from $received\_packet\_list$.

Algorithm 2. The Packet-monitoring-based Neighbor Detection Algorithm

```
Initially nbr_count := N_d and received_packet_list is empty

1:    upon receiving a broadcast packet p_i
2:        if p_i is not in received_packet_list then
3:            add p_i to received_packet_list with an expiration time
4:        else
5:            increase the received_packet_list [p_i].counter by 1
              // record the number of packet p_i received

6:    for every periodic time interval t do
7:        if no broadcast packet p_i is received within t then
8:            nbr_count := received_packet_list [p_i].counter
9:            remove the entry of p_i from the received_packet_list
```

The packet-monitoring-based neighbor detection algorithm takes time to gradually approach the accurate value of the number of neighbors. If the initial value is set closer to the accurate value, the algorithm will converge to the nbr_count faster. Here, we utilize the average network degree to be a basis for initializing the nbr_count. Let A be the area of a MANET, N be the number of mobile nodes in the network, and R be the communication range. The average network degree N_d can be obtained by the following formula:

$$N_d = N(\frac{\pi R^2}{A}) - 1 . \qquad (1)$$

A Three-level Adaptation. The SEER-1(p) uses a fixed rebroadcast probability p for super-critical nodes. According to percolation theory [7], there exists a threshold $P_c < 1$, such that by using P_c as the rebroadcast probability, almost all nodes can receive a broadcast message, and there is no much improvement on reachability for $p > P_c$. Therefore, SEER-1(p) does not work well in various MANET topologies. To give some intuition, we make three observations below.

Observation 1. In a sufficiently large network, only selecting super-critical nodes to rebroadcast received messages suffice to fulfill reachability requirement while achieving higher energy-efficiency, even though all sub-critical nodes decide to sleep to save their precious residual energy.

Observation 2. In a sparse network, a node has fewer neighbors. Some sub-critical nodes are more likely to play a critical role for forwarding messages in order to maintain the connectivity of the network. If they fail to do so, the network is partitioned. Therefore, in addition to super-critical nodes, sub-critical nodes should increase its rebroadcast probability to avoid reachability degradation.

Observation 3. In a dense network, if the neighborhood of a node is crowded enough, we can not only inhibit the sub-critical nodes from forwarding messages but also further decrease the rebroadcast probability of super-critical nodes to reduce redundant transmissions without sacrificing the reachability.

To resolve the dilemma between reachability, the number of rebroadcasting messages, and the system lifetime, we propose a three-level adaptation scheme in which each node can independently adjust its rebroadcast probability according to its residual energy level and the neighborhood status. We extend the fixed probability p into two probability functions Psuper-critical(n) and Psub-critical(n) for super-critical nodes and sub-critical nodes respectively as

$$P_{super-critical}(n) = \begin{cases} 1, & if\ n < n_2, \\ H(n), & if\ n \geq n_2, \end{cases} \quad (2)$$

$$P_{sub-critical}(n) = \begin{cases} 0, & if\ n \geq n_1, \\ L(n), & if\ n < n_1, \end{cases} \quad (3)$$

where n is number of neighbors maintained by our packet-monitoring-based neighbor detection algorithm, $H(n)$ a decrease function within an area $[p_l, 1]$, and $L(n)$ a decrease function within an area $[0,1]$. Following **Observations 1, 2, and 3**, Fig. 1 shows an abstract shape of three-level adaptation. With few neighbors ($n \leq n1$), not only all super-critical nodes need to rebroadcast but

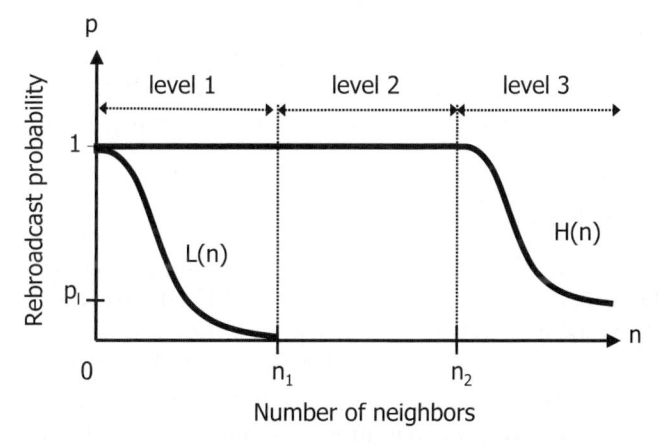

Fig. 1. Abstract shapes of three-level adaptation

sub-critical nodes should gradually increase their rebroadcast probability if n becomes smaller and smaller. When n is close to 0, we force all nodes to participate messages forwarding for the behalf of reachability. Between n_1 and n_2, no sub-critical nodes need to forward received messages. Only super-critical nodes taking over messages forwarding suffices the broadcasting operation to reach equilibrium state, balancing reachability and power saving. After $n \geq n2$, a decrease function H(n) is used to gradually decrease the rebroadcast probability of super-critical nodes to p_l. Note that p_l is a fixed lower bound for the rebroadcast probability of super-critical nodes to guarantee the reachability requirement. This optimization is called SEER-2(n_1, n_2, H(n), L(n)). In section 4.3, we will derive n_1 and n_2 values through experiments.

4 Performance Evaluation

In this section, we first evaluate the performance of our SEER-1(p) scheme and observe the partition ratio of our energy-based diffusion algorithm with different network parameters. Following the experiment results in SEER-1(p) scheme, we derive exact n_1 and n_2 values to set up our SEER-2(n_1, n_2, H(n), L(n)) scheme. We compare our SEER-2 scheme with a simple flooding algorithm and the dynamic probabilistic broadcasting (DPB) algorithm [11]. We implement all the four algorithms and study the following performance metrics, including reachability, saving ratio, the number of message rebroadcasts with different initial energy levels, and extended lifetime.

4.1 Simulation Model

Our simulation is performed in the GloMoSim network simulator [10] (version 2.03). The mobility model used in each of simulations is known as random direction. The transmission range of each node is held constant at 250 meters. The radio frequency at the physical layer is 2.4 GHz of the ISM band. The raw network bandwidth is 2 Mbps and the MAC layer protocol is IEEE 802.11 [3]. One source node is responsible for sending constant bit rate (CBR) flows and each CBR flow consists of 128 byte packets. Our energy consumption model is based on Chen et al. which measured the Lucent 2Mb/s WaveLAN 802.11 cards, observing power consumption cost of 1.4W(transmit), 1.0W(receive), and 0.83W(idle) [1].

4.2 Partition Ratio

Fig. 2 shows the partition ratio vs. the number of diffusion rounds with different node mobility models: 0 km/h, 30 km/h, and 60 km/h. We simulate 200-node networks in a $1500m \times 1500m$ area. Each node has a random initial energy, uniformly distributed over the interval [300 J, 2000 J]. Partition ratio is defined as $\frac{|N_{super}-N_{sub}|}{N}$, where N_{super} and N_{sub} are the number of super-critical nodes and sub-critical nodes respectively after each diffusion round, and N the total number of nodes in the network. We force each node to execute the energy-based

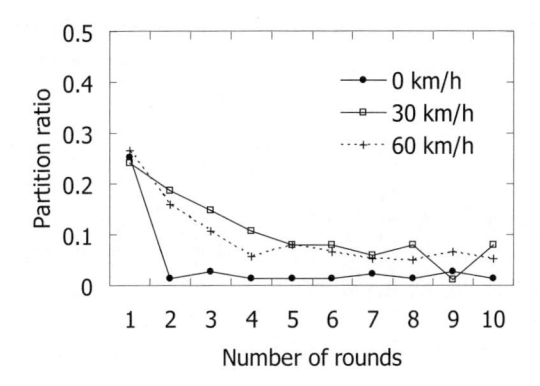

Fig. 2. Partition ratio vs. number of rounds

diffusion operation once in each round. In Fig. 2, we can see that mobility does not affect our energy-based diffusion algorithm very much. After round 5, the partition ratio is very close to 0, especially for static MANETs. In other words, the ratio between the number of super-critical nodes and the number of sub-critical nodes approaches to the desirable ratio 1:1. This implies that, after few diffusion rounds, half of total nodes can independently classify themselves into the sub-critical group and decrease their rebroadcast probability to save precious residual energy.

4.3 Reachability and Forwarding Ratio

Here we study the performance indicated by the following two metrics, of which the first was studied in [8]:

- *REachability (RE)*: the number of mobile node receiving the broadcast message divided by the total number of mobile nodes that are reachable, directly or indirectly, from the source node.
- *Forwarding Ratio (FR)*: The ratio of the nodes that retransmit the packets at least once to the total number of nodes in the network in a broadcast.

We use a fixed area size with different average number of neighbors n. Fig. 3 shows our simulation results for SEER-1(p) with $p = 1$. It can be seen that the results follow the **Observations 1, 2, and 3** discussed earlier. Remember that sub-critical nodes do not forward messages in this scheme. When $n \leq 15$, a situation that a node has fewer neighbors, RE obviously degrades because some sub-critical nodes are more likely to be located in a critical position to maintain the network connectivity. The fact that sub-critical nodes do not forward messages thus incurs the problem of network partition. When $15 < n < 21$, super-critical nodes suffice to achieve high reachability (RE > 0.83). When $n \geq 21$, the chance of receiving the same messages from other neighbor super-critical nodes raises. We can decrease the rebroadcast probability of super-critical nodes to reduce FR. Intuitively, more redundant transmissions can be saved without sacrificing

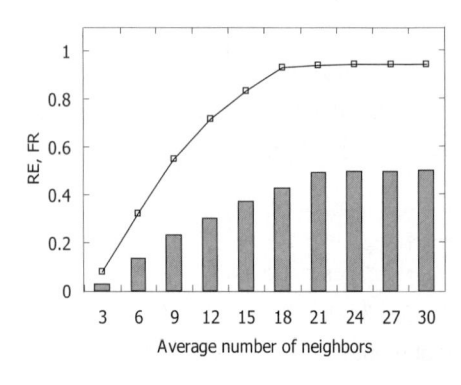

Fig. 3. Performance of SEER-1 scheme: Reachability RE (shown in line) and Forwarding ratio FR (shown in bars) vs. average number of neighbors

Fig. 4. Performance of SEER-2 and DPB schemes: Reachability RE (lines in upper part) and Forwarding Ratio FR (bars in lower part) vs. average number of neighbors

the reachability. From the results in Fig. 3, we let $n_1 = 15$ and $n_2 = 21$ respectively to evaluate the performance of our SEER-2 scheme. As Fig. 4 shows, RE and FR of SEER-2(15, 21, H(n), L(n)) are as good as those of DPB. This demonstrates that utilizing energy-based knowledge to determine rebroadcast probability can produce satisfying broadcast performance.

4.4 Rebroadcasts

In our experiments, the initial power of nodes is set to be a uniform distribution between 300J and 2000J. In Fig. 5, each diamond symbol along the horizontal axis represents an individual node with initial power of various levels. Fig. 5 shows the relationship between the number of relays and the total 200 nodes of different initial energy. In this experiment, a source node generates a total of 12000 broadcast packets at the packet rate of 20 packets per second. In DPB, the number of relays of each node falls roughly between 5000 to 8000 times. This means that even a node with very low energy still has the same rebroadcast probability as a node with very high energy. As expected, our SEER-2 scheme dramatically divides all nodes into super-critical nodes and sub-critical nodes and most rebroadcasting load is shared about evenly by the super-critical nodes. The number of relays of sub-critical nodes is less than 2000 times. It can be noticed that some sub-critical nodes never participate message forwarding. This is because the sub-critical nodes tend to drop the received messages except when the number of its neighbors is less than n_1.

4.5 Network Lifetime

This section shows how much more our SEER-2 scheme can extend network lifetime compared with simple flooding and DPB. We define network lifetime as the time interval from network initialization to the instant of the first node

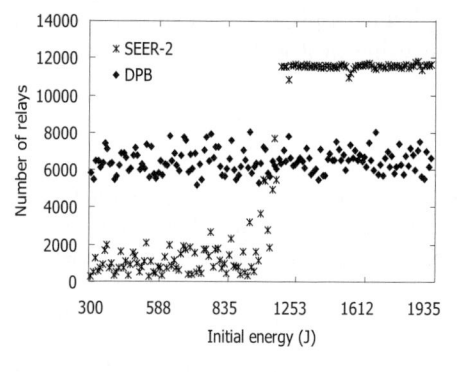

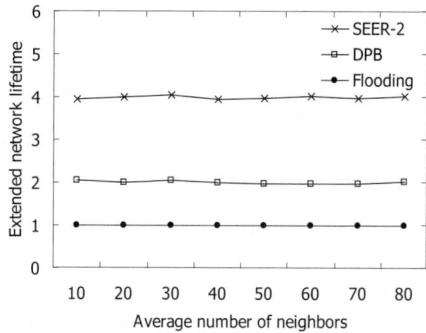

Fig. 5. Number of relays vs. initial energy

Fig. 6. Extended network lifetime vs. average number of neighbors

failure due to battery depletion. We assume that the source node has unlimited energy for generating data traffic, and that the remaining 200 nodes start with random initial energy uniformly distributed over the interval [300 J, 2000 J].

Following the results in section 4.3, Fig. 6 shows that the extend network lifetime of SEER-2 scheme is about a factor of 2 and 4 better than DPB and simple flooding respectively for various node density. This is because we concentrate the load of messages forwarding on super-critical nodes. A sub-critical node in this scheme decreases its rebroadcast probability to save energy and can thus extend the system lifetime. Especially in a denser network when the average number of neighbors is greater than the parameter n_2 of SEER-2 scheme, no sub-critical nodes need to participate in rebroadcast.

5 Conclusion

In this paper, we present two energy-efficient relay schemes namely SEER-1 and SEER-2 respectively. Both schemes utilize a localized energy-based diffusion algorithm to estimate a system energy approximation (SEA), with which each node can independently determine an appropriate rebroadcast probability. To optimize the energy efficiency, and to extend network lifetime without sacrificing the reachability, we also study how to dynamically adjust the rebroadcast probability by using neighbor connectivity information. Simulation results show that the reachability and forwarding ratio of our SEER-2 scheme are as good as those of DPB. This demonstrates that utilizing energy-based knowledge to determine rebroadcast probability can efficiently reduce redundant transmissions without sacrificing reachability. Besides, our SEER-2 scheme dramatically concentrates the greater part of message forwarding load on the nodes with higher residual energy. Following the results, extended network lifetime of SEER-2 is about a factor of 2 and 4 better than that of DPB and flooding scheme respectively. We expect this performance improvement to become even more significant in denser networks. In the future work, we plan to apply these schemes to current MANETs protocols, such as multicast or routing protocols.

References

1. B. Chen, K. Jamieson, H. Balakrishnan, and R. Morris. Span: An Energy-efficient Coordination Algorithm for Topology Maintenance in Ad Hoc Wireless Networks. In *Proc. IEEE/ACM Intl. Conf. on Mobile Computing and Networking (MOBICOM)*, July 2001.
2. Zygmunt J. Haas, Joseph Y. Halpern, and Li Li. Gossip-based Ad Hoc Routing. In *Proc. IEEE INFO-COM*, Jun. 2002.
3. IEEE Computer Society LAN MAN Standards Com-mittee. *Wireless LAN Medium Access Control (MAC)and Physical Layer (PHY) Specifications*. New York, New York, 1997. IEEE Std. 802.11-1997.
4. D. B. Johnson, D. A. Maltz, and J. Broch. *DSR: The Dynamic Source Routing Protocol for Multi-Hop Wireless Ad Hoc Networks*. In Ad Hoc Networking, edited by Charles E. Perkins, Chapter 5, pp. 139-172, Addison-Wesley, 2001.
5. E. D. Kaplan, editor. *Understanding GPS: Principles and Applications*. Artech House, Boston, MA, 1996.
6. Q. Li and D. Rus. Global Clock Synchronization in Sensor Networks. In *Proc. IEEE INFO-COM*, March 2004.
7. R. Meester and R. Roy. *Continuum percolation*. Cambridge University, 1996.
8. S.-Y. Ni, Y.-C. Tseng, Y.-S. Chen and J.-P. Sheu. The Broadcast Storm Problem in a Mobile Ad Hoc Network. In *Proc. IEEE/ACM Intl. Conf. on Mobile Computing and Networking (MOBICOM)*, Aug. 1999.
9. C. E. Perkins, E. M. Royer, and S. Das. Ad Hoc On Demand Distance Vector (AODV) Routing. IETF Internet draft, draft-ietf-manet-aodv-10.txt, March 2002.
10. X. Zeng, R. Bagrodia, and M. Gerla. GloMoSim: A Library for Parallel Simulation of Large-Scale Wireless Networks. In *Proc. 12th Workshop on Parallel and Distributed Simulation*, pp. 154-161, 1998.
11. Qi Zhang and Dharma P. Agrawal. Dynamic Probabilistic Broadcasting in MANETs. *Journal of Parallel and Distributed Computing*, v.65 n.2, p.220-233, February 2005.

Multicost Routing over an Infinite Time Horizon in Energy and Capacity Constrained Wireless Ad-Hoc Networks

Christos A. Papageorgiou[1,2], Panagiotis C. Kokkinos[1], and Emmanouel A. Varvarigos[1,2]

[1] Department of Computer Engineering and Informatics,
University of Patras, 26500 Patras, Greece
[2] Research Academic Computer Technology Institute, Patras, Greece

Abstract. In this work we study the dynamic one-to-one communication problem in energy- and capacity-constrained wireless ad-hoc networks. The performance of such networks is evaluated under random traffic generation and continuous energy recharging at the nodes over an infinite-time horizon. We are interested in the maximum throughput that can be sustained by the network with the node queues being finite and in the average packet delay for a given throughput. We propose a multicost energy-aware routing algorithm and compare its performance to that of minimum-hop routing. The results of our experiments show that generally the energy-aware algorithm achieves a higher maximum throughput than the minimum-hop algorithm. More specifically, when the network is mainly energy-constrained and for the 2-dimensional topology considered, the throughput of the proposed energy-aware routing algorithm is found to be almost twice that of the minimum-hop algorithm.

1 Introduction

In this work we study the dynamic one-to-one communication problem in energy- and capacity/interference-constrained wireless ad-hoc networks. In the model we consider, packets are generated at each network node according to a random process, over an infinite time horizon. All packets have equal length, and require one slot in order to be transmitted over a link. Each packet transmission consumes an equal amount of energy E. Time is slotted, and a new packet is generated at each node with probability p during a slot. Packet destinations are uniformly distributed over all nodes. In addition to the usual capacity and interference constraints, the network is also assumed to be energy constrained. More specifically, we assume that energy is generated at each node of the network at a *recharging rate* of X units of energy per slot, over an infinite time horizon. We propose a multicost energy-aware algorithm for routing the packets in an ad hoc network, and compare its performance to that of minimum-hop routing.

During our comparisons, we are interested in two performance criteria: a) the maximum stability region, which is defined as the maximum throughput that can be sustained by the network with the node queues being finite and

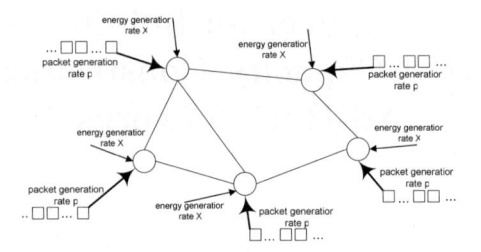

Fig. 1. The infinite-time horizon problem. Packets are generated at each node of the network with probability p during a slot, and have uniformly distributed destinations. Energy is also generated at each node at a rate X units of energy per slot.

b) the average delay suffered by the packets for a given throughput, which is defined as the average time that elapses between the generation of a packet at a node and the time it is received at its destination. We obtain results on the way the maximum stability region of the routing protocols examined changes as a function of the energy generation rate X at steady-state. We also obtain results for the average packet delay as a function of the packet generation rate, when the network is both energy and capacity/interference constrained. Figure 1 summarizes the definition of the problem.

Most previous works [7] studied the performance of ad-hoc networks in the context of the *evacuation problem*, where the network starts with a certain number of packets that have to be served and a certain amount of energy per node, and the objective is to serve the packets in the smallest number of steps, or to serve as many packets as possible before the energy at the nodes is depleted. This is different from the *dynamic one-to-one communication problem* considered in this paper where packets and energy are generated at each node continuously.

In the simulations performed for a specific network topology, we find the maximum packet generation probability p_{max} at the network nodes for which the network is stable, and the average delay for a given packet generation probability $p < p_{max}$ in the stability region. In our experiments we examined two routing algorithms: a multicost energy-aware routing algorithm and the traditional minimum-hop algorithm. The results obtained show that the multicost energy-aware algorithm outperforms the minimum-hop algorithm, achieving larger maximum throughput p_{max} for all recharging rates tested, and a smaller average delay for a given $p < p_{max}$. More specifically, we found that for the 2-dimensional topology considered and in the region where the network is energy-constrained, the throughput of the energy-aware algorithm is almost twice that of the minimum-hop routing algorithm. We also obtain results on the way the average packet delay changes as a function of the traffic load for energy and capacity/interference limited ad hoc networks. We find that the average delay increases with the traffic load more abruptly when the traffic reaches its maximum limitation due to the energy constraint, while it increases more smoothly when the traffic reaches its maximum limitation due to the capacity/interference constraint. We also discuss the effect certain network characteristics, such as the node density, the geographical distance, and the transmission range play on network performance. We argue, for example, that the transmission range of the nodes plays a more important role on performance for energy-limited networks than it plays for capacity/interference-limited networks.

The remainder of the paper is organised as follows. In Section 2 we discuss the impact of the capacity and energy constraints on network performance. In Section 3 we describe the routing algorithms tested in our experiments. In Section 4 we outline the environment under which our experiments were conducted. Section 5 presents the simulation results obtained.

2 Capacity and Energy Limitations

The traffic load that can be inserted in a network is restricted by capacity and interference limitations, and by the energy recharging rate at the nodes. Several works have examined the effect these limitations have on the maximum achievable throughput, for a variety of assumptions on the network topology, the routing algorithm, and the traffic pattern [8],[5]. Energy and its best use has also been the subject of several works; see e.g. [10][2] and [4] where the energy reserves at the nodes are among the criteria that the routing algorithms consider.

Capacity/Interference Limitation: According to the IEEE 802.11 protocol under the RTS/CTS mechanism a node before transmitting using a transmission range R, reserves a transmission floor of area at least equal to πR^2 and at most equal to $\frac{4}{3}\pi R^2$ around it (depending on the relative distance of the transmitter and the intended receiver) and the nodes located in this area cannot transmit. Ad hoc networks that do not use 802.11 often use busy tones [3] to avoid the hidden terminal problem. If the node density is high, then all nodes at a distance of approximately $2R$ from a transmitting node (therefore a total area of $4\pi R^2$) are prevented from transmitting. Therefore, the number of other nodes forbidden from transmitting when a given transmission takes place is similar (within a constant factor) when a busy tone mechanism or an RTS/CTS mechanism is used, and is proportional to R^2 (Fig. 2).

Following [1], we define a collision free set as a set of links that can be used simultaneously without causing collisions or excessive interference at the receiving nodes. The number of simultaneous transmissions the network structure permits, is upper bounded by the maximum cardinality C of the collision-free sets. From the preceding discussion and a simple "sphere packing" argument we infer that C is upper bounded by $\frac{A}{kR^2}$, where A is the area covered by the network and k is a constant between π and 4π that depends on the MAC protocol used and the relative location of the nodes.

Assume now that packets are generated at each node of an N-node network with probability p during each slot, and a packet requires an average of $h(p)$ transmissions to arrive at its destination. All transmissions have a transmission range R and require energy E. The mean number of transmissions per slot is given by the product $N \cdot p \cdot a(p) \cdot h(p)$, where $a(p)$ is the ratio of the total number

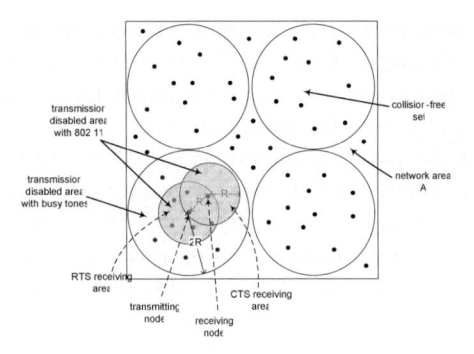

Fig. 2. The division of the network into collision-free sets

of packet transmissions over the number of successful transmissions required to get the packets to their destinations over the paths chosen. Therefore, for the network to be stable the following inequality must hold:

$$N \cdot p \cdot a(p) \cdot h(p) \leq C \leq \frac{A}{kR^2} \qquad (1)$$

The number of hops of the paths $h(p)$ is roughly inversely proportional to the transmission range R of the nodes, and we have $h(p) \geq \frac{L}{R}$, where L is the average physical source-destination distance (with the inequality being closer to equality for dense networks and shortest distance routing). Assuming we are in the stable region and there is no buffer limitation, no packets are lost, and we have $a(p) \geq 1$. Consequently, a limit on the packet generation rate p posed by the capacity/interference constraints is given by

$$p \leq \frac{A}{kRNL} = \frac{1}{k\rho L} \cdot \frac{1}{R}, \qquad (2)$$

where $\rho = N/A$ is the area node density

Energy Limitation: For a wireless ad-hoc network with energy rechargeable nodes to be stable, the mean energy expended at each time slot must be at most equal to the energy inserted in the network in the same period. The average energy expended in each slot is equal to $N \cdot p \cdot h(p) \cdot a(p) \cdot E$. We assume that all nodes use the same transmission radius R and expend energy equal to E for each packet transmission. The average energy inserted in the network during each slot is equal to $N \cdot X$, where X is the energy recharging rate at each node per slot. Consequently a necessary condition for the network to be stable is

$$N \cdot p \cdot h(p) \cdot a(p) \cdot E \leq N \cdot X \qquad (3)$$

The energy expended E for a packet transmission can be expressed as $k'R^\alpha$, for some constant k' (which depends on the channel, the sensitivity of the receiver, and the desired BER), where α is between 2 and 4 depending on the power-loss model. Working in the same manner as in the previous paragraph, and using the inequality $h(p) \geq \frac{L}{R}$, we find that a necessary condition for stability due to the energy constraint is

$$p \leq \frac{X}{k'L} \cdot \frac{1}{R^{\alpha-1}} \qquad (4)$$

The inequalities (2) and (4) show that the energy limitation depends more strongly on R than the network capacity/interference limitation. The stability region shrinks as R increases, showing that using small transmission range is beneficial both for capacity/interference-constrained and energy-constrained ad hoc networks. That is, the amount of traffic that can be served by the network increases when we decrease the transmission range of the nodes, both due to increasing network capacity (better reuse factor) and due to lower spending of the energy reserves. Since in most wireless environments $a > 2$ (a is close to 4 for urban environments), we conclude (at least for dense networks) that for R sufficiently small the network throughput is mainly constrained by capacity/interference limitations, while for R sufficiently large it is constrained by energy limitations.

Equations (2) and (4) show that the energy limitation and the capacity/interference limitation depend in similar ways on the average physical distance L in the network, with the achievable throughput per node falling as L increases. Another conclusion drawn from the above discussion is that even though the capacity/interference limitation decreases as the area node density ρ increases, the energy limitation is independent of ρ. In summary, we expect networks that are sparse or that have a small recharging rate X, or that use a large transmission radius R to be mainly energy-limited as opposed to capacity/interference limited.

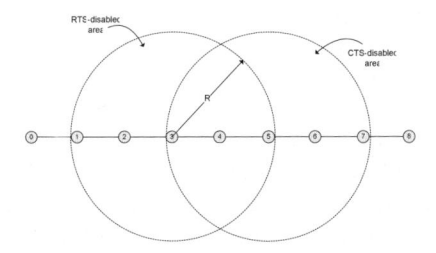

Fig. 3. The case of a linear ad hoc network using the RTS/CTS mechanism

Case of linear ad hoc networks: The preceding discussion assumes a 2-dimensional network. It is worth also studying briefly the case of linear (1-dimensional) ad hoc networks. Depending on whether busy tones or the RTS/CTS mechanism is used, each transmission prevents other nodes in a segment of length $k \cdot R$ from transmitting, where k is a constant between 2 and 4, depending on the MAC scheme used and the distance between the transmitter and receiver (Figure 3 illustrates the case where the RTS/CTS mechanism is used). If L is the length of the linear network, at most $\frac{L}{kR}$ transmissions can take place simultaneously during a slot and a necessary condition for stability is

$$N \cdot p \cdot a(p) \cdot h(p) \leq \frac{L}{kR} \qquad (5)$$

Using $h(p) \geq \frac{L}{R}$ and $a(p) \geq 1$ and defining $\rho = \frac{N}{L}$ on the linear node density of the 1-dimensional network, we infer that

$$p \leq \frac{1}{kN} = \frac{1}{k\rho L} \qquad (6)$$

for some constant k.

The energy limitation can be formed by arguing in a similar way to the 2-dimensional case obtaining again (4).

From (4) and (6) it can be seen that the capacity/interference constraint for linear networks is largely independent of the transmission range R used by the network nodes. Thus, networks of this kind that use a large R are expected to be energy-limited. The dependence of the throughput upper bounds on the physical dimension L of the 1-dimensional network is similar to that of the 2-dimensional case. We also expect, as in the case for 2-dimensional networks sparse linear networks (small ρ) to be mainly energy-limited.

3 Routing Strategies

The behavior of the network in the context of the infinite-time horizon problem is evaluated under two routing algorithms: the traditional *minimum-hop* algorithm and a multi-cost routing algorithm, to be referred to as the *energy-aware* algorithm, which takes energy considerations into account. The multi-cost routing approach is fully described in [6].

Multi-cost Routing: The multi-cost energy-aware routing algorithm considered in this paper uses two cost metrics: The residual energy R_i, and the transmission power T_i at the transmitting node i of a link (i, j). These cost metrics are combined using the "min" and the "+" operators, to obtain the minimum residual energy $R = \min_{i \in P} R_i$ on the nodes of path P and the total energy $T = \sum_{i \in P} T_i$ consumed on path P, respectively. The optimization function f used in order to produce the final scalar path cost is

$$\text{Energy-Aware:} \quad f(T, R) = \frac{\sum_{i \in P} T_i}{\min_{i \in P} R_i}, \qquad (7)$$

where the index i runs over all the nodes on path P.

4 Simulation Environment

In our experiments we used the Network Simulator [9] to simulate a wireless multihop network of 49 nodes arranged in a 7x7 grid topology. Neighboring nodes at the grid were placed at a distance of 50m from each other. The transmission range of the nodes is variable and follows a uniform distribution between 50 and 100 meters. We assume Bernoulli arrivals, where a packet is generated at each node during each slot with probability p. The duration of the slot is 0.08 seconds, while the packet transmission time is 0.016576 seconds, for the 2000 bytes sized packets we use in our experiments. We chose this slot time in order for the RTS/CTS handshake mechanism and packet transmission to have been completed by the time the next packet is generated. Each node has zero initial energy, and the recharging rate X is the same for all nodes. Finally we assume that every node has full knowledge of the network topology and all other information needed for the route computation.

Furthermore, we define a threshold on the residual energy of a node, and when the energy at a node falls below this threshold, the node stops forwarding packets and starts storing them in its queue. The same holds when the next-hop node's residual energy is below this threshold. Each node periodically checks its energy reserves and those of its neighbors, and if they both exceed the threshold the node starts forwarding its packets, decreasing its queue.

5 Results

The performance of the minimum-hop and the energy-aware algorithms was evaluated in the context of the infinite-time horizon problem, for varying recharging rates and packet generation probabilities. We are interested in the steady-state performance of the proposed schemes; the network was assumed to be in steady state when the variance in the packet delivery delay was below some threshold.

The performance metrics of interest were the largest packet generation probability p_{max} for which the network remains stable (maximum throughput) and the average packet delivery delay for a given packet generation probability. By stability we mean that the volume of the incoming traffic can be served appropriately: with small average packet delivery delay and high packet delivery ratio. When either of these conditions is broken, the network is assumed to enter an unstable region, so there is no point in further studying it.

In Fig. 4 the average packet delay is depicted for $X = 5 \cdot 10^{-3}$ and $X = 9 \cdot 10^{-3}$ Joules per slot[1] with respect to the packet generation rate p, for both the minimum-hop and the energy-aware routing algorithms. For both recharging rates, the energy-aware algorithm outperforms the minimum-hop algorithm, by enabling the network to remain stable for heavier traffic loads. For the 2-dimensional topology considered, the traffic generation probabilities that the energy-aware algorithm is able to handle with adequately small packet delivery delay are nearly twice those of the minimum-hop algorithm for both recharging rates considered. Figure 5 illustrates the received-to-sent packets ratio for both recharging rates and routing schemes.

---

[1] To be more specific energy equal to 0.005 joules and 0.009 joules was offered every 10 seconds in the experiments.

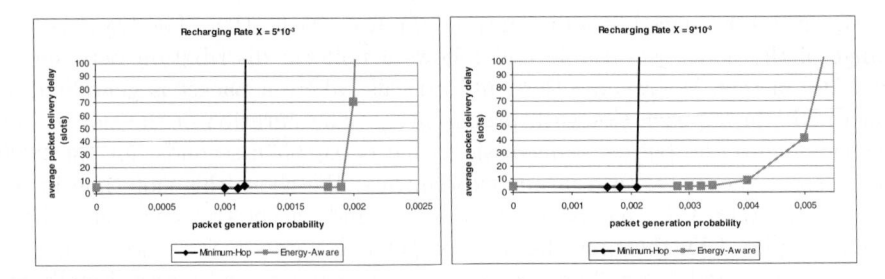

Fig. 4. The packet delay (in slots) for recharging rates $X = 5 \cdot 10^{-3}$ and $X = 9 \cdot 10^{-3}$ Joules per slot

The transition of the network to the unstable region as indicated by the rise in the average packet delay in Fig. 4 is extremely steep for the minimum-hop algorithm for both recharging rates $X = 5 \cdot 10^{-3}$ and $X = 9 \cdot 10^{-3}$ Joules per slot: from values of the delay around 4 or 5 slots in the stable region, there is an almost instant increase to practically infinite values above 100 slots. This is because when the minimum-hop algorithm is used, the network for both values of the recharging rate X is energy constrained; when the energy at some nodes gets depleted, the energy of many other nodes also start getting depleted soon afterwards, and the rise in the delay is very abrupt. In this state the connectivity of the network is weakened and the delivery of the incoming packets becomes difficult (large delays) or impossible (dropping of packets).

When the energy-aware algorithm is used and for $X = 5 \cdot 10^{-3}$ Joules per slot the network is again energy-constrained, but because it uses energy more efficiently, the rise in the delay is less abrupt than with the minimum-hop algorithm. When the energy-aware algorithm is used and the recharging rate is relatively high, $X = 9 \cdot 10^{-3}$ Joules per slot, the network is mainly capacity-constrained and the rise in the delay is rather smooth.

Figure 5 shows the number of received packets with respect to the number of packets that were sent, for $X = 9 \cdot 10^{-3}$ and $X = 15 \cdot 10^{-3}$ Joules per slot[2]. It can be observed that the energy-aware algorithm achieves a higher throughput than the minimum-hop algorithm, since the degration of the received to sent packets ratio begins later than with the minimum-hop algorithm. For both algorithms, the number of packets delivered to their destination grows linearly, initially, with the number of packets that enter the network, since for relatively light traffic they are nearly identical. For probabilities greater than p_{max}, however, there is a steep decline in the ratio. The number of packets successfully delivered to their destinations not only stops increasing as the number of incoming packets grows, but it even declines after the network enters the unstable region.

Figure 6 illustrates the maximum packet generation probability (that is, the maximum throughput) p_{max} for which the network remains stable as a function of the recharging rate X at the network nodes, for both the minimum-hop and

---

[2] To be more specific energy equal to 0.009 joules and 0.015 joules was offered every 10 seconds in the experiments.

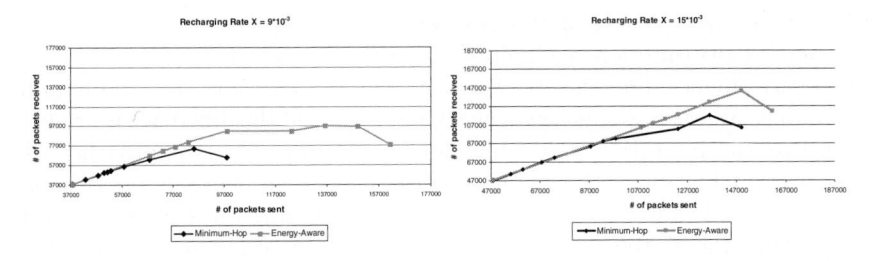

Fig. 5. The number of the packets received versus the number of packets sent for recharging rates $X = 9 \cdot 10^{-3}$ and $X = 15 \cdot 10^{-3}$ Joules per slot

the energy-aware routing algorithm, along with a detail of the figure for smaller recharging rates. p_{max} is taken to be the highest packet generation probability for which the network manages to serve the incoming traffic appropriately, meaning with small average packet delivery delay and high packet delivery ratio. The thresholds set for these two metrics used for detecting experimentally when the network enters the unstable region (above 100 slots for the average packet delivery delay and under 80% for the delivery ratio) are not important qualitatively for the results obtained, since we found that a different setting of the thresholds only causes a small shifting in the values presented without altering any of the conclusions drawn.

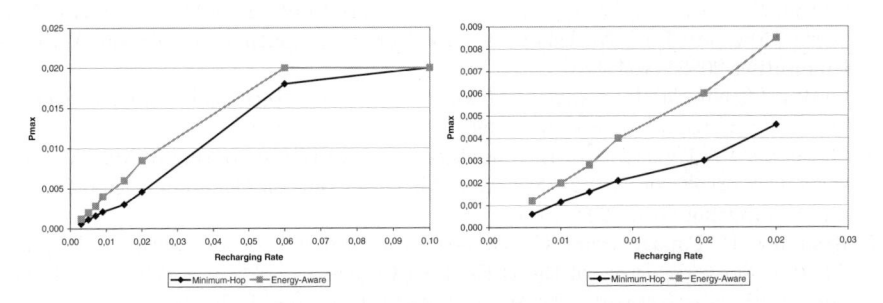

Fig. 6. The maximum traffic generation probability p_{max} versus network nodes' recharging rate X (Joules per second) for the Minimum-Hop and Energy-Aware algorithms and a detail for smaller recharging rates

Figure 6 shows that the energy-aware algorithm outperforms the minimum-hop algorithm, achieving significantly larger p_{max} for all recharging rates considered. The maximum throughput p_{max} seems to depend on the recharging rate almost linearly until the very end, for both routing algorithms. This verifies that the network in this region is energy-constrained since its performance, expressed by p_{max}, increases proportionally with the energy that is offered to it. When the recharging rate increases beyond some point, the network starts getting constrained by capacity/interference limitations, and the rate at which p_{max} grows with respect to the recharging rate is slowed, until it reaches a plateau indicating that the capacity/interference limitation has been reached.

The performance difference between the energy-aware and the minimum-hop algorithm is larger for low energy recharging rates, and the difference is gradually reduced as the limitation posed by the network capacity is approached. The detail part of Fig. 6 highlights the difference between the two algorithms. It can be observed that for the whole range of recharging rates presented in the detail part of Fig. 6, the p_{max} achieved by the energy-aware algorithm is nearly twice that of the minimum-hop algorithm. This is because the further away the network is from the capacity-constrained region, the more important energy efficiency becomes. When energy is the factor defining the ability of the network to serve incoming traffic, the energy-aware algorithm performs better. However, as energy becomes abundant and the capacity limitation starts constraining network performance, the performance gap between the energy-aware and the minimum-hop algorithm is narrowed.

References

1. Bertsekas, D., Gallagher, R.: Data Networks Englewood Cliffs, NJ: Prentice-Hall
2. Chang, J-H., Tassiulas, L.: Maximum lifetime routing in wireless sensor networks IEEE/ACM Trans. Netw. **12** no 4, 2004 609–619
3. Deng, J., Haas, Z.: Dual Busy Tone Multiple Access (DBTMA): A New Medium Access Control for Packet Radio Networks IEEE ICUPC'98, Florence, Italy, 1998
4. Gupta, N., Das, S.R.: Energy-Aware On-Demand Routing for Mobile Ad Hoc Networks Proc. 4th Int'l Workshop on Distributed Computing, Mobile and Wireless Computing 2002), 164–173
5. Gupta, P., Kumar, P.: Capacity of wireless networks Technical report, University of Illinois, Urbana-Champaign, 1999
6. Gutierrez, F.J., Varvarigos, E., Vassiliadis, S.: Multi-cost routing in Max-Min fair share networks Proc. Vol.2. 38th Ann. Allerton Conf. on Communication, Control and Computing 2000 1294–1304
7. Kokkinos, P., Papageorgiou, C., Varvarigos, E.: Energy Aware Routing in Wireless Ad Hoc Networks Proc. of the IEEE Int'l Conf. on a World of Wireless, Mobile, and Multimedia Networks (WoWMoM 2005) Taormina, Italy 2005
8. Li, J., Blake, C., De Couto, D.S.J., Lee, H.I., Morris, R.: Capacity of Ad Hoc wireless networks Proc. 7th Ann. Int'l Conf. on Mobile Computing and Networking, 2001, 61–69
9. NS - Network Simulator: http://www.isi.edu/nsnam/
10. Rodoplu, V., Meng, T.: Minimum energy mobile wireless networks Proc. IEEE Int'l Conf. on Communications 1998 1633–1639

An Adaptive Self-organization Protocol for Wireless Sensor Networks

Kil-Woong Jang[1] and Byung-Soon Kim[2]

[1] Dept. of Mathematical and Information Science, Korea Maritime University
1 YeongDo-Gu Dongsam-Dong, Busan, Korea
jangkw@bada.hhu.ac.kr
[2] Dept. of Computer Education, Andong National University
388 Songchon-Dong, Andong, Korea
byungsoon_kim@hotmail.com

Abstract. This paper proposes a new self-organization protocol for sensors with low-power battery in wireless sensor networks. In our protocol, sensor networks consist of a hierarchical architecture with a sink, which is a root node, using the spanning tree algorithm. Our protocol utilizes some control messages to construct a hierarchical architecture, and by exchanging the messages between nodes, maintains adaptively the network topology by reorganizing the tree architecture as the network evolves. We perform the simulation to evaluate the performance of our protocol over wireless sensor networks. We provide simulation results comparing our protocol with the conventional approaches. The results show that our protocol outperforms other protocols over a broad range of parameters.

1 Introduction

Sensor technology is one of the most challenging technical issues in ubiquitous networks. As sensors have currently the functions of sensing, data processing and communication, a wide range of monitoring applications, such as temperature, pressure, noise and so on, has been commonly studied in the literature [3-6]. Wireless sensor networks have characteristics as follows. A wireless sensor network consists of a large number of sensors, which may be very close to each other, and has a multi-hop wireless topology. Sensors are able to communicate directly in the transmission range. However, to send data to the destination beyond the transmission range, they have to communicate through some intermediate sensors. In addition, they have the constrained batteries, which cannot be recharged or replaced. That is, wireless sensor networks have the different environment than other wireless networks. Therefore, the conventional network protocols are not well applied to the application of wireless sensor networks. To accomplish the above functions, sensors must have capabilities to perform signal processing, computation, and network self-organizing capabilities to achieve scalable, robust and long-lived networks [1,2,7]. Specifically, the low power consumption of sensors is one of the most important requirements of network protocol in wireless sensor networks. The conventional wireless network protocols focus on high quality of services, but wireless sensor network protocols aim primarily to achieve power conservation owing to consisting of sensors with limited, irreplaceable batteries.

Several network protocols have been studied to extend network lifetime in wireless sensor networks. Direct communication protocol [5] is that each sensor transmits directly data to the sink. In this type of protocol, transmit power of each sensor is different due to the distances from the sensor to the sink. If the sensor is far away from the sink, it will quickly dissipate its power. On the other hand, it close to the sink can transmit data using a relative low transmit power. Minimum transmission energy (MTE) routing protocol [4] is designed each sensor to send data for next node by using minimum power. In this type of protocol, the sensors route data destined for the sink through intermediate sensors. Intermediate sensors act as routers for other sensors' data and are chosen such that the transmit power is minimized. The drawback of using this protocol is, due to relaying data, sensors closest to the sink die out first, whereas sensors furthest from the sink die out latest. Another power aware communication protocol is clustering protocol [5,6], where sensors are organized into clusters. In each cluster, a head exists in order to aggregate data from sensors and transfer them to the sink. The head can be decided by static or dynamic methods. For relaying other sensors' data, the head consumes its power more than others, so it would die quickly out.

In this paper, we propose a new self-organization protocol to extend the system life by solving the draws of the conventional protocols and reducing the power consumption of all sensors in the network system. Our protocol is designed by being based on the spanning tree algorithm and makes use of extra control messages to maintain the tree architecture. As all sensors send data destined for the sink by consuming minimum power and dissipating uniformly the power of all sensors, the lifetime of sensor networks is extended.

2 Our Protocol

For analyzing and evaluating the fundamental performance of our protocol, we first describe the power conserving behavior of the protocol. In wireless sensor networks, all sensing nodes have the maximum transmission range, and they send data to the sink directly or through intermediate nodes in this range. In our protocol, if the sink exists in the maximum transmission range of nodes, the nodes directly send data to the sink. Otherwise, they send data to the sink through intermediate nodes closest to the sink.

We first define notations used in our protocol before describing the detail operation of our protocol. A sink and all nodes have a level. The level of the sink is initially set to zero and that of all nodes is set by an infinite value. To distinguish a node with others, each node has a unique identification (ID), which is generally its MAC address. The operation of our protocol consists of the join and rejoin phases. The join phase is a process that each node joins the network when it powers on. In our protocol, each node is initially set by a

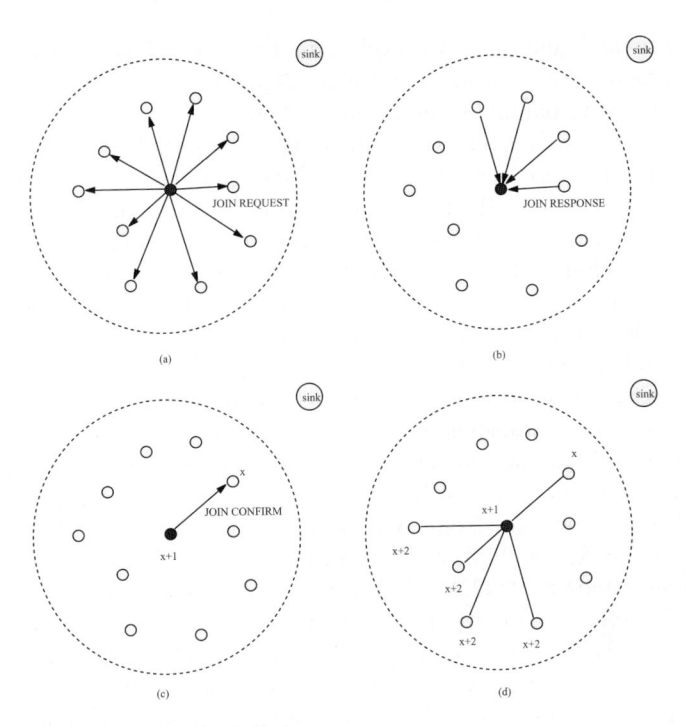

Fig. 1. An example of the join phase in our protocol. Node A is marked with •, and a dashed line is the maximum transmission range of node A. Labels (x, x+1 and x+2) on nodes are their level. (a) node A sends a *JOIN REQUEST* message to neighbors. (b) neighbors reply node A with a *JOIN RESPONSE* message. (c) node A sends a *JOIN CONFIRM* message to the parent. (d) network links after the join phase for node A.

constant threshold value, and if its energy is less than this value, it is unable to act as intermediate node for other nodes. Suppose a node sends data to the sink through an intermediate node. If energy of the intermediate node is less than the threshold value, the intermediate node cannot relay data to the sink. To maintain the connection with the sink, the node must join the network again by choosing new intermediate nodes, which is called the rejoin phase.

Initially, when each node powers on, it operates the following procedure to join a wireless sensor network, as shown in Fig. 1. To simply describe the operation, suppose a node A powers on right now. Node A first sends a *JOIN REQUEST* message to the neighbors, which are sensors or the sink. This message includes node A's ID and level. Upon receiving the message, the neighbors compare their level with node A's level. If their level is lower than node A's level, they send a *JOIN RESPONSE* message to node A. This message includes the sender's ID, node A's ID, the sender's level and the number of children of sender. If node A receives the *JOIN RESPONSE* messages, node A chooses the node sending the message with the lowest level as its parent, and replaces its level by parent's level plus one. For example, node A receives the message from the sink, the sink is parent of node A and node A's level becomes to one. If more than nodes with the lowest level can be chosen, node A chooses a node, whose number of children is the lowest, as its parent. In addition, if there are chosen some parent, then node A randomly chooses one of the nodes as its parent. The main reason using this approach is because each node can send data to the sink using minimum hop count and we can balance the number of child of intermediate nodes. Therefore, the advantages of our protocol is that it minimizes the energy consumption in each node, so prolong network lifetime. On selection of the parent, node A sends a *JOIN CONFIRM* message to the parent. This message includes node A's and parent's IDs. Finally, the parent records node A's ID and increments the number of children in its memory. However, if node A does not receive any response messages from neighbors, node A fails to join the network because no neighbor exists in the transmission range of node A.

After completely joining the network, if node's energy is lower than threshold, the rejoin phase is operated, as shown in Fig. 2. Suppose the energy of node A's parent is below threshold. Node A's parent sends a *RELEASE REQUEST* message to node A, its child. Node A receiving the message, in order to find a new parent, sends a *PROBE REQUEST* message its own ID and level to neighbors. Neighbors, except parent and childen of node A as well as the nodes with energy lower than threshold, send a *PROBE RESPONSE* message to node A. The message includes the sender's ID, node A's ID, the sender's level and the number of child of sender. Upon receiving the message, node A chooses a node with lowest level as new parent, and replaces its level by the parent's level plus one. If more than nodes with the lowest level can be chosen, node A chooses a node having the lowest number of child as new parent. If more than nodes with the lowest number of child can be chosen, then node A randomly chooses one of the nodes as new parent. On selection of new parent, node A sends a *PROBE CONFIRM* message to the parent. This message includes node A's and parent's ID. Finally, the parent records node A's ID and increments the number of child in its memory, and finally the rejoin phase is accomplished. However, if node A does not receive any *PROBE RESPONSE* messages from neighbors, node A reset its own level as infinite and must carry out the join process again. In addition, node A sends a *RELEASE REQUEST* message to children.

During this phase, if node A's level is changed by other value, node A must inform children. Thus it sends a *CHANGE LEVEL* message with its own ID and level to children. Upon receiving the message, each child replaces its level by node A's level plus one, and if it has children, it also sends a *CHANGE LEVEL* message to children.

Intermediate nodes relaying data between nodes and the sink play a role in our protocol such like the head in the clustering protocol. In the clustering protocol, the head is generally chosen without respect to a distance from the sink and aggregates data from nodes, thus the head dies out quicker than other nodes. However, in our protocol, each node sends data to the intermediate node closest to the sink, and changes the intermediate node having low energy with a new intermediate node having highier energy. Thus, our protocol can reduce energy

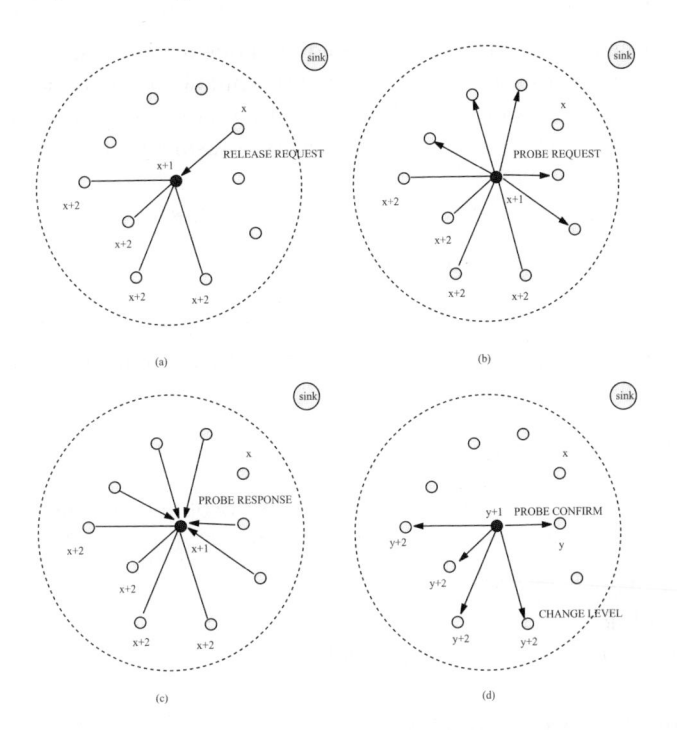

Fig. 2. An example of the update phase in our protocol. (a) node A's parent sends a *RELEASE REQUEST* message to node A. (b) node A sends a *PROBE REQUEST* message to neighbors. (c) neighbors reply a *PROBE RESPONSE* message to node A. (d) node A sends a *PROBE CONFIRM* message to new parent. If node A's level is changed, node A sends a *CHANGE LEVEL* message to children.

dissipated from nodes and solve the problem, that specific nodes die out quickly. In addition, in the direct communication protocol, nodes furthest from the sink die out first, but this problem also can be solved by using our protocol.

As a media access control protocol to transmit data and control messages, we use the IEEE 802.11 power saving mechanism [8]. In IEEE 802.11 power saving mechanism, power management is done based on Ad hoc Traffic Indication (ATIM). Time is divided into beacon intervals, and every node in the network is synchronized by periodic beacon transmissions. However, since our protocol maintains the hierarchical tree structure, after synchronization between parent and child is carried out only once at first, then all nodes in this structure do not need to synchronize more. So every node will start and finish each beacon interval almost at the same time. At the start of each beacon interval, there exists an interval called ATIM window, where every node should be in awake state and be able to exchange messages. If a node A joins or rejoins neighbor, it sends a corresponding message to the node during this interval. On the other hand, if node A has data destined for sink, it sends an ATIM packet to intermediate node during this interval. If the intermediate node receives this message, it will reply back by sendig ATIM-ACK to node A, and both nodes will stay awake for that entire beacon interval. If the intermediate node has not sent or received any messages or ATIM packets during the ATIM window, it enters doze mode and stays until the next beacon time. This allows the radio component of each node to be turned off at all times except during its transmit times, thus minimizing the energy dissipated in the individual sensors.

3 Performance Evaluation

We carried out the computer simulation to evaluate the performance of our protocol. In this section, we describe performance metrics, simulation environment and simulation results.

3.1 Performance Metrics

In order to compare the performance of our protocol and the conventional protocols, the performance metrics that we are interested in are

a) network lifetime (T),
b) number of nodes alive (N), and
c) total energy dissipated in the network (E).

3.2 Simulation Environment

The network model for simulation consists of randomly placed nodes in a constant size square area. Let s denote the network diameter and n denote the total number of nodes in the network. We assume that there are n nodes distributed randomly in a $s \times s$ region, and a sink is positioned at the center of the network. For example, if the network has a 100×100 meter area, the coordination of the sink is positioned at (x=50, y=50). Simulations are performed in wireless LAN environment. The data rate is 11Mbps, and the transmission range of each node is 15 m. In addition, beacon interval is set to 100 ms, and ATIM windows are 200 ms. We assume that each node has 2000 bits data packets and 160 bits control packets, and energy and threshold of each node were assigned 0.5 J and 0.5, respectively. We also assume that if a node dies out, it is not recharged or replaced by a new battery, and the event sensing by nodes is exponentially distributed with rate λ.

In a wireless sensor network, the energy of nodes is mainly dissipated in transmit and receive modes. In order to measure the energy dissipation of nodes, we use a radio model developed in [5]. In this model, nodes have the transmitter and receiver circuitries, which operate independently. The transmitter circuitry consists of a transmit electronics and a transmit amplifier, and the receiver circuitry consists of a receive electronics. Let E_e be the energy dissipated in transmit and receive electronics and E_a be the energy dissipated in transmit amplifier. We assume that $E_e = 50\ nJ/\text{bit}$ and $E_a = 100\ pJ/\text{bit}/m^2$. We also assume that the energy loss happens according to a distance between source and

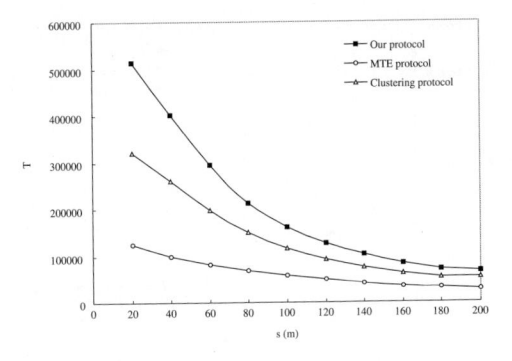

Fig. 3. T for various values of s

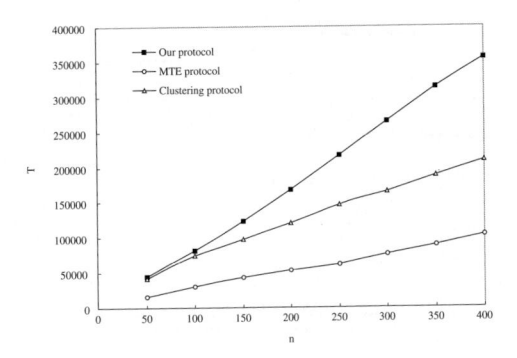

Fig. 4. T for various values of n

destination. Therefore, transmit energy, $E_t(k,d)$, and receive energy, $E_r(k)$, dissipated to send a k bit data packet to a destination apart a distance d are as follows:

$$E_t(k, d) = E_e * k + E_a * k * d^2. \qquad (1)$$

$$E_r(k) = E_e * k. \qquad (2)$$

3.3 Simulation Results

Using some results obtained by simulation, we compare our protocol with the MTE and clustering protocols. In the simulation, according to s, n and λ over a network maintained by a sink, we obtained some performance results, which are T, N and E.

We first experiment on different diameters of the network. We measure the network lifetime for our protocol, the MTE and clustering protocols. The result is in Fig. 3. In this simulation, there are a sink and 200 nodes in the network. In this figure, we can see that our protocol performs better than other protocols over all ranges of parameters. If $s<60\ m$, the lifetime of our protocol is on average 1.5 times longer than that of the clustering protocol and average 2 times longer

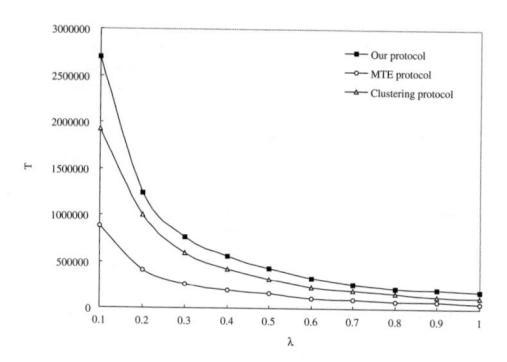

Fig. 5. T for various values of λ

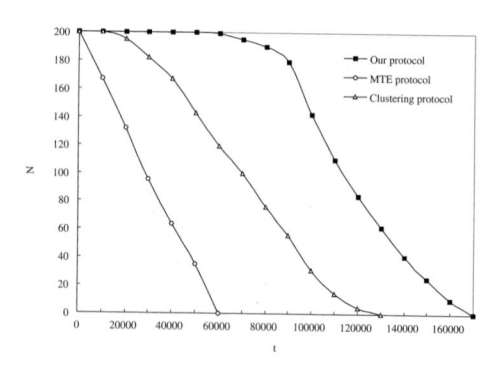

Fig. 6. N for various values of t

than that of the MTE protocol. If $s>60\ m$, the lifetime of our protocol is also on $20 \sim 50$ % longer than that of both. The reason is because all nodes constantly dissipate their energy and they use the optimized transmission path in order to send data to the sink.

Fig. 4 shows the network lifetime according to number of nodes in the network. In this simulation, we used the network with a $200\ m$ diameter. For all the cases we clearly see that as n increases, T for all protocols increases accordingly. Specially, our protocol achieves 2 or 3 times extension in network lifetime compared with other protocols.

Fig. 5 shows network lifetime as we increase λ of nodes for $s=200\ m$ and $n=200$. As like the above results, our protocol outperforms more than other protocols. Note that increase of λ causes to increase an amount of data destined for the sink. As shown in Fig. 5, as λ increases, T for all protocols exponentially decreases but our protocol outperforms other protocols by a slight difference.

Fig. 6 shows the number of nodes alive for various simulation times, t. In this simulation, 200 nodes are deployed in 200×200 meter environment with rectangular topology. In this figure, we see nodes using the MTE protocol die out earlier than other protocols. To send data to the sink, the number of nodes in the MTE protocol need more than in other protocols. Therefore, since the

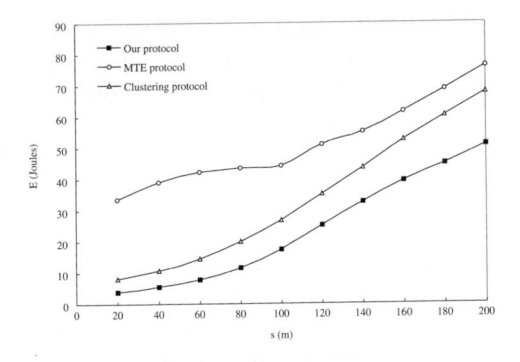

Fig. 7. E for various values of s

intermediate nodes dissipate their energy more, the nodes become to die out quickly. Similar to the above reason, in the clustering protocol, the head in each cluster dies out quicklier than other nodes. In this simulation, we do not consider the battery recharge or replacement of nodes. Note that as the number of die-out nodes increase, remained active nodes will dissipate their battery more than previous. Therefore, in Fig. 6, the MTE and clustering protocols decrease faster than our protocol.

Fig. 7 shows the total energy dissipated in the network for different values of s and $n=200$. In this figure, we see all the protocols increase accordingly the dissipated energy. If $s<100$ m, plot of the MTE protocol increases the different pattern than other protocols. Although a distance between nodes and the sink is close, they use an optimal path to send data to the sink. Thus the hop count is increased and the dissipated energy of the nodes on the path will be increased. On the contrast, if the network size is small, our protocol sends data to the sink directly. In addition, if the network size increases, our protocol establishes the optimal path for all nodes and thus can increase the energy efficiency.

4 Conclusions

In this paper, we have presented a new self-organization protocol to maintain efficiently the energy of sensors in wireless sensor networks. The basic idea of our protocol is to establish an optimal path of each node destined for a sink, and balance the energy of all nodes in wireless sensor networks. To achieve our protocol, each node has its level related to a distance between the node and the sink, and chooses a node with the lowest level and the lowest number of child as its parent. Each node also adaptively changes the parent with another node using the threshold and node's level, If intermediate node, which relays data destined for the sink, has its energy less than the threshold, another node, which has energy more than the threshold, is chosen as new parent. Therefore, we are able to maintain the optimal path of nodes continuously and reduce the number of die-out nodes. Moreover, we can extend network lifetime to balance the energy of all nodes without respect to a distance between the node and the sink. Using the simulation, we evaluated the performance of our protocol in terms of the network lifetime, the number of node alive and the energy dissipated in the network. The simualtion results illustrated that our protocol outperformed the conventional protocols over various range of parameters.

References

1. I. Akyildiz, W. Su, Y. Sankarasubramanian and E. Cayiraci.: Wireless sensor networks: a survey. Computer Networks, No. 38, (2002) 393–422
2. K. Sohrabi, J. Gao, V. Ailawadhi and G. J. Pottie.: Protocols for Self-Organization of a Wireless Sensor Network. IEEE Personal Communications, Oct. (2000) 16–27
3. A. Cerpa and D. Estrin.: ASCENT: Adaptive Self-Configuring sEnsor Networks Topologies. Proceedings of IEEE Infocom, (2000)
4. X. Lin and I. Stojmenovic.: Power-Aware Routing in Ad Hoc Wireless Network. In SITE, University of Ottawa, TR-98-11, Dec. (1998)
5. W. R. Heinzelman, A. Chandrakasna and H. Balakrishnam.: An Application-Specific Protocol Architecture for Wireless Microsensor Networks, IEEE Transaction on Wireless Communications, Vol. 1, No. 4, Oct. (2002) 660–669
6. M, Younis, M. Youssef and K, Arisha.: Energy-aware management for cluster-based sensor networks. Computer Networks, No. 43, (2003) 649–668
7. E. H. Callaway.: Wireless Sensor Networks - Architectures and Protocols, Auerbach publications, (2004)
8. IEEE Standard 802.11 for Wireless Medium Access Control and Physical Layer Specifications, Aug. (1999)

COPRA – A Communication Processing Architecture for Wireless Sensor Networks

Reinhardt Karnapke and Joerg Nolte

BTU Cottbus
{karnapke, jon}@informatik.tu-cottbus.de

Abstract. Typical sensor nodes are composed of cheap hardware because they have to be affordable in great numbers. This means that memory and communication bandwidth are small, CPUs are slow and energy is limited. It also means that all unnecessary software components must be omitted. Thus it is necessary to use application specific communication protocols. As it is cumbersome to write these from scratch every time a configurable framework is needed. COPRA provides such an architectural framework that allows the construction of application specific communication protocol stacks from prefabricated components.

1 Introduction

Sensor networks are collections of small sensor nodes with wireless neighbourhood broadcast facilities. Since sensor networks shall be deployed in large scales (possibly thousands of nodes [1,2]), the overall cost dictates the use of cheap but simple radio transceivers for communication. The latter lack most of the common capabilities of WLAN or bluetooth networks. Even typical tasks like medium access control or the addressing of individual nodes in the direct radio neighbourhood are entirely left to software layers [3]. To make things worse the scarce CPU/memory resources of the sensor nodes do not allow to waste much space and processing power to process complex communication protocols [4]. Thus the designer of the communication software is stuck between a hard place and a rock: the simplicity of the radio requires much more work to be done by the CPU while the processing resources that are needed for this job are scarce. Consequently, communication protocols must be designed as close as possible to their intended use and the processing of the protocol stack must be dedicated to a specific user profile. However, designing application specific protocol stacks from scratch is always cumbersome and error prone.

This paper introduces COPRA[1], an architectural framework for the construction of application specific communication protocols in wireless networks. In COPRA often recurring protocol processing tasks are encapsulated in reusable components (so-called Protocol Processing Stages, PPSs) that can be composed to application specific protocol processing engines (PPEs). Thus application specific protocols do not need to be designed from scratch but can be composed from prefabricated elements.

---

[1] COPRA is part of the COCOS Project which is supported by the German Research Foundation (DFG) in the SPP 1140.

The following sections are structured as follows: section two looks into CO-PRA's structure, section three shows implementation details and section four briefly outlines other attempts in this area. We finish with a look at current status and future work in section five and the conclusion in section six.

2 The COPRA Framework

COPRA is a library of protocol processing stages (PPSs) and a few already defined protocol processing engines (PPEs). A PPS is a special task in communication such as medium access, a PPE is a concatenated set of PPSs. By concatenating only the needed PPSs into a special PPE a lot of memory is saved. Each PPE can again be a part of a larger PPE. In figures one to three you will see examples of PPEs. The PPE seen on Figure 1 includes transceiving, medium access control and error checking, which normally is done by hardware. In case of sensor networks this part must also be managed by COPRA because the cheap radios do not supply such functionality. The example on Figure 2 uses the broadcast PPE from Figure 1 as basis and adds address management. Note that the type of address is entirely configurable. It can be a number or a geographic location or even some property on the node, e.g. the value of the last temperature sampling. The third example on Figure 3 adds multi hop functionality to the PPE.

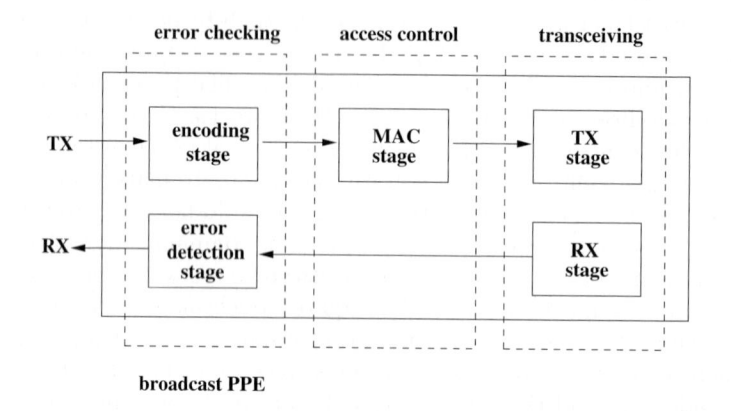

Fig. 1. A single hop broadcast PPE

In the following sections we take a closer look at the two important parts of PPS and PPE.

2.1 A Protocol Processing Stage (PPS)

Most of COPRA's PPSs have a predecessor and a successor, with the exception of the end pieces of a PPE which have only one of them. PPSs normally consist of two parts which represent the direction the data flows: From the upper layers to the lower ones which is the transmitting (Tx) path and the opposing,

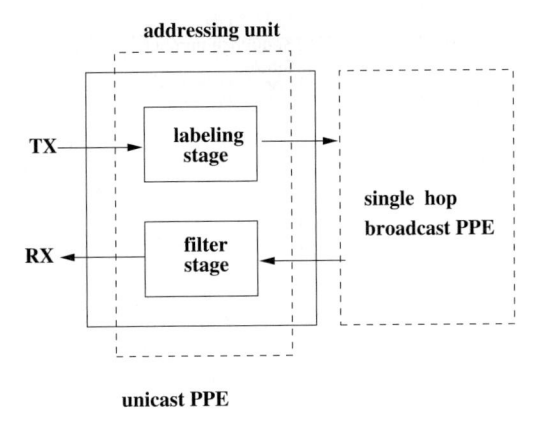

Fig. 2. A single hop unicast PPE

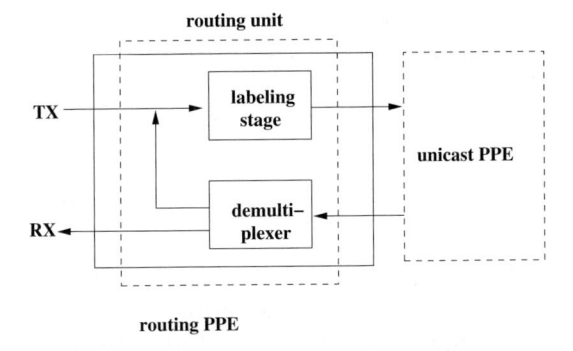

Fig. 3. A routing PPE

receiving (Rx) path. To take this into account we provide the classes `RxStage` and `TxStage` from which a PPS has to be derived. An example for an end piece is the radio which does not have a successor because it transmits the data via hardware drivers. The data is represented as the data structure stack with the well known methods a stack supplies, the type of the stack is configurable as template parameter.

2.2 The Protocol Processing Engine (PPE)

A protocol processing engine consists of a number of PPSs or other PPEs which are linked together. These links represent the transmitting and receiving chains which were already mentioned. Note that the layout is freely configurable. The end pieces of a PPE connect to the application on one side and the hardware drivers on the other. The `Radio` stage does not have a successor in the `TxChain` but uses the interfaces provided by the hardware drivers to transmit the data packets to another node. On the other node the `Radio` stage is the beginning of

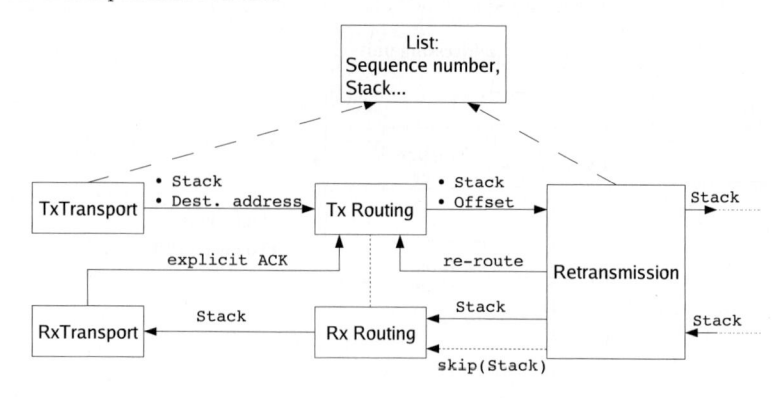

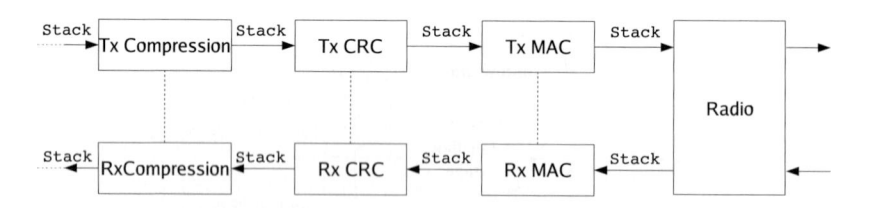

Fig. 4. A complex PPE which is used in our project

the `RxChain` and fills a stack with the data it receives from the hardware. The radio then forwards the stack along the `RxChain`.

Figure 4 shows the largest PPE we have constructed yet. You might notice that the lower half of the picture which contains physical (radio), mac, error correction (crc) and compression follows the scheme mentioned before, where only `rxForward` and `txForward` are used. The upper half splits with the general concept as cross-layer issues arise. The `Retransmission` stage for example shares a data structure with the `Transport` stage. This is necessary because they use the same sequence numbers. The cross-layer issue between the `Retransmission` stage and the `Routing` stage arises from the fact, that retransmissions may fail repeatedly. Then, the `Routing` stage is informed that it has to find new routes. Because of all these issues we tend to see the upper half of the picture as a single entity.

3 Applying the COPRA Framework

When we want to use COPRA we configure it for a specific application. Lets assume that for this application we need to create a new PPE as none of the existing PPEs fits. Lets also assume that there is one particular PPS we need that does not exist either. For this reason we will now take a look at how a PPS is build.

3.1 Implementing a PPS

As example for a PPS the `FilterStage` is discussed here. Its job is only to forward incoming data packets on the `RxChain` if they are addressed to this node (including broadcast). Note that we are using reference counting to determine if the memory can be reused so we only decrement the reference count if the stack is unwanted. Please note also that the address is configurable as template parameter. This way it is up to the user to decide whether to use numbers, geographical identities ore even sensor values for addressing. As the `FilterStage` is a member of the `RxChain` it has to be derived from `RxStage`. In the `accept()` method the address of the destination is taken from the stack and compared to this node's id and the broadcast address. Only if one of these matches the stack is forwarded along the RxChain.

```
template <typename Address>
class FilterStage : public RxStage<Stack> {
  ...
  // called by previous stage in the RxChain
  virtual void accept(Stack* stack)
  {
    Address id;
    // get destination address
    stack->pop(id);

    // test if the packet is addressed to this node or the
        broadcast address
    if((id == myID) || (id == broadcastID))
      rxForward(stack);  // send stack to the next stage
    else
      stack->downRef(); // free memory
  }
}
```

In this example it is easy to see what a user has to do to construct a PPS. To build a member of the `Rx-/TxChain` the PPS has to be derived from `Rx-/TxStage`. The method in which all the work is done is called `accept()` in the RxChain and `deliver()` in the TxChain. This is the only method the designer of the PPS has to fill. When all work is done the method `rx-/txForward()` has to be called, which delivers the stack to the next stage by calling `accept()` (`deliver()`) on it. The forwarding methods are inherited so there is no need for the designer to touch these. They hide the identity of the succeeding stage.

Now that we have build the PPS lets take a look at how a PPE is constructed.

3.2 Composing a PPE

To build a PPE we need to have PPSs. As we have already build these we now have to connect them in the desired order. The following example is a datagram network (`DtgNet`). In this example you will notice that there are not a `RxRadio` and a `TxRadio` but only one `Radio` that works as both. The rxMac is omitted, because all a receiving MAC-layer would do is removing the MAC Header and we do not use any. This is because the data sampled by sensor nodes is normally small and we do not want to waste bandwidth and energy on unnecessary overhead. This PPE enables the application to use the standard way of sending by simply giving an address, a pointer and a length to the PPE's `send()` method. It also provides the method `receive()`, which allows the application to receive messages in the standard form. To receive a message the application supplies a buffer which should be filled with the message. After this is done, the number of received bytes is returned.

The connecting of the PPSs is done in the constructor of the PPE. First the receiving chain is built, then the corresponding transmitting chain follows. The methods `receive()` and `send()` are called by the application and offer the services mentioned above. They take care of memory management by selecting stacks from a pool and returning them once they are not needed anymore.

For simplicity reasons we omitted a few details, e.g. the check whether the buffer is big enough.

```
class DtgNet {

// the elements of the PPE
Pool pool; RcxRadio radio;
TxMac txMac; RxCRC rxCRC; TxCRC txCRC;
LabelingStage<Address> labeling;
FilterStage<Address> filtering;
MessageQueue msgQueue;

// Constructor.
// Here all parts of the PPE are assembled.
DtgNet()
{

  // build receiving chain
  radio.rxConnect(&rxCRC);
  rxCRC.rxConnect(&filtering);
  filtering.rxConnect(this);

  // build sending chain
  labeling.txConnect(&txCRC);
  txCRC.txConnect(&txMac);
  txMac.txConnect(&radio);
}

int receive(char* buf, int size)
{
  Stack* stack = msgQueue.get()
  if(!stack) // no message in the queue
    return 0;
  int used = stack->used(); // determine needed memory
  memcpy(buf, stack->tos(), used); // copy message
```

```
  stack->downRef(); // free memory
  return used; // return size of message
}

void send(char* msg, unsigned size, Address address)
{
  // try to get a new stack from pool
  Stack* stack = new (pool) Stack();
  if (stack) {
    void* buf = stack->alloc(size); // allocate memory
    memcpy(buf, msg, size); // copy message
    labeling.deliver(stack, address); // forward stack
  }
}
```

As you see it is very easy to construct a PPE. By calling `rx-/txConnect` on a PPS we connect it with its successor on the receiving (transmitting) chain. These methods are inherited from `Rx-/TxStage` so again there is no need to care for them. Also, in this example the great benefit of COPRA's modularity can be seen. Lets assume that the MAC Layer used above uses TDMA. Now we may need a different MAC for a different environment but all the rest should stay the same. We then replace the `txMac` with `txCSMAMac`. Now all we have to do is connect this stage instead of the original one and we are done. Another possibility to change this PPE would be to remove one unit, e.g. the addressing unit as seen in figure 2. All this is up to the user to configure. By supplying a variety of stages for all Layers we give the users an easy way to configure individual PPEs according to their needs.

3.3 Writing an Application

Now that we have PPSs and a PPE lets take a final look at the application. What the application does is of course up to the user but the easiest way to use a PPE will be discussed here. There are in fact two ways for an application to use a PPE. One possible way is for the application to be the end piece of the receiving chain or the beginning of the transmitting chain. This way the application needs to inherit from `RxStage` or `TxStage` or both. This may seem a little drawback but it enables the application to use `txForward()` and work with the `accept()` method. It also has another advantage which will be seen when the second way is discussed. The second way is for the application to use a PPE with a special end piece, which allows the usage of standard communication interfaces. This end piece would offer a `send()` method which gets a pointer to the message and its size. In this method it would allocate a stack, copy the data and forward the stack. The advantage of this method is clear. The application does not need to worry about stacks, it does not even need to know it is using a PPE. The disadvantage lies in the end piece of the PPE. It has to copy the message to a stack which takes time. It also costs additional memory on the sensor nodes. An application would use the PPE seen above like this:

```
DtgNet net(myID);
Message msg;
...
// sending
net.send(4711, &msg, sizeof(msg));
...
// receiving
int size = net.receive(&msg, sizeof(msg));
...
```

Please note again that while in this example the address is a number it is entirely up to the user what type of address is being used.

Now that we have seen how the COPRA framework can be used, lets take a look at the cost of using it.

3.4 Code Size

As mentioned above sensor nodes are limited in memory and have slow CPUs. In this section we take a closer look at the size of our framework. There are two figures which go into the code size. First, the size of the code which is independent of COPRA as it would exist even if the framework was not used. Second, the overhead of using the framework. This overhead can be determined as follows:

Each stage has a pointer to its successor, the connecting method and the forwarding one. Also a `vtable` is needed for the inherited functions and the calls to the connecting methods must be made. Finally the call to the constructor of the PPE in which the connections are made needs to be considered.

Two things are included for every PPS, the pointer to the next stage and the `vtable`. The size of these depends upon the CPU in use. In our experiments we use Lego RCX robots [5] which include a Renesas H8/300 processor. This is a 16 Bit processor with a clock frequency of 16 MHz. On a 16 Bit processor the size of a pointer is two bytes which means that the overhead for one PPS includes 2 bytes for the pointer to the next stage, 2 bytes for the pointer to the `vtable` and 6 bytes for the `vtable` itself. Altogether this means an overhead of 10 bytes per PPS.

There are also the inlined connecting and forwarding methods and the constructor of the PPE. As these exist only once for the framework they are not taken into account here.

The next figure shows code sizes of two selected PPEs. The sizes were measured on the RCX robots we used for our experiments. As these sizes are dependent on the CPU in use they may vary on different systems.

PPE	buffer pool	radio	mac	crc	addressing	size (bytes)
broadcast	x	x	x	x		3400
unicast	x	x	x	x	x	3848

4 Related Work

In sensor networks the communication cost is reduced by replacing part of the communication with local computation. While this is a great improvement in battery lifetime it also means that the communication must be done in an application specific way. The authors of [6] call for a family of protocols for general purpose sensor nets. With COPRA such a family exists, as the framework represents a lot of different communication protocol stacks that can be configured according to the applications needs.

COPRA is partly inspired by CORBA and .NET. The channel sink chains in .NET are configurable, meaning that the user can insert whatever sink he needs. These chains are reflected in COPRA's `Rx-/TxChains`. An important difference is however, that COPRA's chains starts where .NETs sinks end. The lowest of .NETs sinks is the `TransportSink`, whereas COPRA is a communication framework. The portable interceptors in CORBA were also an inspiration, as it is possible to insert additional interceptors into a chain. This is reflected in COPRA's PPSs which are connected in a PPE. While CORBA has a predefined order, the PPSs in COPRA can be inserted anywhere in a PPE.

In [7,8] the lack of an overall sensor network architecture is remarked. The authors describe the need for a sensor network protocol which should be located lower than the IP-Layer in the internet. While this so called SP should provide a set of functionalitys it should still stay configurable and be open to cross-layer issues. COPRA offers the configurability and openness required.

5 Current Status and Future Work

At the moment we have 14 different PPSs and 8 PPEs. While this number may not seem very large, it is not necessary for it to become much larger. We are experimenting with some of our PPSs and PPEs on modified RCX robots. These Robots have been additionally equipped with an easy radio ER400TRS radio module which we use instead of the included infrared module (IR). To enable this, a serial port has been inserted which allows us to connect either the IR or the radio module. The IR is still needed to boot the RCX robots but once they are booted we switch to the radios. COPRA is independent of the operation system used, but we decided to use our self developed miniature OS `Reflex`[9] as basis. `Reflex` supports pre-emptive scheduling and provides hardware drivers which we use in some of or PPSs.

In the near future we will need to implement a few more different PPSs for each layer. Once we have these there could be more PPEs and application examples. But it is not our focus to find new applications for sensor networks, only to offer an easier way to build them. Also it is not our goal to build lots of PPEs. That is not necessary as the users will build their own ones. Right now we are using the RCXs only but we are going to equip these with ScatterWeb[10] sensor nodes. This is necessary because the RCXs have only three input channels and the additional serial port, which are connected to touch sensors and the radio.

When we connect the ScatterWeb sensor nodes with the serial port we will be able to use their radio and have their additional sensors.

6 Conclusion

COPRA is an easy to use framework which allows a user to plug and run communication protocols for sensor nodes without having to rewrite the application each time a different hardware is used or the environment is different. Developers can now focus their attention entirely on the application. COPRA performs well in our experimentation environment and we are positive that it will work equally well in the next experiments using the ScatterWeb sensor nodes.

References

1. Chatzigiannakis, I., Nikoletseas, S., Spirakis, P.G.: Efficient and robust protocols for local detection and propagation in smart dust networks. Mob. Netw. Appl. **10**(1-2) (2005) 133–149
2. Chatzigiannakis, I., Nikoletseas, S., Spirakis, P.: Smart dust protocols for local detection and propagation. In: POMC '02: Proceedings of the second ACM international workshop on Principles of mobile computing, New York, NY, USA, ACM Press (2002) 9–16
3. Kahn, J.M., Katz, R.H., Pister, K.S.J.: Next century challenges: mobile networking for smart dust. In: MobiCom '99: Proceedings of the 5th annual ACM/IEEE international conference on Mobile computing and networking, New York, NY, USA, ACM Press (1999) 271–278
4. Sohrabi, K., Ailawadhi, V., Gao, J., Pottie, G.: Protocols for Self Organization of a Wireless Sensor Network. IEEE Personal Communication Magazine **7** (2000) 16–27
5. Patterson-McNeill, H., Binkerd, C.L.: Resources for using lego mindstorms. In: Proceedings of the seventh annual consortium for computing in small colleges central plains conference on The journal of computing in small colleges, , USA, Consortium for Computing Sciences in Colleges (2001) 48–55
6. Heidemann, J., Silva, F., Estrin, D.: Matching data dissemination algorithms to application requirements. In: SenSys '03: Proceedings of the 1st international conference on Embedded networked sensor systems, New York, NY, USA, ACM Press (2003) 218–229
7. Culler, D., Dutta, P., Ee, C.T., Fonseca, R., Hui, J., Levis, P., Polastre, J., Shenker, S., Stoica, I., Tolle, G., Zhao, J.: (Towards a sensor network architecture: Lowering the waistline)
8. Polastre, J., Hui, J., Levis, P., Zhao, J., Culler, D., Shenker, S., Stoica, I.: A unifying link abstraction for wireless sensor networks. In: SenSys '05: Proceedings of the 3rd international conference on Embedded networked sensor systems, New York, NY, USA, ACM Press (2005) 76–89
9. Walther, K., Hemmerling, R., Nolte, J.: Generic trigger variables and event flow wrappers in reflex. In: ECOOP - Workshop on Programming Languages and Operating Systems. (2004)
10. Schiller, J., Liers, A., Ritter, H., Winter, R., Voigt, T.: Scatterweb - low power sensor nodes and energy aware routing. In: Proceedings of the 38th Hawaii International Conference on System Sciences. (2005)

DAEDALUS – A Peer-to-Peer Shared Memory System for Ubiquitous Computing

Peter Ibach[1], Vladimir Stantchev[2], and Christian Keller[1]

[1] Lehrstuhl Rechnerorganisation und Kommunikation, Institut für Informatik,
Humboldt-Universität zu Berlin
Unter den Linden 6, 10099 Berlin
{ibach, keller}@informatik.hu-berlin.de

[2] Net Business Center, Fachgebiet Systemanalyse und EDV, Technische Universität
Berlin, Franklinstrasse 28/29, 10587 Berlin, Vladimir
Stantchev@sysedv.tu-berlin.de

Abstract. Data sharing in a large scale and for high volatility tolerance requires peer-to-peer solutions where traditional multiprocessor shared memory systems are not applicable. Efficiency of those P2P shared memory systems depends, in particular, on scale, dynamics, and concurrent write accesses. We have developed a P2P shared memory solution, DAEDALUS, based on SUN's JXTA framework, and integrated an efficient stochastic locking protocol, proper resource clustering, and semi-hierarchical grouping of nodes. We evaluated the applicability under heavy load, scale, and node mobility. Here, DAEDALUS outperformed a client/server system and solved its inherent scalability problem.

1 Introduction

Shared memory systems provide the foundation for efficient development of distributed applications. A lot of mature shared memory solutions for multiprocessor systems exist. However, data sharing in a large scale and for high volatility tolerance – typically occurring in ubiquitous computing scenarios – is still unaccomplished and has become an active field of research. Under such conditions, peer-to-peer architectures provide advantages over traditional distributed architectures with classical shared memory approaches. On the other hand there are numerous shared memory systems that are well designed for large amount of write accesses, but those are usually intended for supercomputers or cluster computing. However, none of these systems fully cover issues arising in ubiquitous computing scenarios where network topology and quality of service parameters are subject to frequent changes. Providing a synchronized and consistent view on the shared data for all participants with reasonable communication overhead, accordingly, is challenging and requires proper utilization of caching, routing, grouping, data compression, cryptography, forwarding, and consensus.

Therefore, we have developed DAEDALUS, a platform-independent and lightweight framework for peer-to-peer communication. It enables mobile/embedded devices to easily and efficiently share their data. Data may be distributed among thousands of peers and subjected to permanent updates, while further dynamics induced by mobility or environmental changes remain transparent to users or application developers. Devices may join or leave groups in an ad-hoc manner and can be members of an arbitrary number of groups at the same time, while our framework keeps the data *stochastically* in sync and consistent among all members of any group even in case of numerous concurrent write accesses. Our approach therefore uses *stochastic locking* and *semi-hierarchical grouping*. The implementation is based on SUN's JXTA Java classes for peer-to-peer communication. As a case study, we have integrated it in MagicMap, a cooperative WLAN positioning system. The client/server communication here did not scale well and required reliable connectivity. Both problems could be successfully solved using DAEDALUS, which achieved significant improvements regarding dependability, scalability and performance.

2 The MagicMap Application Scenario

MagicMap is a cooperative context aware computing application we introduced in [1,2,3]. Every node senses its environment and uses the observed data to calculate its location and situation. From that, location/situation specific actions can be triggered. The system works cooperatively, i.e., nodes exchange their measurements among each other. Calculations can be done redundantly on multiple nodes to improve fault tolerance, in particular, to prevent a minority of malicious nodes to affect system stability. In our current implementation we use WLAN equipped Laptops, PDAs, and Smartphones that exploit WLAN signal strength to sense the environment and calculate their positions (see Fig. 1). Nodes sense the WLAN received signal strength (RSSI) of neighboring nodes (access points, other clients, or previously measured reference points) and estimate the physical distance. A *spring layout* algorithm moves the nodes with unknown positions such that length of edges best match the calculated physical distance. Thus, the graph converges to a "magic map", where nodes are located approximately at their true physical position.

Since different nodes may calculate devices positions, the calculating nodes need access to signal strength measurements. Consider the following scenario shown in Fig. 2. Node C wants to know the position of node B. Node C, as well as node B, has low processing capabilities. Node A has high processing capabilities and therefore calculates the position of node B. Node E, being sufficiently capable as well, does that calculation too for redundancy reasons. All WLAN-aware nodes sense signal strength (1) and forward it to the nodes where calculation is done (3).

Note, that in this scenario we assume the mobile clients A, B, and C to sense the signal in a symmetric manner, i.e., A senses signal strength from B and symmetrically, B can sense signal strength from A. Some nodes may not sense the signal, in our case node D, which might be an access point or a peer node without MagicMap installed. However, given D is using its WLAN interface, its radio signals can be sensed by other nodes (2).

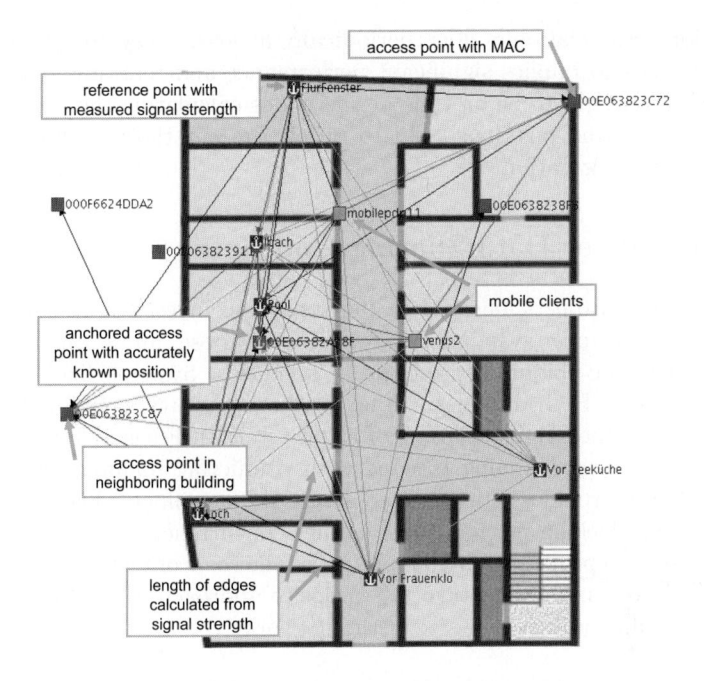

Fig. 1. MagicMap screenshot

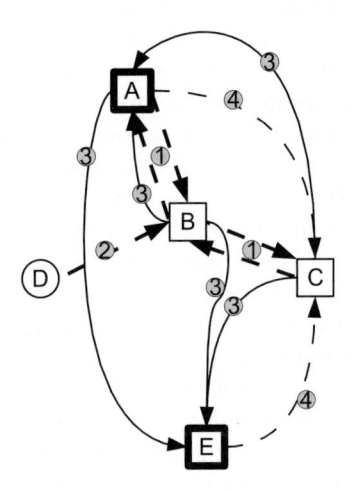

Fig. 2. Example scenario with high-performance nodes A and E and low performance nodes B and C

Finally, the calculated positions are sent to node C (4) who then can use, for example, the mean value of both calculations as best position estimation. In case C receives three or more independent position estimations, it could employ elaborated voting algorithms for improved fault tolerance and resilience against malicious behavior. To provide a real-time picture, signal strength measurement and position recalculation is done periodically at least every 10 seconds. Obviously, this scenario implies significant performance and real-time demands: all measured values need to be on time at the nodes calculating the positions, and finally, all calculated positions need to be on time at those nodes, that have interest in this information.

3 Peer-to-Peer Data Sharing Concepts

Several research projects emerged in the last years investigating efficient data updates in peer-to-peer systems. However, they impose limitations that reduce their usefulness in ubiquitous computing scenarios. Systems like Freenet [4], OceanStore [5], or P-Grid [6] assume no conflicting writes, going as far as limiting updates to the original author of a data item in Freenet. Ivy [7] requires application-level programming to cope with conflicting manipulations of data objects and only provides some tools to detect those conflicts. These systems do not provide any locking mechanisms or other concurrency protocols since their main purpose is to provide high scalability – at the costs of sacrificed consistency. Systems such as JuxMem [8] take the opposite approach: they provide locking mechanisms while limiting the size of the network.

3.1 Concurrency Control – Pessimistic and Optimistic Approaches

There are two opposed approaches for concurrency control, the pessimistic and the optimistic one. The first assumes that conflicting write accesses to a data item might cause intolerable inconsistency and thus have to be avoided anyway (conflict prevention). To guarantee that no other node is performing a concurrent write access to any replica of a data item, a node has to lock that item to prevent it from other concurrent manipulations. In a distributed scenario, this requires two-phase locking, i.e., the node has to wait for all item replicas to confirm the lock request. After the write has been performed, all replicas have to be updated accordingly to obtain a consistent state. Meanwhile, since the data might be temporarily inconsistent, additional write or read accesses to it are not allowed.

Pessimistic locking, hence, is not applicable in highly dynamic networks where typically presumed latencies cannot be guaranteed.

Therefore, an optimistic approach, in contrast, assumes that temporary inconsistency resulting from concurrent writes to a data item is tolerable. It employs conflict resolution instead of the above conflict prevention. Optimism is accounted for the assumption that the number of actual conflicts and resolving them will be manageable and temporary inconsistencies will be rare. If, however, a conflicting update occurs, nodes have to use roll back or roll forward mechanisms to resolve inconsistency and recover a consistent system state. An example is Ivy, which stores the history of operations that have been performed on the items. It does not resolve conflicting updates, but it detects them and provides application-level means to resolve them.

Instead of pessimistic or optimistic conflict handling, its also possible to create a disjoint global storage space, such that conflicts cannot occur. Freenet, for example, combines keys for files with a private key, specific to a user, and thereby creates a global name space with private subspaces. This however, would result in unmanageable network traffic and does not fit the MagicMap scenario where every device can publish estimates of other devices' positions.

Since none of the above approaches seem appropriate for our purpose, we have employed a hybrid approach (see Section 4.2).

3.2 Considering Different Node Capabilities

All above systems assume the peers to possess comparable capabilities. This assumption, although it might be acceptable in workstation environments, is unrealistic in heterogeneous networks of ubiquitous computing. Therefore, caching a snapshot of the overall storage system as required by Ivy is only feasible for very small distributed file systems. OceanStore does allow multiple nodes to change a single data item. To prevent faulty nodes from publishing wrong version information, a Byzantine agreement is formed between all primary replicas. OceanStore however, as well as P-Grid, does not offer means to prevent or resolve conflicting write accesses to the same data item. Since MagicMap updates position information rather frequently, such peer-to-peer systems are likewise not appropriate.

4 The DAEDALUS Peer-to-Peer Shared Memory System

The system architecture is divided into platform dependent and platform independent components, see Fig. 3. Measurements of signal strength and collection of other sensor data is highly dependent on particular hardware, operating system, and drivers. The platform independent components – in particular the DAEDALUS shared memory and the normalization and calculation of position estimations – are written in Java. All components are freely available via our website www.magicmap.org.

4.1 Peer Groups

The basic idea of our shared memory system is to assign every data item a specific peer group. Peers that have interest in this data item join the related group and serve as a replica. The advantage of this approach is scalability. Thereby, the amount of messages send does not depend on the number of nodes participating in the entire system, instead it depends on the number of peers interested in this data item. While the number of nodes in a network could become rather big in real world scenarios, the number of peers interested in a specific data item is limited. The idea, however, has a downside: once no peer is interested in a data item, it will be lost. To prevent this, nodes having enough resources to join multiple groups in parallel will be asked to join this group, in case the number of member nodes is decreasing below certain threshold. As these groups still can grow rather big, a further differentiation is needed. A percentage of all nodes in this group acts as a manager. Managers act as replicas, vote on locking requests, and keep track of the group size.

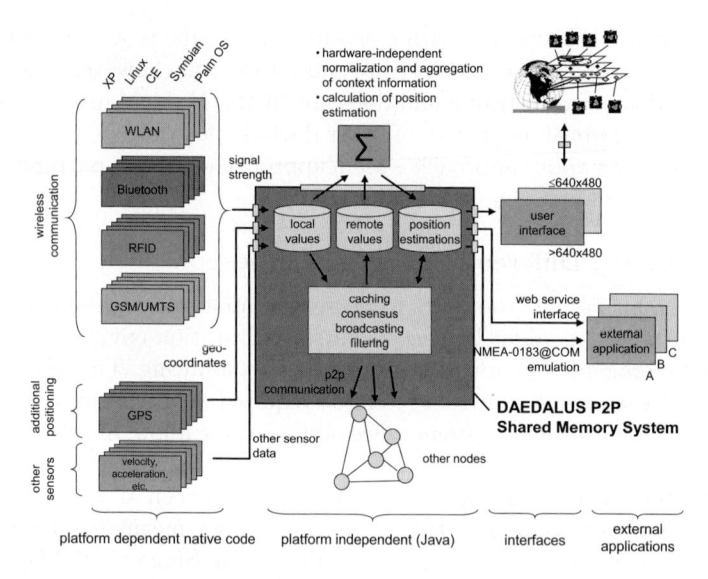

Fig. 3. System architecture with platform dependent and platform independent components communicating via the DAEDALUS peer-to-peer shared memory

4.2 Stochastic Locking – A Hybridization of Pessimistic and Optimistic Concurrency Control

Since both, pessimistic and optimistic approaches are not feasible in our scenario, we pursue a hybrid approach. We use a locking mechanism but we do not require all nodes to answer a lock request. Instead, only a relatively small number of nodes has to answer and broadcast their decision to all managing nodes in a group as shown in Fig. 4. The requesting node has successfully locked a data item, if a majority of those answers is positive. This approach is optimistic, as it assumes that enough nodes receive the lock request messages and there are only a few faulty nodes that give an insane answer regarding a request. It is as well pessimistic to a certain degree, as it reduces the number of conflicts by locking a resource before updating it. While this *stochastic locking* cannot *guarantee* that no conflict occurs, it does provides a high probability of conflict prevention.

4.3 Scalability Considerations

As only a fixed number of managers is required to answer a client request, the expected traffic for each update process is limited and known. However, in order to ensure that the number of managers answering a request does not exceed the threshold, the managers have to keep track how much of them are in a group. Therefore, every peer joining a group broadcasts a hello packet to all managers. A fixed number of managers will provide the new member with all necessary information, such as group size and a list of managers. Every peer node joining the group starts as a manager. If the peer later discovers that there are already

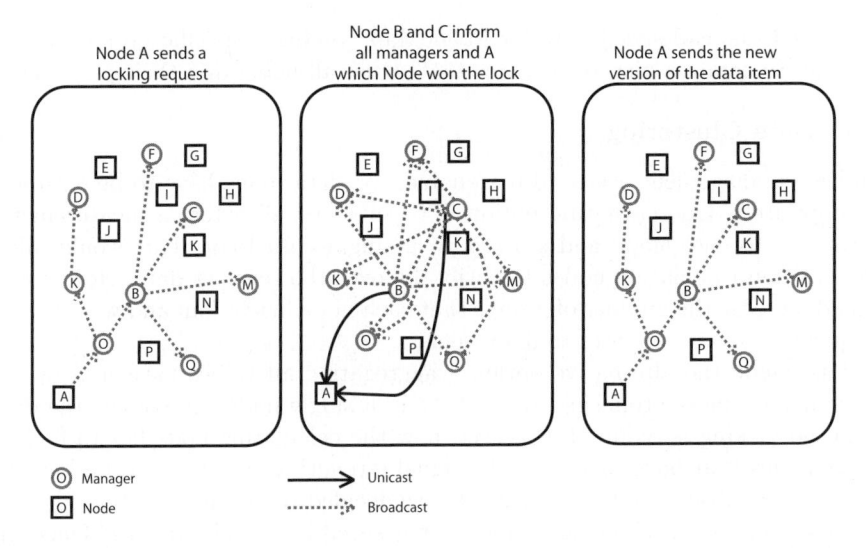

Fig. 4. The locking process. Node A sends a lock request to all manager nodes. Of those manager nodes B and C respond and broadcast their decision to all other managers.

enough managers it can alter its status and become a regular client. Additionally, managers check whether there are still enough managers in the group, and ask clients of the group to become managers, if the number falls below the threshold. On the other hand, if a manager detects that there are not enough members in its group, they call other nodes that still have enough resource capacity available to enter the group.

While the load of a client is independent from the number of nodes in the group, the load of managers does grow with the size of the group. For a single process the load is constant. However, as only the request for data items can be balanced over all managers, the load for writing, locking and counting is not. Therefore, the number of messages a manager has to process increases linear with the number of nodes in the group. This however does not compromise the original goal of low load for small computing devices. As the chance for such a device to be a manager decreases with group size, the load for small devices will not grow beyond a point which depends on the ratio of small and large nodes within it.

4.4 Integration into MagicMap

We implemented the peer-to-peer shared memory system as a Java application which communicates with any local application via UDP datagrams. It supports calls to read the data item of a given name, to store a new version and to lock and unlock the data item. Additionally, we included calls to search for groups and peers. As group names are the same as their data item's name, a search for all groups will result in a list of all available data items. By applying a name scheme an application can easily search for all data items it needs. We have developed one call specifically for MagicMap: joining a specific group. We use this call to create a hierarchical tree that stores all nodes and their positions.

4.5 Data Clustering

To keep management overhead reasonable, the data items have to be clustered appropriately. One clustering option is to subsume all external measurements according to each node and store it in a single data item. This would allow the position calculating nodes to easily discover the relevant data. However, it would increase the number of groups that each node has to join and would cause frequent locks and updates to data items.

This made the alternative option – aggregating all values measured by the same node – most promising to us. As only a single node will change the data item, no locking is required. However, now the calculating node has to find all other nodes that have measured the signal strength of the node to be located. To make the discovery process feasible, we decided to add a data item for each node to store a list of the nodes having measured its signal strength. Thus, the calculating node can scan the list and find all the data items required to calculate the node's position. As this node list has to be updated by different nodes, locking is required. Fortunately, the number of updates to the list typically remain in a manageable amount.

The position values for each node are stored in a single data item. As there are typically less than five nodes actually updating this data item, this does not cause heavy load. We end up with three data items for every MagicMap node. For a node A there are A-*Measurements* where this node stores all signal strengths it sensed, A-*See* stores all nodes that sense signals from A and A-*Position* contains the calculated position of this node. A node that wants to know the position of node A accesses A-*Position*. If no other node has yet calculated the position and the data item is empty, this node may want to calculate the position itself. To do so the node first reads A-*See* and then accesses all measurement data items of the nodes in this list.

5 System Evaluation

We conducted our tests using the MagicMap application as a case study and compared the delay of data item updates in the client-server setup to the DADALUS peer-to-peer setup at different numbers of participating nodes (see Fig. 5).

In the client-server setup, updates were done via a centralized server using Web Service communication. The peer-to-peer setup utilized JXTA broadcast/unicast and comprises locking the data item, updating it, and finally releasing the lock.

For both setups we employed 8 Dell PDAs as "low capable" nodes and 8 desktop computers as "high capable" nodes and simulated further nodes. The ratio of low to high capable nodes was kept at constantly 1:1. We tested each setup for a period of 6 hours and repeated the measurement three times at different days. While we consider the obtained result quite realistic, true real

#Nodes	C/S Avg. Delay	C/S Standard Deviation	P2P Avg. Delay	P2P Standard Deviation
10	2.5 s	11 s	1.0 s	461 s
20	3.5 s	12 s	1.6 s	810 s
40	7.5 s	13 s	2.9 s	1,103 s
80	12.5 s	14.5 s	4.1 s	1,221 s
120	-	-	4.2 s	1,069 s

Fig. 5. Comparing the data update delay of the standard client-server and the DAEDALUS peer-to-peer setup

world measurement with heterogeneous devices in a magnitude of hundreds or even thousands of nodes have to remain for future work.

6 Conclusion and Outlook

We have proposed a peer-to-peer shared memory system designed for ubiquitous computing scenarios. It provides stochastic locking and data clustering to arrive at reasonable performance even at high scale and dynamics. In our WLAN positioning case study implementation we used relatively well equipped Dell PDAs and measured performance parameters. Using these measurements, we further investigated scalability and other quality of service issues by simulation. The results indicate that, regardless of group size, 95% of all data updates will not take longer than 6 seconds, provided that no conflicting writes occure.

Future work may integrate a way to preserve multiple versions of a single data item. Also a privacy scheme has to be developed to protect data and improve system acceptance – since user locations are definitely very sensitive information.

References

1. Ibach, P., Hübner, T., Schweigert, M.: MagicMap - Kooperative Positionsbestimmung über WLAN. In: Chaos Communication Congress, Berlin, Germany (2004)
2. Ibach, P., Schreiner, F., Stantchev, V., Ziemek, H.: Ortung drahtlos kommunizierender Endgerte mit GRIPS/MagicMap. In: 35. Jahrestagung der Gesellschaft für Informatik, Bonn, Germany (2005)
3. Ibach, P., Stantchev, V., Lederer, F., Wei, A., Herbst, T., Kunze, T.: WLAN-based Asset Tracking for Warehouse Management. In: IADIS International Conference e-Commerce, Porto, Portugal (2005)
4. Clarke, I., Sandberg, O., Wiley, B., Hong, T.W.: Freenet: A distributed anonymous information storage and retrieval system. Lecture Notes in Computer Science **2009** (2001)
5. Rhea, S., Eaton, P., Geels, D., Weatherspoon, H., Zhao, B., Kubiatowicz, J.: Pond: The OceanStore prototype. In: Proceedings of the Conference on File and Storage Technologies. (2003) 1–14

6. Aberer, K., Cudré-Mauroux, P., Datta, A., Despotovic, Z., Hauswirth, M., Punceva, M., Schmidt, R.: P-grid: a self-organizing structured p2p system. SIGMOD Record **32**(3) (2003) 29–33
7. Muthitacharoen, A., Morris, R., Gil, T., Chen, B.: Ivy: A read/write peer-to-peer file system. In: Proceedings of the 5th USENIX Symposium on Operating Systems Design and Implementation (OSDI '02), Boston, Massachusetts (2002)
8. Antoniu, G., Boug, L., Jan, M.: Juxmem: An adaptive supportive platform for data sharing on the grid. Scalable Computing: Practice and Experience **6**(3) (2005) 45–55

Context Awareness: An Experiment with Hoarding

João Garcia, Luís Veiga, and Paulo Ferreira

Distributed Systems Group, INESC ID Lisboa, Portugal
http://www.gsd.inesc-id.pt/
{jog, lveiga}@gsd.inesc-id.pt, paulo.ferreira@inesc-id.pt

Abstract. Computer mobility allows people to use computers in varied and changing environments. This variability forces applications to adapt thus requiring awareness of the computational and physical environment (e.g. information about power management, network connections, synchronization opportunities, storage, computation, location-based services, etc.).

An important application for mobility is hoarding, i.e. automatic file replication between devices. To be accurate and not obstructive to the user, the hoarding mechanism requires both context awareness (e.g. amount of usable storage) and estimation of future environment conditions (e.g. network connection, tasks to be performed by the user in the near future, etc.). However, making applications context-aware is hindered by the complexity of dealing with the large variety of different modules, sensors and service platforms, i.e. there is no middleware supporting such applications and their development in a uniform and integrated way.

This paper presents the architecture for an environment awareness system (EAS) and how it applies to hoarding. EAS is a middleware component that acts as an intermediary between applications and all mechanisms that assess the surrounding environment. It lets applications query and combine environment properties in a standardized way. Crucial for the success of automatic file hoarding is the EAS's capability of supporting environment prediction based on simple reasoning and pattern detection. Thus, applications may advise users accordingly or even make decisions on their behalf.

1 Introduction

Mobile computer technology has led people to use their computers in a wide variety of environments, e.g.: a PC at the office, a laptop at the airport, a PDA in a taxi, etc. Users want to work continuously in this data ubiquitous world taking advantage of available resources and not worrying about any system problem that may occur (such as missing files).

Achieving such ubiquity is hard and depends on many applications. Automatic file replication between devices, i.e. hoarding, is a solution for the fundamental problem of data availability in mobile environments. To be accurate while not obstructive to the user, the hoarding mechanism requires both context awareness (e.g. knowing the amount of available memory) and estimation of future environment conditions (tasks to be performed in the future, etc.).

However, creating or adapting applications, like file hoarding systems, is encumbered by the variety of different modules, sensors and service platforms. This is due to the absence of middleware supporting such applications and their development in a uniform and integrated way.

Furthermore, users want to take advantage of any resources as they move around (for example cheap wireless connections). Therefore, the increasing geographical mobility of devices and the mobility of data among devices require applications to be aware of the environment around them and its risks and opportunities. An optimal evaluation of the context should include not only current conditions but also conditions that may be found in the future. For example, if a user is leaving her office, she wants to take along on her laptop all files needed in order to keep on working on her ongoing tasks. But, if her personal computer knew that she is leaving for a meeting related to a specific task, it would only transfer files related to that task.

Power management is another domain where knowing resource availability is highly relevant: if a device were aware of how long its batteries are supposed to last, it could adjust its energy consumption accordingly.

Fulfilling the above mentioned users expectations raises two major problems for application development and execution:

– The heterogeneity of networks, sensors, platforms and services increases the difficulty of building such applications, making a middleware layer supporting applications and their development in a uniform way clearly desirable;
– Current approaches don't take into account users' past habits and future actions. Thus, developers of context-aware applications and users in general must take a large number of decisions concerning the best usage of computational and physical resources.

Existing methods to provide applications with structured context information [1,2], either limit the information provided to applications to specific domains (relative location, user/device identification) or require programmers to modify their applications each time they want to query a new environment property.

This paper presents the architecture for an environment awareness system (EAS) and how it applies to automatic file hoarding. EAS is a middleware component that supports the interaction between applications and any computer-based mechanism able to provide clues regarding the surrounding environment. It lets applications query and combine environment properties in a standardized way by means of an API providing access to the device's context sensors in a uniform way. Each sensor is represented by an environment perception module (EPM), which is a software component capable of polling and/or forecasting a specific environmental property.

In addition to providing a framework for existing and future EPMs, the EAS enables many synergies between EPMs by aggregating and/or applying logic expressions over data from several EPMs. Most important, an EAS is able to forecast future conditions and/or user actions by detecting patterns in the time series of past conditions and maintains a history of past, scheduled and forecast environment conditions. The EAS analyzes personal information manager data and the history of past events to estimate future location, future user activities, etc... For instance, in the case of automatic file hoarding, the EAS is designed to compare the subject of a user's meetings and the file access patterns during previous meetings with the keyword on her files to determine which files need to be replicated to a user's laptop for the meetings on her immediate schedule.

In summary, managing personal files effortlessly is one of the major problems raised by an environment where users have multiple mobile devices. Automatically replicating files among devices, hoarding, can be greatly aided by an EAS because it enables a comparison between users' file accesses and specific environment conditions. Additionally, since the EAS provides an estimation of future conditions, this can be used to decide which data will be most useful in the predicted future.

In the rest of this paper, we begin by presenting the architecture of the EAS. Then, we list a number of EPMs that could currently be integrated into an implementation of an EAS. We show how the EAS is applied to file hoarding and compare our design with existing systems. Finally, we present some of our conclusions and future work.

2 Architecture

The environment awareness system (EAS) is a middleware component, which provides applications with a simple and structured mechanism to query a device's computational and physical environment.

Applications can perform queries on the current situation or request callbacks when certain conditions are met. Queries return an indication of whether an environment property is within a certain range of values.

An EAS event is a timestamped set of property-value pairs representing a change in environment properties. Each property, such as "AbsoluteLocation" or "NetworkConnectivity", is detected by a particular sensor. Each attribute has a domain describing the values it can assume, e.g. "NetworkConnectivity" can be "None", "Poor", "Medium" or "Good". A value describes the status of a property such as "443.23N 217.98W" (a possible value for "AbsoluteLocation") or "45 min" (a "BatteryTime" value). Queries and callbacks can be directed to the local device or to a remote device.

Additionally, the EAS lets programmers specify that certain conditions are to be associated with a particular label: for example, assigning GPS positions to known locations ("home", "office", etc.) or particular circumstances to specific activities (being in room 13 on Monday morning means the person is in a "Staff Meeting"). Users can submit these labeled situations, which are stored by the system and can be referred to in subsequent queries. The hierarchical organization of environment properties and events combines on one hand a simple way of manipulating information about the hardware and, on the other hand, an expressive mechanism to describe situations to which applications want to respond.

The EAS (Fig. 1) is composed of:

- An API layer accessible to applications,
- A callback registry,
- A schedule of EPM probe actions,
- A repository of past, scheduled and forecast events,
- A forecasting model for calculating probable future conditions,
- Several environment perception modules (EPMs),
- A remote invocation interface.

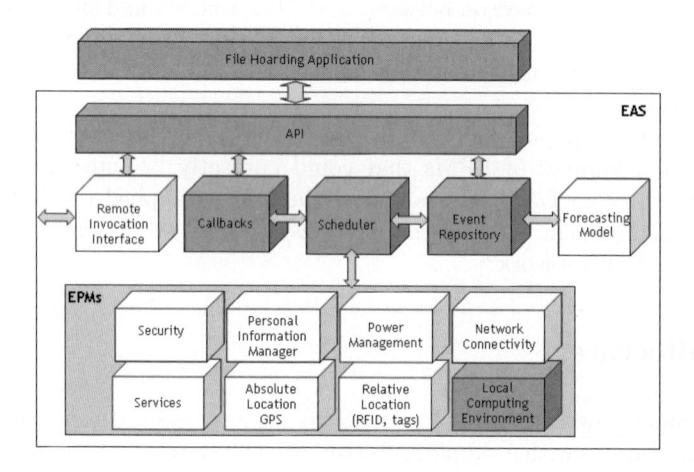

Fig. 1. A file hoarding application on top of the EAS (dark gray boxes were implemented in the prototype of Sec. 3)

Callbacks. Callbacks allow applications to be notified of future conditions. They are associated to changes in the values of one or more environment properties. Callbacks are stored in a data structure indexed by environmental property and remain active until they expire. Whenever a change occurs in an environment property, e.g. "PowerSupply", all callbacks that refer to that property are reviewed to check whether all necessary conditions have been met and the application, which submitted them should be notified.

Probe Scheduler. The EAS periodically probes the physical devices that provide environment property values. Many environment properties are machine characteristics that don't change frequently, if at all. Therefore, they can be stored in the EAS and only be recalculated when the local hardware is reconfigured. There is a schedule of the next moment when each of the available devices must be probed in order to update the properties it detects. Probe actions may be disabled during periods for which there are no registered callbacks.

Event Repository. The event repository stores all changes in environment properties that are detected by the EAS. Observed events are the result of actual probes of EPMs (see Sec. 2.1) whereas forecast events are calculated by the forecasting model. Scheduled events are explicitly inscribed in personal organizer information.

Forecasting Model. A forecasting model was included in order to be able to forecast future situations and to allow applications to perform proactive actions regarding future environment conditions. Many future conditions can be deducted from personal organizer schedules but frequently that leaves a significant amount of time with unknown occupation. This can be complemented with future events forecast based on past recurring conditions. This is performed by an ARIMA[3] forecasting model for discrete variables, which periodically analyzes the event log and tries to detect temporal patterns for each of the detected environment variables and stored labeled situations and inserts new events, adequately tagged as "forecast" into the event repository.

Remote Invocation Interface. An EAS provides a remote service that enables queries between different devices. This opens up the possibility of sharing information among a group of devices so that in some cases other trusted devices can work as extensions of a device. In general, the remote interface is used for queries that refer to neighbouring devices and are most useful to allow applications to take advantage of available resources around it.

2.1 Environment Perception Modules

Environment Perception Modules (EPMs) are the EAS components that interact with physical devices and assess current environment properties. Building them is the greatest challenge in implementing an EAS due to hardware and OS heterogeneity. Each EPM has to include code to assess its properties by probing different devices in different OSs. Currently many of the OS modules and devices that provide environment properties are accessible through standard APIs, which may simplifies the EPM code greatly.

Personal Information Manager. Many computer users run personal organizer software, which can be a source of valuable information regarding the device's future environment. This information isn't presently taken into account by computing systems.

For example, Microsoft's personal organizer, Outlook, provides automation objects APIs, which can be used to programmatically acquire information and store it into a future location table in the EAS. This information could also be acquired from other similar software with a specific EPM.

Once integrated in the EAS, this information can be used to feed the forecasting model. A schedule can be further processed to provide applications with hints and to determine, of those events stored in the event repository, the ones whose corresponding callbacks should be invoked invoked. Thus applications can, asynchronously, send results to hardware not present, assured that data will be sent to it when it connects to the network and becomes available.

Whenever a location is provided by a personal organizer appointment, the situation hierarchy is scanned to check whether that location has been submitted as a situation label. If so, for each of the attributes contained in that situation, scheduled events are inserted into the event repository. Naturally, events scheduled in the personal organizer override forecast events.

Location is a particularly important characteristic of the environment. Therefore, a index of previous locations, and of the corresponding values of the environment properties, is kept in order to derive more information about future locations which are known from scheduled events (see 3.7). Moreover, events, which have a high probability of occurring at a scheduled events location, are added to the event repository as probable events.

Local Computing Environment. This EPM allows applications to obtain information about the software environment in the device, such as, which applications are running, which is the current foreground application and which files and folders are being accessed. Knowing which files and applications are accessed by a user at each device is fundamental in order to configure devices which will be used in the future. A good example of this form of adaptation is the hoarding application described in Sec. 3.

Other. Naturally, the EPMs that were presented above are a fraction of those that will exist and be of interest to applications in the future. Important environment properties we did not discuss are, for instance:

- Absolute location: This EPM would provide GPS coordinates to applications or confirm previously provided labelled locations;
- Relative location and tagging: Currently, there are many ways to provide relative location within a restricted space (WiFi, Bluetooth, RFID). The EAS can uniformly inform applications of their location or of the presence of other device or persons.
- Services: EAS can also be used to integrate and standardize existing service location infra-structures;
- Processing power: CPUs can be benchmarked and classified for applications using a simple description domain;
- Power supply: Time and percentage estimates of battery time are available on most computers today. They can be easily represented as an environment property within the EAS and be used to adjust energy consumption and/or CPU speed.

We can envisage many situations were other environment information and services become relevant: assessing lighting to manage solar charged batteries, using voice processing services at nearby devices, etc. As the EAS' implementation evolves it will incorporate such novelties as new attributes in its event repository.

3 Context-Aware File Hoarding

We chose to use the EAS to address the problem of hoarding, i.e. automatically select the relevant files to be replicated when a user moves from one computer to another (as described in Sec. 1). Automatic file management is an relevant requirement in mobile environments because more and more users have several devices (PC, laptop, mobile phone, PDA) and the storage and bandwidth between them is not constant and unlimited.

The ability to keep a user's files updated on her current device requires first of all that the relevant files be present at that device. It isn't feasible to transfer all of a person's files because many devices have limited storage and bandwidth is often a bottleneck for large transfers. It is impossible to rely on transferring files on demand because network connections aren't constantly available. And finally, selecting files to be replicated manually is time consuming and error prone.

SEER[4] has shown that hoarding files based on the history of most recently used files is the best known heuristic. However, it should be pointed out that this results from a small scale exploration of the parameter space of their algorithms and from assuming that people use only one type of computing device. There are many situations where more sophisticated information is needed. For example, many users do different tasks depending on where they are (at home, commuting, in the office, at a meeting, etc.) and which device they are using (PC, laptop, mobile phone, etc.). Letting users' habits and access patterns determine which files will be hoarded hasn't been tried and can be achieved using the EAS. This would, for example, enable a user to leave her office to do a presentation elsewhere, without worrying about her slides, because the EAS on her PC would have detected that it was necessary to transfer them to the PDA she carries with her.

The first step of hoarding is clustering files so that semantically related files are moved together. Detection of file accesses is performed by the Local Computing Environment (LCE) EPM. Detecting past patterns automatically by correlating file access and environment properties is the job of EAS's forecasting model. Forecasting future actions is essential for a hoarding algorithm that is more sophisticated than just hoarding the most recently used files.

We have implemented a file hoarding application on top of a EAS prototype(Fig. 1). This application monitors user activity and clusters and selects data that should be hoarded in case a user decides to move to another device. Users can assign folders, extensions filenames and applications to any given task. Using that information, the hoarding applications clusters accessed files according to the user's preferences. Only when it is unclear which task a file belongs to, the application then asks whether the file is to be assigned to the current or some other task. This EAS prototype was implemented in C# on Microsoft Windows XP. The local computing environment EPM is composed by a Windows installable file system that intercepts all file system accesses and a monitor of GUI events. So far our experience has shown that the disturbance caused by asking the user to organize her files, quickly fades away. Currently, we log all accesses to selected folders enabling the detection of time patterns in the history of file (and consequently task) accesses. Additionally, personal calendar data allows the EAS to compare the subject of scheduled activities with task and file keywords, thereby improving our hoarding estimates.

4 Related Work

We have been witnessing a trend towards integrating several kinds of devices in an increasingly network-centric manner. Research work developed to deal with the issues raised by this shift have been divided in a number of fields. The most relevant efforts have been made in the areas of location awareness, device/object identification and service location.

Regarding location awareness, in [5] routing efficiency and quality of service is maximized in ad-hoc wireless networks, i.e., networks composed of dynamically repositioning mobile hosts. In this work, location awareness is used at a lower level than in our work. It is used to improve routing algorithms and packet forwarding. Our approach aims at providing applications broader information about their computing environment, its host and its neighbourhood i.e., information about every capability their execution environment supports, and noticing them when these capabilities are subject to temporary or permanent changes.

In the Xerox ParcTab[6] experience, broad work about location and context-awareness, for proximate selection and automatic reconfiguration, ranging from hardware design, user interface customization based on context, and location information is also presented. Context information is based on information about neighbouring hosts. A specially developed predicate language was included for programming context-triggered actions. The system is highly dependent on outside information like responses to homing beacons and positioning devices.

Naturally, these notifications about the execution environment, in the case of network centric applications running in mobile hosts, must include information about nearby devices that may be consulted to obtain such information. This must be implemented with some kind of distributed event processing. There are several approaches to this issue[7,8,9,10]. Bates[7] defends a framework for a federation of heterogeneous components connected, transparently, by distributed events in a publisher-subscriber model. There is an event taxonomy based on event sub-classing. There is an event composition algebra that allows some degree of control over dependency checks between different sets of events and to enforce certain event sequences. Events are logged for future replay or querying.

In Jini[8] the emphasis is put on resource and service discovery. There is only one event class so it lacks expressiveness although each event may contain an arbitrary data object.

This architecture was further refined and extended in Rio[9] with extended event description, capabilities detection for proxy execution, operational strings to represent resources and services and quality of service matching between device capabilities and application requirements.

The Universal Plug and Play[10] approach aims at very similar goals than the previous two but is also centered on remote device and service detection, and data exchanging without any sort of code download. It is supported in a series of standards that makes it more platform independent though less flexible in the dynamic code download aspect. It supports dynamic IP addressing, device and service discovery; control actions are based on SOAP URL invocations and received events are implemented in XML messages defined in GENA.

All these technologies try to take advantage of some form of awareness about the computing environment in some specific way. None of them, though, comprehensively attends to all the properties mentioned in this paper or aims to standardize the representation of environment properties in an extensible manner.

There have also been attempts to create generic context-awareness platforms [11,2,1,12,13,14]. Our decision to design the EAS, came from the realization that some of these platform either were aimed at specific context properties (relative location and user/device identification as in [11,2]), while others require programmers write specific code for each new environment property [1] and that none of them considered knowledge of future conditions as relevant input for device adaptation[13,14]. For example, Gaia[2] results in modified applications that interact with an omni-present infra-structure whereas we would simply like applications to become aware of encircling resources.

5 Conclusions and Future Work

This paper presents an architecture for an integrated environment awareness system (EAS) that allows applications to assess and adapt to the computational, network and physical environments. The system can anticipate future user behaviour based on past patterns in order to take full advantage of the resources available in the future.

We also demonstrate how an EAS can aid the task of automatic file management, in particular file hoarding. Making well informed hoarding decisions requires complex information about users' habits and patterns and these can be acquired and structured by an EAS. We present a prototype of a hoarding application based on a EAS which collects information from the local computing environment in order to perform file clustering and estimates future user file needs by comparing the keywords of accessed user tasks and the subject of calendar scheduled activities.

As future work, we are currently obtaining performance and user experience [15] results on the EAS main features (queries, callbacks) and on the hoarding prototype. We are also refining the design of the EAS architecture modules that weren't implemented for the prototype.

References

1. Salber, D., Dey, A.K., Abowd, G.D.: The context toolkit: Aiding the development of context-enabled applications. In: CHI. (1999) 434–441
2. Shankar, C., Al-Muhtadi, J., Campbell, R., Mickunas, M.: A middleware for enabling personal ubiquitous spaces. In: Workshop on System Support for Ubiquitous Computing (UbiSys '04) at the Sixth Annual Conference on Ubiquitous Computing (UbiComp 2004), Nottingham, UK (2004)

3. Makridakis, S., Wheelwright, S., Hyndman, R.: Forecasting: methods and applications. third edn. John Wiley and Sons, New York (1998)
4. Kuenning, G., Ma, W., Reiher, P., Popek, G.: Simplifying automated hoarding methods. In: Proceedings of the 5th ACM International Workshop on Modeling, Analysis and Simulation of Wireless and Mobile Systems (MSWiM'02), Atlanta, Georgia, USA, ACM (2002)
5. Tseng, Y.C., Wu, S.L., Liao, W.H., Chao, C.M.: Location awareness in ad hoc wireless mobile networks. Computer **34**(6) (2001) 46–52
6. Want, R., Schilit, B.N., Adams, N.I., Gold, R., Petersen, K., Goldberg, D., Ellis, J.R., Weiser, M.: The parctab ubiquitous computing experiment. Technical report, Xerox Corporation Palo Alto Research Center (1995)
7. Bates, J., Bacon, J., Moody, K., Spiteri, M.: Using events for the scalable federation of heterogeneous components. In: EW 8: Proceedings of the 8th ACM SIGOPS European workshop on Support for composing distributed applications, Sintra, Portugal, ACM Press (1998) 58–65
8. Waldo, J.: The Jini Specifications. Addison-Wesley Longman Publishing Co., Inc., Boston, MA, USA (2000)
9. Sun Microsystems: Rio Architecture Overview. (2001)
10. Microsoft Corporation: Universal Plug and Play Device Architecture. (2001)
11. Ferscha, A., Vogl, S., Beer, W.: Context sensing, aggregation, representation and exploitation in wireless networks. Scalable Computing: Practice and Experience **6**(2) (2005) 71–81
12. Verissimo, P., Cahill, V., Casimiro, A., Cheverst, K., Friday, A., Kaiser, J.: CORTEX: Towards Supporting Autonomous and Cooperating Sentient Entities. Proceedings of European Wireless (2002) 595–601
13. Chan, A., Chuang, S.: MobiPADS: A Reflective Middleware for Context-Aware Mobile Computing. IEEE Transactions on Software Engineering **29**(12) (2003) 1072–1085
14. Capra, L., Emmerich, W., Mascolo, C.: CARISMA: Context-Aware Reflective mIddleware System for Mobile Applications. IEEE Transactions on Software Engineering **29**(10) (2003) 929–944
15. Garcia, J., Ferreira, P.: Operating system support for task-aware applications. In: Conference on Mobile and Ubiquitous Systems, Guimarães, Portugal (2006)

A Client-Server Approach to Enhance Interactive Virtual Environments on Mobile Devices over Wireless Ad Hoc Networks[*]

Azzedine Boukerche, Richard Werner Nelem Pazzi, and Tingxue Huang

SITE - University of Ottawa, Canada
PARADISE Research Laboratory
{boukerch, rwerner}@site.uottawa.ca

Abstract. Interactive 3D environments have been studied for years and represent an important application area of computer graphics. However, high quality virtual environment interaction requires powerful computers equipped with expensive graphics accelerator cards. The high 3D data volume and the dynamic nature of bandwidth pose significant challenges when providing a smooth virtual navigation on thin mobile devices over wireless ad hoc networks. In this paper, we show that it is possible to provide a virtual environment walkthrough on mobile devices through a client-server approach. Although mobile devices have low processing power and memory, they can still render images with relative ease. Based on this fact, instead of using traditional geometry-rendering techniques and locally rendering complex scenes, we employ an image-based mechanism on the client that uses images, which are provided by a remote server through an interactive streaming transport protocol. In this paper, we propose a bandwidth feedback algorithm together with a rate control and virtual user path prediction to better adapt the system to the changing bandwidth. We also discuss our ideas and show an extensive set of simulations in order to evaluate the performance of our solutions.

1 Introduction

The fast developments of computer graphics technologies such as fast geometry rendering algorithms and hardware implementation of graphics primitives, and the advances in communication networks and protocols have enabled the creation of a vast number of interesting applications related to navigating in a remote virtual environment, e.g. games, virtual tours, training, virtual shopping, etc. However, complex 3D models have been created for powerful computers, not for thin mobile devices such as cell phones and PDAs. In addition, downloading complex virtual environments requires both high bandwidth and storage capability. The high volume of 3D data and the dynamic nature of bandwidth pose significant challenges in terms of providing smooth virtual navigation on thin mobile devices over wireless ad hoc networks: a mobile device can only hold a

---

[*] This work is partially sponsored by Grants from NSERC Canada Research Chair Program.

fraction of the entire virtual environment; the 3D rendering engine is not able to process complex scenes in real-time, the bandwidth is always changing and because the wireless communication channel is highly susceptible to error. A possible approach to these problems is rendering the complex 3D geometry on a graphics workstation server and transmitting only images to the remote client, depending on the user's position within the virtual environment. Instead of employing the traditional 3D geometry rendering mechanism on the client side, the approach can make use of an inexpensive image-based rendering (IBR) technique. IBR uses images as input and the rendering cost does not depend on the scene complexity, but on the final image resolution. As mobile devices usually provide small displays, IBR methods perfectly fit on these types of devices as the image size and therefore the bandwidth required to transmit the images will be small. Lately, there has been a great deal of interest in IBR algorithms lately. For instance, the view morphing [1] technique, which requires low processing as it renders novel views based on a collection of sample images. IBR algorithms are based on the plenoptic function [2]. View morphing is the simplest IBR algorithm as it relies on a certain amount of geometric information about the scene, whereas lumigraph [3] and lightfield [4] use implicit geometry or no geometry at all; this requires more processing. Basically, as the user walks through the 3D environment, the client device sends its position and orientation to the server, which will update the virtual camera position, render a reference image, and send this image back to the client. The client can use certain reference images to render novel views through the IBR while it is waiting to receive new reference images from the server.

Providing a less expensive rendering technique to the client device is not sufficient to solve all of the problems related to remote interaction on mobile devices. In order to cope with the dynamic bandwidth, efficient transport protocols, which take into consideration the user's behavior in a virtual environment, must be developed. In this paper, we propose a client/server architecture to enhance the user experience in a remote virtual environment through a hybrid rendering system, which uses traditional 3D geometry rendering on the server and an IBR method, such as view morphing [1], on the client. In order to improve the frame rate, or the number of images that the device can display per second, a virtual user path prediction algorithm is proposed, allowing the server to pre-fetch certain images to the client when enough bandwidth is available. The bandwidth feedback mechanism and rate control are designed to optimize the pre-fetching scheme, as its goal is to avoid starvation of images at the client side. The interactive streaming protocol is designed over the Real-Time Protocol (RTP) [5] and the Real-Time Streaming Protocol (RTSP) [6], which provide end-to-end delivery services for data with real-time constraints. For instance, audio and video.

This paper is organized as follows: Section 2 gives an overview of related work. Section 3 presents the proposed system architecture. The algorithms are described in Section 4. Simulation experiment results are shown and discussed in Section 5. Finally, in Section 6 the reader can find our conclusions and future work.

2 Related Work

In this Section we discuss some of the existing 3D rendering mechanisms on mobile devices. We also give the reader a brief review of image-based rendering and provide a discussion about interesting similar solutions to remote virtual environments.

OpenGL ES API [7] is used by several 3D applications on mobile and embedded devices. However, the rendering quality is still poor, or has a very low level of detail. The Mobile 3D Graphics API (M3G), defined in Java Specification Request (JSR 184) [8], is another industry effort to create a standard 3D API for Java-enabled thin devices. A solution to visualize more complex 3D scenes on mobile devices is made possible through Image-Based Rendering (IBR). IBR methods are categorized based on the geometry information they require to render novel views. Some image-based rendering techniques do not require geometric information. For instance, Lightfield [4] renders a new view by interpolating a set of samples without any geometric information such as a depth map. The problem with IBR methods that do not rely on geometric information is the huge storage capacity required to hold all of the pre-acquired image samples.

Our remote interactive system is based on View Morphing [1], which is able to render any novel image by morphing two or more reference images. The basic principle is depicted in Figure 1. Morphing parallel views is the simplest image morphing algorithm. As depicted in Figure 1, images I_0 and I_1 are acquired at points C_0 and C_1 respectively, with focal lengths f_0 and f_1. Novel image I_n, with focal length f_n, at point C_n, is rendered by the interpolation of images I_0 and I_1. There is also an image cache on the client in order to reuse a previously received image, significantly improving system performance and reducing network traffic.

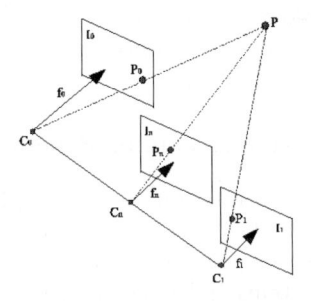

Fig. 1. View morphing with parallel views

QuickTime VR [9] is the most popular image-based rendering system. However, it is limited to panoramic scenarios, and the client device must download the entire environment in order to start the navigation. Our solution offers a higher level of freedom, as a user can walk through the environment and there is no need to download the entire environment as it is being rendered while the user moves through different areas.

A client-server approach to image-based rendering on mobile terminals is presented in [10]. The objective of this solution is to make it possible to render complex scenes on mobile devices through an IBR method and a client/server architecture. However, its main contribution concerns how to place the cameras in order to avoid problems such as exposure and occlusion when using IBR. Thus, the camera placement solution works only for urban scenes, thereby limiting the applications of this solution.

The work presented in [11] aims at the protection of copyrighted 3D models when manipulated by remote users. The server owns the entire 3D environment and sends certain images to the client on demand. The client renders a low-polygon model of the scene as the user manipulates it. When the user stops, the client sends a request to the server, which will send back a high-resolution image of the scene. This approach is different from ours as the client is capable of 3D geometry rendering. This feature is not well suited to low-capacity devices such as PDAs. The reader can refer to [12,13,14] for detailed information on other solutions to remote virtual environments using IBR methods.

Unlike the presented solutions, we propose a system that can make better use of the available bandwidth, which is crucial to applications involving wireless communication and thin devices. To the best of our knowledge, the solutions found in the literature do not address the problem of dynamic bandwidth. Our approach uses a virtual user's path prediction together with bandwidth monitoring and rate control algorithms to adapt the protocol and pre-fetch images to the client when bandwidth permits.

3 The Proposed Streaming System

Our proposed system is organized in the following modules: a modified JPEG codestream, new RTP payload format for view morphing, streaming protocol, bandwidth feedback mechanism, rate control scheme, and path prediction and pre-fetching algorithms.

3.1 The Packetization and Streaming Schemes

We specified a new JPEG codestream to cope with wireless channel errors. Figure 2 shows this new JPEG codestream. A packetization scheme was developed to avoid the errors that a corrupted packet propagate to other packets; it keeps the packetization items independent from one another. A packetization item is an atomic component such as the main header, the layer header, or a pixel block. The image codestream is split into packetization items and is encapsulated in RTP packets, and is then sent to the client.

Streaming multimedia compressed data over wired or wireless networks over RTP requires new payload formats such as H.26x over RTP [15,16] and G.7xx over RTP [17,18]. We introduced a new payload format for view morphing over RTP, as well as a new packet header for view morphing, which are depicted in Figures 3(a) and 3(b) respectively.

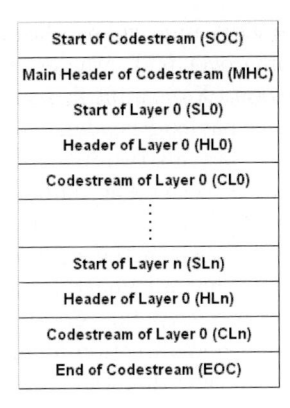

```
┌─────────────────────────────────────┐
│    Start of Codestream (SOC)        │
├─────────────────────────────────────┤
│  Main Header of Codestream (MHC)    │
├─────────────────────────────────────┤
│     Start of Layer 0 (SL0)          │
├─────────────────────────────────────┤
│     Header of Layer 0 (HL0)         │
├─────────────────────────────────────┤
│    Codestream of Layer 0 (CL0)      │
├─────────────────────────────────────┤
│              ⋮                       │
├─────────────────────────────────────┤
│     Start of Layer n (SLn)          │
├─────────────────────────────────────┤
│     Header of Layer 0 (HLn)         │
├─────────────────────────────────────┤
│    Codestream of Layer 0 (CLn)      │
├─────────────────────────────────────┤
│     End of Codestream (EOC)         │
└─────────────────────────────────────┘
```

Fig. 2. Structure of the new JPEG codestream

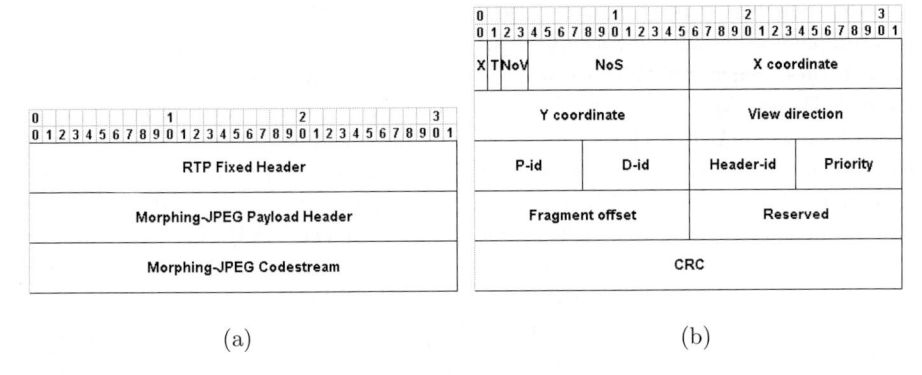

Fig. 3. (a) Structure of a RTP packet for View Morphing. (b) View Morphing payload header.

The modified RTP fields are shown below.

Payload Type: according to the RTP standard, this field specifies the format of the RTP payload and determines its interpretation by the application [17]. Because our payload type is not specified by the RTP profile, the payload type for the View Morphing codestream is not assigned through RTP means; the upper layer defines the payload code;

Number of Video Unit (NVU): If the current packet is the first fragment of a video unit or if it is the whole video unit, then check if this is the first packet for that session. If it is the first packet, the field NVU and timestamp in the RTP header receive a random value; this will be the first value of a sequence number.

Timestamp: The View Morphing stream has no strict sampling instance. Unlike other media types, the timestamp for View Morphing does not indicate the sampling instance. Nevertheless, it is significant for calculating synchronization and jitter when other media streams are associated with View Morphing.

All RTP packets for the same video unit will set the same timestamp. The packet header fields for view morphing over RTP are described below:

The payload header extension (**X**) is a bit flag that is activated when supplementary information follos the payload header. The twin images (**T**) bit field is set to inform the renderer to process two images; otherwise the rendering algorithm can render the image directly. The number of images in a video unit (**NoV**) field informs the renderer whether the packet contains zero, one, or two images. The number of an image in the streaming (**NoS**) field works as a sequence number to make sure that all image fragments were received by the client. The (**X,Y**) fields are the viewpoint coordinates and (**VD**) is the view direction angle. Viewpoint and view direction identification (**P-id and D-id**) help identify the viewpoint and view direction when data is corrupted or lost. The (**header-id**) field helps recover the main header when data is corrupted. The field (**Priority**) indicates the layer priority when sending different quality layers. The fragment **offset** field is used to reassemble the codestream. We also have the **reserved** field for future use. Finally, the **CRC** field detects whether or not the payload header is corrupted, which part is corrupted and tries to correct certain bits. Basically, the assembly of an RTP packet begins with the assignment of the payload type field for the Morphing-JPEG. If the current packet is the first fragment of a video unit or if it is the entire video unit, then check if this is the first packet for that session. If it is the first packet, the field NVU and timestamp in the RTP header receive a random value; this will be the first value of a sequence number. If it is not the first packet, NVU is incremented by one. All other fields are set according to the specification. For instance, field X is set to 1 if an optional payload follos the payload header; the field NoV is set according to the images the client will have to process; for instance, it will be set to 00 so as to instruct the client to process the first image on that video unit, 10 for the second, and 11 for the final one.

On the client side, the algorithm is a simple parser for the server codestream. First it checks if the payload type field is set to Morphing-JPEG, then checks the NVU field to see if it is different from the last one. If it is different, it instructs the application layer to render the image at that moment. The algorithm then gets the timestamp to calculate the synchronization and jitter. If field X is set to 1, the algorithm will locate the optional payload header, parse the codestream, and send it to the application. For the CRC scheme, the algorithm verifies if the viewpoint and direction are correct. If they are correct, the algorithm proceeds by checking the priority and the remaining fields. If the viewpoint and direction are not correct, algorithm approximate values from P-id and D-id, and runs the CRC on them. The final step is to check the offset field. If the offset is equal to the last offset plus the length of the packetization data, then it merely appends the packetization data to the previous one. If this is not the case, the algorithm waits for a short timeout period. If the delayed packet does not arrive during this period, the algorithm informs the server that an RTP packet was lost. This will adjust the parameters according to the bandwidth feedback mechanism.

3.2 The Pre-fetching Mechanism

Based on the path prediction mechanism, the server will pre-fetch certain images to the client. The pre-fetching algorithm takes advantage of available bandwidth to send images in advance to the client, saving some requests and network traffic, and improving the image quality and perhaps the frame rate on the client because the reference images will be available in the client's local cache. The periodic transmission of control packets conducted by RTCP is enough to control the adaptive encodings and the speed of data distribution in wireless network scenarios. Our proposed bandwidth feedback mechanism involves sending ACKs for every received RTP packet. The server can establish the network status by keeping track of these ACKs. An ACK packet is composed of its type, a sequence number for ordering, and a timestamp. We use the timestamp to calculate the round trip time (RTT) of RTP packets. Missing acknowledgments are interpreted as dropped RTP packets. When the server receives an ACK, it will parse the RTP packet and extract the sequence number and timestamp. Then, the server adds the sequence number to a list of successfully transmitted packets. The server then calculates the RTT and adds it to the list of recently transmitted packets. For each recently transmitted packet, if the sequence number does not exist in the list of successfully transmitted packets, and if the RTT of the previous packet minus the RTT of the next packet is greater than the acceptable variation value, the number of packets lost is incremented by one. If the number of packets lost is greater than 0, the network status is set to congested. Otherwise, if the RTT of the last received packet minus the RTT of the first received packet is greater than the threshold, the status is congested. If the RTT of the first received packet minus the RTT of the last received packet is greater than a threshold, the status is unloaded. Otherwise network status is constant.

The path prediction mechanism is based on the virtual user's previous and recent movement within the environment. There are two navigation modes: linear and rotational. For instance, if the user is moving along a straight line, the path prediction can determine that based on his/her previous movements, the user will continue along the same path. In the case of rotation within the virtual environment, the path prediction will obtain the nearby positions based on a threshold angle.

Our rate control mechanism is based on the reports from the bandwidth feedback mechanism. If the network status is congested, the rate is decreased. If the status is unloaded, the rate is increased; and if the status is constant, the rate is left unchanged. The increment value is crucial for the performance of our protocol. Upon receiving a request, the server will determine the network status. If the network is unloaded or congested, the rate is increased or decreased respectively by an amount corresponding to one image.

4 Simulation Experiment Results

We have implemented and simulated our approach on the NS-2 [19] network simulator. We performed a set of simulations in different ad hoc network scenarios.

They consist of one server and 15 to 60 mobile nodes moving at 5m/s to 20m/s in an area of 500x500 m^2. An 802.11b MAC layer was utilized during the simulations. We used the following metrics to evaluate our proposed interactive streaming system: system throughput, end-to-end delay, burst length and burst duration. The average number of images in the cache versus the number of requests was used to evaluate the performance of the pre-fetching algorithm.

As depicted in Figure 4, when only one client is connected to the server, the amount of images in the clients cache is 315 for only 150 requests. The rate control was aware of the bandwidth status, and the server could utilize the available bandwidth to pre-fetch additional reference images. With the pre-fetching feature turned off, the number of images in the clients cache would not exceed 150 (1 image per request). As we increased the number of clients, the bandwidth dropped. The rate control mechanism was able to adjust to the new scenario and reduce the streaming rate. For instance, the server sent 181 images for 150 requests.

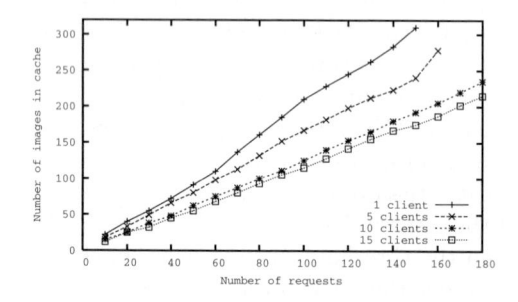

Fig. 4. Pre-fetching mechanism performance

The number of successfully sent packets per second is represented by the system throughput. As depicted in Figure 5(a), the system shows a reasonable scalability. Increasing the number of nodes leads the number of packets per second to increase from 25 to 45.

End-to-end delay is critical to our interactive system, because the system relies on quick response times. As can be seen in Figure 5(b), delay is less than 10ms when 35 nodes are employed and increases along with node density because the number of nodes in a path will most likely be higher.

Burst is a temporary connection lost that occurs when packets cannot reach the destination due to broken paths. Burst length is the number of packets that are dropped during a burst duration. As can be seen in Figures 5(c) and 5(d), burst length and duration depend on node density. With higher densities, routing paths are quickly restored. When a burst occurs, the bandwidth feedback mechanism is aware of the network status and immediately decreases the packet rate. Thus, the client renders novel views based on reference images stored on its cache, with a significant image quality depreciation, until it receives new reference images.

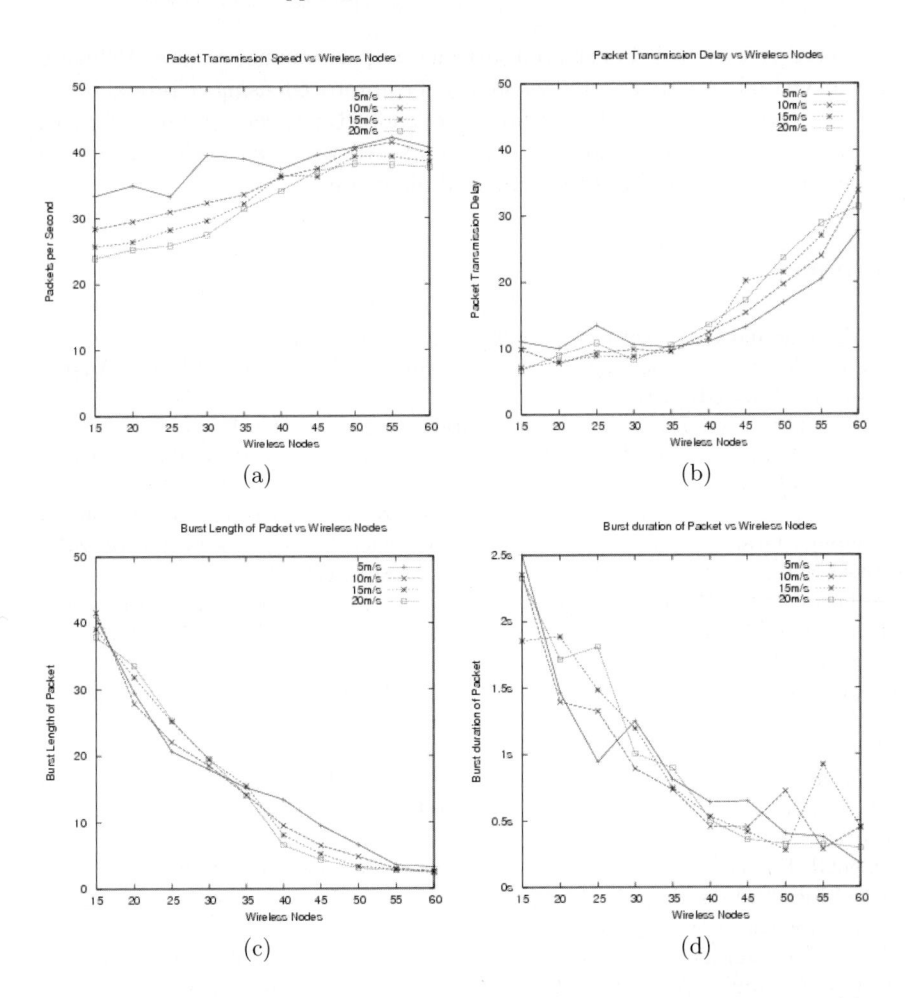

Fig. 5. (a) System throughput. (b) End-to-end delay. (c) Burst length. (d) Burst duration.

5   Conclusions and Future Work

In this paper, we proposed a client/server approach, which consists of a new payload format for RTP, an interactive streaming algorithm, the packetization scheme, and the pre-fetching mechanism, for remote interaction in virtual 3D environments for mobile devices over ad hoc networks. We addressed some issues of remote interactive virtual environments on mobile devices and focused on optimizing bandwidth usage and maximizing user experience. To the best of our knowledge, the related work did not tackle the dynamic bandwidth issue of mobile ad hoc networks. Pre-fetching images when there is available bandwidth has proven to be beneficial to the rendering system because the client does not need to request and wait for the images, which implies long delays. The simulation experiments demonstrated satisfactory performance results. As future work, the virtual user path prediction will be improved using a probabilistic virtual path prediction so as to optimize the tradeoff between sending reference images and the real use of them at the client side. We are also working on system prototype in order to evaluate the proposed algorithm.

References

1. S. M. Seitz and C. M. Dyer. View morphing. In Computer Graphics Proceedings, Annual Conference Series, pages 2130, Proc. SIGGRAPH96 (New Orleans), August 1996. ACM SIGGRAPH.
2. E. H. Adelson and J. Bergen. The plenoptic function and the elements of early vision. In Computational Models of Visual Processing, pages 320. MIT Press, Cambridge, MA, 1991.
3. S. J. Gortler, R. Grzeszczuk, R. Szeliski, and M. F. Cohen. The lumigraph. In Computer Graphics Proceedings, Annual Conference Series, pages 4354, Proc. SIGGRAPH96 (New Orleans), August 1996. ACM SIGGRAPH.
4. M. Levoy and P. Hanrahan. Light field rendering. In Computer Graphics Proceedings, Annual Conference Series, pages 3142, Proc. SIGGRAPH96 (New Orleans), August 1996. ACM SIGGRAPH.
5. H. Schulzrinne, S. Casner, R. Frederick, and V. Jacobson. Rtp: A transport protocol for real-time applications. Standards Track, Network Working Group, January 1996.
6. H. Schulzrinne, A. Rao, R. Lanphier, M. Westerlund, and A. Narasimhan. Real time streaming protocol (rtsp). Inernet Draft, Internet Engineering Task Force, February 2004.
7. OpenGL Embeded System. http://www.khronos.org/opengles/ 2006
8. Java Specification Request 184 (2005) - Mobile 3D Graphics API for J2ME http://www.jcp.org/en/jsr/detail?id=184
9. S. E. Chen. QuickTimeVR an image-based approach to virtual environment navigation. Computer Graphics (SIGGRAPH95), pages 2938, August 1995.
10. G. Thomas, G. Point, and K. Bouatouch. A client-server approach to image-based rendering on mobile terminals. Technical Report, ISSN 0249-6399, France, January 2005.
11. Koller, D., Turitzin, M., Levoy, M., Tarini, M., Croccia, G., Cignoni, P., and Scopigno, R. 2004. Protected interactive 3D graphics via remote rendering. ACM Trans. Graph. 23, 3 (Aug. 2004), 695-703.
12. Biermann, H., Hertzmann, A., Meyer, J., Perlin, K., Stateless Remote Environment Navigation with View Compression, NYU Technical Report 1999-784. April 22, 1999.
13. Chang, C. and Ger, S. 2002. Enhancing 3D Graphics on Mobile Devices by Image-Based Rendering. In Proceedings of the Third IEEE Pacific Rim Conference on Multimedia: Advances in Multimedia information Processing (December 16 - 18, 2002). Y. Chen, L. Chang, and C. Hsu, Eds. Lecture Notes In Computer Science, vol. 2532. Springer-Verlag, London, 1105-1111.
14. J. Li, Y. Tong, Y. Wang, H.-Y. Shum, Y.-Q. Zhang, "Image-based Walkthrough over the Internet", International Workshop on Very Low Bitrate Video Coding (VLBV01), October 2001, Athens, Greece.

15. T. Turletti and C. Huitema. Rfc2032: Rtp payload format for h.261 video streams. Stardards Track, Network Working Group, Octorber 1996.
16. C. Zhu. Rfc2190: Rtp payload format for h.263 video streams. Standards Track, Network Working Group, September 1997.
17. H. Schulzrinne and S. Petrack. Rfc2833: Rtp payload for dtmf digits, telephony tones and telephony signals. Standards Track, Network Working Group, May 2000.
18. R. Zopf. Rfc3389: Real-time transport protocol (rtp) payload for comfort noise (cn). Standards Track, Network Working Group, September 2002.
19. The Network Simulator. http://www.isi.edu/nsnam/ns/

Topic 15: Peer-to-Peer and Web Computing

Henrique J. Domingos, Anne-Marie Kermarrec,
Pascal Felber, and Mark Jelasity

Topic Chairs

Peer-to-peer (P2P) systems have become a major area of research in the past few years. Their potential was first revealed by the hugely popular P2P file sharing applications, which allow any computer (as a peer), anywhere in a large scale distributed computing environment, to share information and resources with others. The computing environments promoted by P2P systems and technology are decentralized in nature, exploring a symmetric pairwise interaction model. They are self- organized and self-coordinated, dynamically adapted to peer arrivals and departures, and highly resilient to failures. As P2P research becomes more mature, new challenges emerge to support complex and heterogeneous distributed environments for sharing and managing data, resources, and knowledge, with highly volatile and dynamic usage patterns. This topic provides a forum for researchers to present new contributions on P2P technologies, applications, and systems, identifying key research issues and new challenges.

Eleven papers were submitted to this topic and four were accepted. These papers address various aspects of P2P overlays and search protocols. In *"Top k RDF Query Evaluation"*, the authors describe a P2P backtracking search strategy for finding the largest k values in large RDF databases. The second paper, *"Roogle: Supporting Efficient High-Dimensional Range Queries in P2P Systems"*, presents a mechanism to index data in a distributed hash table (DHT) and look it up using high-dimensional queries. In *"Creating and Maintaining Replicas in Unstructured Peer-to- peer Systems"*, the authors describe an approach for optimal replication of data in unstructured P2P systems based on square-root replication. Finally, *"DOH: A Content Delivery Peer-to-Peer Network"* presents a scalable content distribution scheme for Web sites, which involves a load balancing component and a content retrieval mechanism based on DHT lookup and caching.

Top k RDF Query Evaluation in Structured P2P Networks

Dominic Battré, Felix Heine, and Odej Kao

University of Paderborn, Fürstenallee 11, 33012 Paderborn, Germany
{battre, fh, okao}@uni-paderborn.de
http://www.upb.de/pc2

Abstract. Berners-Lee's vision of the Semantic Web describes the idea of providing machine readable and processable information using key technologies such as ontologies and automated reasoning in order to create intelligent agents.

The prospective amount of machine readable information available in the future will be large. Thus, heterogeneity and scalability will be central issues, rendering exhaustive searches and central storage of data infeasible. This paper presents a scalable peer-to-peer based approach to distributed querying of Semantic Web information that allows ordering of entries in result sets and limiting the size of result sets which is necessary to prevent results with millions of matches. The system relies on the graph-based W3C standard *Resource Description Framework* (RDF) for knowledge description. Thereby, it enables queries on large, distributed RDF graphs.[1]

1 Introduction

The Semantic Web [3] envisions to make the huge information resources of the Web available for machine-driven evaluation. Electronic agents are supposed to locate information necessary for their objectives, process them, generate conclusions and new information, and finally present the results to either a human user or to other electronic agents.

The Resource Description Framework (RDF, [11]) has been proposed by the W3C in order to formally describe resources. In combination with RDF Schema (RDFS, [4]) it provides sufficient expressibility to describe taxonomies of classes and properties and to infer information from taxonomies described with different schemas.

Query languages like SPARQL [15] with implementations like ARQ of Jena [2] (see http://esw.w3.org/topic/SparqlImplementations for other implementations) allow to query and infer information from RDF databases. These implementations, however, assume that all RDF triples are located in a central data repository, which is a questionable assumption given the growth of information available on the web.

---

[1] Partially supported by the EU within the 6th Framework Programme under contract 001907 "Dynamically Evolving, Large Scale Information Systems" (DELIS).

In [9] we have presented a scalable P2P based RDF querying strategy, which allows for distributed storage of information and selective collecting of RDF triples necessary to answer RDF queries. As current search engines of the web demonstrate, it is conceivable that RDF queries on real-world data will sometime deliver millions of results as well. Therefore, an evaluation procedure which retrieves every matching subgraph is neither desirable nor scalable. Typically, the user is not interested in an exhaustive collection of every matching result, but rather seeks for some matches which are good for her.

In this paper, we restrict to problem to finding the k best results (called *Top k* results) with respect to a single optimization criterion. I.e., the user might specify single variables in an `ORDER BY` clause of the query, but no complex expressions, and limit the number of results k by the `LIMIT` clause.

A simple solution would be to adapt the exhaustive search so that the results are ordered and filtered after the evaluation. However, a main goal of the Top k search is to enhance the scalability. Thus, we have to move away from the strategy to exhaustively collect all candidates before starting the final evaluation. We will rather start the final evaluation immediately, and fetch the candidates from the network step by step as needed. This avoids retrieving parts of the model graph which are never used during the query evaluation, when only k matches are of interest.

In the following section, we will provide an overview of the Top k query algorithm using an example. After that, section 3 describes the algorithm in detail and explains the caching strategies. The evaluation is given in section 4. Section 5 presents related work and section 6 concludes the paper.

2 Overview

In order to locate the RDF triples of various sources that are relevant to a query, we insert each triple three times into a distributed hash table (DHT) which is realized with Pastry [16]. Each triple is inserted into the hash map using the subject, the predicate, and the object as a key to the actual triple. That way, it is possible to look up all triples with a common subject for example and to perform the query processing by starting with one triple that has a URI in either subject, predicate, or object and to proceed from there on, fetching new triples and checking that their values to not contradict previous variable assignments.

Figure 1 shows an example query which consists of the following three triples: $t_1 = \langle v_1, U_1, v_2 \rangle$, $t_2 = \langle v_1, U_2, v_3 \rangle$, $t_3 = \langle v_3, v_1, v_2 \rangle$. U_1 and U_2 are fixed URIs, whereas v_1, v_2, and v_3 are variables. Assume that the user expects the value of v_2 to be a floating point value, and that she looks for matches with values of v_2 to be as large as possible.

As t_1 and t_2 have a bound value (the predicate) – a precondition for DHT lookups – either one can serve as a start for the query evaluation and we choose t_1 arbitrarily. By using U_1 as a DHT index and asking the responsible node to send triples with predicate U_1, we receive candidates for the variables v_1 and v_2. As the number of possible values of v_1 and v_2 can be very large, it is not desirable to

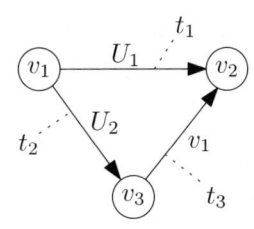

Fig. 1. Example RDF Query

fetch a complete list of matching triples. We rather need a way to query chunks of triples so that we can retrieve more candidates later, in case the first chunk did not deliver a sufficient number of matches. Thus, we ask the node to deliver the triples *in order*, so that we can later specify the highest known candidate to retrieve the next chunk. Because of scalability reasons, the target nodes do not store any state information. Therefore, the client node has to know the current chunk position and send it later to the other node to fetch the next chunk.

Assume the first chunk of five candidates gets retrieved and stored in a table as depicted in figure 2a. By selecting the first candidate, v_1 gets bound to A and v_2 to 10. As t_2 has a bound subject and predicate now, we can choose this as the next triple to proceed recursively. Here, we have to fetch candidates which respect the current variable bindings. As the predicate is bound to U_2, we can use U_2 as DHT index, and retrieve all triples with predicate U_2 which have the subject A. We sort the results by ascending order of v_3.

Assume that the result are two triples, $\langle A, U_2, X \rangle$ and $\langle A, U_2, Y \rangle$. These triples are stored as candidates for t_2 and the first one is selected in the backtracking procedure.

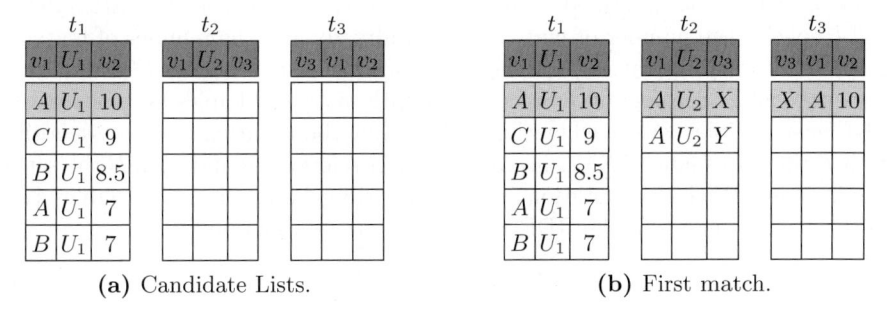

t_1			t_2			t_3			t_1			t_2			t_3		
v_1	U_1	v_2	v_1	U_2	v_3	v_3	v_1	v_2	v_1	U_1	v_2	v_1	U_2	v_3	v_3	v_1	v_2
A	U_1	10							A	U_1	10	A	U_2	X	X	A	10
C	U_1	9							C	U_1	9	A	U_2	Y			
B	U_1	8.5							B	U_1	8.5						
A	U_1	7							A	U_1	7						
B	U_1	7							B	U_1	7						

(a) Candidate Lists. (b) First match.

Fig. 2. Query evaluation

Finally, we fetch candidates for the last triple. As it consists only of variables, we have to use the current binding of one of the variables as DHT index. We choose v_1, and thus use A as DHT index. The remaining two variables are already bound to $v_3 = X$ and $v_2 = 10$. That means that we ask for the *existence* of the triple $\langle X, A, 10 \rangle$. As the triple exists, we have generated the first match (see figure 2b). Afterwards, we backtrack to t_2, select the second candidate, and ask for the existence of $\langle Y, A, 10 \rangle$ (see figure 3). By this procedure, we generate the top matches step by step, only retrieving the candidates as needed. After having generated the requested number of matches, the procedure stops.

However, as we can see e.g. from the last step, it might be useful to have a kind of look-ahead when fetching the candidates. We have contacted the node for URI A twice in short succession to ask for the existence of triple $\langle X, A, 10 \rangle$ and then $\langle Y, A, 10 \rangle$. During the first lookup, the second candidate for t_2 was already known, and therefore, it should have been possible to ask directly for the existence of the second triple in order to save one communication step.

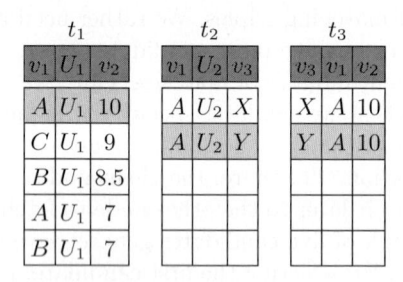

t_1		
v_1	U_1	v_2
A	U_1	10
C	U_1	9
B	U_1	8.5
A	U_1	7
B	U_1	7

t_2		
v_1	U_2	v_3
A	U_2	X
A	U_2	Y

t_3		
v_3	v_1	v_2
X	A	10
Y	A	10

Fig. 3. Second match

3 Top k algorithm and Caching Strategy

In this section we present an evaluation strategy with look-ahead caches that efficiently reduce the amount of information transferred over the network and the number of messages passed between nodes.

The evaluation function, as we can see in figure 4, resembles the basic backtracking strategy as described in the previous section. Its parameters $nr\_matches$ and k specify the number of matches found already and the number of matches to be delivered in total respectively. The ordered list of triples (T_i) describes the query and i represents the recursion depth, as query triples are matched to RDF graph triples with backtracking. We follow the notation of [9] by denoting with $\mathcal{L}$ the set of labels (XML literals and URI references) and with $\mathcal{B}$ the set of blank nodes. With this notation, we can describe the binding of variables $\{v_i\}$ to their actual nodes in the RDF graph as a partial function

$$B : \{v_i\} \to \mathcal{L} \cup \mathcal{B}. \qquad (1)$$

Similarly, we maintain a set of candidates that can possibly be assigned to variables, with

$$C : \{v_i\} \to \text{Pow}(\mathcal{L} \cup \mathcal{B}). \qquad (2)$$

Finally, we define a set of caches that allow retrieving candidates for the RDF query triples. Each triple in the query has one individual cache, but we employ different kinds of caches as we will see later on.

The general idea of the evaluation function is to iterate over all possible assignments to triples (variable j serves as an index for the iteration), assume one,

```
function eval(nr_matches, k, (T_i), i, B, C, Caches)
    if i = |(T_i)| + 1 then
        record B as a match found;  /* all triples are bound */
        return nr_matches + 1;
    end if
    j := 0;  /* counter of inspected candidate triples for T_i */
    loop
        t := Caches.getCache(T_i).getNextCandidate(B, C, j);
        /* the candidate will respect the bindings in B */
        j := j + 1;
        if t = null then
            break;  /* no more candidates available */
        end if
        /* update bindings and candidates: */
        B' := B ∪ Bindings of t;
        C' := C ∪ Candidates of t;
        nr_matches := eval(nr_matches, k, (T_i), i + 1, B', C', Caches);
        if nr_matches ≥ k then
            break;
        end if
    end loop
    return nr_matches;
end function
```

Fig. 4. Evaluation algorithm

and proceed to a recursive evaluation until we encounter contradictions, find a complete match, realize that we have found a sufficient number of matches, or until we cannot assign any more triples. The caches allow to perform this search efficiently even though data are distributed among the peers of the network.

We will describe the strategy of the caches with the example of fetching candidates for triple $t_3 = (v_3, v_1, v_2)$ in figure 2b. As described before, we use the predicate $v_1 = A$ as the DHT key and have two remaining components of the triple to look up; v_3 and v_2. These variables are already bound to the valued $v_3 = X$ and $v_2 = 10$. If the cache knows whether the resulting triple $\langle X, A, 10 \rangle$ exists or does not exist, it can return this answer. Otherwise it has to retrieve the information of the node that is in charge of triples with predicate A. Instead of fetching just one triple it queries this and a chunk of additional queries, anticipating that they will be requested later. As we know from the first occurrences of v_1 and v_3 in columns of figure 2b, the candidates of v_1 are $\{A, C, B\}$, and the candidates of v_3 are $\{X, Y\}$. The cross product of both candidate sets defines a super set of values of possible interest. The cache asks not only whether the triple $\langle X, A, 10 \rangle$ exists but also asks for additional $c - 1$ unknown triples of the cross product, where c is the chunk size. That way, we hope to retrieve information that will be requested later.

Each triple can be split into one component which defines the key of the DHT and two remaining components. If possible, we choose a fixed URI as DHT key, otherwise we iterate over all possible candidates of a variable. In the previous example, the predicate served as DHT key, and subject and object served as the remaining components. The latter ones were both bound variables, but this does not need to be the case. In general we can encounter six different cases, where the two remaining components are:

1. two unbound variables,
2. an unbound variable plus a bound variable,
3. an unbound variable plus an fixed URI or literal,
4. two bound variables,
5. a bound variable plus an fixed URI or literal, and
6. two URIs / literals

A component is bound if it consists of a variable that was seen before in a higher recursion level, unbound with a variable that occurs for the first time, or fixed if it is a URI or literal. The binding of a variable is the known candidate set, where the variable was found the first time. For each of these cases we define a specially optimized type of cache. These caches are depicted in figure 5.

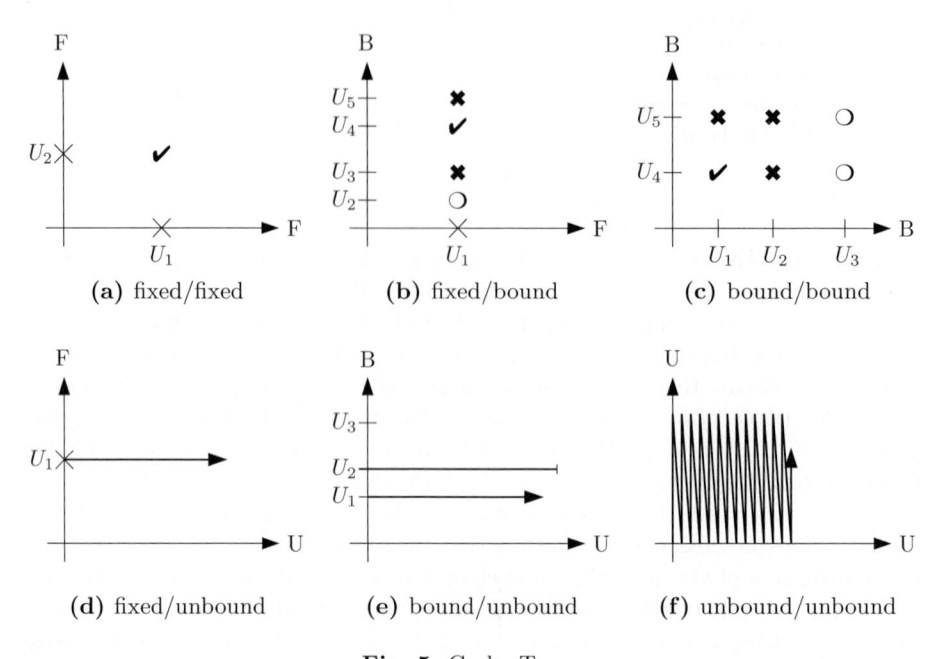

(a) fixed/fixed
(b) fixed/bound
(c) bound/bound
(d) fixed/unbound
(e) bound/unbound
(f) unbound/unbound

Fig. 5. Cache Types

The caches have to query the next chunk of up to c triples for a query. They deliver these chunks triple by triple. For scalability reasons, the peer who will process the query, does not store any state information, so the requesting peer is in charge of submitting the state along with the actual request. The state can consist of the set of triples we want to gather information about (first three cases below) or of a set of markers, which define the last triples for which we know information already (last three cases below).

The simplest cache for fixed/fixed components (see fig. 5a), which occur if a RDF query contains a triple with three URIs, does a simple lookup without lookahead. The state of the cache can be "triple exists in RDF graph" (represented by a check mark in the figure), "triple does not exist in RDF graph" (cross), or "unknown whether triple exists in RDF graph" (circle).

For fixed/bound component pairs (see fig. 5b) the caching is simple as well. A peer requests a chunk of triples by specifying the fixed component and a set of candidates for the bound component for which it wants to retrieve the state.

For bound/bound components (see fig. 5c) we build up a request containing a set of unknown combinations of already known values for the bound variables.

The fixed/unbound cache (see fig. 5d) is similar to the fixed/bound cache, except that it is sufficient to request the next c elements starting after a given position. Therefore, we submit the fixed element and the last inspected value for the unbound element as the request.

The bound/unbound cache (see fig. 5e) extends this by storing and submitting markers for the last known elements in several rows. The peer who processes a request starts sending triples at the first marker until c triples have been sent or continues at the next marker if the row (candidates) do not provide c triples.

For the unbound/unbound cache (see fig. 5f) it is again sufficient to submit a single marker which determines the next triples to be delivered.

4 Evalution

For the evaluation of the strategy described we have generated random resource descriptions that follow the data guide of the JSDL specification [7]. The generation was based on rough but arbitrary estimations (e.g.: of the many operating systems available, the first three will account for the majority of offers and requests). The data generation approach is very similar to the Lehigh University Benchmark (LUBM), see for example [8]. The resource descriptions consist of 11.3 RDF triples on average (standard deviation: 2.6). Queries are less specific than resource descriptions and consist of 3.9 RDF triples on average (s.d.: 1.8). This ensures big result sets which are focus of the Top k strategy.

We have evaluated the Top k strategy for $k = 10$ with a look-ahead of 10 triples against an optimized exhaustive search described in [9]. This exhaustive search employs sophisticated means based on Bloom filters to fetch a minimal set of triples necessary to do a full evaluation locally. For the evaluation, we have processed 100 queries on databases of 100,000, 500,000, and 1,000,000 RDF triples, spread on a P2P network of 64 nodes.

Our main goals were to reduce the number of triples sent over the network and to reduce the number of messages sent over the network.

Figure 6a shows the aggregated number of queries each individual peer had to sent to process all 100 queries on the database of 100,000 triples. We see that some nodes were not involved at all, when trying to find just the first 10 matches. On average, each peer sent 634.7 triples in 308.2 messages for a Top 10

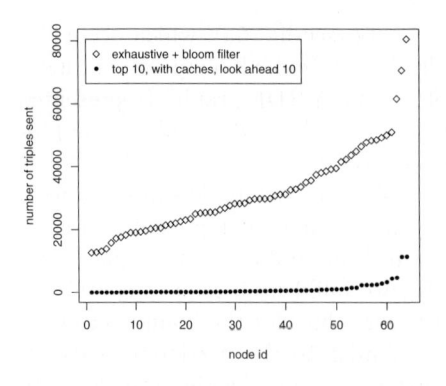

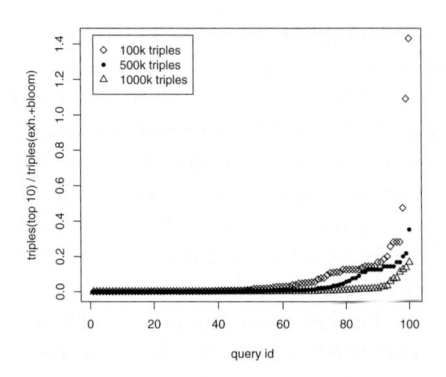

(a) Total number of triples sent by peers to process exhaustive and Top 10 search

(b) Ratio of triples sent by peers for Top 10 search divided by triples sent for exhaustive search

Fig. 6. Empirical analysis

evaluation, while the exhaustive evaluation sent an average of 31485 triples in 18886 messages per node.

We further hypothesize that restricting the number of results gets increasingly important the larger the database is. Therefore, we have analyzed the ratio of triples sent over the network with Top k strategy devided by the number of triples sent with optimized exhaustive search for all three databases. In two of one hundred test queries, the exhaustive search was slightly faster than the Top k search on 100,000 triples because the triple ordering strategy in the Top k search is less optimized. These two queries had very small result sets. In all 98 other test queries, the Top k strategy was superior. In order to disregard these outliers, we describe the median values for the ratios instead or the mean values.

The experimental results support our hypothesis (see fig. 6b). The median ratio of triples sent over the network for Top k divided by the the number of triples sent for an exhaustive search of 100 queries was just 0.89% for a database of 100,000 triples, 0.13% for 500,000 triples, and 0.05% for 1,000,000 triples. The mean ratios were 5.7%, 2.8%, and 0.87% respectively. We see that the Top k is on average significantly faster than the optimized exhaustive search if result sets are big.

Decreasing the look-ahead from 10 triples to 1 increases the total number of triples sent over the network by a factor of 1.61 and the total number of messages by a factor of 1.95 on the database of 100,000 triples.

5 Related Work

The general idea of the semantic web [3] paints the vision of a web where information can be automatically processed by software. The Resource Description Framework (RDF) together with RDF Schema [11,4] is one of the upcoming standards which will help to make this vision a reality.

Kokkinidis and Christophides describe in [12] a P2P based middleware for evaluating queries in the RDF Query Language (RQL) using RDF Schema knowledge. They focus on the construction and optimization of query plans. Their basic approach is different from ours as they require mandatory schema information encoded in RDFS. In our approach, schema information is not required for query processing.

In [5], dynamic query execution for schema-based P2P networks in the context of the Edutella project [14] is described. This work focuses on dynamic query planning and execution. Queries are evaluated in a distributed fashion; the optimizer tries to evaluate operators local to the data.

Kokkinidis et al. do not address Top k evaluation explicity. Nejdl et al. propose a Top k evaluation strategy in [13] but this is fundamentally different from our approach as it is based on a P2P network with super-peer architecture.

The idea of using URIs as the key to distribute information over an DHT-based P2P network has been described in several papers. We have used it in [10] to distribute knowledge based on Description Logics and it has been used in BabelPeers [9], the GridVine project [1], and RDFPeers [6] to distribute RDF triples. The distribution of triples used in this paper is similar to these ideas.

6 Conclusion

In this paper, we have focussed on querying large amounts of distributed RDF-based knowledge. While the Semantic Web is a prominent use case for our algorithm, we argue that other applications like Grid resource discovery are important as well. In most use cases, only a small fraction of the results are relevant for the user. Thus, we devised a Top k query algorithm which delivers only the k best results according to a sorting attribute. The algorithm operates on RDF data distributed over an DHT-based P2P network. It uses caching and look-ahead strategies to reduce both the number of messages and their size.

In the evaluation, we showed that the algorithm indeed reduces network usage significantly compared to an exhaustive evaluation. We further showed that the positive effect increases the larger the underlying RDF knowledge-base grows. Thus our strategy is efficiently increasing scalability of RDF-based P2P data management systems.

References

1. Karl Aberer, Philippe Cudré-Mauroux, Manfred Hauswirth, and Tim Van Pelt. GridVine: Building Internet-Scale Semantic Overlay Networks. In Sheila A. McIlraith, Dimitris Plexousakis, and Frank van Harmelen, editors, *International Semantic Web Conference*, volume 3298 of *Lecture Notes in Computer Science*, pages 107–121. Springer, 2004.
2. ARQ - A SPARQL Processor for Jena. URL http://jena.sourceforge.net/ARQ/.
3. Tim Berners-Lee, James Hendler, and Ora Lassila. The Semantic Web. *Scientific American*, May 2001.

4. Dan Brickley and Ramanathan V. Guha. RDF Vocabulary Description Language 1.0: RDF Schema. http://www.w3.org/TR/rdf-schema, 2004.
5. Ingo Brunkhorst, Hadhami Dhraief, Alfons Kemper, Wolfgang Nejdl, and Christian Wiesner. Distributed Queries and Query Optimization in Schema-Based P2P-Systems. In *Databases, Information Systems, and Peer-to-Peer Computing, First International Workshop, DBISP2P, Berlin Germany, September 7-8, 2003, Revised Papers*, pages 184–199, 2003.
6. Min Cai, Martin Frank, Baoshi Pan, and Robert MacGregor. A Subscribable Peer-to-Peer RDF Repository for Distributed Metadata Management. *Journal of Web Semantics: Science, Services and Agents on the World Wide Web*, 2(2), 2005.
7. Andreas Savva (editor). Job Submission Description Language (JSDL) Specification, Version 1.0, 2005.
8. Yuanbo Guo, Jeff Heflin, and Zhengxiang Pan. Benchmarking DAML+OIL Repositories. In *International Semantic Web Conference*, pages 613–627, 2003.
9. Felix Heine. Scalable P2P based RDF Querying. In *First International Conference on Scalable Information Systems (IN FOSCALE06)*, to appear, 2006.
10. Felix Heine, Matthias Hovestadt, and Odej Kao. Towards Ontology-Driven P2P Grid Resource Discovery. In *GRID*, pages 76–83. IEEE Computer Society, 2004.
11. Graham Klyne and Jeremy J. Carroll. Resource Description Framework (RDF): Concepts and Abstract Syntax. http://www.w3.org/TR/rdf-concepts, 2004.
12. Giorgos Kokkinidis and Vassilis Christophides. Semantic Query Routing and Processing in P2P Database Systems: The ICS-FORTH SQPeer Middleware. In *Current Trends in Database Technology - EDBT 2004 Workshops, EDBT 2004 Workshops PhD, DataX, PIM, P2P&DB, and ClustWeb, Heraklion, Crete, Greece, March 14-18, 2004, Revised Selected Papers*, pages 486–495, 2004.
13. Wolfgang Nejdl, Wolf Siberski, Uwe Thaden, and Wolf-Tilo Balke. Top-k Query Evaluation for Schema-Based Peer-to-Peer Networks. In Sheila A. McIlraith, Dimitris Plexousakis, and Frank van Harmelen, editors, *ISWC 2004: Third International Semantic Web Conference*, volume 3298, pages 137–151, jan 2004.
14. Wolfgang Nejdl, Boris Wolf, Changtao Qu, Stefan Decker, Michael Sintek, Ambjörn Naeve, Mikael Nilsson, Matthias Palmér, and Tore Risch. EDUTELLA: a P2P networking infrastructure based on RDF. In *WWW2002, May 7-11, 2002, Honolulu, Hawaii, USA*, pages 604–615, 2002.
15. Eric Prud'hommeaux and Andy Seaborne (Editors). SPARQL Query Language for RDF. URL http://www.w3.org/TR/rdf-sparql-query/, Nov 2005.
16. Antony Rowstron and Peter Druschel. Pastry: Scalable, decentralized object location and routing for large-scale peer-to-peer systems. In *Proc. of the 18th IFIP/ACM International Conference on Distributed Systems Platforms (Middleware 2001)*, 2001.

Roogle: Supporting Efficient High-Dimensional Range Queries in P2P Systems

Di Wu, Ye Tian, and Kam-Wing Ng

Department of Computer Science & Engineering
The Chinese University of Hong Kong
Shatin, N.T., Hong Kong
{dwu, ytian, kwng}@cse.cuhk.edu.hk

Abstract. Multi-dimensional range query is an important query type and especially useful when the user doesn't know exactly what he is looking for. However, due to improper indexing method and high routing latency, existing schemes cannot perform well under high-dimensional situations. In this paper, we propose Roogle, a decentralized non-flooding P2P search engine that can efficiently support high-dimensional range queries in P2P systems. Roogle makes improvements on both indexing and routing. The high-dimensional data is indexed based on the maximum or minimum value among all dimensions. This simple indexing method performs rather well under high-dimensional situations and tolerates data points with missing values or different dimensionality. To speed query routing, Roogle is built on top of our proposed structured overlay - Aurelia, which has better routing performance by exploiting node heterogeneity. Aurelia also guarantees the data locality and efficiently support range queries. Experimental results from simulation validate the scalability and efficiency of Roogle.

1 Introduction

To fully exploit the gigantic amount of structured data (e.g., multimedia data, computation resources, scientific dataset, etc) shared in the P2P systems, which are inherently spatial, it is expected that advanced query types could also be efficiently supported.

Among advanced queries, the multi-dimensional range query [1] is an important query type and particularly useful for discovery purposes. When the user doesn't know exactly the properties of desired objects, range queries on multiple attributes can be issued to perform search. For instance, in order to locate the movies published in recent two months and with the size smaller than 400MB, one 2-dimensional range query can be issued: *(20060101≤ PublishDate< 20060201) AND (0M< FileSize < 400MB))?*. However, it should be noticed that the query dimensionality varies with different data types, and may be very high in some scenarios (e.g., color histogram, stock dataset).

To date, there have been quite a few schemes (e.g., [1],[2],[3],[4],etc) being proposed to enable multi-dimensional range queries in P2P systems, but most

---
[1] Also known as multi-attribute range query.

of them can only function well under the low-dimensional situations. In case of high dimensionality, their performance will deteriorate greatly due to the "*curse of dimensionality*" [5]. Additionally, their query routing is also inefficient. Many approaches are built on top of traditional structured overlays (e.g., Chord, CAN), which can only provide $O(\log N)$-hop routing performance. Considering the large query range of high-dimensional queries, the latency will be intolerable for users.

In this paper, we address the above issues by proposing Roogle, a decentralized non-flooding P2P search engine that can efficiently support high-dimensional range queries. Due to the characteristics of high-dimensional data, it is safe to index data simply based on the maximum or minimum values among all dimensions. Such indexing method performs rather well under high-dimensional situations, and also tolerates data points with missing values or different dimensionality. At the same time, a locality-preserving structured overlay called *Aurelia* is proposed as the underlying overlay. The routing performance of Aurelia is greatly improved by exploiting node heterogeneity. The size of routing table on each node is proportional to the node capacity, and multicasting is adopted for scalable routing table maintenance. The size of data index on each node is also adaptive to node capacity, and the data key is allowed to be registered into multiple nodes to guarantee the reliability. Finally, the performance of Roogle is evaluated via simulation, and the experimental results confirm the scalability and efficiency of our design.

The rest of the paper is organized as follows. Section 2 provides an overview of related work. Section 3 describes the detailed design of Roogle. Experimental results are presented and analyzed in Section 4. Section 5 concludes the paper.

2 Related Work

In unstructured P2P systems, flooding-based approaches are widely adopted for multi-dimensional range queries, but huge traffic will be incurred. Therefore, most of the existing non-flooding schemes are built on top of structured P2P systems in order to limit the traffic.

In [1], a wide-area resource discovery engine called SWORD is implemented to answer multi-attribute range queries so as to locate computation resources. In SWORD, nodes participate in multiple DHTs, one per attribute. A query is routed in the DHT overlay corresponding to its most selective attribute. In MAAN [2], order-preserving hash is adopted to preserve data locality, and query selectivity is used to identify the relevant DHT for query routing. Similar schemes also include Mercury [4], PHT [6], etc. They all require prior knowledge of attribute selectivity, which is a kind of drawback in the design.

In Squid [3] and SCRAP [7], Space-Filling Curves(SFC) is introduced for dimension reduction. Multi-dimensional data is mapped to single-dimensional data and then is range-partitioned across peers. However, the data locality of SFC will become worse with the increase of dimensionality. To improve the data locality, MURK [7] and SkipIndex [8] adopt KD-tree to partition the multi-dimensional space into hypercuboids, each of which is assigned to a node. They use hashless CAN [9] and SkipGraph [10] as the underlying overlay respectively. However, in dynamic environments, both of them suffer from the problem of load balancing. Additionally, it is hard for them to index data points with missing values in some dimensions, or with different dimensionality.

Different from the above approaches, our proposed Roogle doesn't require prior knowledge of attribute selectivity, and works efficiently under the high-dimensional situation. It avoids the poor performance of range queries caused by awkward DHT-based designs. By exploiting node heterogeneity and guaranteeing data locality, both indexing robustness and routing performance are much improved in Roogle.

3 System Design of Roogle

In this section, we will introduce Roogle from four aspects: overlay structure, data indexing, query processing and load balancing.

3.1 Overlay Structure

The underlying overlay provides the basic routing function for uplevel applications. To improve routing performance, Aurelia is proposed as the underlying overlay structure to support Roogle.

Like Chord, Aurelia organizes nodes into a circular ring that corresponds to the ID space $[0, 2^{128} - 1]$. Initially, each node is assigned a unique node ID by uniform hashing, which consists of three parts: *RangeID*, *LevelID* and *RandomBits*. For example, for the node $\underline{001}010...10\underline{10}$, its *RangeID* and *LevelID* are $\underline{001}$ and $\underline{10}$ respectively, and the bits in between are *RandomBits*.

The *RangeID* is the first r-bit prefix of the node ID, which defines the value range that the node is responsible for. Aurelia abandons the using of hashing to distribute the data objects across the nodes, for hashing destroys the data locality. Instead, Aurelia simply maps the data object to the node according to its normalized value. All the data keys whose first r-bit prefix is the same as the *RangeID* will be placed on that node. The length of *RangeID* determines the size of data index if the objects are uniformly distributed. The nodes can choose a suitable length of *RangeID* based on its capacity. Formally, supposing the range of the ring is mapped to $[0, 1]$, the node with *RangeID* of $b_0 b_1 \ldots b_{r-1}$ will take charge of the range:

$$Range(b_0 b_1 \ldots b_{r-1}) = \left[\frac{\sum_{i=0}^{r-1} b_i \times 2^{r-i-1}}{2^r}, \frac{\sum_{i=0}^{r-1} b_i \times 2^{r-i-1} + 1}{2^r} \right)$$

The *LevelID* is the l-bit suffix of the node ID, which determines the routing table size together with "*Routing Regions*". Here, "*Routing Region*" refers to the region where the routing pointers take effects. Under uniform node distribution, there is only one routing region, i.e., the whole ID space. In case of non-uniform node distribution, the whole ID space is divided into multiple routing regions based on the node density. The node can have different *LevelID*s for different routing regions. For a node A, only the nodes whose node ID is within the routing region and the last l bits are the same as A's *LevelID* appear in the routing table of node A. The shorter the *LevelID*, the bigger the routing table.

In case of uniform node distribution in the ring, the routing scheme of Aurelia will be similar to SmartBoa [11]. However, with node dynamism or load balancing reasons(e.g., under-loaded nodes quit their current position and rejoin the heavy-loaded region), the node distribution in the ring may become non-uniform. In such scenarios, SmartBoa's routing scheme will become inefficient. For the region with dense node distribution, more hops will be required than expected(as illustrated in Fig. 1)). This drawback leads to the design of Aurelia.

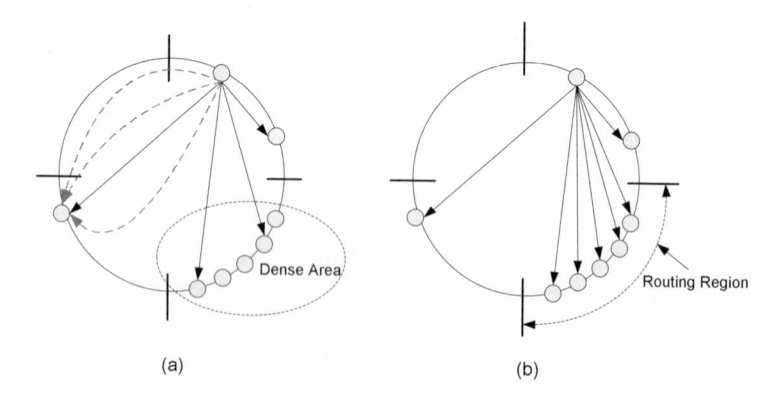

Fig. 1. Routing under non-uniform node distribution (a) SmartBoa's routing pointers; (b) Aurelia's routing pointers

The basic idea of Aurelia is to adjust the density of routing pointers so as to adapt to the node distribution. In the sparse routing region, fewer routing pointers are allocated; while in the dense region, more routing pointers will be allocated for fast routing. The density of routing pointers is controlled by adjustment of *LevelID*. The node seems like hosting multiple virtual nodes with *LevelID* in different lengths, each of which corresponds to a different routing region. By adjusting the length of *LevelID* in different routing region, we can make the density of routing pointers congruous to the node distribution. Note that all these virtual nodes share the same node ID.

As to the routing strategy, Aurelia performs in a greedy-like style. Given a target key, the node always selects the nearest one in the routing table for forwarding. For the powerful nodes, the large routing table enables them to complete the routing even in 1 hop.

To build such kind of routing table, a big challenge is to maintain the node-count distribution in a decentralized way. Aurelia adopts the technique of sampling to collect the statistical information and build the approximate node-count histogram locally. In addition to producing local estimate, the node also periodically makes sampling uniformly in the ring. According to the collected information, the node chooses the most recent statistical data to produce the node-count histogram. After the histogram is built and normalized, the node can adjust its routing pointers in the following way:

Each block in the histogram represents a continuous region with the size of $1/2^k$ of the whole ID space. Based on the ratio between the block's density level and the average density level, the length of virtual nodes' *LevelID* will be increased or decreased accordingly. Such adjustment will impact the density of routing pointers in different routing regions. For sparse routing region, fewer routing pointers are needed, so the *LevelID* of the virtual node in that region can be extended. On the contrary, for dense routing region, more routing pointers are included by shortening the *LevelID*. However, the extension and reduction of the *LevelID* cannot deviate too much from its real level.

To maintain the routing pointers in a scalable fashion, Aurelia adopts multicasting to distribute the events of node join, leave or status change. The multicast tree doesn't require explicit management. It is based on the *LevelID* and routing region to disseminate the event from high-level nodes to low-level nodes. The details of the multicast algorithm are as follows: Each node maintains a list of "*top nodes*", whose *levelID* is the shortest suffix of current nodes's *levelID*. "*Top nodes*" are often powerful nodes that hold more routing pointers. When a node joins or changes its status, it first forwards the event with the node ID to one of its top nodes randomly. Then this top node disseminates the event to all the nodes whose *levelID* is the suffix of the reported node ID. But for the virtual node, if an announced node ID is not within the routing region it wants to be notified, the event is ignored. In every step of the multicast process, the node that receives the event first sends the event to the next lower level, then notifies the other nodes with the same *levelID*. By this approach, we can guarantee the event to be exactly delivered from the high-level nodes to the low-level nodes. When a node leaves, its predecessor will detect the event and help to notify the top nodes and propagates the change information to all the related nodes.

3.2 Data Indexing

Generally, for a multi-dimensional range query, all values of all dimensions must satisfy the query range along each dimension. If any of them fails, the data point will not be qualified. Therefore, a straightforward approach is to index on a small subset of the dimensions. However, the effectiveness of such an approach depends on the data distribution of the selected dimensions and requires prior knowledge.

To avoid this drawback, we index the data based on the maximum or minimum value among all dimensions of the data point. As the proportion of data points with a very big or small value in one dimension will increase with dimensionality, so it is safe to index the data points based on their edges. The feasibility of such indexing method has been validated by [12] in traditional database field.

The transformation process is a simple mapping, which is computationally inexpensive. Let x_{min} and x_{max} be respectively the smallest and biggest values among all the d dimensions of data point $< x_1, x_2, ..., x_d >$, $x_j \in [0,1]$, $1 \leq j \leq d$. Let the corresponding dimensions of x_{min} and x_{max} be d_{min} and d_{max}

respectively, then the high-dimensional point x can be mapped to a point y in the single-dimensional space through the following function,

$$y = \begin{cases} d_{min} \times c + x_{min}, & \text{if } x_{min} + \theta < 1 - x_{max}; \\ d_{max} \times c + x_{max}, & \text{otherwise.} \end{cases}$$

where c is a positive constant to stretch the range of index keys(normally assigned the value 1) and θ is the tuning parameter to adjust the data distribution. In case of distributed systems, the value θ should be negotiated among peers. By random sampling, the "top nodes" collect the information of data distribution. Periodically, they negotiate the value of θ through distributed voting [13] and then multicast the decision to the low-level nodes.

Through the above transformation, we get an 1-dimensional value y, $y \in [1, d+1]$, which can be further normalized as a binary string $b_0 b_1 ... b_{n-1}$ (n is the maximum ID length). The string $b_0 b_1 ... b_{n-1}$ is the key of the data object to be published in the Aurelia overlay.

The key is registered into all of the Aurelia nodes whose *RangeID* is a prefix of that key. In this way, one data object has multiple replicas and makes the indexing service more robust. Even when some nodes that host the data key leave the system, the lookup may still be successful.

The detailed publishing process is as follows: Besides "top nodes", every node also maintains a list of "top index nodes", whose *RangeID* is the shortest among nodes whose *RangeID* is a prefix of the current node's *RangeID*. During data publishing, the registration message is firstly routed to the node whose node ID is nearest to the data key; then the node will select a "top index node" randomly and forward the message to this "top index node", which is responsible for multicasting the message to all the nodes whose index range is covered by it. The multicast algorithm is similar to the above-mentioned multicast algorithm for routing table maintenance, except that it is based on the *RangeID* for multicasting.

During implementation, although the index key is only single-dimensional, the index entry can contain the complete d-dimensional data point. Accordingly, if the query also contains the full original query, then the results can be directly filtered at the side of index nodes. Only the qualified results are returned to the query initiator. In this way, the network traffic is further reduced.

3.3 Query Processing

To perform query, the d-dimensional query in the original data space should be transformed into 1-dimensional queries first. For a d-dimensional range query $q = ([x_{11}, x_{12}], [x_{21}, x_{22}], ..., [x_{d1}, x_{d2}])$, it will be mapped to d subqueries $sq_j = [l_j, h_j]$, $1 \leq j \leq d$, in the single dimension, with:

$$sq_j = \begin{cases} [j + max_{i=1}^d x_{i1}, j + x_{j2}], & \text{if } min_{i=1}^d x_{i1} + \theta \geq 1 - max_{i=1}^d x_{i1}; \\ [j + x_{j1}, j + min_{i=1}^d x_{i2}], & \text{if } min_{i=1}^d x_{i2} + \theta < 1 - max_{i=1}^d x_{i2}; \\ [j + x_{j1}, j + x_{j2}], & \text{otherwise.} \end{cases}$$

The union of the answers from all subqueries provides the candidate answer set from which the query answers can be obtained. Among the d subqueries, some subqueries are not necessary to be evaluated. The subqueries will be eliminated from the query set if any of the following conditions is satisfied:

(i) $min_{i=1}^{d} x_{i1} + \theta \geq 1 - max_{i=1}^{d} x_{i1}$ and $h_j < max_{i=1}^{d} x_{i1}$

(ii) $min_{i=1}^{d} x_{i2} + \theta < 1 - max_{i=1}^{d} x_{i2}$ and $l_j > min_{i=1}^{d} x_{i2}$

In this way, at most d subqueries are required to be evaluated. Suppose that the subquery $[l_j, h_j]$ corresponds to the ID range $[lkey_j, hkey_j]$. To perform range query, the query initiator should first compute the common prefix l_r of $lkey_j$ and $hkey_j$, and then check whether there exists a node in the routing table whose RangeID is a prefix of l_r. If existing, the range query is directly forwarded to that node. To avoid causing too much traffic on the powerful nodes, the prefix matching is based on the best-matching rule. If no entry exists, the node selects the nearest node ID to $lkey_j$ for forwarding.

The node that receives the query checks whether its range intersects with the query range, if its responsible index range covers the whole query range, it returns the results directly; otherwise, the node answers the part it knows and forwards the left part to the node whose ID is nearest to the lowest bound of the range.

In case that the query initiator wants to get the complete answer, all the subqueries should be executed. There are two approaches for execution: sequential or parallel. In the former, the subqueries are executed sequentially from the lowest bound to the highest bound. With each range query being answered, the whole range is increasingly reduced. In the latter, the subqueries are sent in parallel. For efficiency, before issuing subqueries, some subqueries can be combined into one if one node whose index range can answer them all is found.

3.4 Load Balancing

The problem of load balancing is a big issue in case of non-uniform query distribution. Roogle achieves system-wide load balancing through node self-adaptation.

When a node feels overloaded, it randomly selects one node from the uplevel index entries and forwards the later incoming queries to that node. Since the index range of up-level index node covers the range of low-level nodes, the queries can also be resolved. If the node itself is already the "*top index node*", it will firstly try to probe an underloaded node, and request this underloaded node to leave its current position and rejoin the "hot" region, so that the load can be partitioned; however, in case that the probe fails within a time limit, the node reduces its responsible range by extending the *RangeID*. After making changes, the node should notify its predecessor and successor. At the same time, it should also multicast the event to all the low-level nodes. The low-level nodes then change their top index entries accordingly.

4 Experimental Evaluation

Currently, the performance of Roogle is validated by simulation. The hardware platform is a Sun Enterprise E4500 server with 12 UltraSPARC-II 400MHz CPUs, 8GB RAM and 1Gbps network bandwidth.

Two datasets are used to evaluate the query performance of Roogle: one is Corel Image Features dataset [14], which contains 32-dimensional color histogram vectors extracted from 68,040 photo images; another is Movies dataset [15], which contains 11,435 30-dimensional movie records. In the Movies dataset, missing values are common. The node capacity distribution follows the measurement results of Gnutella in [16]. To simulate system dynamics, each node is associated with a lifetime satisfying Pareto distribution ($\alpha = 2.1, \beta = 2$) and the peer arrival follows a Poisson process.

The metric in use is the average number of nodes visited per query. For the purpose of comparison, we also simulate another three schemes: MAAN [2], Squid [3] and Roogle with SkipGraph [10] as the underlying overlay. The simulation results are illustrated in Fig. 2.

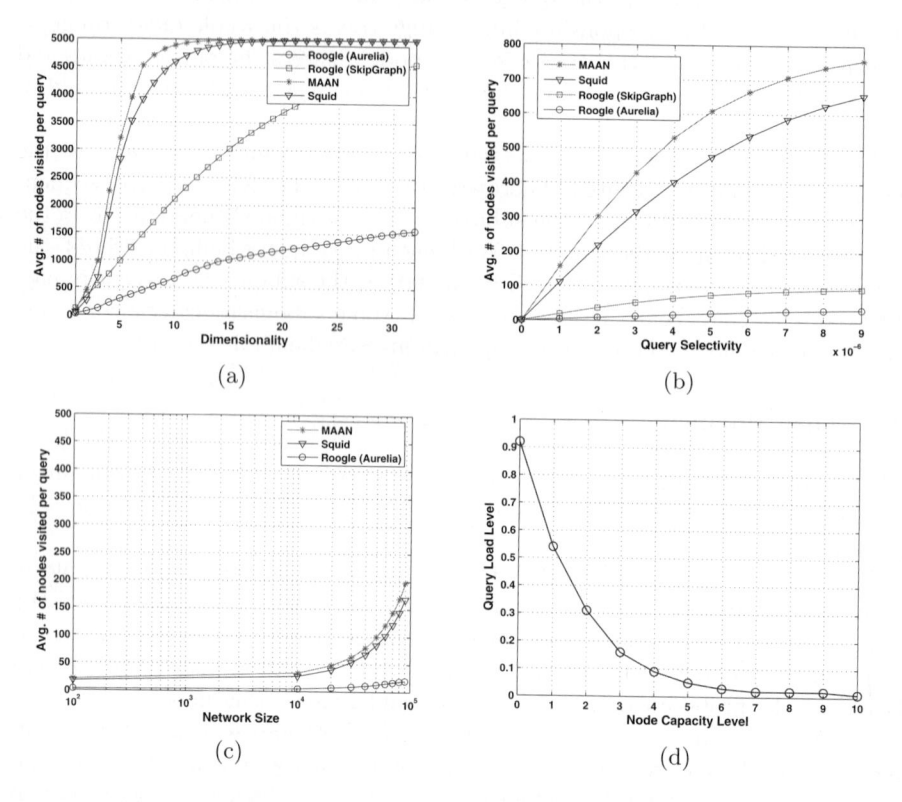

Fig. 2. (a) Effect of dimensionality on query performance; (b) Effect of query selectivity on query performance; (c) Effect of network size on query performance; (d) Query load level of nodes with different capacities

Fig. 2(a) depicts the effect of the dimensionality on query performance. All the schemes perform worse with increasing of dimensionality. When the dimensionality is bigger than 10, the performance of MAAN and Squid is almost degraded to sequential scan, while Roogle has a slower decreasing rate no matter what kind of overlay structure is adopted. But Aurelia further improves the performance by exploiting the node heterogeneity.

In Fig. 2(b), we show the effect of query selectivity on query performance. By varying the query selectivity, we can observe that Roogle has better data locality compared with MAAN and Squid. The locality of MAAN and Squid deteriorates greatly with the increasing of query selectivity. The poor locality of MAAN/Squid is caused by bad indexing performance of locality-preserving hashing and Space-Filling Curve (SFC), in which nearby points in original data space are dispersed to a large region in the single-dimensional space.

We also measure the effect of network size on query performance (as shown in Fig. 2(c)). For a given query, it is observed that Roogle scales well with the increase of network size. On the contrary, for MAAN and Squid, the number of nodes required to visit in order to solve the query increases almost linearly with the network size. When the system size is bigger than 10,000, their performance degrades in exponential speed.

Fig. 2(d) depicts the query load distribution on nodes with different capacity levels. we find that, due to the large routing table and index range, high-level nodes are likely to have more query traffic. But they also enjoy quicker routing and querying than low-level nodes, thus the above cost gets well compensated. The results also show that load balancing is mostly achieved in the system, and all the nodes take a fair portion of load corresponding to their capacity.

5 Conclusion

Our focus in this paper is to efficiently support high-dimensional range queries in P2P systems. The problem is addressed from two aspects: indexing and routing. To overcome the curse of dimensionality, we index P2P shared data simply based on the maximum or minimum value along all dimensions. By exploiting the node heterogeneity, we speed up the query routing. Compared with previous schemes, the performance of our design is much improved.

The next step is to investigate additional query optimization and system resilience to failures, and deploy Roogle in PlanetLab [17] to verify its performance under more realistic environments.

References

1. Oppenheimer, D., Albrecht, J., Patterson, D., Vahdat, A.: Scalable wide-area resource discovery. In: UC Berkeley Technical Report UCB CSD-04-1334. (2004)
2. Cai, M., Frank, M., Chen, J., Szekely, P.: Maan: A multi-attribute addressable network for grid information services. In: Proc. 4th International Workshop on Grid Computing (Grid'03), Phoenix, Arizona (2003)

3. Schmidt, C., Parashar, M.: Enabling flexible queries with guarantees in p2p systems. Internet Computing Journal **Vol.8, No.3** (2004)
4. Bharambe, A.R., Agrawal, M., Seshan, S.: Mercury: Supporting scalable multi-attribute range queries. In: Proc. ACM SIGCOMM'04. (2004)
5. Weber, R., Schek, H.J., Blott, S.: A quantitative analysis and performance study for similarity-search methods in high-dimensional spaces. In: Proc. VLDB'98, St. Johns, Canada (1998)
6. Ramabhadran, S., Ratnasamy, S., Hellerstein, J.M., Shenker, S.: Brief announcement: Prefix hash tree. In: Proc. ACM PODC'04, St. Johns, Canada (2004)
7. Ganesan, P., Yang, B., Garcia-Molina, H.: One torus to rule them all: Multi-dimensional queries in p2p systems. In: Proc. WebDB'04. (2004)
8. Zhang, C., Krishnamurthy, A., Wang, R.Y.: Skipindex: Towards a scalable peer-to-peer index service for high dimensional data. In: Princeton Technical Report, Submitted for publication. (2004)
9. Ratnasamy, S., Francis, P., Handley, M., Karp, R., Shenker, S.: A scalable content-addressable network. In: Proc. ACM SIGCOMM'01, San Diego, CA (2001)
10. Aspnes, J., Shah, G.: Skip graphs. In: Proc. 4th Annual ACM-SIAM Symposium on Discrete Algorithms. (2003)
11. Hu, J., et al.: Smartboa: Constructing p2p overlay network in the heterogeneous internet using irregular routing tables. In: Proc. 3rd International Workshop on Peer-to-Peer Systems (IPTPS'04), San Diego, CA, USA (2004)
12. Yu, C., Bressan, S., Ooi, B.C., Tan, K.: Querying high-dimensional data in single-dimensional space. VLDB Journal **13(2)** (2004) 105–119
13. B., H., Kwiat, K., Upadhyaya, S.: Secure and fault-tolerant voting in distributed systems. In: IEEE Aerospace Conference. (2001)
14. http://kdd.ics.uci.edu/databases/CorelFeatures/CorelFeatures.data.html.
15. http://www-db.stanford.edu/pub/movies/main.html.
16. Saroiu, S.: Measurement and analysis of internet content delivery systems. Doctoral Dissertation, University of Washington (2004)
17. http://www.planet-lab.org/.

Creating and Maintaining Replicas in Unstructured Peer-to-Peer Systems*

Elias Leontiadis[1], Vassilios V. Dimakopoulos[2], and Evaggelia Pitoura[2]

[1] Department of Computer Science, University College London, United Kingdom
[2] Department of Computer Science, University of Ioannina, Ioannina, Greece

Abstract. In peer-to-peer systems, replication is an important issue as it improves search performance and data availability. It has been shown that optimal replication is attained when the number of replicas per item is proportional to the square root of their popularity. In this paper, we focus on updates in the case of optimal replication. In particular, we propose a new practical strategy for achieving square root replication called pull-then-push replication (PtP). With PtP, after a successful search, the requesting node enters a replicate-push phase where it transmits copies of the item to its neighbors. We show that updating the replicas can be significantly improved through an update-push phase where the node that created the copies propagates any updates it has received using similar parameters as in replicate-push. Our experimental results show that replicate-push coupled with an update-push strategy achieves good replica placement and consistency with small message overhead.

1 Introduction

The popularity of file sharing systems (such as Napster and Gnutella) has resulted in attracting much current research in peer-to-peer (p2p) systems. Peer-to-peer systems offer a means for sharing data among a large, diverse and dynamic population of users. An issue central in such systems is resource location, i.e. given a user query for data, to discover the peers with matching data items.

There are two basic approaches for building p2p systems for efficiently locating data. In structured p2p systems, data items are assigned to specific peers using some form of distributed hashing. Locating peers with matching data is then guaranteed to take place by visiting a bounded number of peers, normally logarithmic to the total number of peers in the system. In unstructured p2p systems, there is no assumption about the placement of data items. New nodes connect to some other nodes in the p2p system randomly. When compared with structured p2p systems, unstructured p2p systems usually provide no guarantees for search performance but do not suffer from the cost induced from maintaining the structure and from load balancing procedures necessary in structured p2p systems.

In this paper, we focus on the problem of replication in unstructured p2p systems. Replication improves the performance of search as well as data availability. Availability issues are especially critical in p2p systems, since peers leave the system very often,

---

* Work partially supported by the Integrated Project IST-15964 AEOLUS.

thus making their data unavailable. Previous work on the topic [1,2] showed that optimal (with respect to search performance) replication is achieved when the number of copies per data item is proportional to the square root of their popularity. Here, we propose a new practical strategy for achieving square root replication called pull-then-push replication (PtP). With PtP replication, after a successful search for a data item, the node that posed the query enters a replicate-push phase during which it pushes copies of the item to its neighbors.

We also propose consistency maintenance protocols for copies created using the optimal replication strategy. We show that updating the copies can be significantly improved through an update-push phase where the node that created the copies propagates any updates it receives to its neighbors. Although, replica consistency protocols have been previously proposed (e.g., in [3]), our main contribution is that we study the problem in conjunction with the strategy used to create the copies. Our experiments show that the best results are achieved when update-push uses similar parameters with replicate-push.

2 Optimal Replication

Suppose there are in total m different data items in the network, and that, collectively, the peers have capacity for storing R items[1]. Also, assume that the query rate for item i is q_i, $i = 1, \ldots, m$. Cohen and Shenker [1] developed a theory for optimally replicating the data items in unstructured peer-to-peer networks, given the restriction of R. In particular, they studied different replication strategies and showed that the expected search cost is minimized when the ith item has r_i replicas, where r_i is proportional to $\sqrt{q_i}$.

In their analysis the authors assumed a theoretical random probes (RP) search method: the inquiring node repeatedly probes peers in random and asks for the item, until the item is found. As the authors argued, the RP method captures the essential behavior of the blind search strategies (such as flooding) usually employed in p2p systems because in unstructured networks the topology is unrelated to the location of data. The problem with square-root (SR) replication is that it requires knowledge of the query rate for each item. To alleviate this, the following scheme was proposed: after each successful search, the item is copied to a number of nodes equal to the number of probes. It was shown that, with an analogous rate of item removals, this scheme leads to SR replication.

However, even this scheme is not easily implementable. Keeping track of the number of queried nodes is simply impractical when the usual flooding-based search algorithms are used, due to the excessive number of messages required. But even if a practical way of counting the queried nodes existed, this number would not be equal to the number of random probes that would have been required. The reason is that the theoretic RP strategy stops immediately after locating the item. All practical strategies, however, unleash parallel search paths — if the item is found in one of the search paths, the rest might continue querying nodes until, for example a time-to-live (TTL) parameter was exhausted.

In conclusion, practical strategies for approximating the number of probes are required. In [2], the authors examined a number of such algorithms, namely

---

[1] Data items can be actual copies of the data or just pointers to them.

owner-replication, *path-replication* and *random-replication*. In owner-replication, the inquiring node is the only one that makes a copy of the resource — leading clearly to suboptimal replication. In the other two strategies, the node that provides the resource creates a number of replicas, equal to the distance (in hops) between the inquiring and the offering node. The last two strategies differ only in where the replicas are placed. Path and random replication approach SR replication but not quite accurately. The reason is that if the distance between the inquiring and the offering node is t hops, the RP strategy may not have located the item within just t probes, unless a single path was used for the search. The authors used multiple random walkers, which naturally visit a multiple of t nodes. We next propose a simple but effective scheme.

2.1 Pull-Then-Push Replication

The proposed scheme is based on the following idea: the creation of replicas is delegated to the *inquiring* node, not the providing node. The scheme consists of two phases. The *pull* phase refers to searching for a data item. After a successful search, the inquiring node enters a *push* phase, whereby it transmits the data item to other nodes in the network in order to force creation of replicas. We call this the *Pull-then-Push* (PtP) replication. One can conceive variations of the PtP strategy by utilizing different algorithms for the pull and push phases. Path replication as suggested in [2] could be considered as a type of PtP replication, where the pull phase uses multiple random walkers, while the push phase uses a single path.

In order to reach SR replication, we need to create a number of replicas equal to the number of probed nodes. Consequently, one should utilize the *same* algorithm for the push and the pull phases, so that the push phase visits approximately the same nodes the pull phase visited. For example, if a random BFS search algorithm is used for the pull phase, the same algorithm should be used to broadcast the item during the push phase.

All practical search strategies produce multiple search routes, and utilize some form of TTL to limit the search space (and the resulting message overhead). If during the pull phase the item was found at distance t hops from the inquirer, then the push phase should also stop after t hops. This means that the TTL utilized for the push phases should not be set according to the TTL used during pull, but rather according to t.

However, because of the multiple search routes produced, the tth step may contact quite a large number of nodes. In [4], it was shown that for pure flooding, the number of messages grows exponentially with the TTL; most of those messages are sent in the last step of the search. For example, assume a random network with each peer connected to d other nodes, and a pure flooding strategy, where each peer propagates the query to all its neighbors. If a search returned an item at the 3rd step, approximately $d + d^2 + d^3$ different peers would have been visited, although only one node at distance 3 had the item. This means that $d + d^2 + 1$ probes could be enough and as a result, the best strategy for the push phase would be to use a TTL of 2, not 3. In general, the TTL used for the push phase should be equal to the hop distance at which the item was found minus one.

Recapping, our proposed PtP strategy adheres to the following rules: **(a)** After a successful search, the requester pushes the item back to the network; **(b)** The same algorithm is used for both pull (search) and push; **(c)** The TTL for push is equal to $t - 1$, where t is the hop distance where the resource was found; **(d)** All peers receiving the push message create a replica of the item. In the next section we provide simulation results which confirm that this simple PtP strategy does indeed lead to SR replication.

2.2 Experimental Results

The PtP strategy has been evaluated through extensive simulations. In our simulator, we construct a network of peers/nodes, where each peer is connected to d other peers in random, called its neighbors. Each peer offers a number of data items and also has a fixed number of slots for replicating other items. Initially, all replica slots are empty. Then, we continuously perform searches originating at random peers, for random items. After each search, a push phase occurs, where replication is forced according to the strategy used. If a peer has to replicate an item and has no available slot, a uniformly random slot is emptied so that room is created for the new replica. Results are collected after a sufficiently large number of searches; the single most important metric we extract is the number of replicas, r_i, for each item.

The simulator is capable of utilizing a number of different search (pull) strategies. In all these strategies, a peer that receives a query for a data item, first checks whether it knows about the item; if not, it propagates the query to its neighbors. The strategies differ in the set of neighbors where the queries are propagated, and include [2,4,5]:

- *Pure flooding.* Peers propagate the query to all their neighbors.
- *Random walkers or random paths.* For a single random path, each peer propagates the query to exactly one of its neighbors, in random. Multiple walkers searching in parallel is a variation to decrease the average number of hops: the inquiring node sends the query to a number of its neighbors, each one unleashing a random walker.
- *Random BFS or teeming.* Peers propagate the query to each of their neighbors with some fixed probability ϕ. A *decay* parameter may be utilized so that ϕ decreases with the distance from the inquiring node. If a node is in distance t from the inquiring peer, then the probability of contacting a neighbor is given by: $\phi_t = \phi(1-c)^t$, where $\phi_0 = \phi$ and c is the decay parameter. For $c = 0$ we have simple teeming, while if in addition $\phi = 1$, the strategy is pure flooding.

The same algorithms are used for the push phase. Of course, in this case the peers do not receive queries but just items to propagate immediately to some of their neighbors.

In Fig. 1, we present results for a random network of 1000 peers, each with 4 neighbors on average. A peer has storage space for 10 items, out of a total of $R = 100$ different items. The replication strategies employed are owner, path and PtP replication. For PtP we experimented with all the algorithms presented above and with different parameters. In Fig. 1, we show the results for two of them, one with 5 random walkers and TTL = 10 and one with teeming, TTL = 5 and a decay parameter of $c = 0.4$. The other algorithms exhibited the same behavior, and were omitted for clarity. The plot shows the normalized number of replicas (r_i/R) for each of the items. To make the square-root trend clearer, for this particular plot, we have assumed query rates proportional to the id of the item, so the x-axis could also be named 'query rate'. The plot includes the optimal square-root distribution (SR), drawn with a thick line. We have also experimented with other query rates, including Zipf-like ones, and the results were identical.

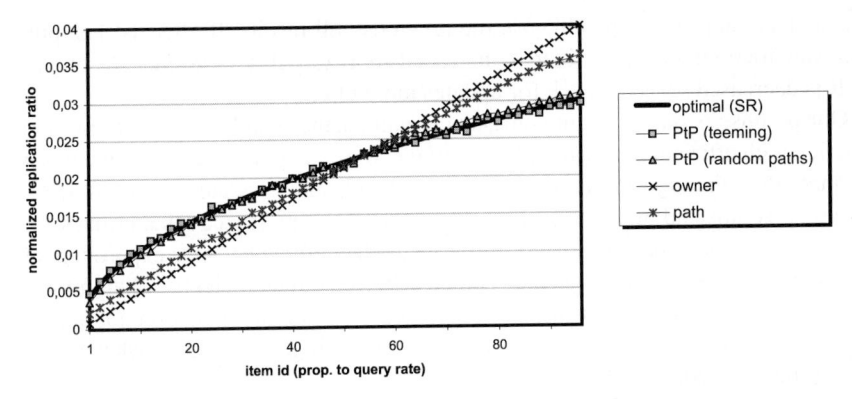

Fig. 1. Distribution of replication ratios under various replication strategies

It should be clear from Fig. 1, that owner replication is far from the optimum. Path replication is better, but does not result in SR replication. Both PtP strategies, although different by nature, led to almost perfect SR replication. This also comes to confirm our intuition that the exact strategies used for the pull/push phases of PtP are not very important, as long as they are the same in both phases. Here we only show PtP's ability to approximate SR replication. Results on PtP's performance and the achieved search gains can be found in [6].

3 Consistency Maintenance

Replication induces the need for consistency maintenance, that is, keeping the replicas up to date whenever changes occur. For the discussion that follows, we assume that each data item has a single *owner*, which is also the single peer that is allowed to modify the item. Upon modification, the replicas which have been spread over the network must be made consistent with the most recent version of the data item.

The problem of consistency maintenance appears in many contexts [7,8]. In [3,9], various strategies were proposed in the context of peer-to-peer systems. In general, updates of a data item are broadcast by the owner and/or are searched for by the peers that have the replicas. Thus, solutions to the consistency maintenance problem utilize **(a)** owner-initiated update *push*, so that peers with replicas are communicated the update, **(b)** replica holder-initiated *pull*, either when needed or periodically, so as to discover new updates, if any, or **(c)** a combined push/pull scheme.

It has been shown that usually a combined push/pull strategy (P/P for short) constitutes the best tradeoff between consistency levels and message overhead [9,5]. The owner performs a limited push of the updates and the peers pull periodically, just in case the owner-initiated push did not reach them.

A basic problem in these P/P protocols is when should a peer pull. Pulling too often creates substantial message overhead. Pulling infrequently may result in missing important updates. Adaptive pull strategies try to minimize the communication overhead, while maintaining good consistency levels by having each replica holder pull at specific intervals. These intervals are determined by a time-to-refresh (TTR) parameter, which is adaptively adjusted depending on the previous pull results. If after the last pull the item was found unchanged, TTR is increased so as to pull less frequently; otherwise, TTR is decreased so as to check for updates more often.

Our premise is that efficient consistency maintenance can be achieved only in conjunction with efficient replication. If the number of replicas and their placement is well-planned, then the algorithms for maintaining them under updates can be much more effective. To this end, we propose a novel push/pull update strategy that utilizes knowledge about replica creation so as to improve update efficiency. Our experiments have shown that consistency maintenance can be achieved quite efficiently when replication is done in the optimal way, using the PtP strategies. Optimal replication not only minimizes the average search costs but also reduces the average update costs when combined with a suitable update strategy.

3.1 Updates Under Optimal Replication

From now on we assume that items have been replicated in the network and that replication has been done using the PtP strategy. As discussed earlier, the PtP strategy requires that, after a successful search, the peer that found the item creates a number of replicas, through a *replicate-push* phase, or R-push for short, with an appropriate TTL value. The basic idea now is to let this peer be held "responsible" for updating the replicas it created, as explained next. With respect to a particular data item, the nodes in the network fall into one of the following three categories:

- *owner*: the single peer that produces new versions of the data item
- *responsible*: a peer that searched for the item in the past (and thus forced the creation of replicas of the item)
- *indifferent*: a peer that was forced to hold a replica of the item.

The strategy, which we call PtPU, is a combination of push/pull. The owner broadcasts new updates to the network, through an update-push, or U-push for short. Whenever a "responsible" peer receives a new version of the item (either through an *update-pull* that it itself performed or an U-push that the item owner initiated), it undertakes the task of updating the replicas it created. In other words, it performs a U-push itself for the new version of the item. Moreover, this U-push should employ the same TTL parameter as the one used in the R-push, thereby reaching approximately the same nodes that were previously reached in order to create replicas.

This scheme has the potential of reducing the overhead of consistency maintenance significantly. A peer that is "responsible" for a resource should check (pull) frequently for newer updates of the item, using a smaller TTR value. Peers which were forced to have replicas of this item ("indifferent" peers) do not need to pull (or, they could pull quite infrequently; *cf* the discussion in Section 4), relying on some "responsible" peer to provide an update for them. Summarizing, our strategy behaves as follows:

- The owner pushes the new versions of the item
- "Responsible" peers pull periodically, and push any updates they become aware of to their neighborhood exactly as when they created the replicas (i.e. with the same parameters as in the push phase of PtP).

– The other peers do nothing; they rely on "responsible" peers to keep them updated.

For the periodic pulls of the "responsible" peers, we follow an adaptive scheme [9], whereby the time-to-pull-next (TTR) is decreased or increased according to the perceived version of the item. If the last pull did not return a newer version, the estimate for the next TTR will be increased by some constant: $TTR_e = TTR + C$. If, on the other hand, a more recent version of the item was found, the next TTR should be decreased. It should be decreased in proportion to the difference, D, in versions between the pulled item and the one the peer had — the higher the difference D, the more the missed updates, and hence the more frequent the pull should be. Thus, the estimate for the new TTR is: $TTR_e = TTR/(D + \beta)$, where β is a parameter that provides some reduction in TTR in the case of $D = 1$. The next TTR is a weighted average of the current TTR and the estimate:

$$TTR \longleftarrow wTTR_e + (1-w)TTR,$$

where, w is a parameter determining the rate of change — smaller values of w make TTR change very slowly, while larger values make TTR adapt quickly to variations.

3.2 Experimental Evaluation

We have evaluated the performance of both the P/P and the PtPU strategies through extensive simulations. The network of peers is constructed and the data items are replicated using the PtP strategy as described in Section 2.2. After creating the replicas, we initiate simulation sessions. Each session runs for a number of rounds (turns). During each turn, the owner of an item creates a new version of the item with a given update probability p_u (update rate) and pushes it to the network. In the P/P strategy, all peers with replicas pull for new versions using adaptive pull. With PtPU, only the "responsible" peers pull using, again, adaptive pull. In addition, the "responsible" peers push any received updates to their neighbors using exactly the same strategy used when the replicas were created (for example, using teeming with the same decay and TTL values).

We evaluate the performance of the update strategies with respect to two parameters: the achieved consistency and the associated message overhead. The consistency level is measured as the percentage of replicas that are up-to-date. We experimented with different strategies for propagating the updates (i.e., pure flooding, random walkers, teeming and teeming with decay). The results attained were qualitative the same, thus, we report here only the results obtained when using teeming with decay, which is the method that gives us the most flexibility in terms of tuning the extend of the propagation. In particular, we present results when using three variations of teeming as summarized in the table that follows. Wide teeming visits more peers, while narrow teeming produces smaller message overhead.

Extend of teeming	c (decay)	TTL
Wide	0.1	5
Medium	0.3	5
Narrow	0.4	4

Regarding the adaptive pull, the tuning of its parameters is beyond the scope of this paper. A set of values that were found to work well in adapting the TTR is: $w = 0.8$,

$b = 0.5$, and $C = 10$ turns, and those are the values that were used in all the experiments presented here. The reader is referred to [10,7] for a detailed discussion of the topic.

Performance with respect to the update rate. The goal of the first set of experiments is to depict the behavior of plain P/P and PtPU under different update rates. We consider two cases: frequent updates ($p_u = 0.1$), and infrequent updates ($p_u = 0.025$). The owner pushes the updates using narrow teeming. The reason for using such a rather limited push is to make the effect of pull more clear. To discover a general trend, we let both strategies utilize exactly the same pull characteristics (i.e. the same variations of teeming) and see how they compare with each other.

The results are shown in Fig. 2 for high update rates and in Fig. 3 for infrequent updates. Each strategy is simulated for pulling with wide, medium and narrow teeming. In the case of high update rates, peers are forced to a high pull overhead in the P/P strategy so as to be frequently updated. In the PtPU case, though, pull is limited. Push messages are more since the "responsible" peers also propagate any updates they receive. For a low update rate, it is easier for any strategy to keep good consistency levels, utilizing fewer messages. Even in this case, though, PtPU achieved consistency levels above 92%, while plain P/P is, at best, a little above 80%. PtPU consistently outperforms P/P by any measure. It results in better consistency levels and, at the same time, fewer messages.

Comparison of the two update policies. In this set of experiments, we compare further the two methods. In particular, we show (i) the level of consistency achieved when the two methods produce the same number of messages and (ii) the number of messages required by each method for achieving the same consistency level. Here, we consider a

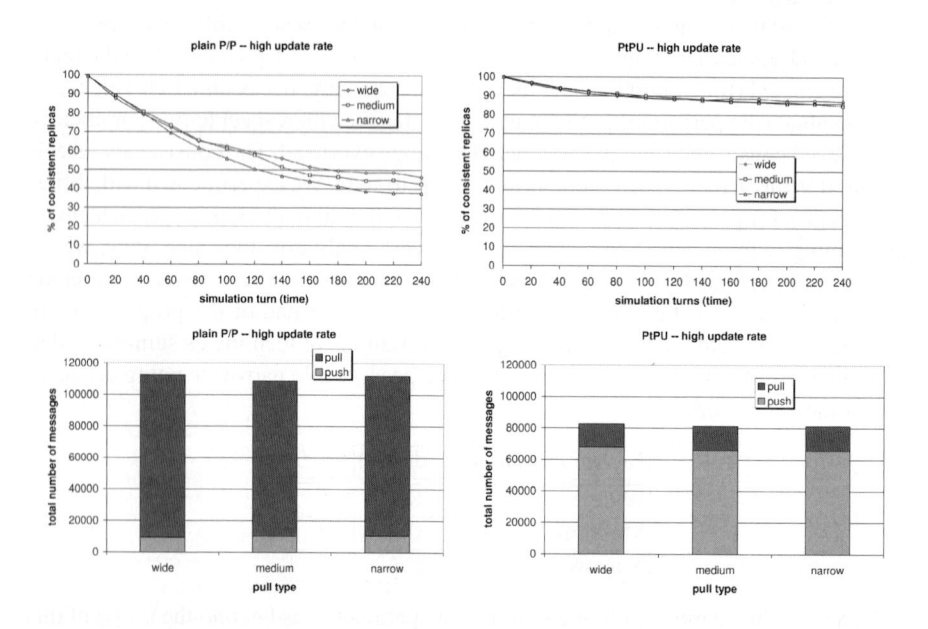

Fig. 2. Performance of the two strategies under high update rates

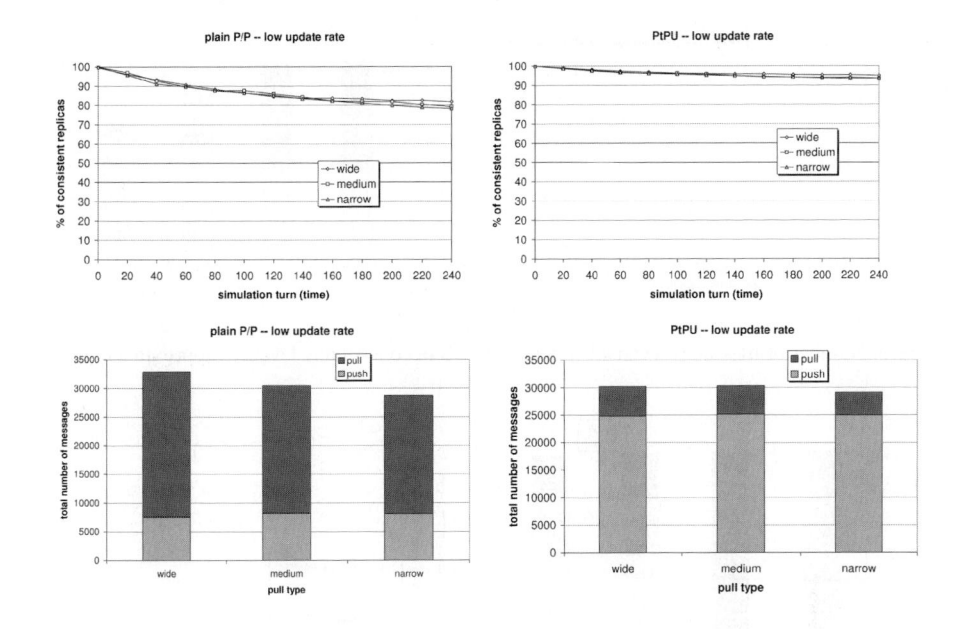

Fig. 3. Performance of the two strategies under low update rates

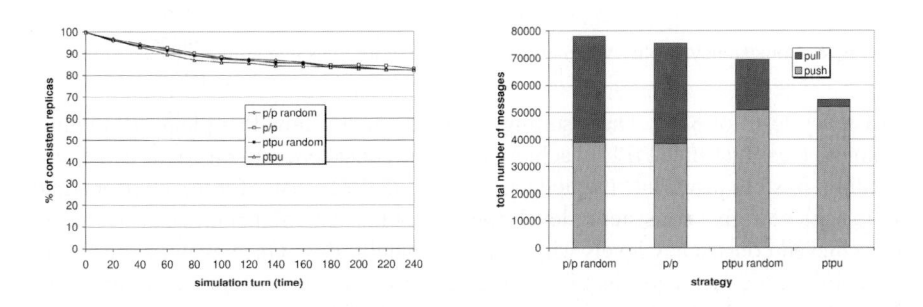

Fig. 4. Number of messages when all strategies result in consistency levels of approximately 82%

medium update rate ($p_u = 0.05$). For each strategy we repeatedly alter the pull parameters until we achieve the same value for the metric of interest (i.e. the consistency level or the number of messages) among all strategies.

The results are presented in Figures 4–6. In the plots, we also consider the performance of P/P and PtPU, for the case where the creation of replicas does not follow the PtP strategy. Instead, after the replication phase, the replicas get scattered across the network. Our goal is to show that loosing the locality induced by the PtP strategy results in worsening the performance of both the P/P and PtPU strategies. Note that the number of replicas is kept the same; what differs is their placement in the network. The strategies under random placement of the replicas are marked with an "(R)" in the plots.

In Fig. 4 the owner uses a narrow push to propagate the updates. We run the simulator tuning the pull parameters until all strategies achieved approximately the same

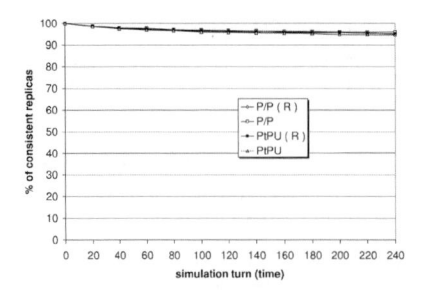

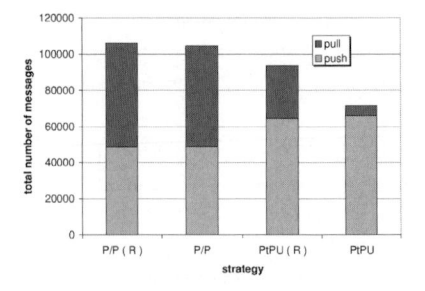

Fig. 5. Number of messages when all strategies result in consistency levels of approximately 95%

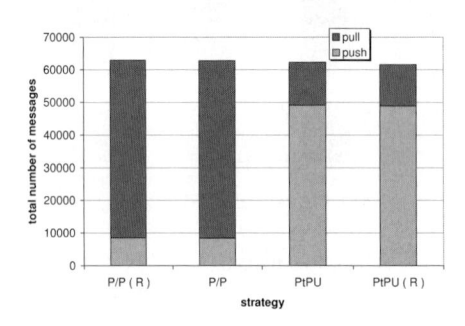

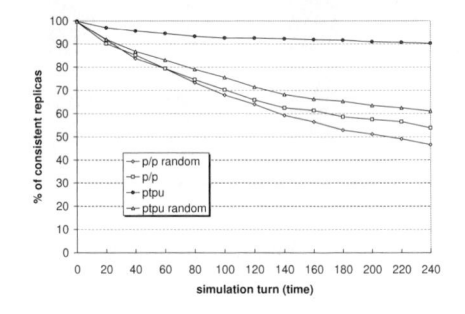

Fig. 6. Consistency quality when all strategies generate the same number of messages

consistency level of 82%. The resulting message counts show that plain P/P required 43% more messages than PtPU to achieve the same consistency. In Fig. 5 the owner uses a medium push to propagate the updates, so as to make it easier for the inferior strategies to achieve higher consistency levels (but, of course, with higher message overhead). The achieved consistency levels where approximately 95%. Once again, plain P/P required 46% more messages than PtPU. In Fig. 6 all strategies generated approximately 62000 messages. PtPU required a narrow pull while P/P's adaptive pull resulted in a wider teeming. The superiority of the PtPU strategy is shown vividly, as it managed to achieve more than 90% consistency.

Another conclusion from these plots is that, indeed, the random placement of replicas makes the performance of P/P and PtPU worse. This validates our intuition that the inherent locality of replica creation through PtP results in more efficient updates.

4 Discussion

In this paper, we consider replication in unstructured p2p systems. The idea behind our approach is that developing protocols for consistency maintenance which utilize knowledge about the strategy used to create the copies increases the efficiency of such protocols. Based on this, we develop a simple strategy for achieving square-root replication, which was previously proved to be optimal for unstructured peer-to-peer systems, and a consistency maintenance protocol that is tuned for our replication strategy.

Our experimental results show that our protocols achieve significantly better consistency for a smaller communication cost than protocols that do not exploit knowledge of the underlying replication strategy. A more detailed version of this work can be found in [6].

In our experiments we have assumed that the network does not change during the replication and update phases. We are currently studying the behavior of our strategies in more dynamic settings where peers enter or leave the system at will. In such environments the PtPU strategy may encounter the following problem: a "responsible" peer could depart from the network, leaving thus a number of "indifferent" nodes without anybody to update their replicas for them. Thus, it is almost imperative that "indifferent" peers should pull, too, just in case the "responsible" node is not near them anymore.

References

1. Cohen, E., Shenker, S.: Replication Strategies in Unstructured Peer-to-Peer Networks. In: Proc. ACM SIGCOMM'02. (2002)
2. Lv, D., Cao, P., Cohen, E., Li, K., Shenker, S.: Search and Replication in Unstructured Peer-to-Peer Networks. In: Proc. ICS'02, 16th ACM Int'l Conference on Supercomputing, New York, USA (2002)
3. Datta, A., Hauswirth, M., Aberer, K.: Updates in highly unreliable, replicated peer-to-peer systems. In: Proc. of ICDCS 2003, 23rd Int'l Conference on Distributed Computing Systems, Providence, Rhode Island (2003) 76–85
4. Dimakopoulos, V.V., Pitoura, E.: Performance analysis of distributed search in open agent system. In: Proc. IPDPS '03, Int'l Parallel and Distributed Processing Symposium, Nice, France (2003)
5. Leontiadis, E., Dimakopoulos, V.V., Pitoura, E.: Cache Updates in a Peer-to-Peer Network of Mobile Agents. In: Proc. P2P2005, 4th Int'l Conference on Peer to Peer Computing, Zurich, Switzerland (2004) 10–17
6. Leontiadis, E., Dimakopoulos, V.V., Pitoura, E.: Creating and Maintaining Replicas in Unstructured Peer-to-Peer Systems. Technical Report TR2006-01, Univ. of Ioannina, Dept. of Computer Science (2006)
7. Srinivasan, R., Liang, C., Ramamritham, K.: Maintaining temporal coherency of virtual data warehouses. In: Proc. RTSS '98, 19th Real Time Systems Symp., Madrid, Spain (1998)
8. Urgaonkar, B., Ninan, A., Raunak, M., Shenoy, R., Ramamritham, K.: Maintaining mutual consistency for cached web objects. In: Proc. ICDCS 2001, 21st Int'l Conference Distributed Computing Systems, Phoenix, AZ, USA (2001)
9. Lan, J., Liu, X., Shenoy, P., Ramamritham, K.: Consistency maintenance in peer-to-peer file sharing networks. In: Proc. of WIAPP'03, 3rd IEEE Workshop on Internet Applications, San Jose, CA, USA (2003) 76–85
10. Lan, J.: Cache Consistency Techniques for Peer-to-Peer File Sharing. Technical report, MSc Thesis, Dept. of Computer Science, Univ. of Massachusetts (2002)

DOH: A Content Delivery Peer-to-Peer Network

Jimmy Jernberg[1], Vladimir Vlassov[1], Ali Ghodsi[1,2], and Seif Haridi[1,2]

[1] School for Information and Communication Technology (ICT), Royal Institute of Technology (KTH), Stockholm, Sweden
[2] Swedish Institute of Computer Science (SICS), Kista, Sweden

Abstract. Many SMEs and non-profit organizations suffer when their Web servers become unavailable due to flash crowd effects when their web site becomes popular. One of the solutions to the flash-crowd problem is to place the web site on a scalable CDN (Content Delivery Network) that replicates the content and distributes the load in order to improve its response time.

In this paper, we present our approach to building a scalable Web Hosting environment as a CDN on top of a structured peer-to-peer system of collaborative web-servers integrated to share the load and to improve the overall system performance, scalability, availability and robustness. Unlike cluster-based solutions, it can run on heterogeneous hardware, over geographically dispersed areas. To validate and evaluate our approach, we have developed a system prototype called DOH (DKS Organized Hosting) that is a CDN implemented on top of the DKS (Distributed K-nary Search) structured P2P system with DHT (Distributed Hash table) functionality [9]. The prototype is implemented in Java, using the DKS middleware, the Jetty web-server, and a modified JavaFTP server. The proposed design of CDN has been evaluated by simulation and by evaluation experiments on the prototype.

1 Introduction

The major focus of our research presented in this article is to design and evaluate a scalable Content Delivery Network (CDN) built on top of a structured P2P system that provides the Distributed Hash Tables (DHT). Such CDN can be used as a Web-hosting environment that allows improving the overall performance and storage capacity, scalability and availability of hosted Web sites.

As a motivational scenario, assume a small company has a web server on a 10 Mbit broadband line, which usually serves it well. One day a large news portal reviews and recommends the company site to the portal users. Since the site becomes a "hot object", it starts generating a huge amount of hits. Subsequently, the company's web server will not be able to cope with the strain, and, eventually, its bandwidth will be totally consumed, making the company's Web-pages unavailable. The situation described above is called the flash crowd effect (also known as the SlashDot effect[1]), when a sudden increase in traffic makes a web site completely unavailable.

One solution for a company to survive a flash crowd is to pay for joining a proprietary CDN like the one owned by Akamai[2] that offers services in distributing the load of heavily trafficked web sites for companies with an extensive web presence. For SMEs and organizations without the need for a CDN on a daily basis, the incurred costs of placing their web-sites on a proprietary CDN might be considered too high.

In our view, one of the cost-efficient approaches to building high-performance and scalable web-sites, CDNs and Web-hosting systems, is to integrate several (open-source) web-servers in a scalable structured P2P system with DHT functionality. A (part of the) URL of a Web-page can be used as a key to determine a web-server on which the page is (to be) stored. The P2P system of web-servers should support content replication in order to improve performance and availability of hosted Web-sites. We believe that this approach should make it possible for SMEs and small organizations to obtain at an affordable price the same hosting services that have been available to big companies for years. CoralCDN[10] is an existing P2P CDN that are already deployed, but while CoralCDN is designed to be an overlay network for handling the flash crowd effect DOH aims for more: to be a low cost, transparent, web-hosting service with a built in ability to handle a flash crowd.

Extensive research has been done in building efficient DHTs on top of structured P2P overlay networks, see e.g.[3], [11], [19], [20], and [21]. A DHT provides a distributed indexing service based on hashing and like an ordinary centralized hash-table, a DHT, whose buckets are distributed among peers, can be used for storing of different kind of information. Note that the DHT should be "open": in the case of a hash collision, when different entries are hashed to the same bucket, a single bucket can contain multiple entries, which should be searched sequentially.

In this paper, we present our approach to building a scalable Web Hosting environment as a CDN on top of a structured P2P system with DHT functionality. In our design, several Web-servers are organized in a structured P2P system in order to share their load and to improve the overall system performance, scalability and storage capacity, as well as availability of hosted web-sites. The underlying P2P overlay network provides an efficient and scalable lookup mechanism needed for DHT, replication and ability to automatically self-organize when nodes join/leave the network.

The DHT is used for fetching and storing web pages. Each of the web-servers is responsible for a region of DHT buckets used to store Web pages, referenced by URLs. Even though the worst-case lookup latency in a structured P2P system with N peers is $O(\log N)$, building a Web-hosting environment as a structured P2P system allows improving overall performance and scalability of Web-hosting due to multiple access points, well-balanced load distribution and content replication, increase in overall storage and computational capacity of the P2P CDN. In our design, we use a sophisticated content replication mechanism, called symmetric replication [13], in order to even more improve the system performance, availability and reliability.

To validate our approach, we have developed, implemented and evaluated a system prototype called DOH (DKS Organized Hosting) that is a content delivery network implemented on top of the DKS (Distributed K-nary Search) structured P2P system with DHT functionality [9]. DOH provides the same features as a corporate CDN at the same cost as a regular low-end web server. The system prototype is implemented in Java, using DKS[3], the Jetty[12] web server, and a modified JavaFTP[6] server package. We have evaluated our proposed CDN design by simulation and by performing evaluation experiments on the developed prototype.

The remainder of the paper is organized as follows: Section 2 describes the DOH architecture. Section 3 presents our DOH prototype. Section 4 presents results of preliminary performance evaluation. Section 5 discusses some related work. Conclusions and future work are given in Section 6.

2 DOH Design

When designing the DOH content delivery network, two different types of users should be considered: a regular user (called User) browsing the Web; and a content provider (called here Publisher) publishing content of a web-site in DOH. As shown in Figure 1, DOH consists of two types of nodes: Translators and DOH-Nodes (or shortly Nodes).

The DOH-Nodes are connected by the DKS P2P middleware in a structured P2P network. Each DOH-Node contains an FTP server, a web server, and is connected to the DKS overlay network (see Figure 1). It serves HTTP requests submitted by Users; confirms identity, inserts, and removes content provided by Publishers.

Translator nodes handles interaction between the User and the system before an HTTP request is sent to a DOH-Node. A Translator redirects the User's browser to DOH-Nodes based on a load-balancing strategy. Each Translator maintains a cache for storing information about other Translators and DOH-nodes including their load status and RTT times, referred to as the Translator-cache. This information is used for redirection decisions, and when Nodes join. Servicing of an HTTP request arrived to one of the DOH Translator nodes, passes the following steps:

1. Redirection from the old home to a Translator. This step is performed when the DNS entry for the requested web page has been updated;
2. Redirection from a Translator to a DOH-node based on current load of the DOH nodes and network congestion;
3. Retrieval of the requested file (replica) from the Translator-cache of the node or, if the cache misses, from the DHT of the DKS P2P system.
4. Unwrap, assemble, and write the file to the disk (cache) of the requested Node;
5. Sending the requested file to the requesting client.

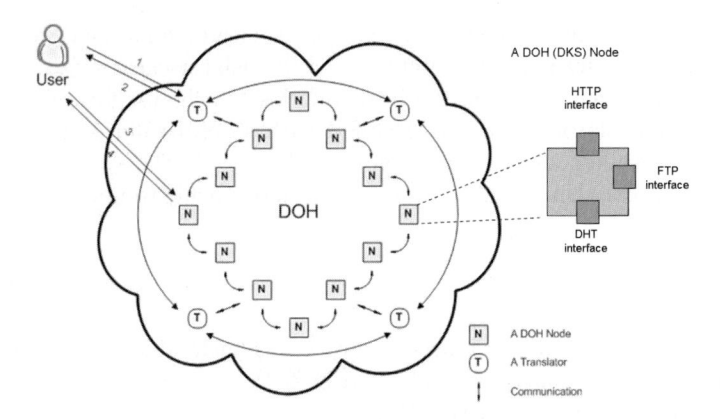

Fig. 1. Architecture of the DKS-based Hosting (DOH) P2P Content Delivery Network

2.1 Translator

A web-hosting system like DOH allows storing (and replicating) content of several web sites in one hosting system. As the content is referenced to by URLs, the system needs to direct a HTTP request to one of DOH-Nodes that serve as access points to the content, i.e. it needs to translate a requested URL to a new URL that redirects a requesting client to one of the DOH-Nodes. To perform this URL-to-URL translation, the DOH system includes Translator nodes that are Users' initial access points to the DOH content delivery network. Translators serve as mediators that redirect web clients to one of the DOH-nodes based on their current load and network congestion. Thus, the URL-to-URL translation performed by Translators aims at load balancing in order to improve availability and performance of the DOH content delivery network.

To perform load balancing, each Translator maintains a cache of information on DOH-nodes: IP-addresses, load, and RTT values; and information on other Translators it knows. This cache is called Translator-cache. When a Translator receives an HTTP request, it checks the Translator-cache to find the currently "best" DOH-node that can service the request. To redirect the client to a DOH-node, the Translator issues an HTTP code 302 message that is used to respond on requests for temporarily moved pages, and adds the IP address of the Node to the new URL when forming the 302-code response to redirect the requesting client. E.g., if the riginal URL is `http://www.url.com/a/b/index.html`, then the translated URL is `http://192.168.2.23/www.url.com/a/b/index.html`, where `192.168.2.23` has been chosen as the currently "best" DOH node to service the request.

Information in the Translator-cache is periodically updated by the Nodes. The data collected in the Translator-cache are used to calculate Nodes' load and network congestion over time. The Translator-cache is also used for bootstrapping, as it contains information on Nodes and Translators that are known to be up and running. There are three levels of the caching structure: (1) level 1 keeps a list of other known Translator-caches, (2) level 2 is the local Translator-cache level, and

(3) level 3 is a Translator-cache-entry that stores data on an actual Node. When a new Node joins DOH, it first contacts the Translator-cache (if any) it used last time. If that cache is down, the Node queries a cache list for online caches. The queried Translator-cache may respond with several valid entries, and the new (booting) Node contacts one of these Nodes to join the system. The same mechanism is used when a Publisher wants to find a Node to upload its content on DOH.

2.2 DOH-Node

In DOH, web content and its replicas are stored in a hash table distributed among the DOH-nodes. The content is replicated in DHT according to the symmetric replication mechanism used in DKS[13].

Each DOH-Node includes three subsystems which are the DKS middleware that connect the Node to the P2P DKS overlay network, an FTP server, and a web server. When a client requests a file, the web server searches the file locally, and if the file does not exist locally, or the local replica is considered too old, the web server will perform a lookup operation in the DKS DHT to retrieve the requested file. The FTP server of a DOH-node is used when Publishers upload content.

Granularity of content stored in DHT

To store contents (or content references) of hosted web sites, the DOH CDN uses a Distributed Hash Table (DHT) provided by the DKS P2P middleware[3]. When content of a web site is stored to or fetched from the DHT, either the entire URL or a part of the URL is considered as a key. The hashed key value determines a DOH-node responsible for the DHT-bucket in which the content is (to be) stored.

There are three strategies of placement of web-site content identified by URLs to the DHT, that differ in the granularity of a web-site content stored in a bucket of the DHT: (1) file-wise placement (uses the entire URL as a key, e.g. http://www.url.com/a/b/index.html); (2) directory-wise placement (uses the directory part of the URL as a key, e.g. http://www.url.com/a/b/); (3) site-wise placement (uses the web-sire part of the URL as a key, e.g. http://www.url.com/).

With the file-wise placement, files that belong to the same web-site can be stored in different DHT buckets and distributed among the DOH-nodes.

With the directory-wise placement, all files of the same directory are hashed to the same bucket, i.e. stored on the same DOH-node. Even though the files are hashed directory-wise it does not mean they have to be returned directory-wise on a DHT get request. The DKS API allows a file name to be sent along with the DHT get request as an additional parameter to retrieve specific entry (the requested file).

The web-site-wise placement is a coarse-grained placement and is similar to the directory-wise placement described above. The site-wise placement causes the entire content of a web-site to be stored in the same DHT bucket.

The coarser the placement is, the lower is the level of content distribution. In our DOH prototype, we support all three different levels of granularity. Preliminary evaluation of file-wise and directory-wise distribution shows that the system performance is not very sensitive to the granularity of the content distribution in the DHT but rather to prefetching and caching. One can expect that the file-wise and directory-wise distribution allows improving performance in the case of intensive concurrent accesses to a web-site, as well as improving its availability in the case of node failures. We leave more detailed evolution of the distribution strategies to our future work.

Symmetric Replication and Adaptive Caching
DKS builds on symmetric replication, which is built on-top of the DHT layer. Symmetric replication of DHT content enables parallel lookups, which increase the responsiveness of the system, while keeping the number of messages needed for restoring the replication degree after dynamism low. In addition to the symmetric replication for reliability and higher performance, DOH also implements an adaptive caching of requested content at DOH-nodes. The caching algorithm has been devised based on a combination of the Directory scheme defined in [14] with the entry caching scheme of DNS. We assume that whenever an object (a file) is requested, it is likely to be requested again from the same or another access point. Therefore it makes sense to cache the object on its way to the node that originates the lookup operation. In DKS, the return path of passing the object to the requesting node is recursive; therefore the object can be cached in the nodes along the return path. Consistency of copies is weak and can be kept by using the if-modified-since field built-in to the header of the HTTP protocol. When a cache is full, the Least Recently Used (LRU) algorithm or some other caching policy, could be used for deciding which objects to evict.

3 A System Prototype

We have implemented the DOH prototype in Java, using the DKS P2P middleware with the DHT API[3], the Jetty[12] web server, and a modified JavaFTP[6] server.

In the DOH-Node prototype, the web server functionality has been implemented by modifying the Jetty web server, which is licensed under the Apache license[4]. For uploading, downloading and removing content in DOH, we use the modified JavaFTP server package, also licensed under the Apache license. The DOH-Nodes are peers in the P2P DKS network, and the DKS DHT API is used to store and retrieve content of web sites hosted in DOH. In order to integrate the Jetty server in our prototype, we have extended the server so that it creates a special handler that searches the DKS DHT if a requested file is not found locally in the web server's cache.

JavaFTP server is a package that implements the FTP standard We have modified the server so that it allows a content provider to upload the files in the DHT rather than to the host's local file system. In order for the DOH system prototype to handle large objects, it uses data fragmentation that, we believe, allows better control over memory usage and to avoid running out of memory on a low-end server.

When a file is stored to the DHT the following steps are performed:

1. The key (it can be either the file name or the directory name or the web-site name) is hashed to get a hash table index using SHA-1, shortened to 64 bits.
2. If necessary (it depends on the file size), the file is fragmented;
3. Each of the fragments or the entire file is put in the DHT entry defined by the hash-table index. The put operation is based on the DKS lookup operation that finds a DOH node responsible for the target bucket. placed.

When retrieving a file from DHT, a DOH node performs the following steps:

1. The key is hashed to obtain a hash table index;
2. A DHT get operation based on the DKS lookup operation is performed that returns an array of entries (all the files) stored in the target DHT entry;
3. If fragmented, the requested file is assembled by combining the fragments. Copies of the file are stored in the web-caches of the nodes involved in the operation.

3.1 Implementation of a Translator

A Translator is a stand-alone node that serves as a web server for web-clients accessing web-sites hosted in DOH. Translator receives HTTP requests and redirects the clients to DOH-nodes. Translator provides the following functionality: (1) maintains a cache of information on DOH nodes (IP addresses, load and RTT times) and on other Translators (to retrieve information from their caches); (2) provides load balancing so that it redirects HTTP requests to DOH-Nodes based on their load and network congestion; (3) displays Node information to Publishers in a human-readable format.

Each Translator redirects the clients to DOH nodes as described above except of a special case when the requested URL refers to `doh_webcache.xml` indicating that a Publisher asks for an IP address of a Node to upload content. In this case, Translator replies with an XML page that contains information on Nodes from its own Translator-cache. From this file the Publisher can choose a node to connect to.

4 Preliminary Evaluation

In this paper, we present results of preliminary evaluation of the approach, leaving more detailed evaluation to our future work. The DOH prototype has been used to evaluate small-scale configuration mostly in order to verify the design, whereas a specially developed DOH simulator has been used to evaluate impact of different design choices (such as the use of content caches, the granularity of content distribution) on the system performance and reliability of DOH with varies (large) configurations.

In our experiments and simulation, the performance is measured as a service time that is the time from receiving a request to sending a reply. A single standalone Jetty web-server (without DHT), has been chosen as a baseline. We assume the synthetic *workload* formed of streams of HTTP requests issued by the number of concurrent independent clients, each of which sends a random sequence of requests to retrieve different randomly selected files from different randomly selected sites with a specified intensity. To generate random sequences of requests, we assume the Zipf distribution, as suggested in [8] for the distribution of incoming page requests in the Web.

4.1 Preliminary Evaluation of the DOH Prototype

We used the prototype mostly in order to validate the DOH design. The evaluation testbed included several Pentium 3, 500 MHz computers with 256 MB RAM running Linux (Red Hat 9.3). We report results of two series of experiments. In the first series, 50 files of the mean size of 10Kb of 6 web-sites were uploaded to a DOH-node. In the second series, the content was "heavier": 18 domains, 47 directories, and 503 files (mean size is still 10Kb). The number of nodes varied from 1 to 6 nodes.

As expected, our experiments have shown that the DOH performance is very sensitive to the use of file caches in DOH-nodes when increasing the number of nodes, i.e. increasing the level of distribution of web-sites in DOH. These results suggest that it is worth to make more efforts to find an caching strategy. Evaluation of the prototype has also shown that the performance of DOH heavily depends on the performance of the underlying DKS network: over 90% of the time used by the system consumed by DKS-related activities when the file size is increased to 4Mb.

We have also preliminary evaluated three different strategies of storing files in DHT: file-, directory-, and site-wise - in order to see whether the system performance is sensitive to the strategy used, and which of the strategies is the best with respect to performance. Remind that the three placement strategies use different parts of a URL as a key to determine a DOH-node responsible for the content pointed to by the URL. Unfortunately, the evaluation results for small-scale system configurations show that there is no clear best candidate to use in all cases studied. If published web pages are changing rapidly or if the load of the network is small, then the file-wise approach yields the best results. If there are seldom changes in the stored sites or the load is high, then the directory-wise or even the site-wise approaches are better to use. The system can perform even better if it can support a combination of at least two of the placement strategies, and DKS indeed supports this kind of flexibility.

A full-scale evaluation of the prototype should be done on configurations larger than the setups we could afford for now. In our future work we intend to evaluate large more realistic configurations of the prototype (with large number of nodes and clients). Our future plans also include further performance optimization of the prototype.

4.2 Performance Evaluation Using the DOH Simulator

As we continue improving and optimizing performance of the DOH prototype and, in particular, the DKS P2P middleware, we have developed an accurate simulator of DOH in order to evaluate the impact of different design choices (such as the use of caches in nodes, different strategies of storing files in DHT: file-, directory-, and site-wise, prefetching, different replication schemes and the number of replicas) and system changes (nodes leave the network, new nodes join the network) on the overall system performance and reliability. The simulator is based on timing estimates obtained from experiments on the DOH prototype and a stand-alone Jetty web-server.

In our simulation, we assume the following workload: a random sequence of requests with the predefined rate (1000-5000 requests/sec) is issued by several clients to retrieve different randomly selected files from different randomly selected sites; the content stored in DOH includes 18 sites, i.e. about 100 directories with about 10 files in each directory; the average file size is assumed to be 30 Kb (concurring with the average file size in the Web, as shown in [5]); the ratio of TTL for files in node caches varies from 0% (no cache) to 100% (always in the cache) of the simulation time. The cache TTL defines how long a file stays in a cache before it's removed from the cache.

The average service time has been computed based on timing estimates obtained from experiments on the DOH prototype with smaller configurations and the Jetty web-server. We assume that there are three major factors that affect the service time: (1) the current load of the server, (2) the size of the requested file, and (3) whether the file is cached or not. The service time was computed as follows:

$$T_s = 2 + Load \times 0.85 + Miss \times (15 + 2 \times fileSize + H \times 100 \times \log_2 N)$$

Here T_s is the service time in ms; $Load$ is the number of parallel requests served; $Miss \in 0, 1$ indicates whether the cache misses ($Miss = 1$) or hits ($Miss = 0$); $fileSize$ is the size of the requested file in Kbytes; $H \in 0, 1$ indicates whether the file is stored locally in one of the local DHT buckets ($H = 0$) or remotely ($H = 1$) and a number of hops is required to find and fetch the file from DHT - the probability that $H = 1$ is f/N, where f is the number of replicas per file in DKS; N is the number of nodes. Numeric constants (in ms) in the formula are average times obtained from experiments on the prototype and the stand-alone Jetty web-server.

We have evaluated the effect of different design choices on the performance of DOH. Figure 2 shows plots of the service time as the function of the number of nodes for different TTL of cached content and different strategies of placement of content to the DHT. As expected, it has been observed that the service time is sensitive to the use of caches: the service time is shorter if cached content stays longer in the cache (i.e. the higher cache hit ratio). The service time degrades as the number of nodes increases because of the increase in the DKS lookup latency. However the service time degrades slower when TTL of cached content is high. This result suggests that it is worth to make more efforts to

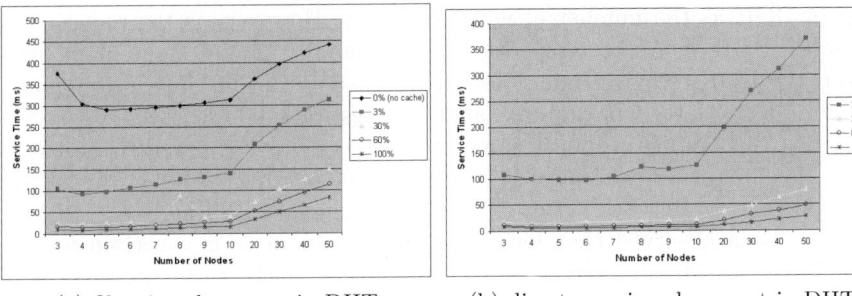

(a) file-wise placement in DHT
(b) directory-wise placement in DHT

Fig. 2. Effect of the use of caches on performance of DOH with different number of nodes, different TTL ratio and different DHT placement strategies. Request rate is 2500 req/sec.

find (more) efficient caching strategies. Plots in Figure 2 also show that DOH with the directory-wise placement (Figure 2 (b)) serves faster than DOH with the file-wise placement (Figure 2 (a)) because the directory-wise placement is combined with prefetching: when a file is fetched from DHT the entire directory is prefetched to the cache of the requesting node.

We have compared performance of DOH with different number of nodes and a stand-alone web server (indicated in plots as cases where the number of nodes is 1). Figure 3 shows plots of the service time for different request rates in DOH with the directory-wise placement and different number of nodes. The TTL of cache content is assumed to be 30 sec. Even though the DOH performance degrades as the number of nodes increases, the service time of DOH with the large number of nodes scales better than the service time of a single server for high request rates. As expected, in the case of low workload, DOH with the small number of nodes performs slower than a stand-alone Jetty server because of an extra overhead introduced by the DKS middleware, that causes increase of the average service

(a)

(b)

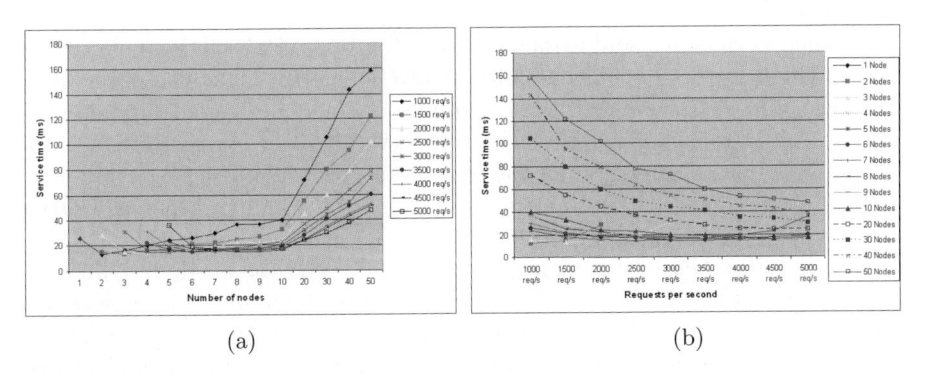

Fig. 3. Performance of DOH with the directory-wise DHT placement strategy. The cache TTL ratio is 30% of 100 sec of the simulation time (i.e. a file stays in the cache 30 sec).

time in DOH as the number of nodes increases. However, as the request rate increases, DOH shows better performance scalability than a stand-alone Jetty server: the DOHs service time does not increase as fast as the service time of the stand-alone Jetty. For example, at the request rate 600 req/sec (100 parallel requests), the average response time for Jetty is 87ms compared to 110ms for DOH. However at service rate of 1170 req/sec (200 parallel requests), the average response time for Jetty is 171ms compared to 135ms for DOH.

It has been observed that for each request rate there is a certain number of nodes, at which the system shows minimum response time, i.e. there is no improve in service time when increasing the number of DOH nodes beyond a certain value. We believe that this effect depends on the distribution of content in DHT and on how the content is cached in DHT nodes. We intend to study this in our future work.

The simulator suggests that DOH will do well with service times under 300ms, with request rates smaller than approximately 1200 requests per second per node.

Thus, our preliminary evaluation has shown that DOH would be able to handle a flash crowd; however the price users would pay is that the page retrieval under normal workloads would be sligthly slower than in the case of a stand-alone web server.

5 Some Related Work

Many P2P systems, like those proposed in [18], [20], [15], [19], and [21], have been used for creating DHTs. There are two main arguments for choosing DKS. First, DKS provides local atomic joins and leaves that guarantees that the DHT will never be in an inconsistent state. Second, DKS uses symmetric replication that allows to improve lookup time as well as reliability of the DHT. It also allows the client to get more than one result when doing a lookup. This feature can be used in a voting protocol, making sure that the retrieved object not has been tampered with. To our best knowledge, no other P2P overlay network provides these features.

Globule[16], SCAN[7] and CoralCDN[10], all propose P2P CDN similar to the one presented in this paper. The authors of Globule[16] make the observation that local web space is cheap, and therefore it could be traded for non-local space, creating replicas on different other servers (called slaves [16]) over the world. In Globule, negotiation for the replication space, configuration, and management are not handled automatically but rather by a human, whereas DOH is autonomous and has the ability to self-organize when a node joins/leaves. SCAN, which is a P2P CDN proposed in [7], uses Tapestry[21] as an underlying P2P network. One of the main goals of SCAN is to keep the number of replicas at a minimum to reduce overhead. This may cause sites to be unavailable whenever the master copy is unavailable. CoralCDN[10] uses the Coral[11] implementation of a Distributed Sloppy Hash Table to keep references to the master copy (or valid cached copies) on different nodes. In CoralCDN, like in SCAN, if the master copy of a site becomes unavailable for Coral, the site will soon become unreachable. In contrast to SCAN and CoralCDN, DOH uses symmetric replication to improve availability of hosted web-sites. Furthermore, in CoralCDN the URLs need to be "coralized" (see [10]) to be a part of the overlay network, i.e. CoralCDN is not even initally transparent to the end-users, which is one of the major design goals of DOH. DotSlash[22] is described by the authors as being a rescue system for web servers during hotspots. The authors of DotSlash do share the same motivation for developing a P2P CDN as in this paper: to help web servers survive a flash crowd. DotSlash does not store content globally (i.e. does not distribute and/or replicate a web-site among nodes as it is done in DOH) but all servers will store their own content. When a flash crowd occurs, an overlay network with rescue servers will be created, and the "hot objects" will be cached at these servers during the flash crowd. This network will be abandoned when workloads are back to normal. In contrast to DotSlash, DOH allows to distribute content of a web-site and its replicas among nodes making the web site more available for intensive concurrent requests.

6 Conclusions and Future Work

In this paper we have presented an approach to building a content delivery network as a structured P2P system of web-servers. This approach allows improving availability and scalability of a web site due to the load distribution, multiple access points, and replication. Such content delivery P2P network of collaborative web-servers can be used as a Web-hosting environment to host several web-sites. This approach can also be considered as one inexpensive solution for surviving of a flash crowd[1].

To validate, and preliminary evaluate our approach we have developed a system prototype called DOH (DKS Hosting system) based on the DKS P2P middleware that integrate the Jetty web-servers in a scalable content delivery network (web-hosting system). Each node in the DOH network is a DKS-node, i.e. is a part of the DKS overlay network, has a web server (to retrieve files) and an FTP server (to download/upload files). The network also includes Translators which are client contact points to the DOH network. Translators are used to distribute requests among DOH nodes to achieve load balancing. A Translator redirects web clients to DOH nodes based on the nodes load and network congestion. Each Translator maintains a cache of information about nodes (including their load and RTT times) that are known to the system. The Translator-cache is also used to help a content provider to find nodes , when uploading content using FTP.

DOH stores files in a Distributed Hash Table (DHT) provided by DKS so each node is able to retrieve the requested files from the DHT and cache them locally for future requests. Thus when a sudden traffic surge occurs, there will not only be one server (DHT-node) serving all the requests but a network of cooperating web servers helping each other by dividing the load.

We have evaluated the prototype for small-scale DOH configurations. To preliminary evaluate medium- and large-scale setup we have developed a DOH simulator based on timing estimates obtained from the performance experiments on the prototype. Evaluation results show that DOH performs better than a stand-alone web-server in the case of the high request rate (a large number of simultaneous requests) and the response time in DOH scales better than the response time in a single web-server, so the approach is valid for solving the intended problem of a flash crowd. However, as expected, with a low request rate, the single web-server outperforms DOH. Experiments also show that performance of DOH is very sensitive to the use of caches. We can also expect that explicit replication supported in DKS will improve performance of the prototype.

Two scenarios of performance has thus been identified: during low and high request rate, and both needs to be addressed in our future work which also includes more detailed performance evaluation, including assessment of impact (if any) of replication on service time; improving the caching mechanism in order to improve system performance; extending the system design to support dynamic contents (web-based applications). There are many issues to be considered when extending the system for dynamic contents, e.g. how to deploy and replicate applications, how to handle transactions, states, and failures; how to store the state to achieve failover. We leave answering all these questions to our future work.

References

1. Adler, S.: The Slashdot Effect: An Analysis of Three Internet Publications [Online] http://ssadler.phy.bnl.gov/adler/SDE/SlashDotEffect.html (2005)
2. Akamai Technologies, Inc. [Online] http://www.akamai.com/
3. Alima, L.O.,El-Ansary, S., Brand, P., Haridi, S.: DKS(N, k, f): A Family of Low Communication, Scalable and Fault-Tolerant Infrastructures for P2P Applications, In *The 3rd Int workshop CCGRID2003*, Tokyo, Japan, (2003)
4. Apache License, Version 2.0 [Online] http://www.apache.org/licenses/LICENSE-2.0
5. Arlitt, M. F., Williamson, C. L.: Internet web servers: Workload characterization and performance implications. In *IEEE/ACM Trans on Networking*, **5(5)** (1997) 631-645
6. R. Bhattacharyya [Online] http://www.mycgiserver.com/ranab/ftp/ (2005)
7. Chen, Y., Katz, R., Kubiatowicz, J.: SCAN: A dynamic, scalable, and efficient content distribution network. *Proc of Int Conf on Pervasive Computing*, Zurich (2002)
8. Crovella, M. E., Taqqu, M. S., Bestavros, A.: Heavy-tailed probability distributions in the World Wide Web. In *A Pract. Guide To Heavy Tails*, Chapman & Hall (1998) 3-26
9. Distributed K-ary System (DKS), [Online] http://dks.sics.se/
10. Freedman, M. J., Freudenthal, E., and Mazi'eres, D: Democratizing Content Publication with Coral. In *Proc of the 1st Symp on Networked Systems Design and Implementation (NSDI 2004)*, San Francisco, USA (2004)
11. Freedman M., Mazi'eres, D.: Sloppy hashing and self-organizing clusters. In *2nd Int Peer To Peer Systems Workshop*, Berkeley, USA (2003)
12. Jetty Java HTTP Servlet Server [Online] http://jetty.mortbay.org/jetty/index.html

13. Ghodsi, A., Alima, L.O., Haridi, S.: Symmetric Replication for Structured Peer-to-Peer Systems. In *The 3rd Int Workshop on Databases, Information Systems and Peer-to-Peer Computing*, Trondheim, Norway (2005)
14. Iyer, S., Rowstron, A., Druschel, P.: Squirrel: A decentralized, peer-to-peer web cache. In *Proc of the 21st Ann ACM Symp on Principles of Distributed Computing*, ACM (2002)
   jimmy:
15. Maymounkov, P., Mazi'eres, D.: Kademlia: A peer-to-peer information system based on the xor metric. In *IPTPS02*, Cambridge, MA (2002)
16. Pierre, G., van Steen, M.: Design and implementation of a user-centered content delivery network. In *Proc. 3rd Workshop on Internet Applications*, San Jose, USA (2003)
17. Pierre, G., van Steen, M., Tanenbaum, A. S.: Dynamically selecting optimal distribution strategies for Web documents. *IEEE Trans on Computers*, **51(6)** (2002) 637651
18. Ratnasamy, S., Francis, P., Handley, M., Karp, R., Schenker, S.: A scalable content-addressable network, In *Proc of the 2001 Conf on Applications, Technologies, Architectures, and Protocols for Computer Communications*, San Diego, USA (2001) 161-172
19. Rowstron, A., Druschel, P.: Pastry: Scalable, distributed object location and routing for large-scale peer-to-peer systems. In *IFIP/ACM Int Conf on Distr Systems Platforms (Middleware)* (2001) 329-350
20. Stoica, I., Morris, R., Karger, D., Kaashoek, M. F., Balakrishnan, H.: Chord: A Scalable Peer-to-Peer Lookup Service for Internet Applications, In *Conf on Applications, Technologies, Architectures, and Protocols for Comp. Communications* (2001) 149-160
21. Zhao, B. Y., Kubiatowicz, J. D., Joseph, A. D.: Tapestry: An infrastructure for fault-tolerant wide-area location and routing. TR UCB/CSD-01-1141, UC Berkeley (2001)
22. Zhao, W., Schulzrinne, H.: DotSlash: A selfconfiguring and scalable rescue system for handling web hotspots effectively. In *Int Workshop on Web Caching and Content Distribution (WCW)*, Beijing, China (2004)

Topic 16: Applications of High-Performance and Grid Computing

Simon J. Cox, Thomas Lippert, Giovanni Erbacci, and Denis Trystram

Topic Chairs

The use of high performance and grid computing has spread rapidly, revolutionising the ability of scientists and engineers to tackle the challenges they face. Driven by commoditisation and open standards: the widespread availability of parallel computers, large data storage, fast networks, maturing Grid middleware, and distributed service-oriented technologies have led to the development and deployment of large scale distributed simulation and data analysis solutions in many areas. The papers in this topic highlight recent progress in applications of all aspects of distributed computing technologies with an emphasis on successes, advances, and lessons learned in the development, implementation, and deployment of novel scientific, engineering and industrial applications on high performance and grid computing platforms. Today's large computational solutions often require access to or generate large volumes of data- indeed today seamless data access and management can be as important to the underlying computational algorithm as raw computing power. Papers in the sessions highlight data intensive applications which couple together High Performance/ Grid computing with large-scale data access/ management.

Task Pool Teams Implementation of the Master Equation Approach for Random Sierpinski Carpets

K.H. Hoffmann[1], M. Hofmann[2], G. Rünger[2], and S. Seeger[1]

[1] Department of Physics, Chemnitz University of Technology
[2] Department of Computer Science, Chemnitz University of Technology

Abstract. We consider the use of task pool teams in implementation of the master equation on random Sierpinski carpets. Though the basic idea of dynamic storage of the probability density reported earlier applies straightforward to random carpets, the randomized construction breaks up most of the simplifications possible for regular carpets. In addition, parallel implementations show highly irregular communication patterns. We compare four implementations on three different Beowulf-Cluster architectures, mainly differing in throughput and latency of their interconnection networks. It appears that task pool teams provide a powerful programming paradigm for handling the irregular communication patterns that arise in our application and show a promising approach to efficiently handle the problems that appear with such randomized structures. This will allow for highly improved modelling of anomalous diffusion in porous media, taking the random structure of real materials into account.

1 Introduction

Random fractal structures are used to model the random structural properties found in many real materials such as aerogels, porous rocks or cements. There we find a fractal structure on certain length scales (spanning about two or three decades) [1], while on larger scales the structure looks rather homogeneous. One feature of these materials is that diffusion is anomalous and the behaviour is very well modeled by random walk processes on regular fractals like Sierpinski carpets [2]. But, these regular fractal structures do not exhibit the transition to normal diffusion found in the real materials. This transition could be captured by performing the Sierpinski carpet construction only to some finite stage and repeating the resulting structure, thereby obtaining a crystal like structure with fractal unit cells. However, this does not capture the randomness of the local fractal structure present in real materials, which has quite an influence on the diffusion properties [3]. This randomness in local structure can be modeled by using newly generated carpets instead of repeating one randomly generated unit cell. While in regular (crystal) structures added disorder usually leads to a decrease in diffusion or transport properties, we find here that disorder can also enhance diffusion on these structures. This is also observed in experiments on ionic conduction in solid electrolytes [4].

In this paper we report how a master equation approach to simulating random walks on random Sierpinski carpet structures may be implemented efficiently. This method is an elegant way to calculate the evolution of the probability densities of random walkers on such structures from a given initial distribution. This initial distribution is assumed to have finite support, usually chosen to be a delta distribution. Though we can apply some concepts developed for regular Sierpinski carpets [2] in a straightforward manner, the randomness of the resulting structures poses some challenges not apparent when considering the simpler regular case. This article describes strategies that can be used for an efficient parallel implementation. An important problem that needs to be solved in order to obtain an efficient parallelization is to develop a strategy for handling the irregular communication patterns that arise due to the random, dynamic growth of the carpet structure covered by the probability distribution. We show that the concept of *task pool teams* [5] provides a suitable framework for handling these issues.

2 Random Sierpinski Carpets

Given a set of $M \times M$ black-and-white patterns (the generators) and a probability distribution for the choice among these patterns, the algorithm to construct random Sierpinski Carpets described by Reis [6] and ben-Avraham [7] is as follows:

1. start from a square (level 0).
2. divide each square into $M \times M$ subsquares.
3. choose a generator pattern at random (according to its probability)
4. remove the subsquares corresponding to white markings in the selected generator
5. for the next level, repeat steps 2 – 4 for each remaining subsquare.

Figure 1 shows an example of the first two refinement steps for a set of three different generators. Note that with just a single generator we obtain regular Sierpinski carpets as a subset of random Sierpinski carpets. The construction procedure can be repeated ad infinitum, where the resulting structure is a random Sierpinski carpet [6]. If we stop at some level l, the resulting pre-carpet pattern of size $M^l \times M^l$ is referred to as an *iterator* of level l.

These pre-carpet structures give a good model for the (in a statistical sense) self-similar micro-structure of porous materials. We therefore use iterators as basic unit in our algorithm to build larger structures by connecting single iterators. For instance, repeating a given iterator in all directions, we obtain a 'crystal' with random unit cell. Extending the carpet in all directions by appending newly created random iterators, we obtain a structure with the same properties as real porous materials. The last method is certainly the most difficult to implement, as virtually no savings can be made in the description of the structure. We therefore discuss an algorithm that allows efficient simulation

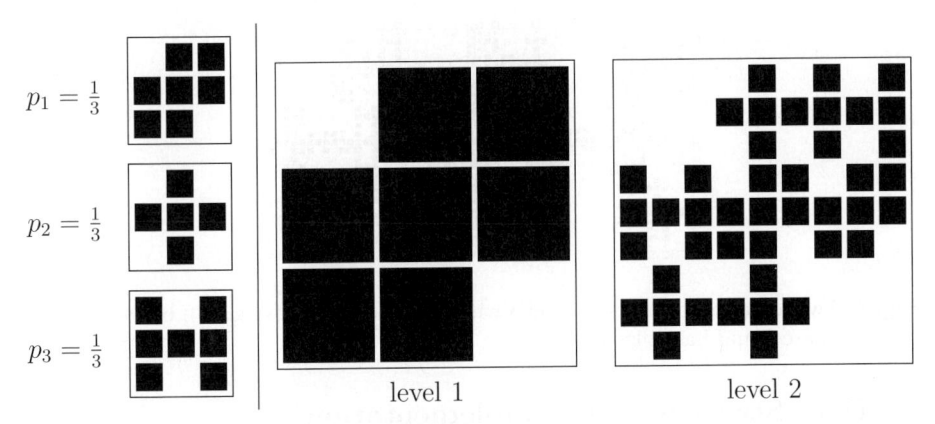

Fig. 1. The first two construction steps for a random Sierpinski carpet constructed from three different generators with equal probability

for the last, most demanding case. However, it can easily be modified to handle other cases well.

In order to iterate the master equation on the resulting structure, we introduce the following terms and notations: Consider a random walker is allowed to hop between the midpoints of the *tiles* (black subsquares) in a carpet. In one discrete time step, the walker can move to one of the neighbouring tiles. De Gennes [8] introduced the analogy of a random walker as an "ant in a labyrinth" and with the so called *myopic ant* or *blind ant* algorithms, we obtain the probabilities W_{ij} for a walker to arrive at tile i coming from tile j. Given some probability $p(t,i)$ to find a walker on tile i at time t, we can calculate the probability $p(t+1,i)$ by accounting for the gain and loss of probability by walkers crossing the boundaries as

$$p(t+1,i) = (1 - L_i)\, p(t,i) + \sum_{j \in <i>} G_{ij}\, p(t,j) \,. \qquad (1)$$

The sum is over the set of all neighbours $<i>$ of tile i, $G_{ij} = W_{ij}$ are the gain factors and $L_i = \sum_{j \in <i>} W_{ji}$ is the overall loss of tile i. By iterating the master equation (1) starting with a delta distribution at the starting point we obtain a new distribution for every time t. This distribution determines the mean square displacement accurately, free of the fluctuations pertinent to direct simulation methods. From this, not only the random walk dimension of the fractal can be determined, but also can this probability distribution be compared with theoretical descriptions of anomalous diffusion, e.g. by fractional diffusion equations.

Iteration of the master equation, however, requires a large amount of computer RAM, as for every point in the carpet that can be reached by a random walk in the considered time t, memory to store two probability values need to be allocated. This memory requirement grows considerably with simulation time t, thus an efficient way of storing and updating these probability values is needed.

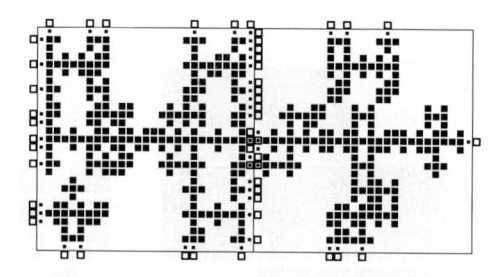

Fig. 2. Two adjacent iterators of level 3 with their body tiles (squares), boundary tiles (small squares) and halo tiles (outlines)

3 Data Structures and Implementation

In [2] we have already reported on an efficient algorithm for regular carpets. Though the basic idea of dynamic storage of $p(t,i)$ applies straightforward to random carpets, the randomized construction breaks up most of the simplifications possible for regular carpets. For instance, with a regular carpet an iterator pattern of some level determines the whole carpet structure. This is not the case for randomized carpets where each iterator is different. Also, with a dynamically growing data structure the connections to neighbour iterators cannot be predicted in advance of the simulation from analysis of the iterator. Instead, it can only be determined once the carpet has actually been constructed and all neighbour iterators are known.

Our basic unit of processing remains an iterator of level l. We start with one iterator that contains the tile with the non-zero part of the initial delta distribution. The carpet is described by a linked list of iterator descriptions, that store topological information and the probability values at the current and next time step. In every time step this list is traversed once in order to calculate the probability values for the next time step. Within an iterator we have to distinguish the following types of black tiles: *body tiles* are inside the iterator but not adjacent to a boundary, *boundary tiles* are inside the iterator and adjacent to a boundary, *halo tiles* are outside the iterator adjacent to a boundary tile. Figure 2 illustrates this situation showing two iterators with their body (■), boundary (·), and halo (□) tiles.

For body tiles we can perform the update calculation without any additional information other than that stored for the iterator required. For boundary tiles, we do not know the surrounding carpet topology in the beginning. Furthermore we need to know the probability value(s) at the adjacent halo tile(s) in order to perform the update for a given tile. Fortunately, the corresponding terms in (1) vanish initially because we have zero probability that walkers are at those positions. Only as soon as the master equation predicts a non-zero probability value at a boundary tile for the next iteration step we need to make sure the neighbouring iterators are present and the data structures are consistent. Halo tiles are not updated according to (1) but by copying the values after updating the corresponding boundary tiles from the neighbour iterator. Doing so allows the task of iterating the master equation to be distributed among multiple processes by distributing the iterators.

4 Parallel Implementation

For the parallel implementation we use a master-worker scheme. The master is responsible of overall program control as well as to keep track of the global carpet topology. With the data structures described above, the workers receive a number of iterator descriptions for the iterators they have to process. Thus the global list of iterators to process is split among the workers and each worker has its local list. A load balancing mechanism is implemented by assigning new iterators to the least-busy workers where load is determined by the number of tiles that need to be updated per iteration.

The processing of one time step is organized in three phases:

1. The master informs all workers to start processing their local list of iterators for updating the probability values of the body and boundary tiles. However, it may happen that workers arrive with non-zero probability at boundary tiles, thereby making a carpet extension necessary. If this happens, the worker reports this event to the master and processes the next iterator until it has finished traversing its local list. The master collects messages about carpet extensions necessary.
2. After all workers have finished processing their local iterator list, the master extends the carpet as necessary by assigning newly created iterators to the workers and notifying the workers of the changed carpet topology.
3. Finally, as the last phase in every iteration the boundary values are exchanged. Once this has been finished, results may be collected or a new iteration is started.

For the simulation of about 32000 time steps, the runtimes of the three phases for a straightforward implementation are shown in Figure 3. The carpet increases up to about 2300 iterators, each of size $5^3 \times 5^3$. The implementation uses MPI to send the various control and data messages. The master and every worker process is assigned to a single cluster node. Measurements have been performed on three different Beowulf-type clusters: (A) the Chemnitzer Linux Cluster CLiC with 512 nodes with single Pentium III/800MHz CPUs, 512MB RAM and FastEthernet interconnect and a Xeon cluster with dual Xeon/2GHz CPUs, 1GB RAM and either (B) GigabitEthernet or (C) SCI interconnects. For the Fast- and GigabitEthernet interconnects, the LAM-MPI implementation and for SCI interconnect the optimized SCAMPI implementation has been used.

As can be seen from Figure 3, the amount of wall-time spent in the first phase decreases as the number of nodes is increased. The carpet extension phase has a fairly constant and rather small amount of execution time, because the carpet extension is handled by the master only. The longest time, however, is spent in the third phase performing the boundary update. While with the SCI interconnect (cluster C) a slight speedup can be observed, the amount of time

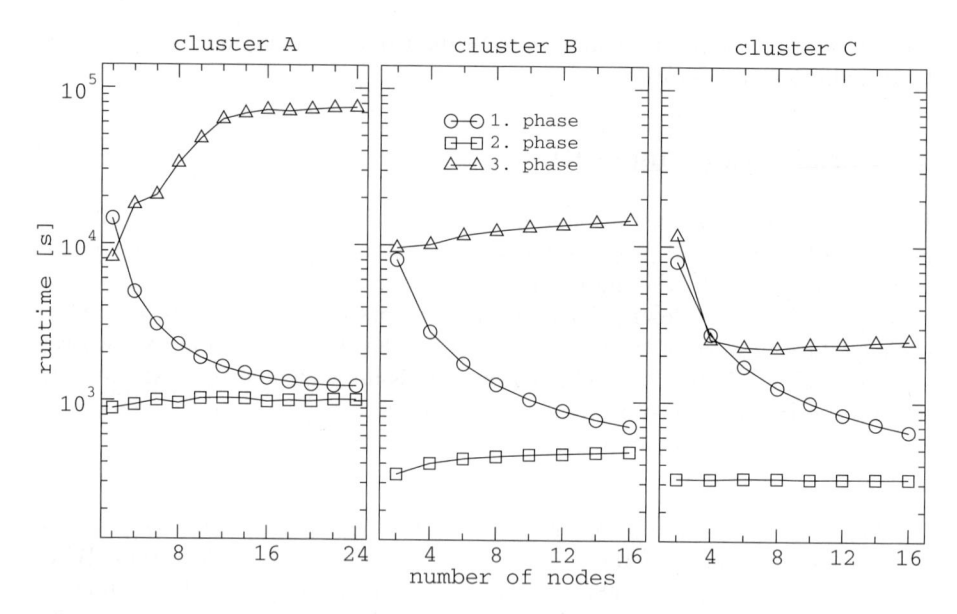

Fig. 3. Runtimes (in wall-time seconds) of the three phases of the main loop: 1) iteration of (1) for each iterator, 2) carpet extension and fixup where new iterators are appended, 3) update of boundary values (possibly between processors)

spent in boundary update remains fairly constant for GigabitEthernet (cluster B). For FastEthernet (cluster A) communication time actually increases with the number of nodes. This is because the communication scheme used in the reference implementation results in many short messages, resulting in high latency times adding up. Another drawback is the highly irregular communication scheme arising from sending and receiving the boundary updates. Because for a single-threaded worker the resulting irregular communication protocol cannot be proven deadlock-free, the third phase is serialized: each worker either sends messages to other workers or waits for incoming messages. The best improvement can therefore be achieved with a better implementation of the boundary update phase.

4.1 Optimized Boundary Updates

For parallelizing the boundary update phase by handling the irregular communication we use *task pool teams*. The *task pool* concept uses a decomposition of the computational work into tasks. A task pool stores the tasks and threads are responsible for the execution of tasks. *Task pool teams* are an approach for extending the idea of task pools to the use of parallel platforms with distributed memory. They combine task pools on single cluster nodes with explicit communication. We use the implementation of task pool teams for SMP cluster presented in [5], which uses Pthreads for SMPs and MPI for communication between SMP nodes. A specific communication thread and a number of worker threads run on each SMP node. Thus, each worker of the master-worker scheme is now actually realized as a collection of internal worker threads and one communication thread. An advantage of task pool teams is to support irregular communication requirements.

In order to speedup the boundary updates we focused on three additional implementations:

- As a first implication from the strong impact of the latency, we start to collect single boundary update messages for each worker until a sufficiently large message can be sent. This avoids many small messages in favour of larger messages thereby reducing the high impact of the latency to start communication. We will refer to this as the boundary collect mechanism.
- To achieve a parallel update with task pool teams we use the communication thread to handle update requests from other workers. At the same time a worker thread is able to process the local iterator list performing the boundary updates. The messages are sent using the specific asynchronous communication which is mapped to MPI operations by the task pool teams implementation. This provides individual point-to-point communication between pairs of workers whenever messages need to be transferred. We will refer to this as the asynchronous parallel update.
- Another method for parallel update with task pool teams uses the specific communication for notifying the workers to perform a boundary update. After this notification all workers participate in sending their messages synchronous by all-to-all communication operation. We refer to this method as the synchronous parallel update.

Both parallel update methods use the boundary collect mechanism for sending larger messages instead of many small ones.

As can be seen from Figure 4, the boundary collect mechanism provides a saving in runtime of about an order of magnitude. This is caused by avoiding many small messages between nodes handling adjacent iterators. Especially for the high latency Fast- and GigabitEthernet (on clusters A and B) this provides the most substantial savings. The additional use of the parallel update scheme leads to different results with the different architectures. For the uniprocessor cluster (A) using only a small number of nodes the runtimes remain fairly unchanged. However, with an increasing number of nodes there is a slight saving in runtime. These rather fair improvements can be attributed to the use of multi-threaded programming on uniprocessor architectures. Using the asynchronous and synchronous method makes no difference. Much better results are obtained with the SMP cluster (B and C). Using the parallel update we observe a gain of another order of magnitude in execution time. This is achieved by using the task pool teams concept for handling the irregular communication. Additional benefits are achieved by overlapping of communication and computation through the parallel execution of communication and worker thread. The results for the asynchronous parallel update are shown only for the SCI interconnect (cluster C). In comparison with the synchronous method the savings in execution time

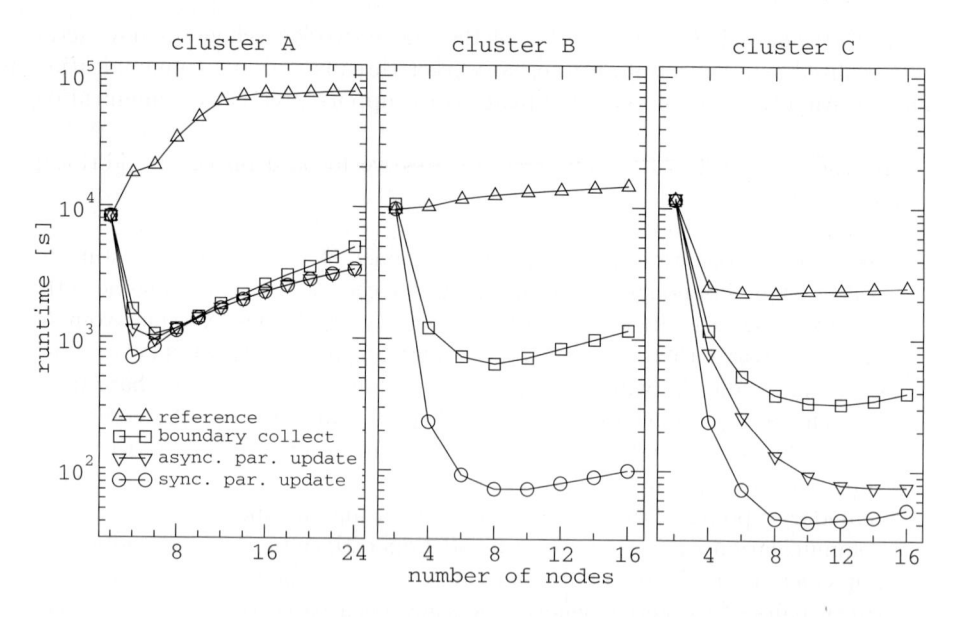

Fig. 4. Runtimes of various implementations of the boundary update phase: reference implementation, boundary collect mechanism and task pool teams with asynchronous and synchronous parallel boundary updates

are rather small. The runtime results for the GigabitEthernet interconnect with the LAM-MPI implementation in the multi-threaded environment are diverse and not shown in the diagram.

4.2 Optimized Iterator Updates and Overall Runtimes

Due to the good results using the task pool teams with the SMP cluster, we extend their usage to another computational expensive part of the simulation. The processing of the local iterator list in the first phase can easily be split into independent tasks. These tasks are executed in parallel by different worker threads.

As can be seen from Figure 5, the multi-threaded implementation of the task pool teams leads to another saving in runtime. For the multiprocessor architecture this is the expected behaviour. However, for the uniprocessor cluster there appears also a slight decrease in execution time as the number of nodes increases. On the multiprocessor cluster no additional benefits are achieved using more worker threads than CPUs per node available.

Finally, in Figure 6 we compare the overall runtimes of the optimizations using task pool teams and the reference implementation. On three different clusters the reference implementation shows a different behaviour in parallel execution. Savings in runtime with an increasing number of nodes are only achieved with the SCI interconnect (cluster C) while using Fast- and GigabitEthernet (cluster A and B) the runtimes increase or remain constant. With the optimizations this behaviour completely change and first of all becomes more independent from the

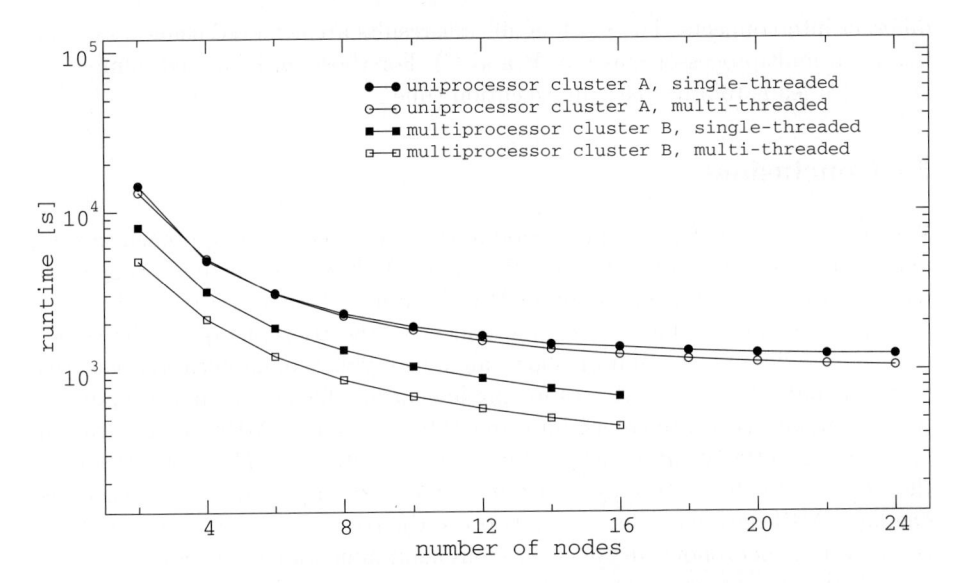

Fig. 5. Runtimes of the first phase updating the iterators according to (1) with single-threaded implementation and multi-threaded using task pool teams with 2 worker threads

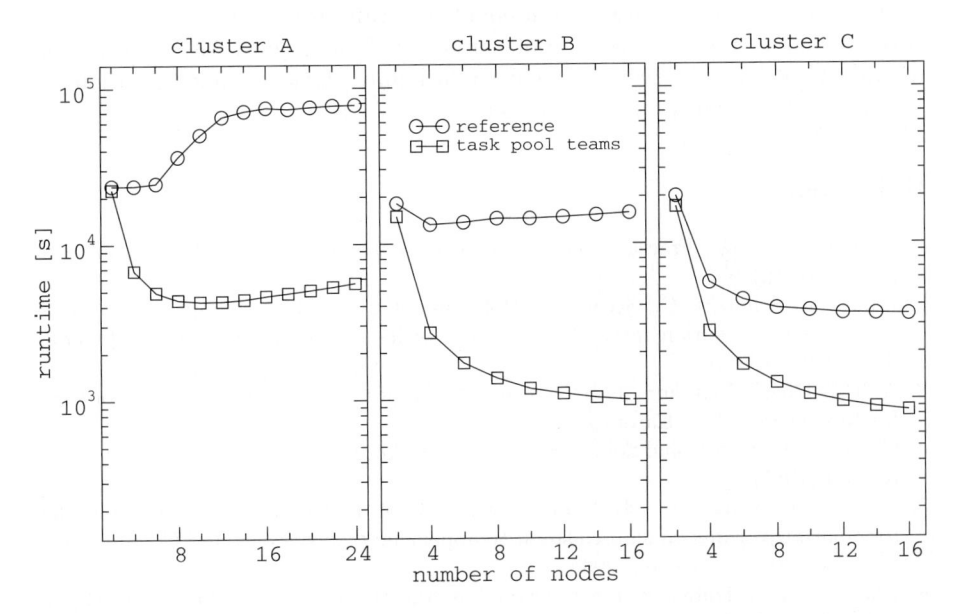

Fig. 6. Overall runtimes of the reference implementation and with the optimizations using task pool teams

different interconnects. The most significant results are achieved using task pool teams on multiprocessor clusters (B and C). For those an increased number of nodes still leads to a decrease in execution time.

5 Conclusions

We have considered an implementation of the master equation approach to simulating diffusion on random Sierpinski carpets. As iterating the master equation requires a huge amount of computer RAM, we have favoured a parallel implementation. However, due to the randomness in the construction of the structures, a parallel implementation shows highly irregular communication patterns that demand adequate strategies for implementing efficient boundary updates. In comparison with a reference implementation that uses MPI communication operation directly, we have analyzed four implementations. The first introduces the boundary collect strategy, collecting small messages and sending them as one large MPI message. The second two use the concept of task pool teams together with synchronous and asynchronous communication operations. The last extends the use of task pool teams to a more computational expensive part of the algorithm. We observe that on high latency communication networks, such as Fast- and GigabitEthernet, the savings due to the boundary collect strategy are most important. However, with an increasing number of nodes and taking SMP clusters into account, the use of task pool teams can result in a further reduction of the boundary update time of about an order of magnitude. Altogether, using the task pool teams concept we achieved a highly efficient implementation for the utilization of multiprocessor clusters.

References

1. Mandelbrot, B.B.: Fractals - Form, Chance and Dimension. W. H. Freeman, San Francisco (1977)
2. Franz, A., Schulzky, C., Seeger, S., Hoffmann, K.: An efficient implementation of the exact enumeration method for random walks on Sierpinski carpets. Fractals **8**(2) (2000) 155–161
3. Anh, D.H.N., Hoffmann, K.H., Seeger, S., Tarafdar, S.: Diffusion in disordered fractals. accepted by Europhys. Lett. (2005)
4. Chandra, S.: Superionic Solids: Principles and Applications. North-Holland, Amsterdam (1981)
5. Hippold, J., Rünger, G.: Task pool teams: A hybrid programming environment for irregular algorithms on smp clusters. to appear in: Concurrency and Computation: Practice and Experience (2006)
6. Reis, F.D.A.A.: Diffusion on regular random fractals. J. Phys. A: Math. Gen. **29**(24) (1996) 7803–7810
7. ben Avraham, D., Havlin, S.: Diffusion and Reactions in Fractals and Disordered Systems. Cambridge University Press, Cambridge, UK (2000)
8. de Gennes, P.G.: La percolation: Un concept unificateur. La Recherche **7**(72) (1976) 919–927

A Preliminary Out-of-Core Extension of a Parallel Multifrontal Solver

Emmanuel Agullo[1], Abdou Guermouche[2], and Jean-Yves L'Excellent[3]

[1] LIP-ENS Lyon, France
[2] LaBRI, Bordeaux, France*
[3] INRIA and LIP-ENS Lyon, France

Abstract. The memory usage of sparse direct solvers can be the bottleneck to solve large-scale problems. This paper describes a first implementation of an *out-of-core* extension to a parallel multifrontal solver (MUMPS). We show that larger problems can be solved on limited-memory machines with reasonable performance, and we illustrate the behaviour of our parallel *out-of-core* factorization. Then we use simulations to discuss how our algorithms can be modified to solve much larger problems.

1 Introduction

The solution of sparse systems of linear equations is a central kernel in many simulation applications. Because of their robustness and performance, direct methods can be preferred to iterative methods. In direct methods, the solution of a system of equations $Ax = b$ is generally decomposed into three steps: (i) an analysis step, that considers only the pattern of the matrix, and builds the necessary data structures for numerical computations; (ii) a numerical factorization step, building the sparse factors (e.g., L and U if we consider an unsymmetric LU factorization); and (iii) a solution step, consisting of a forward elimination (solve $Ly = b$ for y) and a backward substitution (solve $Ux = y$ for x). For large sparse problems, direct approaches often require a large amount of memory, that can be larger than the memory available on the target platform (cluster, high performance computer, ...). In order to solve increasingly large problems, *out-of-core* approaches are then necessary, where disk is used to store data that cannot fit in physical main memory.

Although several authors have worked on sequential or shared-memory *out-of-core* solvers [1,2,3], sparse *out-of-core* direct solvers for distributed-memory machines are less common. In this work, we aim at extending a parallel multifrontal solver (MUMPS, for MUltifrontal Massively Parallel Solver, see [4]), in order to enable the solution of larger problems, thanks to *out-of-core* approaches. Recent contributions by [5] and [6] for uniprocessor approaches pointed out that multifrontal methods may not fit well an *out-of-core* context because large dense matrices have to be processed, that can represent a bottleneck for memory; therefore, they prefer left-looking approaches (or switching to left-looking approaches). However, in a parallel context, increasing the number of

---

* This work was done during an INRIA post-doctoral position at ENSEEIHT-IRIT, Toulouse, France.

processors can help keeping such large frontal matrices in-core. Note also that another type of approach is based on virtual memory and system paging, that can be controlled by low level mechanisms [7] in relation with the application and provide better performance than default LRU mechanisms. However, such approaches are very closely related to the operating system and are not adapted when designing portable codes.

This paper is organized as follows. After a quick description of the memory management in multifrontal methods (Section 2), we present in Section 3 an approach to store the sparse factors L and U to disk. We will observe that this allows us to treat larger problems with a given memory, or the same problem with less memory. In Section 3.4, both a synchronous approach (writing factors to disk as soon as they are computed) and an asynchronous approach (where factors are copied to a buffer and written to disk only when the buffer is full) are analyzed, and compared to the in-core approach on a platform with a large amount of memory. Finally, in order to process much larger problems, we present in Section 4 simulation results where we suppose that the active memory of the solver is also stored on the disk and study how the overall memory can further be reduced. This study is the basis to identify the bottlenecks of our approach when confronted to arbitrarily large problems.

2 Memory Management in a Parallel Multifrontal Method

In multifrontal methods, the task dependencies are represented by a so-called assembly tree [8,9], that is processed from bottom to top during the factorization. At each node of the tree is associated a so-called *frontal matrix*, or *front*, and a task consisting in the partial factorization of the frontal matrix. The partial factorization produces a Schur complement, or *contribution block*, which will be used to update the frontal matrix of the parent node (see [10], for example, for more details). This leads to three areas of storage, one for the factors, one for the contribution blocks, and another one for the current frontal matrix [10]. The active memory (as opposed to the memory for the factors) then corresponds to the sum of the contribution blocks memory (or stack memory) and the memory for the current active matrix. During the factorization process, the memory required for the factors always grows while the stack memory that contains the contribution blocks varies: when the partial factorization of a frontal matrix is performed, a contribution block is stacked which increases the size of the stack; on the other hand, when the frontal matrix of a parent is formed and assembled, the contribution blocks of the children nodes can be discarded and the size of the stack decreases[1].

From the parallel point of view, the parallel multifrontal method as implemented in MUMPS uses a combination of static and dynamic scheduling approaches. Indeed, a first partial mapping is done statically (see [11]) to map some of the tasks to the processors. Then, for parallel tasks corresponding to large frontal matrices of the assembly tree, a master task is in charge of the elimination of the so-called fully summed rows, while dynamic scheduling decisions are used to select the processors in charge of updating the rest of the frontal matrix (see Figure 1). Those decisions are taken to balance workload, possibly under memory constraints (see [12]).

---

[1] In parallel, the contribution blocks management may differ from a pure stack mechanism.

3 Out-of-Core Multifrontal Approach

3.1 Preliminary Study

In the multifrontal method, the factors produced during the factorization step are not reused before the solution step. It then seems natural to first focus on writing *them* to disk. Thus, we present a preliminary study which aims at evaluating by how much the in-core memory can be reduced by writing the factors to disk during the factorization. To do so, we simulated an *out-of-core* treatment of the factors: we free the corresponding memory as soon as each factor is computed. Of course the solution step cannot be performed as factors are definitively lost, but freeing them allowed to analyze real-life problems on a wider range of processors (in this initial study).

We measure the size of the new peak of memory (which actually corresponds to the *active memory* peak) and compare it to the one we would have with an *in-core* factorization (*i.e.* the *total memory peak*). In a distributed memory environment, we are interested in the maximum peak obtained over all the processors as this value represents the memory bottleneck.

For a small number of processors, we observe that the active memory is much smaller than the total memory. In other words, if factors are written to disk as soon as they are computed, only the active memory remains *in-core* and the memory requirements decrease significantly (up to 80 % in the sequential case).

On the other hand, when the number of processors increases, the peak of the active memory decreases more slowly than the total memory as shown in Figure 2. For example, on 64 processors, the active memory peak reaches between 50 and 70 percent of the peak of total memory. In conclusion, on platforms with small numbers of processors, an *out-of-core* treatment of the factors will allow us to process significantly bigger problems; the implementation of such a mechanism is the object of Section 3.2. Nevertheless, either in order to further reduce memory requirements on platforms with only a few processors or to have significant memory savings on many processors, we may have to treat both the factors and the active memory with an *out-of-core* scheme. This will be studied in Section 4.

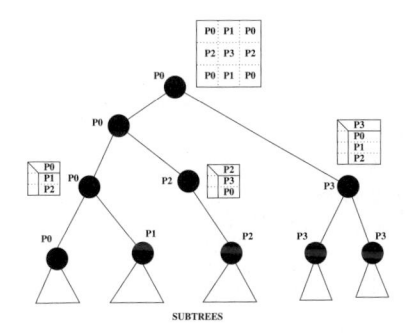

Fig. 1. Example of the distribution of an assembly tree over four processors

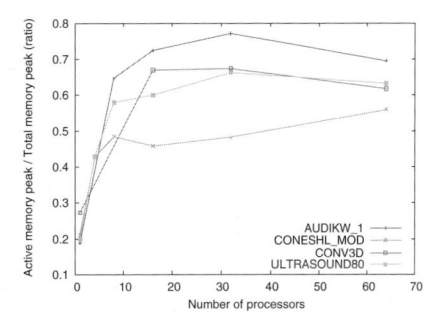

Fig. 2. Ratio of active and total memory peak on different number of processors for several large problems (METIS is used as the reordering technique)

3.2 Out-of-Core Management of the Factors

The performance of I/O mechanisms are essential and impact directly the performance of the whole application. Neither MPI-IO [13] (because files are not shared by processors in our case) nor FG [14] (our I/O threads do not interfere with each other) match our purpose. Both AIO, an asynchronous I/O mechanism optimized at the kernel level, and the recent Fortran 2003 asynchronous I/O layer were not available on our target platform (see Section 3.3). We finally used the standard C I/O routines *fread/fwrite* and *read/write* (or *pread/pwrite* when available) which are known to be efficient low-level kernels.

In the synchronous I/O scheme, the factors are directly written with a synchronous scheme using the standard I/O subroutines (either *fread/fwrite* or *read/write*). In the asynchronous I/O scheme, we associate with each MPI process of our application an I/O thread in charge of all the I/O operations. This allows us to overlap the time needed by I/O operations with computations. The I/O thread is designed over the standard POSIX thread library (pthread library). The communication and the synchronization between the computational thread and the I/O thread are designed using semaphore mechanisms. The communication scheme between the two threads is described in Figure 3. Each time an I/O operation has to be performed, the computational thread posts an I/O request and inserts it into the *queue of waiting requests*. Concerning the I/O thread, it treats the I/O requests in the *queue of waiting requests* using a FIFO strategy. Once an I/O request is finished, it is inserted in the *queue of finished requests* by the I/O thread. The computation thread can then remove it from this queue when checking for the completion of the request.

Together with the two I/O mechanisms described above, we designed a buffered I/O scheme (that can be either synchronous or asynchronous). This approach relies on the fact that we want to free the memory occupied by the factors as soon as possible without necessarily waiting for the completion of the corresponding I/O. Thus, and in order to avoid a complex memory management in a first approach, we added a buffer where factors are copied before they are written to disk. The buffer is divided into two parts so that while an asynchronous I/O operation is occurring on one part, factors that are

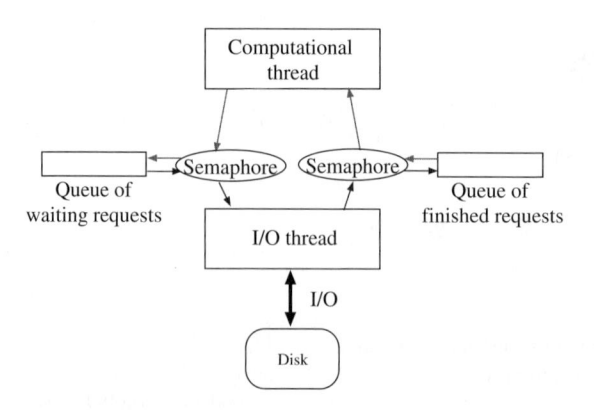

Fig. 3. Thread communication scheme

being computed can be stored in the other part (double buffer mechanism allowing the overlap of I/O operations with computation).

3.3 Experimental Environment

In order to study the impact of the proposed mechanisms, we now experiment with them on several problems (see Table 1) extracted from either the PARASOL collection[2] or coming from other sources. The tests have been performed on the IBM SP system of IDRIS[3] composed of several nodes of either 4 processors at 1.7 GHz or 32 processors at 1.3 GHz. On this machine, we have used from 1 to 128 processors with the following memory constraints: we can access 1.3 GB per processor when asking for more than 128 processors, 3.5 GB per processor for 17-64 processors, 4 GB for 2-16 processors, and 16 GB on 1 processor.

Table 1. Test problems

Matrix	Order	NZ	Type	$nnz(L|U) \times 10^6$	Description
AUDIKW_1	943695	39297771	SYM	1368.6	Automotive crankshaft model (PARASOL)
CONESHL_mod	1262212	43007782	SYM	790.8	provided by SAMTECH; cone with shell and solid element connected by linear constraints with Lagrange multiplier technique
CONV3D64	836550	12548250	UNS	2693.9	provided by CEA-CESTA; generated using AQUILON (http://www.enscpb.fr/master/aquilon)
ULTRASOUND80	531441	330761161	UNS	981.4	Propagation of 3D ultrasound waves, provided by M. Sosonkina, larger than ULTRASOUND3

By default, we used the METIS package [15] to reorder the matrices and thus limit the number of operations and fill-in arising in the subsequent sparse factorization. The results presented in the following sections have been obtained using the dynamic scheduling strategy proposed in [12].

The I/O system used is the IBM GPFS [16] filesystem. With this filesystem it was not possible to write files on disks local to the processors and some performance degradation was observed when several processors write/read an amount of data simultaneously to/from the filesystem: we observed a speed-down between 5 and 50 from 2 to 64 processors when each processor writes a block of 800 MBytes. Finally, it is important to note that we chose to run on this platform because it allows us to run large problems *in-core* and thus compare *out-of-core* and *in-core* approaches (even if the behaviour of the filesystem is not optimal for performance).

3.4 Experiments

First, we have been able to observe that for a small number of processors we use significantly less memory with the *out-of-core* approach: the total memory peak is replaced by the active memory peak, with the improvement ratios of Figure 2. Thus the factorization can be achieved on limited-memory machines.

---

[2] http://www.parallab.uib.no/parasol
[3] Institut du Dveloppement et des Ressources en Informatique Scientifique.

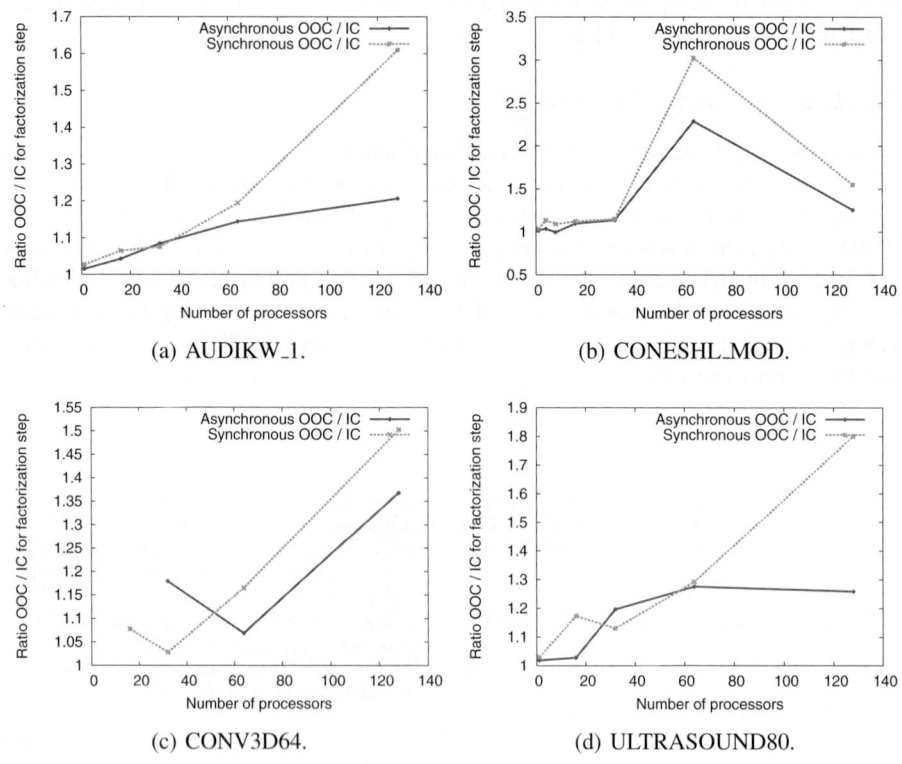

Fig. 4. Execution times (normalized with respect to the *in-core* case) of the synchronous and asynchronous I/O schemes

We now focus on performance issues and report in Figure 4 a comparative study of the *in-core* case, the synchronous *out-of-core* scheme and the asynchronous buffered scheme, when varying the number of processors.

Note that for the buffered case, the size of the I/O buffer is set to twice the size of the largest factor block (to have a double buffer mechanism). As we can see, the performance of the *out-of-core* schemes is indeed close to the *in-core* performance for the sequential case (note that we were not successful in running the CONV3D64 matrix on 1 processor with the *in-core* scheme because the memory requirements are larger than 16 GB). The *out-of-core* schemes are at most 20% slower than the *in-core* case while they need an amount of memory that can be 80 percent smaller as shown in Figure 2 for one processor. Concerning the parallel case, we observe that with the increase of the number of processors, the gap between the *in-core* and the *out-of-core* cases increases. The main reason is the performance degradation of the I/O with the number of processors that we mentioned at the end of Section 3.3. In order to avoid this problem, we have experimented with the smallest of our large test problems on a machine with local disks. In this case, we do not have such a performance degradation, as shown in Figure 5; on the contrary, the *out-of-core* schemes perform as well or even better than

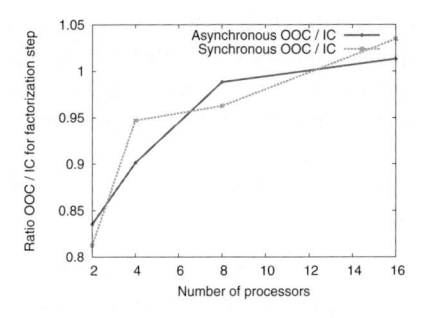

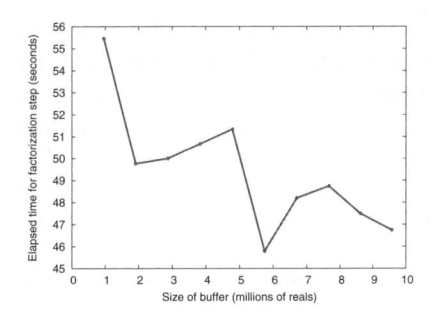

Fig. 5. Performance of the *out-of-core* factorization on a machine (CRAY XD1 system at CERFACS) with local disks for the CONESHL_MOD matrix on different number of processors

Fig. 6. Performance of the *out-of-core* factorization on 32 processors for the CONESHL_MOD matrix with respect to the size of the I/O buffer

the *in-core* one (cache effects resulting from freeing the factors from main memory and using always the same memory area for active frontal matrices). Finally, concerning the comparison of the *out-of-core* schemes, we can see that the asynchronous buffered approach performs better than the synchronous one. However, it has to be noted that even in the synchronous scheme, the system allocates data in memory that also allows to perform *I/O* asynchronously, in a way that is hidden to the application. Otherwise, the performance of the synchronous approach would be much worse.

We artificially decreased the size of the I/O buffer on the matrix CONESHL_MOD on 32 processors (default size was 9.5 million reals for this matrix). We can see from Figure 6 that the factorization time decreases when the size of the buffer increases. Indeed, in our strategy, the nodes that cannot fit into the buffer are written synchronously to disk, slowing down the factorization. (Note that in all cases the size of the buffers ensures a sufficient granularity for the performance of I/O.)

Concerning the solution phase, the size of the memory will generally not be large enough to hold all the factors. Thus, factors have to be read from disk, and the I/O involved increase significantly the time for solution. Note that we use a basic demand-driven scheme, relying on the synchronous low-level I/O mechanisms from Section 3.2. We have observed that the performance of the *out-of-core* solution step is often more than 10 times slower than the *in-core* case. Although disk contention might be an issue on our main target platform in the parallel case, the performance of the solution phase should not be neglected; it becomes critical in an *out-of-core* context and prefetching techniques in close relation with scheduling issues have to be studied. This is the object of current work by the MUMPS group in the context of the PhD of Mila Slavova.

4 Simulation of an *out-of-core* Stack Memory Management

In Section 3, we presented a first *out-of-core* approach for the parallel multifrontal factorization, consisting in writing factors to disk as soon as possible. The results obtained have shown the potential of the approach and how larger problems can be treated.

However this approach also has certain limitations and the stack memory now becomes the limiting factor. Therefore, the next step is to manage the stack of contribution blocks with an *out-of-core* scheme, where a contribution block may be written to disk as soon as it is produced, and read from disk when needed (either with a prefetching mechanism or with a demand-driven scheme).

With the objective to assess the potential of such an approach, we perform in this section simulations with various scenarios for the stack management:

- All-CB *out-of-core* **stack memory.** In this scheme, we suppose that during the assembly step of an active frontal matrix, all the contribution blocks corresponding to its children have been prefetched in memory. Thus, the assembly step is processed as in the *in-core* case.
- One-CB *out-of-core* **stack memory.** In this scheme, we suppose that during the assembly step of an active frontal matrix, only one contribution block corresponding to one of its children is loaded in memory, while the others stay on disk. Thus we interleave the assembly steps with I/O operations.
- Only-Parent *out-of-core* **stack memory.** In this scheme, we suppose that during the assembly step of an active frontal matrix, no contribution block is loaded in memory. Thus, the assembly step is done in an *out-of-core* way. Note that the implementation of such a strategy will not be efficient at all since the assembly steps are not very costly and there is no way to overlap I/O operations with computations. This strategy corresponds to an ideal scenario concerning the size of the in-core memory.

Note that for the three scenarios, we suppose that a contribution block is written to disk as soon as it is computed. In addition, we assume that all the active frontal matrices remain in memory until the end of their factorization.

Results and discussion. Although we experimented with several matrices, we only illustrate in Figure 7 the memory behaviour using the different *out-of-core* memory management strategies and *in-core* case for two test problems on different numbers of processors.

As expected, we see that the strategies for managing the stack *out-of-core* provide a reduced memory requirement. We also observe that the Only-Parent *out-of-core* stack memory management is the one that best decreases the memory needed by the factorization. Although this strategy might not be good for performance, it is here to provide some insight on the best we can do with our assumptions and with the current version of the code. One interesting phenomenon we observed is that the *out-of-core* stack memory management strategies give better results with symmetric matrices (see Figure 7(a)) than with unsymmetric ones (see Figure 7(b)). For unsymmetric matrices and on large numbers of processors, the bottleneck is very often due to the treatment of master tasks (holding the variables that need to be factored when the frontal matrix is parallelized) that are bigger for unsymmetric matrices (see [4]). Since we prefer to keep these tasks in core, a variant of the splitting algorithm of [4] could be applied in a parallel context, to limit the size of those tasks. In addition, we have observed that with our assumption that an active frontal matrix (or part of it if it is distributed over several processors) has to stay in memory while being factored, it would be beneficial to

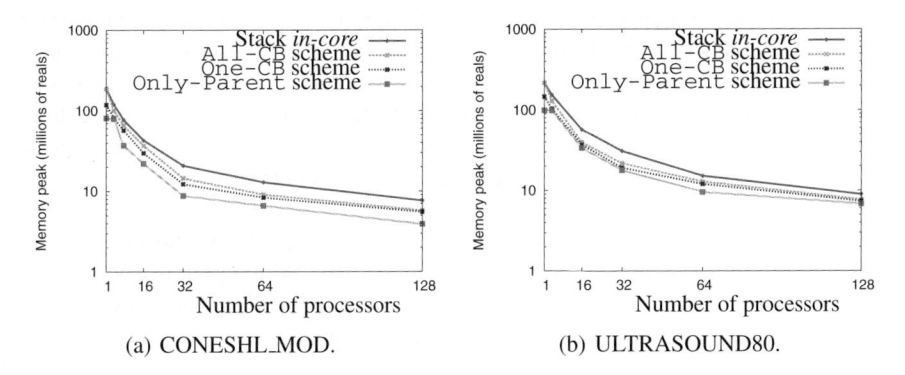

Fig. 7. Memory behaviour with different memory management strategies on different numbers of processors for two large problems (`METIS` is used as reordering technique)

reduce as much as possible the number of simultaneous active tasks on a processor. This can be done by modifying the scheduling strategies currently existing in the parallel multifrontal method.

These results illustrate that the `One-CB` approach could be a good way to design an *out-of-core* stack memory management strategy with reasonable performance. With the modifications discussed above to further decrease the memory peaks, it seems that the intrinsic limits of the sequential multifrontal method become much less critical thanks to parallelism.

5 Future Work

We presented in this paper a first implementation of an *out-of-core* extension of the parallel multifrontal solver `MUMPS`. The selected approach was to drop factors from memory as soon as they are computed and to overlap the I/O operations as much as possible with computations. We illustrated the good behaviour of this approach on a small number of processors and its limitations on larger ones, while first experiments on machines with local I/O showed no significant I/O overhead during the factorization. Nevertheless we noticed that low-level *I/O* mechanisms have to be designed with care as the system is not tuned to *I/O*-intensive and large memory applications.

One key point that must be studied is the design of efficient *out-of-core* stack memory management schemes based on the results presented in Section 4. In this context, the contribution blocks can be considered as read-once/write-once data accessed with a near-to-stack mechanism (for the parallel case the accesses are more irregular). With asynchronous I/O, prefetching algorithms have to be designed. In addition, the number of contribution blocks (for the parallel case) that a processor has in memory is closely related to the scheduling decisions made; both the static and dynamic aspects of scheduling could limit the I/O volume that each processor has to perform and drive some dynamic decisions with the data that are available in memory (for example, give a priority to tasks that depend on/consume contribution blocks already in memory).

In order to treat larger problems where both the factors and the stack memory are *out-of-core*, we have to determine more accurately which type of tasks are responsible for the peak of memory and then to limit their size and/or the number of such tasks that are active at the same time. We have already identified some critical cases in Section 4 and should now modify our algorithms when memory usage becomes a strong priority. Furthermore, adapting the techniques described in [17] could further reduce the stack memory requirements.

We believe that in a parallel context, this study shows that there is still room before reaching intrinsic memory limits of multifrontal methods. Although it is true that large frontal matrices can be problematic in sequential (need for an out-of-core assembly and factorization), this is less the case in a parallel environment.

Acknowledgements

We are grateful to P. R. Amestoy, I. S. Duff and S. Pralet for their remarks on a preliminary version of this paper.

References

1. (The BCSLIB Mathematical/Statistical Library) http://www.boeing.com/phantom/bcslib/.
2. Dobrian, F., Pothen, A.: Oblio: a sparse direct solver library for serial and parallel computations. Technical report, Old Dominion University (2000)
3. Toledo, S.: Taucs: A library of sparse linear solvers, version 2.2 (2003) Available online at http://www.tau.ac.il/~stoledo/taucs/.
4. Amestoy, P.R., Duff, I.S., Koster, J., L'Excellent, J.Y.: A fully asynchronous multifrontal solver using distributed dynamic scheduling. SIAM Journal on Matrix Analysis and Applications **23**(1) (2001) 15–41
5. Rothberg, E., Schreiber, R.: Efficient methods for out-of-core sparse Cholesky factorization. SIAM Journal on Scientific Computing **21**(5) (1999) 129–144
6. Rotkin, V., Toledo, S.: The design and implementation of a new out-of-core sparse Cholesky factorization method. ACM Trans. Math. Softw. **30**(1) (2004) 19–46
7. Cozette, O., Guermouche, A., Utard, G.: Adaptive paging for a multifrontal solver. In: Proceedings of the 18th annual international conference on Supercomputing, ACM Press (2004) 267–276
8. Ashcraft, C., Grimes, R.G., Lewis, J.G., Peyton, B.W., Simon, H.D.: Progress in sparse matrix methods for large linear systems on vector computers. Int. Journal of Supercomputer Applications **1**(4) (1987) 10–30
9. Duff, I.S., Reid, J.K.: The multifrontal solution of indefinite sparse symmetric linear systems. ACM Transactions on Mathematical Software **9** (1983) 302–325
10. Amestoy, P.R., Duff, I.S.: Memory management issues in sparse multifrontal methods on multiprocessors. Int. J. of Supercomputer Applics. **7** (1993) 64–82
11. Amestoy, P.R., Duff, I.S., Vömel, C.: Task scheduling in an asynchronous distributed memory multifrontal solver. SIAM Journal on Matrix Analysis and Applications **26**(2) (2005) 544–565
12. Amestoy, P.R., Guermouche, A., L'Excellent, J.Y., Pralet, S.: Hybrid scheduling for the parallel solution of linear systems. Parallel Computing (2005) To appear.

13. Takhur, R., Gropp, W., Lusk, E.: On implementing mpi-io portably and with high performance. In: Proceedings of the 6th Workshop on I/O in Parallel and Distributed Systems, ACM Press (1999) 23–32
14. Cormen, T.H., Davidson, E.R., Chatterjee, S.: Asynchronous buffered computation design and engineering framework generator (abcdefg). In: 19th International Parallel and Distributed Processing Symposium (IPDPS'05). (2005)
15. Karypis, G., Kumar, V.: METIS – A Software Package for Partitioning Unstructured Graphs, Partitioning Meshes, and Computing Fill-Reducing Orderings of Sparse Matrices – Version 4.0. University of Minnesota. (1998)
16. Schmuck, F., Haskin, R.: GPFS: A shared-disk file system for large computing clusters. In: Proc. of the First Conference on File and Storage Technologies. (2002)
17. Guermouche, A., L'Excellent, J.Y.: Constructing memory-minimizing schedules for multifrontal methods. ACM Transactions on Mathematical Software (2005) To appear.

A Parallel Adaptive Cartesian PDE Solver Using Space–Filling Curves*

Hans-Joachim Bungartz, Miriam Mehl, and Tobias Weinzierl

Technical University Munich, 85748 Garching, Germany
{bungartz, mehl, weinzier}@in.tum.de
http://www5.in.tum.de

Abstract. In this paper, we present a parallel multigrid PDE solver working on adaptive hierarchical cartesian grids. The presentation is restricted to the linear elliptic operator of second order, but extensions are possible and have already been realised as prototypes. Within the solver the handling of the vertices and the degrees of freedom associated to them is implemented solely using stacks and iterates of a Peano space–filling curve. Thus, due to the structuredness of the grid, two administrative bits per vertex are sufficient to store both geometry and grid refinement information. The implementation and parallel extension, using a space–filling curve to obtain a load balanced domain decomposition, will be formalised. In view of the fact that we are using a multigrid solver of linear complexity $\mathcal{O}(n)$, it has to be ensured that communication cost and, hence, the parallel algorithm's overall complexity do not exceed this linear behaviour.

1 Introduction

An important issue of a finite element code is to implement it in an efficient way. We want to examine four different aspects of efficiency: First of all the numerical efficiency covering all mathematical aspects, from modelling and discretization up to the solver. Second, there is the process integration efficiency, representing classical front– and back–end application integration tasks, such as adding a geometry input or embedding a flow solver into a fluid–structure interaction application. Furthermore, we distinguish between the implementation efficiency, regarding everything influencing the actual execution speed of a given program on a given platform, and parallel efficiency. The latter three often suggest the usage of cartesian grids, since then several implementation tasks are simplified. However, cartesian grids are not competitive for any real world application if they do not support adaptivity. On the other hand, with adaptivity the development of a well–suited traversal order, appropriate data structures, and a data access scheme is not a trivial task anymore.

In fact, many multigrid — i.e. numerically efficient — codes suffer from an inefficient implementation, integration, and parallelisation. We want to address

---

* This work has partially been funded by DFG's research unit FOR493 and the DFG project HA 1517/25-1/2.

this problem and, in the following, will derive a traversal and data management algorithm working on adaptive cartesian grids alike [7,12]. This algorithm then is parallelised using a domain decomposition approach based on [6]. Although the results are presented for a three–dimensional Poisson problem on an a priori refined grid only, we are able to solve any d–dimensional problem that can be discretised by a 3^d–point stencil. This is an important subtask of many more complex problems (the pressure Poisson part in the Navier–Stokes equations, e.g.) and starting point for the implementation of more difficult operators, such as the diffusion–convection operator or the diffusion operator for jumping material parameters.

The remainder is organised as follows: In Section 2, we introduce the adaptive cartesian grid our algorithm is based on. Section 3 is concerned with defining a traversal order (a linearisation) for the cells of this grid and exposing a vertex handling scheme, proving that two extra administrative bits per vertex are sufficient, both to store the complete grid structure including the geometry and to solve the equation system. Afterwards, in Section 4, we apply a hierarchical domain decomposition technique to end up with an algorithm whose communication data scales linearly with regard to the maximum number of vertices on the boundary of any partition. In Section 5, we present an upper bound for the corresponding constant, showing it is quasi-optimal. Finally, in Section 6, some numerical results for the Dirichlet Poisson problem are given, showing the efficiency with respect to both memory access and the parallelisation. Some final remarks in Section 7 conclude the discussion.

2 The Adaptive Grid

We create our grid using a hypercube $[0,1]^d$ and embed the computational domain into it. Then, the grid is refined in a recursive way, splitting up each cell into three parts along every coordinate axis. The depth of recursion and, hence, the resolution depends on both the boundary approximation and the numerical accuracy to be obtained. Following the notion of a spacetree (e.g. [1]) for a binary substructuring, we call these trees Peano spacetrees. A more formal definition as well as a reason for the division into three will be given later on.

On this grid, we use a nodal generating system [4] for the operator evaluation, that is a nodal basis on every grid level. Hereby the support of any shape function (hat), suitably scaled and dilated on a level k, shall be $[0, \frac{2}{3^k}]^d$. Consequently, a strictly element–wise assembly of the operators [1] is feasible, whereas within every geometric element only the element's vertices are needed. Since one degree of freedom is assigned to every vertex in this paper, the terms vertex and degree of freedom are used equivalently. Before implementing a solver on such a grid, one has to mention five important facts:

- If the values of an approximation are stored as hierarchical coefficients of the generating system $\hat{u}$ on the vertices, the inverse hierarchical transform (mapping from the hierarchical representation into a nodal basis representation

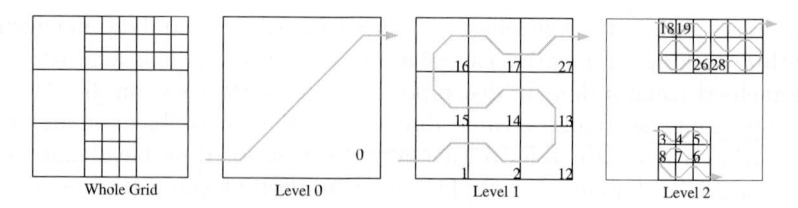

Fig. 1. An adaptive Peano grid and Peano spacetree of height three with corresponding cell order in two dimensions

of the finest level) $u = P\hat{u}$ can be done within one top–down traversal of the cell tree.
- If a value r is given on a vertex of the fine grid, the Galerkin hierarchical transform (mapping the other way round) $\hat{r} = P^T r$ of this value may happen during one bottom–up traversal.
- If a matrix–vector operation $Au = r$ with A generated by a 3^{d-1} stencil is given on any grid level, the result can be computed element–wise. Thus, all elements of this level have to be traversed once. Furthermore, the result value can be stored within the vertices directly, such that an explicit setup of matrix A is not needed at any time.
- Because of the last issue, both a residual computation and a Jacobi update step on any level can be done traversing all geometric elements of this level only once:

$$u^{(n+1)}_{level\ k} = u^{(n)}_{level\ k} + \omega\ diag^{-1}(A)\left(b - Au^{(n)}_{level\ k}\right). \quad (1)$$

- Combining equation (1) with the top–down–bottom–up arguments given above, one is able to implement an additive multigrid scheme with additive smoother [4], doing one depth–first sweep on the cell tree per iteration:

$$\hat{u}^{(n+1)} = \hat{u}^{(n)} + \omega\ diag^{-1}(P^T A P) P^T \left(b - AP\hat{u}^{(n)}\right)$$
$$=: \hat{u}^{(n)} + \omega\ diag^{-1}(P^T A P)\hat{r}^{(n)}. \quad (2)$$

A detailed description of the actual realisation of such a solver can be found in [7,13]. In the following, we will focus on the development of a well–suited depth–first traversal of the grid, on the vertex management, and on the parallelisation. Thereby, regarding the operator evaluation, we focus on a strict element–wise evaluation scheme, where only the 2^d vertices of the current element have to be available at any time. As a result, every vertex is used 2^d times per iteration.

3 Grid Traversal Using a Peano Curve

Space–filling curves [15] are well known to simplify a lot of different tasks, due to their good locality properties ([3,5,6,7,8,10,12,13] e.g.). Their recursive, self–similar definition implies a depth–first traversal of the corresponding cell tree and, therefore, an enumeration of the cells of all levels. We are using the Peano curve as illustrated in Figure 1.

A Peano spacetree is a tree corresponding to a d–dimensional adaptive cartesian grid, where each node has either 0 or 3^d children. There is an order on the tree nodes (i.e. the geometric elements / grid cells) defined top–down by the Peano space–filling curve. Note that, if one inverts the Peano curve on the root level, the order on the child nodes on every level also is inverted. The resulting tree is again a Peano spacetree, which means this set of trees is closed under the invert–traverse operation.

Now, as a result of choosing this hierarchical grid and the Peano traverse, we have to provide a data structure such that the traversal algorithm is able to access the elements' vertices within every node for element–wise operator evaluation. This is not a new problem, e.g. [5] uses a hash function derived from the space–filling curve to access the vertices and shows some nice properties of such a scheme with respect to parallelisation and load balancing. We chose a different approach, exploiting the properties of the curve as well. Here, this idea is explained for the two–dimensional case. The recursive extension to arbitrary dimensionality is very technical, but is based on exactly the same ideas [7,9]. Where necessary, the basic construction ideas for $d > 2$ are presented:

First of all, one can observe that every continuous traverse splits up all vertices of the grid into left and right ones (in terms of their position with respect to the Peano traverse). This is formalised by a left–right classifier function

$$c_{LR_2} : vertices \mapsto \{L, R\} = \{0, 1\} \quad \text{in } \mathbb{R}^2. \tag{3}$$

Given a $d \geq 2$ there are $d - 1$ mappings of both the vertices and the space–filling curve onto the planes $(x_1, x_2), (x_1, x_3), \ldots$. For the Peano curve, the projection property holds [15], i.e. every projection onto the subplanes given before is a Peano curve again (see Figure 2). Thus, on each plane one can evaluate c_{LR_2} and combine the $d - 1$ classifiers resulting in a $d - 1$–dimensional left–right classifier function

$$c_{LR} : vertices \mapsto \{0, 1\}^{d-1}. \tag{4}$$

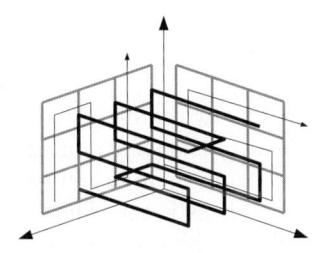

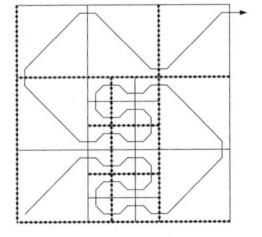

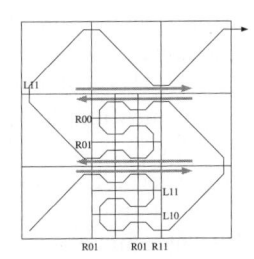

Fig. 2. On the left–hand side one can see the projection property, i.e. the projections of the Peano curve are again Peano curves. In the middle the alternating edge colouration is illustrated (even/odd indicated by dotted/solid), whereas on the right–hand side one can observe the palindrom / stack property (fat arrows). To some vertices their classifier value c is added.

The second idea is to have a look at the chronology a vertex is needed on a two–dimensional grid: If a "left" vertex is needed before another "left" vertex, the next time, the vertex is needed for operator evaluation, it is the other way round. Figure 2 shows this fact using grey arrows. Our idea is to use stacks, since they meet the resulting requirements: put a record (vertex) on the top of a stack after using it the first time, and pop a record from the stack when it is needed the second time. In addition to the left and right stack, one has to add a third idea, the stack colouring, within a hierarchical grid as pointed out first by [6], to avoid access conflicts due to the top–down bottom–up steps of the traverse in the generating system since there might be more than one degree of freedom per vertex.

So, the third idea is to colour the edges of a two–dimensional grid alternating along every axis in an even–odd manner. For example the left and the bottom edge of the root element are coloured. On any refinement level first of all the colours of the edges of the parent element are inherited, then the other edges are coloured again alternating (see Figure 2). Since every vertex is element of two edges, we get an additional qualifier

$$c_{col} : vertices \mapsto \{0,1\}^2 \quad \text{in } \mathbb{R}^2 \tag{5}$$

defining the colour of a vertex. Combining (4) and (5), we end up with a classifier function

$$c : vertices \mapsto \{0,1\}^3 \quad c = c_{LR} \circ c_{col} \quad \text{in } \mathbb{R}^2 \tag{6}$$

for every vertex.

Lemma 1. *For $d = 2$ one is able to implement the whole vertex handling using $2^3 + 2$ stacks only.*

Assume there is a vertex stream, the vertices being ordered according to the very first vertex access. The first time a vertex v is required, it is read from the input stream. After the first usage, the vertex is stored on a stack $c(v)$. Next time it is needed for element–wise evaluation, it lays on top of stack $c(v)$. After the fourth usage, it is written to an output stream. As soon as one iteration is done, you can invert the Peano spacetree traversal order and switch input and output stream using them as stacks. For $d > 2$ this access scheme is extended in a recursive way regarding the axes, and the whole vertex handling can be done using $2^{2d-1} + 2$ stacks [9].

Implementing this algorithm, every vertex is augmented by two administrative bits: The first bit describes whether the vertex is inside or outside the domain. On the second bit we define an or–refinement semantic: An element is refined, if at least the refinement bit of one of the 2^d element's vertices is set. Thus, if n' is the number of vertices of the Peano spacetree, only $2n'$ bits are required for both the geometry and the grid description. During depth–first traversal, the whole traversal order can be reconstructed, evaluating the bits of the vertices of the current element and the current traversal state.

Using stacks is the first key for the high cache efficiency reported in [7,9], since for them the data access is highly local (no jumps within the memory), which is often named spatial locality. Using the Peano curve, the spatial locality [11] of the traverse results in very small temporary stacks and, therefore, good temporal locality of the stack access, which is the second key.

4 Domain Decomposition

Space–filling curves are well known within the parallel community for their good load–balancing and good spatial locality properties, i.e. ratio of surface divided by partition volume. For the Peano spacetree, we define a tree partition, based on an existing fine–grid domain decomposition, in a bottom-up manner.

We assume that we have a Peano domain decomposition of a fine grid into disjoint partitions. A Peano spacetree partition is a Peano spacetree minimal with respect to the number of tree nodes. Within this tree, all nodes of the given fine–grid partition as well as their fathers are contained and are called active. All vertices adjacent to the active elements are called active, too. Besides the active elements, the Peano spacetree partition contains all elements of the global tree, which are adjacent to the active vertices. The additional elements are called passive. As a result, every vertex a degree of freedom is assigned has again 2^d adjacent geometric elements within each spacetree partition it is contained.

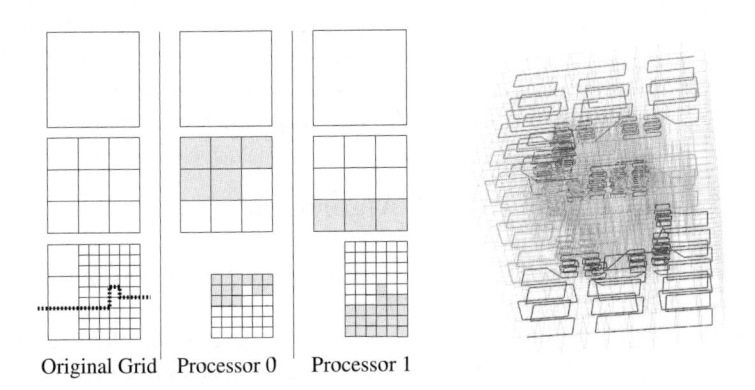

Fig. 3. The left–hand side shows a domain decomposition into two hierarchical partitions. The grey cells are held on a processor, but not evaluated since they do not belong to the processor (passive elements). The example on the right–hand side just gives an idea how a three–dimensional partitioning might look like.

Figure 3 shows two Peano spacetree partitions belonging together: Only the sets of active fine–grid elements are disjoint. For coarse grid elements this may not hold. Now, every processor has to traverse its Peano spacetree partition, whereas on the passive elements no calculation is done. In our additive multi-grid algorithm, the restriction part of equation (2) is done on every processor autonomously without any master process. As we added the passive elements to the partition, the vertex management does not have to be modified and all the vertices, even those for which the process computed only a part of the residual, are transferred to the output stack:

$$\begin{aligned}\hat{r}^{(n)} &= P^T(b - AP\hat{u}^{(n)}) \\ &= \underbrace{P^T(b_{p0} - (AP\hat{u}^{(n)})_{p0})}_{\text{processor 0}} + \underbrace{P^T(b_{p1} - (AP\hat{u}^{(n)})_{p1})}_{\text{processor 1}} + \ldots \end{aligned} \quad (7)$$

When implementing the algorithm, we split up the output stream into two streams, one holding only vertices other processors are interested in. Either of them contains a subset of the global vertex stream that would correspond to a single processor run, and the global order of the vertices is preserved on all the output streams. Every vertex, with at least two processors interested in, has got a set of processors needing its residual contribution. This contribution might be sent to the other processors immediately, before the vertex is stored on the output stack, resulting in an asynchronous communication scheme. It is shown in [12] how to compute the set of interested computers on the fly. Furthermore, it is a good idea to buffer the elementary messages, depending on the hardware used.

After one iteration, all the residual contributions received and the own data, stored on one output stream, have to be merged. Since the order on the vertices is preserved, this can be done in $\mathcal{O}(s)$, where s is the number of vertices that had to be sent. Furthermore, this does not have to be done within a dedicated merge phase, but can be done during the next top–down traversal.

5 Efficiency of the Parallel Algorithm

Prior sections have shown how to implement an algorithm, linear in the number of unknowns n, in a (technically) efficient way on a parallel machine with p nodes without any major intrinsic serial part. According to [5], the performance of our algorithm, where the results have to be synchronised after every iteration, solely depends on the amount of data s'_k to be sent by a node k, such that the computational time per iteration is given by

$$t(n,p) = C_{solver}\frac{n}{p} + C_{startup} + C_{comm}\max_k\{s'_k\}, \quad (8)$$

if one is not able to do the communication in an asynchronous way. The algorithm becomes quasi–optimal [5] for

$$\max_k\{s'_k\} \leq C\left(\frac{n}{p}\right)^{1-1/d}, \quad C \geq \sqrt[d]{\frac{2d^{d-1}\pi^{d/2}}{\Gamma(d/2)}}, \quad (9)$$

reflecting the continuous Hölder continuity with parameter $\frac{1}{d}$ of a continuous space–filling curve [5,15].

In our case, the amount of data depends on the tree's height and the surface s of the fine–grid partition. There is a lot of published work on the interfaces of space–filling curves' partitions (e.g. [3,8]). Since most of this work deals with Hilbert and Lebesgue curves only, the proofs given there have to be transfered into the Peano curve case and have to be augmented by the tree issue.

In the following, we examine regular refined grids with $n \in (3^d)^{\mathbb{N}}$ geometric elements using p processors. The workload (number of geometric elements) is distributed equally among them.

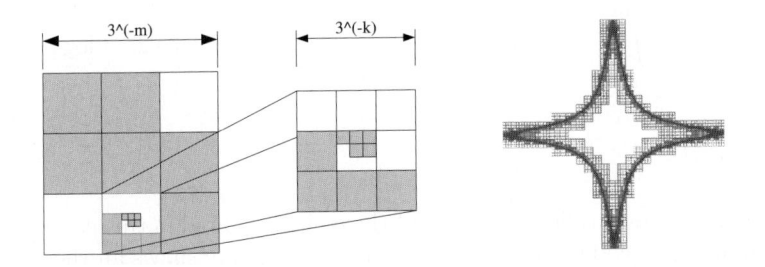

Fig. 4. Construction of a trivial upper bound of the surface of a partition induced by the Peano space–filling curve (grey), and the star shaped domain used in Section 6

Lemma 2. *The number of boundary vertices — vertices adjacent to passive geometric elements — on the fine grid of any partition is bounded by*

$$s' \leq \frac{4d}{1 - 3^{1-d}} 3^{d-1} \left(\frac{n}{p}\right)^{1-1/d}. \tag{10}$$

Proof. The proof follows the argumentation of [8]: Let M be the maximal tree depth, and m be the maximal tree level one would be able to embed the $\frac{n}{p}$ cells of the partition into one geometric element. On level m, the partition is contained in at most two elements, such that the bounding box of the two neighbouring geometric elements s_m is an upper bound for the continuous surface, if the domain was represented in the level's resolution (compare to Figure 4). On the finer levels $k > m$, there might be at most two appendices (cells containing not only active subcells), since the space–filling curve used is compact and continuous (therefore, all the children of a node are visited, before the next node is processed). Their boundary box surface is already considered, but the possibly resulting concave surface parts s_k have to be added to the result. This surface is bounded by the bounding box of an element of level k. Finally, the continuous surface $s(n, p)$ is divided by the fine grid element face size, which is $3^{-M(d-1)}$, giving the number of fine–grid vertices up to a small constant:

$$\left(3^d\right)^M = n \qquad \left(3^d\right)^{M-m-1} \leq \frac{n}{p} \leq \left(3^d\right)^{M-m} \tag{11}$$

$$s(n,p) \leq 2 \sum_{k=m}^{M-1} s_k \leq 2 \sum_{k=m}^{M-1} 2d \left(\frac{1}{3}\right)^{(d-1)k} \leq \frac{4d}{1 - 3^{1-d}} 3^{(1-d)m}$$

$$s' \leq \frac{4d}{1-3^{1-d}}3^{(1-d)m+M(d-1)} \leq \frac{4d}{1-3^{1-d}}3^{d-1}\left(\frac{n}{p}\right)^{1-1/d}. \quad (12)$$

The amount of data sent by one processor is bounded by a geometric series with argument $\left(\frac{1}{3}\right)^{d-1}$ for the grid levels m to M scaled by s', as the number of vertices decreases with this factor for each coarsening step. For the levels $0\ldots-1$ the number of active cells enclosing the partition is bounded by two. Therefore, the number of boundary vertices is bounded by $3 \cdot 2^{d-1}$ for each level.

$$s = s'\sum_{k=m}^{M}\left(\frac{1}{3}\right)^{(d-1)k} + m\frac{3}{2}2^d \leq \underbrace{\frac{3^{d-1}}{1-3^{1-d}}p^{1/d-1}}_{\leq 1} s' + 3 \cdot 2^{d-1}\log_3 \sqrt[d]{p}. \quad (13)$$

6 Results

Figure 5 gives the parallel behaviour of the code presented in this paper for three dimensions. This code is not optimised yet, but already shows all the properties stated in this paper for a Dirichlet–Poisson problem on the cube, a sphere, or a star domain (see Figure 4), as well as the excellent cache behaviour (see [6,7,9,10,12,13], e.g.). The star domain experiment suffers from the lack of dynamic load balancing not implemented yet: Since the ratio of inner cells to cells outside the domain, where no operator evaluation is necessary, is unfavourable, a simple equidistant curve partitioning fails. The same reasoning holds for the sphere.

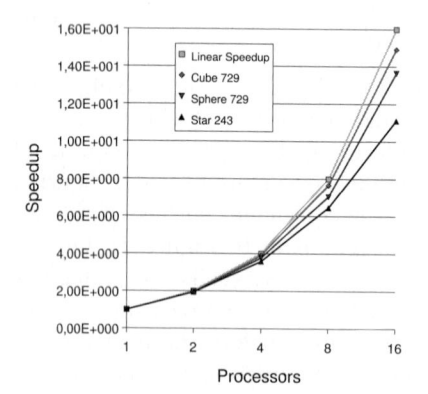

	cube 729^3	sphere 729^3	star 243^3
real dofs	$\approx 4.0 \cdot 10^8$	$\approx 2.4 \cdot 10^7$	$\approx 4.3 \cdot 10^5$
S(2)	1.95	1.95	1.94
S(4)	3.9	3.77	3.58
S(8)	7.66	7.04	6.45
S(16)	14.92	13.65	11.11
L2 CHR	99.96%	99.94%	99.95%

Fig. 5. Some parallel performance results for $d=3$ on a Myrinet cluster of Dual Pentium III 800 MHz with 2GByte RAM per node [12]. $S(p)$ denotes the speedup on p processors, L2 CHR abbreviates level 2 cache–hit rate.

7 Concluding Remarks

In this paper, we have presented a parallel multigrid PDE solver based on the Peano spacetree, handling all the vertices solely using stacks. Since this approach has proven to be of value with respect to memory requirements, parallelisation, and cache efficiency, it is our strategy to use this algorithm within a more complex environment. In fact, we have already used exactly the same approach to prototype a Navier–Stokes solver [14]. Furthermore, it has been shown that our algorithmic approach is well–suited for a posteriori refinement [13]. Since we are working on trees, dynamic load balancing can be implemented in a very natural way by forking trees [10]. Right now we are integrating all these aspects into one d–dimensional PDE solver, embedded into a fluid–structure interaction application framework [2].

It is work in progress, how to extend the scheme to higher order stencils and to provide better estimates on the amount of data to be communicated. Furthermore, the behaviour on a massively parallel cluster and different load balancing strategies have to be evaluated.

Special thanks to Markus Pögl and Markus Langlotz, for doing a first implementation of the algorithm presented and solving many implementation issues.

References

1. Bader M., Frank A.C., Zenger Ch.: An Octree-Based Approach for Fast Elliptic Solvers. High Performance Scientific and Engineering Computing. Springer-Verlag, Berlin Heidelberg (2001)
2. Brenk M., Bungartz H.J., Mehl M., Neckel T.: Fluid–Structure Interaction on Cartesian Grids: Flow Simulation and Coupling Environment. Fluid-Structure Interaction, LNCS (to appear)
3. Gotsman C., Lindenbaum M.: On the Metric Properties of Discrete Space–Filling Curves. IEEE Transactions on Image Processing vol. 5 (**5**) (1996) 794-797
4. Griebel M.: Multilevel algorithms considered as iterative methods on indefinite systems. SFB-Bericht **342/29/91** (1991)
5. Griebel M., Zumbusch G.: Hash–Storage Techniques for Adaptive Multilevel Solvers and Their Domain Decomposition Parallelization. Proceedings of Domain Decomposition Methods 10, DD10 **218** (1998) 279–286
6. Günther F., Krahnke A., Langlotz M., Mehl M., Pögl M., Zenger Ch.: On the Parallelization of a Cache-Optimal Iterative Solver for PDEs Based on Hierarchical Data Structures and Space-Filling Curves. Recent Advances in Parallel Virtual Machine and Message Passing Interface, LNCS **3241** (2004) 425-429
7. Günther F., Mehl M., Pögl M., Zenger Ch.: A cache-aware algorithm for PDEs on hierarchical data structures based on space-filling curves. SIAM Journal on Scientific Computing (to appear)
8. Hungershöfer J., Wierum J.M.: On the Quality of Partitions based on Space–Filling Curves. International Conference on Computational Science 2002, LNCS **2331** (2002): 31-45
9. Hartmann J.: Entwicklung eines cache–optimalen Finite–Element–Verfahrens zur Lösung d-dimensionaler Probleme. Diploma thesis, Technical University Munich (2004)
10. Herder W.: Entwicklung eines cache–optimalen Finite–Element–Verfahrens zur Lösung d-dimensionaler Probleme. Diploma thesis, Technical University Munich (2005)

11. Kowarschik M., Weiß C.: An Overview of Cache Optimization Techniques and Cache-Aware Numerical Algorithms. Proceedings of the GI-Dagstuhl Forschungseminar: Algorithms for Memory Hierarchies, LNCS **2625** (2003) 213-232
12. Langlotz M.: Parallelisierung eines Cache–optimalen 3D Finite–Element–Verfahrens. Diploma thesis, Technical University Munich (2005)
13. Mehl M., Weinzierl T., Zenger C.: A cache–oblivious self–adaptive full multigrid method. Numerical Linear Algebra With Applications (to appear)
14. Neckel T.: Einfache 2D–Fluid–Struktur–Wechselwirkungen mit einer cache–optimalen Finite–Element–Methode. Diploma thesis, Technical University Munich (2005)
15. Sagan H.: Space-Filling Curves, Springer-Verlag, Berlin Heidelberg (1994)

Load Balanced Parallel Simulated Annealing on a Cluster of SMP Nodes

Agnieszka Debudaj-Grabysz[1] and Rolf Rabenseifner[2]

[1] Silesian University of Technology, Department of Computer Science
Akademicka 16, 44-100 Gliwice, Poland
agrabysz@polsl.pl
[2] High-Performance Computing-Center (HLRS), University of Stuttgart
Nobelstr. 19, D-70550 Stuttgart, Germany
rabenseifner@hlrs.de
www.hlrs.de/people/rabenseifner

Abstract. The paper focuses on a parallel implementation of a simulated annealing algorithm. In order to take advantage of the properties of modern clustered SMP architectures a hybrid method using a combination of OpenMP nested in MPI is advocated. The development of the reference implementation is proposed. Furthermore, a few load balancing strategies are introduced: time scheduling at the annealing process level, clustering at the basic annealing step level and suspending—inside of the basic annealing step. The application of the algorithm to VRPTW—a generally accepted benchmark problem—is used to illustrate their positive influence on execution time and the quality of results.

Keywords: Simulated annealing, parallel processing, load balancing, MPI, OpenMP, hybrid parallelization.

1 Introduction

The paper presents a time scheduled algorithm for parallel simulated annealing—a heuristic method of optimization—that is intended to run on modern clusters of shared-memory (SMP) nodes. While clusters of SMPs with numbers of processors ranging into hundreds are becoming more and more popular, the question of how to use them efficiently for parallel simulated annealing, knowing its sequential character, is still open. One of popular programming styles for clustered systems uses different communication environments for their separate components, combining the benefits of both shared and distributed memory systems at the same time. The communication method discussed in the paper adopts such a hybrid approach for simulated annealing and is called a *hybrid communication method* (HC).

The research described in this work is a continuation of the efforts reported in [9], where the reference HC method was introduced. It proved to be the most effective compared with the other tested methods, when solving a bicriterion optimization problem. The paper presents a modification of the reference method, namely the *hybrid communication method with a single data exchange*, which is the way to improve quality of results for the second optimized criterion. The time scheduling aspects for both methods are discussed. A constrained cost is assumed, which means that searching for the optimal solution is performed with a given pool of processors available for a specified period of time. This approach produces linear speed-up.

Simulated annealing (SA) is a heuristic optimization method used when the solution space is too large to explore all possibilities within a reasonable amount of time. The vehicle routing problem with time windows (VRPTW) is an example of such a problem. Other examples are school bus routing, newspaper and mail distribution or delivery of goods to department stores. Optimization of routing lowers distribution costs and parallelization allows a better route to be found within given time constraints.

The SA bibliography focuses on the sequential version of the algorithm (e.g., [2,17]), however parallel versions are investigated too, as the sequential method is considered to be slow when compared with other heuristics [18]. In [1,3,10,12,13] and many others, directional recommendations for parallelization of SA can be found. VRPTW, formally formulated by Solomon [16], who also proposed a suite of tests for benchmarking, also has a rich bibliography [18]. Additionally, a few works discussing parallel SA to solve the VRPTW are known, namely [6,7,4,8]. Nevertheless, in contrast to the constraints applied in the current research, i.e., limited time, the first two take advantage of the parallel algorithm to achieve higher accuracy of solutions, while the others define different stopping criteria for the algorithm.

The plan of the paper is as follows: section 2 presents the theoretical basis of the sequential and parallel SA algorithm. Section 3 describes the two variants of the hybrid communication method, while section 4 presents practical issues, leading to load balanced execution. The results of the experiments are described in section 5. Conclusions follow.

2 Sequential and Parallel Simulated Annealing

In simulated annealing, one searches for the optimal state, i.e., the state that gives either the minimum or maximum value of the *cost function*. It is achieved by comparing the current solution with a random solution from a specific *neighbourhood*. With some probability, worse solutions could be accepted as well, which can prevent convergence to local optima. However, the probability of accepting a worse solution decreases over the process of annealing, in synchronisation with the parameter called *temperature*. An outline of the SA algorithm is presented in Figure 1, where a single execution of the innermost loop step is called a *trial*. The final solution which is returned is the best one ever found. Simulated annealing can be also modelled by using the theory of Markov chains. The algorithm is formed by a sequence of Markov chains where each chain consists of a sequence of trials for which the acceptance criterion with a fixed value of temperature was applied.

```
01  S ← GetInitialSolution();
02  T ← InitialTemperature;
03  for i ← 1 to NumberOfTemperatureReduction do
04      for j ← 1 to EpochLength do
05          S' ← GetSolutionFromNeighbourhood();
06          ΔC ← CostFunction(S') − CostFunction(S);
07          if (ΔC < 0 or AcceptWithProbabilityP(ΔC, T))
08              S ← S';    {i.e., the trial is accepted}
09          end if;
10      end for;
11      T ← λT;    {with λ < 1}
12  end for;
```

Fig. 1. SA algorithm

Since in SA each new state contains modifications to the previous state, the process is often considered to be inherently sequential and its parallelization is not trivial. However, a few strategies for designing a parallel SA algorithm exist, e.g., based on different types of applied decomposition. In the research the creation of trials is decomposed among processors. Additionally, the chain length is fixed, meaning that the number of trials performed within the chain is the same for both the sequential and parallel algorithms.

3   Hybrid Communication—Nesting OpenMP in MPI

Clustered SMP systems support two parallelization levels: the outer parallelization for communication between SMP nodes and the inner parallelization for the shared memory environment within nodes. The HC method tries to exploit the features of the parallel SA approach that can be supported by the architecture: intensively communicating parts can be realised inside the inner level with OpenMP [15], while parts with infrequent communication can be realised at the outer level with MPI [11,14].

3.1   The Reference Method

Outer-level parallelization. Following previous research [9], in the algorithm for the outer level each Markov chain of SA optimization is divided into sub-chains. Their length is equal to the length of the original chain divided by the number of sub-chains. The main idea is to assign a separate sub-chain to each individual cluster node and thus to let nodes generate different sub-chains simultaneously. In this way the computation for generating a Markov chain is divided over all the available nodes. After generating the first Markov chain, the process of generating every consecutive chain is performed without communication between nodes. For each node the outcome of the last trial of the preceding sub-chain is the starting point for the subsequent sub-chain. At the end, the best solution found is picked up as the final one. The usage of multiple sub-chains allows intensive exploration of the search space. However, excessive shortening of the sub-chain length negatively affects the quality of results, so the maximum number of nodes used is limited by reasonable shortening of the sub-chain length.

Inner-level parallelization. Within a node a few threads can communicate to build one sub-chain of the length determined at the outer level. Negligible deterioration of quality is a key requirement for the inner-level algorithm. The idea of parallelization is to divide the total number of trials of each sub-chain into short *sets of trials*. The size of the sets equals the number of threads, so each thread generates one trial at a time, independently of the others. After completing a set, the master thread selects one solution among all the accepted ones and the others are discarded. The selected solution is common for all threads and becomes the starting point for further computation.

3.2 The Method with a Single Data Exchange

Tracing the process of finding solutions one can conclude, that incorporating lightweight communication between nodes could improve the quality of results. During the optimization all processes working on cluster nodes explore the search space, but after the first stage, which is characterized by "long jumps" and large changes of position, it is likely that only a few processes will be working in the "right" area of the global minimum. The rest of them may perform useless computations. One can speculate that global selection of the best result found during the stage of heavy exploration would let all the processes move into these "right" areas, leading to significant improvement of the quality of results. The ratio between durations of the two, above mentioned stages should be carefully selected. The optimization process should be able to use an adequate period of time during the first stage to explore the search space precisely enough to reach the area of the global minimum. On the other hand, the duration of the second stage should be long enough to let the processes exploit the promoted area and further approach the minimum.

4 Load Balancing in the Hybrid Communication Methods

4.1 Outer Level Load Balancing

The major drawbacks to obtain balanced computational load and acceptable speed-up are differences in the execution times of the trials, because the effort of performing them depends on the current configuration. This leads to substantial idle times when stopping the algorithm after generating a number of sub-chains. This is shown in Figure 2, where the times for generating 8 separate sub-chains are presented, based on an example run of the investigated problem (see section 5). To overcome this difficulty a real time limit is set for computation. It derives from the average time needed by the sequential algorithm and is calculated so as to assure linear speed-up.

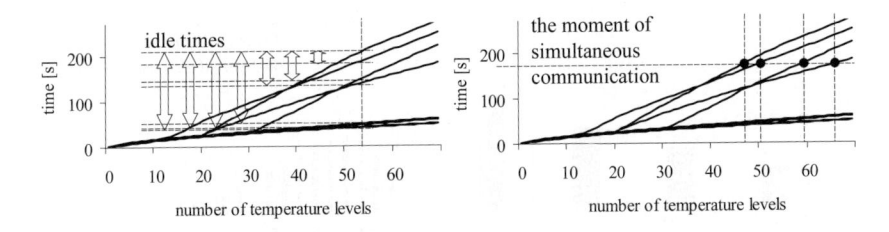

Fig. 2. The times for generating 8 sub-chains based on a run for the investigated problem. Left: scheduling by a fixed number of temperature levels with idle times marked with arrows. Right: time-based scheduling, where communication is scheduled after the same amount of time on all processes.

A time-based scheduling is also suitable when defining the way for announcing the moment of a single data exchange. Specifying the time limit for the computation by measurements of the elapsed time, gives a new opportunity to determine the exact moment of data exchange. Setting the number of data exchanges is straightforward as well. Therefore the proposed method forces one data exchange when a specific percentage of time limit (e.g., after 50%, 70% or 90%) elapses. After selecting and broadcasting the best solution found so far, all processes starts their computation from this agreed solution. The method results in much better balancing than an alternative with a fixed number of temperature levels (i.e., sub-chains) (Figure 2 (left)) and makes more efficient use of the given time limit. In Figure 2 (right) the moment of simultaneous communication is marked on the time axis. This can be individually determined by working processes irrespectively of the number of performed trials.

4.2 Inner Level Load Balancing

To achieve an acceptable inner-level speed-up a few optimization stages were necessary for the OpenMP parallelization with loop worksharing. The need for optimization stems from extremely varying execution time for each trial. In the presented example these differences were within a factor of 100. Consequently, where the number of trials equals the number of threads, i.e., each thread generates one trial at a time, a theoretically calculated speed-up, based on a comparison of the execution times, does not exceed 2 with the use of 4 threads. The average execution times of trials within a set, accumulated throughout the whole example run, are presented in Figure 3 (left). A case of one trial per a thread is marked white. The distance from the average is visible. The white bars show the average execution time of the fastest (left most bar) trial in a set of 4 trials, up to the slowest trial (right most bar). In Figure 3 (right) the histogram of trials with timings that falls into 25 ranges is presented. It also proves their imbalanced distribution.

The first optimization step was to increase the size of a set of trials to make each thread generate a few trials at a time without any communication. In this way the load imbalance was substantially decreased. Nevertheless, in order

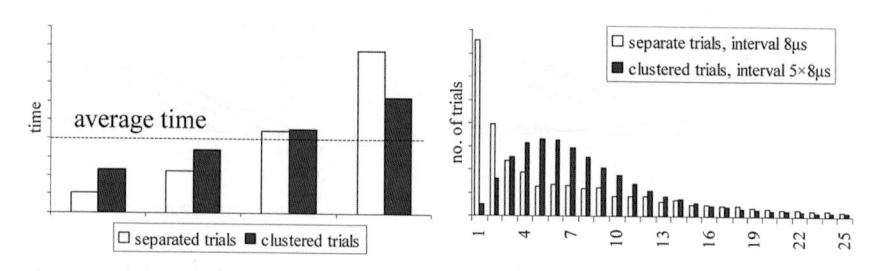

Fig. 3. Trials before the reconfiguration. Left: average execution times of the trials within a set. Right: the histogram of execution times. Example: Solving VRPTW, test R108 from Solomon's benchmark set.

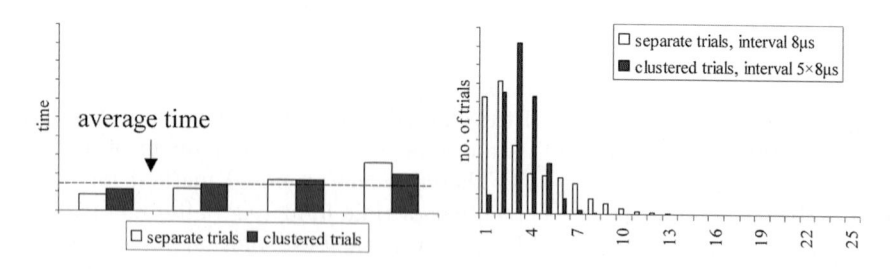

Fig. 4. Trials after the reconfiguration. Left: average execution times of the trials within a set. Right: the histogram of execution times. Example: Solving VRPTW, test R108 from Solomon's benchmark set.

to maintain quality, the size of the set of trials should be as small as possible, because it affects the number of accepted but discarded trials. The average execution times of so called "clustered" trials within a set of the size 20 (= 4 threads × 5 trials/thread), as well as the histogram of execution times are also presented in Figure 3, but marked black. The scale for "clustered" trials was normalised to the scale of "separate" trials to indicate more clear their smaller deviations from the average value and changed distribution.

The second optimization step was the redefinition of a trial, in order to improve load balancing and simultaneously to decrease the number of discarded trials. Hence, the speed-up as well as a quality of results can be increased. As the execution time of each trial is determined by the time for finding a new valid solution S' in the neighbourhood of S, one can limit the number of actions taken within GetSolutionFromNeighbourhood() (see Figure 1). If after only a few disturbances of the current configuration no new solution can be created, the algorithm produces a transitional status "no answer" and suspends the process of completing a trial. After the selection has been made over the set by the master thread, all uncompleted trials are continued. The influence of the redefinition can be seen in Figure 4, where the time scale is the same as in Figure 3. The average execution time of a trial is shortened but better balanced than before the redefinition.

The third optimization step was to choose an appropriate moment for forking as well as for joining parallel threads. As the execution time for a set of trials can be short compared to the OpenMP fork-joined overhead, the parallel loop should comprise a wider region, i.e., the whole temperature step.

5 Experimental Results

In the vehicle routing problem with time windows it is assumed that there is a warehouse, centrally located to customers. There is a road between each pair of customers and between each customer and the warehouse. The objective is to supply goods to all customers at the minimum cost. The solution with fewer route legs (the first goal of optimization) is better then a solution with smaller total distance travelled (the second goal of optimization). Each customer as well as the warehouse has a time window. Each customer has its own demand level and should be visited only once. Each route must start and terminate at the warehouse and should preserve maximum vehicle capacity. As already mentioned, previous work [9] focused on the first goal of optimization, while this paper focuses on the second one, i.e., optimizing the final distance when the minimum number of route legs is already achieved. The sequential algorithm from [5] was the basis for parallelization.

Experiments were carried out on a NEC Xeon EM64T Cluster installed at the High Performance Computing Center, Stuttgart (HLRS). Additionally, for tests of the OpenMP algorithm, a NEC TX-7 (ccNUMA) system was used. The numerical data were obtained by running the program 100 times for Solomon's [16] R108, R111, RC105 and RC108 tests with 100 customers and the same set of parameters. The number of OpenMP threads was 4 and the size of the set of trials was 20, this giving the best combination of efficiency and quality. Due to the lack of access to a genuine clustered SMP machine with 4 CPUs per each node (the NEC Xeon EM64T consists of dual CPU nodes), the usage of 4 OpenMP threads per cluster node was emulated. The emulation was carried out by extending the applied time limit by the speed-up factor coming from a separate set of experiments. Such an extension can be thought as undoing the speed-up to be observed on a cluster of nodes having 4 CPUs instead of 2.

Time results. At the outer level both versions of the hybrid algorithm give linear speed-up, since a real time limit is applied. However, in case of the method with a single data exchange one should consider the additional communication overhead. In investigated examples the time needed for selecting and broadcasting the best solution between nodes was between 0.4ms (2 nodes) to 0.8ms (30 nodes), which is substantially shorter than the execution time. At the inner level the average speed-up factor obtained empirically was 2.7, which gives the efficiency of OpenMP parallelization as 67%.

Quality results. A few parameters for controlling the data exchange were investigated, namely after reaching 50%, 70% and 90% of the time limit. The results

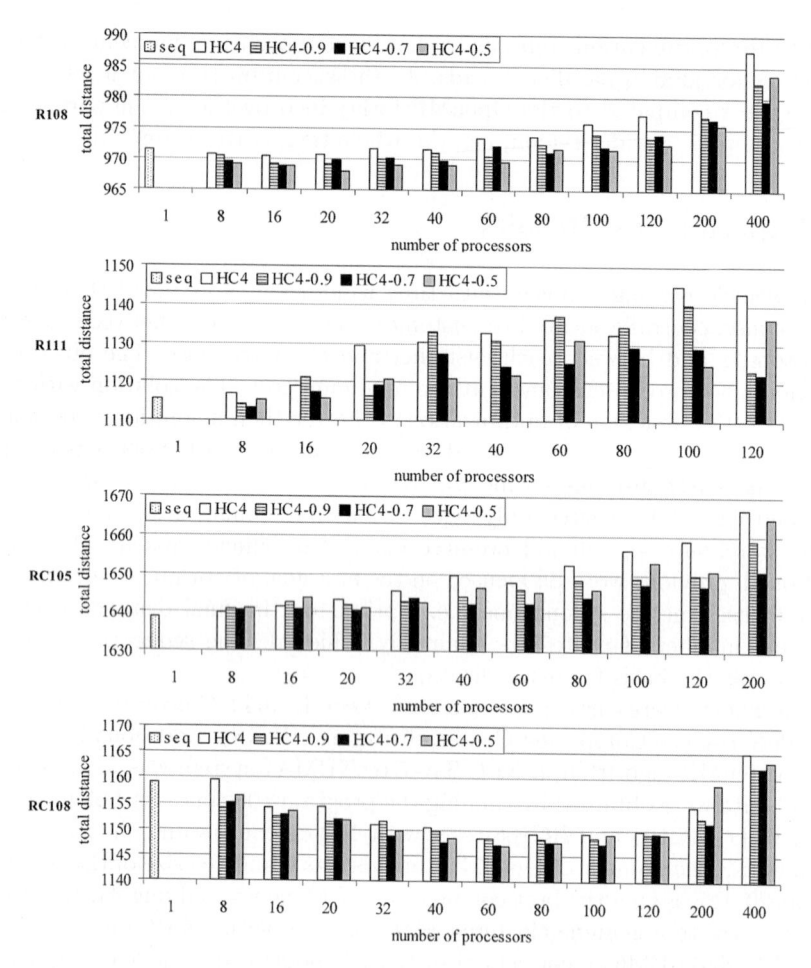

Fig. 5. Comparison of quality results for hybrid communication methods

of experiments are presented in Figure 5. Generally, selecting a common solution after 50% (HC4-0.5) or 70% (HC4-0.7) of the time limit was much better than other tested possibilities. It should be noted that both HC4-0.5 and HC4-0.7 give better results than the reference method (HC4) almost for all investigated numbers of processors (with only one exception). Nevertheless, when compared to the sequential results (SEQ) it can be observed that the quality of the hybrid parallelization depends on a test. E.g., for R108 up to 40 processors, R111 with 8 processors and RC108 (excluding 400 processors) the results of HC4-0.7 are better than for the sequential version, but in other cases, i.e., RC105, R108 with more than 40 processors and R111 (excluding 8 processors) they are worse.

To verify these observations one can incorporate test statistics. Statistical hypotheses $H_0 : x_i = x_j$ versus alternative hypotheses $H_1 : x_i < x_j$ or $H_1' : x_i \neq$

x_j can be tested, where x denotes the mean value of the total travel distance, i, j - populations that are compared (HC4-0.7 with HC4, HC4-0.7 with SEQ, HC4-0.5 with SEQ, respectively). Let s denote the standard deviation, and n, the population size, then u, the test statistic is given by:

$$u = \frac{x_i - x_j}{\sqrt{\frac{s_i^2}{n_i} + \frac{s_j^2}{n_j}}}$$

The significance level is set as 0.05. When comparing HC4-0.7 with HC4 the calculated values u indicate that H_0 should be rejected in favour of H_1 in 64% of tested cases. This means HC4-0.7 gave statistically shorter total distance than HC4. Besides, when comparing HC4-0.7 with SEQ, although for the tests R111 and RC105 H_0 can not be rejected in favour of H_1' up to 20 processors, for R108 it can not be rejected up to 100 processors. In other words there is no evidence that HC4-0.7 gave statistically different results for these cases, compared to the sequential algorithm. Additionally, application of similar reasoning indicates, that for test RC108, statistically HC4-0.7 allowed to achieve solutions with smaller travel distances than its sequential equivalent for numbers of processors up to 200. For test R108, HC4-0.5 compared favourably to the sequential version with up to 40 processors.

6 Conclusions

In this study the implementation of parallel SA algorithm that is intended to run on clusters of SMP nodes is considered. The development of the reference method, based on performing time scheduled data exchange was proposed. Additionally, the paper provides detailed analyses of factors influencing the speedup and efficiency, e.g.: defining the moment for terminating the optimization process, as well as time dependencies between randomly generated trials of SA algorithm. A few optimization strategies were introduced, that resulted in better balancing of the algorithm.

Based on experiments one can conclude that the quality of results for the modified method outperforms the reference one. When compared to sequential results, it needs to be stated that with proposed load balancing strategies it is possible in many cases to achieve better or comparable quality, always with linear speed-up. This observation is valid even up to 200 processors.

Acknowledgement

This work was supported by the EC-funded project HPC-Europa (contract No RII3-CT-2003-506079) and by the State Committee for Scientific Research grant 3T 11F 00429. Computing time was also provided within the framework of the HLRS-NEC cooperation.

References

1. Aarts, E., de Bont, F., Habers, J., van Laarhoven, P.: Parallel implementations of the statistical cooling algorithm. Integration, the VLSI journal (1986) 209–238
2. Aarts, E., Korst, J.: Simulated Annealing and Boltzman Machines, John Wiley & Sons (1989)
3. Azencott, R. (ed): Simulated Annealing Parallelization Techniques. John Wiley & Sons, New York (1992)
4. Arbelaitz, O., Rodriguez, C., Zamakola, I.: Low Cost Parallel Solutions for the VRPTW Optimization Problem, Proceedings of the International Conference on Parallel Processing Workshops, IEEE Computer Society, Valencia–Spain, (2001) 176–181
5. Czarnas, P.: Traveling Salesman Problem With Time Windows. Solution by Simulated Annealing. MSc thesis (in Polish), Uniwersytet Wrocławski, Wrocław (2001)
6. Czech, Z.J., Czarnas, P.: Parallel simulated annealing for the vehicle routing problem with time windows. 10th Euromicro Workshop on Parallel, Distributed and Network-based Processing, Canary Islands–Spain, (2002) 376–383
7. Czech, Z.J., Wieczorek, B.: Frequency of cooperation of parallel simulated annealing processes, Proceedings of the 6th International Conference on Parallel Processing and Applied Mathematics (PPAM'05), Poland (in print)
8. Debudaj-Grabysz, A., Czech, Z.J.: A concurrent implementation of simulated annealing and its application to the VRPTW optimization problem, in Juhasz Z., Kacsuk P., Kranzlmuller D. (ed), Distributed and Parallel Systems. Cluster and Grid Computing. Kluwer International Series in Engineering and Computer Science, Vol. 777 (2004) 201–209
9. Debudaj-Grabysz, A., Rabenseifner, R.: Nesting OpenMP in MPI to implement a hybrid communication method of parallel simulated annealing on a cluster of SMP nodes, in Di Martino B., Kranzlmuller D., Dongarra J.,(ed.), Recent Advances in Parallel Virtual Machine and Message Passing Interface, Springer-Verlag Berlin Heidelberg, LNCS 3666, (2005) 18–27
10. Greening, D.R.: Parallel Simulated Annealing Techniques. Physica D, 42, (1990) 293–306
11. Gropp, W., Lusk, E., Doss, N., Skjellum, A.: A high-performance, portable implementation of the MPI message passing interface standard, Parallel Computing 22(6) (1996) 789–828
12. Lee, F.A.: Parallel Simulated Annealing on a Message-Passing Multi-Computer. PhD thesis, Utah State University (1995)
13. Lee, K.-G., Lee, S.-Y.: Synchronous and Asynchronous Parallel Simulated Annealing with Multiple Markov Chains, IEEE Transactions on Parallel and Distributed Systems, Vol. 7, No. 10 (1996) 993–1008
14. Message Passing Interface Forum. MPI: A Message-Passing Interface Standard, Rel. 1.1, June 1995, www.mpi-forum.org
15. OpenMP C and C++ API 2.5 Specification, from www.openmp.org/specs/
16. Solomon, M.: Algorithms for the vehicle routing and scheduling problem with time windows constraints, Operation Research 35 (1987) 254–265, see also http://w.cba.neu.edu/~msolomon/problems.htm
17. Salamon, P., Sibani, P., Frost, R.: Facts, Conjectures and Improvements for Simulated Annealing, SIAM (2002)
18. Tan, K.C, Lee, L.H., Zhu, Q.L., Ou, K.: Heuristic methods for vehicle routing problem with time windows. Artificial Intelligent in Engineering, Elsevier (2001) 281–295

A Grid Computing Based Virtual Laboratory for Environmental Simulations

I. Ascione[1], G. Giunta[1], P. Mariani[2], R. Montella[1], and A. Riccio[1]

[1] Dept. of Applied Sciences at University of Naples "Parthenope" - Italy
[2] Danish Inst. for Fisheries Research,
Dept. of Marine Ecology and Aquaculture - Denmark

Abstract. The grid computing technology permits the coordinate, efficient and effective use of (geographically spread) computational and storage resources with the aim to achieve high performance throughputs for intensive CPU load applications.

In this paper we describe the development of a virtual laboratory for environmental applications. The software infrastructure, and the related interface, are developed for the straightforward use of shared and distributed observations, software, computing and storage resources. The user can design and execute his experiments building up and assembling data acquisition procedures, numerical models, and applications for the rendering of output data, with limited knowledge of grid computing, thereby focusing his attention to the application.

Our solution aims at the goal of developing black-box grid applications for earth observation, marine and environmental sciences.

1 Introduction

Numerical modeling plays a main role in the earth sciences, filling in the gap between experimental and theoretical approach. Now, the computational approach is widely recognized as the complement to the today scientific analysis. Meanwhile, the huge amount of observed/modeled data, and the need to store, process and refine them, often makes the use of high performance parallel computing the only effective solution to ensure the real usability of numerical applications, as in the case of the atmospheric/oceanography field, where the development of the Earth Simulator supercomputer is just the edge [1].

The grid computing is a key technology in the field of the computational sciences, allowing the use of inhomogeneous and geographically-spread computational resources, shared across a virtual laboratory. Moreover, this technology offers several invaluable tools, ensuring the security, the performance and the availability of applications [2].

A great amount of simulation models have been successfully developed in the past, but a lot of them are poorly engineered and built following a monolithic programming approach, unsuitable for a distributed computing environment. The use of the grid computing technologies is limited to domain specialists, because of the complexity of grid itself and of its middleware complexity. Another source of complexity resides on the use of coupled models, as, for example, in the case of atmosphere/sea-wave/ocean dynamics. The grid enabling approach could be hampered by the grid software and hardware infrastructure complexity. In this context, the buildup of a grid-aware virtual laboratory for environmental applications is a "grand challenge" for computer scientists.

In this paper we describe the implementation and application of a grid-enable virtual laboratory for environmental simulations. This application is built on the componentization of different environmental models: atmospheric circulation, air quality and ocean related models. The grid-enabling approach is described in the next section, while in section 3, we give an example of a grid application providing on-demand or operational weather and sea forecasts.

2 The Grid-Enabling Approach

For our grid infrastructure development, we use the middleware Globus Toolkit [3] version 4.x (GT4), developed within the Globus Alliance and the Global Grid Forum (GGF) with a wide support of institutions belonging to the academia, the government and the business area. The GT4 has been chosen because it exposes its features via web services using common W3C standards as the Web Service Description Language (WSDL), the Standard Object Access Protocol (SOAP), and the Hyper Text Transfer Protocol (HTTP). Complex features, as the service persistence, the state and stateless behavior, the event notification, the data element management and the index services tools are implemented in the respect of this standards. The GT4 also offers support to pre-web services features as the GridFTP protocol, an FTP enhanced version, capable of massive parallel striping and reliable file transfer.

In our grid virtual laboratory we coupled several environmental models: the MM5 (Mesoscale Model 5) [4], the STdEM (Spatio-temporal distribution Emission Model) [5], and the PNAM (Parallel Naples Airshed Model) air quality model [6]. Our grid enabling approach also integrates marine-related environmental models, such as the POM (Princeton Ocean Model) [7], and the WW3 (WaveWatch III) sea-wave propagation model [8]. We enhanced the computational capabilities of the POM model, developing a parallel version with nesting capabilities (POMpn) [9]. Moreover, we recently integrated the WRF (Weather and Research Forecasting model) [10] and the CAMx (Comprehensive Air quality Model with eXtensions) air quality models [11], while we are working on the integration of the sea-wave propagation model SWAN [12].

The grid-enabled version of each model is based on three files: the model package, the launching script and the RSL job description file.

We configured and packaged each model, in order to be independent on the software and hardware configuration of the local machine. A framework approach was used to abstract different model configuration, by exploiting an object oriented programming-like methodology. We standardized the model packing/unpacking, configuring and setup defining which methods, implemented as shell scripts or Java class code, have to be called to perform operations as namelist placeholder, processor configuration and packing and unpacking of input/output data. Each model runs in a private custom environment, so that several instances of the same software from the same or different user can be concurrently executed. This approach is based on a repository where model packages are stored and from where they can be retrieved as instance template at each run.

A job launcher, invoked as a grid job on a remote machine set up the virtual private environment doing all needed data file stage in and stage out operations. In this way all implementation details are hidden and a Resource Specification Language (RSL) file can be used to describe each grid operation to the globusrun-ws job submitter. The launching script is deployed with a stage in file transfer operation on the target machine and represents the job executable file to be run on the remote machine. This scripts unpack the model package eventually downloaded from a repository or copied from a local directory, unpack and inflate the input data, run the model and the pack and deflate results. The script communicates with the job submitter via the standard output and the standard error.

A RSL (Resource Specification Language) file [13] describes the job to be submitted in a very detailed way, specifying the executable path, the current working directory, the files to be staged in before the execution and staged out after the job run, and any additional argument. Job submission is managed through the Grid Resource Allocation Manager (GRAM) tool. All files are named using URL, with protocol details from the target machine point of view and specifying the gridFTP high performance parallel striping transfer protocol when referring to a remote machine.

In the RSL evolution from the Globus Toolkit version 3 to version 4 some operations were simplified, making the RSL less verbose and more expressive; on the other hand, some features, as the automatic management of scratch directories, disappeared, so that we implemented a custom RSL pre-processor for the easy and straightforward definition of jobs, introducing an advanced method of labeling and placeholders parsing and evaluation, macro-based code explosion and late binding capabilities.

The Globus Toolkit 4.x grid middleware provides job submission tools via web-services and pre-webservices infrastructure without any kind of support for job flow scheduling and resource broking, while different grid technologies, such as the Condor [14] and Unicore [15] middleware offer a full support of direct acyclic graph job workflow with conditional branches, recovery features and graphical user interfaces. Our custom software solution was developed with the aim to provide domain scientists of a full configurable, really straightforward grid computing tool minimizing the impact of the grid infrastructure.

As in many grid applications, the final result is obtained by assembling different components executed as jobs on remote machines. Each component could be related to its previous/next component as data producer or data consumer, defining the so-called computational pipeline in which we have one job for one component. For example, consider this simple application: a regional-scale atmospheric model waits until data can be downloaded from a specified service, then acquires the boundary and initial conditions from a global-scale forecast, runs for a specified time period; when data are ready to be processed, another job, encapsulating an ocean circulation model, uses the atmospheric data as boundary conditions; when this second job finishes, the produced data is consumed by another job simulating the wind-driven sea wave propagation and forecasts the wave height/period and direction fields. At last, the user retrieves all pipeline outputs produced by all models. This simple grid application can be implemented via shell scripts and RSL files specifying the target submitting machine in the script itself.

This approach, though operatively correct, presents many disadvantages. The user needs to know the details about the script programming language, the job submission technical details related to a specific middleware and the to system environment setup. The developed code is tightly coupled to its application: any change to the job behavior or model configuration affects the entire application. In case of complex job fluxes, like in a concurrent ramification context, for example when the weather simulation model forces both wave propagation and oceanic circulation models, control code grows in complexity and data consuming/production relationships could be hard to implement, since synchronization issues may arise. Moreover, this kind of approach is potentially insecure because the user must be logged-in to the system to run a script, and this scenario is not applicable in the case of an interactive application on web portal.

In order to enhance the flexibility and to minimize the impact on the grid configuration, we implemented a custom job flow scheduler (JFS). Using this tool the entire complex, multi branch, grid application could be configured through a XML file. The JFS takes care of submitting jobs to computing nodes. JFS integrates itself in the Globus Toolkit environment both as a web service and a command tool with very few configuration needs. It uses a customized version of the Job Description Language (JDL), developed under the Condor project. In this way, every job is described through its RSL file and built in a XML file, which describes the activation order and relationships between jobs. The description language implemented has been defined as Job Flow Definition Language (JFDL) with a suitable XML schema. The following JFDL file implements the coupled use of the MM5/WW3 models:

```
<jfdl:jfs project="experiment01">
  <!-- Job definition -->
  <jfdl:jobs>
    <jfdl:job name="downloadConditions"
         target="dgric.uniparthenope.it"
         rsl="downloadConditions.rsl"/>

    <jfdl:job name="runMM5"
         target="dgbeobi.uniparthenope.it"
         rsl="runMM5.rsl"/>
```

```
    <jfdl:job name="runWW3"
        target="dgbeobe.uniparthenope.it"
        rsl="runWW3.rsl"/>
  </jfdl:jobs>

  <!-- Job Relationship Definition -->
  <jfdl:nodes>
    <jfdl:node job="downloadConditions">
      <jfdl:next>runMM5</jfdl:next>
    </jfdl:node>
    <jfdl:node job="runMM5">
      <jfdl:prev>downloadConditions</jfdl:prev>
      <jfdl:next>runWW3</jfdl:next>
    </jfdl:node>
    <jfdl:node job="runWW3">
      <jfdl:prev>runMM5</jfdl:prev>
    </jfdl:node>
  </jfdl:nodes>
</jfdl:jfs>
```

The file describing the experiment could be divided in two parts: inside the element ⟨jfdl:jobs⟩ each job belonging to the grid application is described specifying its symbolic name, the computing node where it will be submitted and the name of the RSL file specifying all needed resources. Inside the element ⟨jfdl:nodes⟩ the jobs activation order is described using a direct acyclic graph. In this section, each job node is characterized by the reference to all previous jobs, the ⟨jfdl:prev⟩ element, by the way the jobs that have to be finished before the start of the current job, and by a reference to all next jobs which will be submitted after the current job finishes using the ⟨jfdl:next⟩ element.

The described experiment is a typical example of a simple virtual laboratory grid application, but our JFS could submit very complex application graphs, thanks to its Java multithread implementation (Fig. 1).

The Jobflow Scheduler was implemented using the Java language using a class framework encapsulating all described features including XML file parsing based on the StaX [16] package, graph setup and application runtime support. The most interesting class is Job derived from Thread, implementing the job submission in its run method using a clear, effective and efficient algorithm: if the job is to be started, make a join to each thread-related jobs using the previously defined dependence graph. In this way the thread waits until all prerequisite data are successfully produced. Then the job is submitted to the grid using the globusrun-ws service specifying the target factory and the job RSL file. The class Job is an item of the collection Jobs composing the JobFlow class, providing methods for graph setup, management and run. The Jobs run method starts all jobs belonging to the collection with no previous job dependence. For example, more data providers have to download initialization data to feed a consuming job.

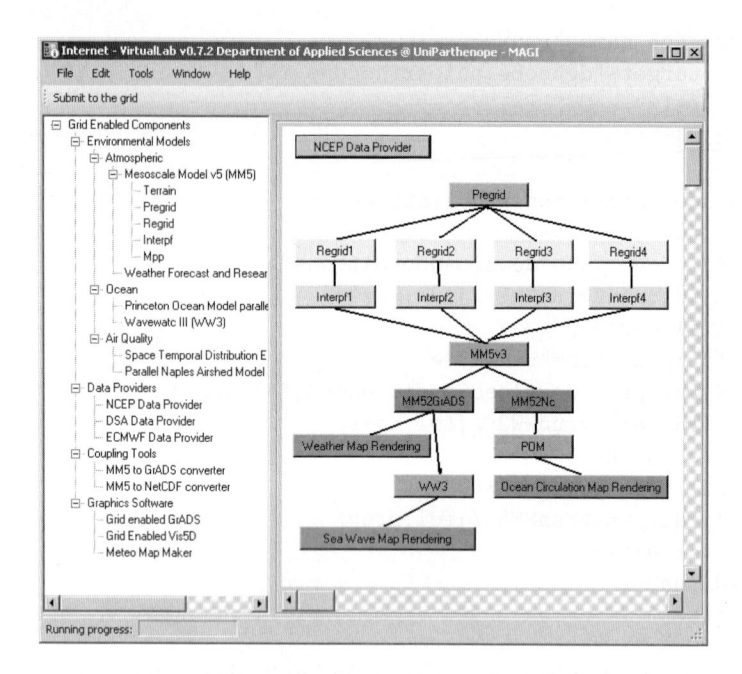

Fig. 1. A GUI for interactive JFDL files editing with direct grid interfacing capabilities via MyProxy

The described grid application is classifiable as a grid-enabled application because it uses the grid to submit a job to the best computing node, but this association is statically performed at the design time. On the other hand a grid-aware application could be adapted in relation to the grid status using a Resource Broker [17] component, designed to submit a job to the best fitting node, based on needed computational and storage requirements. Our JFS automatically activates this feature if no target machine is specified in the ⟨jfdl:job⟩ element. The Resource Broker algorithm is straightforwardly configurable, changing the behavior of the implementation class in a properties file.

3 Laboratory Components

The Jobflow scheduler and the Resource Broker implement the core of the grid based virtual laboratory. The domain scientist can configure and run his experiments using the JFDL and RSL files, or through a web portal, or an under development Java user interface, selecting and assembling each component from a palette.

Actually our virtual laboratory provides several grid components for data acquisition: the NCEPDataProvider performs the data download from the NOAA-NCEP [18] for the initialization of the meteorological model, thanks to a daemon component, completely decoupled from the grid; the ECMWFDataProvider [19] performs the on-demand download of historical data for scenario and "what if" analysis; the DSADataProvider performs the on-demand download of processed data.

Numerical models are grouped in atmospheric circulation models, such as the gWRF and the gMM5 suites, whose components (gTERRAIN, gPREGRID, gREGRID, gINTERPF and gMPP) have been ported to our grid environment; air quality related models as gSTdEM, gPNAM, gCAMx and ocean related models as the gPOMpn, gWW3, gSWAN. We provided our virtual laboratory with a suite of tools for model coupling, data conversion, classification and graphics rendering software. Thanks to our packaged framework for grid enabling legacy software components, adding more grid components is straightforward.

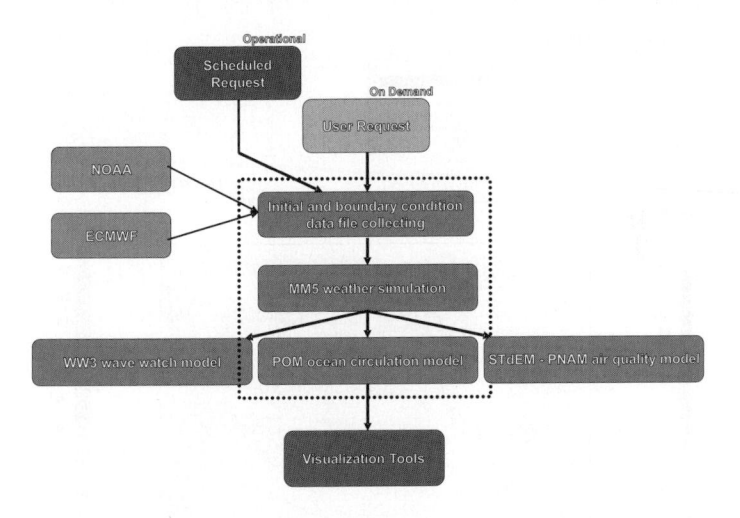

Fig. 2. The grid application building blocks

We used the Jobflow Scheduler (JS) and the Resource Broker (RB) to develop a grid application aiming at producing weather and marine forecasts in both operational and on demand mode, by coupling several simulation models, data acquisition, conversion, and visualization software (Fig. 2). The application workflow is easy to understand: the starting event is produced by the on demand user request, or by the availability of initial data in the case of an operational production environment. Then, the weather forecast model is initialized, and the output data is rendered by a presentation software and concurrently consumed by other models, as ocean dynamics, sea wave propagation or air quality models. Each application branch proceeds on separate thread. This workflow could be represented by an acyclic direct graph into a JFDL file, while each job to be submitted is described by the RSL file and its launching script. Our JS permits the implementation using a single XML self describing file, while the RB makes grid-aware the application with any kind of constrain and without the need to use a storage element as intermediate files repository because of our late binding reference approach. This application run in operational mode with a few maintenance operations, except components or grid middleware upgrades. All performed results are interactively published at the Department web portal and used by several scientists, local institutions and citizens [20] (Fig. 3).

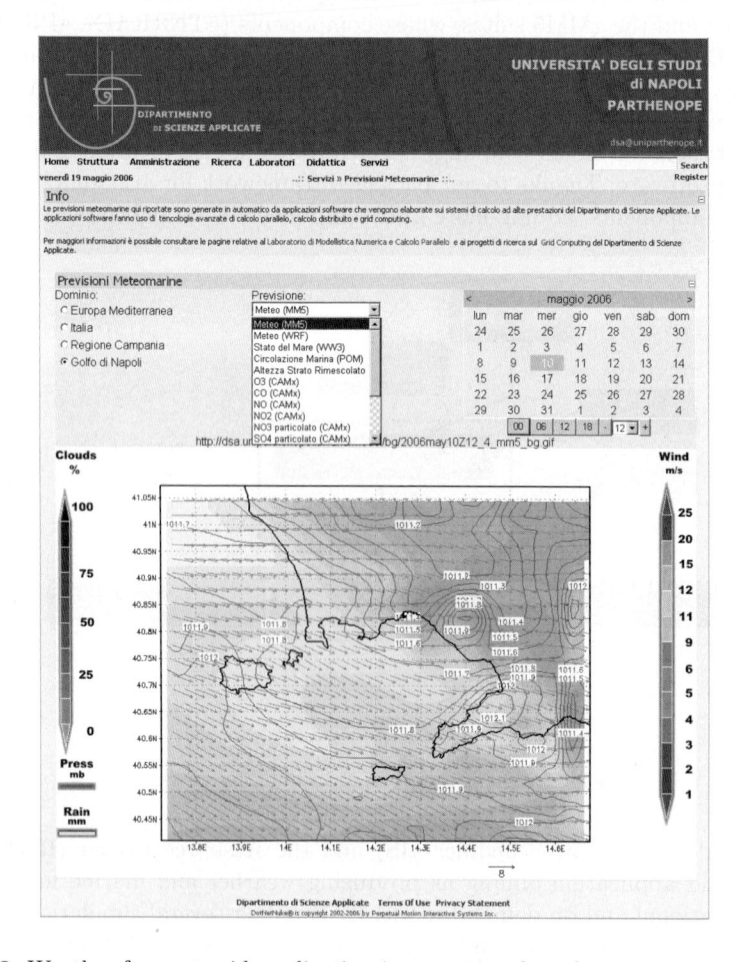

Fig. 3. Weather forecast grid application in operational mode: an output example

4 Conclusions and Future Development

In this paper we described some of our results in the field of grid computing research. The virtual laboratory for earth observation and computational environmental sciences based on the grid computing technology is a tool used both for research and application-oriented uses, running a complex grid application dedicated to operational weather, marine and air quality forecasts on nested domains from the Mediterranean Europe to the Bay of Naples area. Comparison tests between a grid and non-grid implementation, performed using a simple benchmark weather forecast application, affected by networking capabilities, demonstrates that with the number of simulated hours increasing from 72 to 144 the efficiency of the grid implementation rise with clear evidence.

The JS and the RB realized the primary goal of our research providing the power of the computing grids and the high performance computing with the simplicity and the flexibility of a local XML configurable application demonstrating the grid technology features.

The Globus Tooklit middeware version 4 is stable enough to perform production activities in the range of our needs, but some points have to be improved. The JS engine works very well, but it have to be enhanced offering more expression power to the JFDL especially regarding conditional branches and resume features. The RB algorithm have to be well tested and improved.

References

1. Lin, S., Atlas, R., Yeh K.: Global weather prediction and high-end computing at Nasa. Computing in Science & Engineering, **6** (2004)
2. Foster, I., Kesselman, C., Tuecke, S.: The Anatomy of the Grid: Enabling Scalable Virtual Organizations. International Journal of High Performance Computing Applications, **15** (2001) – 200-222
3. The Globus Toolkit, The Globus Alliance, http://www.globus.org
4. Michalakes, J., Canfield, T., Nanjundiah, R., Hammond S., Grell, G.: Parallel Implementation, Validation, and Performance of MM5. Parallel Supercomputing in Atmospheric Science. World Scientific, River Edge, NJ 07661 (1994)
5. Barone, G., D'Ambra, P., di Serafino, D., Giunta, G., Montella, R., Murli, A., Riccio, A.: An Operational Mesoscale Air Quality Model for the Campania Region. Annali Istituto Universitario Navale (2000) – 179-189
6. Barone, G., D'Ambra, P., di Serafino, D., Giunta, G., Murli, A., Riccio, A.: Parallel software for air quality simulation in Naples area. J. Environ. Manag. & Health (2000) 209–215
7. Blumberg, A.F., Mellor, G. L.: A description of a three-dimensional coastal ocean circulation model. Three-Dimensional Coastal ocean Models, edited by N. Heaps, American Geophysical Union. (1987)
8. Tolman. H.L.: A third-generation model for wind waves on slowly varying, unsteady and inhomogeneous depths and currents. J. Phys. Oceanogr. , **21** (1991) 782–797
9. Giunta, G., Montella, R., Mariani P., Riccio, A.: pPOM: A nested, scalable, parallel and Fortran 90 implementation of the Princeton Ocean Model, accepted by Environmental Modelling & Software
10. Michalakes, J., Dudhia, J., Gill, D., Henderson, T., Klemp, J., Skamarock, W., Wang, W.: The Weather Research and Forecast Model: Software Architecture and Performance. 11th ECMWF Workshop on the Use of High Performance Computing in Meteorology. 25–29 October 2004, Reading, U.K.
11. CAMx Comprehensive Air Quality Model with eXtensions. Version 4.20. ENVIRON International Corporation (2005)
12. Booij, N., Holthuijsen, L.H., Ris, R.C.: The SWAN wave model for shallow water. Int. Conf. on Coastal Engineering., Orlando, USA (1996) 668–676
13. Resource Specification Language (RSL). http://globus.org/toolkit/docs/4.0/execution/wsgram/

14. Thain, D., Tannenbaum, T., Livny, M.: Distributed Computing in Practice: The Condor Experience. Concurrency and Computation: Practice and Experience, **17** (2005) 323–356
15. Streit, A., Erwin, D., Lippert, T., Mallmann, D., Menday, R., Rambadt, M., Riedel, M., Romberg, M., Schuller, B., Wieder, P.: UNICORE – from Project Results to Production Grids
16. Stream API for XML, http://dev2dev.bea.com/xml/stax.html
17. Mavilio, C., Montella, R.: A resource broking algorithm for grid computing applications, Technical Report 2005/11, Dept. of Applied Sciences – University of Naples "Parthenope"
18. National Centre for Environmental Prediction. http://www.ncep.noaa.gov
19. European Centre for Medium-Range Weather Forecasts. http://www.ecmwf.int
20. Giunta, G., Montella, R., Mariani P., Riccio, A.: Modeling and computational issues for air/water quality problems: A grid computing approach. Il Nuovo Cimento, **28** 2005

Exploiting Throughput for Pipeline Execution in Streaming Image Processing Applications*

F. Guirado[1], A. Ripoll[2], C. Roig[1], A. Hernàndez[3], and E. Luque[2]

[1] Universitat de Lleida
f.guirado@diei.udl.es, roig@diei.udl.es
[2] Univ. Autònoma de Barcelona
ana.ripoll@uab.es, emilio.luque@uab.es
[3] Computer Vision Center. Univ. Autònoma de Barcelona
aura@cvc.uab.es

Abstract. There is a large range of image processing applications that act on an input sequence of image frames that are continuously received. Throughput is a key performance measure to be optimized when executing them. In this paper we propose a new task replication methodology for optimizing throughput for an image processing application in the field of medicine. The results show that by applying the proposed methodology we are able to achieve the desired throughput in all cases, in such a way that the input frames can be processed at any given rate.

1 Introduction

There is a large range of emerging applications in which data generated in a given external environment is pushed asynchronously to servers that process this information. These applications are characterized by the need to process different instances of an input data stream in a timely and responsive fashion. Hereafter, we refer to such applications as streaming applications [1] [2] [3] [4].

There are two distinct criteria for judging the quality of an execution for these streaming applications: latency and throughput. Latency is the time taken to process individual data, while throughput is the aggregate rate at which the instances of the input data stream are processed. Throughput can also be measured in terms of its inverse, the Iteration Period (IP), which corresponds to the interval of time existing between the execution of two consecutive instances of data.

In this paper, we deal with image-processing applications executing in a streaming manner. These applications are typically composed of a set of computation stages performing different functions. The usual computations in the stages of these kinds of applications are blurring, filtering, interpolation, etc. As they are independent functions, they can be arranged in a set of consecutive stages that at a given time can be simultaneously processing different image

---

* This work was supported by the MEyC-Spain under contract TIN 2004-03388.

frames in pipeline fashion. The sequence of input frames is continuously received. Thus, throughput is a key performance measure to be optimized in these executions [2] [5].

In this context, we address the problem of maximizing the throughput of an image processing application in the field of medicine. The application under study is devoted to the detection of the real arterial structure from a sequence of Intra-Vascular Ultrasound (IVUS) images, captured by a transducer at a specific rate. Real-time constraints must be met in order for images to be processed at the same rate as that at which they are captured [6].

We first define the task model of the IVUS application by capturing its salient computational features. Based on this model, we propose a methodology that performs an innovative task replication technique for those tasks that can process independent input frames in such a way that a given throughput can be achieved. Then, the replicated application is executed in a simulation framework using a task mapping mechanism that considers its iterative behaviour.

We show through experimentation that the task replication technique allows us to reach the given throughput constraint in all cases. Additionally, we show the effectiveness of the whole strategy of replication and mapping in the optimization of processor utilization, as well as the speedup that is achieved.

The rest of the paper is organized as follows. Section 2 exposes the main characteristics and the steps that are performed in the IVUS imaging application. Section 3 describes the proposed methodology of task replication to exploit throughput. Section 4 outlines the main contributions of the literature in relation with the optimization problem that is undertaken in this paper. Section 5 shows the experimentation results that are obtained for the application under study. Finally, Section 6 outlines the main conclusions.

2 The IVUS Imaging Application

This is a study of an image processing application, in the field of medicine, for the detection of the real arterial structure (called adventitia) [7]. The input of the application is a sequence of IVUS frames that are captured by a radio-frequency transducer installed in a catheter. The captured data are sent out to be processed and then converted to images.

Tissue characterization is a fundamental tool for studying and diagnosing the pathologies and lesions associated to the vascular tree. This is an arduous job that requires specialists to manually identify the tissues and visualise them. IVUS imaging is a highly suitable visualization technique for the task as it provides a cross-section of the coronary vessel, revealing its histological properties and tissue organization [6].

As it is so time consuming and due to the subjectivity of the classification depending on the specialist, there is an increasing interest among the medical community in using automatic tissue characterization procedures. These automatic

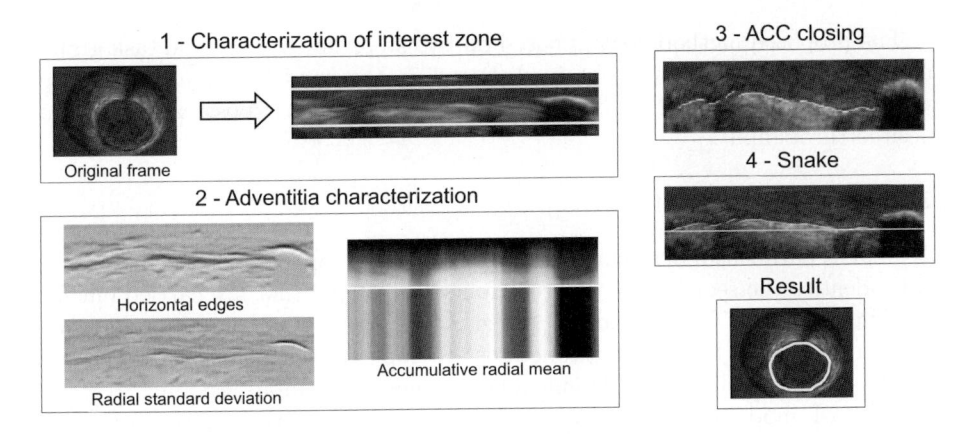

Fig. 1. Stages of the IVUS imaging application

procedures are time-critical, and consequently should provide answers in a minimum time. Figure 1 shows the main stages of the process that are explained below.

1. *Characterization of the interest zone.* Because the original image of adventitia is circular, it is transformed to polar coordinates. In this coordinate system, the adventitia appears as a dark horizontal line. By means of a diffusion method, the image is de-noised and the target structure is enhanced. Then, the band of interest is determined in order to reduce computational cost.
2. *Adventitia characterization.* The three following filters are applied to the image: horizontal edges, radial standard deviation and mean accumulative radial. The three filtered images provide the necessary information for discriminating between four different sets: adventitia, calcium, fibrous structures and the remaining pixels.
3. *Anisotropic contour closing (ACC).* The previous step characterizes the adventitia with a collection of fragmented curve segments. These segments are interpolated using ACC to join them.
4. *B-Snake.* Since the above interpolation process still presents gaps in side branches and calcium sectors, a parametric B-snake is used on the ACC closure in order to close it and obtain a compact and explicit representation. Finally, the identified adventitia is returned to cartesian coordinates to visualize its original circular shape.

3 The Optimization Problem

One of the key problems that arise when executing image-processing applications that act on an input sequence of image frames is having enough throughput to permit their processing at a given rate. We address the optimization of the throughput of these applications by considering the definition of computation stages that can run in pipeline.

The proposed methodology is based on two steps: in the first, the task graph model of the application is obtained. Based on this model, in the second step we apply the convenient replication of tasks that makes it possible to reach a desired throughput.

3.1 The Task Model

To exploit parallelism in streaming image processing applications, the main issue is to identify the sequence of different functions (steps) that are carried out for each input frame. Then, a parallel design of the application can be undertaken in such a way that the different functions are implemented as tasks that can run overlapped in pipeline for different image frames of the input stream.

The task model that we extracted for the IVUS imaging application is composed of 12 tasks that can be modeled using the directed acyclic graph (DAG) structure, $G(V,E)$, illustrated in Figure 2(a). The graph is composed of a set of nodes V, each node representing a task. Each task $Ti \in V$ has an associated computation time $\mu(Ti)$. E is the set of arcs representing task precedences. Each arc $(Ti,Tj) \in E$ has an associated communication volume $c(Ti,Tj)$, in bytes, to be transferred between tasks.

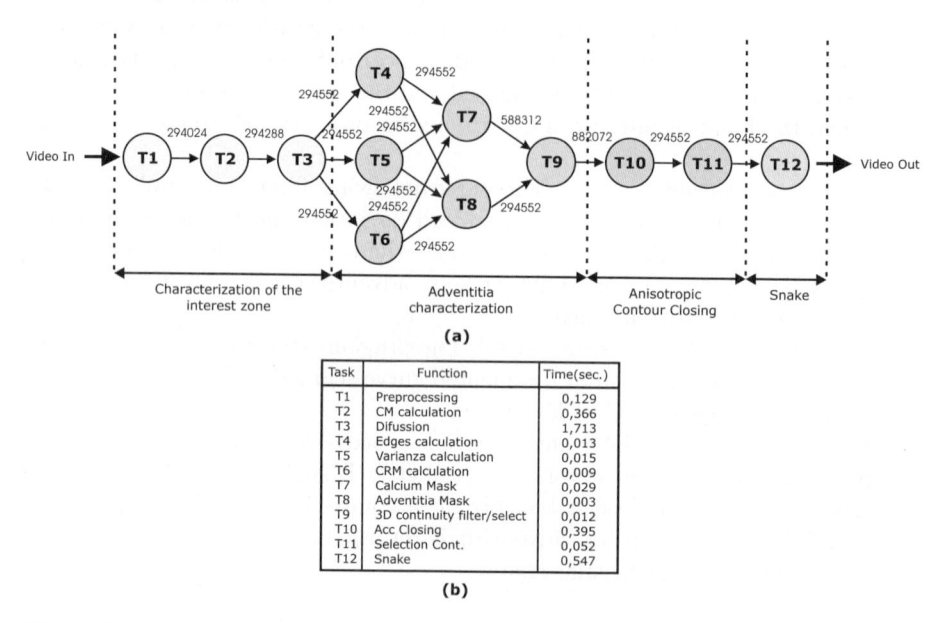

Fig. 2. (a) Task graph model of IVUS imaging application. (b) Functionality and computation time of each task.

Each step of the application can have several tasks performing different functions as indicated in the graph. Figure 2(b) shows the different functions that were identified and their correspondence with the tasks in the graph, together with the task execution time, in seconds. To obtain these task execution times we profiled the execution of the different functions that conform a task in the IVUS sequential algorithm. We also computed the data structures that are shared between functions to determine the communication volume of data that has to be transferred between tasks. The sequential application was programmed in Matlab and executed on a Pentium IV processor at 3GHz with 512 Mb of RAM running windows.

3.2 Task Replication Method

In this subsection, we expose the methodology that is proposed in this paper to achieve a given throughput for streaming applications running in pipeline. We assume that the instances of the input stream have no temporal dependencies between themselves, as is the case with IVUS imaging. Thus, different image frames can be processed concurrently in the same step with replicated tasks. Consequently, throughput is improved, since multiple images are processed in parallel. We propose this methodology instead the typical data parallel approach at the application level, in order to exploit the task parallel capacity without penalizing latency.

Given an IP to be achieved, the replication problem consists of determining the tasks that should be replicated and the number of replications for each. It is established in the literature that the optimum IP in a pipeline execution is given by the maximum computation time $\mu(Ti)$, considering that communications are performed concurrently [4][8]. We enhance this model by taking into account the fact that on several platforms communications cannot be overlapped. Thus, the optimum IP is also influenced by the global amount of communications that are transferred from one task. With these considerations in mind, we propose a replication methodology that consists of the two following steps:

1. *Determine the tasks to be replicated*
   All the tasks are evaluated to decide whether they have to be replicated or not. Each specific task Ti∈V with a communication to Tj∈V, will be replicated if it has accomplished one of the following two conditions: (a) The computation time of Ti is greater than IP or, (b) the communication time to transfer volume c(Ti,Tj) is greater than IP. Taking these conditions for replication into account, the algorithm shown in Figure 4 determines the tasks to be replicated, joined in subgraphs (replicable subgraphs). Starting from the DAG graph of the application, for each task Ti∈V it calls the recursive procedure *group(Ti)*, which returns the set of successor-replicable tasks that form the replicable subgraph to which Ti belongs.

   From the identified subgraph, it ascertains whether there is an intersection with the previously found subgraphs that are stored in *subgraph_set*. Should the subgraph have a task in common with another, these are joined into a single subgraph. Then, the subgraph of replicable tasks is the chosen entity to which replication will be applied in the next step.

2. *Calculate the number of replications*
   In this step the number of copies is calculated for each replicable subgraph and the number of copies for each task inside the subgraph. This provides

us with the most appropriate number of replications for each task, instead of replicating the same number of copies for all the tasks, which, in some cases, would be an excessive replication.

To illustrate this step consider the example of Figure 3(a) where we can see a replicable subgraph with tasks Ti and Tj, along with 2 and 4 replications respectively. Figure 3(b) shows the result that would be obtained if the tasks were replicated individually, which leads to a large amount of dependencies. This could lead to an excessive overhead when communication from the same tasks is serialized as we have considered. Figure 3(c) illustrates the replication result applied on the subgraph level as proposed.

Figure 5 shows the algorithm to be applied to each replicable subgraph, which proceeds as follows. For each task Ti in the subgraph, the corresponding number of replications is calculated as the maximum between computation and communication, divided by the given IP. To calculate the number of replications of the graph we identify its initial tasks. Among these initial tasks, we chose the one with the lowest number of replications assigned. This determines the number of the replications of the whole subgraph.

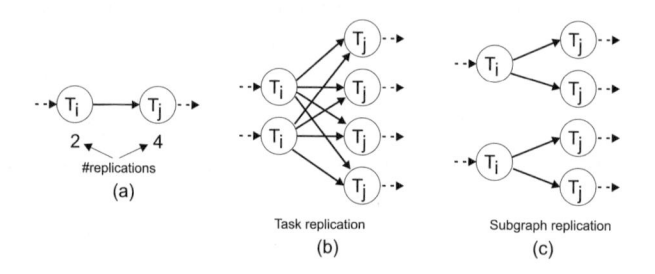

Fig. 3. (a) Replicable subgraph. (b) Result of replication if it was applied on a task level. (c) Result of replication with our method.

```
subgraph_set=∅                          function group(Ti)
non_evaluated_tasks={Ti; Ti∈V}           adjacent_set=∅
for each Ti ∈ non_evaluated_tasks        non_evaluated_tasks=non_evaluated_tasks-{Ti}
  subgraph=group(Ti)                     if is_replicable(Ti)
  if subgraph ≠ ∅                          adjacent_set={Ti}
    for each H ∈ subgraph_set              for each task successor task Tj ∈ V and
      if H ∩ subgraph ≠ ∅ then              is_replicable(Tj)
        subgraph = subgraph ⋃ H              adjacent_set=adjacent_set ⋃ group(Tj)
    end_for                                end_for
    subgraph_set=                        end_if
    =subgraph_set ⋃ subgraph             return adjacent_set
  end_if                                end_function group
end_for
```

Fig. 4. Algorithm for identifying replicable subgraphs of tasks

```
function replication
    for each subgraph ∈ subgraph_set
        for each Ti ∈ subgraph
            number_replications[Ti]=⌈ max(μ(Ti),max(∀Tj successor_of Ti comm(Ti→Tj))) / iteration_period ⌉
        end_for
        Tinit = initial task of subgraph with lowest number_replications[Ti]
        number_replications_subgraph = number_replications[Tinit]
        for each Ti ∈ subgraph
            number_replications[Ti]= ⌈ number_replications[Ti] / number_replications[Tinit] ⌉
        end_for
    end_for
end_function replication
```

Fig. 5. Algorithm for determining the number of replications of the subgraphs and their internal tasks

Table 1 shows the development of the replication method when it is applied to the task graph of the IVUS application for a desired throughput of 214 ms. In this case, we identify three replicable subgraphs (two with a single task). Figure 6 graphically shows the result of this replication method in IVUS application.

Table 1. Development of the replication method for IVUS application

Rep.subgraphs	n_rep.[Ti]	n_rep. inside subgraph	n_rep_subgraph
T2 → T3	T2:2	2/2=1	2
	T3:8	8/2=4	
T10	T10:2	2/2=1	2
T12	T12:3	3/3=1	3

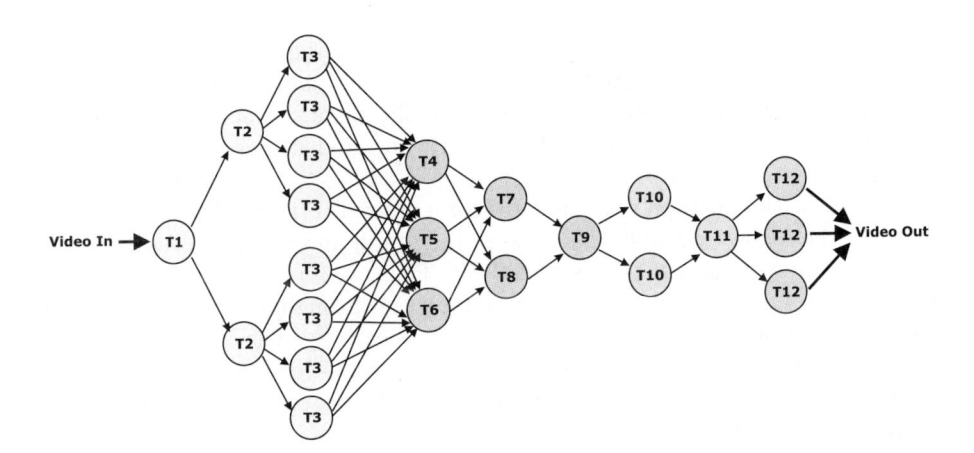

Fig. 6. Ivus replicated graph

4 Related Work

The optimization of throughput in streaming applications running in pipeline has been undertaken by several proposals in the literature. Some authors provide generic solutions to exploit throughput under different constraints without considering replication. Hoang and Rabey in [2] propose an algorithm that solves a resource-optimization problem and maximizes throughput. Hoang and Rabey's proposal was later improved by Yang et al in [3]. This starts the task assignment with the ETF algorithm that is based on a classic DAG heuristic [9]. From the obtained assignment, it processed additional reassignment steps in order to exploit the iterative behaviour of applications. In all these approaches, the maximum throughput is given by the maximum computation time of the tasks in the application, which also indicates the minimum IP achievable.

In order to improve throughput there are approaches that introduce the concept of replication to their techniques, dividing the input frame into several parts and applying data parallelism to each [4] [8]. Lee et al in [5] apply a replication technique where the tasks of the same stage have to be identical. Unlike these previous works, our approach considers the possibility of replication on the task level for applications with arbitrary task structure and without constraints on the kind of tasks within each stage. The replicated tasks perform the same computation for different frames of the input stream. This facilitates the applicability of the replication method as the code of the task does not need to be modified and provides a more feasible solution for improving throughput.

5 Experimentation Results

In this section, we conducted an experiment to show both the applicability and the benefits provided by the proposed approach of task replication when used to execute the IVUS imaging application in a cluster environment.

From the obtained IVUS task graph model, we executed the application in the simulation framework pMAP [10], which simulates the execution of message-passing applications in distributed systems. The underlying system was modelled by defining a set of homogeneous nodes with the same characteristics as those used in the sequential execution. The network was modelled as a Gigabit Ethernet.

The replication method was evaluated for a desired IP that is based on $\mu(T3)$, which is the highest computation time in the graph and consequently indicates the maximum throughput that is achievable. Thus, we replicated the application tasks using our methodology in order to increase throughput by 2, 3, 4, 8 and 16 times, indicated as x2, x3, x4, x8 and x16 respectively. The tasks of the replicated graph were assigned to the processors using a specific mapping mechanism for pipeline applications [11] that makes it possible to exploit throughput and optimize the number of processors.

Figure 7 shows the throughput, in frames/second, that was obtained for the application using our methodology, compared with the maximum throughput that can be theoretically obtained for the different number of replications. As can be observed, we obtained significant similarities for both values. Thus, the replication method is able to achieve the given throughput constraints for the IVUS imaging application.

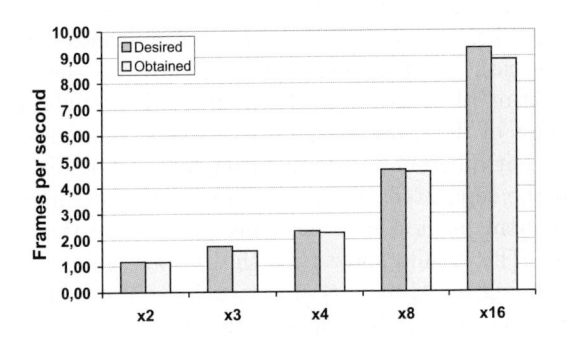

Fig. 7. Throughput

As the number of processors that are used is increased due to task replication, we evaluated the processor utilization in the executions. For each experiment, Figure 8(a) shows the number of required processors with the corresponding average utilization. As can be observed, the worst case is an average utilization of 68% when 4 replications were applied. In all the remaining cases, the percentage utilization is greater that 70%. Thus, the mapping mechanism applied after replication is able to optimize resource utilization.

Finally, we evaluated the speedup in order to analyse the influence of the increase in the number of tasks and processors due to replications. As shown in Figure 8(b), the obtained speedup has the same tendency as the maximum, and the difference between both becomes more significant only when 36 processors were used in the x16 experiment.

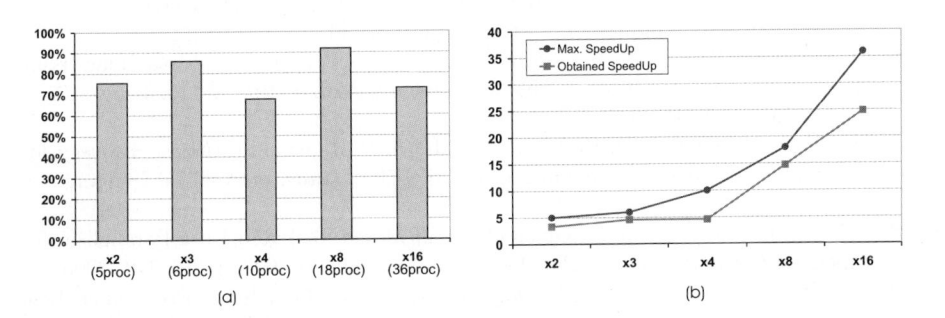

Fig. 8. (a) CPU utilization. (b) Speedup.

6 Conclusions

Optimization of throughput is a key performance measure for optimization in image processing applications that act on an input sequence of image frames. In this work, we addressed the exploitation of throughput based on a real application, IVUS imaging, in the field of medicine.

We have proposed a task replication methodology that consists of two steps: (a) obtaining the task graph model of the application and, (b) applying the convenient replication of tasks to enable the desired throughput.

The effectiveness of the proposed approach was evaluated for the IVUS imaging application. For different values of throughput to be obtained, we applied replication, and the replicated application was executed through simulation in a cluster environment using a task mapping mechanism that considers its iterative behaviour. The results show that the proposed replication method followed by the mapping mechanism is able to achieve the desired throughput for the application under study while maintaining good utilization of resources.

References

1. M. Lee, W. Liu and V. K. Prasanna: A Mapping Methodology for Designing Software Task Pipelines for Embedded Signal Processing. Proc. Workshop on Embedded HPC Systems and Applications of IPPS/SPDP 1998. Pp. 937-944.
2. P. Hoang and J. Rabey: Scheduling of DSP Programs onto Multiprocessors for Maximum Throughput. IEEE Trans. Signal Processing. Vol. 41. N. 6. Pp. 2225-2235. June 1993.
3. M-T. Yang, R. Kasturi and A. Sivasubramaniam: A Pipeline-Based Approach for Scheduling Video Processing Algorithms on NOW. IEEE Trans. on Par. and Distr. Systems. Vol. 14. N. 2. Pp. 119-130. Feb. 2003.
4. A. Choudhary, W. K. Liao, D. Weiner, P. Varshwey, R. Linderman and M. Linderman: Design Implementation and Evaluation of Parallel Pipelined STAP on Parallel Computers. Proc. 12th Int. Parallel Processing Symposium. Florida. Pp. 220-225. April 1998.
5. M. Lee, W. Liu and V. K. Prasanna: Parallel Implementation of a Class of Adaptive Signal Processing Applications. Algorithmica. N. 30. Pp. 645-684. 2001.
6. O. Pujol and P. Radeva: Supervised texture classification for intravascular tissue characterization. Handbook of Medical Imaging. Vol. 2. Pp. 57-110. Kluwer Academic/Plenum Pub. 2005.
7. D. Gil, A. Hernàndez, O. Rodríguez, J. Mauri, P. Radeva: Statistical Strategy for Anisotropic Adventitia Modelling in IVUS. IEEE Trans. on Medical Imaging. (in press)
8. Subhlok J. and Vongran G.: Optimal Use of Mixed Task and Data Parallelism for Pipelined Computations. J. Par. Distr. Computing. Vol. 60. Pp 297-319. 2000.
9. J.J. Hwang, Y. C. Chow, F. D. Anger and C. Y. Lee: Scheduling Precedence Task Graphs in Systems with Interprocessor Communication Times. SIAM J. Comp. Vol. 18. N. 2. Pp. 244-257. April 1989.

10. F. Guirado, A. Ripoll, C. Roig and E. Luque: Performance Prediction Using an Application Oriented Mapping Tool. IEEE Proc. Euromicro Conf. on Parallel, Distributed and Network-based Processing. (PDP) Pp. 184-191. Feb. 2004.
11. F. Guirado, A. Ripoll, C. Roig and E. Luque: Exploitation of Parallelism for Applications with an Input Data Stream: Optimal Resource-Throughput Tradeoffs. IEEE Proc. Euromicro Conf. on Parallel, Distributed and Network-based Processing. (PDP) Pp. 170-178. Feb. 2005.

dCache, Storage System for the Future

Patrick Fuhrmann and Volker Gülzow*

Deutsches Elektronen Synchrotron
Notkestrasse 85, 22607 Hamburg

Abstract. In 2007, the most challenging high energy physics experiment ever, the *Large Hardon Collider(LHC)*, at CERN, will produce a sustained stream of data in the order of 300MB/sec, equivalent to a stack of CDs as high as the Eiffel Tower once per week. This data is, while produced, distributed and persistently stored at several dozens of sites around the world, building the LHC data grid. The destination sites are expected to provide the necessary middle-ware, so called Storage Elements, offering standard protocols to receive the data and to store it at the site specific Storage Systems. A major player in the set of Storage Elements is the *dCache/SRM* system. dCache/SRM has proven to be capable of managing the storage and exchange of several hundreds of terabytes of data, transparently distributed among dozens of disk storage nodes. One of the key design features of the dCache is that although the location and multiplicity of the data is autonomously determined by the system, based on configuration, cpu load and disk space, the name space is uniquely represented within a single file system tree. The system has shown to significantly improve the efficiency of connected tape storage systems, by caching, 'gather & flush' and scheduled staging techniques. Furthermore, it optimizes the throughput to and from data clients as well as smoothing the load of the connected disk storage nodes by dynamically replicating datasets on the detection of load hot spots. The system is tolerant against failures of its data servers which enables administrators to go for commodity disk storage components. Access to the data is provided by various standard protocols. Furthermore the software is coming with an implementation of the *Storage Resource Manager* protocol (SRM), which is evolving to an open standard for grid middleware to communicate with site specific storage fabrics.

1 Contributors

dCache/SRM is a joined effort between the Deutsches Elektronen-Synchrotron[1] in Hamburg and the Fermi National Accelerator Laboratory[2] near Chicago with significant distributions and support from the University of California, San Diego, INFN, Bari as well as from the GridPP people at Rutherford Appleton Laboratory, UK[4] and CERN[3].

---

* For the dCache team.

2 The LHC Computing Grid and the Storage Element

The worlds largest installation of a High Energy Physics Particle accelerator, using superconducting magnets, is the *Large Hardron Collider*[5] at CERN, next to Geneva in Switzerland. A 27 Km tunnel holds two ring pipes equipped with supercooling magnets, accelerating bunches of protons to nearly the speed of light and letting them collide at an energy of 14 TeV. This is the highest energy achieved ever by any accelerator in the world as well as the most intense beam. At four locations within the ring structure, huge detectors are placed, detecting particles produced during beam collisions. Knowing that those collisions will happen at a rate of 800 million times a second and that one bunch crossing may produce up to 20 physical events, it becomes clear that computer science faces the challenge of processing and storing data two orders of magnitude larger than they did for known physics experiments. This in mind, the *LHC Computing Grid Group, LCG*[6] was formed, targeting computing challenges common to all LHC experiments. Although other computing patterns may have solved the upcoming challenges as well, a Tier approach had been chosen. Within this design, the raw data source, namely CERN, builds the Tier 0 centre surrounded by only very few Tier 1 centres per country. Those, in turn, deliver data to some dozens of Tier 2 centers. Most of those centres have already been in place far before agreeing to join the LCG Tier tree and consequently are running their own compute farms and storage fabrics. So, sufficiently flexible interfaces to compute and storage systems had to be defined, allowing interoperability of the tier tree without forcing the local sites to change their existing software stack. In LCG terms, the abstraction of a storage system is called a Storage Element, SE if it complies with a certain set of interfaces allowing interoperability with the LCG middle ware [32].

3 Technical Overview

The intention of this publication is to describe features, behaviour and applications of a storage middleware system, called the dCache/SRM[12][10][27][31].

The core part of the dCache has proven to combine heterogenous disk storage systems in the order of several hundred tera bytes and let its data repository appear under a single filesystem tree. It takes care of data hot spots, failing hardware and makes sure, if configured, that at least a minimum number of copies of each dataset resides within the system to ensure full data availability in case of disk server maintenance or failure. Furthermore, dCache supports a large set of standard access protocols to the data repository and its namespace.

If dCache is connected to a Tertiary Storage System, it optimizes the access to such a system by various technics. Currently Enstore[7], the Open Storage Manager (OSM), the High Performance Storage System (HPSS) and the Tivoli Storage Manager (TSM)[9][29] are supported by the dCache middleware.

Moreover, dCache/SRM supports all interfaces of the LCG storage element definition.

4 Technical Specification

4.1 File Name Space and Dataset Location

dCache strictly separates the filename space[33][24] of its data repository from the actual physical location of the datasets. The filename space is internally managed by a database and interfaced to the user resp. to the application process by the nfs2[16] protocol and through the various ftp filename operations. The location of a particular file may be on one or more dCache data servers as well as within the repository of an external Tertiary Storage Manager. dCache transparently handles all necessary data transfers between nodes and optionally between the external Storage Manager and the cache itself. Inter dCache transfers may be caused by configuration or load balancing constrains. As long as a file is transient, all dCache client operations to the dataset are suspended and resumed as soon as the file is fully available.

4.2 Maintenance and Fault Tolerance

As a result of the name space and data separation, dCache data server nodes, subsequently denoted as pools, can be added at any time without interfering with system operation. Having a Tertiary Storage System attached, or having the system configured to hold multiple copies of each dataset, data nodes can even be shut down at any time. In both setups, the dCache system is extremely tolerant against failures of its data server nodes.

4.3 Data Access Methods

In order to access dataset contents, dCache provides a native protocol (dCap), supporting regular file access functionality. The software package includes a c-language client implementation of this protocol offering the posix *open, read, write, seek, stat, close* as well as the standard filesystem name space operations. This library may be linked against the client application or may be preloaded to overwrite the file system I/O. The library supports pluggable security mechanisms where the GssApi (Kerberos) and ssl security protocols are already implemented. Additionally, it performs all necessary actions to survive a network or pool node failure. It is available for Solaris, Linux, Irix64 and windows. Furthermore, it allows to open files using an URL like syntax without having the dCache nfs file system mounted. In addition to this native access, various FTP dialects[27] are supported, e.g. GssFtp (kerberos)[15] and GsiFtp (GridFtp)[14]. An interface definition is provided, allowing other protocols to be implemented as well.

4.4 Tertiary Storage Manager Connection

Although dCache may be operated stand alone, it can also be connected to one or more Tertiary Storage Systems. In order to interact with such a system, a dCache external procedure must be provided to store data into and retrieve data from the corresponding store. A single dCache instance may talk to as many storage systems as required. The cache provides standard methods to optimize access to those systems. Whenever a dataset is requested and cannot be found on one of the dCache pools, the cache sends a request to the connected Tape Storage Systems and retrieves the file from there. If done so, the file is made available to the requesting client. To select a pool for staging a file, the cache considers configuration information as well as pool load, available space and a *Least Recently Used* algorithms to free space for the incoming data. Data, written into the cache by clients, is collected and, depending on configuration, flushed into the connected tape system based on a timer or on the maximum number of bytes stored, or both. The incoming data is sorted, so that only data is flushed which will go to the same tape or tape set. Mechanisms are provided that allow giving hints to the cache system about which file will be needed in the near future. The cache will do its best to stage the particular file before it's requested for transfer. Space management is internally handled by the dCache itself. Files which have their origin on a connected tape storage system will be removed from cache, based on a Least Recently Used algorithm, if space is running short. Less frequently used files are removed only when new space is needed. In order to allow site administrators to tune dCache according to their local tape storage system or their migration and retrieval rules, dCache provides an open API to centrally steer all interactions with Tertiary Storage Systems.

4.5 Pool Attraction Model

Though dCache distributes datasets autonomously among its data nodes, preferences may be configured. As input, those rules can take the data flow direction, the subdirectory location within the dCache file system, storage information of the connected Storage Systems as well as the IP number of the requesting client and the data transfer protocol, the client is able to support. The cache defines data flow direction as getting the file from a client, delivering a file to a client and fetching a file from the Tertiary Storage System. The simplest setup would direct incoming data to data pools with highly reliable disk systems, collect it and flush it to the Tape Storage System when needed. Those pools could e.g. not be allowed to retrieve data from the Tertiary Storage System as well as deliver data to the clients. The commodity pools on the other hand would only handle data fetched from the Storage System and delivered to the clients because they would never hold the original copy and therefore a disk resp. node failure wouldn't do any harm to the cache. Extended setups may include the network topology to select an appropriate pool node. Those rules result in a matrix of pools from which the load balancing module, described below, may choose the most appropriate candidate. The final decision, which pool to select out of this set, is based on free space, age of file and node load considerations.

4.6 Load Balancing and Pool to Pool Transfers

The load balancing module is, as described above, the second step in the pool selection process. This module keeps itself updated on the number of active data transfers and the age of the least recently used file for each pool. Based on this set of information, the most appropriate pool is chosen. This mechanism is efficient even if requests are arriving in bunches. In other words, as a new request comes in, the scheduler already knows about the overall state change of the whole system triggered by the previous request though this state change might not even have fully evolved. System administrators may decide to make pools with unused files more attractive than pools with only a small number of movers, or some combination. Starting at a certain load, pools can be configured to transfer datasets to other, less loaded pools, to smooth out the overall load pattern. At a certain point, pools may even refetch a file from the Tertiary Storage System rather than an other pool, assuming that all pools, holding the requested dataset are too busy. Regulations are in place to suppress chaotic pool to pool transfer orgies in case the global load is steadily increasing. Furthermore, the maximum numbers of replica of the same file can be defined to avoid having the same set of files on each node.

4.7 File Replica Manager

The Replica Manager Module[26] enforces that at least N copies of each file, distributed over different pool nodes, must exist within the system, but never more than M copies. This approach allows to shut down servers without affecting system availability or to overcome node or disk failures. The administration interface allows to announce a scheduled node shut down to the Replica Manager so that it can adjust the N ¡ M interval prior to the shutdown.

4.8 Data Grid Functionality

In order to comply with the definitions of a LCG Storage Element the storage fabric must provide the following interfaces :

There must be a protocol for locally accessing data. dCache provides this by nfs mounting a server for file name operations but transferring the actual data via faster channels. Local Storage Elements, including dCache, hide this mechanism by being integrated into a local filesystem wrapper software provided by CERN, the *Grid File Access Layer, GFAL*[20].

A secure wide-are transfer protocol must be implemented which, at the time being, is agreed to be GsiFtp, a secure Ftp dialect. Furthermore dCache offers kerberos based FTP as well as regular and secure http access.

To allow central services to select an appropriate Storage Element for file copy or file transfer requests, each Storage Element has to provide sufficient information about its status. This includes its availability as well as its total and available space. Currently this information is provided via the ldap protocol but this, for scalability reasons, is in process of being redesigned. In order to be independend of the actually distribution mechanism, dCache provides an interface to the *Generic Information Provider, GIP*. GIP[23] is responsible to make this information available to the connected grid middle ware.

The forth area, defining a LCG Storage Element, is a protocol which makes a storage area a manageable. The interface is called the *Storage Resource Manager, SRM*[10]. Beside name space operations, it allows to prepare datasets for transfers directly to the client or to initiate third party transfers between Storage Elements. SRM takes care that transfers are retried in case they didn't succeed and handles space reservation and management. In addition, it protects storage systems and data transfer channels from being overloaded by scheduling transfers appropriately. The SRM doesn't do the transfer by itself, instead it allows to negotiate transfer protocols available by the data exchanging parties.

5 Performance Considerations and Future Plans

The core design of dCache has been avoiding central components to be involved in data transfers. Therefor, because of the fact that CPU speed is increasing faster than disk system access speeds or even network transfer speeds, dCache data mover components are always limited by either the performance of the underlying RAID system or by the network components. Consequently dCache data transfer performance turned out to be as good as the hardware it's build upon. This is different for name space operations and the initial open time for datasets. These tasks are processed within central components. Further evaluation on dCache systems beyond 100 TByes of disk space and a frequency of opening files above 10 Hz let us believe that the file system name space simulation software builds the actual bottleneck. To overcome this limitation, the name space module has been revised and will be replaced by Chimera[33][24], a fully database based system specially tuned for this kind of access. Chimera is currently in the extended testing phase.

6 Dissemination

At the time of this publication, dCache is in production at various locations in Europe and the US. The largest installation is, to our knowledge, the CDF system at FERMI [2]. More than 150 Tbytes are stored on commodity disk systems and in the order of 50 Tbytes have been delivered to about 1000 clients daily for more than a year. FERMI dCache installations are typically connected to ENSTORE[7], the FERMI tape storage system. CDF is operating more than 10 tape-less dCache installations outside of FERMI, evaluating the dCache Replica Manager. The US devision of the LHC CMS[19] experiment is using the dCache as Grid Storage Element and large file store in the US and Europe. At DESY, dCache is connected to the Open Storage Manager (OSM) and serving data out of 100 Tbytes of disk space. The German LHC Grid Tier 1 center in Karlruhe (GridKa,[18]) is in the process of building a dCache installation as Grid Storage Element, connected to their Tivoli Storage Manager[9] installation. End of

2005 and beginning of 2006 the majority of sites participating in the LCG data challenges have been transferring and storing their data under the control of dCache/SRM storage elements.

Furthermore dCache is a component of the german D-Grid[21][25] e-science initiative.

References

1. DESY : http://www.desy.de
2. FERMI : http://www.fnal.gov
3. CERN : http://www.cern.ch
4. Rutherford Appleton Laboratory : http://www.cclrc.ac.uk/
5. Large Hadron Collider : http://lhc.web.cern.ch/lhc/
6. LHC Computing Grid : http://lcg.web.cern.ch/LCG/
7. Fermi Enstore http://www.fnal.gov/docs/products/enstore/
8. High Performance Storage System : http://www.hpss-collaboration.org/hpss/
9. Tivoli Storage Manager : http://www-306.ibm.com/software/tivoli/products/storage-mgr/
10. SRM : http://sdm.lbl.gov/srm-wg
11. CASTOR Storage Manager : http://castor.web.cern.ch/castor/
12. dCache Documentation : http://www.dcache.org
13. dCache, the Book : http://www.dcache.org/manuals/Book
14. GsiFtp http://www.globus.org/ datagrid/deliverables/gsiftp-tools.html
15. Secure Ftp : http://www.ietf.org/rfc/rfc2228.txt
16. NFS2 : http://www.ietf.org/rfc/rfc1094.txt
17. Fermi CDF Experiment : http://www-cdf.fnal.gov
18. GridKA : http://www.gridka.de/
19. Cern CMS Experiment : http://cmsinfo.cern.ch
20. Grid GFAL http://lcg.web.cern.ch/LCG/peb/GTA/GTA-ES/Grid-File-AccessDesign-v1.0.doc
21. D-Grid, The German e-science program : http://www.d-grid.de
22. Patrick Fuhrmann et al. dCache, the Upgrade. Spring 2006, CHEP06, Mumbai, India
23. Lawrence Field et al. Grid Deployment Experiences: The path to a production quality LDAP based grid information system. Spring 2006, CHEP06, Mumbai, India
24. Tigran Mkrtchyan et al. Chimera. Spring 2006, CHEP06, Mumbai, India
25. Lars Schley, Martin Radicke et al. A Computational and Data Scheduling Architecture for HEP Application. Spring 2006, CHEP06, Mumbai, India
26. Alex KULYAVTSEV et al. Resilient dCache: Replicating Files for Integrity and Availability Spring 2006, CHEP06, Mumbai, India
27. Timur Perelmutov et al. Enabling Grid features in dCache Spring 2006, CHEP06, Mumbai, India
28. Abhishek Sinh Rana et al. gPLAZMA : Introducing RBAC Security in dCache Spring 2006, CHEP06, Mumbai, India
29. Patrick Fuhrmann et al. The TSM in the LHC Grid World Sep 2005, TSM Symposium , Oxford, UK
30. Patrick Fuhrmann, dCache, the commodity cache. Spring 2004, Twelfth NASA Goddard and Twenty First IEEE Conference on Mass Storage Systems and Technologies. Washington DC, USA

31. Timur Perelmutov, Storage Resource Managers by CMS,LCG. Spring 2004, Twelfth NASA Goddard and Twenty First IEEE Conference on Mass Storage Systems and Technologies. Washington DC
32. Michael Ernst et al. Managed Data Storage and Data Access Services for Data Grids. Sep 2004, CHEP04, Interlaken, Switzerland
33. Tigran Mkrtchyan et al. Chimera, the commodity namespace service. Sep 2004, CHEP04, Interlaken, Switzerland
34. Patrick Fuhrmann et al. dCache, LCG SE and enhanced use cases. Sep 2004, CHEP04, Interlaken, Switzerland
35. Michael Ernst, Patrick Fuhrmann et al. dCache. March 2003, CHEP03, San Diego, USA
36. Patrick Fuhrmann et al. dCache. Sep 2001, CHEP01, Bejing, China

Computing the Diameter of 17-Pancake Graph Using a PC Cluster

Shogo Asai, Yuusuke Kounoike, Yuji Shinano, and Keiichi Kaneko

Tokyo University of Agriculture and Technology, Tokyo 184-8588, Japan
asai@al.cs.tuat.ac.jp, {kounoike, yshinano, k1kaneko}@cc.tuat.ac.jp
http://opt.cs.tuat.ac.jp/

Abstract. An n-pancake graph is a graph whose vertices are the permutations of n symbols and each pair of vertices are connected with an edge if and only if the corresponding permutations can be transitive by a prefix reversal. Since the n-pancake graph has $n!$ vertices, it is known to be a hard problem to compute its diameter by using an algorithm with the polynomial order of the number of vertices. Fundamental approaches of the diameter computation have been proposed. However, the computation of the diameter of 15-pancake graph has been the limit in practice. In order to compute the diameters of the larger pancake graphs, it is indispensable to establish a sustainable parallel system with enough scalability. Therefore, in this study, we have proposed an improved algorithm to compute the diameter and have developed a sustainable parallel system with the Condor/MW framework, and computed the diameters of 16- and 17-pancake graphs by using PC clusters.

1 Introduction

In this paper, let us consider a problem in which a stack of pancakes whose sizes are completely different is rearranged so that the pancakes form a pile where the sizes of pancakes increase from the top to the bottom. As operations of rearrangement, reversing several pancakes from the top of the stack is possible. The problem to obtain the largest number of operations to rearrange the worst-case stack of n pancakes as a function of n is called the pancake sorting problem[1]. This problem is also called the prefix reversal problem.

A pancake graph is a graph whose vertices are the permutations of n symbols from 1 to n and its edges are given between permutations transitive by prefix reversals. Since the graph topology is dependent on n, it is called an n-pancake graph. An n-pancake graph is a regular graph that has $n!$ vertices and its degree is $n-1$. The pancake sorting problem and the problem to obtain the diameter of the pancake graph is equivalent. Since the pancake graphs have many merits such as the symmetric and recursive structures, and the small degrees and diameters against the sizes, much attention is paid to them as a model of interconnection networks for parallel computers[2,3,4]. When we regard the pancake graphs as the model of the interconnection networks, the diameter of the graph is a measure that represents the delay of communication[5,6].

Table 1. The diameters of n-pancake graphs

n	1	2	3	4	5	6	7	8	9	10	11	12	13	14	15
Diameters	0	1	3	4	5	7	8	9	10	11	13	14	15	16	17

To obtain the diameter of an n-pancake graph, it is sufficient to obtain the shortest distances from one vertex to all the vertices. However, the algorithms that depend on the numbers of vertices and/or edges cannot solve the problem practically because the computational time and the memory space increase exponentially. Hence, Kounoike et al.[7] proposed a method that restricts the number of vertices for which the shortest distances must be obtained by taking advantage of the recursive structure of the pancake graphs. This method is based on the method by Heydari et al.[8] to obtain the diameter of the 13-pancake graph and is extended not to execute the unnecessary search. Kounoike has applied the method to give the diameters of 14- and 15-pancake graphs that were unknown so far. Table 1 shows the known diameters of the pancake graphs. Some attentions are paid to the sequence of diameters mathematically, and the sequence up to $n = 13$ is listed in the 'On-Line Encyclopedia of Integer Sequences'[9] as 'Sorting by prefix reversal.' However, no sequence for $n \geq 14$ is listed there. Hence, obtaining the diameters of the larger pancake graphs also contributes the study of the sequences.

In this study, we have improved the method by Kounoike et al. when they obtained the diameter of 15-pancake graph so that it computes the diameters of the larger pancake graphs and implemented it as a parallel computing system. In addition, we made use of the implemented system to obtain the diameters of 16- and 17-pancake graphs that have been unknown.

2 Definitions of Terminology and Symbols

In this section, we define the terminology and symbols used in this paper. Refer [7] for the detailed explanations.

Let S_n be the set of all the permutations of n symbols from 1 to n, and let the symbols 1 to n correspond to the smallest size of pancake to the largest one. Then assume that a permutation $\pi \in S_n$ which is obtained by arranging the symbols from the top pancake to the bottom pancake represents a stack of n pancakes. Let e_n be the permutation $(1, 2, \ldots, n)$ that corresponds to the sorted stack. Let $\sigma \in S_n$ be a permutation that is obtained by reversing the preceding k $(2 \leq k \leq n)$ symbols in $\pi \in S_n$. Then the transformation from the permutation π to the permutation σ is called the prefix reversal of k symbols for the permutation π, and it is denoted $\pi^k = \sigma$. Since we use only the prefix reversals of permutations in this paper, we mention reversals to mean the prefix ones. The successive reversals $(\pi^{x_1})^{x_2}$ of a permutation π are also denoted $\pi^{(x_1,x_2)}$. Moreover, if $x = (x_1, x_2, \ldots, x_m)$ then let π^x represent a successive reversals with $x_1, x_2, \ldots, x_m$ symbols. If $\pi^x = e_n$ then x is called a sorting sequence of π. For a given permutation $\pi \in S_n$, let the function $f(\pi) = \min\{|x| : \pi^x = e_n\}$

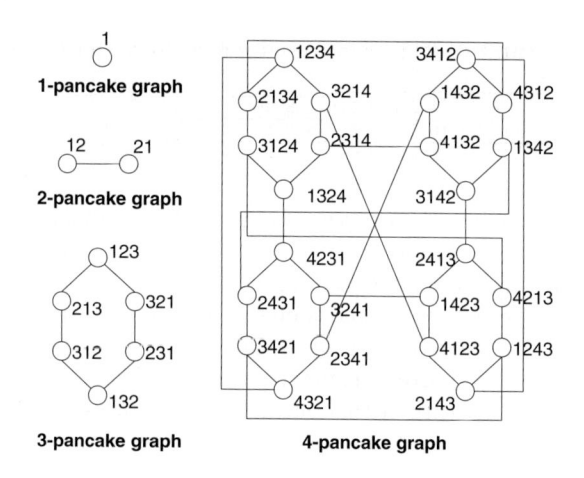

Fig. 1. The pancake graphs

represent the smallest number of reversals to sort the permutation. In addition, let the function $f(n) = \max\{f(\pi) : \pi \in S_n\}$ represent the largest number of reversals to sort the stacks of n pancakes. Conventionally, the same letter f is used for the functions. Note that the meaning of f depends on its argument.

A pancake graph is a graph whose vertices are $\pi \in S_n$, and whose edges are between vertices π and σ where $\sigma = \pi^k$. Since pancake graphs are different depending on n, each pancake graph is called n-pancake graph and denoted by P_n. Figure 1 shows P_1 to P_4. In general, between two vertices in a graph, the path that has the smallest number of edges is called the shortest path between the two vertices, and the number of edges included in the path is called the shortest distance. For arbitrary pair of two vertices in a graph, the longest shortest distance is called the diameter of the graph. By selecting e_n as one of the pair of vertices to which we compute the shortest distance, computing the diameter of an n-pancake graph is equivalent to computing $f(n)$. In this paper, the shortest distance between a vertex $\pi \in S_n$ and the vertex e_n is simply mentioned the distance of π.

3 Basic Method

We took the method by Kounoike et al.[7] by which they obtained $f(15)$ as the basic method to obtain the diameters. The method obtains the dependency between vertices based on the symmetric and recursive properties of pancake graphs and restricts the vertices whose distance computation is necessary.

First, for a permutation $\pi = e_{n-1}^x \in S_{n-1}$, we define a permutation $\sigma_k \in S_n$ $(1 \le k \le n)$ by expression (1). Then, for $f(\sigma_k)$, expression (2) holds.

$$\sigma_k = \begin{cases} ((e_n)^n)^x & k = 1 \\ ((e_n)^{(k,n)})^x & 2 \le k \le n-1 \\ e_n^x & k = n \end{cases} \quad (1)$$

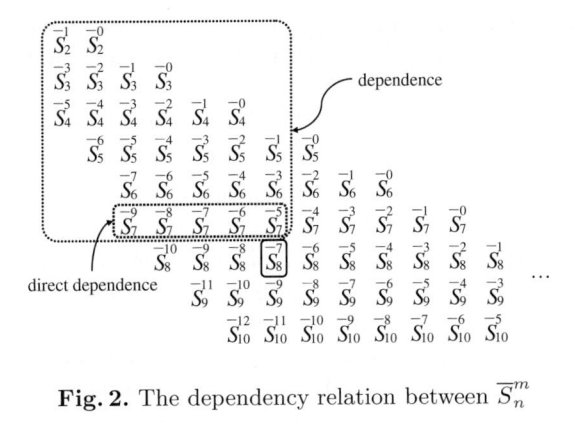

Fig. 2. The dependency relation between $\overline{S}_n^m$

$$f(\sigma_k) \leq \begin{cases} f(\pi) + 1 & k = 1 \\ f(\pi) + 2 & 2 \leq k \leq n-1 \\ f(\pi) & k = n \end{cases} \quad (2)$$

For $\pi \in S_{n-1}$, let $T_k(\pi)$ be the transformation that obtains $\sigma_k \in S_n$, and let $T_k(S)$ be a set of $T_k(\pi)$ for all the elements of the set $S \subseteq S_{n-1}$. In addition, let S_n^m be a set of $\pi \in S_n$ such that $f(\pi) = m$ where S_n^k be empty for k such that $k < 0$ or $k > f(n)$. Then define the set $\overline{S}_n^m$ by expression (3).

$$\overline{S}_n^m = T_1(S_{n-1}^{m-1}) \cup T_2(S_{n-1}^{m-2}) \cup \cdots \cup T_{n-1}(S_{n-1}^{m-2}) \cup T_n(S_{n-1}^m) \quad (3)$$

This is the set of vertices in S_n whose upper bounds are equal to m. Then, from expression (2), $f(\pi) \leq m$ holds for $\pi \in \overline{S}_n^m$. The following relation holds among $\overline{S}_n^m$, S_n^m and S_n:

$$S_n = \bigcup_{k=0}^{f(n-1)+2} \overline{S}_n^k, \quad (4)$$

$$S_n^m \subseteq \bigcup_{k=m}^{f(n-1)+2} \overline{S}_n^k. \quad (5)$$

From expression (4), we can see that

$$f(n) \leq f(n-1) + 2. \quad (6)$$

In Figure 2, a set depends on the sets that are just above or upper left of it, and the lower left and upper right blank parts represent empty sets. We cannot judge if the below part of a set is empty or not until its diameter is computed. Based on the relationship, we can obtain an arbitrary $\overline{S}_n^m$ by repeating transformation and distance computation recursively from $S_1 = S_1^0 = \{e_1\}$. To obtain the diameter $f(n)$, we first obtain $f(\pi)$ for all of $\pi \in \overline{S}_n^{f(n-1)+2}$. Then, depending on the existence of π that satisfies $f(\pi) = f(n-1) + 2$, $f(n)$ is classified as follows:

- In case that π which satisfies $f(\pi) = f(n-1)+2$ exists: $f(n) = f(n-1)+2$ holds. We can finish computation just after such π is found (See expression (6)).
- Otherwise: $f(n) \leq f(n-1)+1$ holds. We can finish computation with the result $f(\pi) = f(n-1)+1$ by showing π which satisfies the equation.

For searching shortest paths, we use A* algorithm. Refer [7] to see the detail of the algorithm.

The implementation by Kounoike et al. fixes the elements of the sets in Figure 2 from the leftmost column by performing transformation and distance computation. The diameters are also obtained in the process. This search method makes it possible to skip the searches of vertices that are known to be unnecessary for diameter computation based on dependency among the sets. However, as the size of the pancake graph increases, the number of elements in the sets becomes very large, and we cannot manage the pancake graph only with the main memory. Their implementation stores all the results of distance computation for later use. The results increase exponentially, and it occupies 21GB of the disk as a compressed file after the computation of $f(14)$. Hence, their implementation has the limitation for diameter computation of the larger pancake graphs.

4 Our New Implementation

In the previous implementation, it is impossible to compute the diameter of the larger pancake graphs because of memory restriction. Hence, we changed the searching method to decrease the number of nodes drastically during the search process. In addition, by devising the representation of each node, we decreased the amount of the memory used. Moreover, we proved that distance computation is unnecessary in some cases and accelerated the search.

4.1 Depth-First Search

If we consider the process of computing diameters the tree search, the search method in the previous implementation corresponds to the breadth-first search inside a specific column in Figure 2. If we can replace it with the depth-first search, much memory space can be saved. However, the simple depth-first search will also search the vertices that have no relation to diameter computation.

Then, we used a method in which the vertices are judged if they have relation to diameter computation or not by using the incumbent diameter value. For a vertex π, if its upper bound value u is known, to judge if the vertex can be discarded or not, it is necessary to know to which column the vertex belongs in Figure 2. Then, let the number obtained by the following expression of $n = |\pi|$ and u be the column number in the figure of dependency relationship.

$$column = n \times 2 - u \tag{7}$$

If the column number calculated by substituting u with the incumbent diameter is less than or equal to the column number of the vertex which we are focusing on, we can discard the vertex.

If we perform the depth-first search by using this method, while the incumbent diameter is smaller than the true diameter, our implementation may search some vertices that are not searched by the previous implementation. However, once the incumbent diameter becomes equal to the true diameter, this situation never occurs. Empirically, we can easily find the vertices that attains $f(n-1)+1$ during $f(n)$ computation. Hence, this method is efficient enough.

4.2 Elimination of Unnecessary Distance Computations

Up to now, we have computed the distance of the transformation even if it does not increase the upper bound, that is, $\sigma_n = T_n(\pi)$ for $\pi = e_{n-1}^x$. This transformation generates a permutation obtained by just adding n at the final position of the permutation $\pi = e_{n-1}^x$. However, it looks impossible to sort this kind of permutations with less operations than $f(\pi)$. Then, we guessed that $f(\pi) = f(\sigma_n)$ and proved it. Hence, there is no need to compute the diameter for $f(\sigma_n)$. By using this, we can improve A* search. By applying this improvement, we could accelerate the program by 5 to 8%.

Proof of $f(\pi) = f(\sigma_n)$. Let $\pi = e_{n-1}^x$ and $\sigma_n = T_n(\pi)$, respectively. In general, to sort the permutation σ_n, it is necessary to execute multiple prefix reversals. Here, we abstract the operation sequence necessary to sort and denote it with an operation sequence $\boldsymbol{X}$.

First, we assume that there exists an operation sequence $\boldsymbol{X}$ for which $|\boldsymbol{X}| < f(\pi)$ holds. Then, let $\boldsymbol{Y} = (y_1, y_2, ..., y_m)$ be the operation sequence where y_i obtained by transforming each element x_i in $\boldsymbol{X} = (x_1, x_2, ..., x_m)$ as follows:

$$y_i = \begin{cases} x_i & x_i < n_{pos} \\ x_i - 1 & x_i \geq n_{pos} \end{cases} \quad (8)$$

where n_{pos} represents the position of n when the operations just before x_i are applied to σ_n.

Each y_i that is constructed by this transformation has the following features:

- In case that n is at the final position, it is just a reversal of no more than $n-1$ symbols.
- Order of the symbols except for n is same as that of the result of operation before transformation.

σ_n has n at the final position in the initial status. Therefore if we use $\boldsymbol{Y}$ instead of $\boldsymbol{X}$, we can sort π without performing the reversal of n symbols. In this case, the number of operations is $|\boldsymbol{Y}| = |\boldsymbol{X}|$. That is, if σ_n can be sorted by $|\boldsymbol{X}|$ operations, $f(\pi)$ can be also sorted by no more than $|\boldsymbol{X}|$ operations. This leads to contradiction. Hence, there does not exist $\boldsymbol{X}$ that satisfies $|\boldsymbol{X}| < f(\pi)$. From this, we can say $f(\pi) = f(\sigma_n)$.

Improvement of A* Search. In the part of the diameter computation by A* search, one vertex which attains the least estimated distance is taken from enumerated elements, and the estimated distances for all of its neighbor vertices are computed. However, if the permutation corresponding to the vertex has n in its final position, then from the proof above, we can see that the shortest distance is obtained without checking the vertex generated by the prefix reversal of n symbols. That is, we can see that there is a shortest path which does not include the vertex. Hence, in case that the final position has n, we can lessen the paths to be searched by ignoring the vertices obtained by the prefix reversal of n symbols. In addition, generalizing this idea, in case that the final part of the permutation is sorted, we can lessen more vertices to be searched by not operating them.

5 Parallelization

We have implemented the proposed system as a parallel system that works based on the Master-Worker method by using the MW framework[10]. By using MW, the number of Workers can be coped with automatically because Condor[11] performs the resource management. In addition, according to the function of MW, in case that some failures on the Worker side are detected, the executed tasks are migrated into normal Workers automatically.

Master fulfills the distribution of child problems and the collection of results, and Workers compute the given child problems. There is a variance among the sizes of child problems (the number of vertices for which distance computations are necessary) and the size of each child problem cannot be expected in advance. Therefore, if a Worker simply solves all of the child problems and returns the result, then it would be inefficient because the Worker that has finished its computation earlier must wait until the completion of computation of other Workers. Hence, we introduced a mechanism in which a Worker will suspend computation after a constant time and divide the suspended situation into multiple child problems. From this, we can maintain a constant number of tasks on the Master side all the time, and the Worker that has completed its computation can start its next computation immediately.

In addition, in parallel execution, we conducted a benchmark task in the initialization process of each Worker to measure the power of the machine on which the Worker runs. As the benchmark task, we selected the computation of $f(15)$. After a minute has passed, computation of the benchmark task on the Worker is stopped and we regard the number of vertices searched per second as the benchmark value of the Worker. Based on this value, we can estimate the execution time when we use other machines.

6 Computations of the Diameters of P_{16} and P_{17}

By using the implemented system, we actually computed the diameters of 16- and 17-pancake graphs. In both cases, we set both of the parameters in execution, the check pointing interval and the interval of the interruption of Worker computation to be 10 minutes. We also set the number of child problems which

Table 2. PC clusters configurations

Computation	Master/Worker	CPU	Memory	No. PCs	Connection
$f(16)$	Master	Pentium2 400MHz	256MB	1	100BASE-TX
	Worker	Pentium3 1GHz dual	1GB	16	
		Pentium2 400MHz	256MB	17	
$f(17)$	Master	Opteron 1.8GHz dual	2GB	1	1000BASE-T
	Worker	Opteron 1.8GHz dual	2GB	107	

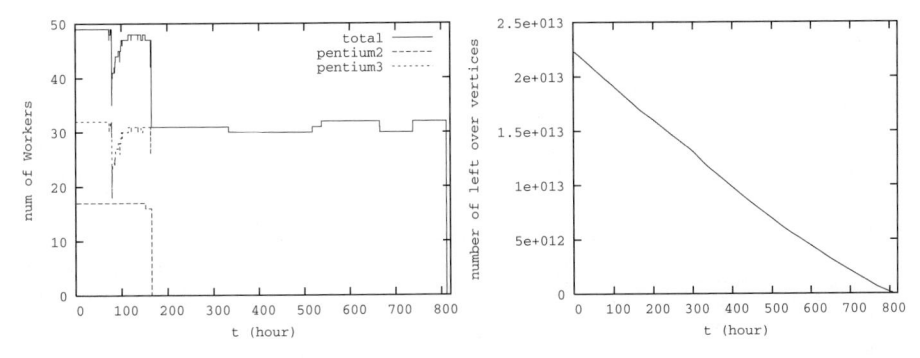

Fig. 3. Total number of Workers ($f(16)$) **Fig. 4.** Number of left over vertices ($f(16)$)

Master holds to be 1024. For these parameters, the optimal values are unknown. However, the values we set are proved to provide the sufficient performance based on preliminary experiments. Table 2 shows the PC clusters configurations that are used for the computations.

In computation of $f(16)$, by the computation during 33 days and 19 hours under the environment with 49 Workers at most, we have obtained the result $\boldsymbol{f(16) = 18}$. From expression (6), it is known that $f(16) \leq f(15) + 2 = 19$. Hence, we have checked that there is no vertex whose distance is 19. Figure 3 shows the change of the number of Workers in the process of computation. In this figure, the number of Workers of Pentium3 decreases rapidly around $t = 80$. This is because the MW framework found an ordinary user's job and a part of computation is automatically interrupted. In addition, around $t = 170$, Pentium2 machines are all stopped because of maintenance. Figure 4 shows the change of the number of remaining vertices in the process of the computation. Here, the number of vertices is the number in case all the vertices are assumed to be necessary for search. From this figure, we can see that the remaining vertices decrease almost linearly. Hence, we could find that the remaining computation time can be expected on the way of the computation process.

In computation of $f(17)$, by the computation during 38 days and 19 hours under the environment with 214 Workers at most, we have checked that there is no vertex whose distance is no less than 20 and obtained the result $\boldsymbol{f(17) = 19}$. Figure 5 shows the change of the number of Workers in the process of computation. In Figure 5, the number of Workers are rapidly increasing around $t = 90$,

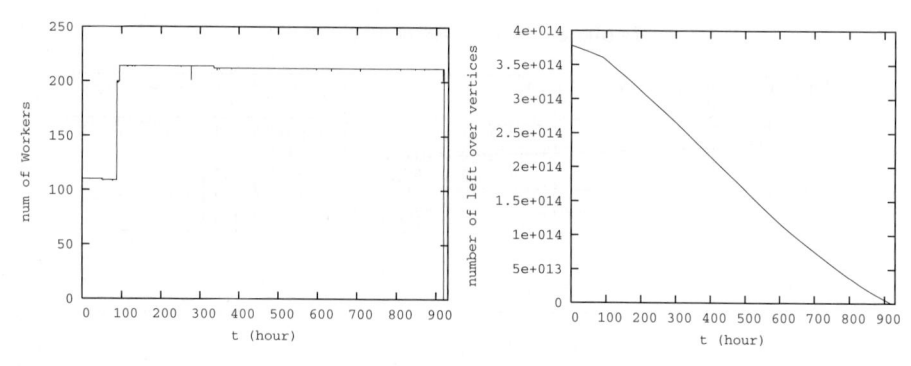

Fig. 5. Total number of Workers ($f(17)$) **Fig. 6.** Number of left over vertices ($f(17)$)

Table 3. Examples of the permutations that attain the diameters of P_{16} and P_{17}

n	Permutation	Sorting Sequence
16	(1, 15, 9, 11, 8, 10, 12, 7, 13, 5, 2, 16, 4, 14, 6, 3)	(10, 12, 16, 3, 5, 12, 3, 2, 4, 3, 5, 6, 8, 12, 3, 13, 15, 2)
	(6, 10, 4, 14, 2, 13, 16, 12, 8, 11, 7, 9, 5, 1, 3, 15)	(10, 8, 12, 5, 6, 2, 4, 14, 4, 15, 10, 2, 16, 15, 13, 2, 5, 3)
	(13, 9, 15, 2, 6, 4, 7, 11, 8, 12, 10, 14, 1, 16, 5, 3)	(11, 4, 3, 10, 6, 8, 9, 6, 13, 11, 14, 16, 3, 4, 2, 12, 14, 2)
17	(1, 4, 2, 7, 13, 3, 5, 17, 10, 15, 9, 14, 8, 12, 6, 16, 11)	(17, 8, 6, 10, 3, 8, 2, 12, 14, 3, 5, 8, 17, 2, 4, 3, 12, 6, 12)
	(7, 13, 2, 4, 1, 3, 5, 17, 10, 15, 9, 14, 8, 12, 6, 16, 11)	(12, 10, 2, 17, 10, 8, 12, 3, 10, 4, 5, 8, 17, 4, 3, 2, 12, 6, 12)
	(11, 15, 4, 2, 3, 1, 5, 8, 6, 17, 13, 16, 12, 14, 10, 7, 9)	(14, 7, 15, 16, 2, 7, 14, 12, 13, 11, 4, 12, 17, 6, 14, 4, 3, 2, 3)

Table 4. Statistical information

n	16	17
Number of (different) workers	49	214
Wall clock time for this job (sec)	2921931.4774	3309757.6983
Overall Parallel Performance	0.9993	0.9994
Equivalent Run Time	103371746009.5473	2375697871296.6587

because we augmented the number of Workers assigned to the computation. The change of the numbers of remaining vertices is shown in Figure 6. Because the ratio of change varies around $t = 90$, we can see the effect of the augmentation of the assigned Workers.

We show some of the permutations and sorting sequences that attain the diameters in Table 3. The statistical information of the computations is shown in Table 4 where Overall Parallel Performance is a ratio of the total time of computation of Workers over the total working time of Workers. Though this value is ideally equal to 1, it is usually a smaller value practically, because of the overhead by communication and the unbalanced task granularity. However, in our system, the values are nearly equal to 1. Therefore we can see that it works very efficiently. We consider that this is because tasks are interrupted in constant time, and at least a constant number of tasks are maintained on the Master side all the time, hence all the Workers can execute the tasks all the time. In addition, Equivalent Run Time is the total sum of the multiplication of the execution time of each Worker and the benchmark value. This is the expected execution time when it is executed on the machine whose benchmark value is 1.

The benchmark value of Pentium3 machine is about 1100 per one CPU. Hence, if all the Workers work all the time, then $f(16)$ can be computed in about 34 days in case of executing it on Pentium3 machines only. In addition, if we compute $f(17)$ by using 16 Pentium3 machines, which are used for computation of $f(16)$, then no less than 4 years would be necessary as the computation time.

In this computation, we counted the number of discarded vertices as well as the number of searched to verify the correctness of the results of computation of the diameters. As a result, the sum of numbers of the discarded and the searched vertices matched the number of total vertices. Hence, we can conclude that the computation is correct. Since computation for each vertex is fulfilled in one CPU, we can also be fully confident in the correctness of the result of computation.

7 Conclusions

In this study, we have improved the method by Kounoike et al. to obtain the diameter of P_{15} so that it is applicable to compute the diameters of the larger scales of pancake graphs and implemented as a parallel computing system. In addition, we applied the system and obtained 16- and 17- pancake graphs by PC clusters. By conventional implementations, it has been impossible to compute the diameters of the larger pancake graphs because of memory restriction. However, our improved method can complete the computation if sufficient time is supplied and the computation time is shorten.

By using the implemented system, we have obtained the diameters of the pancake graphs up to $n = 17$. We want to obtain the diameters of larger pancake graphs. In addition, the known diameters so far satisfy $f(n) = f(n-1) + 2$ only when $n = 3, 6$ and 11 and no n has been found for $n > 11$ which satisfies the equation. We are also interested in such n's.

Acknowledgements

We would like to express our thanks to Prof. Mitsunori Miki and Prof. Tomoyuki Hiroyasu for permission to use the PC cluster on Doshisha University. A part of this research is supported by Japan society for the promotion of sciences, the grant-in-aid(No.16510105).

References

1. Dweighter, H. Amer. Math. Monthly **82** (1975) 1010
2. Akl, S.G., Qiu, K.: Fundamental algorithms for the star and pancake interconnection networks with applications to computational geometry. Networks **23** (1993) 215–225
3. Bass, D.W., Sudborough, I.H.: Pancake problems with restricted prefix reversals and some corresponding cayley networks. Journal of Parallel and Distributed Computing **63**(3) (2003) 327–336

4. Berthomé, P., Ferreira, A., Perennes, S.: Optimal information dissemination in star and pancake networks. IEEE Transactions on Parallel and Distributed Systems **7**(12) (1996) 1292–1300
5. Kumar, V., Grama, A., Gupta, A., Karypis, G.: Introduction to Parallel Computing: Design and Analysis of Algorithms. Benjaming/Cummings (1994)
6. Quinn, M.J.: Parallel Computing: Theory and Practice, second edition. McGraw-Hill (1994)
7. Kounoike, Y., Kaneko, K., Shinano, Y.: Computing the diameters of 14- and 15-pancake graphs. In: Proceedings of the International Symposium on Parallel Architectures, Algorithms and Networks. (2005) 490–495
8. Heydari, M.H., Sudborough, I.H.: On the diameter of the pancake network. J. Algorithms **25**(1) (1997) 67–94
9. AT&T: On-Line Encyclopedia of Integer Sequences http://www.research.att.com/~njas/sequences/.
10. MW project: MW Homepage http://www.cs.wisc.edu/condor/mw/.
11. Condor Team: Condor Project Homepage http://www.cs.wisc.edu/condor/.

Topic 17: High-Performance Bioinformatics

Craig A. Stewart, Michael Schroeder,
Concettina Guerra, and Konagaya Akihiko

Topic Chairs

High performance computational biology and bioinformatics are increasingly required to extract valuable biological and biomedical knowledge from ever-increasing biological data. New computational techniques and new theoretical models are required to simulate complex biological behavior of biological systems. Topic 17 focuses on high-performance and high-throughput computing necessary for management of biological data, extraction of meaning from biological data and using such data in modeling and simulation of biological systems.

Five papers were accepted for Topic 17 this year:

Multidimensional Dynamic Programming for Homology Search on Distributed Systems by Shingo Masuno, Tsutomu Maruyama, Yoshiki Yamaguchi, and Akihiko Konagaya describes a computation method for multidimensional dynamic programming on distributed systems. This paper makes use of FPGA systems in novel and interesting ways, demonstrating the utility that many people expect to see in FPGA-based systems in biocomputing in the future.

Load balancing and Parallel Multiple Sequence Alignment with Tree Accumulation by Guangming Tan proposes a load balancing strategy for parallelizing tree accumulation in progressive alignment in the widely used package ClustalW, reducing overall running time and achieving reasonable speedups.

ZIB Structure Prediction Pipeline: Composing a Complex Biological Workflow through Web Services by Patric May, Hans-Christian Ehrlich, and Thomas Steinke presents status of their efforts for the realization of an automated protein prediction pipeline as an example for a complex biological workflow scenario in a Grid environment based on Web services. As grid computing evolves, many leading experts believe the use of standards-based web services to be of particular value in bioinformatics.

Evaluation of Parallel Paradigms on Anisotropic Nonlinear Diffusion by S. Tabik, E.M. Garzn, I. Garca, and J. J. Fernndez discusses the parallel implementation of Anisotropic Nonlinear Diffusion, a powerful noise reduction technique in the field of computer vision. This technique is applied to the problem of analysis of 3D images, an important problem in high performance computational biology and bioinformatics.

Improving the Research Environment of High Performance Computing for Non-Cluster Experts Based on Knoppix Instant Computing Technology by Fumikazu Konishi, Manabu Ishii, Shingo Ohki, Yusuke Hamano, Shuichi Fukuda, and Akihiko Konagaya presents an approach for instant computing using Knoppix technology that can allow even a non-computer specialist to easily construct and operate a Beowulf cluster. The application InterProScan (from the European Bioinformatics Institute) is used as a demonstration of the value of Knoppix Instant Computing Technology in bioinformatics.

These papers represent the very high-quality submissions received for the topic High-Performance Bioinformatics. The contribution of Shingo Masuno, Tsutomu Maruyama, Yoshiki Yamaguchi, and Akihiko Konagaya is in particular noted as a distinguished contribution. The organizers of the High-Performance Bioinformatics topic would like to thank all authors who submitted papers, the paper review committee, and the Euro-Par 2006 conference organizers.

Multidimensional Dynamic Programming for Homology Search on Distributed Systems

Shingo Masuno[1], Tsutomu Maruyama[1], Yoshiki Yamaguchi[1], and Akihiko Konagaya[2]

[1] Systems and Information Engineering, University of Tsukuba,
1-1-1 Ten-ou-dai Tsukuba Ibaraki, 305-8573, Japan
masuno@darwin.esys.tsukuba.ac.jp
[2] RIKEN Genomic Sciences Center, 1-7-22 Suehiro-cho Tsurumi-ku
Yokohama Kanagawa, 230-0045, Japan

Abstract. Alignment problems in computational biology have been focused recently because of the rapid growth of sequence databases. By computing alignment, we can understand similarity among the sequences. Dynamic programming is a technique to find optimal alignment, but it requires very long computation time. We have shown that dynamic programming for more than two sequences can be efficiently processed on a compact system which consists of an off-the-shelf FPGA board and its host computer (*node*). The performance is, however, not enough for comparing long sequences. In this paper, we describe a computation method for the multidimensional dynamic programming on distributed systems. The method is now being tested using two nodes connected by Ethernet. According to our experiments, it is possible to achieve 5.1 times speedup with 16 nodes, and more speedup can be expected for comparing longer sequences using more number of nodes. The performance is affected only a little by the data transfer delay when comparing long sequences. Therefore, our method can be mapped on any kinds of networks with large delays.

1 Introduction

Alignment problems in computational biology, namely homology search, have been focused recently because of the rapid growth of sequence databases[1,2,3]. By computing alignment, we can investigate similarity among the sequences. Dynamic programming is a technique to find optimal alignment among sequences. In dynamic programming, all causal connections to the final result are stored, and back-traced in order to obtain the optimal alignment. Its computational complexity, however, is very large (order L^N to compare N sequences of length L), and it is not realistic to use algorithms based on dynamic programming even for alignment between two sequences on desk-top computers. In order to reduce the computation time, many heuristic algorithms[6,7,8] or hardware systems [9,10,11,12,13,14,15] have been proposed. Most of them, however, are designed for two-dimensional alignment (alignment between two sequences) because of the complexity to calculate alignment among more than two sequences under limited hardware resources. We have already proposed computational methods for more than two sequences [16,17], and shown that high performance can be achieved on a compact system which consists of an off-the-shelf FPGA board and its host computer (*node*). The performance is, however, not enough for comparing long sequences.

In this paper, we describe a computation method for the multidimensional dynamic programming on distributed systems, which consist of the nodes connected as a ring. The communication pattern between the nodes in our approach is very simple and regular. Each node receives data from its predecessor, and sends its results to its successor. This data transfer can be overlapped with the computation of the dynamic programming. The method is now being tested using two nodes connected by Ethernet.

This paper is organized as follows. Section 2 introduces the outline of dynamic programming for homology search, and our computation method for more than two sequences are described in Section 3. The parallel computation method on distributed systems are given in Section 4, and the estimated performance based on the experimental results is given in Section 5. The current status and future works are given in Section 6.

2 Dynamic Programming for Homology Search

In the dynamic programming for homology search, sequences are compared inserting gaps with extra costs. Figure 1 shows an example of alignment of two sequences by dynamic programming (two-dimensional). In Figure 1(A), scores on each node on the search space ($M \times N$) are calculated using the equation in Figure 2. Scores for each matching between two elements ($M_S[a[x], b[y]]$) and inserting gaps (GC()) are given by *score matrices* [4,5]. In each node, there are three candidates of its score (from the left-upper node, upper node and left node) in two-dimensional search, and the maximum of them is chosen. The paths which give the maximum values are stored, and after calculating scores of all nodes, the paths are back-traced from the last node to the start node to obtain the alignment of the two sequences (Figure 1(B)).

To obtain an alignment of more than two sequences, the same procedure is applied to the sequences. The search space of N-dimensional dynamic programming

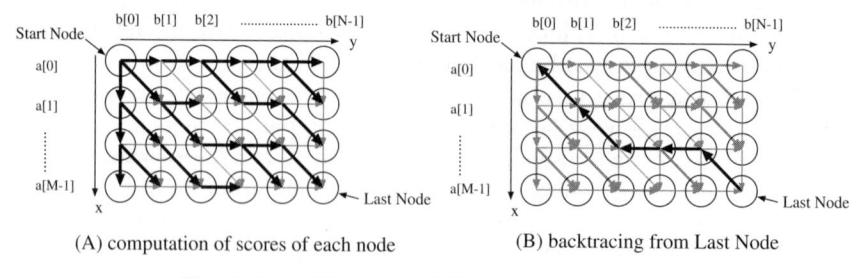

(A) computation of scores of each node   (B) backtracing from Last Node

Fig. 1. Two Dimensional Dynamic Programming

Two-Dimensional Search:
score(x, y) =

$$\max \begin{cases} score(x\text{-}1, y\text{-}1) + M_S[a[x], b[y]] \\ score(x, y\text{-}1) + GC(x, \text{-}) \\ score(x\text{-}1, y) + GC(\text{-}, y) \end{cases}$$

Three-Dimensional Search:
score(x, y, z) =

$$\max \begin{cases} score(x\text{-}1, y\text{-}1, z\text{-}1) + M_S[a[x], b[y], c[z]] \\ score(x, y\text{-}1, z\text{-}1) + M_S[\text{-}, b[y], c[z]] + GC(x, \text{-}, \text{-}) \\ score(x\text{-}1, y, z\text{-}1) + M_S[a[x], \text{-}, c[z]] + GC(\text{-}, y, \text{-}) \\ score(x\text{-}1, y\text{-}1, z) + M_S[a[x], b[y], \text{-}] + GC(\text{-}, \text{-}, z) \\ score(x\text{-}1, y, z) + GC(\text{-}, y, z) \\ score(x, y\text{-}1, z) + GC(x, \text{-}, z) \\ score(x, y, z\text{-}1) + GC(x, y, \text{-}) \end{cases}$$

Fig. 2. Equations to calculate Scores

becomes L^N (when N sequences have length L). As indicated by the equations in Figure 2,

1. the number of candidates of the score for each node is $2^N - 1$ in N-dimensional dynamic programming, and
2. the size of score matrices is k^N (k is the number of type of elements in the sequences), which becomes very large for larger N.

Figure 3 shows the maximum parallelism in dynamic programming. As shown in Figure 3, nodes on a diagonal line (plane) can be processed in parallel. The maximum parallelism in N-dimensional search is the product of the size of N-1 sequences (in the maximum case). When $N=2$, the maximum parallelism is Y, and it takes $X \times Y - 1$ steps to calculate the alignment.

3 Multidimensional Dynamic Programming on an FPGA

In the dynamic programming, we need to store paths to each node to backtrace. The total size of the paths becomes L^N(the number of the nodes in the search space) $\times N$(data bit width of a path), which becomes very large for larger N. However, if the given sequences are not apparently similar, we do not need the alignment. Therefore, in our approach, two types of circuits are configured on FPGA[15,17]. With the first type circuits, the similarity among sequences are checked by computing only the scores. Then, the second type circuits are

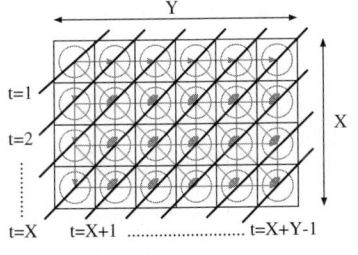

(A) Parallel Processing of two dimensional dynamic programming

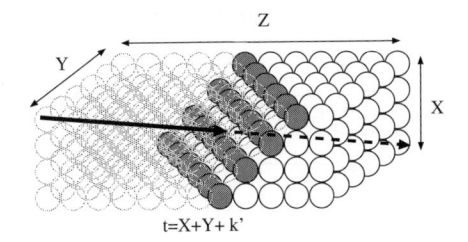

(B) Parallel Processing of three dimensional dynamic programming

Fig. 3. Parallelism in Dynamic Programming

configured on the FPGA, and the alignments are calculated for the sequences with high similarity (score) by storing all causal connections. In the following discussion, we focus on the first type circuits.

In our approach, N-dimensional dynamic programming is achieved by repeating two-dimensional dynamic programming along other dimensions in order to reduce the size of the score matrices which have to be cached on the FPGA (for the protein sequences ($k=24$), the total size of the score matrix becomes 324K words when $N=4$). Suppose that we repeat the following procedure for four-dimensional dynamic programming (a four-dimensional score matrix $\mathrm{M_S}[a[x], b[y], c[z], d[t]]$ is used).

1. Calculate the alignment between two sequences (a and b) without changing other two sequences ($c[z] = C_k$ and $d[t] = D_l$; C_k and D_l are constants).
2. Increment z, and then t ($c[z]$ or(and) $d[t]$ is changed).

Then, we need only a part of the four-dimensional matrix, which is a two-dimensional score matrix ($\mathrm{M_S}[a[x], b[y], C_k, D_l]$) in the first step of the procedure. However, we need different two-dimensional score matrix when the value of $c[z]$ or $d[t]$ is changed. In our implementation, two-dimensional score matrices are implemented using dual-port RAMs in FPGA, and score matrices for next $b[z]$ or/and $d[t]$ (namely next parts of the four-dimensional score matrix) are downloaded from external RAMs on the FPGA board in parallel with the computation of scores. The number of score matrices which are download during the computation becomes 2^{N-2}. Thus, with a certain value of N, the downloading time of the next score matrices exceeds the time of the computation of the two-dimensional dynamic programming, and becomes the bottleneck of this approach.

In the following discussion, suppose that X, Y, Z and T are length of sequences placed along x, y, z and t axes, and W_x, W_y, W_z and W_t are part of sequences which can be processed continuously without extra input/output for boundary data. Figure 4 shows how three-dimensional dynamic programming is executed by the repetition of the two-dimensional dynamic programming. In Figure 4, processing of $W_x \times W_y$ nodes (two-dimensional dynamic programming) in the *scan window* (gray square in the figure) is scanned along z axis (the black arrow shows the *scan line*). When the scan window reaches at the end of z axis,

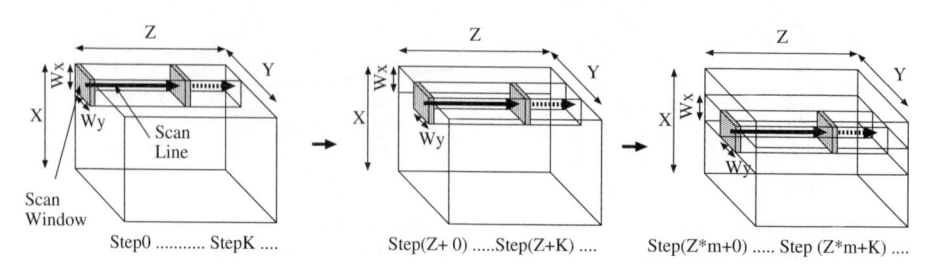

Fig. 4. Three-Dimensional Dynamic Programming

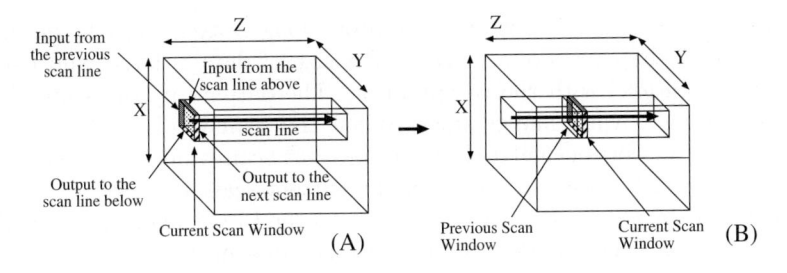

Fig. 5. Boundary Data for Three-Dimensional Dynamic Programming

it is shifted along y axis by W_y, and is scanned along z axis again. After processing $W_x \times Y \times Z$ nodes, the scan window is shifted down along x axis by W_x, and the same procedure is repeated.

Figure 5 shows the data input/output for the three-dimensional dynamic programming. In Figure 5(A), two dark gray rectangles show the inputs to the scan window (light gray square), and two rectangles with slanted lines show the output by the scan window. The outputs are stored, and used for the computation of other scan lines. In Figure 5(B), in order to calculate scores in the current scan window, data in previous scan window are also necessary (those data are not necessary in Figure 5(A), because the scan window is placed at the boundary on the search space, and boundary conditions are given instead of those data). Therefore, the data in previous scan window are held on FPGA.

Figure 6 shows the *scan cube* for four-dimensional dynamic programming (a cube is used instead of the window). Processing of nodes in the cube (size is $W_x \times W_y \times W_z$) is scanned along t axis, changing positions of the scan line. In order to calculate scores of the nodes in the cube, the scan window in the cube (light gray square in Figure 6(A)) is scanned along z axis. Suppose that current cube is on $(x, y, z, t=C_k)$. In order to start the calculation of the scan window (Figure 6(B)(1)), we need scores in dark gray parts and scores in the previous cube along t axis $((x, y, z, t=C_k-1)$ which are temporally held on the FPGA (not shown in the figure) as boundary data. Among these data, two dark gray rectangles in the figure can be obtained while calculating the scores of the nodes in the scan window. However, data in the dark gray square (the

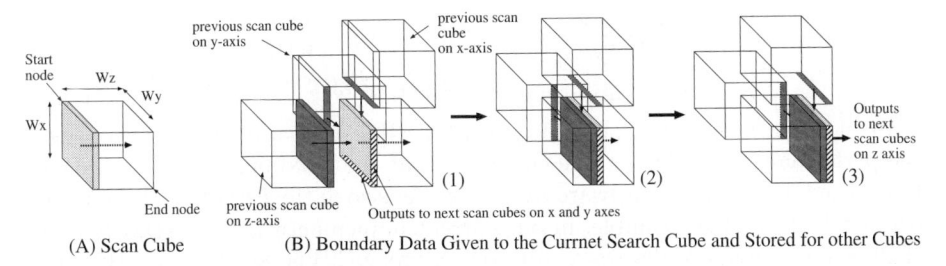

Fig. 6. Four-Dimensional Dynamic Programming

last scan window in the previous scan cube along z axis) need to be loaded before starting the calculation, because the size of data is large, and can not be loaded in parallel with the computation. The outputs by the scan window are two rectangles with slanted lines. When the scan window is in the cube (figure 6(B)(2)), scores calculated in the previous scan window are held on the FPGA, and used for the calculation of the current scan window (the scores in the previous cube along t axis which are held on the FPGA are also used). When the scan window reaches at the end of the cube, scores in the current window are stored for later processing (figure 6(B)(3)). In this processing of the scan cube, there are two types of data;

1. data which can be loaded, and output in parallel with the computation of the scores of the nodes in the scan window (two dark rectangles in Figure 6(B)(1,2,3)), and
2. data which have to be loaded before the computation (dark gray square in Figure 6(B)(1)) and which have to be stored after the computation (dark gray square in Figure 6(B)(3)).

The total clock cycles by our approach can be estimated as follows, when the data width of each element in score matrices is 16 bits, and the external memory banks run at the same speed as the circuit on the FPGA. In the following equations, the first term chooses the maximum of the computation time of the scan window ($W_x + W_y$) and the time to update score matrices which is executed in parallel with the computation. In other terms, constant values show the time to download score matrices, and other values show the time to input/output boundary data (some matrices can not be loaded in parallel with the computation, and we need to download them when $c[z], d[t]$ and so on are changed).

Three-Dimensional:
$$\max\left\{\begin{array}{c} W_x + W_y \\ 24^2/4 \end{array}\right\} \times \frac{XYZ}{W_x W_y}$$

Four-Dimensional:
$$\max\left\{\begin{array}{c} W_x + W_y \\ 24^2/8 \times 2 \end{array}\right\} \times \frac{XYZT}{W_x W_y} + \max\left\{\begin{array}{c} 24^2/2 \\ W_x W_y \times 2/5 \end{array}\right\} \times \frac{XYZT}{W_x W_y W_z}$$

Five-Dimensional:
$$\max\left\{\begin{array}{c} W_x + W_y \\ 24^2/16 \times 4 \end{array}\right\} \times \frac{XYZTU}{W_x W_y} + \max\left\{\begin{array}{c} 24^2/4 \times 2 \\ W_x W_y \times 2/5 \end{array}\right\} \times \frac{XYZTU}{W_x W_y W_z} +$$
$$\max\left\{\begin{array}{c} 24^2/2 \\ W_x W_y W_z \times 2/5 \end{array}\right\} \times \frac{XYZTU}{W_x W_y W_z W_t}$$

Six-Dimensional:
$$\max\left\{\begin{array}{c} W_x + W_y \\ 24^2/32 \times 8 \end{array}\right\} \times \frac{XYZTUV}{W_x W_y} + \max\left\{\begin{array}{c} 24^2/5 \times 4 \\ W_x W_y \times 2/5 \end{array}\right\} \times \frac{XYZTUV}{W_x W_y W_z} +$$
$$\max\left\{\begin{array}{c} 24^2/5 \times 8 \\ W_x W_y W_z \times 2/5 \end{array}\right\} \times \frac{XYZTUV}{W_x W_y W_z W_t} + \max\left\{\begin{array}{c} 24^2/5 \times 7 \\ W_x W_y W_z W_t \times 2/5 \end{array}\right\} \times \frac{XYZTUV}{W_x W_y W_z W_t W_u}$$

In the equations above, $\{W_x, W_y, W_z, W_t, W_u\}$ are parameters which decide the performance, and have to be chosen so that the maximum performance can be realized under given hardware resources (the size of the FPGA, and the memory bandwitdh). For example, in our current implementation on ADM-XRC-II (FPGA board byits Alpha Data) with one Xilinx XC2V6000, $\{W_x, W_y, W_z, W_t\}$ are $\{10, 64, 6, 3\}$ for five-dimensional dynamic programming, and it takes about 1.35×10^4 seconds to calculate the alignment, when the length of the sequences

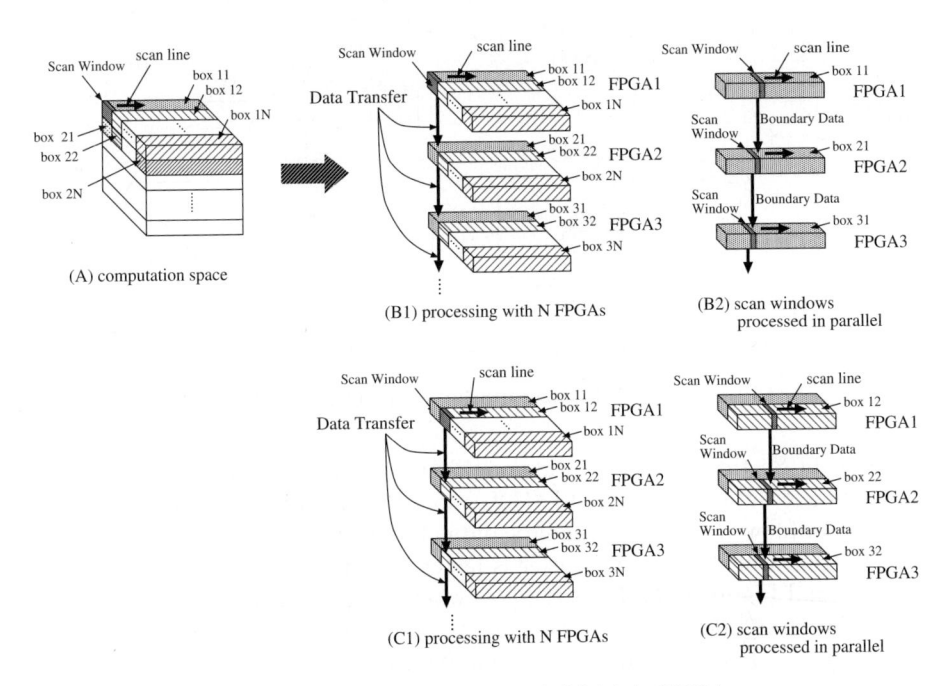

Fig. 7. Parallel Processing with Multiple FPGAs

is 256. This performance is more than 100 times of Pentium 4 2GHz[17], but is still too slow for comparing longer sequences.

4 Multidimensional Dynamic Programming on a Distributed System

Figure 7(A) shows the search space in three-dimensional dynamic programming. With one FPGA, the computation of the scan window is started from the left-hand side of *box 11*, and the scan window is scanned along z axis (*scan line*). After finishing *box 11*, the scan window moves to *box 12*, and the computation of the scan window is repeated. Figure 7(B1) shows how to divide the search space. In Figure 7(B1), $FPGA_k$ processes *box k1 - kN* sequentially. When the first scan window in *box 11* is processed by FPGA1, the boundary data on its bottom are transferred to FPGA2. Then, FPGA2 starts the computation of the first scan window in *box 21*. In the same way, FPGA3 starts the computation of the first scan window in *box 31* as soon as the boundary data for the scan window arrive from FPGA2. Figure 7(B2) shows only the boxes which are processed in parallel. In this parallel processing, data transfer can be overlapped with the computation of scan windows. After finishing the computation of *box 11*, FPGA1 starts the computation of *box 12*, and FPGA2 also starts the computation of *box 22* (Figure 7(C1)(C2)).

Figure 8 shows when the computation of the scan window can be started on FPGA1 and FPGA2. The gray boxes in Figure 8 shows the first term of the

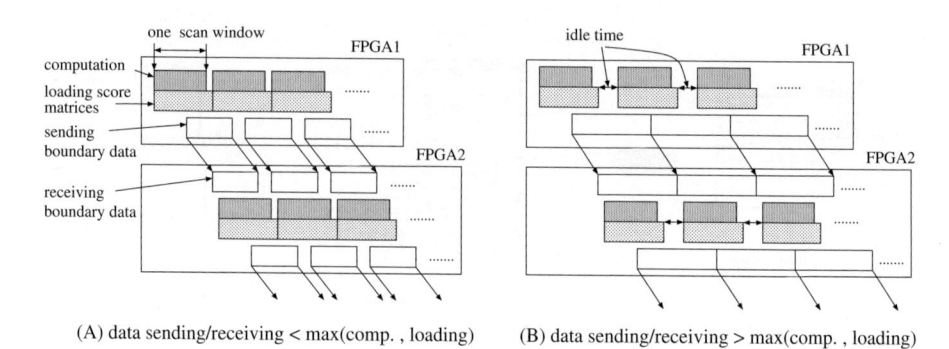

(A) data sending/receiving < max(comp. , loading)   **(B)** data sending/receiving > max(comp. , loading)

Fig. 8. Flow of the computation on FPGA1 and FPGA2

equation in Section 3. During the computation of a scan window in FPGA1, its boundary data are sent to FPGA2, and FPGA2 starts the computation of its scan window using the boundary data. Figure 8(A) shows the flow of the computation when the data transfer is faster than the computation of the scan window, and Figure 8(B) shows the flow when it is slower. In Figure 8(B), each FPGA becomes idle to wait for sending its boundary data to its successor, and for the arrival of the boundary data from its predecessor.

Data transfer delay is not important in our computation method. The reason is as follows. FPGA1 can continue its computation until it finishes all the computation assigned to FPGA1, and the data transfer can be overlapped with the computation of the scan windows. FPGA2 becomes idle when waiting for the first arrival of the boundary data because of the data transfer delay, but after that, FPGA2 can continue its computation as far as the boundary data arrive within a certain delay. Therefore, the increase of the computation time by the data transfer delay is only

the data transfer delay × (the number of FPGAs - 1)

in the total computation time.

Figure 9 shows a distributed system for our computation method. In our approach, the search space is divided along x axis as shown in Figure 7(B1). When the number of FPGAs(N) is smaller than the number of the divided search spaces, some FPGAs have to process several of them sequentially (for example, FPGA$_i$ processes the i-th, $(N+i)$-th, $(2N+i)$-th spaces, and so on). Therefore, the nodes are connected as a ring. Each node on the system consists

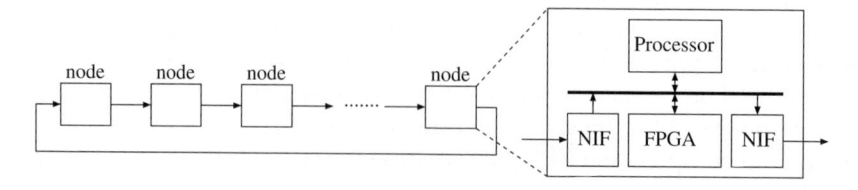

Fig. 9. A distributed system

of an FPGA board with one FPGA, its host processor, and two network interface cards. With two network interface cards, each node receives boundary data from its predecessor, and send new boundary data to its successor.

5 Estimated Performance

We have implemented two circuits (four-dimensional and five-dimensional homology search) on XC2V6000, and they run at 36.6MHz and 31.0MHz respectively. The main reason of the low operational frequency is selectors to choose the maximum $2^N - 1$ candidates.

We are now testing the computation method using two nodes (two FPGA boards and their host processors) connected by Ethernet (100Mbps). Figure 10 shows the performance of the computation method which is estimated based on our experiments (five-dimensional, and legth of all sequences is 256). Boxes with slanted lines correspond to the second and the third terms of the equation shown in Section 3, and grey boxes correspond to the first term (the computation time, and the downloading time of the score matrices which can be executed in parallel with the computation). The size of the scan cube is {10,64,6,3} for non-distributed processing by one FPGA, and {10,32,14,3} for the distributed processing by more than one FPGA. These sizes are dicided so that the maximum performance can be achieved in each case. In the five-dimensional dynamic programming, the time to download score matrices is larger than the computation time. Therefore, we need to minimize the downloading time when processing by one FPGA. However, the downloading time can be hidden by the idle time caused by the slow data transfer on the distributed system, which allows us to focus to minimize the computation time. Because of the lack of the throughput for data transfer, the idle time occupies more than half of the total computation time when the number of FPGAs is larger than one. The computation time with two FPGAs is larger than one FPGA. However, we can obtain performance gain as the number of FPGAs increases. The performance gain becomes 5.1 times with 16 FPGAs, and about 10 times with 26 FPGAs. With 26 FPGAs,

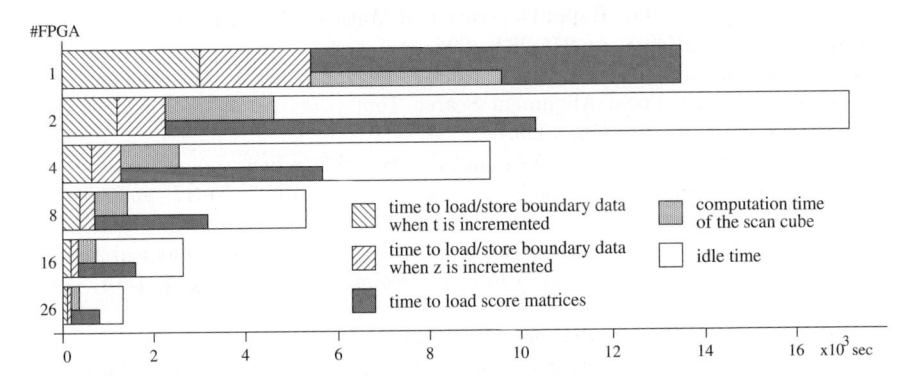

Fig. 10. Estimated performance on the distributed system

each FPGA processes only one divided search space, because the search space is divided to 26 sub-spaces ($X/W_x = 256/10$).

The data transfer delay is not important in our computation method as described in Section 4, when the computation time by each FPGA is large enough. When the number of FPGAs is N, the increase of the total computation time is about $N \times d$ seconds if the data transfer daley becomes d second longer. This increase is very small compared with the total computation time.

6 Conclusions and Future Works

In this paper, we described a computation method for the multidimensional dynamic programming on distributed systems. The method is now being tested using two nodes connected by Ethernet. The data transfer speed of Ethernet (100 Mbps) is not enough, but according to our experiments, it is possible to achieve 5.1 times speedup with 16 nodes. The performance is affected only a little by the data transfer delay when comparing long sequences. Therefore, our method can be mapped on any kinds of networks with large delays.

We still have two major works. First, we need to evaluate the method using more FPGA boards, and then using more FPGA boards placed at distant places. Second, the size of boundary data can be compressed less than half, because two continuous data on the boundary have same values with high probability. We need to implement circuits to compress and uncompress the boundary data on FPGAs.

References

1. National Center for Biotechnology Information (NCBI), "NCBI-GenBank Flat File Release 137.0", http://www.ncbi.nlm.nih.gov/, Aug 2003.
2. European Molecular Biology Laboratory (EMBL), http://www.ebi.ac.uk/embl/.
3. DNA Data Bank of Japan (DDBJ), http://www.ddbj.nig.ac.jp/.
4. Henikoff, S. and Henikoff, J.G.: "Amino Acid Substitution Matrices from Protein Block", Proc. Natl. Acad. Sci. 89, pp.10915-10919, 1992.
5. Jones, D. T. et. al: "The Rapid Generation of Mutation DataMatrices from Proten Sequences", CABIOS 8, pp.275-282, 1992.
6. Stephen F. Altschula, Warren Gisha, Webb Millerb, Eugene W. Meyersc, and David J. Lipman, "Basic Local Alignment Search Tool", Journal of Molecular Biology, Vol.215, Issue 3, pp.403-410, 1990.
7. Stephen F. et al, "Gapped BLAST and PSI-BLAST: a new generation of protein database search programs", Nucleic Acids Research, Vol.25, No.17, pp.3389-3402, 1997.
8. William R. Pearson and David J. Lipman "Improved tools for biological sequence comparison", Proceedings of the National Academy of Sciences of the USA, Vol.85, pp.2444-2448, 1988.
9. PARACEL, "GeneMatcher2", http://www.paracel.com/.
10. Dominique Lavenier, "SAMBA Systolic Accelerators For Molecular Biological Applications", Technical Report RR-2845, 1996.

11. C. Thomas White, et al, "BioSCAN: A VLSI-Based System for Biosequence Analysis", *IEEE International Conference on Computer Design: VLSI in Computer & Processors*, Vol.147, pp.504-509, (1991).
12. TimeLogic Corporation, "Decypher bioinformaticsacceleration solution", *http://www.timelogic.com/products.html*, 2002.
13. Kiran Puttegowda, William Worek, Nicholas Pappas, Anusha Dandapani, and Peter Athanas, "A Run-Time Reconfigurable System for Gene-Sequence Searching", *International VLSI Design Conference*, pp.(to appear), 2003.
14. Steven A. Guccione and Eric Keller, "Gene Matching Using JBits", *International Conferenece on Field-Programmable Logic and Applications*, pp.1168-1171, 2002.
15. Yoshiki Yamaguchi, Yosuke Miyajima, Tsutomu Maruyama, Akihiko Konagaya, "High Speed Homology Search with Run-time Reconfiguration", *International Conferenece on Field-Programmable Logic and Applications*, pp.281-291, 2002.
16. Yoshiki Yamaguchi, Tsutomu Maruyama, Akihiko Konagaya, "Three-Dimensional Dynamic Programming for Homology Search", *International Conferenece on Field-Programmable Logic and Applications*, pp.505-514, 2004.
17. S. Masuno, T. Maruyama, Y. Yoshiki and A. Konagaya, "Multidimensional Dynamic Programming for Homology Search", *International Conferenece on Field-Programmable Logic and Applications*, pp.173-178, 2005.

Load Balancing and Parallel Multiple Sequence Alignment with Tree Accumulation

Guangming Tan[1,2], Liu Peng[1,2], Shengzhong Feng[1], and Ninghui Sun[1]

[1] Institute of Computing Technology, Chinese Academy of Sciences
[2] Graduate School of Chinese Academy of Sciences
{tgm, pengliupl, fsz, snh}@ncic.ac.cn

Abstract. Multiple sequence alignment program, ClustalW, is time consuming, however, commonly used to compare the protein sequences. ClustalW includes two main time consuming parts: pairwise alignment and progressive alignment. Due to the irregular computation based on tree in progressive alignment, available parallel programs can not achieve reasonable speedups for large scale number of sequences. In this paper, progressive alignment is reduced to tree accumulation problem. Load balancing is ignored in previous efficient parallel tree accumulations. We proposed a load balancing strategy for parallelizing tree accumulation in progressive alignment. The new parallel progressive alignment algorithm reducing to tree accumulation with load balancing reduced the overall running time greatly and achieved reasonable speedups.

1 Introduction

Algorithms for multiple sequence alignment [1] are routinely used to find conserved regions in biomolecular sequences, to construct family and superfamily representations of sequences, and to reveal evolutionary histories of species. Conserved subregions in DNA/protein sequences may represent important functions or regulatory elements. The profile or consensus sequences obtained from a multiple alignment can be used to characterize a family or superfamily of species. Multiple sequences alignment is also closely related to phylogenetic analysis. From a mathematical point of view, the multiple sequences alignment is a more complex combinatorial problem which is NP hard. There has been a lot of interest in finding efficient approximation algorithms (PTAS)[2] for these problems. However, the PTAS algorithms have high time complexity so that they become impractical for many long sequences. Some popular heuristic approaches such as progressive alignment [1] that work reasonably well in practice have been proposed. The most widely used algorithm is the progressive alignment algorithms and its typical implementations are ClustalW [1] and DFALIGN [3]. Although the running time has been reduced, the time complexity of the progressive alignment algorithms is $O(n^2m^2)$, where n is the number of sequences and m is the maximum length of all sequences. Since the best known progressive alignment programs is ClustalW, we focus on the parallelization of ClustalW. The basic algorithm behind ClustalW proceeds in three steps.

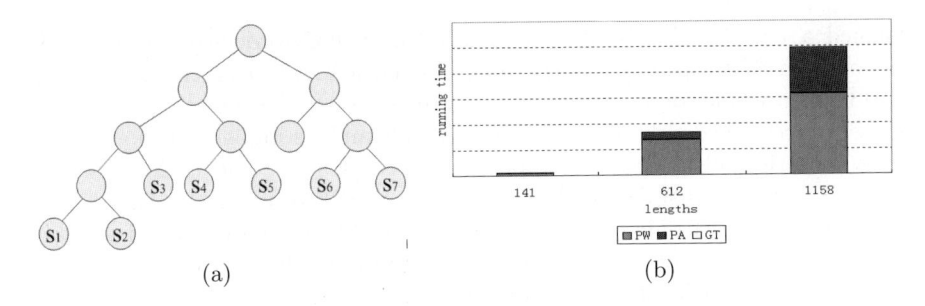

Fig. 1. a). A guide tree. Each leaf represents a sequences, the internal nodes represent the partial alignment from their children. b). The running time distribution of three parts in ClustalW. The number of protein sequences is 781, 1158 and 1770. In most cases the CPU times spent on building the guide tree is less than 1 percent (almost cannot be seen in this figure). The pairwise alignment occupies the most of the overall running time, however, the running time of the progressive alignment significantly increases with the larger number of sequences.

1. **Pairwise alignment(PW):** Compute the optimal alignment cost for each pair of sequences using standard dynamic programming. This results in a distance matrix whose entries indicates the degree of divergence of each pair of sequences in evolution. In fact, this step can be very time consuming and become the bottleneck of the whole process because it has to align $n(n-1)/2$ pairs, where n is the number of sequences. Since each alignment is independent of the rest, the parallelization is a problem of allocating time-independent tasks to parallel processors and can achieve linear speedups [5][6][7][8].

2. **Guide tree(GT):** Compute an evolutionary tree from the distance matrix using some phylogeny reconstruction method. This tree will be used as the guide tree (See Figure 1(a)) which guides the final multiple alignment process are computed from the distance matrix by first using a popular distance based phylogeny reconstruction method, the Neighbor-Joining method [4]. In general, this step can be completed very fast.

3. **Progressive Alignment(PA):** The basic procedure of progressive alignment is to use a series of pairwise alignments to merge larger and larger groups of sequences, following the branching order in the guide tree. Each merger involves aligning two multiple alignments using a dynamic programming algorithm similar to that for the alignment of a pair of sequences. It contains a profile-profile/sequences alignment implemented by dynamic programming algorithm with linear space. In this way, sequences that are highly divergent from the rest of sequences are given due consideration in the alignment process.

Because ClustalW program is widely used and time consuming, there exist some contributions to parallelizing ClustalW algorithm. Mikhailov et al. [5]

designed a parallel ClustalW for shared-memory multiprocessor machines. It runs only on SGI computers with OpenMP and achieves a maximum speedup of 10 for the whole alignment process on 16 processors machine for some protein sequences. Duzlevski [6] used Posix threads and its implementation can be run symmetric multiprocessor computers. Jamse et al. [7] and K. Li [8] implemented a parallel ClustalW for PC cluster using MPI, respectively. They report a fine linear speedup only for pairwise alignment, but the speedup and scalability for the whole alignment are poor because those parallel programs ignore the significant to parallelize progressive alignment.

For the small number of sequences, the efficient parallelization of the step 1 is enough because the running time of progressive alignment is not significant(See figure 1(b)). However, when the number of sequences becomes larger, the poor performance of parallelization in progressive alignment becomes a bottleneck because of the linear speedup in pairwise alignment. Because of the irregular structure based on tree in progressive alignment, it is difficult to efficiently parallelize step 3. The previous parallel programs focuses on the small scale problem, thus the performance of parallel progressive alignment is not important to the overall parallel program for the small number of sequences (less than few hundreds of sequences). However, when aligning the larger number of sequences, the progressive alignment becomes a bottleneck because of the linear speedup in step 1 and the poor performance for parallel progressive alignment. In this paper, we proposed a fast parallel algorithm for multiple sequences alignment program (ClustalW) using load balancing strategy.

2 Parallel Progressive Alignment

2.1 Reducing to Tree Upward Accumulation

There are generally two kinds of accumulations on trees with bounded maximum degree: upward accumulations and downward accumulations[9]. Consider a tree of n nodes, each containing an operation drawn from a set S, and a binary associative operation $\otimes : S \times S \rightarrow S$. Let s_v denote the operation at node v, and $u_1, u_2, ..., u_k$ be an ordered list children of v. Without loss of generality, the upward accumulation problem is to compute $A(v)$ for each node v in the tree where

$$A(v) = \begin{cases} s_v & \text{if v is a leaf} \\ A(u_1) \otimes A(u_2) \otimes ... \otimes A(u_k) & \text{otherwise} \end{cases} \quad (1)$$

If the binary operator is commutative, we can simply write the upward accumulation as:

$$A(v) = \bigotimes_{u \in subtree(v)} s_u \quad (2)$$

Progressive alignment is a profile/sequences alignment progress basing on the guide tree that is a complete binary tree. The leaves are sequences and the internal nodes are profiles. Basing on the guide tree, progressive alignment performs the profile/sequences alignment from leaves to root. Thus, progressive alignment is reduced to the tree accumulation problem. If the binary operator $\otimes$ represents profile/sequences alignment, progressive alignment is reduced to tree accumulation naturally.

contractl(u):
push(u.right.stack, u, u.operator);
u.right.operator = u.operator
(u.left.opertor, u.right.opertor);
u.right.parent = u.parent;
if u.left != **NULL**
 u.parent.left = u.right;
else
 u.parent.right = u.right;
u.right.left = u.left;
if root == u
 root = u.right;

(a)

distribution:
for each node u do **in parallel**
 wait until u.val is computed;
 while u.stack != **NULL do**
  (v, operator) = pop(u.stack);
  while dependency in operator do
   block;
  end
  v.val = operator(u.val);
 end
end

(b)

Fig. 2. Pseudocode procedure for contractl and distribution

2.2 Parallel Tree Accumulation

To get round this problem, the PRAM tree accumulation algorithm operates in two phases [9]: a *contraction* phase in which the tree is reduced to a single leaf and some nodes are put aside on stacks, and a *distribution* phase in which the stacked nodes receive their final values. Each contraction operation removes a leaf node v and its parent (an internal node) by connecting v's sibling directly to its grandparent. Although the final value to be assigned to the internal node is still unknown, yet it is the certain known function (binary operator) of the final value that is to be assigned to the siblings. The deleted internal node and its binary operator are put aside on a stack belonging to all the siblings. When the final value to be assigned to the sibling is computed, the value for the deleted parent can be computed in turn.

Contraction. The contraction operations each remove two nodes, at least one of which is a leaf (See Figure 3). Assume that all leaf nodes of tree are numbered from left to right. Mark all even/odd numbered leaves. For every marked leaf that is left child of their parent u, perform the contraction operation: contractl(u), and then for every marked leaf that is right of their parent u, perform the contraction operation: contractr(u). This guarantees that parents of the leaves contracted are not adjacent [9]. The primitive operation is contraction, which is only called so when u is the internal node and its one of its children is a marked leaf. contractl(u) and contractr(u) a pair of symmetric operation, contractl(u) is defined as fig. 2(a).

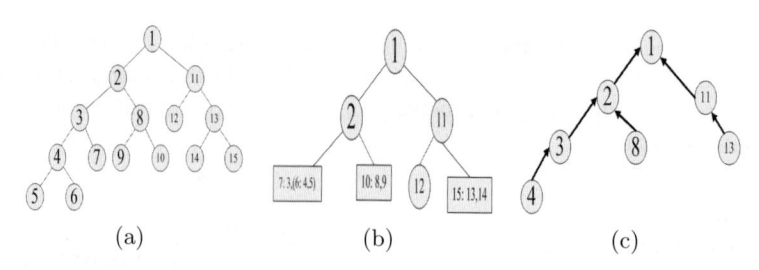

Fig. 3. An illustration of the contraction phase and mapping tree accumulation to task graph. (a). the original tree. The number in each node is number by preorder traversal. (b). the partial contracted tree. The internal node 3, 4, 8 and 11 are removed. The information are stored in their right child. c). The DAG task graph. The direction implicit the order of task dispatched.

Distribution. The contracted nodes are expanded and accumulations at all nodes are accomplished during the distribution phase. The premise is that before a leaf node is expanded, its siblings have correct accumulation. The information stored in each leaf node and the accumulations in its siblings are used to compute the final values of the node. Each node u has stack $u.stack$ with the data structure of (node, function). If (v, h) is in $u.stack$, then operator $(u.val)$ should be assigned to $v.val$, once $u.val$ is computed (See fig. 2(b)).

2.3 Load Balancing Strategy

For a tree accumulation problem, tree contraction has been proven to be efficient if the operations associated with the internal nodes require $O(1)$ time[9]. The available parallel algorithms for tree accumulation, which rest on a common assumption that all binary operators are equal, that is, the computing time of all binary operators is the same, the order has no bearing on the runtime. On the other hand, if the each binary operator on nodes consumes different time, or at least two operations require different time for executing, different orders of operations might well lead to variety, even to the extent of great difference in runtime. Have a deeper analysis of this issue. In the parallel upward accumulation, $A(v)$ in the same level of the tree can be computed in parallel. However, if the binary operators need different running time, then the processor which has completed its computational task will have to wait until all the computational tasks of its brother nodes have finished, which causes poor load balancing, and consequently results in low processor utilization. Moreover, the topology of the tree also has influence on the performance of the parallel algorithm in that the critical path of accumulating from the leaves to the root determines the running time. Unfortunately, previous parallel algorithms hardly focus on the effect of tree topology and almost all of them start accumulating from all leaves, let alone contrive efficient policy to map accumulation to proper processor, which gravely diminish the efficiency of tree accumulation because the processors which compute the shorter branches are left idle most of the time if the tree is unbalanced which is just the case in most practical applications.

Progressive alignment reducing to tree upward accumulation is an exact example for the shortcomings described above. The time of each pairwise alignment is proportional to the product of the length of two sequences. The length of sequences is different, thus each pairwise alignment has different running time. Because of the divergent of all sequences, the guide tree based on the distance matrix may not be a balancing tree. So a naive implementation of previous parallel tree accumulation algorithm can not promise good load balancing and high processor utilization.

Many scheduling algorithms have been proposed and two good surveys on static and dynamic scheduling algorithms can be found in [10], where a parallel program can be described as a directed acyclic graph (DAG). A weighted DAG task graph can be used to represent the problem of tree accumulation on the basis that a task is defined as an operator at a node. The task graph can be constructed in the contraction phase. Each stacked node corresponds to a certain node in the task graph. The weight of a node is an estimated running time of the operator while the weight of an edge is an estimated size of messages from the child task to the parent task. And how to calculate the two weights is determined according to real applications. An compelling example of mapping from the original tree to task tree is shown in Figure 3(c).

Define the b-level of a task as the length of the longest path from the task to the root task, where the length of a path is the sum of all the node and edge weights along the path. Further, the number of internal nodes from each node to root is added to calculate the b-level in order to consider the factor of tree topology. The b-level of a node is bounded from above by the length of a critical path, which is the longest path from the temporal node to the root node in the DAG. The *b-level* of a node is assigned to the node on the stacks in the contraction phase. In the distribution phase, a dynamic priority queue is maintained in order to schedule the tasks for processors. When any popped operator can be computed, it is inserted into the priority queue according to its *b-level*. And if there is any idle processor, remove the task at the head of the queue and schedule it to the processor. The modified distribution algorithm employs a coordinator-worker model. The coordinator maintains the priority queue and schedules and dispatches tasks to workers who execute the real operators. The algorithm of coordinator in the distribution phase are as follow:

Distribution with load balancing:
for each node u **do**
    wait until u.val is computed;
    while u.stack != **NULL do**
        (v, operator) = pop(u.stack);
        while dependency in operator **do**
            block;
        end
        insert(u, v, operator, queue);
        while (queue != **NULL** && idle_procs != 0) **do**

```
              dispatch(queue, processor);
        end
    end
end
```

The coordinator also maintains an idle processors pool. In the beginning, all processors are idle and the pool is full. When one task is assigned to one idle processor, the processor is deleted from the pool. And after one processor has completed its task, it sends a message to coordinator and coordinator adds this processor to the pool.

3 Performance Evaluation

Because the previous parallel programs can not get speedups for progressive alignment, we only evaluated the performance of load balancing. The experiment implemented the load balancing parallel algorithm in cluster systems—distributed memory parallel computers connected by networks. Each node of the cluster system is composed of Xeon 2.8Ghz SMP processors, 4GB memory, while all the nodes are connected via gigabit Ethernet switch. And the parallel program is written using C with MPI library. Moreover, the test data sets are downloaded from PDB bank [11], and they are five different protein family or domain(TROW, WOLPM, WIGBR, ZYMMO, YERPS). For simplicity, some notations are used in the evaluation: lb denotes the parallel algorithm with load balancing and na denotes the naive parallel algorithm.

Speedup: The performance of a parallel algorithm is measured by speedup or efficiency. The speedup of a parallel algorithm using p processors is defined as $Speedup = \frac{T_{serial}}{T_{parallel}(p)}$ and the efficiency is $Efficiency = \frac{Speedup(p)}{p}$. Strictly speaking, T_{serial} is the running time of the fastest known serial algorithm on one processor for the same problem. Figure 4(a) and 4(b) show the speedups as the number of processors and the size of problem size are increased for algorithms with load balancing strategy and without load balancing strategy, respectively. The speedups of lb are much higher than that of na. When the number of processors is less than 16, the parallel program with load balancing strategy can achieve approximate linear speedup. While the number of processors is larger, the speedup of both algorithms increases slowly. The highest speedup 18 of lb occurred when the number of sequences is 3998 and the number of processors is 32, while the highest speedup 8 of na occurred when the number of sequences is 781 and the number of processors is 32.

Time: The most important contribution of load balancing strategy is the reduction of overall running time. The tree accumulation process in progressive alignment presented above comprises computation, communication and other overhead such as scheduling and idle. Table 1 demonstrates the overall running time for two parallel algorithms with the different number of processors and different size of problems. The overall running time of parallel algorithm lb are reduced mostly 3 times as that of na. Because the relative time distributions

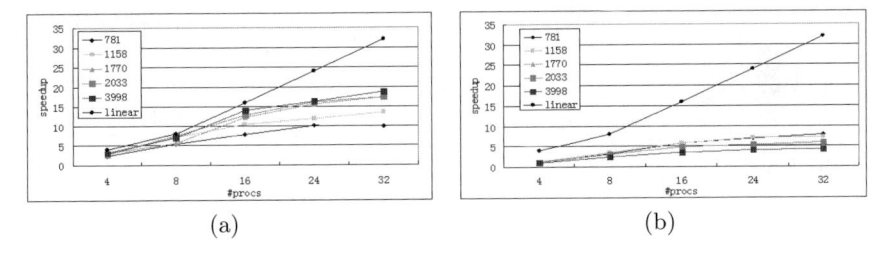

Fig. 4. Speedup for naive parallel algorithm with/without load balancing

Table 1. The overall running time of two parallel algorithms. The number of sequences of 5 data sets are 781, 1158, 1770, 2033 and 3998, the number for processors are 4, 8, 16, 24 and 32. Time: second

		4	8	16	24	32
781	lb	163	73	51	40	40
	na	377	127	68	58	51
1158	lb	199	91	62	55	48
	na	502	189	113	93	91
1770	lb	311	173	79	63	56
	na	765	321	203	183	183
2033	lb	393	146	87	69	64
	na	998	373	226	206	185
3998	lb	448	200	100	86	74
	na	1394	609	406	345	327

of computation, communication and overhead for the different problem size are almost the same, we only analysis the experiment results of overhead in the case of 1158 sequences alignment.

The parallel algorithms are implemented using coordinator-worker model, while the coordinator only performs scheduling and communication. It is the workers who execute the real computation and send/receive message from coordinator. Figure 5(a) shows that there is minor difference of the communication times between two algorithms. However, the communication distribution among all slave processors for *lb* is even more than the distribution for *na*. Figure 5(b) demonstrates communication time distribution among all workers in 32 processors. In fact, the unbalanced communication is relative the reflection of the unbalance computation in each worker. In the presented algorithms, there only exist communications between the coordinator and workers, so the more computation load in one worker, the more communications are needed in the worker. Due to the different computing load and unbalanced binary tree, some slave processors may be idly waiting for another computation task that depends on some other computation tasks running on slower processors. Although the communication and computation load is unbalanced for the parallel algorithm *na*, neither of the time cost are higher than the cost of the parallel algorithm *lb* as

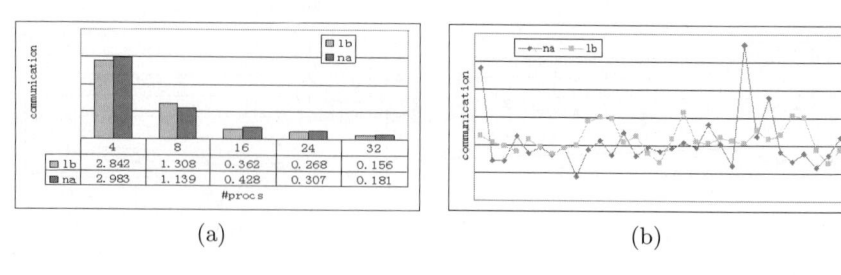

Fig. 5. a) The maximum communication time in seconds with the different number of processors. b) The communication time distribution in 32 processors for two parallel algorithms.

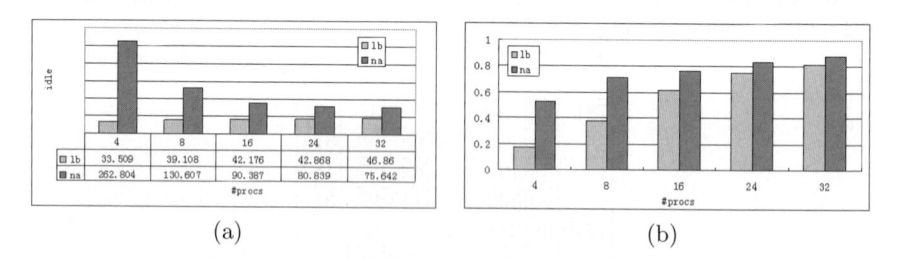

Fig. 6. a) The maximum idle time in seconds for the different number for processors. b)The time proportions of idle to the overall running time.

shown in above analysis. Thereby the communication and computation time pale in terns of their influence on the time reduction. Measure the idle time in each worker processor. Figure 6(a) shows the maximum idle time for different number of processors. The parallel algorithm *lb* mainly focuses on the factors of computation weight and branch length in the task tree to schedule the computation task, and it proved to have reduced the idle time in each processor greatly. With the number of processors increasing, the computation loads in each processor become less, that is, the overall computation is decreasing, so the proportions of idle time to the overall running time become higher (See Figure 6(b)). However, the larger the number of processors, the more the overhead of scheduling task among more processors is, and correspondingly the more the idle time is for the parallel algorithm *lb*.

4 Conclusions

In this paper, a new parallel implementation of progressive alignment through tree accumulation with load balancing is presented. And in the proposed implementations, the load balancing strategy is used in order to take advantage of both weighted tree contraction and tree topology. Moreover, a test for the performance of the algorithm and a comparison with the naive PRAM implementation on a 32-processors Linux cluster system is shown and analyzed in the context, which shows that the parallel tree accumulation algorithm achieves not only reasonable speedups for the data sets used in the evaluation, but also higher speedups than the naive parallel algorithm using load balancing.

This work is supported by National Natural and Science Foundation (90412010) and Youth Foundation of ICT.

References

1. D. T. Julie, G. H. Desmond and J. G. Toby, Clustal W: improving the sensitivity of progressive multiple sequence alignment through sequence weighting, position-specific gap penalties and weight matrix choice, Nucleic Acids Research, 1994, Vol. 22, No. 22, pp. 4673-4680.
2. D. Henikoff, Approximation Algorithms for NP-hard Problems, PWS publishers, 1996.
3. D. Feng and R. F. Doolittle, Progressive sequence alignment as prerequisite to correct phylogenetic trees, Journal of Molecular Evolution, 1987, Vol. 25, pp. 351-360.
4. N. Saitou and M. Nei, The neighbor-joining method: A new method for reconstructing phylogenetic trees. Molecular Biology and Evolutoin, 1987, Vol. 4, pp. 406-425.
5. D. Mikhailov, H. Cofer and R. Gomperts, Performance optimization of ClustalW: Parallel ClustalW, HT Clustal and MULTICLUSTAL. White papers, SGI, 2001.
6. O. Duzlevski, SMP version of ClustalW 1.82, http://bioinfor.pbi.nrc.ca/clustalw-smp.
7. J. J. Cheetham, F. Dehne, S. Pitre, A. R. Chaplin and P. J. Tailon, Parallel CLUSTALW for PC Clusters. Proceedings of International Conference on Computational Science and its Applications, Montreal, Canada , May 18-21, 2003.
8. K. Li, ClustalW-MPI: Clustalw analysis using distributed and parallel computing, Bioinformatics, 2003, Vol. 19, no. 12, pp: 1585-1586
9. J. Gibbons, W. Cai, D. Skillicorn, Efficient parallel algorithms for tree accumulations, Sci. Comput. Programming, 1994, 23, pp. 1-18
10. Y. K. Kwok, I. Ahmad, Static scheduling algorithms for allocating directed task graphs to multiprocessors, ACM Computing Surveys, 1999, Vol. 31, No. 4, pp. 406-471
11. http://www.rcsb.org/pdb/

ZIB Structure Prediction Pipeline: Composing a Complex Biological Workflow Through Web Services

Patrick May, Hans-Christian Ehrlich, and Thomas Steinke

Zuse Institute Berlin, Takustr.7,14195 Berlin, Germany
{patrick.may, ehrlich, steinke}@zib.de

Abstract. In life sciences, scientists are confronted with an exponential growth of biological data, especially in the genomics and proteomics area. The efficient management and use of these data, and its transformation into knowledge are basic requirements for biological research. Therefore, integration of diverse applications and data from geographically distributed computing resources will become a major issue. We will present the status of our efforts for the realization of an automated protein prediction pipeline as an example for a complex biological workflow scenario in a Grid environment based on Web services. This case study demonstrates the ability of an easy orchestration of complex biological workflows based on Web services as building blocks and Triana as workflow engine.

1 Introduction

In the post-genomics era protein structure prediction is still one of the major challenges in bioinformatics research, because the full understanding of the biological function of proteins requires knowledge about its three-dimensional (3D) structure [1]. Although experimental methods are providing high-resolution structure information, they are still expensive in costs and duration. On the other hand, fully automated computational structure prediction tools have made rapid progress over the last years (Critical Assessment of Structure Prediction, CASP [2] and CAFASP [3]). Protein structure prediction is a process which typically involves multiple data processing and decision steps, iterations, as well as the parallel execution of time-consuming applications. In comparison with sequence homology searches with, for example, Blast [4] structure prediction is a much more complex scenario.

Web services provide a well-defined, standardized access to methods independent from its implementation and programming platform. As pointed out in [5], Web services are an emerging technology paradigm for distributed computing. Problem solving environments with standardized workflow description languages (e.g. BPEL4WS [6]) are providing solutions to these problems. Suitable workflow engines support the orchestration [7] of workflows with Web services as building blocks. Complex workflows contain compute and/or storage intensive tasks. Regarding compute intensive tasks, the support of parallel execution models, e.g. task farming or MPI parallelized programs, are therefore an imperative prerequisite. We selected Triana [8] for the following reasons: it easily integrates Web services, and provide a graphical user interface allowing an easy workflow orchestration. Furthermore, it can represent workflows as non-DAG and the workflow engine can be interfaced with selected Grid services which is an important pre-requisite for the next step towards the realization of our workflow in a Grid environment.

There are many initiatives pushing biological applications towards the use of workflow, Grid and/or Web service technologies. Gao et. al. [9] describe a drug discovery data-mining system using Web services. Mattoso et. al. [10] built MHOLline, an automated workflow for comparative modelling with legacy applications using Web service technology. They used BPEL4WS for defining the workflow and IBM BPWS4J 1.0.1 [11] as workflow engine. PROSPECT-PSPP [12] is a fully automated structure prediction pipeline using SOAP for remote procedure calls. Hence, the problem of consistency in data integration projects, which combine common information from different data sources, is still a major obstacle for obtaining unique information sets and data quality in secondary biological databases. There are successful data integration (data warehouse) projects, for example, MSD [13] or Columba [14] with their focus on structural data. The Helmholtz Open BioInformatics Technology initiative (HOBIT) [15] is dedicated to build a technology platform for concatenating applications and resources together with an efficient communication tier for bioinformatics resource access based on Web services.

The scope of this paper is to show that fully automated workflows with Web service components are able to integrate heterogeneous applications and data into a standalone, demanding biological application scenario. One can expect that Web services are one starting point for the realization of a collaborative e-science infrastructure in Grid environments. In this paper we use protein structure prediction as a paradigm for complex biological problems.

The organization of the article is as follows. In the next section we describe the ZIB structure prediction pipeline. Section 3 presents the Web services, followed by workflow definition in Triana. As the most interesting result, we compare the overhead timings of the traditional monolithic workflow using a PERL implementation of the workflow engine with the Web service based workflow using the Triana workflow engine in section 4. Finally, section 5 gives a summary and an outlook towards future work.

2 ZIB Structure Prediction Pipeline

The ZIB structure prediction pipeline has been designed and implemented for the 6th CASP experiment in 2004 [2,16]. In order to provide a fully automated protein prediction tool, the pipeline integrates various prediction and analysis steps. The whole pipeline is designed modular, so that improved methods can be substituted in, as they become available. Fig. 1 shows the global pipeline architecture.

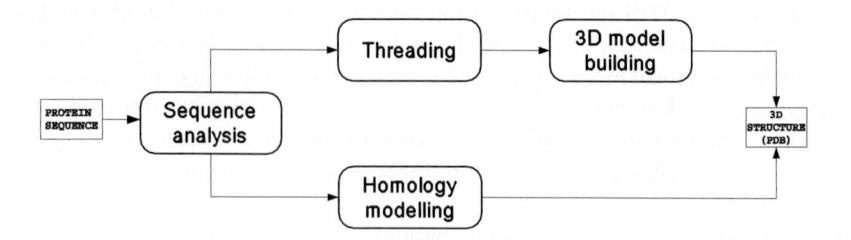

Fig. 1. Schematic representation of the ZIB structure prediction pipeline

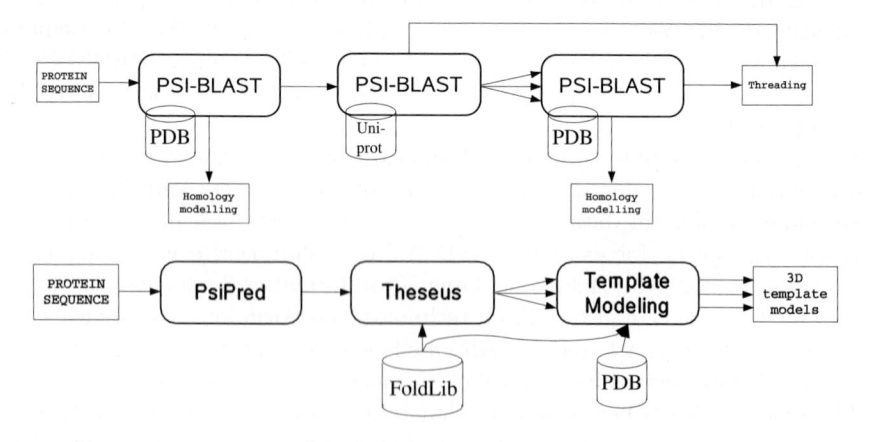

Fig. 2. Sub-Workflows: (top) Sequence analysis, (bottom) Threading

The first step in the workflow is the identification of suitable template structures for homology modelling (Fig. 2, top). A sequence analysis sub-workflow is passed to search for homologous sequences with known structures. Successive PSI-Blast searches are performed in order to find suitable templates. If no template structure has been found in the PDB (Protein Data Bank [17]) database, a second PSI-Blast search in the Uniprot [18] database is initiated followed by parallel PSI-Blast searches in the PDB database starting from the Uniprot hits. If a structural template has been found, an atomic structural model will be generated with MODELLER [19]. If no suitable structural template is detectable, the structure will be predicted by our protein threading implementation. The threading procedure (Fig. 2, bottom) starts with a secondary structure prediction using PsiPred [20]. PsiPred provides a 3-state prediction (helix, strand, loop) together with a reliability score for every sequence position. THESEUS [21] is an MPI-parallelized implementation of a protein threading based on a multi-queue branch-and-bound search algorithm to find the optimal sequence-to-structure alignment through a library of template structures [22]. From the highest scoring template structures the most probable template is selected and submitted to the loop modelling procedure where different 3D models are generated in parallel. Here, MODELLER is used to model the loop regions and the sidechain atoms of the given template structure. At the end, a full atom structure for the target sequence is provided.

The most time consuming step in the sequence analysis procedure is the PSI-Blast search against the Uniprot database (minutes to one hour of CPU time). The prediction of a 3D protein model by threading typically takes many minutes to hours, the modelling steps with MODELLER some minutes to few hours. The types of data to be exchanged and processed are protein sequences, structures and alignments. Data formats are either application specific, e.g. the PDB format for protein structures or Blast-XML for PSI-BLAST, or in-house developed XML schemes for a standardized data exchange.

Fold recognition by threading can be parallized by assigning each of a subset of template structures to a different process. Our parallel threading core is implemented in C++ and uses either MPI for message passing or POSIX threads. Two kind of parallel architectures are designed: a Master-Slave (MS) version, and a Single-Program-Multiple-Data (SPMD) version. In the MS architecture the central component is the MySQL database. A master process or POSIX thread distributes each outstanding template structure to a slave process waiting for work. Based on a first-come-first-serve protocol a dynamic load balancing scheme can be realized. In the SPMD architecture the content of the MySQL template structure database is dumped into a binary file which is cloned on each compute node on a Linux cluster. The template structures are distributed in a static scheme amongst the MPI processes, i.e., each MPI process performs its own subset. Having all template structures processed, one MPI process gathers all results from the remaining concurrent MPI processes. The SPMD approach is significant faster over the MS architecture (shown in Figure 3: the red line

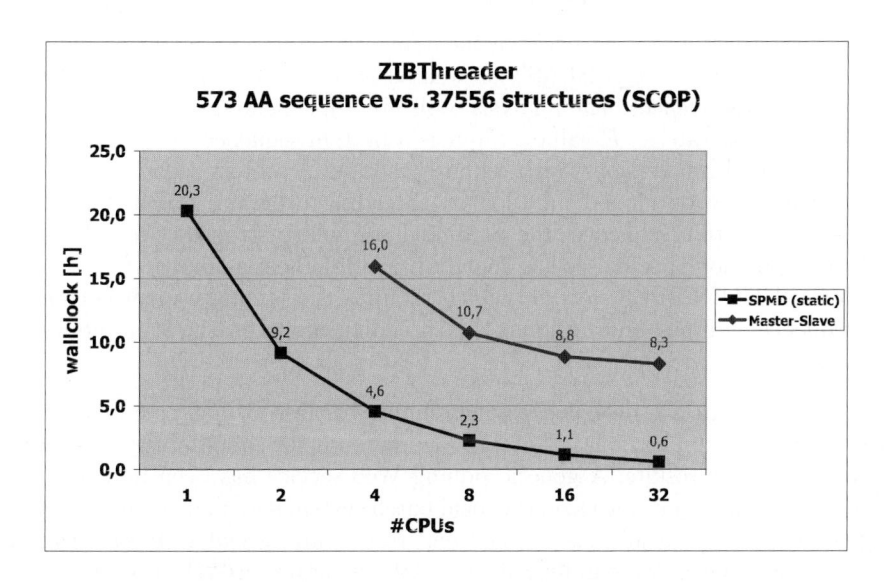

Fig. 3. Performance of the two parallel threading architectures

indicates the MS and the blue line the SPMD architecture). The drawback of the MS architecture is the time determing database connections: the central database server can not timely satisfy the requests from all the slave processes. The SPMD architecture has the extra advantage of parallel I/O. To show the time efficiency of our implementation, we can process a protein sequence consisting of 573 amino acids against 37556 templates structures representing the whole SCOP template database in about 36 minutes on 32 cpus on a IA32 Myrinet Linux cluster .

3 Workflow Implementation with Web Services

3.1 Compute Environment

The implemented pipeline runs on compute resources locally available at our site. Web service applications can either run on a compute cluster complex consisting of an IA32 Myrinet Linux cluster and a Cray XD1 system, or on desktop machines. The resources of the compute cluster complex are managed by a job management system providing a single point of control (job submission and job control). The Triana workflow engine runs either on local desktop machines or on the cluster front-end node. More technical details of the hardware and software configuration can be found elsewhere [23].

3.2 Web Service Implementations

For the ZIB structure prediction pipeline the following applications, part of them are legacy codes, were wrapped into Web services:

- A local Blast program package including the standard sequence analysis tools BLAST and PSI-BLAST as well as FastaCMD for retrieving FASTA formatted sequences. The analysis tools are implemented with standard options (e.g. database, E-value). Input is a protein sequence.
- A local PsiPred version which requires as input a protein sequence.
- The in-house developed parallelized threading program THESEUS which needs a protein sequence, the predicted secondary structure from PsiPred and a position-specific scoring matrix from PSI-Blast as input.
- A local MODELLER version, which requires the template identifier and a sequence as input and optional the threading model in the loop modelling case.

The Web services are designed asynchronous, because of high computational demands of the applications. They provide methods for submitting the job and for collecting the results. A generic polling Web service has been implemented which monitors the job status on the local batch system and informs the workflow engine process that a job has finished and results are available. Parallelization over data is achieved by handling lists as data structure in Web services.

Data, either XML or unfiltered file contents, are transferred through the body of the SOAP message. The Web services were implemented using two different languages: the Blast, PsiPred, and the MODELLER Web services are written and deployed with Java and Apache Axis [24], the THESEUS and the polling Web service are written in Python.

3.3 Workflow Definition with Triana

Our structure prediction workflow is defined and executed by Triana [25]. Triana allows the user to build and execute workflows consisting of Triana units and Web services. Triana is written in Java and supports the implementation of self-written Triana units easily. The Triana GUI provides a Unit Wizzard for generating a skeleton Triana unit code, an editor and an interface for compiling the code. Fig. 4 shows as an example the source code snippet of our `makeFastaCMDrequest` unit:

- The unit has one input port (line 4).
- The unit has two output ports (lines 19 and 20).
- The input for the unit is a Blast result in a XML document.
- The XML document is parsed for possible Blast hits (line 12:`BlastXML.Hit`).
- The output ports send a request string to the FastaCMD Web service with a list of corresponding hit identifiers that are needed by FastaCMD to fetch the corresponding protein sequences, together with the input XML document, which is needed in the further workflow.

Web services can directly be imported into Triana canvas from its WSDL description. By specifying the URI of the WSDL document the Web service is known to Triana and usable as Triana unit. Input and output object types are given by the WSDL description.

Triana supports the loop as control element in its workflow description. Every resource-intensive application has to be submitted to the local batch system.

```
1  /* provides a list of sequence ids for FastaCMD  */
2  public void process() throws Exception {
3    //get the input from the triana module node
4    BlastXML.BlastFtObj BlastXmlObj = (BlastXML.BlastFtObj) getInputAtNode(0);
5
6    StringBuffer IDs = new StringBuffer(); // array of sequence ids
7    Iterator hitIter = BlastXmlObj.HitStorage.keySet().iterator();
8.....
9    while (hitIter.hasNext()){
10     String tmpKey = (String) hitIter.next();
11     //get each hit
12     BlastXML.Hit tmpHit = (BlastXML.Hit) BlastXmlObj.HitStorage.get(tmpKey);
13     //extract the hit accession and generate FastaCMD request string
14     if (requestCount != BlastXmlObj.HitStorage.size()-1)
15       IDs.append(tmpHit.getHit_accession()).append("\n");
16     else
17       IDs.append(tmpHit.getHit_accession());
18   }
19   outputAtNode(0, IDs.toString());
20   outputAtNode(1, BlastXmlObj);
21 }
```

Fig. 4. Example of a Triana unit: `makeFastaCMDrequest`

Then, our polling Web service method `pollStatus` determines the status of a given job (queued, running, finished). The orchestrated Triana sub-workflow unit `pollStatusLoop(PDB)` includes Triana's `LOOP` unit that initiates either the next polling cycle or exits the loop. The decision is made depending on the output of the `pollStatus` Web service: if the stop condition is send (meaning job is finished) the job identifier (being the input of the `pollStatus` Web service) is passed to the next step in the workflow, which is usually a `getResults` Web service method.

The data driven parallelization of sub-workflows (high-throughput computations) like the invocation of a series of PSI-Blast searches against the PDB database in the sequence analysis sub-workflow (see Fig. 2) is implemented through lists. The Blast Web service works on lists of protein sequences as input data, i.e. the Web service method `runBlast` submits all input sequences to the batch system and returns a list of job identifiers that can be handled by the polling Web service.

4 Web services vs. "Scripting"

In this section we compare the performance of two different implementation scenarios of our protein structure prediction pipeline focussing on the associated overhead costs. The "traditional" approach uses a specifically written workflow engine implemented in PERL (scripting approach). The second implementation is based on the Triana workflow engine with Web services as described in the previous sections. In both implementations the time-consuming bioinformatics applications ran as jobs scheduled via the local batch system, i.e. in both cases the steps (1) job submission to compute nodes of the cluster, (2) monitoring the job status (polling), and (3) delivery of results after job termination were identical. Data analysis steps either implemented into Web services or as Triana units have their counterparts in the PERL implementation as well. The Triana workflow engine process was either started on the cluster front-end (TRIANA/Linux, in Table 2) or on a desktop machine connected through a switched 100 Mb/s Ethernet network to the cluster front-end (TRIANA/Windows). The PERL script ran on the front-end only (PERL/Linux).

Additionally, we have implemented a pipeline version **ZIB-jws** including the BLAST Web service from DNA Data Bank Japan (DDBJ)[1] for searches against the PDB instead of our local PSI-BLAST installation (see Fig. 2). The DDBJ provides their Web services also synchronous as well as asynchronous. Therefore, the DDBJ BLAST Web service could be directly plugged into our request and response with polling architecture. For the PERL implementation we used the SOAP::Lite library.

Three experiments were performed to estimate the overhead costs:

1. PDB sequence analysis + homology modelling (Homology(PDB)),
2. PDB sequence analysis + UNIPROT sequence analysis + parallel PDB sequence analysis + homology modelling (Homology(UNIPROT)),

---
[1] http://xml.nig.ac.jp/wsdl/index.jsp

3. PDB sequence analysis + UNIPROT sequence analysis + PsiPred prediction + Threading + loop modelling (Threading).

To maintain the desired partial workflow, for every experiment a specific protein target sequences was used:

1. a sequence that had a significant identity (> 40%) to a sequence in the PDB,
2. a sequence that showed no significant identity to a PDB sequence but where sequences with sufficient similarity were detectable in the PDB through iterative search against the Uniprot database sequences with, and
3. a sequence with no similarity detectable in the PDB and Uniprot.

All experiments were repeated 10 times for the normal ZIB prediction pipeline (ZIB) as well as for the ZIB-jws version to have a minimal representative set of results. Timing information is based on the `gettimeofday` system call.

In all experiments, the total workflow execution times (wall-clock time) as well as the time spent in the execution of non-application steps, i.e. the workflow overhead execution time, were recorded. In the later case, the wall-clock is fetched before and after the invocation of an asynchronous Web service. The mean values for the total workflow times (Table 1), and the mean values of the overhead times (Table 2) over all approaches are summarized. Note that the total workflow execution time depends heavily on the resource usage of the compute complex, since it includes the job waiting time in a batch queue. Fortunately, for that study these numbers are more of a formal interest since our main focus is to estimate the overhead costs of our workflow design with Web services compared to the scripting approach. The 3D structures obtained by the different approaches were validated to manually obtained reference data in order to ensure the correctness of any workflow implementation.

Table 1. Typical total workflow execution times (wall-clock, in seconds) for the two workflow versions

experiment	ZIB	ZIB-jws
Homology(PDB)	134.1	133.5
Homology(UNIPROT)	358.1	358.0
Threading	503.2	503.1

As expected, the overhead times in the Triana/Web service implementation is by an order of magnitude larger compared to the monolithic PERL approach (Table 2). Overall, the total execution times for the ZIB-jws pipeline are slightly better than those for the in-house version. This is because the DDBJ Web service can only execute simple Blast runs, whereas our in-house implementation uses the more time consuming PSI-BLAST. This implies that any workflow engine invokes the Blast result polling services less frequently than in the in-house scenario. Furthermore, the overhead execution time did not include the time for data transfers between the workflow engine process and a Web service. Compared

Table 2. Mean workflow overhead execution times (in seconds) for the two workflow versions

Implementation/Platform	Homology				Threading	
	PDB		UNIPROT			
	ZIB	ZIB-jws	ZIB	ZIB-jws	ZIB	ZIB-jws
TRIANA/Linux	0.120	0.120	0.300	0.300	0.301	0.300
TRIANA/Windows	0.123	0.120	0.305	0.302	0.305	0.303
Perl/Linux	0.012	0.009	0.080	0.078	0.075	0.071

to the total workflow execution (wall-clock) times, the overhead times in both workflow versions of the Web services based workflow implementation are about four orders of magnitude lower and therefore practically negligible (less than 0.1%).

5 Summary and Outlook

We have presented the implementation of a protein structure prediction pipeline as Web service-based workflow using Triana. We have demonstrated that Web services are a versatile technology to integrate various, heterogeneous methods into one stand-alone, fully automated and biological demanding application scenario.

The design of such complex workflows with Web services as buildings blocks are well supported by the Triana problem-solving environment. Additionally, Triana supports the workflow design and the development of self-written Triana units. Within the Triana framework, the processing of workflows with Web services is characterized by an additional, but expected performance overhead. Fortunately, these additional "costs" are usually negligible for workflow scenarios where the time-dominating factors are compute-intensive tasks. Such a coarse-grain segmentation of workflows is the appropriate approach for taking the advantages of Web service technology in real-world scenarios. Moreover, we see today the overall benefit of using the Web service approach in the modular design of the workflow, the improved maintainability, and the more intuitive plug-in of new modules accessible as Web services. Those modules may run locally or are provided by external service providers. The "only" concern of the end user is the functional interface and the corresponding input and output data.

Having a Web service based workflow in place fulfills an important precondition for moving the application scenario into a Grid environment. As long as all services are statically defined in the workflow any flexibility for improving the throughput performance is missing. The next step is to apply brokering services at runtime to select appropriate compute and storage resources for compute and/or storage intensive workflow steps. This approach will allow the transparent use of geographically distributed resources for the workflow processing. It enables the implementation of high-throughput pipelines for solving complex biological questions.

This work constitutes the base for further developments towards a workflow system for protein structure prediction based on Grid services in a Grid environment. Several additional pre- and post-processing steps to further improve the quality of the predicted models will enhance the ZIB structure prediction pipeline. This development is also part of the German MediGRID [26] project.

Acknowledgement

This work is partly funded by BMBF (Germany), grant no. 031U209A (Berlin Center for Genome Based Bioinformatics, BCB) and BMBF (Germany), grant no. 01AK803F (D-Grid/MediGRID-Ressourcenfusion für Forschung in Medizin und Lebenswissenschaften). We would like to thank Falko Krause and Jonas Maaskola for their implementation of local Web services, and the Triana group at Cardiff University for their support.

References

1. Baker, D., Sali, A.: Protein structure prediction and structural genomics. Science **294** (2001) 93–96
2. Moult, J.: A decade of CASP: progress, bottlenecks and prognosis in protein structure prediction. Curr. Opin. Struct. Biol. **15** (2005) 285–289
3. Fischer, D., Barret, C., Bryson, K., Elofsson, A., Godzik, A., Jones, D., Karplus, K., Kelley, L., MacCallum, R., Pawowski, K., Rost, B., Rychlewski, L., Sternberg, M.: CAFASP-1: critical assessment of fully automated structure prediction methods. Proteins **3** (1999) 209–217
4. Altschul, S.F., Madden, T.L., Schaffler, A.A., Zhang, J., Zhang, Z., Miller, W., Lipman, D.J.: Gapped BLAST and PSI-BLAST: a new generation of protein database search programs. Nuc. Acids Res. **25** (1997) 3389–3402
5. Majithia, S., Shields, M., Taylor, I., Wang, I.: Triana: A graphical web service composition and execution toolkit. In: IEEE International Conference on Web Services (ICWS'2004). (2004)
6. Curbera, F., Andrews, T., Dholakia, H., Goland, Y., Klein, J., Leymann, F., Liu, K., Roller, D., Smith, D., Thatte, S., Trickovic, I., Weerawarana, S.: (Business Process Execution Language for Web services, V.1.0) Available via http://www-106.ibm.com/developerworks/webservices/library/ws-bpel.
7. Leymann, F.: (Web Service Flow Language (WSFL), version 1.0)
8. (Triana) Available via http://www.trianacode.org.
9. Gao, H.T., Hayes, J.H., Cai, H.: Integrating biological research through web services. Computer (2005) 26–31
10. Cavalcanti, M.C., Targino, R., Baião, F.A., Rössle, S.C., Bisch, P.M., Pires, P.F., Campos, M.L.M., Mattoso, M.: Managing structural genomic workflows using web services. Data Knowl. Eng. **53**(1) (2005) 45–74
11. (IBM BPWS4J) Available via http://www.alphaworks.ibm.com/tech/bpws4j.
12. Guo, J., Ellrott, K., Chung, W.J., Xu, D., Passovets, S., Xu, Y.: PROSPECT-PSPP: an automated computational pipeline for protein structure prediction. Nucleic Acid Res. **32**(Web Server Issue) (2004) W522–W525

13. Velankar, S., McNeil, P., Mittard-Runte, V., Suarez, A., Barrell, D., Apweiler, R., Henrick, K.: E-MSD: an integrated data resource for bioinformatics. Nucleic Acids Res. **33**(Database issue) (2005) D262–265
14. Trissl, S., Rother, K., Muller, H., Steinke, T., Koch, I., Preissner, R., Froemmel, C., Leser, U.: Columba: an integrated database of proteins, structures, and annotations. BMC Bioinformatics **6**(1) (2005) 81–92
15. (HOBIT (Helmholtz Open Bioinformatics Technology) project) Available via http://hobit.sourceforge.net.
16. Michalsky, E., Goede, A., Preissner, R., May, P., Steinke, T.: A distributed pipeline for structure prediction. In: CASP6 Methods Abstracts, 6th Meeting on the Critical Assessment of Techniques for Protein Structure Prediction, Gaeta, Italy (2004) 112–114
17. Berman, H., Westbrook, J., Feng, Z., Gilliland, G., Bhat, T., Weissig, H., Shyndyalov, I., Bourne, P.: The protein data bank. Nucl. Acids Res **28** (2000) 235–242
18. Bairoch, A., Apweiler, R., Wu, C., Barker, W., Boeckmann, B., Ferro, S., Gasteiger, E., Huang, H., Lopez, R., Magrane, M., Martin, M., Natale, D., O'Donovan, C., Redaschi, N., Yeh, L.: The universal protein resource (uniprot). Nucleic Acids Res. **1**(33) (2005) 154–159
19. Marti-Renom, M., Stuart, A., Fiser, A., Sanchez, R., Melo, F., Sali, A.: Comparitive protein structure modeling of genes and genomes. Annu. Rev. Biophys. Biomol. Struct. **29** (2000) 291–325
20. McGuffin, L., Bryson, K., Jones, D.: The PSIPRED protein structure prediction server. Bioinformatics **16** (2000) 404–405
21. May, P., Steinke, T.: THESEUS - protein structure prediction at ZIB. ZIB Report **06-24** (2006)
22. Lathrop, R.H., Sazhin, A., Sun, Y., Steffen, N., Irani, S.S.: A multi-queue branch-and-bound algorithm for anytime optimal search with biological applications. Genome Informatics **12** (2001) 73–82
23. (BCB-Cluster) Available via http://elfie.bcbio.de.
24. (Apache Axis) Available via http://ws.apache.org/axis.
25. Taylor, I., Wang, I., Shields, M., Majithia, S.: Distributed computing with triana on the grid. Concurrency and Computation:Practice and Experience **17** (2005) 1–18
26. (MediGRID) Available via http://www.medigrid.de/.

Evaluation of Parallel Paradigms on Anisotropic Nonlinear Diffusion*

S. Tabik, E.M. Garzón, I. García, and J.J. Fernández

Dept. Computer Architecture and Electronics. University of Almería, Almería 04120, Spain
{siham-ester-inma-jose}@ace.ual.es

Abstract. Anisotropic Nonlinear Diffusion (AND) is a powerful noise reduction technique in the field of computer vision. This method is based on a Partial Differential Equation (PDE) tightly coupled with a massive set of eigensystems. Denoising large 3D images in biomedicine and structural cellular biology by AND is extremely expensive from a computational point of view, and the requirements may become so huge that parallel computing turns out to be essential. This work addresses the parallel implementation of AND. The parallelization is carried out by means of three paradigms: (1) Shared address space paradigm, (2) Message passing paradigm, and (3) Hybrid paradigm. The three parallel approaches have been evaluated on two parallel platforms: (1) a DSM (Distributed Shared Memory) platform based on cc-NUMA memory access and (2) a cluster of Symmetric biprocessors. An analysis of the performance of the three strategies has been accomplished to determine which is the most suitable paradigm for each platform.

1 Introduction

In many disciplines, raw data acquired from instruments are substantially corrupted by noise and sophisticated filtering techniques are then indispensable for a proper interpretation or post-processing. In general terms, smoothing techniques can be classified into linear and non-linear. Standard linear filtering techniques based on local averages or Gaussian kernels succeed in reducing the noise, but at expenses of poor feature preservation. In other words, they may severely blur the features as their edges are attenuated. However, nonlinear filtering techniques achieve better feature preservation as they try to adaptively tune the strength of the smoothing to the local structures found in the image. Anisotropic nonlinear diffusion (AND) is currently one of the most powerful noise reduction techniques in the field of computer vision [1]. This technique takes into account the local structures found in the image to filter noise, preserve edges and enhance some features, thus considerably increasing the signal-to-noise ratio (SNR) with no significant quantitative distortions of the signal. Pioneered in 1990 by Perona and Malik [2], AND has grown up to become a well-established tool in the last decade [1,3,4]. AND has already been successfully applied in different disciplines, such as medicine [5,6,7] or biology [8,9,10], for denoising multidimensional images. AND has actually been crucial to achieve some recent breakthroughs [11,12,13,14].

---

* This work was supported by the Spanish Ministry of Education and Science through grant TIN2005-00447.

The mathematical basis of AND is a partial differential equation (PDE) tightly coupled with a massive set of eigensystems [10]. The computational cost of AND may be very high, depending on the size of the images. There are some disciplines where the requirements may be so huge –much more than 1 Gbyte in size [15,16]– that parallel computing proves to be essential.

The standard numerical scheme for solving PDEs is based upon an explicit finite difference discretization. More efficient schemes have been specifically designed for nonlinear diffusion [17], though. However, they are complex to implement and, despite their efficiency, they still require to be parallelized [18].

In this work we address the parallelization of AND for its application to denoising of large three-dimensional (3D) volumes in biomedicine and structural cellular biology. We make use of the standard explicit numerical scheme for the discretization. This scheme is commonly used in other fields where PDEs are involved [19] and, as a consequence, the parallel approaches that are presented and discussed here may be valuable for them too.

2 Review of Anisotropic Nonlinear Diffusion

AND accomplishes a sophisticated edge-preserving denoising that takes into account the structures at local scales. AND tunes the strength of the smoothing along different directions based on the local structure estimated at every point of the multidimensional image. Conceptually speaking, AND can be considered as an adaptive gaussian filtering technique in which, for every voxel in the volume, an anisotropic 3D gaussian function is computed whose widths and orientations depend on the local structure [20]. This section presents local structure determination via structure tensors, the concept of diffusion, a diffusion approach commonly used in image processing and, finally, details of the numerical implementation.

2.1 Estimation of Local Structure

The *structure tensor* is the mathematical tool that allows us to estimate the local structure in a multidimensional image. Let $I(\mathbf{x})$ denote a 3D image, where $\mathbf{x} = (x, y, z)$ is the coordinate vector. The structure tensor of I is a symmetric positive semi-definite matrix given by:

$$\mathbf{J}(\nabla I) = \nabla I \cdot \nabla I^T = \begin{bmatrix} I_x^2 & I_x I_y & I_x I_z \\ I_x I_y & I_y^2 & I_y I_z \\ I_x I_z & I_y I_z & I_z^2 \end{bmatrix} \quad (1)$$

where $I_x = \frac{\partial I}{\partial x}$, $I_y = \frac{\partial I}{\partial y}$, $I_z = \frac{\partial I}{\partial z}$ are the derivatives of the image with respect to x, y and z, respectively.

The eigen-analysis of the structure tensor allows determination of the local structural features in the image [1]:

$$\mathbf{J}(\nabla I) = [\mathbf{v_1}\ \mathbf{v_2}\ \mathbf{v_3}] \cdot \begin{bmatrix} \mu_1 & 0 & 0 \\ 0 & \mu_2 & 0 \\ 0 & 0 & \mu_3 \end{bmatrix} \cdot [\mathbf{v_1}\ \mathbf{v_2}\ \mathbf{v_3}]^T \quad (2)$$

The orthogonal eigenvectors $\mathbf{v_1}$, $\mathbf{v_2}$, $\mathbf{v_3}$ provide the preferred local orientations, and the corresponding eigenvalues μ_1, μ_2, μ_3 (assume $\mu_1 \geq \mu_2 \geq \mu_3$) provide the average contrast along these directions. The first eigenvector $\mathbf{v_1}$ represents the direction of the maximum variance. Therefore, $\mathbf{v_1}$ represents the direction normal to the local feature (see Fig. 1).

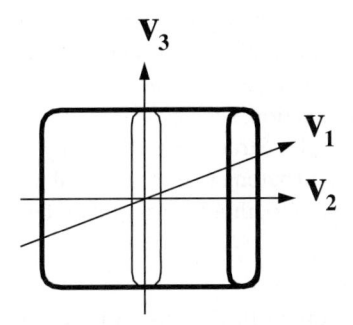

Fig. 1. Local structure found by eigen-analysis of the structure tensor. $\mathbf{v_1}$, $\mathbf{v_2}$, $\mathbf{v_3}$ are the corresponding eigenvectors. $\mathbf{v_1}$ is the direction normal to the local structure.

2.2 Concept of Diffusion in Image Processing

Diffusion is a physical process that equilibrates concentration differences as a function of time, without creating or destroying mass. In image processing, density values play the role of concentration. This observation is expressed by the *diffusion equation* [1]:

$$I_t = \text{div}(\mathbf{D} \cdot \nabla I) \tag{3}$$

where $I_t = \frac{\partial I}{\partial t}$ denotes the derivative of the image I with respect to the time t, ∇I is the gradient vector, $\mathbf{D}$ is a square matrix called *diffusion tensor* and div is the *divergence* operator:

$$\text{div}(\mathbf{f}) = \frac{\partial f_x}{\partial x} + \frac{\partial f_y}{\partial y} + \frac{\partial f_z}{\partial z}$$

In AND the smoothing depends on both the strength of the gradient and its direction measured at a local scale. The diffusion tensor $\mathbf{D}$ is therefore defined as a function of the structure tensor J:

$$\mathbf{D} = [\mathbf{v_1} \ \mathbf{v_2} \ \mathbf{v_3}] \cdot \begin{bmatrix} \lambda_1 & 0 & 0 \\ 0 & \lambda_2 & 0 \\ 0 & 0 & \lambda_3 \end{bmatrix} \cdot [\mathbf{v_1} \ \mathbf{v_2} \ \mathbf{v_3}]^T \tag{4}$$

where $\mathbf{v}_i$ denotes the eigenvectors of the structure tensor. The values of the eigenvalues λ_i define the strength of the smoothing along the direction of the corresponding eigenvector $\mathbf{v}_i$. The values of λ_i rank from 0 (no smoothing) to 1 (strong smoothing).

Therefore, this approach allows smoothing to take place anisotropically according to the eigenvectors determined from the local structure of the image. Consequently, AND allows smoothing on the edges: Smoothing runs along the edges so that they are not only preserved but smoothed. AND has turned out, by far, the most effective denoising method by its capabilities for structure preservation and feature enhancement [1,8,9,10].

2.3 Edge Enhancing Diffusion

One of the most common ways of setting up the diffusion tensor **D** gives rise to the so-called Edge Enhancing Diffusion (EED) approach [1]. The primary effects of EED are edge preservation and enhancement. Here strong smoothing is applied along the preferred directions of the local structure, (the second and third eigenvectors, $\mathbf{v_2}$ and $\mathbf{v_3}$). The strength of the smoothing along the normal of the structure, i.e. the eigenvector $\mathbf{v_1}$, depends on the gradient: the higher the value is, the lower the smoothing strength is. Consequently, λ_i are then set up as: $\lambda_1 = g(|\nabla I|)$, $\lambda_2 = 1$ and $\lambda_3 = 1$, with g being a monotonically decreasing function, such as $g(x) = 1/\sqrt{(1 + x^2/K^2)}$, where $K > 0$ acts as a contrast parameter [1]; Structures with $|\nabla I| > K$ are regarded as edges, otherwise they are considered to belong to the interior of a region. Therefore, smoothing along edges is preferred over smoothing across them, hence edges are preserved and enhanced.

2.4 Numerical Discretization of the Diffusion Equation

The diffusion equation, Eq. (3), can be numerically solved using finite differences. The term $I_t = \frac{\partial I}{\partial t}$ can be replaced by an Euler forward difference approximation. The resulting explicit scheme allows calculation of subsequent versions of the image iteratively:

$$\begin{aligned} I^{s+1} = I^s + \tau \cdot (& \tfrac{\partial}{\partial x}(D_{11}I_x) + \tfrac{\partial}{\partial x}(D_{12}I_y) + \tfrac{\partial}{\partial x}(D_{13}I_z) \\ & + \tfrac{\partial}{\partial y}(D_{21}I_x) + \tfrac{\partial}{\partial y}(D_{22}I_y) + \tfrac{\partial}{\partial y}(D_{23}I_z) \\ & + \tfrac{\partial}{\partial z}(D_{31}I_x) + \tfrac{\partial}{\partial z}(D_{32}I_y) + \tfrac{\partial}{\partial z}(D_{33}I_z)) \end{aligned} \quad (5)$$

where s is the iteration index, τ denotes the time step size, I^s denotes the image at time $t_s = s\tau$, the terms I_x, I_y, I_z are the derivatives of the image I^s with respect to x, y and z, respectively. Finally, the D_{mn} terms represent the components of the diffusion tensor $\mathbf{D}^s$. The standard scheme to approximate the spatial derivatives ($\frac{\partial}{\partial x}$, $\frac{\partial}{\partial y}$ and $\frac{\partial}{\partial z}$) is based on central differences.

In this traditional explicit scheme for solving the partial differential equation Eq. (3), the stability is an issue [1]. The maximum time step that is allowed is $\tau \leq 0.5/N_d$, where N_d is the number of dimensions of the problem. In our case, we are dealing with a three-dimensional problem, so $N_d = 3$. In the experiments carried out in this work, we used a conservative value of $\tau = 0.1$. As far as the number of iterations is concerned, a range of 60-100 iterations is typically used in 3D problems [1,8,9,10] with that value of τ.

For illustration purposes, Fig. 2 shows the result of the application of 60 iterations of AND to a volume of a mitochondrion, a cell organelle, that was obtained by electron microscope tomography [16]. The enhancement in visualizing a slice of the volume is apparent (left: slice from the original volume; right: slice from the filtered volume).

2.5 The Algorithm of the Diffusion Approach

In this work, we propose an optimized algorithm for solving the PDE in Eq. (5) that computes the volume by z-planes, where –without loss of generality– the z-axis is the

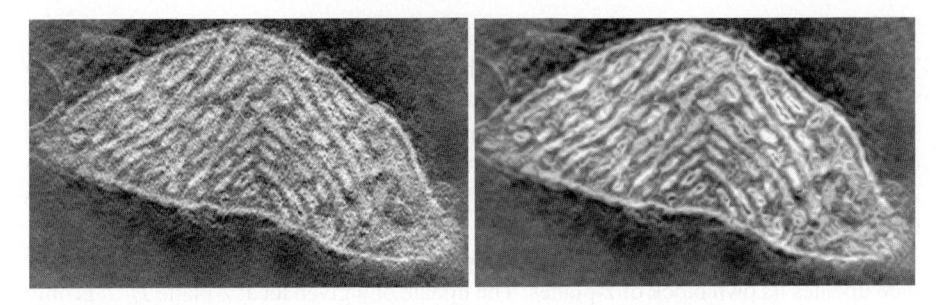

Fig. 2. Left: a slice from a volume of a mitochondrion obtained by electron microscope tomography; Right: the same slice from the volume filtered with anisotropic nonlinear diffusion

direction of the larger image dimension $N_z >= N_x, N_y$. The proposed sequential algorithm consists of the following steps:

Do $s = 0 \ldots n-1$
    Do $k = 1 \ldots N_z$ /* processing the volume by z-planes */
      1. Compute the structure tensor $\mathbf{J}_k^s$ (Eqs. (1) and (2)).
      2. Compute the diffusion tensor $\mathbf{D}_k^s$ from $\mathbf{J}_k^s$ (Eq. (4)).
      3. Compute the resulting z-plane of the image $\mathbf{I}_k^{s+1}$, at step $(s+1)$ from step s by means of Eq. (5). The resulting z-plane of the image corresponds to the diffusion time $t_{(s+1)} = (s+1)\tau$
    End Do
End Do

where s and k denote the index of the iteration and the index of the z-plane respectively. The algorithm is executed iteratively for a number of iterations n. The final image is obtained after a total diffusion time $T = n\tau$. Note from Eq. (5) that $\mathbf{I}_k^{s+1}$ is only a function of $\mathbf{I}_{k-2}^s, \mathbf{I}_{k-1}^s, \mathbf{I}_k^s, \mathbf{I}_{k+1}^s, \mathbf{I}_{k+2}^s$. Our implementation minimizes the memory usage by allocating and computing only the necessary data for updating each single z-plane. Hereinafter, the body of this nested loop is denoted as $\mathbf{I}_k^{s+1} = AND(\mathbf{I}_k^s)$.

3 Parallel Implementation of AND

In this work, AND have been implemented using three parallel programming models: (1) shared address space model based on Pthreads,(2) message passing model, where MPI is applied for message passing between different processors; and (3) hybrid model that uses Pthreads at the node level while MPI is only applied for message passing between processors from different and/or the same nodes. Essentially, the parallel strategies are based on domain decomposition. They consist in distributing the input 3D volume among the processors by blocks of consecutive z-planes, and every processor then applies the AND algorithm to its own block. At the end of every iteration, depending on the specific implementation, boundaries planes must be updated from neighbor processors for their processing in the subsequent iteration. Next, the main characteristics of every parallel code are described:

-Pure Pthreads code. A single thread is mapped onto each processor of the system. The shared address space model allows all the processors to access the shared whole 3D volume. The thread running in each processor updates its corresponding z-planes of the shared volume. Transparently, the neighbor threads then have their boundary z-planes updated thanks to the shared memory. To ensure consistency of the data throughout the algorithm, an additional structure has been defined to hold the boundary z-planes before the neighbors modify them.

-Pure MPI code. Here, one process is spawned on each processor. Each processor then updates its own block of z-planes. The update of a given local z-plane I_k^{s+1} is only a function of I_k^s and its four neighbor z-planes, $I_{k-1}^s, I_{k-2}^s, I_{k+1}^s$ and I_{k+2}^s. Updating the boundary z-planes of the block would imply many communications during one update step. To avoid excessive communications, each processor allocates four additional z-planes to hold the two neighbor z-planes of the two boundaries. At the end of each iteration, the processor then exchanges the updated four boundary z-planes with the immediate neighbor processors by MPI point-to-point communications.

-Hybrid code. The hybrid strategy has been designed in such a way that one MPI process is spawned on each node, and the MPI process then creates as many Pthreads processes as the number of processors in the node. The block of z-planes assigned to the node is shared by all the threads running in the node. Every thread updates its own subset of the block of z-planes, similarly to the shared address space strategy above described. At the end of the iteration, all the threads running in a node are joined. The boundary z-planes are then exchanged among the immediate neighbor nodes by MPI point-to-point communications. The outline of the hybrid code would be as follows:

1. Distribute I^0 among nodes, $N_z^{nd} = \lceil N_z/P \rceil + 4$ planes are assigned to each node.
2. Do $s = 0 \ldots n-1$
   (a) Each thread initializes its auxiliary data structures.
   (b) Do $k = 1 \ldots N_z^{thr} = \lceil (N_z^{nd} - 4)/T \rceil$ /* each thread */
       $I_k^{s+1} = AND(I_k^s)$
       End Do
   (c) Interchange boundary z-planes between neighbor nodes.
3. End Do
4. Collect the image.

where N_z^{nd} and N_z^{thr} denote the local number of z-planes of the volume in every node and in every thread respectively, P and T denotes the number of nodes and processors inside one node respectively, and I^0 denotes the original 3D volume, n denotes the number of iterations, and $AND()$ represents the diffusion algorithm.

In this strategy, it is necessary to control the data distribution at two levels: (1) at the node level, since the total number of z-planes is distributed among nodes, and (2) at the processor level inside the node, as each thread updates its subset of z-planes by applying the AND process.

4 Evaluation of the Parallel Implementation of AND

In this section, we evaluate the performance of three parallel implementations of the AND method: (1) Pure MPI AND-code, (2) Pure Pthreads AND-code and (3) Hybrid AND-code. The evaluation has been carried out on two parallel platforms:

-Distributed Shared Memory (DSM) platform SGI Altix 3700 Bx2 of 8 processors 1600 MHz Intel Itanium 2 Rev with 128 GB RAM. The Altix 3700 computer system is based on a Distributed Shared Memory architecture and uses a cache-coherent Non-Uniform Memory Access (NUMA) where the latency of processors to access to local memory is lower than the latency to access to global memory (or remote memory) [21].

-Cluster of symmetric biprocessors of Intel(R) Xeon(TM) 3.06 GHz with 2 GB RAM, 512 KB cache. Nodes are interconnected via two Gigabit Ethernet networks, one for data (NFS) and the other for computation. The architecture of this cluster is based on a UMA access, where all processors have equally fast (symmetric) access to the memory in the node.

Dimensions of volumes in biomedicine and structural cellular biology usually range between 256x256x256 and 640x640x640. Typical values for n are around 60-100 iterations with $\tau = 0.1$, where n is the number of iterations needed to denoise the volume for an acceptable result [9,10]. In this work, two test volumes with cubic symmetry of sizes 256x256x256 and 640x640x640 have been selected to carry out the evaluation process. Hereinafter, these volumes will be referenced by the size of their edges.

4.1 Distributed Shared Memory Platform: Altix 3700

Let *mpi* be the number of MPI processes and *pt* the number of Pthreads processes. To evaluate the hybrid implementation for a fixed number of processors p, several combinations of values of mpi and pt are possible. Experimental performance results were measured for several combinations of mpi and pt for a fixed p, obtaining similar behavior. In the results shown here, we focus on the case with $pt = 2$ Pthreads processes, and we only increment the number of MPI processes. Fig. 3 shows the speedup achieved by the pure MPI, pure Pthreads and hybrid implementations, for the two volumes, on a 8-processor SGI Altix 3700 Bx2. As it can be seen, in general the three parallel implementations have very good performance. They all approach the ideal linear speedup, with slightly better behavior for the message passing implementation. For the volume 640, some curves exhibit slight levels of superspeedup. Finally, the volume size has turned out to have a very low influence on the speedup. The excellent behavior shown by the message passing version may be thanks to the high speed interconnection technology used in this computer [21]. In summary, the three parallel strategies present very good levels of scalability on this computer platform.

4.2 Cluster of Symmetric Biprocessors

On the cluster of symmetric biprocessors, the performance has been evaluated only for two models: message passing model and hybrid model with $pt = 2$, since on this platform the evaluation of the shared address space model is limited to two processors into one node. The scalability of both models has been analyzed by means of the speedup measurements.

Traditionally, the speedup is only referred to the sequential runtime. Recently, a general concept of speedup has been introduced [22], where the parallel runtime is used as a reference instead. In this evaluation, this concept is taken into account to evaluate the performance of AND on the cluster of SMPs. Specifically, for the volume 640, the parallel runtime with four processors is considered as a reference, since it was not possible

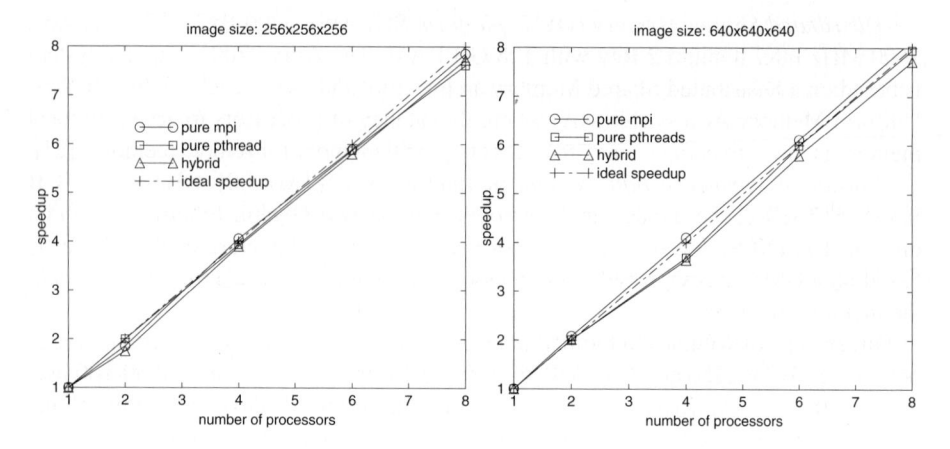

Fig. 3. Speedup on a SGI Altix 3700 Bx2 of the pure MPI, Pure Pthreads and hybrid codes for the volumes 256 and 640

to run the codes on fewer than four processors with these volume sizes. Meanwhile, the sequential runtime is used as reference for the smaller volume 256.

Fig. 4 shows the speedup achieved by the pure MPI code and the hybrid code, for the test volumes, on the cluster of SMPs described above. In general terms, both strategies yield good results, with better performance for the hybrid strategy. It is evident from these figures that the hybrid strategy yields better scalability than the strategy based on message passing, specially for increasing number of processors.

In order to explain the better behavior of the hybrid strategy compared to the message passing one, additional measures of communication times have been obtained as well (results not shown here). Clearly the penalty due to the communications is stronger on the message passing implementation than on the hybrid one, specially as the number of

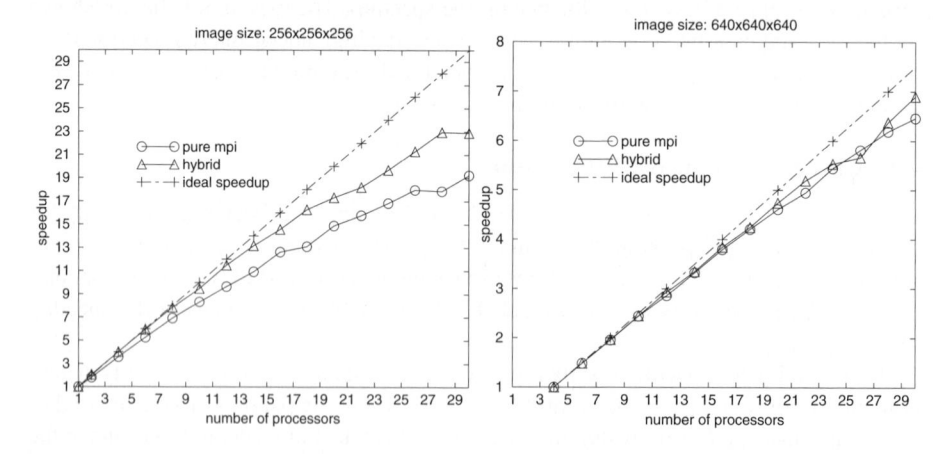

Fig. 4. Speedup on a cluster of SMPs of the pure MPI and hybrid codes for the volumes 256 and 640

processors increases since only communications between pairs of symmetric processors are involved for the hybrid code.

The influence of the problem size on the performance was also analyzed, and the conclusion is that the volume size proves to be relevant on this platform. This behavior is justified by two factors. First, the computational complexity depends linearly on the volume size whereas the amount of communications is proportional to the size of a single z-slice. Therefore, for small volumes the penalties from communications are relevant, specially for the pure MPI code. Second, the local memory hierarchy management improves for larger volume sizes and has a stronger impact in the scalability of both strategies. Therefore, any increase in the problem size is expected to imply a direct improvement in the speedup of both strategies.

5 Conclusions

In this work, we have presented parallel implementations of AND, using three strategies based on: (1) shared address space, (2) the message passing paradigm and (3) a hybrid approach. The evaluation has been carried out on two different architectures: (1) a Distributed Shared Memory platform based on cc-NUMA access and (2) a cluster of Symmetric Biprocessors based on UMA access. In view of the results, we can conclude that the parallel algorithms present good levels of scalability. Furthermore, the evaluation allows us to draw the conclusion that for DSM platforms like the Altix 3700 Bx2, all paradigms yield better and similar speedup. Consequently there is no favorable paradigm for this platform. However, for clusters of SMPs the hybrid paradigm (Pthreads+MPI) is more suitable than a strategy based solely on the message passing paradigm.

Acknowledgments

The authors wish to thank Dr. Guy Perkins (National Center for Microscopy and Imaging Research, San Diego, USA) for kindly providing the mitochondrion dataset.

References

1. Weickert, J.: Anisotropic Diffusion in Image Processing. Teubner (1998)
2. Perona, P., Malik, J.: Scale space and edge detection using anisotropic diffusion. IEEE Trans. Patt. Anal. Mach. Intel. **12** (1990) 629–639
3. Weickert, J.: Coherence-enhancing diffusion filtering. Int. J. Computer Vision **31** (1999) 111–127
4. J. Weickert: Coherence-enhancing diffusion of colour images. Image and Vision Computing **17** (1999) 201–212
5. Gerig, G., Kikinis, R., Kubler, O., Jolesz, F.A.: Nonlinear anisotropic filtering of MRI data. IEEE Trans. Med. Imaging **11** (1992) 221–232
6. Bajla, I., Hollander, I.: Nonlinear filtering of magnetic resonance tomograms by geometry-driven diffusion. Machine Vision and Applications **10** (1998) 243–255

7. Ghita, O., Robinson, K., Lynch, M., Whelan, P.F.: MRI diffusion-based filtering: A note on performance characterisation. Comput. Med. Imaging Graph. **29** (2005) 267–277
8. Frangakis, A.S., Hegerl, R.: Noise reduction in electron tomographic reconstructions using nonlinear anisotropic diffusion. J. Struct. Biol. **135** (2001) 239–250
9. Fernandez, J.J., Li, S.: An improved algorithm for anisotropic nonlinear diffusion for denoising cryo-tomograms. J. Struct. Biol. **144** (2003) 152–161
10. Fernandez, J.J., Li, S.: Anisotropic nonlinear filtering of cellular structures in cryo-electron tomography. Computing in Science and Engineering **7**(5) (2005) 54–61
11. Medalia, O., Weber, I., Frangakis, A.S., Nicastro, D., Gerisch, G., Baumeister, W.: Macromolecular architecture in eukaryotic cells visualized by cryoelectron tomography. Science **298** (2002) 1209–1213
12. Grunewald, K., Desai, P., Winkler, D.C., Heymann, J.B., Belnap, D.M., Baumeister, W., Steven, A.C.: Three-dimensional structure of herpes simplex virus from cryo-electron tomography. Science **302** (2003) 1396–1398
13. Beck, M., Forster, F., Ecke, M., Plitzko, J.M., Melchior, F., Gerisch, G., Baumeister, W., Medalia, O.: Nuclear pore complex structure and dynamics revealed by cryoelectron tomography. Science **306** (2004) 1387–1390
14. Cyrklaff, M., Risco, C., Fernandez, J.J., Jimenez, M.V., Esteban, M., Baumeister, W., Carrascosa, J.L.: Cryo-electron tomography of vaccinia virus. Proc. Natl. Acad. Sci. USA **102** (2005) 2772–2777
15. Fernandez, J.J., Lawrence, A., Roca, J., Garcia, I., Ellisman, M., Carazo, J.: High performance electron tomography of biological specimens. J. Struct. Biol. **138** (2002) 6–20
16. Fernandez, J.J., Carazo, J., Garcia, I.: Three-dimensional reconstruction of cellular structures by electron microscope tomography and parallel computing. J. Paral. Distr. Computing **64** (2004) 285–300
17. Weickert, J., ter Haar Romeny, B.M., Viergever, M.A.: Efficient and reliable schemes for nonlinear diffusion filtering. IEEE Trans. Image Processing **7** (1998) 398–410
18. Bruhn, A., Jakob, T., Fischer, M., Kohlberger, T., Weickert, J., Bruning, U., Schnorr, C.: High performance cluster computing with 3D nonlinear diffusion filters. Real-Time Imaging **10** (2004) 41–51
19. Press, W.H., Flannery, B.P., Teukolsky, S.A., Vetterling, W.T.: Numerical Recipes: The Art of Scientific Computing. Cambridge University Press (1992)
20. Barash, D.: A fundamental relationship between bilateral filtering, adaptive smoothing and the nonlinear diffusion equation. IEEE Trans. Patt. Anal. Mach. Intel. **24** (2002) 844–847
21. Dunigan, T., Vetter, J., Worley, P.: Performance evaluation of the SGI Altix 3700. In: Proceedings of the IEEE Intl. Conf. Parallel Processing, ICPP. (2005) 231–240
22. Akl, S.G.: Superlinear performance in real-time parallel computation. The Journal of Supercomputing **29**(1) (2004) 89–111

Improving the Research Environment of High Performance Computing for Non-cluster Experts Based on Knoppix Instant Computing Technology

Fumikazu Konishi[1], Manabu Ishii[2], Shingo Ohki[1], Yusuke Hamano[3], Shuichi Fukuda[2], and Akihiko Konagaya[1]

[1] RIKEN Genomic Science Center (GSC)
Advanced Genome Information Technology Research Group
Bioknowledge Federation Research Team
{fumikazu, ohki, konagaya}@gsc.riken.jp
[2] Tokyo Metropolitan Institute of Technology
[3] VSN Inc.

Abstract. We have designed and implemented a new portable system that can rapidly construct a computer environment where high-throughput research applications can be performed instantly. One challenge in the instant computing area is constructing a cluster system instantly, and then readily restoring it to its former state. This paper presents an approach for instant computing using Knoppix technology that can allow even a non-computer specialist to easily construct and operate a Beowulf cluster . In the present bio-research field, there is now an urgent need to address the nagging problem posed by having high-performance computers. Therefore, we were assigned the task of proposing a way to build an environment where a cluster computer system can be instantly set up. Through such research, we believe that the technology can be expected to accelerate scientific research. However, when employing this technology in bio-research, a capacity barrier exists when selecting a clustered Knoppix system for a data-driven bioinformatics application. We have approached ways to overcome said barrier by using a virtual integrated RAM-DISK to adapt to a parallel file system. To show an actual example using a reference application, we have chosen InterProScan, which is an integrated application prepared by the European Bioinformatics Institute (EBI) that utilizes many database and scan methods. InterProScan is capable of scaling workload with local computational resources, though biology researchers and even bioinformatics researchers find such extensions difficult to set up. We have achieved the purpose of allowing even researchers who are non-cluster experts to easily build a system of "Knoppix for the InterProScan4.1 High Throughput Computing Edition." The system we developed is capable of not only constructing a cluster computer environment composed of 32 computers in about ten minutes (as opposed to six hours when done manually), but also restoring the original environment by rebooting the pre-existing operating system. The goal of our instant cluster computing is to provide an environment in which any target application can be built instantly from anywhere.

1 Introduction

Over the last decade, high performance computing has become a fundamental technology essential for large-scale scientific research. The Beowulf[1,2,3] parallel workstation that consists of commercial PC components achieves a balanced low-cost architecture for an environment of single-user scientific workstations. However, as Philip Papadopoulos points out, the economics of clusters have changed due to additional and ongoing personnel costs related to the "care and feed" of the machine.[4,5,6]

This paper presents an image-based approach for a light-load deploying system using Linux-based Live CD technology. The image-based system adapts well to temporary usage. The original environment can be easily rolled back as well. We have designed and implemented a new portable system that can rapidly construct a computer environment where high-throughput search applications for protein analysis can be performed instantly. One challenge in this instant computing area is making a target system with a reasonable configuration to enable instant construction, and then easy restoration to the former state. The advantage of instant computing is its demonstrated available technology to solve a given problem without the need for special technical knowledge in order to build a system that can perform the intended application. Consequently, end users have practical needs for instant computing.

2 Related Work

Related work in instant computing technology can be divided into two groups: install-based systems and image-based systems.

2.1 Install-Based Deploying System

NPACI Rocks toolkit. The NPACI Rocks toolkit developed by the University of California at San Diego (UCSD) is designed to address a large-scale, cluster-work support infrastructure for applications that scientists can build and manage by themselves. Rocks has achieved the setup of a system consisting of hardware and software, and which is unified by prescribed system configurations. The Rocks cluster architecture inherited by the Beowulf project was defined as consisting of minimal components for which there are large mean-time-to-failure specifications. [4] The conventional Beowulf parallel workstation intended for scientific applications requiring the repetitive use of large data sets and large applications is composed of a front-end node and work-nodes with dual channel Ethernet networks. Therefore, the presumed Rocks hardware system offers both simplicity and a high degree of practicality. UCSD also developed a robust set of OS installation tools known as NPACI Rocks with RedHat Kickstart. UCSD had deployed all software by using RPM-based automatic configuration technology, and also supported a reinstallation mechanism for forcing the base OS on the computing nodes as well. As far as possible in the Rocks world, such technologies may facilitate the easy building of cluster systems by end users. In addition, similar efforts have been made to realize a light-speed deploying system for cluster computing the Real World Computing Partnership, Scyld Beowulf, Scalable Cluster Environment, Open Cluster Group, VA Linux, and Extreme Linux. These install-based management systems are fitted into a uniform cluster system that can be used for a long time. For weekend computing, end users may want to temporarily construct a cluster computer system, and will not accept a destructive reinstallation because the PC must be restored to its original condition.

2.2 Image-Based Deploying System

Knoppix. Knoppix is a collection of GNU/Linux and features one CD live file system (iso9660) that can be customized as a full-featured portable computer system. [7] The key technologies of Knoppix are automatic hardware detection and configuration, and a compressed loop-back device. The loop-back device allows us to mount a file as a block device, thus reducing the system file image and enabling the file to be read on the-fly decompression. After the CD has been mounted on the loop-back device, additional memory disks (tmpfs) are mounted with a writable ext2 file system for an application program as a normal Linux distribution system. The tmpfs size is adapted from the available size of real memory.

ClusterKnoppix. ClusterKnoppix is a Linux kernel extension for single-system image clustering that adopts Knoppix distribution using the OpenMosix kernel, and is designed to activate a cluster without having to install it on the hard disk. The MOSIX multi-computer system, which is improved by kernel algorithms for sharing the scalability of PC cluster resources, has a feature of preemptive process migration for dynamic load-balancing and memory-ushering, due to the management mechanism required for cluster-wide dynamically distributed resources in a time-sharing parallel execution environment for multiple users. MOSIX offers a general-purpose environment infrastructure for executing large scale, demanding sequential and parallel applications.[8] The MOSIX infrastructure includes such MOSIX File Systems as the Global File System (GFS)[10] and Parallel Virtual File System (PVFS)[11] that provide a unified view of all files on all mounted file systems in all nodes of the MOSIX cluster. The system consists of several functional components for easily building a cluster system. For booting up the work nodes via a network, ClusterKnoppix has the OpenMosix terminal server that integrates the Pre-Boot Execution Environment (PXE), Dynamic Host Configuration Protocol (DHCP), and tftp. Moreover, the system allows new nodes to join the cluster automatically by using the auto discovery feature for improved convenience.[12]

3 Application

In the present bio-research, bioinformatics applications that typically represent a data-oriented approach are utilized with a public and/or in-house database from which a researcher can obtain new findings about topics of interest. To prepare a research environment, bioinformatics applications usually require a long time to perform, and are not easy for researchers who are not experts on information technology. Furthermore, it is very troublesome to build and maintain a system for large-scale computation. We were assigned the task of proposing a way to build an environment where a cluster computer system can be instantly set up. This technology is expected to accelerate the pace of bioinformatics research. Building such a temporary system is not very appealing in view of existing research and development systems. Therefore, there is an urgent need to address this nagging problem.

We have chosen InterProScan [13] as a reference application, which represents data-oriented characteristics. InterProScan is an integrated application prepared by the European Bioinformatics Institute (EBI) that utilizes many databases and scan methods for protein signatures. These well-maintained databases include protein families, domains, and functional sites in which identifiable features found in known proteins can be applied to unknown protein sequences. InterProScan allows a protein science researcher to simultaneously scan several member databases, such as PROSITE patterns, PROSITE profile, PRINTS, PFAM, PRODOM, SMART, and TIGRFAMs. InterProScan is capable of scaling workload with local computational resources, though biology researchers and even bioinformatics researchers find such extensions difficult to set up.

4 System Design and Implementation

We have been addressing the difficulties of instantly constructing a high-performance cluster computing system, and improving the research environment to make it easy for non-cluster experts to do so. Our approach entails two main domains in the deploying environment. First, we will decide on a target application. Secondly, our second domain is remastering a specific service. This section describes the system design and implementation for remastering typical bioinformatics applications on Live-OS.

InterProScan consist of several functional scripts: system configure, pre-procedure, job submitting, status checker, post-procedure, and member database. The database contains 11,972 entries, representing 3079 domains, 8597 families, 228 repeats, 27 active sites, 21 binding sites, and 20 post-translational modification sites. Overall, there are 7,521,179 InterPro hits from 1,466,570 UniProt protein sequences in release 10.0. [14] Thus, the database contains 5.3G-byte file sets comprising 38,391 directories and 38,433 files. InterProScan is a well-known application with abundant directories.

As for why there are so many directories, there is an issue regarding how a protein family model file of the HMMER program is stored in each directory.

Thus, the file system must store much structure information as metadata. Moreover, InterProScan will also submit 12 jobs per sequence file. Thus, storage space greater than the file size of a member database is required when adding the file size of meta-data and the results.

A Live-OS offers the advantage of easily setting up the most suitable environment, but the technical issue of creating more than 6G bytes of data storage space for member databases and results must be addressed. Specifically, a single Live-OS node without a hard disk drive cannot be expected to obtain RAM-DISK space exceeding 2G byte. Therefore, to make our proposed method complementary, a parallel file system has been chosen to integrate RAM-DISK storage with a clustered Live-OS computer. To realize our method of an integrated RAM-DISK, we have designed KnoppixCluster using a traditional architecture for high-performance computing environments such as the Beowulf parallel workstation, which has been defined for single-user multiple computers. In order to develop InterProScan service on KnoppixCluster, we have developed a series of setup scripts: htc_hop, htc_step, and htc_jump. The htc_hop script executes a setup procedure to deploy the back-end image. First, the front-end is booted from a local CD-image. Then the front-end executes htc_hop to activate the features with our configuration, which includes the network card settings, DHCP IP ranges, and client-side NIC drivers. To enable these features, we have chosen the Knoppix terminal server, which allows thin clients such as diskless workstations. [15] After htc_hop is completed, the system is ready to start a back-end-node booting sequence. The back-end nodes must support a PXE prepared by the front-end node for network booting from the terminal server.

The htc_step script can then be executed on the front-end node, provided that the necessary number of back-end-nodes are booted up, thus allowing us to automatically create a configuration file for PVFS2 and Condor [16]. Our configurations that focus on instant computing can instantly create an on-memory-parallel-file system using PVFS2 to integrate a specified memory disk (tempfs) that is mounted with a writable ext2 file system on the back-end nodes. In order to build a service environment for InterProScan4.1 with database release 10.0, additional capacity of 1.2G bytes is necessary for that data structure, although a capacity of 5G bytes should be sufficient to store a database. Thus, the quantity necessary for this structure information can be obtained through experimental observation beforehand. Therefore, this system must use PVFS2 to create a total capacity greater than 6.2G bytes. This capacity thus becomes the condition on which to maintain the minimum system configuration. This condition is evaluated based on the run-time system capacity on the back-end nodes, then the possibility of said system configuration is evaluated, and the script provides information for the end-user. The back-end node serves an important role in providing the on-memory-parallel-filesystem. The back-end node also functions as a work node to perform a given task at the same time. In order to utilize the back-end nodes, the condor scheduling system allows us to deal with parallel jobs involving large-volume processing. The condor is set up by running a condor setup program (condor_configure) on each back-end node in the on-memory-parallel-filesystem through serial processing. In order to collect the software and hardware of "headth" on all nodes, we have chosen Ganglia, which is a lightweight, distributed, multicast-based monitoring system. Ganglia allows us to indicate the number and speed of the CPUs, the kernel version, the amount of RAM installed, and more useful information about all nodes. After all setups are completed, a test job is performed in the htc_jump script to confirm whether all setups can be properly executed, and with the results being verified.

5 Performance

5.1 The Evaluation of Application Performance by the Boot Memory Model

We have evaluated the effects of dividing main memory in the Knoppix Instant Computing System on an application because rewritable space is important for practical use. This problem dictates how much main memory size can be assigned to RAM-DISK to execute typical bioinformatics applications on KNOPPIX.

For example, a computer with 1G byte of memory is able to assigned 400M bytes of memory for operating system use; thus, the remaining 600M bytes of memory can be assigned for RAM-DISK use. When consisting of ten cluster nodes, this system can create a rewritable capacity of 6G bytes . The capacity that can be used for the on-memory-parallel-file system is reduced when too much quantity is allocated to the operating system. The system thus assigns the necessary and sufficient conditions of system memory for a bioinformatics application.

Table 1 shows the test equipment that we used to build Knoppix Cluster for this performance evaluation. The front-end node has 4G bytes of memory to ensure the stability of system operation, and the back-end nodes have 2G bytes of memory. Table 2 shows the experimental conditions for the different boot memory models. As for the experiment, we measured the execution time of

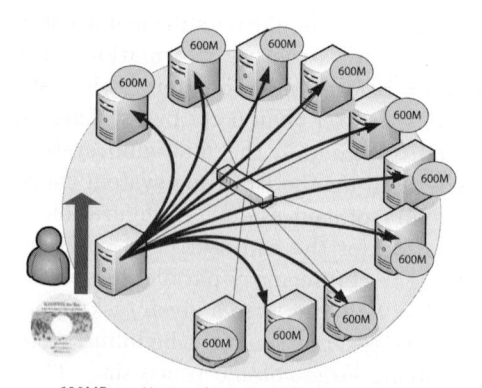

600MBytes X 10 nodes = 6000MBytes = 6GBytes

Fig. 1. A boot model for Knoppix Cluster

Table 1. Test Equipment

	Front-end node	Back-end node
CPU	Pentium4 2.4 GHz	Pentium4 2 GHz
Main Memory	4G bytes	2G bytes
NIC	1000 Base-T	100 Base-T
node	1	10

Table 2. Experiment Conditions for boot memory models

item	parameter
Memory Model	high (800 MB) low (400 MB)
File System	PVFS ver.1
application	Parallel Blast
Database Size (sequences)	1000,10000,10000
Query (sequences)	1,2,4,8,10,100
Number of Proc. (CPU)	1,2,4,8,10
Rep.	5

parallel BLAST extended from NCBI BLAST [18] with a combination of query and database size on the on-memory-parallel-filesystem in both High mode and Low mode. The experiment was repeated five times to evaluate repeatedly, and a total trial experiment count of 1350 times was enforced.

Figure 2 shows the homology search throughput, which is the capability of each memory model in units of time. There were no differences between the two memory models based on the results of the wide-range parameter sweep experiment. Therefore, it was shown that similar performance could be expected for a special configuration of the on-memory-parallel-filesystem.

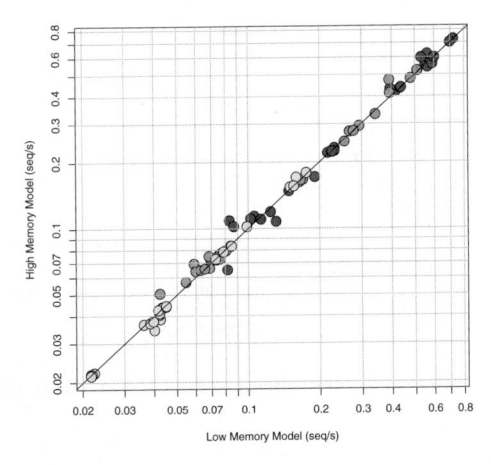

Fig. 2. Homology Search Throughput based on differences in boot memory proportion

5.2 The Evaluation of Instant System Setup Performance

Knoppix-Cluster allows us to instantly reproduce a final system configuration by embedding a construction script in a boot image. An end user without special expertise can use this mechanism to build a cluster system. Table 3 shows the setup time for each step in our instant Knoppix Cluster. It took about ten minutes (on a 30-node scale) to build a cluster computer on the instant system. Even when compared with the reinstallation time stated in the reference paper about ROCKS [4], performance equivalent to that above is realized for this setup time.

Table 3. The setup time for an instant cluster

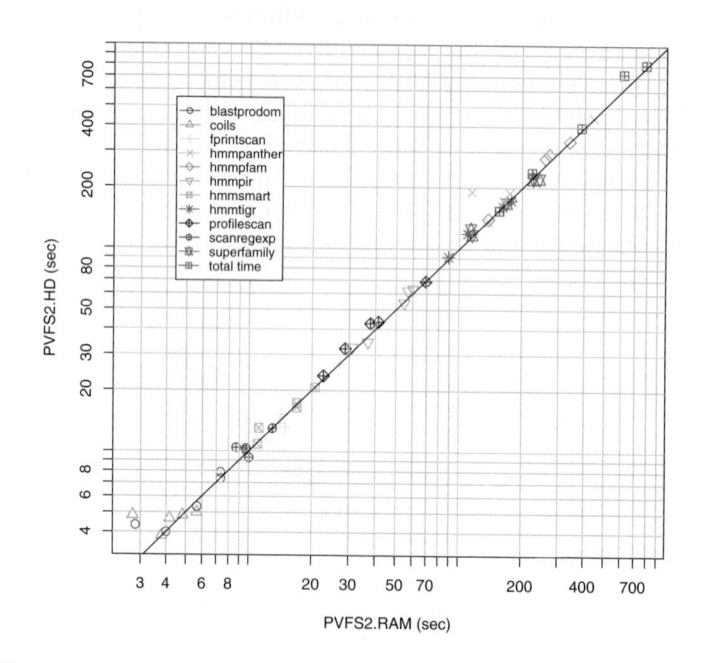

Script		Time (sec)		
Work nodes size		10	20	30
HTC_hop		22	21	19
HTC_Step		214	419	619
File System	(PVFS2)	46	82	118
Scheuler	(Condor)	156	299	445
Monitor	(Ganglia)	12	21	32
Total Time		234	440	638

Fig. 3. The execution time of InterProScan versus RAM-DISK (RD) and Hard Disk (HD)

5.3 The Evaluation of InterProScan4.1 Performance for Instant Computing

An experiment was conducted to verify practical use of the high-throughput application environment instantly provided for non-cluster experts. InterProScan4.1 is a well-known integrated application that can perform a search using 12 programs and a database. Each application has a different executive time distribution. Figure 3 shows the total execution time difference between RAM-DISK and Hard DISK. Since Knoppix Cluster is built by using the on-memory-parallel-filesystem (which is a case of special use), we had to observe the difference in practical execution time.

6 Conclusion

We have presented and implemented a new approach to image-based instant computing technology on Knoppix Cluster for improving the research environment of high performance computing for non-cluster experts. Our work represents the first step in exploring design and implementation issues regarding instant computing technology. We have been very particular about restoring the original condition as held before. We expanded instant computing by using the on-memory-parallel-filesystem for image-based technology from install-based technology, and this technology enabled us to handily build a cluster system. Consequently, we considered its adaptation to practical bioinformatics applications and succeeded in building an InterProScan4.1 environment and distributing images for Knoppix using the InterProScan4.1 High Throughput Computing Edition. [19] The results can then be used as one infrastructure for deploying application service.

Acknowledgments

The authors would like to thank Kuniyasu Suzaki for his valuable contributions regarding Knoppix.

References

1. Donald J. Becker, Thomas Sterling, Daniel Savarese, John E. Dorband, Udaya A. Ranawak, Charles V. Packer, "BEOWULF: A PARALLEL WORKSTATION FOR SCIENTIFIC COMPUTATION", Proceedings, International Conference on Parallel Processing, (1995)
2. Thomas Sterling, Daniel Savarese, Donald J. Becker, Bruce Fryxell, Kevin Olson, "Communication Overhead for Space Science Applications on the Beowulf Parallel Workstation", Proceedings,High Performance and Distributed Computing, (1995) 23–30
3. Thomas Sterling, Donald J. Becker, Daniel Savarese, Michael R. Berry, and Chance Res. ,"Achieving a Balanced Low-Cost Architecture for Mass Storage Management through Multiple Fast Ethernet Channels on the Beowulf Parallel Workstation", Proceedings, International Parallel Processing Symposium, (1996) 104–108

4. Philip M. Papadopoulos, Mason J. Katz, and Greg Bruno, "NPACI Rocks: Tools and Techniques for Easily Deploying Manageable Linux Clusters", Cluster 2001: IEEE International Conference on Cluster Computing, Oct. (2001)
5. Mason J. Katz, Philip M. Papadopoulos, and Greg Bruno, "Leveraging Standard Core Technologies to Programmatically Build Linux Cluster Appliances", Cluster 2002: IEEE International Conference on Cluster Computing, Apr. (2002)
6. Philip M. Papadopoulos, Caroline A. Papadopoulos, Mason J. Katz, William J. Link, and Greg Bruno, "Configuring Large High-Performance Clusters at Lightspeed: A Case Study", Clusters and Computational Grids for Scientific Computing 2002, Dec (2002)
7. Klaus Knopper, "Building a self-contained autoconfigurarion Linux system on an iso9660 file system", 4th Annual Linux Showcase & Conference Atlanta, (2000)
8. Barak A. and La'adan O., "The MOSIX Multicomputer Operating System for High Performance Cluster Computing", Journal of Future Generation Computer Systems (13) 4-5, pp. 361-372, March (1998)
9. Amar L., Barak A. and Shiloh A., "The MOSIX Parallel I/O System for Scalable I/O Performance", Proc. 14-th IASTED International Conference on Parallel and Distributed Computing and Systems (PDCS 2002), pp. 495-500, Cambridge, MA, Nov. (2002)
10. Global File System, "http://www.redhat.com/en_us/USA/home/solutions/gfs/"
11. P. H. Carns, W. B. Ligon III, R. B. Ross, and R. Thakur, "PVFS: A Parallel File System For Linux Clusters", Proceedings of the 4th Annual Linux Showcase and Conference, Atlanta, GA, pp. 317-327, Oct (2000)
12. ClusterKnoppix, "http://bofh.be/clusterknoppix/"
13. Evgeni M. Zdobnov and Rolf Apweiler, "InterProScan–an integration platform for the signature-recognition methods in InterPro. Bioinformatics", 17(9):847-8, Sep (2001)
14. Mulder NJ, et al. , "InterPro, progress and status in 2005", Nucleic Acids Res. 33, Database Issue:D201-5 (2005)
15. Linux Terminal Server Project, http://www.ltsp.org/
16. Michael Litzkow, Miron Livny, and Matt Mutka, "Condor - A Hunter of Idle Workstations", Proceedings of the 8th International Conference of Distributed Computing Systems, pages 104-111, June, (1988)
17. Todd Tannenbaum, Derek Wright, Karen Miller, and Miron Livny, "Condor - A Distributed Job Scheduler", in Thomas Sterling, editor, Beowulf Cluster Computing with Linux, The MIT Press, (2002)
18. Altschul, S.F., Gish, W., Miller, W., Myers, E.W. & Lipman, D.J. ,"Basic local alignment search tool." J. Mol. Biol. 215:403-410 (1990)
19. Knoppix for InterProScan4.1 High Throughput Computing Editon http://big.gsc.riken.jp/index_html/Members/fumikazu/htc/

Topic 18: Embedded Parallel Systems

Jürgen Teich, Stefanos Kaxiras, Toomas Plaks, and Krisztián Flautner

Topic Chairs

Multi-processor systems implemented in System-on-a-Chip technology (MPSoC) are emerging for processing embedded applications such as consumer electronics, mobile phones, computer graphics, and medical imaging, to name a few. Contrary to cluster and grid processing, their design and required compilation techniques are driven by multiple conflicting design objectives simultaneously such as power consumption, speed, monetary cost, and physical as well as memory size. Here, new specification techniques, special parallelization and mapping techniques are needed in order to embed computations optimally into the parallel architecture. Various architectural concepts ranging from fine-grain to coarse-grain parallel SoC architectures with focus on dynamic programmability or reconfigurability are currently emerging in academia and industry.

On account of the outlined importance of MPSoCs in todays and future embedded systems, the topic *Embedded Parallel Systems* is included in the program of Euro-Par for the first time. Unfortunately, the number of submitted papers was comparatively small but nevertheless after a rigorous review process we accepted three papers which form one session at the conference. In the following, we provide a brief outline of the topics addressed in these contributions.

The paper titled "Optimal Localization of Data Dependencies in Algorithm Partitioning Under Resource Constraints" by S. Siegel and R. Merker deals with the communication in dedicated processor arrays. The authors propose an integer linear program in order to minimize the number of necessary channels and the amount of local memory.

The work "FPGA implementation of a Prototype Hierarchical Control Network for Large-Scale Signal Processing Applications" by J. Lemaitre and E. Deprettere presents a prototypical FPGA implementation of a hierarchical control network coupled with a distributed dataflow network. For modeling the network, communicating Finite State Machines and Kahn Process Networks are used.

The third paper "An Embedded Systems Programming Environment for C" by B. Burgstaller, B. Scholz, and A. Ertl presents a programming environment for mixed-mode execution, i.e. code is either executed on the CPU or in a virtual machine. Trade-offs between highly compressed byte-code and the speed of machine code are discussed.

The Topic Committee would like to sincerely thank all the authors submitted papers and the referees who helped with the reviewing process. In particular, we would like to thank Frank Hannig for his valuable assistance in the organization of this topic. Finally, we would like to thank the Euro-Par 2006 Organizing Committee for their support to establish this topic.

Efficient Realization of Data Dependencies in Algorithm Partitioning Under Resource Constraints

Sebastian Siegel and Renate Merker

Dresden University of Technology, Institute of Circuits and Systems,
01062 Dresden, Germany
{siegel, merker}@iee1.et.tu-dresden.de

Abstract. Mapping algorithms to parallel architectures efficiently is very important for a cost-effective design of many modern technical products. In this paper, we present a solution to the problem of efficiently realizing uniform data dependencies on processor arrays. In contrary to existing approaches, we formulate an optimization problem to consider the cost of both: channels and registers. Further, a solution to the optimization problem assigns which channels shall be implemented and it specifies the control for the realization of the uniform data dependencies. We illustrate our method on the edge detection algorithm.

1 Introduction

Many modern technical products need to cope with fast digital signal, image and video processing under real-time requirements. Massively parallel data processing on processor arrays (PAs) is known to accelerate compute-intensive algorithms. The semiconductor industry presents more and more solutions for implementations of PAs in portable and other embedded systems. These solutions range from ASICs, reconfigurable systems in FPGAs, arrays of CPU-cores to platforms such as DRP from NEC [1] and picoArray [2].

PAs are mainly characterized by their processing elements (PEs). Some PEs are connected to the periphery, e. g. via a memory hierarchy (Fig. 1 (b)). The PEs are characterized by functional units and local memory. This paper focuses on the communication within the PA which is realized by regular local interconnections between the PEs.

To exploit the processing performance of PAs, we apply a new design flow which consists of mainly **two steps: 1)** partitioning the algorithm in order to match the PA parameters such as shape of PA, number of PEs, communication to a memory hierarchy [3] and **2)** realizing the data dependencies of the algorithm on the PA. This paper deals with the second step.

The first step can be summarized as follows: We consider compute-intensive algorithms e. g. described as systems of uniform recurrence equations (SUREs) [4]. They consist of many elementary computational tasks, the so-called *iterations*, that are aligned in an iteration space. Using our parameterized partitioning method [5], we map algorithms to PAs, i. e. each iteration to a PE (allocation) and determine the corresponding time of execution (schedule). The optimal use of the data path of each PE for the operations of the algorithm is addressed in [6,7,8,9].

This paper considers the second step: the realization of the data dependencies of an algorithm on the PA. The data dependencies cause data transfer within the PA. This data transfer is realized using *channels* between PEs and *registers* within PEs. Other existing works consider only one or the other. The optimal use of channels for the realization of data dependencies is considered in [10] for one-dimensional PAs and in [11] embedded in the traditional design flow which applies linear space-time mappings. The optimal use of registers is regarded in [12,13] for a single processor machine or in [14] for the design of PAs based on the traditional design flow.

We present an approach which addresses the optimal use of both: channels and registers for the realization of the uniform data dependencies of the algorithms. It is important to consider channels and registers in one model because channels with a delay (e. g. with a pipeline structure) can reduce the usage of registers.

Given several channels and their implementation cost, we formulate and solve the *communication problem* by integer linear programming (ILP). A solution to the communication problem specifies which channels shall be implemented and the control of the data dependencies, i. e. when a channel or a register is used to realize a data dependency. This solution also includes the specification of an inner schedule for all computational tasks (given by the statements of the SURE) within an iteration. Our approach can be extended by existing methods concerning the optimal use of the data path.

To select the channels with minimum cost, our method can be applied in several ways. E. g. in reconfigurable computing where different algorithms shall be implemented on the same PA, we can determine which channels would best realize the communication within the PA for each algorithm. This information combined with the reconfiguration cost can lead to an efficient solution for the communication. In [15] we consider savings in the communication cost by avoiding redundancy. There it is necessary to know the channel selection (binding of data dependencies to channels) a priori. We apply the method presented in this paper to determine an efficient channel selection.

This paper is organized as follows. We describe the notation of algorithms as SUREs and we summarize partitioning in Sect. 2. Section 3 is the main contribution of this paper. In this section we derive a method to formulate and solve the communication problem. We give a solution to the communication problem for an example application in Sect. 4. Finally we draw some conclusions in Sect. 5.

2 Algorithm Coding and Partitioning

To demonstrate our methods for the communication problem, we consider systems of uniform recurrence equations which are defined as follows:

Definition 1 (System of Uniform Recurrence Equations (SURE)).
A system of uniform recurrence equations consists of a set of J statements $(1 \leq j \leq J)$ of the form

$$S_j: \quad y_j[\mathbf{i}] := f_j\left(\ldots, y_i\left[\mathbf{i} - \widetilde{\mathbf{d}}_{j,i}^r\right], \ldots\right), \forall \mathbf{i} \in \mathcal{I}_j, i, j \in \mathbb{N}$$

where $\mathcal{I}_j$ denotes the iteration space of statement S_j, $\mathcal{I}_j$ is a polyhedral subset of a $\mathbb{Z}$-module and f_j denotes a single-valued function.

A variable y_i that is computed by statement S_i is a dependent variable if it is input to some (other) statement S_j. Vector $\widetilde{\mathbf{d}}_{j,i}^r$ denotes the corresponding uniform data dependency. Upper index "r" is used only if more that one data dependency exists between statements S_j and S_i.

With the embedding given by $\mathcal{I} = \text{conv}(\bigcup_j \mathcal{I}_j) \subset \mathbb{Z}^n$ we determine the iteration space $\mathcal{I}$ of the SURE. In Algorithm 1 we show the edge detection algorithm (EDA) in the notation of a SURE which we use throughout this paper as an example.

Algorithm 1. Edge detection algorithm (EDA)

$S_1: \quad q\begin{bmatrix}x\\y\end{bmatrix} = 2 \cdot p_i\begin{bmatrix}x\\y\end{bmatrix}, \quad \begin{pmatrix}x\\y\end{pmatrix} \in \mathcal{I}_1 = \{\begin{pmatrix}x\\y\end{pmatrix} \in \mathbb{Z}^2 \mid \begin{smallmatrix}0 \le x \le N-1\\ 0 \le y \le M-1\end{smallmatrix}\}$

$S_2: \quad h_1\begin{bmatrix}x\\y\end{bmatrix} = p_i\begin{bmatrix}x\\y-1\end{bmatrix} + p_i\begin{bmatrix}x\\y+1\end{bmatrix}, \quad \begin{pmatrix}x\\y\end{pmatrix} \in \mathcal{I}_2 = \{\begin{pmatrix}x\\y\end{pmatrix} \in \mathbb{Z}^2 \mid \begin{smallmatrix}0 \le x \le N-1\\ 1 \le y \le M-2\end{smallmatrix}\}$

$S_3: \quad h_2\begin{bmatrix}x\\y\end{bmatrix} = h_1\begin{bmatrix}x\\y\end{bmatrix} + q\begin{bmatrix}x\\y\end{bmatrix}, \quad \begin{pmatrix}x\\y\end{pmatrix} \in \mathcal{I}_3 = \mathcal{I}_2$

$S_4: \quad v_1\begin{bmatrix}x\\y\end{bmatrix} = p_i\begin{bmatrix}x-1\\y\end{bmatrix} + p_i\begin{bmatrix}x+1\\y\end{bmatrix}, \quad \begin{pmatrix}x\\y\end{pmatrix} \in \mathcal{I}_4 = \{\begin{pmatrix}x\\y\end{pmatrix} \in \mathbb{Z}^2 \mid \begin{smallmatrix}1 \le x \le N-2\\ 0 \le y \le M-1\end{smallmatrix}\}$

$S_5: \quad v_2\begin{bmatrix}x\\y\end{bmatrix} = v_1\begin{bmatrix}x\\y\end{bmatrix} + q\begin{bmatrix}x\\y\end{bmatrix}, \quad \begin{pmatrix}x\\y\end{pmatrix} \in \mathcal{I}_5 = \mathcal{I}_4$

$S_6: \quad h_3\begin{bmatrix}x\\y\end{bmatrix} = h_2\begin{bmatrix}x-2\\y-1\end{bmatrix} - h_2\begin{bmatrix}x\\y-1\end{bmatrix}, \quad \begin{pmatrix}x\\y\end{pmatrix} \in \mathcal{I}_6 = \{\begin{pmatrix}x\\y\end{pmatrix} \in \mathbb{Z}^2 \mid \begin{smallmatrix}2 \le x \le N-1\\ 2 \le y \le M-1\end{smallmatrix}\}$

$S_7: \quad h_4\begin{bmatrix}x\\y\end{bmatrix} = |h_3\begin{bmatrix}x\\y\end{bmatrix}|, \quad \begin{pmatrix}x\\y\end{pmatrix} \in \mathcal{I}_7 = \mathcal{I}_6$

$S_8: \quad v_3\begin{bmatrix}x\\y\end{bmatrix} = v_2\begin{bmatrix}x-1\\y-2\end{bmatrix} - v_2\begin{bmatrix}x-1\\y\end{bmatrix}, \quad \begin{pmatrix}x\\y\end{pmatrix} \in \mathcal{I}_8 = \mathcal{I}_6$

$S_9: \quad v_4\begin{bmatrix}x\\y\end{bmatrix} = |v_3\begin{bmatrix}x\\y\end{bmatrix}|, \quad \begin{pmatrix}x\\y\end{pmatrix} \in \mathcal{I}_9 = \mathcal{I}_6$

$S_{10}: \quad s\begin{bmatrix}x\\y\end{bmatrix} = h_4\begin{bmatrix}x\\y\end{bmatrix} + v_4\begin{bmatrix}x\\y\end{bmatrix}, \quad \begin{pmatrix}x\\y\end{pmatrix} \in \mathcal{I}_{10} = \mathcal{I}_6$

$S_{11}: \quad p_o\begin{bmatrix}x-1\\y-1\end{bmatrix} = \min(255, s\begin{bmatrix}x\\y\end{bmatrix}), \quad \begin{pmatrix}x\\y\end{pmatrix} \in \mathcal{I}_{11} = \mathcal{I}_6$

Next we briefly describe locally parallel, globally sequential (LPGS) partitioning [5] and we use it as a parameterized method to directly map an algorithm to a PA. LPGS-partitioning separates an iteration $\mathbf{i} \in \mathcal{I}$ into $\widehat{\boldsymbol{\kappa}}$ (denoting a partition) and $\boldsymbol{\kappa}$ (representing the position within a partition) by the tiling step as follows [5]:

$$\mathbf{i} = \Theta\widehat{\boldsymbol{\kappa}} + \boldsymbol{\kappa}, \quad 0 \le \kappa_k < \vartheta_k, 1 \le k \le n, \widehat{\boldsymbol{\kappa}} \in \widehat{\mathcal{K}} \subset \mathbb{Z}^n, \boldsymbol{\kappa} \in \mathcal{K} \subset \mathbb{N}^n \quad (1)$$

where $\Theta = \text{diag}(\vartheta_1 \cdots \vartheta_n) \in \mathbb{N}^{n \times n}$ is a square matrix whose diagonal elements represent the size of the partitions in each of the n directions of the iteration space $\mathcal{I}$. The size of the partitions corresponds to the size of the PA. Only two elements of Θ may be greater than one to obtain a two-dimensional PA. The PA consists of $\prod_{i=1}^{n} \vartheta_i$ PEs.

We extend the scheduling function from [5] to determine the time of execution for each statement S_j of an iteration given by $\widehat{\boldsymbol{\kappa}}$ and $\boldsymbol{\kappa}$ as follows:

$$t_j(\widehat{\boldsymbol{\kappa}}, \boldsymbol{\kappa}) = \lambda \boldsymbol{\tau} \widehat{\boldsymbol{\kappa}} + \boldsymbol{\tau}^{\text{offs}} \boldsymbol{\kappa} + b_j \quad \text{with} \quad \boldsymbol{\tau}, \boldsymbol{\tau}^{\text{offs}} \in \mathbb{Z}^{1 \times n}, b_j \in \mathcal{L} . \quad (2)$$

The first term in (2) determines the starting time for each partition. The second term allows to shift the schedule within a partition according to τ^{offs} to avoid data dependency conflicts or to change the time behavior of the I/O of the PA. With b_j we determine the starting time for each statement S_j within the iteration interval λ which is defined as follows:

Definition 2 (Iteration Interval λ). *The iteration interval λ denotes the number of time steps between the beginning of two successive iterations. The set $\mathcal{L} = \{0, 1, 2, \ldots, \lambda - 1\}$ consists of these time steps.*

Note that the parameters b_j will be specified by a solution to the communication problem.

In Fig. 1 (a) we show LPGS-Partitioning for the EDA. Each circle denotes an iteration. The numbers within each circle describe the beginning and the end of the corresponding iteration interval. In Fig. 1 (b) we illustrate the obtained PA with a memory hierarchy. Memory L_0 denotes local registers whose cost will be determined in the communication problem. Some boundary PEs are connected to the memory hierarchy whose size depends on the tile size and the size of the image (M and N). We refer to [3] where we describe how to determine this memory hierarchy. The local interconnections are not depicted in Fig. 1 (b). They will be determined by solving the communication problem.

We introduce the set $\widetilde{\mathcal{D}}$ which comprises all uniform data dependencies with $\forall \tilde{\mathbf{d}} \in \widetilde{\mathcal{D}}: \tilde{d}_k < \vartheta_k$ where $1 \leq k \leq n$. For a data dependency $\tilde{\mathbf{d}} \in \widetilde{\mathcal{D}}$ there exists at least one iteration within a partition serving as the source of the data dependency $\tilde{\mathbf{d}}$ and there exists at least one iteration in the same partition serving as the corresponding drain.

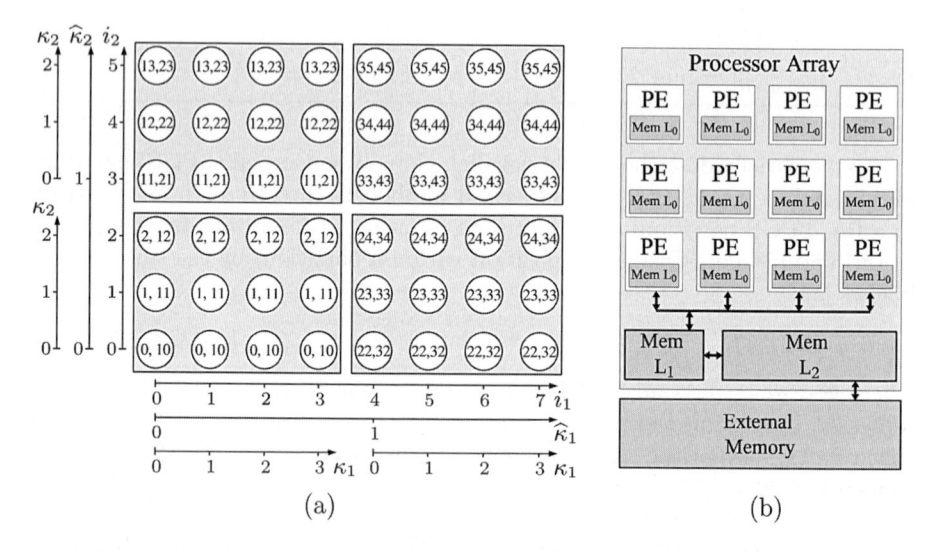

Fig. 1. (a) LPGS-Partitioning of the EDA with an iteration space of size $M \times N = 8 \times 6$ according to $\Theta = \text{diag}(4\ 3)$ and $t_j(\widehat{\boldsymbol{\kappa}}, \boldsymbol{\kappa}) = 11 \cdot (2\ 1) \cdot \widehat{\boldsymbol{\kappa}} + (0\ 1) \cdot \boldsymbol{\kappa} + b_j$, (b) Corresponding PA with a 2-level memory hierarchy

Let $\vartheta_{k_1} > 1$ and $\vartheta_{k_2} > 1$ represent the two dimensions of the partitions. Then, only the elements $\widetilde{d}_{k_1}$ and $\widetilde{d}_{k_2}$ of any vector $\widetilde{\mathbf{d}} \in \widetilde{\mathcal{D}}$ can be greater than zero. To make things easier (especially for $\dim(\mathcal{I}) > 2$), we introduce the set $\mathcal{D} \in \mathbb{Z}^2$ as follows: Each vector $\mathbf{d} \in \mathcal{D}$ corresponds to one and only one vector $\widetilde{\mathbf{d}} \in \widetilde{\mathcal{D}}$ with $d_1 = \widetilde{d}_{k_1}$ and $d_2 = \widetilde{d}_{k_2}$. Hence there exists a bijective mapping between the sets $\widetilde{\mathcal{D}}$ and $\mathcal{D}$. Note that one-dimensional partitions can be regarded as a special case where w. l. o. g. we set $\vartheta_{k_2} = 1$.

For the example of the EDA we consider partitions with $\vartheta_1, \vartheta_2 \geq 3$. Therefore, all the 13 data dependencies which can be extracted from Algorithm 1 belong to the set $\mathcal{D}_{\text{EDA}}$:

$$\mathcal{D}_{\text{EDA}} = \{ \mathbf{d}_{3,1} = \binom{0}{0},\ \mathbf{d}_{3,2} = \binom{0}{0},\ \mathbf{d}_{5,1} = \binom{0}{0},\ \mathbf{d}_{5,4} = \binom{0}{0},\ \mathbf{d}_{6,3}^1 = \binom{2}{1},\ \mathbf{d}_{6,3}^2 = \binom{0}{1},\ \mathbf{d}_{7,6} = \binom{0}{0},$$
$$\mathbf{d}_{8,5}^1 = \binom{1}{2},\ \mathbf{d}_{8,5}^2 = \binom{1}{0},\ \mathbf{d}_{9,8} = \binom{0}{0},\ \mathbf{d}_{10,7} = \binom{0}{0},\ \mathbf{d}_{10,9} = \binom{0}{0},\ \mathbf{d}_{11,10} = \binom{0}{0} \} \ .$$

3 Communication Problem

The uniform data dependencies given by the set $\mathcal{D}$ need to be realized by a conflict free organization of the data transfer they cause. The communication problem consists in minimizing the implementation cost in terms of channels and registers for these data dependencies. In order to determine this cost, we introduce a model which allows a description of the communication.

3.1 Modelling the Communication

A set $\mathcal{W} = \{\mathbf{w}_1, \mathbf{w}_2, \ldots, \mathbf{w}_{|\mathcal{W}|}\}$ of channels $\mathbf{w}_i \in \mathbb{Z}^2$ between PEs is supposed to be given. The elements of $\mathcal{W}^{\text{bi}} \subseteq \mathcal{W}$ denote the channels that may be used bidirectional. To each element $\mathbf{w}_k$ of $\mathcal{W}$ corresponds a delay $l_k^{\mathbf{w}} \in \mathbb{N}$ which represents the time it takes to transfer an instance of a variable on that channel. Each channel with $l_k^{\mathbf{w}} = 0$ denotes a broadcast. Channels with $l_k^{\mathbf{w}} > 0$ represent a pipeline structure which may only be used onedirectional. Hence the delay of channel $\mathbf{w}_k$ may only be non-zero if $\mathbf{w}_k \in \mathcal{W} \setminus \mathcal{W}^{\text{bi}}$.

The realization of data dependency $\mathbf{d}_{j,i}$ consists of two parts. First, it realizes the transfer of an instance of variable y_i from its source to the relative position given by $\mathbf{d}_{j,i}$. Second, the realization is responsible for the storage of that instance until it is input to a data path which executes statement S_j.

We model the transfer of each instance of a dependent variable y_i by a *sequence of moves* which describes the path from its source to its drain. A sequence of moves m where $0 \leq m \leq M_{j,i}$ is characterized by a mapping $\{0, 1, 2, \ldots, M_{j,i}\} \longrightarrow \mathcal{W} \cup \{\mathbf{w}_0\}$ which determines the order in which the channels are used for a realization of data dependency $\mathbf{d}_{j,i}$. With $M_{j,i}$ we denote the maximum number of moves it may take to realize a data dependency $\mathbf{d}_{j,i}$. We determine each $M_{j,i}$ a priori according to [16]. The element $\mathbf{w}_0$ represents "no transfer". With $m = 0$ we describe an initial "move" which will always be mapped to $\mathbf{w}_0$.

Example 1. Suppose $\mathcal{W} = \{\mathbf{w}_1 = \binom{1}{0}, \mathbf{w}_2 = \binom{0}{1}, \mathbf{w}_3 = \binom{1}{1}\}$, $\mathbf{d} = \binom{1}{1}$ and $M = 2$. The following four different mappings would realize data dependency $\mathbf{d}$:

1)
$0 \to \mathbf{w}_0$
$1 \to \mathbf{w}_1$
$2 \to \mathbf{w}_2$

2)
$0 \to \mathbf{w}_0$
$1 \to \mathbf{w}_2$
$2 \to \mathbf{w}_1$

3)
$0 \to \mathbf{w}_0$
$1 \to \mathbf{w}_3$
$2 \to \mathbf{w}_0$

4)
$0 \to \mathbf{w}_0$
$1 \to \mathbf{w}_0$
$2 \to \mathbf{w}_3$

To determine this mapping is one task of the communication problem. We add binary variables $\beta_{j,i,k,m}$ to the communication problem which parameterize this mapping as follows:

$$\beta_{j,i,k,m} = \begin{cases} 1 & \text{if channel } \mathbf{w}_k \text{ is used at move } m \text{ to realize data dep. } \mathbf{d}_{j,i} \\ 0 & \text{otherwise} \end{cases} \quad . \quad (3)$$

In order to assure that each move is mapped to one and only one channel, the variables $\beta_{j,i,k,m}$ are subject to:

$$\forall \mathbf{d}_{j,i} \in \mathcal{D} : \sum_{k=0}^{|\mathcal{W}|} \beta_{j,i,k,m} = 1, \quad 0 \leq m \leq M_{j,i} \quad \text{where} \quad \beta_{j,i,0,0} = 1 \ . \quad (4)$$

To avoid mappings which represent similar paths as given in Example 1 by realizations 3) and 4), we force mappings to channel $\mathbf{w}_0$ to be placed as far to the end as possible in the sequence of moves for $m \geq 1$. Therefore we add the following constraints:

$$\forall \mathbf{d}_{j,i} \in \mathcal{D} : \beta_{j,i,0,m+1} \geq \beta_{j,i,0,m}, \quad 1 \leq m \leq M_{j,i} - 1 \ . \quad (5)$$

To distinguish between the direction in which a bidirectional channel $\mathbf{w}_k \in \mathcal{W}^{\text{bi}}$ is used we add binary variables $\beta^{\text{bi}}_{j,i,k}$ to the communication problem which parameterize the direction of a potential use of channel $\mathbf{w}_k$ by data dependency $\mathbf{d}_{j,i}$ as follows:

$$\beta^{\text{bi}}_{j,i,k} = \begin{cases} 0 & \text{if channel } \mathbf{w}_k \text{ would be used in the direction given by } \mathbf{w}_k \\ 1 & \text{if channel } \mathbf{w}_k \text{ would be used in the direction given by } -\mathbf{w}_k \end{cases} \quad (6)$$

where $\mathbf{w}_k \in \mathcal{W}^{\text{bi}}$. The path of a dependent variable from its source to its drain is fixed once the sequence of moves is determined. To ensure that this path leads to the correct final destination, we add the following constraints:

$$\forall \mathbf{d}_{j,i} \in \mathcal{D} : \mathbf{p}^{\text{rel}}_{j,i,M_{j,i}} = \mathbf{d}_{j,i} \quad (7)$$

where $\mathbf{p}^{\text{rel}}_{j,i,m}$ describes the relative position where an instance of variable y_i is located within the PA (relative to its source) after the mth move of the realization of data dependency $\mathbf{d}_{j,i}$. We determine $\mathbf{p}^{\text{rel}}_{j,i,m}$ as follows:

$$\mathbf{p}^{\text{rel}}_{j,i,m} = \sum_{m'=1}^{m} \left(\sum_{\{k \mid \mathbf{w}_k \in \mathcal{W}^{\text{bi}}\}} \beta_{j,i,k,m'}(1 - 2\beta^{\text{bi}}_{j,i,k}) \mathbf{w}_k + \sum_{\{k \mid \mathbf{w}_k \in \mathcal{W} \setminus \mathcal{W}^{\text{bi}}\}} \beta_{j,i,k,m'} \mathbf{w}_k \right) \ . \quad (8)$$

In the following, we will regard the time behavior of the realization of the data dependencies. The causality of all data dependencies of the set $\mathcal{D}$ is ensured by the following constraints:

$$\forall \mathbf{d}_{j,i} \in \mathcal{D}: t^{\mathrm{d}}_{j,i} = \underbrace{b_j - (b_i + l_i)}_{\text{①}} + \underbrace{\boldsymbol{\tau}^{\mathrm{offs}} \cdot \mathbf{d}_{j,i}}_{\text{②}} - \underbrace{\sum_{m=1}^{M_{j,i}} \sum_{k=1}^{|\mathcal{W}|} \beta_{j,i,k,m} \cdot l^{\mathbf{w}}_k}_{\text{③}} \geq 0 \ . \quad (9)$$

The delay $t^{\mathrm{d}}_{j,i}$ equals the number of time steps for which an instance of the dependent variable y_i needs to be stored until it is used by statement S_j. Of course, this amount of time may not be negative (causality constraint). In (9), term ① determines the time between the availability of an instance of the dependent variable y_i and its use (disregarding the scheduling offset). With l_i we describe the number of time steps after which the result of statement S_i is available at the output register of the data path. We assume that the input of the data path can be connected directly to the end of any local channel or to any local register. Term ② takes the scheduling offset into account as it represents the relative time difference between the iteration serving as the source and the iteration serving as the drain of data dependency $\mathbf{d}_{j,i}$. And term ③ denotes the time it takes to transfer an instance of the dependent variable y_i from the source to the drain.

If $t^{\mathrm{d}}_{j,i} > 0$, then it is necessary to store an instance of the dependent variable y_i along its path for $t^{\mathrm{d}}_{j,i}$ time steps. We introduce variables $t^{\mathrm{r}}_{j,i,m} \in \mathbb{N}$ in the communication problem to parameterize the storage of an instance of a dependent variable y_i along its path from the source to the drain of data dependency $\mathbf{d}_{j,i}$. The value of variable $t^{\mathrm{r}}_{j,i,m}$ gives the number of time steps for which an instance of variable y_i is stored in a local register **after** the mth move of the realization of data dependency $\mathbf{d}_{j,i}$. The variables $t^{\mathrm{r}}_{j,i,m}$ are subject to:

$$\forall \mathbf{d}_{j,i} \in \mathcal{D}: t^{\mathrm{d}}_{j,i} = \sum_{m=0}^{M_{j,i}} t^{\mathrm{r}}_{j,i,m} \ . \quad (10)$$

Equation (10) ascertains that an instance of the dependent variable y_i is stored for as many time steps as given by $t^{\mathrm{d}}_{j,i}$ along its path. Each variable $t^{\mathrm{r}}_{j,i,m}$ with $m = 0$ denotes the time of storage for dependent variable y_i at the source of data dependency $\mathbf{d}_{j,i}$, i. e. before y_i is transported anywhere. This explains why we always map move $m = 0$ to channel $\mathbf{w}_0$. Note that our model above is also valid for data dependencies with $\mathbf{d}_{j,i} = \mathbf{0}$ where we use $M_{j,i} = 0$.

3.2 The Objective Function

The values of variables $\beta_{j,i,k,m}$, $\beta^{\mathrm{bi}}_{j,i,k}$, and $t^{\mathrm{r}}_{j,i,m}$ fully describe the realization of all data dependencies on the PA. In the following, we will determine the objective function of the communication problem. Hence we need to derive the cost for channels and registers. This cost can only be determined indirectly from variables $\beta_{j,i,k,m}$, $\beta^{\mathrm{bi}}_{j,i,k}$, and $t^{\mathrm{r}}_{j,i,m}$. Therefore we introduce further variables in the communication problem through which the cost can be described.

With $t^{\mathrm{in}}_{j,i,m}$ we describe the time at the PE at the relative position $\mathbf{p}^{\mathrm{rel}}_{j,i,m}$ at which an instance of variable y_i arrives as the mth move of the realization of data dependency $\mathbf{d}_{j,i}$. And with $t^{\text{out}}_{j,i,m}$ we describe the time at which that instance of variable y_i leaves the PE to perform move $m+1$ of the realization of data dependency $\mathbf{d}_{j,i}$. Variables $t^{\text{in}}_{j,i,m}$ and $t^{\text{out}}_{j,i,m}$ are determined as follows:

$$t^{\text{in}}_{j,i,m} = \underbrace{b_i + l_i}_{\text{\textcircled{1}}} + \underbrace{\sum_{m'=1}^{m}\sum_{k=1}^{|\mathcal{W}|} \beta_{j,i,k,m'} \cdot l^{\mathbf{w}}_k}_{\text{\textcircled{2}}} + \underbrace{\sum_{m'=0}^{m-1} t^{\text{r}}_{j,i,m'}}_{\text{\textcircled{3}}} - \underbrace{\tau^{\text{offs}} \cdot \mathbf{p}^{\text{rel}}_{j,i,m}}_{\text{\textcircled{4}}} \,, \quad (11)$$

$$t^{\text{out}}_{j,i,m} = t^{\text{in}}_{j,i,m} + t^{\text{r}}_{j,i,m} \,. \tag{12}$$

In (11), term ① determines the time when the source of a data dependency $\mathbf{d}_{j,i}$ is available. Term ② represents the delay caused by the transfer until the mth move. Term ③ gives the delay caused by the storage along its path. And term ④ accounts for the time difference between the source and the position after the mth move of the realization of the data dependency according to the scheduling offset.

Note that for $m = 0$, the time $t^{\text{in}}_{j,i,0}$ denotes the time when an instance of variable y_i is fetched from the data path at the relative position $\mathbf{p}^{\text{rel}}_{j,i,0} = \mathbf{0}$. After the last move ($m = M_{j,i}$), an instance of variable y_i arrives at the drain of data dependency $\mathbf{d}_{j,i}$. And time $t^{\text{out}}_{j,i,M_{j,i}}$ denotes the time when that instance serves as an input to the data path at the relative position $\mathbf{p}^{\text{rel}}_{j,i,M_{j,i}} = \mathbf{d}_{j,i}$ for the computation of statement S_j.

For $\tau^{\text{offs}} = \mathbf{0}$ in (2), it would be sufficient to consider one iteration interval λ of one PE to describe the communication problem [11]. For the general case where the scheduling offset may also be non-zero, we have to use an extended approach to solve the communication problem.

As a consequence we determine a priori for each data dependency $\mathbf{d}_{j,i} \in \mathcal{D}$ a set of time steps $\mathcal{L}'_{j,i}$. The set $\mathcal{L}'_{j,i}$ is defined in a similar way to the set $\mathcal{L}$ (see Def. 2) with the only difference that it takes the scheduling offset into account so that it may consist of some different time steps. We refer to [16] for a detailed derivation of how to determine the set $\mathcal{L}'_{j,i}$.

We introduce binary variables which account for the use of channels and registers caused by the realization of data dependency $\mathbf{d}_{j,i} \in \mathcal{D}$. Binary variable $\gamma_{j,i,k,m,l'}$ denotes whether a channel $\mathbf{w}_k$ is used at time step $l' \in \mathcal{L}'_{j,i}$ at the mth move of the realization of that data dependency. And binary variable $\delta_{j,i,m,l'}$ denotes whether a register is used at time $l' \in \mathcal{L}'_{j,i}$ **after** the mth move of the realization of that data dependency. The binary variables $\gamma_{j,i,k,m,l'}$ and $\delta_{j,i,m,l'}$ are determined as follows:

$$\gamma_{j,i,k,m,l'} = \begin{cases} 1 & \text{if } l' = t^{\text{out}}_{j,i,m-1} + \beta^{\text{bi}}_{j,i,k} \tau^{\text{offs}} \mathbf{w}_k \wedge \beta_{j,i,k,m} = 1 \\ 0 & \text{otherwise} \end{cases}, \; l' \in \mathcal{L}'_{j,i}, \; m > 0 \,, \tag{13}$$

$$\delta_{j,i,m,l'} = \begin{cases} 1 & \text{if } t^{\text{in}}_{j,i,m} \leq l' \wedge l' < t^{\text{out}}_{j,i,m} \\ 0 & \text{otherwise} \end{cases} \quad \text{with} \quad l' \in \mathcal{L}'_{j,i} \,. \tag{14}$$

Note that for a channel with a pipeline structure, the binary variable $\gamma_{j,i,k,m,l'}$ takes the value of one only for the first time step during which the channel is used for the realization of data dependency $\mathbf{d}_{j,i}$. Hence, another communication can begin to use the same channel at the next time step.

The number of channels $\mathbf{w}_k$ used at time $l \in \mathcal{L}$ caused by the realization of data dependency $\mathbf{d}_{j,i}$ is given by variable $c^{\mathbf{w}}_{j,i,k,l}$. The corresponding number of registers that is used at time l is given by $c^{\mathrm{r}}_{j,i,l}$. We determine variables $c^{\mathbf{w}}_{j,i,k,l}$ and $c^{\mathrm{r}}_{j,i,l}$ as follows:

$$c^{\mathbf{w}}_{j,i,k,l} = \sum_{\{l' \in \mathcal{L}'_{j,i} \mid l' \bmod \lambda = l\}} \sum_{m=1}^{M_{j,i}} \gamma_{j,i,k,m,l'} \quad \text{with} \quad l \in \mathcal{L}, \quad (15)$$

$$c^{\mathrm{r}}_{j,i,l} = \sum_{\{l' \in \mathcal{L}'_{j,i} \mid l' \bmod \lambda = l\}} \sum_{m=0}^{M_{j,i}} \delta_{j,i,m,l'} \quad \text{with} \quad l \in \mathcal{L}. \quad (16)$$

In (15) and (16), the first sum considers all time steps of the set $\mathcal{L}'_{j,i}$ that will be mapped to time $l \in \mathcal{L}$ by modulo arithmetics. And the second sum adds over the moves that realize data dependency $\mathbf{d}_{j,i}$.

Next we determine the total cost for the realization of all data dependencies of the set $\mathcal{D}$. Variable $c^{\mathbf{w}}_k$ denotes the maximum number of channels $\mathbf{w}_k$ that is used at an arbitrary PE. And variable c^{r} denotes the maximum number of registers used at an arbitrary PE. The values of variables $c^{\mathbf{w}}_k$ and c^{r} are determined as follows:

$$c^{\mathbf{w}}_k = \max_{l \in \mathcal{L}} \sum_{\mathbf{d}_{j,i} \in \mathcal{D}} c^{\mathbf{w}}_{j,i,k,l} \quad \text{and} \quad c^{\mathrm{r}} = \max_{l \in \mathcal{L}} \sum_{\mathbf{d}_{j,i} \in \mathcal{D}} c^{\mathrm{r}}_{j,i,l}. \quad (17)$$

Finally, the objective function of the communication problem is determined as follows:

$$\min \left(\eta^{\mathrm{r}} \cdot c^{\mathrm{r}} + \sum_{k=1}^{|\mathcal{W}|} \eta^{\mathbf{w}}_k \cdot c^{\mathbf{w}}_k \right) \quad (18)$$

where η^{r} denotes the cost for a register and $\eta^{\mathbf{w}}_k$ denotes the cost for channel $\mathbf{w}_k$.

4 Experimental Results

For the EDA (Algorithm 1) we discuss two different target architectures (PA1 and PA2) as given in Table 1. The cost for a register is $\eta^{\mathrm{r}} = 1$ for PA1 and $\eta^{\mathrm{r}} = 1.5$ for PA2. In both cases, we assume one data path with a latency of one for solving each statement S_j within each PE. Hence, we search for a solution to the communication problem with $\lambda = 11$. The scheduling offset is $\tau^{\mathrm{offs}} = 0$.

In Fig. 2 we illustrate the optimal solution to the communication problem. The starting time of each use of a channel and/or register within the iteration

Table 1. Available channels $\mathbf{w}_k$, their latency $l^{\mathbf{w}}_k$ and their cost $\eta^{\mathbf{w}}_k$

(PA1)						(PA2)					
k	1	2	3	4	5	k	1	2	3	4	5
$\mathbf{w}_k$	$(1\ 0)^T$	$(0\ 1)^T$	$(1\ 1)^T$	$(1\ -1)^T$	$\pm(1\ 0)^T$	$\mathbf{w}_k$	$(1\ 0)^T$	$(0\ 1)^T$	$(1\ 1)^T$	$(1\ -1)^T$	$\pm(1\ 0)^T$
$l^{\mathbf{w}}_k$	3	3	1	2	0	$l^{\mathbf{w}}_k$	2	2	1	2	0
$\eta^{\mathbf{w}}_k$	1.5	1.5	1	2.5	2	$\eta^{\mathbf{w}}_k$	1	1	1.5	3	1.5

interval is shown. The gray background in the use of channel $\mathbf{w}_5$ denotes a use in the negative direction. The inner schedule is given by the succession of the statements S_j.

ILPs were generated to solve the communication problem. Both ILPs consist of 3838 constraints, 2001 binary and 220 integer variables. The ILPs were solved using ILOG CPLEX v. 9.1 on an Athlon 64 Processor 3800+. It took 40 sec. and 72 sec. for PA1 and PA2 respectively to find an optimal solution of each ILP (including the verification that the solution is optimal).

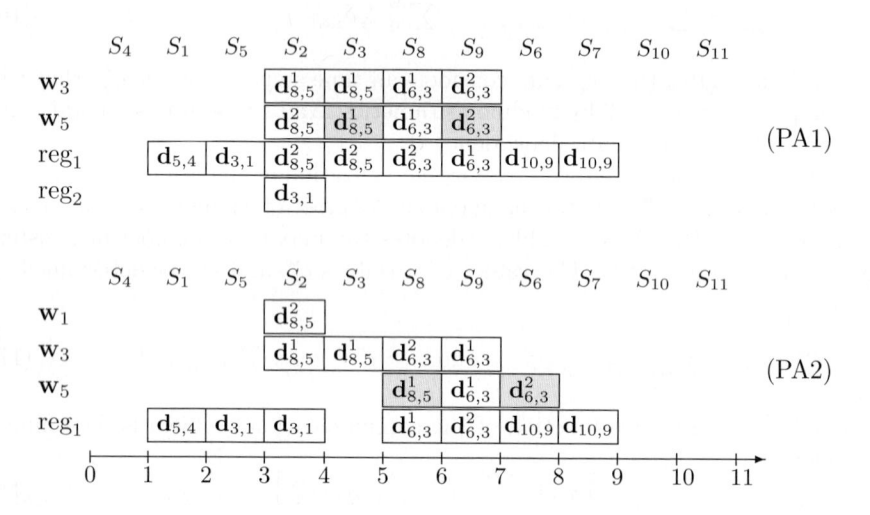

Fig. 2. Bar chart denoting the usage of channels and registers for the communication caused by all data dependencies of the set $\mathcal{D}_{\text{EDA}}$ within an iteration interval $\lambda = 11$

5 Conclusions

In this paper we have formulated and solved the communication problem to realize uniform data dependencies within a PA using ILP. Our method takes the cost of channels and registers of the target architecture into account. A solution to the communication problem determines a selection of channels and the control of the communication caused by the uniform data dependencies. Further, it specifies the inner schedule for all computational tasks within an iteration.

Our method can also be used to find suitable realizations of the uniform data dependencies of algorithms on a given PA with fixed local interconnections. Future work includes an extension of our approach to non-uniform data dependencies (e. g. the realization of input/output) or to multi-level partitioning (e. g. co-partitioning).

References

1. Motomura, M.: A Dynamically Reconfigurable Processor Architecture. In: Microprocessor Forum. (2002)
2. Duller, A., Towner, D., Panesar, G., Gray, A., Robbins, W.: picoArray Technology: The Tool's Story. In: Design, Automation and Test in Europe (DATE'05). Volume 3. (2005) 106–111
3. Siegel, S., Merker, R.: Optimized Data-Reuse in Processor Arrays. In: Proc. IEEE Int. Conf. on Application-Specific Systems, Architectures, and Processors (ASAP 2004). (2004) 315–325
4. Karp, R.M., Miller, R.E., Winograd, S.: The organisation of computations for uniform recurrence equations. JACM **14**(3) (1967) 563–590
5. Siegel, S., Merker, R.: Algorithm Partitioning including Optimized Data-Reuse for Processor Arrays. In: Proc. IEEE International Conference on Parallel Computing in Electrical Engineering (PARELEC 2004). (2004) 85–90
6. Derrien, S., Rajopadhye, S., Sur-Kolay, S.: Combining Instruction and Loop Level Parallelism for Array Synthesis on FPGAs. In: International Symposium on System Synthesis. (2001)
7. Feautrier, P.: Fine-grain Scheduling under Resource Constraints. In: Proc. 7th Int. Workshop on Languages and Compilers for Parallel Computing (LCPC '94). Volume 892 of LNCS., Springer (1994) 1–15
8. Thiele, L.: Resource constraint scheduling of uniform algorithms. Journal of VLSI Signal Processing **10** (1995) 295–310
9. Teich, J., Thiele, L.: A New Approach to Solving Resource-Constrained Scheduling Problems based on a Flow-Model. Technical Report 17, TIK, Swiss Federal Institute of Technology (ETH) Zürich (1996)
10. Dion, M., Risset, T., Robert, Y.: Resource-constrained scheduling of partitioned algorithms on processor arrays. Integration, the VLSI Journal **20** (1996) 139–159
11. Fimmel, D., Merker, R.: Localization of Data Transfer in Processor Arrays. In: Parallel Processing: 5th International Euro-Par Conference (Euro-Par '99). Volume 1685 of LNCS., Springer (1999) 401–408
12. Eisenbeis, C., Sawaya, A.: Optimal loop parallelization under register constraints. Technical Report 2781, INRIA (1996)
13. Müller, J., Fimmel, D., Merker, R.: Optimal Loop Scheduling with Register Constraints Using Flow Graphs. In: Proc. of the 7th Int. Symposium on Parallel Architectures, Algorithms, and Networks (I-SPAN), Hong Kong, China (2004)
14. Fimmel, D.: Optimaler Entwurf paralleler Rechenfelder unter Verwendung ganzzahliger linearer Optimierung. PhD thesis, Dresden University of Technology, Institute of Circuits and Systems (2002)
15. Siegel, S., Merker, R.: Minimum Cost for Channels and Registers in Processor Arrays by Avoiding Redundancy. In: IEEE 17th Int. Conf. on Application-Specific Systems, Architectures, and Processors (ASAP 2006). (2006)
16. Siegel, S., Merker, R.: Optimal Realization of Uniform Data Dependencies in Algorithm Partitioning Under Resource Constraints. Technical report, Dresden University of Technology, Institute of Circuits and Systems (2006, http://www.iee.et.tu-dresden.de/~siegel/paper/SM06a.pdf)

FPGA Implementation of a Prototype Hierarchical Control Network for Large-Scale Signal Processing Applications

Jérôme Lemaitre[1] and Ed Deprettere[2]

[1] ASTRON, Oude Hoogeveensedijk 4, 7991 PD Dwingeloo, The Netherlands
[2] LIACS, Leiden university, NielsBohrweg 1, 2333 CA Leiden, The Netherlands

Abstract. The performance of a high throughput and large-scale signal processing system must not be compromised by the control and monitoring flow that is inherently part of the system. In particular, the interfacing of data flow and control flow components should be such that control does not obstruct the signal flow that is of higher priority. We assume that the signal processing is modeled as a distributed hierarchy of data flow networks, and that the control and monitoring is modeled as a distributed hierarchy of communicating Finite State Machines. The interfaces between leaf-nodes of the control and monitoring network, and the signal processing nodes in the dataflow networks are specified in such a way that the semantics of both network types are preserved. In this paper, we present the prototyping of a control network and its interfacing with a data flow network in a FPGA-based platform, and we analyze the performance of the interfacing in a case study. The HDL code that is involved in the interfaces is generated in a semi-automated way.

1 Introduction

Large-scale signal processing systems such as phased array radio telescopes [15] typically comprise of a hierarchically distributed data flow network (DFN), a hierarchically distributed control network (CN), and an interfacing between these two networks. Depending on the nature of the astronomical source that is observed, the system must be able to operate in modes that range from spectroscopy, pulsar observation or searches for transients. Moreover, disturbances in the high throughput streaming data paths, which are mainly due to radio frequency interferences and changes in the ionosphere, must be monitored and mitigated [14] by re-configuring the dataflow processing at run-time. Thus, to each operational mode corresponds a different set of high-level dataflow processing parameters (e.g. frequency resolution and integration time) and control parameters (e.g schedule to update the number of beams and blanked channels).

We assume that signal processing tasks such as filtering, FFT, beamforming and correlation in the DFN are modeled as nodes of Kahn Process Networks (KPN [12]). The control data for re-configuring and monitoring the processes in the KPNs and/or the components onto which they are mapped is sent over the

CN that has a tree or lattice-like structure, until they reach CN leaf-nodes which interact with KPN nodes through specific interfaces. As shown schematically in Figure 1, the interaction between these nodes is synchronized by means of periodic pulse trains that are distributed to control nodes in a synchronization network. The periods of the pulse trains are so chosen that a command that is sent over the CN to a dataflow process reaches this process during the period that precedes its execution. The process will then execute this command concurrently and complementarily to the DFN data processing.

This paper focuses on the way the CN is modeled and reports on a prototype FPGA implementation. In [7], the interfaces between CN leaf-nodes and KPN processing nodes have been so modeled that the two networks that are designed separately can work together without compromising their individual semantics. This paper also demonstrates that these interfaces can be implemented in a semi-automated way, taking IP re-use and scalability constraints into account, and avoiding error prone and time consuming handcrafting of FPGA implementations [10].

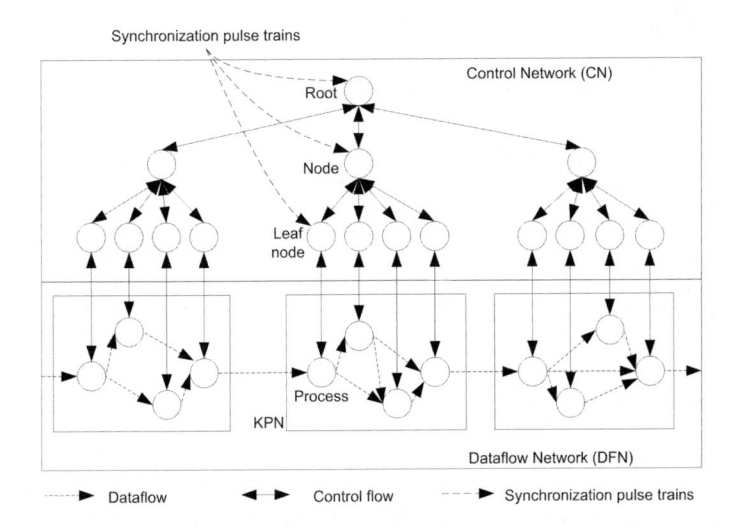

Fig. 1. Interface between a control network (CN) and a data flow network (DFN). The root, nodes and leaf-nodes receive periodic pulse trains in a synchronization network.

In the remaining of this section we give our problem statement, solution approach and related work. The rest of the paper is organized as follows. In Section 2, we explain how to map nodes of the CN onto soft-cores and how to interface CN leaf-nodes with KPN processing nodes that wrap hardware IPs. In Section 3, we present a case study for the control of processes in a KPN in the DFN from leaf-nodes in the CN, and discuss the results concerning the Hardware Description Language (HDL) semi-automation, the IP re-use, and the scaling in the design. Finally, conclusions and future work are drawn in Section 4.

1.1 Problem Statement

The high throughput large-scale systems we are concerned with comprise of two networks that transport and process signal data and control data, respectively. The signals propagate and are processed in a hierarchy of distributed KPNs that together make up the Dataflow Network (DFN). The control data are passed and processed in a tree or lattice like structure of communicating Finite State Machines that make up the Control Network (CN). Signal processing tasks in the KPNs as well as the system components onto which they are mapped are re-configured and their behavior is monitored from the leaf-nodes in the CN for the system to operate in a particular mode. Because the two networks have different semantics as shown in Table 1, their interfaces have to be defined and modeled judiciously to avoid semantic corruption. Their definition and model can be found in [7]. The problem that is addressed in this paper is whether these interfaces can be implemented in a semi-automated way such that, indeed

– The semantics of the two models of computation remain respected, and
– The inclusion of control interfaces does not obstruct the performance of the DFN in terms of throughput and resource usage.

Table 1. Characteristics of the data flow network and control network to interface

	Dataflow network (DFN)	Control network (CN)
Behavior	Deterministic	Sporadic
I/O data type	Streams	Messages
Scheduling	Local	Global
Synchronization	Blocking write and read	Periodic
Timing	Self-timed	Synchronous

1.2 Solution Approach

In the CN, all nodes are connected to other nodes above through a single port, and all nodes, except the leaf-nodes are connected to other nodes below through a single output port. They receive and send control packets from and to these ports, respectively. The leaf-nodes have three ports to below: a configuration-data output port, a command output port, and a monitoring-data input port as shown in Figure 2. Control packets that are received from above are unpacked (possibly after some processing) to separate commands from configuration data before being send downwards. By the same token, incoming monitoring data is packed in control packets that are sent upwards, possibly after some processing of the packets. The DFN is mapped onto networks of platforms that implement the KPNs in the DFN, and each platform has a single entry point for control data. CN leaf-nodes are internal to platforms and are interfaced to platform components onto which processes of a KPN in the DFN are mapped. As a result, all platform components must have a configuration-data input port, a command input port, and a monitoring data output port. A CN leaf-node that is associated

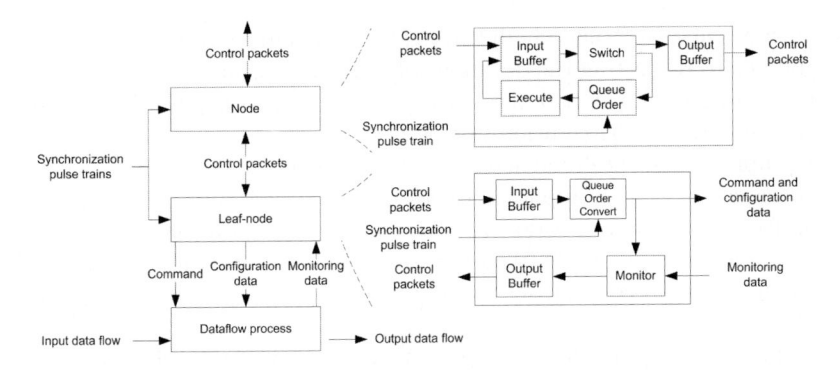

Fig. 2. Approach to interface CN nodes with CN leaf-nodes, and CN leaf-nodes with DFN processes

with a dataflow component therefore unpacks component-specific configuration data, releases component-specific commands, and packs component-specific incoming monitoring data.

In our approach to prototype the hierarchical CN we start from Figure 1. Root is mapped on a PC, and the rest of the figure is mapped on an FPGA-based platform as if it were a DFN platform onto which a KPN was mapped. Thus Node is the platform's single entry point. Node and Leaf-node are mapped onto soft-cores (e.g., the microblaze from Xilinx, or the nios II from Altera) of the FPGA-based platform [16] [17]. Leaf-node is controlling a specific DFN platform component. The control of a dataflow component is kept separated from the dataflow and from the synchronization mechanism by means of three concurrent FSMs. A first FSM synchronizes the execution of commands issued from a leaf-node with the (periodic) execution of the signal processing function. A second FSM controls the dataflow Read and Write ports in a SBF (Stream Based Functions [8]) dataflow model of computation. A third FSM controls the IP Execute function repertoire in the SBF dataflow model of computation. These FSMs are generated automatically in HDL from a high-level specification.

1.3 Related Work

The PSDF model [1] separates the dataflow specification from the control specification, with the objectives of staying within one model of computation and modeling parameterized dataflow in such a way that it remains possible to derive quasi-static schedules [2]. We are dealing with with a large DFN and a large CN instead of a single process network. The DFN is a network of distributed KPNs and the CN is a network of communicating FSM. The interfacing of the two is not done as in [1] because the behavior of the CN is sporadic. Thus we cannot schedule and we cannot afford blockings read at the interface.

The FunState model [3] unifies many models of computation and allows to verify scheduling constraints as well. However, a state transition must take place in a FunState component before a function starts a new execution. Waiting for this transition may obstruct a high throughput streaming dataflow processing. In our approach to respecting the dataflow integrity, pre-defined control procedures are executed in a FSM in a process, while a complementary FSM concurrently controls the dataflow distribution in this process.

In [4], applications are modeled using Process Networks and SBF with non-static parameters. These applications are also mapped onto a FPGA and get configuration data from outside. However, the re-configuration is only possible after a complete network cycle. We want to be able to re-configure each individual periodic dataflow process during any period at run-time, without stopping the entire DFN.

An approach to dynamically reconfiguring a streaming application in a hierarchical SoC with a multiprocessor subsystem is presented in [5]. Processing tasks can be reconfigured through inserting reconfiguration tokens in the data streams. We avoid such insertions by physically separating dataflow and control paths in our implementations. Nevertheless, combining the generic services offered by the shell described in [5] with a standard task-level specification as in [6] would lead to optimized SoC implementations and re-usable modules.

2 Interface Mapping

In this Section we first briefly review the modeling paradigm to abstract platform-specific processing, communication, and synchronization mechanisms to specifying Control and Dataflow processing applications in the large-scale system. Then we detail the mapping of the CN and the DFN into soft-cores and hardware of an FPGA-based platform, respectively, and the interfacing of the two networks.

2.1 Modeling Paradigm

From a separation of concerns viewpoint it is interesting to specify the DFN and CN models independently from each other and to progressively refine them for HW/SW implementation on networks of platforms as shown in Figure 3. Thus, optimized implementations of the separated networks can be re-used from a high-level specification. The first step consists of refining the mechanisms that are involved in the communication between nodes in the models based on generic, platform-independent services of an abstract HW/SW task transaction level interface [6]. These services may be provided with functions (**Read, Write, Execute**) that manipulate vectors of arbitrary data types. These data types may be dataflow tokens in the DFN, or control packets in the CN, which consist of two parts: a header that indicates which command should be executed at what time by which node or leaf-node, and eventually control data information that comes with a specific command.

The second step converts task-level representations to implementations through services that abstract platform-specific intricacies. When targeting a FPGA, each control node is assigned to a soft-core for which we generate code.

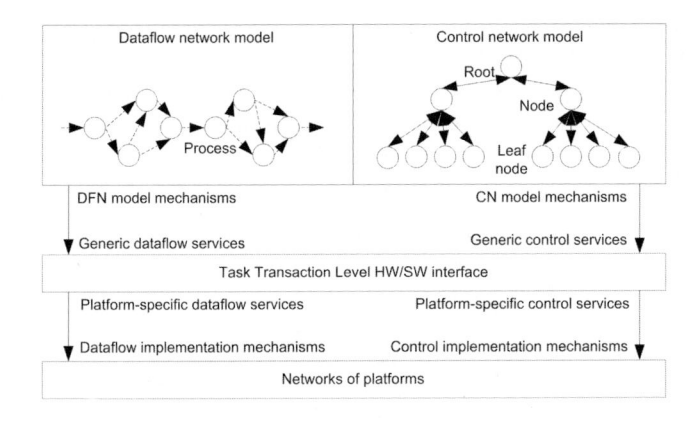

Fig. 3. Modeling Paradigm. The DFN and CN models are separated and gradually refined to be implemented in networks of platforms through a task transaction level interface.

Synchronization pulse trains are handled as interrupts through a real-time operating system. Control packets are defined in structures whose elements can easily be manipulated individually. These soft-cores communicate through embedded memories. Each dataflow process is assigned to a re-configurable processor that is generated in HDL. These dataflow processors exchange tokens through embedded memories, which are not shared with the memories that are used in the control network. Hardware IPs are wrapped in these dataflow processors because we do not want to get involved in low level functions design.

2.2 Node Mapping

The state diagram of a control node executed in a soft-core is shown on the left-hand side in Figure 4 and the corresponding platform-independent code is given on the right-hand side (lines 1-10). The default state is represented in grey (`INIT`, lines 11-15) and corresponds to the definition of a packet and initialization of the communication channels in the CN. These channels are first checked for the presence of packets (`READ&CHECK`). When a packet is received, there are two possibilities (`SWITCH`): it is either sent to the appropriate destination in the hierarchy (`ROUTE`), or it is inserted in a priority queue (`QUEUE&ORDER`, lines 16-22). Finally, there are again two alternatives (`CHECK`): the command that is in the packet on top of the queue must be executed during the current period (`EXECUTE`), else the communication channels are checked again until a new packet enters the node or a new command must be executed.

As detailed in [9], the hardware realization of a KPN node (process) is made of four components: a `Dataflow Read Unit` that gets tokens from dataflow input channels, multiplexes and transmits them to an `Execute Unit`, which consumes these tokens, performs computation using an IP `Function Repertoire` and produces output tokens towards a `Dataflow Write Unit`. This unit demultiplexes

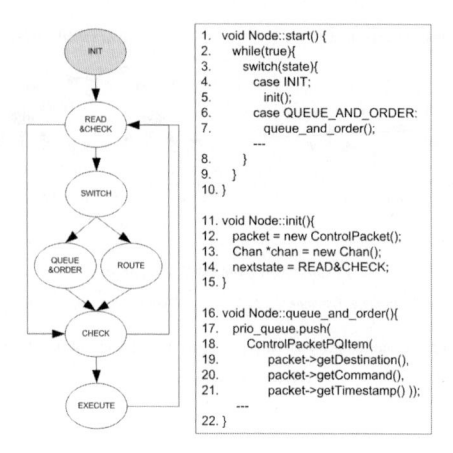

Fig. 4. Mapping a node as a FSM onto a soft-core: state diagram and corresponding platform-independent pseudo code

the output tokens and sends them to output dataflow channels. The fourth component is the `Controller` that supervises the execution of the three other units. All these units are shown in Figure 5.

2.3 Interfacing a Leaf-Node with a Process

In the interface between a CN leaf-node with a DFN process, the additional configuration-data port is connected to a `Control Read Unit` as shown in Figure 5. The Dataflow Read Unit, Dataflow Write Unit and Function Repertoire get their own configuration parameters under the supervision of the Controller. The additional command port is connected to the Controller and the monitoring-data port is connected to a `Control Write Unit`, which probes the dataflow in the Dataflow Write Unit, as well as the state of the node in the Controller. Zooming in into the Controller, three distinct FSMs are executed. The behavior of these FSMs is depicted in Figure 6 (initial states are represented in grey).

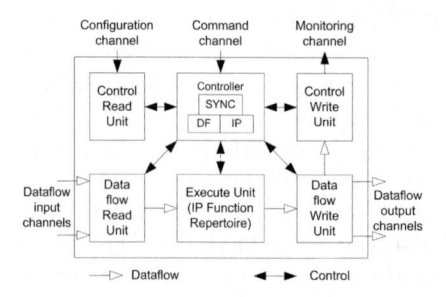

Fig. 5. Hardware implementation of the interface between a leaf-node and a process

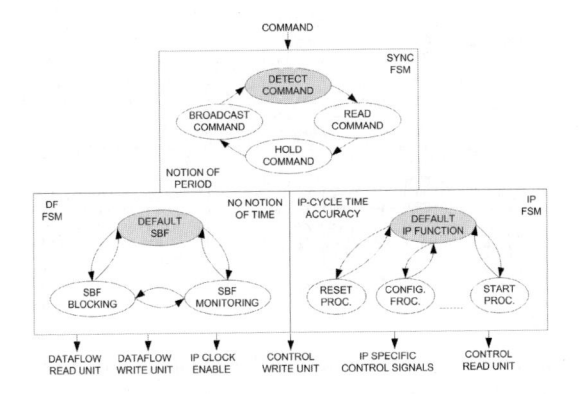

Fig. 6. Separation of concerns with three FSMs in a controller (SYNC for synchronization, DF for dataflow distribution and IP for IP-function control)

A SYNC FSM gets a command issued from a CN leaf-node and synchronizes its broadcasts to the two other FSMs when starting a new period. A DF FSM implements the behavior of an SBF [8] dataflow model of computation in a DFN process, without any notion of time, but a notion of order. It controls the Dataflow Read Unit, the Dataflow Write Unit and the clock enable signal of the IP Function Repertoire. The only command it can execute is a monitoring command. In this case it permits the Control Write Unit getting data from the Dataflow Write Unit. Concurrently, an IP FSM executes commands in states that encapsulate the corresponding IP-specific pre-defined control procedures, with an IP-cycle accurate notion of time. It generates control signals for the IP Function Repertoire and controls the flow of re-configuration data.

3 Case Study

In Section 2 we detailed our approach to map a hierarchical CN and the interface with a DFN onto a FPGA. In this Section we present such an implementation and discuss the results concerning the separation of the two networks, the semi-automation, IP re-use and design-scaling.

3.1 Application

In this case study, the DFN is limited to a single KPN with two processes as shown in Figure 7. The first process generates periodic dataflow patterns (e.g. impulses, ramps or sinewaves) that can be re-configured (e.g. amplitude, frequency) by a leaf-node. The second process wraps an 8-taps FIR filter IP whose taps can be re-configured (e.g. low-pass, band-pass, high-pass characteristics depending on the operational mode) from a leaf-node as well. The two leaf-nodes and the node above in the hierarchical CN are executed in nios II soft-cores in a Stratix II FPGA [16], and support a portable real-time kernel (MicroC/OS-II [13]) to handle the synchronization pulse trains as interrupts. The node is the

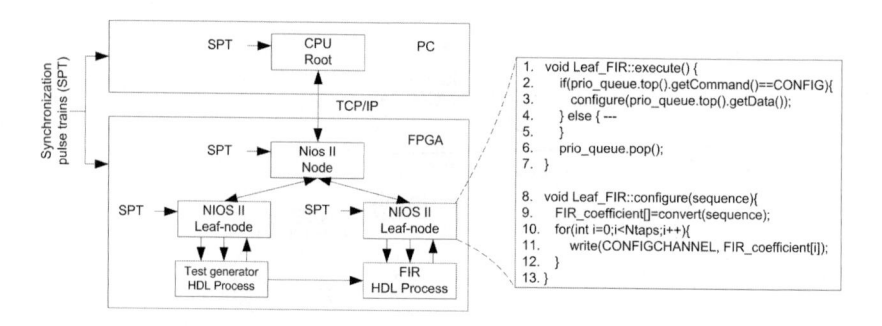

Fig. 7. FPGA implementation of a hierarchical control network to control and monitor a test generator and a FIR filter

single entry point of the FPGA-based platform. It supports a portable lightweight TCP/IP stack from Opencores [18] to communicate with the root that is mapped on a CPU in a PC.

Re-configuring the filter coefficients requires converting the new coefficients to the filter-specific format because coefficients are stored in partial order and distributed in embedded memory segments. This is done in the Execute state of the leaf-node that controls the filter as shown in the pseudo-code in Figure 7. On the occurrence of a synchronization pulse train, a control packet is read from the priority queue. If the command requests re-configuring the filter, then a conversion program is called (lines 2-4) that converts the configuration data to the filter-specific sequence. This sequence is sent to the re-configuration channel (lines 10-12) and the control packet is removed from the queue after its execution (line 6). The leaf-node may then send a command to activate the re-configuration of the process as detailed in Section 2.

3.2 Results

Each nios II soft-core has been implemented in approximately 1,000 (Adaptive) Look-Up Tables (LUT) and 64kB of memory in this application. The middleware library took approximately 1.3MB of external SDRAM for each processor. These results could be improved by mapping nodes and leaf-nodes onto Application Specific Instruction-Set Processors (ASIP [11]) and by sharing the implementation of the middleware between all soft-cores, respectively. Figure 8 shows the impact of the FIR function-scaling on throughput (maximum frequency sustained by the dataflow in the process after synthesis of the process, on the left-hand side) and resource usage (LUT, on the right-hand side). Results are given for the interface presented in this paper (CN-DFN) and for a manufacturer-dependent dataflow only interface (Atlantic [16]). Our interface needs a few more resources since it includes both dataflow and control, and the loss in the throughput is still acceptable (less than 10%) in this case study.

Portable and speed-optimized HDL has been generated from high-level graphical specifications in StateCAD [17] for the three FSMs that are executed in the

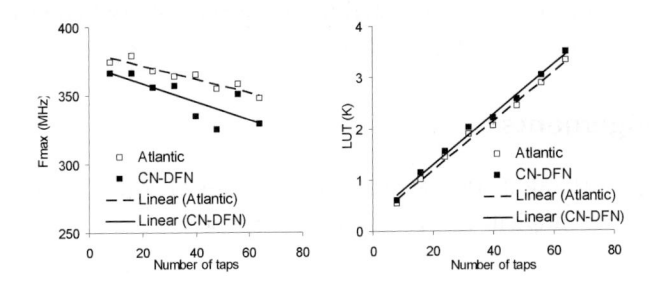

Fig. 8. Impact of design-scaling on throughput and resource usage

controllers of the DFN processes to wrap the hardware IPs. Thus, we avoided time consuming and error prone HDL handcrafted development. Our interfaces permit de-coupling low-speed clock domain(s) in the CN (soft-cores hardly run faster than 150MHz) from the high-speed clock domain(s) in the DFN. However, finding optimal buffer sizes to minimize the chance of blocking in the dataflow communication channels remains problematic. Moreover, the granularity of the dataflow processes in our prototype is large enough for the control network not to be critical since dataflow processes use more resources than a soft-core. Nevertheless their periods should not be shorter in future implementations so that soft-cores in the CN are not overwhelmed handling synchronization pulse trains as interrupts.

Although dataflow processes can be re-configured without the loss of data as in the case of the filter, re-configuring functions may induce transients in the data itself. Thus, the current practise is still to discard the data that is temporarily corrupted due to the transition.

4 Conclusion

We presented a prototype FPGA implementation of a hierarchical control network (modeled as communicating Finite State Machines) and its interfacing with a distributed dataflow network (modeled as communicating Kahn Process Networks). Nodes of the control network have been mapped onto soft-cores and interfaced with the nodes of the dataflow network, without sharing hardware resources, and based on HDL FSMs that isolate the synchronization mechanisms from the dataflow distribution and from the monitoring and control of the dataflow processing tasks, which are executed in hardware IPs. The performance of the interfaces in term of speed and resource usage allowed not to obstruct the performance of the (dominant) dataflow network in a case study. In addition, we anticipated design scaling and design re-use constraints by semi-automating HDL code generation in the mapping of the interfaces.

In the near future we would like to map such interfaces onto a network of re-configurable platforms, mainly consisting of FPGAs and CPUs, so as to evaluate its adaptiveness to larger systems. This mapping could be facilitated by keeping the signal processing separated from the control and monitoring in the architecture exploration of these large systems.

Acknowledgements

We would like to thank J. Bol, M. Lammertink and M. van der Kooij for their contribution to the implementation of the prototype.

References

1. B. Bhattacharya and S. Bhattacharyya, "Parameterized Dataflow Modeling of DSP Systems", In Proc. of the Intl. Conf. on *Acoustics, Speech, and Signal Processing (ASSP 2000)*, Istanbul, Turkey, June 2000.
2. B. Bhattacharya and S. Bhattacharyya, "Quasi-static Scheduling of Reconfigurable Dataflow Graphs for DSP Systems", In Proc. of the Intl. Workshop on *Rapid System Prototyping (RSP 2000)*, Paris, France, June 2000.
3. K. Strehl, L. Thiele, M. Gries, D. Ziegenbein, R. Ernst and J. Teich, "FunState - An Internal Design Representation for Codesign", *IEEE Trans. on VLSI Systems*, 2001.
4. H. Nikolov, T. Stefanov and E. Deprettere, "Modeling and FPGA Implementation of Applications using Parameterized Process Networks with Non-Static Parameters", In Proc. of the Symposium on *Field-Programmable Custom Computing Machines (FCCM'05)*, Napa, California, USA, Apr. 2005.
5. M. Rutten, E-J Pol, J. van Eijnhoven, K. Walters and G. Essink, "Dynamic reconfiguration of streaming graphs on a heterogeneous multiprocessor architecture", *SPIE Electronic Imaging: Embedded processors for Multimedia and Communications II*, vol. 5683, San Jose, CA, USA, January 2005.
6. P. van der Wolf, E. de Kock, T. Henriksson, W. Kruijtzer, G. Essink, "Design and Programming of Embedded Multiprocessors: An Interface-Centric Approach", *CODES+ISS'04*, Stockholm, Sweden, september 2004.
7. J. Lemaitre, S. Alliot and E. Deprettere, "Behavioral specification of control interface for signal processing applications", In Proc. of the Intl. Conf. on *Application-Specific Systems, Architectures, and Processors (ASAP'05)*, Samos, Greece, July 2005.
8. B. Kienhuis and E. Deprettere, "Modelling stream-based applications using the SBF model of computation", *Journal of VLSI Signal Processing-Systems for Signal, Image, and Video Technology*, 34(3):291-300, 2003.
9. S. Derrien, A. Turjan, C. Zissulescu, B. Kienhuis and E. Deprettere, "Deriving Efficient Control in Process Networks with Compaan/laura", *International Journal of Embedded Systems*, vol.1, issue 7, 2005.
10. J. Lemaitre, S. Alliot and E. Deprettere, "Requirements for interfacing IP-components in re-configurable platforms", *Journal of VLSI Signal Processing-Systems for Signal, Image and Video Technology*, vol.43, number 2, 2006.
11. L. L'Hours, "Generating Efficient Custom FPGA Soft-Cores for Control Dominated Applications", in Proc. of the *IEEE conference on Application-Specific Systems, Architectures, and Processors (ASAP'05)*, Samos, Greece, July 2005.
12. G. Kahn, "The semantics of a simple language for parallel programming", in Proc. of the *IFIP Congress 74*, North-Holland Publishing Co., 1974.

13. J. Labrosse, "MicroC/OS-II, The Real-Time Kernel, 2nd edition", CMP Books, 2002.
14. A.J. Boonstra, "Radio Frequency Interference Mitigation in Radio Astronomy", PhD thesis, T.U. Delft, The Netherlands, June 2005.
15. SKA, The Square Kilometre Array radiotelescope: www.skatelescope.org
16. Altera: www.altera.com
17. Xilinx: www.xilinx.com
18. Opencores: www.opencores.org

An Embedded Systems Programming Environment for C⋆

Bernd Burgstaller[1], Bernhard Scholz[1], and Anton Ertl[2]

[1] The University of Sydney
[2] Technische Universität Wien

Abstract. Resource constraints are a major concern with the design, development, and deployment of embedded systems. Embedded systems are highly hardware-dependent and have little computational power. Mobile embedded systems are further constrained by their limited battery capacity. Many of these systems are still programmed in assembly language because there is a lack of efficient programming environments.

To overcome or at least alleviate the restrictions, we propose a lightweight and versatile programming environment for the C programming language that offers mixed-mode execution, i.e., code is either executed on the CPU or on a virtual machine (VM). This mixed-mode execution environment combines the advantages of highly compressed bytecode with the speed of machine code.

We have implemented the programming environment and conducted experiments for selected programs of the MiBench suite and the Spec 2000. The VM has a footprint of 12 KB on the Intel IA32. Initial results show that the performance of the virtual machine is typically only 2 to 36 times slower than the binary execution, with compressed code occupying only 36%–57% of the machine code size. Combining sequences of VM instructions into new VM instructions (superinstructions) increases the execution speed and reduces the VM code size. Preliminary experiments indicate a speedup by a factor of 3.

1 Introduction

Mobile devices powered by batteries constitute a major share of today's embedded systems market. Mobile devices have embedded intelligence, which needs to be programmed. Due to the limitations in terms of power consumption, memory size, and computational power, programming mobile devices is still a difficult problem. To overcome or at least alleviate the problem of programming embedded systems, we introduce a programming environment for C. The C programming language is still the language of choice for mobile and embedded systems, with more than 78% of all surveyed embedded systems firmware and application developers employing it [1].

⋆ This project has been supported by the ARC Discovery Project Grant "Compilation Techniques for Embedded Systems" under Contract DP 0560190 and the ARC Discovery Project Grant "Distributed Data Processing for Wireless Sensor Networks" under Contract DP 0664782.

Our programming environment provides a seamless integration of VM and machine code execution as outlined in Fig. 1. The program is stored as an image which contains bytecode[1] and machine code. Depending on whether the code is bytecode or machine code, it is executed on the VM or on the CPU respectively. Both, CPU and VM, share the same memory and the thread of execution can either jump from the VM to the machine code realm or vice versa.

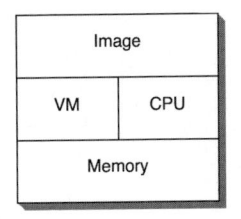

Fig. 1. Model

In this model rarely executed code is run on the VM. Frequently executed code is run on the CPU. This mixed-mode execution combines the advantage of both worlds: machine code is fast however has limited compression potential. Bytecode is stored highly compressed though the execution is slower. This execution model results in small image sizes, which reduces memory footprint and therefore devices will save energy. Also, the costs per device will decrease. Further advantages of VMs are hardware independent execution of C-programs and the fast deployment of programs by downloading them via an inter-network communication.

The contribution of this paper is the implementation of a light-weight programming environment for the C programming language. This programming environment offers mixed-mode execution, i.e., a seamless integration of VM code and machine code. The footprint overhead of the VM is small. The current footprint of the VM on an Intel IA32 architecture is 12 KB.

The paper is organised as follows: in Sec. 2 we discuss the compilation path of the programming environment. In Sec. 3 we discuss the design of our VM. In Sec. 4 we present experimental results. In Sec. 5 we survey related work. We draw our conclusions in Sec. 6.

2 Compilation

Fig. 2 depicts the compilation path of our embedded systems programming environment. Therein an application consists of a set of C source files containing code that can be compiled to either bytecode or to machine code. To allow the programmer to select between the two, we extend the C programming language with two storage class specifiers (cf. [2]), namely vm and mc. Furthermore, we use a command line parameter with the C-compiler to select a default storage class for unassigned entities. (Unassigned entities are entities that have not been assigned one of the above storage class specifiers). With this mechanism we partition the set of entities of a given application into the set of entities assigned to the realm of the VM, and those assigned to machine code.

It is the purpose of the *splitter* to preprocess an application and separate each source file into a corresponding vm and mc file that reflects the programmer's choices with respect to compilation to vm or mc code. The splitter has to achieve a

---

[1] In the context of this paper the word "bytecode" does not denote Java bytecode. Instead, it denotes the instruction code format of our VM.

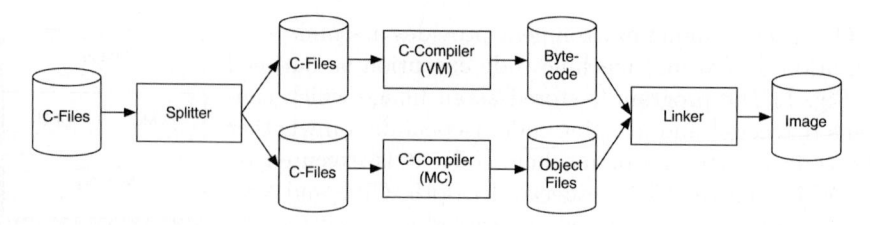

Fig. 2. Compilation Path

```
1   #define MAX 1024
2   mc static void fft_float (float *R_In, float *I_In, float *R_Out, float *I_Out);
3   vm char Buffer[MAX];

4   vm int main(void){
5       float R_In[MAX], I_In[MAX], R_Out[MAX], I_Out[MAX];
6       fft_float (R_In,I_In,R_Out,I_Out);
7       return 0;
8   }

9   mc static void fft_float (float *R_In, float *I_In, float *R_Out, float *I_Out)
10  { /* perform FFT */ }
```
(a) Application

```
1   #define MAX 1024
2   extern void fft_float (float *R_In, float *I_In, float *R_Out, float *I_Out);
3   char Buffer[MAX];
4   int main(void){
5       float R_In[MAX]; float I_In[MAX]; float R_Out[MAX]; float I_Out[MAX];
6       fft_float (R_In,I_In,R_Out,I_Out);
7       return 0;
8   }
```
(b) Application, vm Realm

```
1   #define MAX 1024
2   extern char Buffer[MAX];
3   void fft_float (float *R_In, float *I_In, float *R_Out, float *I_Out)
4   { /* perform FFT */ }
```
(c) Application, mc Realm

Fig. 3. Example: Splitting of Application Sources

clear semantic separation between vm and mc code to enable separate compilation by the vm and mc compilers.

The example in Fig. 3 is a simplified version of the FFT benchmark from the MiBench embedded benchmark suite [3]. Figure 3 (a) denotes the application which, for the sake of simplicity, consists of only one source file. Line 2 and lines 9–10 define a C function fft_float, which, due to the storage class specifier mc, is meant to be compiled to machine code. The main function (lines 4–8 is to be compiled to bytecode; main calls fft_float. Line 3 declares a global buffer variable that is kept in bytecode as well. Figure 3 (b) shows the vm file as output by the splitter. Therein the code for function fft_float has been removed and a corresponding external declaration has been inserted to keep the file compileable. This contrasts the declaration of the global buffer that is kept in the vm file (cf. line 3). All vm and mc storage class specifiers have been removed, because the occurrence of a declaration in the vm file already implies the vm storage class specifier. Likewise for the mc-file of Figure 3 (c). It contains only the definition of the MAX constant, an external declaration for the global buffer, and the code for function fft_float. As can be derived from Fig. 2, the separated files are then compiled by the C compilers for machine- and bytecode.

We employ LCC [4,5] for both bytecode and machine code compilation. LCC comes already equipped with a backend for bytecode, which we extended to facilitate the architecture of our VM (cf. Section 3).

Bytecode and machine code files of a given application are combined by the linker to a so-called *fat binary*. In this linkage step all references are resolved; this includes cross-references between bytecode and machine code to allow for seamless execution between the two. The fat binary can then be downloaded and executed on the embedded device.

3 Virtual Machine

The instruction set of our stack-based VM is closely related to the bytecode interface that comes with LCC [4], with the main deviations being induced by the requirements of the seamless integration of vm and mc execution. Table 1 depicts the instructions provided by our VM. The instruction opcodes cover the leftmost column whereas the column headed "IS-Op." lists operands derived from the instruction stream (all other instruction operands come from the stack). The column entitled "Suffixes" denotes the valid type suffixes for an operand (F=float, I=signed integer, U=unsigned integer, P=pointer, V=void, B=struct).[2] In this way instruction ADDRG receives its pointer argument p from the instruction stream and pushes it onto the stack. Instructions ADDRF and ADDRL receive an integer argument literal from the instruction stream; this literal is then used as an offset to the stack framepointer to compute the address of a formal or local variable. Instruction BADDRG uses its instruction stream argument as an index into a lookup table to derive the address of an mc-entity. The lookup table itself is created by the linker (cf. Section 2). For the remaining instructions of our VM we refer to the descriptions in Table 1.

To make bytecode interpretation acceptable for embedded systems, the performance of the interpretive system must be within reasonable limits compared to the performance of machine code. Due to the large design space for interpreters the achieved performance can vary drastically, with slowdowns between a factor of 10 and more than a factor of 1000 reported in the literature [6].

We used vmgen [7,8] for the implementation of our VM. Vmgen takes VM instruction descriptions as input and generates C code for execution, VM code generation, disassembly, tracing, and profiling. Vmgen already incorporates advances in interpreter technology such as threaded code (representing a VM instruction as the address of the routine that implements the instruction [9]), top

---

[2] Operators contain *byte size* modifiers (i.e., 1, 2, 4, 8), which we have omitted for reasons of brevity.

Table 1. VM Instruction Set

Instruction	IS-Op.	Suffixes	Description
ADD SUB	—	FIUP..	integer addition, subtraction
MUL DIV	—	FIU...	integer multiplication, division
NEG	—	FI....	negation
BAND BOR BXOR	—	.IU...	bitwise and, or, xor
BCOM	—	.IU...	bitwise complement
LSH RSH MOD	—	.IU...	bit shifts and remainder
CNST	a	.IUP..	push literal a
ADDRG	p	...P..	push address p of global
ADDRF	l	...P..	push address of formal parameter, offset l
ADDRL	l	...P..	push address of local variable, offset l
BADDRG	index	...P..	push address of mc entity at index
INDIR	—	FIUP..	pop p; push *p
ASGN	—	FIUP..	pop p; pop arg; *p = arg
ASGN_B	a	.....B	pop p, pop q; copy the block of length a at *q to p
CVI	—	FIU...	convert from signed integer
CVU	—	.IUP..	convert from unsigned integer
CVF	—	FI....	convert from float
CVP	—	..U...	convert from pointer
LABEL	—	....V.	label definition
JUMP	target	....V.	unconditional jump to target
IJUMP	—	....V.	indirect jump
EQ GE GT LE LT NE	target	FIU...	compare and jump to target
ARG	—	FIUP..	top of stack is next outgoing argument
CALL	target	....V.	vm procedure call to target
ICALL	—	....V.	pop p; call procedure at p
INIT	l	....V.	allocate l stack cells for local variables
BCALL	—	FIUPVB	mc procedure call
RET	—	FIUPVB	return from procedure call
HALT	—	....V.	exit the vm interpreter

of stack (TOS) caching (keeping the topmost stack element in a register), and superinstructions (combining frequently occurring patterns of VM instructions).

Figure 4 depicts a refined view of the execution architecture introduced in Sec. 1. The VM comprises a frontend, an interpreter, and stacks. The purpose of the frontend is to parse the bytecode (cf. Table 1) and to issue calls to the interpreter to build the *internal representation* that vmgen uses to store threaded instructions. Once this internal representation has been generated from the instruction stream, the interpreter is started. Our VM employs three stacks: the VM stack is used as the evaluation stack, the Arg stack holds procedure call arguments, and the Prog stack is used for machine code execution. The separate argument stack is due to LCC's ordering of bytecode instructions which intersperses procedure call arguments with other stack operands. The separate Arg

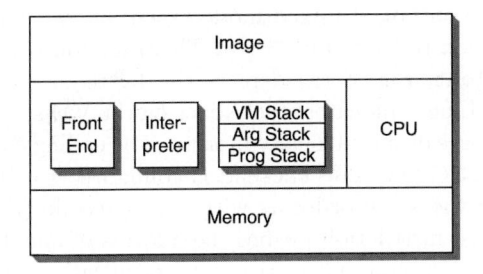

Fig. 4. Refined Execution Model

```
1 baddrg_p4 ( #ul -- p )        1 add_i4 ( l1 l2 -- 1 )
2 p = getsymbol_ptr(ul);        2 1 = l1+l2;

      (a) BADDRG_P4                    (c) ADD_I4          1 bcall_v ( l1 p -- )
                                                           2 indirect_call_v(p,argsp,l1);
1 arg_p4 ( p -- ARGp )          1 addrl_p4 (#1 -- p)       3 argsp = argsp+l1;
2 /* moves p from VM            2 p = (void *)(fp+1);
  stack to Arg stack */                                         (e) BCALL_V

      (b) ARG_P4                     (d) ADDRL_P4
```

Fig. 5. Vmgen Instruction Specifications

stack provides an efficient way to collect procedure call arguments and arrange them in a stack frame (we will elaborate on procedure calls in the following).

Vmgen provides a mechanism to specify the semantics of the instructions provided by the interpreter. As an example, consider Fig. 5 (c) which depicts the specification of the ADD instruction with this mechanism.[3] Therein line 1 describes the stack effect of the instruction: it pops the arguments l1 and l2 from the VM stack and pushes argument 1. Line 2 contains C code that describes how argument 1 is actually computed. The overall semantics for the ADD instruction is to pop l1 and l2 from the VM stack, execute the C code, and push 1 back on the VM stack.

```
1  proc main            9  ADDRL_P4 0
2  INIT 4096           10  ARG_P4              12.1 BADDRG_P4 0
3  ADDRL_P4 12288      11  CNST_I4 4           12.2 BCALL_V
4  ARG_P4              12  CALL_V fft_float        (b) after linking
5  ADDRL_P4 8192       13  CNST_I4 0
6  ARG_P4              14  RET_I4              1 void *st[]={
7  ADDRL_P4 4096       15  endproc             2   (void *)&fft_float,
8  ARG_P4                                      3   0L};

       (a) main.s                                  (c) address table
```

Fig. 6. Call from Bytecode to Machine Code

---

[3] Note that, unlike Table 1, the VM instructions in Fig. 5 include the type and size specifiers.

To illustrate the concept of mixed mode execution, we consider the vm code generated for function main of Fig. 3 (b). This function contains a binary call to function fft_float. Fig. 6 (a) depicts the bytecode for function main as generated by LCC. Line 2 allocates stack space on the VM stack to accommodate the four arrays of floats declared locally in main. The calling concept is illustrated in lines 3–12; our LCC bytecode backend is configured to evaluate parameters of function calls in the same order as with the bytecode (right to left, in this case). Each ADDRL_P4-instruction pushes the address of one float-array onto the VM stack (cf. the corresponding instruction specification in Fig. 5 (d), where the address in p is computed relative to the VM stack framepointer fp). Each subsequent ARG_P4-instruction moves this address from the VM stack to the Arg stack (cf. Fig. 5 (b), where the Arg-prefix denotes the argument stack). The purpose of line 11 is to push the number of stack cells covered by the arguments onto the VM stack. Line 12 contains the call to fft_float.

Once our linker (cf. Fig. 2) generates the fat binary, we employ a scan of the bytecode to collect all references that cannot be resolved within the bytecode itself. For these references we generate a machine code address table. (The address table for our example is shown in Fig. 6 (c), it contains just the address of function fft_float.)

It is only at link time that the actual address of fft_float can be resolved. The linker replaces line 12 of Fig. 6 (a) by the code depicted in Fig. 6 (b) in order to account for the fact that this is a binary call. In line 12.1 the index of function fft_float with respect to the address table is pushed onto the VM stack (cf. Fig. 5 (a) for the corresponding instruction specification). This index is used by instruction BCALL_V (cf. Fig. 5 (e)) to perform the binary call.

```
1    void indirect_call_v(void (*f)(void),void *arg, long arglen) {
2      void *p=alloca(sizeof(Cell)*arglen);
3      memcpy(p,arg,sizeof(Cell)*arglen);
4      return (*f)();
5    }
```

Fig. 7. Binary Call

The binary call mechanism itself is illustrated in Fig. 7. Function alloca allocates space on the program stack to account for the arguments of the call. Thereafter the arguments are copied from the Arg stack to the Prog stack (arg corresponds to the framepointer of the argument stack) and the binary call itself is carried out. To clean up after the call, the current argument frame is removed from the argument stack (cf. Fig. 5 (e)).

Calls of bytecode functions from machine code are carried out via a trampoline (similar to the approach in [10]). The trampoline code sets up the VM and Arg stacks and starts VM execution at the first bytecode instruction of the called function.

4 Experiments

As a testbed we used selected C programs of the MiBench benchmark suite [3] and the Spec CPU 2000 [11] benchmark suite targeting specific areas of the embedded market. We performed our experiments on the Intel IA32 platform to determine

1. the slowdown of programs executed as bytecode on our VM,
2. the VM performance improvement due to superinstructions, and
3. the best possible compression rate by using simple Huffman coding.

4.1 Performance of the Virtual Machine

We compared the performance of the VM to native code on the IA32 platform. To make a fair comparison, LCC was used to generate the bytecode and the machine code of the benchmark programs. In Table 2 the runtimes of the benchmark programs are shown.

Table 2. Performance, Machine Code (s) $\cdot \lambda$ = Bytecode (s)

	Benchmark	Machine Code (s)	Byte Code (s)	λ
Spec2k	gzip	85	2943	34.6
	bzip2	321	11463	35.7
	mcf	55.9	483	8.6
MiBench	basicmath			
	small	0.1	0.35	3.5
	large	2.1	5.2	2.5
	bitcount	0.07	2.16	30.9
	FFT	0.16	4.3	26.9
	adpcm			
	rawcaudio	1.2	32.1	26.8
	rawdaudio	1.2	24.7	20.6
	CRC32	1.0	19.7	19.7

In Table 2 the time measurements are given in seconds. All programs are executed on a Pentium 4 with 1.8GHZ under Linux. The execution time of the benchmark programs vary from 0.1 to 321 seconds when compiled as machine code. If the programs are compiled as bytecode the execution increases varying from 0.35 to 11463 seconds. These results were expected since the execution of bytecode is slower than machine code. The slowdown (Column λ of Table 2) ranges from 2 to 36. Note that the slowdown increases if extensive computations are performed inside of the VM. If machine libraries are called as in basic math, the slowdown is much smaller. This result is not surprising and it goes in line with the expected slowdown factors reported in [7]. Note that this virtual machine already uses the fastest known techniques such as threaded code and an advanced dispatching mechanism, but we have not employed superinstructions in the above experiments.

Latest experiments with superinstructions enabled indicate that significant further improvements with respect to execution times are possible. In profiling the gzip program and using just the top 7% of the most frequently executed bytecode sequences we experienced a reduction of the slowdown from a factor of 36 to a factor of 12. In allowing more superinstructions further improvements can be expected. However, there is a clear tradeoff between the code size and performance of the VM. By converting 7% of the most frequently bytecode sequences the codesize of the VMs increases by 55.5%, i.e., to nearly 19 KB instead of 12 KB.

4.2 Code Compression

Bytecode has properties that allow high compression rates. In this experiment we compared the size of binary executables with Huffman encoded bytecode. As a compression method we split the bytecode stream into three portions for Huffman coding: op-code stream, number stream, and symbol stream. This is a well known technique [12] to improve the compression rate. In this experiment we did not apply a dictionary approach (such as superinstructions or LZW) that stores re-occurring sequences only once. By adding a dictionary approach, even higher compression rates are possible. In Table 3 the results of this experiment are shown.

Table 3. Codesize

	Benchmark	Op-Code (bits)	Number (bits)	Symbol (bits)	Total (bytes)	IR (bytes)	Object (bytes)
Spec2k	gzip	115312	46614	31194	24140	148872	42412
	bzip2	65380	33196	15816	14299	94328	28093
	mcf	26091	12417	3393	5238	39324	10325
MiBench	basicmath						
	small	4728	1965	793	936	6756	2612
	large	5970	2373	1116	1183	8648	3224
	bitcount	5702	1897	289	986	7152	1952
	FFT	6998	2926	529	1307	8756	2467
	adpcm						
	rawcaudio	2774	1309	316	550	4384	1069
	rawdaudio	2766	1310	316	549	4384	1067
	CRC32	1348	400	60	226	1772	528

Columns Op-Code, Number, and Symbol show the number of bits required to store the op-codes, numbers, and symbols of the bytecode. We used different Huffman codes for op-codes and arguments. Column Total gives the number of bytes to store Huffman encoded bytecode of a benchmark program. In Column

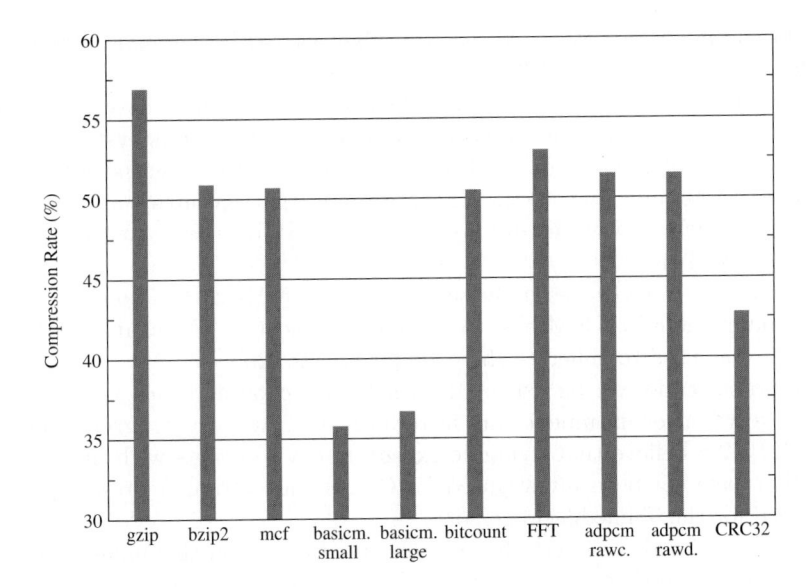

Fig. 8. Compression Rate

Object we show the number of bytes of the Intel IA32 machine code. As shown in Fig. 8 the compression rate of compressed bytecode is very high. Here we compare the size of the compressed bytecode (Column Total) with the size of the machine code (Column Object). Compression rates range from 36% for very small programs to 57% for larger programs. Compression rates can be further improved by employing dictionary based approaches in combination with Huffman codes. This initial result is very motivating; it shows that a high compression rate is achieved by using virtual machine technology.

The number of bytes used for the internal representation is quite expensive, as shown in Column IR of Table 3. The internal representation of the bytecode is bigger than the IA32 machine code. This is attributed to the use of threaded code techniques [9] in which op-codes are replaced by function pointers. In the Intel IA32 architecture threaded code techniques waste roughly 3 bytes per bytecode instruction, i.e., four bytes for a function pointer minus one byte for an opcode. This result indicates that a buffer technique should be applied to keep most frequent executed portions of code in its internal representation. Rarely executed code should be left in its compressed form until it is needed.

5 Related Work

Instead of translating the source code of a high-level language to assembly code, quite often a VM is used. A VM abstracts the properties of the underlying hardware and, therefore, makes the execution of programs hardware independent. In comparison to other implementation techniques of programming languages, VMs have the advantages of (1) portability, (2) ease of implementation, and (3) fast edit-compile-run cycles. VMs are very light-weight, which makes them suitable for embedded systems [13,14].

VM code consists of a sequence of VM instructions, which have many similarities to real machine code. In such a design, the interpretive system consists of two components: (1) a front end, that is a compiler that translates the input language to VM code, and (2) the VM interpreter that executes VM code. Good examples of such an architecture are Java's JVM [15], Prolog's WAM [16], and Smalltalk's VM [17].

Several tools [7,18,19] assist the development of VMs. A VM compiler generates an interpreter for a VM based on a VM specification. For example, the tool vmgen [7] was used to generate the code for Gforth [20].

Interpreted code can be executed with binary code and vice versa. Such a mixed execution environment was introduced for the Java programming language [21]. We believe that dynamic execution environments with mixed-mode execution have not been investigated for C, although a similar project [22] was developed for the Trimedia processor.

Low-end embedded systems have strong restrictions on the amount of available memory, which severely limits the size of the applications. Memory is a scarce commodity for several reasons: available physical space is limited, and power consumption and production costs must be minimised. Therefore, a lot of effort was taken to minimise program sizes of embedded systems applications. Especially in the realm of Java, compression rates of up to 85% of the original program size are not rare [23,24,25]. Instead of using sophisticated compression schemes, alternative representations of the VM code such as trees have been investigated [26]. For binary code several techniques have been introduced [12,27,28,29]. The main technique is to split the code into various portions and to compress them with different compression schemes. Even the instructions are split in op-codes and operands. This gives further opportunities to remove redundancies. Recently an interesting approach was introduced to incorporate compression into the instruction fetch inside a VM using Huffman codes [30]. It has to be investigated to what extend such an approach would affect the performance of our VM.

6 Conclusion

In this paper we have introduced a light-weight programming environment for the C programming language that alleviates the resource constraints present in embedded systems. Our programming environment provides seamless integration of VM and machine code execution.

In our compilation model the programmer assigns storage classes to C functions in order to decide whether they are compiled to bytecode or machine code. Our programming environment uses the LCC compiler, for which we have implemented a bytecode backend. The VM itself was developed with the vmgen specification tool.

We have conducted experiments with selected programs from the MiBench and Spec 2000 benchmark suites. Experiments show that the compressed bytecode occupies only 36%–57% of the corresponding machine code. The bytecode executed on the VM is only 2–36 times slower than machine code. Experiments indicate that superinstructions will further boost the performance by a factor of 3. However, superinstructions increase the footprint of the VM.

References

1. eMedia Asia Ltd. and Gartner, Inc.: Embedded Systems Development Trends: Asia. http://www.eetasia.com (2005)
2. Kernighan, B.W., Ritchie, D.M.: The C Programming Language. Prentice Hall Press, Upper Saddle River, NJ, USA (1988)
3. Guthaus, M.R., Ringenberg, J.S., Ernst, D., Austin, T.M., Mudge, T., Brown, R.B.: MiBench: A Free, Commercially Representative Embedded Benchmark Suite. In: Proceedings of the IEEE 4th Annual Workshop on Workload Characterization. (2001)
4. Hanson, D.R., Fraser, C.W.: A Retargetable C Compiler: Design and Implementation. Addison Wesley (1995)
5. Fraser, C.W.: A Retargetable Compiler for ANSI C. SIGPLAN Not. **26** (1991) 29–43
6. Romer, T.H., Lee, D., Voelker, G.M., Wolman, A., Wong, W.A., Baer, J.L., Bershad, B.N., Levy, H.M.: The Structure and Performance of Interpreters. In: ASPLOS-VII: Proceedings of the 7th International Conference on Architectural Support for Programming Languages and Operating Systems, New York, NY, USA, ACM Press (1996) 150–159
7. Ertl, M.A., Gregg, D., Krall, A., Paysan, B.: vmgen — A Generator of Efficient Virtual Machine Interpreters. Software—Practice and Experience **32** (2002) 265–294
8. Ertl, M.A., Gregg, D.: Building an Interpreter with vmgen. In: Compiler Construction (CC'02), Springer LNCS 2304 (2002) 5–8 Tool Demonstration.
9. Bell, J.R.: Threaded Code. Communications of the ACM **16** (1973)
10. Bruno Haible: Foreign Function Call Libraries. http://www.haible.de/bruno/packages-ffcall.html (2006)
11. Standard Performance Evaluation Corporation: Spec CPU 2000 (2000) http://www.spec.org/.
12. Debray, S., Evans, W.: Profile-Guided Code Compression. In: Proceedings of the ACM SIGPLAN 2002 Conference on Programming Language Design and Implementation, New York, NY, USA, ACM Press (2002) 95–105
13. Levis, P., Culler, D.: Mate: A Tiny Virtual Machine for Sensor Networks. In: International Conference on Architectural Support for Programming Languages and Operating Systems, San Jose, CA, USA. (2002)
14. Various: TinyVM () http://tinyvm.sourceforge.net/.
15. Lindholm, T., Yellin, F.: The Java Virtual Machine Specification. Second edn. Addison-Wesley (1999)
16. Aït-Kaci, H.: Warren's Abstract Machine: A Tutorial Reconstruction. MIT press, Cambridge (1991)
17. Goldberg, A., Robson, D.: Smalltalk-80: The Language and Its Implementation. Addison-Wesley Longman Publishing Co., Inc., Boston, MA, USA (1983)
18. Kelsey, R.A., Rees, J.A.: A Tractable Scheme Implementation. Lisp and Symbolic Computation **7** (1994) 315–335

19. Folliot, B., Piumarta, I., Riccardi, F.: A Dynamically Configurable, Multi-Language Execution Platform. In: Proc. of the 8th ACM SIGOPS European Workshop. (1998) 175–181
20. Various: GForth () http://www.jwdt.com/~paysan/gforth.html.
21. Muller, G., Moura, B., Bellard, F., Consel, C.: Harissa: A Flexible and Efficient Java Environment Mixing Bytecode and Compiled Code. In: Proceedings of the 3rd Conference on Object-Oriented Technologies and Systems, Portland, OR, USA, Usenix (1997) 1–20
22. Hoogerbrugge, J., Augusteijn, L., Trum, J., Wiel, R.V.D.: A Code Compression System Based on Pipelined Interpreters. Softw. Pract. Exper. **29** (1999) 1005–2023
23. Pugh, W.: Compressing Java Class Files. In: SIGPLAN Conference on Programming Language Design and Implementation. (1999) 247–258
24. Clausen, L.R., Schultz, U.P., Consel, C., Muller, G.: Java Bytecode Compression for Low-End Embedded Systems. ACM TOPLAS **22** (2000) 471–489
25. Bradley, Q., Horspool, R., Vitek, J.: JAZZ: An Efficient Compressed Format for Java Archive Files (1998)
26. Kistler, T., Franz, M.: A Tree-Based Alternative to Java Byte-Codes. International Journal of Parallel Programming **27** (1999) 21–33
27. Cooper, K.D., McIntosh, N.: Enhanced Code Compression for Embedded RISC Processors. In: SIGPLAN Conference on Programming Language Design and Implementation. (1999) 139–149
28. Ernst, J., Evans, W.S., Fraser, C.W., Lucco, S., Proebsting, T.A.: Code Compression. In: SIGPLAN Conference on Programming Language Design and Implementation. (1997) 358–365
29. Lekatsas, H., Wolf, W.: SAMC: A Code Compression Algorithm for Embedded Processors. IEEE Transactions on CAD **18** (1999) 1689–1701
30. Latendresse, M., Feeley, M.: Generation of Fast Interpreters for Huffman Compressed Bytecode. In: IVME '03: Proceedings of the 2003 Workshop on Interpreters, Virtual Machines and Emulators, New York, NY, USA, ACM Press (2003) 32–40

Author Index

Agrawal, Gagan 360
Agullo, Emmanuel 1053
Akihiko, Konagaya 1125
Alam, Sadaf R. 65
Alfaro, F.J. 884
Alves, Carlos Eduardo Rodrigues 831
Angel, Eric 157
Argollo, Eduardo 78
Asai, Shogo 1114
Ascione, I. 1085
Asenjo, R. 323
Ayguadé, Eduard 459

Badía, José M. 710
Bagci, Faruk 909
Baille, Fabien 519
Bal, Henri E. 765
Bampis, Evripidis 157
Banino-Rokkones, Cyril 167
Battré, Dominic 995
Baumgartl, Robert 124
Beaumont, Olivier 167
Bednarski, Andrzej 461
Bender, Michael 155
Benner, Peter 710
Bergdorf, M. 730
Bermúdez, Aurelio 853
Bernard, Julien 841
Berrendorf, Rudolf 415
Bilas, Angelos 851
Birkner, Marcel 415
Blachot, Florent 289
Blin, Lelia 519
Bos, Herbert 765
Boukerche, Azzedine 981
Boutammine, Salah-Salim 188
Brehm, Matthias 1
Brzeziński, Jerzy 530
Bungartz, Hans-Joachim 677, 1064
Burgstaller, Bernd 1204
Buyya, Rajkumar 425

Cáceres, Edson Norberto 831
Camargos, Lasaro 549

Cappello, Franck 393
Casado, Rafael 853
Castillo, R. 323
Chan, Hiu Ning (Angela) 655
Chapman, Barbara 277
Chapman, Barbara M. 645
Chatelain, P. 730
Chatelier, Pierre 740
Chen, Liang 360
Chen, Liming 371
Chevalier, Cédric 243
Chiang, Yi-Kai 919
Christopoulou, Artemis A. 405
Čiegis, Raimondas 679
Collard, Jean-Francois 459
Colmenar, J.M. 495
Corbera, F. 323
Cores, F. 777
Cortes, Ana 15
Cosnard, Michel 677
Coulson, Geoff 763
Cox, Simon J. 371, 1041
Cunha, José C. 603

Damaschke, Peter 811
De Bosschere, Koen 459
de Dios, A. 473
de Supinski, Bronis R. 1, 645
Debudaj-Grabysz, Agnieszka 1075
del Cuvillo, Juan 134
Deprettere, Ed 1192
DeRose, Luiz 1
Dimakopoulos, Vassilios V. 1015
Domingos, Henrique J. 993
Douillet, Alban 311
Duato, J. 864, 884
Dumitrescu, Catalin L. 448
Dupont de Dinechin, Benoît 289

Ehrlich, Hans-Christian 1148
Eicker, Norbert 874
Eitrich, Tatjana 350
Emoto, Kento 605
Erbacci, Giovanni 1041

Ertl, Anton 1204
Eterovic S., Yadran 559

Fakler, M. 571
Feitelson, Dror 155
Felber, Pascal 993
Feng, Shengzhong 1138
Fernández, Enol 383
Fernández, J.J. 1159
Ferrari, Gianluigi 897
Ferreira, Paulo 971
Ferscha, Alois 897
Feschet, Fabien 740
Fettweis, G. 299
Flautner, Krisztián 1179
Flich, J. 884
Fraigniaud, Pierre 799
Frenz, S. 571
Frings, Wolfgang 350
Fuhrmann, Patrick 1106
Fukuda, Shuichi 1169

Gabriel, Edgar 517, 700
Gallagher, Andrew J. 655
Gallopoulos, Efstratios 677
Gao, Guang R. 134, 311
Garbey, Marc 700
García, Eva M. 853
García, I. 1159
Garcia, João 971
García, Luis 821
García, P.J. 884
Garnica, O. 495
Garzón, E.M. 1159
Gaudiani, Adriana 78
Gaudiot, Jean-Luc 485
Ghodsi, Ali 1026
Gilabert, F. 864
Giné, Francesc 177
Girkar, Milind 253
Giunta, G. 1085
Glaser, S.J. 751
Goeckelmann, R. 571
Gómez, M.E. 864
Gómez-Villamor, Sergio 3
Gorlatch, Sergei 603
Goscinski, Andrzej 517
Gottlieb, Allan 155
Goundan, Appu S. 655
Gradl, T. 751

Gregorio, Jose A. 851
Guermouche, Abdou 1053
Guerra, Concettina 1125
Guirado, F. 1095
Gülzow, Volker 1106
Guo, J. 299

Ha, Phuong Hoai 811
Hackenberg, Daniel 145
Hamano, Yusuke 1169
Hanzich, Mauricio 177
Haridi, Seif 1026
Heine, Felix 995
Hermanns, Marc-André 47, 415
Hernàndez, A. 1095
Hernández, Porfidio 177, 777
Heuveline, Vincent 689
Heymann, Elisa 383
Hidalgo, J.I. 495
Hieber, S.E. 730
Hoefler, Torsten 124
Hoffmann, K.H. 1043
Hofmann, M. 1043
Hu, Weiwu 506
Hu, Zhenjiang 605
Hu, Ziang 134
Huang, Lei 645
Huang, Tingxue 981
Huard, Guillaume 289
Huckle, T. 751
Hwang, Shyh-In 919

Ibach, Peter 961
Ibáñez, P. 473
Ishii, Manabu 1169
Issarny, Valerie 897
Ivars, Vicente 15

Jalby, William 277
Janda, Rick 89
Jang, Kil-Woong 941
Jeannot, Emmanuel 211
Jelasity, Mark 993
Jernberg, Jimmy 1026
Jiao, Zhuoan 371
Jin, Haoqiang 645
Jost, Gabriele 645
Jung, Hyungsoo 539

Kaneko, Keiichi 1114
Kao, Odej 763, 995

Karagiorgos, Gregory 222
Karl, Wolfgang 25, 459
Karnapke, Reinhardt 951
Kaufmann, Michael 799
Kaxiras, Stefanos 1179
Keen, Aaron W. 655
Kejariwal, Arun 253
Keller, Christian 961
Kelly, Paul 277
Kermarrec, Anne-Marie 993
Kessler, Christoph 461
Kim, Byung-Soon 941
Kim, Namgi 899
Klemm, R. 299
Kobusińska, Anna 530
Kokkinos, Panagiotis C. 931
Konagaya, Akihiko 1127, 1169
Konishi, Fumikazu 1169
Korch, Matthias 720
Kosch, Harald 763
Kosowski, Adrian 592
Kotsalis, E.M. 730
Koumoutsakos, P. 730
Kounoike, Yuusuke 1114
Kral, Stefan 279
Kranzlmüller, Dieter 381
Krizanc, Danny 799
Kühnal, Andrej 47
Kuszner, Lukasz 592

Labarta, Jesús 63
Laforenza, Domenico 381
Laforest, Christian 519
Lanchares, J. 495
Lang, Bruno 350
Larriba-Pey, Josep-L. 3
Lehner, Wolfgang 335
Lemaitre, Jérôme 1192
León, Coromoto 821
Leontiadis, Elias 1015
Leopold, Claudia 635
Lérida, Josep L. 177
L'Excellent, Jean-Yves 1053
Li, Li 35
Limberg, T. 299
Lippert, Thomas 874, 1041
Llabería, J.M. 473
López, P. 864
Lopez, S. 495

Ltaief, Hatem 700
Luchangco, Victor 801
Luque, Emilio 78, 177, 777, 1095
Lysne, Olav 851

Madeira, Edmundo R.M. 549
Malgouyres, Rémy 740
Malony, Allen D. 35, 99
Man, Billy Yan-Kit 655
Margalef, Tomás 1
Mariani, P. 1085
Martínez, A. 884
Martins, Vidal 337
Maruyama, Tsutomu 1127
Masuno, Shingo 1127
Matsuzaki, Kiminori 605
Matus, E. 299
May, Patrick 1148
Mayo, Rafael 710
Megson, Graham 788
Mehl, Miriam 1064
Mehlan, Torsten 124
Mennenga, B. 299
Merker, Renate 1181
Meyerhenke, Henning 232
Mietke, Frank 124
Millot, Daniel 188
Miñana, G. 495
Miranda, Gara 821
Missirlis, Nikolaos M. 222
Mocito, José 582
Mohr, Bernd 47, 63
Montella, R. 1085
Moreau, Luc 381
Morin, Christine 517
Morlier, Tangui 393
Morris, Alan 99
Muntés-Mulero, Victor 3

Nagel, Wolfgang E. 89, 145
Nataraj, Aroon 99
Natvig, Lasse 167
Navarro, A. 323
Neri, Vincent 393
Ng, Kam-Wing 1005
Nicolau, Alexandru 253
Nijhuis, Maik 765
Nolte, Joerg 951
Nussbaum, Dan 801

Ohki, Shingo 1169
Olsson, Ronald A. 655
Orellana, Christian F. 559

Pacitti, Esther 337
Papageorgiou, Christos A. 931
Papatheodorou, Theodore S. 666
Parrot, Christian 188
Pascual, Fanny 157
Pazzi, Richard Werner Nelem 981
Pedone, Fernando 549
Pellegrini, François 243
Peng, Liu 1138
Pérez-Casany, Marta 3
Petzold, Jan 909
Pflüger, Stefan 145
Pinter, Shlomit S. 265
Pitoura, Evaggelia 1015
Placek, Martin 425
Plaks, Toomas 1179
Plata, Oscar 277
Polasek, B. 730
Polychronopoulos, Constantine 253
Polychronopoulos, Eleftherios D. 405
Priebe, Steffen 615

Quiles, Francisco J. 853, 884
Quinlan, Daniel 603
Quintana-Ortí, Enrique S. 710

Rabenseifner, Rolf 1075
Rauber, Thomas 720
Rees, Steve 3
Rehm, Wolfgang 124
Reinefeld, Alexander 381
Rex, Robert 124
Rexachs, Dolores 78
Rezmerita, Ala 393
Riccio, A. 1085
Rilling, Louis 437
Ripoll, A. 777, 1095
Ro, Won W. 485
Robles-Gómez, Antonio 853
Röblitz, Thomas 198
Roch, Jean-Louis 841
Rodrigues, Luís 582
Rodríguez, Casiano 821
Roig, C. 1095
Rong, Hongbo 311

Rui, Hou 506
Rünger, G. 1043
Rünger, Gudula 517
Ruz, Cristian 559
Rzadca, Krzysztof 198

Saad, Yousef 677
Sahelices, B. 473
Saito, Hideki 253
Sánchez, J.L. 884
Sbalzarini, I.F. 730
Schamberger, Stefan 232
Schill, Alexander 897
Schnor, Bettina 851
Schoettner, M. 571
Scholz, Bernhard 1204
Schöne, Robert 145
Schroeder, Michael 1125
Schulte-Herbrüggen, T. 751
Schulthess, P. 571
Schwiegelshohn, Uwe 155
Seeger, S. 1043
Seidel, Jan 415
Seinstra, Frank 763
Senar, Miquel Àngel 15, 383
Shavit, Nir 801
Shende, Sameer 99
Shinano, Yuji 1114
Siegel, Sebastian 1181
Sips, Henk 111
Snavely, Allan 63
Song, Siang Wun 831
Spörl, A. 751
Stantchev, Vladimir 961
Steinke, Thomas 1148
Stewart, Craig A. 1125
Süß, Michael 635
Sun, Ninghui 1138
Sung, Lin-Fei 919
Suppi, R. 777
Szychowiak, Michał 530

Tabik, S. 1159
Takeichi, Masato 605
Talia, Domenico 335
Tan, Guangming 1138
Tao, Jie 25
Teich, Jürgen 1179
Tian, Xinmin 253
Tian, Ye 1005

Tineo, A.   323
Torchinsky, Matías   177
Tran, John   3
Traoré, Daouda   841
Trenkler, Bernd   89
Triska, Markus   279
Trumler, Wolfgang   909
Trystram, Denis   1041
Tsigas, Philippas   811
Tzaferis, Filippos   222

Ueberhuber, Christoph W.   279
Ungerer, Theo   909

Valduriez, Patrick   335
van Gemund, Arjan   111
Varbanescu, Ana Lucia   111
Varvarigos, Emmanouel A.   931
Veidenbaum, Alexander   253
Veiga, Luís   971
Venetis, Ioannis E.   666
Vernier, Flavien   211
Vetter, Jeffrey S.   63, 65
Viñals, V.   473

Vlassov, Vladimir   1026

Walther, Andrea   689
Walther, J.H.   730
Watson, Paul   335
Weinzierl, Tobias   1064
Welch, Peter H.   603
Wolf, Felix   47
Wu, Cheng-Lin   919
Wu, Di   1005
Wu, Yunsong   788

Yamaguchi, Yoshiki   1127
Yang, X.Y.   777
Yeom, Heon Y.   539

Zalmanovici, Marcel   265
Zapata, E.L.   323
Zaroliagis, Christos   799
Zhang, Longbing   506
Zhu, Qian   360
Zhu, Weirong   134
Zrour, Rita   740
Zumbusch, Gerhard   625

Lecture Notes in Computer Science

For information about Vols. 1–4032

please contact your bookseller or Springer

Vol. 4162: R. Královič, P. Urzyczyn (Eds.), Mathematical Foundations of Computer Science 2006. XV, 813 pages. 2006.

Vol. 4153: N. Zheng, X. Jiang, X. Ian (Eds.), Advances in Machine Vision, Image Processing, and Pattern Analysis. XIII, 506 pages. 2006.

Vol. 4146: J.C. Rajapakse, L. Wong, R. Acharya (Eds.), Pattern Recognition in Bioinformatics. XIV, 186 pages. 2006. (Sublibrary LNBI).

Vol. 4144: T. Ball, R.B. Jones (Eds.), Computer Aided Verification. XV, 564 pages. 2006.

Vol. 4139: T. Salakoski, F. Ginter, S. Pyysalo, T. Pahikkala, Advances in Natural Language Processing. XVI, 771 pages. 2006. (Sublibrary LNAI).

Vol. 4138: X. Cheng, W. Li, T. Znati (Eds.), Wireless Algorithms, Systems, and Applications. XVI, 709 pages. 2006.

Vol. 4137: C. Baier, H. Hermanns (Eds.), CONCUR 2006 – Concurrency Theory. XIII, 525 pages. 2006.

Vol. 4133: J. Gratch, M. Young, R. Aylett, D. Ballin, P. Olivier (Eds.), Intelligent Virtual Agents. XIV, 472 pages. 2006. (Sublibrary LNAI).

Vol. 4130: U. Furbach, N. Shankar (Eds.), Automated Reasoning. XV, 680 pages. 2006. (Sublibrary LNAI).

Vol. 4129: D. McGookin, S. Brewster (Eds.), Haptic and Audio Interaction Design. XII, 167 pages. 2006.

Vol. 4128: W.E. Nagel, W.V. Walter, W. Lehner (Eds.), Euro-Par 2006 Parallel Processing. XXXIII, 1221 pages. 2006.

Vol. 4127: E. Damiani, P. Liu (Eds.), Data and Applications Security XX. X, 319 pages. 2006.

Vol. 4121: A. Biere, C.P. Gomes (Eds.), Theory and Applications of Satisfiability Testing - SAT 2006. XII, 438 pages. 2006.

Vol. 4119: C. Dony, J.L. Knudsen, A. Romanovsky, A. Tripathi (Eds.), Advanced Topics in Exception Handling Components. X, 302 pages. 2006.

Vol. 4117: C. Dwork (Ed.), Advances in Cryptology - Crypto 2006. XIII, 621 pages. 2006.

Vol. 4115: D.-S. Huang, K. Li, G.W. Irwin (Eds.), Computational Intelligence and Bioinformatics, Part III. XXI, 803 pages. 2006. (Sublibrary LNBI).

Vol. 4114: D.-S. Huang, K. Li, G.W. Irwin (Eds.), Computational Intelligence, Part II. XXVII, 1337 pages. 2006. (Sublibrary LNAI).

Vol. 4113: D.-S. Huang, K. Li, G.W. Irwin (Eds.), Intelligent Computing, Part I. XXVII, 1331 pages. 2006.

Vol. 4112: D.Z. Chen, D. T. Lee (Eds.), Computing and Combinatorics. XIV, 528 pages. 2006.

Vol. 4111: F.S. de Boer, M.M. Bonsangue, S. Graf, W.-P. de Roever (Eds.), Formal Methods for Components and Objects. VIII, 447 pages. 2006.

Vol. 4109: D.-Y. Yeung, J.T. Kwok, A. Fred, F. Roli, D. de Ridder (Eds.), Structural, Syntactic, and Statistical Pattern Recognition. XXI, 939 pages. 2006.

Vol. 4108: J.M. Borwein, W.M. Farmer (Eds.), Mathematical Knowledge Management. VIII, 295 pages. 2006. (Sublibrary LNAI).

Vol. 4106: T.R. Roth-Berghofer, M.H. Göker, H. A. Güvenir (Eds.), Advances in Case-Based Reasoning. XIV, 566 pages. 2006. (Sublibrary LNAI).

Vol. 4104: T. Kunz, S.S. Ravi (Eds.), Ad-Hoc, Mobile, and Wireless Networks. XII, 474 pages. 2006.

Vol. 4099: Q. Yang, G. Webb (Eds.), PRICAI 2006: Trends in Artificial Intelligence. XXVIII, 1263 pages. 2006. (Sublibrary LNAI).

Vol. 4098: F. Pfenning (Ed.), Term Rewriting and Applications. XIII, 415 pages. 2006.

Vol. 4097: X. Zhou, O. Sokolsky, L. Yan, E.-S. Jung, Z. Shao, Y. Mu, D.C. Lee, D. Kim, Y.-S. Jeong, C.-Z. Xu (Eds.), Emerging Directions in Embedded and Ubiquitous Computing. XXVII, 1034 pages. 2006.

Vol. 4096: E. Sha, S.-K. Han, C.-Z. Xu, M.H. Kim, L.T. Yang, B. Xiao (Eds.), Embedded and Ubiquitous Computing. XXIV, 1170 pages. 2006.

Vol. 4094: O. H. Ibarra, H.-C. Yen (Eds.), Implementation and Application of Automata. XIII, 291 pages. 2006.

Vol. 4093: X. Li, O.R. Zaïane, Z. Li (Eds.), Advanced Data Mining and Applications. XXI, 1110 pages. 2006. (Sublibrary LNAI).

Vol. 4092: J. Lang, F. Lin, J. Wang (Eds.), Knowledge Science, Engineering and Management. XV, 664 pages. 2006. (Sublibrary LNAI).

Vol. 4091: G.-Z. Yang, T. Jiang, D. Shen, L. Gu, J. Yang (Eds.), Medical Imaging and Augmented Reality. XIII, 399 pages. 2006.

Vol. 4090: S. Spaccapietra, K. Aberer, P. Cudré-Mauroux (Eds.), Journal on Data Semantics VI. XI, 211 pages. 2006.

Vol. 4089: W. Löwe, M. Südholt (Eds.), Software Composition. X, 339 pages. 2006.

Vol. 4088: Z.-Z. Shi, R. Sadananda (Eds.), Agent Computing and Multi-Agent Systems. XVII, 827 pages. 2006. (Sublibrary LNAI).

Vol. 4085: J. Misra, T. Nipkow, E. Sekerinski (Eds.), FM 2006: Formal Methods. XV, 620 pages. 2006.

Vol. 4079: S. Etalle, M. Truszczyński (Eds.), Logic Programming. XIV, 474 pages. 2006.

Vol. 4077: M.-S. Kim, K. Shimada (Eds.), Geometric Modeling and Processing - GMP 2006. XVI, 696 pages. 2006.

Vol. 4076: F. Hess, S. Pauli, M. Pohst (Eds.), Algorithmic Number Theory. X, 599 pages. 2006.

Vol. 4075: U. Leser, F. Naumann, B. Eckman (Eds.), Data Integration in the Life Sciences. XI, 298 pages. 2006. (Sublibrary LNBI).

Vol. 4074: M. Burmester, A. Yasinsac (Eds.), Secure Mobile Ad-hoc Networks and Sensors. X, 193 pages. 2006.

Vol. 4073: A. Butz, B. Fisher, A. Krüger, P. Olivier (Eds.), Smart Graphics. XI, 263 pages. 2006.

Vol. 4072: M. Harders, G. Székely (Eds.), Biomedical Simulation. XI, 216 pages. 2006.

Vol. 4071: H. Sundaram, M. Naphade, J.R. Smith, Y. Rui (Eds.), Image and Video Retrieval. XII, 547 pages. 2006.

Vol. 4070: C. Priami, X. Hu, Y. Pan, T.Y. Lin (Eds.), Transactions on Computational Systems Biology V. IX, 129 pages. 2006. (Sublibrary LNBI).

Vol. 4069: F.J. Perales, R.B. Fisher (Eds.), Articulated Motion and Deformable Objects. XV, 526 pages. 2006.

Vol. 4068: H. Schärfe, P. Hitzler, P. Øhrstrøm (Eds.), Conceptual Structures: Inspiration and Application. XI, 455 pages. 2006. (Sublibrary LNAI).

Vol. 4067: D. Thomas (Ed.), ECOOP 2006 – Object-Oriented Programming. XIV, 527 pages. 2006.

Vol. 4066: A. Rensink, J. Warmer (Eds.), Model Driven Architecture – Foundations and Applications. XII, 392 pages. 2006.

Vol. 4065: P. Perner (Ed.), Advances in Data Mining. XI, 592 pages. 2006. (Sublibrary LNAI).

Vol. 4064: R. Büschkes, P. Laskov (Eds.), Detection of Intrusions and Malware & Vulnerability Assessment. X, 195 pages. 2006.

Vol. 4063: I. Gorton, G.T. Heineman, I. Crnkovic, H.W. Schmidt, J.A. Stafford, C.A. Szyperski, K. Wallnau (Eds.), Component-Based Software Engineering. XI, 394 pages. 2006.

Vol. 4062: G. Wang, J.F. Peters, A. Skowron, Y. Yao (Eds.), Rough Sets and Knowledge Technology. XX, 810 pages. 2006. (Sublibrary LNAI).

Vol. 4061: K. Miesenberger, J. Klaus, W. Zagler, A. Karshmer (Eds.), Computers Helping People with Special Needs. XXIX, 1356 pages. 2006.

Vol. 4060: K. Futatsugi, J.-P. Jouannaud, J. Meseguer (Eds.), Algebra, Meaning, and Computation. XXXVIII, 643 pages. 2006.

Vol. 4059: L. Arge, R. Freivalds (Eds.), Algorithm Theory – SWAT 2006. XII, 436 pages. 2006.

Vol. 4058: L.M. Batten, R. Safavi-Naini (Eds.), Information Security and Privacy. XII, 446 pages. 2006.

Vol. 4057: J.P.W. Pluim, B. Likar, F.A. Gerritsen (Eds.), Biomedical Image Registration. XII, 324 pages. 2006.

Vol. 4056: P. Flocchini, L. Gąsieniec (Eds.), Structural Information and Communication Complexity. X, 357 pages. 2006.

Vol. 4055: J. Lee, J. Shim, S.-g. Lee, C. Bussler, S. Shim (Eds.), Data Engineering Issues in E-Commerce and Services. IX, 290 pages. 2006.

Vol. 4054: A. Horváth, M. Telek (Eds.), Formal Methods and Stochastic Models for Performance Evaluation. VIII, 239 pages. 2006.

Vol. 4053: M. Ikeda, K.D. Ashley, T.-W. Chan (Eds.), Intelligent Tutoring Systems. XXVI, 821 pages. 2006.

Vol. 4052: M. Bugliesi, B. Preneel, V. Sassone, I. Wegener (Eds.), Automata, Languages and Programming, Part II. XXIV, 603 pages. 2006.

Vol. 4051: M. Bugliesi, B. Preneel, V. Sassone, I. Wegener (Eds.), Automata, Languages and Programming, Part I. XXIII, 729 pages. 2006.

Vol. 4049: S. Parsons, N. Maudet, P. Moraitis, I. Rahwan (Eds.), Argumentation in Multi-Agent Systems. XIV, 313 pages. 2006. (Sublibrary LNAI).

Vol. 4048: L. Goble, J.-J.C.. Meyer (Eds.), Deontic Logic and Artificial Normative Systems. X, 273 pages. 2006. (Sublibrary LNAI).

Vol. 4047: M. Robshaw (Ed.), Fast Software Encryption. XI, 434 pages. 2006.

Vol. 4046: S.M. Astley, M. Brady, C. Rose, R. Zwiggelaar (Eds.), Digital Mammography. XVI, 654 pages. 2006.

Vol. 4045: D. Barker-Plummer, R. Cox, N. Swoboda (Eds.), Diagrammatic Representation and Inference. XII, 301 pages. 2006. (Sublibrary LNAI).

Vol. 4044: P. Abrahamsson, M. Marchesi, G. Succi (Eds.), Extreme Programming and Agile Processes in Software Engineering. XII, 230 pages. 2006.

Vol. 4043: A.S. Atzeni, A. Lioy (Eds.), Public Key Infrastructure. XI, 261 pages. 2006.

Vol. 4042: D. Bell, J. Hong (Eds.), Flexible and Efficient Information Handling. XVI, 296 pages. 2006.

Vol. 4041: S.-W. Cheng, C.K. Poon (Eds.), Algorithmic Aspects in Information and Management. XI, 395 pages. 2006.

Vol. 4040: R. Reulke, U. Eckardt, B. Flach, U. Knauer, K. Polthier (Eds.), Combinatorial Image Analysis. XII, 482 pages. 2006.

Vol. 4039: M. Morisio (Ed.), Reuse of Off-the-Shelf Components. XIII, 444 pages. 2006.

Vol. 4038: P. Ciancarini, H. Wiklicky (Eds.), Coordination Models and Languages. VIII, 299 pages. 2006.

Vol. 4037: R. Gorrieri, H. Wehrheim (Eds.), Formal Methods for Open Object-Based Distributed Systems. XVII, 474 pages. 2006.

Vol. 4036: O. H. Ibarra, Z. Dang (Eds.), Developments in Language Theory. XII, 456 pages. 2006.

Vol. 4035: T. Nishita, Q. Peng, H.-P. Seidel (Eds.), Advances in Computer Graphics. XX, 771 pages. 2006.

Vol. 4034: J. Münch, M. Vierimaa (Eds.), Product-Focused Software Process Improvement. XVII, 474 pages. 2006.

Vol. 4033: B. Stiller, P. Reichl, B. Tuffin (Eds.), Performability Has its Price. X, 103 pages. 2006.